KU-493-581

A Treatise on the

STEAM ENGINE

Historical, Practical and Descriptive

Volume I

A
Treatise on the
STEAM ENGINE

Historical, Practical and Descriptive (1827)

by JOHN FAREY

Volume 1

A Reprint

DAVID & CHARLES REPRINTS

ISBN 0 7153 4945 7

This work was originally published in 1827
by Longman, Rees, Orme, Brown and Green
Paternoster Row, London

Reprinted by the present publishers 1971

COUNTY
COPY

PUBLISHER'S NOTE

Volume 2 of this work was in course of preparation at the time of the author's
death in 1851, but was never published. David & Charles will shortly be pub-
lishing a reproduction of the page proofs of this work (in book form uniform
with this reprint of Volume 1), complete with corrections in the author's own
hand. The documents from which this reproduction will be taken are deposited
in the National Reference Library of Science and Invention.

HERTFORDSHIRE
COUNTY LIBRARY
621.1
5338296

Printed in Great Britain by
Latimer Trend & Company Limited Whitstable
for David & Charles (Publishers) Limited
South Devon House Newton Abbot
Devon

A

TREATISE

ON THE

STEAM ENGINE,

Historical, Practical, and Descriptive.

A

TREATISE

ON THE

STEAM ENGINE,

𝕳istorical, 𝕻ractical, and 𝕯escriptive.

BY

JOHN FAREY, ENGINEER.

Voilà la plus merveilleuse de toutes les Machines; le Mécanisme ressemble à celui des animaux. La chaleur est le principe de son mouvement; il se fait dans ses différens tuyaux une circulation, comme celle du sang dans les veines, ayant des valvules qui s'ouvrent et se ferment à propos; elle se nourrit, s'évacue d'elle-même dans des tems réglés, et tire de son travail tout ce qu'il lui faut pour subsister. Cette Machine a pris sa naissance en Angleterre, et toutes les Machines à feu qu'on a construites ailleurs que dans la Grande Bretagne ont été éxécutées par des Anglois. BELIDOR, *Architecture Hydraulique.*

ILLUSTRATED BY

NUMEROUS ENGRAVINGS AND DIAGRAMS.

LONDON:

PRINTED FOR LONGMAN, REES, ORME, BROWN, AND GREEN,
PATERNOSTER-ROW.

1827.

PREFACE.

THE high state of wealth and civilization which the English people have attained within the last half century, has been greatly promoted by the application of the power of the steam-engine to various purposes of the useful arts, in aid of manual labour. In an uncultivated state of society, all that the higher classes consume must be obtained by the degradation and slavery of the mass of the population; for unless the industry of the working class is systematically applied, and aided by the use of machines, there can be but little surplus wealth to maintain an educated class in society, and produce that state of general affluence which is conducive to the progress of civilization, and the developement of intellect.

It is only in comparatively modern times, that the natural powers of currents of wind and water have been applied to facilitate laborious operations; but at the commencement of the last century, in addition to the general application of wind and water powers, a new element was subjected to the laws of mechanics in England, and with the greatest success. The productiveness of labour has been so greatly increased by this gigantic auxiliary, and by the improved system of manufactures and commerce to which it has given birth, that those conveniences of life, and attributes of wealth, which were formerly considered to be one of the distinctions of the first classes of the community, are now acknowledged to belong equally to that middle class which may be said to consist of labourers, who apply their minds to useful industry, instead of their hands.

This change has produced a great advancement in the state of society. The desire of supplying the artificial wants thus created, has excited a great energy of mind, and an active exercise of knowledge amongst that middle class, which has influenced the whole mass of society, so as to have produced a high state of civilization; and in its consequences has induced a refinement of habits, manners, and sentiments, which can only result from a general diffusion of wealth and knowledge. The condition of the labouring class has been greatly improved at the same time, and a more rapid improvement will follow, from a perseverance in that system of education which is now in active operation amongst the common people.

The history of the origin and progress of so important and useful an invention, cannot fail to be an object of interest to every intelligent inquirer. The principles upon which useful results are to be obtained from the application of steam power, should be studied in detail, by all those who take any part in the direction of productive industry, and it will be found, that although much has been done, there are frequent occasions for new applications of those machines.

The object of the present treatise is to give a historical, practical, and descriptive account of the steam-engine, and its application to useful purposes.

The historical part is intended to form a complete account of the invention, from its first origin, to its present state of perfection, from which the statesman and political economist may observe the influence that the adoption of steam power has exercised upon our national prosperity and advancement, during a century past, and may form an opinion of the future advantages to be expected from more extended applications of the same principles.

The practical and descriptive part is intended to form a course of instruction for professional students, in the practice and principles of making and using steam-engines. At present, such students are left to form and digest their own crude and imperfect observations; and for want of a scientific guide, their conclusions are liable to be tinctured with many erroneous notions, and false assumptions. The same part is also intended to form a manual, which, by the aid of tables and theorems for calculation, will facilitate the practice of experienced professional engineers; and will tend to perfect the practice of those engineers and others, who require to employ steam-engines, and apply them to various purposes; but who do not possess a complete knowledge of their construction, and therefore require information, which it has hitherto been difficult to attain. The principles of mechanical and chemical action, upon which the operation of steam-engines depends, and the application of those principles to practice, will be also embodied in this part.

Another part will contain a record and brief explanation of all the speculative projects which have been proposed for the improvement of steam-engines, so as to exhibit to mechanical inventors the various ideas which have been suggested for that object; the instruction that they may obtain from other parts of the work, together with the history of the circumstances from which the great points of invention have originated, may enable them to make some further improvements.

To attain these objects is not an easy task; for it is only in the course of an active professional practice, that sufficient information is to be obtained; and all competent engineers are too much engaged to find leisure for a literary occupation. The author could not have undertaken such a work, if he had not formed the plan, and collected materials for its execution, at his first entrance into business; and as it became known that he had such an object in view, he has continually received contributions of information from other engineers, who wished to promote the undertaking; and, with very few exceptions, this feeling has been general in the profession. In this way the author became personally acquainted with two great authorities in this branch of mechanics, the late Mr. Watt, and Mr. Woolf, and received from them a full knowledge of the origin and progress of their respective inventions, and of the principles which they followed in applying those inventions to useful practice.

At the commencement of his professional studies in the years 1805 and 1806, the author felt the want of a guide of this kind; and after carefully studying Dr. Robison's article in the Encyclopædia Britannica, and M. Prony's Architecture Hydraulique, and finding them insufficient for that purpose, he determined to preserve notes of all the observations and investigations, by which he should acquire his own knowledge of the construction and principles of operation in steam-engines, and other machines, and their various applications; in the expectation that at some future period a useful publication might be formed from those materials. Professional avocations have long since occupied all the time which might have been devoted to such an object, but it has never been abandoned; and if the sale of the present work should prove sufficient to induce the publishers to undertake others of a similar nature, the author has an accumulation of notes, which in the course of years he may find time to arrange in a corresponding form with the present work.

In the year 1815 the author drew up a descriptive article on the steam-engine for Dr. Rees' Cyclopædia; but the plan of that work, and the limited number of engravings, rendered it necessary to avoid details, which must constitute the great value of a practical treatise. Since the publication of that article in 1816, the want of a correct manual has been still more felt, from the great and increasing extension of the use of steam power, and the author was advised by his friends in the profession to resume his original project; this he undertook to do in 1820, and the bulk of the historical part was written, and most of the plates were engraved by the late Mr. Lowry in the next year; but the author being obliged to reside at a distance from London, by engagements which left no leisure for this object, the impression has been carried on at intervals, and the publication has been unavoidably protracted until the present time.

To instruct students, it is necessary to state such elementary propositions in mechanics, as have a direct application to the subject of steam-engines; for this purpose a series of definitions are given in an introduction to the present volume. These definitions have been formed from a full examination of the works of the best writers on the theory of mechanics, viz., Belidor, Emerson, Smeaton, Hutton, Banks, Gregory, Robison, Young, and others. The author has endeavoured to preserve their modes of reasoning, and the mathematical accuracy of their conclusions, without employing the language of geometrical or algebraical analysis; but all quantities are represented in numbers, and their proportions established by the ordinary processes of arithmetic; this plan has been adopted in order to render the principles very apparent to those who are not accustomed to any other mode of calculus. This part of the work is intended to give practical men an exact knowledge of the true principles upon which their operations ought to be conducted; and other parts to show the means of applying those principles to their daily practice, in the construction and use of steam-engines.

To readers who are conversant with mathematical investigations, the mode of stating the propositions will appear to leave them without sufficient demonstration; but the principles which it is intended to explain and define (rather than to demonstrate) have been so well established by mathematical evidence as to leave no doubt of their truth; and such readers may be referred to the mathematical writers whose principles have been adopted.

Dr. Gregory's Treatise of Mechanics, in two volumes, octavo, contains the best collection of mathematical investigations; and the fullest and clearest exposition of the principles will be found in Dr. Young's Course of Lectures on Natural Philosophy, in two volumes, quarto; also Dr. Robison's articles on Mechanical Philosophy, in the Encyclopædia Britannica, reprinted by Dr. Brewster, in four volumes, octavo. To the young student, a previous mathematical course is strongly recommended; the best guides are, Martin's System of Mathematical Institutions, in two volumes, octavo; Dr. Hutton's Course, in three volumes, octavo; and Dr. Gregory's Mathematics for Practical Men.

One great object of the present work is to furnish practical engineers with a series of rules for calculating all proportions, and quantities, which can be required to be known for the construction and use of steam-engines; these rules have been deduced from very numerous observations made upon steam-engines and mills of all kinds and of all magnitudes. In each case the observations have been very carefully compared, and assorted in series, according to the similarity of circumstances, and then such formulæ have been deduced from them, as would give results corresponding equally well with all parts of the series. The construction of these formulæ has been a work of great labour, of which very little appears, because only the results of the investigations are retained in the form of an arithmetical rule. The greater part of these rules have been formed by the author for his own use, in professional practice; and having undergone the test of continual application, during a course of several years, and received frequent corrections, he is justified in claiming some confidence in their accuracy.

The principles which regulate the proportions of the different quantities, which are to be computed by each rule, are stated in the most concise terms which could be chosen, without using algebraical substitutions; these have been avoided throughout the work, because the methods of algebra and fluxions are only necessary to investigate the formulæ whereby computations may be performed in numbers, by the processes of ordinary arithmetic ; and it is sufficient for practical use to have rules which will give the required results.

The method of performing each calculation by the sliding rule is added, and will tend to facilitate the computations. This valuable instrument was introduced into considerable use amongst engineers by Mr. Watt, and only requires a good collection of formulæ to become of universal application. The author hopes that what he has done will contribute to extend the use of that excellent mode of computation amongst the profession.

The history of the invention of the steam-engine and its application is divided into chapters. The first of which contains an account of the various projects and attempts which were made, during the seventeenth century, to obtain a moving power from fire, and a description of the first working engine which was invented by Mr. Savery for raising water, but it never came into extensive use.

The second chapter is on the invention, principle, and construction of the fire-engine of Newcomen, which was the first engine brought into important use, and it is still very extensively employed : this subject is treated at length, and rules are laid down for the proportions of its parts.

The third chapter is on the various applications of Newcomen's invention, which were made during the first half century after its invention.

The fourth chapter is on the introduction of cast iron into the construction of machinery, and the application of the fire-engine to the manufacture of iron.

The fifth chapter is a history of the origin and progress of Mr. Watt's invention of his first steam-engine for pumping water ; with an account of that engine, and of the rules which he established for the proportions of its parts.

The sixth chapter gives an account of the first application of the steam-engine to give continuous circular motion to mills ; with a complete description of the principle, operation, and structure of Mr. Watt's rotative engine for that purpose ; and the dimensions of several standard engines made by himself, which have been in use for years, and which perform as well, as any modern engines, which depend upon the same application of steam.

The seventh chapter is a treatise on the construction, use, and application of the sliding rule, to the purposes of calculations relative to steam-engines.

The eighth chapter is a collection of rules, for calculating the proportions and dimensions for all the parts of Mr. Watt's rotative steam-engine.

The ninth chapter is in continuation of the history of the invention of the steam-engine, and describes those improvements upon Mr. Watt's engine, which were proposed and executed by his cotemporaries.

The present volume concludes at that part of the history of the invention of the steam-engine, when it had been brought to that degree of perfection, in which all its principles of action were fully developed, and realised in practice ; and although the art of constructing these machines has been very greatly improved since Mr. Watt's time, and their applications widely extended, there have been no important inventions established in practice, except high pressure engines, and particularly those of Mr. Woolf, which are still used in the same form as he first made them. The remainder of the subject will consist of technical descriptions of the structure of such steam-engines, as are now in use, and as they are made by the best engineers ; this with their applications to various purposes, and the principles which should be followed in making such applications, will form the subject of another volume.

CONTENTS.

Introduction.

THE steam-engine defined, 1; is composed of the mechanical powers; and actuates secondary machines, 2. Importance of the invention, 3; and of its various applications, 4, 5; which are enumerated, 6. Number of steam-engines in use, 7.

On steam and the manner of its application; its expansive force, 9; its condensibility by cold, 10. Action of the atmospheric pressure, 11. Successive inventions of Savery, Newcomen, Watt, and Woolf, 12.

DEFINITIONS OF MECHANICAL PRINCIPLES. Matter, body, mass, volume, 13. Weight, density, rarity, specific gravity. Motion, place, time, velocity, measure of velocity, 14. Laws of motion; inertiæ, force, equilibrium, statical and dynamic, 15. Pressure, and its measure. *Mechanical power,* and its measure, 16. Energy of moving bodies, and its measure, 17. Distinction between the measure of power and of energy, 18. Impetus, and its measure, 19. Measure of quantities of matter; power requisite to give motion to bodies, 20.

Law of accelerated motion, 21; and standard measure for it, 22. Table of the motion of falling bodies, and rules for calculating, 23. Motion of bodies rolling on inclined planes. Centre of gravity of different bodies, 24. Rectilinear motion; curvilinear motion, 25; free motion; constrained motion, 26. Circular motion; centripetal force. *Centrifugal force,* and its measure, 27. Rules for calculating it, 28. Efficient velocity of bodies moving in circles, 29. Centre of gyration,

and general rule for calculating it, 30. *Mechanical power resident in moving bodies,* and rules, 31; when moving with circular motion, 32; and rules, 33. Rules for the centre of gyration in levers, 34, 35; in wheels, &c., 36 to 38. Pendulums, 39. Centre of oscillation, 40. *Centrifugal pendulums,* 41; with rules for governors, 42 to 44.

Mechanical powers, 45. Lever, 46. Wheel and axis, 47. Pulley, 48. Inclined plane; Wedge, 49. Screw; Funicular action, 50. Hydrostatical action, 51; hydrostatic press, 52. Pneumatical action, 53. *Combinations of mechanical powers,* 54; machine, engine, mill, 55. Machinery. *Motive forces,* animals, wind, water, and steam, 56; pressure of the atmosphere; gunpowder; laws of motive forces, 57. *Resisting forces,* gravitation, cohesion, elasticity of solids and of fluids, 58. Combinations of resisting forces, 59. *Friction,* and its principles of action, 60 to 62. Preponderance, 63. Velocity of machines, 64. *Mechanical effect,* and its measure; manpower, horse-power, and horse-power in steam-engines, 65. Maximum effect of machines, 66.

ON STEAM, definition, quantity of heat, 67. *Latent heat,* 68. Standard for the quantity of heat, 70. Density of steam, 71; *Elastic force of steam,* and rule for calculating, 72. Tables of its elastic force, 73, 74. *Standard for the density of steam,* 75. Expansibility of air, rule for calculating it, 77. Rule for calculating the density of steam, 78; and table, 79. Explanatory examples, 80.

PART I.

Historical Account of the Origin and Progress of the Invention of the Steam-Engine, from the Discovery of its Principle, to its present State.

CHAP. I. ON THE ORIGIN OF MACHINES TO BE MOVED BY FIRE: Æolipile, 81. Philosophical discoveries in the 17th century, 82. Salomon de Caus on steam, with a figure, 83, 84. Giovanni Branca, and Bishop Wilkins, 85. Discovery of the atmospheric pressure, by

Galileo, Pascal, 87; Otto Guericke, and figure of his machine, 88. *The Marquis of Worcester's fire water-work,* 89, 90. Sir Samuel Morland's principles of the force of fire, 91. *Papin's inventions:* his digester, 93; his project for raising water, 94; and for transmitting force to a distance, with a figure, 95, 96; his project for using gunpowder, with a figure, 97; and for using steam, 98.

MR. SAVERY'S FIRE-ENGINE, shewn to the Royal Society in 1699, 99; his description of it, Plate I., 100; and manner of working it, 101 to 104. Another engine by Savery, 106 to 108. Safety valve invented by Papin, 108; his fire-engine, 109. Desagulier's improvement on Savery's engine, with a figure, 111; its operation, 114; and advantages, 115. *Applications of Savery's engine*, 117; improvements in its construction, 119. Savery's engine to raise water for a water-wheel, with a figure, 122. Consumption of fuel in Savery's engine, 125.

CHAP. II. NEWCOMEN'S ATMOSPHERIC FIRE-ENGINE, 1710. Principle of its operation, with a figure, 128; Description of it, with the improvements of Mr. Beighton and Mr. Smeaton, Plates II. and III., 133; its operation, 139; particulars of different parts of it, 145.

Power of the atmospheric engine, and rules for the proportions of its parts, 155. Tables of the weight of columns of water, and the pressure upon pistons, 161. Rules for proportioning pumps to the atmospheric engine, 162. Mr. Curr's table of cylinders and pumps, 164.

Performance of different atmospheric engines. Of Mr. Smeaton's experimental engine, one-horse power, 166; of his engine of 40-horse power, at Long Benton, 172. *Mr. Smeaton's table* for the proportions and parts of the atmospheric engine, 183; and explanation, 184. Directions for adjusting an engine, 187; cataract for regulating its motion, 189.

Mr. Smeaton's large atmospheric engine of 76-horse power, at Chase Water mine, 190; Plates V. and VI.; its working-gear, Plate VII., 198; adjustment of the engine, 202; weights of the parts and prices, 204.

Atmospheric engine of 63-horse power, by Mr. Curr, Plate VIII., 205. Mr. Curr's tables of the proportions for the parts of the atmospheric engine, 206; and explanation, 210.

CHAP. III. APPLICATIONS OF THE ATMOSPHERIC ENGINE TO DIFFERENT PURPOSES, 212. Description of the pit work, and pumps for draining a mine, Plate IV., 214. *Mr. Curr's tables of dimensions of the parts of pump work, and of clacks and buckets,* 220.

Application of the atmospheric engine to drain collieries, 225. Cost of the materials of an atmospheric engine, in 1727, 230; and in 1775, 232. Performance of atmospheric engines at Newcastle, in 1769, 234; and in Cornwall, 237. Mr. Smeaton's method of supplying the boilers of a fire-engine with soft water, 238.

Application of the atmospheric engine for public waterworks, 242. Engine at York Buildings waterworks, with a figure of its boiler, 244;

and of its pumps, 247. Engine at London Bridge water works, with figure, 255. Small atmospheric engines at York, and at Hull, 256. Mr. Smeaton's portable fire-engine, Plates XIV. and IX., 257.

Application of the fire-engine to drain dry docks, 263. Mr. Smeaton's engine for the docks at Cronstadt, with figure of its boiler, 264. *Application of the fire-engine to drain fen lands,* 266.

CHAP. IV. ON THE MANUFACTURE OF IRON; AND THE APPLICATION OF CAST IRON IN ENGINES AND MILL WORK, 269. Old method of making iron, 270. New method begun at Colebrook Dale, and at Carron, 273. Mr. Smeaton's blowing machine for the furnaces at Carron, by a water-wheel, 276. Small atmospheric engines and blowing machines, with water-wheels, by Mr. Smeaton, 279, 280.

Application of the atmospheric engine to blow furnaces by the direct action of a forcing pump, 281. Blowing machine, with lifting pump, 285. Water regulator for blowing machines, 286. Air-chamber for a blowing machine, 288.

Machine for boring cylinders, 290. Boring mill at Carron, by Mr. Smeaton; with figure, 291.

Application of the atmospheric engine to raise water for working mills by water-wheels, 296. An atmospheric engine and machine for drawing coals, by Mr. Smeaton, 297. Other similar machines, 305. Conclusion respecting Newcomen's fire-engine, 306.

CHAP. V. MR. WATT'S STEAM-ENGINE, 1769, Memoirs of Mr. Watt, 309. Origin of his invention, 310; his first model, 315. Mr. Watt's first patent, 316. Act of parliament to extend its term, 318. Partnership with Mr. Boulton, 319. Progress of the invention, 320. Description of Mr. Watt's first steam-engine, with a figure, 321; progressive improvements on it, 325. Mr. Wilkinson's boring machine, 326. Several engines erected by Messrs. Boulton and Watt, 329. On applying Mr. Watt's condenser to the atmospheric engine, 329.

Advantages of Mr. Watt's engine over Newcomen's, 330. Improved form of Mr. Watt's engine, with a figure, 333; further improvements, with a figure, 334. *Performance of Mr. Watt's first engines,* 336. Coal measure, 337. Mr. Boulton's rule for calculating the consumption of fuel, 338. *Mr. Watt's expansion steam-engine,* 339. Table of the operation of the expansion, 341. Dr. Robison's investigation, figure, and table, 342. Table of hyperbolic logarithms, 345. Hornblower's patent, 346. Mr. Watt's second patent, 346; and his third patent for the expansion engine, with figure, 347 to 352.

Description of Mr. Watt's single engine for draining mines, 1784, Plate X., 353; the operation, 357; and management, 361; particulars of different parts, 365. Rules for the quantity of water evaporated from the boiler ; for the surface of the boiler ; the consumption of coals ; and power of the engine ; effect of its expansion, 366. Apparatus connected with the boiler, 369 ; steam-gage, with figure, 372. *Valves*, with a figure, 373. Condensing apparatus, 374 ; rules for the quantity of injection, 376 ; barometer-gage, 378 ; air-pump, 379 ; and steam case for the cylinder, 380. Joints ; Great lever, 382.

Mr. Hornblower's double cylinder expansion engine; with a figure, 384 ; its performance, 387. Messrs. Boulton and Watt's statement to parliament, 389. Operation of the steam in Mr. Hornblower's engine, 391, compared with Mr. Watt's expansion engine, 393.

Mr. Watt's experiments on steam, 394 ; its elastic force, 395 ; figures of the apparatus, 396, 397. Tables of the elastic force of steam, 396 and 398. Tables of the elastic force of the vapours of salt and water, and of spirits of wine, 399. Quantity of water in a given volume of steam, by Mr. Watt and Mr. Smeaton, 401. Quantity of heat required to form steam, by Mr. Watt, 401 to 405.

CHAP. VI. APPLICATION OF THE STEAM-ENGINE TO TURN MILLS, WITH CIRCULAR MOTION, 406. Origin of the invention, 408. Atmospheric engine with a crank and fly-wheel, with a figure, 410 ; its operation, 413 ; calculation of the action of the fly-wheel, 414 ; and of the crank, 415. Table of the operation of a steam-engine turning a crank and fly-wheel, 417. Rule to calculate the effective leverage of a crank when in different positions, 418 ; action of the crank, 419.

Mr. Watt's second patent, for applying the steam-engine to turn mills, 423 ; his sun and planet wheels, 425. Double acting engines, 426 ; by Mr. Watt, 427. Mr. Watt's fourth patent, containing, his rotative engine with mercury, 429 ; his parallel motions with figures, 430 ; his methods of working pumps, rolling-mills, and hammers, 432 ; new working-gear, and locomotive engine, 433. First application of Mr. Watt's rotative engines at the London Breweries, 434 ; his throttle-valve and governor, with figures, 436.

Estimation of the power of steam-engines by horse-power, 438. Rules for calculating the power of mills and engines in horse-power, 440. *Mr. Watt's fifth patent for smoke-burning furnaces*, 440. First establishment of steam mills, 442 ; the Albion flour mills, and other steam mills and manufactories, 443, 444.

Mr. Watt's rotative double acting engine for turning mills, 1785, 443 ; general description of Plates XI., XII., and XIII., 445 ; action of the piston, 447 ; of the steam, 448. Moving parts of the engine, 448 ; action of the sun and planet wheels, 449 ; and the fly-wheel,

450 ; the working-gear, 451 ; and valves, 452. Particular parts of the engine, 453 ; its fixed framing, 455 ; regulation of its motion, 457. Operation of Mr. Watt's rotative engine, 459 ; blowing through ; starting ; action during the descending stroke, and during the ascending stroke, 460 ; action of the air-pump ; regulation of the motion, 461 ; to stop the motion, 462. The governor and throttle-valve, 465. Regulation of the valves and working-gear, 467 ; adjustment of the working gear ; 468 ; joints, 471. Merits of this engine, 473 ; the different sizes, 474. Management of the piston, 475 ; manner of packing the piston, 477 ; proof of the fitting of the piston and valves, 479.

Mr. Watt's Indicator, 481 ; example of the use of it, 483. Indicator proposed by Dr. Robison, 484. *Performance* of Mr. Watt's double rotative engine, according to the indicator, 486.

Dimensions of Mr. Watt's rotative engine of 10-horse power, 489. Multiplying-wheel and pinion for the fly-wheel, 492. Dimensions of a 20-horse rotative engine, 495. A 20-horse engine without sun and planet wheels, 499. A 30-horse engine, 500. A 40-horse engine, 502 ; with a rolling-mill, 505. Dimensions of Mr. Watt's 50-horse engine at the Albion mills, 509 ; its performance, 515. Another 50-horse engine, 517.

Mr. Watt's counter, to ascertain the number of strokes made by an engine, 520. Construction of the nosles of Mr. Watt's double engine, 521 ; and of the working-gear, with a figure, 524 ; other varieties of the working-gear, 527, and 529 ; weights to the working-gear, 530.

CHAP. VII. *Application of the sliding-rule for calculating the dimensions for the parts of steam-engines*, 531. Method of performing calculations by the sliding-rule, 533. Nature of logarithms, and their use, 534. Division of lines into logarithmic spaces, with a figure, 535. Gunter's scale, 536. The Soho sliding-rule, with figure. Notation on the sliding-rule, 537. Multiplication, 539. Inverted sliding-rule, with figure, 541. Division by the sliding-rule. Multiplication and division at one operation, 543. Proportion or the rule of three, 544 ; increasing and diminishing proportion, 545. Reduction of vulgar fractions to decimals, with tables, 547. Squares and square roots. Reciprocals of numbers, 548.

Uses of the line of single radius on the sliding-rule. To multiply the square of a number by another number, 549. To find the corresponding diameters and lengths of square prisms or cylinders, having the same solidity, 550. To find the areas of circles, 551. Table of the circumferences and areas of circles, 552 to 555. Table of divisors for mensuration by the sliding-rule, 556 to 559. Gage points, 559. Rules for calculating the solidities of rectangular prisms, and of pyramids or cones, 560; of frustums of square pyramids or cones, 561; of ellipsoids or spheroids, 562. Cubes and cube roots of numbers, *ibid.* Twelve rules for performing various compound calculations by the sliding-rule at single operations, 563 to 565. To find the hypothenuse of a right-angled triangle, 566. Directions to engineers for the choice of a sliding-rule, 566. A new sliding-rule, by the author, 567; table of gage-points for it, 569; table of specific gravities for it, 570. Use of the line of one and a half radius, 571. To calculate the quantity of water which will flow in a cascade over the edge of a weir, and a table, 572.

CHAP. VIII. RULES FOR CALCULATING THE DIMENSIONS OF THE PARTS OF MR. WATT'S ROTATIVE ENGINE, 573. *Table and rule for calculating the dimensions of the cylinders,* 574. Rule for the expenditure of steam by each horse power. Actual expenditure of steam, 576. Quantity of water evaporated from the boiler, 577. Rule for the dimensions of the hot water pump, 578. Modes of expressing the performance of engines in respect to fuel, and rules, 578, 579.

Dimensions for boilers of Mr. Watt's rotative engines, 579. Rule to find the surface to be exposed to the heat, 581. Dimensions of the fire-grate, 583; of the furnace and flues, 584; and of the chimney, 585. Dimensions of a boiler for a 30-horse engine, 585. Rule for computing the surfaces of boilers, 588. Dimensions of the steam pipe and rule, 589. Rules for the safety-valve and its load, 590, 591. Rules for the apertures of the induction and eduction valves, 591. Dimensions of the air-pump, and of the cold water pump, 592. Quantity of cold water required, 593. Aperture through the injection cock, 594.

Proportions for the lengths of the moving parts. Length of stroke, 594; of the great lever and its vibration, 596. Proportions of the parallel motion, 597; rule for calculating them, 598.

Rules for calculating the strength requisite in the moving parts, 599. The piston-rod, 600; links to suspend the piston, 601. Proportions and rules for screw-bolts, 602. The connecting rod, 603. Main joints, 604. Axis of the great lever, 605. Strength of gudgeons, 605; cast iron gudgeons, 606; wrought iron gudgeons, 607; axletrees of carriages, 608. Oak beams for great levers, 609; strength and stiffness of oak beams, 611. Cast iron levers, 612; strength of cast iron beams, 613.

Sun and planet wheels, 614. Pitch and strength of their teeth, 615; number of revolutions, and radial link, 617. Rotative axes, strength of necks, 618; cast iron necks, 619; wrought iron necks, 620. Multiplying wheels, 621. Strength of the teeth for wheels, and the necks for axes for ordinary mill-work, 623; for the strongest proportion, 625; for water-mills, and for watch-work, 627; for cranes, 628. Rule to proportion the necks of axes to the teeth of wheels. Wooden teeth, 629; proportions for toothed wheel-work, 630. Pitch, breadth, and length of teeth, and their form, 632. Rules for the diameters of wheels and the pitch of their teeth, 633; and Mr. Donkin's table, 635.

Proportions for fly-wheels, 636; rules for computing their effects, 638; rules for adapting the fly-wheels of engines, 640 to 642; rules for computing the weight of fly-wheels. Remarks upon the rules in Chap. VIII., 643. Conclusion respecting Mr. Watt's inventions, 645; his method of copying writings, 646; Law-suits respecting his patent right, 648. Law of patents, 649. Extent of his invention, 650. His character as an inventor, 651. On the constitution of mind which produces inventive genius and judgment, 652. Number of Mr. Watt's engines in use, in 1805, 654. Engines in France and in Russia, 655.

CHAP. IX. *Improvements on Mr. Watt's engine, by his contemporaries.* Mr. Cameron's, 656. Mr. Symington's, 657. Double acting atmospheric engines, 658; by Mr. Thomson, with a figure, 659; by Messrs. Sherratt, with a figure, 662. Mr. Cartwright's engine, with a figure, 665; his metallic piston, 668. Mr. Sadler's engine, 669. Atmospheric engines with condensers, 671. Rotative engines operating by a direct action of steam, without reciprocating pistons, Plate XV., 671; Mr. Watt's, 672; Mr. Hornblower's, 673; Mr. Murdock's, 676.

Independent engines, 677. Murdock's bell crank engine, Plate XVI., 677; his sliding valve, 679. Murray's independent engine with White's parallel motion, Plate XVII., 682; and bark mill, 687. Murray's engine with a great lever under the cistern, 688; Murray's independent engine, with box slide valve, Plate XVIII., 691.

Boulton, Watt, and Co.'s 36-horse rotative engine, with multiplying wheel, Plates XIX. and XX., 695; its working gear, with a figure, 697. 20-horse engine of the same construction, and rolling mill for sheet lead, Plate XXI., 708. 10-horse engine and horizontal sugar-mill, Plate XXII., 712. Boulton, Watt, and Co.'s single engine for waterworks, Plates XXIII. and XXIV., 715. Boulton, Watt, and Co.'s double engine for blowing furnaces for smelting iron, Plate XXV., 723.

LIST OF THE ENGRAVINGS.

PLATE I.   Mr. Savery's Patent Engine for raising water by fire, 1698, p. 99, 100. and 106.

II. and III.   Newcomen's Atmospheric Steam Engine for drawing water from Mines, as constructed by Mr. Smeaton in 1772, p. 134 to 155, and 172 to 182.

IV.   Pit Work and Pumps of a Steam Engine for draining a Colliery, p. 214.

V. VI. and VII.   Large Atmospheric Steam Engine, constructed at Chasewater mine in Cornwall, by Mr. Smeaton in 1775, p. 190. 192. 198 and 238.

VIII.   Atmospheric Steam Engine, constructed by Mr. Curr in 1790, p. 205.

IX. and XIV.   A Portable Atmospheric Engine for draining temporary excavations, proposed by Mr. Smeaton in 1765, p. 257 and 259.

X.   Mr. Watt's Single Engine for pumping water for draining Mines, as constructed by Messrs. Boulton and Watt in 1788, p. 353 to 383.

XI. XII. and XIII.   Mr. Watt's Patent Rotative Steam Engine for turning Mills, as constructed by Messrs. Boulton and Watt, Soho, from 1787 to 1800, p. 445 to 530.

XV.   Projects for Rotative Steam Engines, Mr. Watt's, 1782, p. 672; Mr. Hornblower's, 1798, p. 673; and Mr. Murdock's, 1799, p. 676.

XVI.   Independent Rotative Engines, Mr. Murdock's Bell Crank, 1802, p. 677, and Messrs. Fenton, Murray, and Wood's, 1806, p. 688.

XVII.   Independent Rotative Engine with White's Parallel Motion, by Messrs. Fenton, Murray, and Wood, 1802, p. 682, and Bark Mill, p. 687.

XVIII.   Independent Rotative Engine of six-horse power, by Messrs. Fenton, Murray, and Wood, 1810, p. 691.

XIX. and XX.   Mr. Watt's Rotative Engine of thirty-six-horse power, as constructed by Messrs. Boulton, Watt, and Co., in 1808, p. 695 to 707.

XXI.   A Twenty-horse Engine of the same construction as the last, applied to a mill for rolling sheet lead, constructed by Messrs. Loyd and Ostell in 1810, p. 708.

XXII.   A Ten-horse Independent Engine applied to a horizontal sugar mill, p. 712.

XXIII. and XXIV.   Mr. Watt's Single Engine for pumping water for the supply of towns, as constructed by Messrs. Boulton, Watt, and Co. in 1803, p. 715 to 723.

XXV.   Mr. Watt's Double Engine for blowing air for furnaces, for smelting iron, constructed by Messrs. Boulton, Watt, and Co., 1807, p. 723 to 728.

N. B.   *The Binder is to place the plates at the end of the volume.*

Introduction.

THE STEAM-ENGINE is a compound machine which exerts a moving force, and is the *primum mobile* or first moving power to give motion to other machines, mills, or engines, by which various useful operations in the arts and manufactures are performed. The steam-engine is very extensively used in Britain for all those purposes to which the power of horse-mills, wind-mills, or water-mills can be applied; and some operations which cannot be effected by those agents, are performed with facility by steam power.

The mechanical force or power of the steam-engine, is obtained from the expansion or rarefaction of water into elastic steam by heat, and, from the subsequent contraction or condensation of that steam into water, by cold. It was formerly called the fire-engine, because it is in reality actuated by the fire, which causes the water to boil and become steam.

In treating of the steam-engine as a first moving machine or power, it must be considered separately from the secondary machines which are put in motion by it. A first mover must, in all cases, be endowed with that mechanical energy or force, which will impel some secondary machine, and give it the motion and power, requisite to overcome the resistance occasioned by the operation which is to be performed by it : and here it should be remarked, that the first mover does not actually produce the force with which it operates, but is adapted to collect and concentrate the force which arises from some natural cause, so as to derive motion from that cause ; and it must be provided with parts to modify such motion and force, and transmit it in a suitable manner and direction, to the secondary machine with which it is connected.

A common hand-pump, which is fixed over a well to raise water for domestic purposes, is a familiar instance of a secondary machine and its first mover. The man who works the lever or handle of the pump by the strength of his arms is the first mover, because, by his muscular force, he gives the power and motion necessary to actuate the pump, which is only the secondary machine, though it performs the required operation of raising the water. If a steam-engine were to be applied for this purpose, it would be substituted for the man, and, instead of his muscular strength, the expansive and contractile properties of the steam of boiling water would be applied in the steam-engine, in such manner as to produce motion in its parts, and those moving parts would be adapted to communicate their motion to the handle of the pump, to elevate and depress it alternately, and raise the water.

For another instance, suppose a laborer to be employed to turn a grindstone, or malt-mill, or a turner's wheel, with continuous circular motion, he must apply the force of his hands and arms, to a winch-handle on the end of the axis of the stone or wheel, so as to turn it round, and a quicker motion may be communicated from the revolving wheel, by an endless band or strap, to a lathe or other machine : in this case, also, the man is the first mover, and the grindstone, or the mill, or lathe, is the secondary machine. If a steam-engine were to be substituted for the man, it would be applied to the handle or crank of the wheel, in such manner as to give it a continuous circular motion.

B

Men who row a boat, or horses employed to draw a carriage, and wind-mill sails or water-wheels, which give motion to the millstones of corn mills, may be cited as examples of first moving powers, actuating operative machines, which can be, and are every day, impelled by the steam-engine with equal effect.

The first moving powers must be distinguished from what are usually denominated mechanical powers, and which are described as the elements or constituent parts of all machines (a). These mechanical powers are, the lever, the pulley, the wheel and axle, the wedge, the inclined plane, the screw, and two others of modern invention; but, however these mechanical powers may be combined or varied, they merely modify the action of the moving power, and transmit its force in an efficient manner to the matter which is to be operated upon; whereas the moving power must contain some natural cause of motion and force, by which the operation is to be effected, although the agency of a mechanical power, or a combination of them, may be necessary to transmit the power, in a suitable direction, and with the requisite force and velocity to produce the required effect. For instance, in the case of the pump just mentioned, a lever is employed to transmit the force of the man's arms to the bucket of the pump, and, by the intervention of this mechanical power, the direction of the motion given by the man is reversed, and the force which he exerts is greatly increased; but the quantity of motion being diminished, just as much as the force is increased, the mechanical effect (that is, the quantity of water raised by the pump) is not augmented by the application of the lever; the man is still the first moving power; and if he were to apply his strength immediately to the bucket of the pump, and draw it up by his hands as a saw is worked, without the intervention of the lever, he could produce the same effect of raising water, though not so conveniently, or with so little fatigue, as in the ordinary mode of pumping by a lever.

The more complicated first movers as horse-wheels, wind-mills, water-wheels, and steam-engines, are machines composed of the different mechanical powers, variously combined, for the purpose of transmitting the first moving force of the animal, the wind, the water, or the steam, in a suitable manner to the secondary machines, such as pumps, grindstones, millstones, saws, and a variety of others, which perform the desired operations.

The present work is intended to treat of the steam-engine as a first moving machine, complete in itself, and adapted to produce the requisite power and motion to impel other operative machines; but, with the exception of some of the most striking examples of the application of steam power, it is not intended to enter into the description of those secondary machines, by which mechanical power is made to perform useful operations.

The steam-engine is an invention highly creditable to human genius and in-

(a) Machines are described by writers on the theory of mechanics, as organs or tools interposed between the workman, or natural agent, and the task to be accomplished, in order to facilitate the performance of that work, which, under the circumstances proposed, would have been difficult, if not impossible, without the intervention of some mechanical contrivance. Machines are thus interposed chiefly for three reasons: 1st. To accommodate the direction of the moving force, to the direction of the resistance which is to be overcome. 2d. To enable a power, which has a certain velocity, to perform work with a different velocity. 3d. To enable a natural power, having a certain intensity, to balance, and overcome another power, obstacle, or resistance, of greater intensity. Each of these purposes may be accomplished in different ways: i. e. either by machines, which perform motions about some fixed and supported centre, as the lever, the pulley, and the wheel and axle; or by machines having a fixed path or road, along which the resistance, or body to be moved, is impelled, as the inclined plane, the wedge, and the screw. Compound machines are peculiar combinations of these six mechanical powers.—*Gregory's Mechanics, II. p.* 1.

dustry, for it exhibits the most valuable application of philosophical principles to the arts of life, and has produced greater and more general changes in the practice of mechanics, than has ever been effected by any one invention recorded in history.

The axe, the saw, and other simple tools used by carpenters and smiths, as well as the spade, the plough, and the application of horses and oxen to draw burdens, were invented in such early ages, that they were considered the production of the demigods ; but for a long time after the simple implements and machines were invented, men were obliged to perform all labour by their own personal strength. The most degrading labour of hewing wood, and drawing water, fell to the lot of the slaves, whilst thrashing and grinding corn, as well as spinning and weaving, were the constant employment of the female sex. The next advance towards our present state of improvement, was the employment of horses and oxen ; According to Diodorus Siculus, Minerva was worshipped under the name of Boormia, for having first taught the yoking of oxen to a plough, and horses to the levers of mills, for grinding of corn.

The natural elements were next called in, to aid men in their labours, and the powers of wind and water were employed to turn mills. Antipater, of Thessalonica, addresses the female sex thus : " Women ! you who have hitherto been employed to grind corn, for the future let your arms rest. It is no longer for you that the birds announce, by their songs, the dawn of the morning. Ceres has ordered the river nymphs to move the heavy millstones, and to perform your labour." But these inventions, though highly useful, left much more to be accomplished, for falls of water are only to be found in certain situations, and the " wind bloweth where it listeth." It was reserved for the inventors of the steam-engine, in the eighteenth century, to discover a power that could be at the command of man; and the progress of this wonderful invention has been so rapid, that, in less than a century from its first origin, we find it applied to all purposes of art and manufacture. In one place we find the miner employing it to drain water from the deepest chasms of the earth, whilst in another it sets the wind's uncertainty at defiance, and conveys our packets across the ocean with a precision that would formerly have been deemed chimerical. The water we drink, the bread we eat, the clothes we wear, the furniture we use, nay even the paper and printing of the books and newspapers we read, are alike the produce of its versatile power.

All other inventions appear insignificant when compared with the modern steam-engine. A ship, with all her accessaries, and the extent of knowledge requisite to conduct her through a distant voyage, are such striking instances of the intellectual power of man, and of his enterprising disposition, that, if we reflect upon the number of sciences which are applied to practice in the construction and management of a vessel, we must be struck with admiration, particularly when we consider the advantages that mankind have derived from this invention, and the improvements which it has effected in the state of civilization, by uniting all the inhabitants of the globe, as it were, into one society, and enabling them to supply their mutual wants, by a general diffusion of the bounties of nature.

The steam-engine follows next to the ship in the scale of inventions, when considered in reference to its utility, and as an instance of the persevering ingenuity of man, to bend the powers of nature to his will, and employ their energies to supply his real and artificial wants; but when we consider the steam-engine as a production of genius, it must be allowed to take the lead of all other inventions.

The art of navigation, is the result of the combined ingenuity and experience of all nations, from the earliest period of history, to the present time ; and the successive and almost imperceptible improvements, by which it has arrived at its present state of perfection, have, in most instances, originated from accident, and been improved by continual practice : We do not know to whom we are indebted for the most important inventions, such as the mariner's compass, gunpowder, the telescope, &c., and many other most useful servants to human weakness and ingenuity have been first discovered by chance ; but the steam-engine is the invention of a few individuals, in the first origin it was the result of philosophical inquiry, and the production of very ingenious minds ; and almost every material alteration in its principles of action has been an application of scientific knowledge.

The natives of Britain will more readily grant this pre-eminence to the steam-engine, from the circumstance of its having been invented, and brought into general use, by their countrymen within a century ; and particularly as it has been one of the principal means of effecting those great improvements, which have taken place in all our national manufactures within the last thirty years. That amazing increase of productive industry, which has enabled us to extend our commerce to its present magnitude, could never have been effected without the aid of this new power. In fact, there is every reason to suppose that, if the steam-engine had not been brought into use, this country, instead of increasing in wealth and prosperity during the last century, would have retrograded greatly ; because the mines of coals, iron, copper, lead, and tin, which have, in all ages, formed so considerable a portion of the wealth of England, were, at the beginning of the last century, nearly exhausted and worked out to the greatest depths to which it was practicable to draw off the water by aqueducts and simple machinery ; and, without the aid of steam-engines, it is probable that fuel, timber, and all the common metals, would long since have become too scarce in England to have supplied the necessities of a numerous population.

For more than half a century after the first introduction of the steam or fire-engine, as it was then called, it was essentially a hydraulic machine, its powers being limited to the operation of raising water. It was then chiefly applied to the draining of mines, or supplying towns with water ; and, in some few instances, the water raised by a fire-engine was used to turn a water-mill, in lieu of the natural fall of a river. During the last forty years, the steam-engine has been very greatly improved, and its power is now applied to give continuous circular motion to mills of every description, and to propel vessels and carriages, in lieu of the strength of men or animals, or the natural powers of the wind, or streams of water. In consequence, the steam-engine is become a common mechanical agent for a great variety of useful purposes in the arts and manufactures, and has given rise to an entirely new system of manufacturing, by self-acting machines, in lieu of human labour. New applications of its powers are continually made with advantage ; and it is impossible to foresee to what extent the useful arts may be still further improved by its assistance.

The old fire-engine for raising water was a most valuable invention, as it afforded means of draining mines, which could not have been worked without it. That single application of steam power was productive of immense riches to this country during the whole of the last century. The modern steam-engine is still more efficacious for draining mines, and being also applied to all purposes of manufacture, it has given a magnitude of effect to the industry of the inhabitants of this country which is unknown elsewhere, and which would have been impossible without this powerful auxiliary to human labor.

The steam engine, in the hands of many of our enterprising countrymen, has been made to realize some of the Oriental fables of those beneficent and laborious genii, who, at the request of some favored mortal, would raise populous cities in the midst of deserts, excavate subterraneous palaces, create fountains and cascades of water, and transport their favorites from place to place (*b*). With the assistance of a lively imagination, and the licence usually granted to a poet in the exercise of it, the operations which are every day performed by steam power, in some of our large mines and manufactories, might be described in terms which would appear equally incredible and fabulous to those who are unacquainted with this wonderful invention. There are several instances of steam engines having a force of 120 horse-power, which is equal to the strength of more than a thousand men acting together (*c*). One of these steam giants, if we may be allowed the expression, will be found at a copper mine, laboring night and day, without any intermission, to draw a great quantity of water perpendicularly from the bottom of a pit 1200 feet deep. Let us for a moment suppose, that 1000 men were employed instead of such an engine, they could not produce the same effect as the single machine: indeed, if they attempted to climb up the ladders and carry the water in buckets, or to hand such buckets from one to another, without the aid of any other machinery, 5000 men, acting at the same time, would scarcely be sufficient. But allowing that 1000 men were so well disciplined as to apply their united strength to equal advantage with the engine, and thereby draw up the required quantity of water, they would soon be fatigued with their exertion, and other men must be employed to relieve them; for which reason three sets of men must be kept to carry on the

(*b*) An able writer in Scotland has spoken of the modern steam engine as "stupendous alike for its "force and its flexibility; for the prodigious power which it can exert, and the ease, precision, and "ductility with which its force can be varied, distributed, and applied. The trunk of an elephant, that "can pick up a pin, or rend an oak, is as nothing to it. It can engrave a seal, and crush masses of "obdurate metal before it; draw out, without breaking, a thread as fine as gossamer, and propel a "ship of war like a bubble in the water. It can embroider muslin, and forge anchors; cut steel into "ribands, and impel loaded vessels against the fury of the winds and waves.

"It would be difficult to estimate the value of the benefits which these inventions have conferred "upon the country: there is no branch of industry that has not been indebted to them; and in all "the most material, they have not only widened most magnificently the field of its exertions, but "multiplied a hundred-fold the amount of its productions. It is our improved steam-engine that "has fought the battles of Europe, and exalted and sustained, through the late tremendous contest, "the political greatness of our nation. It is the same great power which enables us to pay the interest "of our debt, and to maintain the arduous struggle in which we are still engaged, with the skill and "capital of countries less oppressed with taxation. But these are poor and narrow views of its "importance: it has increased indefinitely the mass of human comforts and enjoyments, and rendered "cheap and accessible all over the world the materials of wealth and prosperity: it has armed the "feeble hand of man, in short, with a power to which no limits can be assigned; completed the "dominion of mind, over the most refractory qualities of matter, and laid a sure foundation for "all those future miracles of mechanic power which are to aid and reward the labours of after "generations. It chiefly is to the genius of one man* that all this perfection is owing; and certainly "no man ever before bestowed such a gift on his kind. The blessing is not only universal, but "unbounded; and the fabled inventors of the plough and the loom, who were deified by the gratitude "of their rude contemporaries, conferred less important benefits on mankind than the inventors of our "present steam engine."—*Memoirs of Mr. Watt.*

(*c*) The strength of an ordinary man is stated by the best authorities to be equal to the raising of a weight of 3750 pounds avoirdupois, or 60 cubic feet of water, one foot high per minute; this force he can continue to exert during several hours per day.

Engineers estimate the force of steam-engines by a measure which they term the horse-power; it is 33 000 pounds, or 528 cubic feet of water, raised one ft. high per minute; therefore a HP. is equal to the force of $(528 \div 60 =)$ 8·8 men. At this rate, 114 HP. is equal to 1000 men.

The strongest horses cannot exert the force thus denominated horse-power for any continuance. At an average, the strength of a good horse may be taken at 22 000 lbs. or 352 cubic feet of water, raised one ft. per minute; hence a horse is equal to two-thirds of a HP. or $(352 \div 60 =)$ 5·866 men; but a horse can only work at this rate during eight hours per day, and a steam-engine will work all the twenty-four hours, if we require it.

* Mr. Watt.

work incessantly; and at least 500 more would be requisite to replace those who would be occasionally indisposed, and to give every man one day of rest in each week. Thus we have a laborious and indefatigable servant, doing as much work as 3500 men could do, and so docile, that it requires no other government or assistance than that of two men to attend and feed it occasionally with fuel. This statement is the exact truth of that which is done every day at some of the large copper mines in Cornwall, where there are instances of three and four large engines acting in concert to drain one mine, besides smaller engines to draw up the ore out of the pits.

In visiting any extensive iron work, we shall find similar engines at work to drain the water out of the mines of coal and iron ores; and another engine of 80 or 100 horse-power will be found employed in blowing air by vast bellows into furnaces, in which the iron is extracted from the ore by excessive heat. This engine will be found surrounded by several others of smaller dimensions; viz. 10, 20, 30, and 40 H.P.; some employed in drawing up the coal and ore out of the pits, others in beating out the red-hot iron by rapid blows of ponderous hammers, which would require 20 men to lift them up to make a single blow; and other engines in rolling or flattening the iron bars into long ribands or into flat plates, with as much seeming ease as the dough for pastry is extended by a rolling-pin.

In all great towns in Britain, water for domestic purposes is forced through pipes by the power of steam-engines of from 20 to 60 horse-power, and distributed by small branches, which ramify like the arteries in the human body, and extend through every street, so as to supply water in abundance to every house.

Numerous steam vessels are now navigating the rivers and coast of England and America; some of which contain steam-engines of 60, 80, 100, and 120 horse-power, acting with as much force as the united strength of 600 or 800, and even 1000 men. These engines propel the vessels much more steadily than any galleys can be rowed by slaves, and they move through the water (without any assistance from the wind) as quick as coaches can travel on the best roads.

Small steam-engines, of six and eight horse-power, are also applied instead of horses, to draw waggons upon rail-roads, for the conveyance of coals and other heavy goods. Attempts have been made to propel coaches by similar means, and recently a steam-engine has been used to break the stones for making roads.

In all our populous towns, a multitude of steam-engines, of all sizes, are continually at work for a great variety of purposes; such as pumping water, grinding corn, sawing timber and stone, rasping logwood, expressing oil from seeds, grinding cutlery, forming lead or copper into sheets or hollow pipes, fulling and scouring woollen cloth, twisting ropes and cables, drawing wire, and for every description of laborious employment. We find them also in all extensive breweries and distilleries, in tanneries, soap manufactories, iron-foundries, and in the national establishments of dock-yards and arsenals. Their number is daily increasing, and they are continually applied to new purposes. There are many extensive manufactories of these useful machines, where, under the direction of skilful mechanicians, one steam-engine is employed to assist the workmen in all the difficult and laborious operations of making other steam-engines, as well as the operative machines to be impelled by them. They are made by this powerful assistant with as much ease as watches, and finished in the most elegant manner; their parts are combined in various forms, suitable for the different purposes to which they are to be applied; and in power they are made from that of one man to that of several hundreds.

Steam-engines are still more extensively applied to perform many simple operations, which are so easily effected by the unaided strength of the hands, that at first sight it would appear absurd to call in the aid of this gigantic power; such as spinning the delicate fibres of cotton into threads, weaving those threads into cloth, then bleaching, washing, dying, and printing ornamental figures in colours upon it, and lastly glazing, pressing, and packing up the printed cloth in bales ready for the market.

Steam power is also applied for twisting silk, for spinning flax and worsted into thread, and winding sewing thread into balls; coining money, cutting diamonds, grinding plate glass, grinding spectacle glasses, making lace, printing books and newspapers, making pins, bending card wires and button shanks, dressing hats, making weavers' reeds, grinding snuff, grinding drugs of all kinds, making mustard, grinding chocolate, making lozenges, making soda-water, and even for chopping sausage-meat.

The refinement of modern invention is such, that the most trifling operations in all the manufactures of articles for clothing, whether woollen, cotton, worsted, flax, or silk, are now performed by curious machines, which are put in motion without human labor, either by the powers of water-falls or steam-engines(d). The latter power is often preferred, because a manufactory by steam power may be established in any convenient situation where fuel can be procured, and it is not liable to interruption from occasional failure of the power; but water power can only be obtained in particular situations, which are frequently unfavourable in other respects; and the supply of water is subject to diminution in dry weather, or to total stoppage in frost, and to excessive accumulation in rainy seasons, so as to suspend the works.

A steam-engine may be set up any where, and if increase of power is afterwards wanted, other engines can be added; but a water-work has its natural limits. In many districts of England, where extensive manufactories have been long established with water-power, steam-engines have since been added at every considerable work to assist in dry seasons, frosts or floods, when the water cannot act, or can only act partially. Again, natural falls of water are mostly found on rivers in the open country; but steam-engines can be placed in the centres of populous towns, where laborers are easily procured.

Steam-power is frequently preferred, as a first mover for those mills which consist of a number of small machines, each performing some delicate operation; such machines require considerable assistance from work-people to direct their actions, and supply them with the materials upon which they are to operate. As all manufactories of this nature, require many work-people, they are more advantageously carried on by steam-power in populous towns, than by water-power in the country: this is fully proved by the number of large manufactories in London, Manchester, Leeds, and Glasgow (e).

(d) M. Biot, of the French Academy of Sciences, in 1818, when narrating the voyages he had made, during the operations for determining the figure of the earth, has the following passage: " I " next visited the most industrious counties of industrious England; I saw there the powers of nature " employed in the service of man, under every supposable shape, and man himself reserved for those " operations which mind alone can direct or perform."

(e) In London there are about 290 steam-engines for water-works, small manufactories, and steamboats: they amount to about 5460 horse-power, or about equal to the strength of 48 000 men working continually.

In Manchester there are about 240 large steam-engines in manufactories, exerting about 4760 HP., or 42 000 men.

In Leeds about 130 steam-engines in manufactories, exerting about 2330 HP., or 20 500 men.

In Glasgow there are 80 or 90 steam-engines. At Birmingham, Sheffield, and particularly in all the smaller towns of Lancashire and Yorkshire, steam-mills are very numerous, and many of them are on an extensive scale.

An extensive cotton-mill is a striking instance of the application of the greatest powers to perform a prodigious quantity of light and easy work. A steam-engine of 100 horse-power, which has the strength of 880 men, gives a rapid motion to 50 000 spindles, for spinning fine cotton threads : each spindle forms a separate thread, and the whole number work together in an immense building, erected on purpose and so adapted to receive the machines, that no room is lost. Besides these spindles, the engine gives motion to an extensive suite of preparing machines which work the cotton by many successive operations, beginning with the cotton-wool in its raw and dirty state as it comes from abroad in bags, beating out the dirt, carding, or combing out and disentangling the fibres, till they are all laid straight and parallel to each other, and drawn out into long minute bands, ready to be twisted into threads by the spindles. Although spinning is not an operation of main force, the advantages of machinery in this case are still greater than in laborious work ; for if the threads were to be spun by the distaff and spindle in the simplest manner which was in use in Queen Elizabeth's time, each person could spin but one thread at a time, and the most diligent and expert spinner could not produce one-fourth as much thread as one of the spindles which are turned by this engine. Seven hundred and fifty people are sufficient to attend all the operations of such a cotton-mill ; and by the assistance of the steam-engine they will be enabled to spin as much thread as 200 000 persons could do without machinery, or one person can do as much as 266. The engine itself only requires two men to attend it, and supply it with fuel. Each spindle in a mill will produce between two and a half and three hanks (of 840 yards each) per day, which is upwards of a mile and a quarter of thread in twelve hours ; so that the 50 000 spindles will produce 62 500 miles of thread every day of twelve hours, which is more than a sufficient length to go two and a half times round the globe.

There are many instances, in the North of England, of cotton-spinning establishments of even much greater extent than this, but they usually employ two or more steam-engines of 40, 50, and 60 horse-power. A similar system of self-acting machines, moved by steam-power, is followed in the manufactories of woollen and flax, and many of those establishments are of equal magnitude with the cotton-mills.

In short, the invention and introduction of the steam-engine within the last forty years, has changed the system of industry for the useful arts as completely as the invention of gunpowder and fire-arms changed the system of warfare two centuries before.

After defining terms, and explaining certain principles of action, the present work will be divided into three parts, with an Appendix. First, *An historical account* of the origin and progress of the invention of the steam-engine, from the first discovery of its principles to its present state of perfection. Secondly, *Technical descriptions* of steam-engines now in established use, and their application to various useful purposes. Thirdly, *Theory of the principles* of action in steam-engines, the nature of combustion and fire, the properties of heat, and its operation in producing mechanical force in elastic fluids. The Appendix will be speculative, or a *record of projects* for the improvement of the steam-engine, from its origin to the present time.

ON STEAM, AND THE MANNER OF ITS APPLICATION.

THE varieties of steam-engines in common use are so numerous, and they differ so much in their construction and operation, that an inquirer who begins to examine the subject for the first time, is confused by its complexity. As an introduction to this study, a brief account must be given of the properties of steam, which is the agent employed in all cases, and then it must be shown how these properties are brought into action, in each class of steam-engines, before we can enter into the details of their construction.

The power of the steam-engine is derived from that physical property whereby water, in common with other liquids, is disposed to rarefy, dilate, or expand by heat, so as to become an elastic fluid, or vapour, which is called steam : it is also the property of that elastic steam to condense and contract in bulk by cold ; that is, by the abstraction of the heat which occasioned its expansion. In consequence of these properties, a few drops of water, rarefied by heat, will expand into, and form a sufficient volume of steam to fill a hollow vessel of considerable size ; but by the application of cold water, or in other words by abstracting and withdrawing the heat from that steam, it will suddenly condense or contract in bulk, so as to return again into the same few drops of water from which it originated.

The steam-engine derives its force from one of these two principles of expansion and condensation, or from the combination of the two. To have more precise ideas of the nature of steam, and of the manner in which it is produced by heat, we may refer to some facts with which every person is familiar.

The expansive force of steam may be shown by a common tea-kettle, which has a close lid fitted on very tight, and a spout proceeding from one side, suppose this kettle is partly filled with water, so as to cover the bottom, but not to rise up to the part from which the spout proceeds, and that it is placed over a fire and heated until the water boils. When the heat is raised a little above the boiling point, a hot vapour, which we call steam, will begin to issue from the orifice of the spout ; and, as more heat is applied, it will issue with noise and violence, in a continued stream. This vapour, or steam, is nothing more than water, rarefied by heat into an aërial and elastic fluid, and a very small quantity of water, with a sufficiency of heat, will produce abundance of such steam ; for, although the issuing current of steam is very rapid, the water in the kettle is found to waste and boil away very gradually : it has been determined by experiment, that one measure of water will form about seventeen hundred measures of steam, such as issues at the spout of the kettle.

As this steam is an exceedingly light and elastic fluid, a great quantity may be retained by force within the kettle, even two or three times as much as it would hold without any confinement ; but the steam which is thus shut up, will become very hot, and will exert a proportionate force to burst open the sides and lid of the kettle, and escape. This may be shown, by stopping up the spout of the kettle with a cork, so as to prevent the escape of the steam ; if the heat is continued, steam will still be produced from the water, and will accumulate within the kettle for some time, until there is so much confined, that it can no longer be retained by the strength of the vessel, but will force its way out, either by blowing up the lid or by driving the cork out of the spout. The force of steam, thus confined, may become so great, that it will burst the strongest vessels, if sufficient heat is applied, and continued long enough. This effect is shown by those small glass bubbles which are sold by the glass-blowers, and are called crackers : they are hollow, about the size of peas, with a small stalk to each, and contain a little water, but are hermetically sealed, so that it cannot escape ; by sticking the stem of one of these crackers into a candle, with the bulb close to the flame, the contained water is heated, and converted into steam, which will increase in force with the heat, till it bursts the glass into hundreds of pieces, with a report as loud as that of a pistol, and generally the wick of the candle is destroyed.

c

The effort which accumulated steam makes to escape from its confinement, is called the expansive force of steam; this force may be applied, either to expel and drive up water from any vessel into which the confined steam is suddenly admitted, or if the vessel is cylindrical, the steam may give motion to a moveable stopper, or piston, which is so accurately fitted to the interior capacity of the vessel, that the steam cannot escape without pushing the piston before it.

The condensation of the elastic steam may be easily shown, by a simple experiment with any hollow vessel, which is close on all sides; for instance, a small barrel or cask, with a stop-cock fixed in each end of it; let both these cocks be open, and insert the spout of one of them, into the spout of the boiling kettle, the steam therefrom will be received into the cask, and will drive out the contained air in a stream, through the cock at the other end of the cask, and the steam will take its place, until the cask is quite filled with hot steam, which will then issue in a stream at the other cock, the same as it did from the spout of the kettle in the first instance. When this has continued some time, let both cocks be shut, and let the cask be removed from the vicinity of the fire, to allow it to become cold, or this cooling may be hastened by pouring cold water on the outside of the cask; when all the heat has thus passed away from the steam, the principle which occasioned the expansion being gone, the steam contracts again into the same bulk which it occupied before it was expanded by the heat. This bulk is only about one seventeen-hundredth part of that of the steam; therefore the steam, which quite filled the cask when it was hot, condenses into a small quantity of water when it is cold, leaving the remaining capacity of the cask nearly exhausted and empty: this is called a vacuum, there being no air, and only a very small portion of rare vapour in it. In this state the surrounding air, which always presses on the outside of the cask, is no longer resisted or balanced by the steam withinside, and therefore the sides and ends of the cask must bear the whole pressure of the atmosphere or surrounding air; and, if they are not strong enough, they will be crushed inwards. The pressure will be shown by opening one of the cocks, and the air will suddenly rush into the cask, with great force and velocity.

This force can be applied to mechanical purposes in different ways. The pressure of the atmosphere, which tends to fill up the vacuum occasioned by the condensation of the steam, is capable of raising water by what is called suction, to any height less than 34 feet, as may be shown by immersing the spout of the cock of the vacuous cask in a vessel of water, before it is opened, and the instant the cock is turned, the water will be sucked into the cask with as much violence as the air was in the former instance: now, if the vessel of water were placed 33 feet below the cask, with a leaden pipe of that length joined to the spout of the cock, to convey the water up to it (supposing every part to remain air-tight), the water would still be sucked up into the cask.

The pressure of the atmosphere may also be applied to give motion to a stopper or moveable piston, which is fitted so closely into a cylinder inserted into the end of the cask, that the air cannot escape by it, or get into the cask without pushing the stopper or piston into the cylinder before it: this it will do with great force, which may be applied to mechanical purposes.

The action of the atmospheric pressure is a consequence of the weight of the air bearing upon the earth; for although air is very light, in comparison with solid substances, it is, nevertheless, matter in a fluid and elastic state, and therefore it has weight, which is very sensible. The air of the atmosphere surrounds the whole earth, and extends to a height of many miles above its surface; the weight of all this air rests upon the earth, and surrounds every thing which is within our reach, so as to pervade all such spaces as are not occupied by some other matter : for instance, a vessel, which, in common language, we call empty, is in reality full of air, and that air is not easily extracted, except by putting some other, more solid matter into the place of it ; as, when we pour water into the vessel, it will displace the air. This surrounding air presses with great force upon every thing which it touches, and in all directions, upwards and sideways equally as downwards.

We may, by mechanical means, extract the air from a hollow vessel, and leave it empty; but it requires considerable force to effect such exhaustion. This can be experienced in some degree, by applying the mouth and sucking out part of the air from a bottle ; into which, when removed from the lips, the air will return with violence and noise to fill up the exhausted space. By a pump we can exhaust the air more completely from the interior of a close vessel, so as to make a vacuum therein. As soon as any portion of the air is extracted, the external air appears to begin to press with force on the outside surface of the vessel, and tends to enter into it : in fact, it did press with equal force whilst it was full of air, but then the pressure or elastic force of the contained air, against the inside of the vessel being equal to the pressure of the external air against its outside, the opposite pressures balanced each other. The air within being removed, either wholly or in part, the pressure on the outside is no longer balanced, but tends to crush the vessel inwards and to fill up the void space.

The weight with which this pressure acts, varies somewhat with the weather, as is shown by the barometer ; but at a mean it may be taken at $14\frac{3}{4}$ pounds Avoirdupois, upon every square inch of surface. To explain this : suppose a hollow vessel to have a hole very nearly an inch square in the upper side of it, and that this hole is covered with a piece of soft wet leather exactly an inch square ; also that a thin flat plate of metal of the same size is laid over the leather ; if the vessel be completely exhausted of air, by a pump, or by condensing steam, as before described, so that there is a perfect vacuum within it, the external air will then press down the plate with a force of $14\frac{3}{4}$ lbs., because a column of air one inch square, and several miles perpendicular height from the earth's surface upwards, will be of that weight, and must be borne by the plate ; for there is nothing beneath the plate to support it. When the air is admitted freely into the vessel, the internal air will press as much upwards beneath the plate, as the external air presses downwards on the upper side, and these opposite pressures balancing each other, the plate will only be borne down by its own weight.

The weight or pressure of the atmosphere upon the surface of the earth is about equal to the pressure which would be occasioned by covering it with a shell of water of $33\frac{3}{4}$ ft. thick spread over all parts of its surface, or a shell of mercury 30 inches thick, would press with the same force.

In speaking of the atmosphere as extending many miles high above the earth's surface, it must be considered, that any given bulk of air is lighter, rarer, and thinner, in proportion as it is situated higher above the earth's surface ; because the air near the earth is pressed upon with the weight of all the air which lies above it ; and being an elastic and compressible fluid, the lower regions thereof are condensed by the weight of all the upper regions and crowded into a less space, occasioning it to be denser and heavier in the lower than in the upper regions.

The real weight of the air varies with the weather, but it may be taken at a mean, in dry weather when the barometer stands at 30 inches, and Fahrenheit's thermometer at 60 degrees, that the air, at the surface of the earth, is 830 times lighter than an equal volume of water. Now, as a cubic foot of water weighs $62\frac{1}{2}$ lbs. or 1000 oz. Avoirdupois, or 437 500 grains, a cub. ft. of air will weigh $1\frac{1}{5}$ oz. Avoirdupois, or 527 grs. and 100 cubic inches $30\frac{1}{2}$ grs.

The subject of the nature and properties of steam, will be resumed and more fully treated in its proper place; but for the present, having enumerated the principal facts, we may proceed to state the manner in which the operation of each kind of steam-engine is produced by those properties of steam already enumerated.

The first steam-engine was invented by the *Marquis of Worcester* in 1663. It operated by the expansive force of steam only, and was adapted for forcing up water in a jet or fountain. It was never brought into use.

The next inventor was *Captain Savery*, who produced an engine in 1699 for raising water by fire. It operated both by the expansion and condensation of steam, the two principles being employed alternately: the pressure of the air forced up water into a vacuous space obtained by condensation. This was a valuable invention, and was used until more perfect engines were brought forward.

The third engine is called after *Mr. Newcomen*, the person who first put it in practice, and in fact invented the engine, about the year 1710, although the first idea of such an engine, and its principle of action, had been given some years before by M. Papin. It is also called the atmospheric engine, because it operates by the atmospheric pressure, without the expansive force of steam, for the steam is only used to fill a hollow cylinder, and expel the air from it: this steam being cooled or condensed by throwing in cold water, a vacuum is produced within the cylinder, and then, the pressure of the atmosphere gives motion to a piston or plug, which is exactly fitted into the cylinder, with liberty to move freely therein, but the air cannot get into the cylinder without pushing the piston before it. Newcomen's engine was very extensively used for raising water by means of pumps; and, during half a century, his invention was a source of great wealth to this country, by draining valuable mines; it was also employed for supplying towns with water, and it is now used for turning mills with continuous circular motion; for the force which the atmosphere exerts upon the moveable piston, may be communicated by a lever, and crank, and fly-wheel, to other machines, but the former engines were only adapted to raise water. Newcomen's engine is still in use, although it has been, in a great degree, superseded by more improved steam-engines.

The modern steam-engine was invented by *Mr. Watt* in 1769. It operates by the expansive force of steam, and also by its condensation, both principles being employed at once, to give motion to a piston moveable in a cylinder. A vacuum is first obtained by the condensation of the steam, but instead of allowing the atmospheric air to press upon the piston, and give it motion, elastic steam is made to act upon the piston; hence it is really an engine moved by steam. The condensation is also performed in a vessel separate from the cylinder in a more perfect manner than in Newcomen's engine, and a greater effect is obtained. Mr. Watt brought the steam-engine to great perfection about 1775, and made it operate with a much less consumption of fuel than his predecessors. At first he only applied it to raise water, by pumps, but by a series of subsequent inventions he adapted the motion of the piston to produce *continuous circular motion*, and thereby made his engine applicable to all purposes of manufacture. He completed this grand invention in 1784; an immense number of steam-mills have since been established in this country, and their use is every day extending over all parts of the world.

Of late years *high-pressure steam-engines* have been brought into use. This kind of engine operates by the expansive force of very hot and highly elastic steam pressing against a moveable piston: the steam is not condensed to produce a vacuum, but is allowed to escape into the open air, after it has performed its office. They were first made by *Mr. Trevithick*, in 1806, but the principle of such an engine was proposed early in the last century. The high-pressure engines are simple, and, for some particular purposes, very useful, but the boilers are liable to explode by the violence of the confined steam, and in general they have not been found to answer so well as Mr. Watt's engines. High-pressure engines are in great repute in some parts of America, particularly for steam-boats : they were introduced in that country by Mr. Oliver Evans.

The last and most improved steam-engine was invented by *Mr. Woolf* in 1805, and has since been brought into very extensive use by him. It is also a high-pressure steam-engine, but the high pressure steam, after exerting its force upon a moveable piston, is used over again, to act on a second larger piston, and is then condensed to obtain a vacuum the same as in Mr. Watt's engine. The improvement is very important, as the power is obtained by this double use of the steam, with a much smaller consumption of fuel than in any other engine.

The first three inventors, the Marquis of Worcester, Savery, and Newcomen, succeeded in giving this wonderful machine great force, and, in their hands, it was capable of overcoming great resistances ; but it wanted that activity and celerity of motion which distinguishes the modern engine, its power was not so much under the controul of its attendant, and it required abundance of fuel. Mr. Watt and the modern improvers have obviated these defects, and at the same time that they have really augmented its force, they have rendered it capable of very rapid movements, and put its powers so completely under controul, that it is now the most tractable, as well as the most active, laborer we can employ.

These inventors, and many others, have conferred great benefits on mankind by their labors, and the history of their progressive discoveries merits our attention. We will, therefore, go through a detailed account of the invention of the steam-engine, from its first origin to the present time. It is not intended to enter into all the speculative projects which have occupied the minds of ingenious mechanicians, but only to describe such inventions as have been brought into real use, with occasional notice of those philosophical discoveries and mechanical projects, which have led the way to important improvements. The terms used in mechanical disquisitions must be first defined, and several established principles of action must be stated.

DEFINITIONS, AND OUTLINES OF THOSE PRINCIPLES OF MECHANICAL AND CHEMICAL ACTION WHICH GIVE EFFECT TO THE OPERATIONS OF THE STEAM-ENGINE.

MATTER is the general term for that substance of which every thing in nature is composed : its most essential mechanical qualities are, 1. Extension, or the property of occupying space ; 2. Weight, or the property of being attracted by the earth ; and 3. Inertia, or the property of passive indifference to motion or rest.

Body is any collective quantity of matter, which occupies some defined portion of space, and therefore exists in some form, figure, or shape ; bodies may either be in the solid, fluid, or aeriform state.

Mass is a body containing a determinate quantity of matter, but without regard to extension, or the space it may occupy.

Volume is the measure of extension of a body ; that is, the portion of space it occupies. *Capacity* is also a determinate portion of space, but without contemplating any matter included in that space. Volume and capacity must be determined by combining the three dimensions of length, breadth, and thickness ; thus the space of one foot in length, one foot in breadth, and one foot in thickness, is called a cubic foot, and may be used as a measure of capacity ; thus, we may say, the volume of a certain mass of cast iron, or water, or air, is equal to a cubic foot ; and that the capacity or solid content of a certain vessel is equal to a cubic foot.

WEIGHT is the result of that continual attraction of gravitation which the earth exerts to draw all bodies towards its centre. We conclude that two weights are equal, when they produce exactly the same effect in the same circumstances, as when they counterpoise one another at the extremities of a balance, whose arms are precisely alike. The *mass* or quantity of matter in any body is always estimated by its weight; for if the earth attracts two bodies equally, it is assumed that they contain equal masses or quantities of matter; for instance, $62\frac{1}{2}$ pounds of cast-iron, $62\frac{1}{2}$ lbs. of water, or $62\frac{1}{2}$ lbs. of air, are considered as equal masses of matter.

Density is a term of comparison to express the relative quantities of matter contained in different bodies, having the same volume; for instance, a cubic foot of cast-iron weighs 450 pounds, and a c. f. of water weighs $62\frac{1}{2}$ lbs.; therefore the density of cast-iron is $7\cdot2$ times the density of water.

Rarity is the reverse of density; thus water is $7\cdot2$ times rarer than cast-iron.

The specific gravity of any substance, is its density compared with that of water, or the weight of any volume thereof compared with the weight of an equal volume of water; thus, the specific gravity of cast-iron is $7\cdot2$, that of water being 1. The specific gravity of aeriform bodies, is usually stated in comparison with air, instead of water.

When we find that the several parts of a body are alike in every other respect, we may safely conclude that they are alike in weight: such bodies are said to be *homogeneous* or of uniform density, their weight being proportional to their volume; hence by ascertaining the weight of some determinate volume of a certain homogeneous substance as a standard (a cubic foot of water for instance), we may take multiples and submultiples of that volume, and thus obtain convenient measures of all masses of water from their volumes, without the necessity of a continual recurrence to the operation of weighing. And like masses of any other substance (whereof the density or specific gravity is known) may be determined by comparing their volumes with that of known masses of water.

MOTION is a compound idea, which involves the ideas of time and place. A body in motion may be described as being in the act of changing its place, or position, in respect to other bodies with which we compare it.

PLACE is that portion of space which a body occupies in respect to other bodies; the only way in which we can identify place, as a portion of space, is, by reference to some bodies occupying other adjacent portions thereof.

TIME is that portion of duration or existence, or of the succession of events, which occur during some defined motion of a body: we can best conceive portions of time as indicated by motion, or as the operation of a body passing successively through different portions of space. Thus, our standard *measures of time*, are the motions of the heavenly bodies; the period in which the earth revolves round the sun determines the year; the circuit of the moon round the earth measures the month; and the rotation of the earth on its axis marks the day; and, by the motion of pendulums and clocks, the day is artificially divided into hours, minutes, and seconds.

VELOCITY is the expression for the rapidity or celerity of motion; that is, the extent of motion, or the space, a body passes through in a given time. A body is said to move with *uniform velocity* when its rate of motion continues always the same, or when it passes through equal spaces in every succeeding equal interval of time. A body is said to move with *accelerated velocity* when the rapidity of the motion is continually increasing, so that the body moves through a greater space, at every succeeding equal interval of time. *Retarded Velocity* is the reverse of accelerated velocity, the spaces moved through being less in each succeeding equal interval of time.

The measure of Velocity is the space or distance through which a body moves uniformly in a determinate portion of time; thus we say a coach, or a steam-boat, travels with a velocity of ten miles per hour, or 880 feet per minute, or $14\frac{2}{3}$ ft. per second: these are all different terms of expression for the same velocity. In varying velocities, the rate of increase or diminution of the velocity must be expressed. When the increase or the diminution of velocity continues the same in successive equal intervals of time, such are said to be *uniformly accelerated*, or *uniformly retarded* velocities. For instance, the motion of a body falling freely by gravity, is uniformly accelerated; for it increases at the rate of $8\frac{1}{24}$ ft. during every succeeding quarter of a second of time; hence it acquires a velocity of $16\frac{1}{12}$ ft. per second when it has been falling for half a second; $24\frac{1}{8}$ ft. per second when it has fallen three-fourths of a second; $32\frac{1}{4}$ ft. per second after falling a whole second, and so on increasing $8\frac{1}{24}$ ft. at the end of each succeeding quarter-second.

THE THREE MECHANICAL AXIOMS, or LAWS OF MOTION, established by Sir Isaac Newton, are—

I. *The Law of Inertiæ.* Every body continues in its state of rest, or of uniform motion in a right line, until that state is changed by the exertion of some force.

II. *The Law of disturbing Forces.* Any change which is effected in the quiescence or motion of a body, will observe the direction of the disturbing force; and the change of quiescence or motion will be proportionate to that force.

III. *The Law of Reaction.* Reaction is in all cases equal to, and contrary to action; for the mutual actions of two bodies upon each other, are always equal, and they operate in contrary directions.

INERTIÆ is a property inherent in matter, whereby it has a tendency to continue in its existing state, whether of rest or of uniform motion, and to resist any change of that state. Thus a body being at rest, requires an exertion of force to give it motion; and a body which is moving with a certain velocity, requires an exertion of force to give it either a greater or a less velocity.

The term *vis inertiæ* was formerly used, but inertiæ is not a force, for there is no active power resident in matter, but it is absolutely passive, and indifferent to motion or rest; hence the largest bodies are put in motion by the action of the smallest forces.

FORCE is any action or cause, which can produce motion in bodies at rest, or increase, diminish, or modify the motion of moving bodies. It is any effort which one body may exert upon another, to put it in motion from a quiescent state, or to make any change, in the direction, or the velocity of its motion.

The action of a force upon a body, may be so counteracted by another equivalent force, acting in an opposite direction, that the contending forces will neutralize each other, and then the quiescence of the body will not be disturbed. A force so circumstanced is termed *a quiescent force or pressure.* And a force which gives motion to a body, is called *an impelling or moving force.*

The intimate nature of forces can only be examined by the effects they are capable of producing upon bodies, without regard to their origin. We can form no conception of a quiescent force or pressure, which is so counteracted that it does not produce motion, except by considering what motion, or change of motion, it would produce by acting upon a body, if it were not counteracted.

When a body has been put in motion by the exertion of any force, which continues to act on the moving body, it will give it a uniformly accelerated motion; such is the operation of the force of gravitation upon falling bodies. To continue the motion of a body uniformly, no exertion of force is required beyond that which was communicated to put it in motion. Bodies move uniformly, either when the action of the original impelling force has ceased to be exerted upon them; or else when some reacting force becomes operative in an opposite direction, with an intensity equivalent to that which the impelling force continues to exert; so that the acting and reacting forces neutralize each other, and cease to affect the body, which then continues its uniform motion by its own inertiæ.

EQUILIBRIUM is that state of equality and opposite action between contending forces, whereby they neutralize and counteract each other, and cease to affect the quiescence or motion of the bodies upon which they act. Equilibrium may be distinguished into two kinds; that of quiescent bodies called statical equilibrium; and that of bodies moving uniformly, called dynamic equilibrium.

Statical equilibrium is the equality of all the opposing forces which act upon bodies, when they are at rest; for if a body is urged in opposite directions, by different forces which are equal to each other, it will have no tendency to obey either, or to move in any direction; such is the state of a scale beam, when each scale is equally loaded. And such is the state of bodies which are urged by some force tending to produce motion, but are retained at rest, by the *passive resistance or strength of bodies;* that is the resistance which solid matter opposes to change of form.

Dynamic equilibrium is the equality of all the impelling and resisting forces, which act upon bodies when in motion; so that they will preserve an uniform velocity by their own inertiæ, neither being accelerated by any predominance of the impelling force, nor retarded by any excess of the opposing force. Uniform motion can only result from such a state of dynamic equilibrium.

PRESSURE, *or quiescent force*, is any force exerting a continual effort to move a body which is withheld from motion, either by an equivalent reacting force; or by the passive resistance of other bodies with which it is in contact.

* c 4

Measure of quiescent force or pressure. The force of gravitation is made the standard by which all other forces are compared, and their proportions represented by numbers. To measure an impelling force which operates upon a body, to give it motion, a weight must be applied in an opposite direction, to counteract the force, and prevent any motion ; and then that weight becomes the measure of the force or pressure in question. By weighing any heavy body in a balance, we ascertain what pressure it will exert upon the supports which are to prevent its motion by their passive strength.

The pressure exerted by elastic steam, when it is confined in a close vessel, may be measured by applying a loose cover over a hole in the vessel, and loading as much weight upon the cover, as will hold it down, and stop the steam, so as to prevent its escape ; that weight added to the pressure exerted by the atmospheric air upon the top of the loose cover (see p. 11) will measure the pressure that the confined steam exerts against so much of the interior surface of the vessel.

In this work the pound avoirdupois will be commonly used for the unit of weight by which forces are to be measured ; and in some cases the cubic foot of water, which weighs $62\frac{1}{2}$ pounds.

MECHANICAL POWER, *or power*, is an expression for the quantity of forcible motion which is communicated to bodies by the active operation of any impelling force upon them ; being the compound of the force exerted upon bodies, and the motion produced by that exertion.

A moving body which is impelled by one force, and at the same time urged in a contrary direction by an equivalent resisting force, will by its own inertiæ continue to move uniformly, with the velocity originally given to it ; because all the impelling force that is exerted upon the body, will be neutralized and expended in overcoming the resisting force. This is the state of dynamic equilibrium.

Whenever the impelling force which urges a body, exceeds the resisting force, its motion will accelerate ; or if the resisting force exceeds the impelling force, the motion will be retarded. But so long as the motion continues to be uniform, the impelling and resisting forces must be exactly equivalent to each other. The circumstances which attend uniform motion must be carefully distinguished from those which occasion accelerated or retarded motion.

Measure of mechanical power or forcible motion, is the product which is obtained by multiplying the impelling force, by the space through which that force acts. Or the product of the resisting force, into the space through which it acts, will give the same result. It is most convenient to estimate the impelling force in pounds, and the space through which it acts in feet.

Example. Suppose a horse draws a rope which passes over a pulley, to suspend a bucket of water, of 80 pounds weight, in a well $137\frac{1}{2}$ feet deep. The mechanical power exerted to draw up the bucket may be expressed by (80 lbs. × 137·5 ft. =) 11 000 pounds raised one foot high. Or if the bucket had been 110 pounds weight, and was raised 200 feet, the mechanical power would be (110 lbs. × 200 feet =) 22 000 pounds raised one foot high ; that is double the former exertion of power. The friction of the pulley, and the bending of the rope over it, tends to increase the resistance that the horse must overcome, beyond the mere weight raised.

The mechanical power exerted by air, or steam, when acting against a moveable piston which is fitted into a cylinder, may be determined by applying such a counteracting weight to the piston, as will prevent its motion ; and the mechanical power exerted during motion, will be represented by the product obtained by multiplying the equivalent weight, by the space through which the piston moves.

Example. Suppose a piston is fitted into a cylinder 28 inches diameter, and 6 feet long ; let the interior be exhausted, by filling it with steam to displace the common air, and condensing that steam by cold water (see p. 10) ; the external air will then press against the piston, to force it into the exhausted cylinder. If the pressure against the piston is 44 000 pounds, and it moves 5 feet ; then the mechanical power exerted will be (44 000 lbs. × 5 feet =) 22 000 lbs. raised one foot. *Note.* If the pressure against the piston varies during its motion, an average of the variations must be taken.

Mechanical power may be estimated by the product obtained by multiplying the force exerted, into the uniform velocity with which it acts, during a given time ; but this expression must be restricted to the case of uniform motion. The time which is occupied in the exertion, is not a necessary particular to determine the amount of mechanical power, except when the velocity of action is taken to represent the space through which the force acts. For instance, when a horse, or the piston of a steam-engine, exerts a power equal to that of raising 22 000 lbs. one foot high, that power will be the same, whether it was exerted in one minute, or in five seconds, or in any other space of time.

If we are considering the mechanical power exerted, it is best to estimate it according to the vertical height in feet, through which a given weight in pounds descends ; or by the height that the weight is raised, if we consider the resistance overcome. For gravitation in a vertical direction is a determinate force, provided the descent, or ascent, is performed with slow and uniform motion. But the mechanical power exerted by a moving mass is the same, whether it moves through a vertical or a horizontal space, provided that its motion is produced in opposition to a counteracting force, as stated in the definition of mechanical power. Uniformity of motion is a proof that such counteracting force, is exactly equivalent to the mechanical power, or that the motive force which is the cause of the motion, and the force of the resistance overcome, which is the effect of such motion, bear an uniform relation to each other. This is the state of dynamic equilibrium. But variable motion ensues whenever the relation between them changes.

When any sensible change of velocity takes place, the operation of the inertiæ of the moving body, to resist change of motion, must be taken into the account; because mechanical power is always concerned in such changes, either as the cause, or as the effect thereof. The power communicated to a body to produce its motion, or an increase of its motion, is called its Energy; or when the motion of a body is destroyed, or impeded, it restores that power, by what is called its Impetus.

In all cases of variable motion, the energy or impetus of the moving matter must be considered, but the mechanical power which is exerted, may still be measured by the space actually described by the moving body, during some interval of time which we adopt for our observation, although it is usually by the velocity that our observations are taken.

Example. Suppose that a mass of 88 cub. ft. of water, weighing 5500 lbs. moves with an uniformly accelerating motion, which is at the rate of 40 ft. per minute, at the commencement of the interval during which we observe it, and at the rate of 80 ft. per minute at the end of that interval, which we will suppose to be one minute. Now, the space passed over in the minute will be 60 ft.; for it would have been 40 ft. without any acceleration, and to this, half the increase of velocity must be added. Therefore, the mechanical power really exerted during that minute is (5500 × 60 =) 330 000 lbs. moved 1 ft.

Note. In measuring the mechanical effect produced by this power, the resistance overcome, would appear to be less than 330 000, because part of the power exerted, is not expended in overcoming resistance, but is communicated to the moving mass, to increase its velocity, from 40 ft. per minute, to 80 ft. per minute. The quantity of power so communicated will be stated in another article.

ENERGY OF MOVING BODIES is the mechanical power communicated to them, to produce their motion from a state of rest, in opposition to the inertiæ of matter. The power so communicated continues to reside in such moving bodies, so long as they preserve their motion undiminished, and constitutes what is called their energy, or inherent force.

Energy cannot be an active force in any case of bodies moving with uniform velocity; for where there is no change, the inherent force or energy, can have no other operation than to continue the motion. The mechanical power originally communicated to a body, and which occasioned its motion, resides in that body in a latent or inactive state, so long as the velocity continues uniform, for it produces no effect, and undergoes no diminution: but that latent force is always ready to be called into action, and to exert mechanical power against any obstacle which occasions a diminution of the velocity. Any power so elicited from a moving body, and rendered active, is called its Impetus.

Measure of energy is the weight, or mass of moving matter, multiplied into the square of the velocity with which it moves; this product expresses the mechanical power which is accumulated in the moving body, being that which was originally communicated to it, to produce the motion which it possesses. The energy, therefore, expresses the whole power which resides in the moving mass, but which can only be rendered active by destroying all the motion, and bringing it to rest. This force is entirely latent, or inactive, in the case of uniform velocity; for it is only during some change of velocity that the impetus of bodies has any operation, or that the property of inertiæ in matter can be said to have any relation to force.

Note. In computing energy, we must take the last acquired velocity with which the body is actually moving, whether uniformly or not; for the body must necessarily have moved with acceleration, from a state of rest, to have acquired that velocity with which it is moving uniformly, in consequence of acceleration having ceased; but during the continuance of uniform motion, no further force can be required to continue the motion.

Example. The energy of a mass of one cubic foot of cast-iron, which moves with a velocity of 40 ft. per second, may be represented by 1 cub. ft. × (40 × 40 =) 1600 = 1600 energy. This is equivalent to the energy of another mass of 4 cub. ft. of cast-iron, moving with a velocity of only 20 ft. per second, for 4 cub. ft. × (20 × 20 =) 400 = 1600 energy. The same obstacle or resistance, would destroy the motion of either of these masses, and put them at rest.

D

All proportions of this kind may be found with the greatest facility by the sliding rule (*f*), with the two graduated lines usually marked C and D. Thus, upon the line C, find the number which expresses the mass, and move the slide so as to bring that number opposite to 1, on the line D. The rule being thus set, any velocity of the mass being represented by a number on the line D, the corresponding number on the line C will express the energy of the mass when moving with that velocity, thus :—

Sliding Rule. $\left\{ \begin{array}{l} \text{C Mass moving} \\ \text{D \quad 1.} \end{array} \right.$ Energy   *Example.* $\dfrac{\text{C 4 cub. ft.}}{\text{D \quad .1.}}$ 1600 energy
Velocity. 20 ft. per sec.

The energy, therefore, expresses the whole power which resides in the moving mass, but which can only be rendered active by destroying all the motion, and bringing it to rest. This force is entirely latent or inactive in the case of uniform velocity; for it is only during some change of velocity that the impetus of bodies has any operation, or that the property of inertiæ in matter, can be said to have any relation to force.

Distinction between the measure of mechanical power, and the measure of Energy. The amount of mechanical power expended in overcoming resisting force, and that which constitutes the energy of moving masses, bear very different relations to the velocity of motion; for the mechanical power exerted by moving bodies, is measured by the simple velocity, whilst their energy (which is also mechanical power) is measured by the square of that velocity; it is therefore requisite to have a clear discrimination of the cases to which these two different measures apply.

The measure to which we must in all cases refer the amount of mechanical power, is the mass multiplied into the space described, and therefore it has no direct relation to time; but for convenience, we commonly infer the space described, from the velocity of the motion. Now in some cases, the velocity indicates the rate at which mechanical power is expended, and in other cases, the velocity indicates what power is accumulated in the moving body, ready to be expended when called for.

Uniformity of motion is certain evidence that the power is really expending, and not accumulating; for whenever any accumulation takes place, the velocity must of necessity increase. In uniform motion, the space described, is in proportion to the velocity: hence, if we multiply the mass by the uniform velocity, we obtain the same result as if we multiply the mass by the space described; but the latter must always be looked to, as the real quantity involved.

From these properties of uniform motion, we are certain, that in all cases where it obtains, the mass multiplied by the simple velocity, will express the quantity of mechanical power uniformly exerted, and expended in the production of an uniform effect, of overcoming a continual succession of resistances.

The energy is that mechanical power which was originally communicated to the moving body to change its state from rest to motion, in opposition to the natural inertiæ of matter. All power thus communicated, is accumulated in the moving body, and though it is inactive, so long as the body continues to move uniformly, it must be punctually given out again, when the motion diminishes, and as the body returns from its state of uniform motion to that of rest. To measure the power thus accumulated, we must multiply the moving mass into the space through which it moved, whilst the power was actually communicated to it.

For convenience, we usually observe the ultimate velocity which the mass has acquired; and to find therefrom the space it has described, we must square that velocity, because we know that the mass moved with accelerated velocity, when describing the space in question: and the space a body passes through with accelerated velocity from rest, is as the square of velocity acquired from rest, by that acceleration. It is true, that at the time we observe the ultimate velocity, it may be uniform, and usually is so; but the real quantity involved, is the space described by the moving mass, when it was actually producing the motion from a state of rest; and it must necessarily have moved with acceleration from its original state of rest, to have acquired that uniform motion, at which we take our observation.

The velocity acquired, is certain evidence that there has been an accelerated motion; and the mechanical power communicated in producing it, must have been as the square of the velocity acquired.

From these properties of inertiæ and of accelerated velocity, we are certain, that the mass

---

(*f*) The sliding rule is an instrument of such extensive use in performing ordinary calculations, that all engineers and mechanicians should be well versed in the practice of it. A brief description of the use of the sliding rule will be given at the end of these definitions.

multiplied into the square of the last acquired velocity, will express the quantity of mechanical power which has been communicated in the first instance, to move the body, in opposition to its inertiæ, from a state of rest to that of motion, with the velocity then acquired.

Therefore the two cases of expenditure, or of accumulation of power, may always be distinguished by the nature of the motion; in bodies moving with uniform velocity, the mechanical power communicated to them must be expended, and is proportionate to the simple velocity; and in accelerated velocities, the mechanical power communicated to moving bodies must be accumulated in them, and is proportionate to the square of the last acquired velocity.

It has been already stated that the mechanical power exerted by steam, or elastic fluids, is measured by assuming, that a weight, which is equivalent to the pressure, is multiplied into the space, through which the pressure acts. But to measure the energy or power communicated to an elastic fluid to produce its motion, we must only consider the actual weight of matter contained in that fluid, and multiply it by the square of the velocity with which it moves. The weight of elastic fluids is so small, that the energy of a quantity which exerts a very great mechanical power is but trifling.

Example. The quantity of steam which must be introduced into a cylinder of 6·18 inches diameter, to move the piston through a space of 110 feet per minute is 22·9 cubic feet, and assuming that a cub. ft. of steam is only a cub. inch of water in an aeriform state, the weight of 22·9 cub. ft. will be only ·826 of a pound; therefore the energy will be only that of a small weight, moving with a velocity of 1·83 feet per second, although the mechanical power exerted by the same steam is a pressure of 300 lbs., moving with the velocity of 1·83 feet per sec. The energy of the piston and other moving parts of the apparatus, is necessarily connected with that of the steam, but is a separate consideration.

IMPETUS, *or force of Impetus*, is the mechanical power which is elicited from moving bodies, and exerted upon any obstacles which occasion a diminution of their velocity. It is a consequence of the inertiæ of moving matter, and is derived from that mechanical power which always resides in moving bodies, and which is termed energy, whilst in a latent state; but when it is brought into activity it is called impetus.

For instance, the force which is elicited from a bullet when it strikes against a fixed obstacle, and is thereby brought from rapid motion to a state of rest, is called its impetus; or the impetus of the fly-wheel of a machine is the power which it exerts against any resistance, by which its motion is impeded, and its velocity diminished. The blow of a hammer, which a blacksmith swings round his head with great velocity, is another instance of impetus.

The force of impetus is the consequence of an abstraction from the motion with which a moving body is endowed; for every loss of motion is necessarily attended with a restoration of all that mechanical power which was originally communicated to the body, to occasion the motion which it has so lost.

Measure of Impetus. The impetus of a moving body is derived from its energy, or the force inherent in it; impetus will therefore be represented by the difference between the energy before and after the diminution of velocity during which the impetus is operative.

If the motion is totally destroyed, and the body brought to rest, then the impetus will be equal to all the energy it possessed.

In most cases which occur in practical mechanics, only a part of the motion of a moving body is abstracted from it; leaving it still some motion remaining. The quantity of its energy, or the force of impetus, which will be brought into operation upon any obstacle, which occasions a certain diminution of the velocity of a moving body, will be represented, by the difference between the squares of the velocities, before and after, such diminution respectively.

Example. If a mass of one cub. ft. of cast-iron, moving with an uniform velocity of 40 ft. per second, is reduced to an uniform velocity of 20 ft. per second, that part of the energy, or the force of impetus which becomes operative to resist such change, may be represented by 1200, which is the difference between (1 cub. ft. × 40 × 40 =) 1600, the impetus with the first velocity, and (1 × 20 × 20 =) 400 the impetus with the last velocity. Now, suppose that the other mass of 4 cub. ft. which moves with a velocity of 20 ft. per second, is reduced to move only 10 ft. per second, then the impetus with the first velocity is (4 cub. ft. × 20 × 20 =) 1600, and the impetus with the last velocity is (4 × 10 × 10 =) 400, and the difference between these two is 1200. Therefore in both these cases, the change of velocity being equal to one half of all that the moving body possessed, the same quantity of power is brought into operation by the impetus, during that change.

MEASURE OF MASSES OR QUANTITIES OF MATTER. It has been already men-
D 2

tioned that masses are measured by their weights; but *inertia* is as universal a property of matter as weight, and is not subject to certain causes of variation which affect the weight of bodies; for the latter is only the operation of the earth's attraction, and depends on the form and proximity of the earth; it varies at different parts of the earth, and diminishes at a distance from its surface. The force, or mechanical power, requisite to give a body a certain velocity of motion, is also a measure of the quantity of matter it contains, and is free from the variations above-mentioned.

Suppose that a body is so suspended, as to be free to move in a certain direction, and that the resistance of the air, and of friction, and the operation of its own gravity, are rendered insensible, by the mode of its suspension; still a moving force will be required to overcome its inertiæ, and urge it into motion; but, in the case supposed, the smallest force will be sufficient, and the same force, by acting similarly upon different bodies, will give them different degrees of velocity, according to the mass of matter in each body.

A body thus circumstanced may be represented by two equal weights so suspended at the two ends of a line passing over a pulley, as to balance each other perfectly, but by adding a small weight, to one of them, they will both be put in motion. Now if several pairs of such balanced weights. be tried in succession, the mechanical power of the weight which must be added to each pair, to give them a certain velocity, by acting similarly through a certain space, will be found proportionate to the quantity of matter so put in motion.

Estimating the quantity of matter in bodies, in this way, it is found to be proportional to the weight of those bodies, when the experiment is tried at the same part of the earth's surface. Hence weight may generally be taken for a measure of the quantity of matter, as before stated.

The falling of bodies in vacuo, and the vibrations of pendulums, demonstrate the same fact, for all kinds of bodies, fall with the same velocity, when similarly circumstanced; and pendulums of the same length, vibrate alike, whatever be their weight or substance. But the same pendulum, removed to different parts of the earth, is found to vibrate differently, because the force of gravity alters, whilst inertiæ continues in all places the same.

Mechanical Power requisite to give the same Body different Velocities. Supposing a body to be, as in the preceding case, free from all resistance, except its own inertness, the smallest force applied to it, will put it in motion; and if the force is continued uniformly, the velocity of the body will increase uniformly. This is called uniformly accelerated velocity. Now the mechanical power, requisite to communicate different velocities, to the same body, will be found to be, as the squares of the velocity acquired. This is the measure of energy before stated.

A given amount of mechanical power will always produce a given velocity in the same body, however it may be applied; for whether the power is so applied to a body, as to operate thereon with great force for a short time, or with smaller force during a longer time, the same effect will be produced when the same power has been communicated to the body; because the real measure of mechanical power, depends upon the space through which a force acts, and not upon the time, or manner of its action.

When the power is so applied as to act on the body, with an uniform force, but with an increasing velocity, the velocity communicated to the body will be simply as the time during which the force has so operated; but the quantity of power communicated, will be as the square of the time, because of the accelerated velocity with which the force has acted; for the space through which a body passes whilst its motion is uniformly accelerating, is always as the square of the time.

Note. This measure by the square of the velocity, only applies to the energy acquired by means of accelerated motion, in cases when the only resistance to the motive power is the inertia of matter; in such cases, all the power exerted, will be found accumulated in the energy which the moving body possesses.

But when the motion of a body is resisted by any opposing force, which the mechanical power must overcome, the motion will become uniform, after it has once been accumulated by acceleration, to a certain point, at which the dynamic equilibrium obtains, and all the further exertion of power will be really expended to overcome the resisting force.

In cases of uniform motion, there can be no accumulation of power going on, as in the case of acceleration. Hence, though the mechanical power communicated, to generate motion by acceleration, is as *the square of the velocity*, generated (or rather accumulated), the mechanical power expended to overcome resistance, and continue any motion uniformly, after it is generated, will be only *as the simple velocity* so continued, in opposition to resisting force.

In the case of velocity generated by acceleration, the mechanical power communicated, will be found accumulated in the energy of the moving body, and is there in reserve, ready to be faithfully accounted for. But in the case of motion continued uniformly, there can be no accumulation; for the mechanical power has been regularly expended in overcoming the resisting force.

LAW OF ACCELERATED MOTION IN FALLING BODIES. The force of gravitation being made the standard for comparing and measuring other forces, it is important to state the quantity of motion which it will generate in bodies.

The force of gravitation is shown by that constant effort, which all bodies exert to approach towards the centre of the earth. This force is greatest at the surface of the earth; for either above, or beneath, that surface it diminishes; and it varies at different parts of the surface of the earth.

As we ascend above the surface of the earth, the force of its attraction diminishes, as the square of the distance from the centre; so that a body situated at the surface of the earth, which is about 3960 English miles distant from its centre, will be attracted with four times the force, that it would be, if removed to a distance of 3960 miles from the earth's surface, or 7920 miles from its centre.

The force of attraction on bodies situated within the earth, and beneath its surface, is simply, or directly, proportionate to the distance from the centre. Therefore, at a distance of 1980 miles from the centre, the force of attraction would be only half as great, as at the surface of the earth.

The force of attraction also varies at different parts of the earth's surface; being greatest at the poles, and least at the equator. This arises, partly, from the surface at the equator being more distant from the centre than the surface at the poles, and partly from the effect of centrifugal force at the equator.

In any of those situations to which we can gain access, these variations in the force of attraction, are not sufficiently great, to produce any sensible differences in the motions of falling bodies; and for practical purposes, the force of gravitation may be considered as uniform, and constantly acting.

Any body, or mass of matter, which is withheld from approaching towards the centre of the earth, will, by the force of gravitation, exert a quiescent force, or uniform pressure, upon that body which sustains it at rest, or withholds it from motion; and this force, which is called weight, is proportionate to the quantity of matter, as before stated.

A body which is not so withheld from approaching towards the centre of the earth, will be urged into rectilinear motion, by the force of gravitation, or, in common language, it will fall. If the body is quite unresisted, and at liberty to obey this force, without any opposition, except what arises from the natural inertiæ of matter, it will move in a straight line, or fall, with a velocity continually increasing, from the time it begins to move, until it reaches the earth, or until its motion is stopped by some other obstacle.

For gravitation being an uniform and constant action, continues to urge the falling body, after it has begun its motion; and whatever velocity it may have already acquired by falling, the force of gravitation will continually give it a greater velocity as it falls farther.

The motion of a falling body is said to be *uniformly accelerated*, for in equal portions of time, equal increments of velocity are gained; consequently the velocity acquired, is proportionate to the time it has been so increasing from the beginning of the motion; but the space described during such increasing velocity, being the product of both time and velocity, must be as the square of the time from the beginning, or as the square of the last acquired velocity; Thus,

The velocity acquired by a falling body is proportionate to
$\Big\{$ The time expended in falling from the beginning of the motion; or to
The square root of the height through which it has fallen.

The height or space through which it has fallen, will be proportionate to
$\Big\{$ The square of the velocity acquired by the fall; or to
The square of the time expended in the fall.

Therefore, supposing equal portions of time, to be represented by the numbers 1, 2, 3, 4, 5, &c., then the velocities which a body will have acquired by falling, will be represented by the same numbers; but the heights, or spaces through which it has descended from the point of rest, will be represented by the numbers 1, 4, 9, 16, 25, &c.

The space through which a body passes, with accelerated velocity from its state of rest, is in all cases as the square of the last acquired velocity.

The velocity which a body acquires by falling a given height, is termed the velocity due to that height; and conversely, it is called the height due to the velocity.

It has been ascertained by experiment, that a body falling freely from rest, will descend $16\frac{1}{12}$ ft. in the first second of time (g), and will then have acquired a velocity, which, being continued uniformly, will carry it through $32\frac{1}{6}$ ft. in the next second. For the height which a body will fall from rest with an uniformly accelerating motion, is, in all cases, half the space it will pass through in the same time, with the last-acquired velocity continued uniformly. The rate of increase of velocity whilst a body is falling, is $8\frac{1}{24}$ ft. in every quarter of a second, and the progressive effect of gravitation in generating motion in a falling body during the first three seconds is as follows:

(g) Experiments have been made by letting bodies fall in tall exhausted receivers; but as this method is not susceptible of great accuracy, the height which a heavy body will fall in a second, has been inferred, from the length of a pendulum vibrating once in a second.

It has been determined by Captain Kater, that, in London, a pendulum, suspended in vacuo, at the level of sea, and performing a vibration in every second of mean time, measures 39·1386 inches, by Sir George Shuckburgh's standard brass scale, in a temperature of 62 degrees of Fahrenheit's thermometer. The length of the pendulum being measured from the centre of suspension to the centre of oscillation. It may be demonstrated, that during the time in which a pendulum vibrates once, in a small circular arc, a heavy body will fall 4·9348 times the length of the pendulum, measured from its centre of suspension to its centre of oscillation.

The reason of this proportion is, that the time in which any pendulum vibrates, is to the time in which a heavy body will fall through half the length of the pendulum, as the circumference of a circle is to its diameter; or as 3·14159 to 1. Now the height that a body falls, being as the square of the time of falling, the height which a body will fall during the time of a vibration of any pendulum, is to half the length of that pendulum, as the square of the circumference of a circle is to the square of its diameter, or as 9·8696 to 1. Consequently the height fallen is to the whole length of the pendulum as (the half of 9·8696, or) 4·9348 to 1. Now, 4·9348 times 39·1386 inches is 193·141 inches, or 16·095 ft.; but it is usually stated at $16\frac{1}{12}$ feet, the decimals of an inch being neglected.

Motion of a heavy body falling in vacuo.		
Height fallen. Feet.	**Time elapsed. Seconds.**	**Velocity acquired. Feet per second.**
0	0 rest.	No velocity.
$1\frac{1}{192}$	$\frac{1}{4}$	$8\frac{1}{24}$
$4\frac{1}{48}$	$\frac{1}{2}$	$16\frac{1}{12}$
$9\frac{3}{64}$	$\frac{3}{4}$	$24\frac{1}{8}$
$16\frac{1}{12}$	$1'$	$32\frac{1}{6}$
$25\frac{25}{192}$	$1\frac{1}{4}$	$40\frac{5}{24}$
$36\frac{3}{16}$	$1\frac{1}{2}$	$48\frac{1}{4}$
$49\frac{49}{192}$	$1\frac{3}{4}$	$56\frac{7}{24}$
$64\frac{1}{3}$	2	$64\frac{1}{3}$
$81\frac{27}{64}$	$2\frac{1}{4}$	$72\frac{3}{8}$
$100\frac{25}{48}$	$2\frac{1}{2}$	$80\frac{5}{12}$
$121\frac{121}{192}$	$2\frac{3}{4}$	$88\frac{11}{24}$
$144\frac{3}{4}$	3	$96\frac{1}{2}$

The velocities of falling bodies are so frequently required to be known, that the following rules must be stated, for calculating all the cases which can occur.

I. *To find the velocity a body will acquire by falling for any given time.*

RULE. Multiply the time, in seconds, by $32\frac{1}{6}$, and it will give the velocity acquired, in feet per second.

Example.—In 4 seconds: $4 \times 32\frac{1}{6} = 128\frac{2}{3}$ feet per second velocity.

Sliding Rule $\begin{cases} \text{A } 6 & \text{Time elapsed in seconds} \\ \text{B } 193 & \text{Velocity feet per second.} \end{cases}$ *Example.* $\quad \begin{array}{ll} \text{A } 6 & \text{4 seconds elapsed} \\ \text{B } 193 & \text{128·6 ft. per second.} \end{array}$

II. *To find the velocity a body will acquire by falling any given height.*

RULE. Multiply the height, in feet, by $64\frac{1}{3}$, and the square-root of the product, will be the velocity acquired per second, in feet. Or multiply the square root of the height, by 8·021, and it will give the same result.

Example.—In falling $257\frac{1}{3}$ feet; $257\frac{1}{3} \times 64\frac{1}{3} = 16\,555\frac{1}{9}$, the square root of which is $128\frac{2}{3}$ feet, per second velocity. Or the square root of $257\frac{1}{3}$ is 16·04 which $\times$ 8·021 $= 128\frac{2}{3}$ as before.

Sliding Rule $\begin{cases} \text{C } 14 & \text{Height fallen in feet.} \\ \text{D } 30 & \text{Velocity feet per sec.} \end{cases}$ *Example.* $\quad \begin{array}{ll} \text{C } 14 & \text{257·3 feet fallen.} \\ \text{D } 30 & \text{128·6 ft. per sec.} \end{array}$

III. *To find the height a body will fall in any given time.*

RULE. Multiply the square of the time, in seconds, by $16\frac{1}{12}$, and it will give the height fallen in feet.

Example.—In 4 seconds; 4 square $= 16 \times 16\frac{1}{12} = 257\frac{1}{3}$ feet, the height fallen.

Sliding Rule $\begin{cases} \text{C } 700 & \text{Height fallen in feet.} \\ \text{D } 6·6 & \text{Time elapsed seconds.} \end{cases}$ *Example.* $\quad \begin{array}{ll} \text{C } 700 & \text{257·3 ft. fallen.} \\ \text{D } 6·6 & \text{4 seconds.} \end{array}$

NOTE—All the above rules may be used conversely; as I. To find the time a body must fall to acquire a given velocity. II. To find the height a body must fall to acquire a given velocity: and, III. To find the time a body will require to fall a given height.

The velocity with which bodies fall by gravity, is not influenced by their weight, figure, density, or size; for, as every particle of matter in a body, is equally attracted by the earth, it can make no difference whether a single particle falls by itself, or a great number of particles united into one body. As it is supposed that the bodies fall in vacuo, all resistance to their motion is removed, except the inertiæ; and it is found by experiments in letting bodies fall within the exhausted receiver of an air-pump, that the heaviest metals, and the lightest feathers, fall with exactly the same velocity. When bodies fall through the air, it causes a resistance to their motion, which is nearly as the squares of their velocity: and in addition to this, the air deducts something from the weight of the body, by buoying it upwards, in the same manner, but in a less degree, as when bodies fall in water. Therefore, when bodies fall in air, their velocities will be less than above stated, and the deficiency of velocity will depend upon their density, figure, and other circumstances.

If a body is projected perpendicularly upwards by mechanical force, with the velocity which it would have acquired by falling from any height, it will ascend with a motion uniformly decreasing in velocity, until it reaches that height from which it may be supposed to have fallen. In this case, the time and velocity of the ascending motion, is precisely the same as that of the descending motion, but the changes of velocity take place in an inverted order. This is retarded velocity.

The motions of bodies rolling down inclined planes. All the laws of falling bodies before stated are in the simple case of bodies descending vertically, or in a direct line towards the centre of the earth. Bodies descending inclined planes observe the same law of acceleration, and may all be referred to the height which a body would fall vertically in the same time, in the following manner: Suppose a vertical line let fall on the lowest end of the inclined plane, and a line being drawn from the upper end of the plane, at right angles thereto, the intersection of this last line, with the vertical line, is the height from which a body will fall vertically, to the lower end of the inclined plane, in the same time, that another body will roll down from the upper, to the lower end of the inclined plane, and the velocity acquired in both cases, will be the same.

THE CENTRE OF GRAVITY is that point, in relation to any solid body, where its whole weight may be supposed to be collected, and, if it is suspended by that point, the weight of all its parts will balance each other in all positions; for instance, in a globe of homogenous matter, the centre of gravity is in its centre; for it is evident, that if it is suspended by its centre, it will turn any way, and yet will be always in equilibrium: in a straight rod, a staff, or beam, of uniform size and substance, the centre of gravity will be in the middle of its length.

The centre of gravity is not always within the body itself; for instance, in a ring or in a hollow cylinder, globe, or cup, the centre of gravity will fall in the hollow space: all bodies, however, have a centre of gravity, and also all systems, consisting of several bodies united together, at invariable distances.

When a body performs motion about its centre of gravity, without any motion of that centre, only the parts of the body are affected, but when the centre of gravity moves, it is to be considered as an actual motion of the whole mass; for instance, if the centre of gravity of a body is immoveably supported, whilst it is turned about upon it, there can be no descent of the mass towards the earth; for, though certain parts of it have descended, other parts precisely equivalent have ascended.

In all cases of rectilinear motion, the velocity with which a body is said to move, is, in reality, the velocity with which its centre of gravity moves along the right line; because, in rectilinear motion, all parts of the body are assumed to move with the same velocity, unless the contrary is stated; as in the cases of rolling bodies, which have a motion about their own centre; and, even in those cases, the rectilinear motion of the body, is still according to the progress of the centre of gravity along the right line.

The position of the centre of gravity being known, a very great simplification of mechanical disquisitions can be made, for it enables us to state the conditions of equilibrium in whole masses of matter, with ease and simplicity, because the weight may be conceived to act wholly at the centre of gravity. A few cases will make this clear.

If the centre of gravity of a body, be supported by a force equal to the weight of the body, it will remain in equilibrium in every possible position.

If the centre of gravity be in the same vertical line as the point of support, the body will be in equilibrium. But it will not now, as before, be indifferent to any position; for, when the centre of gravity is vertically above the point of support, the smallest inclination will altogether destroy the equilibrium. This is called the *unstable equilibrium*; hence the difficulty of balancing a body on a point or edge, or an egg on its end. When the centre of gravity is vertically below the point of suspension, the equilibrium being disturbed, it will, after some oscillation, be restored : this is the case in a pendulum, or in a plumb line.

When a body rests upon a horizontal plane, if the vertical line, passing through the centre of gravity, falls within the base, the body will stand ; but if the line falls without the base, the body will fall or overset ; and, when that line does fall within the base, the farther it is within the circumference of that base, the greater will be the stability of the body in its position.

A body, whose centre of gravity is high above the base whereon it stands, is overset by a smaller .inclination, and is therefore more frequently overset by accident, than one having the same base, and whose centre is lower, or nearer to the base.

Method of determining the centre of gravity in various bodies of uniform density.

In a right line.—Suppose a single row of equal particles connected by the force of cohesion, the centre of gravity is obviously in the middle of the length.

In the surface of any triangle.—Draw a straight line from any of the angles, bisecting the opposite side, the centre of gravity will be in that line, at one-third part of its length from the base.

In the surface of a parallelogram.—The centre of gravity is the point in which the diagonals intersect one another ; or it may be found by drawing straight lines across it, bisecting the opposite sides.

In the surface of any regular polygon.—If the number of sides be even, draw two diagonals joining opposite angles ; if it be odd, draw from two of the angular points perpendiculars to the opposite sides. In either case the point of intersection of the lines so drawn, will be the centre of gravity.

In prisms and cylinders.—In these solids, the centre of gravity, is in the middle of the axis, or straight line, joining the centres of gravity of their opposite ends. For a line so situated will pass through the centres of gravity of all the laminæ (parallel to the ends), into which the solids may be conceived to be divided.

In pyramids and cones the centre of gravity is also in the axis or centre line, joining the vertex, and the centre of gravity of the base ; but this line must be divided into four equal parts, and the centre of gravity will be at one-fourth part of the whole, from the base.

Hence in all pyramids and cones, of the same vertical height, their centres of gravity will be at the same height above the base.

Every solid body bounded by plane surfaces, may be divided into pyramids, and the centre of gravity of each one, may be found by the process now described.

In a hemisphere the centre of gravity is in the axis, or line passing through the centre of the base, perpendicular to the plane thereof, and is situated at $\frac{3}{8}$ of the whole height (or radius of the sphere) above the centre of the base.

When the figure of a body or its density, is irregular, the centre of gravity may be found experimentally thus : Let the body be suspended in equilibrio, by a string attached successively to several different points ; and supposing the several directions of the string, produced through the body, the point in which they all intersect each other, is the centre of gravity.

When we merely have occasion to learn the distance of the centre of gravity, from one of the extremities of a body, it may be balanced upon any sharp edge, and the centre will then be vertically above that edge.

RECTILINEAR MOTION is when a body describes a straight line in its motion. This is the natural motion of bodies, for they always move in straight lines, unless some extraneous force acts, to draw them aside from the straight path, and then, by their inertiæ, they resist such forces as tend to divert them from the straight line.

In rectilinear motion, every part of the moving body proceeds with the same velocity, except in the cases of rolling bodies, which revolve on their own centres. When a body, after describing a given extent of rectilinear motion, returns again, upon the same right line, it is called *reciprocating motion.*

CURVILINEAR MOTION is when a body describes a curve in its motion. A moving body will, by its inertiæ, continue in motion with the same velocity, and in the same direction, until its perseverance in that state is prevented by obstacles or adverse forces. Consequently a body cannot describe a curve-line, unless it is subjected to the influence of some constantly acting force, or meets with a continued succession of obstacles, which at every instant change the direction of its motion and deflect it from the rectilinear course.

Accordingly we observe, on throwing a stone horizontally, or obliquely upwards in the air, that

E

the force of gravity acts continually, to change the direction of the motion which was originally communicated to it; but its inertiæ tending to carry it forwards in a straight line, modifies the action of gravity, so that it does not fall down all at once, but there is a perpetual conflict between the inertiæ and gravitation, whereby the direction of the motion is continually changed, and a curvilinear motion produced, its path being a portion of a parabola. Also, when we have a stone or other weight suspended in a sling, and whirl it round with a circular motion, we feel sensibly that it makes an effort to escape from the circular path in which it is compelled to move; and, accordingly, when we release the stone from the sling, it flies off in a line, which is straight at its commencement, and would continue so, if the force of gravitation did not operate to draw it continually towards the earth.

In all curvilinear motions, the different parts of the moving body must move with different velocities, accordingly as the parts are situated on the concave, or on the convex side of the curve described.

FREE MOTION is when a body is so circumstanced, as to be at liberty to move in any direction without impediment. A body absolutely free being once put in motion, could only move in a straight line, and with uniform velocity; but all bodies with which we are acquainted, are influenced by certain forces, by which this natural motion is modified, and curvilinear motion, or variable velocity is produced. A body is said to have free motion, when it is detached from contact with all other bodies, and is therefore free to move in such direction and with such velocity as to maintain a constant equilibrium between its own natural tendency to uniform rectilinear motion, and all the different forces which operate upon it, to change that natural motion.

This is the case with the celestial bodies: the moon revolves round the earth, and the earth and the other planets revolve round the sun in elliptical orbits, with varying velocities, which result from the combined effect of the centripetal force of attraction, or gravitation, towards the central body, and the centrifugal force produced by the force originally communicated, which tends always to continue the motion in a right line.

Also, a projectile, or body which is thrown obliquely upwards in the air, moves from the earth, and then returns in a parabolic curve, resulting from the projectile force originally impressed upon it, combined with the attraction of the earth. But as this motion is modified by the resistance of the air through which it passes, such projectiles are not so free as the planetary bodies, because they are not completely free from contact with other bodies. A body falling perpendicularly towards the earth in a vacuous space, is an instance of free motion.

CONSTRAINED MOTION is when a body is compelled, by actual contact with other bodies, to move in any particular direction, or with any stated velocity. For instance, in a wheel which is fixed upon an axis, and turned round by mechanical power, every part is constrained to revolve in circles, described about the common centre; and the parts which are at different distances from that centre must move with different velocities: and although the matter thus constrained to move in circles may be solicited by the force of gravity, and by its inertiæ and centrifugal force, to quit the circular path, it is not at liberty to do so.

All the operations of machines belong to the class of constrained motion; and it is only in the case of bodies falling in vacuo, and in the motions of the heavenly bodies, that instances of free motion can be found. It is, therefore, to the consideration of constrained motion that our attention must be particularly directed: nevertheless, the preceding definitions relate entirely to free motion, because the laws of motion in bodies so circumstanced, as to freely obey their influence, must be our standard; which being established, we may consider the effect of the particular constraint under which the artificial motions of bodies are performed.

We find frequent instances of moving bodies which are nearly free in some respects, and nearly or quite constrained in others. For instance: a ship sailing on the ocean, is nearly free to take any direction and velocity of horizontal motion, which results from the force of the wind and the resistance of the water, but it is so far constrained by its own gravity and its contact with the water, that it cannot be said to be free to move upwards or downwards. A spherical ball, rolling upon a horizontal plane, is a still more obvious case; for by its contact with the plane, it is absolutely constrained to horizontal motion, though it is free to move in any direction horizontally: or a weight, suspended by a string, is constrained to move in the surface of a portion of sphere, whereof the point of suspension is the centre, but it is free to vibrate in any direction in that surface.

CIRCULAR MOTION is that particular case of curvilinear motion, when a body is constrained to describe a circle in its motion. If it continues to move forwards, so as to describe the entire circle, it is called *rotatory motion*, or continuous circular motion, and the body is said to *revolve* in a circle: the centre of the circle is called the *centre of motion*, or *the axis of motion*.

In all circular motions, the deflection or deviation from a rectilinear motion continues the same at every part of its course, and the revolving body must preserve a constant distance from the axis, or fixed centre of motion.

CENTRIFUGAL FORCE, in a body describing curvilinear motion, is the result of a constant tendency which the body has, to depart from the curve, and assume the rectilinear motion, which is natural to bodies. (See the law of inertiæ, p. 14.)

This is very obvious in the case of whirling a stone in a sling; for so long as we oblige the stone to travel in the circular path, it makes a considerable effort to recede, and fly off from the centre of motion. This effort is called the centrifugal force. The reason of the force is also very apparent in the case of the sling, for the instant we release the stone, and allow that centrifugal force to operate, we observe that the stone departs from the circle with rectilinear motion.

CENTRIPETAL FORCE is the force requisite to counteract centrifugal force, and retain a body in its curvilinear path; the attraction of gravitation is the centripetal force, which operates on the celestial bodies, to retain them in their orbits.

In the instance of the sling, the strength of the string counteracts the tendency of the stone to fly out from the centre. *Note.* In all cases of free circular motion, the centripetal force must be exactly equal, and contrary in its direction to the centrifugal force; for if the tendency to recede from the centre is not equal to the tendency to approach it, an invariable distance from that centre cannot be maintained. In machines circular motion is almost always of the nature of constrained motion, the revolving bodies being forcibly retained in their circular paths; consequently, centrifugal force, in such cases, is of the nature of a pressure, for it is a force, so balanced by counteracting force, that no motion is produced by it.

In all cases of constrained circular motion, the centrifugal force is not permitted to operate, to remove the body farther from the centre of motion; therefore the centrifugal force can occasion no alteration in the law of motion, or of force with which the body moves forward in its circle; and hence all the laws of rectilinear motion, apply equally to circular motion. The only difference is, that in rectilinear motion, the velocity with which the centre of gravity moves along the right line, is to be taken for the velocity of the whole mass; but in circular motion, a different centre, called the centre of gyration, must be taken for that purpose, because those parts of the mass which are farthest from the axis, or centre of motion, move with greater velocity than the parts nearer to the centre of motion.

Measure of Centrifugal Force of Bodies in Circular Motion. A body which is compelled to revolve in a circular orbit, will, by its centrifugal force, describe the largest circle it is allowed to do, because the curvature of a large circle differs less from a straight line, than that of a small circle. The law of inertiæ is more violated, when a body describes a small circle, than when it describes a larger one; hence all other circumstances of weight and velocity being alike, the centrifugal force is proportional to the circumference of the circle, in which the body revolves.

The centrifugal force of a body which revolves in any circle in a given time, is proportionate to the diameter of the circle, provided the revolution is performed in a given time.

Example. The circular rim of a fly-wheel, of 20 feet diameter, being of the same weight as that of another wheel of 10 feet diameter, and both wheels revolving in the same period, the rim of the large wheel, will have twice as much centrifugal force as that of the small one.

Sliding Rule. { A Centrifu. force. Centrifu. force. *Example.* A 2 centr. force. 1 centr. force.
B Diam. of circle. Diam. of circle. B 20 feet diam. 10 feet. diam.

Note. In estimating the velocity with which a body moves in a circular orbit, we must always take the velocity wherewith the point called its centre of gyration, moves in its circle, as the mean velocity of the whole mass; and not the centre of gravity, which only refers to rectilinear motion, without rotation.

The centrifugal force of a body moving with different velocities in the same circle, is proportional to the square of the velocity with which it moves in that circle; or, to the square of the number of revolutions performed in a given time, is the same thing.

Example. The centrifugal force of the rim of any fly-wheel will be four times as great, when it makes 40 revolutions per minute, as when it makes only 20 revolutions per minute.

Sliding Rule. $\begin{cases} C & \dfrac{\text{Centr. force.}}{\text{Revolutions.}} & \dfrac{\text{Centr. force.}}{\text{Revolutions.}} \end{cases}$ *Example.* $\dfrac{C}{D} \dfrac{\text{1 force.}}{\text{20 revo.}} \dfrac{\text{4 force.}}{\text{40 revo.}}$

General rule for finding the centrifugal force of a body of a given weight, revolving with a certain uniform velocity, in a circle of a given diameter, by the Marquis de l'Hôpital.

It is founded upon the height, from which the body must have fallen by gravity, to have acquired the velocity with which its centre of gyration moves in the circle which it describes. Then as the radius of that circle, is to double the height due to the velocity, so is the weight of the body to its centrifugal force. Hence the following rule.

RULE. Divide the velocity in feet per second, by 4·01 and the square of the quotient, is four times the height in feet, due to the velocity. Divide this quadruple height by the diameter of the circle, and the quotient is the centrifugal force when the weight of the body is 1. Consequently multiplying it by the weight of the body, gives us the actual centrifugal force.

Example. Suppose the rim of a fly-wheel which is 20 ft. diameter, moves with a velocity of $32\frac{1}{6}$ ft. per second. Now $32\cdot16 \div 4\cdot01 = 8\cdot02$, the square of which is $64\cdot33$ feet for the quadruple height, due to the velocity, and divided by 20 ft. diameter, gives $3\cdot216$ times the weight of the rim for its centrifugal force; that is, the centrifugal force is nearly $3\frac{1}{4}$ times as great as the weight of the revolving body.

The calculation may be performed by the sliding rule, at two operations, viz.

To find the height due to the stated velocity.

Sliding Rule. $\begin{cases} C & 14 & \text{Height fallen in feet.} \\ D & 30 & \text{Velocity, ft. per sec.} \end{cases}$ *Example.* $\dfrac{C}{D} \dfrac{14}{30} \dfrac{16\frac{1}{12}\text{ feet fallen.}}{32\frac{1}{6}\text{ feet per sec.}}$

To find the proportion between the weight of the body and its centrifugal force.

Sliding Rule. $\begin{cases} A & \dfrac{\text{Radius of circle.}}{\text{Double the height.}} & \dfrac{\text{Weight of body.}}{\text{Centrifugal force.}} \end{cases}$ *Example.* $\dfrac{A}{B} \dfrac{\text{10 ft. rad.}}{32\frac{1}{6}\text{ feet.}} \dfrac{\text{10 lbs. weight.}}{32\frac{1}{6}\text{ lbs. force.}}$

Another method of finding the ratio between the centrifugal force, and the weight of a body revolving in a circle, the diameter of the circle being given in feet, and the number of revolutions it makes in a minute.

RULE. Multiply the square of the number of revolutions per minute, by the diameter of the circle in feet, then divide the product by the constant number 5870, and the quotient is the centrifugal force of the body, in terms of its weight, which is supposed to be 1.

Example.—Suppose a stone of 2 lbs. weight is placed in a sling, and whirled round in a circle of 4 feet diameter, at the rate of 120 revolutions per minute, what will be the centrifugal force, or tendency to fly off from the centre?

Now, 120 revolutions per minute squared is 14 400, × 4 ft. diam. = 57 600 ÷ 5870 = 9·81, for the ratio of the centrifugal force to the weight; which latter being 2 lbs., the centrifugal force, to break the sling and escape, is 19·6 lbs.

The same question may be solved by the former rule, thus: The velocity of the stone in its circle will be (4 × 3·1416 = 12·566 ft. circumference, × 2 revolutions per second =) 25·133 ft. per second. The height due to that velocity is 9·81 ft.; and as double that height or 19·62 ft. is to 2 ft. the radius of the circle, so is the centrifugal force of the body, to its weight; that is, 9·81 times.

The calculation may be performed with great facility by the sliding rule; for, by using all the four lines, the squaring, and the multiplication and division are performed at once. The constant number must be found on the line A of the rule, and is then called a Gage point, and in this particular case, the slider of the rule must be drawn out of the groove, and replaced in a reversed position, or end for end; this state of the rule is called the inverted slide, for the line C in the slider acts inverted thus, ꓛ against A, and the line B acts inverted thus, ꓭ against D. To set the rule for this case, find the number of revolutions made per minute on the line D, and move the slider till the number on the inverted line ꓭ, expressing the diameter of the circle in feet, is brought to correspond with the former number; then find the Gage point or constant number 5870, on the line A, and the corresponding number on the inverted line ꝺ, is the centrifugal force of the body, its weight being 1. Thus,

Sliding Rule, slide inverted, all four lines.	A	˙5870 Gage point.	Example.	A	5870 Gage point.
	ꝺ	Centrifugal force.		ꝺ	9˙8 Centrifu. force.
	ꞵ	Diameter in feet.		ꞵ	4 feet diameter.
	D	Revolutions per min.		D	120 revo. per min.

When the velocity is taken by the revolutions per second, the constant number or Gage point is 1˙63.

Note. It is not absolutely necessary to use all the four lines of the rule in the above case; for if the constant number 242 is found on the line D, the number corresponding thereto, will be the same as that corresponding to the gage point 5870 on the line A; Thus,

Sliding Rule, slide inverted.	ꞵ	Diameter in feet.	Centrifugal force.	Example.	ꞵ	4 ft. diam.	9˙8
	C	Revolu. per min.	242		C	120 per min.	242

The weight of the revolving body multiplied by the centrifugal force, as determined by the preceding rules, will be the actual force or pressure which it exerts to recede from the centre of motion.

To find the rate at which a body must revolve in any circle, that its centrifugal force may be equal to its weight. Having given the diameter of the circle in feet, to find the number of revolutions it must make per minute.

Rule. Divide the constant number 5870 by the diameter of the circle in feet, and the square root of the quotient is the number of revolutions it will make per minute, when the centrifugal force is equal to the weight.

Example. In a circle 6˙5 ft. diameter, a body must revolve 30 times per minute, for 5870 ÷ 6˙5 ft. = 903, the square root of which is 30˙05 revolutions per minute.

It is done by the sliding rule with the slide inverted, by the two lines B and D; and when once set, as below, it forms a complete table of the number of turns per minute, proper for each diameter.

Sliding Rule, slide inverted.	ꞵ	9˙4 feet diameter.	Example.	ꞵ	10 feet diameter.	and	20
	D	25 revo. per min.		D	24˙2 revo. per min.		17˙1

METHOD OF ESTIMATING THE EFFICIENT VELOCITIES OF BODIES MOVING WITH CIRCULAR MOTION. In the definitions of energy and impetus (pp. 17 and 19), the moving bodies are supposed to describe right lines; but the same definitions will apply to circular motions, when we know the real velocity with which the bodies move in circles.

In rectilinear motion, all parts of the body are assumed to have equal motion, (the body being without any rotation,) and then the centre of gravity is the centre of inertiæ: but in circular motion, all those parts of the body which are at different distances from the centre of motion, must have different velocities of motion. Hence we require some means of reducing all the different velocities of the parts of the body, to one common or standard velocity, which we may take for the common velocity of the whole mass.

This is done by means of an assumed point, called the *centre of gyration*, which is the centre of inertiæ, in bodies moving in circles; in the same manner as the centre of gravity is the centre of inertiæ in bodies moving in right lines.

If a body moves about an axis, or fixed centre of motion, so that the several parts can have no other motion than in circles about that centre, then the particles by their inertiæ, resist the communication of motion to any given point, with forces which are as the weight of the particles themselves multiplied into the squares of their distances from the centre of motion. And if a body is in motion, its inertiæ acts in the same manner, to preserve or continue that motion uniformly.

CENTRE OF GYRATION IN BODIES WHICH REVOLVE IN CIRCLES, *or move about fixed centres of Motion,* is a point where the effect of the inertiæ, or the energy of the whole mass, may be supposed to be collected; for the different parts of bodies which describe circles, or move about a fixed centre, must have different velocities, according to their distances from such fixed centre. And as the energies of moving bodies is measured by the product of their masses, multiplied into the squares of their velocities, it follows, that the energies of those parts of the body which are near to the centre of motion, will be much less than that of the parts more remote therefrom.

Now, the centre of gyration is a point, into which, if all the matter of a revolving body could be collected, it would continue to have the same energy to revolve in the circle, as it had when the parts were disposed in their respective places, at different distances from the centre of motion.

Therefore, suppose a given force to act upon a body at its centre of gyration, in the direction of a tangent to the circle which that centre must describe round the fixed centre of motion, such a force will move the centre of gyration with the same velocity in its circle, as it would move an equal mass of matter in a right line, by acting at its centre of gravity; such mass being supposed to be free and unconstrained in its motion.

Consequently, if all the matter of the revolving body could be collected into its centre of gyration, the energy would be represented by multiplying the total weight of the body, into the square of the velocity with which the centre of gyration moves in the circular orbit which it describes.

The use of the centre of gyration is, to enable us to compare the energy of a body moving in a circle, with that of a body of the same weight moving in a right line, the energy of the latter being represented by the multiplication of the whole weight, into the square of the velocity with which the centre of gravity proceeds along that right line.

To find the distance of the centre of gyration of any revolving body from the centre or axis of motion. GENERAL RULE.—Multiply the weight of each particle of the body by the square of its distance from the axis, and divide the sum of all these products, by the weight of the whole mass; the square root of the quotient, will be the distance of the centre of gyration from the axis of motion.

Example. Suppose three cannon balls to be fixed on a straight lever, which is assumed to be without weight. One ball weighing 2 lbs. is fixed at a distance of 10 inches from the axis of motion; another which weighs 4 lbs. at 6 inches distance; and the third, which weighs 6 lbs. is placed at 4 inches distance; required the distance of the centre of gyration from the axis of motion.

Now, 2 lbs. × 10 inch. × 10 = 200; and 4 lbs. × 6 inch. × 6 = 144; and 6 lbs. × 4 inch. × 4 = 96. The sum of these three products is (200 + 144 + 96 =) 440; which divided by (2 + 4 + 6 =) 12 lbs. (the sum of the three weights) is 36·66; the square root of this 6·05 inch. is the distance of the centre of gyration from the axis of motion. Therefore a single ball of 12 lbs. weight placed at 6·05 inch. from the axis of motion, and revolving in the same time, would have the same impetus or energy as the three balls in their respective places.

The calculation may be performed by the sliding rule at two operations, viz.

To find the product of each weight into the square of its distance from the axis:

Sliding Rule, slide inverted.	B	Weight.	Product		Example.	B	4 lbs.	144 prod.
	D	Distance.	1			D	6 inch.	1

Then to divide the sum of the products, by the sum of the weights, and to extract the square root of the quotient which gives the distance of the centre of gyration from the axis:

Sliding Rule, slide inverted.	B	Sum of products.	Sum of weights.		Exam.	B	440 prod.	12 weights.
	D	1	Dist. of cent. of gyrat.			D	1	6·05 dist.

The distance of the centre of gyration depends upon the figure of the body or system of revolving bodies. The preceding rule will apply to all cases of bodies, whereof the weight can be ascertained as being collected in particular points; but for other bodies different rules must be used. The origin of these rules will appear from the following article.

QUANTITY OF MECHANICAL POWER REQUISITE TO COMMUNICATE DIFFERENT VELOCITIES OF MOTION TO HEAVY BODIES, which are so circumstanced as to be free from all impediment to motion, except that which arises from the inertiæ of the matter they contain.

This is shown by a body falling towards the earth by its own gravity, in a vacuous space, for then the mechanical power exerted in communicating motion, is the whole weight of the body, multiplied by the height from which it has fallen, so that the height fallen, is a measure of the mechanical power communicated, and is proportioned to the square of the velocity acquired, as has been already shown, p. 22. In speaking of the velocity acquired, it must be supposed that every particle of the body moves with the same velocity; and with this condition, the law of falling bodies applies generally to all cases of motion communicated to bodies in opposition to their inertiæ, whether the direction of their motion is free or constrained. The power so communicated is termed the energy of the moving body.

GENERAL RULE. Multiply the mass of matter moved, by the height due to the velocity it has acquired, the product is the mechanical power communicated in generating that velocity of motion in the body, and is termed the energy of the body when moving with that velocity. (h)

Example. Suppose a waggon on a railway to weigh 2500 pounds, what mechanical power must be communicated to it, to urge it from rest into motion, with a velocity of three miles per hour; that is 4·4 feet per second?

Now 4·4 feet per sec. ÷ 8·02 = ·5487, the square of which is ·301 feet, the height that a body must fall to acquire the velocity which will carry it with uniform motion through a distance of three miles per hour; therefore, the mechanical power communicated is (2500 lbs. × ·301 feet =) 752·5 lbs. descending one foot high.

Note. This exertion of power is only required to set the waggon going in the first instance, and does not include any resistance from the friction of its wheels. The inertiæ once overcome requires no further accession of power; but friction, which is continually operating to retard the motion, is a separate consideration.

The same result may be obtained by one operation with the sliding rule, thus—

Sliding Rule. $\begin{cases} C & \text{Mass moved.} & \text{Power communicated.} \\ D & 8\cdot02 & \text{Velocity, feet per sec.} \end{cases}$ *Exam.* $\begin{array}{ccc} C & 2500 \text{ lbs.} & 752\cdot5 \text{ lbs. descend 1 ft.} \\ D & 8\cdot02 & 4\cdot4 \text{ ft. per sec. veloc.} \end{array}$

Note. The power communicated, as determined by this rule, is the mass of matter, which, by descending 1 foot, with slow and uniform motion, would exert the mechanical power in question. Therefore, if the mass moved is stated in pounds weight, the result obtained, will be the number of pounds weight, which must descend 1 foot; or if the mass is given in cubic feet of water, or other matter, the result will be the number of such cubic feet, which must descend 1 foot, to generate the motion of the mass.

Example. Suppose a canal boat displaces 400 cubic feet of water, what power must be communicated to it, to urge it from rest into motion, with a velocity of three miles per hour (or 4·4 feet per second) supposing the friction and resistance of the water to be neglected; and what power must be expended to give it a velocity of 6 miles per hour, or 8·8 feet per second?

Sliding Rule. $\begin{cases} C & 400 \text{ C. F.} & 120\cdot4 \text{ C. F. descending 1 foot.} \\ D & 8\cdot02 & 4\cdot4 \text{ feet per second.} \end{cases}$ or, $\begin{array}{c} 481\cdot6 \text{ C. F. desc. 1 ft.} \\ 8\cdot8 \text{ feet per second.} \end{array}$

Answer. 120·4 cubic feet of water, descending 1 foot, would generate a velocity of 3 miles per hour; and 481·6 cub. ft., descending 1 foot, would produce a velocity of 6 miles per hour.

(h) *To find the height a body must fall to acquire a given velocity.* The second rule before given, p. 23, may be stated conversely, viz.: RULE. Divide the square of the velocity in feet per second, by 64⅓, and the quotient is the height fallen in feet. Or divide the velocity in feet per second, by 8·021, and the square of the quotient is the height fallen in feet.

Sliding Rule. $\begin{cases} C & 14 & \text{Height fallen in feet.} \\ D & 30 & \text{Velocity, feet per sec.} \end{cases}$ *Example.* $\begin{array}{ccc} C & 14 & \cdot301 \text{ feet fallen.} \\ D & 30 & 4\cdot4 \text{ feet per sec.} \end{array}$

In all cases of moving bodies having acquired any given velocity of motion, the energy or the mechanical power exerted in communicating such motion will be the same, without regard to the time or manner of its action. (See p. 20.)

For instance, whether the railway waggon, or the boat, is urged into motion from rest, by a great force acting during one second, or whether its motion is generated by a small force, producing a very gradual and imperceptible acceleration during an hour, the same quantity of mechanical power must, in either case, be communicated to produce the same velocity.

Note. In referring the quantity of mechanical power to the weight which must descend through a given height, it must be always understood that such descent is performed with slow and uniform motion.

The power thus exerted, in communicating motion, is not lost or expended, but only laid up in store in the moving body, and constitutes its energy of motion, for the moving body will restore all or part of that mechanical power by its force of impetus, when any obstacle is so opposed to the continuance of its motion, as to occasion a cessation, or diminution thereof. (This has been fully explained at pp. 18 and 19.)

The law of falling bodies is directly applicable to those cases, where all the particles of the moving body have the same velocity, and therefore includes all cases of rectilinear motion, whether free or constrained, or whether the direction is vertical or horizontal; but bodies rolling on planes must be excepted, because they have a revolving motion on their own centres, independent of their progress in the right line; and in bodies moving with curvilinear motion, the particles must move with different velocities, therefore if these differences are considerable, it must be taken into the consideration.

In cases of bodies revolving about immoveable centres or axes of motion, the different particles can have no motion except in circles described about such centres or axes, and the velocities of particles situated at different distances from the centres, must be greater or lesser according to such distances.

To apply the law of falling bodies, we must therefore imagine the particles composing such revolving bodies to be divided and collected into several small bodies, situated at different distances from the centre, and therefore moving with different velocities, and then we can determine the energy or the power which must be communicated to each of the supposed separate bodies, to give it the velocity it actually possesses, when it forms part of the entire revolving body. The sum of all the powers so determined, is the total energy, or the power which must be communicated to the entire body, to generate the motion it actually possesses.

The most simple case of revolving body is that of a straight rod of uniform size, moveable about one of its extremities as a centre, or axis of motion; therefore one extremity will not move at all when it is turned round, and the other will describe a circle.

To determine the power which must be communicated to such a rod, to give it a certain velocity of motion, suppose it to be 20 inches long, and that its weight is divided into 10 balls, or equal weights, like bullets, which are threaded upon a fine straight wire 20 inches long, like a string of beads. Suppose this wire to be inflexible, and without weight in itself, and that one end of it is the centre of motion, and that a ball is placed upon it at one inch therefrom, and then a ball at every two inches, viz. at 3, 5, 7, 9, 11, 13, 15, 17, and 19 inches, respectively from the centre.

When this assemblage of weights is whirled round like a sling, the centre of each ball will describe a circle, and the ten circles thus described will be 2, 6, 10, 14, 18, 22, 26, 30, 34, and 38 inches diameter respectively. If the assemblage of balls makes 114·6 revolutions per minute, each ball will move in its circular orbit with a velocity of 1, 3, 5, 7, 9, 11, 13, 15, 17, and 19 feet per second respectively.

The height due to the velocity with which each ball actually moves, being calculated by the preceding rule, and multiplied by the mass of matter contained in each ball, the product will be the power communicated to each ball, and the sum of these products is the mechanical power communicated to the whole assemblage, to produce the motion it actually possesses, for it is the common product of all the separate masses, into the heights due to the respective velocities with which they move.

Velocity, feet per second.	Height fallen feet.	Mass of each ball.	Mechanical power communicated.
1	·015	1	·015
3	·140	1	·140
5	·389	1	·389
7	·762	1	·762
9	1·260	1	1·260
11	1·882	1	1·882
13	2·633	1	2·663
15	3·500	1	3·500
17	4·510	1	4·510
19	5·620	1	5·620
		10	20·711

The extremity of the assemblage of balls, that is, the extremity of the wire, moves with a velocity of 20 feet per second, and the height due to that velocity is 6·22 feet: now, if we divide the sum of the products 20·711 by 6·22 ft., we shall have 3·3 of the balls (that is, one-third of the whole) for the mass, which the same power could put in motion, in a straight line, with the velocity of 20 ft. per second: from this the following conclusion is drawn.

The mechanical power which must be communicated to a straight uniform rod or lever, to put it in motion about one of its extremities, as a fixed centre or axis, is the same as that which must be communicated to one-third of its weight of matter to give it motion in a straight line, with the same velocity as the extremity of the lever moves in its circle.

RULE. Multiply one-third of the weight of the lever, by the height due to the velocity, with which the extremity of the lever moves in its circle, the product is the mechanical power communicated.

Example. Suppose a round rod or bar of iron, weighing 52·7 pounds, to be 2 inches diameter, and 5 feet long, and to project out from an axis like a lever, or like one of the arms of a fly-wheel, which makes 35 revolutions per minute. The extremity of the lever describes a circle 10 feet diameter, or (10 × 3·14 =) 31·4 ft. circumference, and it moves with a velocity of (31·4 × 35 revolutions =) 1099·5 feet per minute, or (÷ 60 =) 18·32 ft. per second. The height due to this velocity is 5·23 feet; which, multiplied by 17·57 lbs. (one-third of the weight), gives 91·8 lbs. descending 1 ft., for the mechanical power communicated (*i*).

In case the end of the lever does not reach to the axis, it must first be con-

---

(*i*) All calculations on these subjects will be facilitated by the following rules:

To find the velocity with which any revolving body moves in the circle which it describes.

RULE. Multiply the diameter of the circle in feet by the number of revolutions made per minute, and divide by 19·1, the quotient is the velocity in ft. per second.

Example. Suppose a revolving body describes a circle 10 ft. diameter, and makes 35 revolutions per minute. 10 × 35 = 350 ÷ 19·1 = 18·33 feet per second.

Sliding Rule, slide inverted.	A Diam. of circle, feet.	19·1
	Ɔ Revolutions per min.	Velocity ft. per sec.

Example.

	A	10 feet.	19·1
	Ɔ	35 rev.	18·33

To find the height due to the velocity with which a body revolves in a circle.

RULE. Multiply the diameter of the circle in feet, by the number of revolutions per minute, and divide the product by the constant number 153·2, the square of the quotient is the height in feet, due to the velocity with which the body moves in its circle.

Example. The circle in the above case is 10 feet diam. × 35 revolutions per min. = 350 ÷ 153·2 = 2·287, the square of which is 5·23 ft. for the height as before.

This cannot be done at one operation by the ordinary slide rule; but if we multiply the diameter of the circle in feet, by the number of revolutions per minute, then we may employ that product on the slide rule; viz.

Sliding Rule.	C 1	Height due to velo. feet.
	D 153·2	Diam. in ft. × revo. per min.

Exam.

	C 1	5·23 feet height.
	D 153·2	350 diam. × revo.

F

sidered as if it did reach the centre, and then the effect of the part which is wanting must be determined and deducted.

Example. Suppose an uniform rod 2·5 feet long, weighing 26·35 pounds, revolves 35 times per minute, one of its ends 5 ft. from the centre of motion, and the other 2·5 ft. The preceding case is the calculation of a rod 5 ft. long, reaching quite to the centre, and we must now calculate another of only 2·5 ft. long, weighing 26·35 lbs. The velocity of the extremity will be 9·16 ft. per second, and the height due to this velocity is 1·3075 ft.; which, multiplied by 8·783 lbs. (one-third of the weight 26·35 lbs.), gives 11·47 lbs. for the power; this being deducted from 91·8 lbs., the power of the lever 5 ft. long, leaves 80·33 lbs. descending 1 ft., for the mechanical power communicated to the rod in question.

Note. This calculation may be abridged upon the following principle.

The mechanical power which must be communicated to straight levers of different lengths, but of the same size and density, to make them revolve in the same period, about one of their extremities, as fixed centres or axes, is proportionate to the cubes of the distances of their other extremities from those axes.

To find the mechanical power which must be communicated to any central portion of an uniform lever (that is, a portion extending from the centre), having given, the power communicated to the entire lever.

RULE. Divide the length of the entire lever, by the length of the central portion, and cube the quotient; then divide the power communicated to the entire lever, by that cube, and the quotient is the power due to the central portion.

Example. 5 feet long ÷ 2·5 ft. long = 2, the cube of which is 8. The power of the entire lever of 5 ft. is 91·8 lbs., which ÷ by 8 gives 11·475 lbs. for the mechanical power which must be communicated to a rod or lever extending 2·5 ft. from the centre.

Note. All the above proportions are adapted to cases of very slender rods or bars, whereof the size is uniform, and bears but a small proportion to their length; but it may be used without sensible error, for almost all kinds of levers or arms which project from an axis or centre of motion.

In cases where the lever is larger, and heavier at one end than the other, we must recur to the supposition of an assemblage of balls, which are not of equal weight; but the mass of each ball being properly stated, and multiplied into the height due to its velocity, as in the case above, we may adapt the rule to any case.

To find the centre of gyration in a straight uniform rod or lever, we must resume the consideration of the supposed assemblage of balls above mentioned; and if we divide the sum of the products, 20·711, by 10, the sum of the masses, we shall have 2·07 feet for the space which the whole mass must have descended, to have exerted all the power communicated. The velocity due to that height is 11·55 feet per second; that is (11·55 ÷ 20 =) ·577 of the velocity (of 20 feet per second) with which the extreme end of the assemblage of balls moved in its circle; whence the following conclusion may be drawn.

The mechanical power which must be communicated to a straight uniform rod or lever, to put it in motion about one of its extremities as a fixed centre or axis, is the same as that which must be communicated to an equal weight of matter, to give it motion in a straight line, with ·57735 of the velocity with which the extremity of the lever moves in its circle.

The point in the revolving lever which moves with that velocity, is called the centre of gyration.

In any revolving lever, of uniform size, moving about one of its extremities, as a centre or axis, the centre of gyration will be at ·57735 of its whole length from the centre of motion.

RULE. Multiply the whole length of the lever by ·57735, and it will give the distance from the centre of motion to the centre of gyration; it is very nearly $\frac{5 \cdot 2}{9 \cdot 0}$ of the whole length of the lever.

Example. In a lever of 5 feet long, the centre of gyration is at (5 ft. × ·577 =) 2·88 ft. from the centre of motion.

Sliding Rule. $\left\{\begin{array}{ll} \text{A} & 45 \\ \hline \text{B} & 26 \end{array}\right.$ $\dfrac{\text{Length of the lever.}}{\text{Dist. of cent. of gyration.}}$ *Example.* $\dfrac{\text{A} \quad 45}{\text{B} \quad 26}$ $\dfrac{\text{5 ft. length.}}{\text{2·8 ft. dist.}}$

To find the centre of gyration in a straight uniform rod revolving about a centre which is at a distance beyond the end of the rod. Having given the distances of each extremity of the rod from its centre of motion; to find the distance of the centre of gyration therefrom.

RULE. Divide the difference between the cubes of the distances of the two extremities, by three times the difference between those distances, (that is, three times the length of the revolving rod,) the square root of the quotient, is the distance of the centre of gyration from the axis.

Example. Suppose a straight uniform rod 5 ft. long, one of its ends 10 ft. distant from the axis, the other 5 ft. distant; the difference between the cube of 10 ft. (or 1000) and the cube of 5 (or 125) is 875; divide this by 15, which is three times 5 ft. the length of the rod, the quotient is 58·33, and the square root of this = 7·637 ft. is the distance of the centre of gyration from the axis. The calculation may be performed by the slide rule at two operations, viz.

To find the cubes of the distances.

Sliding Rule. $\left\{\begin{array}{ll} \text{C} & \text{Distance.} \\ \hline \text{D} & 1 \end{array}\right.$ $\dfrac{\text{Cube of distance.}}{\text{Distance.}}$ *Example.* $\dfrac{\text{C} \quad \text{5 dist.}}{\text{D} \quad 1}$ $\dfrac{\text{125 cube}}{\text{5 dist.}}$

To divide the difference between the cubes of the distances by three times the difference between the simple distances, and to extract the square root of the quotient.

Sliding Rule. $\left\{\begin{array}{ll} \text{C} & \text{Three times differ. of dist.} \\ \hline \text{D} & 1 \end{array}\right.$ $\dfrac{\text{Differ. between the cubes.}}{\text{Dist. of centre of gyration.}}$ *Exam.* $\dfrac{\text{C} \quad 15}{\text{D} \quad 1}$ $\dfrac{875}{7·637}$

To find the centre of gyration in a straight uniform lever, which reaches both ways from the axis or centre of motion, so as to form a short arm at one end, and a longer arm at the other end. Having given the distance of the extremity of each arm from the axis, to find the common distance of their centres of gyration from the axis.

RULE. Divide the sum of the cubes of the distances of the two extremities, by three times the sum of those distances; the square root of the quotient, is the distance of the centre of gyration.

Example. Suppose a lever or beam 21 ft. long, poised on an axis at 14 ft. from one extremity and 7 ft. from the other. The cube of 14 ft. (= 2744) added to the cube of 7 ft. (= 343) is 3087; divide this by 63, which is the sum of three times 14 ft. (= 42), and three times 7 ft. (= 21), the quotient is 49, and the square root of this = 7 ft. is the distance of the centre of gyration from the axis of motion.

Sliding Rule. $\left\{\begin{array}{ll} \text{C} & \text{Three times sum of dist.} \\ \hline \text{D} & 1 \end{array}\right.$ $\dfrac{\text{Sum of the cubes.}}{\text{Dist. of cent. of gyration.}}$ *Exam.* $\dfrac{\text{C} \quad 63}{\text{D} \quad 1}$ $\dfrac{3087}{7}$

To find the common distance for the centres of gyration of a lever with several arms, revolving about a common axis or centre of motion. If the arms of the lever are all alike, their centres of gyration will be at the same distance from the axis; but if the centres of gyration of each arm, being found separately, are not alike, then all the weight of each arm must be supposed to be collected into its centre of gyration, and a common distance must be found between those centres, by the following statement of the general rule first given (p. 30,) viz.

RULE. Multiply the weight of each arm into the square of the distance of its centre of gyration from the axis; divide the sum of all the products, by the weight of all the arms, and the square root of the quotient will be the common distance.

Example. Suppose a lever has three arms of uniform size, proceeding from the common axis, and that one arm, which weighs two pounds, is 17·32 inches long from the axis; another, which weighs 4 lbs. is 10·38 inches long; and the third, which weighs 6 lbs. is 6·92 inches long. The result is the same as that of the example of three cannon-balls before stated (p. 30).

The next simple case of a revolving body, is that of a solid wheel, or circular plate, revolving on an axis which passes through its centre, like a grindstone or millstone.

To determine the power which must be communicated to such a body to give it a certain velocity of motion. Suppose a solid wheel, 40 inches diameter, of uniform density and thickness, so that its weight will be proportionate to its area, if it revolves 114·6 times per minute, the circumference will move 20 feet per second.

Now we must imagine a revolving mass, of the same weight as the wheel, to be constructed, with ten circles like metallic hoops, placed concentric one within the other, at equal distances, leaving considerable spaces between them; suppose these circular hoops to be 2, 6, 10, 14, 18, 22, 26, 30, 34, and 38 inches diameter respectively, and that they are firmly united together, by small radial wires, which we must suppose to be inflexible and without weight.

When this mass is whirled round at the rate of 114·6 times per minute, each circle will move with the same velocity as the corresponding ball in the preceding case (p. 33), and hence the two cases may be compared; the only difference between them is, that the balls were all of the same weight; but the concentric circles are heavier as they are larger. The mass of matter contained in each circle may be represented by its radius, because the circumferences of circles are proportionate to their radii. The height due to the velocity with which each circle moves, being multiplied by the mass of matter contained in that circle, the product will be the power communicated to each circle, and the sum of these products, is the mechanical power communicated to the whole assemblage, to give it the motion it actually possesses; for it is the common product of all the separate masses, into the heights due to the respective velocities with which they move.

Velocity, feet per second.	Height fallen feet.	Mass of each circle.	Mechanical power communicated.
1	·015	1	·015
3	·140	3	·420
5	·389	5	1·945
7	·762	7	5·334
9	1·260	9	11·340
11	1·882	11	20·702
13	2·633	13	34·229
15	3·500	15	52·500
17	4·510	17	76·670
19	5·620	19	106·780
	Sums.	100	309·935
			3·099

The circumference of the wheel of 40 inches diameter moves with a velocity of 20 feet per second, and the height due to that velocity is 6·22 feet. Now if we divide 309·9, the sum of the products, by 6·22 ft. we shall have very nearly 50 (that is, one half of the whole) for the mass which the same power could put in motion in a straight line, with the velocity of 20 feet per second; whence the following conclusion.

The mechanical power which must be communicated to a solid circular wheel of uniform thickness and density, to make it revolve upon an axis passing through its centre, is the same as that which must be communicated to one-half of its weight of matter, to give it motion in a straight line, with the same velocity as the circumference of the wheel moves in its circle.

RULE. Multiply one half of the weight of the revolving wheel, by the height due to the velocity with which the circumference of the wheel moves; the product is the mechanical power communicated.

Example. Suppose a grindstone 4·375 feet diameter, weighing 3500 lbs. makes 270 revolutions per minute, what power must be communicated to it to give it that motion. By the rule formerly given (p. 33), the velocity of the circumference will be 61·83 ft. per second, and the height due to this velocity is 59·4 ft., the mechanical power is (half the weight 1750 lbs. × 59·4 ft. =) 103·950 lbs. raised 1 ft.

In case the revolving wheel is not an entire or solid circle, but only a ring or annulus, like the rim of a fly-wheel, it must first be considered as an entire circle, or solid wheel, and then the effect of the part which is wanting, must be calculated and deducted.

Example. Suppose the rim of a cast-iron fly-wheel is 22 ft. diameter outside, and 20 ft. diameter inside, the thickness of the rim being 6 inches, and that it makes 36 revolutions per minute; what power must be communicated to the rim, to give it that motion, the weight of the arms being left out of the consideration?

The mass of cast-iron in this ring is 33 cubic feet; because a solid wheel, 22 ft. diameter and 6 inches thick, would be 190 cub. ft.; from which deduct 157 cub. ft. for the mass of a solid wheel 20 ft. diam. and we have 33 cub. ft. for the rim alone.

Now supposing a solid wheel 22 ft. diam. the mass would be 190 cub. ft.; the velocity of the circumference would be 41·47 ft. per second; and the height due to that velocity would be 26·8 ft. which multiplied by 95 cub. ft. (or half the mass) gives 2546 cub. ft. of cast-iron, raised 1 foot, for the power communicated.

Then supposing another solid wheel, 20 ft. diam.; the mass 157 cub. ft.; the velocity of the circum. 37·7 ft. per sec.; and the height due to that velocity 22·1 ft. which multiplied by 78·5 cub. ft. (or half the mass), gives 1735 cub. ft. raised 1 foot, for the power.

This being deducted from 2546, the power due to the entire wheel leaves 811 cub. ft. of cast-iron, or (× 450 lbs. per cub. ft. =) 364 959 pounds weight, raised 1 foot, for the mechanical power which must be communicated to the rim of the wheel in question, to give it a velocity of 36 revolutions per minute. *Note.* This calculation may be abridged upon the following principle.

The mechanical power which must be communicated to solid wheels of different diameters, but of the same thickness and density, to make them revolve in the same period, is proportionate to the fourth powers of their diameters.

Example. Suppose five grindstones of 1 foot, 2, 3, 4, and 5 feet diameter respectively, but all of them being of the same thickness, and composed of the same kind of stone. The power which must be communicated to each stone, to make them all revolve the same number of times in a minute, may be represented by (1 × 1 × 1 × 1 =) 1; (2 × 2 × 2 × 2 =) 16; (3 × 3 × 3 × 3 =) 81; (4 × 4 × 4 × 4 =) 256, and (5 × 5 × 5 × 5 =) 625.

Again: the mechanical power communicated to the two supposed solid wheels, of 20 and 22 feet diameter, in the above case, being represented by 1735 and 2546, the proportion between those two numbers will be found the same, as between 160 000, which is the fourth power of 20, and 234 256, which is the fourth power of 22.

The reasons for this proportion are, I. That the velocities of the circumferences of such wheels are proportionate to their diameters. II. The energies of moving bodies are proportionate to the squares of their velocities. III. The masses of such solid wheels are proportionate to the squares of their diameters.

To find the mechanical power which must be communicated to any central portion of a solid circular wheel (that is, to a smaller solid wheel, within the larger one), having given, the power communicated to the entire wheel.

RULE. Divide the diameter of the entire wheel, by the diameter of the central portion, and raise the fourth power of the quotient; the power communicated to the entire wheel, being divided by this fourth power, gives the power communicated to the central portion.

Example. The power communicated to a solid wheel of cast-iron 22 feet diameter, and 6 inches thick, to make it revolve 36 times per minute, being 2546 cubic feet of cast-iron raised 1 foot, what power must be communicated to a central portion or smaller solid wheel, 20 feet diameter, contained within the other? Now 22 ft. ÷ 20 ft. = 1·1, the fourth power of which is 1·1 × 1·1 × 1·1 × 1·1 =) 1·464. And 2546 cub. ft. ÷ 1·464 gives 1735 cub. ft. for the power which must be communicated to the central portion.

Note. The above proportions are applicable to all cases of solid circular wheels, of uniform thickness and density, revolving on axes passing through their centres.

In cases where the thickness or density of the revolving wheel, is not uniform, we must resume the supposition of an assemblage of circles of suitable masses to represent the revolving wheel in question, and the mass of each circle, being multiplied into the height due to the velocity with which it moves, the rule may be adapted to the particular case.

To find the centre of gyration in a solid circular wheel, we must recur to the supposed assemblage of circles above-mentioned. And if we divide the sum of the products 309·9, by 100 the sum of the masses, we shall have 3·099 feet for the space the whole mass must have descended, to have exerted the whole power communicated. The velocity due to that height is very nearly 14·14 ft. per second, that is (14·14 ÷ 20 =) ·707 of the velocity (of 20 ft. per sec.) with which the periphery of the circle moved; whence the following conclusion.

The mechanical power which must be communicated to a solid circular wheel, to make it revolve upon an axis passing through its centre, is the same as that which must be communicated to an equal weight of matter, to give it motion in a straight line, with ·7071 of the velocity with which the periphery of the wheel moves in its circle.

Any point in the wheel, which moves with that velocity, is called a centre of gyration.

Note: in this case, there will be an infinite number of centres of gyration, all situated at the same distance from the axis of motion, and therefore forming the periphery of a circle, which may be called the circle of gyration. The diameter of this circle will in all cases be ·707 of the diameter of the solid wheel.

RULE. Multiply the diameter of the solid wheel by ·7071, and it will give the diameter of the circle, in which the centres of gyration are situated.

Sliding Rule. $\begin{cases} A & 75 & \text{Diameter of solid wheel.} \\ B & 53 & \text{Diam. of circle of gyration.} \end{cases}$ *Example.* $\begin{array}{lll} A & 75 & \text{20 ft. diam. of wheel.} \\ B & 53 & \text{14·14 ft. diam. of circle.} \end{array}$

To find the velocity with which the centres of gyration revolve in their circular orbits, in all cases of solid wheels: having given the diameter of the wheel in feet, and the number of revolutions per minute, to find the velocity of the motion of the centres of gyration, in feet per second.

RULE. Multiply the diameter of the wheel in feet, by the number of revolutions per minute, and divide the product by 27, the quotient is the velocity of the centres of gyration in feet per second.

Example. Suppose a millstone, 5 feet diameter, making 100 turns per minute, 5 ft. × 100 turns = 500 ÷ 27 = 18·5 feet per second for the velocity of the centres of gyration in their circle.

Sliding Rule, slide inverted. $\begin{cases} A & \dfrac{\text{Diam. solid wheel ft.} \qquad\qquad 27}{\text{Revo. per minute.} \quad \text{Velo. of cent. of gyra. ft. per sec.}} \end{cases}$ *Exam.* $\dfrac{\text{22 ft.} \qquad 27.}{\text{36 rev.} \quad 29·3 \text{ ft.}}$

To find the centres of gyration in a circular ring or annulus: having given the radius of the outside of the ring, and the radius of the inside thereof, to find the radius of the circle in which the centres of gyration are situated.

RULE. Divide the difference between the fourth powers of the two radii, by twice the difference between the squares of those radii; and the square root of the quotient is the radius of the circle, in which the centres of gyration of the ring are situated.

Example. A ring 11 feet radius outside and 10 feet radius inside. The difference between 14 641 (the fourth power of 11) and 10 000 (the fourth power of 10) is 4641. Twice the difference between 121 (the square of 11) and 100 (the square of 10) is (21 × 2 =) 42. Then 4641 ÷ 42 = 110·5, the square root of which, 10·519, is the radius of the centre of gyration.

Proof. When this ring revolves 35 times per minute, the centres of gyration will describe a circle 21·04 feet diam. and will move therein with a velocity of 39·65 feet per second; the height due to this velocity is 24·5 feet, and if the mass of matter in the ring was 33 cubic feet, its energy would be (24·5 × 33 =) 808·5 cubic feet descending 1 foot.

Distances of the centres of gyration from the centre of motion, in different revolving bodies.

In a straight uniform rod revolving about one end, the length of the rod × ·5773
In a circular plate revolving on its centre,   the radius of the circle × ·7071
In a circular plate $\begin{cases} \text{revolving about one of} \\ \text{its diameters as an axis} \end{cases}$ radius × ·5
In a thin circular ring
In a solid sphere $\begin{cases} \text{revolving about one of} \\ \text{its diameters as an axis} \end{cases}$ radius × ·7071
In a thin hollow sphere   radius × ·6325
                                                        radius × ·8164
In a cone revolving about its axis,   the radius of the circular base × ·5477
In a right angled cone revolving about its vertex,   the height of the cone × ·866
In a paraboloid revolving about its axis,   the radius of the circular base × ·5773

Note. The weight of the revolving body, multiplied into the height due to the velocity with which the centre of gyration moves in its circle, is the energy of that body, or the mechanical power which must be communicated to it, to give it that motion.

A PENDULUM is a heavy body, so freely suspended by a rod, or line, from a fixed support, as to be at liberty to perform alternate motions in the arc of a circle described about the centre of suspension.

The natural quiescent position of a pendulous body, is when the suspending rod, or line, is vertical; because the centre of gravity of the body is in its lowest position when it is situated exactly beneath the centre of suspension.

If, by the exertion of any force, a pendulum is removed or drawn aside from the vertical position, the centre of gravity must rise, and consequently the pendulum will return, or fall towards the vertical line by the action of gravity, when that disturbing force ceases to act. The moving mass acquires a considerable energy of motion, by the descent of its centre of gravity in the circular arc, during the return to the vertical line, and as this energy is not destroyed or counteracted when it arrives at that position, it will carry the pendulum forward in the circular arc, on the opposite side of the vertical line, and then the centre of gravity rising, counteracts the energy, until it is quite expended, and consequently the motion in that direction ceases, leaving the pendulum to return to the vertical line by its gravity; but in descending, during such return, it acquires a new energy, which carries it forward in the arc as before.

In this manner a pendulous body continues to vibrate or oscillate backwards and forwards, on each side of the vertical line, by virtue of the power originally communicated to it; and if it were not for the resistance of the air, and the friction of the centre of suspension, such vibrating motion would continue for ever.

The motion of a pendulum is kept up by the compound action of gravity and inertiæ. For gravity causes the pendulous body to move towards the vertical line; but that motion is at first impeded by the inertiæ, until the motion being gradually generated, the acquired energy is sufficient to carry the pendulum beyond the vertical line on the opposite side, and occasion its rise, until the continued retarding force of gravity, destroys the energy and causes its return.

When a body which is suspended as a pendulum is removed from the vertical position, the tendency of the weight of each particle, to restore it to the vertical position, is proportional to the horizontal distance of such particle from the vertical line passing through the point of suspension; and the united forces of all the particles, may be expressed by the product of the whole weight into the horizontal distance of the centre of gravity, from that vertical line.

The mass which is to be moved in the circular arc by this force, is to be estimated by the sum of all the products obtained by multiplying the weight of each particle, into the square of its distance from the centre of motion.

The motion of a pendulum in one direction, from its state of rest, until it begins to return in an opposite direction, is termed *a vibration* or *an oscillation;* and the time which elapses during such motion, is called the *time of vibration,* or *of an oscillation.*

A *simple pendulum* is, when the suspending rod or line is supposed to be inflexible, and without weight; and when the weight, or pendulous body, is collected into a mathematical point, the distance from the centre of suspension, to that point is *the length of the pendulum.* A small leaden bullet, suspended by a long and fine hair, will represent a simple pendulum, and possesses the following properties:

I. If the pendulum vibrates in the arc of a circle, the velocity acquired by the ball when it arrives at the lowest point, will be proportionate to the chord of the arc which it describes in its descent.

II. The force which accelerates the motion of the ball, in its circular arc, is proportioned to the whole force of gravity, as the horizontal distance of the ball, from the vertical line, is to the length of the pendulum.

III. The number of vibrations made in a given time, by any pendulum is nearly the same, whether the vibration be longer or shorter. *Note.* This is true, when the ball describes a curve called a cycloid, but is not quite exact when the ball describes a circular arc, unless that arc is so short, that the cycloid and the circle do not differ perceptibly.

IV. The time of vibration in a short circular arc, is proportioned to the time in which a body would fall by gravity through half the length of the pendulum, as the circumference of a circle is to its diameter.

V. The number of vibrations made in a given time, by pendulums of different lengths, is inversely as the square roots of their lengths; provided the force of gravity continues the same.

VI. The lengths of pendulums vibrating in the same time, is directly as the force of gravity.

To find the height a body will fall, in the time that a pendulum makes one vibration, having given the length of the pendulum.

RULE. Multiply the length of the pendulum by 4·9348, and it will give the height through which a body will fall in the time whilst the pendulum makes one vibration.

Example. The length of a pendulum, which vibrates in vacuo once in a second, is 39·1386 inches; which, multiplied by 4·9348, gives 193·141 inches for the height through which a body will fall in vacuo, in one second.   (See page 21.)

Sliding Rule. $\left\{\begin{array}{ll} \text{A} & 81 \\ \hline \text{B} & 400 \end{array}\right.$ Length of pendulum.
Height a body will fall.   *Example.* $\begin{array}{ll} \text{A} & 81 \\ \hline \text{B} & 400 \end{array}$ 3·2615 ft. length.
16·095 ft. fallen.

To find the length of a pendulum, which will perform a given number of vibrations in a minute.

RULE. Divide the constant number 375·36 by the number of vibrations made per minute, and the square of the quotient is the length of the pendulum in inches.

Example. To make 60 vibrations per minute. 375·36 ÷ 60 = 6·256, the square of which is 39·138 inches for the length of the pendulum.

Or conversely. Divide the constant number 375·36 by the square root of the length of the pendulum, in inches, and the quotient is the number of vibrations it will make in a minute.

Example. A pendulum of 22 inches long. The square root of 22 is 4·69; and 375·36 ÷ 4·69 = 80 vibrations per minute.

Sliding Rule, slide inverted. $\left\{\begin{array}{ll} \text{C} & 22 \\ \hline \text{D} & 80 \end{array}\right.$ Length of pendulum inches.
Vibrations made per min.   *Example.* $\begin{array}{ll} \text{C} & 22 \\ \hline \text{D} & 80 \end{array}$ 39·13 inches length.
60 vibr. per minute.

In most cases the weight of the pendulous body is not collected in one point; for instance, when the rod by which it is suspended forms a considerable portion of its weight; or the whole weight may be disposed in that rod.   In all such cases, the efficient length of the pendulum must be measured from the centre of suspension, to an imaginary point termed the centre of oscillation, and which is at such a distance from the centre of motion as is equal to the length of a simple pendulum, which will vibrate in the same time as the pendulous body in question.

THE CENTRE OF OSCILLATION IN A PENDULOUS BODY is a point in the line passing through the centre of suspension, and the centre of gravity.   If all the matter of the body could be collected in that point, any force applied there, would generate the same angular velocity, in a given time, as the same force would generate in the same time, by acting similarly at the centre of gravity of the pendulum, when all the parts thereof are situated in their respective places.

The whole weight of the body may be considered as collected in its centre of gravity, and the force of gravity may be considered as acting entirely at that point.   The inertiæ of the pendulous body may also be considered as acting entirely at its centre of gyration, which has been before explained.   Hence the position of the centre of oscillation of a pendulous body, has a constant relation to the position of its centre of gyration, and its centre of gravity; viz. the distance from the centre of motion to the centre of gyration, is a mean proportional between the distances of the centres of oscillation, and gravity, from the centre of motion.   Whence the following rule:—

To find the distance of the centre of oscillation of a pendulous body from its centre of motion.

Having given the distance of the centre of gyration, and the distance of the centre of gravity from the centre of motion:

RULE.   Divide the square of the distance of the centre of gyration, by the distance of the centre of gravity, the product is the distance of the centre of oscillation.

Example.   A straight uniform rod 20 inches long, being suspended by one end, the centre of gyration will be at 11·55 inches from the centre of motion (see p. 34), and the centre of gravity will be at 10 inches therefrom.   The square of 11·55 is 133·3 ÷ by 10 = 13·33 inches from the centre of suspension to the centre of oscillation.   It is situated at (13·3 ÷ 20 =) ·66, or exactly two-thirds of the whole length.   If the centre of gyration is not known, then,

RULE.   Multiply the weight of each particle of the pendulous body, into the square of its distance from the centre of motion, and divide the sum of all these products, by the product of the sum of all the weights, multiplied into the distance of the centre of gravity from the centre of motion. The quotient is the distance of the centre of oscillation.

Example.   Suppose a straight rod to be represented by 10 equal bullets, threaded upon a wire 20 inches long, at distances of 1, 3, 5, 7, 9, 11, 13, 15, 17, and 19 inches; the squares of these distances will be 1, 9, 25, 49, 81, 121, 169, 225, 289, and 361, the weight of each ball being 1 lb. the same numbers will also represent the products, and their sum is 1330.

The distance from the centre of motion to the centre of gravity is 10 inches; which, multiplied by 10 weights = 100. And lastly, 1330 ÷ 100 gives 13.3 inc. for the distance between the centre of oscillation and the centre of motion, as before.

By this method, the centre of oscillation of any regularly formed body may be found; for we have only to imagine a similar body to be composed of several detached balls or masses connected together, in the same manner as before directed for finding the centre of gyration (p. 33).

To find the distance of the centre of oscillation of a pendulous body from its centre of gravity.

RULE. Multiply the weight of each particle of the pendulous body, by the square of its distance from the centre of gravity, and divide the sum of all these products, by the product obtained by multiplying the whole weight of the body, by the distance between its centre of gravity and its centre of motion. The quotient is the distance of the centre of oscillation, beneath the centre of gravity.

Example. Suppose ten equal bullets threaded on a wire 20 inches long, as in the former example; the centre of gravity will be in the middle of its length, and the centres of the balls will be respectively at 1, 3, 5, 7, and 9 inches' distance on each side thereof; the squares of these distances will be 1, 9, 25, 49, and 81 on one side of the centre of gravity, and the same on the opposite side; the weight of each ball being 1, the same numbers will also represent the respective products, and their sum is 330. The distance between the centre of motion and the centre of gravity is 10 inc., which × by 10 weights = 100; and lastly, 330 ÷ 100 gives 3·3 inches for the distance of the centre of oscillation beneath the centre of gravity; which being at 10 inches' distance from the centre of motion, the centre of oscillation is (10 + 3.3 =) 13.3 inches therefrom, as before.

In a uniform rod suspended by one extremity, the centre of oscillation will be at two-thirds of the length of the rod from its centre of motion. Or if the same rod be suspended with one-third of its length above the centre of motion, and two-thirds beneath it, the centre of oscillation will be at the lower extremity, and consequently the rod will vibrate in the same time as when it was suspended by the upper extremity.

The centre of oscillation and the centre of motion are reciprocal; for if a body which is so suspended by any point as to vibrate thereon, be inverted and suspended by its centre of oscillation, the former point of suspension will become the centre of oscillation, and the vibrations in both positions will be performed in equal times. From this property, the centre of oscillation in any irregular body may be determined experimentally; for the number of vibrations which it performs in a given time, when suspended on any point, being ascertained, it may then be inverted, and suspended by some other point, which may be found by repeated trials to be so situated, that the vibrations will be exactly the same as before, then the distance between these two points is the length of a simple pendulum, which will vibrate in the same time. It was upon this principle that Captain Kater determined the length of a pendulum vibrating seconds to be 39·1393 inches of the new imperial standard measure (a).

Example. Suppose a rod or bar, of uniform size and weight, 58.709 inches long, to be suspended by one of its extremities; it will perform a vibration in a second, for the centre of oscillation will be at two-thirds of the length of the rod from the centre of motion, or 39.1393 inches. If the rod is inverted, and suspended by its centre of oscillation, one-third of its length will be above the centre of motion, and two-thirds below, but it will vibrate in exactly the same time as in the first instance.

A REVOLVING OR CENTRIFUGAL PENDULUM, is a heavy ball suspended by a string or rod from a fixed point, in such manner as to revolve about a vertical axis, and describe a horizontal circle. It is also called a *conical pendulum*, because the string or rod of suspension describes the surface of a cone during its revolution.

In such a pendulum the centrifugal force of the revolving ball tends to remove it from the vertical axis, so as to keep the rod of suspension extended, at an angle with the vertical line; but the force of gravity tends to draw the ball towards the vertical axis. If the ball of the pendulum is caused to move with uniform velocity in the circle which it describes, these contending forces will maintain a

---

(a) The length of the seconds' pendulum has been stated (in pp. 22 and 40) at 39.1386 inches of Sir George Shuckburg's brass standard scale. Since that article was printed, an Act of Parliament has been passed, declaring a brass standard scale, made in 1760, to be the genuine and original measure, which is now denominated the Imperial Standard Yard; and that the seconds' pendulum is 39.1393 inches of this Imperial Standard Measure.

G

constant equilibrium, whereby the ball will preserve a constant distance from the vertical axis. If the velocity of the revolving motion is increased, the centrifugal force of the ball will increase, and will overbalance its gravitating tendency, so as to cause it to recede from the vertical axis, and describe a larger circle ; but in receding from the vertical axis, the gravitating tendency is also increased until it balances the increased centrifugal force, and then the ball will retain its new distance from the vertical axis, and will continue to describe that larger circle, until some other change takes place in its revolving motion. For instance, if the velocity is diminished, the gravitating tendency will exceed the diminished centrifugal force, and will cause the ball to approach the vertical axis, so as to describe a smaller circle. The apparatus called a *governor*, for regulating the velocity of the motion of a steam-engine, is constructed on this principle.

Dr. Young has remarked, that the motion of a centrifugal pendulum, may be imagined to be compounded of two different vibrations of a simple pendulum, taking place at the same time, but in directions at right angles to each other ; for whilst a simple pendulum is under the influence of a force which causes it to vibrate from north to south, it is capable, at the same time, of receiving the impression of another force, which, if it operated alone, would cause it to vibrate from east to west. And assuming the two forces to be equal, but acting in directions at right angles to each other, their combined action on the ball of the pendulum will cause it to revolve with uniform motion in a horizontal circle, at a constant horizontal distance from the vertical axis, and at a constant perpendicular distance beneath the point of suspension.

One revolution of a centrifugal pendulum, will be performed in the same time as two vibrations of a simple pendulum, whereof the length is equal to the vertical height of the point of suspension, above the plane of the circle described by the centre of the revolving ball : or, in other words, half the period of a revolution of a conical pendulum is equal to the time of one vibration of a simple pendulum whereof the length is equal to the altitude of the cone described by the rod of suspension. It follows, that the period of the revolution of a centrifugal pendulum, is not so much determined by the length of the pendulum, or the size of the circle described by its ball, as by the vertical height of the point of suspension above the plane of that circle.

To find the diameter of the circle described by the centre of the ball of a centrifugal pendulum. Having given the number of revolutions the pendulum makes per minute; and its length in inches, between the centre of the ball and the point of suspension ;

RULE. Divide the constant number 187·68, by the number of revolutions made per minute, and the square of the quotient will be the vertical height in inches, of the point of suspension, above the plane of the circle described by the centre of the ball (*a*). Then, to find the radius of that circle, deduct the square of the vertical height in inches, from the square of the length of the revolving pendulum in inches, and the square root of the remainder is the radius in inches, of the circle described by the centre of the ball ; and twice that square root, is the diameter of the circle in inches.

Example. Suppose the length to be 28 inches from the point of suspension to the centre of the ball, and that the pendulum revolves 37 times per minute. 187·68 ÷ 37 revolutions = 5·07, the square of which is 25·7 inches for the vertical height, or altitude of the cone ; the square of this is 660·5, and the square of 28 inches length is 784. Deducting 660·5 from 784, the remainder is 123·5, and the square root of this is 11·1 inches for the radius of the circle described by the centre of the ball ; or 22·2 inches for its diameter.

(*a*) To do this by the sliding rule, it must be set as follows, and will then form a table :

Sliding Rule, slide inverted.

B	22	Altitude of cone inches.	*Exam.*	B	22	25·7 inc. vertical height.
D	40	Revolutions per minute.		D	40	37 revolutions per min.

In many cases it is convenient to compute the effect of a revolving pendulum from the centrifugal force of the ball, rather than by comparison with a simple pendulum, because the point of suspension does not always coincide with the vertical axis of rotation. This is frequently the case in governors for steam engines; and then the point of suspension is not a fixed centre of motion, but it describes a small horizontal circle about the vertical axis, and consequently the rod of suspension does not describe a complete cone, but only a frustum of a cone.

If we find what tendency the ball will have to approach towards the vertical axis, by its own gravity, and consider that as the centripetal force, we may then proceed (by the converse of the rule given in p. 28) to compute what velocity of circular motion must be given to the ball, in order that its centrifugal force may balance and counteract that centripetal force.

The force with which the ball tends to approach towards a vertical line passing through its point of suspension, bears the same proportion to the weight of the ball, as the horizontal distance from the centre of the ball to the vertical line, bears to the vertical height of the point of suspension above the centre of the ball. Therefore if the horizontal distance is divided by the vertical height, the quotient will be the force with which the ball will tend towards the vertical line, expressed in terms of its own weight, which is supposed to be 1 (*a*). For instance, in the preceding case, 11·1 inc. horizontal ÷ 25·7 inc. vertical = ·432 of the weight of the ball.

Having thus determined the centripetal force, the rate of revolution requisite to produce an equivalent centrifugal force may be found by the rule p. 28, stated conversely, viz. Multiply the centripetal force of the ball, expressed in terms of its own weight, by the constant number 70 440; and divide the product by the diameter of the circle described by the centre of the ball in inches; the square root of the quotient is the number of revolutions it must make per minute.

Example. Suppose that the centripetal force of the ball is ·432 of its own weight, and that the centre of the ball is to describe a circle of 22·2 inches diameter: then 70440 × ·432 = 30430 ÷ 22·2 inc. dia. = 1370, the square root of which is 37 revolutions per minute, the same as stated in the preceding method.

This calculation will require two operations on the sliding rule, thus. If we know the angle in degrees, that the rod of suspension makes with the vertical, then set the slider for the tangent of that number of degrees, on the line t at the back of the slider (*a*); or otherwise, we can set the rule in the same position, by means of the vertical height of the point of suspension above the centre of the ball, and the horizontal distance of that centre from the vertical line passing through the point of suspension, thus:

I. Sliding Rule. { Angle in degrees on the line t behind the slide. }

A	Vertical height inches	70 440 constant multiplier.
B	Horizontal distance inc.	Product for 2d operation.

The product thus obtained is that of the centripetal force into the constant multiplier; then,

II. Sliding Rule. {

C	Diam. of circle inches.	Product of 1st operation.	*Exam.*	C	22·2	3043
D	1	Revolutions per minute.		D	1	37R.

———————————————————————————

(*a*) The tangent of the angle that the rod of suspension makes with the vertical line, will represent the force with which the ball tends towards the vertical, expressed in terms of its own weight.

If the angle is expressed in degrees, its tangent may be found in tables of sines and tangents; or else by means of a line marked t, on the back of the slider of the Soho Sliding Rule; the slide must be drawn out towards the right hand, so as to expose to view the divisions of the line t from under the rule, until that division which represents the number of degrees of the angle in question, corresponds with the end of the rule; the slider being thus set for the angle, its tangent will be found on the line B, opposite to 1 on the line A. For instance, the angle in the above case is about 23½ degrees, then,

Soho Sliding Rule. { 23½ deg. line t behind the slide, }

A	1. Radius	and	Vertical height	25·7 inc.
B	·423 Tangent		Horizontal dist.	11·1 inc.

Note. The tendency of the ball to approach the vertical will be the same in all cases when the suspending rod makes the same angle with the vertical, whatever may be its length or the distance of the ball from the point of suspension.

When the centripetal force is thus computed separately from the centrifugal force, we are enabled to take into account any other tendency which the ball may have to approach the axis, in addition to that of its own weight, viz. the weight of the suspension rods, or of the parts connected with them ; in such cases the additional force must be expressed in terms of the weight of the ball, and added to the centripetal force of the ball itself, as determined by the preceding method.

If the suspension rod of a revolving pendulum is required to expand to some given angle from the vertical position, the constant number 70 440 being multiplied by the tangent of that angle, will give other constant numbers suitable for such cases, of which the two following are the most common.

To find the rate at which the ball of a centrifugal pendulum must revolve in any given circle, to cause the rod of suspension to assume an angle of 30 degrees with the vertical line.

RULE. Divide the constant number 40 670 by the diameter of the circle described by the centre of the ball in inches; the square root of the quotient is the number of times the ball must revolve per minute.

Example. Suppose the centre of the ball is to describe a circle of 30 inches diameter, then, $40\,670 \div 30 = 1355 \cdot 66$, the square root of which is $36 \cdot 8$ revolutions per min.

Sliding Rule, slide inverted.	B	60	Diam. of circle inches.	*Example.*	B	60	30 inc. diam.
	D	26	Revolutions per min.		D	26	36·8 revolut.

The rule thus set, forms a complete table of the number of revolutions suitable for each diameter.

To find the rate at which the pendulum must revolve, that its rod may assume an angle of 45 degrees. In this case, the centripetal force is equal to the whole weight of the ball, therefore the rule must be in effect the same as that in p. 29.

RULE. Divide the constant number 70 440 by the diameter of the circle in inches; the square root of the quotient is the number of times the ball must revolve per minute.

Sliding Rule, slide inverted.	B	44	Diam. of circle inches.	*Example.*	B	44	52 inc. diam.
	D	40	Revolutions per min.		D	40	36·8 revolut.

The rule thus set forms a complete table of the number of revolutions suitable for each diameter.

Note. In all the preceding rules relative to centrifugal pendulums, the action of the ball alone is considered, the rod of suspension being assumed to be without weight; but when the weight of the rod, and other parts connected with it, bears a sensible proportion to that of the ball, an equivalent allowance must be made.

THE MECHANICAL POWERS are the simple elements of which all machines are composed ; they are usually stated as six in number, viz. *the lever ; the wheel and axis ; the pulley ; the inclined plane ; the screw ; and the wedge.* There are also two others of modern invention. All these are the material agents, by means of which contending or opposing forces are brought to operate upon one another. The simple mechanical powers admit of being combined, so as to act in concert, for the attainment of a common object, and such combinations are usually termed machines.

The mechanical powers are used to transmit the force of some first mover, and direct its action to the resistance which is required to be overcome by the exertion of that force. The office of the interposed mechanical power may be, either to multiply or increase the force with which the first mover acts, so as to enable it to overcome the resistance of a greater force ; or else to augment the motion which the first mover possesses, and cause it to act on the resisting force through an increased space ; and in some cases a mechanical power is introduced merely to change the direction in which the force of its first mover acts, and to transmit that force in a new direction suitable to that of the resistance which is to be overcome.

The mechanical powers are therefore the organs by which a given force may be accumulated to any required degree of intensity; and whereby the motion resulting from the exertion of a given force, may be modified in its velocity and direction, so as to become adapted to any required purpose. It is usual to employ the term *power*, to express the force which is communicated to any mechanical power, and the term *weight* or *resistance* to denote the contending or opposing force, which the mechanical power is required to counteract or overcome.

Mechanical powers are not always required to transmit motion; for in many cases it is only required to produce a statical equilibrium between the power and the weight, by means of some mechanical power which is interposed between them.

In considering the mechanical powers, and their various combinations, as the means of increasing either the force or the motion which is communicated to them by their first moving powers, it must always be kept in mind, that the resulting force cannot be increased, without a corresponding diminution of the original motion; nor can the resulting motion be increased without a corresponding diminution of the original force; consequently no application of mechanical powers can augment the amount of moving force or mechanical power, as defined in p. 16. For the utmost effect of any mechanical power is to decompose any moving force which is communicated to it, into the constituent elements of force and motion, and then to recompose them again into a new moving force, having a different proportion of those elements, but forming the same amount as the original.

The product obtained by multiplying the force which is communicated to any mechanical power, by the space through which such force acts, will in all cases be equal to the product of the weight or resistance into the height or space through which it is raised or overcome by the aid of that mechanical power (*a*); this is an essential reservation which must never be overlooked in considering the properties of the mechanical powers, and should be preserved in the memory, by a maxim that " *whatever is gained in force, is lost in space.*" The truth of this maxim will become apparent when the following reason for the augmentation of force is understood.

All the mechanical powers operate by means of certain fixed and motionless supports, which are so disposed, as to sustain part of the opposing force, in order to relieve the moving organs from the opposition of such part of the force; in consequence, the motive force which is communicated to the moving organs, is caused to act with all its efficacy upon such proportion of the resisting force, as those organs really do sustain; but it is evident, that the portion of resisting force which is by this expedient thrown upon the fixed support will not receive any motion. The result is, that although a great opposing force may be overcome by a smaller active force, this effect is attained by a proportionate sacrifice of the motion of the smaller force, because only part of that motion is transmitted to the greater force which is overcome.

The increase of velocity, which may be attained by means of a mechanical power, s only the converse of the same proposition. A force capable of acting through a small space is communicated to the moving organs, certain parts of which are confined and restrained from motion, by means of the fixed supports, in such manner, that the motion which is thus prevented from taking place in those parts, shall be accumulated in other parts of the moving organs to which the resistance is applied; the space through which the resisting force is overcome, will then be greater than the space through which the original force acts, because the motion, which might have been distributed amongst several moving parts, is accumulated into one part; but the force which can be exerted with such accumulated motion, is proportionably diminished, because part of the original force is thrown upon the fixed support.

(*a*) This is on the assumption that the mechanical power acts without friction; but in practice there will always be an actual loss of moving force, by every mechanical power which is employed to transmit or modify it, in consequence of the friction of its moving parts; and in complicated combinations the loss from friction is a very considerable proportion of the whole power applied.

The most important property of all the mechanical powers being included in the preceding statement, it is not necessary to enter very minutely into the description of each variety.

A LEVER is an inflexible bar, which is capable of moving freely about a fixed point or centre, or axis of motion, so that every part of the lever will describe an arch of a circle about the common centre, and the motion of all those different parts will be proportionate to their distances from the common centre. When any force is applied to act upon a lever at a point situated at any given distance from the centre of motion, that distance may be called the real length or radius of the lever, because it is the radius of the circle which such point must describe during the motion of the lever. Let us suppose that the force acts in a direction at right angles to the radius of the lever, that is, at right angles to a line drawn from the point of action to the centre of motion. The effect of such a force to move or turn the lever about its centre, may be represented by the product obtained by multiplying the acting force by the radius, or distance at which the force acts from the centre of the lever; and any other force so applied as to produce the same product, will have an equal and similar effect on the lever.

Example. Suppose a straight inflexible lever to be poised horizontally on a centre in the middle of its length, in the manner of a scale beam; and that its length is 24 feet; if a weight of 6 tons is suspended from one end of this lever, at a radius of 12 feet, or at a point 12 feet distant from the centre, so as to pull at right angles to that radius, then the effect of the weight to move the lever may be represented by (6 × 12 =) 72. And suppose another weight of 4 tons to be suspended from the same lever, at a radius of 9 feet, but on the opposite side of that centre to the first weight, the effect of this second weight will be (4 × 9 =) 36, or half the effect of the former. Again, if we suppose another weight of three tons to be suspended at a radius of 12 feet, and at the same side of the centre as the last weight, its effect would be also 36; so that the combined effect of both the latter weights would be 72, or equal to that of the first one.

The sliding rule is very convenient to show what weights must be applied to a lever at different radii or distances from the centre, to produce the same product, or effect to give it motion: thus,

Sliding Rule, slide inverted. { A Distances from centre.   1   C Weights or forces.   Product. }   *Ex.*   A 12 feet   1   C 6 tons   72   or   8 feet. / 9 tons.

Note. This supposes that there is no resistance from friction, and that the lever itself is devoid of weight, or else that the effect of the weight of one part of the lever counterpoises the weight of the other part, so as to produce an equilibrium between them. The weight of the lever acts at its centre of gravity, and has the same effect as if a weight equal to that of the lever were suspended from the centre of gravity of the lever; the position of this point will of course depend upon the form of the lever, and may be determined by the methods stated in p. 25.

The above rule assumes that the force acts upon the lever in a direction at right angles to its radius; that is, at right angles to a line drawn from the point of action to the centre of motion. When the force does not act at right angles to the radius, we must take the distance from the centre, to the line of direction in which the force acts, measured upon a line drawn at right angles to that line of direction. This distance may be called the effective radius, or effective length of leverage. And if we assume this to be the distance at which the force is applied from the centre, we may proceed to calculate the effects of such force to move the lever, in the same manner as above.

In all cases, whether the force does or does not act at right angles to the real radius of the lever, the effective radius or length of leverage will be the radius of such a circle as, being described about the centre of motion, will have for its tangent the line of direction in which the force acts.

When a lever is moved by a force applied to a point at any given distance from its centre, the effective radius of leverage will change perpetually during the motion, and will become less than the real radius of the lever, because the line of direction in which the force acts, must continually alter in its direct distance from the centre of motion. The effective radius will be in all cases, the sine of the angle which the real radius of the lever makes with the line of direction in which the force acts; for instance, when that is a right angle, as supposed in the preceding rule, the sine of a right angle being equal to the radius, the effective radius will be the same as the real radius.

In other cases, when the angle between the real radius and the direction of the action is known in degrees, the effective leverage may be determined, by taking the sine of that angle from a table of sines, and multiplying that sine by the real length or radius of the lever, the product will be the effective radius.

Example. Suppose the real radius of the lever is 8 feet, and that a force of 3 tons acts in such a direction as to make an angle of 60 degrees with that radius. The sine of an angle of 60 degrees is ·866 × 8 feet radius = 6·928 feet effective radius, or length of leverage × 3 tons' force = 20·784 will represent the force to move the lever.

Soho Sliding Rule. $\left\{\begin{array}{l}\text{Line s behind the slide}\\\text{angle of action on radius.}\end{array}\right\}$ $\dfrac{\text{A Real radius of lever.}}{\text{B Effective radius.}}$ *Ex.* 60 deg. $\dfrac{\text{A 8 feet radius.}}{\text{B 6·9 ft. effective.}}$

When it is not known what angle the line of action makes with the real radius; but if it is known how far the point of action is distant from a line drawn through the centre, at right angles to the line of action, then the square of that distance being deducted from the square of the real radius of the lever, the square root of the remainder will be the effective radius.

Example. Suppose the real radius of a lever is 9 feet, and that the point of action is removed 3 feet from that position, in which the real radius would be at right angles to the direction of the action. The square of 3 feet is 9, which deducted from (the square of 9 =) 81 leaves 72, the square root of which is 8·4 feet effective leverage.

A WHEEL AND AXIS is a cylindrical axis or roller, mounted upon pivots at each end, so as to be capable of turning round thereupon, in order to wind up a rope in coils round about its cylindrical surface; the resistance which is to be overcome, or weight which is to be raised, is applied to this rope. A circular wheel is also fixed fast upon the axis, and the power to give motion to the wheel and axis, is applied to a cord wrapped round the circumference of that wheel, so that by drawing off and unwinding the cord from the wheel it shall be turned round together with the axis, in such manner as to wind up the rope around the axis. But in as much as the diameter of the wheel is greater than that of the axis, so the length of cord drawn off the wheel, will be greater than the length of rope wound up round the axis. The resistance overcome by winding up the rope, will exceed the force applied to draw off the cord, in the same proportion as the space through which that force acts, exceeds the space through which the resistance is overcome.

The wheel and axis may be considered as being composed of an infinite number of equal and similar levers, radiating about a common centre, so as to form a circle; the radius of the wheel and the radius of the axis, being the distances at which the force and resistance are applied from the centre of motion of such levers.

It is most convenient in calculations to take the diameters of the wheel and axis; and the diameter or thickness of the cord, and of the rope, should be added to the real diameters of the wheel and of the axis.   The diameter of the wheel added to that of the cord, which is applied upon its circumference, being multiplied by the force which is applied to draw that cord, the product will represent the effect of the force to turn the wheel and axis round; if this product be divided by the diameter of the axis, added to that of the rope which is applied to its circumference, the quotient will be the force or resistance which that rope will overcome, or keep in equilibrium.

Sliding Rule, slide inverted.	A	Diam. of wheel + diam. of cord.	Diam. of axis + diam. of rope.
	?	Force applied to that cord.	Resistance overcome by rope.

Note.   In the wheel and axis the effective radius or length of leverage is always the same, and is equal to the real radius or length of leverage, because in all cases the direction in which the force acts, is a tangent to the same circle. This is true, provided that the wheel is a circle, with the axis passing through its centre, and that the axis is a cylinder movable about its central line or imaginary axis. But either the wheel or its axis, or both, may be of any other curved figure than a circle, so as that in turning round, the forces may continually act at different distance from the centre of motion; such cases may be considered as a series of levers of different lengths, arranged about a common centre of motion as radii, and coming successively into action.

Wheels are very commonly formed with projecting teeth on their circumferences; such teeth being adapted to enter into the spaces between similar teeth, formed on the circumferences of other wheels of greater or lesser diameter, or upon the edges of straight racks.   The similitude of all cases of toothed wheelwork with those of simple levers is very obvious, as each projecting tooth may be considered to be the extremity of one of the series of levers which radiate from the common centre of motion.

A PULLEY is a small circular wheel adapted to turn freely round upon a pin or axis which passes through its centre; the circumference is hollowed out with a groove, to receive a rope or cord, which is bent over the pulley, in order to change its direction; and the pulley, by turning round, will allow the rope to move freely endways when it is drawn over the pulley, on the same principle that the wheels of a carriage facilitate its motion on a roadway: the pulley is usually fitted into a cell in a frame called the block, and the pin or axis of the pulley is supported by the block, which may have two, or three or more, pulleys placed side by side on the same pin.

In some cases a block is suspended from a fixed support, and the rope which passes over the pulley has the force or power applied at one end, and the resistance or weight is applied at the other end: in any case where the power and the resistance are applied to the same rope, the power cannot be augmented by conducting the rope over a pulley; but the direction in which the force acts may be changed.

To apply a pulley so as to overcome a great resistance by a smaller force, the resistance must be applied to the centre pin of the pulley, or to the block in which it is mounted, one end of the rope must be made fast to a fixed support, and the force must be applied to the other end; then if the rope passes round one half of the pulley, so that the two parts of the rope are parallel, the resistance overcome by the block, will be twice as great as the force which is applied to the rope; because one half of the resistance is borne by the fixed support to which the end of the rope is fastened.   The motion communicated to the block, will be only half as much as the motion with which the force acts upon the rope.

In all combinations of pulleys over which only one rope circulates, the augmentation of any force which is applied to that rope, will be according to the number of pulleys in the moving block, to which the resistance is applied; but whatever stationary pulleys may be combined in the same system with pulleys in movable blocks, such stationary pulleys will have no effect to augment the force. One moving pulley doubles the force; two moving pulleys augment the force four times; three moving pulleys six times; and four moving pulleys eight times. The resistance which may be overcome by the moving block, is in all cases equal to the product obtained by multiplying the force which is applied to the rope, by twice the number of pulleys upon which the rope is applied, in that moving block.

An inclined plane is any path or roadway which makes an angle with the horizon; if a heavy body is rolled or moved up such an inclined path, its weight will be raised upwards, and this weight is the resistance overcome. As the inclined plane sustains a large proportion of the weight of the body, only the remainder of that weight can act against the force by which the body is drawn up the plane. The force which is required to draw the body up the inclined plane, will be less than the resistance which is overcome, in the same proportion as the length of the inclined plane is less than its vertical height or departure from a horizontal line. The length of the plane, or so much of it as the body is moved thereupon, is the space through which the force acts, and the vertical height that the body has been raised in so moving, is the space through which the resistance is overcome.

If any force which is applied to draw a body up an inclined plane, be multiplied by the length of the plane, and the product divided by the vertical height that the plane ascends in that length, the quotient will be the resistance or weight of the body, which may be moved up the plane by that force.

Sliding Rule, { A Length of inclined plane. Vertical ascent of plane. *Exam.* A 24 feet. 3 feet.
slide inverted. { ⊃ Force applied to move body. Resistance overcome. ⊃ 150 lbs. 1200 lbs.

When it is known what angle the inclined plane makes with the horizontal in degrees, the force which must be applied to move a body up the plane, may be represented by the sine of that angle, the weight of body being radius or 1.

Soho Sliding Rule. { Line s behind the slide { A Length of inclined plane. *or* Resistance overcome.
{ angle of plane with horiz. { B Vertical ascent of plane. Force applied to body.

A wedge is a piece of metal made thin and sharp at one end or edge, and tapering or increasing to some thickness at the other end; it is usually driven with the point forwards, by the blows of a mallet or hammer, in order to force it into masses of wood or stone, which it is required to separate or split into pieces; or, in some cases, a wedge is driven beneath any heavy mass, in order to raise it up to a very small height from the support on which it rests.

The principle of the wedge is the same as that of the inclined plane, for the two sides or surfaces of the tapering wedge are two planes inclined to each other; the difference of application is, that the wedge is moved or forced forwards beneath the resistance; but the inclined plane is stationary, and the resistance is drawn up or moved along the same. This makes no difference in the effects, or in the mode of calculating them; for the length of the wedge being taken as the length of the inclined plane, and the thickness of the wedge at its thick end being taken as the vertical ascent of the plane, the preceding rules may be applied to any cases of the wedge.

A screw is a cylinder with a thread or fillet projecting from its circumference, and winding spirally about it; the screw must be fitted into a cell or cylindrical hole, having a corresponding spiral thread projecting inwards from the interior circumference of the hole, and exactly filling the spaces between the spiral

H

thread of the screw.   If the screw is turned round in its cell, the spiral direction of the thread will cause the cylinder to advance slowly endways in the direction of its axis; and if a motive force is applied so as to turn the screw round, its advancing motion may be applied to overcome a resistance.   The screw may be turned by affixing a bar or lever to the cylinder of the screw, and applying the force to some point in that lever, at a distance from its centre of motion, so as to move that point round in a circle.   In this way the action of the screw is compounded with that of the lever.

A screw is often applied to act upon the teeth of a wheel, to turn it round, and it is then called an endless screw or worm.   The screw is placed in the direction of a tangent to the wheel, and the teeth of the wheel are adapted to fit into the spaces between the threads of the screw, which being turned round, its threads, by their continual advancement in the direction of its axis, push the teeth of the wheel round in their circular path.

The screw of itself is only a modification of the inclined plane or wedge, the inclined path being so formed around the cylinder of the screw, that by turning the cylinder round, the inclined plane will be moved endways, or advanced beneath the resistance which is to be overcome.   The length of spiral thread, which is wound about the cylinder, is the length of the inclined plane; and the space which that spiral advances along the cylinder in a direction parallel to its axis, is the vertical height or ascent of the plane.

If the cylinder were turned round by a force applied to a cord, wound about the spiral groove, or spaces between the spirals of the thread of the screw, then the effect of the screw would be the same as that of the inclined plane; but when the force to turn the screw is applied by means of a lever at a greater distance from the centre, than the circumference of the cylinder, then that force will be augmented, in as much as the diameter of the circle described by the point at which the force is applied to the lever, exceeds the diameter of the cylinder.

If the force applied to the lever of a screw, be multiplied by the circumference of the circle described by the point at which the force acts, and the product be divided by the distance which the spiral thread advances parallel to the axis of the screw, in making one turn, the quotient will be the resistance which may be overcome by the screw in moving endways.

Note. Screws are frequently made with two, three, or more spiral threads, winding spirally round the cylinder, independently of one another,—these are termed double, triple, or quadruple threaded screws; but in all such cases, the distance which any one of the spiral threads advances parallel to the axis, must be taken as the data for calculation.

Two other mechanical powers, which are not commonly noticed in that character by writers on mechanics, deserve especial notice, because they are extensively used as mechanical powers in practice.   One is called the funicular principle, or the action of stretched cords, when bended from their straight lines; and the other is the hydrostatical principle, or the action of water or other liquids upon moveable pistons; it is also called the pneumatical principle, when air or elastic fluids are the active agents, instead of water or liquids.

THE FUNICULAR ACTION is that force which is exerted by a stretched rope or cord, to draw together the two supports between which it is extended, when any force is applied laterally to the middle of the cord, so as to bend or deflect it from a right line.   This is the action of a bow-string, whereby it causes the two extremities of the bow to approach nearer towards each other, whenever the string is drawn, in preparation for throwing an arrow by it.

A cord or rope cannot operate as a mechanical power, or produce any augmentation of force, so long as that force is applied to act in the direction of the length of the cord, for then it can only transmit the force unaltered to the resistance.   But when a force acts to deflect a stretched cord from its natural straight line, the space through which such deflecting force acts, is much greater than the space through which the supports for the ends of the cord are moved towards each other;

and the force with which the supports are moved through such diminished space, will be greater than the original deflecting force, in proportion as that force acts through a greater space.

Instead of a simple cord, two straight inflexible bars or rods may be united together by a joint, in the manner of a carpenter's rule; and the two ends being extended between, and jointed to two fixed supports, the funicular action will be exerted upon those supports, whenever the middle joint, which connects the two bars, is moved to or from the straight line between the two extreme joints. The action is the same with inflexible bars, as with a simple string, but as the bars admit of acting in both directions, they may be used to force the supports farther asunder from each other, when their middle joint is moved towards the straight line, as well as to draw them towards each other, when the joint is removed from that line.

The augmentation of force, which can be attained by the funicular action, varies in every position of the bended string, being greatest when it is nearest to the straight line, and diminishing as the deflexion becomes more considerable. Suppose that the deflecting force is applied in the middle of the string, and that it acts at right angles to the straight line between its two extreme supports, then the deflecting force being multiplied by half the length of the string, and the product divided by the distance that the middle of the string is deflected from the straight line; the quotient is the force which the string must endure in the direction of its length, tending to break it. The force exerted by the string, to move the supports for its two ends towards each other, along the straight line between them, will be rather less than the force which the string endures in the direction of its length; because each half of the bended string makes an angle with that straight line, and consequently that force must be somewhat diminished, when it acts on the supports; but this difference is so small as to be nearly insensible in practical application.

The HYDROSTATICAL ACTION of water, or the pneumatical action of elastic fluids upon moveable pistons or stoppers, which are fitted into hollow cylinders, filled with such fluids, is a most important mechanical power; and being the foundation of the operation of the steam-engine, and of pumps, demands a full explanation. It derives its efficacy from a leading property of fluids, which may be thus explained. When any fluid mass is so contained within any close vessel, as to completely occupy all the interior space or capacity thereof, and form contact with its interior surfaces of boundary on all sides; if the boundary is complete and perfect at every part, a force of compressure may be exerted upon the fluid mass, by moving any part of the interior boundary, so as to encroach upon the included space; but the contained fluid will resist such motion, in consequence of its incompressibility, or of its elasticity; and it will transmit and distribute such force equally, and in all directions, to every part of the interior surface of boundary with which the fluid is in contact, so as to press such boundary outwards with a force proportionate to the surface of each part respectively.

Suppose, for instance, that the interior capacity of a hollow cylinder, into which a moveable piston is accurately fitted, is completely filled with water, and that the piston is pressed, in order to force it farther into the hollow cylinder, so as to diminish the space occupied by the water; but that space being closely bounded on all sides, and the water being an incompressible fluid, will not permit any sensible motion of the piston so circumstanced, unless there is some opening or deficiency of the boundary, whereby part of the water can escape out of the cylinder; but whatever force of compressure is exerted by the piston upon the water, will be transmitted thereby, to act equally against every part of the interior surface with which the water is in contact; and the water will press against every part of that surface, to force the same outwards in a direction at right angles to that surface, and with an equal intensity of pressure, whatever that direction may be, whether upwards, downwards, or sideways; and consequently the force exerted by the water, against any particular portion of surface with which it is in contact, will be proportionate to the extent of that surface, compared with the surface of the moveable piston, to which the compressing force is applied.

For instance, the flat bottom, or close end of the cylinder, opposite to the movable piston, being of the same size therewith, will be pressed by the water, with a force equal to that which the piston exerts upon the water; for the water merely transmits that force; and so if we consider the force exerted upon any other portion of the interior surface against which the water acts, that force will bear the same proportion to the compressing force applied by the piston, as the surface in question bears to the surface of the piston itself.

If we suppose that instead of a close bottom to the cylinder, another movable piston is fitted into it, opposite to the first mentioned piston, and the space included between them being entirely filled with water, it is very obvious, that if one piston is forced, and moved further into the cylinder, then the included water acting against the other piston, must move it further out of the cylinder with an equal force, and with equal motion. This transmission of force and motion will take place, however long the cylinder may be, and if that intermediate part of the cylinder into which the pistons are not required to pass, is altered in form, or contracted in size, there will be no difference, provided that it is sufficiently open to allow the water to pass freely from one piston to act upon the other; or the two pistons may be fitted into two distinct cylinders placed at any required distance from each other, with a pipe or open passage communicating freely from the interior of one cylinder to that of the other, then if either of the pistons is moved, or forced into its cylinder, it must displace the water from that cylinder, and expel it through the pipe, which will convey it into the interior of the other cylinder, where it will displace and move the other piston through an equal space, and with the same force as the first piston acts upon the water, which merely transmits the force unaltered.

In the above cases, the two cylinders are assumed to be of the same size, but if one piston is larger than the other, then the motion of the small piston, will be as much greater than that of the large piston, as the area of the former is smaller than that of the latter; because all the water which is displaced from one cylinder by the motion of its piston, being transferred to the other cylinder, must displace and move its piston so much, as to make room for the reception of the water so transferred; consequently, the extent of the motion thus produced in either piston, will be inversely as the area of that piston, compared with the area of the other piston.

The Hydrostatic Press, which is the most forcible of all the mechanical powers, is an application of the above principle. A large cylinder is fitted with a moveable piston; and to this piston, the resistance which is to be overcome is applied; the piston is moved by forcibly intruding water by a pump into the interior capacity of the cylinder, so as to displace and move the piston out of that cylinder. The pump is in reality a smaller cylinder, similar to the large one, and fitted with its moveable piston, which being thrust into that cylinder, drives out the water therefrom, and forces it through a pipe, into the large cylinder.

The space through which the large piston is moved, is as much less than the space through which the small piston of the pump is moved, as the area of the small piston is less than that of the large piston; consequently, the force which the large piston will exert upon the resistance that it is required to overcome, is greater than the force which the small piston exerts upon the water, in the same proportion, as the area of the large piston is greater than the area of the small piston.

The force exerted by the small piston, being multiplied by the area of the large piston, and the product divided by the area of the small piston, the quotient will be the augmented force exerted by the large piston,

Sliding Rule. $\left\{ \dfrac{\text{A Force on small pist. Force of large pist.}}{\text{B Area of small pist. Area of large pist.}} \right.$ *Ex.* $\dfrac{\text{A 60 lbs. force. 735 lbs. force.}}{\text{B 4 square inc. 49 square inc.}}$

Or, as it is most convenient to use the diameters of the pistons instead of their area,

Sliding Rule. $\left\{ \dfrac{\text{C Force on small pist. Force of large pist.}}{\text{D Diam. of small pist. Diam. of large pist.}} \right.$ *Ex.* $\dfrac{\text{C 60 lbs. force. 735 lbs. force.}}{\text{D 2 inc. diam. 7 inc. diam.}}$

The relative motions of the large and small pistons in respect to each other, will be the inverse of the above proportion, viz. if the motion of the small piston be multiplied by the area thereof, and the product divided by the area of the large piston, the quotient will be the motion communicated to the large piston.

The same proportion will apply in cases of pumps, wherein there is but one piston employed to expel water out of a cylinder, and force it through a pipe; to find what velocity, or extent of motion, will be communicated to the water in passing through the pipe, relatively to the velocity or extent of motion with which the piston is moved in its cylinder; the velocity of the piston being multiplied by the area thereof, and the product divided by the area of the pipe or orifice, through which the water is forced, the quotient will be the velocity with which the water will pass through such pipe or orifice.

Sliding Rule, $\left\{\dfrac{\text{q Velocity of piston. Velocity of water.}}{\text{D Diam. of piston. Diam. of orifice.}}\right.$ *Ex.* $\dfrac{\text{q 36 ft. per min. 1760 ft per min.}}{\text{D 7 inc. diam. 1 inch diameter.}}$
slide inverted.

Note. In the hydrostatic action, the water or other fluid which transmits the force and motion from one piston to another, is assumed to be incompressible, so that it will not diminish in volume, in consequence of the pressure to which it is subjected, but that it transmits all the motion which it receives.

The pneumatical action of elastic fluids. The gasses and air, and steam, will yield, when a competent force of compressure is exerted upon them, and they will diminish in volume so as to occupy less space; but the density being increased in proportion as the volume is diminished, the elastic force increases, in consequence of a greater mass being accumulated into the same space; and provided that the temperature of an elastic fluid remains unaltered, its elasticity will be proportionate to its density, or inversely as its volume.

When an elastic fluid is made the medium to transmit force, in order to give motion to a piston, the force with which the fluid will press against the piston, will depend upon its elastic force, and the intensity of that force will depend upon the space that the fluid is permitted to occupy. If the fluid is compressed into such a space, that its elastic force acting against a piston, is equal to the resistance which is opposed to the motion of that piston, then the action of the fluid will be the same as if it were incompressible; for in fact it will not yield any more to that force of compressure to which it is subjected, and consequently any encroachment which is made upon the space occupied by the fluid, must produce an equivalent motion of the piston, according to the proportions already stated for the hydrostatic action.

In computing the action of elastic fluids, the intensity of their elastic forces must be always considered, and referred to some standard measure, such as the force exerted by the elastic fluid against one square inch of surface, reckoned by pounds avoirdupois weight. Or the vertical height in inches of a column of mercury, whereof the weight will be equal to the elastic force of the fluid. Or the vertical height of a column of water measured in feet. Or the elastic force of any fluid may be compared with that of the atmospheric air when in its ordinary state, see p. 11.

All elastic fluids have an unlimited tendency to expand in volume, whereby any given mass is capable of occupying any assignable extent of vacant space which is allotted to it; but in all situations which are accessible to man, the air of the atmosphere occupies every portion of space from which it is not excluded by solid or liquid bodies, or by some elastic fluid possessing a force of elasticity equal to that of the atmosphere; that is, possessing an equal tendency to dilate itself and occupy a greater space. The atmospheric pressure forms a common measure for the elastic force of all elastic fluids, which are not shut up in close vessels, so as to be insulated from any communication with the atmosphere; for if a mass of any elastic fluid has an open communication with the air of the atmosphere, the pressure thereof, will limit the space occupied by such mass of elastic fluid, to such an extent, as will cause its elastic force to become precisely equal to that of the atmospheric air. This force varies in different states of the weather, but may be assumed at a mean to be equal to the weight of a column of mercury 30 inches high, as is shown by the ordinary barometer or weather-glass. The equivalent column of water is 13·55 times as high, or 33·87 feet. And the pressure against a square inch of surface, is 14·7 pounds avoirdupois.

All these are different terms for expressing that intensity of elastic force which will be assumed by any elastic fluid occupying a space which has a free communication with the atmospheric air; and provided that no change takes place in the temperature of the fluid, it will require an exertion of mechanical power to cause any alteration in the elasticity which thus establishes itself; for any alteration, whether to increase or diminish the elastic force, can only be produced by compressing the fluid mass into a less space, or by dilating it into a larger space than the atmospheric pressure would permit it to occupy.

The force required to compress or dilate any mass of any elastic fluid, so as to produce any given alteration of its elastic force, will be proportionate to the alteration produced in its volume, provided that the temperature of the fluid continues unchanged. Suppose for instance, that a space of 10 cubic feet is filled with ordinary atmospheric air, which exerts a pressure of 14·7 pounds against each square inch of the interior surface of boundary to that space; then if the space of 10 cubic feet is diminished to 5 cubic feet, supposing none of the included air to have escaped, and that its temperature remains unaltered, that air, in its diminished volume of one half, will have acquired a double elasticity, so as to press with a force of 29·4 lbs. per square inch against the boundary. Or if the space of 10 cubic feet had been increased to 20 cubic feet, supposing no more air to have entered, and no change of temperature to have taken place, then the included air in its increased volume of double, would possess only half the original elasticity, and its pressure on each square inch of its boundary would be only 7·35 lbs.

The elastic force of any mass of elastic fluid, which occupies a given extent of space, being multiplied by that space, and the product divided by the space which it occupies when compressed or dilated, the quotient is the elastic force in the dilated or compressed state.

| Sliding Rule, slide inverted. | A Elasticity. ○ Volume. | Elasticity. Volume. | *Exam.* | A 14·7 lbs. per squ. inc. ○ 24 cubic feet. | 4·9 lbs. per squ. inc. 72 cubic feet. |

Force may be transmitted from one piston to another, by means of an elastic fluid included within the two cylinders into which such pistons are fitted, provided that the fluid can pass freely from the capacity of one cylinder to the other, and that all the space which the fluid occupied in the two cylinders, and their passage of communication, is closely bounded on all sides, so that the fluid cannot escape. The force which will be transmitted from one piston to another so circumstanced, will depend upon the elastic force of the contained fluid, and that is regulated by the space the fluid occupies as above stated. If the pressure of the elastic fluid against that piston, to which the resistance is applied, is not equal to that resistance, then the piston will not be moved, in consequence of the motion of the other piston: nevertheless all the force which is exerted to move one piston, will be transmitted to the other, by the increased elastic force which the fluid will acquire from its compression; but the force so transmitted cannot produce any motion of the piston, unless it equals the resistance which is opposed to its motion.

Combinations of the mechanical powers. All machines will be found upon investigation to be combinations of the mechanical powers variously modified; the greatest number of such modifications will be found to consist of parts which have motions of partial or entire rotation round fixed centres, and which derive their efficacy from levers virtually contained in them: thus the pulley, and the wheel and axis, and all toothed wheelwork, are cases of levers; the screw is compounded of the lever, with a variety of the inclined plane or wedge. The number of absolutely simple mechanical powers may be reduced to a few principles already stated, which assume an indescribable variety of forms and motions, according to the manner in which they are applied, and combined together in machinery.

In all combinations of the simple mechanical powers, the conditions of equilibrium may be computed, by calculating the effects of each separate mechanical power, and considering the effect produced, or the resistance overcome by one power, as the motive force which is applied to actuate the next power.

The proportion that the motive force exerted upon any combination of mechanical powers, bears to the force of resistance which may be overcome thereby, may in all cases be determined by the same general proportion that has been already given for any mechanical power; viz. the product obtained by multiplying the motive force, by the space through which it acts, will be equal to the product obtained by multiplying the resisting force by the space through which it is overcome. This implies that the forces act with motion, and in cases where no motion is produced, the proportion of the forces should be taken according to the relative space through which they would act, if they did move.

Note respecting all the preceding statements of the action of the mechanical powers. The rules already laid down, will merely determine the conditions of statical equilibrium between the motive force, and the resisting force ; the expression " resistance overcome" means in reality the resistance counteracted, for no motion will result from such a state of equilibrium, as the forces will have when they are apportioned by those rules. To overcome the resisting force, and actually produce motion, some addition must in all cases be made to the motive force, beyond what is required to counteract or produce a state of statical equilibrium between them : this additional force is termed *preponderance*, and its amount does not bear any settled proportion to the motive force, or to the resisting force, but it depends upon circumstances. A great part of the preponderance is exerted to overcome the friction, with which the moving parts of the mechanical power rub upon their fixed parts ; and the remainder is exerted in urging those moving parts into motion from rest, in opposition to their inertiæ. It is evident that both these resistances must entirely depend upon the number of the moving parts, the manner of their action against their fixed supports, their masses, or weights, &c.

That part of the additional force, or preponderance, which is required to overcome friction, is absolutely lost and expended, and must be continually renewed, as long as motion is to be kept up ; but that part of the preponderance which is exerted to put the parts in motion, cannot be lost, for it will continue to reside in the moving masses, constituting what is termed their energy (see p. 17), so long as the motion continues ; and whenever the motion ceases, all that resident force or energy must be elicited from the moving masses, in the act of their coming to rest ; the force so elicited is called impetus (see p. 19). So long as any part of the force of preponderance continues to be exerted to urge the parts into motion, such motion will be with accelerated velocity ; but as soon as the parts acquire a uniform velocity, that uniformity is evidence that all further exertion of the preponderance is expended to overcome the friction. In continuous uniform motions the effect of inertiæ is a nullity, for inertia is only operative at the commencement or conclusion of the motion, viz. whilst it is accelerating from rest, to its state of uniformity, or whilst it is retarding from its uniformity to a state of rest.

MACHINE or *engine* is, properly speaking, a tool, but either of these terms implies a complicated construction, adapted to expedite labour, by performing it according to certain invariable principles. The term *machine* should only be applied to such tools or instruments, as are employed to perform intricate or forcible operations, and in which the simple mechanical powers are conspicuous. Machine is nearly synonymous with engine ; but the latter is a modern term which was intended to convey an honourable distinction, being bestowed only on mechanical contrivances, in which unusual ingenuity and skill are manifest.

The terms Machine, Engine, and Mill, are frequently used without a proper discrimination ; they all signify the practical application of the science of mechanics to the purposes of the arts, and all consist of combinations of the simple mechanical powers.

Machine should be a general term for any mill, engine, instrument, or apparatus having moving parts, and which is employed to augment the intensity of moving forces, or to regulate their action. *Machinery* should also be used as a general term, signifying the moving and operative parts of any machine or engine whatever, and therefore it includes the simple mechanical powers and their combinations. The synonymous term *mechanism* should be applied to machinery on a small scale, such as the parts of watches, clocks, and mathematical instruments, or to the smallest parts of any other machine. Thus we may say the machinery of a flour-mill, or sawing-mill ; and the mechanism of a clock, watch, orrery, &c.

The term *Engine* should be restricted to those machines which apply some principle of Hydraulics or Pneumatics, to the useful arts, their operations depending upon, or actuating fluids ; such as a steam engine, a water engine, pumping engine, blowing engine, pressure engine, and fire extinguishing engine.

Mill should be applied to large and powerful compound machines, or systems of machines ; including their first mover, as a cotton mill, which contains a number of different machines, and also the water wheel, or steam engine, which actuates them all ; so likewise a rolling mill, fulling mill, grinding mill, logwood mill, sawing mill, &c. &c.

Note. There are exceptions to this use of the term Mill, which is derived from a Latin word signifying to grind, and hence machines for grinding or reducing solid bodies to powder are called mills, as corn mill, bark mill, coffee mill, &c.; though according to the above arrangement these should be termed machines; and some curious mechanical tools which are called engines, as dividing engine, clockmaker's engine, &c. should also be called machines.

MACHINERY is a general term for all mechanical organs, which can be combined together so as to form the limbs or moving parts of machines; the term commonly includes the fixed parts of machines, as well as the moving parts, and may be considered to designate the instruments or organs by which the principles of mechanics are carried into execution, and rendered applicable to all the purposes of art and manufactures.

The common object of all mechanism, or machinery, is to convey and modify the force and motion, or power, of the first mover of the machine, and communicate it in a proper manner to the matter to be operated upon:

Thus, by means of the machinery of toothed wheels, cranks, and levers, the slow rotative motion of a water-wheel may be converted either into a rapid reciprocating motion, to work sawing, or other machines; or into a slow reciprocating motion to work pumps, &c. The velocity of the motion of the first mover may be increased or diminished by the machinery of wheelwork, according as the occasion may require either great velocity, or great power to be exerted.

In like manner, by the machinery of a parallel motion, great-lever, connecting rod, crank and fly-wheel, the rectilinear reciprocating motion of the piston-rod of a steam engine is converted into a continuous circular motion; and again this motion can, by the machinery of toothed wheel-work, be adapted, either in velocity or power, to work various machines, such as grinding-stones, circular saws, spinning machines, &c. which require great velocity; or rolling-mills, sugar mills, boring-machines, rasping machines, and other machines which require great power to give them motion, and which must therefore operate with less velocity.

Dr. Robison observes, that " the contrivance and direction of such works constitute the profession of the *Engineer;* a profession which ought by no means to be confounded with that of the mechanic, the artisan, or manufacturer. It is one of the liberal arts, as deserving of the title as medicine, surgery, architecture, painting, or sculpture. And whether we consider the importance of it to this flourishing nation, or the science that is necessary for giving eminence to the professor, it is very doubtful whether it should not take place of the three last named, and go *pari passu* with surgery and medicine."

MOTIVE FORCES are those natural forces which admit of being collected and accumulated in a suitable manner to act upon machines, so as to give them forcible motion, or mechanical power; that is, motion in opposition to some resistance which is overcome. The motive forces which are commonly used in practical mechanics, may be classed as follows :

I. *The muscular strength of men* is applied to actuate all kinds of small machines; and the strength of brute animals is applied to move carriages, and to turn the wheels of mills.

II. *The wind or natural current of the atmospheric air* is applied to propel ships at sea, and to turn the sails of windmills.

III. *The natural current of water in rivers and brooks* is used to turn the wheels of water mills.

IV. *The elastic force of steam* is the motive force in steam engines. Steam is water rarefied and expanded by heat to a vast augmentation of volume, and in such manner as to become an elastic fluid, possessing a constant tendency to further expansion or increase of volume; consequently steam will exert a force of pressure against the surfaces of boundary to any space which it occupies (see p. 9). This force of pressure can be applied to give forcible motion to a piston fitted into a cylinder, by admitting the steam into the cylinder, in order to displace the piston from the internal capacity of such cylinder. And after the steam has produced this effect, it may be deprived of nearly all its elastic force, and very greatly reduced in volume, by withdrawing from the water of which the steam is composed, that heat which occasioned its augmented volume and its elasticity; consequently any space which is occupied by elastic steam may be made nearly void, or may be greatly exhausted, by cooling and condensing that elastic steam into liquid water (see p. 10).

This method is used to exhaust the interior capacity of a cylinder into which a moveable piston is fitted, and hot elastic steam being admitted to press against the opposite surface of such piston, will force it into the exhausted space, with nearly all the elastic force of the steam, because that force is not materially opposed or counteracted by the remnant of the steam which previously occupied that space, but which has been condensed by withdrawing its heat from it (see p. 12).

The pressure of the atmosphere is employed in some steam engines as a secondary force; the steam being admitted into any space to displace the air therefrom by its elastic force, may then be cooled or condensed, so as to leave such space exhausted and nearly void; the pressure which the atmospheric air will exert, to return into and re-occupy the void space, may be used to raise water, or to impel a moveable piston before it, see p. 10.

V. *The elastic force of the gases* which are generated in the sudden combustion of fired gunpowder, is employed as a motive force, for some purposes, such as to break up hard rocks, and to project bullets and shells. The great force of these gases is occasioned by the violent heat which is generated during the combustion of the compound of inflammable matters.

The intensity or activity of any motive force may be measured by considering it as a quiescent force or pressure (see p. 16), and applying such a weight as will counteract and produce statical equilibrium with the force, so as to prevent motion in the body on which it operates. The quiescent force thus ascertained, being multiplied by the space through which that force acts on the body to which it gives motion, the product will represent the mechanical power possessed by the motive force. To find how much of this power is realized in impelling the body, we must inquire whether the intensity of the moving force remains the same, when it acts with motion, as when it was quiescent; for different motive forces are differently affected by motion, and except gravitation, they all lose a part of their activity as the velocity of their action increases; and each has some velocity of action, which it cannot exceed.

The force of gravity does not appear to be sensibly impaired by motion; for at any velocity that is ever produced in machines, the effect of gravity appears to be just as great to increase the motion of a falling body (whatever velocity it may have acquired), as it had to commence the motion from rest, in the first instance (*a*). Hence the uniformly accelerated motion of falling bodies (see p. 21); and on this account gravitation is well qualified to be a measure of force (see p. 16). When a motive force derives its efficacy from gravitation (as in the case of a natural current of water acting to turn a water-wheel), the amount of mechanical power which can be realized by a machine which is impelled by a given expenditure of such a motive force, will diminish according to the velocity with which the gravitating body acts upon the machine. Because part of the power must be communicated to the gravitating body to give it motion from rest; the power so communicated constitutes its energy, and is as the square of the velocity attained by it. This power the gravitating body retains after it has ceased to act upon the machine, and hence the slower it acts, the more power it will communicate to the machine, in proportion to the expenditure of motive force, because less will be retained in the energy of the body itself.

The mechanical power which men and animals can exert, diminishes with the velocity of their action, according to some law, which has not been sufficiently established by experiments.

The elastic force of steam, or the pressure of the atmosphere when used as motive forces, must also exert less power on machines when they act with quick motion, than when they act slower, because some power must be communicated to the steam or air, to put it in motion; this difference is very small, for the power is only diminished by that which constitutes the energy of the fluid which acts upon the machine (see p. 19). Hence steam-engines may be so constructed as to act with great rapidity of motion, without a corresponding loss of the motive force. When a column of water is applied, instead of steam, to impel a piston in a cylinder, similar to that of a steam-engine, the same force may be attained, but not with the same velocity of action; or if water is made to impel a piston with the same velocity as steam usually does, then the force upon the piston becomes greatly impaired, and only a part of the mechanical power is realized, because a great part of the motive force is communicated to the water to produce its own motion or energy, and all that is lost, to the piston. Steam is so very light that its energy is imperceptible, even when its elastic pressure is very great, and when it is acting with a considerable velocity.

The force of fired gunpowder admits of a greater velocity of action than any other motive force, and is capable of giving a prodigious velocity of motion to bodies upon which it acts, but the mechanical power which it exerts becomes lessened when it acts with its greatest velocity.

(*a*) The force of gravitation diminishes at different distances from the centre of the earth (see p. 21), and as a body must continually approach that centre when it falls by gravity, the activity of gravitation must be continually increasing during the motion; but any effect of this kind is absolutely insensible in all cases which can occur in practical mechanics.

I

RESISTING FORCES are all those natural forces which oppose themselves to any motions of the matter upon which our artificial machines are required to operate. The common object of constructing machines is to overcome such resisting forces, and to give motion to bodies, or the parts of bodies, in opposition to them ; or in other words to produce changes, either in the place which bodies occupy, or in the form which they possess.

To effect such purpose, an adequate motive force must be provided, and machinery must be so combined as to constitute a machine possessing parts of three descriptions ; viz., parts which are adapted to receive the action of the bodies in which the motive force resides ; and other parts adapted to act upon the bodies which it is required to impel with motion, or whose forms are to be altered, in opposition to the resisting forces; also intermediate parts so combined as to acquire the properties of mechanical powers, and which being interposed between the parts to which the motive force is applied, and those to which the resisting forces are applied, will transmit the force and motion from one to the other, so that it may operate with a suitable direction of action and intensity of force, to overcome the resisting forces, and produce that rapidity of motion, in opposition to them, which is required. The rapidity will depend upon the proportions given to the parts which act as mechanical powers, to enable them to transmit the motive force, with such modification of its intensity of action, as to exceed the intensity of the resisting force with a sufficient preponderance, whilst the machine is at rest.

Resisting forces are measured by applying a sufficient weight to overcome them, and which by its descent will produce motion in opposition to them (see p. 57).

The resisting forces which are opposed to the motion of machines, are as various as the objects to which machines are applied, but they may be arranged under the following heads.

I. *The force of gravitation* which must be overcome in all machines for raising water, and for raising weights of any kind ; this being the most common resisting force, and the most determinate, all the others are measured, and compared by reference to it.

II. *The force of cohesion* which occasions the particles of matter to adhere together, and form solid bodies, capable of resisting change of form. Cohesion must be overcome, in all operations of grinding, pulverizing, sawing, boring, punching, &c. Whenever the force of cohesion is destroyed in bodies, heat is generated. In liquid bodies there is also an adhesion between the particles, whereby they resist change of form.

III. *The force of elasticity in solid bodies*, is that which restores them to their original form, upon the cessation of any force, which has produced a temporary change, of form. The reaction, or return to the original form, results from the cohesive force of the particles, when cohesion is not destroyed, but has only yielded, during the action of some superior force. The elasticity of solid bodies is a resisting force, which must be counteracted, in all operations of pressing elastic substances, such as hemp, cotton, cloth, paper, oleaginous seeds, &c.

Note. The force of elasticity, is to a certain extent of the same nature as that of cohesion, for all solid bodies, are more or less elastic ; and the resisting forces of cohesion, and elasticity, are so combined in a great variety of operations, that they cannot be distinguished as distinct forces. For instance in the operations of hammering, rolling, and wire-drawing masses of metal, they resist a certain degree of compression by their elasticity, but if the compression is carried beyond certain limits, the internal arrangement of the particles is permanently changed, and then the body will not recover its primitive form, but it will retain the effect of the compression.

IV. *The elastic force of aëriform fluids*, is their tendency to enlarge in volume, and occupy more space, at the same time that they may by superior force, be condensed, and compressed into less space than they really do occupy ; this is the resisting force, in the operation of condensing air to blow furnaces, in bottling up soda water, and portable gas for illumination.

Machines must encounter the resisting forces of cohesion and elasticity, whenever their operations occasion any change of figure in solid or liquid bodies, or any alteration in the internal organization of their particles ; or whenever they act to increase the density of any aëriform fluid, by compressing it into less space, so as to diminish its volume.

In addition to these resistances, the friction with which the moving parts of machines rub upon their fixed parts, always tends to retard and impede their motions; and the inertiæ of those moving parts, and of the bodies which are put in motion by them, has an influence upon the operations of machines, whenever any changes take place in the rapidity of such motions; that is, in all cases when their motions are not uniform. The resistance of the air is a considerable impediment to the continuance of rapid motions in the parts of machines; and the motion of such bodies as are immersed in water, or other liquids, is retarded by the resistance which the particles of the liquid oppose to their separation. Some of these resistances will occur as inevitable attendants on the operations of all machines, and will deduct from the useful effects which might otherwise be expected from them; hence these incidental resistances should be considered separately from the resisting forces which machines are intended to overcome for some useful purpose.

The motions of machines are greatly influenced by the kind of resistances they are applied to overcome, compared with the kind of motive forces by which they are impelled, because different kinds of resisting forces and of motive forces, are differently affected by motion.

The resistance of friction is not altered by motion, for it requires the same force to overcome it at all velocities. The resistance of gravitation is not altered by motion; but when it is overcome with accelerated or retarded velocity, the effect of inertiæ is superadded to that of gravity. The resistance of cohesion, or the force requisite to separate the particles of solid bodies, probably increases with the velocity of the motion with which that resistance is overcome, but this has not been proved by experiment; and the resistance which elastic bodies oppose to compression, is greater when the motion is rapid, than when it is slow, but the law has not been ascertained, and probably varies in different bodies. The resistance which fluids oppose to the motion of bodies which are immersed in them, increases according to the squares of the velocities of the motion.

The forces which must be exerted to overcome different resistances at different velocities being as above, the expenditure of mechanical power in each case will be represented by the product obtained by multiplying the resistance by the velocity with which it is overcome. For instance, the power which must be expended to overcome the resistance of fluids, and cause bodies to move through them, will be as the cubes of the velocities of the motions; so that to produce a double velocity, eight times the power must be expended. The power which must be communicated to bodies to give them motion in opposition to their inertiæ, will be as the squares of the velocities which have been generated, with accelerated motion from rest. The power which must be expended to overcome friction, and gravitation with uniform motion, is simply proportionate to the velocity of the motion which is produced in opposition to those resistances.

Combinations of resisting forces. The simple cases of resisting forces rarely occur in practice, for several causes are usually combined to resist the motions of machines. It is sometimes very difficult to reduce the force of combined resistances to any common measure, because the different resistances may observe different laws of increase.

For instance, when any machinery is used to raise up a heavy body in opposition to its gravity, the inertiæ of the heavy body will occasion an additional resistance to the motion during the transition from the quiescent state to the state of motion; but when the intended velocity of motion is acquired, and for so long as the motion continues uniformly with that velocity, the inertiæ ceases to have any effect either to oppose or to increase the motion. The gravity of the body which is raised, and the friction of the moving parts of the machinery, and the resistance of the air, will continue to oppose the motion under all circumstances; and whenever the velocity of the motion is changed, the inertiæ of the moving parts will again oppose a resistance to such change (see p. 17).

When a hammer is raised with rapidity, in preparation to strike a blow, its gravity and inertiæ will both be operative to resist the motion.

In all compound machines friction has a very considerable operation to retard their motion; and in many mechanical operations, such as polishing, grinding, sawing, boring, &c. the ultimate task to be performed is very nearly the same as friction, and follows the same law of increase. All machines being thus subject to friction, must require a greater expenditure of mechanical power to give them motion than can be accounted for, in the sensible motions they have communicated to the bodies which have been subjected to their action.

In a steam-engine for raising water by pumps, the compound resistance will consist of the gravity of the water; its inertiæ when it is changed from rest to motion; the adhesion of its particles,

whereby the water resists change of form, when squeezed through the small passages of the valves;
and as all these changes of form are attended with great change of velocity, the resistance of adhesion
in fluids is inseparably combined with that of inertiæ. Lastly, the friction of all the moving parts of
the engine, which are in contact with fixed parts, including also the friction of the water itself, against
the interior surfaces of the pipes, through which it moves.

In a steam-boat or other vessel, which is to be impelled through water, by the mechanical power
of steam-engines, the resistance is very complicated. In the passage of the vessel through the fluid,
new masses of it are continually put in motion, from a state of rest, and therefore resist that
motion by their inertiæ. Part of the water so put in motion, is thrown up before the vessel into an
elevated heap above the level surface, and therefore resists by its weight. From its adhesion the
water resists that change of form which must take place to admit the vessel to pass through it.
The friction of the water against the bottom of the vessel is a great resistance, and the friction of the
moving parts of the machinery must also be considered.

FRICTION is that resistance which is occasioned to the motion of bodies by
their contact with other bodies which do not partake of the motion, so that a relative motion of sliding or rubbing must take place between the surfaces which are
in contact. Friction is the result of a natural attraction and force of adhesion
which exists between the surfaces of bodies when they are in close contact, and
which gives them a tendency to unite together, with more or less force. There is
also a mutual penetration or interlocking, of certain imperceptible prominences or
asperities, into corresponding pores or cavities, which may be supposed to exist in
all surfaces however smooth; consequently when any motion takes place between
two surfaces which are forcibly retained in contact with each other, small particles
of the prominences which interlock, must be broken off and separated from the
mass; or else the minute prominences of one surface must mount upon, and pass
over those of the other surface, so as to remove the bodies to a greater distance
from each other, in opposition to the force which tends to keep them in contact.

The resistance of friction partakes in some measure of the nature of the resistance of cohesion,
and heat is produced by the friction of rubbing surfaces in the same manner as in other cases
when the cohesion of bodies is overcome by force. The greatest heat is excited by friction when
the wearing and abrasion of the particles is greatest. Oil and grease, or animal fat, being interposed between rubbing surfaces, diminishes their friction, by filling up their pores, so as to lessen
the adhesion and mutual penetration, and thus diminish the wearing away of the particles: water
or tar, or any liquid matter which is not corrosive, has a similar tendency, though in a less degree,
than unctuous matter.

Friction cannot properly be termed a force, because it cannot in any case produce motion, but
always retards it, so as to diminish the velocity which would otherwise be produced by the exertion
of a given motive force. The amount of the resistance must be measured during the motion between
the rubbing surfaces, and from experiments of this kind, the following facts appear to be ascertained.

I. The force which must be exerted to overcome friction, and produce relative
motion between the smooth surfaces of bodies which are in contact, is nearly proportionate to the pressure by which the contact is preserved; and within certain
limits it appears that the greater or lesser extent of the surface in contact, makes
no sensible difference in the amount of the friction.

For instance, suppose a loaded sledge requires a certain force of traction to draw it, or slide it
along upon a smooth and level floor, if the same sledge is loaded so as to press on the same floor with
a double weight, then it will require a double force of traction to slide it; but the weight of the
sledge being the same, it will require the same force to slide it when the surfaces which bear upon
the floor are narrow, as when they are broad.

This proportion appears to be correct in all cases when the pressure upon a given rubbing surface
is so proportioned to the hardness and smoothness of the two surfaces, that their mutual prominences
and asperities can rise up out of the cavities of the opposite surface, and slide one over the other, without
breaking off and separating the particles of matter which form those prominences. The force required to overcome this cause of friction will be proportionate to the pressure.

The limitation to this proportion is, when the rubbing surfaces are so forcibly pressed together, that the prominent asperities of one surface will imprint into, and form corresponding cavities in the other surface; the motion between the surfaces will then occasion the interlocked particles to be cut and broken away from the solid masses to which they belong, and the resistance which must be overcome, is that of the cohesion of the particles of matter which are separated. In such cases heat is always excited in the rubbing surfaces; this may be particularly observed when a rough hard body is rubbed over a soft body, however smooth it may be; as in the operations of grinding, or filing, whereby visible particles are cut, and broken off with violence, and noise and heat. This cause of the resistance of friction is always more or less operative, as is shown by the continual wear and abrasion which takes place between rubbing surfaces, and by the heat which is produced.

In all cases of rapid cutting and wearing, that action is diminished, by increasing the surface in proportion to the pressure; consequently the friction will bear a less proportion to the pressure, which retains the surfaces in contact, when that pressure is exerted on a larger surface, than on a smaller; but if the surface is sufficient to render the cutting and wearing and heating imperceptible, then the friction is nearly as the pressure. The actual force of pressure which a given surface will sustain without sensibly cutting, when they are rubbed together, depends upon a variety of circumstances, such as the hardness of the substances, and the smoothness of their surfaces; the degree of mutual attraction between them; the kind of fluid which is interposed; and the rapidity of the rubbing motion. In practice a sufficient surface must always be given, to prevent rapid wearing, otherwise the heat which is excited will accumulate, so as to dry up the interposed fluid, and by softening the surfaces, and impairing their polish, the friction and heat will be increased beyond all bounds.

II. The force which must be exerted to overcome friction, and produce relative motion between the smooth surfaces of bodies which are in contact, is independent of the velocity of that motion, in all cases when the wear or cutting and heating is insensibly slow, for then the same force will produce any velocity of motion, according to the velocity with which that force acts.

For instance, the force of traction by which a loaded sledge is drawn along a smooth floor, is the same whether it is drawn quicker or slower. It follows that the amount of mechanical power which must be expended to overcome friction, will be proportionate to the velocity of the motion which is produced by that expenditure; because a constant force must be exerted with a varying velocity of action. For instance, whatever amount of mechanical power may be lost by the friction of the parts of any machine, whilst it is moving with a given velocity (supposing the pressure on the rubbing parts to remain the same) a double power will be lost by friction, when the machine moves with a double velocity, and so on in proportion.

It also follows, that the amount of mechanical power lost by friction, will in all cases be proportionate to the space passed over by the surfaces in contact, without regard to time; for instance, the amount of mechanical power which must be expended to move a carriage for a given distance along a level railway, will be the same whether it travels quickly or slowly over that distance.

III. The actual proportion between the pressure by which the contact of rubbing surfaces is preserved, and the force which will overcome their friction, and produce relative motion between the surfaces, depends upon the hardness and smoothness of the substances of which the surfaces are composed, for this determines the extent of the mutual penetration and interlocking of the insensible asperities; the degree of mutual attraction and adhesion which exists between the two substances of which the surfaces are composed, has a great influence upon the friction. In general friction is greater when both the rubbing surfaces are of the same kind of substance, than when the surfaces are of different substances, provided that they are suitably paired, so as not to act chemically on each other; for two dissimilar substances may be chosen, which will have less mutual attraction and adhesion, than that which exists between two surfaces of the same substance; and when two surfaces of the same substance are applied together, the asperities and cavities of one surface, will correspond and fit more exactly into those of the other, than if the substances were unlike; and therefore a greater force will be requisite to cause them to rise one out of the other. The amount of friction also varies according to the kind of liquid matter which is interposed between the surfaces. These different causes have different degrees of influence upon the amount of friction according to circumstances.

The most important circumstance is hardness of substance, to resist any sensible imprinting of one surface upon the other, but the degree of hardness requisite must be relative to the pressure upon a given surface; because two softer bodies may imprint less into each other by a small pressure, than two harder bodies by a great pressure. Hard bodies also bear the finest polish, because their particles cohere more firmly to the mass; and therefore in the operation of polishing, the particles break away from the surface individually, or in smaller clusters than happens in softer substances, whose particles come away in clusters or masses of more sensible magnitude, so as to leave larger cavities, and higher prominences on the polished surface, than in harder bodies. This may be easily understood by considering the degree of polish which could be produced on a surface of soft coarse limestone, compared with that of a surface of marble.

If one of the rubbing surfaces is a hard body well polished, the other surface may be a soft body, because the asperities of the soft body cannot penetrate or imprint into the hard surface, and by the polish given to that surface, its own asperities are greatly reduced; if both surfaces are hard bodies and well polished, there will be still less friction. Experiments are wanting upon the actual friction of the parts of machines; the following are the results of M. Coulomb's experiments.

A smooth surface of oak wood may be slided slowly over another similar surface of oak wood (the motion being in the direction of their grains) by a force of traction equal to one-tenth of the pressure which holds the surfaces in contact. If the surfaces remain for some time in contact under the pressure without motion, they will imprint and adhere, so as to require a force equal to ·43 hundredths of the pressure to begin the motion from rest, although that motion may be continued by a force of one-tenth. When the grain of one surface of oak crosses that of the other, the friction is ·26 of the pressure to begin the motion from rest, and one-tenth to continue that motion. Oak sliding against fir, in the direction of the grain ·65 to start, and ·16 to continue. Fir against fir ·56 to start, and ·17 to continue. Elm against elm ·47 to start, and ·1 tenth to continue the motion.

These proportions are when the surfaces are dry; but if fresh grease be interposed, the friction of oak against oak, in the direction of the grain is ·38 to begin the motion, and ·038 to continue it: but when the grease has been worked some time, and become soft ·24 to begin motion, and ·08 to continue it.

When a polished surface of iron slides upon oak, in the direction of the grain, the surfaces being dry, the friction is ·2 tenths of the pressure to begin motion, and ·17 to continue it; or with soft tallow between the surfaces ·11 to start from rest, and ·07 to continue the motion.

The friction of a polished surface of iron, sliding over another polished surface of iron, is about ·28 of the pressure when the surfaces are dry; or ·1 tenth, when oil or soft tallow grease is applied. The friction between metallic surfaces, is nearly the same to begin the motion as to continue it.

The friction of a polished surface of iron sliding over a polished surface of brass is about ·26 of the pressure when dry; or ·09 (about one eleventh) when oiled or greased.

Note. When grease is interposed between rubbing surfaces, although the friction is greatly diminished, yet the adhesive and clammy nature of the grease, occasions some resistance to the motion between the surfaces; this resistance is in proportion to the surface in contact; but the friction of the surfaces being nearly proportionate to the pressure, and not to the surface, the effect of interposing grease will be greater in some cases than in others, and the kind of grease has also an influence. The most fluid grease or liquid oil is the best, but it must be continually renewed, because it is more rapidly consumed by the friction than hard tallow.

When the motion is quick, if the pressure is not too great, a gentle heat is produced by the friction, but it will not increase, so long as the surfaces are well supplied with grease or oil; the friction will then be less than if the motion were slower, probably because the heat and rapid motion renders the grease more fluid, and obviates the effect of its adhesive quality, without softening or affecting the polish of the surfaces. Under these circumstances, which are the most favourable, if the supply of grease fails, the surfaces come into close contact, the friction increases, and generates a greater heat, by which the polish of the surfaces is impaired, and then they begin to cut and wear rapidly; the friction and heat increase prodigiously after the surfaces begin to cut each other, and may even produce fire; metallic surfaces often become red hot, and solder or melt themselves together into one solid mass. The forcible separation of the particles of the rubbing surfaces from each other, in opposition to their force of cohesion, generates heat, which, if it accumulates, will soften the surfaces and dry up the unctuous matter, so as to promote adhesion and increase the friction.

The pressure must therefore be so moderate, as not to force the surfaces into closer contact than their hardness can resist without imprinting into each other, or causing such adhesion as will make them cut or wear rapidly. This caution is indispensable when a rapid motion is required; otherwise the friction will produce an accumulation of heat, and go on increasing beyond bounds. With a slow motion, even if the surfaces do cut each other in some degree, and produce heat, it will be carried off by surrounding bodies, so as not to accumulate.

PREPONDERANCE is that portion of the motive force of a machine, which is an excess beyond the resisting force, and which is supposed to be lost in the operation of the machine, or expended in producing its motion. If a correct estimate is formed of the force requisite to overcome the friction of the moving parts of a machine, and all those incidental resistances which operate before the motion begins, it will appear that the moving force exceeds the resisting forces.

In reality, when a uniform motion is attained, all the preponderance of the motive force must be expended in overcoming friction and the resistance of the air, or other incidental resistance which arises or increases during the motion, and which, combined with the useful resisting forces, forms a total exactly equal to that of the motive forces, so as to leave no preponderating force; this is the state of *dynamic equilibrium* which constitutes uniform motion. So long as there remains any preponderance of the motive forces over the resisting forces, that preponderance must produce accelerated motion; and uniform motion cannot be attained, until the preponderance which existed at the commencement of the motion becomes nul; this will happen when some certain velocity is attained, and then the acceleration ceases, and the motion will continue uniform.

For instance, suppose two equal weights to be suspended at the opposite ends of a line, which passes over a pulley, they will be in equilibrium, and will have no tendency to motion; but if a sufficient weight is added to one of them, it will cause that weight to descend with it, and thus draw up the other weight. The weight so added is called the preponderating force, and is that which produces the motion, for its force is expended in overcoming the friction of the pulley and the line, the friction of the air, and the inertiæ of both weights. The inertiæ cannot oppose the motion, except whilst some change is taking place in the rate of motion, and it must cease to oppose any resistance whatever, when the motion becomes uniform. The friction offers a constant resistance to the motion at all velocities. The resistance of the air to motion increases as the square of the velocity.

If the experiment were performed in vacuo, the motion of the weights would go on accelerating, because the descending force of the preponderating weight is opposed by the friction, just in the same degree when the motion is rapid, as when it is slow. If the force of preponderance were only equal to overcoming the friction, then it could produce no motion in the weights, but only an indifference to motion or rest. If the preponderance is greater than is necessary to overcome the friction, then that excess will tend to produce motion, and as that tendency will continue to act after motion is produced, the result will be uniformly accelerated motion the same as that of a body falling freely in vacuo (see pp. 14 and 21), though performed with a slower rate of increase.

The velocity of motion which two weights thus circumstanced will acquire in any given time, will bear the same proportion to the velocity which a body would acquire in the same time, by falling freely in vacuo, as the accelerating force bears to the whole weight of matter moved. The accelerating force is the preponderating weight, minus the friction, but in a body falling freely, it is the whole weight of that body; and the masses to be moved being the same, the velocities produced in the two cases will bear the same proportion as the accelerating forces bear to each other.

For instance, suppose that the two equal weights, and the line, weigh 28 pounds, and that a weight of 4 pounds is added to one of them, then the mass to be moved would be 32 pounds; supposing that the friction of the pulley, and of the line, required two pounds to overcome it, and produce indifference to motion or rest, then there would remain 2 pounds' accelerating force, to move a mass of 32 lbs., or the accelerating force would be one-sixteenth of the weight of the mass to be moved; in this case the velocity produced (or the space described) in any given time, would be one-sixteenth of the velocity acquired (or the space described) by a body falling freely in vacuo during the same time (see p. 23).

The principle is the same, when a body falls through a fluid medium, by the weight of which a part of its gravity is counteracted; for instance when a stone sinks in water, if it were twice as heavy as an equal volume of water, then the accelerating force would be only one half of the weight of the mass moved; but in all such cases, the resistance which the fluid opposes to the motion must be taken into consideration.

If the two weights above mentioned move through the air, or through water, the resistance that the fluid offers to their motion will increase as the square of the velocity; and therefore by degrees as motion is attained, the resistance will increase so as to counteract the progress of the acceleration, and the rate of acceleration will become continually slower, until it ceases altogether: this will happen when that velocity is acquired, at which the resistance of the air will become equal to all that part of the preponderance which is not already employed in overcoming the friction, and which was active at first in producing the acceleration; but when a sufficient velocity is attained, the acceleration will cease, because there is no longer any preponderance remaining; for the resistance in its augmented state becomes exactly equal to the motive force, and the dynamic equilibrium is established, so that the motion will go on uniformly; the effect of inertiæ to oppose the motion also ceases, but the inertiæ will continue the motion which is already acquired, and will oppose any tendency to alteration of the velocity.

The common notion of preponderance being expended in producing motion, is only true whilst the motion is accelerating, and even then it is only accumulated or laid up in store in the moving mass constituting its energy (see p. 17), and ready to be given out again. When uniform motion is attained, it is evidence that the preponderance has ceased, either because the resistance has increased with the motion, or else that the motive force has diminished with the motion, or both these causes may have operated to produce the dynamic equilibrium.

In all cases when the preponderating force is gravitation, uniform motion cannot be attained, unless some part of the resistance which is overcome has the property of increasing with the velocity; neither friction nor inertiæ have this property, but the resistance of fluids has.

As another instance of preponderance, suppose the average pressure of the steam upon the piston of a steam-engine, to be 2200 pounds, when that piston is moving with a velocity of 200 feet per minute. The mechanical power exerted by the motive force will be 440 000 lbs. raised one foot per minute. If this engine is employed to raise water by pumps, it will probably be found to raise only 1650 lbs. at the rate of 200 feet per minute, therefore the mechanical effect of its performance will be only 330 000 lbs. raised one foot per minute.

The preponderance or excess of the motive force beyond the resisting force is 550 lbs. acting with a velocity of 200 feet per minute, which is a mechanical power of 110 000 lbs. raised one foot per minute. This power is expended in overcoming the friction of the moving parts of the engine, and various other incidental resistances, which come into operation when the motion of the piston is 200 feet per minute, so as to be equal to the preponderance of 550 lbs. The causes of such incidental resistances have been already stated, but as they form no part of the useful effect which the engine is required to produce, they are not usually submitted to calculation, and are accounted as loss.

CONDITIONS WHICH DETERMINE THE VELOCITY OF THE MOTIONS OF MACHINES. This depends upon the preponderance or excess of the motive forces beyond the resisting forces, supposing the preponderating force to be measured when the machine is at rest; but the motion which will be generated by its action, cannot be predetermined without considering the alteration which will take place in intensity of the motive force, and of the resisting forces, when they act with motion.

Almost all the useful operations of machines are performed with uniform motion, and it has been already stated, that in such cases all the preponderating force must be expended in overcoming certain resistances; but as motion cannot be commenced from rest, except by the preponderance of the motive force, the commencement of the motion of a machine by acceleration, and the uniform continuance of that motion when attained, must observe different laws. The preponderance must be an active force at first starting, but it must either become gradually passive, or become neutralized in the course of the acceleration; or else uniform motion could not be attained.

In some cases the resistance, or a part of it, increases with the velocity, so that at some velocity the augmented resistance will become equal to the motive force, leaving no preponderance, and then the dynamic equilibrium being established, the motion will continue uniform. In other cases the motive force diminishes with the velocity, so as to become equal to the resistance, and leave none of that preponderance which must have existed in the first instance to have commenced the motion.

In practice both these causes frequently concur, and uniform motion in machines is the result of diminished motive force on the one hand, and increased resistance on the other, whereby they come to an equality at some velocity. Hence it is obvious that there can be no determinate proportion between the motive force and the resisting forces, or any settled amount of preponderance which will cause the machine to move with any particular velocity, but it must depend upon the combination of various circumstances.

MECHANICAL EFFECT is a term implying the performance of any operation which requires an expenditure of mechanical power for that purpose. All mechanical effects are resolvable, into cases of forcible motion produced in bodies, in opposition to some resisting force ; that of gravitation being usually chosen for the measure.

To estimate the mechanical effects which are produced by machines, we should take into account all the resisting forces which oppose their motions ; and having found the weight requisite to overcome each force respectively, with the required rapidity of motion, the mechanical power may be calculated by multiplying such weight by the space through which it acts ; or by the velocity with which it descends uniformly during a given time (see p. 16.) The motive forces being estimated in the same manner, the mechanical power exerted by them, must be equivalent to that exerted by the resisting forces, in all cases of uniform motion or dynamic equilibrium.

In practice, the real amount of the various resisting forces cannot always be ascertained, particularly that of friction and inertiæ, which are only operative to resist motion, but are not forces in a quiescent state ; and most of those complicated resistances, which arise from a combination of causes, are not reducible to any rules, at present known.

Hence the term mechanical effect is usually confined to the useful result, which is attained by any given expenditure of power, accounting all other resistances as loss : for instance, the mechanical effect of an engine for raising water, is calculated according to the quantity of water which it raises to a given height, in a given time ; and the loss of power from incidental resistances is not included ; or if it is ever calculated, it is stated as an addition to the mechanical effect.

Thus we say that a steam-engine raises 528 cubic feet of water per minute, to a height of 10 feet, besides overcoming the friction of its own parts, working its own air-pump, &c. Or that a flour-mill grinds 10 bushels of wheat per hour, into flour, independently of the friction of its machinery. In the case of a locomotive engine which draws loaded waggons along a level railway, no mechanical effect is produced, according to the above expression, for all the power is exerted to overcome the friction of the machinery, which varies with the circumstances, and the amount can only be ascertained by actual trial.

STANDARD MEASURES OF THE MECHANICAL POWER EXERTED BY MACHINES. The force of men and horses being most familiar to common people, have been adopted as standards of comparison, but are very indefinite, and subject to great variation in different individuals. The following are determinate measures for the quantities of mechanical power exerted by any machine in a given time ; the power being expressed by the product which is obtained by multiplying the weight which is equivalent to the resisting force, by the space through which it acts with a uniform motion, during a given time.

Note. It has been already stated (p. 16) that time is not an essential part of the consideration of the measure of mechanical power ; but, in the present instance, the question is what amount of mechanical power is exerted in a given time by an individual first mover, whether it is an animal, or a machine ; and therefore the time in which it can complete a given task is an essential particular.

The power of a man may be assumed equal to that of raising 60 cubic feet of water (or a weight of 3750 pounds avoirdupois) to a height of one foot, in one minute. Or to any other height in feet a proportionate weight in pounds, so as to produce the same product of 3750 pounds. A stout labourer will continue to work at this rate during eight hours per day. A day's work for such a labourer may be taken at 28 800 cubic feet of water raised one foot high.

The power of a horse may be assumed, on an average, equal to that of raising 352 cubic feet of water (or a weight of 22 000 pounds) to a height of one foot in a minute. This is equal to the power of 5·867 men. A good horse will continue to work at this rate during 8 hours per day. A day's work for such a horse may be taken at 168 960 cubic feet of water raised one foot high.

The horse power in steam-engines is a measure adopted by engineers, and is equal to that of raising 528 cubic feet of water (or a weight of 33 000 pounds) to a height of one foot in a minute. This horse power is denoted by the letters HP in this work.

A horse power is equal to the power of 1½ ordinary horses, or of 8·8 men ; and as the power of steam-engines is always denominated by this measure, they are equivalent in power to half as many more horses, as the specified number of horse power : for instance, the power of a 10 horse steam-engine is equal to that of 15 horses acting together ; and if the engine works night and day, when each horse can only work during 8 hours out of the 24, it will really perform the work of 45 horses ; for it would require that number of horses to be kept to execute the same work.

In all these cases, the weight raised is to be considered as the useful mechanical effect produced,

K

independently of any losses of power by friction, or other accidental causes; consequently the mechanical power necessarily expended, to produce these useful effects, will be much greater.

To find the mechanical power which is exerted by any engine, or machine, in horse powers.
RULE. Multiply the uniform force or pressure in pounds, with which the motive force acts, by the space in feet, through which that force acts in one minute; and divide the product by 33 000 lbs. The quotient is the horse power exerted by the engine or machine.

Example. Suppose the pressure on the piston of a steam-engine to be 1650 lbs. and that it moves at the rate of 200 feet per minute. Then $1650 \times 200 = 330\,000 \div 33\,000 = 10$ horse power.

Sliding Rule, $\left\{ \dfrac{\text{A Pressure in pounds} \qquad 33\,000}{\text{O Motion feet per min.} \qquad \text{Horse power.}} \right.$ *Ex.* $\dfrac{\text{A } 1650 \text{ lbs.} \qquad 33\,000}{\text{O } 200 \text{ ft.} \qquad 10 \text{ HP.}}$
slide inverted.

To find the mechanical power which is exerted, in horse powers, by a force which acts in a circular path.
RULE. Multiply the uniform force or pressure, in pounds by the diameter in feet, of the circle in which it moves; multiply the product by the number of revolutions which it makes per minute; and divide the last product by 10 504. The quotient is the horse power exerted.

Example. Suppose that the circumference of the multiplying wheel of a steam-engine is 5 feet diameter, and that it is impelled with a force of 840·32 pounds, and makes 25 turns per minute; then $840\cdot32 \text{ lbs.} \times 5 \text{ ft.} = 4201\cdot6 \times 25 \text{ revolut.} = 105\,040 : 10\,504 = 10$ horse power.

Sliding Rule, $\left\{ \dfrac{\text{A Force or pressure in pounds} \qquad 10\,5}{\text{O Diam. in ft.} \times \text{revol. per min.} \quad \text{Horse power.}} \right.$ *Ex.* $\dfrac{\text{A Pressure} \quad 840 \qquad 105}{\text{O } (5\times25=)125 \quad 10\text{HP.}}$
slide inverted.

THE MAXIMUM OF EFFECT IN MACHINES. The conditions under which machines may produce their utmost effect, is to be considered either in respect to the amount of mechanical power that they are capable or deriving from a given exertion of the moving force by which they are actuated; or in respect to the amount of mechanical power which any given machine is capable of exerting, without reference to the expenditure of impelling force. These considerations are both important, and they are rarely separable; but to attain a maximum in one sense, will sometimes require different conditions to those which will produce it in the other; for it does not follow that a machine which is acting so as to exert its utmost power, will also afford that power at the least expense.

To obtain the maximum of effect from a given expenditure of motive force, a machine must be adapted to receive all the power of that motive force, without permitting any of its activity to escape, or be expended in waste, so as to fail of exerting its full effect, to impel the machine. This depends chiefly upon the manner in which the machine is constructed; its acting parts must be proportioned to the forces which are to act upon them, according to the velocities, intensities, and directions of those actions. These circumstances must be considered in reference to some particular velocity with which the motive force is required to impel the machine; because each kind of motive force has some particular velocity, which will enable it to act with the greatest advantage upon machinery, so as to communicate the greatest mechanical power thereto; for the activity of all those motive forces which are applied to give motion to machines, diminishes with the velocity of their action, and on the other hand, the intensity of most of the resisting forces increases with the velocity of the motion with which they are overcome, (see p. 64).

For instance, men or animals cannot move their limbs quicker than some certain velocity, even if they have no resistance to overcome; and, on the other hand, if an excessive resistance is applied they cannot move at all; in either case no mechanical power is realized. In the same manner the natural currents of wind and water, are limited in the rapidity of their motions, and will act most efficaciously on machines with some certain velocities. If the parts of the machine which receive the action of the wind or water, move very quickly, then they can only realize part of the force of the current; because after the wind or water has quitted, and ceased to act upon the moving parts, it will continue to move with the same velocity as those parts, and will therefore retain a proportion of the power of the motive force, instead of communicating the whole to the machinery.

To attain the maximum effect in any given machine, the resisting force must be so proportioned to the motive force, that they will come to an exact dynamic equilibrium, when the machine is impelled with that velocity at which the greatest product will be obtained by multiplying the resisting force overcome by the machine, by the space through which it acts.

The elastic force of steam is limited in the velocity of the motion with which it can act; but its utmost velocity would be so great, that the pistons of steam-engines would admit of a much quicker motion than that with which they usually act, before the maximum of effect would be attained. In most steam-engines the power is regulated by contracting the aperture through which the steam enters into the cylinder, to act upon the piston; and the motive force is not permitted to act freely upon the piston, but only with a limited activity. The size of the aperture being determined, there will be some particular velocity of the piston, at which that quantity of steam which will pass through the aperture, into the cylinder, will produce a maximum effect, according to the conditions already stated. In other cases the power is limited by the quantity of steam which the boiler can generate.

The motion of water wheels is also regulated by restraining the flow of water upon the wheel, through the aperture of a sluice; but the extent of their power must be limited by the quantity of water which the current of the river will afford, and the height through which it falls.

In the action of gunpowder in fire-arms the case is different from that of steam; for it is usually required to throw bullets with a greater velocity than that which would produce a maximum of effect from the consumption of a given quantity of powder; hence the mechanical power of gunpowder is much greater when it is applied to blow up rocks, with a comparatively slow motion, than when it is used to throw bullets with a great velocity.

STATEMENT OF THE PROPERTIES OF STEAM UNDER DIFFERENT CIRCUMSTANCES.

This subject has not been sufficiently investigated by experiment, to give us confidence in the exactitude of the quantities which should be established as standards for the chief properties of steam. The following statement contains such points as appear to be tolerably settled, and the most probable conjectures respecting other conclusions, which require to be verified by more exact experiments than those which have yet been made. It is extremely difficult to conduct such experiments, so as to attain results which can be depended upon.

Steam is water so impregnated with heat, as to assume the state of an aeriform or elastic fluid. In common language the word steam denotes that hot white vapour which proceeds from the surface of boiling water, and ascends into the air in the form of a cloud; but the opacity and cloudy appearance of steam is merely owing to the admixture of common air; for when steam is contained in close vessels so as to have no communication with the atmospheric air, it is a transparent elastic fluid, which possesses a greater or lesser degree of elastic force, and of density or rarity, according to its heat and its temperature. When steam is very rare, and has but a small elastic force, and a low temperature, it is common to call it vapour.

To obtain clear ideas of the nature of steam, we must consider it as water, which contains a much greater charge of heat than it contains in its usual liquid state; and that the water is expanded or rarified in volume by that surcharge of heat, so as to occupy a prodigiously greater space than it occupied in the liquid state; also that the extension in volume is attended with the property of aeriform elasticity, whereby the steam always tends to occupy more and more space, and consequently it will exert a force of pressure against those surfaces which form the boundaries of any space which is allotted to it. (See p. 53.)

This view of the subject leads us to consider the following properties of steam. The quantity of heat in steam, compared with the quantity of water of which it is composed. The quantity of water that is contained in a given volume of steam; or the space that a given weight of steam will occupy, under given conditions of elasticity and temperature; or in other words, the density of the steam. Also the elastic force of steam, in different circumstances of temperature and density; or the pressure that it will exert in consequence of its tendency to dilate, and

K 2

occupy a greater space. These different quantities are subject to great variations according to circumstances, but in all cases they bear certain relations to each other, which must be well understood, in order to give us correct notions of what takes place in the interior of the vessels of a steam engine, during its operation.

The quantity of heat in steam is much greater than is apparent to the sense of feeling, or than can be shown by a thermometer. All matter contains or possesses heat as an inherent property, and though there are no satisfactory means of estimating the quantity of heat that is thus inherent in bodies, yet that quantity of heat can be increased or diminished to a certain extent; and by observing the effects attendant upon such alterations, some safe inferences may be drawn. One of these is, that heat is capable of disappearing in bodies as hotness, so as to become insensible or hidden, and latent in the matter, and then it ceases to affect the thermometer.

When any accession of heat which may be communicated to liquid bodies ceases to give them an increase of hotness, it gives them the new property of aeriform elasticity. It is by a portion of heat thus becoming insensible or latent in bodies, that they are changed from the state of solids, to that of liquids; and that liquid bodies are converted into aeriform or elastic fluids. The portion of heat which produces these effects may be termed the latent heat of liquidity, and the latent heat of elasticity; and whenever heat thus becomes latent in liquid bodies, it must in the act of becoming latent give to such liquid bodies the property of aeriform elasticity.

Experiments have been made, by mixing known quantities of steam, with known quantities of cold water, and the results prove that steam contains a prodigious quantity of heat, which it retains in a concealed or latent state, ready to be faithfully accounted for, and communicated to any colder body. From the scalding power of steam, we should be disposed to think that it is much hotter than boiling water, but this is a mistake; for it will raise the thermometer no higher than the water from which it proceeds. And yet if we make the steam from the spout of a tea-kettle pass into and mix with a mass of cold water, that steam will be condensed or changed into water; and when one pound of water in the state of steam, has been condensed in this manner, it will have heated the mass of cold water with which it is mixed, as much as if 960 pounds of boiling hot water had been thrown in, instead of one single pound of water in the form of steam.

The heat thus denominated latent in steam, is that quantity of heat which is contained in any given weight of elastic steam, over and above the quantity of heat which is contained in an equal weight of liquid water, having the same sensible heat or temperature as that steam; this may be called its *latent heat of elasticity.*

For instance, the steam which rises from water, when it is boiling in an open vessel, so as to be exposed to the pressure of the atmosphere, when the mercury in the barometer stands 30 inches high, must have the same elasticity as the atmospheric air, and it will have a temperature of 212 degrees of Fahrenheit's thermometer (a); but in addition to the heat thus sensibly manifested, the steam contains as much more heat concealed within it, in a latent state, as would raise the temperature of 960

(a) All temperatures will be expressed in degrees of Fahrenheit's mercurial thermometer, in this work, because that thermometer is most commonly used in England. The 32d degree of Fahrenheit's scale is the freezing temperature, or that at which ice begins to melt into water; and the 212th degree is the temperature of boiling water; or of the steam which rises from water which is boiling in an open vessel, so as to be exposed to the pressure of the atmosphere, when the column of mercury in the barometer is 30 inches high, and the temperature of the air is 62 degrees of Fahrenheit. The tube of the thermometer being of uniform bore, the space between these two ascertained points is divided into 180 equal divisions, and forms the scale of Fahrenheit's thermometer; the same equal divisions are extended above 212 degrees, as far as is required to express the highest temperature which the mercurial thermometer can indicate; that is, about 680 degrees, at which temperature mercury will boil into vapour.

times its own weight of water one degree, (say from 211 deg. to 212 deg.). Or it would heat 96 times its own weight of water 10 degrees; say from 202 deg. to 212 deg.

Nevertheless, the boiling water from which that steam rises into the open air, is 212 deg. or of the same temperature as the steam; and the difference between the steam and the boiling water, viz. its aeriform quality, and its prodigious augmentation of volume (equal in this case to 1700 times) is solely owing to the heat which became latent in the boiling water, at the moment when it assumed the elastic form.

That heat could not have become latent, unless the water had at the same time assumed the aeriform state; and as the change from the liquid to the aeriform state is necessarily attended with a great increase of volume, such change cannot take place, whilst the water is confined and retained within a limited space, so as to be prevented from expanding in volume.

For instance, suppose that heat is communicated to water, when it is so confined in a large close vessel, that the steam which is produced from it cannot escape, but is retained within a small limited space; the elasticity of that steam will exert a force of compression upon the water, whereby it will be restrained from enlargement of volume, until the disposition to such enlargement is able to overcome the force of compression. Under these circumstances, the heat which is communicated to the compressed water, will raise its temperature, but will not become latent in it, except in that small portion of water which is converted into steam; and the tendency of the heat to become latent and convert more of the water into steam, will cause the water to exert the same pressure against the surfaces forming the boundaries of the space which it occupies, as is exerted by the steam, which is of the same temperature, as that to which the confined water is raised. The quantity of heat necessary to give that temperature, and force to the water, is far less than that which an equal weight of the steam of the same force requires; because there is only the sensible heat in the water, but in the steam there must be latent heat in addition to its sensible heat.

The effect produced by communicating an accession of heat to water, is either to produce elastic fluidity, and extension of volume in becoming latent; or else if the disposition to extension is forcibly restrained, then the accession of heat will produce pressure, and an increase of temperature, without becoming latent. In most cases both these effects are combined, and they take place at the same time. Part of the communicated heat becomes latent, and produces extension, whilst another part remains sensible, and produces elastic force.

If a portion of water is heated in a large close vessel, which allows only a certain space for steam, so that the steam which rises from the water, when heat is communicated to it, will be confined to accumulate in the vessel; then the elastic force and density, of the confined steam, and its temperature, will increase according to a particular progression. The increase of elasticity and density, results from the confinement and accumulation of more and more steam in the same space, wherein it cannot acquire that volume which the heat would give to it, if permitted, by becoming latent. Under these circumstances of restrained action, only part of that heat which is communicated to the water, when it is converted into steam, can become latent in that steam. The portion of the heat which does become latent, causes the water into which it enters, to expand and form as great a volume of steam, as the confined space will permit; the remaining portion of the heat which does not become latent, remains sensible, so as to increase the temperature and elasticity of the steam, because there is not room for it to exist in the steam in a latent state, unless the steam could receive a further augmentation of volume than the space will allow.

The concealment of heat in a latent state in steam, may be illustrated by comparing it to that of water contained in a damp sponge, which is so forcibly grasped in the hand, as to confine it to a certain volume. If water is communicated to the sponge in this state, the latter will absorb some water, which it will retain internally without appearing wet, until more and more water being added, it becomes overcharged, and then the surplus of water will show itself in sensible wetness on the surface. But if this wet sponge is released from its compression, it will enlarge in volume, and then its capacity for absorbing water will increase, so that all the visible wetness will enter into the substance, and the surface will appear dry again.

So steam, when retained under compressure, can only have a certain quantity of heat latent in it, and the remainder of the heat it contains will be sensible; but if the steam is released from the compressure, and allowed to expand in volume, part of the sensible heat will become latent during the expansion; whence we may reasonably infer, that the expansion is occasioned by the heat thus changing its character from sensible to latent.

Standard for the quantity of heat in steam. The best experiments seem to prove that any given weight of steam, contains such a quantity of heat, as would be sufficient to raise the temperature of 1172 times its own weight of liquid water, one degree of Fahrenheit's thermometer; provided that the heat could be abstracted from the steam, until its temperature fell to the zero of Fahrenheit's scale; but as this is never the case, the temperature of the water, which is converted into steam (or the temperature of the water into which the steam is condensed), must be deducted from 1172; and the remainder will express the weight of liquid water, which can be warmed one degree of Fahrenheit, by that quantity of heat, which is contained in one weight of steam, in addition to the quantity of heat which is naturally inherent in an equal weight of liquid water, when it is at that temperature which has been so deducted from 1172.

For instance, to convert one pound of water at 62 degrees into one pound of steam at any required temperature, as much heat must be thrown into it, as would warm $(1172 - 62 =)1110$ pounds of water one degree. Or to condense and convert one pound of steam at any temperature into one pound of water at 62 degrees, as much heat must be withdrawn from that steam, as will raise the temperature of 1110 pounds of water one degree; for instance from 61 degrees to 62 deg.

If the quantity of water which is thus used for the measure of heat be less, it is assumed that the increase of its temperature will be proportionably greater; for instance, 555 pounds would be heated two degrees, say from 60 to 62 degrees; or 111 pounds of water would be heated 10 degrees, say from 52 to 62 degrees, and so on.

The sum of the sensible and latent heat of steam, is supposed to be in all cases a constant quantity, which may be represented by the number 1172 as above; thus steam at 212 degrees of temperature has 960 of latent heat in it; but steam of 275 degrees of temperature has only 897 of latent heat; again, steam of 62 degrees temperature has 1110 of latent heat. Therefore, as the temperature and elasticity of steam increase, the latent heat diminishes, and *vice versa*.

According to this statement, the quantity of heat contained in a given quantity, by weight, of steam, is assumed to be the same, whatever may be the temperature of that steam; for instance, one pound of steam at 212 degrees of temperature, must lose as much heat, in order that it may be condensed and reduced to one pound of water at 62 degrees, as if the pound of steam had been at 275 degrees, or at any other temperature (*a*). The experiments which have been

---

(*a*) This is the conclusion that Mr. Clement has drawn from his experiments, and Mr. Watt had a similar idea; it is ascertained that the truth is not widely different from the above statement, but there is reason to suppose that it is not strictly true, from the circumstance that the capacity for heat is greater in the same bodies when they are dilated, than when they are more dense; or, in other words, that the quantity of heat required to produce a given change of temperature, in a given weight or mass of matter, will be greater when that mass occupies a greater space.

According to the above statement the same weight of water should have the same capacity for heat, under all circumstances, whether it be in the state of liquid water, or in the state of steam, and whatever the density of that steam may be; but the fact is that steam has a less capacity for heat, when its density is greater. It has never been determined what is the capacity of steam at one state of density, elasticity, and temperature, compared with an equal weight of water, or of steam, at other states; but steam is known to obey the general law, that the capacity of bodies for heat is greater when they are rarer, or occupy more space; so that a given mass of any elastic fluid will acquire a greater capacity for heat, when it expands in volume. For when very hot and highly elastic steam

made upon this subject are not so satisfactory as could be wished ; and a complete series of accurate experiments on the heat, and density of steam, at different elasticities and temperatures, is very much wanted.

The density of steam depends upon the quantity of water that is contained in a given volume of steam. The heat which is communicated to water to convert it into steam, only produces expansion of volume, and elastic force in the steam, without making any alteration in the weight of the water, of which that steam is composed. We may therefore inquire what is the extent of expansion, or augmentation of volume, that any mass, or given weight of water undergoes, by being converted into steam, under given conditions. When we speak of a given weight of steam as a mass or quantity, we should consider it as a given weight of water in the state of steam, because the heat has no weight in itself.

It is obvious, that very hot and elastic steam must consist of a greater quantity of heat and water, crowded into a given space ; and cool and weak steam of a smaller quantity of heat and water, contained in an equal space ; this difference of circumstances constitutes the two relative states of *plenum* and *vacuum*, as those terms are used in reference to steam-engines. A plenum is a space so filled with water and heat, in the form of steam, that it is crowded therewith, and consequently the excess of steam makes an effort to escape ; whilst the term vacuum denotes a space containing so little water in the form of steam, that it will not very strongly resist the entrance of other matter into the space which it occupies. These terms are only relative to each other, like hot and cold.

As steam is an elastic fluid, a given mass of it may have any assignable volume, according to the space which it is allowed to occupy ; or to the force by which it is compressed and confined within that space. We must attend to the elastic force which steam exerts against the surfaces which form the boundaries of the space that it occupies ; because the intensity of the elastic force, indicates the manner in which the steam occupies the space. The elasticity is also dependent upon the temperature, and therefore we must take that circumstance into account ; so that the three considerations of density, elasticity, and temperature, must be taken altogether, in relation to each other.

On the elastic force of steam at different temperatures. The law by which steam increases in elasticity by accession of heat, when it is allowed to saturate itself with water, and thereby to increase in density according to its own progression, has been ascertained by numerous experiments made by Mr. Watt, Dr. Robison, M. Bettancourt, Mr. Southern, and others ; and they agree so nearly as to give us great confidence in their accuracy. The experiments made by Mr. Southern have been adopted for this work because they are the most complete and were made with a very correct thermometer.

To acquire an idea of the nature of these experiments, we may suppose that a syphon barometer or an inverted glass syphon-tube, like the letter U, is partly filled with mercury at the lower part of its loop ; suppose that one of the legs is sealed at top, so as to exclude the atmospheric air, and that there is a perfect vacuum in that leg ; then the mercury that it contains, can have no other pressure upon it, except that of its own weight. Suppose that the other leg of the syphon communicates with the interior of a close vessel, or boiler, in which steam is produced from water, by communicating heat to that water ; the steam will therefore press upon the surface of the mercury in

is allowed to rush out into the open air, and expand itself, until it comes to the same elasticity as the atmosphere, its temperature will sink greatly below that of boiling water, which is the temperature due to steam of that elasticity. And air of a mean temperature being forced into a close vessel, becomes warmer by the compression ; or if it is allowed to rush out from its confinement, the issuing blast of expanded air becomes much colder than the surrounding air.

that leg with which it communicates, whilst there is no counteracting pressure on the mercury in the other leg; consequently the steam will depress the mercury in one leg, and raise it in the other in opposition to its weight, and the difference of the two levels to which the mercury rises in the two legs, will measure the elasticity of the steam.

Suppose, for instance, that there is a perfect vacuum in the steam boiler, then the mercury in both legs would stand at the same level, not being pressed on either side; but suppose that the water is heated and produces steam, which is retained and accumulated, until the temperature rises up to 212 degrees, and the steam acquires the same elasticity as the atmospheric air, then it would press down the mercury in the leg exposed to its action, and cause it to mount up in the other vacuous leg, until the difference of level in the two legs would be 30 inches, the same as a common atmospheric barometer would indicate. In all cases the difference of level between the surfaces of the mercury in the two legs, being measured in inches, will express the elastic force that the steam in the boiler possesses, when it is heated and accumulated to different temperatures.

The common notion of steam is, that it is always scalding hot; because that is observed to be the case, so long as the pressure of the atmosphere acts upon the water to retain it in the liquid state; for it will then require the temperature to be raised to 212 degrees, before its tendency to rarefy into steam can overcome that pressure. When a portion of water is shut up in a close vessel, if the vacant space above that water is exhausted of air, so as to produce a vacuum, and thus relieve the water from all other pressure except that of its own weight, then some steam will rise from the water and fill up the void space, and the elastic force and density of that steam will depend upon its temperature.

Whatever the temperature of the water may be, it will produce steam of a corresponding elasticity; if it is as low as freezing, the elasticity of the steam will then be very small; but it will increase with the temperature, according to the progression indicated by the first table. Hence it is impossible to have a perfect vacuum in any close vessel which contains liquid water in it; for that water will yield steam of more or less elasticity according to its temperature.

From the comparison of a great number of his experiments, Mr. Southern invented a method of calculating the elasticity of steam, at different temperatures, when saturated with water. Mr. Southern's method is embodied in the following rule, which will give results very nearly corresponding with the experiments:

To find the elasticity of steam of any given temperature; that temperature being expressed in degrees of Fahrenheit's thermometer; and the elasticity being expressed by the height in inches, of the column of mercury that the steam will support.

Rule. To the given temperature in degrees of Fahrenheit, add the constant temperature 51·3 degrees; and take out the logarithm of the augmented temperature, from a table of logarithms. Multiply that logarithm by the constant number 5·13; and from the product, deduct the constant logarithm 10·94123; then by the table of logarithms, find the number corresponding to the remainder, and that number is one-tenth of an inch less than the height required; therefore, by adding one-tenth of an inch to the number, we have the proper height in inches, of the column of mercury that the steam will support (a).

Example. What is the elasticity of steam at 212 degrees of temperature? 212 deg. + 51·3 deg. = 263·3 deg.; the logarithm of that number is 2·42045, which × 5·13 = 12·4169; from this deduct the constant logarithm 10·94123, and the remainder is 1·47567; the number corresponding to this logarithm is 29·1 inches; and adding ·1 inc. we have 30 inches of mercury for the required elasticity.

The rule may be used conversely to find the temperature of steam of any given elasticity. Deduct one-tenth of an inch from the height in inches, of the column of mercury which expresses the elasticity; take out the logarithm of the diminished height, and add to it the constant logarithm 10·94123; then divide the sum of these logarithms by the constant number 5·13; and find, by the table of logarithms, the number which corresponds to the quotient. That number is 51·3 degrees more

---

(a) The effect of multiplying the logarithm by 5·13, is to raise the 5·13th power of the temperature, when augmented by its constant addition of 51·3 degrees; and then by deducting the constant logarithm 10·94123, that power is divided by the constant number (87 344 000 000) viz. eighty-seven thousand, three hundred and forty-four millions; the quotient resulting from this division, with the constant addition of one-tenth of an inch, is the required elasticity in inches of mercury.

than the required temperature; therefore, by deducting the constant number 51·3 from the quotient, we have the proper temperature in degrees.

Example. What is the temperature of steam of an elasticity of 120 inches of mercury?— 120 inches − ·1 = 119·9 inc.; the logarithm of this is 2·07882 + the constant logarithm 10·94123 = 13·02005 for the sum of logarithms, which ÷ 5·13 constant number = 2·53802 quotient; the number corresponding to this logarithm is 345·2 degrees; from which deduct the constant temperature 51·3 degrees, and we have 293·9 degrees for the required temperature.

The following tables have been calculated by Mr. Southern's method. The first column of each table expresses the temperature of the steam in degrees of Fahrenheit's thermometer, and the other six columns express the elastic force that the steam will possess, when it is accumulated and heated up to those different temperatures. The elasticity is expressed in different corresponding terms for the convenience of its application to different purposes.

The second column is the height in inches, of a column of mercury that the steam will support. The numbers in the first and second columns have been calculated by the rule. The third column is the corresponding height of a column of water, assuming that mercury is 13·548 times heavier than water; and therefore that a column of one inch of mercury is equivalent to a column of 1·129 feet of water. The fourth column is the pressure in pounds avoirdupois, which is exerted against every square inch of the interior surface of the vessel containing the steam. A cubic foot of water weighs 62¼ pounds, and therefore a column of water one inch square and one foot high, weighs ·434 of a pound; or a column of mercury one inch square and one inch high weighs ·49 of a pound.

The three remaining columns show the difference between the elasticity of steam at different temperatures, and the pressure of the atmosphere, which is supposed to press into the vessel containing the steam, with a constant force of a column of mercury 30 inches high.

If the elasticity of the steam is less than the atmospheric pressure, the air will exert a force to enter into the space occupied by the steam; so as to tend to crush the sides of the vessel together with a force which is shown in different terms by the three last columns of the first table.

Table of the Elastic Force of Vapor, or Steam at different temperatures, below boiling.

N. B. The steam is supposed to be saturated with water.

Temperature by Fahrenheit's mercurial Thermometer.	Elasticity of the vapour or the pressure that it will exert to enter into a void space.			Excess of Atmospheric Pressure, or force exerted by the external air to enter into a space filled with the vapour.		
	Column of mercury.	Column of water.	Pressure on a square inch.	Column of mercury.	Column of water.	Pressure on a square inch.
Degrees.	Inches.	Feet.	Pounds avoirdupois.	Inches.	Feet.	Pounds.
32 freezing	0·18	0 20	0·09	29·82	33·67	14 61
42	0·25	0·28	0·12	29·75	33·59	14·58
52	0·35	0·39	0·17	29·65	33·48	14·53
62	0·50	0 56	0·24	29·50	33·31	14·46
72	0·71	0·80	0·35	29·29	33·07	14·35
82	1·01	1·14	0·50	28·99	32·73	14·20
92	1·42	1 60	0·70	28·58	32·27	14 00
102	1·97	2·22	0 97	28 03	31·65	13·73
112	2·68	3·02	1·31	27·32	30·85	13·41
122	3 60	4 06	1·76	26·40	29 81	12·94
132	4·76	5·37	2·33	25·24	28·50	12·37
142	6·22	7 02	3·05	23·78	26·85	11 65
152	8·03	9 06	3·93	21·97	24·81	10·77
162	10·25	11·57	5 02	19·75	22·30	9·68
172	12·94	14 60	6·34	17·06	19 27	8 36
182	16·17	18 25	7·92	13·83	15·62	6·78
192	20 04	22·62	9 82	9·96	11·25	4·88
202	24·61	27·78	12 06	5·39	6 09	2·64
212 boiling	30·00	33 87	14·70	The steam equal to the atmosphere.		

L

When the force of the steam exceeds that of the atmosphere, the contained steam will exert a force to escape from the vessel into the open air; so as to have a tendency to burst open the vessel, with a force which is shown in the three last columns of the second table.

Table of the Elastic Force of Steam at different temperatures, above boiling.

N. B. The steam is supposed to be saturated with water.

Temperature by Fahrenheit's mercurial Thermometer.	Elasticity of the steam, or the pressure that it will exert to enter into a void space.			Pressure of the steam above the atmosphere, or force exerted by the steam to escape from a close vessel into the open air.		
	Column of mercury.	Column of water.	Pressure on a square inch.	Column of mercury.	Column of water.	Pressure on a square inch.
Degrees.	Inches.	Feet.	Pounds.	Inches.	Feet.	Pounds.
212 = 1 atmo.	30·0	33·87	14·70	The steam equal to the atmosphere.		
222	36·32	41·00	17·78	6·3	7·13	3·08
232	43·6	49·22	21·36	13·6	15·35	6·66
242	52·2	58·93	25·57	22·2	25·06	10·87
250 2 = 2 at.	60 0	67·74	29·40	30·0	33·87	14·70
252	61·9	69·87	30·33	31·9	36·00	15·63
262	73·0	82·40	35·77	43·0	48·53	21·07
272	85·8	96·87	42·04	55·8	63·00	27·34
275 = 3 at.	90·0	101·61	44·10	60·0	67·74	29·40
282	100·3	113·25	49 16	70·3	79 38	34·46
292	116·7	131·75	57·18	86·7	97·88	42·48
293·9 = 4 at.	120·0	135·48	58·80	90·0	101·61	44·10
302	135·2	152·60	66·22	105·2	118·73	51·52
309·2 = 5 at.	150·0	169·35	73·50	120·0	135·48	58·80
312	156·0	176·10	76·44	126·0	142·23	61·74
322	179·3	202·40	87·85	149·3	168·53	73·15
322·3 = 6 at.	180·0	203·22	88·20	150·0	169·35	73·50
332	205·4	231·80	100·60	175·4	197·93	85·90
333·7 = 7 at.	210 0	237·09	102·90	180·0	203·22	88·20
342	234·4	264·60	114·80	204·4	230·73	100·10
343 8 = 8 at.	240·0	270·96	117·60	210·0	237·09	102·90

	Degrees.	Degrees.	Degrees.	
From	32	to 212	is 212	for the first atm.
	212	250·2	38·2	second.
	250 2	275	24·8	third.
	275	293·9	18·9	fourth.
	293·9	309·2	15·3	fifth.
	309·2	322·3	13·1	sixth.
	322·3	333·7	11·4	seventh.
	333·7	343·8	10·1	eighth.

The above tables show that the elastic force of confined steam increases more rapidly than its temperature, as measured by a thermometer; for the number of additional degrees of temperature, which will produce an additional atmosphere of elasticity, is continually diminishing as the temperature and elasticity become greater.

This circumstance has misled many projectors, who have not distinguished between temperature and quantity of heat, or between a mere force of pressure, and forcible motion, or mechanical power, but have supposed that a less quantity of heat would produce a greater mechanical effect, when applied in high pressure steam, than in low pressure steam.

Standard for the density of steam. When water is boiled in an open vessel, the steam which rises from it, must have the same elastic force as the atmospheric air ; this varies with the weather, but it is assumed, at a mean, to press with a force of 14·7 pounds avoirdupois, upon each square inch of surface, or equal to the weight of a column of mercury, of 30 inches vertical height. The temperature of steam, under the above circumstances, is made one of the standard points for the scales of thermometers, and is called 212 degrees on Fahrenheit's scale.

The volume of steam, whereof the elastic force is 30 inches of mercury, and the temperature 212 degrees of Fahrenheit's thermometer, is supposed to be 1700 times greater than the volume of an equal weight of pure liquid water. The temperature of that water, or the pressure to which it is subjected, are not very important conditions, but it is supposed to be at 62 degrees of Fahrenheit and to bear the pressure of the atmosphere.

The absolute weight of a cubic foot of pure water at 62 degrees of temperature, is 62·321 pounds according to the new Imperial standard ; but a cubic foot of common water may be assumed to weigh 62¼ pounds avoirdupois, or 437 500 grains, which ÷ 1700 gives 257·4 grains for the weight of a cubic foot of steam circumstanced as above, or ·03676 of a pound. And 27·2 cubic feet of such steam will weigh one pound avoirdupois.

This being taken as a standard for the density of steam, when at a standard elasticity and temperature, we may proceed to consider the law by which that density varies at other elasticities, and other temperatures ; but further experiments on this subject are wanted, to enable us to state that law with precision.

Probable law of the density of steam under different circumstances. It has been assumed from the results of experiments on air, that the elasticity of all elastic fluids is proportionate to their density, or to the quantity of matter contained in a given space, provided that their temperature remains unchanged ; and it is supposed that steam will obey the same law.

To explain this hypothesis, suppose a cubic foot of steam of the temperature of 212 degrees, and of an elasticity capable of sustaining the pressure of a column of mercury 30 inches high, it will contain one seventeen hundredth part of a cubic foot of water, or 257·4 grains of water as above stated. If this steam is allowed to pass into an additional space of one cubic foot, it will then occupy a space of two cubic feet, and its density will necessarily be just half as much as it was in the first instance, because there is the same quantity of matter diffused in a double space.

During the expansion of the steam, its temperature will subside of itself, much below 212 degrees, because the same quantity of heat cannot keep up a constant temperature, in the same mass of matter, when it occupies a greater space. If more heat is communicated to the steam so as to restore its temperature to 212 degrees, then according to the above assumption, the elastic force of that steam should be exactly half of what it was in the first instance, or it would be capable of sustaining a column of mercury 15 inches high. This has not been proved by any experiments which have come to the author's knowledge, but there is no reason to doubt the fact.

The converse of the above supposition would be as follows : suppose that the cubic foot of steam were to be compressed into a space of half a cubic foot, so as to give it a double density, then the temperature of the steam would rise of itself far above 212 degrees ; but it could not be restored to that temperature, without diminishing the density, for if any heat were withdrawn from the steam, in order to lower its temperature, a corresponding portion of the water contained in the steam, would immediately condense, or precipitate, and return to the liquid state ; because, nearly all the heat which the steam originally contained, is absolutely essential to the continuance of all the water in the aeriform state, whatever the circumstances of density or elasticity may be.

It is well ascertained that all permanently elastic fluids, are equally and similarly affected by the same changes of temperature ; and whether such changes of temperature affect the elastic force of the fluid, or its volume, still any permanently elastic fluid, whether common air, carbonic acid gas, or hydrogen gas, will be equally expanded in volume, or equally increased in elasticity, by making the

L 2

same alteration in its temperature. It is probable that steam obeys the same law, so long as it preserves the property of a permanently elastic fluid, that is, whilst it continues to retain the same quantity of water in an elastic state, without precipitating or condensing any portion into liquid water. This will be the case when more heat is added to steam than what the water has taken up, in consequence of its own natural disposition to receive heat; such steam then becomes like a permanently elastic fluid, and that additional heat must be all withdrawn from it, before any liquid can be condensed from it.

When heat is communicated to water in order to convert it into steam, each portion of that water may be supposed to assume the aeriform state, the instant that it has acquired that dose of heat, which by becoming latent, will give the aeriform state to such portion of water. In this state, steam is of the nature of a condensible vapour, for any abstraction of heat will cause some condensation into liquid.

Any additional quantity of heat may be communicated to steam after it has risen from water, without adding any more water to it, and if the volume of the steam remains unchanged, such additional heat will increase its temperature, and its elasticity without increasing its density. On the other hand, we cannot communicate any more water to steam, beyond that quantity which it will naturally take up with it, when it rises from water, unless a corresponding accession of heat is also communicated; because steam in the state in which it rises naturally from water, may be said to be saturated therewith, so that it can hold no more water in solution. There is no intermediate state between the liquid water and elastic steam, consequently, any portion of heat being withdrawn from steam which is saturated with water, a corresponding portion of the steam must be precipitated or condensed into water.

The elasticity and temperature of steam which is saturated with water, must of necessity increase at the same time with its density, because when more heat is communicated, more water must be converted into steam, and if that increased quantity is confined to the same space, it must become more dense and more elastic, and its temperature must rise. But when steam is surcharged with heat so as to be of the nature of a permanently elastic fluid, an accession of heat will only alter its elasticity, but cannot affect its density, so long as the volume remains unchanged.

To obtain clear ideas of all the circumstances which govern the elasticity and density of steam, we must consider both cases, viz. the most simple case of steam which is overcharged with heat, and which is therefore of the nature of a permanently elastic fluid; when the volume remaining the same, the increase of its elasticity, by accession of heat, is not attended with any increase of density. Also the compound case when the increase of elasticity by accession of heat, is attended with an increase of density, as is the case in steam which is allowed to take up as much water as it is disposed to do, and which is therefore saturated with water. The simple case is contained in the compound case, and should therefore be considered first; it has never been subjected to direct experiment, but there is every reason to suppose that steam surcharged with heat, observes the same law of expansion as air, and all other elastic fluids do; and in the absence of any direct evidence we may assume that for the fact.

Law of the expansion of common air by heat. It has been determined by accurate experiments of M. Gay Lussac, Mr. Dalton, and others, that a mass of dry air at any elasticity, which occupies a space of 100 cubic feet, or cubic inches, when it is at the freezing temperature, or 32 degrees, will expand so much in volume as to occupy $137\frac{1}{2}$ cubic feet, or cubic inches, when the temperature is raised to boiling, or 212 degrees; provided that the elasticity of the air remains the same at the boiling, as it was at the freezing temperature.

By the same law, any given volume of air which has a sufficient elasticity to support a column of mercury 30 inches high, when it is at the freezing temperature, or 32 degrees, will increase so much in elasticity, when the temperature is raised to boiling, or 212 degrees, as to support a column of mercury (1·375 times 30 inc.=) 41·25 inches high; provided that the volume of the air, or the space that the given mass occupies, remains the same at the boiling, as it was at the freezing temperature.

During the expansion of any given mass of air, of a constant elasticity, the increments of its volume are very nearly proportionate to the increments of temperature by which they are produced; that temperature being measured by a mercurial thermometer with a uniformly divided scale. The expansion produced by every additional degree of temperature on Fahrenheit's scale, is nearly at the rate of one 480th part of the volume that the air had, when it was at the freezing temperature.

The above proportion is very near the truth, in all the range of temperatures between freezing and boiling. But Messrs. Dulong and Petit made a series of experiments on the expansion of air at high temperatures, which show that the increments are not quite so rapid in the higher temperatures as in the lower; still the difference is very small, and Mr. Tredgold has formed a rule which will give results sufficiently near to these experiments for practical use.

To find the volume of a given mass of air at any temperature. Having given the volume of air at one temperature, expressed in degrees of Fahrenheit, to find the volume that the same air will have, when it is heated to any other temperature; the elasticity remaining the same in both cases.

RULE. Add the constant number 459 to the two temperatures respectively. Then multiply the given volume, by the sum of 459+ the temperature due to the unknown volume; and divide the product by the sum of 459+ the temperature due to the given volume. The quotient is the unknown volume required.

Example. Suppose 136·66 cubic feet of air at 212 degrees, what space will it occupy if it is heated up to 572 degrees? Thus 212 deg. + 459 = 671. And 572 deg. + 459 = 1031. The known volume 136·66 × 1031 = 140896 product, which ÷ 671 = 209·98 cubic feet, will be occupied at 572 degrees.

The same rule will serve to find the elasticity of a given mass of air at any temperature. Having given the elasticity at one temperature, to find how much that elasticity will be increased by heating the air to any other temperature; the volume remaining the same in both cases.

We have only to substitute the word elasticity for volume in the above rule.

Example. Suppose air at 62 degrees to have such an elasticity as to support a column of mercury 30 inches high, what elasticity will it have, if it is heated up to 242 degrees, the volume of the air remaining unchanged?

Thus, 62 degrees + 459 = 521. And 242 degrees + 459 = 701. The known elasticity 30 inc. × 701 = 21030 product, which ÷ 512 = 41·09 inches of mercury is the required elasticity.

Law of the expansion of Dry Air by heat.			
Temperature. Degrees of Fahrenheit.	Volume of Dry Air.		
	Experiments.	Calculated.	Calculated.
—33	0·865	0·868	0·865
32	1·000	1·000	1·000
212	1·375	1 366	1·373
302	1·558	1·550	1·562
392	1·739	1·733	1·747
482	1 919	1·916	1·934
572	2·098	2 010	2 120
680	2·312	2·320	2·344

The annexed table contains the experiments of Messrs. Dulong and Petit, in the two first columns; and in the third the results of the calculations made by the above rule.

The fourth column contains other calculations made by the same rule, but with a constant number 450, instead of 459 as above. These correspond best with the observations made between the freezing and boiling temperatures, but not so well with the higher temperatures.

The density of steam at different temperatures and elasticities. This has not been sufficiently established by direct experiment, and different opinions are held respecting it; Mr. Southern concluded from the result of his experiments, that the density of steam which is saturated with water, is exactly proportionate to its elasticity at all temperatures.

For instance, if steam of the same elasticity as the atmospheric air (= 30 inches of mercury and 212 degrees of temperature) is 1700 times lighter than an equal volume of cold water; then steam of two atmospheres (= 60 inches of mercury and 250·2 degrees of temperature) would be 850 times lighter. And steam of four atmospheres would be 425 times lighter, and so on. Or steam of half an atmosphere (= 15 inches of mercury and 178·6 degrees of temperature) would be 3400 times lighter than an equal bulk of cold water.

It is most probable that this simple law is modified by the circumstance, that the elasticity of steam must be augmented by the increase of temperature beyond what is merely due to its increased density, and that this augmentation takes place in the same manner and in the same proportion as the elasticity of air or any other elastic fluid is augmented by increase of temperature when there is no alteration of density. According to this hypothesis the following rule is formed from that before given for air.

To find the volume of steam at any given elasticity, compared with the volume of an equal weight of cold water. The steam is supposed to be saturated with water; and its temperature is supposed to be known, either from experiment, or from a calculation according to the preceding rule and table.

RULE. To the temperature of the steam in degrees of Fahrenheit's mercurial thermometer, add the constant number 459; multiply the sum by the constant multiplier 76; and divide the product by the elasticity of the steam in inches of mercury. The quotient will be the required volume of the steam, compared with the volume of an equal weight of cold water.

Note. If the elasticity is expressed in pounds per square inch, then use 37·24 for the constant multiplier, instead of 76.

Example. What will be the volume of steam which has an elasticity of 3 atmospheres or 90 inches of mercury; its temperature being 275 degrees. Thus 275 degrees + the constant addition 459 = 734 sum, × constant multiplier 76 = 55784 product ÷ 90 inches = 619·6 volumes.

Sliding Rule, $\begin{cases}$ A Temp. deg. + 459   Times the volume of water.   *Exam.*   A 734   619·6 vol.

slide inverted $\supset$ 76   Elasticity inches of mercury $\supset$ 76   90 inches.

Note. This rule assumes that the volume of steam is 1700 times the volume of an equal weight of cold water, when the steam has the same elasticity as the atmosphere; so as to support a column of 30 inches of mercury, or 33·87 feet of water; or to exert a pressure of 14·7 pounds per square inch, and when its temperature is 212 degrees.

The constant numbers used in the above rule are thus obtained: 459 is the constant addition which is used in the rule for the expansion of air. The above calculation requires the same number 459 to be added to the 212 degrees, the standard temperature of steam equal to the atmosphere, making 671, which should be used for a constant divisor; the standard number 1700 volumes should be used as a constant dividend; and the standard elasticity of 30 inches of mercury, should also be used as a constant multiplier. In the above rule all these numbers are concentrated into one constant multiplier (1700 × 30 = 51 000 ÷ 671 =) 76, by the use of which the calculation is much abridged.

Note. If the constant number 450 is adopted for calculating the expansion of air instead of 459 (see p. 77), then the constant multiplier for the above rule will be 77 instead of 76. Thus 212 + 450 = 662 and 51 000 ÷ 662 = 77 very nearly.

According to this rule the volume of any given weight of steam in different states of temperature and elasticity, will be greater than in the inverse proportion to that elasticity, as follows:

Table of the Elasticity and Volume of a given weight of Steam at different Temperatures.

N. B. The steam is supposed to be saturated with water.

Temperature. Degrees Fahrenheit.	Elasticity of the steam.			Volume of a given weight of steam.
	Atmospheres.	Inches of mercury.	Pounds per square inch.	
82	$\frac{1}{30}$	1	·49	41116
133·8	$\frac{1}{6}$	5	2·45	9010
161	$\frac{1}{3}$	10	4·9	4712
178·6	$\frac{1}{2}$	15	7·35	3230
192	$\frac{2}{3}$	20	9·8	2474
212	1	30	14·7	1700
233·7	1½	45	22·05	1170
250·2	2	60	29·4	898
263·8	2½	75	36·75	732
275	3	90	44·1	620
285	3½	105	51·45	538
293·9	4	120	58·8	471
301·9	4½	135	66·15	428
309·2	5	150	73·5	389
316	5½	165	80·85	357
322·3	6	180	88·2	330
328·1	6½	195	95·55	307
333·7	7	210	102·9	287
338·9	7½	225	110·25	270
343·8	8	240	117·6	254

The numbers in the last column express the volume of steam which may be produced, or the space which may be filled, by converting one volume of water into steam of different temperatures and elasticities.

The quantity of heat required for that purpose will be nearly the same in all cases; because the quantity of water to be converted into steam is always the same (a); and when steam of a higher temperature and greater elasticity is produced, it will occupy a less space, and *vice versa*.

The table, therefore, exhibits all the consequences which result from charging one cubic foot of water at 62 degrees of temperature, with such an additional quantity of heat as will convert it into steam. That additional quantity of heat is supposed to be as much as would be sufficient to raise the temperature of (1172−62=) 1110 cubic feet of water one degree, say from 61 degrees to 62 degrees. Or 555 cubic feet of water 2 degrees, say from 60 to 62 degrees.

The above table is a representation of the progressive changes which take place in the state of the steam which is produced in a high pressure steam boiler, as the temperature is raised from the state of cold water, and the steam is accumulated, until it acquires an elasticity of eight atmospheres. The first column of the table shows the increments of temperature; the second, third, and fourth, the corresponding elasticities; and the fifth, the corresponding rarities of the steam, or the number of volumes of steam, of the description stated in the other columns, which will result from the vaporization of one volume of liquid water.

(a) It has been hinted, p. 70, that this is not strictly true, although it is probably near enough to the truth for all practical conclusions. The capacity of steam for heat diminishes as it becomes more dense, so that a less quantity of heat will be required to effect a given augmentation of the temperature of a given weight of steam, when its density, elasticity, and temperature are greater, than when they are less. It would be very desirable to have these facts determined by more exact experiments than any which have yet been made.

To explain this more fully, suppose that 20 cubic feet of cold water are enclosed in a strong close boiler, which is capable of containing 30 cubic feet; then 10 cubic feet of space will be left vacant above the water, to receive the steam which rises from the water, when it is heated by fire. Suppose that vacant space to be entirely exhausted of the common air, so as to relieve the water from the pressure of the atmosphere; and that the temperature of the water is raised to 82 degrees; then the very small quantity of steam, which would be evolved at that temperature, would fill the space of 10 cubic feet, and its temperature would be 82 degrees, the same as that of the water; but the steam would be so rare, that it could only exert a pressure of ·49 of a pound, against each square inch of the interior surface of the boiler; or equal to that which would be produced by a column of mercury one inch high; and as the atmospheric air is continually pressing against the outside surface of the boiler, with a force of 14·7 pounds per square inch, the unbalanced pressure of the air, tending to crush the boiler and enter into the space, would be 14·21 lbs. per square inch, or $\frac{29}{30}$ ths of the whole pressure of the atmosphere.

According to the above table, this rare steam would be only one 41116th part of the weight of an equal volume of water; or one (1700 × 30 =)51 000th part, according to Mr. Southern's statement. As a cubic foot of water (at 62·5lbs.) weighs 437 500 grains, each cubic foot of such rare steam would weigh only (437 500 ÷ 41 116 =) 10·63 grains; consequently the weight of the steam which fills the space of 10 cubic feet, would be only 106·35 grains.

The quantity of heat contained in that steam, above that quantity of heat which is naturally contained in an equal weight of water at 82 degrees, would be sufficient to add one degree to the temperature of (1172 − 82 =) 1090 times its weight of water. That would be 1090 × 106·3 grains = 107 300 grains of water warmed from 81 degrees to 82 degrees. 107 300 grains (÷ 7000) is equal to 15·47 pounds, or (109000 ÷ 41116 =) ·245 of a cubic foot of water.

The heat of the fire, which is supposed to be applied to the boiler, being communicated to the contained water, will convert more and more of the water into steam; and supposing that steam is retained in the boiler without drawing it off, the steam must accumulate in the space of 10 cubic feet, so as to become progressively denser, and more elastic, and of a higher temperature, in the manner shown by the different numbers in the table.

For instance, when so much steam is accumulated that it will exert a pressure equal to the atmosphere, or 14·7 pounds per square inch, its temperature will be 212 degrees, and that steam will contain one 1700th part of its own volume of water; therefore 10 cubic feet of that steam will weigh (10 times 62·5 = 625 ÷ 1700 =) ·3675 of a pound. The quantity of heat in that steam, more than that which is contained in an equal weight of water at 82 degrees, is as much as would warm (1172 − 82 =) 1090 times its weight of water one degree; so that the heat of 10 cubic feet of such steam, would warm (1090 × ·6375 =) 643 pounds of water from 81 to 82 degrees.

As a last instance, suppose that the steam is progressively accumulated within the boiler until its elasticity becomes 8 times as great as that of the atmospheric air, so that it will exert 117·6 pounds against each square inch of the internal surface of the boiler; and as the atmospheric air presses against the outside, at the rate of 14·7 lbs. per square inch, the force which the steam exerts to burst open the boiler, and escape into the air, will be 102·9 lbs. per square inch. The temperature of the steam thus retained will be 343·8 degrees, and its density, according to the above table, will be one 254th part of the density of an equal volume of liquid water, or one (1700÷8 =) 212·5th part according to Mr. Southern. Hence 10 cubic feet of steam would weigh (625 lbs.÷254 =) 2·46 pounds.

The quantity of heat contained in this hot steam will be as much more than that which is contained in an equal weight of water at 82 degrees, as would add one degree to the temperature of (1172 ÷82 =) 1090 times its weight of water; so that (2·46 lbs. × 1090 =) 2830 pounds of water at 81 degrees, might be warmed to 82 degrees, by the heat contained in 10 cubic feet of such steam; or 1415 pounds of water might be raised from 80 to 82 degrees, &c.

In all these cases the water from which the steam rises will be of the same temperature as the steam which is in contact with it; and the water being compressed by the elastic force of the steam, it will be prevented from expanding, so as to assume the aeriform state, excepting so much of it as can be supplied with the necessary latent heat of elasticity, without abstracting any of that sensible heat from the water, which is necessary to keep it at the same temperature as the steam with which it is in contact.

The above instances explain all the most essential properties of steam, as accurately as they admit of being defined, according to the best information the author has been able to obtain; but in the absence of sufficient experiments to determine the truth with accuracy, the preceding calculations cannot be recommended to practical men with the same confidence as the other statements contained in this introduction.

PART I.

AN HISTORICAL ACCOUNT OF THE ORIGIN AND PROGRESS OF THE INVENTION OF THE STEAM-ENGINE, FROM THE FIRST DISCOVERY OF ITS PRINCIPLE, TO ITS PRESENT STATE OF PERFECTION.

CHAPTER I.

On the Origin of Machines to be moved by Fire or Steam.

The expansive force of steam was known to the ancients by the operation of a simple instrument called the Æolipile : it is a hollow ball of thin metal, with a long slender pipe, or spout proceeding from it, and terminating in a very small orifice : it is, in fact, a metal retort.   The ball of the æolipile being partly filled with water, it is placed upon a fire to make the contained water boil rapidly, steam then rises from it, and, after a little time, the hot steam will rush out of the orifice in a continuous current, with great violence and noise.   If greater heat is applied, the efflux of steam will become more violent, for the orifice being very small, the steam cannot escape so fast as it is produced, it therefore accumulates in the ball until its elasticity becomes sufficient to force it through the orifice with increased rapidity, and thereby discharge the steam, as fast as it is produced from the water.

The æolipile, when in action, exhibits the expansive force of steam ; and the condensation of steam by cold, so as to form a vacuum, is also experienced in the use of the same instrument ; for the usual method of filling the ball with water is to plunge it into cold water, when it is strongly heated and filled with steam. The cold then condenses the steam contained within the ball, so that it contracts into a very small space, leaving a vacuum within the ball ; and the cold water is immediately forced into the ball, by the pressure of the external air, although the orifice of the spout is so small, that water could not be introduced by any other means.

The true principle of the action of this instrument was so little understood by the ancient authors who have described it, that the steam which issued from it, when placed on the fire, was commonly supposed to be air produced by the decomposition of the water, and some have proposed to use the æolipile for blowing furnaces.

The first idea of employing the force of steam to produce motion, was by the current issuing from an æolipile.   The writings of the Philosopher Hero of Alexandria, who lived 130 years before the birth of Christ, give us some ideas of the knowledge the ancients possessed on the subject of mechanics.   One of his books, entitled " Spiritalium," describes many machines which operate by the pressure of air and water ; but in general they are curious rather than useful.   Two instruments,

M

which are described in propositions 50 and 71, are to be moved by an issuing blast of confined steam or heated air; In the first of these, the steam is generated in a vessel shaped like a vase or urn, and the steam rises through an upright pipe, the extremity or spout of which turns horizontally, and has a revolving wheel, so poised as to be capable of turning round upon the spout as an axis. The steam being thus introduced into the central part of the wheel, passes through two projecting arms, and escapes into the air through the extremities (which are bent sideways), and the wheel is caused to revolve by the efflux of the steam, in the same manner as the wheels in fireworks. As these contrivances are not of a nature to be of any real use, they must be considered as philosophical toys.

During the dark ages of ignorance and superstition which followed the decline of the Roman empire, all the arts and sciences were neglected in Europe; a small remnant of the writings of the ancient philosophers was preserved by monks, who read and copied them, without comprehending their real meaning, but which they frequently perverted to suit their own reveries of magic, astrology, alchemy, logic, &c. Hence arose absurd dogmas of schools, which cramped the exercise of human intellect during many ages, and the researches of students being limited to the discovery of the opinions which the ancients held upon different subjects, they neglected the investigation of facts. After a long lapse of barbarous ignorance, the useful arts and trades, which were practised by vulgar artizans, were improved, and many discoveries were imported into Europe from Arabia, where science was first cultivated. By degrees, the useless learning and philosophy of the schoolmen began to fall into disrepute, and the progress of civilization was promoted by several important inventions, which were made by those who were unlearned; such was the origin of the mariners' compass, fire-arms, and the arts of navigation, building, painting in oil, engraving, printing, &c.; these inventions have produced a great and general revolution in the state of Europe, and have wonderfully improved the condition of mankind, in all parts of the world.

The revival of the arts and sciences in Europe may be dated from the time of the emperor Charles V. in Germany and Spain; Francis I. in France; Henry VIII. in England, and pope Sixtus V. in Italy. Mathematics, astronomy, and navigation were cultivated with great success from this time, and mechanics became a favourite study with many ingenious men. The principal writers on this subject, during the sixteenth century, were Agricola in Germany, who died in 1555; Jacob Besson in France, 1578, and Augustino Ramelli in Italy, 1588. Their works contain descriptions and engravings of a vast variety of machines, which evince great readiness of invention, but are very deficient in the mathematical principles of mechanics. From a careful examination of these writings, and others of less note, it does not appear that they had any idea of machines to be moved by fire or steam.

PHILOSOPHICAL DISCOVERIES IN THE SEVENTEENTH CENTURY.

From the beginning of the seventeenth century, experimental philosophy was prosecuted with great ardour in every country in Europe, and the possibility of deriving mechanical power from the atmospherical pressure, and from steam, seems to have engaged the attention of several ingenious philosophers, who began by inquiring into the principles of the different phenomena of nature; and when they had detected powerful agencies, set themselves to contriving expedients by which their operation could be rendered useful to mankind. The principles laid down by these philosophers, and their first ideas and hints, though insufficient of themselves to produce any useful machines, gave employment to a number of industrious speculators of an inferior class, who, by combining practice with theory, have made their labors equally useful with those of philosophers of a higher order.

The first who deserves mention is Solomon de Caus, an engineer and architect to Louis XIII. King of France. He came to England in 1612, in the service of the Elector Palatine, who married the daughter of King James I. De Caus was employed by the Prince of Wales in ornamenting the gardens of his house at Richmond; and whilst there, he composed a work on perspective, which was published at London, 1612; and another, on the construction of sun-dials, was printed at Paris, 1624. His most important work is dated Heidelberg, 1615, and was afterwards published in French at Paris; it is entitled—" *Les Raisons des* " *Forces Mouvantes, avec diverses Machines tant utiles que plaisantes, par Salomon* " *de Caus, Ingenieur et Architecte du Roy.* Paris, 1623, *folio.*"

This is a curious work, which begins with definitions of the four elements— fire, air, water, and earth. Air is defined to be a cold, dry, and light element, which can be compressed, and rendered very violent; and he afterwards says :— " The violence will be great, when water exhales in air, by means of fire, and that " the said air is enclosed; as, for example, Take a ball of copper, of one or two " feet diameter, and one inch thick, which being filled with water by a small hole, " which shall be strongly stopped with a peg, so that neither air nor water can " escape, it is certain that if we put the said ball upon a great fire, so that it will " become very hot, that it will cause a compression so violent, that the ball will " burst in pieces, with a noise like a petard."

Theorem I. is, that the parts of the elements mix together for a time, and then each returns to its place.—" Upon this subject here is an example : Take a " round vessel of copper, soldered close on every side, and with a tube, whereof " one end approaches nearly to the bottom of the vessel, and the other end, which " projects on the outside of the vessel, has a stop-cock; there is also a hole in the " top of the vessel, with a plug to stop it. If this vessel will contain three pots of " water, then pour in one pot of water, and place the vessel on the fire, about three " or four minutes, leaving the hole open; then take the vessel off the fire, and a " little after, pour out the water at the hole, and it will be found that a part of the " said water has been evaporated by the heat of the fire. Then pour in one pot of " water, as before, and stop up the hole, and the cock, and put the vessel on the fire " for the same time as before; then take it off, and let it cool of itself, without " opening the plug, and after it is quite cold, pour out the water, and it will be " found exactly the same quantity as was put in. Thus we see that the water, " which was evaporated (the first time that the vessel was put on the fire), is " returned into water the second time, when that vapour has been shut up in " the vessel and cooled of itself.

" Another demonstration of this is, that after having put the measure of water " into the vessel, and shut the vent-hole and opened the cock, put the vessel on the " fire, and put the pot under the cock; then the water of the vessel, raising itself " by the heat of the fire, will run out through the cock : but about one-sixth or " one-eighth part of the water will not run out, because the violence of the vapour " which causes the water to rise, proceeds from the said water; which vapour goes " out through the cock after the water with great violence. There is also another " example in quicksilver, or mercury, which is a fluid mineral, but being heated " by fire, exhales in vapour, and mixes with the air for a time; but after the said " vapour is cooled, it returns to its first nature of quicksilver. The vapour of " water is much lighter, and therefore it rises higher," &c. &c.

M 2

Theorem II. is, that there is no vacuum known to us : in which he follows the opinion of Aristotle, in opposition to that of Epicurus.

" There are five different methods of raising water higher than its level, and " there are several different machines by each method : 1. By want of vacuity. " 2. By its own proper means. 3. By the aid of fire. 4. By the air. 5. By " machines, variously composed, and moved by the strength of men or horses :" and of each of these he gives an example, illustrated by a figure.

Theorem III. " When water rises by want of vacuity, it is to descend lower " than its level."

Theorem IV. " Water cannot rise by its own proper means, unless it " is to descend lower than its level."—These two theorems are demonstrated by examples of syphons.

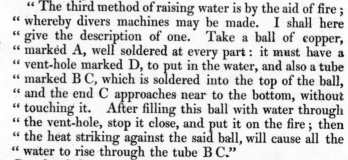

Theorem V. Water may be raised by the aid of fire higher than its level.—On this head the author gives us the following demonstration :

" The third method of raising water is by the aid of fire ; " whereby divers machines may be made. I shall here " give the description of one. Take a ball of copper, " marked A, well soldered at every part : it must have a " vent-hole marked D, to put in the water, and also a tube " marked B C, which is soldered into the top of the ball, " and the end C approaches near to the bottom, without " touching it. After filling this ball with water through " the vent-hole, stop it close, and put it on the fire ; then " the heat striking against the said ball, will cause all the " water to rise through the tube B C."

This is all that De Caus has left us upon the use of steam : the remainder of his book contains theorems on the mechanical powers, and many problems, describing various machines for raising water, clepsydra, self-acting musical instruments, and organs. The 12th problem is, to make a machine which shall move of itself. This is to operate by the heat of the sun's rays striking upon the outside of a vessel which contains some water, and the remaining space is filled with air, which being expanded by the heat, expels the water from the vessel, and raises it up in a cylindric tube, in which a light copper ball floats so as to be put in motion, and this float, by a fine thread, is made to turn a small axis, with an index on the end, to point to divisions painted on a circular dial, like that of a clock. This machine is proposed to indicate the heat of the weather, by the motion of the index; it is in fact, an air thermometer.

De Caus also shows the application of the same principle, in a machine to raise water by the heat of the sun : this he calls a continual fountain, and says it will have a great effect in hot countries, like Spain and Italy. The machine consists of four cubical vessels of copper, about a foot square : they are placed in a row, and a pipe passes over them, with a branch descending into each one, nearly to the bottom ; this pipe is to conduct the water to the elevated basin into which it is to be raised, and the extremity has a valve, to allow water to pass upwards, but not to return. The vessels are all filled about one-third of their depth with water, which is supplied to them by a pipe passing beneath all the four, with a branch turning up, into the bottom of each, the extremity of this pipe is immersed in the water of the spring from which the water is to be drawn or raised, and is provided with a valve opening upwards, so as to allow the water to rise upwards, but to prevent its return. The action of the machine is like the former, the air in the four vessels being rarified by the heat of the sun in the day, expels the water from them, through the branched pipe and its valve, into the elevated basin of the fountain; during the night the air cools, and returns to its former bulk, and a fresh supply of water is drawn into the vessels from the spring, through the lower branched pipe and valve. In another problem he proposes to increase the effect of the sun's rays by burning glasses.

This machine, though ingenious, would not be possessed of sufficient power to be of any real use, but it is deserving of notice, because the apparatus is very well arranged; and if the inventor had thought of applying steam, according to his first theorems, instead of air, and the heat of a fire, instead of that of the sun's rays, it would have formed a good engine for raising water, but many years elapsed before such an application was made (*l*), though both the principle of action, and a suitable form of apparatus are described in this work.

Solomon De Caus appears to have been a man of invention, and very correct in his notions, considering the time in which he lived: his other works show an intimate acquaintance with geometry, astronomy, and music, as well as mechanics and hydraulics. He should be distinguished from Isaac De Caus, also an engineer and architect, a native of Dieppe, who wrote a book in folio intitled "*Nouvelle Invention de Lever l'Eau plus haut que sa Source, avec quelque Machines mouvantes par le Moyen de l'Eau, et un Discours de la Conduite d'Icelle.*" This work has no date, but from the language, it appears to be older than Solomon's book of 1623; indeed the latter has all the character of a second and improved edition of the former, for the drawings and the machines described in both are exactly the same: there is a chapter on moving forces, and both contain the continual fountain, by the solar heat; but it is only in Solomon's work, that the definitions and theorems, on raising water by fire, are to be found. The plates in the latter are much better engraved, and the whole is executed in a very superior style to that of Isaac's.

The next proposal on record, to employ the force of steam for useful purposes, was by Giovanni Branca, an engineer and architect of Loretto, who contrived different kinds of mills to be worked by steam issuing from a large æolipile, and blowing forcibly against vanes fixed on the circumference of a wheel. Branca was the author of a number of ingenious inventions, which he dedicated to M. Cenci, governor of Loretto, in 1628, and published them in a work printed at Rome in the year following: it is a thin quarto volume, entitled "*Le Machine volume nuovo,* " *et di molto artificio da fare effetti maravigliosi tanto Spiritali quanto di Animale* " *Operatione, arichito di bellissime figure. Del Sig. Giovanni Branca, Cittadino*

(*l*) In Sanderson's edition of Rymer's Fœdera, vol. xix. page 239, is a copy of a patent or especial privilege granted by King Charles I. in 1630, to David Ramseye, Esq. one of the grooms of the privy chamber, to practise the following inventions, which he is stated to have invented:

1. To multiply and make saltpeter in any open field, in fower acres of ground, sufficient to serve all our dominions; 2. To raise water from low pitts by fire; 3. To make any sort of mills to goe on standing waters by continual motion, without the help of wind, waite, or horse; 4. To make all sorts of tapistrie without any weaving loom or waie ever yet in use in this kingdome; 5. To make boates, shippes, and barges to goe against strong wind and tide; 6. To make the earth more fertile than usual; 7. To raise water from low places, and mynes, and coal-pitts, by a new waie never yet in use; 8. To make hard iron soft, and likewise copper to be tuffe and soft, which is not in use within this kingdome; And 9. To make yellow wax white verie speedilie.

The exclusive privilege was for the term of fourteen years, and the patentee was to pay a yearly rent of 3*l*. 6*s*. 8*d*. to the king.   Dated 21st January, 1630.

This Ramseye appears to have been a wholesale projector, and he had other grants from Charles I. for inventions. It is not on record how he intended to raise water from low pits by fire, but it is very probable that it was something borrowed from the book, written by Solomon De Caus in 1615, and published in 1623.

" *Romano.   In Roma*, 1629." This work contains sixty-three engravings, with their descriptions in Italian and in Latin.   The æolipile, which is exhibited in the twenty-fifth plate, is a copper vessel or boiler, placed upon a brasier filled with burning charcoal : the upper part of this vessel is formed like a human head, with a pipe proceeding from the mouth, towards the circumference of a large horizontal wheel, which is surrounded with buckets like an overshot water-wheel.   This being turned rapidly round by the blast of steam, communicates its motion by toothed wheels to the different machines.   Several different applications of his proposed new power are described ; viz. mortars and pestles for pounding the materials to make gunpowder, and rolling stones for grinding the same ; machines for drawing water by buckets, and for sawing timber, driving piles, &c. &c.   The author appears to have been a man of genius, and his inventions are intended to perform useful operations : some have since been brought into use ; but the force which he could have thus obtained from steam, would have been found altogether inconsiderable, if he had ever put it in practice.

Bishop Wilkins, in his Mathematical Magic, printed in 1648, says that an æolipile may be placed in a chimney corner, so as to blow upon the sails of a wheel, and give motion to a spit for roasting meat ; but he afterwards describes a common smoke-jack, and calls it a much better machine.

Steam has so little density, that the utmost effect it can produce by percussion is very trifling, notwithstanding the great velocity with which it moves.   The blast of steam which issues from an æolipile, or from the spout of a boiling tea-kettle, appears to rush out with so much force, that at first sight it would be supposed that its power, on a larger scale, might be applied in lieu of the natural current of the wind, to give motion to machinery; but, on farther examination, it will be found, that the steam being less than half the specific gravity of the common air, its motion is greatly resisted and impeded by the surrounding air ; and as the steam contains so little matter or weight, it cannot communicate any considerable force by its impetus or concussion when it strikes a solid body.   (See p. 19.)   It must also be considered, that the force of the steam is wholly dependent upon the heat it contains, and therefore, in proportion as that heat dissipates itself in the air, or on surrounding bodies, the steam condenses and is lost.   This may be observed in the tea-kettle, which has been before instanced ; for the steam, which seems to have such an active force at the orifice of the spout, becomes a mere mist at a few inches distance, without any remaining motion or energy; and if the issuing current of steam were directed to strike upon any kind of vanes, with a view of obtaining motion from it, the condensation of the steam would be still more sudden, because the substance of such vanes would absorb the heat of the steam more rapidly than the air.

For these reasons, nothing can be expected from the motion of steam in the open air; and in reality, all the useful applications of that agent are by causing it to operate within close vessels.

DISCOVERY OF THE ATMOSPHERIC PRESSURE, 1643.

THE operation of the modern steam-engine depends principally upon making a vacuum or empty space ; this the old philosophers and schoolmen pronounced an impossibility, for they supposed all space to be filled with matter of some kind, and asserted that all our ideas of space are inseparable from those of matter.   To create a vacuum, they gravely maintained would require the hand of Omnipotence, transcending the utmost power of men, or even devils; for nature, they said, had a horror of a vacuum : such was the kind of reasoning from the time of Aristotle till within two centuries of the present age. Whilst this was the current opinion of the best informed philosophers, nothing but absurdities could be expected amongst their followers. But when we trace the steps by which true philosophy was established, about the middle of the seventeenth century, upon the wrecks of the Aristotelian tenets, it is pleasing to observe how soon these principles were brought into important use in the steam-engine.

The justly celebrated Galileo first suspected nature's abhorrence of a vacuum to be partial, or confined within certain limits. A pump having been constructed to draw water from a very deep well, at a villa of the Grand Duke Ferdinand II., at Florence in Italy, it was found that the water could not be raised higher than thirty-two feet by suction, but that the upper part of the tube, if it was higher, remained empty ; this incident being reported to Galileo, threw him into a train of reflections upon the cause of such an unexpected result, and he made many experiments at Florence, about the year 1640, which served to confirm his suspicions, that the doctrine of the schools was unsound. From the course of inquiry he had taken just before his death, in 1642, there is every reason to suppose that he would have discovered the real cause of that continual pressure, or effort which the air exerts, to enter into and fill up every space which is not previously filled with some more solid matter ; for that which the schoolmen called an abhorrence or disgust in nature, is only the effect of the weight of the mass of air, resting upon the earth.

Galileo's investigations were continued by his pupil, Torricelli, who, in the year 1643, discovered the pressure of the external atmosphere, and fully demonstrated its effects by means of the barometer, which he had just then invented. He died in 1647, but the news of his discovery having spread through all the countries in Europe, this interesting subject began to engage the attention of all philosophers, and, in the course of a few years, became generally understood by all men of education.

In France, the barometer was used with the greatest success by Blaise Pascal, a genius of the highest order, and the most original that France has ever produced. He succeeded in establishing the truth of the Torricellian discovery, in spite of a very illiberal opposition from the Jesuits, who maintained the doctrines of Aristotle with the most malignant obstinacy. From his investigations on the equilibrium of fluids in 1647, Pascal invented the mechanical power which has since been applied to such valuable use by Mr. Bramah, and called the hydrostatic press. Pascal wrote an account of his discoveries and experiments, in an excellent work entitled " *Traitez de l'Equilibre des Liqueurs, et de la pésanteur de la Masse de l'Air. Par M. Pascal. Paris*, 1698, 12mo." It was not published till after his death.

In Germany, great discoveries were made by Otto Guericke, a wealthy magistrate of Magdeburg, who amused himself by constructing curious pieces of mechanism. He conceived the idea of extracting the air from a wooden cask by a pump ; and after many failures, he succeeded in obtaining a vacuum. By that means he found out the weight of the atmosphere about the year 1650, although the previous experiments of Torricelli were unknown to him. By his various experiments with the air-pump, Guericke acquired a very complete knowledge of the atmospheric pressure, and being of a more mechanical turn than any of his predecessors, he invented and made several powerful machines which operated by that force. These he exhibited in 1654 before the Princes of the German empire, when assembled at the diet of Ratisbon.

The principal agent in all these machines was a small air-pump, by which the air could be extracted from any hollow vessel to which it was applied; and he had also a large hollow cylinder A, 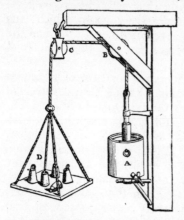 with a piston accurately fitted into it, which performed wonders, by the weight of the atmosphere pressing upon the piston, when the air was drawn out from the cylinder by the small pump. In one case, a strong rope was fastened to the piston of this cylinder, and conducted over pullies, B C, and a scale D being tied to the end of the rope, was filled with great weights, which were all lifted up from the ground by the strength of one man working the air-pump, which was applied to a cock, or spout, at the lower part of the cylinder, to extract the air from beneath the piston, and then the weight of the incumbent air bearing upon the upper surface of the piston, forced it down into the cylinder, and by the rope, drew up the scale loaded with weights.

This experiment was varied by employing a large glass globe, or bottle, in the neck of which a stop cock was cemented, to shut or open the mouth at pleasure. This globe being exhausted of air by the air-pump, the cock was closed, and the globe detached from the pump. The spout of the cock was then connected with the spout in the lower part of the large cylinder before-mentioned; and on opening the cock, part of the air contained in the cylinder, was suddenly transferred to the glass globe, so as to partially exhaust the cylinder in an instant, and raise the weights by the pressure of the atmosphere on the piston. In this way, only a partial vacuum could be obtained, because the whole of the air could not be extracted from the cylinder, and therefore the weight raised was less than in the former instance; but as the effect was produced instantaneously, it was more striking. In another case, the weights and scale were removed from the rope, and 20 men were employed to pull the end of the rope with all their strength, but the force of the piston overcame their united efforts, as soon as the cylinder was exhausted.

Another experiment, which was exhibited by Guericke at the same time, has ever since been termed the Magdeburg experiment: it is performed with two hemispheres of copper, which fit together, to make a hollow ball, and the edges where they join, are very exactly fitted together, with a piece of wet leather put between their junction, to make it quite tight, and prevent any air entering; but there is nothing to fasten the two hemispheres together, they are simply laid one upon the other, and the air pumped out from the space within the ball; the pressure of the atmosphere on the outside of the hemispheres, being then unbalanced by any pressure within, fastens the two hemispheres together so firmly that they cannot be separated without a great effort. The hemispheres which the inventor exhibited at Ratisbon, sustained the force of two teams of twelve horses each, without separating: the horses were harnessed to the two hemispheres, so as to pull in opposite directions, but they were unable to detach them so long as the vacuum continued; but when a small pin-hole was opened, and the air admitted, the two hemispheres separated by their own weight. An account of these experiments was published by Gaspard Schottus, in his *Technica Curiosa*, 1664. And the author afterwards published his own inventions, with many excellent engravings, in a folio volume in Latin, entitled " Ottonis de Guericke, Experimenta Nova ut vocantur Magdeburgica, de Vacuo Spatio, Amsterdam, 1672."

After this time, we find philosophers reasoning correctly upon the pressure of fluids and of the air. The machines which were made by Pascal and Guericke must have shown what a prodigious accumulation of force could be made by cylinders with pistons; at the time when they were new discoveries, they were much more talked of than they have been since, and no doubt contributed to the invention of the steam-engine, when combined with the principles laid down by Solomon de Caus.

The study of experimental philosophy had been commenced in England by Lord Bacon, who died in 1626; but it does not appear to have been generally cultivated, till the time of Mr. Boyle and Dr. Hooke, who made an air-pump at Oxford, in 1658, and by their various improvements and discoveries soon raised a school of English philosophers, which gave rise to the Royal Society in 1663.

THE MARQUIS OF WORCESTER'S 'FIRE WATER-WORK, 1663.

THE first real steam-engine was invented by Edward Somerset, Marquis of Worcester, who, in the reign of King Charles II., and in the year 1663, published a small pamphlet, entitled " A Century of the Names and Scantlings of the Marquis of Worcester's Inventions," written in 1659.

This little work was addressed to the king and parliament, and published with a view to obtain parliamentary encouragement for the prosecution of one hundred projects, which it enumerates in a mysterious style, but without sufficient descriptions to enable any person to put the inventions in execution. His account of the fire-engine, though too concise to give us any distinct notions of its structure, is exact as far as it goes, and agrees so nearly with what had been published forty years before, by Solomon de Caus, in his Raisons des Forces Mouvantes, that it is very probable the marquis had read that book.

No. 68. A fire water-work.—" An admirable and most forcible way to drive " up water by fire; not by drawing or sucking it upwards, for that must be as " the philosopher calleth it, *intrà sphæram activitatis*, which is but at such a " distance. But this way hath no bounder, if the vessel be strong enough; for I " have taken a piece of whole cannon, whereof the end was burst, and filled it " three-quarters full of water, stopping and screwing up the broken end, as also " the touch-hole, and making a constant fire under it; within twenty-four hours " it burst, and made a great crack; so that having a way to make my vessels, so " that they are strengthened by the force within them, and the one to fill after the " other, I have seen the water run like a constant fountain stream forty feet " high: one vessel of water, rarefied by fire, driveth up forty of cold water. And " a man that tends the work is but to turn two cocks, that one vessel of water " being consumed, another begins to force and refill with cold water, and so " successively; the fire being tended and kept constant, which the self-same person " may likewise abundantly perform in the interim, between the necessity of turning " the said cocks."

This passage certainly contains a description of an engine for raising water by the expansive force of steam; and from his expression, that one vessel of water, converted into steam, may force up forty vessels of cold water to the height of forty feet, it is probable that he had actually tried the experiment, by a working model.

N

The marquis concluded his Century of Inventions by a promise to leave to posterity a book, wherein under each head, the means of putting his inventions in execution were to be described, with the assistance of plates; but as this work never appeared, we can only judge of his abilities by this specimen. He appears to have been a person of much knowledge and ingenuity; but his obscure and enigmatical account of these inventions forms a most striking contrast to the writings of De Caus, Pascal, and Otto Guericke. The marquis seems not to have intended to instruct the world, so much as to raise wonder; and his encomiums on the utility and importance of his inventions are, in the highest degree, extravagant, more resembling the puffs of an advertising tradesman, than the patriotic communications of a nobleman. The Marquis of Worcester was indeed a projector, and very importunate and mysterious in his applications for public encouragement. From some expressions in the preface to his book, it may be concluded that he had already ruined his fortune by his schemes, and hoped to retrieve his affairs by the assistance of Parliament.

It does not appear that the noble inventor succeeded in his object, or that any public encouragement was given to his propositions; although he speaks in his preface of being rewarded by an act for some water commanding engine, of which he published a pompous account in a small quarto volume of twenty-two pages, entitled "An exact and true definition of the most stupendous water commanding engine, " invented by the Right Honourable (and deservedly to be praised and admired,) " Edward Somerset, Lord Marquis of Worcester, and by his lordship himself pre- " sented to his most excellent Majesty King Charles the Second." His character as a projector, and the many failures to which persons of that turn of mind are every day exposed, probably excited prejudice against him, which prevented all attention to his projects.

It would seem surprising that an invention, by which the steam of boiling water is stated to be capable of exerting a power equal to that of gunpowder, should have been neglected; yet when we consider that many things in this Century of Inventions are in the style of legerdemain, and others of them absolutely impossible, and contrary to all established rules of science, we are not to be surprised at the neglect which the whole experienced. For example, the ninety-ninth number of the Century is " How to make one pound weight to raise an hundred as high as one pound falleth, and yet the hundred pounds descending doth, what nothing less than one hundred pounds can effect." There are three or four others of the same character.

It must be also considered, that these projects were published at a time when true science was beginning to take place of empiricism; for the Century of Inventions appeared very soon after the establishment of the Royal Society, in the time of Mr. Boyle, Dr. Hooke, Dr. Wallis, Sir Christopher Wren, Sir Isaac Newton, and others equally skilled in calculations, and in the inventive parts of mechanics.

These circumstances explain satisfactorily why the Century of Inventions should have been coldly received; and we may conclude that the steam-engine was condemned to obscurity, on account of the wonders and fallacies with which it was accompanied.

It is but justice to the memory of the Marquis of Worcester to state, that the fire-engine is not the only invention of merit contained in his Century; on the contrary, several have been reinvented and brought into use since his time; for example, the art of writing short-hand, telegraphs, floating baths, speaking statues, carriages from which the horses can be disengaged if unruly, combination locks, secret escutcheons for locks, candle-moulds, his conceited door, rasping-mill, gravel-engine, &c. It is also probable that others may yet be brought to perfection, but the greater part are so marvellous, that for the reputation of our noble author, it is to be wished he had published nothing but an explicit account of his fire water-work; he would have rendered his name immortal by that means, without any other inventions.

SIR SAMUEL MORLAND'S PRINCIPLES OF THE FORCE OF FIRE, 1682.

THE next person, in succession, who deserves to be recorded among the inventors of the steam-engine is Sir Samuel Morland: his father was a zealous partizan of Charles II., and was created a baronet for services performed during that king's exile. The son was a man of celebrity in his time, and a number of ingenious inventions are attributed to him; such as the drum-capstan for ships, the speaking-trumpet, plunger-pump, &c. In 1680, he was appointed Master of the Works to King Charles the Second, and in the following year was sent to France to execute some water-works for the French king, Louis XIV.

In 1683, whilst in France, he wrote a small book in French, entitled "Elevation " des Eaux, par toute sorte de Machines, reduite à la Mesure, au Poids, et à la Ba- " lance. Presentée à sa Majesté très Chrestienne, par le Chevalier Morland, Gen- " tilhomme Ordinaire de la Chambre Privée, et Maistre des Méchaniques du Roi " de la Grande Brétaigne, 1683." This book is preserved in manuscript in the Harleian Collection at the British Museum; it is written on vellum, and consists of only thirty-eight small pages. It contains tables of measures and weights, theorems for calculating the contents of cylinders, and the weights of different columns of water, the requisite thickness of lead for pipes, to sustain the pressure of different columns of water; also, a sketch and brief description of the hydro-pneumatic machine called Hero's fountain, or the Chremnitz machine, where a descending column of water compresses air, and drives it out of one vessel into another, from which it expels and forces up another column of water, to a still higher level than the original supply.

The chapter on steam-engines occupies only the last four pages, of which the following is a translation:—" *The Principles of the new Force of Fire,* " *invented by the Chevalier Morland, in the Year* 1682, *and presented to his* " *Christian Majesty,* 1683." " Water being evaporated by the force of fire, " these vapours immediately require a greater space (about two thousand times) " than the water occupied before, and too forcible to be always imprisoned, will " burst a piece of cannon. But being governed according to the rules of statics, " and reduced by science, to measure, weight, and balance, then they will peaceably " carry their burden (like good horses), and thus become of great use to mankind, " particularly to raise water according to the following table, which shows the " number of pounds which can be raised 1800 times per hour, to six inches in " height, by cylinders half filled with water, as well as the different diameters and " depths of those cylinders."

N 2

Cylinders.		Pounds weight to be raised.
Diameter in feet.	Depth in feet.	
1	2	15
2	4	120
3	6	405
4	8	960
5	10	1875
6	12	3240
Number of cylinders of 6 feet diameter, and 12 feet in depth.	1	3240
	2	6480
	3	9720
	4	12960
	5	16200
	6	19440
	7	22680
	8	25920
	9	29160
	10	32400

This account is not very intelligible, nor does it appear how the cylinders were to be employed; but it is most probable Sir Samuel built upon the same foundation as the Marquis of Worcester, and that the water was intended to be introduced into the steam cylinder, and to be expelled therefrom by the expansive force of the steam. He gives no hint of using a piston in his cylinder, nor is the pressure of the atmosphere alluded to. His cylinders in the above table appear to be twice as deep as their diameters, until they become 6 ft. diameter, and 12 ft. deep, which is the largest size he mentions. The height of 6 inches raised 1800 times per hour, would be 30 strokes per minute of half a foot each, or 15 ft. French = 16 ft. English, for the motion per minute. The cylinder of 2 ft. diameter, and 4 ft. in depth, is stated as raising 120 lbs.; which, at 16 ft. per minute, is 1920 lbs. raised 1 ft. per minute. This is only half as great as the power of one man; the other cylinders are all calculated by the same proportion: hence, whatever his method may have been, it was capable of but very trifling effects, compared with our present engines.

As Morland held a place under Charles II. at the time he proposed this engine in France, it is probable his schemes did not meet with the encouragement he expected at home. Whilst he remained in France, he published a work bearing nearly the same title as the small manuscript, but it contains no mention of the force of fire :—" Elevation des Eaux, par toute sorte des Machines, reduite à la " Mesure, au Poids, à la Balance, par le moyen d'un nouveau Piston et corps de " pompe, et d'un nouveau mouvement Cyclo Elliptique. Par le Chevalier Mor- " land. Paris, 1685." 4to.

The author states that he was sent by his master, King Charles II. of England, to Louis XIV. of France, in 1681, to direct the execution of water-works. He relates many experiments made at St. Germain, on the weight of the water of the Seine, and he gives tables of the weights of different columns of water, the contents of cylinders, &c.

He describes his new piston and pump-barrel, which he says he exhibited at St. Germain in 1683. It is a solid cylinder or plunger, made smooth on the outside, and fitted into a collar of leathers at the top of the pump-barrel, which need not then be bored inside, but the motion of the solid plunger up and down through the collar of leathers will produce the same effect as that of a piston, fitted into a bored barrel, or hollow cylinder. This kind of plunger-pump is now used for draining mines by steam-engines, and for many other useful purposes. His cyclo-elliptical motion, is an elliptical wheel fixed on a revolving axis, to give an alternating motion to the piston of the pump, in lieu of a crank.

Soon after this publication, Morland returned home, and resided near the court till his death, in 1696.   The celebrated John Evelyn, in his Diary, mentions a visit he paid to Sir Samuel at his house at Hammersmith, in 1695, when he had become aged and blind, but still retained his ingenuity (*l*).   Sir Samuel Morland also invented other machines besides this fire-engine, and his plunger forcing-pump; viz. a speaking-trumpet, and two arithmetical machines, of which he published a description under the title of " The Description and Use of Two Arithmetic " Instruments, together with a short Treatise, explaining the ordinary Operations " of Arithmetic, &c.   Presented to His Most Excellent Majesty Charles II. by " S. Morland, in 1662."   This work, which is exceedingly rare, is illustrated with twelve plates, in which the different parts of the machine are exhibited.

PAPIN'S INVENTIONS, 1688 AND 1690.

Among the philosophers who applied themselves to the invention of machines to be actuated by the force of steam, the celebrated Denys Papin deserves most honourable mention ; and his projects being all published, are more on record than those of his predecessors.   Papin was born at Blois in France, and was educated as a physician.   After obtaining a degree of doctor in medicine in his own country, and making some new experiments at Paris, he travelled into England, and taking an active part in the new philosophy, he was elected a fellow of the Royal Society in December, 1680.   He passed some years in London, and assisted the celebrated Mr. Boyle in various experiments with the air-pump, of which an account is given in the History of the Royal Society, and in Mr. Boyle's " Continuation of New " Experiments Physico-Mechanical, 1682."

Whilst Papin remained in London, he invented the apparatus which is still known by the name of Papin's Digester.   In 1681 he published a description of it, with various experiments, by his new method of dissolving bones and other animal solids in water, by confining them in close vessels, which he called digesters, and which he made sufficiently strong to retain the steam and prevent all evaporation, so as to accumulate a great degree of heat.   It was also published at Paris under the title of " *Le manière d'amolir les os, et de faire cuire toutes sortes de* " *viandes en peu de temps, et à peu de fraix.   Avec une description de la machine* " *nouvellement inventée par M. Papin.   Paris*, 1682."

About the same time Dr. Hooke, the most inquisitive experimental philosopher of that inquisitive age, observed that water could not be heated above a certain temperature in the open air ; for as soon as it begins to boil, its temperature remains fixed, and an increase of heat only produces a more violent ebullition, and a more rapid waste.   But by confining the steam, Papin found that new effects could be produced.   In the course of these experiments, he became very familiar with the expansive force of steam, when heated considerably above the boiling point.   The great strength which his digesters required, and the means he was obliged to use to keep the covers down, and prevent them being blown off by the force of the confined steam, must have shown him what a powerful agent he was operating with ; but it does not appear that he thought, at that time, of employing this force in mechanics.

(*l*) " October 25th, 1695.—The Archbishop and myself went to Hammersmith, to visit Sir Samuel Morland, who was entirely blind ; a very mortifying sight.   He showed us his invention of writing, which was very ingenious ; also his wooden kalender, which instructed him, all by feeling ; and other pretty and useful inventions of mills, pumps, &c. ; and the pump he had erected that serves water to his garden and to passengers, with an inscription, and brings from a filthy part of the Thames near it, a most perfect and pure water."

The construction of the digester was afterwards improved, and it has since been greatly employed in chemical and philosophical experiments, as well as for culinary purposes.

Papin appears to have been employed by the Royal Society to make experiments on philosophical subjects, during the years 1683 to 1687; and he succeeded in demonstrating many curious facts in pneumatics. Being a Calvinist, he was prevented from returning to his native country by the revocation of the edict of Nantes; but he left England in 1687, being appointed by the Landgrave of Upper Hesse to be professor of mathematics at Marburg in Germany; in which situation he continued his philosophical researches for many years, and published occasional papers in the Acta Eruditorum of Leipsic, and others in the Philosophical Transactions of London.

In 1688 he published in the Acts, a method of prolonging the motion of water-wheels to a great distance, by drawing air through pipes; also a project for a new use of gunpowder. Papers on both these subjects had been read to the Royal Society in 1686 and 1687. In the Acts for 1690, is a further addition to his former paper on the use of gunpowder. In this he proposes to use steam, and here is the origin of the steam-engine. These three papers, which are in Latin, contain the sum of Papin's inventions, and are certainly of great merit. He afterwards republished them in French, with some additions, in a small book, entitled " Recueil de diverse Pièces touchant quelques nouvelles Machines, par D. Papin, " Cassel, 1695." It is an interesting publication, and does the author great credit. These papers deserve particular notice, as illustrative of the progress of his speculations towards a useful result.

Papin's method of transmitting the action of a water-wheel to a distance, by drawing air through pipes, is a very valuable invention, though it has been but little used: he first proposed the principle enigmatically, in the Philosophical Transactions for 1685, No. 173, as a new way of raising water; and, after many different solutions of the problem had been published by the English academicians, he showed the real application and use of the plan, to raise water out of a mine, or to force water up a high tower, by the power of a running stream, situated at a considerable distance. It is described in the Philosophical Transactions, No. 178.

The machine at Marly, in France, which pumps up the water of the river Seine, to supply the water-works in the Royal Gardens at Versailles, was then new, and was the wonder of the age; it consists of a number of water-wheels placed in the river, and by means of leading rods their power is transmitted up a high hill, to work a suite of pumps, situated nearly a quarter of a mile distant from the river. Papin proposed to use the following method to transmit the power from the water-wheels in the Seine, to raise water at Versailles instead of the cumbersome machinery which had just then been made for communicating the motion to the pumps.

The water-wheel which was placed in the river to be turned by the force of the current, was to be provided with cranks formed upon its axis, to give alternate motion to the pistons of two large air cylinders, made like forcing-pumps, but without any valves, and from the lower end of each cylinder an air pipe was to be conducted to the tower where the water was to be raised, and which, he thought, might be situated at a great distance from the water-wheel; the extremities of these air-pipes were to be divided into several branches, in order to communicate with as many chests, or receivers, which were to be fixed in the tower at different stages, and the water was to be raised by successive lifts, from one receiver to the next. For this purpose, each receiver was intended to communicate by an ascending suction-pipe, with the receiver immediately beneath it, and by another forcing-pipe, with a receiver immediately above it : the pipes were all to be furnished with valves, which would open to allow the water to rise up, but would shut to prevent the descent of the water.

The intention was to alternately rarify and compress the air in the receivers, by the motion of the pistons of the air cylinders : for instance, when one of the pistons was drawn up, the air would be drawn out of those receivers with which it communicated, and in consequence they would draw up water by their suction-pipes, and fill themselves from the receivers immediately beneath them ; also on the return of the same piston, the air would be forced back again, into these receivers, so as to expel the water from them, and raise it through the forcing-pipes into the receivers next above.　The different branches of the air-pipes from the two air-cylinders, were to be so connected with the different receivers, that whilst one of the cylinders operated, to draw the air out of one set of receivers, and make them suck up the water from the other set of receivers immediately beneath, the other cylinder would force the air into the last mentioned receivers, so as to drive the water up out of them ; therefore the actions of sucking and forcing would mutually assist each other to raise the water.

The ingenious inventor did not foresee that the elasticity of the air which he employed, would wholly defeat his end ; for when the piston of the air cylinder was forced down, it could not compress the quantity of air contained in a long pipe, so as to produce a sufficient condensation to force or raise water out of the receivers ; nor, on the other hand, could a sufficient rarefaction of the air be made by the ascent of the piston, to produce any efficient suction in the receivers, when placed at a considerable distance from the cylinders.

Papin afterwards made some alterations in his project, to obviate the objections which were urged by Dr. Hooke and other English philosophers, and in 1686 he proposed another machine to the Royal Society, and afterwards published the description in the Acta Eruditorum for 1688, p. 644, with a figure.

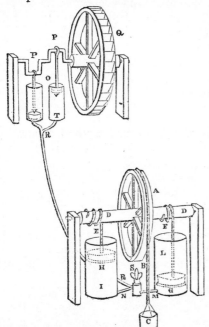

In this method, the two pumps O, which are worked by alternate cranks P P on the axle of the water-wheel Q, are provided with valves, similar to those of a sucking-pump, and they act alternately, to draw the air through the conveyance pipe R R, which leads to the mine, so as to exhaust that pipe.　At the mine are two large cylinders, I and L, with pistons G and H, fitted into them, and ropes E and F, which are fastened to each of the pistons, are wrapped several times round a horizontal axis D D, which is placed over both cylinders I and L.

Upon the middle of the axis D is a large wheel A, to wind up a cord B, which descends into the mine, and has a bucket C attached to each end.　Now by turning the axis D, and wheel A, first one way round, and then the other, the full and empty buckets are alternately drawn up, or lowered down in the mine, to draw up either water or ore therefrom, like the buckets in a well.

This alternate motion is given to the wheel and axle by the pistons, in the following manner. The ropes E and F from the piston of the two cylinders, are wrapped round the axis in opposite directions, so that when one piston is pressed down, it will draw the rope and turn the axle, which winding up the other rope, will draw up the piston of the other cylinder. A single conveyance-pipe R R, leads from both the air-pumps O at the water-wheel; but when the pipe arrives at the mine it divides into two branches, N and M, one for each cylinder, and at the intersection of these branches a double-passaged cock S is placed, which will admit the air from either of the cylinders, I and L, into the conveyance-pipe R R, which leads it to the air-pumps, or it will admit the atmospheric air into either of the cylinders; these passages are opened alternately by turning the cock, so that whilst the air from one cylinder is drawing off through the conveyance-pipe by the air-pumps, the atmospheric air will have free entrance to the other cylinder.

The air-pumps O being kept in continual action by the water-wheel, draw the air away from the cylinders I and L, by a continual suction through the long conveyance-pipe R R, and the air will be exhausted from either of those cylinders I or L, according as the cock S is turned, so as to form a vacuum under the piston of that cylinder, and therefore the pressure of the atmosphere will act, to press it down into its cylinder; and by the rope E or F, which is attached to it, and wound round the axle D of the wheel A, the axle and wheel will be turned round, and will draw up the cord B which passes over the wheel, so as to raise up one bucket in the mine, and lower down the other. During this descent of one piston, the other piston is freely at liberty to be drawn up in its cylinder, because the cock S admits the atmospheric air into the same. When the piston, under which the vacuum has been made, is pressed down to the bottom of its cylinder, the other piston will be drawn up to the top of its cylinder, by its rope winding upon the axis. In this state the cock S is turned the other way, in order to exhaust the air from that cylinder in which the piston is at the top, and to admit fresh air into the other cylinder, in which the piston is at the bottom. This will cause the other piston to be pressed down, and turn the axis and wheel round in an opposite direction to what it did before, and will draw up the opposite bucket from the mine.

In this way a constant reciprocating motion of the axis is kept up, and the power of the water-wheel is transmitted by the conveyance-pipe, to any required distance, where, by using a larger or smaller cylinder, it may be made to act with any required force and velocity. The motion of the pistons may easily be applied to turn the cock at the desired moment, and then the machine becomes self-acting.

It is rather surprising, that so simple and advantageous a method of exerting power, at a distance from the first mover, should have remained neglected and unnoticed so long as it has been.

Papin's method may be applied to other useful purposes; but when the pistons are required to pull with a sudden motion, as in the case of stamping coins, the air-pipe must terminate in a receiver, or air-chamber near the cylinders. This will be kept exhausted by the pumps, and being of sufficient capacity, the air will rush into it, and be taken away from beneath the piston the instant the cock is opened; whereas, without such a receiver, the air would be drawn off gradually by the pumps. If the conveyance pipe is made of large dimensions, it will effect the same end most completely.

Returning to the subject of Papin's claims to the invention of the steam-engine; we find that his experiments with his digester, whilst in England in 1682, had rendered the elastic power of steam very familiar to him; he also made many attempts to employ the force of gunpowder in mechanics, and for raising water. He showed experiments on this subject to the Royal Society in 1687, immediately before he left England; but being unable to effect his object by gunpowder, he afterwards proposed to employ steam instead.

Papin's project of 1687, for using gunpowder, is of no real utility : indeed, he was sensible that he had not succeeded, and said he published it only to excite the learned of other countries to perfect it. (See the Acta Eruditorem for 1688, p. 497.) He proposed to employ a cylinder with a piston fitted into it; and to the rod of this piston the weights which were to be raised were applied by a rope conducted over pulleys, and they were to be lifted by the pressure of the atmospheric air upon the piston, when the air was exhausted from the cylinder. Thus far he followed Otto Guericke ; but instead of exhausting the cylinder by an air-pump, Papin proposed to drive out the air by firing off a small quantity of gunpowder in the inside of the cylinder. The piston had a large aperture in it, covered by a valve opening upwards ; and he thought that the sudden blast of flame produced by the explosion of the gunpowder would lift up the valve, and drive out the air at the aperture ; but when the flame ceased, the valve would fall, and prevent the return of the air into the cylinder, so as to leave it exhausted, and then the pressure of the atmosphere on the upper surface of the piston not being balanced by any air beneath, would press the piston down with great force.

Papin speaks of two models of this machine which he tried ; one of them five inches diameter and 16 inches high ; but he says he could never succeed in expelling the whole of the air from the inside of the cylinder : about one-fifth always remained, and that fifth, as he says, reduced the force to one-half of what it would have been if he could have expelled the whole of the air.

In the Acts for 1690, p. 410, Papin published an addition to his former paper. He says, that finding it impossible to make a complete vacuum in the cylinder by means of gunpowder, he had endeavoured to obtain his object by means of water, which, he says, " has the property when changed into vapour to spring like the " air and afterwards to recondense itself so well by cold, that there remains

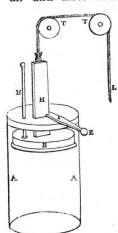

" no appearance of this force or spring." The machine which he proposed is a cylinder, A A, made of thin metal, and fitted with a piston B, which can slide freely up and down in the cylinder. A small quantity of water is put into the bottom of this cylinder, and the piston B put down so as to touch the water, the air being expelled through a hole in the piston, which is afterwards plugged up by a plug M. A fire is applied beneath the bottom of the cylinder to heat the water within, and make it boil, which will soon be done, the bottom being made of very thin metal. The water is then changed into vapour, which exerts so strong a pressure beneath the piston, that it surmounts the pressure of the atmosphere, and pushes the piston upwards to the top of the cylinder, and then a latch E falls into a notch in the stem H of the piston, to prevent it from going down again. The fire must now be taken away from beneath the cylinder, and the vapours in this thin cylinder soon recondense themselves into water by cold, and leave the cylinder entirely empty of air. In this state the machine is ready to exert its force, for by only releasing the latch E, the piston will be pressed down into the cylinder by the weight of the atmosphere which presses upon its upper surface, whilst there is

o

no air beneath it to resist its descent, and a rope L being fastened to the stem of the piston H, may be conducted over pulleys T T, and applied to raise weights.

The principle of the atmospheric steam-engine is suggested in this rude machine, which was afterwards perfected; the author states that he tried the experiment with a cylinder $2\frac{1}{2}$ inches diameter, which he found capable of raising 60 pounds, (that is $= 12\frac{1}{4}$ pounds per square inch) and that it could be made to repeat its action once a minute; from this data he calculated, that a cylinder of little more than two feet diameter, and four feet in height, would have sufficient force to raise a weight of 8000 pounds every minute to a height of four feet; this is 32,000 pounds raised one foot per minute, which is nearly a modern horse-power, or equal to the force of $8\frac{1}{2}$ men.

Papin states his invention to be applicable to draw water from mines, to throw bombs, and to row vessels against wind and tide; for the latter purpose he proposed to place revolving rowers, or paddle wheels, at the sides of the vessels, and by means of three or four of his new invented cylinders, to give a continual motion to the axis, whereon the paddle wheels were fixed; to communicate this motion to the wheels, the stems of the pistons of the cylinders were to be toothed to engage with small toothed wheels fitted upon the axis in such manner as to turn those wheels, and the axis with its paddle wheels, whenever the pressure of the atmosphere made the pistons descend into their respective cylinders; by having several cylinders they could be made to operate in succession, and give an uninterrupted motion to the axis and rowers, one cylinder acting whilst the others were heating, in preparation to repeat their action; the toothed wheels upon the axis of the rowers, were to be provided with ratchets and clicks similar to the winding up part of a watch, so as to be capable of turning round freely, in an opposite direction to the axis, when the pistons were rising up in their cylinders; but when the pistons were pressed down into the cylinders, the clicks would catch in the teeth of the ratchet wheels, and carry the axis round with the toothed wheels.

In the reprint of this project in 1695, the author also describes a new invented furnace, and revolving bellows, which he had invented for the purpose of boiling water by an internal fire-place surrounded on all sides by the water, and explains how it may be applied to heat the cylinders of his proposed machine with such increased rapidity as to make four strokes per minute. Every part of this ingenious project has been realized in the modern steam-boat; but a cylinder of the size he mentions is now made to produce more than twenty times the force he calculated.

An account of Papin's recueil de diverse pieces is printed in the Philosophical Transactions for 1697, No. 226, vol. 19, p. 481. The part which relates to the steam-engine is noticed in the following words: " The fourth letter shows a " method of draining mines where you have not the conveniency of a near river " to play the aforesaid engine (with air-pumps and cylinders connected by an " air pipe); where, having touched upon the inconveniency of making a vacuum in " the cylinder for this purpose with gunpowder, (according to his first scheme of " 1687) he proposes the alternately turning a small surface of water into vapour, " by fire applied to the bottom of the cylinder that contains it, which vapour " forces up the plug in the cylinder to a considerable height, and which (as the " vapour condenses as the water cools when taken from the fire) descends again " by the air's pressure, and is applied to raise the water out of the mine."

This invention is highly creditable to Papin; and though much more remained to be invented in order to put the plan in execution, and make an efficient machine, the philosophical principle is fully pointed out. It is not on record, that Papin ever succeeded in applying his invention to any useful purpose; he was not a practician; and at that time workmen and practical mill-wrights and engineers were not competent to take up an idea so different from their ordinary practice, and involving so many new considerations. He speaks of the difficulty of making cylinders as a great bar to the general adoption of his project, and recommends a manufactory of such cylinders to be established.

In a subsequent publication, dated 1709, Papin informs us that he made a machine in 1698, but before he had made any satisfactory trial, it was destroyed by ice floating down the river wherein it was fixed. He afterwards abandoned the principle of operating with cylinders and pistons by the pressure of the atmosphere, in favour of the expansive force of steam acting immediately upon the water which is to be raised; in this he followed the Marquis of Worcester; but some years before Papin proposed this machine of 1709, Captain Savery in England had invented and made a very complete engine for raising water by steam, which operates by both principles acting alternately, viz. a vacuum being formed by the condensation of the steam, water is raised by suction, and the expansive force of the steam is then used to drive the same water upwards to a greater height. This was the first steam-engine which was brought into real use, though Papin's proposed cylinder was afterwards perfected and improved by the English mechanicians so as to supersede it.

CAPTAIN SAVERY'S STEAM-ENGINE, 1698.

MR. THOMAS SAVERY, commonly called Captain Savery (m), obtained a patent in 1698 for a new invention " for raising water, and occasioning motion to all " sorts of mill-work, by the impellent force of fire." This patent bears date the 25th July, 1698, in the tenth year of the reign of William III. It is stated that the invention will be of great use for draining of mines, serving towns with water, and for working all sorts of mills.

At that time it was not the custom for patentees to give specifications of their inventions, and there is no official record of what Savery's plan was at the time of the patent being granted; but from the philosophical publications of the day we can collect the following particulars.

In June 1699, he showed a working model of his engine to the Royal Society, and in their transactions for that year, viz. No. 253, vol. xxi. there is the following register:

" Mr. Savery, June 14th, 1699, entertained the Society with showing his " engine to raise water by the force of fire. He was thanked by the V. P. for " showing the experiment, which succeeded according to expectation, and was ap- " proved of."

The above is accompanied with a copper-plate figure, (see Plate I. fig. 1.) and the following references by way of description.

A the furnace.

B the boiler.

C C two cocks, which convey the steam by turns to the vessels D D.

D D the vessels which receive the water from the bottom in order to discharge it again at the top.

(m) He is said to have been called Captain by the miners in Cornwall, in consequence of his being employed to drain the water for them; it is still their custom to give the title of Captain to engineers: Mr. Savery was treasurer to the commissioners for recovering the sick and wounded.

E E E E valves.

F F cocks which keep up the water, while the valves on occasion are cleansed.

G the force-pipe.

H the sucking-pipe.

I the water.

From this description, although very defective, it appears that the engine then exhibited by Mr. Savery was for raising water not only by the expansive force of steam, like the Marquis of Worcester's, but also by the condensation of steam, the water being first raised by the pressure of the atmosphere to a given height from the well into the receivers, and then forced out of the receivers up the remaining height, by the expansive force of steam, in the same manner as proposed by the Marquis.

This action was performed alternately in two receivers, so that while the vacuum was formed in one, to draw up water from the well, the pressure of the steam was operating in the other, to force up water into the reservoir ; but both receivers being supplied by the same suction-pipe and the same forcing-pipe, the engine could be made to keep a continual stream, or so nearly uniform as to suffer very little interruption.

Desaguliers relates, that Savery represented his discovery of the power of steam to be owing to an accident ; for having drank a flask of Florence wine at a tavern, and thrown the empty flask upon the fire, he called for a basin of water to wash his hands, and perceiving that the little wine left in the flask had filled up the flask with steam, he took the flask by the neck, and plunged the mouth of it under the surface of the water in the basin, and the water was immediately driven up into the flask by the pressure of the atmospheric air.   This was the same action as that of filling an æolipile with water, as before mentioned ; and it also appears, that Papin had already suggested the idea of a still more useful application of this principle.

Savery afterwards published a very complete account of his engine in a small book, of 84 pages, 12mo. entitled " The Miner's Friend, or an Engine to raise " Water by Fire described ; and the Manner of fixing it in Mines, with an Account " of the several Uses it is applicable unto, and an Answer to the Objections made " against it.   Printed at London, in 1702, by Thomas Savery, gentleman."   This little book was separately addressed to King William III., to whom the engine had been shown at Hampton Court ; to the Royal Society ; and also to the Mine Adventurers of England, who were invited to adopt the invention.

This engine displays much ingenuity, and is nearly as perfect in its contrivance as the same kind of engine has ever been made since that time : it is very clearly described, and its advantages pointed out, in modest language, but with forcible arguments to induce the Mine Adventurers to adopt it.   The copperplate is copied in Plate I. figures, 2, 3, and 4.   The original, which is handsomely engraved, is marked, B. Lens, delin. J. Sturt, sculpt.   The following is Savery's own description, with some few additions, taken from Dr. Harris's Lexicon Technicum, vol. i. article Engine.

A Description of the Draft of the Engine for raising Water by Fire ; by Thomas Savery.   See Plate I. fig. 2, 3, and 4.

" A A the furnaces which contain the boilers.

B 1, B 2, the two fire-places.

C the funnel, or chimney, which is common to both furnaces.

In these two furnaces are placed two vessels of copper, which I call boilers, the one large, as L, the other small, as D.

D the small boiler contained in the furnace, which is heated by the fire at B 2.

E the pipe and cock to admit cold water into the small boiler to fill it.

F the screw that covers and confines the cock E to the top of the small boiler.

G a small gauge cock at the top of a pipe, going within eight inches of the bottom of the small boiler.

H a larger pipe, which goes the same depth into the small boiler.

I a clack or valve at the top of the pipe H (opening upwards).

K a pipe going from the box above the said clack or valve, into the great boiler, and passing about an inch into it.

L L the great boiler contained in the other furnace, which is heated by the fire at B 1.

M, fig. 4, the screw with the regulator, which is moved by the handle Z, and opens or shuts the apertures at which the steam passes out of the great boiler into the steam-pipes O O.

N a small gauge cock at the top of a pipe which goes half way down into the great boiler.

O 1, O 2, steam-pipes, one end of each screwed to the regulator (see fig. 4), the other ends to the receivers P P, to convey the steam from the great boiler into those receivers.

P 1, P 2, copper vessels called receivers, which are to receive the water which is to be raised.

Q screw joints, by which the branches of the water-pipes are connected with the lower parts of the receivers.

R 1, 2, 3, and 4, valves or clacks of brass in the water-pipes, two above the branches Q, and two below them; they allow the water to pass upwards through the pipes, but prevent its descent: there are screw-plugs to take out on occasion, to get at the valves R.

S the forcing-pipe which conveys the water upwards to its place of delivery, when it is forced out from the receivers by the impellent steam.

T the sucking-pipe, which conveys the water up from the bottom of the pit, to fill the receivers by suction.

V a square frame of wood, or a box, with holes round its bottom in the water, to enclose the lower end of the sucking-pipe, to keep away dirt and obstructions.

X a cistern with a buoy-cock coming from the force-pipe, so as it shall always be kept filled with cold water.

Y Y a cock and pipe coming from the bottom of the said cistern, with a spout to let the cold water run down on the outside of either of the receivers P P.

Z the handle of the regulator, to move it by, either open or shut, so as to let the steam out of the great boiler into either of the receivers."—*Vide Miner's Friend.*

The Manner of working the Engine.

The first thing is to fix the two boilers of the engine in a good double furnace, so contrived that the flame of the fire may circulate round, and encompass the boilers to the best advantage, as you do coppers for brewing.

Before you make any fire, unscrew the two small gauge-pipes and cocks, G and N, belonging to the two boilers, and at the holes, fill the great boiler L two-thirds full of water, and the small boiler D quite full; then screw in the said pipes again as fast and tight as possible, and light the fire under the large boiler at B 1, to make the water therein boil, and the steam of it being quite confined must become wonderfully compressed, and therefore will, on the opening of a way for it to issue out (which is done by pushing the handle Z of the regulator as far as it will go from you), rush with a great force through the steam-pipe O 1, into the receiver P 1, driving out all the air before it, and forcing it up through the clack R 1, into the force-pipe, as you will perceive by the noise and rattling of that clack; and, when all the air is thus driven out, the receiver P 1 will be very much heated by the steam. When you find it is thoroughly emptied, and is grown very hot, as you may both see and feel, then pull the handle Z of the regulator towards you, by which means you will stop the steam-pipe O 1, so that no more steam can come into the receiver P 1, but you will open a way for it to pass through the other steam-pipe O 2, and by that means fill the other receiver P 2 with the hot steam, until that vessel has discharged its air through the clack R 2 up the force-pipe, as the other vessel did before.

While this is doing, let some cold water be poured on the first-mentioned receiver P 1, from the spout Y, by which means the steam in it being cooled and condensed, and contracted into a very little room, a vacuum or emptiness is created, and consequently the steam pressing but very little (if at all) on the clack R 3 at the bottom of the receiver P 1, there is nothing there to counterbalance the pressure of the atmosphere on the surface of the water at the lower part V of the sucking-pipe T, wherefore the water will be pressed up, and ascend into and fill the receiver P 1, by what is commonly called suction: the water as it rises lifts up the clack or valve R 3, which afterwards falling down again and shutting close, hinders the descent of the water that way.

The receiver P 2 being by this time emptied of its air, push the handle of the regulator from you again, and the force of the steam coming from the great boiler will be again admitted through Q 1, and will act upon the surface of the water contained in the receiver P 1; which surface only being heated by the steam, it does not condense it, but the steam gravitates or presses with an elastic quality like air, and still increasing its elasticity or spring until it counterpoises, or rather exceeds the weight of the column of water in the receiver and pipe S, which it will then necessarily drive up through the passage Q R 1 into the force-pipe S. The steam takes up some time to recover its power, but it will at last discharge the water out at the top of the force-pipe S, as it is represented in Fig. 3. After the same manner, though alternately, the receiver P 2 is filled with water by means of the suction, and then emptied by the impellent force of the steam, whereby a regular stream is kept continually running out at top of the force-pipe S, and so the water is raised very easily from the bottom of the mine, &c. to the place where it is designed to be discharged. I should add, that after the engine begins to work, and the water is risen into and hath filled the force-pipe S, then it also fills the little cistern X, and by that means supplies the spout or pipe Y Y, which I call the condensing pipe, and which by its handle can be turned sideways over either of the receivers, *and is then open;* by this spout cold water is conveyed down from the force-pipe to fall upon the outside of either of the receivers when thoroughly heated by the steam, in order to cool and condense the steam within, and make it suck (as it is usually called) the water out of the well up into that receiver.

It is easy for any one, that never saw the engine, after half an hour's experience, to keep a constant stream; for on the outside of the receiver you may see how the water goes out as well as if the receiver were transparent: for as far as the steam continues within the receiver, so far is that vessel dry without, and so very hot as scarce to endure the least touch of the hand ; but as far as the water is withinside of the said vessel, it will be cold and wet on the outside where any water has fallen on it ; which cold and moisture vanish as fast as the steam in its descent takes place of the water.

But if you force all the water out of the receiver, the steam, or a small part thereof, will go through the clack R 1 or R 2, and will rattle that clack so as to give notice to move the handle of the regulator, and then the steam begins to force out the water from the other receiver P, without the least alteration of the stream, only sometimes the stream will be rather stronger than before, if you pull the handle before any considerable quantity of steam be got up the clack R ; but it is much better to let none of that steam go off, for that is but losing so much strength, and it is easily prevented by pulling the regulator some little time before that receiver which is forcing is quite emptied (*n*).

This being done, turn the cock, or condensing-pipe Y of the cistern X, over the empty receiver, so that the cold water proceeding from X may run down through Y, which is never opened but when turned over one of the receivers, but when it stands between them is tight and stanch.   This cold water falling on the outside of the receiver, by its coolness causes that steam which had such great force just before, to condense and become an empty space, so that the receiver is immediately refilled by the external pressure of the atmosphere, or what is vulgarly called suction, whilst the other receiver is emptying by the impellent force of the steam, which being done, you push the handle of the regulator from you, and thus throw the force into the other receiver, pulling the condensing pipe over the receiver P 2, causing the steam in that vessel to condense, so that it fills while the other empties.

The labour of turning these two parts of the engine, viz. the regulator and condensing water-cock, and tending the fire, being no more than what a boy's strength can perform for a day together, and is as easily learned as their driving of a horse in a tub-gin ; yet, after all, I would have men employed in working of the engine, and those too the most apprehensive, supposing them more careful than boys.   The difference of this charge is not to be mentioned when we consider the vast profit which those who use this engine will reap by it.

The ingenuous reader will here probably object, that the steam being the cause of this motion and force, and that as steam is but water rarefied, the boiler L must in some certain time be emptied, so as the work of the engine must stop to replenish the boiler, or endanger the burning out or melting the bottom of the boiler.   To answer which, please to observe the use of the small boiler D; it is supplied with water from the force-pipe by a small pipe and cock E; when it is thought fit by the person tending the engine to replenish the great boiler (which requires about an hour and a half or two hours' time to the sinking one foot of water), he turns the cock E, so that there can be no communication between the force-pipe S, and the small boiler D, and putting in a little fire under the small boiler at B 2, the water will there grow presently hot, and when it boils, its own steam, which hath no vent out, will gain more strength than the steam in the great boiler.   For the force of the great boiler being perpetually spending and

(*n*) *Note.* Captain Savery was mistaken in this ; for it is necessary to blow some steam occasionally through the receiver into the force-pipe, in order to evacuate the air from the receiver ; otherwise that small portion of air, which is always extricated from water by boiling, as well as any air that might get into the receiver by leakage, would accumulate therein, and prevent the vacuum being formed by the condensation, because air will not condense.

going out, and the other confined and increasing, it is not long before the force in the small boiler exceeds that in the great one; so that the water in the small boiler being depressed by its own steam pressing on its surface, will force the water up the pipe H, through K, into the great boiler L; and so long will it run till the surface of the water in the small boiler D gets to be as low as the bottom of the pipe H, and then the steam and water will run together, and by its noise and rattling of the clack I, will give sufficient assurance to him that works the engine that the small boiler hath emptied and discharged itself into the greater one L, and carried in as much water as is then necessary; after which, by turning the cock E again, you may let fresh cold water out of the force-pipe S into the lesser boiler D, as before, and thus there will be a constant motion and a continual supply of the engine, without fear of decay or disorder. And inasmuch as from the top of the small boiler D to the bottom of its pipe H (which is within eight inches of the bottom of the boiler) there is contained about as much water as will replenish the great boiler L one foot, so you may be certain it is replenished one foot of course.

Also, to know when the great boiler wants replenishing or not, you need only turn the gauge-cock N, and if water come out, there is no need to replenish it, but if steam only come, you may conclude there is want of water; and the like will the cock G do in reference to the small boiler D, showing when it is necessary to supply that with fresh water from S, so that in working the engine there is very little skill or labour required: it is only to be injured by either a stupid or wilful neglect.

And if a master is suspicious of the design of a servant to do mischief, it is easily discovered by those gauge-pipes; for if he come when the engine is at work, and find the surface of the water in the great boiler L below the bottom of the gauge-pipe N, or the water in the small boiler D below the bottom of the gauge-pipe G, such a servant deserves correction, though three hours after that, the working on, would not damage or exhaust the boilers.

In a word, the clacks being, in all water-works, always found the better the longer they are used, and all the moving parts of our engine being of like nature, the furnace being made of Sturbridge or Windsor brick or fire-stone, I do not see it possible for the engine to decay in many years; for the clacks, buckets, and mitre pipes, regulator and cocks, are all brass; and the vessels, made of the best hammered copper, of sufficient thickness to sustain the force of the working of the engine. In short, the engine is so naturally adapted to perform what is required, that even those of the most ordinary and meanest capacity may work it for some years without any injury.—*Vide Miner's Friend.*

After thus describing his engine, the ingenious inventor enumerates the following uses to which it may be applied, and which he describes rather fully, viz. 1st. To raise water for turning all sorts of mills; 2d. Supplying palaces, noblemen's and gentlemen's houses with water, and giving the means of extinguishing fires therein, by the water so raised; 3d. The supplying cities and towns with water; 4th. Draining fens and marshes; 5th. For ships; 6th. For draining mines of water, and preventing damps in the said mines.

The engine above described does not differ essentially from that described in the Philosophical Transactions, but it is more completely put into form, and improved in some of the minor particulars. For instance, the original engine (see fig. 1) had only one boiler, and there was no provision for supplying it with water, to replace the waste occasioned by the evaporation, except by stopping the action of the engine, whenever the boiler was so far emptied as to risk the burning of the vessel. And after the boiler was replenished, the engine could not begin to work again, until that cold water which was introduced was made to boil.

The engine, which is described in the Miner's Friend, has a subsidiary boiler, in which a quantity of water is made hot, in readiness for supplying the great boiler, and the power of the steam raised in the subsidiary boiler is employed to force the water contained in it into the other, or great boiler, which actuates the engine : by this means the transposition of the feeding water is instantly performed, and being already at a proper heat, it is immediately ready to produce steam for carrying on the work.

There is also another improvement in the construction of this engine. His first engine was worked by two separate cocks, which the operator was obliged to turn separately at every change of stroke ; and if he turned them wrong, he was not only liable to damage the engine, but he prevented its effect, and lost a part of the operation ; whereas in this second engine the communications are made by the double sliding-valve or regulator. This is a brass-plate, shaped like a fan, and moving on a centre withinside the boiler, so as to slide horizontally in contact with the under surface of the cover of the boiler, to which it is accurately fitted by grinding, and thus at pleasure opens or shuts the orifices or entries to the steam-pipes of the two receivers alternately. This regulator acts with less friction than that of a cock of equal bore ; and by the motion of a single handle, at once opens the steam-pipe from one receiver, and closes that which belongs to the other receiver.

The contrivance of the regulator afterwards proved of great consequence, for it was universally adopted in all the steam-engines which were used for half a century, although it has since been superseded by more perfect means of opening and shutting the steam-passages.

Dr. Harris copied Savery's description and figure of his engine into the Lexicon Technicum, or English Dictionary of Arts and Sciences. He speaks of the Captain as a person of great merit and ingenuity, with whom he was probably acquainted. Dr. Harris first mentions another machine of Savery's, for rowing a ship in a calm by paddle-wheels placed at the vessel's side, of which the Captain published an account in 1698 ; and it is worthy of remark, that the same kind of wheels, when actuated by improved steam-engines, is the only method, amongst a vast number of others, which has been found to answer for rowing vessels (o). Dr. Harris, in proceeding to the fire-engine, says, " The other engine is for raising water by the force of fire, in which he has shown as great ingenuity, depth of thought, and true mechanic skill, as ever discovered itself in any design of this nature."

Captain Savery died before the year 1717. It is not known how far he succeeded in applying his engine to real use ; but it appears, he actually did make them under his own inspection, for in his address to the Royal Society (in the Miner's Friend), he says, that since the time he exhibited his model to them, " I " have met with great difficulties and expense to instruct handicraft artificers to " form my engines according to my desire ; but my workmen, after much ex- " perience, are become such masters of the thing, that they oblige themselves to " deliver what engines they make me exactly tight and fit for service, and as such " I dare warrant them to every body that has occasion for them."

(o) Papin had also the same idea of rowing vessels by his steam cylinders as before-mentioned, but Savery's proposal was to give motion to his paddle-wheels, by the strength of men applied to the ship's capstan.

P

Also in his address to the gentlemen adventurers in the mines of England, he says, that the frequent disorders and cumbersomeness of water-engines then in use, " encouraged me to invent engines to work by this new force ; that though I " was obliged to encounter the *oddest* and almost insuperable difficulties, I spared " neither *time, pains*, nor *money*, till I had absolutely conquered them."

It is said that Savery made an engine at Wednesbury in Staffordshire, to drain a tract of land which had been inundated from a coal-pit.

Stephen Switzer, in his " Introduction to a General System of Hydrostatics and Hydraulics," in two vols. 4to. 1729, has described the engine for raising water by fire, in vol. ii. p. 325, as the particular contrivance and sole invention of the ingenious Capt. Savery, some time since deceased, but then a most noted engineer, and one of the Commissioners of the sick and wounded. Switzer says he had the honour to be well acquainted with Captain Savery. " The first hint from which it is said he " took this engine was from a tobacco-pipe, which he immersed to wash or cool it, " as is sometimes done. He discovered by the rarefaction of the air in the tube " by the heat or steam of the water, and the gravitation or impulse of the exterior " air, that the water was made to spring through the tube of the pipe in a " wonderful surprising manner ; though others say that the learned Marquis of " Worcester gave the first hint for this raising water by fire."—" I have heard " him say myself, that the very first time he played, it was in a potter's house at " Lambeth, where, though it was a small engine, yet it forced its way through the " roof, and struck up the tiles." Switzer's figures and description of Savery's engine are copied from the Miner's Friend, and from Bradley.

Captain Savery made some small engines with one receiver, for supplying houses with water. In Bradley's " New Improvements of Planting and Gardening," which was printed in 1718, he speaks of one of these engines in the following terms :

" Supposing the situation of a house or garden to be a considerable height " above any pond, river, or spring, and that it has at present no other conveniency " of water than what is brought continually by men or horses to it. In this case, " the wonderful invention of the late Mr. Savery, F. R. S. for raising water by " fire, will not only supply the defect, by flinging up as much water as may be " desired, but may be maintained with little trouble and very small expense.

" It is now about six years since Mr. Savery set up one of them for that " curious gentleman, Mr. Balle, at Cambden-house, Kensington, near London, " which has succeeded so well, that there has not been any want of water since it " has been built ; and, with the improvements since made to it, I am apt to believe " will be less subject to be out of order than any engine whatever.

" For the satisfaction of the curious, I have given a design of it (See Fig 5, " Plate I.) agreeable to that at Cambden-house, which is, I think, the truest " proportioned of any about London : the several parts of it, and the manner of " working it, I shall explain as follows :—

" A, the fire.
" B, the boiler ; a copper vessel of a spherical figure, in which water is boiled " and evaporated into steam, which passes through
" C, the regulator, which opens to let it into
" D, the steam-pipe (of copper) through which it descends into
" E, the receiver, which is a vessel of copper, that at first setting at work is " full of air, which the steam will discharge through
" F, the engine-tree, and up the clack at
" K, (the plug at the said clack to come at and repair the same if need be) ; " and so the air goes up

" L, the force-pipe. After E is void of air, which is to be found by its
 " being hot all over, then stop the steam at C, and throw a little cold
 " water on E, and the sucking-clack will open at
" I, (which is the plug of the said clack) and fill E with water, which will
 " ascend through
" G, the sucking-pipe, from
" H, the pond, well, or river.
" This being done, proceed to raise your water ; viz.
 " 1. Turn C, to let the steam pass from the boiler into E, and it will force
" the water therein through F by K up L, which water cannot descend, because
" of the clack at I. When E is thus emptied, which may easily be perceived by
" its being hot, as before, turn C, and confine your steam in B ; then open the
" cock M, which will let a little cold water upon E, and that, by condensing
" the steam in E, will cause the water immediately to ascend from H, and
" replenish E.
 " Then turn C to let the steam into E, and it will force the water out of it
" up L into a cistern O, placed at the top to receive it. Then confine your steam
" at C as before, and turn M for the space of a second or two of time, and E will
" be refilled, which may be again discharged up L, as before. So this work may
" be continued as long as you please, if you keep water in B.
 " If you turn the cock N, and then only steam comes out of it (without hot
" water) the boiler must be replenished with fresh water ; but one boiler of water
" will last a long while.
 " When you have raised water enough, and you design to leave off working
" the engine, take away all the fire from under the boiler, and open the cock N to
" let out the steam, which would otherwise (was it to remain confined) perhaps
" burst the engine.
 " The proportion of the several parts of this engine as it now stands at
" Cambden-house is thus :—
 " The pipe G, from the surface of the water H, to the engine-tree F, is sixteen
" feet, which is the length it sucks the water, but might be made to draw water
" about twenty-eight feet.
 " From the engine-tree F, up to the great cistern O, which receives the water,
" is forty-two feet ; but might be a hundred feet high, if such a quantity of steam
" be allowed as is proportionable to the length of the pipe.
 " The diameter of the bore, as well of the sucking-pipe G, as of the force-pipe
" L, is three inches ; and of the steam-pipe D about an inch.
 " The receiver holds thirteen gallons of water, and the boiler three times
" that quantity.
 " When this engine begins to work, you may raise four of the receivers full
" of water in one minute ; which is fifty-two gallons raised 58 feet (p): and at that
" rate, in an hour's time, may be flung up three thousand one hundred and twenty
" gallons ; which is above eighty-six barrels : or, if there were two receivers, one
" to suck while the other discharged itself, as has been practised, then we might
" raise six thousand two hundred and forty gallons in the same time.
 " The prime cost of such an engine is about 50l. and the quantity of coal
" required for each working is about half a peck ; so that the expense will be very
" inconsiderable in comparison to what the carriage of water upon horses would

(p) A gallon of water, ale-measure, weighs 10·2 lbs. which × 52 gallons = 530·4 lbs. × 58 ft.
= 30763 lbs. raised 1 foot, which is nearly a horse-power, or about 8¾ men's power.

" amount to ; and in such countries where wood is plenty the expense would be
" much less.

" Now if the quantity of water, which is here said to rise in an hour's time,
" may not be thought enough for the use designed, the same engine may either be
" continued to work for four or five hours, or another of a larger extent might be
" set up.   I have seen one with two receivers, which held each of them a barrel,
" i. e. thirty-six gallons of water, which would furnish so great a quantity, that
" even some tolerable large fountains might be supplied by it."

It does not appear, from these descriptions, that Captain Savery ever made
use of a safety-valve to the boilers of his engine, although such a contrivance is
very necessary to enable any steam-engine to work with safety ; for if the steam is
kept confined and the fire continued, the heat will accumulate, and the expansive
force of the steam will increase beyond all bounds, so as to burst open the strongest
vessel.   This was observed by the Marquis of Worcester, who relates an experiment in which a cannon was burst by the expansion of confined steam.

The safety-valve was invented by Papin, in 1682, for his digester, or culinary
vessel for stewing meat, and digesting bones, by great heat and pressure.   The
use of the safety-valve is to allow the steam to escape from the vessel into the open
air, when its elastic force becomes so great, as to endanger the bursting of the
vessel ; so that the steam is not absolutely confined, but only retained until it has
a certain force, and if it grows stronger than that force it blows away.

The safety-valve is nothing more than a valve opening outwards, and well
fitted to close an aperture which is made in the top of the boiler, and is kept shut
by a weight or a lever, which is loaded with a weight, capable of sliding upon the
lever in the manner of a steelyard : so that the pressure of the weight upon the
valve can be regulated at pleasure, according to the strength of steam which is
required ; but, in all cases, it must be loaded so as to permit the steam to lift it
up and escape, when it arrives at a pressure which would endanger the boiler or
receiver being burst open by the force of the confined steam. Without this simple
contrivance, the working of such an engine must have been a very dangerous
occupation : the attendant must have kept the engine constantly at work, as long
as ever the fire was burning, so as to work off the steam as fast as it was produced.
In the engine described in the Miner's Friend, it is probable that the regulator
was so contrived, that it could never close both the steam-pipes at once, but as
soon as the passage to one receiver was closed the other would be opened, by which
means the steam could never have been confined with any greater force than the
pressure of the column of water in the perpendicular forcing-pipe.   The safety-
valve appears to have been first applied to Savery's engine by Dr. Desaguliers,
about 1717, although it had been previously proposed by Papin for such an engine.

If we make a close comparison between Captain Savery's engine and those of
his predecessors, it will be in every respect favourable to his character as an
inventor ; and, as a practical engineer, all the details of his invention are made
out in a masterly style, all accidents and contingencies are provided for, so as to
render it a real working engine ; whereas De Caus, the Marquis of Worcester,
Sir Samuel Morland, and Papin, though ingenious philosophers, only produced
mere outlines, which required great labour and skill of subsequent inventors to fill
up, and make them sufficiently complete to be put in execution.

From the great difference between Savery's engine and any of the others,
there is every reason to suppose that he invented the whole himself, and is entitled
to the honour of the original discovery of the expansive force of steam, and the
means of making a vacuum by the condensation of that steam, as well as the inven-

tion of a very complete and perfect machine, in which those principles could be applied to a useful purpose.

Nevertheless, Savery has been accused by some of his countrymen of plagiarism from the Marquis of Worcester; and, by the French writers, the invention is usually attributed to Papin, and Savery is mentioned as 'having borrowed his ideas from Papin, and put them in execution. This is not at all probable, for the Marquis of Worcester did not discover the condensation of steam; and Papin, who did make that discovery, in 1689, proposed to employ it in a cylinder with a piston, a project which was afterwards realized in England by Newcomen; but Papin had so little faith in the merits or practicability of that scheme, that he no sooner saw Captain Savery's book, than he abandoned his own plan of condensation, and set about an engine, which is similar to that proposed by the Marquis of Worcester, without any condensation, which is the great merit of Captain Savery.

PAPIN'S STEAM-ENGINE, 1707.

By his own account, it appears that he had made some experiments in 1698, by order of Charles, Landgrave of Hesse, but without effecting any thing. This was about the time of Savery's patent; and nine years afterwards Papin published an account of his invention in a tract entitled " *Nouvelle Maniere pour lever l'Eau par la Force du Feu, mis en Lumiere. Par D. Papin. Cassell, 1707.*"

This machine does not differ in principle from that of the Marquis of Worcester, but is less perfect than Savery's, because it works only by the expansive power of steam; the chief difference from the Marquis of Worcester, in construction, is, that the receiver being made cylindrical, the steam is separated from the cold water by a moveable cover floating on its surface; and the stream of water, which is expelled from the receiver by the force of the steam, is made to flow in some degree constantly, by being thrown into a large air-vessel, from which it is projected in a jet by the elasticity of the compressed air.

In this publication, Papin admits that he had seen the engraving of Savery's engine. He says, that in the year 1698, he made a great number of experiments, by order of his Serene Highness Charles, Landgrave of Hesse, to raise water by force of fire, in a more advantageous manner than that which he had proposed some years before (viz. 1690). In these experiments he did not succeed, but he communicated his ideas to several persons, and particularly to M. Leibnitz, who answered, that the same thought had occurred to himself.

He also acknowledged that Captain Savery was about that time working upon the same subject in England, and that Savery had first published the fruit of his researches; that from 1698 the affair had lain dormant till the year 1705, when he received a letter from M. Leibnitz, then in London, which contained the engraving of Mr. Savery's engine, and desired Papin's opinion upon it. On showing this to the Landgrave, he ordered Papin to resume the work, and perfect the inventions which he had begun before; and which Papin then published, not with a view to make it supposed that Captain Savery had taken the thoughts from him, but to show the world its obligation to the Landgrave, in having *first* formed a design so useful, and in having brought it to its present degree of perfection; and he labours much to show that his engine is preferable to that of Captain Savery.

The great difficulty of this undertaking, and the importance of the object, affords some excuse for the extravagant compliments which Papin lavishes on his patron and himself upon the success they met with, after encountering many unforeseen difficulties, and making experiments which succeeded, as he tell us, *quite contrary* to their expectations; but it cannot be allowed that Papin's experiments in 1698 were the first, because the Marquis of Worcester's publication was earlier by no less than thirty-five years; nor were they probable to have been so early as Savery's beginning, for it cannot be supposed that he would have incurred the expense of a patent, without some previous experiments to confirm his speculation, or that he could have brought his engine to the degree of perfection in which he exhibited it to the Royal Society, on the 14th of June, 1699, in a year, at a period when workmen were not ready or skilful in the execution of such machines as they now are in this country.   This idea is confirmed by what he says in his Miner's Friend.

About the time of this publication, Papin wrote a letter to the Royal Society to inform them of his new invented machine, and proposed, if the Society would bear the expense, to make a machine on his plan to perform nearly double what could be performed by Savery.   In the register of the Royal Society where this proposition is recorded, a note is added by Sir Isaac Newton.   There is also a long letter from Mr. Thomas Savery, which was read to the Society in 1706; it states that in 1705 the young Prince of Hesse (who was afterwards killed in Flanders) came several times to see his engine, and had every part completely explained to him, and proposed to Savery to go to Cassel to make such engines there, but this he could not do.   Again, in 1706, a gentleman came to him from the Prince of Hesse, with a small model of an engine made partly on his plan, but so imperfect, that it would not work, and requested Savery to set them right, which he did.

The fire-engine described by Papin in his publication of 1707, is proposed to give motion to a water-wheel, the water being forced by the repellent power of steam out of the steam receiver into an air vessel; the re-action of the condensed air was to throw the water out again in a continuous jet, which was directed to strike the floats of the water-wheel, in order to give it motion; the steam after having expelled the water from the steam receiver was allowed to escape into the open air.   M. Papin's machine is, on the whole, far inferior to the engine of Captain Savery, as it wants the advantage of the grand principle of condensation, and is only a return to the Marquis of Worcester's idea; it cannot therefore be called an improvement, except in the separation of the hot steam from the cold water by a diaphragm, or floating piston: this is a considerable improvement on the Marquis of Worcester's, and would be also an advantageous addition to Savery's, if the condensing water could be as well applied to run down the outside of a cylindrical vessel as an oval one.   Papin's greatest expectation of advantage was from the use of a red-hot iron heater to be introduced into the steam receiver.

This engine of Papin's, and also a fire wheel proposed in France in 1699, by M. Amontons, were never put into real practice, they do not therefore belong to this part of the history; but an account of both will be given among the projects which are recorded in the appendix to this work.

DESAGULIERS' IMPROVEMENT OF SAVERY'S ENGINE.

IN 1718, Dr. Desaguliers made an engine on Savery's plan, but in an improved form.   He says, that in considering Savery's engine with Dr. Gravesande in 1716, they thought there was a great waste of steam, by its acting constantly in the receivers without intermission; for the steam becoming weakened as it expels the water, must require some time to recover its force, and in this time it must heat the surface of the water in the receiver, and to a certain depth below that surface:

they thought that if an engine were so contrived, that after the steam had pressed up one receiver full of water, instead of being thrown into another, it should be confined in the boiler till the receiver was re-filled by the atmosphere, and then turned upon the water, the steam would have acquired so much force from its confinement, that it would press suddenly upon the surface of the water, and discharge a considerable portion of it even before it became heated below the surface.

In pursuance of this idea, they had a model made which could either be used with one or two receivers, and found, on experiment, that one receiver could be discharged three times in the same time that two receivers could be discharged once each. They also learned that Captain Savery had made an engine at Kensington, for Mr. Ball, with only one receiver, which acted very well: from this they concluded that a simple engine, which would be more easily worked, and would cost little more than half the money, would raise one-third more water.

In consequence of this success, Desaguliers made an engine with a spherical boiler, provided with a safety-valve, and a receiver of about one-fifth of the capacity of the boiler, and of a cylindrical figure, tall, and of small diameter in proportion. The steam was admitted into the receiver at top through a double-passaged cock, the handle of which being turned towards the boiler, admitted steam; or, being turned towards the force-pipe, admitted a jet of cold water to run into the receiver to perform the condensation, instead of throwing cold water upon the outside of the receiver; by this means a more perfect condensation was obtained, and with a less waste of cold water than by the original plan. The cold water was conveyed by a small pipe, which branched out from the great forcing-pipe to the double-passaged cock, which, when its handle was turned that way, would admit the cold water into the receiver. This water fell into a sort of cullender fixed in the top of the receiver, and pierced with holes, so as to disperse the injected water in a shower of drops within the receiver.

This method of injecting cold water into the receiver was a great improvement on Savery's plan, and has been followed ever since; but it must be observed, that the cold water cannot enter through the cock into the receiver at the first instant that it is opened, because the pressure of the column of water in the force-pipe must necessarily be less than that of the steam within the receiver; therefore, the injection will not commence until after the steam-cock is shut, and then the condensation, or loss of heat, which always takes place within the receiver from the coldness of the water, will very soon diminish the heat, and consequently the pressure of the steam, so that it will no longer balance the pressure of the column of water, which it had just before lifted into the force-pipe. This being the case, the injection-water begins to run, and falls in a shower through the steam contained in the receiver. The effect of this shower to produce the condensation is surprising from the rapidity of its action. The injection, being a portion of the same water which had just before quitted the receiver, it would be supposed to have nearly the same temperature as that which was then in contact with the steam; and hence the increased rapidity of the condensation can only arise from the dispersion of the water into drops. When the cold water is contained in the lower part of the vessel, the surface only of the water is exposed to the steam, and soon becomes so heated that it will not condense with that great rapidity which it did at first, and as heat will not descend in fluids, the mass of the water is not affected. On the other hand, a small quantity of water dispersed in drops will be completely exposed to the steam, and will take up therefrom, in an instant, as much heat as will reduce the temperature of the steam, and increase the heat of the injection-water, until they approach to an equality of temperature. From this it follows, that the degree of condensation which can be obtained by dispersing cold water in very small drops within the receiver will be in proportion to the coldness and quantity of the injection-water.

The quantity of cold water required for condensation by injection into the receiver is far less than when it was applied on the outside, because the receiver will not transmit the heat of the steam through it so quickly, and the water therefore ran down the outside of the receiver, and descended into the well, without being much warmed, and without having extracted much heat from the steam within. All this water was lost, having been first raised to the full height by the power of the engine, and then suffered to run down again into the well; there was also a loss of heat from that film of cold water which remained upon the outside of the receiver, and which must be all evaporated by the heat of the fresh steam when it was re-admitted into the receiver; on the other hand, when the cold water is injected internally, it is much less in quantity, and it is not an entire loss of the water which has been raised, but only of the height to which it has been raised by forcing, because it does not run down any lower than into the receiver.

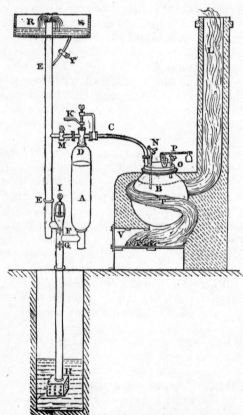

Desaguliers' engine is very complete, and is perhaps the best form of Savery's invention. The figure is copied from his course of Experimental Philosophy, Vol. II. p. 485, with the following description:

A, the cylindrical receiver, made of copper: it communicates at bottom with the sucking and forcing-pipes, between the valves F and G, and at top, by its steam-cock D K, with the steam-pipe C, and the injection-cock M.

B, the spherical boiler, also of copper; it must contain at least five times as much as the receiver: it is fixed in a furnace of brickwork over the fire-place V, and the fire and flame are conducted round the outside of the boiler at T. The boiler has a copper cover screwed on to the top, which cover contains the steam-pipe C, communicating from the boiler to the cock D K at the top of the receiver; also, the lowest gauge-pipe N, and the highest gauge-pipe O; likewise a safety-valve, which is kept down by the steelyard and weight P, to confine the steam to any required degree of elasticity. When the engine is at work, the surface of the water in the boiler must always be lower than the bottom of the short gauge-pipe O, and higher than the bottom of the long gauge-pipe N.

D K, the steam-cock, the key or plug of which is kept down by a perpendicular screw, held over it by a sort of gibbet: the handle K of the cock is either turned towards C, to receive the steam from the boiler, or towards M (as in the figure) to shut out the steam, and admit a jet of cold water from the ascending force-pipe E E, through the injection-cock M, which must always be open during the operation.   When the handle K of the cock is placed between these two positions, both passages are shut.

N. B. At D, in the top of the receiver A, just beneath the cock D K, a cullender, or spreading-plate, is fixed : it is pierced with holes, to divide the steam or the cold water, alternately, into small jets, when admitted by the cock into the receiver.

E F G, the horse or water-pipe, with several elbows and three branches. One of these is soldered to the lower end of the perpendicular forcing-pipe E E, another to the upper end of the upright sucking-pipe G H, and the third to the lower part of the receiver A.   This horse contains the sucking-valve at G, and the forcing-valve at F, which are easily come at by unscrewing I, and removing the plug or stopper beneath it.

I is a screw going through a stirrup, and pressing on the plug which fits into the hole over the valves F and G, and is made tight by double canvas put under it.   The stirrup and its screw I turn down like the bow or handle of a pail, so as to go out of the way, when the plug is to be withdrawn, in order to examine the valves.

R, the elevated cistern into which the water is thrown up.   It is usually placed at the top of a building.   It communicates with the forcing-pipe E by a cock and small branch at Y, to fill the force-pipe upon occasion.

G H, the suction-pipe, to draw the water up out of the well.

H, a box in the well at the lower end of the suction-pipe.   It is perforated with holes, to admit the water, but to keep out dirt from entering the pipe with the water.

T, the flue or passage round the boiler, for the flame and smoke to pass from the fire-place V up to the chimney L.

V, the door of the fire-place.   The fire-grate and the ash-hole are seen beneath V.   The construction of these parts is the same as that of the furnace for a brewer's copper.

The other three figures show the construction of the steam-cock D K, the plug having a cavity at one side, which goes down through the bottom of the cock at D, to lead down into the receiver, in order to admit the steam and jet of cold water

alternately, according as the cavity is turned towards the steam-passage C, or towards the water-passage M.   The plug D K has also a notch behind at Q, to take in the water from the water-passage M, and convey it by the steam-passage C into the boiler, when it is requisite to charge the boiler with fresh water.   The two sections of the cock and its plug show two of its different positions : in one, the cold water from M is admitted to run down through the plug into the receiver, but the steam-passage C is shut.   In the

other position, the plug being turned half round, the passage through the plug down into the receiver is quite closed, but the water is allowed to run from M, through the notch Q, to the steam-passage C, in order to replenish the boiler as before-mentioned.

OPERATION OF SAVERY'S ENGINE, IMPROVED BY DESAGULIERS.

TAKE off, or turn over, the steel-yard P of the safety-valve, and open the cock of the short gauge-pipe O, then having put a long nail or any piece of iron under the safety-valve to keep it up, pour in water at the funnel round that valve, and as the water goes down into the boiler, the air will blow out at O, till the surface of the water comes up above the bottom of the short gauge-pipe O, and then the boiler is sufficiently filled.

Then having turned the handle K, so as to shut the steam-cock D from communicating with the boiler, and likewise shut the two gauge-cocks upon the boiler at N and O, put the steelyard P on the valve, with its weight to the near valve, and light the fire at V. For raising water to a small height, the engine may work when the weight of the steelyard P is very near the valve; but for a great height, it must be placed further off. In trying the boiler at first, that you may be sure your steelyard weight is not too heavy, remove it on the yard from the valve, notch by notch, as you increase your fire, till the steam under the valve causes the weight to rise when it is placed at the notch, farthest from the valve.

N. B. You may fasten a string to pull up the steelyard on occasion, so as to stand at a distance, when you make the strongest trial, that you may be certain the weight is not too heavy, and take care that no additional weight, is ever put to the steelyard weight.

Fill the receiver A with water, from the force-pipe E before you begin. This is done by taking out the key or plug K of the cock D K (the screw which confines it down, being first screwed up a little, and then turned off sideways, with the gibbet which supports it.) Then the cocks Y and M being open, the water will run into the receiver. You may grease the key of the cock as you put it in again, and then the engine will be prepared to work.

All being ready, and the steam beginning to lift the safety-valve, turn the handle K of the steam-cock D K on the receiver, towards C, to let in the steam from the boiler B, along the steam-pipe C, first opening a small passage and then the full bore; the steam spreading through the small holes of the cullender plate at D, will press suddenly on the surface of the water in the receiver A, till that surface is driven down to the forcing-valve F, the water ascending through the forcing-pipe E up to the cistern R, the valve F will be heard to fall, when all the water is gone out of the receiver A, and the outside will feel very hot at that time.

The water from the receiver, being thus suddenly emptied into the cistern R, through the forcing-pipe, the receiver is full of steam; and the cock M being sufficiently open, turn back the handle K towards M, whereupon a jet of cold water will spout into the receiver, through the spreading plate at D, and fall in a shower of drops among the steam, which will immediately condense, and the air, pressing on the surface of the water in the well at H, will force the water up through the suction-pipe G H, and valve G, into the receiver, and will fill it, until the outside of the receiver will feel cold, nearly up to D.

Then, turning the handle K towards C, as before, it will admit the steam into the receiver, to drive the water out through F, and by the force-pipe E, up to R: then return the handle K to M, to admit another injection, and so on.

This engine may work about four or five hours before the boiler is exhausted; which you may know when both the gauge-cocks N and O being opened, both give steam; because then the surface of the water in the boiler, being below the bottom of the longest gauge-pipe N, it is too low, and the boiler is in danger of being burned.

When both cocks, O and N, give water, there is too much water in the boiler, and some must be thrown away, by opening the cock O, and emitting water therefrom, till that cock gives steam.

So that when the cock O, if it is opened, gives steam, and the cock N, if it is opened, gives water, the boiler is in right order.

When you would replenish the boiler, you may either remove the fire first, or let out the steam through the cocks N and O. ·Then turn the handle K from you behind, and when it stands between M and C, it will place the notch Q in the plug of the cock, in the situation shown by the separate figure; and then (the cock M being still open, as well as the cock Y,) the water will run from the cistern R, through the forcing-pipe E, and the cocks M and D, and will pass by the steam-pipe C into the boiler, without going into the receiver, which you may have filled before if you please, the steelyard P being taken off of the valve, and the cock O being open, will let out the steam, or air, as the water comes into the boiler.

To know when you have water enough in the boiler, observe when the cock O ceases to blow air, and the valve (which must not be propped open) will then dance up and down. Turn back the handle K, and shut the cock Y, and the engine will be ready to go on working again. *(Exper. Philos.)*

According to these improvements, Desaguliers caused seven of these fire-engines to be erected, after the year 1717 or 1718. The first was for the Czar Peter the Great, and it was set up in his gardens near St. Petersburgh. The boiler of this engine was spherical (as they all were in this way, the steam being so much stronger than the air), and held between five and six hogsheads; the receiver held one hogshead, and was filled and emptied four times in a minute. The water was drawn up by suction, or the pressure of the atmosphere, 29 feet high out of the well, and then pressed up 11 feet higher. The pipes were all of copper, but soldered to the valve-piece with soft solder, which held very well for that height; but Desaguliers says, he did not venture either upon a greater quantity for that height, or a greater height for that quantity; for if the quantity was larger than above, the boiler must have been larger, and steam of the same force would have had a greater surface to act upon, which might have burst the boiler, or would have required it to have been made much thicker.

Another engine of this sort, which he put up for a friend in 1720, drew up the water 29 feet from the well, and then, by the pressure of the steam, it was forced up 24 feet higher, into a cistern holding about thirty tuns, placed at the top of a tower, from which it ran down again, through a conduit-pipe, and played several jets in the garden. But sometimes, no jets being played, the water was discharged at the height of 6 or 8 feet, out of the force-pipes, to fill ponds and water meadows in dry weather, which it did with a less strength of steam, than would drive the water into the tower; or if the same steam was kept up, it would make eight or nine strokes in a minute, instead of about six, as when the water was driven up into the cistern. Upon the safety-valve there was a steelyard, the place of whose weight showed the strength of the steam, and how high it was capable of raising water; but when the weight was at the very end of the steelyard, the steam, being then very strong, would lift it up, and escape at the valve, rather than damage the boiler.

Twenty-five years after this engine was made, a man, who was entirely ignorant of the nature of the engine, undertook to work it without any instructions; and having placed the weight at the farther end of the steelyard, in order to collect steam, to make the engine work quicker, and also hung a very heavy plumber's iron upon the end of the steelyard, the confined steam not being able to lift the safety-valve and steelyard, when loaded with all this unusual weight, accumulated until it burst open the boiler with a great explosion, and the pieces which flew asunder, killed the unfortunate man who stood near it.

These accounts show how high, and in what quantity, this kind of fire-engine can safely raise the water. About as much fire, as a common large parlour-fire, was sufficient to work this engine, and raise fifteen tons per hour 53 feet high; so that if the cistern was kept full, the jets could be made to play to entertain friends at any time, and then a man being sent to light the fire under the boiler, the engine would raise more water to supply the jets, before the cistern was empty, and thus keep them supplied as long as the fire was kept in.

Desaguliers concludes his description with a comparison between his engine on Savery's plan, and the cylinder atmospheric engine, invented by Newcomen, which will be described in the next chapter. He says, " Savery's engine on my plan " consists of so few parts, that it comes very cheap, in proportion to the water it " raises, but has its limits. Newcomen's cylinder-engine has also its limits the " other way: it must not be too small; for then it will have a great deal of " friction, in proportion to the water that it raises, and will cost too much, having " as many parts as the largest engines, which are the best and the cheapest, in " proportion to the water they raise. The loss by friction is also as the diameter " of the cylinder, whereas the water raised is as the square of the diameter, and " a much greater proportion of the whole power is employed to move the small " machinery in a small engine, than in a great one.

" I had an experimental proof of this at Westminster, in the year 1728 or " 1729, when a Mr. Jones built a working model of Newcomen's engine in my " garden (which model he intended to present to the King of Spain). I had at " the same time, near the place where he erected his engine, one of mine, on " Savery's plan, which raised ten tuns an hour about 38 feet high. He made his " boiler of the exact size of mine, and his cylinder was six inches bore, and about " two feet in length. When his model or cylinder-engine was finished, it raised " but four tuns per hour, into the same cistern as mine. It cost him 300*l.*, and " mine, having all copper pipes, had cost me but 80*l.*"

ON THE APPLICATIONS OF SAVERY'S FIRE-ENGINE, ITS ADVANTAGES AND DEFECTS.

IT appears that a number of small engines were erected, under the authority of the patent, for the supply of gentlemen's houses and gardens with water, in different parts of England, and they succeeded very well; they were also tried for waterworks and for mines.

Desaguliers, in his Course of Experimental Philosophy, which was written in 1743, after more improved steam-engines were brought into use, says that " Captain Savery made a great many experiments to bring his engine to perfection, " and did erect several, which raised water very well for gentlemen's seats, but " could not succeed for mines, or supplying towns, where the water was to be " raised very high, and in great quantities, for then the steam required to be " boiled up to such a strength, as to be ready to tear all the vessels to pieces. " The heat which is sufficient to boil water, will produce steam whose spring is " of the same strength as common air; but that steam having effected the removal " of the atmosphere, and then being condensed, the return of the atmosphere to " press up the water, is only capable of bringing the water up to a little above 30 " feet; and for every 32 feet that the water is to be raised higher, the steam " must be so many times stronger than the air. For example: if the water is to

" be forced up 90 or 100 feet higher than the receiver, where the steam acts upon
" it, the steam must be three or four times stronger than the common air, and
" a great deal stronger than that, (perhaps six times stronger) because the hot
" steam striking upon the surface of cold water in the receiver, condenses itself,
" and thereby becomes ineffectual till the surface of the water, and a small depth
" of it, is made warm enough not to condense any more steam ; and then (and
" not before) the water yields to the pressure of the elastic steam to make it rise.
" I have known Captain Savery, at York Buildings waterworks in London, make
" steam eight or ten times stronger than common air ; and then its heat was so
" great, that it would melt common soft solder, and its strength so great as to
" blow open several of the joints of his machine, so that he was forced to be at the
" pains and charge, to have all his joints soldered with spelter or hard solder."

Mr. Savery proposed at the commencement, to apply his engine to give motion to all sorts of millwork ; but it does not appear that it was attempted in his time, for at that period there were scarcely any mills which could have supported the expense of erecting and maintaining such engines, even where coals were cheapest.

For draining fen lands, Savery's engine was not well adapted ; because, in most cases, the water is required to be raised to only a small height, but in very great quantities ; therefore several engines would be wanted for one drainage, and great part of their power would be lost, because the perpendicular height at which the water would be discharged, would always be less than the height to which the atmospheric pressure can raise the water. To ships, it is probable this engine was never applied, nor has the inventor pointed out the means. Hence it appears that his engines were only applicable in a small number of cases.

They could not be generally employed in mines, for several important reasons. The working part of the engine must necessarily be placed from 26 to 29 feet above the bottom of the mine ; and if, by any accident, the engine should be deranged, and unable to work for a few days, the water would rise above that level, and the engine would be drowned and irrecoverably lost, without some other engines to drain the mine, to that level. As the power of suction in this engine cannot extend much more than 26 feet, the rest of the perpendicular lift, must be obtained by the expansive force of the steam ; and for every 30 or 32 feet of altitude of this column, a pressure, equal to the atmosphere, must be exerted on the inside of the boiler and receiver, tending to burst them open.

It would not be practicable, in constant work, to force the water by steam of more than three atmospheres pressure, or about 64 feet above the engine ; and this limits the power of an engine, on Savery's plan, to about 90 feet lift at the utmost. Hence a separate engine would be required for every 15 fathoms of the depth of a mine, and they must raise the water from one to another ; therefore if any one engine was deranged, the rest must stop until it was refitted.

Another difficulty was in the quantity of water which could be raised with safety : the size of Savery's largest boiler did not exceed 30 inches diameter, and the capacity of the receiver could be but small ; therefore, most mines would require several engines at the same level. The charge, trouble, and difficulty attending such a number, would prevent their introduction, even in cases where they would have been of great service.

The consumption of fuel in Savery's engines was likewise very great compared with the modern engines ; and they were always in danger of blowing up, when they were employed to raise water to any considerable height.

Suppose, for instance, the water is to be raised 90 feet, viz. 26 by suction, and the remaining 64 feet by the force of the steam. To effect this, the pressure within the vessels must be considerably more than three atmospheres; and every square inch of the interior surface of the boiler and receiver will be pressed with a force of more than 30 pounds, tending to burst them open. This moderate height will therefore require very strong vessels, and the joints must be made with the greatest care; for although that pressure is less than is usual in pumps, and other hydraulic machines, yet there is more danger of the vessels being burst by steam of great elasticity, than by an equal pressure of a column of water; because the force of the steam is always accumulating, and is liable to be suddenly increased by any accession of the heat; and also the heat tends to weaken the vessels, particularly the boiler, which sooner or later, must be reduced in thickness at the bottom, and will then burst.

With a view to strengthen the boiler, hoops and internal bars have been proposed, according to the suggestion of the Marquis of Worcester; but this would have been of very little service, because the still greater evil, of the condensation of the steam would remain. It is therefore better to divide the mine into engines, of from 60 to 80 feet lift, according to Mr. Savery's first proposition, than to attempt using steam of such a degree of elasticity as to require those precautions.

It appears from the tables of the elastic force of steam, that it must be heated to a temperature of 273° of Fahrenheit's thermometer, before it can overcome a column of water of 64 feet in altitude; and as this steam must come into immediate contact with the surface of the cold water in the receiver, which is perhaps as low as 50°, the condensation of the steam is excessive at first, and must continue until the surface of the water acquires nearly the same temperature as the steam; which, however, it will soon do, because the heat is transmitted downward very slowly. When the surface of the water is sufficiently heated, the steam, which before was condensed as fast as it came in contact with the water, will begin to accumulate in the receiver, and press upon the surface of the water; the elasticity of the steam increases as this accumulation goes on, until it so much exceeds the pressure of the column of water as to drive it up the force-pipe; but as soon as any of the water, is thus expelled from the receiver, a renewal of the condensation is produced by the cold and wet surface, of that part of the receiver, which was before filled with water; this condensation will be even more rapid than the former, because the vessel, being made of metal, will absorb the heat more rapidly than the water did; this will delay the process of forcing out the water, until the metallic substance of the receiver, is made as hot as the steam.

Mr. Savery seems to have been fully aware of this, for he says in the Miner's Friend, "The steam acts upon the surface of the water in the receiver; "which surface only being heated by the steam, it does not condense, but the steam "gravitates, or presses with an elastic quality like air, and still increasing its "elasticity or spring, until it counterpoises, or rather exceeds the weight of the "column of water in the force-pipe, which then it will necessarily drive up that "pipe; the steam then takes some time to recover its power, but it will at last "discharge the water out at the top of the pipe. You may see on the outside of "the receiver how the water goes out, as well as if it was transparent; for so far "as the steam is contained within the vessel, it is dry without, and so hot as scarcely "to endure the least touch of the hand; but as far as the water is withinside the "said vessel, it will be cold and wet on the outside, where any water has fallen on

" it ; which cold and moisture vanish as fast as the steam takes the place of the
" water in its descent." The heat thus given to the metal of the receiver must be
all dissipated in waste by the cold water which is thrown on the outside to make
the condensation.

IMPROVEMENTS IN THE CONSTRUCTION OF SAVERY'S ENGINE.

THE rapid condensation which must take place, when steam of a great elastic
force is brought into immediate contact with cold water, is an insuperable bar
to the raising of water to any considerable height, on Savery's plan ; for the receiver
must necessarily be heated, before the steam can exert an elastic force in it ; and
then, in order to make the vacuum, it is equally requisite to cool the receiver again.

It is desirable to have all the interior surface of the receiver, and of the water,
as hot as possible when the steam is to act in it, and as cold as possible when the
vacuum is required ; because, in all cases, the elasticity of steam depends upon its
temperature, as may be seen in the table of elasticities.

In reality, a perfect vacuum can never be obtained, for there is always a
vapour in the receiver, more or less elastic, according to its temperature ; and if the
receiver be imperfectly cooled, the vacuum will be proportionably imperfect. For
instance, if the inside of the receiver were only cooled to a temperature of 112 deg.
of Fahrenheit's thermometer, the elasticity of the vapour remaining in the receiver,
would be equal to the pressure of a column of water of three feet in height ; even
in that case, the water might be raised by suction to a height of 26 feet above the
surface of the water in the well, because the whole pressure of the atmosphere
being more than 33 feet, there would still be an ascending force, equal to the
pressure of a column of more than four feet of water, to make the water run
upwards through the suction-pipe and its valves. But suppose the receiver was
cooled down to 80 deg. the remaining vapour would be only equal to a column of
water of one foot ; and therefore in the latter case, the water might be raised two
feet higher by suction than in the former case, and with equal rapidity. Hence
there is a loss of effect from imperfectly cooling the receiver ; and although in either
case, the force of suction would be sufficient to fill the receiver when placed at an
altitude of 26 feet above the well ; yet with an imperfect vacuum the water will rise
slowly through the suction-pipe, and will take a considerable time to fill the receiver :
but when the vacuum is more perfect, or the altitude less, the receiver will be
quickly filled, and the engine will raise more water in the same time.

For these reasons, the heat which is communicated to the surface of the water,
and to the metal of the receiver, is not only a waste of the strength of the steam,
which should force up the water, but is a great evil in preventing the subsequent
condensation, and rendering the vacuum imperfect.

The most obvious remedy for the loss of heat by the contact of the steam
with the water is, to employ a cylindrical receiver with a floating cover on the
surface of the water, to separate the hot steam from the cold water. This was
proposed by Papin in his publication of 1707 ; but it does not appear to have been
put in practice at that time, although it would have been very easily applied to
Desagulier's form of the engine. The best float, is a box of thin copper, hollow
within, and very closely soldered : it should be circular like a grindstone, and made
to fit the inside of the cylindrical receiver as nearly as possible, without actually
touching its sides ; the upper and under surfaces should be covered with varnished

wood, or the inside of the box may be filled with burnt cork, to prevent the heat penetrating through the float to the water. But this would be only a partial remedy for the evil, because the condensation from the cold and wet surface of the receiver would still continue.

When the hot steam is admitted, the heat will pass very quickly through the metal of the receiver, as Mr. Savery observed; and if the receiver is made of thick metal to give it strength, it must absorb a considerable portion of heat in its own substance, as well as what it communicates to the surrounding air. This loss will recur every time the steam is admitted, and all this heat will return into the receiver again, when the cold water is injected, and then it will do harm by raising a vapour and impairing the vacuum.

The waste of heat from this cause, will be rendered very small, if the receiver is made of wood hooped like a cask, and lined with very thin sheet copper; because wood transmits heat so slowly, that the inside surface of a thick wooden vessel may be heated and then cooled again, as often as the operation of this engine requires, before any sensible heat can penetrate far into the thickness of the wood: or a cylindrical receiver may be made of thin sheet copper, soldered up, and enclosed within another cylinder of cast iron, so much larger so as to leave a space of two inches all round between the two; a layer of powder of charcoal being hard rammed into this space, would effectually prevent the passage of the heat, and yet would support the thin copper vessel very firmly.

Therefore, to construct Savery's engine in the best manner, it should have a cylindrical receiver of very thin copper, enclosed within a strong wooden vessel; or within a cast-iron vessel, with a layer of charcoal interposed between them; and the receiver should be fitted with a hollow floating piston, to cover the surface of the water.

Another circumstance deserving attention, is to allow very ample passages for the water, that it may freely enter into the receiver by the suction-pipe, and as quickly make its exit by the force-pipe. An engine so constructed will work with celerity, and the steam will discharge the water in the least possible time, whereby the loss of heat by condensation, will be less: for the same reason, it is not advisable to attempt to raise the water by suction, to its utmost height, which is from 30 to 32 feet, because it would then rise very slowly: but, by reducing the height to 26 feet, the resistance is diminished, whilst the motive force remains the same, and therefore the water is drawn up more quickly.

The rapidity with which the water will rise up into the receiver, depends greatly upon the size of the suction-pipe, and valves, compared with the receiver. The old engines were very faulty in that respect: the water-pipes being made too small, and the passages of the valves still more contracted; hence the water must have risen very slowly in the receiver, even when it passed very rapidly through the small ascending-pipes. By enlarging the pipes, a considerable increase of effect might have been obtained, both in the height to which the water could be raised, and the quantity which could be brought up in a given time, with the same expenditure of steam.

The area of the water-passage, at its most contracted part, should not be less than half the area of the cylindrical receiver; and if still larger than half that area, the engine would act quicker, which is a great advantage to the performance, because there is less time for the heat to be dissipated, and more work is performed.

Savery's engine being well executed, with due attention to these circumstances, will perform as well as an engine can do, whilst the water which is to be raised is permitted to enter into the steam receiver; but it was not until the piston was made to fit exactly into the cylindrical receiver, and the water kept altogether out of it, that the steam-engine was rendered an efficient machine.   This form, which was suggested by Papin in 1690, and perfected by Newcomen in 1710, introduced much complexity in the machine ; for it became necessary to have a separate receiver with its piston, or, in other words, a pump, to raise the water, and also machinery to communicate the motion of the steam-piston to that of the pump.   The simplicity of Savery's engine, and the certainty of its action, renders it a valuable machine for raising water, in situations to which it is applicable ; but these are so few, that the steam-engine would have been of very limited utility, if Savery's principle had not been improved upon.

Since more perfect steam-engines have come into use, some ingenious engineers have proceeded on Savery's plan ; and have endeavoured to improve it so far, that it could be employed for mines, but without success.   Mr. Blakely made many attempts to introduce Savery's engine in an improved form.   He obtained a patent in 1766 ; from the specification of which, it appears that his improvement was to use oil to float upon the surface of the water in the receiver, and form a piston, or disk, between the hot steam and the cold water, so as to avoid condensation of the steam by touching the surface of the water.   Or air was to be admitted into the receiver, for the same purpose : in this case, two receivers were to be used ; one in the same situation as Savery's, which was to receive the air ; and the hot steam, when admitted into it, compressed that air, and forced it to descend by a pipe, to the second receiver placed beneath it, or at the bottom of the well ; and this condensed air was to expel the water therefrom, and drive it up the force-pipe.   By this means he hoped to prevent the steam coming in contact with the water, and thus avoid the condensation.   Mr. Blakely's engine was not found to answer any better than the original plan of Savery ; because air will not make a complete separation between the steam and the cold water, as the inventor expected, and a great power is required, to compress the air sufficiently to make it lift the column of water ; and in order to get rid of this compressed air, it must afterwards be let out into the atmosphere, or forced along with the water up the pipe : in either case, the power required to produce the compression of the air is all lost.   A review of these projects will be given in its proper place in the Appendix ; but to avoid returning to the description of Savery's engine, such improved forms of it as are really useful, will be now described, before proceeding to the account of other steam-engines.

Savery's engine may be usefully employed for raising water to a height of 30 or 35 feet, which can be done principally by suction, with only a very slight pressure for the remainder.   Several small engines have been erected upon this plan ; and where the water which is raised requires to be immediately heated, they are very capital machines ; because all the loss of heat being thrown into the water, warms it, before it enters the boiler, in which it is to be heated, so as to economise the whole of the heat.   For instance, for the purpose of raising water into the evaporating boilers for a salt or alum work, or for a brewery ; they are also particularly applicable, for raising water for warm-baths.

A small engine of this kind was made by Mr. Genjembre, of Paris, in 1820, for raising water for a floating-bath in the river Seine, and answered the purpose very completely.   When circumstances will allow, the most advantageous way is to perform the whole lift by the suction, and even to allow the water a sufficient descent, to empty the receiver by its own gravity, without forcing it in the least by the steam.   In this form of the engine, the water cannot be raised above 20 or 22 feet ; but the advantage is, that the steam need not have much greater elastic force

R

than the atmospheric air, or just as much as is sufficient to make it enter briskly
into the receiver, as the water runs out.   The temperature of the steam will then
be only a little above the boiling point, or about 217 deg. and consequently the
loss by the condensation is not so serious ; and as the steam is not pressed upon
the water, it is not brought into such close contact with the cold surface, as in
the forcing engines.

When the engine operates thus by suction only, the injection-water must be
thrown in by a small force-pump ; and there must be a small air-valve applied
to the receiver, at which the steam when it first enters may blow out that air which
will unavoidably insinuate itself into the engine, and which, without this provision,
would accumulate until it would fill the receiver, and impair the vacuum.   Also,
to render the engine complete, it must be made self-acting, that it may perform all
its functions of opening and shutting its valves, and supplying itself with water,
without any assistance from the attendant, who will then have only to keep up
the fire.

Several engines upon this principle, with various improvements taken from
other steam-engines, were erected by Mr. Joshua Rigley, many years ago, at
Manchester, and in other parts of Lancashire, to impel the machinery of some of
the earliest manufactories and cotton-mills in that district.   The engine usually
raised the water about 16 to 20 feet high ; and the water descending again, gave
motion to an overshot water-wheel.   This is Mr. Savery's original project for
working mills by his engine : but Mr. Rigley contrived his engines to work without
an attendant ; the motion of the water-wheel being made to open and shut the
regulator, and injection-cock, at the proper intervals.   They continued in use for
some years, but were at length given up in favor of better engines.

In Nicholson's Philosophical Journal, 4to. vol. i. p. 419, the following sketch
and description is given of an engine of this kind, which was erected at Mr. Kier's
manufactory, St. Pancras, London, and which worked there for many years, to
turn lathes, &c.

DESCRIPTION OF A STEAM-ENGINE, ON SAVERY'S PRINCIPLE, WHICH RAISES WATER TO TURN A WATER-WHEEL.

" The figure is a section of this engine, taken through the centre. B represents
" a boiler, shaped like a waggon, seven feet long, five feet wide, and five deep : it
" was considered as being of dimensions sufficient to work a larger engine ; a
" circumstance which must, in a certain degree, diminish the effects of the present
" one.   The boiler feeds itself with water from an elevated cistern, by a pipe
" which descends into the boiler, and has a valve in it, at the upper end, which
" shuts downwards, and is connected by a wire with a float on the surface of the
" water within the boiler, so as to open the valve, whenever the water subsides
" below its intended level ; for the float which swims on the water then sinks, and
" by its weight draws the valve up, to allow the water from the cistern to run
" down the pipe and supply the deficiency : but as the water in the boiler rises,
" the float closes the valve.   The boiler, therefore, remains constantly, or nearly at
" the same degree of fulness.

" The steam is conveyed by a pipe C to a box D, through which, by the
" opening and shutting of a valve, it can be admitted to the cylindrical receiver A.
" The axis K serves as a key to open and shut the valve, which is a circular plate,
" formed conical on the edge, and fits in a corresponding aperture in the bottom of
" the box D. H is a cistern, from which the engine draws its water through a vertical

" suction-pipe, in which a valve, G, is placed to prevent the return of the water.
" R is another cistern, into which the water is delivered from the receiver A,
" through the spout F, which is provided with a valve, opening outwards.   W W
" represents an overshot water-wheel, 18 feet in diameter, of which the axis S
" communicates motion to the lathes and other machines used in the manufactory.

" The engine raises the water from the lower cistern H, by suction, into the

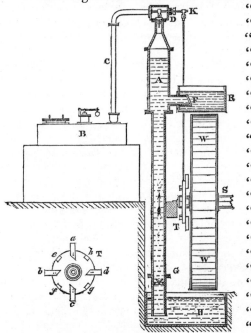

" receiver A ; from which it runs into
" the upper cistern R, and thence
" flows through a sluice into the buckets
" of the water-wheel W, to give it
" motion. The water, as it is discharged
" from the buckets of the wheel, falls
" again into the lower cistern H.   As
" the same water circulates continually
" in both the cisterns, it becomes warmer
" than the hand after working a short
" time ; for which reason, the injection-
" water is forced up by a small forcing-
" pump from a well.   This injection-
" pump is worked by the water-wheel,
" by means of a loaded lever, or pump-
" handle, which is raised up by the
" motion of the wheel, and then left
" to descend suddenly by its weight,
" and force up the water into the re-
" ceiver.   A leaden pipe passes from
" this forcing-pump, to the upper or
" conical part of the receiver A, for the
" purpose of injecting cold water at the
" proper time.   Neither of these could be represented with convenience in the
" present section.

" The manner in which the steam and cold water are alternately admitted
" into the receiver A, remains to be explained.   Upon the extremity of the axis S
" of the water-wheel, a solid wooden wheel T is fixed ; it is about four feet in
" diameter, and turns round with the water-wheel.   It is represented separately,
" as seen in the front : a, b, c, d, are four cleats, all or any number of which may
" be fixed on the wheel at a time.   Each cleat has its correspondent block, e, f, g, h,
" on the opposite surface of the wheel. The use of these is to work the engine. Thus,
" suppose the water-wheel, and this wheel T, with all the revolving apparatus,
" is turning round, one of the cleats a meets in its rotation with a lever, which it
" lifts up, and this opens the steam-valve D, by a rod of communication reaching
" to the handle of the axis K.   The steam consequently passes into the receiver
" A, and the steam-valve shuts again, as soon as the cleat a of the wheel T has
" passed away from the lever by the motion of the wheel.   All this time the cor-
" respondent block e, on the other side of the wheel T, had been operating to
" raise up the loaded lever, which forms the handle of the forcing-pump ; and at
" the same instant that the steam-valve D is shut, as above-mentioned, the block
" e quits the loaded lever, after having raised it up, and leaves it to descend
" suddenly, by its own weight.   This depresses the forcer of the pump, and
" thereby throws a jet of cold water up into the receiver A, and it falls in a shower
" of drops through the steam which fills the receiver, so as to cool and condense

" the steam, and make a vacuum therein.   The pressure of the atmosphere upon
" the surface of the water in the cistern H then causes the water to mount up the
" perpendicular suction-pipe, through the valve G, towards the exhausted receiver.

" When the engine is first set to work, the water-wheel being motionless, the
" steam-valve and injection-pump are moved by hand ; and if the engine has been
" long out of work, two or three strokes may be necessary to raise the water to the
" top of the receiver A, so as to fill it full of water.   As soon as this is the case,
" and the steam-valve is opened to admit steam into the receiver, the whole con-
" tents of water, above the spout and valve F, then flows out of the receiver A,
" by its own gravity, into the upper cistern R.

" The water which is thus raised, is suffered to flow from the cistern upon the
" overshot water-wheel W, through a sluice ; and by that means keeps the wheel
" in motion, and replenishes the lower cistern.   There is no reservoir for the
" injection-water, but the requisite quantity is driven up at each stroke ; and as
" this is done by the sudden descent of the loaded lever of the pump, the water
" is injected very suddenly into the receiver."

Hence we see, that this is in effect the same as Savery's original engine,
except that it is not applied for the sole purpose of raising water, but gives motion
to a water-wheel ; and it does not force the water by the steam at all.   The water
merely runs out of the receiver by its own gravity, and consequently never requires
steam much stronger than the atmosphere.   From the effect of this engine, it may
fairly be concluded, that the immediate action of steam forcing against water, can
never be beneficial, except at places where fuel can be had extremely cheap.

It was found, at the first erection of this engine, that the consumption of
steam, by contact with the water, was very great, but the defect was afterwards
remedied by fixing a small air-valve in the steam-box, which was struck open by the
machinery for an instant, immediately before the admission of steam.   It may be
presumed, that the air occupied a space above the water, and prevented their coming
together.   To get rid of this air, the steam in the boiler was kept up so much
stronger than the atmosphere, as to rush into the receiver with sufficient force
to drive out the air, through the same valve at which it entered, and which opened
outwards for that purpose, but it would shut, to prevent the air entering, when the
vacuum was made, except for an instant before the steam was admitted, and at that
instant it was opened by the machinery.   After this engine had been some years in
use, the air-valve was disused, being found unnecessary, and the engine worked
very well without it, not being so air-tight as at first.

The motion of the water-wheel is regulated by an apparatus called a governor,
invented by Mr. Watt, and which will be fully described in the account of his
engine.   It is a perpendicular axis, which revolves by communication with the
water-wheel, and carries round two pendulums, which move on a joint fixed to the
vertical axis.   When the rotation is very quick, the balls of these pendulums fly
outwards from the perpendicular axis, by the centrifugal force, and this motion
draws down a lever, which is connected with the sluice of the upper cistern ; so
that the sluice is made to fall or rise, according as the velocity of the wheel is
greater or less.   By this disposition, when the wheel moves very speedily from
lightness of work, or any other cause, the quantity of water let down from the
upper cistern is immediately diminished ; and, on the contrary, the quantity of
water is rendered greater, when the slowness of the movement shows, that more is
wanted.

The engine here described continued to work many years.   From the simplicity of its construction it is not liable to wear out for a very long time.

Mr. Kier thought it a profitable engine to himself, and that it would be serviceable for raising water, where coals are cheap.   It was stated to consume six bushels of good coals, in twelve hours' work, when in its best state, or seven bushels when at the worst.   Under these circumstances it was said to make ten strokes per minute, each throwing out seven cubic feet of water at an aperture 20 feet above the water beneath.   This is at the rate of 70 cubic feet per minute raised 20 feet = 1400 cubic feet raised 1 foot, and ÷ by 528 cubic feet, or one horse power, it gives $2\frac{2}{3}$ horse power, for the power of this engine ; but its consumption of coals was probably much greater than above stated, at least when working regularly with the $2\frac{2}{3}$ horse power.

ON THE CONSUMPTION OF FUEL IN SAVERY'S ENGINES.

We have no account of the quantity of water which could be raised by Savery's original engine, to a given height, with a given quantity of fuel.   He tells us in the Miner's Friend, that to lift a three-inch bore of water 60 feet high, would only require a fire-place, for the furnace, of 20 inches deep, or 14 or 15 wide ; the expense of fuel for which, he says, would be inconsiderable, when compared with the advantages to be derived from the use of the engine.

Bradley says, the small engine at Kensington, which he describes, would raise 52 gallons per minute, to a height of 58 feet, and that it burned half a peck of coals each time of working ; but neither of these accounts are sufficiently precise to found any calculation upon them.   Switzer, in repeating this statement in 1729, adds, that a bushel of coals will not cost above twelve-pence in London, and is much cheaper in many other places.

Mr. Smeaton made trial of two engines, which were put up in Manchester in 1774, by Mr. Rigley, to work water-wheels.   They operated by suction only, and were very like that last described.

The first had an upright cylindrical receiver, 16 inches diameter, and about 22 feet high.   It discharged the water at 14 ft. above the water in the well, and about 8 ft. beneath the top of the receiver.   It made 12 strokes per minute, and each stroke filled the receiver about 6 feet high.   Now a cylinder, 6 ft. long and 16 inc. diameter, contains about $8\frac{1}{3}$ cub. ft. × 12 strokes = 100 cub. ft. of water per minute × 14 ft. = 1400 cub. ft. raised 1 foot : this divided by 528 cub. ft. for a horse power, gives $2\frac{2}{3}$ horse power.   This engine consumed 3 hundred-weight of coals in 4 hours, or $\frac{3}{4}$ of a hundred-weight (which is exactly a bushel) per hour.   In that time it exerted a sufficient power to raise (1400 × 60 =) 84,000 cub. ft. to a height of 1 ft.   The weight of this water, at $62\frac{1}{2}$ pounds to a cub. ft. would be 5,250,000, or $5\frac{1}{4}$ million pounds weight, raised to a height of 1 foot by the consumption of a bushel of coals, weighing 84 pounds.

The other engine was larger.   The cylindrical receiver was 2 ft. diameter withinside, and 7 ft. high.   It delivered the water at 19 ft. high above the surface of the water in the well, and it made $7\frac{1}{4}$ strokes per minute, each stroke filling the receiver 6 ft. high.   The quantity is $18\frac{3}{4}$ cub. ft. per stroke, or 136 cub. ft. per minute, raised 19 ft. ; that is, very nearly 5 horse power.

This engine consumed 32 cwt. of coals in 24 hours, or $1\frac{3}{4}$ bushels per hour ; and at that rate, each bushel would raise nearly $5\frac{1}{2}$ million pounds weight 1 ft. high.

As neither of these engines were constructed with the best boilers, such as have since been brought into use in modern steam-engines, it is probable that a greater effect might be produced from the fuel.

CHAPTER II.

Mr. Thomas Newcomen's Fire-Engine, called the Atmospheric Steam-Engine, 1710.

THIS engine is named after its inventor, Mr. Thomas Newcomen, iron-monger, of Dartmouth, in Devonshire. He was an Anabaptist, and appears to have been a person of ingenuity and of some reading : he is said to have been personally acquainted with the celebrated philosopher Dr. Hooke. Mr. Newcomen appears to have grounded his invention upon that of Mr. Savery; for he was in the habit of visiting the tin-mines in Cornwall, where Savery was well known, and had derived his title of Captain from his attempts to introduce his engine for draining the mines, many of which were at that period worked out so deep, as to be unproductive and unprofitable, merely for want of some cheaper and more powerful machines than the hand-pumps or horse-machines, which they then used to drain them of water.

Mr. Savery was not successful in those attempts, because his plan required a direct action of the steam upon the cold water, a principle which either limited him to the height of 26 feet, or compelled him to use steam of a high temperature and elastic force, in which case the boiler and vessels required great strength, and a large quantity of fuel was necessarily consumed, to produce steam sufficiently dense. It is probable that these inconveniences occupied the thoughts of other ingenious men, who had seen Mr. Savery's engine, and become sensible of its importance; and the application of a cylinder and moveable piston had been already proposed, though the difficulties of execution must have retarded its adoption for a considerable time.

The first idea of a steam-engine with a piston was given by Papin, in his publication of 1690, (see p. 97), but it does not appear that his scheme was ever tried ; and indeed his plan was so imperfect, that it required great improvements to render it at all practicable. Papin himself was so far from being sensible of the merit of his own idea, that as soon as he became acquainted with Mr. Savery's engine, he abandoned his own project, and made a plan for an engine, in imitation of Savery's, in which the steam was to press directly upon the water, but with the intervention of a piston to float upon the surface of the water, and diminish the condensation. This engine is described in his publication of 1707, in the preface to which he gave an extract from a letter he wrote to Mr. Lebnitz, in 1698 : " We now raise water by the force of fire, in a more advantageous manner " than that which I had published some years before ; for, besides the suction, " we now also use the pressure which water exerts upon other bodies, in dilating " itself by heat, instead of which I before employed the suction only, the effects " of which are much more limited." His floating piston, though an improvement on Savery's engine, was not sufficient to remedy its defects : in consequence, Papin's last engine was never brought into any use ; for on the whole, it was not so complete a machine as Savery's.

The merit of perfecting Papin's original project of 1690, and putting it into execution, is entirely due to Mr. Newcomen and his associate Mr. John Calley, who was a plumber and glazier at Dartmouth, in Devonshire. According to Dr. Robison's account, it appears that Newcomen was acquainted with at least some of Papin's writings, for he says that among Dr. Hooke's papers, in the possession of the Royal Society, there are some notes of observations for the use of Newcomen, on Papin's method of transmitting the action of a mill to a distance, by drawing air through pipes. (See p. 94.) It would appear from these notes that Dr. Hooke had dissuaded Newcomen from erecting a machine on this principle, which in the first form, as proposed by Papin in 1685, was impracticable, though he afterwards improved it about 1686. Dr. Hooke had exposed the fallacy of Papin's first project in several discourses before the Royal Society, and did not consider his improved edition of it, (see p. 95) as practicable in execution, from the difficulty of making the air-pipes and cylinders sufficiently tight.

Dr. Robison has recorded one passage in these notes of Dr. Hooke's, which is remarkable: " Could he (meaning Papin) make a speedy vacuum under your " second piston, your work is done." The date of these notes is not mentioned, but Papin had been trying to obtain such a vacuum, in 1687, by gunpowder; and then in his project of 1690, he pointed out the means of making a vacuum under the piston, by first filling the cylinder with steam, just so much stronger than the atmosphere as is sufficient to displace the air, and then by cooling and condensing this steam, the cylinder is left void, and the atmospheric pressure acting upon the piston, is capable of lifting very great weights of water, with safety and effect. It does not appear whether Dr. Hooke or Newcomen were acquainted with that particular project of Papin's or not, but most likely they were, for it was published in Latin, in the Acta Eruditorem, for 1690; in French, in Papin's Recueil, at Cassel, 1695; and in English, in the Philosophical Transactions for 1697. Even if they built upon Papin's foundation, the merit of realizing the project, and overcoming all the difficulties of execution, is certainly due to Newcomen and Calley, for Papin merely gave a first idea of the principle; and Dr. Hooke died in 1703, before Newcomen produced his engine.

Mr. Savery made claim to Newcomen's engine as a modification of his own invention; and in consequence of the right he had to the mode of condensation, and making a vacuum according to his patent, he is said to have been admitted by Newcomen and Calley to an association with them in the patent, which was granted in 1705. It does not appear that they completed any engine until 1711; but about the year 1713 they had brought their plan to a degree of perfection not much inferior to the atmospheric engines which are in use at the present day.

Switzer, in his introduction to a system of Hydraulics, 1729, after describing Savery's engine, says, the last improvement of the surprising engine for raising water by fire, by Mr. Thomas Newcomen, is in a great measure summed up in that noble engine erected for the use of the York buildings. " It is generally " said to be an improvement to Mr. Savery's engine, but I am well informed, that " Mr. Newcomen was as early in his invention as Mr. Savery was in his, only the " latter being nearer the court, had obtained his patent before the other knew it; " on which account Mr. Newcomen was glad to come in as a partner to it." It is certain that Newcomen had been occupied with some such invention, about or even before the time that Savery obtained his patent.

Desaguliers, in his account of the invention, makes no mention of Mr. Savery being associated; but says, " that Thomas Newcomen, ironmonger, and

" John Calley, glazier, of Dartmouth, Southams, in the county of Devonshire,
" (Anabaptists,) made several experiments in private about the year 1710, and in
" the latter end of the year 1711 made proposals to drain the water of a colliery
" at Griff, in Warwickshire, where the proprietors employed 500 horses, at an ex-
" pense of 900l. a year; but their invention not meeting with the reception they
" expected, in March following, through the acquaintance of Mr. Potter of
" Bromsgrove, in Worcestershire, they bargained to draw water for Mr. Back of
" Wolverhampton; where, after a great many laborious attempts, they did make
" the engine work: but not being either philosophers to understand the reason,
" or mathematicians enough to calculate the powers and proportions of the parts,
" they very luckily, by accident, found what they sought for.

" They were at a loss about the pumps, but being so near Birmingham,
" and having the assistance of so many admirable and ingenious workmen, they
" came, about 1712, to the method of making the pump-valves, clacks, and
" buckets, whereas they had but an imperfect notion of them before. One thing
" is very remarkable: as they at first were working, they were surprised to see
" the engine go several strokes, and very quick together, when, after a search,
" they found a hole in the piston, which let the cold water in to condense the
" steam in the inside of the cylinder, whereas before they had always done it on
" the outside. They used before to work with a buoy to the cylinder, enclosed
" in a pipe, which buoy rose when the steam was strong and opened the injection,
" and made a stroke; thereby they were capable of only giving six, eight, or ten
" strokes in a minute, till a boy, named Humphrey Potter (q), in 1713, who at-
" tended the engine, added (what he called *scoggan*) a catch, that the beam al-
" ways opened, and then it would go 15 or 16 strokes a minute. But this being
" perplexed with catches and strings, Mr. Henry Beighton, in an engine he had
" built at Newcastle-upon-Tyne, in 1718, took them all away, but the beam it-
" self, and supplied them in a much better manner." Since that time no very
material alterations have been made in this species of engine, except the addition
of the crank and fly-wheel, to produce continuous circular motion by a direct
application of the force of the piston.

PRINCIPLE OF THE ATMOSPHERIC ENGINE.

To have an idea of its principles and mode of operation, suppose a very large
syringe or hollow cylinder E, to be placed in an upright position, and a piston J or
round plug, shaped like a grindstone, accurately fitted into the cylinder, so as to fill
it exactly; the interior surface of the cylinder E, being made smooth and true, that
the piston J may slide freely up and down in it, although it fits so close that neither water nor air can pass by the edge of the piston; this piston is suspended by
a rod and chain, from one end of a lever L p L, which is poised upon a centre p
like a scale-beam, and at the opposite end of the same lever, is another chain and
rod to suspend the stem or rod M, of a sucking pump, such as is used for domestic
purposes, and by which the water is to be raised out of the mine or well, in which
the pump is immersed.

(q) An Englishman, named Potter, went to the continent, about the year 1720, and constructed
an engine at a mine in Hungary; it is described by Leupold in his Theatrum Machinarum Hydraulicarum, 1725 with many encomiums on Potter, who was considered as the inventor.

It is evident that if the piston J is pressed down into the cylinder E, the 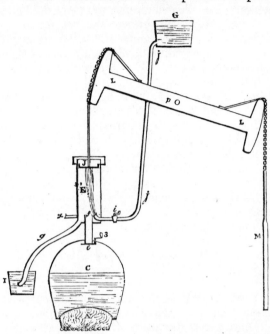 rod M with the bucket of the pump will be drawn up, with a corresponding motion, and will raise the water through the barrel of the pump, till it runs out at the spout. To cause the return of the piston J, when it is released from the pressure which occasioned its descent, a sufficient counter weight is applied to the rod M of the pump, to overbalance the piston, and draw it upwards to the top of the cylinder, as shown in the figure.

To force the piston down into the cylinder E, the air must be exhausted from its interior capacity, and the weight of the atmosphere resting upon the upper surface of the piston, will no longer be balanced by any counter pressure acting beneath it; provided the air could be effectually exhausted from the cylinder, so as to make a perfect vacuum therein, the whole pressure of the atmosphere on the piston, amounting to $14\frac{3}{4}$ lbs. upon every square inch of its upper surface, would act in the same manner, as if a weight of solid matter were heaped upon the piston, to press it down into the empty cylinder, and draw up the pump bucket and the water in the pump (r).

In order to produce the vacuum at pleasure in the interior capacity of the cylinder, hot steam, and cold water, must be introduced into it alternately; for this purpose there are two entrances into the bottom of the cylinder, for the alternate admission of steam and cold water, and two exit passages for the evacuation of useless water, and air, from the cylinder. Each of these four passages is closed occasionally by a valve or cock.

The first entrance is the orifice e of a pipe ef, which descends from the bottom part of the cylinder E, like the spout of a common syringe; the lower end of this pipe communicates with the upper part of a boiler C, which is placed over a furnace A, in the same manner as a brewing copper, but it is a close boiler, being covered with a close dome, or hemispherical top C, to keep in the steam, which rises from the boiling water, contained in the boiler. At the lower end e, of the steam pipe ef, is a valve or regulator, which, when open, permits the steam to pass from the boiler C into the cylinder E, or when shut, stops the communication.

The second entrance into the bottom of the cylinder, is to admit the extremity

---

(r) To form an idea of the amount of the atmospheric pressure on the piston, suppose a number of round cast iron weights, of exactly the same size as the piston to be placed one upon another, like a pile of grindstones, and suspended to the end of the lever in place of the pump rod M: the piston J being suspended to the opposite end of the lever; if the whole height of such a pile of iron weights, amounted to $4\frac{3}{4}$ ft. they would only balance the weight of the atmosphere upon the piston, when all the air was exhausted from the cylinder; or if instead of cast iron weights we suppose they were circular pieces of ice, the whole height of the pile must be more than 34 ft. to make a balance against the atmosphere's pressure on the piston.

S

f of a small pipe *jj*, which conveys cold water from an elevated cistern, G, into the interior of the cylinder, at every interval when the communication from the boiler is cut off; the end of this injection-pipe *j*, turns up through the bottom of the cylinder, with a spout near *f*, in order to throw the cold water upwards in the cylinder, (as shown in the figure,) with a jet or fountain, which, by dispersing in a shower through the steam, cools and condenses it. This injection-pipe *jj*, has a cock *i*, to stop the stream at pleasure.

The third is an exit passage, and leads out of the cylinder: it is the orifice of a drain-pipe *g*, called the eduction-pipe, and its use is, to carry off the condensed steam, and useless injection-water; it therefore proceeds from the very lowest part of the cylinder, and has a considerable descent, to drain off that water. The lower extremity of the eduction-pipe *g*, is covered by a valve opening outwards: this valve is immersed in a cistern I, which receives the hot water.

The fourth passage is a small lateral exit pipe or spout *x*, with a valve opening outwards, to permit the escape of the air, or permanently elastic fluid, which cannot be so condensed by the application of cold water, as to run off through the eduction-pipe *g*. This valve *x*, is called the snifting clack.

The weight of the pump-rod M serves as a counterpoise, and exceeds the weight of the piston J, so much, as to draw up the piston to the top of the cylinder, as in the figure, whenever the steam from the boiler C, or the atmospheric air, is freely admitted into the cylinder.

These are the most essential parts of the engine, and are here represented in their simplest form, merely to give a knowledge of their particular offices, and the propriety of the peculiar forms which are given to them will then be apparent. Let us now consider how this machine is to be put in motion, and what is the manner of its operation.

The fire being made at A, on the grate of the furnace, beneath the boiler C, the water therein is made to boil briskly, so as to afford steam in abundance, and rather more elastic than the atmospheric air. In this state, the steam-regulator *e* being opened by its handle 3, the steam will enter from the boiler, through the steam-pipe *ef* into the cylinder E; and the steam being only half the weight of common air, will rise to the top of the cylinder, and expel the air therefrom, through the snifting valve *x*. When all the air is discharged, the interior of the cylinder is quite filled with hot steam, which will press upwards beneath the piston J, with as much force as the atmospheric air presses downwards upon its upper surface.

The regulator *e* is then shut, and the injection-cock *i* is opened, to admit a stream of cold water, which rises (as per figure) in a jet within the cylinder, and strikes with force against the bottom of the piston J; by which means it is dispersed in a shower of drops, through the steam. This cold water cools the steam, or abstracts so much of the heat from it, that the elastic steam, which so entirely filled the cylinder, as to displace the air therefrom, becomes immediately condensed into a very small quantity of hot water, leaving the interior capacity of the cylinder nearly void; and then the pressure of the atmosphere, on the upper surface of the piston J, being no longer borne up, by the elasticity of the steam beneath it, forces the piston down into the vacuous cylinder, so as to draw up the pump bucket M, which is suspended at the opposite end of the great lever L L, and this raises the water contained in the barrel of the pump.

During this descent of the piston, the injection-cock *i* is closed, and when the piston has arrived nearly at the bottom of the cylinder, the steam-cock *e* is again opened, to admit a fresh supply of elastic steam from the boiler C, into the cylinder.

This steam fills the small space between the piston and the bottom of the cylinder, and assists the piston to rise, by pressing beneath its under surface, with a little more force than the atmospheric air presses on the upper surface ; the steam assists the hot waste water, which remains in the bottom of the cylinder, to run off through the eduction-pipe g. The steam also drives out, through the snifting valve x, any air, or elastic fluid, which was not condensed by the cold water. It is for this purpose, that the steam requires to be kept up rather stronger than the atmosphere ; and therefore it will more than counterpoise the pressure of the atmosphere, on the upper surface of the piston. In consequence of the piston being thus set at liberty, and even lifted by the steam, it is rapidly drawn up, by the action of the counterpoise M, at the opposite end of the lever L L, and quickly regains its original position at the top of the cylinder.

A repetition of this process, namely, of shutting off the steam, and injecting cold water, causes the piston again to descend ; and in this manner the alternations may be continued, as long as the boiler will supply steam, and the cistern cold water.

It is to be understood, that the steam-regulator and the injection-cock are opened and shut by apparatus fixed to the working lever L L, in such a manner as to strike the levers of those cocks, at the precise instant of time, when their effects are required to be produced. A small pump is also connected with the working lever, to raise a continual supply of cold water, up into the cistern G ; and a small portion of the hot water which drains out of the cylinder, through the eduction-pipe g, is introduced from the cistern I into the boiler, to supply the waste of evaporation therefrom : so that the attendant has no other office to perform, but that of keeping up the fire, beneath the boiler.

The action of the engine depends upon a constant supply of hot steam and cold water ; the former to fill the cylinder, and the latter to empty it ; for it must be considered, that all the heat communicated from the fire to the steam, is carried off in the hot water, which runs waste out of the cylinder. In fact, the steam must be considered, as a means of conveying the heat into the cylinder, and diffusing it through the capacity thereof, so as to displace the piston, and the atmospheric air from the cylinder ; and the injection of cold water, as the means of gathering up this heat from all parts of the cylinder, into a small space at the bottom thereof, and thus make room for the piston, and the atmospheric air, to return again into the cylinder.

As an example of the application of Newcomen's engine, to the purpose of draining water out of a mine, suppose that the sucking-pump, which is placed in the pit, is to lift the water 54 yards perpendicular, and suppose that pump is 8 inches bore. The column of water, which must be raised when the rod M, and the bucket of the pump, is drawn up, will weigh 3535 pounds. The chain which is attached to the upper end of the rod M of the pump, being suspended from one extremity of the long lever or working-beam L L, the piston J of the steam-cylinder E, must be suspended from the opposite extremity of the same beam L by another chain, as before described ; and the lever being poised on a centre p, just in the middle of its length, the motion of the piston J, and of the pump-rods M, will be equal, but in opposite directions.

To give this piston a sufficient power to make it draw up the water in the pump at the opposite end of the lever, with celerity, the cylinder must be 24 inches diameter ; the area or surface of the piston will then be 452 square inches, and if each square inch is pressed with a force of 7·8 pounds, it will very nearly balance the weight of water in the pump (for $452 \times 7·8 = 3525$ instead of

3535), but the actual force of the piston, is considerably greater than 7·8 pounds per square inch; for the full pressure of the atmosphere upon the upper surface thereof is about 14¾ pounds per square inch, and from this we must deduct the pressure or elasticity of that vapour or rare steam, which remains in the cylinder after the cold water is injected, not being condensed by that degree of cold.

The elasticity of this vapour will depend upon the temperature to which the interior space of the cylinder is cooled by the injection. This is usually between 140 and 160 degrees of Fahrenheit's thermometer. Let us take 152 deg. as the temperature, and by the table of elasticities it appears, that the force or elasticity of the steam remaining in the cylinder will be nearly 4 pounds per square inch: which force acting upwards beneath the piston, must be deducted from the full pressure of 14¾ lbs., and leaves 10¾ lbs. per square inch for the effective force of the piston, which, multiplied by 452 square inches, gives 4859 lbs. for the unbalanced force of the piston; this is greater than the load in the pumps by 1324 lbs.

One half of this excess, or 662 lbs., must be allowed for the counterweight; for the weight of the pump-rods must exceed the weight of the piston, by that quantity, in order to give the pump-rod a proper preponderance, to draw up the piston, when the steam is admitted into the cylinder. On the other hand, when the cold water is injected into the cylinder, and the steam condensed, the descending force of the piston, will exceed the united resistance of the column of water, and of the counterweight, by 662 pounds.

This extra force, or preponderance of 662 lbs. acting alternately in opposite directions, is necessary to overcome the friction of the parts, and urge them into motion; and the rapidity of this motion will be greater or less, according to the goodness of the engine, in respect to the resistance from friction and the weight of its moving parts, and in respect to the perfection of the vacuum, which is formed within the cylinder by the injection. An engine executed in the best manner, according to the foregoing dimensions, may be expected to make 15 strokes per minute of five feet each; or the piston will pass through 75 ft. per minute when actually raising the water, without counting the returning stroke.

The power or mechanical effect produced by this engine is 3535 lbs. of water raised 75 ft. per minute, which is equivalent to 265125 lbs. raised one foot high per minute; and this quantity, divided by 33000 lbs. (which is esteemed a horse power) shows that the engine is 8 HP. The quantity of water it will raise from the bottom of the pit 54 yards deep, is 1·74 cub. ft. per stroke; and at 15 strokes per minute, 26·1 cub. ft. per minute, or 1566 cub. ft. per hour.

An engine of these dimensions is but a very small one; yet it serves to show the superiority of Newcomen's engine to that of Savery. The latter raised the water, partly by the force of steam, and partly by the pressure of the atmosphere; but Newcomen's engine raises the water entirely by the pressure of the atmosphere; for the steam is employed merely as the most expeditious method of displacing the air, and then producing a void, into which the atmospherical pressure may impel the first mover of the machine; the elasticity of the steam is not the first mover.

To have drained the mine above mentioned, by one of Savery's engines, he would have raised the water about 26 ft. by suction, and must then have forced it 136 ft. higher; he must therefore have employed steam of an elastic force, of nearly 59 lbs. on each square inch when in the receiver, this would have required a temperature of 309 deg. after all the loss of heat, by the condensation of the steam;

but this action would have been so rapid on introducing such hot steam into the same vessel with water of only 50 or 60 deg. of temperature, that it would have been wholly impracticable to have thrown up any considerable quantity of water.

The purpose could only have been effected by two engines, one placed above the other, each raising the water half way, or 81 feet; in which case, 25 ft. being done by suction, the forcing of the water would be only 56 ft. more; and this being but 24·4 lbs. per square inch, would require a temperature of about 267 deg. which would be practicable, but with a great waste of fuel, and with the trouble of attending two engines, instead of one: and they must have been large engines to have raised the required quantity of water, for we have seen (p. 125), that an engine of Savery's, with a cylinder two feet diameter, and making a six-feet stroke, is only 5 HP.

On Newcomen's system there is no need of steam of great or dangerous elasticity; for it operates with very moderate heat, and consequently with a much smaller quantity of fuel than Savery's. The power of Newcomen's engine is not bounded by the strength of the boilers, and vessels, to resist internal pressure, but only by the dimensions which it is practicable, and expedient to make boilers, and cylinders, to contain the requisite quantity of steam, of the ordinary pressure, and the strength which can be given to the working lever, chains, and other parts, which communicate the force of the piston, to the rod of the pump. Newcomen's engine can also be applied to other mechanical purposes, besides that of raising water. For instance, to blow air by bellows or pumps, into a furnace; or, by connecting a crank and fly-wheel with the rod M, which is suspended from the extremity of the great lever L, the reciprocating motion of the great lever, may be changed into a circular motion. Savery's engine is necessarily restricted to the purpose of raising water, and could not be applied to work mills, except by the intervention of a water-wheel, as before shown.

The brief description of the atmospheric engine just given, is very nearly the form in which it was presented to the public by Mr. Newcomen himself about 1712, and when the invention seemed to have justified his confidence in its practicability. But many difficulties still occurred in the execution, and were removed one by one. At first, the valves were opened and shut by hand, and required the most exact and unremitting care of the attendant, to perform those operations at the precise moment: the least neglect or inadvertence might be ruinous to the machine, by beating out the bottom of the cylinder, or allowing the piston to be wholly drawn out of it. Stops were contrived to prevent both of these accidents; then strings were used, to connect the handles of the cocks, with the lever, so that they should be turned, whenever it reached certain positions. These strings were gradually changed, and improved, into detents and catches of different shapes: till at last, in 1718, Mr. Beighton, a very ingenious and well-informed engineer, simplified the whole of these subordinate movements, and brought the machine into the form in which it has continued, to the present day, without any material change.

DESCRIPTION OF NEWCOMEN'S ATMOSPHERIC ENGINE, WITH THE IMPROVEMENTS OF MR. BEIGHTON, 1718, AND MR. SMEATON, 1772.

From the time that Mr. Beighton brought Newcomen's engine to a standard form, it remained during half a century, without any improvement; for, although great numbers were constructed of all dimensions, they were commonly executed by ordinary mechanics, who were incapable of calculating, and paid but little attention to proportions. About the year 1772, Mr. John Smeaton, the celebrated engineer, investigated the subject of the steam-engine, as he had before done that of water-mills, and wind-mills; and, although he did not add any thing to the invention of Newcomen, he established just proportions for engines of all sizes; and the performance of the engines he constructed, greatly exceeded the common sort, as they had been usually made before his time.

The drawings in plates II and III, represent an engine constructed in 1772, after Mr. Smeaton's designs, at Long Benton Colliery, near Newcastle. This engine, which Mr. S. considered as his standard, is in every respect the same machine as Beighton's, but of larger dimensions than was usual in his time.

Plate II, is a section of the whole engine, and the building which contains it; and plate III, a ground plan of the building, with figures of detached parts on a larger scale. The same characters, and figures, of reference are used in all the different views.

A, the fire-place, situated beneath the boiler, with a door opening in front to introduce the fuel.

B, the fire-grate, on which the fire is made, to heat the water in the boiler, and convert it into steam. The space beneath the grate is called the ash-hole; it is open in front, to give free admission to the air, to pass under the grate, and rise up between the bars, to animate the fire.

C C, the boiler, made of iron plates, riveted together in a circular form, covered with a hemispherical dome. It is about half filled with water, leaving the other half for steam; the lower part of the boiler is enclosed in brickwork, but a clear space is left beneath, for the flame of the fire to act under the circular bottom; and a circular channel or flue, extends all round the lower part of the boiler, as shown by the plan of the engine-house in plate III, that the utmost heat of the fire may be communicated to the boiler; the smoke passes from the circular flue, into the chimney.

D, the perpendicular chimney, raised to a great height, to cause a strong draught of air through the fire-place; this chimney is a passage formed in the wall of the building, as shown by the dotted lines, plate II.

Note. This engine being of large dimensions, requires two boilers and furnaces, to supply it with steam; one boiler being placed immediately beneath the cylinder, as shown in the elevation, and the other in a side building, as shown in the plan, plate III. The steam from the outside boiler, is conveyed into the other by a steam-pipe *a*, which forms an open communication between the two boilers.

e is the regulator or steam-cock, placed beneath the flat top of the boiler, and within the dome: it is a brass plate, shaped like a fan, as shown in the enlarged plan of the working gear, in plate III. The regulator *e* moves horizontally about a centre, so as to stop the lower orifice of the steam-pipe *f*, when turned one way, or to open the passage when turned the other way. The perpendicular centre-pin or axis, to which the regulator is fixed, passes up through a socket, in the top plate of the boiler C, and has a handle 3, fixed on, the upper end on the outside of the boiler, to give motion to the regulator within the boiler, in order to open or shut the communication between the boiler and the cylinder at every stroke.

E, the cylinder, made of cast-iron, bored smooth and true within; it has a strong projecting flange in the middle, to suspend it between an assemblage of strong beams, F F, which extend across the building, and their ends are worked into the two side walls thereof. The bottom of the cylinder is a basin or hemisphere, which is united to the lower end of the cylinder, by a number of screw-bolts, passing through two flanges or rims, one projecting on the outside of the cylinder, and the other projecting from the edge of the basin.

f is the steam-pipe, rising up from the top of the boiler, to introduce the steam into the cylinder, whenever the regulator *e* is opened. The steam-pipe rises up some distance within the cylinder bottom, to prevent the injected water, which lodges there, from running down into the boiler.

J J, the circular piston of cast-iron, which is made to fit the cylinder very exactly, but with liberty to slide up and down. It has a flange or circular rim, upon its upper surface, rising four or five inches: a quantity of hemp or oakum is stuffed between this flange and the inside of the cylinder, and is kept down by weights, to prevent the entrance of air or water, or the escape of steam. The under surface of the piston is covered with wood plank, that it may not cool the steam so much as the bare iron would do.

n is the iron shank, or rod, of the piston, and K K the double chain by which it is suspended from the end of the great lever-beam L; which end being formed to the arc of a circle, is called the arch-head.

L L, the great lever or working beam, which is poised on its centre *p*, in the manner of a scale-beam; it is made of two large logs of timber and four smaller ones, bent together at each end, and kept at a distance from each other in the middle, by the iron axis or centre *p*, as represented in the section, plate II. The arch-heads at the ends of the lever are for the chains K to work upon, so as to give a perpendicular direction to the shank of the piston and to the pump-rod M, which are suspended by those chains, from the opposite ends of the great lever.

M M, the pump-rod, which goes down into the mine, and works pumps therein, to draw the water from the bottom of the mine, to some subterraneous adit, or passage where it can run off. The lower part of the rod M, is divided into branches, to work as many pumps as the mine may require. The manner of applying the pumps is shown in plate IV. The pump-rod M being suspended from one arch of the great lever, and the piston rod from the other arch, and both being at the same distance from the centre of motion *p*, the motions of the piston and of the pump-rod will always be equal, but in opposite directions. Each pump in the mine, has a valve in the moveable bucket, and another at the bottom of the pump, both valves opening upwards. The pumps raise the water when their rods and buckets are drawn up, but are inactive when they descend.

S, a deep cistern or pit, divided off from the great pit of the mine, to receive cold water, which is conducted into it by a trough, and the superfluous water is carried off by another trough, so that the cistern is always kept full to the desired height. This cold water, which is for the injection, is collected in ponds, on the surface of the earth, either from rain or from a brook.

R, the jack-head pump, which is a small lifting-pump, worked by an iron rod connected with the great lever by a chain, but being nearer to the centre *p*, than the great pump rod, it makes a short stroke. The jack-head pump stands near the end wall of the building, and raises a supply of cold water from the pit S, by a perpendicular iron-pipe and spout Q Q, into the jack-head cistern G.

G, the jack-head cistern, for supplying the injection-water to the cylinder. It is always kept full by the pump, R; and is fixed on the top of the house, to be so high above the cylinder bottom, as to give the jet of injection a great velocity into the cylinder, when the cock is opened. On the side of this cistern, opposite to the pipe Q, is a waste-pipe, for conveying away the superfluous water.

j j, the injection-pipe, of lead or copper, which descends from the cistern G, to the injection-cock *i*, and the end of it enters the cylinder bottom, and turns up in a curve at the extremity. It has a thin plate of brass, screwed upon the end, which is within the cylinder, with a square adjutage hole in it, to cause the jet of cold water, from the jack-head cistern G, to fly up in a stream, against the under surface of the piston, and condense the steam contained in the cylinder.

Note. Upon the upper end of the injection-pipe *j j*, a leather flap valve is fitted (within the cistern G), and when the engine stands still, this valve is shut, to prevent waste of water by leakage ; but before the engine is set to work, this valve must be lifted up by a cord, and it is kept open all the time the engine continues to work.

d, a small pipe, which branches off from the injection-pipe *j*, and has a cock to run a small stream of water into the cylinder, to cover the surface of the piston, and keep the hemp packing moist, so as to prevent the leakage of air by the piston into the cylinder.

l l, the working plug or rod, by which the steam-regulator *e*, and injection-cock *i*, are opened and shut alternately. It is suspended by a chain from the great lever, and is a heavy piece of timber, with a slot cut vertically down the middle of it. Holes are bored through it horizontally, to receive pins, which give motion to the working-handles of the regulator and injection-cock when the plug works up and down, by the motion of the engine.

3, the handle of the steam-regulator. It is attached to the regulator by a perpendicular spindle, which comes up through a socket in the brass regulator plate, which forms the top plate of the boiler. The regulator itself is a plate of brass, *e*, shaped like a fan, which is moved horizontally by the handle 3, and opens or shuts the orifice of the lower end of the steam-pipe *f*, within the boiler. It is represented in the enlarged plan, Plate III.

14, the stirrup-rod, which communicates motion to the handle 3 of the regulator, one end being jointed thereto, with a pin put through both ; the other end of the rod is open like a stirrup, whence its name. The stirrup is suspended horizontally, by links from the axis of the Y, so as to hang at a proper distance beneath the same : but with liberty to move freely endways, in order to turn the regulator round about its centre.

16, the hammer or tumbling-bob, fixed at the upper end of a lever, called the Y, because it is formed with two forked legs at the lower end, like the letter Y when inverted (thus λ). This hammer, or Y, is fixed on a horizontal axis *t*, which is supported on pivots at its ends, and is provided with two curved levers or arms, one of which, marked 4, enters into the slot in the working plug *l*, and the other, *r*, is close to one side of the plug. The lever *r* has a projecting handle, for the attendant to work the regulator by, occasionally. Both arms, *r* and 4, being fastened on the axis *t* of the Y, are alternately raised and depressed, by two pins in the working plug *l*, so as to throw the hammer or bob, 16, backwards and forwards on either side the perpendicular ; and the two legs of the Y, which enter into the opening of the stirrup, 14, give motion thereto, and open or shut the regulator.

For instance, when the piston J gets to the top of the cylinder, the pin in the slot of the plug *l* raises the arm 4, and turns the hammer 16, towards the cylinder, until it falls over beyond the perpendicular, as represented in the figures. The leg 12 of the Y then strikes against the cross end of the stirrup 14, and draws forwards the handle 3 of the regulator, so as to turn the regulator *e* suddenly, and shut the passage for the steam.

On the other hand, when the piston gets to the bottom of the cylinder, the pin, which projects from the outside of the plug, presses down the handle *r*, and turns the hammer 16, away from the cylinder. When it falls over beyond the perpendicular, the leg 11 of the Y strikes the cross end of the stirrup 14, and suddenly pushes the handle 3 of the regulator *e* backwards, so as to open the passage again.

i is the injection-cock, made of brass, and soldered to the injection-pipe. The turning plug is accurately fitted into the brass socket, by grinding, and the plug is perforated, so as to allow an open passage for the water, when that per-foration corresponds with the direction of the pipe, but when the plug is turned one-fourth round, it closes the passage. A lever or handle is fixed on the top of the plug, to turn it by, and the weight of the plug is borne up by a screw beneath, to avoid jambing in its socket.

s, the handle for the injection-cock. It is moved by a pin projecting out from the plug *l*. The handle *s* is fixed to a horizontal axis, *v*, which has an arm, 14, projecting downwards, and the lower end thereof is connected by a short rod, 6, with the handle or lever, which is fixed to the plug of the cock *i*; so that when the handle *s* is raised up, as in the figure, the passage of the cock will be open, or when the handle *s* is pressed down, the cock will be shut. A loaded lever or hammer, 8, is fixed to the axis *v*, to open the cock *i* by its weight; but when the handle *s* is depressed, and this weight raised, it is held up by a hook or latch 18, which detains it in that position, till the piston reaches the top of its course. A small chain, 22, which is connected with the upper part of the plug *l*, then lifts up the catch 18, and releases the weight 8, which falls, and suddenly opens the injection-cock *i*, to admit a jet of cold water into the cylinder, which, condensing the steam, makes a vacuum within; and the pressure of the atmosphere forcing down the piston J into the cylinder, causes the plug *l* to descend also. The pin in the plug *l*, catches the bent end of the handle *s*, in its descent, and by pressing it down, shuts the injection-cock. The regulator is then opened by the working gear in the manner before described, to admit fresh steam, and so on alternately: when the regulator is shut, the injection-cock is opened, and *vice versâ.*

g, the eduction-pipe, to drain off the water which is injected into the cylinder at each stroke. Its upper end communicates with the lowest part of the cylinder bottom, and its lower end descends into the hot well I, and is covered with a hanging valve *m*, which lets out the injected water when the steam is admitted into the cylinder; but the valve shuts close, to prevent the water returning again, when the vacuum is made in the cylinder.

I, the hot well. It is a small cistern to receive all the waste water which is conveyed from the cylinder by the eduction-pipe *g*, and to keep it in reserve, to feed the boiler, in order to supply the waste occasioned by the continual evaporation of the steam.

T, the feeding-pipe, to replenish the boiler with water from the hot well. It is a perpendicular pipe, which goes down more than half way to the bottom of the boiler; so that the lower end is always immersed in the water, and the upper end, which rises up through the dome of the boiler, is open at top, with a funnel or cup upon it. This upper part has a lateral branch out of the hot well, with a cock, to admit a large or small quantity of water, as occasion requires, to replace what is evaporated.

y y, two gauge-cocks, at the upper ends of two pipes, which descend into the boiler. One descends lower than the other, and both are open at the lower ends. Their use is to show when the boiler contains a proper quantity of water; for, upon opening the cocks, one should give steam and the other should give water, because the intended level of the water in the boiler, is between the lower ends of the two. If they both give water, there is too much, or if both give steam, there

T

is too little water in the boiler, and the cock in the branch of the feeding-pipe T, must be regulated according to these signs.

x is the snifting-valve, covering the orifice of a small spout at the lower part of the cylinder; the valve opens outwards, and when the first puff of steam is admitted into the cylinder, it expels through this valve, any air which may have leaked by the piston, or which may have gained admittance with the injection water. A leaden pipe extends from the snifting-valve *x*, and passes out through the wall of the building, to convey away the air, and the steam which must necessarily follow it; but this pipe is provided with a cock, *k*, to diminish the passage to what is sufficient to discharge the air only, and prevent needless waste of steam.

In some convenient part of the dome of each boiler, a circular hole must be made, large enough for a man to enter, in order to clean the boiler, or repair it occasionally; when the engine is at work, this man-hole is closed by a door, or cover of iron plate, screwed over it, by a number of screw bolts arranged round the margin of the cover; there must also be a drain-pipe from the lowest part of the boiler, to draw off the water from it occasionally, through a cock or a plug.

In the top part of the outside boiler is the safety-valve, also called the puppet-clack: it is an aperture fitted with a brass valve, which is loaded with a lead weight, sufficiently heavy to keep it down, and retain the steam in the ordinary rate of working the engine; but if the engine stops, or if the fire is made too strong, the accumulated steam will lift up the loaded valve, and escape into the open air: as the two boilers are always connected by the steam pipe *a*, one safety valve serves both, or if more convenient, it may be placed on that steam pipe, as at V in the plan, Plate III. The weight of the safety-valve is usually regulated to 1 or $1\frac{1}{2}$ pounds, for every square inch of the under surface of the valve, beneath which the steam acts.

c c is a circular channel or cup, round the top of the cylinder E, to prevent the water upon the piston J dashing over, when it rises too high; and a small lead pipe conveys the superfluous water from this channel, down to the waste water drain. This pipe is omitted in the figures.

u u are catch pins, to limit the motion of the piston; they are two strong iron bars, which pass through the arch heads, at each end of the great lever L, and project out on both sides thereof, so as to strike down upon the fixed spring beams U, and prevent any further motion, when the piston, or the pump-rod, has moved through the full length of the intended stroke. Also, in case the chains at either end should break, these catch pins stop the lever from exceeding its proper motion.

U, the spring beams, which are placed on each side the great lever to meet its catch pins *u*; they are provided with strong wooden springs, so placed as to intercept the catch pins, and moderate the blow which they sometimes make on the spring beams, when the engine works too forcibly.

The building or engine house consists of four walls of very solid masonry; the principal one is called the lever wall, because it forms the fulcrum of the great lever L, which passes through an arched opening at the upper part of the wall, the pivots of its axis *p* being mounted in bearings, which are lodged on the masonry in the opening; one end of the lever is within the house over the cylinder, and the other is out in the open air over the mine or pit; the lever wall has two perpendicular passages in it, DD, to form the chimneys of the two furnaces, and the injection-cistern G is placed on the top of the wall, above the great lever. The lever wall must be built very solid, for it has to sustain the weight of all the moving parts of the engine, and also double the force of the piston, and it is subjected to violent shocks from the motion.

The two side walls of the house are also of considerable thickness, and are worked solid with the lever wall; in the lower part of these side walls are niches or recesses, to leave room for the centre boiler, and the flue which extends round it, as shewn by the dotted line in the plan. The ends of the cylinder-beams are worked into each of the side walls, which therefore sustain the cylinder, and must also resist all the effort the piston makes. The cylinder-beams are very strong, each one being composed of four logs combined by bolts.

The end wall requires less strength than the others, but is worked solid with them. An archway is left in the lower part of the end wall, to introduce the boiler in the first instance, or to renew it when worn out; the arch is closed by a slight brick wall, in which is an entrance door to the engine house, and the ash pit and fire door of the furnace, is opposite to that door.

The auxiliary boiler is contained in a side building (as shewn in the plan, Plate III.), which communicates with the engine house by an archway through the side wall, that the attendant may have ready access to both furnaces; the steam pipe *a* also passes through an opening in that side wall, and the hot well or cistern I, is placed in another opening therein, so as to be equally convenient for feeding both boilers.

The engine-house has three floors, so situated as to give convenient access to all parts of the engine; thus, one floor is level with the top of the brickwork which encloses the boiler, and part of the dome of the boiler rises above this floor; another floor is laid on the top of the cylinder beams, so that the cylinder rises up through it; and the third floor is a little lower than the spring beams. The whole building is covered with a roof, and a strong beam is extended across the top of the side walls, nearly over the cylinder, to afford a suspension for a tackle to occasionally lift the piston out of the cylinder, or to take up the great lever. A strong windlass, such as is used in small ships to heave up the anchor, is mounted between the ends of the spring beams over the cylinder; and from this windlass a rope can be carried to a pulley suspended over the cylinder, from the beam in the roof, to lift the piston.

A recess is formed in the lower part of the lever wall, opposite to the working gear, to leave room for the attendant to stand, when he works the handles to start or stop the engine: there is a window through the wall to give him light, and also to enable him to see the pump rods; a seat is provided for him beneath the window, which is his proper post, when not occupied at the furnaces below. There is usually an entrance door into the house, through the side wall, at the level of this floor, and a passage through into the side building, which is covered by a low roof; (see also an end view of the building in Plate IV. fig. 1).

The spring beams U, are lodged into the lever wall, at one end, and the other ends are fixed on a strong beam, which is extended across the house, between its side walls, nearly over the cylinder. The spring beams at the outer end of the great lever, are supported by upright posts erected upon a strong beam, which lies horizontally across the mouth of the pit of the mine.

OPERATION OF THE ATMOSPHERIC ENGINE.

WHEN the engine is not at work, its resting position or attitude, is such as appears in the drawing, Plate II.; the pump-rods, M, preponderating by their weight, have drawn the great piston to the top of the cylinder; this preponderance is termed the counter-weight.

T 2

To prepare for working, the boilers must be filled about half full of water, and large fires must be made under them ; when the water boils and produces steam heated to about 216 deg. it will exert a pressure of about 1½ pounds beneath each square inch of the safety-valve, and it will lift up that valve and escape.

The machine will still continue at rest, because the steam-regulator and the leather flap valve, at the upper end of the injection-pipe are shut, and the cylinder remains full of air.   To clear the cylinder of air, the man who attends the engine, depresses the handle r of the working gear, so as to throw the hammer 16 away from the cylinder, and the leg 11 of the Y, acting upon the cross end of the stirrup 14, pushes it backwards, and opens the regulator suddenly ; the steam from the boiler will immediately rush into the cylinder, and mixing with the air therein, will expel some air through the snifting-clack x, but steam being a lighter fluid than air, it will ascend by degrees, and occupy the upper part of the cylinder beneath the piston, and drive out the air at the valve x ; at its first entering, the steam will be condensed by the cold surface of the cylinder and piston, and the water produced from it will trickle down the inside, and as soon as any quantity is accumulated, it will drain down into the eduction-pipe g.   This condensation and waste of steam will continue till the whole cylinder and piston are made as hot as boiling water.

When this happens, the steam itself will begin to escape at the snifting-valve x, and issue through the cock k and pipe ; at first slowly and very cloudy, being mixed with much air, the cloudy appearance of steam being always owing to its mixture with common air.   The blast at x will grow stronger by degrees, and become more transparent, as it carries off the common air which filled the cylinder.

We supposed, at first, that the water was boiling briskly, so that the steam was issuing at the safety-valve, which is in the top of the outside boiler.   The opening of the steam-regulator, put an end to this discharge at once, because the cold cylinder drew off the steam from the boiler with astonishing rapidity, until it became so heated as not to condense any more.

When the manager of the engine perceives that the blast at the snifting-valve is strong and steady, and that the boiler is again fully supplied with steam of a proper strength, as appears by the renewal of the discharge at the safety-valve, the engine is ready for starting.

He now lifts up the handle r, till the hammer 16 of the Y falls over the perpendicular towards the cylinder, and its leg 12, striking the cross pin of the stirrup 14, draws it forwards, and shuts the steam-regulator ; at the same instant he lifts up the leather-valve at the upper end of the injection-pipe j, within the cistern G, by its cord ; and the injection-cock, being already open, the pressure of the column of water, which runs down into the pipe j, forces some water through the cock, and it spouts up into the cylinder against the piston, as shewn in Plate II.

The cold water thus coming in contact with some of the pure steam which now fills the cylinder, condenses it, and makes a partial void, into which the more distant steam immediately expands, and being also condensed by the cold water which continues to flow, the same effect is produced on the steam beyond it : and thus it happens, that the abstraction of a small quantity of heat, from an inconsiderable volume of steam, produces a general condensation throughout a cylinder of which the capacity is very extensive.

The instant any condensation takes place, the snifting-valve x, and the valve m of the eduction-pipe g, close themselves, by the external pressure of the atmosphere, and thus the entrance of air or water into the cylinder is prevented.

When the steam that remains in the cylinder, no longer balances the atmospheric pressure on the surface of the water in the injection-cistern G, the water spouts still more rapidly through the hole j, by the joint action of the column, and of the unbalanced pressure of the atmosphere. As the velocity of the injection-water increases, the jet dashes violently against the bottom of the piston, and is scattered through the whole capacity of the cylinder; the condensation of the steam thus becomes universal, till the elasticity of the steam which remains, is reduced to between $\frac{1}{3}$ and $\frac{1}{4}$ of that of the atmospheric air.

The whole pressure of the atmosphere, therefore, continues to be exerted on the upper surface of the piston, while there is but little pressure on its under side; and if the resistance at the outer end of the great lever is inferior to this unbalanced pressure, it must yield to it; the piston J must descend, and the pump-rod M must ascend, bringing along with it, the buckets of the several pumps which are to raise the water of the mine.

But the motion does not begin at the first instant that the injection is made, for the piston having been drawn to the top of the cylinder, by the preponderance of the pump-rods, it must remain there, till the difference between the elasticity of the steam beneath it, and the pressure of the atmosphere above it, exceeds this preponderance. There must be a small space of time between the beginning of the condensation and the beginning of the motion : this interval is very small, not exceeding the half or third part of a second; but it may be very distinctly observed by an attentive spectator, who may perceive, that the instant the injection-cock is opened, if the cylinder has the slightest yielding in its suspension, it will heave upwards a little, by the pressure of the air on the bottom. Its own weight is not at all equal to this pressure; and instead of its being necessary to support it by a strong floor, it must be kept down by large beams, loaded at the ends with heavy walls. This heaving of the cylinder, shows the instantaneous commencement of the condensation; and it is not till after this has passed, that the piston is seen to start and begin its descent.

The motion of the piston must continue, till it reaches the bottom of its course, for it is not like the motion which would take place in a cylinder of air rarefied or exhausted to the same degree. In this latter case, the impelling force would be continually diminished, because the capacity of the cylinder diminishing by the descent of the piston, the air in it, would continually become more dense and elastic, until the piston would stop at a certain height, where the elasticity of the included air, together with the resistance of the pumps, would balance the atmospherical pressure on the piston. But when the contents of the cylinder are pure steam, and the continued injection of cold water, keeps down its temperature nearly the same as it was at the beginning, the elasticity of the remaining steam cannot increase much, during the descent of the piston, for it must correspond to the temperature to which the steam is reduced by the application of the cold water.

For this reason, a great part of the impelling force which urges the piston into motion at the beginning, continues to operate to the end of the descent; and if the resistance of the load which is drawn up at the outer end of the lever, is uniform, the motion of the piston will accelerate in its descent, as heavy bodies do, in falling by the action of gravitation, except that the motion will be slower. If the cylinder has been completely purged of common air, before the steam passage is shut, and if none has entered since, the piston will descend to the very bottom of the cylinder, unless the steam is readmitted in time. To prevent any accident from such descent, the spring-beams U are placed so as to intercept the catch-pins *u* at the ends of the great lever, and stop the piston, whenever it exceeds its intended motion.

When the manager sees the piston as low as he thinks proper, he shuts the injection-cock, by depressing the handle *s ;* and at the same time he opens the steam-passage, by forcing down the handle *r*, which oversets the hammer, and its leg 11, catching the cross-pin of the stirrup 14, opens the regulator.

The steam has been accumulating above the water in the boilers, during the whole time of the piston's descent. The moment, therefore, that the steam-passage is opened, the steam having an elasticity of rather more than one pound per square inch, greater than that of the atmospheric air, rushes into the cylinder, and blows open the snifting-valve *x*, and assists the water which had entered by the former injection, and what resulted from the condensed steam, to descend by its own weight through the eduction-pipe *g*, and through the valve *m*, into the hot-well I.

This water is nearly boiling-hot, or at least its surface ; for whilst it remains in the bottom of the cylinder, it will condense steam, till it acquires this temperature, and it cannot run down, till its surface will condense no more. There is some waste of steam at its first admission, in order to heat the inside of the cylinder, and the surface of the injected water, and of the piston, to the boiling temperature ; but the space being small, and already very warm, it is soon done, and if the engine is properly constructed, but little more steam is required than what will warm the cylinder ; for the eduction-pipe *g* being of large dimensions, it will receive a great part of the injection-water, during the descent of the piston, and this portion will be removed from the contact of the steam : and the piston itself, being covered with wood on the underside, cannot absorb much heat.

The first effect of the entering steam, is of great service to drive out of the cylinder, all the vapour which remains there ; and it passes through the snifting-valve *x*, and cock *k*, and the pipe, into the open air. What is thus expelled is seldom pure steam, or watery vapour, but a mixture of air and steam ; for all water contains a quantity of air in a state of chemical union ; but the union being feeble, a boiling heat is sufficient for disengaging the greatest part of it, by increasing its elasticity. Air is also disengaged, by simply removing the external pressure of the atmosphere ; as is clearly seen when we expose a glass of water in an exhausted receiver. There-fore the small space below the piston, contains watery vapour, mixed with all the air which had been disengaged from the water in the boilers by ebullition, and all that was separated from the injection-water, by the diminution of external pressure, in addition to any which may have entered by leakage, by the piston, or through the joints. By blowing the hot steam through this space, all the air, or mixture of air and steam, is discharged at the snifting-valve, and its place is occupied by pure steam, hot from the boilers.

Let us now consider the state of the piston, when setting out on its return. It is evident that it will recoil, or begin to rise by the counter-weight, the moment the steam-passage is opened ; for at that instant, the excess of atmospherical pressure by which it had been forced down, in opposition to the preponderance of the outer end of the lever, is relieved by the upward pressure of the steam beneath it. At the first instant of the return of the pump-rods, they draw up the piston with great violence ; for the weight of all the water in the pumps, acts in addition to the counter-weight, to draw up the piston. But the lower valves in the pumps, close after an inch of motion, or less, and the further descent of the columns of water is prevented, the valves sustaining the weight of those columns. After this the piston will rise gradually, by the action of the counter-weight, which is moderate, and its operation modified by a combination of circumstances.

The action of the pumps is very different in the two motions of the engine; for whilst the piston was making its working stroke, it was lifting the columns of water in the pumps, and the absolute weight of the pump-rods also, except that a portion of that weight is sustained by the immersion of part of the rods in the water. The wooden rods which are generally used, being soaked in water, and joined by iron straps, are a little heavier than water, and they occupy about one-fifth of the bulk of the water in the pumps, into which they descend. In the returning stroke, the weight of the columns of water is sustained by the valves; but the weight of the pump-rods (after deducting that which they lose by immersion) exceeds the weight of the piston, so much, as to form what is called the counterweight, to draw up the piston.

The piston might be drawn upwards by the counter-weight, even though the steam which is admitted into the cylinder, were not so elastic as common air; for suppose the pressure of the air on a square inch to be, $14\frac{3}{4}$ pounds; and that the counter-weight was equal to $1\frac{1}{4}$ lbs. for every sq. inch of the surface of the piston, in that case the piston might rise, if the elasticity of the steam was equal to $14\frac{3}{4} - 1\frac{1}{4}$, or) $13\frac{1}{2}$ lbs. per sq. inch.

But in practice the steam must be stronger than the air, in order to blow out and discharge the air; it will therefore enter the cylinder without any effort of the piston to draw, or suck it in. Nor can the engine operate, if the counter-weight is too great, so as to draw up the piston too suddenly at first, and make a greater void within the cylinder, than the steam can supply; for that would reduce the pressure of the steam within the cylinder, so as to prevent it from snifting or blowing out the air, which is an essential condition to the working of the engine.

In filling the cylinder with steam, it will require a much more copious supply of steam, than merely to fill up the space left by the ascent of the piston; for as the vacuum which occasioned the descent of the piston, was only a consequence of the interior of the cylinder being sufficiently cooled to condense the steam, this cooled surface must be presented to the steam during the rise of the piston, and will condense steam a second time. Hence the piston cannot rise faster than that part of the cylinder, which the piston has already quitted, can be heated up to the boiling point, and much steam must be expended in so heating it; because the internal surface of the cylinder must not only be raised to the temperature of boiling water, as the piston rises, but must be made perfectly dry; and the film of water left on it, by the ascending piston, must be completely evaporated, otherwise it will continue to condense steam.

On this account, though the steam, on issuing from the boiler, is stronger than the atmosphere, its bulk becomes so much diminished, by the condensation of a great part of it within the cylinder, that, after the first start, it can scarcely be supplied through the steam-pipe, so fast as is required to fill up the space left by the ascent of the piston, and to replace the increasing loss by condensation. At first, when the piston is at the bottom, the space to be filled is small, and the piston has not yet begun to ascend: and also the boiler is very full of steam. All these circumstances concur to give the first puff of entering steam the power of expelling the air at the snifting-valve, and some steam will follow, but very little; for as soon as the piston gets into motion, the cylinder will demand all the supply of steam to fill it; and as the piston rises higher, an increased supply of steam is required, because the steam is exposed to a greater extent of cold surface. In consequence, the issue of steam at the snift-valve, though sharp at first, ceases entirely when the piston has ascended a little.

Hence the steam can give no impulse to the piston, to assist it in rising, except at the first instant; but on the contrary, the piston will afterwards be restrained from rising, with a greater velocity than the steam can be supplied, though it is at liberty to move with that velocity with which the steam is supplied.

The moving force, during the ascent of the piston, results solely from the counter-weight, or preponderating weight of the pump-rods. This force is expended, partly in returning the piston to the top of the cylinder, in overcoming the friction of the rubbing parts of the engine, and the inertia of its several moving parts, and partly in returning the pump-buckets to the bottom of their respective working barrels, that they may be prepared to repeat their working stroke. This latter requires force, for each bucket must be pushed down through the water in the barrel, which must lift up, and pass through the valves in the bucket, with a velocity bearing the same proportion to the velocity of the bucket, as the area of the pump-barrel, bears to the area of the opening of the valves, through which the water must pass; this resistance to the descent of the pump-buckets will increase, as the square of the velocity, with which they descend through the water.

From this general consideration of the ascent of the piston, it is obvious that the motion will differ greatly from the descent. It cannot be supposed to accelerate, because the steam can only be supplied to the cylinder, in a limited quantity through the steam-pipe. In consequence, the motion accelerates till it becomes as quick as the supply of steam will permit, and then the motion becomes uniform. If we observe the working of an atmospheric steam-engine, we may observe that the motion, during the rise of the piston, is extremely uniform, whereas in the working stroke, it is very sensibly accelerated.

These two motions form a complete stroke, which may be repeated by shutting the regulator, and opening the injection-cock, when the piston has reached the proper height.

The first two or three strokes are performed by the attendant working the handles, but when he has ascertained that all the parts are in order, he puts pins into the holes of the plug-beam, and the motion of the engine will then open and shut the steam regulator, and the injection-cock, at the required moments; and so precisely that the piston will never exceed or fall short of its intended course, and the catch pins will very rarely strike the spring beams. The cord of the leather-valve, in the cistern G, at the upper end of the injection-pipe, is tied fast to keep that valve open, and allow the water to pass; but when it is required to stop the engine, this cord is untied, and the valve shutting, cuts off the supply of injection-water, so that the engine cannot proceed.

The jack-head pump R raises a quantity of water, at each stroke by the pipe Q into the injection-cistern G; this supply is rather more than what is injected into the cylinder, that there may be an ample supply, and the surplus runs down a waste-pipe from the cistern, and returns into the pit S. The small branch d, which proceeds from the injection-pipe j, to the top of the cylinder, has a cock at the end, which must be so adjusted that water will run from it, and keep a constant supply of a few inches, above the piston, to keep it tight. Every time the piston comes to the top of the cylinder, it will bring the water along with it, and the surplus of its evaporation and leakage is conveyed away from the engine by a waste-pipe leading down into a common drain.

The cold water which is injected into the cylinder, and which becomes hot by condensing the steam, runs out of the cylinder, through the eduction-pipe *g*, into the hot well I, which is the common receptacle for hot water.

A regulated quantity of hot water runs out of the hot well I, through the perpendicular feeding-pipes T, into the boilers respectively, to supply the evaporation from them in steam. The quantity of hot water required to feed the boilers, is but a very small part of that which comes into the hot well; the surplus runs from the hot well, by a waste-pipe into a common drain, which will convey it quite away from the engine, without mixing with the water in the pit S, which should be kept as cold as possible, in order that a less quantity may serve to condense the steam.

In many situations where a constant supply of cold water cannot be obtained from a river or brook, the same water must necessarily be used over and over again, for injection. In that case, the waste water from the hot well, must be conveyed into a large shallow pond in the open air, and left there long enough to become quite cold, before it is re-admitted into the pit S. It frequently proves difficult to obtain a sufficient supply of cold water for a large engine, as a great quantity is required, and it must be quite cold, or the operation of the engine will be impaired; it should be recollected that almost all the heat communicated by the fire to the water in the boiler, must be carried off in the hot water, which runs away waste, from the hot well.

PARTICULARS OF DIFFERENT PARTS OF THE ATMOSPHERIC ENGINE.

WE must now pay attention to the construction of the parts of this engine, and notice some minute particulars which have not yet been mentioned.

The fire-grate B in the furnace, should not have the bars so close as to prevent the free admission of air, nor so open as to permit the coals to fall between them; if the width of the top edge of each bar is $2\frac{1}{4}$ inches, with $\frac{3}{4}$ of an inch space left for air passage between the bars, the fire will burn very well upon the grate.

The dimensions of the fire-grate and furnace, must be regulated according to the size of the boiler, as will be more fully stated; and the height from the grate to the bottom of the boiler, should be always between 12 and 24 inches.

The boiler may be made of iron, or copper plates, or of cast-iron, the bottom being of such materials as will withstand the effects of the fire, and have sufficient strength; and the top part must be adapted to retain the elastic steam. The sides of the boiler are nearly cylindrical, and the bottom is a little concave on the lower side, so that the flame from the fire, being allowed a space beneath the whole concave surface, rises up against the bottom, and communicates its first heat thereto, and then passes into the circular channel or flue, which encompasses the outside of the lower part of the boiler, and conducts the flame in a circuit round the same, before it reaches the chimney.

The upper part or dome of the boiler is made hemispherical, which is the best form for resisting the elasticity of the steam; any other form will do, provided it is of sufficient strength. Mr. Newcomen's first boilers were made of lead for the dome, and copper-plate for the lower part, but iron-plate is now used for both parts.

The celerity of the motion of the engine, depends greatly upon the capacity

U

of the boiler-top; for if it is too small, the steam must be heated to a greater degree, to give it a sufficient elastic force to work the engine, and then the condensation on entering the cylinder will be greater. If the boiler-top contains eight or ten times the quantity of steam used at each stroke, no more fire will be required to preserve the elasticity of the steam, than is sufficient to keep the water in a proper state of boiling; this is the smallest size for the boiler-top.

In the modern engines it is usual to place a damper, or iron slider, in the chimney, or in the flue leading into the chimney; and this has a chain or lever, by which the attendant can regulate the aperture into the chimney, and consequently the draught of the fire, according to the quantity of fuel on the grate, or according to the working of the engine, so as to keep the steam to a great regularity: for it is evident, that when the engine works slowly, it will require less steam and fuel, than when it works rapidly; without the damper, the engine is constantly subject to have an excess, or a deficiency, in the supply of steam, notwithstanding the greatest care and attention on the part of the attendant, to supply the fire regularly. This contrivance was not in use in Mr. Smeaton's time.

In many of the old engines, the dome or upper part of the boiler was made much larger than the under part, so as to project all round over the flue or circular channel for the flame, in order that the water might be contained over the top of the flue, and receive the ascending heat from the flame therein; this idea was extended, by carrying the flue or circular passage in a large iron-pipe, contained wholly within the water, and making a complete circuit round the inside of the boiler: sometimes the internal flue was formed into a circle, and at other times it was a spiral, beginning in the centre of the boiler, and making two complete turns before it entered the chimney. By this means, after the fire in the furnace had heated the water by its direct action beneath the bottom, the flame heated it again, by the flue being wholly included in the water, and having every part of its surface in contact with it; this was thought to be preferable to carrying it in a flue or passage round the outside of the boiler, because only a third or a half of the surface of the flue could then be in contact with the water, the other surface being brick-work.

In either case, whether the flame is conducted in a flue or passage round the outside of the boiler, or in a pipe round the inside of it, the size of the flue, or aperture through which the flame is to pass, ought to be gradually diminished, from its entrance at the furnace, to its egress into the chimney; and the section of the chimney at that place, should not exceed the section of the flue or pipe, and should also be somewhat less at the chimney-top. This circumstance is rarely attended to, but will tend to produce the best draught.

The perpendicular height of the chimney should not be less than 48 feet, from the level of the fire-grate, to the aperture at the top of the chimney.

In all the old engines the boiler was placed immediately beneath the cylinder, as represented in Plate II; and then the regulator was placed immediately within the boiler, and acted against the under surface of its top, in the same manner as in the first engine of Mr. Savery, who invented the regulator. When the engines were so large as to require two boilers, the additional one was placed in a side building, as described; but it was a subsequent improvement on Newcomen's engine, to remove the boiler entirely from beneath the cylinder into a small shed at the outside of the engine-house: by this means the height of the building is considerably reduced, and as the lever-wall which supports the main centre does not require to be carried to so great a height, it is more able to withstand the violent shocks, to which it is constantly subjected from the working of the engine. Another, and still greater advantage is, that the cylinder may be placed upon a solid pier of masonry, formed on the ground, where the boiler is represented in Plate II; and being fastened down thereto by bolts, it is held much firmer than it can be by suspending it between beams. A boiler is placed at each side of the engine-house, and the steam from them, is conveyed by pipes to a circular box, situated beneath the cylinder, to contain the steam regulator.

All engines for mines should have a spare boiler, in addition to those which are requisite to work them; or at least, a place should be provided for a spare boiler, and when the original boiler requires to be repaired or renewed, it can be replaced by erecting another at one side, and carrying another steam-pipe to the steam-box, so that the working of the engine can be continued without any stoppage; this circumstance is of the greatest importance where water must be constantly drained.

In either case, whether the regulator is placed in the boiler itself, or in the steam-box beneath the cylinder, it is constructed in the manner represented in the enlarged plan, Plate III. It is in two parts, one moveable, the other fixed. The moveable piece is a flat plate of brass, in shape resembling a fan; the upper surface of which applies exactly to the whole circumference of the orifice of the steam-pipe, and completely excludes the steam from the cylinder. The fan is moveable round an upright axis, which is accurately fitted into a conical socket, and passes up through the top of the boiler or steam-box, so that the regulator can be turned aside by a lever, fixed on the upper end of the axis, in order to uncover or open the passage. The orifice of the steam-passage is formed in the fixed part of the regulator, which is also a brass plate, of a sectorial shape, and is fixed into an opening in the top plate of the boiler, or of the steam-box, by a number of rivets all round the margin, so as to form one surface with the top of the boiler or steam-box; from the upper side of this fixed brass regulator-plate, the steam-pipe rises for a few inches, and is united to the iron part of the steam-pipe, which projects downwards from the bottom of the cylinder; a plate of sheet-lead is wrapped round the outside of both pipes, to cover the joint, and is bound fast by a lapping of small cord. To admit of this, the two parts of the steam-pipe are made exactly the same size on the outside, and a piece of canvas, soaked in white lead and oil, is applied immediately on the pipes beneath the sheet of lead. This very insecure mode of making the joint continued to be the universal practice for many years; and though the shaking of the cylinder soon rendered it leaky, the attendant could always renew the cloth and the cord. The orifice of the steam-pipe beneath the fixed brass plate, is not circular, but of a sectorial form, corresponding with the regulator itself, but smaller, so as to allow the regulator to cover it securely, and the margin round the aperture, projects a little beneath the under surface of the regulator-plate. It is to this prominent margin that the upper surface of the regulator is fitted by grinding. The socket through which the upright axis of the regulator passes, is also of brass, in the same piece with the fixed regulator-plate; and the axis, which is rather conical, is fitted into it by grinding, like the turning-plug of a cock. This axis is iron, with a square at the lower end, to which the regulator e is firmly fastened, and another square at the top, on which the lever or handle 3 is fixed, to move it from the outside. In some cases, the regulator is borne upwards by a spring, to keep it in close contact with the fixed regulator-plate. The lower surface of the regulator then has a protuberance in the middle, to rest on a strong flat spring, which is placed below it, across the mouth of the steam-pipe, and presses the regulator towards the steam-pipe, so that it applies very close.

This protuberance slides along the spring, while the regulator turns to the right or left.

The handle 3 of the regulator, and the end of the rod or stirrup 14, are pierced with several holes, and a pin is put through them, to unite them like a joint. The motion of the handle of the regulator may be increased or diminished by choosing for the joint a hole, near to the axis, or remote from it; and the exact position at which the regulator is to stop each way, is determined by pins stuck in a horizontal rail, on which the end of the handle 3 rests.

The tumbling-bob or hammer 16 of the Y, has a long leather-check strap fastened to it, and the end of the strap is fastened to a beam above it; in such a manner that the hammer may be alternately held up, to the right and left of the perpendicular. By adjusting the length of the strap, the Y may be stopped in any desired position. The two legs 11, 12, of the fork of the Y, spread out from each other, and also from the line of the stalk, and they are of such length as to reach the horizontal cross-pin, which forms the cross end of the fork or stirrup 14, which is suspended below the axis by two loose links.

Now, suppose the cross-pin of the stirrup to be hanging perpendicularly beneath the axis of the Y, and the stalk of the Y to be held perpendicular, if it is carried a little outward from the cylinder and then let go, it will suddenly fall farther out by its weight, without affecting the stirrup at first, until the outer leg 11 strikes on the horizontal cross-pin of the stirrup, and then it pushes the stirrup endways towards the cylinder, and opens the regulator. It thus puts the regulator in motion with a smart jerk, which is an effectual way of overcoming the cohesion and friction of the regulator against the mouth of the steam-pipe. This push is adjusted to the proper length by the check-strap, which stops the Y when it has gone far enough. If the stalk of the Y is again raised up to the perpendicular, the width between its legs 11, 12, is such as to permit this motion, and something more, without affecting the stirrup. But when pushed still nearer to the cylinder, it tumbles suddenly towards it by its own weight, and then the other leg of the fork strikes the cross-pin of the stirrup and moves the regulator in the opposite direction, till the hammer is checked by the strap, by these motions of the Y, the regulator is opened or shut suddenly.

This opening and shutting of the steam-passage must be executed a moment before the piston arrives at bottom or top of the cylinder; and by placing the pins in the plug-beam l, which act upon the two handles of the Y, at a proper height, the motion may be regulated exactly. For this reason, the plug is pierced through with a number of holes, that the places of these pins may be varied at pleasure. By this means, and by a proper curvature of the handles, the adjustment may be made to correspond exactly with the intended length of the stroke.

In the same manner the motions of the injection-cock are also adjusted to the precise moment that is proper for them. The different pins are so placed in the plug-beam, that the steam-regulator may be completely shut before the injection-cock is opened. The inherent motion of the machine will give a small addition to the ascent of the piston, after the regulator is shut, and without expending more steam; but by leaving the steam rather less elastic than before, the subsequent descent of the piston is promoted.

The injection-cock is also provided with a hammer weight 8, to make it open suddenly. It has an arm 14, extending from the axis of the hammer, and connected by a rod 6, with the handle of the cock i; and the head of the hammer 8 is a sufficient weight, to open the cock in an instant. When this hammer-weight is lifted up to its utmost, the cock is shut, and in this position the weight is detained by a small latch 18, which is lifted up by a chain connected with the plug-beam, at the moment when the piston arrives at the top of the cylinder, and thus releasing the weight 8, it falls all at once, and opens the cock in an instant; but when the piston descends nearly to the bottom, another pin in the plug-frame takes the handle *s* of the lever, and gradually closing the cock, raises the hammer-weight till the latch 18 detains it, which happens when the piston is quite at the bottom of its motion.

The injection-cock ought to be opened suddenly; but there is much propriety in closing it gradually: for after the first dash of the cold water, against the bottom of the piston, the condensation is nearly complete, and only a little more water is necessary; but a continual accession of some cold water is required, to maintain the condensation as the capacity of the cylinder diminishes, and as the water which is already injected becomes warmed. The continuance of this small injection prevents the vapour in the cylinder increasing in elasticity as the piston descends, or at least, until it has descended nearly to the bottom.

The effect of the injection in condensing the steam in the cylinder is influenced by the height of the reservoir, above the aperture of the spout. Mr. Beighton's rule was, that if the engine makes a six-feet stroke, the jack-head cistern should be at least twelve feet perpendicular, above the bottom of the cylinder. But Mr. Smeaton found that the condensation was more perfect, when the height was 24 feet perpendicular, or 36 feet for the largest engines. The injection-cistern should always be placed as high as the building will admit, so as to give a smartness to the jet. The size of the aperture of the spout must depend upon the capacity of the cylinder, as will be shown by a table; but if the cylinder be very large, it is usual to have three or four small holes rather than one large one, in order that the jet may be more effectually dispersed through the whole capacity of the cylinder. The injection-pipe and cock should be sufficiently large to supply the injection-water freely, but the aperture of the spout or jet, should be considerably less than the passage of the cock or pipe.

It is an advantage to force the injection-water into the cylinder with great velocity; and with this view, Mr. Curr, in 1797, recommended that the cistern which contains the cold water should in all cases be placed 36 feet above the top of the cylinder: the jet of cold water will then be divided into very minute drops, which being dispersed generally through the steam which is contained in the cylinder, will condense it very rapidly, so as to make a better vacuum than if the cistern were placed lower and injected with a feeble jet. It is a common error to suppose that a high column of water is unnecessary, because the pressure of the atmosphere is found to force the water into the exhausted cylinder, with great velocity. But the cold water is most wanted at first, when the cylinder is full of steam of equal elasticity with the atmosphere, and the water can then have no other tendency to enter into the cylinder, except by its fall from the cistern. After the condensation is begun, the atmospheric pressure will give an additional and increasing impetus to the jet, as the condensation becomes more complete; but this assistance to the column will be too late to produce the required effect. The cold water should be thrown in with violence at first, that the condensation may be effected as quickly as possible.

The injection-cap, or jet, according to Mr. Smeaton, should be one square hole through a brass plate, and rounded from the under side, that it may throw up a full bore. The middle of the jet should not be directed to strike the centre of the piston bottom, but the jet should rise perpendicularly, so as to strike the piston bottom at right angles. That part of the injection-pipe which is within the cylinder, should be wrapped round with two or three thicknesses of tarred marline, or small rope, to separate the metal of the pipe from the contact of the steam, or hot water ; which not only saves the condensation of some steam, but by preventing the pipe becoming hot, that portion of injection-water which is contained in the pipe is kept cool, and the stream which afterwards flows through the pipe, will enter in its coolest state; or otherwise the injection-pipe within the cylinder is made of wood, hollowed out within, as shown in the section ; and the brass injection-cap is fastened over the orifice, which turns up. This piece of wooden-pipe is jambed fast in between the steam-pipe in the centre of the bottom of the cylinder, and the internal orifice of the nozel, or iron branch, which joins to the injection-cock.

The jack-head pump R, by which the cold water is raised up into the injection-cistern G, is of the same internal construction as a common sucking-pump, for it has a valve at the bottom of the barrel, and another valve in the bucket, and it raises the water when the bucket is drawn up ; but to enable it to lift that water to a greater height, the top of the barrel is closed by a cover, through the middle of which the rod passes, in a tight collar of leathers; and a forcing-pipe Q is joined to the upper part of the barrel, to convey the water upwards to the cistern.

A leaden pipe is applied beyond the snifting-valve x, with a cock h, which being partially closed, the snifting can be regulated, if it should be found too great, and emit more steam than is requisite to carry out all the air from the cylinder : the pipe conveys this air and steam, through the wall of the house to the outside.

Mr. Smeaton made his engines with a wooden bottom to the piston, as we have before noticed. This was because wood receives and communicates heat, much less rapidly than metals. The piston is kept much cooler, than any other part to which the steam has access, not only from the water which is poured upon it to keep it tight, and prevent the leakage of air into the cylinder, but also because it receives the first, and most direct action of the cold injection-water ; and as the steam, in entering the cylinder through the steam-pipe, first meets the cold surface of the piston, it is thereby condensed in a greater degree than by an equal portion of the internal surface of the cylinder. By covering the bottom of the piston with wood, it will absorb less heat from the steam ; and for the same reason, the cold water, when it is thrown up against the piston, will be less heated by its contact with the piston, the wood acting as a neutral body to the fluids which alternately act beneath it.

The water, which is kept upon the top of the piston, to prevent any entrance of air, necessarily becomes very hot, and it was thought proper to employ its over-plus, for supplying the waste of the boiler. This was accordingly practised for some time ; but Mr. Beighton improved this economical thought, by supplying the boiler from the eduction-pipe g, the water of which, coming from the cylinder, must be still hotter, than that above the piston.

This contrivance required attention to several circumstances, which will be easily understood by considering the section, plate II. The eduction-pipe g comes out of the lowest part of the bottom of the cylinder, in an inclined direction, and descends into the hot-well I, where it turns up, and is covered with a valve. The injected water, accumulating at the bottom of the cylinder, will run into the eduction-pipe g, as soon as the steam is admitted, and opening the valve in the bottom, will flow into the hot-well I. The upright feeding-pipe T, goes through the dome of the boiler, and reaches down a few inches below the water line. The top of this pipe rises about five feet above the surface of the water in the boiler : it is open at both ends, and has a horizontal branch, from near its upper end, communicating with the hot-well. In this communicating branch is a cock, by which its passage may be diminished at pleasure. Now, supposing the steam in the boiler to be very strong, it will cause the boiling water to rise in the feeding-pipe T, and passing along this branch, to rise also in the hot-well, and run over. But the surface of water in the hot-well being 4 or 5 feet, above the surface of the water in the boiler, the steam is rarely strong enough to produce this effect; on the contrary, the water from the hot-well will run through the branch, and go down into the boiler, by the feeding-pipe, as fast as the opening of the cock will admit. By properly adjusting the cock in the branch of the feed-pipe, the boiler may be supplied with water, as fast as the waste in steam requires.

In engines of smaller dimensions than that represented in the plate, the height of the dome of the boiler being less, the hot-well is not sufficiently elevated to cause the water therefrom, to run down into the boiler. It is always desirable to place the hot-well so low, that the water will have a descent of at least $3\frac{1}{2}$ or 4 feet from the bottom of the cylinder into the hot-well, and then it will drain rapidly from the cylinder, as soon as the steam is admitted therein, and produce less waste of heat, than if it required more time to run off. It is also necessary to have at least 4 feet fall from the hot-well into the boiler, or the steam would frequently be so strong, as to prevent the entrance of the water; hence, if the height from the surface of the water in the boiler, to the cylinder bottom is less than 8 feet, the feeding of the boiler must be performed, by carrying the upper end of the feeding-pipe T at least 4 feet above the surface of the water in the boiler, and forming a small funnel or basin upon the upper end of it, into which a supply of warm water must be poured by a small cock, joined to the bottom of the cylinder, at the very lowest place, or else to the upper part of the eduction-pipe, so as to receive part of the water therefrom. This cock is in fact another small eduction-pipe, for the sole purpose of feeding the boiler; and accordingly the orifice must be closed by a small flap-valve opening outwards, and immersed in the funnel at the top of the feed-pipe, which forms the hot-well of this small eduction-pipe. The cock will serve to regulate the quantity of water supplied to the boiler.

The small quantity of water which is necessary to supply the boiler, might be immediately taken from the cold cistern, without greatly diminishing the production of steam; for the quantity of heat necessary to raise the sensible heat of cold water, to that of the usual temperature of the hot-well, is small, when compared with the quantity of heat which must be combined with it, in order to convert it into steam.

The heat which must be expended, in boiling off a cubic foot of water in steam, after it has been heated to a boiling temperature, is about 6 times as much as would raise it to that boiling temperature from the temperature of 50 degrees, and above $9\frac{1}{2}$ times as much, as would raise it from 52° to 152°, which is the ordinary temperature of the hot-well. For this reason, but little difference can be observed in the performance of such engines as are fed with the hot water, and those which have their boilers supplied from a brook, except that about one-tenth more fuel is consumed. The hot water has, however, the advantage of being free from air; and when an engine must derive all its supplies from mine water, the water from the eduction-pipe is far preferable to that from the top of the cylinder, because it has been in a measure distilled and purified.

It is of the greatest importance to have pure soft water, for supplying the boiler; most mines produce water strongly impregnated with mineral salts, and earthy matters, which will not evaporate, but they deposit in the boiler, and form a stony crust on the bottom, which resists the transmission of heat; and then the fire tends to destroy the metal of the boiler, instead of heating the water. The matter deposited, is sometimes of a corrosive quality, in which case the metal is attacked on both sides at once. The clearest water of running brooks is preferable to any other, as it rarely contains saline or earthy impregnations, and it holds less air in a state of solution, than spring water; but water recently brought up from a deep mine contains still more, because the solution was aided in those situations by increased pressure: such water sparkles when poured out into a glass. As it is of consequence to the good performance of an engine, to use water free from air, earth, and salts, it is best to take rain, or fresh river water, if it can only be procured in sufficient quantity for the boiler; but in many situations there is no other than brackish water, in which case, it is advisable to keep the water, for some time before it is used, in a large shallow pond exposed to the air; and to use the same water over and over again for the injection, pouring it out from the hot-well waste-pipe, into shallow ponds till it is cold, and then pumping it up to the injection cistern again. The operation may be carried on in this manner very well, if a sufficient quantity of water is brought into use; for, by the heat and exhaustion which it undergoes in the operation of the engine, it parts with the air it contains, and by cooling in the ponds, it deposits its saline and earthy matters, and becomes much more pure than fresh pit water.

In cases where a small supply of soft water can be obtained for the boiler, but not enough for the injection also, Mr. Smeaton placed a pan of very thin copper in the hot-well, so as to be entirely surrounded by the hot water proceeding from the eduction-pipe. The soft water being conducted by a small pipe into this internal pan, remained there, till it had acquired nearly as much heat as the water of the hot-well, and was then drawn off by the branch and cock of the feeding-pipe to supply the boiler, having received warmth from the impure water without being mixed with it.

Another and better form of this apparatus would be, to use the worm-tub and pipe of a common still, for a hot-well: the eduction-pipe and valve being immersed in the space within the spiral coils of the worm-pipe, would fill the tub with warm water; and the cold fresh water, for the supply of the boiler, being conducted through that spiral pipe, would acquire heat from the surrounding water, without mixing with it.

The cylinders of the first steam-engines were made of brass, but on account of the great expense, cast-iron was substituted, as soon as larger engines came into use. Desaguliers in 1743 speaks of this change, and very strongly recommends the brass cylinder as preferable, both on account of its being smoother withinside, and being thinner in the metal, it occasions less loss of heat to warm and cool it, when the steam and cold water are alternately admitted; and he says the advantage in fuel will compensate for the extra expense.

The interior surface of the cylinder requires to be bored with great exactness; and it must have a sufficient thickness of metal to resist the pressure of the atmosphere, without bending or altering its figure. The piston is made of cast-iron, as nearly as possible to fit the inside of the cylinder, and has all round it, within three inches of the circumference, a circular ledge, or rim projecting upwards from it, which both strengthens the piston, and also leaves a space between it and the interior of the cylinder, to receive the hemp packing, which keeps the piston tight. Mr. Smeaton recommended the lower surface of the piston to be covered with elm or beech plank, about $2\frac{1}{4}$ inches thick. (See Plate III.) The planking consisted of two broad planks, crossing each other at right angles, and halved into each other at the intersection, so as to come to an equal thickness: the remaining parts or sectors between the arms of this cross, were filled up with pieces of the same plank, tongued and well fitted together, and bolted to the cast-iron of the piston with one or two circular iron rings, inlaid into the lower surface, to make it strong. The wood was surrounded on the edge with a circular iron hoop, a quarter of an inch less in diameter than the bore of the cylinder. The cast-iron piston being 6 inches less in diameter than the cylinder, the margin of the wood planking projected all round, and formed the bottom of the groove, to receive the packing, and the edge of the cast-iron formed the side of the groove. The wood was screwed to the iron, with a double thickness of tarred flannel between them, to exclude the air between the iron and the wood. A piston thus constructed is less liable to conduct heat; and the grain of the wood, radiating in all directions from the centre, the planking is not liable to expand by the wet.

The packing round the edge of the piston, consists of a very soft hempen rope; the yarns being very slightly twisted in themselves, and those yarns being loosely twisted together form a strand, and three or four of such strands plaited together form a thick flat rope, called a gasket; this is coiled round in the space between the up-right edge of the piston, and the inside of the cylinder, which space being of considerable width and depth, requires several coils of the gasket to fill it. The interstices between the folds, and plaits of the hemp, should be filled with tallow-grease; and it is better if the whole of the gasket is previously soaked in melted tallow. The coils of gasket being rammed down very tight into the groove, are kept down, by heavy weights of cast-iron, which are segments of a circle, and when they are all put in their places, they form a ring, which fills all the groove round the edge of the piston, and entirely covers the hemp.

The shank of the piston is made with two prongs, to unite it firmly to the piston; or, if the engine is large, it has four prongs; the shank must also have two, or four chains to the arch head of the great lever. These chains are of a similar construction to those used for the fusees of watches; the links are flat plates of iron, with a hole through each end, and they are united by round pins; the alternate links are single and double, each single link being put between two double links, and the joint pin passes through all three; this is shown in Plate III, which also explains how the two chains of the piston and of the pump-rod are harnessed, or united in pairs to the ends of a short horizontal link *o o*, from the middle of which the shank *n* is suspended; by this means the strain of the piston is equally divided between the two chains. For engines of the largest size, four chains are used, and they must be divided into two pairs, each pair with its horizontal link as above; and the middle parts of these two links, must be united to the ends of a longer horizontal link, from the middle of which the shank of the piston is hung: and in this way all the four chains will bear equally, on the same principle as horses are usually harnessed to a carriage.

The upper ends of the chains are jointed to the ends of strong iron bars, let into the upper ends of the arch heads, and projecting sufficiently beyond the circle thereof, to unite with the chains; the other ends of these bars are bolted to the top of the lever, by which means they brace the arch head. These bars also rest upon the catch-pins, which pass through mortices across the upper ends of the arch heads.

The original method of making the great working lever, was to employ a large tree, and place the axis or fulcrum under the middle of it, with proper bands to

X

fasten it.   Mr. Smeaton constructed the great levers of his engines of several beams combined together, in the manner represented in Plate II.   The lever of that engine has six pieces, the two middle ones are whole balks, 12 inches by 24 inches, put together with the gudgeon between them, which is 5 inches thick, and notched into the beams, to keep it in its place : the ends of the beams are then sprung together, and bolted fast.   This being done, another pair of timbers are applied on the outside of the two former, and then others outside these ; for the largest engines, two pairs more were employed, making ten balks in the whole.   All these being firmly united by bolts, several mortices are cut through, between the joints, as shown by the small square marks in the figure ; and into these hard oak wedges are driven, so that they will be half in each beam, and prevent them from slipping or sliding upon each other in the least ; in this case, the outside beams act as ties, by their longitudinal strength rather than their inflexibility.   The great beams which suspend the cylinder, and extend across the house, are compounded of several pieces, in the same manner ; and the cylinder has a projecting flange from the middle of it, to bolt it down to those beams.

The axis of the great lever is cast-iron, flat and wide in the middle part, where it passes between the beams, but it grows thicker and narrower where it projects out at each side ; and these parts are formed to round pins, or gudgeons, on which the lever is poised, like a scale-beam, or like a cannon upon its trunnions.   The gudgeons of the axis, rest in brass sockets, supported in large blocks of wood or stone, solidly fixed into the masonry of the lever wall.   Since Mr. Smeaton's time, this kind of framed beam, has been abandoned in favour of a simple beam of oak, because the joints of the different pieces were found to work loose in the course of time.

The pump-rod, or spear M, which descends into the mine, is made in several lengths of wood, united by iron plates or straps, applied over the joints, and secured to the wood by bolts, see Plate IV.   Fir is the best wood for pump-rods, as it is light, and can be obtained in very long pieces ; and it will bear a great strain endways, if the several lengths are firmly united.   The pumps are of the kind called sucking-pumps : each pump is composed of several lengths of upright iron pipes, united together by screw-bolts, so as to form a perpendicular column as high as the water is intended to be raised.   The lower length, called the wind-bore, is pierced with small holes at the bottom, and is immersed in the water at the bottom of the mine.   A few feet above this, a clack or valve-seat is fitted into the next length of the pipe ; it has two trap-doors, or valves of iron-plate, faced with leather, and with leather hinges, on which they will open upwards, to allow the water to rise up through the pump, but will fall down, and close the passage, to prevent any return.   The next length of the pump, above this clack, is called the working-barrel : it is bored truly cylindrical and smooth, to receive the moveable bucket, which is a circular piston, surrounded with leather on the edge, and exactly fitted into the bored part of the working-barrel, so as to move up and down freely therein, but not to allow any water to pass down, between its leathered edge and the interior surface of the barrel.   The bucket is perforated, and provided with two similar valves or doors, to cover the passage, and prevent the water descending, though they will allow it to ascend.   The bucket is affixed by an iron shank, to the lower end of the wooden spear, which passes down into the pump from the top, through all the water which it contains.

A pump of this kind, raises water when the rod and bucket are drawn up, for then its valves shut, and all the water in the pump is drawn upwards before it, so that part of the contents runs over at the top of the pump, whilst fresh water is drawn in, through the holes at the bottom of the wind-bore; the lower valves open to admit that water to rise up into the working barrel, and fill up the space left by the ascent of the bucket; but, when the bucket returns, no more water is raised than what is displaced from the pump, by the immersion of the pump-rod into the water therein, for the valves in the bucket open, and it descends through the water in the working barrel, without displacing it; the lower valves of the clack being shut, prevent the water running out at the bottom of the pump, when the bucket descends. The moveable valves in the bucket, and the fixed ones in the clack, thus sustain the water by turns; they permit it to rise in the pump, but always prevent it returning.

When a mine is too deep for one pump to raise the water at once, the pit must be divided into two or more lifts; and as many different pumps must be employed, each lifting the water into a cistern, for the supply of that which is above it. Fifty yards is as great a depth as is proper for one lift, but in some very deep mines, they are deeper. It is very difficult, in these cases, to make the pipes sufficiently strong, to bear the pressure of the water, particularly the shock which takes place, when the whole column of water recoils upon the valves in shutting: the blow which they then make, is like the stroke of a forge-hammer, and soon destroys the joints. The only effectual remedy is to add an air-vessel at the side of the pump: but the miners usually make a hole in the suction-pipe of the pump, just below the clack, and insert a cock, with a small valve opening inwards: through this cock a quantity of air is admitted every time the pump draws, and this air, mixing with the water in the barrel, condenses, when the valves shut suddenly, and by its elasticity eases the violence of the shock. When the mine is pumped almost dry, the pumps will draw in a little air, at every stroke, at the bottom of the suction-pipes; and this answers the same purpose.

ON THE POWER OF THE ATMOSPHERIC ENGINE, AND RULES FOR PROPORTIONING ITS DIMENSIONS TO THE WORK IT IS REQUIRED TO PERFORM.

Mr. Newcomen brought forward his engine at a time when the most valuable mines in England could not be pursued, for want of more powerful or cheaper hydraulic machines than were then known. In the course of a few years his invention was put in practice at almost all the mines then existing, and new ones were opened in situations where it would have been impossible to have worked mines before. The cylinder of the first engine which he erected in 1712, at Griff, in Warwickshire, was only 22 inches diameter, and the second and third at Newcastle, were about the same size; the fourth at Austhorpe, was 23 inches (a), and it was some years before any were made so large as 36 inches; but these which we now

(a) Mr. Smeaton made many inquiries into the particulars of a fire-engine, which was erected about the year 1714, by the Patentees, at a mine on the Moor Hall estate at Austhorpe, in Yorkshire, where Mr. S. resided; he was informed, that when this engine was built, there were only three engines in existence. The first which was made, was near Coventry, and had no working gear, but the cocks were opened and shut by hand; there were two others at Newcastle, and the Austhorpe engine was supposed to be the fourth engine which Newcomen made. Mr. Calley, his partner, attended the building of it, and died at Austhorpe in 1717. The patentees had 250l. a year, for working and keeping the engine in order; but they burned out four boilers in the time it was worked, which was only about four years. An old man who had worked the engine in his youth, told Mr. S. that the cylinder was 23 inc. diam. and 6 ft. stroke, and it would make about 15 strokes per minute, when worked by hand, but in general it made about 12 when working itself. The pit was 47 yards call small engines, were so much more powerful than any former means of draining water, that they were found sufficient, until the mines, by growing deeper, required more power. The most obvious means of increasing the force, was to change the cylinder and boiler for others of larger dimensions, and when this expedient had been repeated, until the lever and other parts, would bear no greater strain, a new and larger engine was erected. In this manner the invention proceeded for many years, until, by gradual increase, the cylinders for common use were made 48, 60, and 72 inches diameter.

As it was found impracticable to make them much larger, engineers began to study the means of improving their performance, without increasing their dimensions : and the consumption of fuel in these large engines was so great an expense, as to balance the profits of many mines.

At first the economy of fuel was not much considered, because the steam-engine was found to be cheaper than any other means of draining water. The best engineers were those who made engines which would perform the task assigned to them, and which, in comparison to their dimensions, and expense of erection, would draw the most water, and be the most certain in the continuance of their operation. We have no accounts of the quantity of fuel consumed by any of those early engines, in proportion to the water which they raised to a given height; but the rules by which they apportioned their cylinders to the work to be performed have been preserved.

Desaguliers informs us, that Mr. Newcomen's mode of estimating the power of his engine, was to square the diameter of the cylinder in inches, and cutting off the last figure, he called it long hundred-weights; then writing a cipher on the right hand, he called the number on that side odd pounds: this he reckoned tolerably exact at a mean, or rather when the barometer was above 30 inches, and the air heavy. The effect of cutting off the last figure from the square of the diameter, is to divide the area of the cylinder into portions of 10 circular inches each; and as the pressure on each of these portions is estimated at a long hundred-weight, or 120 lbs., the pressure will be $120 \div 10 = 12$ lbs. per circular inc. or $15 \cdot 3$ lbs. per square inch; but this rule is too great for the full pressure of the atmosphere, even if the vacuum were perfect(b). To compensate for imperfections, and for the friction of the several parts, Newcomen allowed between one-third and one-fourth part of the atmospheric pressure.

Desaguliers says this rule will agree nearly with the work performed at Griff engine, which at every stroke lifted between two-thirds and three-fourths of the weight of the atmospherical column pressing on the piston; i. e. between 10 and 11 lbs. on each squ. inc.; and he gives the following estimation in round numbers: The diameter of the cylinder is 22 inches; this squared is 484; cut off the last figure, and add a cypher, and we have 48 cwt. 40 lbs. for the entire pressure of the

deep, and the pumps, which were in two lifts and nine inches bore, drew the water up 37 yards, into a level, which conveyed it away. The pump which supplied the cold water for injection, was about 4 inc. diam. and 3 ft. stroke. The boiler bottom was set 2 ft. 8 inc. above the fire-grate; it burned 24 or 25 corves of coals in 24 hours.

The steam-engine was first introduced into Scotland in 1720, at Elphinstone coal-work, in Stirlingshire; the cylinder was 28 inc. diam.

In 1723, a steam-engine was erected at a mine at Konigsberg in Hungary, by an Englishman named Potter; the cylinder was about 30 inc. diam.

(b) If instead of the long-hundred of 120 lbs. he had taken the common hundred of 112 lbs. he would have had $112 \div 10 = 11 \cdot 2$ lbs. per circ. inc. or $14\frac{1}{4}$ lbs. per squ. inc. This would have been nearer to the medium pressure of the atmosphere, which is $14\frac{3}{4}$ lbs. per squ. inc. when the barometer stands at 30 inches.

atmosphere. The column of water in the pumps weighs about $27\frac{1}{2}$ cwt., and adding 9 cwt. for the weight of 73 yards of iron-rods, the weight to be lifted at the end of the lever, would be $36\frac{1}{2}$ cwt., but from which we must subtract about 4 cwt. for the piston and other weight, at that end of the lever, reducing the load to $32\frac{1}{2}$ cwt. ; so that the weight of the atmosphere being 48 cwt. 40 lbs. raises a weight of $32\frac{1}{2}$ cwt., (that is very near $\frac{2}{3}$,) with a motion of six feet in two seconds. This requires an effective pressure on the piston of near $10\frac{1}{4}$ lbs. per squ. inc., including friction and counter-weight; but merely to balance the weight of the water in the pump, demands a pressure of only $8\frac{2}{3}$ lbs. per squ. inc. of the piston. Hence in the working stroke, the preponderance of the piston is 5 cwt., or about $1\frac{1}{2}$ lbs. per squ. inc. above the resistance of the load, and the counter-weight. The counterweight, which occasions the motion during the returning stroke, is also 5 cwt. or about $1\frac{1}{2}$ lbs. per squ. inc.; this produces a motion of 15 strokes per minute of six feet, or 180 feet per minute (c). The rate of the motion, or the load in the pumps, is probably overstated in this account, as no atmospheric engines can now be made to do as much. Yet Mr. Beighton calculated and published a table of steam-engines in 1719, which supposes the engine to be loaded with a column of water, amounting to 8 lbs. for each square inch of the piston; and it assumes that the engine will then make 16 strokes per minute, of 6 ft. or 192 ft. motion per minute. As this table was arranged in a very convenient form, for the use of persons unaccustomed to calculation, it came into general use. The largest cylinder, calculated in Mr. Beighton's table, was only 40 inches diameter.

This estimation of 8 lbs. load to each square inch of the piston, continued for many years to be a rule with engineers; and if the engines were of a better or worse construction, they would work with a greater or less rapidity, because all the excess of pressure which could be obtained above the 8 lbs. was appropriated to overcome the friction and inertia of the parts of the machine, and to raise the counterweight. The greater this additional quantity was, the quicker the engine would move, and would raise a greater quantity of water in the same time; but as they never came up to Mr. Beighton's standard of 192 ft. motion per min., succeeding engineers diminished the load to 7 lbs. and even 6 lbs. per squ. inc., in order to obtain a greater celerity of motion.

It has been a common mistake with operative engineers, to reckon merely by the diameter of the cylinder, and the perpendicular height and diameter of the pumps, without attending to the quantity of the motion, or the number of strokes per minute, and their length.

Without these particulars it is impossible to calculate the quantity of water raised to a given height in a given time, which is the true measure of the mechanical power of an engine: it would be like attempting to measure the contents of a solid body, by only two dimensions. The load upon engines has at different times been made to vary from 5 lbs. to 10 lbs. per square inch of the piston; but when working with a light resistance, they must be expected to move with an increased speed. For instance, if a piston loaded to 5 lbs. should move through double the space in the

---

(c) A complete stroke of a steam-engine includes the descent of the piston, which is termed the working stroke; and also the ascent of the piston, which is called the returning stroke. The actual motion of the piston, when it makes 15 strokes per min. of 6 ft. is therefore 180 ft. per min.; but the actual motion of the column of water is only half as much, or 96 ft. per min. because it remains motionless half the time.

In speaking of the motion of the piston, we take the whole motion; but to estimate the power of an engine, we must only consider the motion of the column of water, as its useful or effective motion.

same time, that it would do if loaded with 10 lbs. per squ. inc. then the same quantity of water would be raised to a given height, in the same time, in either case. In the steam-engine, as well as in other machines, there is a maximum of effect, which cannot be exceeded without applying some new principle; and though by imperfect execution, an engine may fall short of what it should do, the best workmanship can only produce a certain effect.

Many experiments have been made by different engineers, to find the resistance which an engine will overcome, so as to produce the maximum effect, but they have not agreed exactly in their results. Some have preferred 8 lbs. per square inc., after Mr. Beighton; others, 7 lbs.; and, perhaps, $7\frac{1}{2}$ lbs. or half the entire atmospheric pressure, is very near the point. Mr. Smeaton, who made great progress in investigating the best proportions for atmospheric steam-engines (d),

(d) Having constant occasion to employ steam-engines in the great works which he executed, he turned his attention to consider the means of improving their effect, and diminishing the consumption of fuel. In calculating the proportions for an engine for the New River Company, in 1767, he considered that there was a great loss of power by the stoppage of the water at every stroke, as well as to put the heavy lever-beam, piston, rods, and chains, from a state of rest into motion, twice at every stroke; he therefore determined to work the engine slowly, and with large pumps, and to put upon the piston all the load it would bear. To reduce the velocity of the column of water still more, he determined to place the fulcrum of the beam out of the centre, making the stroke of the piston nine feet, and that of the pump (which was 18 inc. diam. and lifted 35 ft.) only six feet. This arrangement obliged him to employ a long narrow cylinder, only 18 inches diameter; but from this circumstance he expected to obtain some advantage, because every part of the steam, being near to the surface of the cylinder, would be readily condensed; and, in consequence, a small quantity of injection-water would serve the cylinder, which would itself be more heated. Under all these appearances of advantage, he ventured to burden the piston with a load of $10\frac{1}{4}$ lbs. per inch.

" Having once seen a common engine struggle under this burthen, I thought myself (says this ingenious engineer) quite secure of the result; but how great was my surprise and mortification, to find that, instead of requiring less injection-water than common, the pump, although it was calculated to afford as much cold water as usual (in proportion to the area of the cylinder, with a sufficient overplus to answer all imaginable wants) it was unable to supply the engine with injection; and to keep the engine in motion two men were required to assist, to raise more injection-water by hand: at the same time the cylinder was so cold, that I could keep my hand upon any part of it, and bear it for a length of time in the hot-well. By good fortune, the engine performed the work it was appointed to do, as to the raising of water; but the coals by no means answered to my calculation. The injection-pump being enlarged, the engine was put in a state for doing business, and I tried many experiments, but without any good effect, till I altered the fulcrum of the beam so much, as to reduce the load upon the piston from $10\frac{1}{4}$ lbs. to $8\frac{1}{4}$ per inch. Under this load, though it shortened the stroke in the pump, the engine went so much quicker, as not only to raise more water, but it consumed less coals, took less injection-water, the cylinder became hot, and the injection-water came out at 160° of Fahrenheit; so that the engine, in every respect did its work better, and went more pleasantly. This convinced me that a considerable condensation of the steam took place, in entering the cylinder, and that I had lost more in this way by the coldness of the cylinder, than I had gained by the increase of load. In short, this single alteration seemed to have unfettered the engine; but the extent of this condensation under different circumstances of heat, and where to strike such a medium, as to obtain the best result, was still unknown to me. I resolved, if possible, to make myself master of the subject, and immediately began to build a small fire-engine, which I could easily convert into different shapes for experiments." This experimental engine was set to work at Austhorpe in the winter of 1769; it had a cylinder of 10 inc. diam. and made a stroke of 3 ft. 2 inc.

Mr. Smeaton made a series of experiments with it, so diversified as to give him information on all the important points affecting the performance of an engine. He noted down all his observations with great care in tables, and from comparisons of their results, deduced rules for the proportions of all parts of his engines, on the same system as he had before analysed the powers, and settled the proportions of wind-mills and water-mills.

He afterwards constructed many engines of the largest dimensions, the performance of which fully verified his conclusions. The first of these has been already described: it was made at Long Benton colliery in 1772; the cylinder 52 inc. diam. And soon after he made an engine with a cylinder of 66 inc. diam. for the Empress of Russia, and which is still in use to drain the docks at Cronstadt, near Petersburg; and in 1774, he executed another with a 72 inc. cylinder, for the Mine of Chase-water in Cornwall.

tried many experiments to determine this point; and, from their results, he concluded, that engines work to the greatest advantage when the resistance from the column of water in the pumps, is equal to a column of water 18 feet high pressing upon the piston, this is 7·81 lbs. per squ. inc., and therefore agrees nearly with Mr. Beighton's; but he found, that with this resistance, the motion of the piston was never greater than 168 ft. per min. in the largest engines, and only 132 for the smallest; and even this velocity cannot be attained in engines for draining deep mines, because they are encumbered in their motions, by a great load of pump-rods, and other parts, which must be put into motion without contributing to the effect in raising water.

In estimating the mechanical power of engines by the resistance opposed to each square inch of the area of the piston, it must be considered as the useful effect of the engine, or the column of water it will raise, without any deductions for friction, inertia, counter-weight, &c. To calculate the actual resistance of the different lifts of pumps in the engine-pit of a coal-mine or copper-mine, we find we must take into the account (besides the altitudes and the diameters of the pumps,) the friction of the buckets, and of the water against the insides of the pumps; the opening of strong double-leathered valves, loaded with stones and gravel that enter at the foot of the pumps, the inertia of the pump-rods, the chains, and the great lever, the friction of the great piston, the working gear, &c.; all these resistances are to be overcome by the pressure on the piston, in addition to the load of 7 lbs. per squ. inc. These additional resistances vary in amount in different engines, according to circumstances, but are always considerable, and are in many cases equal to half what is required for the work performed: this will raise the unbalanced pressure of the atmosphere, to $10\frac{1}{2}$ lbs. per squ. inc. When this is the case, the vapour which remains in the cylinder must be equal in pressure to $4\frac{1}{4}$ lbs. per squ. inc.: and this, by the table of elasticities, will be a temperature of about 154° of Fahrenheit.

In general, the water in the hot-well is above this temperature. Mr. Smeaton examined several engines, and found the temperature of the hot-well to vary from 142° to 174°. We have but little information concerning the state of the vacuum in the atmospheric engines, when working in their usual state, but it must be considerably more perfect, than has been suggested by the idea of a load of $7\frac{1}{2}$ lbs. on each square inch of the piston. When the counterweight and the friction and inertia are added, the resistance can rarely be less than 11 lbs. per squ. inc. Mr. Hornblower states in Gregory's Mechanics, vol. II, that he tried the vacuum of several engines in Cornwall with a mercurial barometer, and found that the vacuum in the cylinder raised the mercury to 23, and sometimes 24 inc. instead of 30 inc. at which it would have stood, if the vacuum had been perfect. The mean of these observations, will be $11\frac{1}{2}$ lbs. on each square inch of the piston.

It is obvious, that if the interior of the cylinder were cooled down to a lower temperature, the elasticity of the steam or vapour remaining in it, might be still more diminished, and a greater effect produced; for instance, if it were cooled to 82 degrees, the elasticity of the remaining steam would be only $\frac{1}{2}$ a pound per square inch; but this would require a very great quantity of cold water to be injected, and even if it were practicable, to cool the cylinder so low, the consumption of steam to heat the cylinder again to the temperature of 216° would be excessive, as Mr. Smeaton found in his first attempts. It has, therefore, been found advisable in practice, to throw in no more injection-water, than will produce a sufficient vacuum to enable the engine to raise a load of 7, or $7\frac{1}{2}$ pounds per square inch, independent of all friction, and other unavoidable resistance to motion : with this load one of Mr. Smeaton's best engines made 12 strokes per minute, of 7 feet each, or 168 feet motion per minute.

Mr. John Curr, who wrote a work on the Atmospheric Engine, in 1797, states that he constructed one, which, when loaded with 7 lbs. per squ. inc. made 12 strokes per minute, of $8\frac{1}{2}$ feet length or 204 ft. motion per minute. He tried the same engine with the load increased to $8\frac{1}{2}$ lbs. per squ. inc. ; but it would then only make 9 strokes per minute, of 8 feet, or 144 ft. per minute, which is a much smaller performance ; he also reduced the load, to 6 lbs. per squ. inc. and found the result less advantageous than with 7 lbs., though better than with $8\frac{1}{2}$ lbs. From these statements we may collect, that the load for an atmospheric engine should be calculated at 7 lbs. per squ. inc. of the piston.

It is the usual practice to estimate the pressure on the piston, by the pounds on a square inch, though the pressure in pounds on a circular inch, would be much more convenient for calculations relative to steam-engines ; because all the vessels being cylindrical, the areas of the cylinders in circular inches, can be obtained by only squaring the diameter of the circle in inches; for example : a circle of 10 inches diameter contains $10 \times 10 = 100$ circular inches ; but when we require the area in square inches, the square of the diameter must be multiplied by ·7854 ; for example : a circle 10 inches diameter contains $(10 \times 10 = 100 \times ·7854 =) 78·54$ square inches.

The pressure of any column of water in pounds avoirdupois, upon each square inch, is found by multiplying the height of the column of water in feet, by ·434 lbs. ; for example : the pressure of a column 162 feet high, is $(162 \times ·434 =) 70·38$ lbs. per squ. inc. (*a*).

The weight of cylindrical columns of water, in pounds avoirdupois, is obtained by multiplying the square of the diameter in inches, by ·341 lbs., and multiplying the product, by the perpendicular altitude in feet. For example : a pump 8 inches diameter, which is to lift the water 162 feet perpendicular ; what is the weight of the column of water in pounds ? $8 \times 8 = 64$ circ. inc. $\times ·341$ lbs. $= 21·824$ lbs., for 1 foot, and $\times 162$ ft. $= 3535·5$ lbs. for 162 ft. high.

(*a*) The foundation of these rules is, that a cubic foot of pure water (at the temperature of 57 degrees of Fahrenheit's thermometer) weighs 1000 ounces or $62\frac{1}{4}$ pounds avoirdupois ; and as a square foot contains $(12 \times 12 =) 144$ square inches, a prism 1 inch square, and 1 foot high, weighs $(62·5 \div 144 =) ·434028$ of a pound avoirdupois ; this is called a square inch foot.

Again, a square foot contains 183·346 circular inches (for $144 \div ·7854 = 183·346$) ; therefore, a cylinder one inc. diam., and 1 foot high, weighs $(62·5 \div 183·346 =) ·34088$ of a pound avoirdupois ; it is called a cylindrical inch foot. A cylindrical foot contains 144 of these cylindrical inch feet, and weighs 49·0874 lbs. A cylindrical inch weighs ·0284 of a pound ; and a cubic inch weighs ·03617 of a pound.

Calculations of this kind will be greatly facilitated by the following tables.

Table of the Weight and Contents of Columns of Water in Pumps.

Diam. of the Pump.	Weight of one foot high.	Contents of one foot high
inches.	pounds.	cubic feet.
5½	10·31	·165
6	12.27	·196
6½	14·40	·231
7	16·71	·267
7½	19·18	·307
8	21·82	·349
8½	24·63	·394
9	27·62	·442
9½	30·77	·492
10	34·09	·546
10½	37·59	·601
11	41·25	·660
11½	45·09	·722
12	49·10	·785
12½	53·27	·852
13	57·62	·922
13½	62·14	·994
14	66·83	1·07
14½	71·68	1·15
15	76·71	1·23
15½	81·91	1·31
16	87·28	1·40
16½	92·82	1·49
17	98·53	1·58
17½	104·41	1·67
18	110·47	1·77
18½	116·69	1·87
19	123·08	1·97
19½	129·64	2·07
20	136·38	2.18
20½	143·28	2·29
21	150·36	2·40
21½	157·60	2·52
22	165·02	2·64
22½	172·60	2·76
23	180·36	2·89
23½	188·29	3·01
24	196·39	3·14

Table of the Force of Cylinders at seven pounds pressure per square inch.

Diam. of Cylinder.	Pressure on the Piston.	Area of Cylinder.	Diam. of Cylind.	Pressure on the Piston.	Area of Cylinder.
inches.	pounds.	squ. inches.	inches.	pounds.	squ. inches.
9	445	63·62	49	13 200	1885·74
10	550	78·54	50	13 744	1963·49
11	665	95·03	51	14 300	2042·82
12	792	113·10	52	14 866	2123·72
			53	15 443	2206·18
13	929	132·73	54	16 032	2290·22
14	1078	153·94	55	16 631	2375·83
15	1237	176·71	56	17 241	2463·01
16	1407	201·06	57	17 862	2551·76
17	1589	226·98	58	18 495	2642·08
18	1781	254·47	59	19 138	2733·97
19	1985	283·53	60	19 792	2827·43
20	2199	314·16			
21	2425	346·36	61	20 457	2922·47
22	2661	380·13	62	21 134	3019·07
23	2908	415·48	63	21 821	3117·24
24	3167	452·39	64	22 519	3216·99
			65	23 228	3318·31
25	3436	490·87	66	23 948	3421·19
26	3717	530·93	67	24 680	3525·65
27	4008	572·55	68	25 422	3631·68
28	4310	615·75	69	26 175	3739·25
29	4624	660·52	70	26 939	3848·48
30	4948	706·86	71	27 714	3959·19
31	5283	754·77			
32	5630	804·25	72	28 501	4071.50
33	5987	855·30	73	29 298	4185·39
34	6355	907·92	74	30 106	4300·84
35	6735	962·11	75	30 925	4417·86
36	7125	1017·88	76	31 755	4536·46
			77	32 596	4656·63
37	7526	1075·21	78	33 449	4778·36
38	7939	1134·11	79	34 312	4901·67
39	8362	1194·59	80	35 186	5026·55
40	8796	1256·64	81	36 071	5152·10
41	9242	1320·25	82	36 967	5281·02
42	9698	1385·44	83	37 874	5410·61
43	10 165	1452·20	84	38 792	5541·77
44	10 644	1520·53			
45	11 133	1590·43			
46	11 633	1661·90			
47	12 145	1734·94			
48	12 667	1809·56			

N. B. 7 lbs. pressure per sq. inch is = 5½ lbs. pressure per circ. inch, or = to the pressure of a col. of water 16·13 ft. high.

Y

To find the pressure which is exerted by any column of water, in pounds avoirdupois, upon each square inch of its base. Having given the perpendicular height of the column in feet, above that base.

RULE. Multiply the height in feet, by ·434 lbs.; the product is the pressure in pounds.

Sliding Rule. $\left\{ \begin{array}{ll} A & 30 \quad \text{Perpendicular height in feet.} \\ B & 13 \quad \text{Pressure per sq. inc. in pounds.} \end{array} \right.$ *Example.* $\dfrac{A \quad 30 \quad\quad 162 \text{ feet high.}}{B \quad 13 \quad\quad 70 \cdot 3 \text{ lbs. per sq. inc.}}$

The sliding rule being thus set, becomes a complete table for the pressure, at any height.

To find the weight of any cylindrical column of water in pounds. Having given, the diameter of its base in inches, and the perpendicular height of the column, above that base, in feet.

RULE. Multiply the square of the diameter of the base in inches, by ·341 lbs.; the product is the weight of the column in pounds.

Sliding Rule. $\left\{ \begin{array}{ll} C & \text{Height of col. ft.} \quad \text{Weight of column lbs.} \\ D & 1 \cdot 713 \quad\quad\quad \text{Diameter of base inc.} \end{array} \right.$ *Example.* $\dfrac{C \quad 162 \text{ ft.} \quad\quad 3535 \text{ lbs. wt.}}{D \quad 1 \cdot 713 \quad\quad 8 \text{ inc. diam.}}$

The rule thus set, forms a table for the weight of any diameter, at that height.

To find the area of a circle in square inches. Having given the diameter in inches.

RULE. Multiply the square of the diameter in inches, by ·7854; the product is the area in square inches.

Example. 15 inc. diam.; 15 inc. × 15 inc. = 225 circ. inc. × ·7854 = 176·71 square inches.

Sliding Rule. $\left\{ \begin{array}{ll} C & 43 \quad \text{Area square inc.} \\ D & 7 \cdot 4 \quad \text{Diameter inches.} \end{array} \right.$ *Example.* $\dfrac{C \quad 43 \quad\quad \text{area } 176 \cdot 7 \text{ sq. inc.}}{D \quad 7 \cdot 4 \quad\quad \text{diam. } 15 \text{ inches.}}$

The rule thus set, forms a table for the area corresponding to any diameter.

To find the diameter of the cylinder, proper to give motion to any pump. Having given, the weight of the column in pounds, and the intended load upon the piston in lbs. per square inch.

RULE. Divide the weight of the column of water in pounds, by the intended load or pressure on the piston, in lbs. per sq. inch, the quotient is the area of the required cylinder in sq. inches; divide this by ·7854, and extract the square root of the product; the root is the diameter of the cylinder in inches.

Example. A pump, 8 inc. diam. 162 ft. lift; what must be the size of the cylinder, assuming it to be loaded with 7·8 lbs. per sq. inch? The weight of the column is (8 × 8 = 64 × ·341 = 21·824 × 162 =) 3535·5 lbs. Then 3535·5 ÷ 7·8 = 453·3 sq. inc. ÷ ·7854 = 577 circ. inc. the square root of which is very nearly 24 inc. diam. for the cylinder required.

Note. A still more convenient method is, to first reduce the intended load or pressure upon the piston, into its equivalent column of water, by dividing the pressure in lbs. per sq. inch by ·434.

Example. 7·8 lbs. per sq. inch ÷ ·434 = 17·96 ft. Hence the pressure of 7·8 lbs. upon each sq. inch of the piston, is the same as if it had a column of water 17·96 ft. high, resting upon it.

Sliding Rule, slide inverted. $\left\{ \begin{array}{ll} A & 1 \quad\quad\quad ·434 \\ \mho & \text{lbs. per sq. inc.} \quad \text{Col. of water ft.} \end{array} \right.$ *Example.* $\dfrac{A \quad 1 \quad\quad\quad ·434}{\mho \quad 7 \cdot 8 \text{ lbs. per sq. inc.} \quad 17 \cdot 96 \text{ ft. col.}}$

To find the diameter of the cylinder, proper to give motion to any pump. Having given, the diameter of the pump in inches, the height it is to lift in feet, and the intended load or pressure on the piston, expressed by the height of a column of water in feet.

RULE. Multiply the square of the diameter of the pump in inches, by the height it is to lift, in feet, and divide the product by the height of the columnar pressure on the piston, in feet; the quotient is the area of the required cylinder, in circular inc. and consequently the square root of that quotient is the diameter of the cylinder in inches.

Example. A pump, 8 inc. diam. 162 ft. lift: the load on the piston being a column of 17·96 ft. high. 8 in. × 8 in. = 64 circ. inc. × 162 ft. = 10 368 cylindrical inch feet to be lifted, and ÷ 17·96 ft. = 577 circ. inc. area of the cylinder; the square root of which = 24 inc. is the diam. of the cylinder.

Sliding Rule, slide inverted. $\left\{ \begin{array}{ll} \text{B} & \text{Col. on piston ft.} \quad \text{Height of lift feet.} \\ \text{D} & \text{Diam. of cyl. inc.} \quad \text{Diam. of pump inc.} \end{array} \right.$ *Exam.* $\dfrac{\text{B} \quad 17 \cdot 96 \text{ ft. press.} \quad 162 \text{ feet lift.}}{\text{D} \quad 24 \text{ inc. cylind.} \quad 8 \text{ inc. pump.}}$

The preceding rules will serve to calculate a cylinder for any required load per square inch on the piston; but when the load is regulated to 7 lbs. per sq. in., which appears to be the best burden, the calculation may be much simplified thus :—

To find the diameter of a cylinder, which will work a given pump, the load on the piston being 7 lbs. per sq. inch.   Having given, the diameter of the pump in inches, and the height of the column in feet.

RULE.   Multiply the square of the diameter of the pump in inches, by the height of the column in feet, multiply the product by ·062, and extract the square root of that product; the root is the diameter of the cylinder in inches.

Example.   For a 12 inch pump, to lift water 50 fathoms, what size cylinder is required? $12 \times 12 = 144$ circ. inc. $\times 300$ ft. $= 43\,200 \times ·062 = 2678·4$ circ. inc. for the area of the piston, the square root of this number, is $51·75$ inc. for the diameter of the cylinder.

N. B.   7 lbs. pressure per sq. inch is equal to the pressure of a column of water 16·13 ft. high, (for 7 lbs. ÷ ·434 = 16·13); therefore dividing by 16·13 will give the same result, as multiplying by ·062, for that number is the reciprocal of the other.

Sliding Rule, { ß   16·13 ft. pressure.   Height of lift feet.   *Exam.* ß 16·13 feet.   300 feet lift.
slide inverted. { D Diam. of cyl. inc.   Diam. of pump inc.   D 51·7 inc. cyl. 12 in. pump.

Note.   All the above rules suppose that the length of the stroke is the same in the pump, as in the cylinder; or that the fulcrum or centre of motion, of the great lever, is exactly in the middle of its length, like a balance with equal arms.   But it frequently happens that the arms are not equal, and the injection-pump is always a shorter stroke than the great pump.   In such cases, the column of water in any pump, which makes a shorter stroke than the piston, must be reduced to an equivalent column, having the same stroke as the piston: this is done by multiplying the column of the pump, by the length of its stroke, and dividing the product, by the length of the stroke of the piston.   And if there are several pumps of different lifts, diameters, and lengths of stroke, they must be all reduced to a common measure, and the results added together, to bring them into one.

To find the load which will be occasioned on the piston, by several pumps of different diameters, lifts, and lengths of stroke.   Having given the length of stroke made by the piston in feet, the diameter of each pump in inches, its lift in feet, and its length of stroke in feet.

RULE.   Multiply the height to which each pump lifts its water in feet, by the length of the stroke of that pump in feet, and divide that product by the length of the stroke of the piston, also in feet, then multiply each quotient so obtained, by the square of the diameter of its pump in inches, and the product will be the load occasioned on the piston by each pump, expressed in cylindrical inch ft., and the sum of all the different pumps so calculated, will be the total load on the piston in cyl. inc. ft.

Example.   Suppose an engine, which is to make a 7 ft. stroke, is to work two pumps as follows: one of 8 inc. diam. 162 ft. lift, with a 5 ft. stroke; and another pump, 12 inc. diam. 300 ft. lift, to work with a 6 ft. stroke.   Now 162 ft. lift × 5 ft. stroke = 810 ÷ 7 ft. stroke = 115·7 ft. height of column when reduced, × (8 × 8 =) 64 circ. inc. = 7405 cylin. inc. ft. load on the piston, by the first pump.   And 300 ft. lift × 6 ft. stroke = 1800 ÷ 7 ft. stroke = 257·1 ft. height of column when reduced × (12 × 12 =) 144 circ. inc. = 37 022 cyl. inc. ft. load on the piston by that pump.   The sum of the two is (7405 + 37 022 =) 44 427 cyl. inc. ft. for the total load on the piston by both pumps.

Note.   If products obtained by multiplying the lift of each pump, into the length of its stroke, and then into the square of its diameter, are added together, and the sum divided by the length of the stroke of the piston, the same result will be obtained thus: (162 × 5 =) 810 × 64 = 51 840 added to (300 × 6 =) 1800 × 144 = 259 200; the sum of both is 311 040, and ÷ 7 ft. = 44 434 cyl. inc. ft. total load of both pumps.

To find the diameter of a cylinder to work different pumps.   Having given, the total load occasioned on the piston by all the pumps, in cyl. inc. ft. by the preceding rule, and also the intended load which the piston is to bear, expressed in the height of a column of water in feet.

RULE.   Divide the total load on the piston in cyl. inc. ft. by the columnar pressure on the piston in feet; the quotient is the area of the required cylinder in circ. inc., and, consequently, the square root of that quotient is the diameter of the cylinder in inches.

Example.   44 434 cyl. inc. ft. ÷ 16·13 feet column = 2753 circ. inc., the square root of which 52·5 inc. diam. of cylinder.

Note.   Instead of comparing the different lengths of stroke of the several pumps, with the length of the stroke of the piston, we may obtain the same results, by comparing the distances from the centre of motion of the great lever of the engine, to the parts where each pump-rod is suspended from that lever (that is, the radii of their effective levers respectively) with the radius or distance from the same centre of motion, to the part where the piston-rod is suspended from the lever.

For instance, suppose the piston-rod of the engine, making a 7 ft. stroke, was suspended at $10\frac{1}{2}$ ft. radius from the centre of motion; then the great pump, which makes only a 6 ft. stroke, must have been suspended at a radius or distance of 9 ft. from that centre; and the small pump, which makes a 5 ft. stroke, must have been at $7\frac{1}{2}$ ft. radius.

Pumps of different sizes.

Cylinders proportioned to the different depths from which the water is to be drawn.

The depth from which the water is to be drawn in fathoms is in the top line, and diameters of the cylinders in inches in the columns beneath.

Diameters of the cylinders in inches.

Diam. of pump (inches)	Weight of the column of water in one fathom (pounds)	Quantity of water contained in one foot (Cubic ft.)	5	10	15	20	25	30	35	40	45	50	55	60	65	70	75	80	85	90	95	100
7	100·24	·267	9·6	13·5	16·6	19·1	21·3	23·4	25·3	27·0	28·7	30·2	31·7	33·1	34·5	35·8	37·0	38·2	39·4	40·5	41·6	42·7
8	130·92	·349	10·9	15·4	18·9	21·8	24·4	26·7	28·9	30·9	32·7	34·5	36·2	37·8	39·3	40·8	42·3	43·6	45·0	46·3	47·6	48·8
9	165·70	·442	12·3	17·4	21·3	24·6	27·5	30·1	32·5	34·7	36·8	38·8	40·7	42·5	44·3	45·9	47·5	49·1	50·6	51·2	53·5	54·9
10	204·57	·546	13·7	19·3	23·6	27·3	30·5	33·4	36·1	38·6	40·9	43·1	45·2	47·2	49·2	51·0	52·8	54·5	56·2	57·9	59·5	60·1
11	247·53	·660	15·0	21·2	26·0	30·0	33·5	36·8	39·7	42·4	45·0	47·4	49·8	52·0	54·1	56·1	58·1	60·0	61·9	63·7	65·4	67·1
12	294·58	·785	16·4	23·1	28·3	32·7	36·6	40·1	43·3	46·3	49·1	51·8	54·3	56·7	59·0	61·2	63·4	65·5	67·5	69·4	71·3	73·2
13	345·72	·922	17·7	25·0	30·7	35·5	39·6	43·5	46·9	50·0	53·2	56·1	58·9	61·4	63·9	66·3	68·7	70·9	73·1	75·2	77·3	79·3
14	400·95	1·07	19·1	27·0	33·1	38·2	42·7	46·8	50·5	54·0	57·3	60·4	63·3	66·1	68·9	71·5	74·0	76·4	78·7	81·0	83·2	85·4
15	460·28	1·23	20·5	28·9	35·4	41·0	45·7	50·1	54·1	57·9	61·4	64·7	67·9	70·9	73·8	76·6	79·2	81·8	84·4	86·8	89·2	91·5
16	523·69	1·40	21·9	30·8	37·8	43·6	48·8	53·4	57·7	61·7	65·5	69·0	72·4	75·6	78·7	81·7	84·5	87·3	90·0	92·6	95·1	97·6
17	591·20	1·58	23·2	32·8	40·2	46·4	51·8	56·8	61·3	65·6	69·6	73·3	76·9	80·3	83·6	86·8	89·8	92·8	95·6	98·4		
18	662·80	1·77	24·6	34·7	42·5	49·1	54·9	60·1	65·0	69·4	73·7	77·6	81·4	85·0	88·5	91·9	95·1	98·3				
19	738·49	1·97	25·9	36·7	44·9	51·8	57·9	63·5	68·6	73·3	77·7	82·0	86·0	89·8	93·4	97·0						
20	818·27	2·18	27·2	38·6	47·2	54·6	61·0	66·8	72·2	77·2	81·8	86·3	90·5	94·5	98·4							
21	902·14	2·41	28·6	40·5	49·6	57·3	64·1	70·2	75·8	81·0	85·9	90·6	95·0	99·2								
22	990·11	2·64	30·0	42·4	52·0	60·0	67·1	73·5	79·4	84·9	90·0	94·9	99·5									
23	1082·16	2·89	31·4	44·4	54·3	62·7	70·2	76·8	83·0	88·7	94·1	99·2										
24	1178·31	3·14	32·8	46·3	56·7	65·5	73·2	80·2	86·6	92·6	98·2											

This Table is adapted to load the piston with 7 lbs. per square inch.

N. B.—It is assumed in all cases, that the length of stroke of the pump, is the same as that of the piston.

Calculated by John Curr, 1797.

To obtain the weight of the column of water in pounds, multiply the number in the second column by the depth in fathoms. *Example.*—Suppose an 8 inch pump is to lift 27 fathoms; each fathom weighs 130·92 lbs. per table, × 27 fath. = 3535 lbs. weight of the column.

The third column is useful to ascertain the quantity of water any pump will raise, when working any given number of strokes per minute, of a given length. *Example.* Suppose an 8 inch pump is worked 15 strokes per minute, of 5 ft. = 75 ft. motion per min. Now an 8 inch pump contains, per Table, 349 of a cub. ft. in each foot high and × 75 ft. = 26·17 cub. ft. of water will be raised per minute.

Now if the numbers 10·5 ft. rad., 9 ft. rad., and 7·5 ft. rad. are used in the above calculations, in lieu of 7 ft. stroke, 6 ft. stroke, and 5 ft. stroke, the same results will be obtained in either case.

Calculations of this kind may be greatly facilitated by the use of the sliding rule, each pump being calculated by a separate operation.

First, from the actual heights of the columns of water in those pumps which work with different lengths of stroke, to find the heights that the columns for the same pumps would be, if they worked with the same length of stroke as the piston.

Sliding Rule, slide inverted. { A Stroke of pump ft. Stroke of piston ft. _Example._ A 5 ft. stroke. 7 ft. stroke.
{ ↄ Column of pump ft. Equivalent col. ft. ↄ 162 ft. col. 115·6 feet·

By the same rule, a pump of 6 ft. stroke, with 300 ft. column, will be the same load, as if it made 7 ft. stroke, with 257 ft. column.

Having thus supposed all the pumps brought to work with the same length of stroke, we may proceed to suppose all the several columns collected into one.

Therefore to find the height that each column would have been, if it had been of the same diameter as that of the largest pump (which in the above example is 12 inc.)

Sliding Rule, slide inverted. { ᗺ Height of its colm. feet. Height of its colm. ft. _Exam._ 115·6 ft. col. 51·4 ft. col.
{ D Diam. of small pump in. Diam. of great pump. 8 inc. diam. 12 inc. dia.

Hence the weight of a column 8 inc. diam. 115·6 ft. high, is the same as that of a column 12 inc. diam. and 51·4 ft. high. This may easily be conceived by supposing the water to be poured out of the small pump into the large one. Hence the load on the piston, occasioned by the two pumps above-mentioned, will be equal to that of one pump 12 inc. diam. and (257 ft. + 51·4 ft. =) 308·4 ft. column, making the same length of stroke at the piston, viz. 7 feet.

The rule for finding the diameter of the cylinder in such case, has been before stated; viz. when the load is intended to be 7 lbs. per sq. inch, or 16·13 ft. of water.

Sliding Rule, slide inverted. { ᗺ Height of its col. ft. 16·13 ft. column. _Exam._ ᗺ 308·6 ft. lift. 16·13 feet.
{ D Diam. of pump inc. Diam. of cyl. inc. D 12 inc. pump. 52·5 inc. cyl.

The opposite table is calculated to show the diameters of cylinders corresponding to different pumps; the load being 7 lbs. per sq. inch.

The first upright column under the head of pumps, contains the different diameters of the pump, from 7 to 24 inches; and the numbers at the heads of the different columns of figures, entitled cylinders, are the depths from which the water is to be raised, in fathoms of 6 ft. each, and extending from 5 to 100 fathoms. The diameter of the required cylinder in inches, is found in the line opposite to the diameter of the pump, and beneath the figure expressing the depth in fathoms.

Example. For a 12-inch pump, to lift 50 fathoms. Seek 12 in the first column, and trace that line horizontally across the table, until in the column which is headed 50 at top, 51·8 is found for the diameter of the cylinder.

The rapidity with which an engine will perform its motions, when loaded according to this table, depends upon so many circumstances, that no exact rule can be given. The motion of the piston will vary from 72 feet per minute in small engines to 96 feet, in large ones. The mechanical power exerted by any engine, depends upon the load which it overcomes, and the velocity with which it moves in opposition to that load.

To find the power of an engine, in horse-power, when the load on the piston is 7 lbs. per sq. inch. Having given, the diameter of the cylinder in inches, the length of the stroke in feet, and the number of strokes made per minute.

RULE. Multiply the square of the diameter of the cylinder in inches, by the length of the stroke in feet, and multiply the product by the number of strokes per minute; then divide the product by 6000. The quotient is the power of the engine in horse-power.

Example. A cylinder, 60 inches diameter, making 12 strokes per minute of 8 feet length. 60 × 60 = 3600 × 8 ft. = 28 800 × 12 strokes = 345 600 ÷ 6000 = 57 horse-power.

Sliding Rule. { C Motion feet per min. Horse-power. _Exam._ C 96 ft. motion. 57 HP.
{ D 25 Diam. of cyl. inches. D 25 60 inc. diam.

DIMENSIONS AND PROPORTIONS OF ATMOSPHERIC ENGINES OF DIFFERENT SIZES, AND THE RESULTS OF THEIR PERFORMANCE.

MR. SMEATON had a very extensive practice in the construction of atmospheric steam-engines ; and his observations upon their performances, being very judiciously made, and faithfully recorded, give us a very exact knowledge of this kind of engine.

The smallest size that he constructed, was his experimental engine, which he had in his own house ; it exerted a very little less than 1 horse-power.

The cylinder was 9·9 inch. diam. (= 77 squ. inc.) and the full length of stroke allowed by the catch-pins was 3 ft. 2 inc. ; but in working the stroke was 3 ft. : it made 17½ strokes per min. = 52·5 ft. effective motion per min. when raising water, and as much in returning ; or 105 ft. motion per min. in the whole. This was its performance when in its best state, being loaded with a resistance of 7·89 lbs. per squ. inc. of the piston, = 607 lbs. total load ; and then it consumed at the rate of ·625 of a Yorkshire bushel (of 88 lbs. of coals), = 55 lbs. of coals per hour. The coals were of the kind called Hage-moor.

Now 607 lbs. raised 52·5 ft. per min. is = 31 867 lbs. raised one ft. per min. ; and ÷ 33 000 lbs. for a HP gives ·966 HP for the power of this little engine. Again, as it consumed 55 lbs. of coals per hour, a London bushel (= 84 lbs.) would supply it 91·6 minutes ; and in that time it would raise (31 867 lbs. × 91·6 =) 2 919 017 lbs. one ft. high. Therefore this small engine raised 2·92 million lbs. of water to a height of one foot, by each bushel of coals it consumed. This was the useful mechanical effect of the engine in raising water by its principal pump, and also by its injection pump ; but no allowance being made for the loss of power, by the friction of the moving parts, or for the loss of heat, by communication to the air.

The temperature of the hot well was only 134 deg. when the injection-water was 66 deg. or 68 deg. lower. The quantity of injection-water was found by measurement to be 95 cub. inc. per stroke, or 1663 cub. inc. per min. The quantity evaporated from the boiler was found to be about 8·9 cub. inc. per stroke, or 156 cub. inc. per min. ; from which it appears that the quantity of injection-water was 10·66 times that of the water contained in the steam.

The quantity of water evaporated into steam by each bushel (= 84 lbs.) of coals was 8·25 cub. ft. ; for 156 cub. inc. per min. × 91·6 min., the time that a bushel lasted = 14 267 cub. inc. or (÷ 1728 =) 8·25 cub. ft. ; that is, 6·14 lbs. of water evaporated by each pound of coals consumed.

The temperature of the steam in the boiler was probably about 216 deg., to which adding 960 for its latent heat, and we have the number 1176 to represent the heat of the steam. The heat which must have been abstracted from this steam, in order to condense and reduce it down to 134 deg. the temperature of the hot-well, will be represented by (1176 − 134 =) 1042. As the injection-water gained 68 deg. of temperature from the condensation of the steam, the quantity of injection must have been (1042 ÷ 68 =) 15·33 times that of the water in the steam. Hence, if the above statement is correct, viz. that the injection was only 10·6 times the evaporation, a very large proportion of the heat of the steam was communicated to the air, from the external surface of the cylinder, the temperature of which was found to be 134 deg., that of the air at the time of these observations being 60 deg.

A barometer tube was applied to the cylinder ; the upper end of the tube communicating with the interior of the cylinder, whilst the atmospheric air pressed upon the mercury in the basin or cistern of the barometer. When the engine made its working stroke, the unbalanced pressure of the atmosphere raised this column of mercury to 23·2 inc., that is = 11·35 lbs. per squ. inc. (× 77 sq. inc. =) 873 lbs. pressure on the piston.

At the same time the ordinary barometer, or weather-glass, stood at 29·3 inc.; therefore the difference in the elasticity of the steam remaining within the cylinder, and that of the external air, was equal to a column of mercury of 6·1 inc. high = 3 lbs. per squ. inc.

Now, assuming that the interior of the cylinder was of the same temperature as the hot well, (viz. 134 deg.), then, by the table of elasticities, the steam remaining in the cylinder should have been about 2·4 lbs. instead of 3 lbs. Part of this difference may be attributed to the elasticity of air contained in the cylinder.

The resistance of the load being 7·89 lbs. per squ. inc., the descending force of the piston must have been (11·35 − 7·89 −) 3·46 lbs. per squ. inc. The counterweight, or preponderance of the pump-rod, was 93 lbs. = 1·21 lbs. per sq. inch of the piston; which being deducted from 3·46 lbs. leaves 2·25 lbs. per sq. inch = 173 lbs. for the actual preponderance to overcome the friction, and produce motion in the working stroke.

The resistance from friction was 27 lbs. when the piston moved very slowly during the returning stroke; but it must have been much more during the working stroke, when the engine was at work with its full velocity, and sustained the column of water.

This little engine had very little encumbrance of useless matter to be moved, because there was only a short length of pump-rod; and the pump being 10·4 inc. diam. the water must have met with but little resistance in passing the valves.

The weight of the piston, its rod and chains, amounted only to 92 lbs.; the pump-rod and its bucket were less, but were loaded with ballast, to make up 185 lbs., giving a preponderance of 93 lbs. for the counterweight. The great lever, and all its appurtenances, weighed about 2 cwt., and, according to the principle already stated, (see p. 33,) if a weight of about 38 lbs. had been applied to the piston, and another equal weight to the pump-rod, they would have had the same energy, as the weight of matter contained in the great lever itself.

The total weight of the moving parts of the engine may be stated at 353lbs. and as the column of water in the pump weighed 607 lbs., hence the weight of the moving parts was ·58 of that of the column of water. Again 353 lbs. of moving parts ÷ 77 squ. inc. area of the piston is = 4·58 lbs. of dead load per square inch.

During the working stroke, the total weight suspended at the piston end of the lever, may be estimated at 1003 lbs.; and that at the pump end 830 lbs.: the difference between them, or preponderance of the piston, was 173 lbs.; and rejecting the atmospheric pressure, as being a very trifling mass of matter, and possessing but little energy (see p. 19), the actual mass to be moved was (moving parts 353 + 607 water =) 960 lbs., which was about 5·55 times the preponderance of 173 lbs.

During the returning stroke, the weight depending at the piston end was 130 lbs., and at the pump end 223 lbs.; the preponderance of the counter weight was 93 lbs ,and the sum or mass to be moved 353 lbs. = 3·8 times the preponderance; but this does not allow for any unbalanced pressure of the air on the piston, during its ascent; although it must have been very considerable, because the steam was so rapidly condensed on entering the cylinder, that the steam-pipe would not have supplied it so fast as the ascent of the piston required, unless the piston had been forcibly drawn up by the counter-weight, and the steam thus sucked into the cylinder.

The quantity of steam condensed by the interior surface of the cylinder may be thus estimated. The quantity of water evaporated being 8·9 cub. inc. per stroke, would, according to the usual estimate, produce 1700 times = 15130 cub. inc. (= 8·76 cub. ft.) of steam per stroke. The area of the cylinder being 77 squ. inc. and 36 inc. stroke, it would contain only 2772 cub. inc.; to this must be added the space remaining at the bottom of the cylinder, when the piston was at its lowest. This was 14 inc. deep (× 77 =) 1078 cub. inc., and also allowing 190 cub. inc. for the contents of the steam and eduction-pipes, the whole capacity of the cylinder was (4040 cub. inc. =) 2·34 cub. ft. to be filled with steam every stroke; and this being deducted from 8·76 cub. ft. of steam produced per stroke, leaves 6·42 cub. ft. of steam, lost by condensation every time; that is ·733 of the whole quantity produced.

The interior surface of the cylinder was (9·9 inc. diam. = 31·1 inc. circumf. × 50 inc. long inside =) 1555 squ. inc.; to which adding the surfaces of the cylinder bottom, and of the piston, and also of the injection-pipe, steam-pipe, and eduction-pipes within the cylinder, the whole surface exposed to the steam was (2320 squ. inch. =) 16·11 squ. ft. This surface condensed 6·42 cub. ft. of steam per stroke; that is, at the rate of ·3985 cub. ft. per stroke, or (× 17·5 strokes =) 6·97 cub. ft. of steam condensed per minute, by each squ. foot of internal surface.

The contents of the cylinder was 2·34 cub. ft. and its internal surface was 16·11 squ. ft., that is (2·34 ÷ 16·11 =) ·145 squ. ft. of surface to each cub. ft. of capacity.   But to state this in more intelligible terms ; suppose the cylinder to have been cut open and unrolled, so as to be spread out in a flat surface (of 16·11 squ. ft.), and supposing the (2·34 cub. ft. of) steam contained in the cylinder, to have been distributed over that surface in a layer of uniform thickness, then that layer would have been (·145 of a ft. × 12 =) 1·74 inc. thick ; but the quantity of steam condensed each stroke (6·42 cub. ft.) would have covered the surface to an additional thickness of (·3985 ft. =) 4·78 inc.   This latter quantity may be considered as if it had leaked through the metal of the cylinder : for the heat which formed the elastic quality of the steam, was really absorbed into the metal. Hence the whole quantity of steam produced (= 6·52 inc. thick on the internal surface) was 3·75 times as much as was required merely to fill the capacity of the cylinder.

The construction of this small engine was very similar to the large one already described.   There is an engraving of it amongst the plates of steam-engines, in Dr. Rees's Cyclopædia.

The boiler was of copper, 3½ ft. diam. at the largest part, and 2½ ft. at the bottom part, which was exposed to the fire ; but the boiler was found too small, and it required great attention to the fire to keep up the steam.

The strength of the steam was indicated by a column of mercury, contained in a glass tube bent like an inverted syphon ; and in the usual course of working, the steam raised this column of mercury 5 inc., or 2·45 lbs. per squ. inc.

The spout of the injection-pipe within the cylinder was a square aperture in a brass plate, the area of which was ·078 of a squ. inc., and the surface of the water in the injection-cistern was 12 feet perpendicular above the orifice.   One side of the aperture for the injection was a moveable slider, which could be moved by a fine micrometer screw, the stem of which came through the side of the cylinder to the outside, and had a head fixed upon the end, to enable the manager to turn the screw, and thus enlarge or diminish the size of the aperture at pleasure, so as to give the engine more or less injection-water.

The large pump which this engine worked, was provided with short lengths of wooden tubes, which could be fixed upon the top of the iron working barrel, so as to raise the water to a greater or lesser height, in order to try the performance of the engine under different burdens.

With this engine Mr. Smeaton, during four years, made a course of experiments, of which he recorded more than 130 in a book of tables, with calculations of the performance of the engine during each experiment ; both with respect to the mechanical power exerted by the engine, and also the power it exerted during the time that it was worked by a bushel of coals.   The experiments were in general continued until two or three bushels of coals were consumed.

His practice was to adjust the engine to good working order, and then after making a careful observation of its performance in that state, some one circumstance was altered, in quantity or proportion, and then the effect of the engine was tried under such change ; all the other circumstances, except that one which was the object of the experiment, being kept as nearly as possible unchanged.

The first object of his experiments was to determine the best mode of managing the fire ; and it appeared that the greatest produce was obtained from a thin clear fire, evenly spread over the grate, and affording a clear bright flame, stirring the fire as often as was necessary to keep it burning bright.

When the fire was made with the same coals, heaped thick on the grate, to keep up a large body of fire, and stirring it but very rarely, the performance or mechanical power exerted by the engine, with a given allowance of fuel, was only five-sixths of what it was with a clear bright fire ; though all the other circumstances were kept as nearly as possible the same in both cases.

The fire-grate was fixed at 14 inc. beneath the bottom of the boiler in the central part.   In some experiments it was raised to 9 inc., and in others it was lowered to 23 inc.   The effect with the grate at 12 inc. was found better than when at 22 inc., as 5 to 4.

When the engine was worked slowly, with the apparatus called the cataract, the effect produced by the fuel was found to be less, than when the engine was worked continuously with its full velocity, without any cessation between the strokes.

The cataract is a contrivance to open the injection-cock a certain number of times per minute, and thus regulate the number of strokes the engine shall make. The diminution of effect was a consequence of the loss of heat from the cylinder to the external air, which was more rapid during the cessation than at any other time, because the cylinder was then filled with steam.

Another great object of these experiments was, to determine the amount of the load, or resistance to the motion of the piston, under which it would produce the greatest effect.

The mechanical effect of the steam-engine, may be considered either in respect to the mechanical power exerted by a given engine, in a given time, or in respect to the expenditure of fuel, by which a given mechanical effect is produced. These are separate considerations, but both of them are important, and in particular cases, one or other of them may demand the preference, according to the circumstances.

For instance, if a mine is to be drained of water, and if, by becoming deeper and more extensive, its engine is scarcely equal to the task, the principal object will be to augment the power of the engine, by every possible means, without regard to the consumption of fuel. This is a very common case in coal mines, where fuel can be had free of expense. On the other hand, an engine for a lead mine, or a copper mine, may be fully equal to the task of draining the water, and yet the expense of fuel may be so great, as to take away the profit of the mine. In this case, the main consideration will be to obtain the greatest mechanical power, with the least consumption of fuel. In other cases, both considerations may operate.

Mr. Smeaton found that his small engine produced its maximum effect, both in respect to the power it exerted in a given time, and also in respect to the power exerted by a given consumption of fuel, when it was loaded with a resistance equal to that of a column of water of the same diameter as the cylinder, and 18 feet high ; or 7·81 lbs. per square inch. The motion of the piston was then $17\frac{1}{2}$ strokes per minute, of 3 feet length $= 52\frac{1}{2}$ ft. motion per min.

This he ascertained by adjusting the engine, so that it worked in its best manner, with any given load ; and then by adding or taking off one of the short lengths at the top of the pump barrel, the engine was obliged to raise the water so much higher, or the water ran off so much lower, as to alter the load considerably ; and all the other circumstances being kept the same, the comparison of the results, showed the advantage, or disadvantage, of the change of the load.

When the load was reduced from 7·9 lbs. per squ. inc., to 6·6 lbs., he found the power of the engine was lessened from 100 to 94 ; and the power exerted by a bushel of coals was lessened just in the same proportion.

When the load was still farther reduced from 6·6 lbs. to 5·5 lbs. per squ. inc., the power was diminished from 94 to only 82 ; and the produce from a bushel was lessened from 94 to 80.

On the other hand, when the load was increased from 7·8 lbs. to 8·8 lbs., the power was increased from 100 to 107 ; but the performance of a bushel of coals was lessened from 100 to 97.

When the load was still farther increased from 8 lbs. to 9·1, the power was diminished from 100 to 96, and the effect of a bushel of coals from 97 to 93.

These proportions are not of general application, but are in some measure peculiar to this engine, which had only a limited supply of steam on account of the small size of its boiler, and also because its cylinder condensed so much steam when it entered into the cylinder, that the piston could not move with a proportionably increased rapidity when the load was diminished ; for the loss by condensation, which recurred each stroke, was $2\frac{3}{4}$ times as great as the quantity of steam which really produced the effect of moving the piston ; hence, the rapidity of the motion was limited by the supply of steam from the boiler, rather than by the resistance opposed to the motion.

z

If the boiler had been larger, then the power of the engine would not have been so much diminished when the load was reduced, because the engine might have worked quicker, to have compensated for the lightness of its load. But in that case the diminution of the effect, from a given quantity of fuel, would have been still greater than above, because the condensation of the steam would have been increased in proportion to the increased number of strokes.

It was found advantageous to admit a continual leakage of air into the cylinder; for this purpose a small cock was inserted into the bottom part of the cylinder, and was opened as much as it could be to allow the piston to complete its working stroke. When this cock was shut, the engine required a greater counterweight to return the piston, or else it would move slower in the returning stroke, so that in either case the power of the engine was lessened, when it worked without this leakage of air; and the effect of a given quantity of fuel was lessened by shutting the air-cock, in the proportion of 12 to 11, and in other cases as 12 to 10.

The air thus admitted into the cylinder, would have a beneficial effect, to diminish the condensation of the steam against the internal surface of the cylinder, because the air would have a tendency to collect against that surface, and form a non-conducting lining to the cylinder, so as to keep the steam from actual contact with it.

For instance, when steam is admitted into the cold cylinder, which already contains a small quantity of air, that portion of air which is not expelled by the snifting, must mix with the entering steam as the piston ascends. The steam which comes in contact with the interior surface of the cylinder, is condensed, because its latent heat is absorbed by the metal of the cylinder; consequently that portion of the steam which loses its latent heat, contracts in volume, and returns to the state of liquid water, which adheres to the surface, in a very thin film. The condensation and absence of that steam, would leave a void space all round the interior surface, if it were not instantly filled up by fresh steam, advancing from the more central part of the cylinder, which is always kept supplied with fresh steam from the boiler; but the steam which so advances to the surface of the cylinder, being condensed by the contact therewith, more and more steam must continually radiate from the central parts to the circumference, until the cylinder becomes sufficiently heated, and then it will cease to condense.

During all this process, the air which is intermixed in the steam, must necessarily be carried along with it, from the centre to the circumference. But as the air will not be condensed by contact with the internal surface of the cylinder, it must accumulate against that surface, on the same principle that the water produced by the condensation of the steam, collects and accumulates there; and if the condensation continues long enough, the whole of the air in the cylinder, will at last be collected in contact with its interior surface, forming a complete lining between the metal, and that pure steam which has been supplied from the boiler without admixture of air, and which rises up in the central part of the cylinder and thence proceeds to the surface, as it is drawn thither by the absence of that steam which is condensed.

This may be familiarly illustrated by comparing the situation of the steam mixed with air, and introduced into a cold cylinder, with that of turbid water, containing particles of dirt, when poured into a porous vessel of stone or pottery, in order to be filtered and separated from its impurities, by the pure water leaking through the vessel at imperceptible pores, which are too minute to permit the particles of dirt, or even the finer particles of mud to pass. Now it is evident, that not only the matter which is so minutely diffused in the water, as merely to render it turbid, but also the grosser particles of dirt, which are floating in all parts of the vessel, will by degrees, be carried with the current of water, from the central parts of the vessel to the interior surface thereof; but as they cannot pass through with the fair water, they must be separated from the same, and be left collected on that surface, in a very thin scurf of mud or sediment, covered with a thicker layer of grosser dirt. And it is equally evident, that when these layers of mud and dirt have accumulated to a sensible thickness, they will prevent the water from coming in contact with the surface, and thereby impede or prevent the passage of more water.

The heat contained in the steam may be compared to the pure water in the filtering vessel, for it is mixed with water and with air, and the heat will readily leak or filter through the metal of the cylinder; but neither the water nor the air can pass, and must consequently be left on the interior surface, separated from the heat which has so passed. We may presume that the water of the steam is more intimately mixed or combined with the heat than the air is, because the steam is only half the specific gravity of air, and is probably divided and separated into more minute particles by its solution by the heat; hence, like the finer particles of the mud, which were dissolved in the turbid water, the water contained in the steam will be found collected in a thin layer, in close contact with the interior surface of the cylinder, the same as the mud would be, against the inside of the filtering vessel; and also the air, which was less intimately combined with the heat, will be found collected in a layer over the water, in the same manner as those grosser particles of dirt, which were less intimately mixed with the water, would cover the scurf of mud with a layer of dirt.

The necessity of admitting air into the receiver of an engine on Savery's system, has been already mentioned (p. 124), and in all cases, it may be concluded that if air is admitted along with steam, into a cold vessel, that the air will collect in contact with that cold surface, and tend to preserve the steam from the full extent of condensation, which would otherwise take place.

We may attempt an estimate of the quantity of air contained in the cylinder of the small engine in the following manner:

We have already supposed that the air increased the elasticity of the steam which remained in the cylinder, during the working stroke, from 2·4 lbs. to 3 lbs. per sq. inch. Therefore the elasticity of that air, was equal to ·6 of a pound, when retained under a compressure of 3 lbs. per sq. inch, in consequence of being mixed with steam of 2·4 lbs. elasticity. Under these circumstances, the air must have formed ·6 out of 2·4 ($= \frac{1}{4}$) of the density of all the steam in the cylinder. But at the end of the stroke, when all this air was accumulated into the space left at the bottom of the cylinder, and restored to its natural density, by the admission of steam of an elasticity equal to about 15 lbs. per sq. inch, the air would be under a compressure 5 times as great, and would therefore occupy only $\frac{1}{5}$ of the space it did when rarefied; that is only ($\frac{1}{5}$ of $\frac{1}{4}$ or) $\frac{1}{20}$th part of the contents of the cylinder.

Now if we assume that one-half the air would be expelled by the snifting, then it would leave a quantity of air equal to $\frac{1}{40}$th of the contents of the cylinder to be mixed with the steam. As the steam required to fill the cylinder, was 3¾ times, the mere capacity of the cylinder, it is obvious that the steam first admitted, and mixed with the air, would be condensed and renewed 3¾ times over; its absence being each time replaced by fresh steam from the boiler. Under these circumstances, according to the process before described, the whole of the air would most probably be accumulated on the interior surface of the cylinder.

The capacity of the cylinder being 2·34 cub. ft. $\frac{1}{40}$th would be ·0585 of a cub. ft. of air; and its internal surface being 16·11 sq. ft. the layer of air to be spread over that surface, would be (·0585 ÷ 16·11 =) ·00363 ft. (or × 12 =) ·04356 inc.; that is, $\frac{1}{23}$ of an inch thick.

The thickness of the film of water, or condensed steam, which would be left spread over the inside surface of the cylinder, would be $\frac{1}{181}$ part of an inch; that is $\frac{1}{1700}$ part of the thickness of the layer of steam, which was lost by condensation, and which has already been estimated at 4·78 inc.

As the advantage of admitting air into the cylinder, is only to diminish condensation, it is evident that in those engines which condense the greatest proportion of steam, the advantage of air will be greatest.

Note. Mr. Smeaton made a great many experiments with his experimental engine on the different kinds of coals, which were then used for fire-engines: the principal results were as follows:

The coals in proper sized pieces, termed round coals, were found superior to the small sized coals of the same quality, termed sleck, in the proportion of 100 to 80 nearly; so that the coals and the sleck being of the same quality, except as to the size of the pieces, the produce of the sleck was but four-fifths of that of the coals.

It is usual to wet the small coals with water, to make them cake together in masses on the fire; but this was found to diminish the effect in a small degree.

When the engine was worked with coke, it was found that the effect of the coke was $\frac{5}{6}$ of that of an equal weight of the same coals, from which the coke was made; and 66 lbs. of coke were obtained from 100 lbs. of coals.

Billets of ash-wood moderately dry, and split into pieces about 1 inc. diam., were found inferior to an equal weight of the common Yorkshire coals, called Halton coals, in proportion of 42 to 100; so that 42 pounds of that sort of coals, would do as much as 100 pounds of ash billets: 84 pounds weight of this ash wood, measured $3\frac{1}{2}$ cub. ft. in the wood stack.

The performance of the Newcastle coals of the kind called Team-top, were found superior to the common Yorkshire coals called Halton, in the proportion of 120 to 100. Cannel coal from Wake-field in Yorkshire, was superior to Halton as 133 to 100. Hage-moor coals, and Flockton coals, were superior to Halton as 130 to 100. Middleton-wood coals, and Welsh coals, were superior to Halton as 110 to 100. The coals called Berwick-moor, were inferior to Halton in the proportion of 86 to 100. In all these trials of different kinds of coals, the engine was kept in the same state.

Having now detailed the principal facts relative to the performance of a very small atmospheric engine, we may proceed to give similar particulars of a large engine, as a standard for practice; for it should be observed, that the performance of this small engine is an extreme case, which is very useful to point out principles and relative proportions, but is not to be depended upon for actual quantities.

Dimensions and Proportions of Mr. Smeaton's Atmospheric Engine at Long Benton. (See the description, p. 134, and plates II. and III.)

This was the standard by which Mr. Smeaton proportioned all his other engines. It performed extremely well; and being a medium size, between the largest and the smallest engines, its dimensions and proportions may be taken for a good example of Newcomen's invention, in its most perfect state.

Diameter of the cylinder 52 inc., and ($52 \times 52 = 2704$ circ. inc. $\times \cdot 7854 =$) 2124 sq. inc. or ($\div 144 =$) 14·75 sq. ft. area. The piston was regulated by the catch-pins to move $7\frac{1}{3}$ ft. and it worked 7 ft. stroke; therefore 103·25 cub. ft. of the space within the cylinder, were occupied by the motion of the piston at every stroke. It sometimes made $12\frac{3}{4}$ strokes per minute, but regularly 12; therefore, the piston had 84 ft. motion per min. when drawing water, without reckoning the returning stroke, which was also 84 ft. per min.; or the whole motion of the piston 168 ft. per min.

It worked two columns of pumps, 12·2 inc. bore or diameter, and each $24\frac{3}{4}$ fathoms (=148 ft.) lift, or 296 ft. for the height of the two columns. Their weight was ($12·2 \times 12·2 = 148·84$ circ. inc. $\times \cdot 341$ lbs. =) 50·75 lbs for every foot high; $\times 296$ ft. = 15 022 lbs. total weight.

The jack-head was a forcing-pump 7 inc. bore. It raised the water $70\frac{1}{2}$ ft. and made $5\frac{1}{2}$ ft. stroke. The weight of the column was ($7 \times 7 = 49 \times \cdot 341 = 16·7 \times 70·5 =$) 1177 lbs.; but as it moved only 5·5 ft. whilst the piston moved 7 ft. the resistance occasioned to the piston by this pump was only 925 lbs. It raised the water during the returning stroke.

The whole load of water, or resistance to the motion of the piston, was ($15\,022 + 925 =$) 15 947 lbs., which is more than 142 hundred-weight, or 7 tons 2 cwt. And 15 947 lbs. $\div$ 2124 sq. inc. area of the piston, is a little more than $7\frac{1}{2}$ lbs. effective load per sq. inch.

The quantity of water raised by the main pumps was ($12 \cdot 2 \times 12 \cdot 2 = 148 \cdot 84 \div 183 \cdot 3 = \cdot 81$ sq. ft. $\times 7$ ft. $= 5 \cdot 67 \times 12$ strokes $=$) 68 cub. ft. per min. raised 296 ft.; $= 20\,130$ cub. ft. raised 1 foot per min.; and $20\,130 \div 528$ cub. ft. (which can be raised 1 ft. per min. by a horse-power) $= 38 \cdot 15$ horse-power.   And the quantity raised by the jack-head pump was $17 \cdot 6$ cub. ft. per min. raised $70\frac{1}{2}$ ft.; $= 1241$ ($\div 528$) $= 2 \cdot 35$ horse-power for supplying the injection cistern.

The effective force of the engine was $40\frac{1}{2}$ HP ; that is estimating by the work actually performed, without considering the power lost by the friction of its parts, but including the power necessary to supply it with injection water.

The two furnaces consumed $8\frac{1}{2}$ bushels of Newcastle coals per hour.   Taking the weight of coals at 84 lbs. per bush. it was 714 lbs. per hour for $40\frac{1}{2}$ HP ; that is $17 \cdot 63$ lbs. per hour for each HP.   Or it consumed a bushel of coals in (60 $\div$ $8 \cdot 5 =$) $7 \cdot 06$ minutes $\times$ 12 strokes per min. $= 84 \cdot 7$ strokes $\times$ 7 ft. stroke $= 593$ ft. motion, with one bushel, and 593 strokes $\times$ by $15\,947$ lbs. load $= 9 \cdot 45$ millions pounds weight raised 1 foot high by the consumption of one bushel (or 84 lbs.) of coals.   This effect was exclusive of friction, but the power required to supply the injection water is included as part of the useful effect.

The evaporation of water from the two boilers was $1\frac{1}{2}$ cub. ft. per minute, or 90 cub. ft. per hour, by $8\frac{1}{2}$ bushels of coals $= 10 \cdot 58$ cub. ft. evaporated by each bushel, or 84 lbs., that is ($\cdot 126$ of a cub. ft. or) $7 \cdot 88$ lbs. of water, evaporated by each pound of coals consumed.

The boiler beneath the cylinder was 12 ft. diam. area $= 113$ sq. ft.; the other boiler $13\frac{1}{3}$ ft. diam. area $= 140$ sq. ft.; so that the area of their united surfaces of water was 253 sq. ft.; therefore $1 \cdot 5$ cub. ft. evaporated per minute from this surface, would sink it $\frac{1}{140}$ of a ft. $= \cdot 071$ inc.; that is, nearly $\frac{3}{4}$ of a tenth of an inch per min. or $4 \cdot 27$ inc. per hour evaporated.

Also 253 sq. ft. $\div 40 \cdot 5$ horse-power $= 6 \cdot 25$ sq. ft. of horizontal surface of water in the boilers to each HP.

The capacities of the two boilers were as follow : First, the domes or upper parts of the boilers which contained the steam, was equal, in the 12 ft. boiler, to a hemisphere of $12 \cdot 2$ ft. diam. $= 475$ cub. ft.; and the dome of the $13\frac{1}{3}$ ft. boiler was equal to a hemisphere $13 \cdot 6$ ft. diam. $= 650$ cub. ft. Therefore the space allowed in the domes of the two boilers to contain steam was 1125 cub. ft. for $40 \cdot 5$ HP $= 27 \cdot 75$ cub. ft. of steam space to each HP.

Again, the lower or water parts of the boilers were, in the 12 ft. boiler, $10 \cdot 25$ ft. diameter at the bottom, 12 ft. diameter at the water line, and $4 \cdot 5$ ft. mean depth of water $= 440$ cub. ft. of water. The lower part of the $13\frac{1}{3}$ ft. boiler was $11 \cdot 4$ ft. diam. at bottom, $13 \cdot 3$ ft. diam. at the water line, and $4 \cdot 5$ ft. mean depth of water $= 544$ cub. ft. of water.   Hence the quantity of water in both boilers was 984 cub. ft.

The total capacity of both boilers was 2109 cub. ft. to $40 \cdot 5$ HP $= 52$ cub. ft. of boiler space to each horse-power.

The surfaces at the bottoms of the boilers, which were exposed to the direct action of the flame ascending from the fuel, were as follow ;· Beneath the 12-ft. boiler, a circle 9 ft. diam. $= 63 \cdot 6$ squ. ft.; and beneath the $13\frac{1}{3}$-ft. boiler, a circle of 10 ft. diam. $= 78 \cdot 5$ squ. ft.   Total $142 \cdot 1$ squ. ft. of boiler bottom exposed to the direct action of the fire ; $\div 40 \cdot 5 = 3 \cdot 5$ sq. ft. for each horse-power.

N.B.   The bottoms were concave on the under sides; the crowns or centre parts of the circles rising 10 inches higher than the circumferences.   This circumstance, though it increased the surface, is not taken into account.

The external surfaces of the boilers, which were exposed to the spent flame circulating round them in their flues, were as follow ; Flue round the 12-ft. boiler, $4 \cdot 3$ ft. high ; circumference of the boiler, at a mean between the upper and lower part of the flue, $34 \cdot 8$ ft.; surface 150 squ. ft.   Flue round the $13\frac{1}{3}$-ft. boiler also $4 \cdot 3$ high, by $38 \cdot 8$ ft. mean circumference $= 167$ squ. ft. of surface. Total of surface exposed in the flues 317 sq. ft. $\div 40 \cdot 5$ horse-power $= 7 \cdot 83$ squ. ft. to a horse power.

The chimneys were 47 ft. high, from the fire grates to the tops where the smoke issued, and they were 2 ft. square within.

The fire grates were as follows ; A grate 4 ft. squ. $= 16$ squ. ft. beneath the 12-ft. boiler ; and another grate $4 \cdot 5$ ft. by $4 \cdot 25$ ft. beneath the $13\frac{1}{3}$-ft. boiler $= 19 \cdot 1$ squ. ft.   Total $35 \cdot 1$ squ. ft. $\div 40 \cdot 5$ horse-power $= \cdot 867$ of a squ. ft. of fire grate to each horse-power.   The height from the grate bars to the bottom of the boilers was 2 ft. in the centre, and 14 inc. at the circumference.

The two boilers evaporated $1 \cdot 5$ cub. ft. of water per minute with $11 \cdot 9$ lbs. of coals, burned upon $35 \cdot 1$ squ. ft. of fire grate, the flame acting upon $142 \cdot 1$ squ. ft.

of direct fire surface, beneath the boilers, and 317 squ. ft. of surface exposed in their flues, and 253 squ. ft. of horizontal surface of water.

It has been concluded from the results of many experiments, that 1 cub. ft. of water, when converted by heat into steam, will produce 1728 cub. ft. of steam, of the same elasticity as the air of the atmosphere, when the barometer stands at 30 inches; but as the steam in an engine must be a little stronger, we will assume that one measure of water produces 1700 measures of steam.

At this rate, the two boilers, which evaporated 1·5 cub. ft. of water per min. must have produced 2550 cub. ft. of steam per min.; and as this supplied 40·5 horse-power, it is = 62·95 cub. ft. per min. for each HP.

The space within the cylinder, which was occupied by the piston in its motion, has been already stated at 103·25 cub. ft. per stroke; and at the rate of 12 strokes per minute, this was = 1239 cub. ft. of steam per min. expended by the motion of the piston, and ÷ 40·5 horse-power = 30·58 cub. ft. of steam expended per min. for each horse-power.

But the cylinder was 9 ft. long within, although the piston only descended 7 ft. into it, leaving a useless space of 2 ft. (a) beneath the piston when at the bottom of its course. This space, added to the contents of the steam-pipe, (which was 8·5 inches diam.) contained 30·1 cub. ft. which was lost every stroke, that is = 361 cub. ft. of steam expended per min. without effect.

In this way (1239 + 361 =) 1600 cub. ft. of steam per min. are accounted for: and if the above estimate of 2550 cub. ft. for the quantity produced by both boilers, be correct, then 950 cub. ft. per min. must have disappeared by leakage, and by condensation upon the cold interior surface of the cylinder and piston, when the steam was admitted into it after every stroke. As the leakage must have been very inconsiderable, nearly all this loss must be attributed to condensation.

The internal surface of the cylinder, to which the steam was exposed, was (52 inc. diam. =) 13·61 ft. circumference × 9 ft. long = 122·5 squ. ft. The piston and cylinder bottom were 14·75 squ. ft. each, or 29·5 squ. ft.; and to this, adding the surface of the steam-pipe, and injection and eduction-pipes within the cylinder, the whole is 168 squ. ft. of internal surface, producing the condensation of 950 cub. ft. per min.; therefore each squ. ft. condenses 5·66 cub. ft. of steam per min.: and as the engine made 12 strokes per min. it is (5·66 ÷ 12 =) ·472 cub. ft. condensed per stroke, by each squ. foot of internal surface.

The account of the expenditure of steam per minute is as follows in cubic feet.

	By 40½ horse-power.	By one horse-power.
Steam expended by the motion of the piston	1239	30·58
Steam wasted to fill the extra space in the cylinder	361	8·92
Steam condensed by 168 squ. ft. of internal surface	950	23·45
Steam produced by the boilers	2550	62·95

Note. The extra space at the bottom of the cylinder was double what was requisite. If the engine had been properly constructed, the waste would have been only 180 cub. ft. per min. or 1·4 per horse-power.

The quantity of steam required to fill the cylinder each stroke, was (2550 ÷ 1600 =) 1·594 times the mere capacity of the cylinder.

If we suppose that the interior of the cylinder was cooled down to a temperature of about 152 deg. of Fahrenheit, every time the injection was thrown in to make the working stroke, then the elasticity of the vapour or steam remaining

(a) *Note.* The piston of this engine might have worked with an 8-ft. stroke, and then only half the quantity of steam would have been lost at the bottom.

The cylinder at Long Benton was made with a flat bottom, being an old cylinder. Mr. Smeaton's plan was to make a hemispherical basin at the bottom of the cylinder, as is represented in the Plate.

Also the injection at Long Benton was à forcing-pump, and raised the water during the return-ing stroke of the piston; but a lifting-pump, as is represented in the plate, was most commonly used in Mr. Smeaton's engines.

in the cylinder, would have been nearly 4 lbs. per sq. inch; but to allow for the small portion of air, which might get into the cylinder, say $4\frac{1}{4}$ lbs. in which case, the piston would have been borne upwards by that force, whilst it was borne downwards by the pressure of the atmosphere, acting with $14\frac{3}{4}$ lbs. per sq. inch on its upper surface; so that the real descending force of the piston would have been about $10\frac{1}{2}$ lbs. per sq. inch, which, on 2124 square inches, would be 22 300 lbs.

The mass of matter to be put in motion in this engine was very considerable, because the pumps did not raise the water to the surface of the ground, but discharged it into an adit or level, by which the water ran off at a great depth below the surface.   (See Plate IV.)   In consequence, there was an extra length of $57\frac{1}{2}$ fathoms of the main spear or rod, to reach down from the great lever of the engine to the lowest of the two pumps; this rod was 6 inc. squ., and with its iron joints and bolts, weighed about $2\frac{1}{4}$ tons.

To this main rod were suspended two other smaller rods, which passed down through the water in the two pumps, and had the buckets thereof affixed to their lower ends respectively.   These rods were five inches square, and the two measured 44 fath. in length; their weight, including iron work, was about two tons.

The weight of all these wooden rods, with their iron straps and bolts to unite the different pieces, the two pump-buckets and their iron shanks, together with the main chains at the arch head of the great lever, amounted to about 5 tons, or 11 200 lbs. suspended at the pump end of the great lever.   But the lever was relieved of a part of this weight, by the immersion of the two small rods in the water in the pumps(b).   These rods were 5 inc. squ. and about 44 fath. = 264 ft. length.   The weight of water they displaced would be (5 × 5 = 25 squ. inc. × ·434 lbs. = 10· 85 lbs. per ft. × 264 ft. =) 2864 lbs.   This weight deducted from 11200 lbs. leaves 8336 lbs. actually weighing down at the outer end of the great lever.

To counterbalance this weight, the opposite end of the great lever sustained the weight of the main chains at the arch head, and the piston, with its iron shank and appurtenances, amounting to about 3035 lbs.; and also an iron balance-weight of 1721 lbs. was loaded on the piston, to counterpoise so much more of the weight of the pump-work, as to leave a preponderance of about 3580 lbs. to the pump end of the lever, and 925 lbs. of this was required to depress the forcer of the injection-pump, and the remaining 2655 lbs. was to draw up the piston when the steam was admitted into the cylinder.   This latter part of the preponderating weight, or counterweight, was at the rate of $1\frac{1}{4}$ lb. for each squ. inc. of the piston.

The mass of matter in the great lever, being put in motion and stopped every stroke, must also be taken into account in estimating the quantity of matter in motion.   The two ends of the lever being equally heavy and in equilibrium, its weight would have no effect on the balance or equipoise of the engine, when it was standing still; but the inertia of so great a mass would be considerable when it was passing from rest into motion, and *vice versa*.

The great lever contained about 170 cub. ft. of timber; and, with all its iron work, together with the plug-beam, and the rod and bucket of the injection-pump, must have weighed about four tons, or two tons for each half of its length from the centre.

The energy of a lever, which moves about a centre, at one of its ends (see page 33,) is equal to a mass of one-third of its own weight, moving with the same velocity as its extreme end: therefore, if we suppose that two-thirds of a ton, or 1490 lbs. had been added to the weight which was suspended at each end of the lever, those two weights would have balanced each other when at rest, but when in motion they would have formed an equivalent for the weight of the lever itself.

The total weight of all the moving parts of the engine was equivalent to 18 936 lbs. or 169 cwt. or 8· 45 tons, without including the columns of water,

(b) The effect of the immersion of the rods in the water is, in all cases, to diminish their weight as much as the weight of all the water which they displace.   For instance: when the engine makes its working stroke, and draws up the pump-rods in the pumps, the water which surrounds them rises with them, and therefore the weight of the water actually raised by the engine, is less than the whole weight of the columns of water contained in those pumps, by all the weight of that water which is displaced by the rods: for, in fact, the weight raised by the engine, is partly that of the rods, and partly that of the water in which they are immersed.

And in like manner, when the rods descend into the pumps, the engine is relieved of as much of their weight, as the weight of all the water displaced by them.

which weighed 15 947 lbs., 142·4 cwt. or 7·12 tons. Hence the weight of the moving parts was 1· 19 times the weight of the column of water ; and 18 936 lbs. of moving parts ÷ 2124 squ. inc. area of piston is 8· 91 lbs. of dead weight, for each squ. inch of the piston.

The account of the different opposing forces in this engine may be thus stated :

Balance of the Forces during the Working-stroke.

Forces at the cylinder end of the great lever.		Forces at the pump end of the great lever.	
	pounds.		pounds.
Weight of the piston with its appurtenances - - -	3035	The pump-rods. { Their absolute weight - 11200 Lost by immersion - 2864 } 8336	
Balance weight to the pump-rods	1721	Operative weight - - 8336	
Equivalent for half the weight of the great lever - -	1490	Equivalent for half the weight of the great lever - -	1490
The unbalanced pressure of the atmosphere 10· 5 lbs. per squ. inc. on 2124 squ. inc. - -	22300	Column of water in the main pumps omitting the injection -	15022
Total	28546	Total	24848

The excess of force at the cylinder end, or preponderance, to overcome friction, and produce motion in the working stroke, is 3698 lbs. which is 1· 74 lbs. per squ. inc. of the piston. Weight of matter moved 31102 lbs.

To find the total weight of matter which was put in motion, the weights suspended at both ends of the lever must be added together ; except that the unbalanced pressure of the atmosphere on the piston is not to be included, but only the actual weight of the 103· 25. cub. ft. of air, which entered into the cylinder, when the piston descended ; and this weight is not quite 8 lbs. In the working stroke, the weight of the pump-rods is diminished by all that they lost by immersion in the water ; because that part of their weight was included in the weight of the column of water. The injection column being at rest forms no part of the weight of matter to be moved during the working stroke.

Balance of the Forces during the Returning-stroke.

Forces at the cylinder end of the great lever.		Forces at the pump end of the great lever.	
	pounds.		pounds.
Weight of the piston and its appurtenances - - -	3035	The pump-rods. { Operative weight as before 8336 Deduct the column of the injection-pump - 925 } 7411	
Balance weight to the pump-rods -	1721		
Equivalent for half the weight of the lever - -	1490	Equivalent for half the weight of the great lever	1490
Total	6246	Total	8901

The counterweight at the pump end, or preponderance, to overcome friction, and produce motion in the returning stroke, is 2655 lbs. = 1· 25 lbs. per squ. inc. Weight of matter moved 20585 lbs.

In the returning stroke, the whole weight of the rods must be considered as a part of the moving mass, and an addition must also be made, for that portion of the water in the pumps, which was displaced and put in motion by the rods, when they descended through it ; as the motion given to this water was much less rapid than that of the rods themselves, it will be sufficient to take (one-fourth of the 2864 lbs. of water displaced =) 716 lbs. for the mass of water moved. We may allow 8 lbs. for the air, as before, and also the injection column forms a part of the weight of matter moved during the returning stroke.

The preponderating force of the piston may be estimated at 3698 lbs. during the working stroke, and this force gave motion to a mass of 31 102, or 8· 42 times its own weight ; and in the returning stroke, 2655 lbs. preponderance gave motion to a mass of 20 585 lbs. or 7· 75 times its own weight : but it was only at the first commencement of each motion, that these preponderating forces could operate with their full effect, because the friction and the resistance of the water in the pumps tended to retard the motion, from the time it began, and these resistances increased with the motion.

It is difficult to estimate the amount of force required to overcome the friction and resistance of all the moving parts of the engine, and also to put the mass of matter in motion : but the following computation is an approximation, which will give some idea of what it must have been (c).

The greatest friction was that of the piston and of the pump buckets. The circumference of the piston (52 inc. diam.) was 13· 61 ft., and if the hemp packing had been 4 inc. deep, the surface of hemp which rubbed against the internal surface of the cylinder, would have been nearly 4½ squ. ft. If we assume the resistance by each squ. foot of surface at 144 lbs. the friction of the piston would have been 648 lbs.; that is, it would have required 648 lbs. to have moved it down in the cylinder, with the velocity it actually moved with, in working.   In the returning stroke the friction would be less, from the diminished pressure—say 600 lbs.

The circumferences of the leathered pump buckets being (12· 2 inc. diam. =) 38⅓ inc. circumf. and that of the injection pump bucket (7 inc. diam. =) 22 inc. circumf. the whole would be 8¼ ft.; and if the leathers had been about three inches deep, the surface of wet leather rubbing against the interior surfaces of the three pump barrels would have been about 2 squ. ft.

The friction of these buckets must have been much greater during the working stroke, when the leathers were forcibly pressed against the inside of the barrels, than in the returning stroke ; therefore, suppose the friction during the working stroke to have been 288 lbs. per squ. ft., but in the returning stroke only half as much, this will give 576 lbs. for the friction of the pumps in the working stroke, and 288 lbs. in the returning stroke.

To allow for the friction of the joints of the great chains, the gudgeons of the great lever, the working gear, and the rod of the injection-pump, suppose 200 lbs. in the working stroke, and 150 lbs. in the returning stroke.

The next resistance to be considered, is that of the water passing through the pipes of the pumps ; this is greatly influenced by the velocity of the motion.   The piston (making 12 strokes per min. of 7 ft. or one complete stroke in 5 seconds) must have passed through 168 ft. per min.   But allowing half a second to have been lost at every change, the motion would have been 7 ft. in 2 seconds. Therefore we will assume the velocity of the water in the pumps to have been at the rate of 210 per min.   It appears, from experiments made by Mr. Smeaton, that the pressure of 6· 8 inc. perpendicular column of water, will cause water to run with a velocity of 210 ft. per min. through a pipe 12 inc. diam. and 100 ft. long.   As the two pumps were 296 ft. high, they would require 20 inc. column of 12· 2 in diam.; the weight of which, = 84½ lbs. is the friction of the water moving through the two great pumps.

The ascending pipe of the injection-pump was 5 inc. diam. and 70½ ft. high.   The pump bucket made a 5¼ ft. stroke in 2 sec. or 165 ft. per min.; and the pump-barrel, being 7 inc. diam. (= 49 circ. inc.) whilst the pipe was only 5 inc. diam. (= 25 circ. inc.) the water therein must have moved with a velocity of (⁴⁹⁄₂₅ of 165 =) 323 ft. per min.   To force water with this velocity, through a pipe 5 inc. diam. would require a column of about 34 inc. for 100 ft. long, or 24 inc. for 70½ ft. long ; hence the weight of a column, 24 inc. high, and 5 inc. diam. = 17 lbs. is the resistance of the injection water in passing through its pipe.

The water must have met with a considerable resistance in passing through the narrow apertures of the valves of the great pumps, because the uninterrupted area of those apertures, not being more than ¼ of the area of the pump barrels, the water must have rushed through them, with a velocity of (4 × 210 ft. =) 840 ft. per min.   To urge water with that velocity through a narrow aperture re-

---

(c) It is one of the most difficult problems in mechanics, to determine the motion which an atmospheric engine will have under different circumstances.   A mathematical investigation of the theory of its motions has been attempted by Mr. Bossut, in his Hydrodynamique, but without success, and it has been carried farther by Dr. Robison, in the Encyclopædia Britannica.   Mr. Watt, who revised a modern edition of this memoir, observes, that Dr. Robison's calculations, being founded on principles not sufficiently known to be subjected to rule, are in a great degree unnecessary, but that they contain much ingenious reasoning, and may lead to the formation of more correct formulæ.

quires a column of nearly 87 inc. high; but as the area of this rapid stream of water was only $\frac{1}{4}$ of the area of the barrel, the resistance it occasioned would be equal to the weight of ($\frac{1}{4}$ of 87 =) 22 inc. of the main column of 12·2 inc. diam. = 93 lbs. for the valves of each of the great pumps, or 186 lbs. for both.

The resistance, occasioned by the valves of the injection pump, would be 19 lbs. Thus, 165 ft. motion × 4 = 660 ft. per min. velocity; the column to produce this is 54 inc. $\frac{1}{4}$ of which = 13·5 inc. of the column, 7 inc. diam. weight 19 lbs.

The resistance the water must have met with in entering by suction, at the small holes at the bottom of the pumps, may be taken at one-seventh of that of passing through the valves, or 29 lbs.

In the returning stroke, the water in the pumps was motionless, but as the rods passed down through the water in the pumps, they occasioned some resistance, which may be put down at 97 lbs.

The amount of the several resistances may be summed up as follows:—

	Resistances.		In the working stroke.	In the returning stroke.
Friction of the moving parts.	The great piston, hemp, 4½ sq. ft.	.	648	600
	The three pump buckets, leather, 2 sq. ft.		576	288
	The joints, working gear rods, &c.	.	200	150
Resistance of the water.	In the great pump	. . .	85 ⎫ 102	97
	In the injection pump	. .	17 ⎭	
	Valves of the great pump	. .	186 ⎫ 205	205
	Do. of injection pump	. .	19 ⎭	
	Suction of the water	. .	29	00
	Total	. .	1760 lbs.	1340 lbs.

The sum of the resistances both ways may be estimated at (1760 + 1340 =) 3100 lbs.; and this, with a motion of 84 ft. per minute, would be equal to 260 400 lbs. moved 1 ft. per min.; and ÷ 33 000 lbs. gives 7·9 horse-power, supposed to be expended in overcoming these resistances. It would be almost $\frac{1}{5}$ of (40½ HP, which is) the effective power of the engine.

The resistance during the working stroke is estimated at 1760 lbs. If this is deducted from the preponderating force of 3698 lbs. it leaves 1938 lbs. of unbalanced force remaining to continue the motion of the moving mass, which is 31 102 lbs. or 16·06 times the unbalanced force (a).

And during the returning stroke, 1340 lbs. resistance, deducted from 2655 lbs. preponderance, leaves 1315 lbs. of unbalanced force to continue the motion of the mass, which is 20 585 lbs. or 15·64 times the unbalanced force.

If this estimation is correct, it appears that the working stroke in this engine was produced by a preponderating force, equal at the commencement to 3698 lbs. or between $\frac{1}{8}$ and $\frac{1}{9}$ part of the weight of the matter to be put in motion. But during the motion so produced, the various resistances which opposed themselves to its continuance, reduced the unbalanced preponderating force to 1938 lbs., which is $\frac{1}{16}$ part of the moving weight.

And in the returning stroke the preponderance at the commencement was 2655 lbs. or rather more than $\frac{1}{8}$ of the weight to be moved; but after the motion was produced, the resistances reduced it to 1315 lbs., which is rather more than $\frac{1}{15}$ of the moving weight.

The motion of the piston being 12 strokes per minute of 7 ft. one stroke was completed in 5 seconds; and assuming that an interval of $\frac{1}{2}$ second was lost at

(a) This and the two following conclusions are founded on the assumption, that the preponderating force of 3698 lbs. or 2655 lbs. remained constantly the same, during all the progress of the motion. The preponderating force has been inferred from the temperature of the water in the hot-well, assuming the elasticity of the steam remaining in the cylinder, to have been correspondent to that temperature. This statement must undoubtedly be a correct mean of that elasticity, and preponderance, during the whole descent, but it is uncertain whether it continued to be the same, during all the stroke, or whether it varied.

every change of motion, at the top or at the bottom of the stroke, then the piston would have moved through 7 ft. in two seconds. But it would be rather less than 2 sec. for the working stroke, and rather more for the returning stroke, because it is usual to adjust an engine so as to perform the working stroke with a quicker motion, than the returning stroke, and the estimate shows that the greatest proportion of unbalanced force was in the returning stroke.

The motion of the piston, during the working stroke, appears to accelerate from the time it starts, till it finishes its motion; but it is difficult to judge of the fact, and it is more probable that it accelerates during the first two-thirds, and then becomes uniform, during the latter third of the descent. In the returning stroke it can be seen, that the motion very soon acquires a uniform velocity, probably when it has gone through only one-fifth of the stroke.

According to theory (b), the motion of the piston, under the circumstances above stated, would have been accelerated, from the beginning to the end of the stroke, because more than half of the impelling force continued to act on the moving mass, after the motion was produced. The result would have been an accelerated motion, though not uniformly accelerated (see p. 21). But there are other circumstances, which operated to modify that result.

During the working stroke, the piston descended with an accelerating motion, during a great part of its course; but the injection-cock being closed gradually, diminished the flow of cold water, when the piston approached towards the bottom of its course (see p. 149.) Consequently the vapour, or rare steam, remaining in the small space within the cylinder, beneath the piston, must have increased very rapidly in temperature and density, when the piston descended near to the bottom; and this steam becoming denser and denser, opposed a resistance, if not to the continuance of the motion, at least to the continuance of the acceleration, which had taken place from the commencement of the descent, and which continued until the flow of cold water was diminished; but after that, the motion approached more to uniformity, and it might even begin to be retarded, before the end of the stroke.

In fact the motion must in all cases be retarded, when it comes very near to the end of the stroke, because in ordinary working, the piston does not stop absolutely at once, as it does when the catch-pins strike on the spring beams by accident, but the motion terminates by very rapid retardation, till it ceases quite.

The pins in the working-plug are regulated by experiment, so that they shut the injection-cock, and open the regulator some space before the piston arrives at the end of its course, and the impetus of the moving matter, continues the motion to the full limits of the course. It is during this termination of the course of the piston, after the steam is admitted into the bottom of the cylinder, that the steam operates to heat the interior surface. and produce the snifting (see p. 142).

The great circumstance which regulates the motion of the piston during the returning stroke, is the supply of steam through the steam-pipe from the boiler, compared with the demand for steam in the cylinder, to supply the space left by the ascent of the piston, and by the condensation of steam by its internal surface.

(b) A heavy body falling in vacuo, would descend $64\frac{1}{3}$ feet in two seconds (see p. 23.) As the piston of this engine moved seven feet in two seconds, it was nearly $\frac{1}{9}$ of the motion of a falling body. According to the preceding estimate, the preponderating or impelling force tending to urge the mass of matter into motion, was more than $\frac{1}{9}$ part of the weight of that mass; but as the various resistances to the motion reduced that impelling force to only $\frac{1}{16}$, when the motion was acquired, it ought to have been expected that the rapidity of the motion of the piston would have been less than $\frac{1}{9}$ of that of a falling body.

Accordingly the ascending motion of the piston accelerates very rapidly, till it acquires the full motion which the supply of steam will permit, and then it must continue uniform.

Hence, during the working stroke, the law of the accelerating motion of the piston is influenced by the diminution of the preponderating force of the piston, towards the conclusion of the stroke, owing to the cooling or condensation of the steam not being completely kept up. And during the returning stroke, the motion is influenced by the constant and nearly uniform resistance, of drawing or sucking in the steam through the steam-pipe.

These circumstances are too complicated to be subjected to calculation, unless we had a more exact knowledge of the state of the condensation within the cylinder, during the descent of the piston. In the preceding estimate, the state of the steam remaining in the cylinder, has been inferred from the single fact of the temperature of the hot-well; this may be taken as a fair indication of the internal temperature of the cylinder at a mean of the whole stroke, but it does not inform us what it was, at each different stage of the motion of the piston.

The steam regulator is closed before the piston arrives at the top of its course, and the impetus of the moving matter continues the motion to its full limits: the steam in the cylinder is therefore expanded a little, and in so doing, the motion of the piston is resisted, and it is brought quietly to rest, without making the catch-pins strike; when the ascending course is very nearly completed, the injection-cock is opened and the condensation begins.

The quantity of cold water required to produce the condensation of the steam, may be estimated as follows:

Assuming the injection-water to have been at the temperature of 52 degrees, and the water in the hot-well, to have been 152 deg., then that water would have gained 100 deg. from the steam. Suppose the temperature of the steam to have been 215 deg. in the cylinder, it would also contain as much heat in a latent state, as would raise the temperature of 960 times its weight of water one degree. Therefore, all the heat of the steam may be represented by $(215 + 960 =) 1175$; but as it was cooled down to 152 deg. it must have lost $(1175 - 152 =) 1023$ deg.

A sufficient quantity of water must have been injected to have absorbed all this heat in raising its own temperature 100 deg. The quantity would be $(1023 \div 100 =) 10 \cdot 23$ times the quantity of water contained in the steam. The water evaporated being at the rate of $1 \cdot 5$ cub. ft. per minute, the quantity of water injected must have been $(1 \cdot 5 \times 10 \cdot 23 =) 15 \cdot 35$ cub. ft. per minute. The injection-pump raised $17 \cdot 6$ cub. ft. per minute, which was therefore an ample allowance.

The aperture through which the injection-jet spouted, was a square of $1 \cdot 19$ inc., the area of which was $1 \cdot 42$ square inches. The perpendicular height from this orifice, to the surface of the water in the injection-cistern was $29\frac{1}{4}$ feet, but the velocity of the injection was increased, as soon as the condensation of the steam took place, because the pressure of the atmosphere on the surface of the water in the cistern, would not be opposed by an equal counteracting pressure within the cylinder, and also the jet would be spouting in vacuo. The unbalanced atmospheric pressure was $10 \cdot 5$ lbs. per squ. inc., which $(\div \cdot 434$ lbs.$)$ is equal to a column of $24 \cdot 2$ ft. of water; and this, added to $29 \cdot 25$ ft., gives $53 \cdot 45$ ft. column to produce the jet. A heavy body falling from this height, would acquire a velocity of $58 \cdot 6$ ft. per second (see p. 23), and with this velocity, the discharge through the aperture of $1 \cdot 42$ squ. inc., would be $\cdot 578$ cub. ft. per second.

The injection-cock was opened a little before the piston arrived at the top of its course, and it was closed before the piston reached the bottom, and the whole period of the descent (including the loss of time at the top of the stroke) being $2\frac{1}{2}$ seconds, we may suppose the cock to have been open during 2 seconds. Therefore the discharge would be $(\cdot 578 \times 2 =) 1 \cdot 1 5$ cub. ft. per stroke, or $(\times 12$ strokes per min. $=) 13 \cdot 8$ cub. ft. per min., which is rather less than the previous calculation of the quantity required for condensation.

The aperture for the injection-spout was a square hole, with one of the sides moveable as a slider ; and, by means of a screw, which came through the metal of the cylinder to the outside, the size of the aperture could be regulated, according to the quantity of injection water required in different seasons, and in different states of the engine.

We may conclude this statement of the actual performance of Mr. Smeaton's engine, by observing that the preponderating or impelling force, which appears to have remained unbalanced, after all the resistances to motion were overcome, was not a total loss of power ; because this force (amounting to 1938 lbs. in the working stroke, and 1315 lbs. in the returning stroke) was employed in putting the moving mass into motion, but then the impetus of that moving mass was sufficient to continue the motion, after the impelling force had become too weak to have so continued it.

At the top of the stroke, the preponderance was much greater than the resistance to be overcome, and therefore it urged the parts into rapid motion ; but towards the conclusion of the stroke, the preponderance was diminished, so as to become less than the resistance ; but then the impetus of the moving matter continued the motion, till that impetus was all expended, and thus the moving mass was brought to rest.

It was by this means that the piston finished its stroke both ways, without causing the catch-pins to strike the spring-beams. The pins in the plug-beam being regulated, so that the inherent motion would be completely expended, in carrying the piston to its exact length of 7 ft. stroke, although the regulator and injection-cock were moved, a little before the piston arrived at the termination of its course.

Hence, the excess of impellent force which was given to the piston, beyond what was absolutely requisite to overcome the resistances to its motion, was made available to prolong that motion, beyond the point at which the impellent force ceased to operate ; so that the real loss, or expenditure of power, to produce the motion of the engine, was the sum of the resistances already estimated, or nearly eight horse power, which is less than $\frac{1}{5}$ of the effective power of the engine.

This engine was well proportioned, except in the circumstance of having so much needless space left at the bottom of the cylinder, when the piston was at the lowest part of its course. If the cylinder had been made one foot shorter, the piston would not have approached within one foot of the bottom, which is an ample allowance to avoid striking by accident. In that case, 14·75 cubic feet of steam would have been saved per stroke, or 177 cubic feet per minute ; also, there would have been 13·61 square feet less of internal surface, condensing (5·66 cub. ft. per square ft. or) 77 cubic ft. per minute. On the whole 254 cub. ft. of steam would have been saved out of 2550 per minute, or very nearly $\frac{1}{10}$; therefore $\frac{9}{10}$ of the fuel, and boiler's feed, and injection-water, would have produced just the same effects as above stated.

At this rate an atmospheric steam-engine of 40 horse power, on Mr. Smeaton's plan, would have consumed 7·56 bushels (= 635 lbs.) of coals per hour, or 15·87 lbs. per hour, per HP ; and it would have raised $10\frac{1}{2}$ millions pounds one foot high with a bushel of coals. The water evaporated from its boilers would have been $1\frac{1}{3}$ cubic feet per minute. And the water required for injection about $10\frac{1}{4}$ times as much, or $13\frac{2}{3}$ cubic feet per minute.

The engine at Long Benton also had a greater encumbrance of pump-rods, than those engines which only draw their water to the surface ; for the first 57 fathoms of the large rod, to reach down to the pumps, together with its counterbalance weight, was an extra dead load on the engine of $2\frac{1}{2}$ tons.

It only remains to state the principal dimensions of the moving parts of this engine. The drawings, Plates II. and III. being exact to their respective scales, will give the dimensions of the building (c).

The cylinder beams F F, were 4 feet deep, by 1 foot broad, and their length of bearing between the side walls of the house, $12\frac{1}{3}$ feet. The great lever was 22 ft. long, from the centre of the piston-rod to that of the pump-rod. The dimensions of the timber was 50 inches deep, by 24 inches wide, at the centre, and 40 inc. by 24 inc. at the ends. It was composed of 12 pieces, viz. 6 one above the other, as seen in the elevation, and two such sets, side by side. The whole was bound together by two iron straps at the centre, and twelve iron bolts at each end, at the places shown by the dotted lines; and the pieces were further secured by keys, inserted into mortises, cut out between the joints, as shown in the drawing.

The axis or centre of the lever, was of cast-iron, 5 inc. thick, by 18 inc. broad, in the part between the timbers, and the cylindrical ends or gudgeons were 7 inc. diameter and 7 inc. long. They rested in brass bearings, fitted into large wooden blocks, which were worked in with the masonry of the lever wall. The length of the axis between the bearings was 3 feet.

The whole weight which was poised upon the gudgeons of the axis, was 60 386 lbs. or 27 tons during the working stroke; and 21 802 lbs. or $9\frac{3}{4}$ tons, in the returning stroke.

The iron shank or rod n of the piston, was composed of three rods, each two inches square; the iron pin of the joint, by which they were jointed to the main chains K at o, was $2\frac{3}{4}$ inc. diam. The middle links of each chain, were wrought iron, $1\frac{1}{2}$ inc. by 2 inc.; and two of these links, (containing 6 squ. inc. of section,) bore the whole load, which during the working stroke was 27 056 lbs. or 12 tons. The joint pins of the links of the chains were $1\frac{3}{4}$ inc. diam.

The rod which worked the injection-pump, was connected with the great lever, at $8\frac{3}{4}$ ft. distance from the common centre of motion, the piston and main spear M being suspended at 11 ft. distance therefrom; consequently the pump made a 5-ft. stroke, when the piston, and main pumps, made 7 ft. stroke. It was a forcing-pump with a solid piston, and the barrel was open at top; it raised the water during the returning stroke.

The engine was provided with a small air-cock, inserted into the bottom part of the cylinder, or into the eduction-pipe, and left constantly open, to allow a little leakage of air into the cylinder, with a view of diminishing the condensation of the steam against the internal surface of the cylinder, as before explained. But the advantage of thus admitting air, is not so marked in a large engine, as in a small one, as before stated. In general, an engine which works constantly, becomes so far leaky at the joints of the cylinder bottom, and steam-pipe, as to admit as much air into the cylinder as it ought to have, and then the air-cock may be kept shut; but it is useful to a new engine, when in good order, to open it.

Having now stated every requisite particular of Mr. Smeaton's standard engine, the proper proportions for similar engines of all sizes, are exhibited in the following table.

(c) *Note.* The drawings Plates II. and III. were reduced by the author, from Mr. Smeaton's original plans from which the engine at Long Benton was constructed.

Note. In the drawing, the injection-pump R is represented as a lifting-pump, with a collar of leather at the top of the barrel, for the rod to pass through; such a pump would raise the water during the working stroke. The rod of the pump is also represented as being much nearer to the centre of motion of the great lever, than it really was, because the pump is supposed to be placed in a separate pit S, into which the cold water for the injection is collected: in this case, as the pump would make a shorter stroke, it would require to be proportionably larger in diameter than has been stated.

In the Long Benton engine, the injection-pump was placed in the great pit of the mine, near to the main spear M, and it drew the water from a cistern placed in that pit; the pump itself was placed so low down in the pit, that no part of it would have been seen in the figure, if it had been represented as it really was. The top of the pump barrel should never be more than 12 feet above the surface of the water it is to draw from. The forcing-pump was some advantage in the engine at Long Benton, because the weight of its column, which was 925 lbs. on the piston, counterbalanced so much of the weight of the main spear; whereas if the injection-pump had been a lifting-pump such as is represented in the drawing, and which is the ordinary construction, then the balance weight upon the piston must have been 925 lbs. heavier than it was (p. 176,) or 2646 lbs. instead of 1721 lbs.; and also, as the piston must have raised the column, during the working stroke, the moving mass would have been 1850 lbs. greater during the working stroke, with a lifting injection-pump, than with a forcing-pump.

Mr. Smeaton's Table for the Proportions of the parts of Newcomen's Engines, deduced from actual experiments, 1772.

Horse Power.	Cylinder.			Motion of the Piston.			Boilers.				Injection Water.						Power.		
33 000 lbs. 1 foot per min.	Diameter.	Area.		No. per minute.	Length of Stroke.	Motion per minute.	Centre Boiler, Diameter.	Diam. of the outside boilers.	Fire Surface.	Diameter of Steam-Pipe.	Square Aperture.	Height of the column.	Quantity. Per minute.	Quantity. Per stroke.	Boiler's Feed per stroke.	Newcastle Coals per hour.	Pumpage, or load on the piston.	Great Product per minute, or quantity of water raised 1 foot per min.	Effect per minute of 1 Bushel per hour.
	Inches.	Circ. Inc.	Sq. Inc.		Ft. In.	Feet.	Feet.		Sq. Ft.	Inches.	Inches.	Feet.	Cub. Ft.	Cy. I. Ft.	Cub. In.	Bushels.	Cyl. In. Ft.	Cyl. Inc. Ft.	Cyl. In. Ft.
1⅕	10	100	78	16½	3 ,, 10	63·3	5½		30	2·2	·28	14	1·2	13	9·3	·6	1800	113 940	200 000
1¾	12	144	113	16½	4 ,, 0	66·0	6		37½	2·5	·33	16	1·6	18	12·7	·7	2592	171 072	231 000
2½	14	196	154	16¼	4 ,, 2	67·7	6½		44	2·8	·37	18	2·0	23	15·9	·9	3528	238 881	263 000
3⅓	16	256	201	16	4 ,, 4	69·3	7		51	3·0	·42	20	2·4	28	19·7	1·1	4608	319 488	288 000
4¼	18	324	254	15¾	4 ,, 6	70·9	7½		58½	3·3	·47	20	2·9	34	23·9	1·3	5832	413 372	311 000
5⅓	20	400	314	15½	4 ,, 8	72·3	8		66½	3·6	·51	22	3·5	41	28·5	1·6	7200	520 800	334 000
6⅔	22	484	380	15¼	4 ,, 10	73·7	8½		75½	3·9	·56	22	4·0	48	33·6	1·8	8712	642 162	355 000
8	24	576	452	15	5 ,, 0	75·0	9		84½	4·2	·60	22	4·6	56	39·3	2·1	10 368	777 600	374 000
9½	26	676	531	14¾	5 ,, 2	76·2	9½		94	4·5	·64	22	5·2	65	45·3	2·4	12 168	927 323	393 000
11⅓	28	784	616	14½	5 ,, 4	77·3	10		104	4·8	·69	24	5·9	74	52·0	2·7	14112	1 091 328	410 000
13	30	900	707	14¼	5 ,, 6	78·4	10½		115	5·1	·73	24	6·6	84	59·1	3·0	16 200	1 269 432	427 000
15	32	1024	804	14	5 ,, 8	79.3	11		126	5·4	·78	24	7·3	95	66·8	3·3	18 432	1 462 272	443 000
17¼	34	1156	908	13¾	5 ,, 10	80·2	11½		138	5·7	·82	24	8·0	107	75·0	3·6	20 808	1 669 010	459 000
19⅔	36	1296	1018	13½	6 ,, 0	81·0	12		150	6·0	·87	26	8·8	120	84	4·0	23 328	1 889 568	472 000
22	38	1444	1134	13¼	6 ,, 2	81·7	12½		163	6·4	·91	26	9·7	134	93	4·4	25 992	2 123 806	486 000
24½	40	1600	1257	13	6 ,, 4	82·3	13		176	6·7	·95	26	10·5	148	104	4·8	28 800	2 371 200	498 000
27¼	42	1764	1385	12¾	6 ,, 6	82·9	13½		190	7·0	·99	26	11·4	164	115	5·2	31 752	2 631 606	510 000
30	44	1936	1520	12½	6 ,, 8	83·3	14		204	7·3	1·03	28	12·3	181	126	5·6	34 848	2 904 000	520 000
33	46	2116	1662	12¼	6 ,, 10	83·7	14½		219	7·7	1·07	28	13·3	199	139	6·0	38 088	3 188 346	531 000
36	48	2304	1810	12	7 ,, 0	84·0	15		234	8·0	1·11	28	14·3	218	152	6·5	41 472	3 483 648	539 000
39	50	2500	1963	11¾	7 ,, 2	84·2	11	& 11	252	8·3	1·15	29	15·3	239	167	6·9	45 000	3 789 450	548 000
42½	52	2704	2124	11½	7 ,, 4	84·3	11¼	11½	276	8·7	1·19	29	16·4	261	182	7·4	48 672	4 104 672	555 000
45¾	54	2916	2290	11¼	7 ,, 6	84·4	12	12	300	9·0	1·22	29	17·5	284	199	7·9	52 488	4 427 888	561 000
49¼	56	3136	2463	11	7 ,, 8	84·3	12	13	326	9·3	1·26	30	18·6	310	216	8·4	56 448	4 760 448	567 000
52¼	58	3364	2642	10¾	7 ,, 10	84·2	12	14	354	9·7	1·29	30	19·7	336	235	8·9	60 552	5 099 084	572 000
56¼	60	3600	2827	10½	8 ,, 0	84·0	12	15	384	10·0	1·33	31	20·9	365	255	9·5	64 800	5 443 200	575 000
60	62	3844	3019	10¼	8 ,, 2	83·7	12	& 2 of 11	402	10·3	1·36	31	22·2	396	277	10·0	69 192	5 792 062	579 000
63½	64	4096	3217	10	8 ,, 4	83·3	12	2 11½	426	10·7	1·39	32	23·4	429	300	10·6	73 728	6 144 000	581 000
67	66	4356	3421	9¾	8 ,, 6	82·9	12	2 12	450	11·0	1·42	33	24·7	464	324	11·2	78 408	6 498 455	581 000
70¾	68	4624	3632	9½	8 ,, 8	82·3	12	2 12½	476	11·3	1·45	34	26·0	502	351	11·8	83 232	6 852 768	583 000
74½	70	4900	3848	9¼	8 ,, 10	81·7	12	2 13	502	11·7	1·47	35	27·3	542	379	12·4	88 200	7 206 822	583 000
78	72	5184	4071	9	9 ,, 0	81·0	12	2 13½	530	12·0	1·50	36	28·8	586	409	13·0	93 312	7 558 272	581 000

The surface in the flue is taken at one-half.

N. B. The weight of the column of water in the pumps, including the injection-pump, is to be in all cases 7·81 lbs. per sq. inch of the piston; and it is supposed that the engine is not loaded with any dry pump-rods, or any other dead weight, than its own parts, and so much pump-rod as is actually immersed in the water of the pumps. If there is any more dead load, an allowance must be made.

EXPLANATION OF MR. SMEATON'S TABLE.

Several of the different columns of this table explain themselves sufficiently, with the following observations.

The table is calculated on the supposition, that *the weight of the column of water* in the pumps, including that of the injection-pump, amounts to 7·81 lbs. for each squ. inc. of the piston. This load is equivalent to the pressure which would be occasioned by a column of water, 18 ft. high, resting upon the piston. All the engines, large and small, are proposed to be loaded in the same degree; and the weight of the column of water in the injection-pump, as well as that in the main-pumps, must be included in estimating the performance.

The length of stroke, and number of strokes made per minute, may be varied considerably, without producing any alteration in effect, provided the motion of the piston continues the same, as denoted by the next column. The load being at the same rate in all cases, the different rate of the motion of the piston, in different sizes, is an indication of the superior effect of large engines. The smallest cylinder of 10 inc. diam. will move only $63\frac{1}{3}$ ft. per min., and the largest cylinder of 72 inc. will move 81 ft. The intermediate size of 48-inc. cylinder will move 84 ft. per min.

The diameters of the boilers are taken at their largest part, which is near the water line. The figure and proportion of the boilers is shown in the engravings, and has been already mentioned. The largest size boiler is 15 ft. diam. for a 48-inc. cylinder; but for larger cylinders, two boilers of smaller dimensions are used; and for cylinders of 62 inc. and upwards, three boilers, as shown by the table. The boiler beneath the cylinder is called the centre boiler; the others are placed in outside buildings, on one or both sides of the engine-house, as already described.

The column entitled fire surface denotes the number of square feet of the external surface of the boilers, which receive the action of the flame; the surface exposed in the flues being put down at half, because it was considered to be only a spent flame. *Example.* The engine at Long Benton before calculated, had a 12-ft. centre boiler, the direct fire surface of which was 63·6 squ. ft., and in the flue 150 squ. ft.; half of which (= 75) + 63·6 = 138·6 squ. ft. of fire surface. In the table, a 12-ft. boiler for a 36-inc. cylinder is marked 150 squ. ft. of fire surface.

The dimensions of the fire grates, according to Mr. Smeaton's practice, were as follow:

For a boiler $10\frac{1}{2}$ diam. the fire grate $3\frac{1}{2}$ ft. wide, by 4 ft. long (= 14 squ. ft.); grate bars 2 ft. long, $2\frac{1}{4}$ inc. wide at the top edge, and three-quarters inch space between them; chimney 20 inc. squ. inside, 44 ft. high. For a boiler 12 ft. diam. grate 4 ft. squ. = 16 squ. ft. Boiler 13 ft. diam. grate 4 ft. by $4\frac{1}{2}$ ft. = 18 squ. ft. Boiler 15 ft. diam.; grate $4\frac{1}{2}$ by 5 ft. = $22\frac{1}{2}$ squ. ft.

The diameter of the steam-pipe is measured internally, and for large engines above 48-inc. cylinder it is one-sixth of the diameter of the cylinder; but a rather greater proportion is allowed for the smaller sizes. The aperture of the regulator must be fully equal to that of the steam-pipe. The motion of the regulator about its centre or axis, to be 60 degrees from the position in which it is quite shut, to that in which it is quite open.

The aperture of the snifting valve for a cylinder 30-inch. diam. $1\frac{1}{2}$ inc. diam.; 36-inc. cylinder, $1\frac{3}{4}$-inc. diam.; 48-inc. cylinder, 2 inc. diam.; 60-inc. cylinder, $2\frac{1}{2}$-inc. diam., and for a 72-inc. cylinder, 3 inc. diam.

The aperture of the injection-spout is to be a square orifice in a brass plate; the dimensions of the side of the square, is shown in inches and decimals by the table; and the next column is the perpendicular height of the column which is to produce the jet. It is measured in feet from the level of the orifice, to that of the water in the injection-cistern. The height should never be less than 14 ft. and from that to 36 ft.; but in all cases it should be as high as the building will admit the cistern to be placed.

The bore of the injection-pipe for an engine with a cylinder 12-inc. diam. should be 2 inc. diam.; 24-inc. cylinder, $2\frac{1}{2}$ inc. pipe; 36-inc. cylinder, 3 inc. pipe; 48-inc. cylinder, 4 inc. pipe; 60-inc. cylinder, 5 inc. pipe; and 72 inc. cylinder, 6 inc. pipe.

The diameter of the conical plug, or turning part of the injection-cock, to be the same as that of the injection-pipe, up to 60-inc. cylinder; but all cocks above that to be only 5 inc. diam. The aperture, or water-way of the cock, to be equal in height, to the diameter of the pipe, and in width to be three-fifths of the diameter of the plug. Therefore, all above 60-inc. cylinders will be 3 inc. width of water-way; and a 72-inc. cylinder, will be 3 inc. by 6 inc. water-way, and the plug 5 inc. diam. The plug turns round 80 deg. to open and shut the water-way.

The quantity of injection-water is expressed in two terms, viz. in cub. ft. per min., and in cylindrical inch feet per stroke; the latter is for the convenience of apportioning the jack-head or injection-pump, according to its intended length of stroke. Thus, divide the number of cyl. inc. ft. stated in the table, by the length of the stroke of the injection-pump, in feet, and the square root of the product, is the diameter of the pump in inches. *Example.* A 52-inc. cylinder requires 261 cylind. inc. ft. of injection per stroke; if the injection-pump is to make $5\frac{1}{2}$ ft. stroke, then 261 cyl. inc. ft. $\div$ 5·5 ft. = 47·5 circ. inc., the square root of which is 6·89 inc. for the diameter of the pump: in the engine at Long Benton it was 7 inc. diam.

Sliding Rule, { Ⴔ Cyl. inc. ft. per str.   Stroke inj. pump ft. *Ex.* Ⴔ 261 cyl. inc. ft.   5·5 ft. stroke.
slide inverted. { D   1   Diam. inj. pump inc.   D   1   6·89 inc. diam.

The quantity of water required to supply the boiler at each stroke is stated in cubic inches. For instance, the 52 inc. cylinder requires 182 cubic inc. per stroke, × 11·5 strokes per min. = 2094 cub. inc. or 1·21 cub. ft. per min., which is rather less than was actually evaporated at Long Benton; but there was a considerable waste of steam in that engine, as before explained.

The column entitled Pumpage is the load, or weight of the column of water in the pumps, expressed in cylindrical inch feet, each of which weighs ·341 of a pound, therefore, by multiplying the pumpage number by ·341, we obtain the load in pounds, which the piston must overcome. For instance, a 52-inch cylinder 48 672 cyl. in. ft. × ·341 lbs. = 16 597 lbs., and the area of the piston being 2124 squ. inc., the load is 7·81 lbs per squ. inc.

The pumpage numbers are obtained by multiplying the square of the diameter of the cylinder in inches by 18 ft.; because the pressure of 7·81 lbs. per squ. inc. is equal to a column of water 18 feet high. *Example.* 52-inc. cylinder, square of diam. 2704 cir. inc. × 18 ft. = 48 672 cyl. in. ft. pumpage.

The pumpage numbers in the table, afford a ready mode to find the dimensions of the pumps which the engine can work. *Example.* The length of the stroke of the injection-pump being fixed, divide the cylindrical inch feet of injection per stroke, by that length in feet, and we obtain the area of the injection-pump in circular inches, as before stated. (*Example.* 52-inch. cylinder, 261 cyl. in. ft. $\div$ 5·5 lbs. = 47·5 circ. inc.) Multiply this by the height in feet to which the injection-pump must raise the water, and the product (viz. 70·5 feet × 47·5 = 3349) is the pumpage load in cyl. inc. ft. occasioned by the injection-pump.

Sliding Rule, { A Cyl. in. ft. inj. per str.   Pumpage of inj. pu. *Ex.* A 261 cyl. in. ft.   3349 pumpage.
slide inverted. { Ↄ Lift of inj. pump ft.   Stroke of inj. pu. ft.   Ↄ 70·5 ft. lift.   5·5 ft. stroke.

Deduct this from the pumpage number in the table (viz. 48 672), and we have (45 323) the number of cyl. inc. ft. remaining for the main pumps to lift; then if we divide this by the height in feet to which the water is to be lifted, the quotient will be the area of the main-pump in circular inches, and therefore, the square root of that quotient will be the diameter of the pump in inches.

Example. If the lift is 296 feet; 45 323 cyl. in. ft. $\div$ 296 ft. = 153·1 circ. inc., the square root of which is 12·37 inc. for the diam. of the pump. N. B. The Long Benton pumps were only 12·2 inc. diam., because that engine was only loaded to 7·5 lbs. instead of 7·81 lbs. per squ. inc.

Sliding Rule, { Ⴔ Pumpage cyl. in. ft.   Height of col. ft. *Ex.* Ⴔ 45 323 pumpage   296 feet lift.
slide inverted. { D   1   Diam. of pump inc.   D   1   12·37 inc. diam.

The great product per minute, is the mechanical effect of the engine, expressed by the number of cylindrical inch feet of water which it will raise to a height of one foot in a minute. This is obtained by multiplying the pumpage number by the motion of the piston per minute. Thus, for a 52-inc. cylinder, 48 672 cyl. inc. ft. × 84·33 ft. = 4 104 672 cyl. inc. ft. great product per minute.

In modern steam engines, the mechanical power is estimated by the horse-power, which is 33 000 pounds raised 1 foot per minute, or (33 000 lbs. $\div$ ·341 lbs. =) 968 000 cyl. inc. ft. of water is the same weight. Therefore, if Mr. Smeaton's great product per minute, is divided by 968 000, it will give the horse-power, as denoted by the first column in the table.

Example. Great product 4 104 672 cyl. inc. ft. $\div$ 968 000 cyl. inc. ft. for one HP = 42·4 horse-power.

This mode of estimating by horse-power has been introduced since Mr. Smeaton's time.

The consumption of fuel is stated in bushels of the best Newcastle coals, consumed in an hour's working. Mr. S. estimated the weight of a bushel at 88 pounds, but it is now more usually taken at 84 pounds. He considered the consumption of fuel to be greatly regulated by the extent of the interior surface of the cylinder, with which the steam came in contact; and therefore founded his calculation for coals, upon the proportion between the surface of the cylinder in square inches, and its capacity in cubic inches. The effect of this has been before explained. It is to find how many inches thick the steam contained in the cylinder would cover over all its internal surface, supposing it to be unrolled and spread out in a flat surface, and the steam distributed in a layer upon it.

B B

To find the number of bushels of coals, which an engine will consume per hour, calculate the internal surface of the cylinder, in square inches, (by multiplying the circumference of the cylinder in inches, by its length in inches), and add three times the square of the diameter, to allow for the piston bottom, the cylinder bottom, and the surface of the pipes, which are within the cylinder. Next calculate the solid contents of the cylinder, in cubic inches, (by multiplying the area of the cylinder in square inches, by the length in inches), and then find the proportion between the superficial, and the solid measure of the cylinder. According to the number thus obtained, a divisor must be selected from the following table:

Proportion of the surface of the Cylinder, to its capacity in square and cubic inches.	Proportionate effect of one bushel per hour.	Differences.
1	98	
2	188	90
3	273	85
4	349	76
5	414	65
6	468	54
7	512	44
8	545	33
9	567	22
10	578	11
11	578	0
12	572	8

Lastly, cut off three places of figures, from the great product per minute, and dividing by the above divisor, the quotient is the number of bushels of coals, that the engine will consume per hour.

Example. 52-inch cylinder × 3·1416 = 163 inc. circumference × 96 inc. in length = 15 648 sq. inc. + 8112 (which is 3 times the square of the diameter) = 23 760 sq. inc. of internal surface.

The content of the cylinder is (2124 sq. inc × 108 inc. long) 203 904 cub. inc. ÷ 23 760 sq. inc. = 8·58 cub. inc. to each sq. inch. Therefore the divisor will be between 8 and 9 of the above table, or 558. And the great product, with the last three figures cut off, is 4104 ÷ 558 = 7·36 bushels of coals per hour. In the table it is put down 7·4 bushels. The Long Benton engine consumed 8·5 bushels per hour, but the coals were not of the best quality.

By thus proceeding upon the proportion between the surface, and the content, of the cylinder, a proper allowance is made for the loss of steam, which takes place from condensation, when it enters into the cylinder, at every stroke, after it has been cooled by the injection thrown into it.

The last column of the table, entitled, Effect per minute of one bushel per hour, is the number of cylindrical inch feet, which are raised 1 foot per minute, by every bushel which is consumed per hour. This is obtained by dividing the great product, by the bushels consumed.

Example. In a 52-inch cylinder, 4 104 672 cyl. inc. ft. great product, ÷ 7·4 bush. = 555 000 cyl. inc. ft. effect per min. of 1 bushel per hour.

The number of pounds weight, raised 1 foot high, by the consumption of a bushel of coals weighing 84 lbs. is now commonly used as the standard of comparison, for the performance of steam-engines, in respect to fuel. To reduce Mr. Smeaton's numbers to this standard, they must be multiplied by ·341 lbs. to obtain the pounds weight, and then multiplied by 60 minutes; and then an allowance must be made, for the bushel being accounted 84 lbs. instead of 88 lbs. as Mr. S. used.

All this will be effected, by multiplying Mr. Smeaton's numbers, in the last column, by 19·53.

Example. Effect 555 000 cyl. inc. ft. × 19·53 = 10 839 150 lbs. or 10·83 millions pounds, raised 1 foot high, by every bushel (of 84 lbs.) of the best Newcastle coals, consumed in an engine of 52-inch cylinder.

An atmospheric engine being made according to the foregoing proportions, and its parts being put together in their proper places, the requisite counterweight is roughly determined in the following manner.

The pumps must be filled with water, leaving their buckets without the leathers, and the piston without any packing. In this state, a weight equal to about 1 lb. per squ. inc. being laid upon the piston, the engine is ballasted at either end of the great lever as it may require, until it is found in exact balance. Then the piston being relieved from its weights, will have a counterweight tending to draw it up with a force equal to 1 lb. per squ. inc.

This is for engines of the largest dimensions; but as the proportion of loss by friction of the piston and buckets is greater in small cylinders and pumps, smaller engines must have 1¼ lb., and the smallest engines 1½ lb. per inc., and so on in proportion. When it is not convenient to fill the pumps with water, up to the top, allowance must be made for the difference of the pump-rods, not being immersed in the water.

Mr. Smeaton expected that his engines, when loaded with a neat burden of 7·81 lbs. per squ. inc., and the counterweight as above, would make their returning stroke rather quicker, than the returning stroke, and this he preferred.

The exact proportion of the counterweight depends upon so many contingent circumstances, that it would be impossible to apply any theorem to practice, even if the theory were established; and the final adjustment is easily ascertained by experiment.

The reason for allowing a preponderance to that end of the lever, is simply that the pump-buckets may descend, and that the piston may rise, and allow the steam to fill the cylinder. The manner of adjusting its quantity in an engine first setting to work is as follows. Suppose the water already up to the top of the pumps, the steam being admitted into the cylinder till it has driven out the air, the operator shuts the steam-passage, without supplying any injection, and the engine will make its first stroke, though very quietly, by the external condensation from the surface of the cylinder. He then cautiously allows steam to enter the cylinder, and according to the tendency of the piston to rise, he forms his judgment of the degree of counterweight necessary. If it rises too slow, he puts iron or other ballast, upon the pump end of the lever; and if it rises too quick, he places these weights on the piston end.

There are two important circumstances to attend to in this regulation. First, that the pump-buckets shall descend as quick as they can, but without such force as to occasion a violent shock, to stop their motion at the end of the stroke; and secondly, that the piston shall not be drawn up faster than the steam-regulator (with the degree of opening that is given to it) can supply steam; for that would impede the snifting, to discharge the air and condensing water; and unless these are performed punctually, the engine soon ceases to work.

As neither the air nor the water can be discharged instantaneously from the cylinder, a certain time is requisite, in proportion to the quantity of each, and to the strength of the steam; and the piston must not rise so quick as to prevent the steam acting on the air and condensing water, which will happen when the engine has too great a counterweight, and the steam is low; for then the piston ascending faster than the boiler can supply steam, there can be no proper discharge, and after a few strokes the engine will stop in the middle of the working stroke, not having force to complete it.

This is an extreme case, which can only happen when the steam is let down very low, and the counterweight much too heavy. In the usual course of working an engine, which is properly proportioned to its work, and duly adjusted thereto, the engine-man can retard or accelerate the returning strokes of the engine, in some degree by the regulation of the fire; for if the engine should return too quick, he lets down the damper in the flue of the chimney; and if it returns too slow, he raises the damper. By these means he can vary the action of the steam, on the lower side of the piston, from 1 lb. to $1\frac{1}{2}$ lbs. on the squ. inc. greater than the pressure of the atmosphere, which in a 52-inc. cylinder will amount to 1062 lbs. This is a sufficient latitude to make the engine return very quick, or very slow, but does not alter the rapidity of the working stroke.

This method of proportioning the counterweight, is on the supposition that the engine is to work with its full intended velocity. When an engine is erected on a mine, or pit, which is sinking, the quantity of water to be lifted by the pump being small, the engine must work slowly, and the counterweight must be in proportion. The lever will perhaps require some counterpoise to the weight of the piston, because of the lightness of the pump-rods; but as the pit becomes deeper, successive lengths of pump-rods are applied for the different lifts of pumps, and so much of the weight must then be removed to the piston end of the lever, as to keep the engine under command; for as the motion during the returning stroke depends upon the counterweight, this must be regulated according to the quantity of water which the engine has to draw, or to the number of strokes to be made in a minute.

As the motion must be increased when the quantity of water increases, a greater counterweight must be added; but it is not until the engine works at its intended load, that the counterweight should be apportioned, as before mentioned, at the rate of 1 pound, or $1\frac{1}{2}$ lbs. per squ. inc.

Whilst an engine is working with only a portion of its full load, the injection water must be very sparingly supplied, so as to condense imperfectly within the cylinder, or else the piston would descend with great velocity, and strike upon the spring beams with violence, so as to beat every thing to pieces.

The quantity of the injection is regulated by the size of the aperture through which the jet spouts; and also by the adjustment of that pin in the plug-beam, which presses down the handle, to close the injection-cock, before the piston reaches the bottom of its course, and the other pins in the plug must be regulated so that the engine will work with a shorter stroke, than its proper length.

As a further remedy for this inconvenience it was usual to fix a small air-cock on the upper part of the eduction-pipe, or some other part, having free communication with the cylinder. For the largest engines this was only the size of a small common beer-cock; and the snift being properly regulated, this cock was opened as much as it could be, to allow the piston to come fully down into the cylinder.

The effect of this constant leakage or admission of air into the cylinder, in diminishing the loss of steam by condensation, has been already stated; but in this instance it was to impair the descending power of the piston, when it arrived near the bottom of the cylinder, and thus diminish the excessive acceleration of the motion, which must otherwise have taken place when the engine was too lightly loaded.

The cock in the pipe which proceeds from the snifting-valve, is also very useful to moderate the descending force of the piston, when an engine is not loaded; by closing that cock, and thus diminishing the passage, the snifting will be impeded, so that the air will not be effectually expelled from the cylinder, and consequently the vacuum will be impaired, and the force of the piston moderated.

When a mine is going down, and the engine pit receives all the water from the different parts of the mine, the quickness of the stroke must depend upon the uniform influx of the water, and the engine must be so accurately regulated to this quantity of water, as to sup it up at every stroke. Now if this supping up is violent, the air will be drawn into the pumps at the conclusion of every stroke, and cause the engine to work irregularly; and, on the other hand, if the strokes of the engine are not quick enough, the water will gain on the miners and prevent them working. The rapidity of the motion, must in this case be regulated by the quantity of injection, and by the snift, which will determine the motion of the working stroke, and by the counterweight which will regulate the time of the returning stroke. But a much better regulation of the velocity of the engine can be attained by the use of a cataract, or apparatus usually called by the miners, Jack in the Box.

The cataract was commonly used in Mr. Smeaton's time in the engines for the mines in Cornwall, to regulate the motion of the engine to any given number of strokes per minute, so that the desired quantity of water could be drawn, without wasting steam in working quicker than necessary.

When an engine works by the cataract, it performs both its working and its returning strokes, with a proper celerity of motion ; but after each stroke is completed, and the piston has reached the top of the cylinder, it remains there motionless, for a measured space of time, which being elapsed, it makes another stroke, then waits again, and so on.

The cataract is a small apparatus or water-clock, to measure these intervals of rest, and to so apportion them that the engine will make the required number of strokes per minute.

The cataract is a small tumbling-bob, U Y moving on a centre in the same manner as the χ of the working gear ; but instead of a hammer weight, or bob at the top, it has a small box or cup Y, which is filled with water by a small stream, dripping continually through a cock z, which can at pleasure be regulated to run a greater or lesser stream.

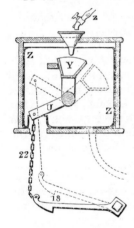

This cataract, enclosed in its box Z Z, is fixed immediately over the working gear, (see Z, Plate II) on a stool, placed on that floor of the engine, which surrounds the cylinder ; and the cock z is soldered into the branch pipe *d*, which supplies the piston.

The lever or upright arm, on the top of which the cup Y is fixed, has a second arm U, and this is connected by the chain 22, with the hook or latch 18, of the injection-handle. This latch serves as a counterweight, which is heavy enough to make the cup Y assume the vertical position, or nearly so, (as in the figure), whenever the cup is empty ; but, in a short space of time, the cup becomes filled with water, by the continual running of the small stream, which flows from the cock z, and runs by the funnel, into the cup. The cup is then of sufficient weight, to make the cataract fall over, as shown by the dotted lines, but in that position, the cup inclines so much, that it discharges its contents, and the counterweight of the latch 18, causes its immediate return to the resting position, as in the figure ; where it will remain, until the cup Y is again filled, by the running stream, and then the cataract will make another stroke.

When the engine works with the cataract, the chain 22 of the injection-latch is detached from the plug *l*, and a short chain which is connected with the arm U of the tumbling cataract, is united to the injection latch 18, instead of the chain 22. The water which is discharged from the tumbling-cup Y, every time it makes its stroke, collects at the bottom of the box in which the cataract is enclosed, and is carried off by a branch, into the drain-pipe from the cylinder top.

It is evident, that by turning the cock z, and diminishing the stream of water, the intervals between these strokes will be lengthened ; for the cataract is in fact, a clepsydra or water-going clock, which performs its motions at regulated periods of time. At every one of these strokes, it lifts up the injection-latch 18, and causes the engine to make a working stroke.

To explain the operation of the cataract, we will suppose the steam-regulator open, the injection-cock shut, and the piston to have just arrived at the top of the cylinder. A pin, in the plug-beam *l*, seizes the handle 4, and overthrows the hammer or tumbling-bob 16, of the Y, towards the cylinder, whereby the leg 12 of the Y, draws the stirrup-rod 14, and shuts the regulator.

In this situation the engine will remain, until the stream of water, which flows from the pipe *d* through the cock *z*, fills the cup Y, at the top of the cataract tumbler, and causes it to fall over; it then suddenly draws up the chain 22, and raises the latch 18. This lets fall the weight 8, and it opens the injection-cock, to throw a jet into the cylinder, which condenses the steam, and causes the piston to descend. When it arrives at the bottom, the pin in the plug *l* depresses the handle *s*, which shuts the injection-cock, and then, by depressing the handle *r*, overthrows the hammer 16, and opens the regulator. This admits steam again into the cylinder, and the counterweight makes the piston return.

The cataract had returned, the instant that its cup had inclined so much, as to throw out the water, and the cup then began to fill again; but it will not act, to discharge the latch of the injection-cock, until it is quite filled; and the injection-cock cannot open till this happens, so that the piston waits at the top of the stroke till the cataract is ready. And this time of waiting can be regulated, by diminishing or increasing the stream which runs down from the cock *z*, into the cup Y, so as to draw up exactly as much water as drains into the mine.

If there is but little water in the mine, so that the engine is only required to make four or five strokes per minute, the interval of rest between the successive strokes will be considerable.

The structure of this apparatus is apparent from the figure. The two levers Y and U are affixed to an axis, which moves freely upon pivots supported in the sides of the box, and a rail is fixed across the box, for the cup Y to rest against, when it is in the position of the figure, waiting to become filled. The box is lined with sheet lead, to receive the water discharged by the cup, and it runs away by the drain-pipe. The chain 22 passes through a small tube fixed in the bottom of the box, and which prevents the water running down by the chain.

DESCRIPTION OF AN ATMOSPHERIC ENGINE OF THE LARGEST DIMENSIONS, ERECTED IN 1775, AT CHASE-WATER MINE, IN CORNWALL, ACCORDING TO THE DESIGNS OF MR. SMEATON.

[See Plates V, VI, and VII.]

THE construction of this engine was the same as that at Long Benton, but it was nearly double the power of the latter. After the full description we have given of Mr. Smeaton's engines, it is unnecessary to enter into any other particulars of this engine, except its general dimensions. It was the most powerful engine which had been then constructed (*a*).

Plate V contains an upright section of the whole engine, and Plates VI and VII contain detached figures of the parts, on a larger scale.

The cylinder was 72 inches diameter, and the utmost length of stroke allowed by the catch pins was $9\frac{1}{2}$ feet; but it usually worked 9 feet, and made 9 strokes per minute, as per table, or 81 feet effective motion per minute. It worked three columns of pumps, $16\frac{3}{4}$ inc. diam. 9 feet stroke, each lifting the water about 17 fathoms, or 51 fathoms lift in all. The injection-pump was 12 inc. diam. 5 ft. stroke, and raised the water 12 fathoms, from the cold water pit, to the injection-cistern at the top of the house.

(*a*) *Note.* There were formerly two atmospheric engines working on this mine; one a 64 inc. cylinder, the other a 62 inc.; both six feet stroke. These two engines were stated to consume $16\frac{1}{2}$ bushels of coals per hour. The quantity of water they raised to keep the mines drained, was 80 cubic feet per minute, in summer, and 100 cub. ft. per min. in winter.

One of these engines worked the lower column of pumps, $18\frac{1}{2}$ inc. diam., and 24 fathoms lift, and the other engine worked the upper column $17\frac{1}{4}$ inc. diam., and 26 fathoms lift; so that one

The weight of the columns of water in the pumps was 31 187 lbs. or 14 tons nearly. This load was at the rate of $7\frac{2}{3}$ lbs. per square inch of the piston. The engine exerted more than 76 horse-power.

Thus the great pumps (16·75 inc. diam. × 16·75 =) 280 circ. inches area × ·341 lbs. = 95·5 lbs. for each foot high × (51 fathoms =) 306 ft. high = 29 223 lbs. for the total weight of the column in the great pumps. And the injection-pump (12 inc. diam. × 12 =)144 circular inches area × ·341 = 49·1 lbs. per foot × (12 fathoms =) 72 feet = 3535 lbs. weight, but as this made only 5 feet stroke, and the piston made 9 feet stroke, the resistance it occasioned to the piston was ($\frac{5}{9}$ of 3535 =) 1964 lbs. Therefore the total resistance to the motion of the piston was (29 223 + 1964 lbs. =) 31 187 lbs.

The area of the cylinder (72 inc. diam.) was 4071 square inches, and 31 187 lbs. ÷ 4071 squ. inc. = 7·66 lbs. pressure per square inch.

And 31 187 lbs. load × 81 feet motion per min. = 2 526 147 lbs. raised one foot per minute ÷ by 33 000 lbs. a horse-power = 76·5 HP for this engine. This is very nearly the same power as the Long Benton engine, in proportion to the diameter of the cylinder.

This engine had three boilers, each 15 feet diam., one placed beneath the cylinder, and two others in low buildings at each side of the engine-house.

The fire-grates were each 4 ft. 9 inc. by 4 ft. 10 inc. = 23 square feet each, or 69 square feet of surface, in the three. There was a separate chimney for each boiler; dimensions 22 inc. by 34 inc. inside, and 52 feet high, from the level of the fire-grate, to the top of the chimney.

The bottom of each boiler was 12 feet diam., and the circular space or furnace for the flame to act in, beneath the bottom, was 11 feet diam.; the fire-grate, which was nearly on a level with the flat floor of this space, was one foot below the bottom of the boiler, but the bottom was concave beneath, and rose one foot higher in the middle, than at the circumference, so that the clear height, from the fire-grate, to the centre of the boiler bottom, was two feet.

The back, or end of the fire-grate, farthest from the fire-door, extended to the centre of the boiler, and the other, or front end of the fire-grate, advanced quite to the circumference of the circular furnace of 11 feet diam. At the farthest extremity of the furnace, diametrically opposite to the fire-door, was a passage $2\frac{1}{2}$ ft. wide by one foot high, to lead the flame from the furnace, into the flue or passage round the circumference of the boiler. This flue was six feet high by $2\frac{1}{4}$ ft. wide at bottom, so that the exterior wall of the flue, was a circle 17 feet diameter. A projecting pier of brick-work was placed in the passage of the flue, near to the part which entered into the chimney: this contracted the aperture of the passage to about three feet high, by 16 inches wide, as is shown in the section Plate V, by the letters r, s, y; the dotted lines o, p, q, being the passage to the chimney.

In all circular boilers, the passage of the circular flue, which encompasses the lower part, is interrupted by a brick-work partition built across it, nearly opposite the fire-door, so as to stop the passage entirely; the entrance into the flue, from the furnace beneath the boiler, is at one side of this partition, and the passage from the flue into the chimney, is at the other side of the same partition.

---

engine raised the water to the other, and that engine raised it up high enough to run away, by the subterranean level or adit, which was 24 fathoms below the surface. In the whole the water was raised 50 fathoms.

This was a common arrangement for draining the large mines in Cornwall, which required more power than one engine usually possessed; this plan was called *shammaling*.

A few years after this engine was erected, another adit was opened on a lower level by 20 fathoms, so as to take away the water at 44 fathoms below the surface.

The works at this, and all the other mines in Cornwall, have been greatly extended since Mr. Smeaton's time.

In consequence, the flame from the furnace, under the boiler, rises up from beneath the same, through the opening at the place farthest from the fire-grate, and enters into the circular flue ; but though that entrance is quite close to the chimney, the brick partition prevents the flame passing into the chimney that way, hence it is compelled to make a complete circuit round the lower part of the boiler, till it arrives again at the same brick partition, but on the other side of it ; and then it escapes into the chimney.

In this long circuit the flame communicates so much of its heat to the boiler, that it is supposed that nothing but hot air is left to pass into the chimney ; but this effect will very much depend upon the size of the opening, from the circular flue, into the chimney where it is contracted, as before mentioned. In modern engines a sliding damper is fitted into that part of the passage, to enable the attendant to regulate the opening at pleasure. The level of the water in the boiler must be three or four inches above the top of the flue round the outside of the boiler, so as to ensure that there will always be water withinside the boiler, opposite to all parts where the fire can act on the outside.

The bars of the fire-grate were made round, of wrought-iron, $2\frac{1}{4}$ inc. diameter, in order that whenever they became bent by the heat of the fire, the curved side might be turned upwards the next time the engine stopped. The spaces between the bars were $\frac{3}{4}$ of an inch wide, and they were supported on racks at each end. N.B. These bars were not found durable, and were afterwards changed for cast-iron bars of greater depth and strength.

The building for this engine was very strong ; the house was 36 feet long, by 20 feet wide, outside of the walls ; the great lever wall was 10 feet thick at the base, and up to the level of the boiler head floor, and then the upper part was 5 ft. thick. The end wall of the house was eight feet thick at bottom. The side walls 2 ft. 10 inc. thick. The inside dimensions of the house were 23 ft. long, by $14\frac{1}{3}$ ft. wide. All the walls were built of granite.

The length of the great lever was $27\frac{1}{3}$ feet, from the centre of the cylinder to the centre of the pump-rod. The size of the pit or shaft of the mine, was 10 feet long, by seven feet wide, inside. The house was not built expressly for Mr. Smeaton's engine, but was the building of one of the old engines. A large archway 18 feet wide, was left in the lower part of each of the side walls, to introduce the boiler ; but the former engine which stood in this house had only one boiler, of an oblong form, 22 feet long, by $8\frac{1}{2}$ feet wide, and its ends passed beneath these arches. The preceding particulars of this engine, are taken from Mr. Smeaton's drawings, and we may subjoin the following papers of his own.

Mr. Smeaton's Remarks upon particular parts of the Engine for Chase Water Mine. (b)

[See Mr. Smeaton's Reports, Vol. II, p. 350.]

It is unnecessary to enter into a minute description of those parts which are to be made in the common manner, and which are obvious to all practitioners ; it is sufficient to explain those that are more particular. In general, the parts being measured off from the drawings, and enlarged according to their respective scales, and the positions of the great parts being first fixed, all the other parts will fit together, and perform their respective offices, without material alterations ; and although many different proportions for the parts, will answer the purpose, yet those laid down in the plans will be found to perform completely, and work with certainty.

(b) These remarks, and the three plates of Chase Water fire-engine, are extracted, with very little alteration, from the printed edition of Mr. Smeaton's Reports, in three volumes quarto, 1812. The drawings for that work were made by the author from Mr. Smeaton's original plans.

Fig. 4, Plate VI, is a plan of the under surface of the piston.   It is a planking of wood, which must be fastened by the bolts and iron-rings A A, and B B, to the under surface of the cast-iron piston plate C D, fig. 3, to form the complete piston ; as is shown in the section, fig. 5, A B C D.

This planking is to be of elm or beech-wood, about $2\frac{1}{4}$ inches thick when worked.   The edges of the two planks forming the cross, are to be grooved with a plough-plane, about $\frac{3}{4}$ of an inch wide, and as much in depth, and the corner-pieces tongued, so as to fit thereto.   A few rivets must be put through the cross planks at their intersection, where they are halved into each other.   This circular wood piston being hooped on the edge, with a good iron hoop, $\frac{1}{2}$ an inch thick, and $2\frac{1}{4}$ broad, the whole will be bound tight together.   The hoop, out and out, to be $\frac{1}{4}$ of an inch less in diameter, than the cylinder.

The flat iron-rings, shown at A A and B B, fig. 4, are to be let into the wood, flush with the under surface of the piston ; and the bolt-heads are to be chamfered or counter-sunk flush into the rings.   The planking is to be applied beneath the cast-iron, C D, with a double thickness of flannel and tar, between the wood and the iron ; and, in case of irregularities, the hollows must be fitted up with additional thicknesses of flannel, so as to exclude the air, between the plate and the wood ; and the bolts, which pass up through the iron piston, and have the nuts screwed on their top ends, must be carefully secured so as to make a water-tight joint from above.   There are 18 of these bolts in the large ring A A, and 8 bolts in the small ring B B.

The planking being thus secured, a sheathing of clean deal-boards, free from sap, $\frac{3}{4}$ of an inch thick, must be nailed on beneath the planking, with tar and hair, or a single thickness of flannel, so as to exclude the air between the planking and sheathing ; and at last, the sheathing is to be dressed off, so as to be flat and smooth.   If any parts of the extremities of the piston-shanks are prominent beneath the cast-iron, they must be let into the planking, and the air excluded from the cavity, by tar and hair, white lead and oil, &c.

The upright feeding-pipe L, Plate V, is designed to answer the double purpose of a lower gauge-cock and a safety-pipe.   Two or three small holes are to be pierced through it, at the depth proper for a lower gauge-cock, that is, at the lowest level which is proper for the water in the boiler, and 3 inches below these holes the feeding-pipe is to be pierced with a hole $\frac{1}{2}$ inch diameter ; and again, at 3 inches lower, the pipe must terminate altogether.   By this means, if the water in the boiler sinks too low, the small holes in the feeding-pipe will emit steam, and give notice by a rackling noise ; but if the water is suffered to subside still lower, the steam will blow out by the $\frac{1}{2}$ inch hole, and the noise will produce a greater alarm ; and when the water sinks so low, as at intervals to be below the bottom of the pipe, the water and steam will issue in such a manner, as to make a very great noise, and call the engine-man to his duty, even if he may have fallen asleep.

The injection-cap, figs. 6 and 7, Pl. VI. is a square hole $1\frac{1}{2}$ inches square, which it is supposed will yield a sufficient injection, when the engine has to work its full lift ; but at first setting the engine to work, this hole should be much less. At first it may be pierced to only $\frac{1}{2}$ inch square, and enlarged as the weight of the column to be lifted, renders it necessary.   The trouble of drawing the piston out of the cylinder, to perform this enlargement gradually, often tempts an engine-keeper to begin with the injection aperture of the full size, and may occasion much mischief at starting a large engine.

C C

194 HISTORICAL ACCOUNT. [PART I.

The injection-cap is fixed by four screws, on the top of a short brass tube, which is driven fast into the end of the wooden injection-pipe, within the cylinder, as shown in fig. 5, the outside of the brass tube being cut rough, so as to stick in the wood. A piece of wood is interposed between the top of the brass tube, and the injection-cap to make the joint, and this may be reduced on one side or the other, as may be requisite, to throw the jet exactly perpendicular, that it may strike the piston bottom at right angles, and be equally dispersed through the cylinder. The under side of the square orifice is chamfered or rounded, that it may throw up a full bore. The cap can be removed and replaced by others, with different apertures.

The wooden injection-pipe is 6 inches bore, and 12 inches diameter outside, one end of it is fitted to the orifice of the injection-pipe, in the hemispherical bottom, and it is fixed by jambing the other end fast, against the steam-pipe in the centre of the hemisphere. The wood preserves the steam from condensation, by the coldness of the injection-water.

It is of consequence to have such tackle for drawing the piston, as to do it readily. For this purpose provide an iron cross of competent strength, the whole length of one bar being about 2 feet, with a shorter bar welded across the middle of it, about 1 foot or 15 inches long ; each end of the long bar to be terminated with an eye or loop. Also provide a rope of suitable strength, and in the whole about two yards long, with a hook at each end.

To draw the piston with this tackle, the two short arms of the cross must be introduced between those two branches of the piston-shank (See fig. 3, Plate VI,) which are united at top, the two long arms of the cross being in a direction parallel to the cylinder beams ; then the two hooks of the rope, must be hooked into the two eyes at the ends of the cross, and the lower tackle-block of pulleys being hooked to the middle of the rope, the purchase of the tackle may be made to heave up the piston: whereby the great chains will be eased of the weight of the piston, without altering their situation, and the great bolt which unites the chains to the piston-shank, may be easily drawn out, and the piston raised up out of the cylinder, into a proper situation to admit a man down into the cylinder. The alteration within the cylinder being completed, the piston may be restored to its place, with little loss of time or difficulty.

As the piston must frequently be rested upon the cup at the top of the cylinder, the weight would be liable to bilge it, if it were made of lead; it is therefore to be made of cast iron, and the piston-head water should be supplied by introducing a pipe through the side of the cup, with a stop-cock to regulate the quantity. Then the top edge of the cup will be clear all round, to rest the piston upon it, without disturbing that pipe and cock. Some precaution in these respects is the more necessary on account of the wooden bottom.

It would have been very convenient to have made the aperture capable of adjustment from the outside. In a correspondence relative to this part, Mr. Smeaton says " the sliding cap at Long Benton was rather intended for an experiment to fix the proper size of the injection-hole, than for general use, for unless it is executed with some nicety it will not answer:" and he adds, that he was obliged to make that cap, and all the apparatus belonging to it, with his own hands. " There are certainly some advantages attending a sliding injection-cap, especially in an engine where the temperature of the injection-water, is liable to vary with the seasons. The screw that regulates the opening of the cap, works in a little conical piece of brass, which being driven (with white lead and oil) into a hole broached taper, through the metal of the cylinder, makes an air-tight joint, and as several of the threads of the screw, are contained in this piece of brass, and on its first application it is dipped in tar, the joint is made tight, without further trouble."

The cast-iron piston CD, figs. 3 and 5, is a flat circular dish, 66 inches diameter outside, and 1½ inc. thick, the rim or raised edge being 5 inches high on the outside ; and through this rim are 18 holes, for the bolts to fasten the wooden bottom.   In the middle part of the flat plate, are four large square holes, to receive the ends of the shanks or prongs, which suspend the piston.   The plate is made thicker round those holes, which are made larger at the lower side, and the ends of the branches are dovetailed, but being smaller than the holes, these can be inserted into them, and then by driving in a key, by the side of each dovetail, it is jambed so fast into the hole, that it cannot draw out again.

Figs. 3 and 5 also explain the harness or parts by which the piston is suspended from the great lever.   The four prongs which are affixed to the piston, are each 2 inches square ; they are formed into two loops at the upper end, and are united by one single joint-pin, 3 inc. diam. to three other long upright links, as shown in the drawing ; and these are connected by another joint-pin, 3 inc. diam. with a short horizontal link, each end of which is connected by a pin of 2½-inc. diam. with a pair of long upright links, which at the upper ends are jointed by similar pins to short horizontal links, to the ends of which, the chains are connected.   By this arrangement, each of the four chains must bear an equal part of the weight.   The middle links of each chain are formed out of wrought iron bar, 2 inc. by 1 inc. welded into loops, as shown in the figure.   Therefore the section of the metal in each link is 4 squ. inc., and four of these, (= 16 squ. inc.) bear all the weight. The outside links of each chain are made of iron 1 inc. squ., but being double the number of the former, they have the same strength ; the joint-pins of the chains are 2 inc. diam. ; the length of each link, or the distance from the centre of one pin to that of the next, 9 inc.

The form of the great lever, or working beam, is sufficiently shown at D D, Plate V, but it is composed of twice as many pieces as appear in that view.   The cast-iron plate for the axis of the lever is 30 inc. wide by 5 inc. in thickness.   The four pieces of timber, which form the middle part of the lever, are made of whole balks of fir, 12 inc. squ. when dressed ; the axis is notched, three-quarters of an inch, into each of these beams, to keep the axis steady ; and when they are fitted, the ends are sprung together, as in the drawing, and the outsides are moulded off towards the extremities, by a fair curve.

It is best to spring these four principal pieces together, before the rest of the beams ; and leaving them a little longer than the others, their ends may be confined together by hoops ; and their curved outsides being made fair, the sixteen external plank pieces, which are all 6 inc. by 12 inc. are to be applied upon the four principals, one at a time, by strong notched pieces of wood, which the carpenters call clams : several of these clams being applied on each side, between the places where the bolts are to be, these, by the aid of wedges, will bring all the pieces together ; then the iron bolts and plates are to be fitted, and all the twenty pieces bound fast together.   Lastly ; the places for the drift-keys are to be mortised out in the joints, between the different pieces ; or, to save trouble in mortising, a little attention to measures, will enable the carpenter to nick the pieces with a saw, and split out the cores, before they are brought together ; and the mortices thus made, being a little too short at first, they are to be very carefully dressed straight and made true, with a heading chisel.   After this, the oak keys are to be adapted to the mortices, with a plane, one by one, easing the parts that bear.   The keys are to be made of heart of oak, perfectly dry and hard.   The keys are to have no drift up and down ; that is, they are to be quite parallel in their thickness, which is

$2\frac{1}{2}$ inches : but they take their drift endways of the mortices, being 6 inches wide at one side of the lever, and 5 at the other.

N. B.—If any key fits well, but drives too far, or, if by tightening occasionally after the engine is set to work, it gets driven too far, it is better to put it in again with a strip of pasteboard of an equal thickness, than to make a new key, as the old one is more sure to fit. The dressing of the ends of the mortices as near as possible to a straight line, across all the breadth of the beams, is the principal nicety of workmanship that is required in framing the great lever (c).

The same system of combining timber is to be used for the cylinder-beams FF, Plate V, between which the cylinder is suspended, and which are kept down, by the ends being entered into the side walls of the house. The keys being for the same use as in the great lever, the same care must be observed in fitting them accurately into the mortices. Each cylinder-beam F is 1 ft. broad by $5\frac{2}{3}$ ft. deep, and is composed of six timbers, united by twelve bolts passing through all the depth of each beam, and with twenty oak keys, inserted between the joints of the timbers, viz. four keys in each joint. The cylinder-beams are the same depth at the ends as at the middle, and their length between the side walls of the house, by which they are supported, is $14\frac{1}{3}$ ft.

As the bolts of the iron plates, which unite the different pieces of the wooden pump-rod or spear, are the most liable to fail, a method is shown in figs. 8, 9, and 10, Plate VI, to give them a very increased strength, without cutting or weakening the wood more than usual. For this purpose, the bolts are not a uniform square, but are $1\frac{1}{2}$ inc. by 1 inc., and are driven through the iron plates and the wood spear, the breadthways upwards. The whole substance of the bolt is well fitted into the plate at the nut end, as well as at the head end of the bolt. If this increase of projection is thought to be objectionable, it may be sufficiently reduced by using drift-keys, instead of screws and nuts, and that will also render the work more simple.

The lower ends of the iron plates for the top of the main-spear, are bent or hooked inwards, into the wood, that they may hold more securely ; and the same method will be proper also for the other spear-plates which unite the different lengths of wood. The hook part should not penetrate more than about five-eighths of an inch into the wood.

The spear is supposed to be a single rod from the engine, down to near the upper pump-head ; then it is proposed to be divided into three separate rods or spears, united flatways by the usual method, to go down to the three lifts of pumps.

(c) The great lever of this engine appears by the drawings to have been of prodigious strength. The depth in the centre was 74 inc., of which the iron axis occupied 5 inc. at the middle parts. The breadth was 24 inc. It consisted of twenty beams of fir, viz. ten one above another, as seen in the figures and two such sets side by side. The depth of the lever at the extremities was 60 inc. by 24 inc. These twenty pieces were united by two large iron bands near the centre, $3\frac{1}{2}$ inc. wide, by $\frac{3}{4}$ inc. thick, with nuts at their upper ends to bind the wood fast ; and also thirty-two iron bolts, 1 inc. diam., each of which passed through the whole depth of the beams, at the places shown by the dotted lines in the figure ; but there were four bolts abreast at each of those places. The upper and lower ends of each of these bolts, passed through sixteen strong iron plates, which were let into the wood at the top and bottom of the lever ; and as each plate extended all across its breadth, they united the two sets of beams, which were side by side.

There were two large oak keys, 7 inc. by 5 inc. inserted between the four principal beams in the middle, and fifty-four smaller keys, $2\frac{1}{2}$ inc. by 5 inc. inserted between the joints of the different beams, towards the ends of the lever, as shown in the plate. The cylindrical gudgeons or pivots, at the ends of the axis, on which this immense lever was poised, were $8\frac{1}{4}$ inc. diam. and $8\frac{1}{2}$ inc. length of bearing ; they rested in brass sockets, let into large blocks of wood, worked into the masonry of the lever wall: the length or distance between these bearings was $3\frac{1}{2}$ ft.

But the uppermost pump should be wrought by the middle of the three parallel spears, in order that the weight of dead spear, on each side of the middle one, may be as nearly equal, as the case admits ; and on that account, the great spear will be less liable to swag in working.   And as the three pumps are in different heights, and will not interfere, the two outside ones may be placed nearer to the middle than usual, if found more convenient.

The upper end of the spear is shown at figs. 8 and 9 to be 7 inc. squ., and is intended to continue of that size, down to the division into three parallel spears. But for greater security in joining the several pieces, it may be more eligible to make it 8 inches by 6 inc. or rather 10 inc. by 5 inc. for the upper length, and the rest of the different pieces to be united in the following manner.

The lower half of the length of the first piece, to be cut half away, reducing its breadth by a shoulder, to a rod 5 inc. square.   Suppose the lengths of spear-wood are 50 ft. this shoulder will be at 25 ft. from the top ; then another piece, of 5 inc. square and 50 ft. long, being applied, it will reach 25 ft. below the end of the first piece.   The jump-joint of the ends of the two pieces, to be secured with three spear-plates, one on each side of the joint, and one opposite to that side where the wood is whole : these plates to be cross-bolted together, with bolts of the kind already mentioned.   To the shoulder formed between the two pieces last joined, a third length of 5 inches square is to be applied, and it will exceed the former by half its length ; the butt-joint to be secured by three plates, as before ; and so on till the required length of spear, 10 inc. by 5 inc., is formed.

By this means, one half of the wood, or 5 inc. squ. will everywhere be whole, and the other half of 5 inches will be secured by spear-plates, in the strongest manner.   To keep the rods which are thus united together, from drawing, in case of any failure in either half, a mortice of 1 inc. by 5 inc. may be cut out, in the joint between the two 5-inch pieces ; and a key of dry oak, of 1 inc. thick, and 5 inches high, being driven into the mortice, as is shown between the different beams of the great lever, one half the thickness of this key will be in one of the 5-inc. pieces, and the other half in the other, so as to prevent them slipping one by the other.   The place of this key, should be between the two last bolts at the ends of the spear-plates, and also two more such keys may be introduced at equal distances between each set of spear-plates, putting two clinch-bolts through both pieces of 5 inches, one above, and the other below each key.   The three small parallel spears, below the division are supposed to be made and secured in the usual way.

As the work is to be executed by experienced engineers, acquainted with the usual practice, it is to be understood that those things for which no particular directions are given, are to be done in the ordinary way, subject to local conveniences, at the discretion of the engineers.

The steam-pipe is 12 inc. diam., the eduction-pipe 8 inc. diam., and the two short necks for the injection, and for the snifting are 6 inc. diam.   All these parts are cast in the same piece with the hemispherical bottom, as shown in fig. 5.

The snifting-clack or valve, is a short cylindrical plug of boxwood, driven tight into the neck in the cylinder bottom, and perforated with an orifice 3 inc. diam., which is fitted with a circular flap-valve of brass, applied to the flat end of the plug, so as to cover the orifice.   It is hung on a hinge-joint at the upper side. A conical copper funnel is screwed to the neck, so as to form a case or cover to the valve; and a stop-cock of $2\frac{1}{2}$ or 3 inc. water-way is soldered to the spout of this funnel, to regulate the aperture of the snift at pleasure.

Explanation of the Working Gear, Plate VII.

Fig. 1, Plate VII, is a plan of the lower side or working surface of the cast-iron regulator-plate, which is fixed in the top or crown of the boiler. Fig. 2 is a plan of the moveable valve, or brass plate, which works against the lower surface of the fixed plate: and fig. 3 is a section of the complete regulator, showing the two parts put together.

Figs. 4, 5, 6, 7, and 8 of the same plate, represent the working gear on a large scale, to show its different positions.

The aperture in the regulator-plate is shown in fig. 1. The prominent margin, or border, round the aperture, to which the moveable plate is fitted, is $\frac{3}{4}$ of an inch wide: the area of the sectorial aperture is equal to that of the circular passage of the steam-pipe, of 12 inc. diam.; and it is brought into that size and form, in the height of the short steam-pipe, fig. 3; so that the upper part thereof is a circular figure, though the lower part conforms to the aperture of the regulator.

The upright axis of the regulator is a square iron pin, having a conical plug of brass cast upon it, which is accurately fitted into the socket of the fixed part, by grinding, as shown in fig. 3. The upper and lower ends of the square iron pin, project above and below the brass plug, to fix on the handle or spanner at top, on the outside of the boiler, and the moveable regulator-plate or valve, fig. 2, at the bottom, withinside the boiler.

In setting the boiler and cylinder, some space ought to be left between the upper end of the regulator steam-pipe, fig. 3, and the lower end of the cylinder-bottom steam-pipe; as shown in Plate V. This space may be filled up with a ring of iron or lead, so as to make a continuation of the two pipes inside, and support the wrapping of canvas, and sheet lead, which is to be bound round it with cord, to make the joint.

The cast-iron regulator-plate, figs. 1 and 3, Plate VII, is to be fixed upon a copper plate adapted thereto, and also to the top of the boiler, so as to close the opening of 3 ft. diam. which is left in the top of the boiler. This plate is to be only one-eighth of an inch thick, and to be set, neither concave or convex, but as flat as possible; so that it will spring up and down, and give way to the working of the engine, without straining the joints.

The cylinder and the boiler, with the regulator, and the great lever, with the plug-beam, being fixed according to the general design, Plate V, the next thing is, to examine the angles which the regulator and injection-cock require to be turned, in order to completely open and shut their respective passages.

The dotted lines in the plan, fig. 8, represent the different positions of the handle or spanner of the regulator, in which it is open and shut. A is the socket, in which the perpendicular conical pin of the regulator turns. (N. B. To bring the figure into less compass, the centre A is brought forwards from its true position towards the working gear.) The line A a is supposed to be at right angles to the position of the great lever, and parallel to the cylinder-beams. The line A b is the position of the spanner, when the regulator is shut, and A C its position when open. The angle b A c will therefore be the whole motion of the regulator, and spanner about their centre. This angle is supposed to be 50 deg.; the angle b A a, being 20 deg., and a A c, 30 deg.

The radius or length of the spanner A b, from the centre, to the joint-pin which unites it to the sliding stirrup-rod, is 12 inc., and the links which suspend the other end of the sliding stirrup-rod from the axis of the hammer Y are 12 inc. length. N. B. The angle b A a is made less than the angle a A c, because the sliding valve requires most force to move it when it is shut. When it is open it is not confined, except by its own friction: but when it is shut it must endure the pressure of the steam beneath it, when there is a vacuum in the steam-pipe above it.

Fig. 6 is a section of the square water-way of the injection-cock; the dotted lines show the relative size of the brass turning plug, and its socket. The mean diameter of the cock-plug being 5 inc. will allow an aperture 3 inc. wide, and 6 inc. high.

A plug of this size is of a considerable weight; but to diminish the friction, a part of the weight is to be borne upon the point of a perpendicular screw, supported by a frame hooked upon the square branches of the cock, on each side of the socket for the plug: d e, figs. 4 and 6, is this screw: fg, fg is the suspended frame, and h is a counter-nut, to prevent the screw moving, by the turning of the cock-plug. This frame hangs loose without any confinement, being kept to its place by the point of the screw, and the hooks are made so slight and thin, as to allow them to spring, and bear up any part of the weight of the plug, till it is found to move with freedom. As this tends to prevent unnecessary wear, it is material to the performance of the engine.

The plug of the injection-cock should be made to take out, in order to grease it frequently with hog's-lard, or other soft grease, and a valve must be fitted at top of the injection-pipe, to stop the water while this is doing. By the incessant motion of the engine, the injection-cock will not remain long tight, but to catch any water which may leak out by the plug, a copper funnel i i k, is supported in an iron ring m m, which is fastened by a branch n, (see also fig. 5) to one of the posts D of the working-gear. The water may be received from the spout k of the tunnel, into a small wooden trough, and conveyed to the waste water.

An injection-cock, made on the above proportions, will completely open and shut the passage by an angular motion of 80°, as is shown in the Plan, fig. 5, where B o is the position of the injection-spanner when the passage is open, and B p the position when it is shut.

The height of the handle or spanner of the regulator being given, as y r, in fig. 4, the height of the tumbler's centre x will be 10 inches higher. And the height of the injection-spanner B, being given, the height of the centre Z, of the lever called the F, will be $13\frac{1}{2}$ inc. higher. The Plan, fig. 5, shows the distance of the two points o and p, fig. 4, where the acting radius of the fork Z o, or Z p meets the injection-spanner, when the cock is quite open, or quite shut. The distance B q of the plane of the F lever, from the centre line of the injection-cock, is $7\frac{1}{2}$ inc., but by bending the fork a little towards, or from the cock, it will cause the spanner of the cock to turn more or less as may be requisite.

DD, fig. 4, is the upper part of the fore-post marked D in the Plan, fig. 5, but the rest of the post is supposed to be removed in fig. 4, to show what is behind it; E E E is the other post, also marked E in the Plan.

To support the F lever, and keep it steady in working, it is fixed upon an axis of a competent length, and its latch above, is also fixed upon an axis; the near end of which is supported by a bracket G H H, from the fore-post D. The arm supporting the near end of the axis of the F lever, is similar, but is shown broken off, as K L L; the far ends of both axes are supported by a back-post MM, because they could not go across to the far pillar E, on account of the tumbler or Y. The pin or stud at S is driven into the pillar M, to support the latch from falling too far, after the F lever is discharged, and the tail $l\,l$ attached to the axis of the latch, serves as a handle to strike off the injection, in working by hand.

In fig. 5, the latch with its axis and arm lying directly above the F, cannot be distinguished, but the handle $l\,l$ of the latch tail is shown.

The only parts which require an exact form are the F, and those two arms, 3 and 7, of the Y which work in the plug-beam; these parts may be moulded by drawing them out upon a board, thus:

For the arms of the tumbler, or Y, first form a square section of its axis, draw a line representing $x\,t$ across the square, and set off the points 8 v w & and t, according to the measures, and at these points cross the line $x\,y$, with the perpendiculars 1 v 4, 2 w 5, 3 & 6, and t 7 setting off the points 1, 2, 3, on one side, according to the distances v 1, w 2 and & 3, measured from the scale, and also set off the points 4, 5, 6, 7, from the distances v 4, w 5, & 6, and t 7, on the other side; this done, draw a fair curve through the points 3, 2, 1, 8, and 7, 6, 5, 4, 8, and lastly, fixing a square piece of wood or iron, upon the square representing the axis, the board is prepared. The arms of the Y being formed in iron, with square holes through them to fit the axis, apply them upon the prominent square on the board; and having heated the iron as much as necessary, each arm must be bended till it answers to its intended curve.

The legs of the Y may be set to a proper opening by drawing a circle $y\,y$, 10, fig. 4, with a radius equal to the length of the suspending links, viz. 12 inches, and from a middle line x 10, setting off 5 inches each way, from 10 towards 9, and from 10 towards 11, then the Y being in like manner applied upon a central square, the legs are to be opened, till the insides touch the two points 11 and 9 respectively.

The figure of the F will be constructed by drawing a line 12, Z, 13; and with a radius of 2 ft. 2 inc. striking the circle 13, 14, 15; then draw the line Z 14, for the middle line of the fork, so that the angle 15, Z, 14, may be 75°; and setting off the arch 14, 16, equal to the arch 14, 15, the line z, 16, will give the proper position for the middle line of the hammer-weight. The angle or bend Z, 12, 17, in the handle of the F is 131°, or 49° from the straight line. Lastly, marking off the proper thicknesses, and the branch to be caught by the latch, the F lever may be formed from this mould.

The positions and shapes of the other parts will be readily determined by measurement from the design; but wherever the measures marked upon the plan differ from those resulting from the scale, the figured measures are to be adhered to.

Fig. 7 is a slider, which may be fixed upon the plug-beam, instead of the pin marked in the plan at N; and the rounded end of the slider at N will perform the office of that pin, which is to depress the handle of the F, to shut the injection-cock. Figure 7 shows a part of the plug-beam in two positions, with the slider affixed to it. One position shows the left-hand half of the plug-beam, as seen in front, the spectator looking towards the cylinder. The dotted line P P represents the middle line of the slot which is cut down the plug-beam.

The other view shows the plug-beam sideways, but on the opposite side to that shown in Fig. 4.

In both positions of fig. 7, Q R shows a part of the plug-beam, N V the slider, and W 20, W 20, two screws for retaining the slider in its place. By this slider, the working stroke of the engine may be adjusted by smaller differences than the distance of the holes would occasion, in the same manner as the length of the returning stroke is adjusted by more or fewer saddle pieces of leather.

I, fig. 4, is a roller which reaches from the post D to the post E, and is furnished with a ratchet and catch, to adjust the check rope for the tumbler of the Y to a proper length. A soft matted rope, soaked in tar, answers well for this purpose, having little elasticity, the same for the check rope to the faller of the F.

A sliding rail for the handle or spanner of the regulator to rest upon is not represented, being a thing of course.

N. B. If the weight of the tumbler falling towards the cylinder, should not sufficiently shut the regulator, the long shank of the Y may be bent from a right line a little over towards the cylinder, which will in effect counteract the weight of the arms, acting in a contrary direction.

Note. In other engines which Mr. Smeaton constructed afterwards, he used a better method of communicating the motion of the F lever, to the injection-cock. This was a toothed sector fixed on the axis of the F, in place of the fork, and acting with a toothed-wheel, fixed on the upper end of the turning-plug of the injection-cock, in place of a simple arm or lever. (See Plate VII. fig. 9.) He says,

" No part of a fire-engine is more apt to get out of order and be troublesome, than the injection-cock, when moved by the common F motion, with a forked lever, which acts upon a single arm on the cock-plug; because it cannot be well made to produce a complete quarter turn of the cock-plug, so that there will remain some parts that never mutually apply themselves, and therefore produce an inequality in the wear. A semicircular toothed-wheel upon the top end of the cock-plug would be sufficient for the purpose; but by making it a whole circle, the plug may be turned round to either side, and worked day for day, to each side, which will further tend to produce an equal wear; and this, with daily greasing, will preserve the cock four times as long, before it becomes equally imperfect with the common one."

The toothed sector, fixed on the axis of the F lever in place of the arm F, was 12 inches radius, and had five teeth, or round pins, projecting from the face of it, parallel to the axis of motion: these worked into the teeth of the wheel o p, fixed on the top of the turning plug, which wheel was 10 inches diameter, and had 12 teeth or pins projecting upwards from its face, that is, parallel to the axis of motion. The operation is the same as that already described.

The injection-pipe was made of lead, and of large diameter; not merely on account of giving a free passage to the water, but to cause the water to move very slowly through the pipe; otherwise the energy of the moving column would have been so great, when the current was suddenly stopped, by shutting the injection-cock, as to have strained the joints, and burst the pipe. The joints of the pipe were soldered.

The joint between the cylinder and its bottom, was made by a ring of lead, screwed between their flanges, and made tight by strong white lead putty, smeared on the lead. The joints of the pumps were also made with lead rings, wrapped round with strips of flannel soaked in tar. The lead rings were smaller than the circle of the screw-bolts by which the flanges were united.

D D

Directions to be observed in adjusting the engine for work.

As a considerable weight will be required, at one or other end of the great lever, to make a proper balance of the pump-rods and piston, a number of cast-iron weights should be provided. They should be adapted one to another, so as to pack into one of the triangular spaces formed under the ends of the great lever, between the arch-heads and their braces or struts. These weights being made in pieces 2 feet in length, will reach through all the breadth of the beams: they should be about 1 cwt. or $1\frac{1}{2}$ cwt. each piece, that as many may be used as is found needful (*a*).

All the parts of the engine being put together, and in their respective places, but the buckets being without leathers, and the piston being without packing, place a weight of 38 cwt. ($=$ 4256 lbs.) upon the piston; that is a little more than 1 pound per square inch; then try, by levers or small tackle, whether the great lever, with all its accompaniments, are in balance; that is, whether the resistance to motion be equal both ways. If it be, it is well; if not, put as many of the balancing weights on one end, or the other, of the great lever as will bring the whole to a balance, or indifference to motion either way; then fix the balancing weights fast to the lever, so that they cannot be displaced in working, and take away the 38 cwt. from the piston. It is expected that the engine, when set to work with a full supply of steam, will go quicker in the returning stroke, than in the working stroke, and if so, the balance must not be altered on that account (*b*). The counterweight of the engine when thus adjusted, will be a little more than 1 lb. per sq. inch of the piston.

Take the mean diameter of the valve of the puppet-clack, or safety-valve, in inches, and multiply it by itself, and the product will be the proper number of pounds weight, for the load of the puppet-clack, including its own weight. For instance, if the puppet-clack is $5\frac{1}{2}$ inches diameter, then $5 \cdot 5 \times 5 \cdot 5 = 30 \cdot 25$; therefore its weight must be $30\frac{1}{4}$ lbs. including the weight of the moving parts of the valve itself. The elasticity of the steam will then be regulated to 1 lb. per circular inch, or ·78 lbs per sq. inch.

A small cock, such as are used for wine or spirit-casks, must be inserted into some convenient part of the upper side of the sloping eduction-pipe, not only to admit air to fill the cylinder when the engine ceases to work, but also to admit a small quantity of air continually while it is working. At first setting the engine to work, this cock must be shut, and the cock in the snifting-pipe must be quite opened, but when the engine is brought to work with a steady motion, diminish the aperture of the snift-pipe by closing its cock, till there is not more appearance of steam issuing from the end of the pipe, in the whole of one stroke, than would form a cloud of the apparent bulk of a hogshead. When this is done,

(*a*) These directions refer to an engine of 66 inc. cylinder, which was sent out to Russia, for draining the docks at Cronstadt, but are here made applicable to the engine at Chase-water.

(*b*) Dr. Robison, as well as Mr. Smeaton, directed that the working stroke should be made slower than the returning stroke; but Mr. Watt was of a different opinion: for he says (in his Annotations to Dr. Robison's Account of the Atmospheric Engine) "it is now generally agreed that an engine to work well, should make the acting stroke in less time than the returning one; or in the engine-keeper's terms, it should go slower out of the house, than it comes in. For if the buckets of the pumps are defective, as is often the case, the quicker the working stroke is taken, the less water will be lost; and the slower the returning stroke is made, the less resistance will be opposed to the descent of the buckets of the pumps, by the throttling of the water, in passing through the narrow apertures of the valves.

give vent gradually by the air-cock, till the piston will but barely bring the catch-pins down to the springs of the cylinder end, when the steam is of a proper strength; or when the steam is weak, adjust the air-cock so that the catch-pins will fall short of the springs, by about six inches.

The best way of making the piston tight, is by packing and ramming down oakum with a mixture of horse-dung, into the groove, till it becomes like pulp, and will hold water upon the top. The packing is not confined down, except by cast-iron weights laid on in segments, so that 8 pieces may complete the circle, and fill up the groove. The weights may be about 3 inches in thickness, which will be of sufficient weight to keep down the packing: each segment should have two handles to lift it out by.

Choose that kind of coal, which affords a bright clear flame, such as the splint coal of Scotland. Small coal, approaching to dust, or round coals in large pieces, are equally to be avoided. Coals in pieces, from the size of a sparrow's egg, to that of a hen's egg, answer best, in proportion to their weight.

Management of the Coals.—Break every coal that is bigger than a goose egg, and the oftener fire is fed the better. Feed the fire a little at a time and often, spreading the fuel equally over the grate. It is no matter how few red coals compose the fire. Keep up the steam, so as to show a little waste at the safety-valve at every stroke; and as soon as ever the engine is found to fall off from its motion, repeat the feed to the fire.

What is wanted is, a bright clear flame, thinly and evenly spread over the whole grate. A heap of red coals is of little use; and when the fire grows too thick by red coals, let it be broken down with the coal-rake, keeping the bars as open, and as evenly open, as possible; not laying on more coals at a time, than what will last five minutes, and spreading them as thinly and evenly over the surface as they can be, observing to give those places most that appear most consumed.

Management of the Fire when Sleck or Small Coal, is used for Fuel.—Open the fire-door when the steam is at the strongest, and with proper toss of the shovel, spread the sleck thinly over the thin places, missing the black ones of the former feed; and keep doing this constantly, repairing the thin places, and keeping as good a body of fire as can be, which will be done by opening the fire-door and repairing oftener than when working with coals. Whenever the engine is stopped from any cause, take the opportunity of cleaning the fire-grate, which must never be suffered to grow foul.

Management of the Cataract.—When miners are at work in the sump, or bottom part of the pit, if the engine draws more water than the drainage requires, put on the cataract to the injection-latch; and it must be adjusted to the exact number of strokes per minute, which will keep up with the influx of water. But if the water accumulates at any time when they are not at work, throw off the cataract, and self work the engine, till the pit is emptied, then apply the cataract, and so on.—N. B. When the cataract is at work, diminish the fire proportionally to the motion, and also take care to lessen the boiler's feed; less water being consumed in steam.

Management of the Air-cock.—After the engine is got steadily to work, without the assistance of the hand, open the air-cock till it begins to flirt a little water at each stroke. If the piston descends rapidly, give more air; in short, give the cylinder all it will take, without spoiling the descent of the piston.

The engine at Chase-water, having three boilers, had three men to attend it; but two would have been sufficient in ordinary working, by calling in other assistance when requisite, to pack the piston, clean the boilers, &c. &c.

The engine at Long Benton had two men, and smaller engines only one man; the common rule was one man to each boiler.

The engine at Chase-water Mine was, at the time it was erected, the most powerful machine in existence; it worked for a few years, and was then altered by Mr. Watt to his improved system, which soon after superseded all the atmospheric engines in Cornwall, where fuel is very expensive, and the mines very deep.

Mr. Smeaton also constructed other engines of the same dimensions as that at Chase-water, for the collieries near Newcastle, and in Yorkshire, and in Scotland; some of which remain in use to the present time.

Particulars of the weights of some of the principal parts of the cast iron-work, for Chase-water Fire-engine; executed at Carron Iron-works, Scotland, 1775.

	Tons.	cwt.	lbs.
The cylinder bored quite through, 72 inc. diam. 10½ ft. long, with a flange round its lower end, and a smaller flange round the top; also a very strong flange near the middle, to suspend it by, and fasten it to the beams	4	16	4
The hemispherical cylinder-bottom, with its flange, to unite it to the cylinder; also a short steam-pipe, and necks for the injection-pipe, snifting-pipe, and eduction-pipe	1	13	36
Cup, or circular channel, to fix on the top of the cylinder	0	13	0
The cast-iron piston	0	14	10
Cast-iron axis for the great lever	0	16	14
Cast-iron part of the regulator, or steam valve	0	1	98

Prices paid to the Carron Company for the above castings, delivered in Cornwall.

For the bored cylinder and working barrels, and for the regulator-valve, 28s. per cwt.
For the clack and buckets, turned, 21s. per cwt.
For the cylinder-bottom, piston, cup, clack door-pieces for the pumps, &c. 18s. per cwt.
For the axis and similar castings, 16s. per cwt.
For the common pipes for the pumps, and other plain pipe-work, and furnace doors, 14s. per cwt.
For grate bars, and bearer bars, and hearth plates, weights, &c. 11s. per cwt.
For wrought-iron screw bolts and nuts, 5d. per lb.
For brass castings for the pump clack-seats, and the regulator-valve, injection-cock, &c. 16d. per lb. without turning or fitting.

Since Mr. Smeaton's time, the improvement of the atmospheric-engine has been much neglected. For the more perfect engine of Mr. Watt, which will be described in the next chapter, having the decided preference in most situations, competent engineers have paid but little attention to Newcomen's engine. But in the districts where coals are cheap, the latter is still in very extensive use for such purposes as draining coal-mines, and for drawing up coals out of the pits, and for some iron-works.

In these situations they can be worked by the small sleck, or waste coals, without expense; and their simplicity and cheapness of construction are strong recommendations, as they are more easily kept in order, and more readily repaired when deranged, than Mr. Watt's engines.

In 1797, Mr. John Curr, an experienced coal-miner, of Sheffield, published a small quarto volume, entitled "*The Coal-Viewer, and Engine-Builder's Practical Companion,*" in which he gives engravings of an atmospheric engine, and copious tables of dimensions for all the parts, proportioned to different sizes of cylinders. It is a useful guide for those who require to construct such engines, being a complete manual for their instruction.

Dimensions and Proportions of an Atmospheric Engine, constructed by Mr. John Curr, at Attercliff Common Colliery, near Sheffield, Yorkshire, 1790.

See Plate VIII.

THE construction and principal dimensions of this engine are sufficiently explained by the engraving, Plate VIII., which contains a side elevation and an end view of the whole engine, showing its two boilers.   After a full description of other atmospheric engines, which has been given, p. 134 and p. 190, it is unnecessary to enter into a detailed description of these drawings, farther than to explain such parts as are different from the preceding engines.

The boiler is not placed beneath the cylinder, as in the original plan of Newcomen, but a boiler and furnace is placed on each side of the engine-house, and the steam is conveyed from them by pipes, to a steam-box or receiver, situated beneath the bottom of the cylinder; and within this box the regulator-valve is placed, in the same manner as it was formerly placed within the boiler.

The space beneath the cylinder being unoccupied by the boiler, two strong cross beams are placed there to fasten the cylinder down to, by four upright legs or pillars, in addition to the two cross beams, by which the cylinder is further supported in the middle of its length.   The ends of all these four cylinder-beams are firmly fixed into the side walls of the engine-house, as is shown in the end view.

The steam receiver for the regulator is placed in the space beneath the cylinder bottom, between its four legs or pillars, and the hot well is situated close to the steam receiver, so as to be partly beneath the cylinder.

The great working lever is composed of a single beam of oak timber, instead of being framed of several pieces, as was practised by Mr. Smeaton; the centre, or axis of motion of the lever, is at the underside of the beam, and works in a socket placed on the top of the wall, immediately beneath the beam; and as the axis does not project more than 4 inches beyond each side of the beam, it is not liable to be broken by the force; but the cylinder which rests in the socket and forms the axis of motion, being small in diameter, has but little friction; the beam is curved downwards at each end, where the arch-heads are fixed on, in order to bring the centre of gravity of the moving mass nearer to the centre of motion.

The injection-cistern is placed at the top of the lever wall, over the centre of the great lever, but at a greater height than in Mr. Smeaton's engines.

The diameter of the cylinder was 61 inches ($=$ 2922 square inches).   The piston was capable of making 9 feet stroke, but usually worked $8\frac{1}{2}$ feet, and made 12 strokes per minute ($=$ 102 feet motion per min.)

It worked five sets of pumps, viz. one set of 13 inches diam. $24\frac{1}{2}$ fathoms lift, weight of the column of water 8480 lbs.   Another pump $13\frac{1}{8}$ inc. diameter, $23\frac{1}{2}$ fathoms lift, weight of the column 8290 lbs.   These two pumps were wrought by the main spear, suspended at $12\frac{1}{2}$ ft. from the centre of motion, viz. at the extremity of the lever, which was 25 ft. long.   Also a pump 15 inc. diam. 5 fathoms lift, weight 2302 lbs. suspended at $7\frac{1}{3}$ ft. from the centre of motion.   A pump $15\frac{1}{4}$ inc. diam. $5\frac{1}{4}$ fathoms lift, weight 2500 lbs. suspended at $6\frac{1}{4}$ ft. from the centre of motion.   And lastly, the injection-pump was 9 inc. diam. 10 fathoms lift, weight 1657 lbs. suspended at 8 ft. from the centre of motion.   The weight of all these columns combined, was equivalent to 20 434 lbs. resistance to the motion of the piston, or 9·12 tons, and at the rate of very nearly 7 lbs. per squ. inc.

The mechanical power exerted by this engine was more than 63 horse-power; for 20 434 lbs. on the piston $\times$ 102 ft. motion $=$ 2 084 268 lbs. raised 1 foot high per min. and ÷ 33 000 lbs. (a horse-power) = 63·16 HP : which is nearly the same power as the 64-inc. cylinder in Mr. Smeaton's Table.

This engine had two boilers, on Mr. Curr's plan, of $14\frac{1}{2}$ ft. diam., the furnaces of which consumed 10 hundred weight (= 1120 lbs.) of sleck, or small coals per hour, or $13\frac{1}{3}$ bushels (of 84 lbs. each) ; and supplied the engine very fully with steam. This is at the rate of (1120 lbs. ÷ 63 HP =) 17·75 lbs. per hour per horse-power.

Or the consumption being 1120 lbs. in 60 minutes, then 84 lbs. (the weight of a bushel) must have worked the engine $4\frac{1}{2}$ minutes, in which time it raised (2 084 268 lbs. × 4·5 min. =) 9·38 millions pounds, 1 foot high, by the consumption of a bushel (= 84 lbs.) of sleck or small coals (a).

Mr. Curr states, that he tried this engine, with an additional pump 7 inches diameter $45\frac{1}{4}$ fathoms lift, weight 4540 lbs. suspended at $12\frac{1}{2}$ feet radius from the centre of motion. This addition increased the load to $8\frac{1}{3}$ lbs. per squ. inc. of the piston ; but with the utmost force of fire, the engine could not then be made to perform more than 9 strokes per minute 8 feet long, or 72 feet per minute : which is far inferior to the power it exerted, when it was loaded as above, with 7 lbs. per squ. inc. because it then went 102 ft. per min.

He also tried the same engine with a load of 6·1 lbs. per squ. inc., and then it performed better, than with $8\frac{1}{2}$ lbs., but not so well as with 7 lbs. ; and hence Mr. Curr concluded, that 7 lbs. per. squ. inc. is the best load for an atmospheric engine (See his Table, p. 164.) (b).

Mr. Curr has given all the requisite dimensions and directions, for constructing engines on his plan (see Plate VIII) of all dimensions, from 25 inc. cylinder, to 70 inc. The most important particulars are contained in the opposite tables, which are abstracts of the originals.

Explanation of Mr. Curr's Tables for the Parts of Atmospheric Engines.

THE grate-bars for the furnace to be cast-iron, 6 inches deep, 3 inc. broad on the top edges, and $1\frac{1}{2}$ at bottom : they are laid with spaces between them $1\frac{1}{4}$ inc. wide. The bars are of the same length as the fire-grate : they are supported at each end, on strong bearing bars, 6 inc. by 4, laid across the width of the ash-pit. At each end of each grate-bar are projecting knobs, which makes it 4 inches wide on the top edge at the ends. So that when the bars are laid close together, side by side, with these wide parts touching, the requisite spaces are left between the other parts of the bars. There are similar projections in the middle of each bar, to keep them from bending sideways, by the heat of the fire.

The fire-door is mounted on hinges, before an iron frame, which is set in the brickwork. The aperture of the door-frame is 13 inc. high by 19 inc. wide.

The boilers are to be made of iron-plates, the edges of the adjacent plates overlapping each other about $1\frac{1}{2}$ inches, and united by rivet-pins $\frac{1}{2}$ an inch diameter, disposed at about 2 inches asunder.

It was formerly the practice to interpose slips of strong brown paper, smeared on both sides with white lead, between the overlapping joints of the plates, to

(a) The author saw this engine at work in 1809, and also some other engines in the neighbourhood of Sheffield, made by Mr. Curr, according to his tables.

(b) This conclusion is very nearly the same as that of Mr. Smeaton, because the latter, in fixing the load at 7·81 lbs. per squ. inc. (p. 169) supposed the engine to have no dead weight of pump-rods, and proposed to make an allowance in cases where there was a dead load, (see p. 183.) But Mr. Curr's estimate of 7 lbs. is intended to include the dead weight of pump-rods. In engines for collieries, as they are usually constructed, the allowance to be made on account of the extra weight of pump-rods will commonly be about $\frac{1}{2}$ or $\frac{3}{4}$ of a pound per squ. inc. ; which being deducted from Mr. Smeaton's standard, will bring it to correspond nearly with that of Mr. Curr.

Mr. Curr's Tables of Proportions for the Parts of Atmospheric Steam-engines, 1797.

Cylinders.	Force.	Boilers. Chimney-top, 48 ft. above grate.				Fire Grates for each boiler.				Injection Cistern, 46 feet high.					Dimensions of Engine-house.				
Diameter in inches.	Pressure on the piston at 7 lbs. per squ. inc.	No. of boilers.	Diam. of the boilers.	Diam. of the chimney pipe.	Diam. of the steam pipe.	Squ. feet.	Width.	Length.	Bars.	Water-way of injection-cock.	Square hole for jet.	Bore of injection-pump.	Bore of injection-pipe.	Bore of eduction-pipe.	Length and breadth inside.	Thickness of Walls. Lever Wall.	Thickness of Walls. Side Walls.	Thickness of Walls. End wall.	Number of Bricks.
	lbs.		Feet.	Inches.	Inches		Ft. In.	Ft. In.	No.	Inches.	Inches,	Inches.	Inches.	Inches.	Feet.	Inches.	Inches.	Inches.	Thousand.
25	3436	1	$8\frac{1}{2}$	16	7	9·7	2 6 by	3 $10\frac{1}{2}$	7	$2\frac{3}{4}$ by $\frac{7}{8}$	$\frac{3}{4}$	$3\frac{3}{4}$	$3\frac{1}{4}$	$5\frac{1}{2}$	$15\frac{1}{2}$ by 7	30	19	15	48
30	4948	1	$10\frac{1}{2}$	18	8	13·7	3 2 ×	4 $4\frac{1}{2}$	9	3 × 1	$\frac{7}{8}$	$4\frac{1}{2}$	$3\frac{1}{2}$	6	16 × 7	35	19	19	55
35	6735	1	$12\frac{1}{2}$	20	9	18·7	3 10 ×	4 $10\frac{1}{2}$	11	$3\frac{1}{8}$ × 1	1	$5\frac{1}{4}$	$3\frac{3}{4}$	$6\frac{1}{2}$	16 × $7\frac{1}{3}$	39	$24\frac{1}{2}$	19	65
40	8796	1	14	22	$9\frac{1}{2}$	22·7	4 4 ×	5 3	13	$3\frac{1}{4}$ × 1	$1\frac{1}{8}$	6	4	$7\frac{1}{4}$	16 × $7\frac{1}{2}$	39	$24\frac{1}{2}$	19	70
45	11133	2	11	19	10	15·0	3 4 ×	4 6	10	$3\frac{3}{8}$ × $1\frac{1}{8}$	$1\frac{1}{4}$	$6\frac{3}{4}$	$4\frac{1}{4}$	$7\frac{1}{2}$	$16\frac{1}{3}$ × 8	44	$24\frac{1}{2}$	$24\frac{1}{2}$	80
50	13744	2	$12\frac{1}{2}$	20	$10\frac{1}{2}$	18·7	3 10 ×	4 $10\frac{1}{2}$	11	$3\frac{1}{2}$ × $1\frac{1}{8}$	$1\frac{3}{8}$	$7\frac{3}{4}$	$4\frac{1}{2}$	8	$16\frac{2}{3}$ × $8\frac{1}{3}$	49	$24\frac{1}{2}$	$24\frac{1}{2}$	90
55	16631	2	$13\frac{1}{2}$	21	11	21·3	4 2 ×	5 $1\frac{1}{2}$	12	$3\frac{7}{8}$ × $1\frac{1}{8}$	$1\frac{1}{2}$	$8\frac{1}{4}$	$4\frac{3}{4}$	$8\frac{1}{4}$	17 × $8\frac{5}{6}$	54	29	$24\frac{1}{2}$	102
60	19792	2	$14\frac{1}{2}$	22	12	23·1	4 6 ×	5 $4\frac{1}{2}$	13	$4\frac{1}{4}$ × $1\frac{1}{4}$	$1\frac{5}{8}$	9	5	$8\frac{1}{2}$	$17\frac{1}{2}$ × $9\frac{1}{4}$	60	29	$24\frac{1}{2}$	120
65	23228	2	16	24	$12\frac{1}{2}$	28·8	5 0 ×	5 9	15	$4\frac{1}{2}$ × $1\frac{1}{4}$	$1\frac{3}{4}$	$9\frac{3}{4}$	$5\frac{1}{4}$	$8\frac{3}{4}$	$17\frac{5}{6}$ × $9\frac{2}{3}$	70	34	29	134
70	26939	2	17	25	13	32·0	5 4 ×	6 0	16	$4\frac{3}{4}$ × $1\frac{3}{8}$	$1\frac{7}{8}$	$10\frac{1}{2}$	$5\frac{1}{2}$	9	$18\frac{1}{4}$ × 10	78	34	29	150

Dimensions and weights of the principal moving parts.

Cylinders.	Size of the great lever, an oak beam 25 feet long in a single piece.			Arch-heads oak $10\frac{1}{2}$ feet long.		Axis or centre of lever.	Piston.		Piston Weights.	Wrought iron shanks for piston.		Size and weight of catch-pins at the ends of the great lever. Wrought iron.		Dimensions and weight of arch-head chains.		
Diam. inc.	Square of the middle.	Square at the ends.	Cubic feet of oak.	Breadth and width.	Cubic feet of oak.	Diam. of bearing.	Thickness at circum. and centre.	Weight.	3-inch square.	Square of one of the four shanks for the piston.	Weight.			Diam. of the pins for the chains.	Size of the links of the chain.	Weight.
	Inches.	Inches.		Inches.		Inches.	Inches.	Pounds.	Pounds.	Inches.	Pounds.	Inches.	Pounds.	Inches.	Inches.	Pounds.
25	22 by 20	19 × 15	70	15 by 10	23	$3\frac{3}{4}$ }	$1\frac{1}{4}$ to 3	{ 297	160	$1\frac{1}{4}$	78	$2\frac{3}{4}$ by $1\frac{3}{4}$	86	$1\frac{1}{4}$	$1\frac{1}{8}$ by $\frac{1}{2}$ or $\frac{3}{4}$	232
30	24 × 22	21 × 16	80	16 × 11	27	4 }		{ 441	168	$1\frac{1}{2}$	114	3 × 2	107	$1\frac{3}{8}$	$1\frac{1}{4}$ × $\frac{5}{8}$ or $\frac{15}{16}$	315
35	26 × 23	22 × 17	95	17 × 12	31	$4\frac{1}{4}$ }	$1\frac{1}{2}$ to $3\frac{1}{2}$	{ 680	231	$1\frac{3}{4}$	142	$3\frac{1}{4}$ × 2	120	$1\frac{1}{2}$	$1\frac{3}{8}$ × $\frac{5}{8}$ or $\frac{15}{16}$	360
40	28 × 25	24 × 18	110	18 × $12\frac{1}{2}$	35	$4\frac{1}{2}$ }		{ 887	252	2	206	$3\frac{1}{2}$ × $2\frac{1}{8}$	142	$1\frac{3}{4}$	$1\frac{1}{2}$ × $\frac{3}{4}$ or $1\frac{1}{8}$	468
45	30 × 27	26 × 19	125	19 × 13	38	$4\frac{3}{4}$ }	$1\frac{3}{4}$ to 4	{ 1262	304	$2\frac{1}{4}$	260	$3\frac{3}{4}$ × $2\frac{1}{8}$	161	$1\frac{7}{8}$	$1\frac{5}{8}$ × $\frac{7}{8}$ or $1\frac{5}{16}$	590
50	32 × 29	28 × 21	140	21 × 14	44	5 }		{ 1553	308	$2\frac{1}{2}$	329	4 × $2\frac{1}{4}$	190	2	$1\frac{3}{4}$ × 1 or $1\frac{1}{2}$	720
55	33 × 30	29 × 22	155	22 × $14\frac{1}{2}$	48	$5\frac{1}{2}$	2 to $4\frac{1}{2}$	2097	377	$2\frac{3}{4}$	398	$4\frac{1}{4}$ × $2\frac{1}{4}$	210	2 of $1\frac{3}{4}$	$1\frac{1}{2}$ × $\frac{3}{4}$ or $1\frac{1}{8}$	936
60	35 × 32	31 × 24	175	24 × 15	54	6		2490	392	3	484	$4\frac{3}{4}$ × $2\frac{1}{4}$	240	2 of $1\frac{7}{8}$	$1\frac{1}{2}$ × $\frac{7}{8}$ or $1\frac{5}{16}$	1114
65	36 × 33	32 × 25	190	25 × $15\frac{1}{2}$	60	$6\frac{1}{4}$ }	2 to 5	{ 3124	450	$3\frac{1}{3}$	520	5 × $2\frac{1}{2}$	287	2 of $1\frac{7}{8}$	$1\frac{5}{8}$ × 1 or $1\frac{1}{2}$	1320
70	38 × 35	33 × 26	210	26 × 16	65	$6\frac{1}{2}$ }		{ 3627	504	$3\frac{1}{4}$	580	$5\frac{1}{2}$ × $2\frac{1}{2}$	323	2 of 2	$1\frac{3}{4}$ × 1 or $1\frac{1}{2}$	1440

In all these engines, the length of the stroke is supposed to be 9 feet, and that the engine works $8\frac{1}{2}$ feet stroke in common work.

make them tight; but by subsequent improvements in the workmanship, the plates are so closely fitted, as to require no paper or white lead in the joints; and they are much better without, for if the overlappings are made at all close, the rusting of the iron will soon fill up all minute crevices, and make the boiler perfectly steam-tight.

Mr. Curr directs the boiler to be set in a furnace of a very different structure to those before described. The boiler itself is made nearly in the same form as Mr. Smeaton's, the bottom being concave, and the fire-grate placed beneath it; the level of the grate 13 inc. below the lowest part of the circumference of the boiler bottom. The weight of the boiler is supported on 12 small piers of brick-work, 15 inc. square, arranged in a circle round the circumference of the bottom, but leaving intervals between them 15 inc. wide, by 13 inc. high, for the flame to pass up, from the bottom in all directions, round the outside of the boiler.

The boiler is surrounded by a dome of brick-work, built in the form of a bottle or decanter, with a tall neck rising from it, over the centre of the boiler to form the chimney. This brick-work does not touch the boiler at any part, but leaves a space of 9 inc. all round between the boiler and the brick-work, for the flame to act in, as it rises up through the intervals between the 12 brick pillars from beneath the bottom, and the flame ascends on the outside of the boiler to the chimney, which is over the centre of the boiler. The internal diameter of the chimney is 22 inc. and the height to the top is about 48 feet above the fire-grate.

Mr. Curr says, this method of placing the boiler upon detached brick-piers, 15 inc. square, and enclosing it within a circular wall, shaped like a bottle, the brickwork being 10 inches thick at the bottom part, and only 5 above, he found to answer better, in seven years' practice, than any other method he had tried, the heat being admitted to the boiler in a very impartial manner. The above dimensions are suitable to a boiler of $14\frac{1}{2}$ feet diameter.

In this way, the flame acts on the boiler in ascending, instead of circulating horizontally round it; and its action is not confined to the lower part of the boiler, containing the water, but the upper part, containing the steam, will also receive a share of the heat.

Boilers constructed according to this plan, must be placed at the outside of the engine-house, and the steam conveyed from them, by steam-pipes, to a circular chest or steam-receiver, placed beneath the cylinder, and containing the regulator.

The cylinder is not suspended between beams, but is placed upon 4 legs, or short hollow iron pillars, which are erected upon strong beams, 14 inc. by 10 inc. extending across the house, and the ends are worked into the side-walls near their foundations. The cylinder is fastened down to these beams, by strong bolts, which pass down through the hollow of the pillars. The space between the pillars allows room for the steam-receiver and the hot-well. The cylinder is also fastened between cylinder-beams, 20 inc. by 18 inc. near its middle, as before described; but they do not require such great strength, as if they bore the whole strain of the cylinder.

The cylinder for a 9 feet stroke, should be 10 ft. long: it should have a cup ring round the top of it, and 4 projecting ears from the middle of it, to fasten it down to the cylinder beams, with bolts $1\frac{1}{2}$ inc. square. Cylinders, of less than 45 inc. may be $\frac{7}{8}$ thickness of metal, when bored, from 45 inc. to 55, 1 inch thick; and above 56 inc. $1\frac{1}{8}$ inc. thick.

For small cylinders, under 36 inc. diam. the flange at the lower end of the cylinder, should be $3\frac{1}{2}$ inc. broad and $1\frac{1}{8}$ inc. thick; for cylinders, from 36 inc.

to 45 inc., $3\frac{1}{2}$ inc. broad and $1\frac{1}{4}$ inc. thick.   From 46 to 55 inc., 4 inc. broad and $1\frac{1}{2}$ thick ; and above 55 inc., 4 inc. broad and $1\frac{3}{4}$ inc. thick.   The bolt-holes to be about 9 or 10 inc. apart : the bolts for cylinders under 40 inc. $1\frac{1}{4}$ inc. diam. and above 40 inc. $1\frac{1}{2}$ inc. bolts.

The cylinder bottom is a flat dish, of the same diameter as the cylinder, and 12 inches deep, with a flange round its upper edge, to unite it to the cylinder ; a ring of lead being interposed in the joint, to make it fit.   The steam-pipe rises up inside in the centre, 3 inches above the flat bottom, to prevent the injection water running down the steam-pipe.   The lower end of the steam-pipe terminates with a flange, by which it is united to a corresponding flange, on the top of a neck or short steam-pipe, rising from the top, or cover of the steam-box or receiver, which contains the regulator.   The cylinder bottom has also three other short necks, with flanges ; one to receive the injection-pipe in front, and another exactly opposite to it, for the snifting-pipe, and a larger neck proceeding downwards, to fix the eduction-pipe to.

The top or cover of the steam-box, forms the regulator plate, beneath which the moveable valve or regulator applies.   The socket or tube, to receive the upright axis of the regulator, is 6 inches deep, and tapers from the upper side downwards ; the moveable regulator or valve should be fixed with a spring, to keep it as tight as possible.   The steam-box has a branch projecting from each side, to join to the steam-pipes from the boilers ; these steam-pipes are 12 inches diameter, and have a little descent towards each boiler ; they are joined to the branches of the steam-box, by lap joints, or by short cast-iron tubes 14 inches long, put over each joint, and wedged all round to fill up the space ; the other end of each steam-pipe is united by an oblique flange, to the top of the boiler.

The discharge or safety valves, to allow the superabundant steam to escape, are adapted to these pipes, and the handles from them, pass through the walls into the engine-house, to enable the engine-man to discharge the steam.

The eduction-pipe proceeds from the bottom of the cylinder, in a sloping direction into the hot-well, which is a box or cistern of cast-iron, of a triangular shape, placed in the space beneath the cylinder.   The extremity of the eduction-pipe is covered by a brass hanging valve, which is suspended by a hinge joint at the upper side ; it is called the horse-foot valve, because the end of the eduction-pipe is cut off obliquely, so as to resemble a hoof in form.   From each side of the hot-well a feeding-pipe proceeds, with a regulating cock to supply the boiler.

The man-holes, to give an entrance into each boiler, are short lengths of cast-iron pipe, 21 inc. diam. inside, and $2\frac{1}{2}$ feet long, with a flange at each end, one of which is screwed to the boiler, and the other reaches through the brick-work, to the outside, to fix on the door.   The gauge-cocks are also fixed at the ends of pipes, which reach from the boiler, through the brickwork.

The injection-pipe, and also the pipe of the injection-pump, and the waste-pipe or overflow, are made of cast-iron, in several lengths, joined by flanges, as shown in the drawing.   The end beyond the injection-cock, is inserted into the neck at the bottom of the cylinder, and turns up at the extremity ; over which an iron plate $\frac{3}{8}$ of an inch thick is fastened, by 4 screws, and the plate is perforated with a square hole, for the jet to issue from.

The injection-cistern is placed 36 feet high above the top of the cylinder, in order to give a great velocity to the injection, and produce a minute division and dispersion of the cold water in the steam.   The orifice of the jet for the 61-inch cylinder is a square hole of $1\frac{5}{8}$ inc. square, and the sizes for other engines are stated in the table.   They are much larger than directed by Mr. Smeaton, because

E E

Mr. Curr intended to throw in a great quantity of injection at first, and thus to produce a sudden condensation, and the working gear was to be regulated so as to close the injection-cock gradually, and diminish the stream, as soon as the piston had descended a little way. Mr. Curr states this as a great advantage; and supposed the increased effect of his engine, of 61 inch cylinder, before quoted, to be entirely derived from that circumstance. The injection-cistern in that engine was placed so high, that the surface of the water in it, was 46 feet vertical height above the orifice of the jet.

The most advantageous system of working an atmospheric engine is, to throw in a very copious injection at the first, and with a great velocity, because the cylinder full of steam should be cooled and condensed as quickly as possible; but the interior of the cylinder, being once cooled, a very reduced stream of injection will keep down the temperature sufficiently as the piston descends, and as the capacity of the space beneath the piston diminishes.

Now supposing that a good vacuum is thus obtained very suddenly at the commencement, the piston must preponderate with much more force than is stated in the estimates (p. 178), whilst the resistances will be nearly inactive. Accordingly the moving mass will be urged into motion, with a very sudden acceleration, during the first part of the descent of the piston; but as motion takes place, and as the velocity increases, the resistances will become more and more active, and thus tend to diminish the rate of acceleration, though it must still go on.

As the piston descends further, and as the capacity of the cylinder diminishes, the small flow of injection will become unable to keep down the temperature; consequently, the elasticity of the vapour, or steam, remaining beneath the piston, must begin to increase, and thereby diminish the descending force of the piston; but this will be no evil, because the impetus of the moving mass will be sufficient to continue the motion it has acquired.

When the injection-cock begins to close, towards the end of the stroke, the elasticity of the steam beneath the piston must increase with rapidity, both from the diminishing flow of water, and from the rapidly diminishing space which that steam has to occupy; consequently the descending force of the piston will be greatly impaired, and the continuance of the descent, must result almost entirely from the impetus of the moving mass: therefore as this energy of motion expends itself, the velocity must diminish, till at last the injection being quite stopped, the regulator is opened, and readmits the steam, which will destroy all the remaining descending force of the piston, and prepare it for returning.

Hence, if a lively motion is given to the piston at first, so as to accumulate a considerable energy in the moving mass, it is not absolutely necessary to preserve the condensation undiminished to the end of the stroke; because the impetus of the moving mass will continue the motion, when the diminishing preponderance of the piston, would not of itself be sufficient to do so, in its enfeebled state.

By thus allowing the steam in the bottom of the cylinder to accumulate its heat, and increase in elasticity towards the end of the stroke, we at least save some of the inevitable waste by the condensation of the steam, when it first enters the small space at the bottom of the cylinder; for the warmer that space is, the less steam and heat will be required to bring it to the boiling temperature.

This method of injecting a sudden affusion of cold water, will in some measure counteract the natural tendency to accelerated motion, during the working stroke; for the motion is brought to a greater uniformity, by urging it most forcibly at first, whilst the resistance of the inertiæ of the moving parts is to be overcome, and diminishing the impelling force as the motion takes place.

The injection-pump is intended to work two-thirds of the stroke of the piston, and the diameter of the cylinder is $6\frac{2}{3}$ times the diameter of the pumps. Hence the capacity of the cylinder is $66\frac{2}{3}$ times as great as that of the pump ; so that one measure of cold water must condense $66\frac{2}{3}$ measures of steam in the cylinder, including that steam which is wasted by the condensation, on entering the cylinder.

The great lever is formed of a single beam of oak, 35 inc. by 32 inc. in the middle ; and 31 by 24 inc. at each end ; the length 25 ft. between the centres of the chains at each end. The beam is curved a little in its length, so that the ends are depressed ; the gudgeon or fulcrum is applied at the under side of the beam, and does not pass through the wood. In this way the bearing will be immediately beneath the beam.

The axis or centre of motion is a portion of a cylinder, 6 inch. diam. which is formed in one piece with a flat square plate of cast iron, $4\frac{1}{2}$ ft. long by 3 ft. wide, and $2\frac{1}{2}$ inc. thick, which is fitted to the underside of the beam and let into the wood 1 inc. ; about two-thirds of the circumference of the cylinder projects out below the flat plate, and this part is fitted into a brass socket, which is received into a cast-iron box, supported on the top of the lever-wall. By this means, though the working part of the axis is small in diameter, it is extremely strong ; because the cylinder, 6 inc. diam. possesses all the strength of the plate, which is $2\frac{1}{2}$ inc. thick, and the weight of the beam is applied immediately over it.

The arch-heads are fixed at each end of the lever, by mortises and tennons, and secured thereto by two iron stays or braces, called martingales, to the extremities of which the main chains are linked. These chains are about $9\frac{1}{2}$ ft. long ; each one is composed of two links, placed side by side, between three links, all the five links being united by the same pin. The iron of the links is $1\frac{1}{2}$ by $\frac{7}{8}$ inc. for the thin links which are the triple ones ; and $1\frac{1}{2}$ inc. by $1\frac{5}{16}$ for the thick links, which are the intermediate or double ones ; the pins are $1\frac{7}{8}$ diam. Engines above 55-inch cylinders have two chains at each end of the lever ; the smaller sizes have but one chain. The weights marked in the table, are those of the set of chains complete for one end of the lever.

The catch-pins are fixed through the arch-heads at each end, to limit the motion of the lever ; their weights and dimensions are given in the table.

The piston is made of cast-iron, convex on the lower side, and thicker in the centre than at the circumference, as is shown by the table ; the piston is about one-eighth of an inch less in diameter than the cylinder. The stuffing-rim, which is raised upon the upper side of the piston, all round its circumference, is about 4 inc. from the inside of the cylinder, and that space is to receive the packing, which is kept down by the six piston-weights, forming a circular ring 3 inc. square in the section. The weights of these parts are shown by the table ; also the dimensions and weights of the four piston-shanks, by which the piston is suspended to the main-chains. These shanks are in four branches affixed into the piston, near the circumference, but are brought together in the centre, to attach to the chains by one joint-pin.

Mr. Curr also describes an atmospheric engine with two stop-valves, to rise and fall perpendicularly into conical apertures or seats, for the alternate admission of steam and cold water into the cylinder, instead of the regulator and injection-cock. This is the modern form of construction for such engines, and will be described in its proper place, because they are improvements adopted from Mr. Watt's engines, which will be described in the fifth chapter.

CHAPTER III.

*On the application of the Atmospheric Engine to different purposes,
during the first fifty years after its introduction.*

THE great object which the first inventors of steam-engines had in view, was
the drainage of mines.   In this Mr. Savery could not succeed, and was obliged to
content himself with other inferior employments for his engine.   But Mr. Newcomen succeeded very well; and for many years after the first introduction of his
engines, the boldest speculations in mining were easily realized, without reaching
the limits of their powers.

The established coal mines about Coventry and Newcastle were the first to
extend their works upon the new system; and as the knowledge of the invention
spread by degrees over other districts of England and Scotland, where coals
abound, new mines were opened in all parts.

The more valuable mines of tin and copper in Cornwall were very soon
deepened by the aid of steam-engines, and those of large size, as the mines had
been previously worked out very deep, by water wheels.   There were also some
engines for draining lead-mines in Derbyshire and in Cumberland.

In London a few engines were set up for supplying water to the houses, in
place of horse engines, which were before used; and some new establishments
were begun with engines.   Other principal towns had also small water-works.

These were nearly all the uses to which the atmospheric engines were applied
in a large way, during fifty years after the first introduction, except a few
instances of pumping up water, by an engine, to supply the water-wheel of a mill,
according to Mr. Savery's original idea.   This was first executed in 1752 for a
manufactory near Bristol, and 25 years afterwards it became a common practice.

A few fire-engines were applied to mines on the Continent, in France, Hungary,
Sweden, and Russia; but they were all made in England, and set up by Englishmen.
This fact is stated by Mons. Belidor, in his *Architecture Hydraulique*,
1734, as a proof that the fire-engine is entirely an English production; for notwithstanding the claims made by the French, in consequence of Papin's suggestions
in 1690, nothing was ever done by them, even after the making of engines was
become a common mechanic trade in Britain (*a*).

In 1736 a patent was granted to Mr. Jonathan Hulls, for a machine for
carrying ships and vessels out of, or into any harbour or river, against wind and
tide, or in a calm.   He published a pamphlet, with an engraving of his machine,
which was a Newcomen's engine, fitted on board a vessel, and adapted, by ropes
passing over pulleys, to give continuous circular motion to rowers, or paddle-
wheels, placed at the stern of the vessel.   The project was not perfected, or
put to any real use, and in fact, was little more than a renewal of Papin's first
scheme of 1690, (see p. 98.)

When the atmospheric engine was applied to a mine, the pumps were always
of the kind called sucking pumps, with valves in the moveable buckets; and they

(*a*) Mr. Martin Triewald appears to have been the first foreigner who studied the fire-engine;
he was a native of Sweden, and travelled to England in 1716; he was employed for some years by
the proprietor of some coal-mines at Newcastle, to superintend an engine; he also attended the
philosophical lectures of Desaguliers, in London.   In 1726 he returned to his native country, where
he set up an engine, the parts for which were constructed in England.   Triewald was afterwards
appointed engineer to the king of Sweden.

could only raise the water up to the level of the engine, but not much higher.   The pumps were placed in the pit, and drew up the water, every time the piston descended.   For small depths, only one pump was used, but as the mines became deeper, it became necessary to use two or more pumps, to lift the water from one to another in succession.

For water-works the established plan was, to force the water up by the engine, through a pipe into a cistern at the top of a high tower, from which it descended again by another pipe, and was distributed through the streets, by branches from that pipe.   In some cases, the cistern into which the water was pumped up, was a pond or reservoir formed on elevated ground, and made large enough to contain a considerable supply of water.   But in all these cases, the water required to be forced up to a considerable height above the engine, and therefore the sucking-pump, such as was used for mines, was not applicable.

The most common mode of forcing water, was by a forcing pump, with a solid piston, which raised the water in descending, that is, during the returning stroke, or ascent of the steam-piston of the engine, the rod of the forcing-pump was therefore loaded with a large counterweight, sufficiently heavy to force up the column of water : this weight was raised by the force of the steam-piston when it descended, and in returning it made the effective stroke in the pump.

Another plan was to use a lifting force-pump, usually called a jack-head pump. It is constructed like a sucking-pump, with valves in the bucket, but a lifting-pipe proceeds sideways from the top of the barrel, which is closed by a cover ; the pump-rod passes through the cover, and a close collar of leathers is fitted to the rod, which is made smooth and polished, so that it may move freely up and down therein, without allowing any leakage of water.   This cover prevents the water from passing out at the top of the barrel, and therefore it must pass up the lifting-pipe to any required height.   The water is raised by this pump when the steam-piston makes its descending stroke, the same as with a sucking-pump.

In some engines for water-works, both kinds of pumps were combined, the two pumps being placed side by side : the force-pump raised water during the ascent of the piston, by the returning force of the counterweight, and the lifting-pump raised water, by the direct action of the steam-piston, during the descending stroke.   The advantage of this arrangement was, that the water was made to flow continually, or with very short intermissions, when the engine changed the direction of its motion.   A large air vessel was sometimes added to receive the water, and by the re-action of the compressed air, it equalized the motion of the water through the pipe, and kept up a continuous stream.

We shall now proceed to give the dimensions and proportions of the best examples of atmospheric engines, applied to each of the purposes for which they were actually employed, previous to the year 1775 ; which may be considered as half a century after they had come into common use.

The engines for draining mines have been fully explained ; but it yet remains to show their application, and the construction of the pump-work, in the pit or shaft of the mine.   A large engine and its pump-work for a public waterworks, to supply a town, will be described, and particulars given, of a small engine for the same purpose.   Also a portable engine, for the purpose of draining foundations, wells, or any similar works for temporary occasions.   Lastly, instances of engines applied to raise water, to turn overshot water-wheels : for machines for blowing air into furnaces for smelting iron, and other instances of machines for drawing up coals out of a mine.   These particulars, in addition to what has been already given, will give our readers sufficient information on the subject of the applications of Newcomen's engine.

DESCRIPTION OF THE PIT-WORK, AND PUMPS, OF AN ATMOSPHERIC ENGINE FOR DRAINING A MINE.

Plate IV.

The pumps for this purpose are generally sucking-pumps, with two valves; viz. one in the bucket, which is moveable up and down in the barrel; and another fixed valve at the bottom of the barrel. The construction is similar to that of the pumps commonly used for domestic purposes, but the mine pumps are made of cast-iron, and have doors at the upper and lower parts of the barrel, which can be opened, to give access to the valve or bucket, when they require repairs or renewal of the leathers. The doors are fastened on with screws.

The pump-rods descend into the pumps, and have the buckets fastened to the lower ends. The upper ends of the rods of all the several pumps in the pit, are united into one main spear, which is suspended by the chains, from the arch-head of the great lever of the engine; so that all the pump-buckets are drawn up together, and all raise their water at the same time.

When the depth of the mine is considerable, the pit is divided into two or three lifts, and as many different pumps are used. The lowest, to raise the water from the bottom of the mine, into a cistern fixed in the pit, at the first lift; and in this cistern the lower end of the second pump stands, and raises the water therefrom into a second cistern, from which the third or upper pump raises it to the surface, or to the level or subterraneous passage, at which it can run off to some lower land than that on which the engine is erected.

It rarely happens that an engine draws the water quite to the surface; for, unless it is placed just at the level of the sea, an adit may, in most situations, be carried up from a brook at a distance, so as to intersect the pit at a considerable depth below the surface, and thereby diminish the height to which the water is to be raised by the engine. There are instances in the mining districts of Cornwall, Derbyshire, and Cumberland, of these subterraneous levels, adits, or soughs, being carried many miles in length from the sea, or from a great river, in order to drain all the mines of a district, to a great depth below the surface; and then only such mines as require to be worked beneath this level, have any engine at all; and all the engines in such a district discharge their water into the common level, which is made to communicate with them all by branches.

Figures 1 and 2, Plate IV, are sketches to explain the usual manner of arranging the pumps in the shaft or pit of a coal-mine; fig. 1 being a section of the upper part of the pit, and fig. 2 a continuation of the same section, to show the lower part. The water is supposed to be raised at twice by two pumps: the lowest raising the water from the bottom of the pit into a cistern, out of which the upper pump draws the water, and raises it up to an adit, level, or subterraneous drain, by which the water runs away into the lowest brook or river in the neighbourhood.

A A, fig. 1, is the level of the ground. The end view of the engine-house shows the lever-wall, and the arch-head at the outer end of the great lever, from which the pump-rod M is suspended by its chain. D D are the two chimneys, and G the injection cistern, placed on the top of the lever-wall, between them. C is the side building, containing the second boiler.

The pit-head frame is also shown in fig. 1: it consists of two legs or posts, erected nearly upright, upon a strong horizontal beam, which lies across the mouth of the pit.   These legs sustain the outer ends of the spring-beams, and are also carried up much higher, to support two horizontal pieces, between which a pulley is fitted, and receives a very strong rope, for occasionally hoisting the pumps and rods in the pit.   The rope being conducted over another pulley, nearly two-thirds down one of the legs, is wound round an upright axis or windlass, which is supported in a frame, and has long horizontal levers fixed into it at the lower end ; it is put in motion by a sufficient number of men walking round in a circle, and pushing the levers before them, like a ship's capstan ; and by this means the rope is wound up, to lift the pumps or rods when requisite.

The pump-rod or spear M descends nearly in the centre of the pit, and is guided at different places between cross-beams, to keep it in its vertical direction. At each of these cross-beams, a stage or resting place is formed, and the ascent from one to the next is by ladders, as shown in the figures.   At the level, or adit, marked L, the main spear terminates in an iron triangle, or fork Y, called the Y, because it resembles that letter inverted, thus λ, see fig. 5 ; and to each of the prongs of the fork, a smaller wooden rod is suspended, as shown in fig. 2 and fig. 5.   These two are the real pump-rods ; one of them goes down into the upper pump P, and the other, which is much longer, descends perpendicularly, by the side of the upper pump, and then goes down into the lower pump.

Each pump consists of a column of iron-pipes, placed one above the other, and united by flanges and screw-bolts : see the enlarged section of the pump, fig. 8. The lowest length of pipe, H, is called wind-bore, or suction-piece ; and the foot, which is immersed in water, has a number of holes to admit the water, but to prevent the entrance of dirt.   The next length I is called the clack-piece, because it has a conical seat bored out in it to receive the clack-box or valve K ; and immediately above this, is an opening sideways at k, and a cover called the clack-door, to take off, and give access to the valves when they require repair.

The next length E E of the pump is the working-barrel : it is bored true, and smooth withinside, for the bucket L to work freely up and down in it.   This part is sometimes made of bell-metal, particularly when the water is of a corrosive nature. The length of pipe P above the working-barrel, has a side opening p, and a cover called the bucket-door, to open when the bucket or its valves require repair. The different lengths of pipes above this, which are called pump-pieces, are only plain cylindrical pipes, with flanges to unite them.

All the joints are made tight by rings of lead, wrapped with flannel soaked in tar.   One of these rings is interposed between the flanges of each joint, and they are forcibly compressed by the screw-bolts.

In this manner a column is formed as high as is required ; and the upper length is joined to a wooden spout, or trough, p, which delivers the water into the cistern r of the upper pump, or into the adit.

The bucket T is shown in figs. 6 and 7, on a larger scale, and the clack is constructed in a similar manner.   It is a circular ring of cast-iron, with a bar diametrically across it.   A circular piece of thick leather, such as the soles of shoes are made of, is placed upon the ring ; and being about an inch larger in diameter than the internal diameter of the ring, it will have about half an inch overlap, or bearing on the ring, so as entirely to cover the apertures.   This leather is fastened down, by a bar which is placed across it, exactly over the cross-bar of the bucket, and fastened down by the two prongs of the pump-rod, which pass down through mortises in this bar, and also through the leather and through the cross-bar of the bucket ; and are keyed across beneath, so as to unite all the parts together.

The leather is fortified by strong iron plates, which are riveted upon the upper side, and smaller plates are also riveted at the under side : these plates form the real valves ; the use of the leather is to form the hinges, or flaps on which they turn up, when they open, and also to make facings for the under sides to fit tight when shut. Double semicircular valves of this kind, having the joint or hinge across the middle, are called butterfly valves, from their resemblance to the wings of that insect, when turned up.

The wooden rod or spear m has an iron stem, united to it at the lower end, and this is fixed by a hand-in-hand joint n to a short length, or shank of the bucket, which divides into two prongs at the lower end, and the extremities of these prongs pass through the cross bar of the bucket, as before mentioned. The two parts of the joint are made to interlock together, like two hands clasped, as shown at n, fig. 6, and they are kept in contact by a strong square hoop, driven down over the two ; but when the hoop is driven up, the joint is easily detached, and is therefore called a take-off joint.

The cast-iron bucket is made conical on the outside circumference, and is surrounded by a hoop or rim of leather, which rises an inch above the top surface of the bucket and valves; and from its conical figure it spreads out at its upper edge to a larger diameter than the bucket. To confine the leather to the bucket, it is first fastened by a few wooden nails driven into small holes in the bucket, and is then surrounded by a thin iron hoop driven upwards : this hoop is conical within to fit the bucket, but the outside is cylindrical as shown in the section, fig. 6, and very nearly fills the working barrel ; the top edge of the leather rim, which is cut thin, and spread out rather larger than the hoop, fills the barrel exactly, when in its place, and the thin edge applied so closely to the inside of the barrel as to prevent any leakage of water ; for the pressure of the water forces the leather in closer contact with the barrel. To keep the external hoop of the bucket in its place, another cross-bar is applied beneath the main cross-bar of the bucket, and its ends catch beneath the lower edge of the bucket hoop, so as to retain it ; the two prongs of the iron-shank pass through this cross-bar, and it is thus made fast by the same cross-keys which are driven through the lower ends of the two prongs.

The clack K, fig. 8, is constructed in the same manner as the bucket, with butterfly valves, and is surrounded with leather on the edge ; but it has no external hoop, because the clack seat is bored out to a cone, which exactly fits the outside of the leather, and the weight of the column of water jams it fast into its place. The two iron prongs of the clack, are united so as to form a strong loop, in order that the clack may be occasionally drawn up, by a hook at the end of a rope, let down through the barrel ; the bucket and rod being first withdrawn.

The pipes or pump pieces are about three quarters of an inch, or an inch, larger in diameter than the working barrel, to allow the bucket and clack to be thus drawn out. The clack-piece and wind-bore are about an inch less in diameter, than the working barrel.

The wooden-rods or pump-spears are made of fir wood, or mast timber, in lengths of from 40 to 60 feet, they were formerly scarfed together at the joints, each piece being cut away to half the thickness, and united by bolts, but as this occasioned waste of timber the rods were frequently composed of two thicknesses or flat pieces, put together as shown in fig. 5, to form a square rod ; the joints of one set of pieces being opposite to the middle of the other pieces, and a long strap, or plate of iron, being applied over each joint, the whole were firmly united by bolts as represented in the figure ; as a security from the ends drawing asunder, small keys of oak were also let into the two adjoining surfaces of the two pieces, to prevent them sliding one by the other. This was the old fashion, and was recommended by Mr. Smeaton, see p. 196 ; for the largest rods three iron plates were applied to each joint.

The two small rods $m\,m$, fig. 5, are connected with the λ, by joint-pins which can be withdrawn, to detach either or both of the rods occasionally. The rod m of the lower pump Q has also a hand-in-hand or take-off joint, n, fig. 8, just at the top or trough of that pump ; and at the lower end of each rod is a similar joint, to unite the shank of the bucket to it, as before described.

These provisions are to facilitate the repair of the pumps. When a valve fails, from the leather being worn out, a spare bucket being kept ready prepared, the pump-rod is detached just at the top of the pump, and drawn up out of the pump, by the rope of the capstan, till the bucket appears above the top of the pump : the shank of the old bucket is then removed at its take-off joint n, and replaced by the new bucket ; and if the clack also requires to be renewed, the capstan-rope is lowered down into the pump, with a hook at the end, which being hooked to the loop of the clack, it is drawn up out of its seat, and the new one let down.

If only the clack requires renewal, but not the bucket, the clack can be lifted up out of its seat through the clack-door k, by an iron lever, without withdrawing the rod and bucket from the barrel ; and sometimes the bucket is made so that it can be replaced through the bucket-door, without drawing the rod, only raising it up, so that the bucket comes opposite to the bucket-door p.

The drawings in Plate IV are supposed to represent the pit work of the engine at Long Benton, represented in Plates II and III, and already described, pp. 134 and 172 ; but as the small figures 1 and 2 do not admit of being strictly proportioned to a scale, they must be considered as explanatory sketches.

The suction-pumps, such as above described, cannot draw the water by suction, above 20 or 25 feet at the utmost, that is, the moveable bucket must never rise above 20 or 25 feet above the surface of the water, in the cistern or well from which it draws ; and therefore all the height which the pump is required to raise the water above 25 feet, must be done by the column of pipes or pump-pieces added above the working barrel ; and as the pump-rod must pass down through the water in those pipes, they are 1 inch more in diameter than the working barrel, to allow for the space that the rods occupy in them.

The usual proportion is to make the lower length of the pump, or wind bore, about $7\frac{1}{2}$ ft. long, the clack-piece also $7\frac{1}{2}$ ft., and the working barrel two feet or more longer than the stroke which the bucket is intended to make in it. This, with a 7-ft. stroke, will cause the bucket to rise from about 14 ft. to 21 ft. above the surface of the water, in which the lower end of the wind-bore is immersed ; and this height of suction should not be greater, to draw the water to a certainty : nor should it be less, because the clack-door should be high enough up in the pit, to avoid any risk of the water rising so high in the pit, as to cover it in a short time, in case of any failure of the bucket or clack, and the consequent stoppage of the engine ; because the water would prevent access to the door, and occasion much more trouble to draw the bucket and clack up out of the pump to repair them from above.

Hence it should be calculated, that when the bucket or clack fails, and the engine ceases to work, the water will take so much time to rise up to the clack door, as will enable the workmen to renew the faulty pieces, without any interruption from the water. But as cases may occur, when the clack-door will be drowned,

F F

full provision must be made, for executing all repairs of that sort from above, as before described.

When the lower end of the wind-bore is intended to stand in a wood cistern, from which it is to draw its water, it usually has a flat close bottom, which stands upon a strong plank of wood, as shown in fig. 8 ; and this must be very solidly supported beneath, by a strong beam fixed across the pit. The water is admitted through several upright slits, or narrow openings made round the lower part of the wind-bore : these openings should not be more than $2\frac{1}{2}$ inc. wide, to prevent the admission of dirt ; but they must be extended so much in height, as to give a water-passage fully equal to the area of the working barrel, so that the free entrance of the water will not be at all impeded ; otherwise the resistance to the engine will be greatly increased, beyond what has been shown at p. 178.

The construction of the foot of the wind-bore is a circumstance of importance, to prevent the admission of dirt, and yet allow an uninterrupted passage for the water. In the construction already described, the lower part of the pipe is in effect cut away to eight small pillars or legs, which support the weight of the pumps, but leave passage for the water between them. When these legs are reduced very small, an additional thickness is given to them on the outside ; so that the outside of the lower part of the wind-bore, forms a larger circle than the outside diameter of the upper part. The width of the passages should be less at the outside, where the water is to enter, than withinside the pipe, and then they will be less liable to choke up with dirt.

It is also a good provision to enclose the lower part of the wind-bore in a cylindrical cage, made of thin iron or copper bars, like a lantern or bird-cage, which will prevent dirt from passing. It should be at least twice the diameter of the out-side of the pipe, so as to surround it on all sides at a considerable distance. With the same view, it is proper to raise the pump-foot on a thick block, that the aper-tures may not be too near the bottom of the cistern, so as to draw in stones or heavy matters which may be at the bottom of the water ; and also the water in the cistern should be a considerable depth above the apertures, that light chips and floating bodies may not be drawn down into the pumps. But the most important point is to allow ample water-passage, and then the current, or draught of water, into these apertures will not be so rapid as to draw in extraneous matters from a distance, and therefore if they are kept at a distance, the pumps will not be interrupted.

The lowest length of pump, which must draw from the sump or bottom part of the pit, where the miners are at work, does not admit of all these precautions to keep out dirt, but there are others of as much importance.

The lower end of a wind-bore for sinking in a mine, is enlarged to a bulb, which is of greater diameter than the pipe. This enlarged part is usually two or three feet long, and the lower end is shaped like the half of an egg, and this bulb is perforated in all parts, with holes about $1\frac{1}{2}$ inc. diam. called snore holes, to admit the water. It is not practicable in this case to make the water-passage fully equal to the area of the working barrel, as before stated ; nor is it desirable that it should be so, because the pump being required to sup up all the water in the bottom of the pit, it must necessarily draw more or less air into the holes at every stroke. This would occasion the engine to work very irregularly, and the catch-pins at the end of the great lever, would strike the spring-beams at almost every stroke, if a very open water-passage were allowed. But by contracting the passages so that the water experiences a great resistance at passing through the small holes, it makes less difference when air is drawn in ; for it enters with great difficulty and violence, and the engine in consequence works more steadily. This is called snoring, and is a most deafening sound.   And with the same view, it is desirable that when a mine is sinking, the lowest length of pumps should be as short a lift as can conveniently be allowed ; because, as that length will necessarily be a very uncertain resistance, it is desirable to make it as small a part of the whole load as possible, that its fluctuations may have less effect upon the motion of the engine.

To regulate the quantity of water that the sinking-pump shall draw from the bottom of the pit, the upper rows of these snore-holes, in the bulb end of wind-bore, are plugged up with wooden plugs, which the miners can open, or stop up at pleasure, as the water increases or diminishes, or according to the going of the engine. When a pit or mine is first sinking, it is necessary to draw the water completely from the bottom, that the miners may not be obliged to stand in the water to work, which, besides the inconvenience, would prevent them from using gunpowder, which they must employ if the bottom is rocky.

Sometimes instead of the plugs of wood, the lower part of the wind-bore, where the holes are, is surrounded by a circular apron, or bag of strong leather, bound fast, and tight round the pipe, above the holes, and descending to the water, so as to cover and stop all those holes which are above the water.   This apron the miners can, from time to time, turn up or down, like the cuff of a coat-sleeve, in order to open more or less of the holes, and thus regulate the access of water into the wind-bore, so as to prevent the pumps drawing too much air, because the free entrance of air would derange the motion of the engine.

In sinking pits, the lower lift of pumps must stand in the deepest hole which has been made in the bottom of the pit ; and therefore when a new hole has been made, either by picking away or blasting up the rock, near to the foot of the wind-bore, it is moved by levers towards such new hole, and dropped into it, so as to draw the water therefrom.   In long lifts, and heavy pumps, this lowest and moveable length of the pump, is wholly or in part suspended by a strong rope, drawn by the great capstan, which is provided at the top of the shaft for drawing the pump-rods, or the pumps themselves in cases of need.

As the shaft is deepened, new lengths of pump-barrels are added at the top of the lower lift of pumps, generally about six feet at a time ; and the pump-rods are lengthened as the pump requires.

It is of the utmost importance in mine work, to have every facility for repairing without loss of time ; and all parts of the work should be made with the greatest solidity, so as to bear severe strains without injury.

The following tables and scale of dimensions for pumps and pit-work is laid down by Mr. Curr, in his Engine Builder's Companion, 1797.

The engine-spears, or pump-rods, to be of fir-wood, and the splicing joints to overlap 4 ft. and be secured by iron plates, bolted to the rods over the joints. These plates are thicker in the middle, and diminish to each end.   They are pierced for the bolt-holes at about 1 ft. asunder.

The method now generally used is, to unite the pieces of wood by the iron plates alone, without any overlapping of the wood, the ends being exactly fitted end to end, and long plates of iron applied at each side, and fastened by bolts going through all ; small pump-rods have two iron plates, and the largest size have four plates.

The square iron bucket-rod is that which is attached to the lower end of the wooden rod, and occupies the working barrel immediately above the bucket-shank. The length in the table is that of the open part, which receives the wooden rod, and is bolted thereto.

Mr. Curr's Tables of the Dimensions of the Pump Work for Engines for draining Mines.

Dimensions of Pump-Spears and Rods.

Diam. of pump.	Size of square wood rod.	Iron joint-plates and bolts.			Square iron bucket-rods, $11\frac{1}{2}$ feet long.		Iron bucket shanks.	
		Length.	Breadth.—Thickness.	Diam.	Size.	Length.	Strength in common.	Strength at the joint.
Inches.	Inches.	Feet.	Inc. Middle. Ends.	Inches.	Inches.	Feet.		
6	3 squ.	6	$2\frac{1}{2}$ by $\frac{3}{8}$ to $\frac{3}{16}$	$\frac{5}{8}$	$1\frac{1}{2}$ squ.	$4\frac{1}{2}$	2 by $1\frac{1}{4}$	2 squ.
8	$3\frac{1}{2}$	$6\frac{1}{2}$	$2\frac{3}{4} \times \frac{7}{16} \times \frac{3}{16}$	$\frac{11}{16}$	$1\frac{3}{4}$	$4\frac{3}{4}$	$2\frac{1}{4} \times 1\frac{3}{8}$	$2\frac{1}{4}$
10	4	7	$3 \times \frac{1}{2} \times \frac{1}{4}$	$\frac{3}{4}$	2	5	$2\frac{1}{2} \times 1\frac{1}{2}$	$2\frac{1}{2}$
12	$4\frac{1}{2}$	$7\frac{1}{2}$	$3\frac{1}{2} \times \frac{9}{16} \times \frac{5}{16}$	$\frac{3}{4}$	$2\frac{1}{4}$	$5\frac{1}{4}$	$2\frac{5}{8} \times 1\frac{3}{4}$	$2\frac{5}{8}$
14	5	8	$3\frac{1}{2} \times \frac{5}{8} \times \frac{3}{8}$	$\frac{7}{8}$	$2\frac{1}{2}$	$5\frac{1}{2}$	$2\frac{3}{4} \times 2$	$2\frac{3}{4}$
16	$5\frac{1}{2}$	$8\frac{1}{2}$	$4 \times \frac{3}{4} \times \frac{1}{2}$	$\frac{7}{8}$	$2\frac{3}{4}$	6	$3 \times 2\frac{1}{4}$	3
18	6	9	$4 \times \frac{13}{16} \times \frac{9}{16}$	$\frac{15}{16}$	3	$6\frac{1}{2}$	$3\frac{1}{8} \times 2\frac{1}{2}$	$3\frac{1}{8}$
20	$6\frac{1}{2}$	$9\frac{1}{2}$	$4\frac{1}{2} \times \frac{7}{8} \times \frac{5}{8}$	1	$3\frac{1}{4}$	7	$3\frac{1}{4} \times 2\frac{3}{4}$	$3\frac{1}{4}$
22	7	10	$4\frac{1}{2} \times \frac{15}{16} \times \frac{11}{16}$	$1\frac{1}{8}$	$3\frac{1}{2}$	$7\frac{1}{2}$	$3\frac{1}{2} \times 3$	$3\frac{1}{2}$
24	$7\frac{1}{2}$	$10\frac{1}{2}$	$4\frac{3}{4} \times 1 \times \frac{3}{4}$	$1\frac{1}{4}$	$3\frac{3}{4}$	8	$3\frac{3}{4} \times 3\frac{1}{4}$	$3\frac{3}{4}$

Dimensions of Buckets and Clacks.

Diam. of pump.	Cast-iron buckets & clacks.			Iron bucket-hoops.		Cross Bars and Bolts.			
	Depth. Middle sides.	Thick-ness top edge.	Holes for Shank.	Depth.	Thick-ness bot-tom edge.	Thickness of Bar. Middle.	Thickness of Bar. Ends.	Breadth of Bars.	Diam. of Bolts.
Inches.	Inches.	Inches.	Inches.	Inches.	Inches.	Inches.	Inches.	Inches.	Inches.
6	3	$\frac{3}{8}$	$2\frac{3}{4}$ by $\frac{3}{8}$	1	$\frac{1}{8}$	$\frac{3}{4}$	$\frac{1}{2}$	$2\frac{1}{2}$	1
8	$3\frac{1}{4}$	$\frac{7}{16}$	$3 \times \frac{7}{16}$	$1\frac{1}{4}$	$\frac{3}{16}$	1	$\frac{3}{4}$	$2\frac{1}{2}$	$1\frac{1}{8}$
10	$3\frac{1}{2}$	$\frac{1}{2}$	$3\frac{1}{4} \times \frac{1}{2}$	$1\frac{1}{2}$	$\frac{1}{4}$	$1\frac{1}{4}$	$\frac{7}{8}$	$2\frac{1}{2}$	$1\frac{3}{8}$
12	$3\frac{3}{4}$	$\frac{9}{16}$	$3\frac{1}{2} \times \frac{9}{16}$	$1\frac{3}{4}$	$\frac{5}{16}$	$1\frac{3}{4}$	1	3	$1\frac{1}{2}$
14	4 & 5	$\frac{5}{8}$	$3\frac{3}{4} \times \frac{5}{8}$	2	$\frac{3}{8}$	2	$1\frac{1}{4}$	3	$1\frac{3}{4}$
16	$4\frac{1}{4}$ 6	$\frac{3}{4}$	$4 \times \frac{3}{4}$	$2\frac{1}{2}$	$\frac{7}{16}$	$2\frac{1}{2}$	$1\frac{3}{4}$	3	$1\frac{7}{8}$
18	$4\frac{5}{8}$ $6\frac{1}{2}$	$\frac{7}{8}$	$4 \times \frac{3}{4}$	$2\frac{5}{8}$	$\frac{1}{2}$	3	$2\frac{1}{4}$	3	2
20	5 $6\frac{3}{4}$	1	$4\frac{1}{2} \times \frac{13}{16}$	$2\frac{3}{4}$	$\frac{9}{16}$	$3\frac{3}{8}$	$2\frac{1}{2}$	3	$2\frac{1}{4}$
22	$5\frac{1}{2}$ $7\frac{1}{4}$	$1\frac{1}{8}$	$4\frac{3}{4} \times \frac{7}{8}$	$2\frac{7}{8}$	$\frac{5}{8}$	$3\frac{3}{4}$	$2\frac{3}{4}$	3	$2\frac{3}{8}$
24	6 $7\frac{3}{4}$	$1\frac{1}{4}$	5×1	3	$\frac{5}{8}$	4	3	3	$2\frac{1}{2}$

The first table shows the dimensions of the rods and joint-plates which go down into the pumps; the strength is adapted to lift a column of 18 fathoms (= 108 ft.), but if the depth exceeds that, the work must be a little stronger. For instance, the engine at Long Benton (see p. 175), had pumps of 12·2 inc. diam. and $24\frac{2}{3}$ fath. lift; and therefore the rods and iron-work were of dimensions suitable for a 14-inc. pump above.

The strength of the bucket-shank, is shown by the size of the holes in the buckets. This supposes the shank is not divided into two prongs, as before described, but that it is in one piece, which passes through the middle of the cross bar of the bucket, and is cross-keyed below. In this case the leather for the valve is fastened down upon the cross bar, by two strong bolts, going down through every thing. The dimensions of these cross-bars, and bolts, is given in the second Table.

The plain pipes for the pumps are always in lengths of 9 ft.; the flanges to be 3 inc. broad. Ten-inch pumps should have six bolts in each flange, and 13-inch pumps eight bolts. The metal should increase in thickness at the lower part, where the great pressure of the water is. For a 12-inch pump, the first four pipes, at the top, need not be more than three-fourths of an inch thick; the next four pipes seven-eighths of an inch thick. The flanges to be one inch thick; the bolt-holes $1\frac{1}{4}$ square, to receive bolts $1\frac{1}{8}$ square. The next four pipes one inch thick, and the flanges $1\frac{1}{8}$; and the two or three pipes below these should be $1\frac{1}{8}$ inc. thick, and the flanges $1\frac{1}{4}$. This makes above 45 yards, which is as great a depth as should be worked at one lift, in a general way. The weight of each pipe should be marked on it, to distinguish those of different thicknesses.

The bucket-piece, which is immediately above the working barrel, should be 6 ft. long, and $1\frac{1}{4}$ inc. thick in the pipe part; the flanges $1\frac{1}{2}$ thick; the chamber opposite the door, and the projecting part $1\frac{1}{2}$ inc. thick. The flange round the door-way 2 inc. thick, with bolt-holes 2 inc. square to receive bolts of $1\frac{7}{8}$ square inc. to fasten the door in its place.

The working-barrel should be 9 ft. long for a 7 ft. stroke, and the metal left $1\frac{1}{4}$ inc. thick when bored; the flanges $1\frac{1}{2}$ inc. thick; the upper end and lower end must be formed a little bell-mouthed, to allow the bucket to pass down into it easily.

All the common pipes of the pumps above the working barrel should be 1 inc. larger in diameter than the working barrel, both for the convenience of drawing up the bucket, and to allow for the space taken up by the pump-rod within the pump.

The clack-piece, above the clack-seat, to be the same size and thickness, as the working-barrel, but one inch less in diameter below the seat. The chamber and door to be the same as that above. The clack-seat to be bored out, for four inches deep, and the bottom diameter one inch less than the top diameter. With this taper the clack will jam fast in its seat.

The wind-bore may be six or eight feet long, and one inch less diameter than the barrel; the metal and flanges $1\frac{3}{8}$ inc. thick. The lower part of the wind-bore may be pierced with upright openings or slits all round, so as to give as much opening for water-passage, as the area of the pipe. But it is better to swell the lower end of the pipe to a larger diameter, terminating at the bottom with an egg-shape: all this part to be perforated with round holes two inches diameter.

The buckets for pumps of more than nine inches diameter, should be $1\frac{1}{4}$ less than the barrel when turned; smaller sizes, one inch or $\frac{3}{4}$ less, to allow room for the leather and the bucket hoop. The outside of the hoop must be cylindrical, and nearly as large as the barrel; the inside should be turned tapering, to fit the outside of the bucket, allowing the space all round for the thickness of the leather; the taper to be one inch in diameter, for 4 inches in depth.

Buckets less than 12 inches to be flat at the top where the valves lie, as shown in fig. 6, but the valves of larger sizes to be inclined, the hinge or joint being lower than the part which opens; and therefore the valves when shut have the same position as those in the figure have when open. In this way, the water passes more freely through the passage, and up the pump.

Mr. Smeaton in his directions for Chase Water engine, says—

" The durability of the leather of the bucket, depends greatly upon a proper proportion and construction of the bucket-hoops. The bucket-hoops on the outside should be a cylinder, that is, the same diameter above and below, their tapering or conical inside must therefore be formed by the different thickness of the metal; their external diameter should not be more than one quarter of an inch less than their respective barrels, that is, they should be as big as possible, so as to allow for the unequal thickness of the leather, without jamming the hoop. The hoop to be made as broad as can be allowed, but the leather need not stand above an inch above the hoop; this being carefully attended to, the barrel will be as little as possible subject to chamber or wear in the working part, either by the hoop or the leather." The bored part of the working-barrel must be considerably longer than is requisite for the mere stroke of the bucket, because there is no certainty of the exact length of the rods, so as to ensure that the bucket shall rise and fall exactly in the bored part.

The butterfly valves, with two semicircular flaps opening back to back on a joint, or hinge fixed across the middle of the bucket or clack, were used for all the common sizes of pumps; but for very large pumps, from 18 inches to 24 inc. diam., it was usual to divide the valves into four flaps instead of two, that is, to make two pairs of butterfly valves.

Thus the ring of the bucket or clack, had one bar diametrically across its centre, and two other bars parallel thereto, dividing the area of the circle into four apertures; upon these two latter bars, the joints or leathern hinges of the flaps, were fastened by cross-bars, and the shank of the bucket was divided into four prongs, or branches, two of which passed down through mortises in each of the extra bars across the bucket, and fastened altogether by keys beneath. Hence each pair of valves opened back to back, like the butterfly valves; but as the two flaps at the middle, opened up in the centre of the barrel, the water had a more free passage, and each of the four flaps being much smaller than it would have been, if there had been only two, they occasioned less shock in falling, when they shut suddenly, by the return of the column of water; and were consequently less liable to be worn out; and also the consequences of the failure of a small valve, are less likely to be serious than that of a large one.

In many mines, the water is of an exceedingly corrosive quality, and quickly destroys the iron-work of the pumps: in such cases the working barrels must be made of brass or gun metal, and the buckets and clacks also; for with iron barrels, if the engine ceases to work for a few days, or if it works short of the full length of stroke, the interior surface of the barrel becomes corroded and rough, at all those parts which are not continually scoured, by the working up and down of the bucket; and then if the engine is afterwards required to work the full length of its stroke, the leathers of the bucket, are quickly cut and worn through, by the rough places in the barrel.

Wooden pump-pipes were also used instead of cast-iron in some mines where the water was very corrosive: they were made in pieces or staves put together and hooped like casks, but very thick and strong.

A short length of about 5 inches, made of brass, was sometimes introduced, between the clack-door piece, and the top of the wind-bore, to form the conical seat for the clack; this brass-piece formed the bottom part of the clack-piece.   When the conical seat was bored out in the iron clack-piece, the corrosion of the metal sometimes fixed the clack so fast in the cone, that it could not be drawn out, when it was necessary to replace it ; or in other cases, if the clack did not get fastened, when the rusting commenced, the seat would become leaky, and the clack would get loose in working.

The fast jamming of the clack into its seat, is a point of great importance in mining pumps, that it may not be subject to rise up, in the working of the engine ; nor to get fixed so fast, that it cannot be drawn up, out of its seat by the tackle from above, in case the water prevents access to it, through the clack door.

The permanency of the leathern valves in the bucket, and clack, is also very important to the working of an engine ; for if any valve fails, the engine is certain to strike very violently on the catch-pins, and frequently some parts are broken or strained.   If the valve in the bucket fails, then the engine will strike at the end of the working stroke, in consequence of the engine losing its load or resistance to the force of the piston.   But if the valve in the clack fails, then the blow will be struck at the end of the returning stroke, in consequence of the weight of the column of water continuing to rest upon the bucket, when there is no force on the piston to counterbalance it.   These evils also take place in part, when any dirt or obstruction gets into the valves, so as to prevent them from shutting quite tight down.

And when air is drawn in at the foot, or lower end of the pumps, the engine is also liable to strike at the end of the returning stroke, in consequence of the air which has been drawn into the barrel beneath the bucket, yielding to the returning stroke, and being at first unable to open the valves which then sustain the column of water above the bucket, so that the weight of that column is left in part dependent upon the bucket, until the air beneath it is so compressed as to open the valves loaded with water ; but as this cannot happen till near the end of the stroke, the engine may strike a violent blow in consequence.

It is very dangerous for a heavy engine to draw air into the large pumps, when it is working with its full power and length of stroke ; because the catch-pins at the ends of the great lever, will then strike with great violence upon the spring beams, and the chains, or other parts are liable to be broken by the shock.

One of the provisions to prevent such accidents is, to fix guard-stops on the main spear, to strike upon strong beams, fixed across the pit, when the spear is at the lowest part of its motion.   These stops are pieces of wood of the same size as the spear, fastened on upon one side of it, by clamp bolts, so as to form a projecting cleat.   The stop beam, which is fixed across the pit, is placed as close to one side of the rod as it can be, not to touch ; but this projecting stop being over the beam, comes down to rest upon it, if the spear ever exceeds the intended working stroke.

In the Long Benton engine, a guard stop of this kind was fixed on the spear, at a few fathoms below the surface ; and as a further security, at the lower end of the great spear, two beams were fixed across the pit, beneath the base or cross piece of the triangular frame called the λ, by which the two small rods were united to the lower end of the spear.   These beams stopped the further descent of the rods, if the spear moved further than was intended.

But with all these provisions, an engine should never be allowed to work with its full length of stroke, when the water is nearly pumped out, and when the pumps are therefore liable to draw air.

A float was also placed in the sump or bottom of the pit, from which the lowest pump drew the water ; and a cord from this float passed up the pit to the surface, and then was conducted over proper pulleys into the engine-house, where the end had a small weight appended to it, which hung down like a bell-pull near to the working gear ; and by its rising or falling as the float subsided, gave the engine-keeper notice of the state of the water in the pit ; and therefore when the water was nearly pumped out, he could attend the working gear to shorten the stroke of the engine, so as to avoid striking the catch-pins, when the pumps drew air, or at least, to avoid dangerous shocks.

The float was a cylinder of wood, included within a piece of iron pipe, such as the pumps were made of ; it was placed on the bottom of the pit, in a vertical position, by the side of the wind-bore of the lowest pump, the lower end being pierced all round with holes, to admit the water freely into it. The float being rather less than the pipe, could freely rise and fall, and the cord from it went up to the engine-house. The top of the pipe was closed by an iron cover, with a hole in the centre, for the cord to pass through. The float had a piece of iron fixed in the lower end of it, and also an iron staple in the upper end, to which the cord was fastened. When the water was nearly drawn out of the pit, the float subsided, and the iron at the lower end of it came to rest upon a piece of iron fixed at the bottom of the pipe. In this state, by snatching the weight at the end of the line, in the same manner as a bell-wire is pulled, the engine-man could easily feel whether the float was aground or afloat, by the knocking of the iron bottom of the float on the iron bottom of the pipe. For when the water was high enough to float it, the cord would return the pull, and oscillate or vibrate ; but it would have a very different motion, when the float struck the bottom. It could also be known when it rose up and floated against the cover of the pipe, because the cord would not then yield to a pull.

Some judgment of the state of the water, could also be formed by the progressive rise and fall of the weight, at the end of the cord ; but this would be uncertain, on account of the variations of the length of a cord of 80 fathoms long. But the indications of the float being afloat or aground, could never be mistaken ; and as soon as it was found to strike on the bottom, the engine-keeper attended at the working gear, and altered the pins in the plug, so as to shorten the length of the stroke a foot at each end, in order that the catch-pins might not strike, even when the pumps drew air, and occasioned sudden snatches in the motion of the engine.

For large engines, which worked three lifts of pumps, as at Chase Water, (p. 197), the Y at the lower end of the main spear was made as shown in fig. 5, but longer, and with a third loop in the middle, proceeding directly down from the spear, to suspend the upper pump-rod : the two others being suspended at *m m*, as in the figure.

In deep mines, such as the copper mines in Cornwall, which require large engines, the pump-rods are of such great weight, as to require a second lever to be placed on the opposite side of the pit to the engine-house ; and being connected by a chain or rod, with the main spear, a balance weight is applied at the outer end of the lever, to counterbalance part of the weight of the rods ; because it would be too great a weight for any lever to bear, if a sufficient balance were put upon the piston end of the great lever of the engine ; and independently of the mere strength, it would be dangerous to have so great a weight on the piston, or on that end of the lever, because when the catch-pins struck on the spring beams (as all engines must be expected to do at times), the blow would be so violent as to break down every thing before it. But the load at the end of the balance lever is so placed, as to descend to the earth, when it has moved its intended space ; and if it moves farther, it strikes upon a solid mass of stone-work, without injury.

The weight is usually a large block of stone, fastened by bolts upon the end of the balance lever ; or, in other cases, a large chest is formed at one end of the lever, and it filled with stones (*a*).

APPLICATION OF THE ATMOSPHERIC ENGINE TO DRAIN COLLIERIES.

THE steam-engine was applied to the draining of mines at its first establishment as a complete invention, by Mr. Newcomen.   It appears that mines and collieries, in England and Scotland, were quite in their infancy at the beginning of the last century, and have been raised into the greatest national importance, by the assistance of the steam-engine.

The mining of coals is most intimately connected with the adoption of the steam-engine ; whether we consider the facilities which that invention has given to the working of coal-mines, or the subsequent advantages derived from the use of coals, as fuel to work steam-engines for the purposes of the arts and manufactures. Without an abundant supply of coals, the use of steam-engines, and the practice of the modern system of manufactures, would be very limited.

The following account of the history of coal as an article of commerce, is taken from Dr. Rees's Cyclopædia, (vol. viii. article Coal).

It is supposed that coals were worked by the Romans, at Benwell, near Newcastle upon Tyne. The first charter for the licence of digging coals, was granted by King Henry III. in the year 1239 : it was then denominated sea-coal : and in 1281, Newcastle was famous for its great trade in this article. The first authentic accounts of coal being wrought in Scotland was in the year 1291, on the lands belonging to the Abbey of Dunfermline.

In 1306, the use of sea-coal was prohibited in London, from its supposed tendency to corrupt the air.   Nevertheless it was, soon after, the common fuel at the king's palace in London.   And in 1325, a trade was opened between France and England, in which corn was imported, and coals exported.   In 1379, a duty of sixpence per ton was imposed upon ships coming from Newcastle with coals.   At this period, the inhabitants of the county of Durham had obtained no privilege to load or unload coals on the south side of the Tyne ; but in 1384, Richard II, on account of his devotion to Cuthbert, the tutelary saint of Durham, granted them licence to export the produce of their mines without paying any duties to the corporation of Newcastle.

In 1421, it was enacted by the English parliament, that the keels or lighters carrying coals to the ships should measure exactly twenty chaldrons, to prevent frauds in the duties payable to the king.   Æneas Sylvius, afterwards Pope Pius II, visited England about the middle of the fifteenth century, and remarked that the poor of Scotland received for alms pieces of stone, which they burnt in place of wood, of which at that time the country was destitute.   About the beginning of the sixteenth century, the best coals were sold in London at the rate of 4*s*. 1*d*. per chaldron, and at Newcastle for about 2*s*. 6*d*.; and in 1563, an act was passed in Scotland to prevent the exportation of coals, which had occasioned a great dearth of fuel in that country.

In 1582, Queen Elizabeth obtained a lease of a great part of the coal-mines of Durham, for 93 years, at the annual rent of 90*l*. which occasioned an advance in the price of coals.   The lease was afterwards assigned to Thomas Sutton, the founder of the Charter-house in London, who assigned it again to the corporation of Newcastle, for the sum of 12,000*l*.; and the price of coals was immediately advanced to seven and eight shillings per chaldron.

About this time, a charter was granted to incorporate a new company, called hostmen or coal-engrossers, for selling all coals to the shipping ; in consequence of which the corporation imposed one shilling per chaldron additional upon this article.

It appears by an order of the hostmen's company, dated 1600, that tram-waggons and waggon-ways, had not then been invented, but that the coals were brought down from the pits to the staiths

---

(*a*) An atmospheric engine with this additional balance lever for the pump-rods is represented in Leupold Theatrum Machinarum Hydraulicarum, Vol. II, tab. 64 ; being the representation of an engine, erected at Konigsbern in Hungary, in 1723, by an English engineer named Potter.

And in Pryces Mineralogiæ 1778, is an engraving of an atmospheric engine, with a balance lever applied in the manner which was then commonly practised for the deep mines in Cornwall, and which is still followed, for the modern engines.

by the side of the river, in wains, holding eight bolls each. About this period, an engine for drawing the water out of the coal-mines, by the power of water-wheels, was invented in Scotland, by a predecessor of the first Earl of Balcarras, who obtained from James VI. of Scotland, a patent for it for 21 years. This improvement was not adopted till some time after, in the neighbourhood of Newcastle.

In 1615, the coal trade of Newcastle employed 400 sail of ships, one-half of which supplied London, the remainder the other part of the kingdom. The French are represented as trading to Newcastle at this time for coals, in fleets of 50 sail at once, serving the ports of Picardy, Normandy, Bretagne, and as far as Rochelle and Bourdeaux; while the ships of Bremen, Embden, Holland, and Zealand were supplying the inhabitants of Flanders.

In 1630, David Ramsey, a great projector, obtained an exclusive charter from Charles I. to raise water from low mines and coal-pits by fire, (see p. 85.) In the same year, the king let to farm an impost of five shillings per chaldron, on coals transported out of England and Wales, to any part beyond the seas, and 1s. 8d. above the 5s. on any coals exported beyond seas by any Englishman; and also 3s. 4d. for every chaldron to be exported, except for Ireland and Scotland.

In 1631, an information was made in the Star Chamber, by the Attorney-General against the hostmen of Newcastle, for mixing 40,000 chaldrons of coals with slates, &c. It seems that they had been before fined and imprisoned, but they still continued to cheat the metropolis, even after those severe measures of government. In 1634, the king, Charles I., solely by his own authority, imposed a duty of four shillings per chaldron on all sea-coal, stone-coal, or pit-coal, exported from England to foreign parts.

In 1637, one shilling per chaldron appears to have been paid, on the foreign vent of coals, to the mayor and corporation of Newcastle. The bishop of Durham wrote to the mayor, 10th January, 1638, commanding an immediate restoration of the exaction. In 1643, the Scots besieged Newcastle; and it is said, that all the coal-mines were ordered to be set on fire, but it was prevented by General Lesley. In 1648, coals were so excessively dear in London, that many of the poor are said to have died for want of fuel.

In November, 1653, articles were again exhibited against the town of Newcastle, concerning the coal trade; and the cause, as usual, was given against them. About this time, the port of Sunderland appears to be rising into importance. In 1658, the customs upon all coals exported were let to Mr. Martin Nowel, at 22,000l. per annum, of which sum 19,783l. was for those of Scotland. Commissioners were appointed by Oliver Cromwell, for measuring of keels.

In 1663, the collieries in Perthshire appear to have been very extensive in Scotland. An act was then passed, constituting the Culross chalder of coals, the standard for Scotland. From the profits of these collieries, the Abbey of Culross is said to have been built. In 1667, coals are said to have been sold in London for above 20s. a chaldron. About 320 keels were at that time employed upon the river Tyne; in the coal trade, each of which carried annually 800 chaldrons on board the ships. Sixteen chaldrons of Newcastle are equal to 31 of London pool measure.

In December, 1667, after the great fire in London, the parliament fixed that the price of coals till the 25th of March following, should not exceed 30s. per chaldron. A duty of one shilling per chaldron was granted to the Lord Mayor, to enable him to rebuild the churches and other public edifices. This was afterwards increased to three shillings, to continue for twenty years. In 1677, Charles II. granted to the Duke of Richmond one shilling per chaldron on coals brought to London; which was continued in the family till the year 1800, when it was purchased by government for the annual sum of 1900l. payable to the Duke and his successors. This duty, at present, produces to government 2500l. annually.

At the end of the 17th century, 1400 ships are said to have been employed in exporting yearly from Newcastle 200,000 chaldrons of coals, Newcastle measure, which was about two-thirds of the whole trade.

About 1725, Newcomen's engine was commonly adopted to drain the collieries, and the coal trade has since been prodigiously augmented in consequence. About 1758, a machine was invented by Mr. Meninzies to draw up coals out of the pits by the descent of a bucket full of water. Horse machines had been always employed before, but this new plan was a considerable improvement.

In 1741, a drawback was granted of the duty on all coals used for the fire-engines employed for working the tin and copper mines in Cornwall.

From 1770 to 1776, the average of the coals annually shipped from Newcastle was 380,000 chaldrons, Newcastle-measure; which in 1800 had increased to upwards of half a million of chaldrons annually.

The first collieries which were worked, were drained of water without machinery by dry levels, or subterranean adits, carried up from the low grounds at a distance, till they intersected the bed of the coals, and thus carried off the water from them. The extent of the coal fields which could thus be drained was very limited, and machines turned by water were next introduced, during the seventeenth century.

Mr. Bald, in his View of the Coal Trade of Scotland, printed in 1812, informs us that, about 1690, water-wheels and chains of buckets were commonly employed to drain collieries in Scotland.   The axle of the wheel extended across the pit mouth, and small wheels were fixed upon the axle to receive endless chains of two or three tires ; which reached down to the coal.   To these chains were attached a number of oblong wooden buckets or troughs, in a horizontal position, which circulated continually with the chains, ascending on one side, and descending on the other, filling at the bottom, coming up full, and discharging at the top, as they turned over the wheels on the great axletree, and then descending empty to be filled again.

When there was abundance of water to turn the water-wheel, the full complement of buckets was placed upon the chains ; but as the water decreased, a proportional number of buckets was taken off. This was the only plan for regulating this machine to the power.   But it was very imperfect ; for although each bucket was full of water when it was lifted from the pit-bottom, none of them remained more than half full, when they arrived at the discharging point above the axletree, this was owing to the swinging of the chains, and the water spilled from the buckets, was constantly pouring down the pit in a deluge.

This machine was very expensive to make and keep in repair.   The chains for a pit of eighty yards deep, cost then 160l., and when a joint-pin gave way, the whole set of chains and buckets fell to the bottom with the most tremendous crash, and every bucket was splintered to pieces.   To supply the water-wheels of these machines, many large artificial ponds and collections of water were formed on the surface at a very great expense.

Where water could not be procured to work these machines, the same sort of chains of buckets were constructed on a small scale, and adapted to be moved with horses.   This was comparatively very expensive, and could only draw water from a small depth, so that those deeper fields of coal, where neither a day level nor water machinery could be employed, remained useless to their owner and to the public : and that to all appearance for ever, as there was no other device for getting clear of water.

In the year 1708, a plan was projected in Scotland, for drawing water from coal-mines by windmills and pumps, but there was at that time no person in Scotland capable of executing the work, except John Young, the millwright of Montrose, who had been sent at the expense of that town to Holland, to inspect the mills in use there. It was suggested, that if this millwright could not be procured, application should be made to the mechanical priest in Lancashire for his advice. Windmills were accordingly erected upon several collieries ; but although they were efficient machines at certain times, they were very irregular ; and in a long period of calm weather, the mines would be drowned, and all the workmen thrown idle.   The contingent expenses of these machines were also very great, and they were only applicable in open and elevated situations.

In the year 1709, John Earl of Mar, who paid the most minute attention to the improvement of his collieries in Clackmannanshire, sent the manager of his works to Newcastle, to inspect the machinery of that district, and learn the mode of conducting colliery operations in every department.

From his report it appears, that the machines then in use, were water-wheels, and horse-engines, with chain-pumps ; the common depth of the pits was from twenty to thirty fathoms, and a few from fifty to sixty fathoms ; the expense of sinking one of these was about 55l., and the machine for drawing the coals cost only 28l.   It appears, that when it was requisite to draw water from the depth of thirty fathoms, two pits were sunk at a little distance from each other ; one pit was made thirty fathoms deep, the other only half that depth. One machine drew the water half way up the deep pit, and then it was poured into a mine, which communicated with the bottom of the other pit. From this, the water was raised to the surface by another machine. In deeper mines, a third pit, with a third machine, was resorted to. But in Scotland, at the same time, the machinery was more powerful; as water was raised at once from the depth of forty fathoms, by the chains of buckets before described.

Barlow's machine is also mentioned as being in use at Newcastle, but the steam-engine is not even hinted at. The colliery which was then upon the greatest scale at Newcastle raised annually 25 000 chaldrons of coals, and made a clear profit of 5000l. per annum; a great amount, when compared with the present value of money. The collieries then going were few in number. At that time waggon-ways were used at Newcastle; but they were not introduced into Scotland till a considerable time afterwards.

In 1710, the Earl of Mar engaged Mr. George Sorocould, engineer from Derby, to inspect his collieries in Clackmannanshire, and give plans for improving his machinery, particularly for drawing water, which was the great desideratum. The specified recompense which the engineer was to have for his trouble was 50l. sterling money. He recommended sucking pumps to be used in place of chains and buckets, but there was no person in Scotland to put his plans in execution at that time: yet the chain and bucket-engine was soon after superseded by the water-wheel, with cranks and levers working with pumps. This machine is the most simple of all the hydraulic engines, and it remains in common use to the present day. It is so easily kept in repair, that any colliery which possesses a supply of water for the wheels and can be drained by it, is nearly upon a footing with a level free colliery. Attempts have been made to improve this machine, but the results have not been very favourable.

The sinking of pits in Scotland at this period, was the most tedious and severe work that can be imagined. An engine-pit required several years for its completion, for the beds of strong hard sand-stone are much thicker, and more frequent than in the English coal-fields. Gunpowder was not then used in Scotland for sinking, although it had been applied in England, from the year 1680.

In this way collieries were carried on, until the invention of the steam-engine. Mr. Newcomen obtained his patent in the year 1705, and the engine was brought to real use for a colliery in 1712, at Griff, near Coventry, in Warwickshire, where, according to Desaguliers, they before employed horses at an expense of 900l. a year. Between this time and the year 1720, they began to be commonly used at Newcastle, and in Yorkshire. In the Statistical Accounts of Scotland, the steam-engine erected at Elphinstone, near Falkirk, in the county of Stirling, and parish of Airth, is stated to be the second which was erected in Scotland, but it is not known where the first was put up.

According to Pryce (Mineralogia Cornubiensis), printed in 1778, the most esteemed means of draining water from the mines in Cornwall, before the introduction of the steam-engine, was by pumps worked by small water-wheels of 12 or 15 feet diameter. These dimensions were thought the best; and, if the fall was considerable, several such wheels were placed one above the other, to be worked by the same fall of water. One mine is mentioned in which there were seven small water-wheels, placed one over the other. About the year 1700, Mr. John Costar, an engineer, of Bristol, went into Cornwall, and taught them to make one large water-wheel of 30 or 40 feet diameter, instead of these small ones; and this was found to be a very great improvement.

Mr. Bald has printed some interesting papers, relative to the erection of a steam-engine, at Edmonstone colliery, in the county of Mid Lothian, which was the second engine that was set up in Scotland. The first is a licence, dated May, 1725, from the committee in London, appointed and authorized by the proprietors of the invention for raising water by fire, to Andrew Wauchope, of Edmonstone, Esq.

This licence states, that as the colliery at Edmonstone could not be wrought by reason of water, liberty is granted to Andrew Wauchope to erect one engine, at his own cost, with a steam cylinder, not to exceed 9 feet long, and 28 inches diameter, according to the manner then used at Elphinstone, in Scotland. For this license, the proprietor agreed to pay a yearly rent of 80*l.* for the term of 8 years. The committee engaged to furnish the cylinder, regulator, and other brass work, at the cost of the proprietor. The deed contains a great many clauses, conditions, and reservations; which show that the erection of an engine was a most arduous undertaking, and that the proprietor had but little confidence as to the result. It is natural to suppose, that engines at this early period would be imperfect in their execution, and that they would frequently go wrong; and in consequence it was provided, that if the engine could not work for three months at one time, in consequence of any accident, a proportionable abatement was to be made from the rent, but in case of non-payment of the rent, the committee were empowered to seize upon and sell the engine.

It appears by a receipt, which was given in 1735, that only 240*l.* was paid, in discharge of the obligation of this agreement, although the sum specified in the licence would have been 640*l.* to have been paid by quarterly payments.

The workmen employed in the execution of this engine were sent from England. The steam-cylinder, the working-barrel, and all the buckets and clacks, were made of brass, somewhere beyond London. The common pumps for the pit were of elm wood, of a bore nine inches diameter, and made out of solid trees, hooped with iron: these were brought from London. The boiler-top was made of lead: it is presumed that this was used on account of the supposition that plates of iron, riveted together, could not be made sufficiently tight to contain steam.

The cost of this engine was about 1200*l.*; which is very great, if compared with the present value of money. This great expense was occasioned by the quantity of brass employed. The substituting of cast-metal, and the great improvements which have been made in the foundry department, have comparatively reduced the cost of these machines so much, that the materials for such an engine, with a complete pile of cast-iron pumps, can at present be furnished for a sum not much exceeding half the sum paid for the Edmonstone machine.

Cost of the Materials for an Atmospheric Engine for draining a Mine in 1727.

The price of engine work, in these times, will be seen by the following items, which are taken from an account of the money paid for the fire-engine, by John Potter, engineer, and repaid to him by Mr. Wauchope in 1727 :—

	L.	s.	d.
A cylinder, 29 inches diameter, with workmanship carried to London, and all other charges and expenses	250	0	0
A piston	9	10	0
A brass barrel, 7 feet long	17	10	0
One brass bucket, and one clack	0	13	0
Paid for elm pumps at London	53	4	6
Two cast-metal barrels, 9 feet long, and 9 inches diameter, and expenses after them	41	16	6
Two brass buckets, and two clacks, 9 inches diameter ; a brass regulator and injection-cock, and other cocks ; sinking-vouls, injection-caps, snifting-vouls, and feeding-vouls	35	5	0
One jeck for the wy	0	12	0
Plates, and rivet-iron for making the boiler	75	10	0
Six Swedish plates	6	3	0
The plumber's bill for lead, and a lead top for the boiler, with sheet lead and lead pipes	78	10	6
Soder	15	10	0
Timber bought in Yorkshire for the engine, with carriage by land and water, and freight to Newcastle	82	16	0
Forty-four cwt. 1 qr. 14 lb. of chains, screw work, and all other iron work about the engine, except the hoops of the pumps, at 5*d*. per pound	103	10	10
Ropes	19	16	0
Cast metal bars for the furnace	16	14	0
A hand-screw	1	17	0
Two brass codds for the regulator-beam	5	14	0
Plank for the plug-holes in the pumps	2	10	6
Leather	18	13	0
Iron hoops for the pumps, with screw-bolts and plates for ditto, 18 cwt. at 4*d*. per pound	33	12	0
Two copper pipes	0	8	0
Carriage of the materials from London and Newcastle to Scotland .	22	2	6
John Potter, pains of going and coming upon account of Edmonstone engine	50	0	0
Travelling expenses of the workmen, and their wages . . .	52	18	0

Six workmen are mentioned as being sent from England. Their expenses were paid, and 15*s*. per week wages. The whole amounts to 1007*l*. for the parts of the engine when arrived at the place, independent of any building, or labour for fixing the engine.

The engine was put up by John and Abraham Potter, engineers, of the bishoprick of Durham, who entered into an agreement with the tacksman or lessee of the colliery, to make it a sufficient and well-going engine, actually drawing water, by pumps of 7 and 9 inches bore. All the expenses of erection were to be advanced by the engineers, but to be repaid to them with a premium of 10 per cent, out of the price and profit of the first coals raised. Also Abraham Potter was to be paid 200*l*. a year salary for his pains, and for keeping servants to attend the engine, and for wear and tear in repairing ; but he was to be furnished with coals laid down at the engine, and was to be allowed a reasonable time for repairing the engine, if it went wrong ; but any accident that might happen to the boiler, cylinder, or engine-shaft by fire, was to be repaired out of the profits of the coal-work. In addition to this, the engineers were to have a half of all the free profits of the coal work, after the first cost of the engine, (which is computed at 1200*l*.,) had been paid, and the annual expense of coals, and repairs and the 200*l*.

yearly salary to the engineers. On the other hand it is stipulated, that if, by any unforeseen accident, the engine was not able to draw the water, and make it a going work, the engineers should be allowed to take away all the materials furnished by them, and to be paid a reasonable allowance for their pains and charges. If this engine had been as perfect in its performance, as the atmospheric engines have since been made, it might have exerted 12 horse-power, but allowing for imperfections, it was probably not more than 8 horse-power.

It appears that brass cylinders were always used for the first steam-engines; for, although cast-iron was used for pipes, it was so hard that it was extremely difficult to bore or cut it. Desaguliers says, in his Experimental Philosophy, 1743, Vol. II: "Some people make use of cast-iron cylinders for their fire-engines; but I would advise nobody to have them; because, though there are workmen that can bore them very smooth, yet none of them can be cast less than an inch thick, and therefore they can neither be heated nor cooled so soon as others; which will make a stroke or two in a minute difference, whereby an eighth or a tenth less water will be raised. A brass cylinder, of the largest size, has been cast under $\frac{1}{3}$ of an inch in thickness; and, at the long run, the advantage of heating and cooling quick, will recompense the difference in the first expense, especially when we consider the intrinsic value of the brass."

In tracing the progressive improvement of Newcomen's engine, we find Desaguliers states, that about the year 1717, he communicated to Mr. Beighton the use of the steel-yard over the puppet-clack or safety-valve, which he applied to some engines. He also says that "the way of leathering the piston for a fire-engine was found out by accident (by Newcomen) about 1713. Having then screwed a large broad piece of leather to the piston, which turned up the sides of the cylinder two or three inches, in working it wore through, and cut that piece from the other, which falling flat on the piston, wrought with its edge to the cylinder, and having been in a long time, was worn very narrow; which being taken out, they had the happy discovery, whereby they found that a bridle-rein, or even a soft thick piece of rope, or match going round, would make the piston air and water tight."

If this account is correct, it shows how very cold the cylinder must always have been, for if the steam had been of a proper temperature within the cylinder, it would have been so hot, as to have destroyed the leather. Desaguliers was probably misinformed in respect to the use of leather for the piston; it might have been tried for the piston, and afterwards adopted for the pumps; but it is very doubtful if it could ever have worked in a complete engine; but that the rope or match of which he speaks, was substituted from necessity. The use of the leaden top or dome for the boiler was also continued for many years, until the workmen were able to rivet iron plates so close as to retain the steam. One of the greatest imperfections in the proportions of all the old engines was the deficiency of power in the boiler, to supply the engine fully with steam; hence they could not work with rapidity, or perform with the full effect due to the size of the cylinder. But this was so little understood, that when an engine became unequal to the task of draining a mine, the cylinder was always changed for a larger one, under the idea of obtaining more power; but frequently without increasing the size of the boiler, so as to give it any increased supply of steam; and therefore, though the enlarged cylinder would lift a heavier column of water in the pumps, it would act with a slower motion, so as to perform but little more work than a smaller cylinder with an adequate boiler. This circumstance of the slowness of the motion was commonly overlooked by engine-makers, who prided themselves upon the great force of their engines; so that, at the time when Mr. Smeaton began his improvements, the generality of engines did not perform above half the work they might have done, with proper proportions, such as Mr. S. laid down.

Cost of the materials for a complete Atmospheric Engine for draining a Mine, 1775.

Cylinder 48 inches diam., the piston made 7 ft. stroke: it would go 12 strokes per min., but usually worked 9 strokes; the piston was loaded to $7\frac{1}{2}$ lbs. per square inch.

The depth of the pit which it drained was 49 fathoms, from the surface of the ground to the bottom; but as the water ran off by an adit or level at $2\frac{1}{2}$ fathoms beneath the surface, the water was lifted $46\frac{1}{2}$ fathoms by the engine: this was done at two lifts of pumps. The lower working barrel was 12 inches diameter, 23 fathoms lift, weight of the column 6760 lbs., and the other $11\frac{7}{8}$ diam. $23\frac{1}{2}$ fathoms lift, weight 6760 lbs.; total weight of both columns 13 520 lbs.

The injection-pump was 8 inches diameter, $3\frac{2}{3}$ feet stroke, and 14 fathoms lift.

I. The Building of the Engine-House.

	L.	s.	d.
Masonry and carpentry of the house, viz. its foundations, walls, floors, roof, doors, windows, and every part ready to receive the engine	420	0	0
Brickwork for the two furnaces, and setting the two boilers	160	0	0
Total cost of building	580	0	0

II. The Engine-Work.

Timber work.

	L.	s.	d.	
The two cylinder-beams to suspend the cylinder	40	0	0	
The four spring-beams, and strong wood frames	6	0	0	
Other wood framing	7	10	0	106 0 0
The jack-head, and the hot-well, wood cisterns	14	0	0	
The great lever, or working-beam, and its arches	35	0	0	
The plug-beam, and the posts for the working-gear	3	10	0	

Cast-iron work.

	L.	s.	d.	
Twenty-four grate bars for the furnaces, 29 cwt. at 12s.	17	8	0	
Two furnace-doors, and frames and arches, $7\frac{1}{2}$ cwt. at 28s.	10	10	0	154 18 0
The cylinder, its bottom, and its piston, 80 cwt. at 30s.	120	0	0	
The injection-pump, working-barrel	7	0	0	

Wrought-iron work.

	L.	s.	d.	
The two iron-plate boilers, with dome tops, 12 ft. diam. 10 tons at 23l.	230	0	0	
Four long bolts and 4 strong stirrups for the cylinder beams	2	10	0	
The shank of the piston, $1\frac{1}{4}$ cwt. at 37s. 4d.	2	16	0	
The axis of the great lever, and cheek plates, $2\frac{1}{4}$ cwt.	4	4	0	
Two catch-pins, 6 screw hoops, and 4 cheek plates	5	0	0	101 9 0
The stirrups, martingales, and bolts for the great lever, 6 cwt.	11	4	0	
The two great chains for the piston and spear, $32\frac{1}{2}$ cwt. at 42s.	68	5	0	
The chains and rods for the injection-pump, and plug-beam	4	10	0	
The iron-work of the working gear	3	0	0	

		L.	s.	d.

Sixty fathoms of wooden pipes for the injection-pump and waste-pipes   30   0   0

Plumbers' work.
- Injection-pipe, steam-pipe ⎫
- Communicating pipe, 3 feeding pipes ⎬ 30 cwt. at 18s.   *L. s. d.* 27 0 0 ⎫
- Snifting-pipe and box, and cylinder-cup ⎭   63 15 0
- Sheet lead for lining the injection-cistern and hot-well, 49 cwt. at 15s.   36 15 0 ⎭

Brass work.

	L.	s.	d.
Two brass bearings for the axis of the great lever, and iron boxes for them	6	0	0
Two brass buckets and 2 brass clacks for the injection-pump	4	0	0
Brass regulator for the steam-passage, 168 lbs. ⎰ at 1s. 6d.	12	12	0
Brass injection-cock, 4 inc. diam. 62 lbs. ⎱	4	13	0
Snifting-valve and cock	1	10	0
Brass pipes for injection and snifting	3	0	0
Safety-valve or puppet-clack for the boiler	0	15	0
Two gauge-cocks and pipes for the boiler	3	10	0
Three feeding-cocks for the boiler and pipes	5	0	0
Eduction-valve flap, brass, 40 lbs. ⎰	4	0	0
Eduction-pipe of copper, 124 lbs. ⎬ at 2s. per lb.	12	8	0
Sundry copper pipes, &c. 20 lbs. ⎱	2	0	0

Brass work total: 59 8 0

Total cost of the engine   745 10 0

III. *The Pit-head Tackle for Lifting the Pump-work.*

	L.	s.	d.
Two large spear poles, and 4 stays to support them	25	0	0
A capstan and capstan-rope	21	0	0
A whim, gin, and rope	10	10	0
Two single-pulleyed blocks, 2 three-fold and 2 five-fold blocks	10	0	0

Total cost of the pit-head tackle   66 10 0

IV. *The Pit-work for the Pumps in the Pit.*

	L.	s.	d.
Forty-six fathoms of main pump pipes, at 6l. per fathom, or 18s. 8d. per cwt.	276	0	0
Two brass working barrels, at 1s. 1d. per lb. 178 0 0 ⎱	191	0	0
Four brass buckets and 4 brass clacks, at 1s. per lb. 13 0 0 ⎰			
Eighty fathoms of wood spear for the pit	10	10	0
Wood cistern and troughs for the pump, timber-work for fixing them	30	0	0
Twelve pairs of plates for the joints of the spear, at 3¼d. per lb.	40	0	0
The Y and its shank, plates, and cross bars, 8 cwt. ⎫	14	18	8
Four short joints, and 2 long ones for the buckets, 9 cwt. ⎬ at 1l. 17s. 4d.	16	16	0
One large off-take joint for the main pump, 3 cwt. ⎭	5	12	0

Total cost of the pit-work   584 16 8

Total cost of the engine, with all its appurtenances, about 2000*l*.

Mr. Smeaton took some pains to ascertain the number of fire-engines actually at work for draining mines in 1769. The great field for them was in the collieries at Newcastle, and he obtained a list of 100 engines which had been erected there; many of these had been worn out, and given up, but those which were then in action, amounting to 57, were of the following dimensions ; but he suspected their performance to be very small, compared with their dimensions, and their consumption of fuel to be very great.

Fire-engines at work at Newcastle, in 1769.

No. of engines.	Diam. of cylinder.	Area in squ. inc.
2	75	8 836
1	73	4 185
2	72	8 143
2	70	7 697
1	66	3 421
2	64	6 434
3	62	9 057
9	60	25 447
1	52	2 124
6	48	10 857
2	47	3 470
6	42	8 312
3	36	3 054
1	32	804
2	28	1 231
6	small	2 280
8	unknown	11 085
57 engines.		116 435

Probably 1188 horse-power.

To ascertain this, he made exact observations upon 15 engines, of different sizes, and which were pointed out to him as forming a fair average of the whole ; and then computed their performance. The results of this computation show, that the powers of the engines bore no settled proportion to the dimensions of the cylinder.

The sum of the great products, of all the 15 engines taken together, was 36 438 667 cylindrical inch feet of water, raised 1 foot high per minute. The sum of the area of all the 15 cylinders was 36 899 square inches ; hence, the great product of each square inch was 988 cyl. inch feet of water ($=$ 337 lbs.) raised 1 foot per min.

The average pressure or load upon each square inch of the piston was 6 ·72 lbs. ; and consequently the average motion of the pistons must have been 50 feet per minute (for 337 lbs. $\div$ 6 ·72 lbs. $=$ 50 ·1).

The modern horse-power, by which engines are now estimated, is 968 000 cylindrical inch feet of water ($=$ 33 000 lbs. weight) raised 1 foot per min., and divided by 988 cyl. in. ft. (which is the great product of each square inch as above) gives us 98 square inches of piston to produce one horse-power.

The united area of the cylinders of all the 57 engines at Newcastle, being 116 435 square inches, divided by 98, will give 1188 horse-power, for the total power of all the engines combined.

Mr. Smeaton's computation of the effect of 15 fire-engines at work at Newcastle, in 1769.

Horse-power.	Diam. of cylinder.	Area in square inches.	Great product Cyl. inc. ft.	lb. per square inch.	Effect of 1 bushel of coals per hour.	Millions pounds raised 1 foot.
37 ·6	75	4418	3 636 633	6 ·17	235 000	4 ·59
42 ·9	74	4301	4 152 730	6 ·36	371 000	7 ·25
35 ·3	72	4072	3 417 750	6 ·54	305 000	5 ·96
18 ·9	66	3421	1 829 520	6 ·08	165 000	3 ·22
26 ·8	61	2922	2 590 596	5 ·42	230 000	4 ·49
30 ·6	60	2827	2 960 048	6 ·86	329 000	6 ·43
40 ·8	60	2827	3 949 004	8 ·61	301 000	5 ·88
22 ·9	52	2124	2 214 337	6 ·34	331 000	6 ·47
22 ·9	52	2124	2 214 337	6 ·34	193 000	3 ·77
27 ·0	52	2124	2 617 839	6 ·34	363 000	7 ·09
26 ·3	52	2124	2 549 319	9 ·01	354 000	6 ·92
18 ·4	42	1385	1 779 840	10 ·9	351 000	6 ·86
16 ·7	42	1385	1 621 069	9 ·25	381 000	7 ·44
4 ·5	26	531	433 251	5 ·66	217 000	4 ·24
4 ·9	20	314	472 394	9 ·86	168 000	3 ·28
376 ·5 Sums.		36899	36 438 667 M.	6 ·72	286 000	5 ·59

The meaning of Mr. Smeaton's great product has been already explained (p. 185) ; and dividing those numbers by 968 000, gives the horse-power of each engine, as put down in the first column.

The sum of all the engines is 376·5 horse-power ; and the united area of all the cylinders is 36899 squ. inc. ; that is 98 squ. inches to each HP.

The average load is 6 ·72 lbs. per squ. inc., ($\times$ 98 squ. inc. $=$ 658 ·6 lbs. load for each HP), and the motion must have been 50 ·1 feet per min. to have made 33 000 lbs.

The effect of 1 bushel of coals per hour, is also explained (p. 186) ; and multiplying these numbers by 19 ·35, gives the number of pounds raised 1 foot high by each bushel ($=$ 84 lbs.) of coals consumed ; as is put down in millions, in the last column ; the average of the whole is 5 ·59 millions.

In general these engines were worked with waste small coals; but some of them, which had a difficulty in raising a sufficiency of steam, were worked with the best coals they could procure; because the engine could then do more work, than with the waste coals.

These observations were made before Mr. Smeaton began his experiments on the fire-engine, and it was in consequence of finding these engines to be so defective, that he was induced to set about those improvements in their construction which he afterwards brought to perfection.

The principal cause of the defective performance of these engines, was their imperfect execution and faulty proportions; the cylinders were very imperfectly bored, and therefore the pistons could not be fitted to move freely in them or made to fit tight, and consequently a constant leakage of water and air took place, as well as great loss of power from friction; the pump work was equally faulty.

The boilers were too small to supply the cylinder properly with steam, and ill-constructed; the fire-grates were placed too low beneath the bottom, which rose up very high in the middle; they had also a most slovenly practice of heaping a great mass of coals on the fire-grate, making a most intense heat beneath the centre of the boiler, and not sufficient heat at the circumference; and very little attention was paid to keeping the boiler free from internal scales or crust. All these deficiencies tended to diminish the supply of steam, and caused the engines to move very slowly.

The engineers who constructed these engines were very ignorant; and so far from seeing these defects, and removing them, they reckoned every thing to depend upon the size of the cylinder, and the load or burden per square inch of the piston, without taking into account the velocity of the motion, or making proper endeavours to gain the greatest mechanical power; all they looked to was to make engines of great force, which would lift great weights of water: hence the parts required to be very large, and strong, but they did not obtain an adequate result.

Mr. Smeaton's engine at Long Benton (see p. 172) carried a load of $7\frac{1}{2}$ lbs. per square inch, and moved 84 feet per minute; or the area of the piston, 2124 squ. inc. $\div$ $40\frac{1}{2}$ horse-power, gives $52 \cdot 4$ square inches of the piston to produce one HP, instead of 98 squ. inc. (b).

In respect to the consumption of fuel, the common engines were equally defective; for all the faults above mentioned, which diminished the effective power of the engine, made little or no diminution in the fuel.

Another great cause of waste of fuel was, from the cylinder being too long for the motion of the piston; or that the working gear was so adjusted, that the piston did not make the full stroke it ought to have done; consequently a great space was left beneath the piston, when it had descended to the bottom of its course; and the steam requisite to fill this space was lost, without producing any effect.

Again, the piston was not usually drawn up so high in the cylinder as it should have been, and the consequence was, that the water which was continually

(b) The engine of 52-inch cylinder and $26 \cdot 3$ horse-power, quoted in Mr. Smeaton's computation, at Long Benton, on the same pit where Mr. Smeaton afterwards built the new engine on his plan; but the old cylinder and pump-work were retained. Its performance appears amongst the best in that list.

In 1772, before Mr. Smeaton began his alterations, an experiment was made upon this old engine when in its best state, in which it was loaded with $8 \cdot 92$ lbs. per squ. inc., and made $7\frac{1}{4}$ strokes per min. of 7 ft. = $54\frac{1}{4}$ ft. motion per min. It exerted $32 \cdot 1$ horse-power; which is at the rate of $66 \cdot 1$ square inches to each horse-power, and burned $12 \cdot 81$ bushels of coals per hour; which is at the rate of $4 \cdot 82$ millions lbs. raised 1 ft. high by each bushel. The new engine afterwards raised $9 \cdot 45$ millions, or very nearly double.

poured upon it, to keep it tight, accumulated to many inches deep upon the upper surface of the piston, and by its coldness, produced an excessive condensation of steam, when it was admitted to fill the cylinder; for the heat passed up through the metal of the piston, very readily to the water, and was carried off in warming that water. The imperfection of the cylinder necessarily required a considerable flow of water upon the piston to supply the leakage, and but little attention was paid to keep the stream of water as small as possible to supply that leakage.

To all these defects we may add, that the steam-pipe and regulator were commonly too small to admit the steam into the cylinder, and that it was rarely opened to the full extent by the working gear; hence, a great counterweight was necessary to raise up the piston, and draw the steam out of the boiler into the cylinder, with the necessary rapidity. The injection-cistern was not placed high enough to give the injection a proper velocity into the cylinder.

The engines at the mines in Cornwall were constructed with a greater attention to economy in fuel than those at Newcastle, but were in reality not much better. Price, in the appendix to his Mineralogia Cornubiensis, 1778, says—

" Mr. Newcomen's invention of the fire-engine, enabled us to sink our mines to twice the depth we could formerly do, by any other machinery. Since this invention was completed, most other attempts at its improvement, have been very unsuccessful; but the vast consumption of fuel in those engines, is an immense drawback on the profit of our mines; for every fire-engine of magnitude consumes 3000*l*. worth of coals per annum. This heavy tax amounts almost to a prohibition.

" We have several fire-engines, that are perhaps the largest in the kingdom: the house cylinders are generally from 54 to 70 inc. diam. Many trials of mechanical skill have been made by our engineers to very little purpose, for the total application of heat, and saving of fuel. The fire-place has been diminished, and enlarged again. The flame has been carried round from the bottom of the boiler, in a spiral direction, and conveyed through the body of the water in a tube; and one, two, or three tubes, before its arrival at the chimney. Some have used a double boiler, so that the fire might act in every possible point of contact, and others have built a moorstone boiler, heated by three tubes of flame passing through it."

Mr. Hornblower has given an account of one of these stone boilers, in Nicholson's Journal, 8vo., vol. viii, p. 169.

Amongst the expedients tried in Cornwall to reduce the consumption of fuel in the engines, it was suggested, that in the several operations of smelting the produce of the mines, much heat must be carried off from the intense fires of their furnaces, which might be employed under the engine boiler, as that required a lower temperature; accordingly an engine was erected on a copper-mine at Camborne in Cornwall, with a set of furnaces so attached to the engine, that their superabundant heat might be rendered available to raise steam.

To effect this, the engine boiler was built of granite, called moor-stone in Cornwall; the masonry well jointed, and put together with a sort of lime, brought from Aberthaw, in South Wales, which has the property of setting under water. To communicate the heat to the water, copper tubes were placed within it, passing horizontally through all the length of the boiler, from end to end: the furnaces were placed at one end of the boiler, and the flame from them was conveyed through the tubes.

At the bottom of the boiler was a cock as usual, to draw off the water when it was required to clean it, and it was discovered that when the fires were lighted, and the water in the boiler made to boil so as to afford steam enough to work the engine, still on drawing water from the bottom part of the boiler, by this cock, it was not scalding hot; the tubes had been situated near the surface of the water, for it was not known at that time how very slowly heat is transmitted downwards in fluids; they were afterwards placed lower down.

By some notes sent to Mr. Smeaton in 1773, it appears that this stone boiler was 20 feet long, by 9 feet wide within, and $8\frac{1}{2}$ feet deep, the fire was applied in three copper tubes 22 inches diameter fixed within the water, and extending through all the length of the boiler. Two of these tubes were fixed within $7\frac{1}{2}$ inches of the bottom of the boiler, and the other one was over the space between them, at 2 feet above the bottom. The flame was conducted from the furnace through one tube, then turned back into the other tube, and then returned through the third, so as to pass three times through the whole length of the water.

It was intended to heat this boiler, by the superfluous flame and spare heat of a roasting furnace for preparing copper ore for smelting; the three copper tubes forming part of the flue, or chimney of the furnace; but it did not answer, in practice.

The failure was thought by some to be owing to the heat lost in the great mass of stone-work, and another boiler was made for Mr. Smeaton's engine at Chase-water mine, to collect the waste heat of the furnaces; but it was made of iron-plates riveted together instead of stone, though the construction in other respects, was nearly the same as the stone boiler. The length was 22 feet, by $8\frac{1}{2}$ feet wide, and $10\frac{3}{4}$ feet deep, with four internal tubes 22 inches diameter; two of them were fixed at 12 inches above the bottom, and the other two at $3\frac{3}{4}$ feet above the bottom. This boiler was intended to be used in addition to the centre boiler of the engine, and in lieu of the two side boilers.

From some minutes collected by Mr. Smeaton, of the engines at work in Cornwall, in 1770, it would appear that there were then about 18 large engines; eight of them are specified as being above 60-inch cylinders, and from that to 70-inc., the others not stated. If we suppose they were 48-inc., the whole when summed up as before, for the engines at Newcastle, will make about 44 300 square inches of piston; and reckoning 90 square inches to a horse-power, it would be 481 HP. This is not professed to be an exact account, like the former. The principal makers of these engines was Mr. Jonathan Hornblower and Mr. John Nancarrow.

For draining mines in which the water can be conveyed away by an adit or level at a great depth beneath the surface, the weight of the dry spear to reach down to the pumps is a great load upon the engine. For instance, the pit at Long Benton colliery (p. 175) was 97 fathoms deep to the bottom, but the pump lifted the water only $49\frac{1}{3}$ fathoms, and it ran off at $57\frac{1}{2}$ fathoms beneath the surface. Also the pit at Chase-water Mine (p. 191) was 75 fathoms deep, and the water was lifted by the engine 51 fathoms at its first erection, but afterwards only 31 fathoms, and then it ran off by a level at 44 fathoms below the surface.

The most notable case of this kind was at some lead mines at Winster in Derbyshire. The workings were commenced on high ground, and after sinking very deep, a subterranean level or sough was excavated, with prodigious labour and expense, from low ground at a distance, so as to intersect the workings of several neighbouring mines, at 100 fathoms below the surface.

In 1777, one of these mines, called Yatestoop, required to pursue the ore still lower beneath this level; and for that purpose an atmospheric engine of 70-inch cylinder was erected at the surface, with a dry spear of 100 fathoms long, to reach down to the level, and there it lifted water 15 fathoms from the bottom of the pit into that level, by two pumps, one 25 inc. diam., the other 14 inc. diam. The water for the injection was obliged to be pumped up from the level to the surface,

100 fathoms, in pumps 9 inc. diam. which were divided into four lifts, and raised the water from one to another.

This engine was so overburdened with the great length of spear, and still more by the raising of its own injection-water from such a depth, that it could not move with sufficient celerity to perform well; and the elevation at which it was placed must also have made some difference in the barometrical weight of the atmosphere.

In consequence, an atmospheric engine, of $64\frac{1}{2}$-inch cylinder, was, some years afterwards, fixed underground, at the level, 100 fathoms beneath the surface, to work the same pumps which were worked by the former engine at the surface. The excavation in the rock to receive the engine-house and boiler cost 300*l.* before the building was begun. It was built of grit stone, let down from the surface: the lever-wall 5 feet thick, and the side walls 3 feet thick. The engine was constructed in the usual form, but it was supplied by one boiler, which was 20 feet diameter. The flue of the chimney was conducted up in the shaft. This work was planned and executed by Mr. Francis Thompson, of Ashover, in Derbyshire; an operative engineer, who had an extensive practice in that district, at the period when its mines were in great activity.

Another instance of a fire-engine working underground, was in a colliery at Whitehaven, in 1776. It was placed 80 fathoms beneath the surface, and worked a series of pumps disposed down the dip, or inclination of the strata of coal, which was very rapid. The pumps lifted 4 fathoms each, from one to another, and were worked by one sliding rod from the engine. The intention of this disposition was, to avoid piercing the floor, or bottom strata of the coals, which must have been done to have made a perpendicular pit down from the surface, and would have let in much more water upon the workings.

In many situations where the bed of coals dips suddenly, and if the strata beneath the coals is of a porous nature, it is of great importance to preserve the water-tight floor of the coal perfect, in order to prevent the passage of the water, and in all such cases the pumps must be placed on the slope of the strata, instead of in perpendicular pits. The same thing happens continually in lead and copper mines, where the veins are not in vertical planes; and the pumps must follow the direction of the vein, because it is too expensive to sink perpendicular shafts in the solid rock to place the pumps in.

Description of Mr. Smeaton's method of supplying the boilers of a fire-engine with soft water, in cases where the mine water is of a corrosive quality, 1779 (a).

It has been the general practice, wherever the situation will admit, to supply the injection-pumps of fire-engines, with the water of a burn or rivulet of soft water, or from a soft water spring; in consequence, the boilers of such engines become supplied with soft water, which will not furr the boilers.

The very great advantage of this practice is well known; but where the situation does not afford any supply from a rivulet or soft water spring, artificial reservoirs should be made of sufficient capacity to collect rain water enough to supply the injection water in the intervals between rains, and great showers; but the quantity of water necessary for the injection of a large fire-engine, is so considerable, that very few situations admit of making such reservoirs of a sufficient

(*a*) From Smeaton's Reports, 4to. Vol. II. p. 362.

extent, particularly at collieries, where the ground is always completely drained beneath. For want of any tolerable certainty of supply by this expedient, and its great expense, in land the injection-water is usually raised from the main pump-heads of the mine, which is in general of a hard nature, and frequently so corrosive and adhesive to the boilers, as to do them more injury in three months than as many years' working would do with soft water.

It will, therefore, be a great improvement to the fire-engine, to be enabled to supply the boilers with soft or rain water in all situations; particularly where, as above mentioned, a supply of soft water has not yet been procured. In the course of experiments which I went through some years ago upon the fire-engine, this idea led me to investigate the quantity of water that was necessary for injection, in proportion to the quantity that was necessary to feed the boiler, and I found that the latter was in most cases less than one-twelfth part of the former (a).

The boiler will not feed unless the hot-well water is let into it by the feeding-cock, from a considerable height above the water in the boiler; but if the small quantity necessary for feeding the boiler can be separately supplied from a reservoir of rain water, that smaller quantity may by a contrivance be warmed in some degree, and the chill taken off by the great quantity of hot water produced from the injection; therefore the engine may be worked by injection-water of any quality provided it is cold, and the boiler may be fed by such water as is of a good quality; without allowing any communication between the two kinds of water except that of heat.

With this view I placed in the hot-well of my experimental engine, a pan made of tinned plate, as large as could be contained therein, leaving a vacancy between the sides of the hot-well, and the sides of the pan, for the injection-water to ascend, and pass off at the hot-well spout as usual. The feeding-pipe to the boiler was connected with this internal pan; which was continually supplied with cold water from a neighbouring well, in such quantity as, by continually running into the boiler, kept up the feed therein. I did not find any sensible difference in the product of work by the engine in proportion to the coals consumed, than it did in its ordinary way of being fed from the hot-well.

The following method is founded upon the above experiment; and the apparatus represented (Plate VI, figs. 1 and 2) is the best and easiest way which has occurred to me, for applying those principles; and which cannot fail of success, on the supposition of a reservoir being made large enough to hold as much water as the boiler feed will require betwixt rain and rain.

Fig. 1 is a section of the hot-well. A B C D, is the hot-well, supposed to be cut through the middle. E F G, part of the sink or eduction-pipe, with the horse-foot valve G, closing the extreme end of it.

H I K L, is a section of the thin copper pan, made to fill as much of the vacant space of the hot-well as possible: it may touch the sink-pipe, and but just clear the joint of the horse-foot valve, the bottom being a little rounding; and about an inch higher at K than at I, in order to induce a part of the injection-water to ascend beneath the sloping part H, which without that advantage it would not be apt to do, the waste-spout being near the part L. Those three sides of the pan which apply to the three sides of the hot-well, are to be about $1\frac{1}{2}$ inc. distant therefrom at the top, and about 2 inches distant at the bottom of the pan; and according to these directions, the dimensions of the pan are to be taken from that of the hot-well.

(a) The author's calculations, founded on Mr. Smeaton's data, show the injection water to be between 10 and 11 times the boiler feed. See p. 166.

The fresh feeding water being brought to the engine-house by a long lead pipe from the feeding reservoir, is to be poured into the internal pan H K I L, by the tunnel Q, and it will be continually receive heat from its immersion in the hot-well water up to the dotted line O P. The water so warmed in the pan must be continually conveyed away by two spouts, one marked M, the other N; from whence it is to be conveyed to the feeding-pipes of the two boilers respectively.

Fig. 2 is a plan of the hot-well, and the pan immersed therein; A B C D is the top of the hot-well; E is a section of the sloping part of the sink-pipe, and R the waste spout for the hot-well water, which is to be conveyed away from the engine; H I K L shows the top border of the copper pan, which has a flange, projecting an inch on all the four sides, and riveted to an iron frame, h i k l, which rests upon the top of the hot-well. This frame is intended to strengthen the pan, which is to be made of slight copper. When any thing is to be done at the horse-foot valve, the frame and pan may be lifted out altogether with unfixing any part.

M is the feeding-spout of the centre boiler, beneath the cylinder, and N the feeding-spout for the off-boiler in the out-building.

The warm water from M falls into a wooden spout, which leads into a funnel upon a pipe of about 1½-inc. bore, which is turned and joined to the upright feeding-pipe of the centre boiler. In like manner, the spout N is provided with a wooden spout to convey the feeding-water for the off-boiler into a funnel and pipe of the same kind, connected with its feeding-pipe.

Q, a branch upon the square frame, with a hole in it to receive the tunnel and its short upright pipe, which receives the fresh water from a cock at the end of the lead pipe which comes from the reservoir. The upright pipe of the funnel reaches down nearly to the bottom of the copper pan, as shown in fig. 1.

To give the water some degree of warmth before it enters the pan, it is brought from the small reservoir in which it is collected, through from 40 to 60 yards of lead pipe, which is immersed in a long wooden lander trough, which conveys the waste hot-well water away from the engine after having left the hot-well.

This lander trough is composed of three deals, and is eight inches wide, and nine deep, inside measure, to convey the waste water from the hot-well, by a small descent of about one foot in 40 yards, into any pond or receptacle from which the water may run away by a ditch.

In this lander trough is laid a lead pipe of two inches bore, made of sheet lead, turned and burnt, and extending all the length of the trough, or from 40 to 60 yards at least. The end of the lead pipe is joined to a wooden pipe, which brings the fresh water from the reservoir containing soft water. The leaden pipe conveys it into the feeding-pan, but being all the way immersed in the current of hot water through the lander trough, the fresh water passing through the lead pipe will have the chill taken off before it enters the pan.

The other end of the lead pipe passes up to the pan through a sloping pipe of wood, which conveys the waste water from the spout R of the hot-well down to the lander trough. To be out of the way, it is proposed to carry this sloping wood pipe under the floor, which makes the boiler-head floor over the stoke-hole, and thence it passes into the lander trough to run away, as before mentioned.

At the distance of every three or four feet, little ribs or bars are to be laid across the bottom of the lander trough, to support the lead pipe from the bottom, that the hot water may pass round it; and these pieces will also impede the water from running so quick through the trough as not to drown and cover the lead pipe completely.

The end of the lead pipe, at the place where the lander trough terminates, being several feet below the level of the top of the hot-well, it is necessary to join the lead pipe well to the wooden pipes which come from the soft water reservoir; and the entry of these wooden pipes from that reservoir must be one foot higher level than the cock at the extreme end of the lead pipe which delivers into the funnel of the copper pan.

The reservoir is supposed to be made artificially, or dug upon such ground that the entry of the wooden pipe from the reservoir shall be about three feet below the surface of the water in the reservoir when it is full.

By this arrangement, the cold fresh water from the reservoir will first descend through the wooden pipes, and then ascend again through the lead pipe in the midst of the waste water, and finally pass by the cock at the end of that pipe into the funnel Q, and thence, descending by its pipe, down to the bottom of the copper pan; but as it becomes warmed, it will gradually ascend towards the surface of the pan, and pass off by the two spouts to the two boilers respectively.

Besides the two-inch stop-cock at the extreme end of the lead pipe, another $2\frac{1}{2}$-inch stop-cock should be fixed in any convenient part of the lead pipe by way of regulation.

This latter cock should be regulated so as to supply a quantity into the pan equivalent to the boiler consumption, which being once found, will rarely require alteration; but when the engine stops working, there will be nothing to do but to shut the stop-cock at the end of the pipe, and when it begins working to open it again—the just regulation of quantity being still preserved by the other stop-cock in the lead pipe. Nevertheless there will be just the same attention necessary in the engine-keeper to see that the water in the boilers keeps right by the gauge-cocks, as in the common way of feeding; and he must give the regulating-cock in the lead pipe a touch accordingly.

It may happen, that one of the feeding-spouts M or N, will run less water from the pan than the other, or that one boiler may raise more steam than the other; in which case a couple of thin wedges must be driven under the iron frame which supports the pan on the opposite side to that which runs too little. The quantity may thus be accurately adjusted between them, whilst the whole quantity is regulated by the cock in the pipe.

If the engine stops working for only half an hour, and at any regular periods, it may be most convenient to regulate the boiler's feed to such a rate as to take in the water continually, without shutting the cock when the engine stops working; for though in the interval the water will go into the boilers almost cold, yet the fire which must necessarily be kept up will keep the whole to a boiling state. But if the feeding were stopped, the same fire only produces a loss of steam at the puppet-clack, and consequently of feeding water.

It is advisable to connect the feeding-pipes of the boilers by branch pipes and cocks with the hot-well in the usual manner. These cocks may be shut when the pan is supplied with fresh water; but if that supply is ever interrupted, then the boilers can be fed from the hot-well in the common way.

Lastly, the snift-pipe, instead of being carried out of doors, may be conducted down into the copper pan in the hot-well; and its end being perforated with several small holes, and being immersed about four inches into the pan's water, this will not be so deep as to prevent its snifting properly; but by this means, the heat of the steam of the snifting-pipe, that otherwise would be lost, will be added to that of the feeding-water in the pan.

I I

Computation of Quantity.—A cubic yard of water will last an engine of a 60-inc. cylinder in boiler's feed full 18 minutes(a); and supposing the engine to work twelve hours in twenty-four, it will be at the rate of 40 cub. yards per day. A reservoir, therefore, of 40 yards square and 2 yards mean depth, will last the engine eighty days; and a reservoir of 76 yards long and 26 yards mean width, will nearly do the same thing at 5 ft. deep.

It will be well to dig the small reservoir a foot deeper near the entry of the wooden pipe, that the water may go in clear of sediment.

Some mineral waters are of a more corrosive nature than others, and the earthy concretions formed within the boiler by deposition from the water, are of different degrees of hardness; sometimes white and soft like chalk, others yellow, brown, and dark brown, and as hard as stone. Some of these hard concretions are laminar like slate. The substances deposited by mineral waters are principally lime, with various compounds of sulphur, salts, and oxyde of iron, and sometimes of copper.

The water of lead mines frequently forms a very hard crust, but thin. That from copper mines not so hard, and more slaty, and of a corrosive nature. The water from iron mines and collieries deposits a most abundant scurf, but of a soft and porous substance. The water from a soft sandy limestone is the worst of all, as to the quantity deposited, but it is not corrosive. Some waters will deposit a scurf upon the interior of the cylinder when injected into it, and on the under surface of the piston, particularly when it is of iron.

APPLICATION OF THE ATMOSPHERIC ENGINE FOR RAISING WATER FOR A PUBLIC WATERWORKS FOR THE SUPPLY OF TOWNS.

THE most considerable work of this kind in existence, in 1775, was at the York Buildings' Water-works, at Villiers-street, in the Strand, London. It was erected about the year 1752, and, for many years afterwards, it was reputed the best engine of its time, being referred to by writers of that period as a standard of Newcomen's engine. Its performance therefore deserves to be recorded.

This establishment appears to have been the first to encourage the inventors of the fire-engine. Mr. Savery made one of his largest engines on this spot, before the year 1710: it is mentioned by Desaguliers, (see p. 117.) And they afterwards had an engine made by Mr. Newcomen: it was the first which was put up in London, and is mentioned by Switzer (see p. 127), and other writers, as being a very complete specimen of Newcomen's invention.

London was at that time chiefly supplied with water by the New River Water-works, the London Bridge, and the Chelsea. The New River being brought to a

(a) According to Mr. Smeaton's table (p. 182) the boiler's feed for a 60-inch cylinder is 255 cubic inches per stroke × 10½ strokes per minute = 2677 cubic inches, or 1·55 cubic feet per min.; or in 18 min. 27·9 cub. ft.

The above plan was made for an engine of 60-inch cylinder, constructed from Mr. Smeaton's designs, at Gateshead Park Colliery, in Northumberland, in 1778. It answered the purpose very well.

A similar apparatus was afterwards applied by him, to an engine made in 1780, at Middleton Colliery, near Leeds, the cylinder 72-inch diam., the same as that at Chase-water. The apparatus was like that represented in figs. 1 and 2, Plate VII, but larger; viz. the bore of the eduction-pipe E F was 8 inches; the hot-well was 4¼ feet long at top A D; and 3 feet at bottom B C, by 2¼ wide; and 3½ feet deep D C. The copper-pan was of corresponding dimensions, and it had three spouts to feed the three boilers, by which this engine was worked. The lead-pipe and cocks, to convey the soft water from the reservoir to the pan in the hot-well, was 2½ inches bore.

reservoir, on high ground at Islington, required but little aid of engines, for the water ran at once into the greater part of their pipes.   At London Bridge, and at Chelsea, the water was pumped up from the river Thames, by the power of water-wheels turned by the current; and there was no expense for power.

At the York Buildings, they raised the water from the Thames, by engines worked by horses, which occasioned them a great expense.   Several other similar establishments were situated on the banks of the river, with horse engines, to supply their immediate neighbourhood, but, except the York Buildings, they were on a very small scale.

In the *Voyages de M. de la Motraye, en diverses provinces et places en Europe, &c.* folio, 1732, Vol. III. is a folio plate engraving, representing Mr. Savery's engine, and also Mr. Newcomen's engine, as standing close together, and raising water into the same cistern.   It is stated to be a representation of those at York Buildings' Water-works.   It is probable that both were kept there together, till the superiority of Newcomen's occasioned the other to be totally disused (*a*). The size of his first engine is not mentioned, but it was probably a brass cylinder, of about 22 or 24 inches diameter, like all the early ones.

It appears from a paper, by Mr. Keane Fitzgerald, in the Philosophical Transactions for 1763, Vol. LIII. p. 139, that there was then an engine at York Buildings, of 45-inch cylinder, 8 feet stroke; which made $7\frac{1}{2}$ strokes per minute, or 120 feet motion.   The pumps were 12 inches diameter, and two of them a lifting-pump and a forcing-pump: they raised the water 100 feet.   The column of water was 9830 lbs. or near 6·18 lbs. per square inch of the piston, (this would be 35·7 horse-power.)   It consumed 4 bushels of coals per hour, having an improved boiler.

This account is likely to be considerably overrated.   The temperature of the hot-well was 180 deg.   The object of Mr. Fitzgerald's paper was to describe some improvements which he had applied to this engine, with a view of lessening the friction, by the application of friction-wheels or quadrants, for the gudgeons of the axis of the great lever to rest upon, and some rollers to guide the plug-beam. These quadrants were $2\frac{1}{2}$ feet radius, and rested upon pivots only $1\frac{1}{2}$ inches diameter; the gudgeons (which it appears had been originally 6 inches diameter) were reduced to only $1\frac{1}{2}$ diameter in the bearing parts, which rested upon the quadrants, but, by the particular form of the new gudgeons, they had the same strength, as if they had been 6 inches diameter.   Two quadrants were applied, side by side, beneath each gudgeon.

The axis of the lever had been originally placed across beneath the wooden beam of which the lever was formed, but Mr. F. changed the axis to the upper

(*a*) M. de la Motraye's remarks on this invention are very flattering to the English, and are worth recording, as coming from a man who was continually travelling on public affairs through all parts of Europe; and who, it is evident from his writings, was well acquainted with the characters of the public of different countries, at that time.

" I never came back to London but I did always find something new and excellent, either as to literature, mathematics, or mechanics—some new production of the vast rich genius of the English, either for the public instruction, or aggrandizing, orning, and the usefulness of that city. It is not to be imagined how far the richness or fecundity of that genius can go: never a nation, how numerous soever it might be, was so universally and so deeply learned, from the scepter to the crook; never carried all the arts and sciences to such high degree of perfection.

" The new fire-engine is placed between the water-engine of London Bridge, and another towards Chelsea; there is one towards Islington, some few years ago moved by a wind-mill, and now I think by horses or water.   That new fire-engine has the appearance of a column or tower."   This is followed by a very brief description of the print of Newcomen's engine.   *Vide* Voyages de M. de la Motraye, Vol. III. p. 360.

side of the beam, and fixed it thereto by proper bands, with screw bolts.   By this means, the centre of gravity of the moving mass was brought below the centre of motion, instead of being above, as it had been before.

The beam was 30 inches deep by 26 inc. broad in the middle, and 24 deep by 22 broad at the ends; length 27 feet; and, together with its arch-heads, weighed about 5 tons.   He says, that when this lever was altered with the axis above the beam, and the gudgeons placed on the four friction quadrants, it could be swung with a thread, and would continue in a state of vibration for several minutes.

These alterations, according to Mr. F. produced so great an improvement in the performance of the engine, that it afterwards went 9 strokes per minute, with the same consumption of fuel as it before required for $7\frac{1}{2}$ strokes; owing, as he states, to the extraordinary regularity of its stroke, "which does not abate of its full length suddenly, as it used to do, when the strength of the fire abated." This Mr. F. takes to be occasioned, in a great measure, from placing the axis above the beam, by which the centre of gravity becomes reversed to what it was before.   It then had a tendency to prevent the return of the lever at the end of the stroke; but after the axis was altered, and the friction of the gudgeons reduced, if either end of the lever was brought down, it had a tendency to return, by the weight of the lever itself, and could therefore work a longer stroke without striking the stop springs.

The boiler of this engine was made of copper plates, and of an unusual construction : the fire-place being within the central part of the boiler, and surrounded by the water.   It was said to be a patent plan, but the name of the inventor or patentee is forgotten.   The boiler was 15 feet in diameter, and was said to consume one-fourth less fuel than any other engine of the same dimensions.

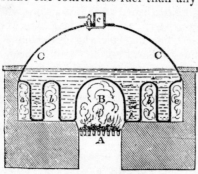

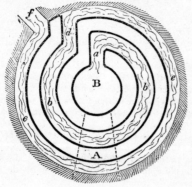

The figure of the boiler was a segment of a sphere C C, with a flat bottom.   The fire A acted beneath the centre of this bottom, within a large cylindrical cavity B, with a dome top, which rose up within the water, so as to be surrounded therewith.   From this cylindrical cavity, a hollow flue $a\,b\,b$, was carried round within the water, in a spiral direction, making one complete convolution, and then it arrived at the outside of the boiler, and communicated with a circular flue $e\,e$, which was formed in the brickwork, and extended all round the circumference of the boiler, and finally passed off into the vertical chimney by a passage f.

The spiral flue was formed of copper plate, and joined to the bottom of the boiler, or rather formed part of the bottom; but as the water surrounded the flue on both sides, and at the top, and was in no part more than 9 inches thickness or depth upon the copper of the flue, the heat was thought to be most effectually applied to the water, having so small a thickness or depth to pass through.

The fire-grate A being situated beneath the hollow vessel B, in the centre of the bottom of the boiler, and the flame rising therefrom into it, gave heat to the surrounding water, and converted it into steam. The flame then passed away through the spiral flue *a b b*, and then it made another complete revolution round the outside of the boiler at *e e*, before it could escape into the chimney at *f*. By this means, the heat was very completely communicated to the water (*a*).

In addition to this engine, there was another of a larger size placed close by the side of it in the same building; the construction very similar. From a note taken by Mr. Watt in 1775, it appears that the cylinder was 49 inc. diam. The piston made a 9-ft. stroke, and went 8¼ strokes per min., or about 74-ft. motion per min.; it worked two pumps of 13 inc. diam. each, which raised the water 102 ft.

One was a lifting pump with a valve in the bucket, but with a cover on the top of the working barrel; and the rod of the pump being smooth and polished, passed through the cover in a collar of leather. This sort of pump was called a jack-head pump, and it raised the water during the working stroke of the engine.

The other was a forcing pump, with a solid piston, and raised the water during the returning stroke of the engine. For this purpose, a heavy counter-weight was applied to the rod of the forcing pump to cause its descent. The two pumps were placed side by side, as close as they could stand, and the two pump-rods were united at the top by a triangular frame, the vertex of which was suspended by the main chains from the arch-head of the great lever, and one of the pump-rods was suspended from each of the angles at the base of the triangle. To insure a steady motion, the extreme ends of the horizontal piece, which formed the base of the triangle, were received in vertical grooves formed in the edges of two upright beams, and moved up and down therein.

The rod of the lifting-pump, which descended from one of the angles of the triangle, was a strong iron bar; the lower part made cylindrical and straight, and polished, to work through the collar of leathers in the jack-head of the pump. The rod of the forcing-pump, which descended from the other angle of the triangle, was formed of two beams or strong planks of timber, firmly united together at the upper and lower ends; but the middle parts were opened out from each other, to leave a sufficient clear space between them to receive pigs of lead or cast-iron, by the weight of which the piston of the pump was depressed and the column of water raised. The lower ends of the two planks where they were united, were also joined to the upright stem or piston-rod of the pump.

(*a*) In the Philosophical Transactions, 1757 (Vol. L. p. 53), is another paper by Mr. Keane Fitzgerald, stating the supposed results of some experiments which he made on the boiler of this engine, to try the effect of the method proposed by Dr. Hales, for purifying salt water, by blowing showers of air through it.

Mr. Fitzgerald's proposition was to introduce air into the boiler, by a pipe which was fixed within the water, near to the flat bottom of the boiler, and bent circularly so as to extend round the central hollow vessel B, in the space between that vessel and the spiral flue *b b*; this pipe was pierced full of holes, and air was forced into it by a pair of bellows, which were worked by the engine, and the air was supposed to rise through the water in bubbles and assist the formation of steam.

On trial it was thought that the introduction of air caused the boiler to raise ⅛th more steam with the same fuel. But the next year the inventor made another communication to the Society, Vol. L. p. 370, in which he candidly acknowledged, that he had found out that he was totally mistaken in his experiment, for that the bellows never forced any air at all into the boiler, having been split withinside, and that on further trial his plan would not answer. He explains the circumstance of the improvement which he had noticed in the production of steam, to be owing to greater care of the fire-man.

The candour of this recantation is a lesson to projectors, who have very frequently made similar mistakes, but have commonly left it to others to discover, and still more to publish, the inefficacy of their project.

Both pumps delivered their water by branch pipes into a large receiver or air vessel, from which the water was conveyed by a vertical pipe to a cistern fixed near the top of a high steeple or octagonal tower, erected for that purpose at the end of the engine-house.   From this cistern a pipe descended to convey the water down again to the main service pipes laid in the streets.   The cistern was not of large dimensions, not being intended as a reservoir to contain any stock of water, but only to preserve an elevated column of water, to form a regular and constant load for the engine; and also to cause the water to run regularly through the descending pipe into the service-pipes, which were laid beneath the pavement in the streets, and which by smaller branches distributed the water to the houses.   There was also a third pipe descending from the upper part of the cistern to convey away the surplus water when the cistern was full, and would otherwise have run over; so that there were three vertical pipes in the tower.

This was the arrangement universally adopted in the water-works in London; and, although it required the water to be always forced much higher than was absolutely necessary, that was supposed to be a necessary evil, to avoid accidents from those continual changes in the motion of the engine which would have taken place if the water had been forced directly from the pumps into the service pipes in the street; because the resistance to the engine would then have been variable, according as the water was given out to a greater or lesser number of houses, and according to the heights at which the water was to be delivered; and these circumstances must change perpetually, and too suddenly for the engine-keeper to accommodate the motion of the engine to them.

The two pumps were placed in a very large well or cistern, which received the water from the river Thames.   In some cases, large square tanks or cisterns, lined with brick walls, were formed in the ground in the neighbourhood of the engine-house.   The water from the river being received into these pits at high-water, was retained in them for the services of the engines when the tide ebbed in the river. By this means the engine had rather less height to lift the water; and the water, by remaining some hours at rest in the tanks, deposited some of its impurities.   This was an advantageous practice; but it required a very great extent of tanks to receive a sufficient quantity of water to serve so large an engine as that at York-buildings during all the ebb-tide.

The engine raised 136 cub. ft. of water per minute by its two pumps, to a height of 102 ft.; which is an effect of $26\frac{1}{4}$ horse-power, or much less than Mr. Smeaton's engines of the same size.   It worked seven hours every day, and therefore raised 57 120 cub. ft. of water, or 2127 cubic yards per day.   There was no separate injection-pump, but the injection-pipe was supplied from the main-pipe.

The consumption of fuel was 6 bushels of coals ($=$ 504 lbs.) per hour; that is, 19·2 lbs. per hour for each horse-power, or 8·67 million lbs. raised 1 ft. high by each bushel, or 84 lbs. of coals.

The weight of the column of water in each pump was 5890 lbs. or the two 11 780 lbs.; which was $6\frac{1}{4}$ lbs. for each squ. inc. of the piston.

The temperature of the water in the hot-well was 170 deg. when that of the injection-water was 46 deg.; so that it gained 124 deg. from the steam.   The quantity of injection was (1400 cub. inc. $=$) ·81 of a cub. ft. per stroke, or 6·68 cub. ft. per minute.

The steam or vapour remaining in the cylinder when it was cooled to 170 deg. would have had an elasticity of about 6 lbs. per squ. inc.; and deducting this from the full pressure of the atmosphere of $14\frac{3}{4}$ lbs. leaves $8\frac{3}{4}$ lbs. of unbalanced pressure acting upon the piston.   The load or resistance to its motion being $6\frac{1}{4}$ lbs.

and allowing about 1¼ lbs. for the counterweight, the preponderance would also be 1¼ lbs. per squ. inc. to overcome friction and produce motion.

The weight applied to the rod of the forcing-pump was equal to the weight of the column of water in that pump ; and this was wholly in addition to the counterweight of 1¼ lbs. per squ. inc. of the piston.

The quantity of water evaporated from the boiler was not noticed.

The boiler of this engine was of copper, with the internal spiral flue the same as the other engine, but of proportionably larger dimensions.

The axis of the great lever of this engine was fixed to the upper side of the beam of timber which formed the lever, and which was suspended to the axis by strong iron straps. The gudgeons of the axis were laid upon friction-wheels, or rather sectors, forming portions of wheels, with a view of diminishing the friction.

As the pumps of both these engines were the standard kind commonly used and most esteemed in engines for water-works at that time, they merit a particular description.

The lifting-pump consisted of a working-barrel M M, mounted upon its wind-

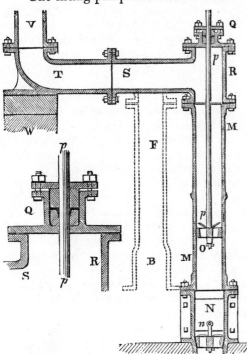

bore and clack-piece N, with a clack n fitted into it, exactly of the same construction as the common mining-pump before described, p. 246. (See also fig. 8, Plate IV.) The bucket O was also of the same construction as there described, but the pump-rod or bucket-shank p p, was made cylindrical and very straight and true, like a polished wire, that it might be closely fitted through a collar of leathers at Q, in the cover which closed the top of the pump. On the top of the working-barrel M, a piece R S was fixed by a flange and screw-bolts : it was called a jack-head piece, and had a pipe S proceeding horizontally from it, to convey away the water. To this pipe another, T, was joined, which turned up at the end with a flange, upon which the perpendicular column of rising pipes V were erected, and they extended up to the cistern at the top of the tower.

The flange at the top of the jack-head piece had a flat cover or circular lid screwed upon it at Q, and in the centre of this was a short tube or nozle, through which the polished pump-rod p passed. The bottom part of this nozle was fitted very exactly to the rod, but the upper part was made larger to receive rings of leather, which were put round the rod, and embraced it so closely as to prevent the escape of any water by the rod. The leathers were confined in their places by a perforated plug of hard wood, which was fitted exactly into the enlarged part of the nozle, and the rod passed through the perforation in the centre of the plug ; which was pressed hard down upon the leather by a flat circular plate, which was also perforated in the centre to admit the rod through it, and was screwed down by three or four screw-bolts, which passed through a small flange projecting at the top of the nozle, Q.

The construction of the collar of leathers is explained more fully by the enlarged section. It was composed of several circular rings of thick soft leather, soaked in melted tallow, the central holes in which were made to fit as exactly as possible to the polished rod. These rings being placed one above another to fill up the nozle, were compressed by screwing down the top cover. This plan answered very well if the collars were very correctly fitted; but as that was rarely the case, and they became leaky, only one or two leathers were put at the bottom of the nozle, and the remainder was filled up with a piece of soft hemp rope, coiled round the rod, another leather was then applied, and the cover screwed down, and by its pressure, it tended to concentrate the hemp round the rod, and thus prevent leakage.

The method recommended by Mr. Smeaton was more complete. A round piece of thick leather was prepared, by soaking in water till it became very soft, and then pressing it into a mould, which gave it the form of a hat, the inside or cylindrical part of which was the same size as the pump rod. The leather being left in the mould, till it became quite dry, would preserve that form. The top, or crown of the hat, was then cut out with a narrow knife, holding the blade in the direction of the axis of the cylindrical inside of the hat, so as to cut out a circle of exactly the same size as the inside of the hat, and leaving the cut edge of the leather sharp.

The leather so formed was put into the nozle, with the cylindrical part surrounding the rod, and fitting it closely. The flat circular rim of the leather was held fast between two plugs of wood fitted into the nozle, one below the leather and the other above it. The perforation, through the lowest of these plugs, was exactly fitted to the pump-rod, at the lower part; but at the upper part was enlarged out, to receive the cylindrical part of the leather, which encompassed the rod. The upper plug was exactly fitted to the rod, and was pressed hard down upon the leather by the screw-bolts and cover.

A single collar of this kind would keep perfectly water-tight, because the cylindrical part of the leather would spring and fit round the rod, and the pressure of the water, in endeavouring to escape, would press the leather in still closer contact with the rod. When the leather or the wood plugs became worn, new ones could be put in very readily.

The forcing-pump was placed close to the side of the lifting-pump; and both pumps were fixed down to the same beams, drew the water out of the same well, and forced or lifted the water up the same rising pipe.

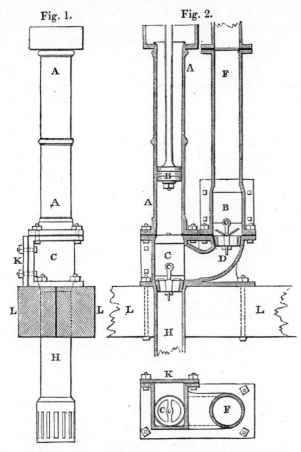

Fig. 1.   Fig. 2.

Fig. 1 is an elevation of the forcing-pump taken endways, of the engine, as it would appear to a spectator standing beneath the great lever of the engine. The barrel of the lifting-pump, if it had been represented in this view, would have stood close by the side of the barrel A A, in the same relative position as fig. 2, but much nearer together.

Fig. 2 is a section of the same forcing-pump, taken sideways, of the engine; and in this direction, if the lifting-pump had been represented, it would have come before the forcing-pump, and have hidden it from the view.

The working-barrel A A is screwed down by a flange at its lower end to a square box C, with which it has a free communication; the upper side of the box has a flange, to which the flange of the barrel is fastened.

This box is square, as shown in the plan, beneath fig. 2, and is pierced at the bottom with a round hole, which forms the orifice of the upper end of the suction-pipe. The upper part of this pipe is enlarged to form a conical seat at the bottom of the box; but beneath that seat, it is the same diameter as the barrel, and continues downwards into the water of the well, being the sucking-pipe of the pump; and the lower end is formed with blast holes.

In the tapering part, or conical seat, a common double clack is inserted, so as to stop the upper end of the suction-pipe, as is seen beneath C. This may be called the sucking-clack. One of the sides of the square box C is carried forward to form the clack-hole door, the cover of which, K, is fastened on by screw-bolts passing through a projecting flange.

The forcer or piston B, is exactly fitted into the working-barrel. It has no perforation or water passage through it, but is surrounded by two cup leathers, facing both ways, and filling the barrel, so as to prevent any water passing by the forcer. When the forcer B is drawn up during the working-stroke of the engine, the water rises up through the clack C, and follows the forcer, so as to fill the barrel A A; but when the forcer B descends, by the weight with which its rod is loaded, the water having no passage through the forcer, and the clack C shutting down, the water cannot escape through the barrel; but as there is a curved pipe D proceeding from one side of the box C, the water will force a passage through it.

The upper part of this curved pipe terminates in a conical seat, adapted to receive another common double clack, exactly like the former. The lids of this clack open to allow the water to pass, when the forcer descends, but they shut, so as to prevent the return of the water, when the forcer rises. This latter may be called the forcing clack.

This forcing clack seat opens in a square flange, forming a continuation of the flange at the top of the box C, on which the working-barrel is fastened; and upon the other part of this large flange, the lowest length B F of the forcing-pipe is fastened over the forcing-clack, so as to stand up close by the side of the working-barrel. The lower part B of this forcing-pipe is a square box of the same size as the box C; and, in like manner, it is provided with a door to give access to the forcing-clack. In case the forcing-pump was to be made independent of the lifting-pump, the forcing-pipe B F, which received the water thus forced up by the pump, might be carried up in a perpendicular column, as high as the water is required to be raised; and to sustain the weight of that column, a large flat flange is formed at the bottom of the box C, and extending beneath the turn-pipe D, with a bracket from the pipe to strengthen it.

The pump is fixed upon two strong beams of timber, fitted together side by side, with a round hole left between them, to receive the suction-pipe H of the pump. These beams being bolted fast together like one piece, and solidly fixed, form the foundation for the pump, the flat flange fitting on the top surface of the compound beam, and being fastened thereto by bolts (g).

When the lifting and forcing-pumps were combined, as in the engine at York Buildings, the lifting-pump was fixed down between the same two foundation-beams, so that the two pumps stood side by side; and the forcing-pipe F of the forcing-pump, was joined to the horizontal branch of the jack-head-piece of the lifting-pump, as shown by the dotted lines B F, in the figure of that pump; a short branch being formed at the lower side of the branch R S, near to the pump, to join the flange at the top of the forcing-pipe B F. By this means the water from the forcing-pump was conveyed up the pipe B F S T V, and one perpendicular column of pipes at V served both pumps. The weight of this column was supported upon a block at w, laying on the top of the wall of the well, in which the pumps were placed, the lower part of the turn-pipe T having a flat flange beneath it to rest upon the block w, with a bracket from the turn T to strengthen the flange.

The pipe R S of the lifting-pump, and the turn-pipe C D of the forcing-pump, are turned as much as they can be towards each other, but still do not pass exactly over the pipe B of the forcing-pump; that pipe, as the pipe R S, must bend backwards in rising from B to F, so as to intersect the pipe R S. The two pumps thus combined do not interfere with each other: the two clack-doors of the forcing-pump at B and C, and the clack-door of the lifting-pump at N, being on the outsides of their respective pumps, permit a free access to all the valves; and, by removing the cover of the lifting-pump barrel, the bucket can be drawn out to repair, and the clack n can also be drawn up through the barrel if required; and so of the forcing-pump, if the forcer is withdrawn, the clack beneath it can be drawn up through the barrel to repair it, in the same manner as before described for the common mining pumps (p. 216).

(g) The forcing-pump above described is in reality copied from the drawing of the injection-pump to the engine at Long Benton (see p. 182), and the lifting-pump is the drawing of the injection-pump for the engine at Chase Water (see p. 191). But the pumps at York Buildings were constructed on exactly the same model on a larger scale.

The forcer or solid piston of the forcing-pump, is furnished with two cup leathers, so called from their resemblance to cups, or rather saucers, being flat circular pieces of leather, turned up all round on the edges, and these edges exactly fitting into the barrel. The forcer consists of three circular plates of iron or brass, fitted on the lower end of the rod, and fastened on by nuts, the flat parts of the two leathers being held between the plates, the middle of these plates is fitted as exactly as possible into the barrel, but the upper and lower ones are rather smaller, to allow space for the cylindrical part of the cup leathers to turn up or down between them, and the inside of the barrel; one of the cups being turned up, and the other turned down, the passage of the water by the forcer is prevented both in ascending and descending.

The application of two pumps to lift and force the water alternately, so as to keep up a continuous flow of water, was advantageous, because the water would then be conveyed equally well by a conduit-pipe, of only half the area which would have been required if all the water had been forced by one pump, and of course with intervals of rest between every stroke.

It was a still further improvement to connect a large air vessel with the pipe which conveyed the water away from the pump: this pipe entered into the lower part of the vessel, and therefore all the water was delivered into it by the pumps, and compressed the air contained in the upper part of the vessel; the ascending pipe also proceeded from the lower part of the same air vessel, and the re-action of the compressed air forced the water out through that pipe, with a nearly uniform force, so as to keep up an even stream, without any intermission, notwithstanding that there would be a momentary cessation of the delivery of water into the air vessel by the pumps, at every change of stroke.

This addition rendered the motion of the engine much more easy, because the column of water was kept in constant motion, without wasting power to start it anew at each pulsation.

In 1777 this engine underwent considerable repairs, and Mr. Smeaton was consulted on the occasion, as he usually was upon every similar subject which occurred in his time: he gave the following directions for its improvement, with a view of augmenting its power, which, as before noticed, was considerably below his standard.

" The under surface of the piston must be covered with a planking of elm, or beech wood, about 1½ inc. thick, applied beneath the iron piston, and fastened thereto with 20 screw bolts of ¾ of an inch diameter. This planking being adapted to the piston, a double thickness of flannel, or a thickness of the strong cloth called fearnought, soaked in tar, is to be interposed between the iron and the wood; the bolt heads are to be in the underside of the wood, which is to be covered with a second sheathing of beech wood, ¾ of an inch thick; a second thickness of fearnought and tar being interposed between the two surfaces of wood. The last wood sheathing is to be fastened to the planking with nails with small heads, and no more of them to be used than is necessary to fasten it, so that as little metal as possible may be exposed in the under surface." (The directions for framing this work are the same as those already given for the Chase-water engine.)

" The rationale of this is, that wood is a much more imperfect conductor of heat than metals; that is, it communicates heat much less readily; and that the piston will necessarily be the coldest part of the cylinder, not only by the water continually poured upon the top to keep it tight, but also from its receiving the first stroke of the cold injection water; and as the steam, when it issues through the steam-pipe from the boiler, must rush violently in the first instance against this cold body, it becomes thereby considerably deadened, contracted, and condensed, so as to take a greater quantity of steam than necessary from the boiler, to fill the cylinder completely full.

" But when the steam strikes against wood, which conducts heat slowly and receives less heat from the steam, a less condensation takes place. Also, by the same reason, the cold injection water which is thrown up against the piston loses but little of its coldness, and gets but little heat from the piston; it is, therefore, better qualified to produce its effect of cooling the steam.

" The wood therefore acting as a neutral body, neither cools nor heats the fluids that alternately strike against it, so much as the iron could do.

" A lesser thickness of tarred flannel between the planking and the sheathing might suffice to make good the joint, and exclude the air; but as tarred flannel is a still less rapid conductor of heat than wood, which by getting soaked with water partakes of the nature of a watery fluid; the heat which penetrates the sheathing is stopped by the interjacent stratum of tar and flannel from getting into the iron-work necessary for fixing the planking; for the same reason, next to beech, which is a close wood, the sheathing should be of good resinous Riga fir.

" Besides the above particulars, the following are also to be attended to.

" The orifice of the injection-cap should be a square hole of $1\frac{1}{10}$ inc. square, and the edges of the hole rounded from the underside, that it may throw up a full bore; the middle of the jet should not (as in common) be directed to the centre of the piston when at top, but it should rise quite perpendicular, and in consequence strike the bottom of the piston, at right angles.

" It must be observed, by going into the boiler, or looking in through the man-hole door, when the regulator is open, that upon letting the injection give a dash as it will do when at work,—whether any water comes down the steam-pipe into the boiler; because if more falls through the regulator than a few drops, the cap of the injection must be a little inclined from, or towards the centre, till as little as possible falls through the regulator. And in order that less of the injection water may get into the steam-pipe, the top of the steam-pipe should be chamfered on the edge from the outside: or if the steam-pipe should be suspected to be too low, or if it is found to rise less than 5 or 6 inches above the cylinder bottom, this end may be answered by driving a short tube of thin copper into the top of the iron steam-pipe.

" That part of the injection-pipe which is inside the cylinder, should be wrapped with a double or treble thickness of tarred cord or marline, so as to form a thick coat, to separate the cold metal from contact with the steam, in order to leave the injection water as cool as possible.

" The injection water ought always to come from the highest reservoir possible, and in this case, as the engine raises water to the top of a tower, the pipe may be branched off from the column of the pipe which conveys the water to the top; using the small cistern at the top of the house, only till the engine is got fairly to work.

" The regulator must be made to open fully, if not so already (which is seldom the case); that it may let the steam ascend from the boiler uninterruptedly by the full bore of the steam-pipe, the least part of which should be the top.

" A small air cock (the least wine cock that can be procured will do) should be put upon the upper part of the eduction-pipe, or any other part where it will freely communicate with the interior of the cylinder.

" A cap and pipe should be put over the snifting clack, with an inch and half cock upon it, by the partial opening of which the snifting can be regulated; but which will not be wanted, unless the valve should snift more than enough when it is left quite open, as it will be apt to do when all the preceding articles are put in execution.

" It is also expected, that when all the above amendments are made, that the weight upon the piston rod of the forcing-pump will bring the lever more freely out of the house than it has hitherto done, that is, it will move quicker in the returning stroke ; but no part of this weight must be removed, because the lever ought to have the preference in going out of the house, rather than in coming in.

" The proper use of the air cock is, that after the engine is got steadily to work, and the snift emits no more than it should do, that the air cock should be opened as much as it can be to let the engine come fairly in, that is, for the piston to come fully down, and if it goes out in preference to that of coming in, the better.

" I expect my alterations will make the engine go quicker, so that it will raise more water, and either serve the customers more amply, or diminish the hours of service.   In case the engine will then raise more water than the main service-pipes will take, a cataract should be applied to open the injection-cock, so as to reduce the number of strokes to exactly what is wanted, and no more; whereby the consumption of coals will be reduced in proportion as the strokes are diminished below the utmost the engine is capable of doing in its improved state."

This engine continued to work constantly till 1805, when a new engine was erected on the same premises, in place of the former engine, of 45 inc. ; but the old one of 49 inc. was still kept as a reserve, and was worked occasionally till 1813, when it was pulled down, a new and more powerful modern engine being put up in the place of it : it was then the last atmospheric engine remaining in London. The whole establishment of these water-works has since been abandoned.

The author frequently visited this engine in 1804 and 1805, when it worked constantly, and made many sketches of its parts, as it formed the subject of his earliest studies on the subject of the steam-engine.   It had undergone no alteration from the state described by Mr. Watt and Mr. Smeaton, nearly 30 years before.

The London Bridge Water-works were the most considerable in London, next to the New River ; they had both water-wheels, and a fire-engine for occasional use.

The old London bridge is built with stone, but the foundations are made according to the ancient method of driving piles into the bottom of the river, and cutting off the tops level with the lowest water line.   Upon these piles the stone piers of the bridge are founded ; but as the original piles were subject to decay, and admitted of no renewal, it became necessary to surround them with a bank of gravel and chalk ; and to retain the chalk, casings of piles were driven in all round the piers, to form enclosures round the foundations, called starlings.   These diminish the space between the arches, so as to occasion a very rapid current of the water in running through them, because the water-passage bears only a small pro-portion of the artificial solids thus placed in the way of the current ; and this reduces nearly all the arches, as it were, to sluices.

Of the twenty arches in this bridge, six were devoted to the water-wheels for the engines, viz. five on the London side, and one on the Southwark side.   In 1763, two of the small arches in the middle of the bridge were thrown into one, by moving the pier and starling between them.   This was a great improvement to the navigation of the river, but it diminished the current of water so much that it became necessary to add another larger wheel to supply the deficiency.

The original machines are described by Mr. Beighton in the Philosophical Transactions for 1731, No. 417, vol. vii. p. 5; and the large machine made by Mr. Smeaton in 1767, is described in his Reports, vol. ii. p. 27. See also Dr. Rees' Cyclopædia, vol. xxxviii. article, Water. These wheels were very powerful at the most rapid period of the tide, but became inactive as the tide slackened, and they were stopped entirely for a considerable time at the turn of high and low water. It was therefore found necessary to add a fire-engine, to continue the service occasionally in neap-tides, when the current was slow; and in the night the engine was prepared for working in case of fire breaking out in the town, whilst the wheels were stopped at the turn of the tide.

The cylinder of this engine was 34 inc. diam. = 908 squ. inc. The piston made a 7-ft. stroke; the pump was 12 inc. diam. and lifted the water to the top of a tower 120 ft. high; weight of the column 5900 lbs. = $6\frac{1}{2}$ lbs. per squ. inc. of the piston.

The engine commonly made eight strokes per min. = 56 ft. motion. This is only 10 horse-power, and it burned three bushels of coals per hour; so that its performance was very greatly inferior to that of the York-buildings engine.

The engine had a forcing-pump, which was worked by a very heavy counter-weight applied to the pump-rod. The descent of the piston raised this weight, which in returning raised the column of water. Mr. Smeaton recommended the engine to be altered, to make it raise the column of water immediately by the direct action of the piston, and thus do away with the useless weight.

" To produce this effect, it will be necessary to change the working barrel, with its valves, and a new one may be introduced of a larger size, which will be more advantageous.

" At present, the whole of the water raised by the fire-engine is raised to the top of the tower, said to be 120 feet high; though in common, one-half, two-thirds, or three-fourths of that height would be sufficient. I therefore propose, as a very material improvement, not only to save fuel, but to raise more water, to unite the main conduit-pipe from the fire-engine with those from the water-engines; by which means the fire-engine will, like the water-engines, never be burdened with a higher column than is sufficient for the service then on.

" To this I am sensible there is an objection, namely, that the column to be lifted by the fire-engine being then variable, it will not work with the requisite degree of steadiness. In the ordinary way of applying the injection this would certainly be the case; but I have a method of applying the injection whereby the engine-keeper is enabled extempore, while the engine is working, to vary the quantity proportionably to the column to be lifted. Hereby the ill effects arising from a variation of the column may be prevented, and a proportionable saving made in the fuel.

" As the engine, at present, when lifting the whole column to the top of the tower, would bear a greater load, and would consequently be considerably under-loaded when the column is lowered, as above recommended, I propose to put in a 13-inch working barrel, instead of the 12-inch, as now; by which means the engine will raise one-fifth more water at every stroke. And by virtue of the other changes, I expect the coals will be reduced from three bushels per hour, when lifting a 12-inc. bore to the top of the tower, as at present, to two bushels at an average when forcing a 13-inc. bore directly into the main conduit-pipe, without first going up the tower.

" The boiler is too small for the cylinder, and not of the very best proportion;

but, as I understand it is in good condition, and may serve some years, I do not at present recommend any alteration in that part."

The boiler of this engine was of the kind called a flange boiler, the upper part being made to project over the flue, which went round the lower part; as will be readily understood from the sketch.

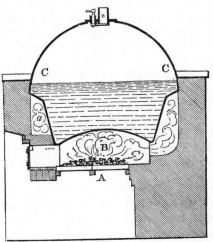

The hemispherical dome at top was $10\frac{1}{4}$ ft. diam., but it was contracted at the flange to $8\frac{1}{2}$ ft., and still more at the bottom to 7 ft. The rise of the bottom was 1 ft., and the fire-grate was placed $2\frac{1}{2}$ ft. below the bottom, in the centre. That ·part of the boiler which was exposed to the fire in the flue was not a vertical surface, but was inclined, and then the projecting flange formed the top of the flue; by which means it was supposed that the flame had a better effect in heating the water, because the natural tendency of heat is to ascend perpendicularly.

This form of boiler appears to have been used by Mr. Newcomen in his earliest engines; it was generally considered to be more economical in the consumption of fuel than the boilers formerly described in Plate II, and which were called tun boilers. The flange boilers were accordingly used for engines in London and in Cornwall, where coals were expensive: but the flange, where it projected over the flue, having but a small depth of water upon it, was liable to be left dry by negligent engine-keepers, and then the boiler would be much injured. They succeeded best when made of copper. For this reason, and the extra expense of making them, and the liability of the mineral water to deposit a crust on the flange, these boilers were not used in the collieries, where coals being cheap, the economy of boilers was a greater consideration than that of fuel.

The other engines at work in London, previous to 1775, were two for the Chelsea Water-works, at Pimlico; one 32-inc. cylinder, the other 28-inc. They took the water from a cut or canal proceeding from the river, and in which the mud was deposited, and forced it up by pipes to a large pond or reservoir on the rising ground in the Royal Park, near Piccadilly. The water which was received into the canal from the river every tide at high water, was allowed to run out into the river again when the tide subsided, and then worked water-wheels, by which water was pumped up as well as by these engines.

There were also two engines, each of 30-inc. cylinder, at Shadwell Water-works; a small one at Lambeth, and another of 24-inc. at Stratford; also a small one at the New River Head at Islington, 18-inc. cylinder. In all, ten engines were employed in the water-works; the area of all the pistons taken together being 8376 square inches. They appear to have performed better than the engines of the same period at Newcastle, chiefly because they were not encumbered with a dead weight of pump-rods. Taking them on an average at 80 square inches to a horse-power, it would be about 105 HP; and probably the water-wheels and horse-machines would amount to as much more power. There were about fourteen water-wheels employed in different places.

A SMALL ATMOSPHERIC ENGINE, 8-HORSE POWER, FOR A PUBLIC WATER-WORKS
IN A PROVINCIAL TOWN.

THIS was for the supply of the city of York : it raised the water of the river
Ouse up into a large cistern, which was made of wood, and lined with sheet-lead,
and held a very considerable quantity of water. The engine was made about the
middle of the last century, on the proportions then usually followed ; and after
working many years, Mr. Smeaton purchased a share in the water-works, and took
great pains to remodel the engine and put the works upon the best footing.

The cylinder was 25 inches diameter = 491 square inches ; the piston
worked 5¼ feet stroke, and went 12 strokes per minute = 63 feet per min. ; the
pump was 10 inc. diam. and raised the water at a mean 72 feet high, from the
surface of the water in the well to that in the reservoir, which was fixed on the
top of the house, and covered the whole area. The weight of the column of water
was 2455 lbs. or 5 lbs. per square inch of the piston.

This weight being raised 63 feet per minute, was 4·7 horse-power. The injection-water was taken from the cistern without a separate injection-pump : the
injection-aperture was ⅝ of an inch diam. ; the temperature of the hot-well was
usually 176 deg. ; the boiler was 7 feet diameter.

This engine worked at the above rate, with 255 lbs. = 3·03 bushels of good
coals per hour, which is very nearly 3 millions pounds raised 1 foot, by a bushel,
or 84 lbs. of coals.

This was the state of the engine before Mr. Smeaton began to improve it.
He first applied a wooden bottom to the piston, and made all the other alterations
directed by him for the York Buildings engine, as before quoted. And by a
similar trial, with the same kind of coals, after the alterations, the engine went 15
strokes per minute, or 78¾ feet per min. (÷ 5·86 HP), with 202 lbs. of coals per
hour ; which is at the rate of 40·83 millions pounds, raised 1 foot by a bushel.

And by a trial of the sleck or small coal, made immediately afterwards, 232
lbs. were consumed per hour, which is ·87 of the effect of the coals, or 4·2 millions.

Three years afterwards, the old boiler being worn out, it was replaced by a
new one of copper, 9 feet diameter at the largest, 7 feet diam. at bottom, and 4
feet deep in the copper part, and the upper dome was made of lead. At the same
time, the length of the working-stroke was increased to 6 feet, and an additional
lifting-pump was applied by the side of the former pump (which was a forcing-
pump), the two being combined, to raise water in the same manner as before
described. This new pump was 8 inc. diam., but worked a shorter stroke than the
principal pump, which was 10 inc. diam. The weight of the column of water was
increased by this additional pump to 3575 lbs. or 7·28 lbs. per square inch on the
piston.

After this alteration, the engine went 13 strokes per minute of 6 feet = 78
feet per min. which is 8·45 horse-power ; and it consumed 208 lbs. of coals per
hour, which is at the rate of 6·75 millions pounds raised 1 foot high by a bushel.
At this time the engine raised 62 cubic feet per minute, or 442 hogsheads per
hour, wine measure.

A cataract was also attached to the engine, and at times, when the service
did not require so much water, the engine was worked at 10 strokes per min. of 6
feet, or 60 feet per min. The effective power was then 6·29 HP, and it raised
328 hogsheads per hour ; 2·2 bushels of coals per hour were then consumed, that
is at the rate of 5·66 millions ; or 2·56 bushels of sleck per hour produced the

same effect, that is 4· 56 millions. Here the effect of the sleck is ·80 of that of the coals.

Hence it appears that there was a loss of effect of the fuel as 6· 75 to 5· 66, or as 100 to 84, by working slowly with the cataract. It was tried to work the engine at its full speed, till the reservoir was full, and then to damp the fire, and stop the engine, till the reservoir was almost empty, and then set it to work again. In this way it worked 48 minutes, and then stopped for 16 minutes, the stoppage being $\frac{1}{4}$ of the whole time : the produce from the fuel was better than with the cataract stopping at every stroke, as 12 to 13.

The town of Kingston-upon-Hull had also its water-works. The engine was the same size cylinder as that·at York, 25 inches diam. It was made by Mr. Hindley, of York, a celebrated clock-maker and mechanist in his time. It worked without a great lever ; the pump, which was 12 inches diameter, being placed immediately beneath the cylinder, and the piston-rod of the engine connected to that of the pump by a frame, which included the cylinder within it, and moved up and down like a window-sash. Thus the piston-rod was affixed to the middle of the top horizontal bar of the frame, and the two upright side-rails of the frame went down on each side of the cylinder, and were united, beneath the bottom of the cylinder, by the lower cross-bar of the frame : to the middle of this lower cross-bar the pump-rod was fixed, being exactly in the line of the piston-rod. It was a forcing-pump, 12 inc. diam. ; and the pressure of the atmosphere on the piston pressed down the forcer of the pump, and raised the column of water about 29 feet high.

It was intended to make the steam in the boiler strong enough to lift up the piston, with its rod, and the sliding-frame, and pump-rod, and forcer, when it was admitted into the cylinder. The inventor died before it was completed ; but, on trial of the elastic force of steam, it could not be made to act, and a counterpoise to the weight of the moving parts was applied, by a chain conducted over pulleys. In this state the engine worked for some years, but produced only a very small effect : it made $13\frac{1}{2}$ strokes per minute of 4 feet, or about $2\frac{1}{4}$ HP, and burned $72\frac{1}{2}$ lbs. of coals per hour.

Mr. Smeaton made a small oil-mill, to be worked with the water raised by this engine during the night, when it was not required to serve the town. It had an overshot water-wheel, 27 feet diam. (see Smeaton's Reports, II. 398), but the engine being found inadequate to work it effectually, a new one, made by Mr. Watt, was afterwards added at Mr. Smeaton's recommendation. This was one of the earliest applications of the fire-engine to work a mill. It was found to do the work of 3 horses acting together.

A PORTABLE FIRE ENGINE FOR DRAINING TEMPORARY EXCAVATIONS.

ALL the engines hitherto described, required to be fixed and supported in a building, the expense of which formed a considerable portion of the cost of the engine ; this circumstance rendered the fire engine inapplicable for temporary purposes, such as draining the foundations of bridges, or other extensive buildings. Mr. Smeaton, having frequent occasion to employ great numbers of men, and at other times horse-machines, to pump out water in such cases, proposed a plan in 1765, for a small portable fire-engine, which could readily be removed from place to place, and quickly fixed up for work ; all the machinery being supported in a wooden frame which stood on the ground, on its own legs, quite independently of the building in which it might be placed, or it could work quite in the open air if required.   The boiler was also contrived with an internal furnace, so as to be complete in itself, without any brickwork-setting for the furnace.

This engine is described with plates in *Smeaton's Reports, Vol.* I. *p.* 223, and one of those plates is given in Plate IX., to explain the boiler and parts of the engine.

To bring the engine into a small compass, the great lever was laid aside, and a circular wheel, or large pulley substituted, the chain from the piston shank being conducted over this wheel, and the shank of the pump rod attached to the other end of the chain, the motion of the piston would be communicated to the pump rod in the same manner as by a lever.

The wood frame was shaped like two of the letters A, placed one behind the other, leaving a space between them ; the vertexes forming supports to the gudgeons of the axis of the pulley-wheel ; and the cylinder bottom was supported upon the two horizontal sills, upon which the two A frames were erected, and which formed the bases of the two triangles ; these sills were supported at a proper height from the ground, upon four upright legs, united by horizontal pieces and cross braces, so as to form a strong stool or table, which became the basement of the engine, and was placed upon brick foundation piers.

The injection cistern was placed over the pulley-wheel, and supported by four upright legs, rising up from the sloping sides of the two A frames, near their upper ends.   A smaller pulley-wheel was fixed on the same axis as the great pulley-wheel, and upon this, two smaller chains were applied, one to suspend the rod of the injection pump, and the other for the plug-beam of the working gear.

The boiler was not placed beneath the cylinder, as in other engines, but at the side, and the steam was conveyed from it by a steam-pipe, to a box beneath the cylinder, containing the regulator ; the boiler was placed on a pier of brickwork, at the side of those piers which supported the legs of the frame or stool, but on a lower level ; the boiler was shaped like a large tea-kettle, (see Plate IX.), and the fire-place was in the centre of it, so as to be surrounded on all sides by the water.   At one side of the boiler was an opening for the fire-door ; and from this, a large tube or pipe led through the water, and joined to a hollow sphere of cast-iron, situated in the centre of the boiler, and in which the fire was made, upon a grate at the lower part of the sphere ; from this sphere, another large tube, or ash-pit, descended perpendicularly beneath the grate, through the bottom of the boiler, and was open below to supply air to the fire ; also, opposite to the fire-door was a third large tube or chimney, proceeding from the central sphere, through the side of the boiler, and it then turned up in the manner of the spout of a tea-kettle, to carry off the smoke into a tall cylindrical pipe or chimney, made of iron-plate.

This engine was intended to have a cylinder 18 inches diameter, and to make 6 feet stroke. The pulley wheel to be $6\frac{1}{4}$ feet diameter.   The triangles or A frames to be $9\frac{1}{2}$ feet base, and 11 feet perpendicular, and the clear space or width between the two parallel A frames, 3 feet inside; the height of the stool upon which the sills of bases of the triangles were supported $6\frac{1}{4}$ feet.   The boiler to be 6 feet diameter, with an internal spherical fire-place in the centre 34 inches diameter ; the openings or pipes for the fire-door and for the chimney 15 inches, and the fire-grate and ash-pit tube 18 inches diameter.

This engine was expected to work at the rate of 10 strokes per minute, of 6 feet = 60 feet motion per minute, when the piston was loaded with a resistance of 8·91 lbs. per square inch, = 2268 lbs.   This would have produced $4\frac{1}{8}$ horse power.   It was afterwards found on trial of another engine of this size, that it would produce that effect and more.

In his computations on the power of this engine, Mr. Smeaton says, " It has been found by experiment, that a horse will raise about 250 hogsheads, wine measure, 10 feet high per hour ;" but, he reckons that the quantity raised by this engine " would be more than six times the quantity raised by one horse; and, consequently, it will act with more than the power of 6 horses at a time; but, in order to keep up this force with horses, night and day, 3 sets will be required, and, consequently, this engine working constantly will be more than equivalent to 18 horses."

A hogshead, wine measure, contains 8·42 cubic feet, and when filled with water, weighs 526·36 lbs. 250 hogsheads per hour, is at the rate of $4\frac{1}{6}$ hogsheads per minute, and is equivalent to 21·9317 lbs. raised 1 foot per minute. This is very nearly $\frac{2}{3}$ of the modern horse power, which is used for estimating the power of steam engines, and which has been before stated as equal to $1\frac{1}{4}$ horses.

Mr. Smeaton's explanation of the Section of the Boiler of his Portable Fire-Engine, 1765.

[See Plate IX.]

Fig. 1 shows a section of the boiler, cylinder, and pipes, with the working-gear; the whole being divested of the framing, in order to render every part distinct. The boiler is supposed to be turned one quarter round from its true position, in order that the most material parts may be brought into one view. It is also to be noted, that every vessel and pipe is supposed to be cut through the middle, in order to show the contents, and that the section is not confined to any particular plane.

" For the more perfect explaining of the principal figure, it will be best to begin with the small plan, fig. 2.

" ABCD is a plan of the cylinder bottom, bolted down upon the two main beams, AB and CD, the dotted circle EF, shows the circumference of the bottom flange of the cylinder, and the circle GH the diameter of the cylinder within; the hole I, answers to the steam-pipe and regulator, and the hole K, to the eduction or sinking pipe, and the circle LM is its flange. The circle NO shows the size and position of the regulator plate, and the dotted circle PQ, the size of the receiver in which the sliding valve of the regulator R works, and which in the position represented is open; ST is the lever by which it is worked, and when that lever is brought forward into the position SV, then the valve will cover the aperture of the regulator.

" *Explanation of Fig. 1.*

" A is the boiler, made of iron plates $\frac{1}{4}$th of an inch thick, riveted together.

" B the fire-place, which is intended to be of a spherical figure of cast-iron, and contained entirely within the water in the boiler; the coals are to be introduced by the large pipe, or opening C, and the smoke carried off by the curved pipe D; and in order to promote a sufficient draught, the iron funnel or chimney E is added; the ashes fall through the grate S, and the wide pipe F, into the ash-hole below; the whole of the fire-place being joined to the boiler by proper flanges, as is shown in the figure, and always covered with water; as will be known by the two gauge cocks, 6, 6.

" Though it is expected that a small force of fire thus applied, will keep the engine going, yet, as that force cannot be wholly exhausted within the compass of the boiler, it is proposed to surround the curved pipe D, with a copper vessel, adapted to the shape thereof, as represented in the figure; into which it is proposed first to introduce the feeding water, and thereby prepare it to a greater degree of heat, than if it were brought immediately from the hot-well into the boiler. The water so heated, in the space round the curved chimney-pipe, is to be introduced into the boiler by a small hole at C.

" The bars of the grate S are intended to be cast in a loose ring, so as to be capable of being taken out and replaced, when occasion requires.

" T is a cork to empty the boiler, when it requires to be cleaned.

" G is a section of the lower end of the cylinder, into which the steam is conveyed from the boiler, by the steam-pipes *d e*, HI, of which the first lengths *d e*, are of copper, and are disunited at *f*, for the sake of taking the engine easily to pieces; and are joined by first bending a piece of sheet-lead round the joint, and then wrapping them with cloths and cord, as is common in great engines.

" H is the receiver for the steam; it is made of cast-iron, and contains the regulator or valve, and I is the steam-pipe, by which the steam passes into the cylinder, when the regulator valve is open, as in the present situation of the engine it is supposed to be; and in that case, the piston K will begin to return, or ascend, by the counter-weight of the pump-rod.

" L is the snifting-clack, fixed in the lower part of the cylinder; at this the steam blows out, when the parts are in the situation represented.

" M is the injection-cock, which is now shut.

" N, the injection-pipe which brings down the water from the cistern, fixed above the great pulley-wheel.

" O is that part of the injection-pipe, which conveys the water into the cylinder, when the injection-cock is opened; the pipe O terminates in a cap, which is perforated on the upper side, so as to direct the jet of water perpendicular; the injection-spout is placed in the centre of the cylinder, though not so drawn, for, as part of the termination of the injection-pipe lies behind the steam-pipe I, it would interrupt the complete view, and, therefore, the pipe O is represented shorter than it really is.

" *Operation of the Engine.*

" It is to be premised that the plug-beam, P, moves up and down with the piston, but with a diminished velocity, according to the diameters of the two pulley-wheels, from which they are respectively suspended, as before stated. The plug-beam is furnished with pins, which lie in four different planes, answerable to four different detents, or arms of the working-gear. The pin 3, lies upon the foreside; the pin *g*, is within the long mortise or slot, which divides the plug-beam into two parts; the pins S, 1, 2, lie upon the far side, and the peg *p*, stands in the face of the plug-beam, upon the farther check.

" When the piston is risen as high in the cylinder, as its stroke is intended, the pin *g* meets, and lifts up the arm *h*, of the tumbler, *i k*, and oversets it into the position *i m ;* so that the point *k*, moving towards the plug-beam, into the place of *m*, the regulator valve is shut, by drawing the rod *k o*, which is jointed to its lever. At the same instant, the peg *p* will have risen so high, as to lift up the latch *q*, whereby the beak *r*, of the faller (or F, as it is commonly called) is disengaged from the latch, and the weight W, by descending into the place *v*, carries the fork *x* along with it, and turns the lever or handle of the injection-cock into the place *y*, whereby that cock is opened, and the jet of cold water played up into the cylinder. The steam is thus reduced into water (or condensed, as it is commonly called), and a vacuum produced within the cylinder.

The pressure of the air then causes the piston to descend, and in descending, the peg S on the far side, meeting the handle of the faller or F, carries it down to *z*, and shuts the injection-cock, and also hooks the beak *r* behind the latch *q*; in this position, the handle, *z*, of the F, becomes parallel to the plug-frame, and the pins 1 and 2, succeeding the pins *s*, retain it in that position ; the piston still descending until the peg 3, on the far side, takes the arm 4, of the tumbler at *n*, and depresses it so as to bring back the regulator into its present position, and the farther descent of the piston is stopped by letting in the steam.

" The cold water thrown in by the injection, at each stroke, and also the water from the condensed steam, is evacuated from the cylinder, while the piston is rising, by means of the eduction or sinking pipe and valve Q; this pipe lying exactly behind the steam-pipe I, is thereby in a great measure hidden. The steam-pipe is carried several inches above the bottom of the cylinder, to prevent the injection-water from running down into the boiler ; but the eduction-pipe rises no higher than the bottom, in order to drain all the water down into it ; and as the bottom of the cylinder is commonly elevated from $1\frac{1}{2}$ feet to 3 feet above the valve Q, at the bottom of the eduction pipe, (commonly called the horse-foot valve,) the pressure of this column of water opens that valve, and makes its way into the cistern R, called the hot well ; but when the vacuum is made in the cylinder, the valve Q shuts ; and being immersed under the surface of the hot well water, is thereby kept air-tight.

" There is a considerable quantity of air contained in water, when in its natural state, but that air separates from the water by boiling, and in some degree by considerable heating ; hence a small quantity of air passes along with the steam into the cylinder, and a farther quantity of air also escapes from the injection water, by being considerably heated with the steam ; and although the air is capable of being considerably expanded by heat, and contracted by cold, yet the degree is so very much less than that of steam, that if the air generated at each stroke was not evacuated, it would accumulate in the cylinder, and render the vacuum so imperfect after a few strokes, that the piston would stop in the middle of its descent.

" When the piston is down, this air will be lodged between the piston and the surface of the injection water, which covers the bottom of the cylinder ; and on the first rush of the steam into the cylinder, when the regulator is opened, the piston not being instantly put into a contrary motion, the steam finds a passage at the snifting clack, and blows out for a small space of time, so as to carry out with it the whole, or the greatest part, of the air generated at each stroke : this prevents its increase beyond a certain degree.

" The surface of the hot well water being in general from 3 to 5 feet higher than the surface of the water in the boiler, this height of column is sufficient to force its way into the boiler, although it is resisted by the action of the steam within the boiler, when sufficiently elastic for use ; the hot water, therefore, finds its way down the upright feeding pipe 5, 6, first passing through the feeding cock 7, which is opened as much as is found necessary for supplying water to the boiler, as fast as it is consumed in steam.

" It is evident, that whenever the repulsion of the steam within the boiler is not too great to suffer the surface of the water in the upright feeding pipe to be below that of the hot well, that the boiler will feed, by the passage of the water through the pipe ; but whenever the repulsion of the steam is such, as to keep the water in the feeding pipe above that in the hot well, then the water will revert into the hot well. As the repulsion of the steam within the boiler is alternately greater and less, accordingly as the regulator valve is shut or open, during each stroke of the engine, it follows, that the boiler may take its feed whilst the cylinder is drawing the steam from the boiler, and revert into the hot well whilst the piston is descending. In order, therefore, to make this matter safe, and at the same time to bring the engine into as small a compass as possible, a valve is placed in a small box, at 8, at one end of the hot well, to prevent such a reversion.

"It is necessary to continue the upright feeding pipe as usual, considerably above the hot well, for as the water is in an oscillating state within this pipe, according to the variable repellency of the steam within the boiler, it must be carried up so high as to prevent its frequently overflowing; and it is necessary to be open at top, otherwise the steam generated in the pipe would prevent its feeding regularly. The hot well in this view is represented considerably longer than it is; for supposing the boiler turned a quarter round, the feeding pipe 5, 6, would fall directly behind the puppet clack 12, and then the hot well will be shorter by half the diameter of the boiler.

"The puppet clack 9, 12, being loaded with a certain weight, in proportion to its size, whenever the repulsion of the steam in the boiler is greater than to lift up this clack, it escapes, without bursting the boiler, which it otherwise might do; and when the engine is stopped, the person attending lifts it up, by pulling the cord 12, 13, in order to discharge the steam.

"A pipe of communication, 10, 11, for steam will be necessary, between the receptacle for feeding and the boiler; for while the water in the boiler is heating, to make the engine work, (after standing some time), the feeding cock 7, being then shut, the water in this receptacle will boil before that in the boiler, and possibly by the sudden expansion of the steam upon boiling, might drive the water out of the receptacle at the small hole C, and being then emptied of water, might, by overheating the copper and cast-iron pipe, break the soldered joints.

"N. B. It is to be understood that in this, as in all other engines, the weight of the main pump spear, is so far to exceed the weight of the piston, as to raise it, and overhaul all the gear; and if it is not so, it must be weighted till it will do it.

"It does not appear that this plan for a portable engine was put into execution by Mr. Smeaton till some years afterwards, as will be described; but the boiler, cylinder, and working gear, represented in plate IX. were put in practice in the engine at the New River Head, in 1769, (before mentioned at p. 158). In other respects that engine was on the common plan, with a great lever, and fixed in a house."

It had two boilers, one on the construction of Plate IX., but $6\frac{1}{2}$ feet diam. and 8 feet high; the other was a common flange boiler, 8 feet diam. in the top, and $5\frac{1}{2}$ feet diam. below the flange; the fire grate 2 feet by $2\frac{1}{2}$ feet = 5 square feet; either boiler could be used to work the engine.

The cylinder was 18 inches diam. = 254 squ. inc., the piston made 9 feet stroke; the pump was also 18 inches diameter, but made only 6 feet stroke, the great lever being 12 feet radius at the cylinder end, and only 8 feet radius at the pump end. The height of the column was 35 feet, and the weight of the column 3869 lbs., or including the weight of the injection column, 10·2 lbs. per square inch, on the piston.

In this state the engine made upon an average 6 strokes per minute, = 54 feet motion of the piston; this was $4\frac{3}{4}$ horse power; but it burned $1\frac{2}{3}$ bushels of coals per hour, which is at the rate of 5 million lbs. lifted 1 foot high by a bushel. It then required ·388 of a cubic foot of injection water each stroke, or $2\frac{1}{3}$ cubic feet per minute; the space in the cylinder being 15·9 cubic feet per stroke, was 41 times the quantity injected. The temperature of the hot well was 153 deg. The portable or urn boiler, with the internal fire, was found to burn more fuel than the other, which was therefore commonly used.

This consumption of fuel being too great in proportion to the work performed, Mr. Smeaton altered the fulcrum of the great lever, so as to make it 13 feet radius at the piston end, and 7 feet at the pump end; therefore the stroke of the piston being 9 feet, as before, that of the pump was reduced to 4·85 feet; and the load on the piston, including the injection, was then 8·3 lbs. per square inch.

The engine then went 8 strokes per minute, and burned $1\frac{1}{3}$ bushels per hour; but afterwards it was improved to make $9\frac{1}{2}$ strokes per minute, = $85\frac{1}{2}$ feet motion per minute, when burning $1\frac{1}{2}$ bushels per hour. This was an effect of 5·47 horse power, and at the rate of 7·22 million lbs. raised 1 foot high by a bushel of coals; which was a great performance for an atmospheric engine of that size; but it had no incumbrance of dead weight of pump-rods. The trials of the coals consumed by this engine were frequently repeated, and always with nearly the same result; the coals were the best sorts of Newcastle.

When both boilers were worked at once, to give an abundant supply of steam, the engine went $11\frac{1}{2}$ strokes per min., or $103\frac{1}{2}$ ft. per min. $= 6\cdot 62$ horse-power. The consumption of both boilers was then under two bushels per hour; which is at the rate of $5\frac{1}{4}$ millions.

APPLICATION OF A FIRE-ENGINE TO DRAIN DRY DOCKS FOR SHIPPING.

This was first applied by Sir Charles Knowles to the magnificent docks constructed by Peter the Great, at Cronstadt in Russia. Cronstadt is the sea-port for the city of Petersburg, and has three capacious docks; one for the Russian ships of war, another for frigates and sloops of war, and a third for merchants' vessels: also the vast canal, or dry docks, for refitting men of war, which were planned and begun by Peter, in 1719; but being neglected under his successors, were not completed until the reign of his daughter Elizabeth.

Ten ships could be put upon the stocks at the same time; for, being floated into the canal, the gates were shut, and the water drawn off by windmills, so as to lay them dry. These docks were provided with three sets of gates, for admitting and letting out the ships; but these entrance-gates, when shut close, prevented the further entrance of water. As there is no advantage of tide in the Baltic sea to run off the water, two large windmills, 100 feet high, were constructed on the walls of the canal, after the plan commonly practised at that day in Holland; and they were constructed by Dutch workmen, who were brought from Holland. But this part of the project proved very inadequate, for it required continual pumping during a whole year, to clear the docks by these two mills.

The canal in the beginning was a most laborious and expensive work, for the whole island of Cronstadt is a flat bank of sand, and the stone walls with which the whole is lined, were all built by Batterdeaux. The canal is 105 feet broad and 40 feet deep; and the greatest canal, which forms the body and head of the cross, and which is called Peter's canal, is 4221 feet in length. They are all paved with stone floors.

The plan of this great work appears to have been very imperfect in other respects, as it was first executed. The docks consist of two canals, intersecting each other in the form of a cross, with a lozenge or quadrangle between the arms and head; but there were no internal divisions to separate the space into different docks, only the three outward gates to keep out the water from the whole; consequently it was necessary to empty the whole space, whether only one ship was put into dock, or whether as many as it would hold were taken in; nor could one ship be removed from the dock before the whole were finished. And as the two wind-mills required a year to clear out the water, the docks could only be used once in that time. Hence it frequently happened that the ships could not be laid dry at the time when they were most wanted.

The works were in this state, when Admiral Sir Charles Knowles was invited into Russia, by the Empress Catherine, to remodel the Russian navy; and the deficiencies of the docks proving a serious impediment to his views of improvement, he determined upon dividing the two arms and head of the cross into three separate dry docks, one of which would hold two ships. This was done by placing large double gates to separate those arms from the great central quadrangle, and also another pair of gates were placed to separate the great canal, or longer leg or body of the cross from the quadrangle. By this arrangement, four ships could be laid dry, to be repaired at once, in the separate docks, and two ships could be lying afloat in the quadrangle, ready to change places with any of those ships in the docks when finished; and at the same time five other ships could lie at once in the long leg of the canal to clean.

Stone drains were laid underground, with proper sluices, to let the water into or out of any one of the docks, without interference with any other. And to pump out the water, it was determined to substitute a fire-engine, it is said, on the suggestion of Dr. Robison, who attended the Admiral to Russia. Accordingly, in 1773, complete plans and descriptions of all the circumstances were sent to Mr. Smeaton, to enable him to fix the dimensions and make the plans for an engine suitable to the work to be performed.

By a computation founded upon these data, Mr. S. found that there must be very considerable leakages of water into the docks and canal, notwithstanding the substantial lining of stone-work, otherwise the two windmills would have evacuated the whole in a few months. He therefore recommended a fire-engine of 66-inch cylinder, which would possess three times the average power of the two windmills; and with this, he expected the whole space of the docks could be laid dry in a fortnight; or that the water could be drained from one of the double docks in one day. He says:—

"An engine of 66-inch cylinder will raise 24 300 tons of water, in 24 hours, to a height of 53 feet: that is rather more than 1000 tons per hour; and if made on the best plan, will consume $7\frac{1}{2}$ chaldrons of the best Newcastle coals, London measure, in 24 hours: that is, equivalent to 4 chaldrons Newcastle measure. And as the supply of this quantity of coals at Cronstadt will be a considerable expense, I recommend the best of the two windmills (viz. that to which the saw-mill is adapted) to be preserved, to work as it does now, to pump the water; but to pull down the other mill which is decayed, and to build the engine upon its present foundation walls,"—which were stated to be good masonry, built up from the bottom of the reservoir 60 feet high.

This advice was acted upon, and the engine was made at Carron, on the same construction as that at Chase-water already described; the principal dimensions were as follow: cylinder 66 inc. diam., working length of stroke $8\frac{1}{2}$ feet, and it worked two pumps of 26 inches diam., which lifted the water out of a well or pit communicating with the docks: the lift being 33 feet perpendicular, when the engine first began to draw the water from the docks, and 53 feet perpendicular when the water was completely drained from them. Also the injection-pump, which was 11 inches diameter, and about 5 feet stroke; it drew the water from the bottom of the same well as the main-pumps, and also lifted it to the cistern, which was placed at the top of the house, 52 feet higher than the main-pump delivery.

The windbores of the pumps were 25 inc. diam. and 12 feet long, to raise the clack-doors as high from the bottom as could be conveniently done, that they might be less subject to be drowned when the water rose high. The working-barrels were of the usual construction, 26 inches diameter, and 10 feet long, and above each of these was fixed a column of pump pipes, 32 feet high and 27 inches diameter, made in 4 lengths each. The two pumps stood close together, side by side, in the well; and the two pump-rods were united at top by a short horizontal bar, the middle of which was suspended, by the shank of the four great chains, which were harnessed together in the manner represented in Plate VI.

The pump-rods were made of cast-iron, and 4 inches square, that their weight might form a balance to the weight of the piston.   At the top of the two pumps was a shallow cistern or trough, to receive and carry off the water, just above the level of the ground on which the engine was erected.

The weight of each column of water was 12 214 lbs. when the water was raised from the bottom of the docks, and the column of the injection pump 4240 lbs.; in the whole, the utmost weight to be lifted was 28 668 lbs., or $8\frac{1}{3}$ lbs. per square inch on the piston.

The links of the great chains were made of cast-iron, 4 inc. broad by $2\frac{3}{8}$ inc. thick, for the single links, 4 of which bore all the strain of the two main pumps. The joint pins to unite these links were wrought iron, 2 inches diameter.   The great lever was of the same dimensions as that at Chase Water.

This engine had three boilers, each 10 feet diameter.   They were made entirely of cast-iron, on a plan which Mr. Smeaton had before recommended to the Carron Company, for their own returning engine.   The upper part of each boiler was a hemisphere, C C, 10 feet diam. cast in one piece, with a projecting

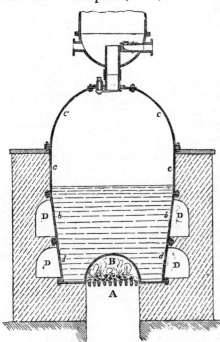

flange all round its lower edge ; and by this it was united to a cylinder $c\,c$, placed with its axis vertical : it was 10 feet diam. and 4 feet deep, cast in one piece, and united by a flange to another piece, $b\,b$, of 4 feet deep, 10 feet diameter at the top, but tapering to 9 feet diam. at the bottom. This cylindrical piece was joined by a flange, to another piece, $d\,d$, $3\frac{1}{3}$ feet deep, 9 feet diameter at top, and 8 feet at bottom.

This lower piece was united to the bottom plate of the boiler, which was flat, except a raised part, B, across the centre, to form the furnace, which rose up within the bottom, in the form of an arched tunnel, passing through and across the bottom part of the boiler.   This tunnel or fire-place, B, was a half cylinder, $4\frac{2}{3}$ feet diam. with its axis horizontal, and extending all across the centre of the flat bottom of the boiler.

Each boiler was therefore $16\frac{1}{3}$ feet high from the top of the hemispherical dome C, to the flat bottom on which it stood : that bottom being 8 feet diameter, and the top a hemisphere of 10 feet diam.   The water line was $7\frac{1}{3}$ feet from the bottom, and the depth of water standing upon the cylindrical fire-place, B, was 5 feet.   The fire-grate, A, which was placed beneath the arched fire-place B, was $4\frac{1}{2}$ feet wide and $4\frac{1}{2}$ feet long, or $20\frac{1}{4}$ square feet, and the furnaces of all those boilers 60 square feet. The flame from the fire-grate was conducted across the bottom of the boiler, into a circular flue or channel, D D, which encompassed the boiler, making two convolutions round the lower part, which contained the water, and then passing away into the chimney.

The cast-iron in each of these boilers, weighed $15\frac{1}{2}$ tons.   They were found to answer very well, and many others were afterwards made on the same plan (a); but for smaller sized engines, the upper cylindrical piece, c c, was omitted, and the hemisphere fastened to the flange on the top of the piece, b b. The piece, c c, was introduced in the boilers at Cronstadt, to give more steam room, without requiring the dome to be more than 10 feet diameter, which was as large a vessel as could be cast in one piece.

This engine was set to work in 1777, and rendered the docks very serviceable, as it was able to clear the water, in a very short time, from such of the separate docks as required it. It worked from 10 to 13 strokes per minute of $8\frac{1}{2}$ feet, according to the height of the water (b).

The building of this engine was an immense mass of masonry, the lever wall being 100 feet high, from the foundation to the top, on which the injection cistern was placed; the upper part for 53 feet high, was 6 feet thick, and the lower part from the foundation to the floor of the engine-house, was 47 feet high and 11 feet thick; the other walls to enclose the pumps, and from the well, were the same height, and corresponded with the walls of the canal and docks.

APPLICATION OF THE FIRE-ENGINE TO DRAIN TRACTS OF FEN LANDS.

THERE are extensive districts of land in the Netherlands, which are lower than the level of the sea; but which are artificially embanked on all sides, and the water raised up into the rivers, by the power of large windmills, so as to render the land capable of cultivation. A sufficient number of mills must be provided to perform the drainage in all seasons of the year. This system has been adopted by the English, on a large scale, in Lincolnshire and Cambridgeshire.

In 1772, a lake near Rotterdam, in Holland, which covered about 7000 English acres of land, was embanked and drained by 34 windmills; the greatest depth of the original lake was 12 feet below the low water level in the river Maese; part of these windmills were appointed to raise the water from the surface of the land, up into a large basin or canal, which was carried across the land between embankments; but the water in this canal being only on the level of the low water in the river, the other windmills were required to raise the water out of this canal over the sea dyke into the river, at all periods of the tide, which rose 5 feet from low to high water.

At the commencement of this work, it had been proposed to the magistrates to employ fire-engines for the drainage, but they determined to follow the old system of windmills; till, about 1776, when a large Newcomen's engine was applied to assist those mills which raise the water of the basin into the river: the execution was entrusted to a M. Van Liender, who procured the iron-work from

(a) Cast-iron boilers of the common form had been frequently used for small fire-engines at an early period. Boilers or pans for brewers and dyers, were first begun in cast-iron, and were then adopted for engines; they were esteemed durable, but took more fuel than plate-iron boilers on account of the thickness of the metal.

It afterwards became a practice in the neighbourhood of Whitehaven in Cumberland, to make the domes or tops of the boilers of cast-iron plates formed in segments or pieces, which were joined together by bolts, passing through flanges, projecting up from the edge or border of each plate; the joints between the flanges being made tight by driving in wooden wedges from the outside; by this means, if any one plate failed, it could be changed without deranging the others; the lower parts of the boilers were of iron-plate.

(b) The author visited the docks at Cronstadt in 1819, and saw the building containing Mr. Smeaton's engine: it was not then at work, but he was informed that it had undergone no alteration since its first erection.

England. This was the first application of the steam-engine to such a purpose, and seemed to be attended with a considerable difficulty on account of the small height and great quantity of water to be raised; and also, from the continual fluctuations of the height to which it was to be raised, it does not appear to have fully succeeded. The following account of this work was written by an eye witness in 1778.

Considering that at or about low tide, the water is only to be raised a few inches, but at high water 5 feet high; they have endeavoured to take advantage of this circumstance in the disposition of their pumps, by working with a greater number of pumps at low, than at high tide. Their cylinder is 52-inch diam., and 9 ft. long; the boiler 18 ft. diam., with a double flue, an inside and an outside one; the boiler is placed on the outside of the house, and there is a receiver or steam vessel below the cylinder, with a regulator in it in the common way; the diam. of the steam-pipe is $8\frac{1}{2}$ inc., and of the injection-pipe $4\frac{1}{2}$ inc. The house is 10 feet wide where the cylinder hangs, and sufficiently long; they have one main working lever beam 27 ft. long, of oak, made of two pieces put together spring fashion, bent above and straight below, each piece 18 inc. deep, and 12 inc. broad; and two others, one on each side, of the same length, but a little slighter made: these beams are connected through their centre by a gudgeon 26 ft. long.

At the end of each beam is a circular pump, of 72 inches diameter; and at 8 ft. from the end, towards the gudgeon of each beam, is another pump, 72 inc. squ., making in all six pumps, three of them circular, and three square. The outer pumps have a 6-ft. stroke, the inner ones about a 30-inc. stroke; the pumps are all of fir-wood, made stave-fashion; the buckets and clacks are also made of wood; and, through a notion they had of avoiding friction in the buckets, they made them each an inch less than the working barrel. The tops of all the pumps are enclosed in a wooden floor, made water-tight, upon the surface of the water in the basin, and upon a level with low-water mark in the river; so that at low tide this floor is dry, and at high tide the water of the river flows on to the floor, and stands five feet above the top of the pumps. The depth of the pump-well is 9 ft., but may be made 12 if necessary, but no more.

At ebb-tide, when the floor is dry, the engine works all the six pumps together, and so continues to do until the tide rises about twelve inches deep on the floor; at which time, one of the inner or short pumps is taken off or disjoined: and when the tide rises a few inches higher, one of the longer pumps is disjoined, and the former short one put on again; and so continually, as the tide rises, more pumps are gradually taken off, until at high tide, the engine works with only two pumps, a long one and a short one; and when the tide begins to ebb, they are joined on again successively, until at last the engine has all the six pumps at work, as at first; and this operation of putting on and taking off the pumps is done without stopping the engine.

This plan has been found to be defective; for whenever one of the pumps is taken off or disjoined, the engine has too much power for some time; and commonly, when a pump is put on at the ebbing of the tide, she has too little power, and sometimes entirely stands for some time.

It appears that this application of the fire-engine was thought to be an object of national importance in Holland; for the Batavian Society at Rotterdam published a programma, in 1778, in which it is stated, that the pumps annexed to the above engine do not answer to expectation, being found too weak to sustain the power applied; and from the great loss of water during the time it works, the power of the engine is in some measure defeated.

In consequence the Society offered a considerable premium " for a plan of an apparatus which, being annexed to a steam-engine, is capable of raising water to all heights under 5 feet, which is the extreme height wanted to raise the water from the basin into the river; but this height diminishes as the water falls in the river by the ebb tide ; the quantity of water to be raised must therefore augment in an inverse ratio to the diminution of height, so as to be double at $2\frac{1}{2}$ feet, quadruple at $1\frac{1}{4}$ ft. &c.   The length of stroke must not exceed 6 ft., and the plans must be adapted to a fire-engine of a cylinder 52 inc. diam. and 7 ft. stroke."

Dr. Robison, in the Encyclopædia Britannica, says that the States of Holland proposed draining the Haerlem Meer, and even reducing the Zuyder Zee, by steam-engines ; and Mr. Watt adds, that a considerable engine was erected by them at Mydrecht, for draining a large extent of country, in which it was successful.

CHAPTER IV.

On the manufacture of Iron, and the application of Cast-iron in the construction of Steam-engines and Mill Work, and the application of the Fire-engine to blow Furnaces and work Water-wheels.

THE progress of the invention of the steam-engine, and of the modern system of manufactories, has been very greatly promoted by the use of cast-iron, as a material for constructing the parts of engines, mills, machines, buildings, &c. This was a new application, made about the middle of the last century, in consequence of an improved system of making iron, with pit-coal for fuel; which was then brought into use, to the great advantage of this country, and which has, in its consequences, tended very considerably to produce the present refined state of mechanical arts. It is a system still scarcely known in other countries, and not practised to any extent; and the want of this valuable material, and of coals, is one reason, amongst many others, of the deficiencies of other countries in those arts in which we so greatly excel.

The working of coal-mines, and the making of cast-iron, are two subjects intimately connected with the history of the steam-engine, and with each other; for the practice of both those arts has been greatly facilitated by the application of engines to drain the mines, and to blow furnaces; and the use of steam-engines for all other purposes to which they are now applied, is entirely dependent upon the facility of procuring coals for fuel, and cast-iron for their construction.

The ancient method of smelting iron from the ore was performed in furnaces or hearths, like blacksmiths' forges, by fires made with wood-charcoal; and it is still the practice to use charcoal in other countries, where wood can be procured: but charred pit-coal, or coke, is now used for the manufacture of iron in Britain. The combustion of the charcoal was excited by a strong blast of air, blown into the furnace by bellows, moved at first by men, and afterwards on a larger scale, by the power of a water-wheel. These bellows were formerly made of wood and leather, on the same construction as ordinary bellows for domestic use; but bellows made entirely of wood were afterwards introduced with great advantage, as they gave a stronger blast. These were found quite sufficient for the charcoal furnaces, and are still in general use on the continent.

The manufacture of iron upon this system was very extensive in England in the time of Queen Elizabeth and King James I. Iron cannon are said to have been first cast in England in 1547, but very few were made until towards the end of the sixteenth century. In 1615, it is said, there were 300 furnaces for smelting iron ore with charcoal of wood; and wood was then the common fuel for domestic use. But in succeeding reigns, the lands were cleared of trees by the progress of agriculture; and the general increase of population, with less frugal habits, occasioned such a great increase in the demand for fire-wood, and also timber for building, and for the construction of ships, that the iron furnaces could not be maintained in such great numbers; and as the forests in their immediate vicinities became exhausted, they were given up, one by one, till, at the beginning of the last century, not above fifty furnaces remained in use, each producing about 300 tons of cast-iron per annum.

The cast-iron thus made with charcoal was of very little use, except as a material for making bar-iron ; and the art of moulding and casting iron was scarcely known, and very little practised, except for the coarsest purposes.

In the Philosophical Transactions, 1677, vol. xii. p. 931, is an account by Mr. Powle, of the iron-works at that time carried on in the Forest of Dean, in Gloucestershire.

The forest between the rivers Wye and Severn is generally a stiff clay soil, naturally inclined to wood, especially hazel and oak, but which is now almost devoured by the increase of the iron-works. Within the forest is found great plenty of coal and iron ore. Their best ore affords a great quantity of iron, but of a short brittle nature if it is melted alone. They therefore mix the ore with a proportion of cinder, or refuse of ore, from which the metal has been extracted by a former fusion. This mixture gives the iron an excellent temper of toughness.

In former times no other bellows were used than such as could be worked by the strength of men ; and as the heat thus excited could only extract part of the metal, a great proportion remained in the cinder or refuse, which is to be found in great quantities in all parts of the country where any former works have stood.

The ore is first calcined without fusion, in kilns like ordinary lime-kilns, filled up with coal and ore, stratum, super stratum. After the calcination the ore is smelted in the furnaces, which are built of brick or stone, about 24 ft. squ. on the outside, and near 30 ft. high, the inside shaped like an egg, and not above 8 or 10 ft. over where widest, but the top and bottom having a narrower compass. Behind the furnace are placed two huge pair of bellows, whose noses meet at a little hole near the bottom. These are compressed together by certain buttons placed on the axis of a very large wheel, which is turned about by water, in the manner of an overshot wheel. As soon as these buttons are slid off, the bellows are raised again by the counterpoise of weights; by which they are made to play alternately, the one giving its blast all the time the other is rising.

The furnaces are filled with ore and cinder, intermixed with charcoal, which being once kindled, and the fire kept up very fierce by the bellows, the materials run together into a hard cake or lump, which is sustained by the fashion of the furnace, and through this the metal as it melts trickles down into the receiver at the bottom, where there is a passage open, by which they take away the scum and dross, and let out the metal as they see occasion.

Before the mouth of the furnace lies a great bed of sand, in which are made furrows, of the shape into which they intend to cast the iron. Into these, when the receiver is full, they let in the metal, which is made so very fluid by the violence of the fire, that it not only runs to a considerable distance, but continues boiling for a good while. The furnaces are kept at work for many months together, never suffering the fire to slacken night nor day ; but supplying the waste of the fuel, and other materials, with fresh poured in at the top.

Several attempts have been made to introduce the use of sea-coal in these works instead of charcoal ; the former being to be had at an easier rate than the latter ; but hitherto they have proved ineffectual—the workmen finding by experience, that a sea-coal fire, how vehement soever, will not penetrate the most fixed parts of the ore, and so leaving much of the metal unmelted.

From these furnaces they bring their sows and pigs of iron, as they call them, to their forges. These are of two sorts, though standing together under the same roof ; one they call their finery, the other the chafery. Both of them are open hearths, on which they place great heaps of sea-coal, and behind them are bellows, like to those of the furnaces, but nothing near so large. They place three or four pigs of iron in the finery with their ends in the fire, which softens them by degrees, and at length runs the metal together in a round lump, which they call a half bloom. This they take out and beat into a thick short square, by a weighty hammer raised by the motion of a water wheel; this they heat again in the finery red-hot, and hammer it again with the same hammer into the shape of a bar in the middle with a square knob at each end ; and lastly, by repeated heating in the other fire or chafery, and working under the hammer, they bring it into iron bars.

For several purposes, as for the backs of chimneys, hearths of ovens, or the like, they have a sort of cast-iron which they take out of the receiver of the furnace in great ladles, and pour it into moulds of fine sand ; but this sort of iron is so brittle, that being heated, with one blow of a hammer it breaks all to pieces.

Also, in the Philosophical Transactions for 1693, Vol. XVII, p. 695, is an account by Mr. John Sturdy, of the iron-works at Milthorpe-forge, in Lancashire.

They have several sorts of iron stones, producing iron of different qualities, and they mix different sorts of stones together. The ore is got at Furnace, 15 miles from Milthorp; some of it lies in beds like coal, others in veins between limestone rocks.

The small dusty part of the charcoal is used for burning the iron stone; and when burnt they throw it into the top of the furnace, with $\frac{1}{17}$ part of unburned limestone, and a due proportion of charcoal. The furnace is built at the side of a hill, the bottom two yards square inside, for a yard or more perpendicular, and lined with a wall of fire stone, to protect the masonry of the furnace from the heat; thence rising up for six or seven yards higher, square wise, but tapering, to about half a yard square at the top hole, the fire is blown at the bottom by very large bellows, played with water, and the furnace is filled to the top with ore, limestone, and charcoal; when it has sunk a yard and a half down into the furnace, they put in more.

The limestone is to make the ore melt freely, and cast the cinder, which is taken off at a hole before the metal is run: nothing remains at the bottom of the hearth but what is melted, either into iron or cinder. Of late they have made much better iron than heretofore, by melting the sow metal over again, and by using turf and charcoal; whereas formerly their fuel was only charcoal. They once tried pit coal, but with bad success.

The forge is like a common blacksmith's, with a hearth made of sow iron, in which they make a charcoal fire, and put in ore, first broken into pieces like a pigeon's egg; it is melted by the blast, which is continued with fresh charcoal for about 12 hours, and the melted glassy cinder is let out, leaving the iron in a lump, which is never in a perfect fusion; this is taken out and beaten under great hammers played with water, and after several heatings in the same furnace it is brought into bars. They get about one hundred weight of metal at one melting, being the product of about three times as much ore. No limestone or any other flux is used.

It is most probable that it was the sow metal or cast-iron, which was thus treated in the forge, and not ore, though it is stated so by this writer.

We may collect from these accounts, as well as from others of the same period, that cast-iron, as then made with charcoal, was unfit for the purposes to which it is now applied; and this is still the case in France, and other parts of the Continent, where the old system of manufacture continues to be practised, exactly as described at the latter part of the 17th century.

The defects of cast-iron, as usually made with charcoal, are, excessive hardness, brittleness, and want of internal soundness or solidity. Most of these defects arise from a want of fusibility, whereby it cannot be made sufficiently fluid to become perfectly incorporated; nor can it be cast with any certainty of filling every part of the mould, so as to form sound castings. It is probable that these defects are not entirely to be attributed to the use of charcoal, but, in a great measure, to that system of treatment with charcoal, which was followed in England, and which is still followed in all other countries, and which has been perfected by long experience, in the view to produce such cast-iron as will make the best wrought or bar iron, without regard to the quality, when in the state of cast-iron.

The object of the modern system of manufacturing iron in England is changed, since cast-iron has become one of the most useful metals; and those changes of the treament of the ore, which became essential when the coke of pit-coal was first substituted for the charcoal of wood, tended to the improvement of the cast-iron, by giving it the qualities of fusibility, cohesion, and softness, in a much greater degree than it was thought possible to attain when charcoal was used; for such coarse articles as were then cast in iron were too hard to be worked with tools to any advantage, and so brittle and unsound in their texture as to be unfit for any purpose in which the metal would be subjected to severe strains.

The facilities which we now have, of forming almost any article of machinery in cast-iron, was not then available, because the art of moulding and casting iron was very imperfectly understood. Heavy cannons were cast hollow, and smoothed out withinside by boring; pump-barrels, for water-works, were made by the same workmen; and pipes, for the conveyance of water, were also cast, but very thick, and in short lengths, and frequently unsound.

By degrees, the iron-founders improved so much as to be able to cast cylinders for the atmospheric engines, which, in the commencement, were made of brass (*a*). If we add to these the articles of shot for cannon, iron railing for fences, ballast for ships, weights, hearth-plates, fire-backs, and grate-bars for furnaces, it will comprise nearly all the uses to which cast-iron was applied, before the introduction of pit-coal as fuel for the blast furnaces,—and those applications were only to a very limited extent.

At the same time that the manufacture of iron was falling to decay in England, from the deficiency of wood fuel, the demand for bar-iron was continually increasing, and was supplied by immense annual importations from Russia and Sweden; where ores of excellent quality are found, and where numerous and extensive forests still furnish abundance of wood for fuel.

It appears that, during the seventeenth century, the use of pit-coal as fuel for household purposes, became common in those districts of England and Scotland which are in the neighbourhood of the coal-mines, and also wherever coals could be conveyed by water-carriage; and, by degrees, most trades, requiring the aid of great fires, such as potters, brewers, distillers, dyers, &c. acquired the art of using pit-coals for fuel, as wood became more scarce and coals more plentiful; but the application of pit-coal to the purposes of the iron manufacture seems to have been attended with great difficulties, and was not established till after a great many fruitless attempts had been made, during more than a century.

James the First granted several patents for the exclusive privilege of manufacturing iron with pit-coals, but none of the projectors succeeded at all, till a Mr. Dudley, in 1619, succeeded so far as to make iron with pit-coals, but only at the rate of three tons per week, from one furnace. This meritorious effort he continued with unabating ardour during a long life, but with very inadequate success, owing to the prejudice and malignant opposition of the trade, the distracted state of the country during the civil wars, and his imperfect success in his process. Dudley had several other patents for his successive improvements; and in 1663, when he applied for his last, he stated that he could then make 7 tons of cast-iron per week, with an improved furnace and bellows, which one man could work for an hour without being tired. Blacksmiths began to use pit-coal during the latter part of the 17th century; and also the chafery forges, for working bar-iron, were frequently worked with this fuel.

After this time, the project of smelting iron, with coke or charred pit-coal, was frequently revived, as is mentioned by Mr. Powle, in 1677, and by Mr. Sturdy, in 1693, though without success; but, even when its practicability was at length proved by experiments and by trials in the large way, it was not established as a real manufacture till about 1740, when the works at Colebrook Dale, in Shropshire, were first commenced under the direction of a Mr. Ford (*b*), and they

(*a*) It is doubtful whether cast-iron cylinders, for fire-engines, were made with the charcoal iron, before the pit-coal iron was introduced. The first mention of iron cylinders is by Desaguliers, who wrote in 1743, but he recommended brass cylinders as being preferable. The Rev. Mr. Mason, in the Philosophical Transactions, 1747, attributes to a Mr. Ford, of Colebrook Dale, the art of making cast-iron with pit-coal so soft and tough as to cast cannon, which would bear turning like wrought iron; but, in the same paper, he speaks of Mr. Ford making a cylinder, for a fire-engine, of spelter, which casts as true as brass, and bores as well, and works better than iron, the rust of which resists the motion of the piston. But whether the iron cylinders thus alluded to were made of this improved cast-iron, or had been made by other founders before, does not appear.

(*b*) In the Philosophical Transactions, 1747, vol. xliv. p. 370, is an account, by the Rev. Mr. Mason, of what he saw at Colebrook Dale :—

"Several attempts have been made to run iron-ore with pit-coal; I think it hath not succeeded anywhere, because we have had no account of its being practised; but I find that Mr. Ford, from iron-ore and coal, both got in the same dale, makes iron brittle or tough as he pleases, there being cannon thus cast so soft, as to bear turning like wrought iron."

soon became established.   They began in a small way, as several others had done before, but had the good fortune to succeed so far at first as to keep working on, till, by continued practice, they improved the process, and brought it to bear.

The coals were charred and reduced to coke by burning in heaps, nearly in the same manner as the wood was formerly burned to charcoal.   The iron stone, or ore, was also prepared by roasting or calcining in heaps, and was then inter-mixed with coke and limestone, and thrown into a large furnace, where it was gradually revived, and melted from the stone, by the heat of an intense fire, kept up by the combustion of the coke, and blown by large bellows, moved by a water wheel.   The process was conducted in every respect in a similar manner to that formerly practised, only with the change of the coke instead of charcoal for fuel. The difficulties which had so long retarded this mode of manufacture, were over-come by employing larger and higher furnaces, and more powerful blowing ma-chines, than had been requisite for the old charcoal furnaces; for coke not having the same inflammability as charcoal, this deficiency must be supplied by accumu-lating the fire in larger masses, and animating it by a more copious and forcible current of air, that the requisite heat may be excited, and that the ore may remain long enough in contact with the burning fuel with which the furnace is filled, to separate the metal from the earthy matters with which it was combined in the ore.

For several years after the establishment of the Colebrook Dale works, it was supposed that their success was owing to some peculiar properties in the coals and iron ore, which they worked; and this notion was strengthened by the failure of some other subsequent attempts on a small scale.   It was not until these first works had grown to a considerable extent, that others found out the secret of following them in their career; at the same time, the excellence of the cast-iron which they made, and the new uses to which they could apply this material, produced them a trade which was an enviable object to others, who had iron ore and coals in plenty, but could rarely procure wood to work the ore with charcoal, nor could they make good cast-iron with it when they did (a).

In 1759 the Carron iron works in Scotland were commenced by Mr. Roebuck, Mr. Garbet, and others; who advanced an ample capital, and chose an excellent situation on the banks of the river Carron, near Falkirk, in the immediate vicinity of the mines of coals and iron ore, and possessing all the water of the river, with a considerable fall, to give motion to the machinery.

Their first furnace was started at the commencement of the year 1760; it was blown by bellows, worked by a water-wheel; but their success was very doubtful at first, until they called in the assistance of Mr. Smeaton, who was then just rising to eminence; he retrieved their affairs by constructing them a more powerful and complete blowing machine for a second furnace.   It operated with four large cylinders, or forcing-pumps, which were worked by an over-shot water-wheel, with alternate strokes to avoid intermission of the blast.   This succeeded very well; for, notwithstanding many imperfections, it gave them the means of exciting a suf-ficient heat, and then they soon found out the proper treatment of their coals and ore, so as to produce cast-iron of a very superior quality to any thing which had been known before.

(a) In the Philosophical Transactions, 1747, Vol. XLIV. p. 370, is an account by the Rev. Mr. Mason of what he saw at Colebrook Dale:—

"Several attempts have been made to run iron ore with pit-coal; I think it hath not succeeded anywhere, because we have had no account of its being practised; but I find that Mr. Ford, from iron ore and coal, both got in the same dale, makes iron brittle or tough, as he pleases; there being cannon thus cast so soft, as to bear turning like wrought iron."

N N

In consequence of this success, Mr. S. afterwards rebuilt their first machine, on his plan, with many improvements; and in a short time they became very celebrated for their cannons, engine-cylinders, iron-pipes : and having a school of ingenious moulders and founders, they succeeded every day in applying their materials to new uses. Mr. Gascoigne was afterwards the principal manager, and with the aid of Mr. Smeaton in the construction of all their machinery, the works were extended with a rapidity, and to an extent, quite unknown in such concerns at that period.

It appears from a report by Mr. Smeaton on the state of their machines in 1769, ten years after their first commencement, that they had then four furnaces in activity, each provided with its own blowing machine, which upon an average blew about 1500 cubic feet of air per minute into each furnace; they had also a boring-mill for cannons, and another for cylinders and pumps, and a third for turning cast-iron; likewise a forge for making their cast-iron into bar-iron, and other appendages. In all there were ten large water wheels, worked by the water of the river, which was sufficient for them all, except in dry seasons; but as it was then inadequate to supply even the blowing machines, large reservoirs had been constructed by the river side, about two miles above the works, to collect water. But these proving insufficient, as the works increased in extent, a fire engine, with a cylinder 72 inches diameter, was added, to return the water of the river in dry seasons; the pump was 52 inches diameter, and it raised the water 24 feet, from the lower pond to the upper.

This engine was erected about 1765, and was either the first, or amongst the first applications of the power of steam to work mills on a large scale, and it was only intended as an auxiliary, to work at particular seasons; but, like most other first trials, the engine proved a very bad one, and notwithstanding its great size, it would only make 6 or 8 strokes per minute, of $5\frac{1}{2}$ feet each, and was subject to such frequent stoppages, as to raise only 440 cubic feet per minute 24 feet high, or only 20 horse power; and the supply of the river in the dryest times, being only at the rate of 660 cubic feet per minute, or 30 HP, they required more power; for when the four furnaces were properly blown, their water wheels expended 3500 cubic feet per minute.

The engine was made before Mr. Smeaton took up the subject, but being consulted upon the deficiencies of their power, and the means of extending their works still farther, he caused all their water wheels to be improved, and made new and more perfect blowing machines, which worked with less water : new reservoirs were also constructed, and the top water was raised to 27 feet fall; and for boring and other work, new water wheels were erected upon a second fall, beneath the principal one.

These different improvements were carried into execution by degrees, and had the desired effect. The works continually increasing in prosperity, two more furnaces were afterwards added, and then the fire engine was rebuilt, on Mr. Smeaton's plan, so as to keep them all going in the dryest times. The details of this subject will be found in Mr. Smeaton's successive reports, (see Smeaton's Reports, 4to. Vol. I.)

Mr. Smeaton promoted the success of this establishment very greatly, not only by the new machinery which he planned for it, but also by extending their foundry business, and continually applying cast-iron to new uses in the various mills which he directed; all these works in cast-iron he recommended to be executed at Carron, where the workmen under Mr. Gascoigne succeeded extremely well in the foundry business. The first cast-iron axis for a windmill had been applied by Mr. S. in 1754, but he found great difficulties in obtaining sound cast-ings which could be depended upon, until the Carron company had established their foundry.

The first cast-iron axis for a water wheel was applied by Mr. Smeaton for one of the blowing machines at Carron in 1769, and they soon after constructed large cog wheels, and axes or shafts, of all dimensions, in cast-iron for their own mills; and as they were found to answer much better than wooden ones, they were con-stantly recommended afterwards by Mr. Smeaton for other mills which he designed, and by degrees came into very general use in the course of twenty years.

The success of the Carron works, following upon that of the Colebrook Dale, stimulated many other enterprising individuals to follow their example, and most of those who began with sufficient power succeeded; the process of converting the coke pig or cast-iron into bar-iron was also brought to bear, but this was done in the old way, with charcoal for the finery, and coals for the chafing.

Mr. John Wilkinson began his works at Broseley, soon after Carron was established, also Mr. Reynolds at Ketley, Mr. Walker at Rotherham in Yorkshire, and others.   The art being now perfected, and workmen trained in these schools, iron works were established by capitalists in all parts of England and Scotland, where coals and iron could be obtained, and still more in South Wales, where establishments were founded by a few enterprising Englishmen, which have since increased to a most immense magnitude.

In 1788 there were 59 of the new furnaces in action in England, Wales, and Scotland; they were said to produce 900 tons of coke pig iron per annum from each furnace; and at the same period only 26 of the old charcoal furnaces re-mained, their produce being about 550 tons annually from each furnace.

All the first iron works were situated on falls of water, for working their ma-chinery, which was generally copied from Carron; but the great success of their trade caused most of them to extend beyond their original views, and then they added fire engines, to return the water upon their wheels, as before mentioned. When it became requisite to work a returning engine on this system constantly, all the year round, night and day, it was found to be more advantageous to connect the piston of the blowing pump, or cylinder, at once to the great lever of the engine, so as to be worked by a direct action of the piston, without the intervention of water and a water wheel.

This application of the fire engine was first tried at some of the new works which began in the neighbourhood of Colebrook Dale, a few years after the cylin-drical bellows were made to answer so well at Carron, but it appears that it did not succeed so completely at first, as the water wheels with four cylinders had done, because the blast was irregular and interrupted at every stroke of the engine; and until the remedy for this inconvenience was found out, the original machines at Carron continued to be the model which was followed in all cases : and if a sufficient natural fall of water could not be obtained for the water wheel, it was supplied by a fire-engine, to pump up water.   These were called returning engines.

Mr. Smeaton being consulted in 1772 to erect blowing machinery for a new furnace, where there was not a sufficient fall of water, gave the following opinion upon the direct application of the engine. " A fire-engine is indeed a reciprocating motion, but not at all suited to the working of cylinders for a constant blast.   It is an impediment to a fire-engine to have any curb upon it, which may prevent its own peculiar motions, and it would require two fire-engines to work at once, that one might be in the middle of its stroke while the other was at the stop; and although it is not impossible to regulate them so as to keep time together, yet it would be an intricate piece of work, and after all would do the business unsteadily."   A water-wheel with four cylinders was in consequence erected for this furnace.

In 1775 the Carron works consisted of five blast furnaces, with four machines for blowing them ; each worked by its own overshot water-wheel, with four cranks upon its axis, to give motion, by means of long levers, to the pistons of four large forcing pumps or blowing cylinders, each of which blew the contained air through a common nose-pipe into the furnace, when the piston was pressed down ; and drew in a fresh supply of air into the cylinder when the piston was drawn up again ; and as the four cylinders all blew into the same nose-pipe, and acted alternately, they kept up nearly a continuous blast of air.

Engravings of these machines are contained in Smeaton's Reports, Vol. I. p. 365 ; and it appears from his computations, that the four machines blew about 6000 cubic feet of common air per minute into the four furnaces, and expended 66 000 tons of water in 24 hours : taking this to be wine measure, it would be 1544 cubic feet per minute, which, at 24 feet fall, is 70 horse-power.

The nose-pipes then used for blast furnaces were not less than $2\frac{1}{2}$ inches diameter, nor greater than $2\frac{3}{4}$ inches diameter, one such pipe to each furnace ; and about 1500 cubic feet of common air was usually forced through each, and with this blast a furnace usually made 20 tons of cast-iron per week.

At first cannons, and heavy castings, were run immediately from the blast furnaces into the moulds ; but it was afterwards found to be an improvement to run the cast-iron into pigs, which were remelted for the foundry use in a reverberatory furnace with pit coals ; this enabled the blast furnace to be worked to more advantage, and with less blast of air than before.

The dimensions of the best of these machines, called No. 2, were as follow :—

Water-wheel 20 feet diameter, and $4\frac{1}{2}$ feet wide inside the buckets, with 60 buckets ; depth of the buckets 10 inches ; the whole fall of the water 24 feet ; surface of the upper pond $3\frac{1}{2}$ feet above the top of the wheel, and 6 inches clearance below.

The axis of the wheel was of cast-iron, 9 inches diameter, $7\frac{1}{2}$ feet long, between the bearings, with necks or gudgeons at each end, 7 inches diameter, and 7 inches length of bearing ; one end of this axis was connected by a square, and a coupling box, with a cast-iron axis 7 inches square, which had two cranks formed upon it, of about 17 inches radius. This axis was prolonged by another similar axis, with two other cranks formed upon it, the whole forming one line of horizontal axis in continuation of that of the water-wheel, with four cranks upon it, which being set at right angles to each other, would operate in exact alternation.

These cranks, by wrought iron connecting-rods, $2\frac{1}{4}$ inches square and $10\frac{1}{2}$ feet long, gave motion to four long levers, or working beams : one end of each lever was placed over one of the cranks, and the other end over one of the blowing cylinders, each lever being poised on a centre near the middle. Those ends of the levers to which the connecting-rods were jointed, were about 8 feet long, from the centres of motion ; and the other ends to which the piston rods of the cylinders were jointed, were 12 feet from those centres.

The great levers were each composed of two beams, 12 inches square, applied one upon the other, with the axis between them, separating them in the middle, and they were sprung together at each end. On the crank end of each lever a large weight of cast-iron was fixed, to counterbalance and draw up the piston, so that the bearings of the cranks and the joints of the connecting-rods were always lifting upwards.

The framing of the machine consisted of eight horizontal beams, or two to each cylinder, all laid parallel to each other, and united by two other long beams, framed across their ends, to form a large horizontal grating ; the bearings for the crank axis being supported on one end of these eight beams, and the four cylinders standing on the other ends, and the air-chest across them all. These beams were 24 inches deep by 12 inches broad ; they formed the foundation, and were supported on brick walls ; the cranks worked in the spaces between them. Upon each of the eight horizontal beams an upright frame was erected, to support the gudgeons of the four great levers ; these frames were each composed of two beams, 12 inches square, placed in a sloping direction, and meeting at top, so as to form a triangle, 15 feet at the base and 12 feet perpendicular ; the eight triangles were all tied together by a horizontal piece fixed into the upper part of the triangle ; and the vertex of each triangle formed the support for the gudgeons of the axis of each of the great levers.

The four cylinders were each 54·2 inches diameter, and the pistons made 4·35 feet stroke ; they were open at top and closed at bottom, and the pistons fitted into them, nearly in the same manner as for the atmospheric fire-engines; in the bottom of each cylinder was an opening covered with double valves, to admit air, when the piston was drawn up, but to shut down when the piston returned. There was also a square curved pipe proceeding downwards out of the cylinder bottom, and turning up again to join to the lower side of a long square trunk, or pipe, 15 inches square and $16\frac{1}{2}$ feet long, which extended horizontally along the front of all the four cylinders, so as to form a common receiver for the air from all of them; the orifices in the bottom of this wind-trunk, where the curved pipes from each cylinder entered into it, were each closed by valves shutting downwards, to allow the air to pass out of the cylinders into the air-trunk, but to prevent it returning; lastly, from the middle part of the air-trunk, one pipe of 7 inches diameter proceeded to the furnace, the extremity diminishing with a regular taper to a blow-pipe, with an orifice $2\frac{5}{8}$ inches diameter, and through this all the air passed.

The usual rate of working this machine was to blow 24 cylinders per minute, the water-wheel then making 6 revolutions per minute; the capacity of each cylinder was ($54\cdot2$ inc. diam. $4\cdot35$ ft. stroke =) $69\cdot75$ cubic feet; but the piston did not go down within 3 inches of the bottom of the cylinder, and including the passages to convey away the air, about $5\frac{1}{4}$ cubic feet of space was left at the bottom, from which the piston could not displace the air. The compression of the air was not noticed by Mr. Smeaton, but if we assume it to have been $2\frac{1}{2}$ lbs. per square inch more elastic than the external air, it cannot be far from the truth; and taking the elasticity of the common air at $14\cdot7$ lbs. per square inch, that of the compressed air would be $17\cdot2$ lbs. per square inch.

The bulk of air being inversely as its elasticity, the volume of the compressed air must have been $\frac{147}{172}$ of that of the common air taken into the cylinders, and as each cylinder contains about 75 cubic feet, when the piston was at the top of its stroke, this air would be condensed to $64\cdot1$ cubic feet, before it would have acquired a sufficient elasticity to be able to open the valves, and force its way into the wind-trunk; but as $5\cdot25$ cubic feet of this condensed air would remain at the bottom of the cylinder when the piston was at the bottom of the stroke, the quantity expelled each stroke could not be more than ($64\cdot1 - 5\cdot25 =$) $59\cdot85$ cubic feet, supposing there was no leakage; and if we suppose $\frac{1}{10}$ of the whole to have escaped by the piston and through the valves, it will leave about 54 cubic feet of condensed air, driven out per stroke, × 24 strokes per minute = 1296 cubic feet per minute supplied to the furnace.

As all this quantity passed through an orifice $2\cdot625$ inches diameter (= $6\cdot89$ circular inches), the velocity of the effluent air must have been 575 feet per second; for 1296 cubic feet × $183\cdot3 =$ 237 700 cylindrical inch feet, ÷ $6\cdot89$ circular inch = 34 500 feet per minute, ÷ 60 = 575 feet per second.

It is very common with engineers to estimate the effects of blowing machines merely by the quantity of air taken into the cylinder, without considering the compression it undergoes, and without making any allowance for leakage. At that rate the present machine would be said to have blown ($69\cdot75 \times 24 =$) 1674 cubic feet of air per minute; and the velocity, supposing all that quantity to have passed through an orifice $2\frac{5}{8}$ inches diameter, would have been stated at 44 500 feet per minute, or 742 feet per second; but this is a fallacious mode of estimation.

This quantity was blown into one furnace, which made pig-iron, at the rate of from 18 to 20 tons per week; but by throwing an additional quantity of water upon the wheel, the machine could be made to work quicker, when required, so as to supply air for another small furnace.

The power required to work this machine may be estimated from the following observations. The water was delivered from a spout in a horizontal direction into the top bucket of the water-wheel, and the shuttle was drawn up $2\frac{1}{4}$ inches, so as to open an aperture of that height, by $45\frac{1}{2}$ inches wide, = $102\cdot4$ square inches, or (÷ 144 =) $\cdot711$ of a square foot. The depth of the bottom of this aperture, beneath the surface of the water in the upper pond, was $41\frac{1}{2}$ inches; and the depth to the top of the aperture $39\frac{1}{4}$ inches.

The velocity a body acquires by falling $41\frac{1}{2}$ inches is 895 feet per minute, and the velocity due to $39\frac{1}{4}$ inches is 870 feet per minute; the mean of these two numbers = $882\cdot5$ feet per minute, may be taken for the mean velocity, with which the water would strike into the buckets of the wheel. But in issuing through an aperture, the water does not at first acquire its full velocity, because it cannot be all at once urged into its full motion; and Mr. Smeaton found by many experiments made upon sluices, similar to those of water-wheels, that the velocity of the water, when it is actually passing through the aperture, is only $\cdot707$ of the velocity that a body would acquire by falling the height from the surface to the aperture, and which velocity it acquires very nearly immediately after having quitted the aperture.

Hence to find the quantity of water poured upon the wheel, the velocity must be reckoned at ($882\cdot5 \times \cdot707 =$) 624 feet per minute, × by $\cdot711$ of a square foot, the area of the aperture = 444 cubic feet per minute, discharged.

The whole height to which the water was raised by the fire-engine, was 24 feet perpendicular, and the expenditure of water being 444 cubic feet per minute, the power requisite to keep the machine going at the above rate, was 444 cub. ft. × 24 ft. (= 10 650 ÷ 528 cub. ft. for a HP) = 20·2 horse-power.

It is evident that there must have been a loss of part of this power in applying the water to the wheel, and the amount of this may be thus estimated. The water delivered into the top bucket of the wheel would be at least 6 inches below the top of the wheel ; the bottom of the wheel was 3 inches clear above the lower water ; and 3 inches more of the fall was lost by the water running from the pit below the wheel to the well for the pump, and in returning from the pump-spout to the trough of the wheel. The buckets would begin to discharge their contents when they arrived within about 3 feet perpendicular from the bottom of the wheel, and they would become quite empty when they were within 2 feet of the bottom ; so that about 2½ feet of the fall would be lost in the water running out of the buckets.

In the whole, the action of the water to turn the wheel by its gravity would only continue whilst it was falling through 17 feet perpendicular ; but to this we must add the force which would be communicated to the wheel by the impulsion of the water when it entered horizontally with force into the top bucket. The water would then strike, with a velocity of 882·5 feet per minute, against the boards of the buckets, at a distance of about 9½ feet from the centre of the wheel ; so that its action would be in the direction of a tangent to a circle 19 feet diameter, the circumference of which being 59·7 feet, and the wheel making 6 revolutions per minute, the velocity would be 358·2 feet per minute ; so that the velocity of the wheel would be (358·2 ÷ 882·5 =) ·406 of the velocity of the water which strikes it.

According to Mr. Smeaton's experiments on undershot wheels, the maximum effect was produced when the velocity of the wheel was $\frac{5}{12}$ (= ·416) of the velocity of the water, (which is very near the above proportion) and in that case the mechanical power communicated to the wheel by impulsion, was only half of that which was possessed by the water, the other half being lost in communicating the stroke. We may therefore take half of the 3½ feet actual fall which produced the velocity of the water, (or 1¾ feet) to have been efficacious in turning the wheel ; which, added to 17 feet, gives 18¾ feet effective fall out of 24 feet, or a little more than ¾ (a).

The power exerted by 444 cubic feet per minute, falling 18·75 feet, is 15·76 horse-power, which was actually exerted in turning the water-wheel, compressing and giving velocity to the air, and overcoming the friction of all the moving parts.

This machine will serve as an example for the standard blowing machine for iron furnaces at that period ; another machine was afterwards added, making five in all (b).

The returning engine for the Carron works was rebuilt by Mr. Smeaton in 1780. The old cylinder of 72 inches diameter was used, but the piston made to

(a) The author has found this to be very nearly the same in many overshot water-wheels which he has examined ; and ventures to state as a fair approximation, that in an overshot water-wheel of the ordinary construction, the loss of fall in applying the water to the wheel, is about one-fourth of the difference of level between the upper and lower ponds ; this is without making any allowance for loss of quantity, by leakage or for friction of the wheel.

(b) The author visited the Carron works in 1819, and saw these machines still at work, just in the same form as in Mr. Smeaton's plans ; but as the modern system of management for the blast furnace requires double the quantity of air formerly blown into a furnace, a large steam-engine has been added to the five machines, and works constantly. The wind-trunks of all the five machines and that of the engine, are connected together by pipes, and blow all the five furnaces by one common system.

work 9 feet stroke; the pump of 52 inches diameter having been found very in-convenient to renew the leathers for such a large bucket and clack, four pumps of 30 inches diameter were substituted. The rods for two of these pumps were suspended by chains from the arch head, at the extremity of the great lever, and made 9 feet stroke; the rods for the other two pumps were suspended by chains from two other arches, one affixed on each side of the lever, but nearer to its centre of motion, so that those pumps made but 6 feet stroke; but all the four pumps made their stroke at once, and raised the water from the level of the lower pond which received the water from the water-wheels, and returned it up to the upper pond, the whole lift being 27 feet.

As there was no danger in this case of the clack doors being drowned, they were placed only 3 feet above the water, which being floored over, a convenient stage was formed to repair the clacks; the leathers for each clack were fastened down by six bolts and nuts, which could be removed, and new leathers put in, without lifting the clack out of its seat; each clack was formed with four leaves or valves, that is, with two pair of butterfly valves.

The working barrels were 10 feet in length, with pump-pipes joined on the tops of them, to raise the water up to the cistern or trough, which communicated with the upper pond, for the water-wheels. The pump-buckets were drawn up out of the barrels, when it was required to repair them; the pump-rods were made of cast-iron of a large size, so as to form a sufficient weight for the counterweight to draw up the piston. The engine was worked with three cast-iron boilers, of the same construction as those before described at Cronstadt.

The construction and dimensions of the other parts of the engine were the same as that at Chase-water; and its power being nearly the same, it was sufficient to return the water for four of the blowing machines. The other water-wheels which were supplied out of the same pond, for the forges and other purposes, took all the water of the river in dry seasons (a).

A small Fire-engine and Blowing Machine for an Iron Furnace, 1779.

All the earlier iron works were placed on natural water-falls, and fire-engines were only called in as auxiliaries when they became indispensable; but when fire-engines became more known and understood, iron works were begun with the in-tention of blowing them by that power.

Mr. Smeaton designed a small fire-engine and blowing machine for an iron furnace, which was erected at Seacroft, in Yorkshire, in 1779. There was an old corn-mill, on the site of the works, which had a small supply of water in the winter, but in dry seasons and in summer the water failed in great part. Mr. Smeaton's report on the power necessary for working the blowing machine for this furnace is printed in the 2d volume of his Reports, p. 373, and is a good example of his mode of computation for such cases.

The utmost fall of the water for the old corn-mill was $33\frac{1}{3}$ feet, but to allow a proper run for the water in a long drain, there was only $31\frac{1}{3}$ feet fall at the place of the furnace; he therefore proposed an overshot water-wheel, 28 feet diameter: and says, that to work a coke furnace roundly, at a middling speed, would require 7294 tons of water per day to be expended upon the wheel, 28 feet diameter.

(a) Mr. Smeaton's returning engine was also remaining to return the water of the river in dry seasons, but the great lever having decayed had been replaced by a lever of cast-iron, with parallel levers, instead of arch heads and chains.

He also calculated, that the same quantity of water expended upon the old corn-mill, (which had a water-wheel only 14 feet diameter) would grind corn at the rate of three bushels, or a load per hour; therefore for so much of the year as the corn-mill had been able to grind at the rate of a load per hour, the natural supply of water might be expected to work the furnace without any foreign aid.

From the information gained on the spot, he inferred that during five months in the year there would be a full supply of water; during three other months there would be $\frac{2}{3}$ of a full supply; and that during the other four months, there would on an average be water enough to work the wheel 6 hours per day. On the whole, the furnace might be expected to have only eight months' supply of water in the year; and therefore without some subsidiary power, the furnace could not be kept working continually.

He therefore advised a fire-engine of sufficient power to supply the wheel, independently of the natural supply; and the engine being set to work, whenever the supply became inadequate to work the machine properly, it would furnish all the water expended by the wheel, leaving the natural supply to accumulate in the mill-ponds, till they became full, and then the engine might be stopped, to save fuel, until the ponds were emptied again.

The required quantity of 7294 tons of water per 24 hours (= 182 cubic feet per minute) could be raised by a fire-engine with a cylinder of 30 inches diameter, which would consume 72 cwt. of good coals per 24 hours, of the quality called Halton Bright, or 86 cwt. of raw sleck, which might cost 3s. 3d. per day; so that if the engine worked four months in the year, the whole cost of the fuel would be less than £20. The proposed engine could be attended by one man, when it could perform its task by working only 12 hours per day, or by two men when it worked 24 hours per day.

The engine and machine which were erected upon this recommendation were as follows: cylinder 30 inches diameter, 6 feet stroke; the pump 21 inches diameter; the proportions of the parts, according to Mr. Smeaton's table, p. 183. The water-wheel was 30 feet diameter, with four blowing cylinders 54 inches diameter, $4\frac{1}{2}$ feet stroke, which were worked by four cranks formed on a cast-iron axis, placed in the line of the axis of the water-wheel and connected therewith; the water-wheel usually made $4\frac{1}{2}$ times per minute, so that about 18 strokes of the cylinders were given, and each containing $71\frac{1}{2}$ cubic feet, the whole quantity of air which was blown by the machine was 1286 cubic feet per minute, without considering the effect of the compression of the air; the nose-pipe was 4 inches diameter. The water was raised 34 feet by the pump, and the quantity being 182 cubic feet per minute would be 11·7 horse-power, which is a much less allowance for one furnace than that at Carron before stated.

This engine and machine performed very well. The author has a sketch which was taken by his father, who saw it at work in 1782; but the work was not successful, for owing to a bad quality in the coals and iron ore, they could never make good iron, until they procured coals from another district. The works were carried on for some years with coals brought from a distance of 100 miles, but the expense proved so great, that it was at length given up.

ANOTHER FIRE-ENGINE AND BLOWING MACHINE FOR AN IRON FURNACE, 1780.

A furnace was established at Beaufort, near Abergavenny, in Wales, in 1780. It was one of the first of those establishments which have since become so numerous in Wales; they had a small and casual supply of water, with 46 feet fall. Mr.

Smeaton gave the plans for a fire-engine of 36-inch cylinder, 6½ feet stroke, to raise 203 cubic feet of water per minute 46 feet high, which is 17·7 horse-power.

This quantity of water fell upon an overshot water-wheel, 42 feet diameter, and 18 inches wide within the buckets; it had 96 buckets, 10 inc. deep; the gudgeon of the axis of the wheel was 8 inc. diameter, the end was connected with an axis formed into 4 cranks, in order to work four blowing cylinders by levers, as before described; the cylinders were 60 inches diameter, and the pistons made 4½ feet stroke = 88 cubic feet content.

The wheel usually made about 6 revolutions per minute, so as to blow into the furnace, 24 cylinders of air, and making no allowance for the loss of effect by leakage and by compression, the quantity would have been 2112 cubic feet of air per minute; this air was forced through a nose-pipe of 3¼ inches diameter.

The boiler for the engine was 12 feet diameter, 11½ feet high, the bottom 10 feet diameter; the fire-grate 4 feet square; the chimney 24 inches square inside, and 43 feet high above the grate. It burned 3 cwt. per hour of the best coals, or 4 cwt. of the ordinary Welsh coals. The steam-pipe was 6 inches diameter; the aperture of the snifting valve 1¾ inch diameter; the injection aperture ⅞ of an inch square; the height of the column 25 feet.

The great lever was 18 feet long, 50 inches deep, at the centre, by 14 inches broad; the gudgeons of its axis 6 inches diameter.

The beams between which the cylinder was suspended were each 36 inches deep, by 12 inches broad, and 12 feet long between the walls. The piston-shank was 2 inches square, and united by one joint-pin, 1⅝ inch diameter, to the two chains for the arch-head; the links of the chain were made of ¾ inch square iron bar, and 8 of these sections (= 4½ square inches) bore the whole strain; joint-pins of the chains 1¼ inch diameter.

APPLICATION OF THE ATMOSPHERIC-ENGINE TO BLOW AN IRON FURNACE BY THE DIRECT ACTION OF A FORCING-PUMP FOR AIR, 1780.

This method was brought into use about 1784, and when the means of keeping up a continuous blast had been attained, it was found a much simpler plan than that of pumping up water to supply a water-wheel. The engine was constructed in the usual form, already described (p. 133 and 205); but in place of the pump, a large blowing cylinder was substituted; this cylinder was fitted with a piston, the shank of which was suspended in the usual manner, by chains, from the arch at the extremity of the great lever; the blowing piston was loaded with a sufficient weight to cause it to descend, and force out the air from the blowing cylinder, during the returning stroke of the engine.

The blowing cylinder was in fact a large forcing pump, open at top and closed at bottom, but there were apertures in the bottom, covered with leather flap valves, which admitted the atmospheric air freely into the cylinder, when the piston was drawn up, during the working stroke of the engine; and there was also a passage or wind-trunk, to convey the air away from the lower part of the cylinder, when it was forced out, by the descent of the loaded piston, during the returning stroke of the engine; this passage was provided with hanging valves, which would allow the air to pass out from the blowing-cylinder, into the wind-trunk, and thence by a conveyance-pipe to the nose-pipe of the furnace.

It is obvious that the air would thus be blown at intervals, with a total cessation during the working stroke of the engine. To equalize the discharge of

air into the furnace, the wind-trunk which proceeded from the bottom of the
blowing cylinder did not join directly to the conveyance-pipe, but it was joined to
the bottom of another larger cylinder, into which a piston was fitted to rise and fall,
and was loaded with a sufficient weight to produce the requisite pressure upon the contained air, to force it out of the large cylinder, through the conveyance-pipe which
proceeded from the bottom thereof to the nose-pipe of the furnace. This was called
the regulating cylinder, and its piston, the floating piston, or fly piston. Its
operation was precisely the same as the upper compartment of the common smith's
bellows; for instance, the blowing cylinder was adapted to blow twice as much air
at every stroke, as the nose-pipe would discharge into the furnace during the time
of that stroke; therefore, in the returning stroke of the engine, when the piston of
the blowing cylinder descended, and forced out the contained air, one half of that
air went to the furnace, but the other half was retained in the regulating cylinder,
where it obtained space for its reception by raising up the piston; and during the
working stroke of the engine, when the blowing cylinder did not force out any air
at all, the floating piston, subsiding in the regulating cylinder by its own weight,
forced out the reserved air, through the conveyance-pipe to the nose-pipe of the
furnace.

By this means the supply to the furnace was kept up without any intermission;
but still the force of the blast was not perfectly regular, because the friction of
the floating piston in its cylinder, and also the inertiæ of the great weight with
which it was loaded, were impediments to its motion; and hence the air within the
regulating cylinder could not raise the floating piston, until its elasticity was
increased beyond what it should have been at a mean. And for the same reason,
when the blowing cylinder ceased to force any more air into the regulating cylinder,
the floating piston would not instantly subside, but it remained stationary until the
elasticity of the air in the cylinder was a little diminished. Hence the efflux of
the air through the nose-pipe was sensibly greater during the returning stroke of
the engine, and less during the working stroke; but the alternations were not so
great as to impair the operation of the furnace in any material degree.

The height to which the floating piston could rise in the regulating cylinder
was limited by a strong stop-spring, which was made of ash wood and fixed in a
horizontal position, with one end projecting over the upright stem of the floating
piston, the other end being firmly fastened. When the floating piston had risen
to its intended height, the upper end of its stem came in contact with the end of
this wooden spring, which by its elasticity resisted, though it did not prevent, the
farther motion of the piston, but it occasioned an increase of pressure upon
the air, and consequently that air was forced with greater velocity through the
nose-pipe. To prevent the floating piston from being forced quite out of the regulating cylinder, a safety valve was provided in the piston, and loaded with sufficient
weight to keep it shut; a lever was fixed on the piston, and one end being connected with this valve, the opposite end was so intercepted by a fixed stop, as to
open the valve and discharge the superabundant air, whenever the piston arrived
at its utmost height.

The greatest defect of this apparatus was the want of an increasing resistance
to the rise of the floating piston, so as to occasion the elasticity of the air contained
in the cylinder, to be gradually and imperceptibly increased, as more air was accumulated; if this had been the case, the air would have been discharged into the
furnace with a velocity suitable to the quantity blown by the blowing cylinder; but
in the ordinary construction above described, the weight or load on the floating
piston was uniform in its pressure on the air, excepting those alternations at every

stroke which were occasioned by the friction of the piston and the inertia of the load; the approach to uniformity of the load was a defect, for as the engine would frequently vary in the velocity of its motion, the load could scarcely be regulated so correctly to the speed of the engine, as to discharge the air into the furnace at exactly the same rate as it was forced into the regulating cylinder, by the blowing cylinder; for instance, when the engine went slow, if the load on the floating piston was too great, it would by degrees subside so low in the cylinder as to touch the bottom; or when the engine went quick, if the load was too small, then the floating piston would rise up to the top of the regulating cylinder, till it was stopped by the spring, or till the waste valve was opened; but in either case the blast would become very irregular, the instant that the free motion of the floating piston was impeded by touching the bottom, or the stop-spring.

As a partial remedy for this defect, it became a practice to make the floating piston very thick and heavy in itself, so as to require only a small additional weight to regulate its action. Two strong beams were laid horizontally across the open top of the blowing cylinder, and from these a heavy ring of cast-iron was suspended horizontally, within the cylinder, at such a height beneath its top edge, that when the floating piston had risen nearly to its greatest height, it would come in contact with this ring so as to lift its weight, which would then form an additional load on the floating piston, and would occasion an increased pressure on the air, whereby it would be blown with a greater velocity, as soon as ever the piston was raised to its greatest height. The ring was suspended from the beams by four or six strong check-straps, which sustained its weight, whenever the floating piston subsided in the cylinder to its proper limit; but when so suspended, the ring would be at all times ready to oppose the rising of the piston above its proper limit. This ring answered the purpose of the wooden stop-spring, but was less abrupt in its action. The cross-beams over the cylinder, prevented the ring rising so high as to allow the floating piston to be forced out of the cylinder.

Two stop-springs were also applied, to limit the descent of the piston into the cylinder: these were firmly fixed at one end to the great cross-beams, and the other ends of the springs were close by the side of the perpendicular stem or rod of the piston, which had an iron catch-pin put through it, and the ends projected out at each side, so as to come in contact with the ends of the springs, whenever the piston subsided to its lowest position. These springs prevented the floating piston from abruptly striking the bottom of the cylinder, in the same manner as the stop-springs, and catch-pins, at each end of the great lever of the engine, limited the descent of its piston into the steam-cylinder. These applications produced a considerable improvement in the operation of the machine, by rendering the blast more continuous and regular.

The following are the dimensions of a blowing-engine of this kind, which was erected at an iron-works in Scotland, about the year 1790, and is recorded by Dr. Robison, in the Encyclopædia Britannica, article Pneumatics.

Steam-cylinder 40 inches diameter (=1257 square inches) (a) the piston making a stroke of 6 feet. The blowing cylinder 60 inches diameter, (= 2827 square inches area) and the piston also 6 feet stroke; it was loaded with a weight of $3\frac{1}{2}$ tons (=7840 lbs.) which is at the rate of 2·77 lbs. per square inch.

(a) The author examined this engine in 1819, when it had long been disused; a modern double engine had been erected on the same spot to blow two furnaces; but the old engine had been preserved to be worked in case of any accident happening to the new one. The steam-cylinder was 44 inches diameter, instead of 40, as stated by Dr. R.; but probably the original cylinder had been replaced by a larger one.

This weight was raised up by the pressure of the atmosphere on the piston of the steam-cylinder, during the working stroke of the engine, and by its descent during the returning stroke, it forced the air out of the blowing cylinder. The area of the steam-cylinder being 1257 sq. inc. the load of 7840 lbs. was at the rate of 6·23 lbs. per sq. inch of the piston.

The air was conveyed by a wind-trunk, or passage, from the bottom of the blowing cylinder into the bottom of the regulating cylinder, which was 96 inc. diam. (= 7238 sq. inch area) and 6 feet long; its capacity was about $2\frac{1}{2}$ times that of the blowing cylinder; the floating piston and the weight with which it was loaded, weighed $8\frac{1}{2}$ tons = 19 040 lbs., which is at the rate of 2·63 lbs. per sq. inch. The floating piston usually rose and fell at every stroke of the engine, through a space of about a foot in the regulating cylinder.

The conduit-pipe which proceeded from the bottom of the regulating cylinder to convey the air to the furnace was 12 inc. diam., terminating with a nose-pipe (b). There were valves in the bottom of the blowing cylinder opening upwards, to admit air into this cylinder, but to prevent its return; and there were also valves in the wind-trunk or passage, to permit the air to pass from the blowing cylinder, into the regulating cylinder, but to prevent its return. The conveyance-pipe, which carried the air away from the bottom of the regulating cylinder to the nose-pipe of the furnace, required no valves, because the air passed constantly through it, either in consequence of the air being forced into the regulating cylinder, by the descent of the piston of the blowing cylinder, or else in consequence of the subsidence of the floating piston in the regulating cylinder, by its own weight, during the inaction of the blowing cylinder.

The blowing cylinder of 60 inches diam. and 6 feet stroke, would take in 117·8 cubic feet of air at every stroke, and it usually made from 15 to 18 strokes per minute; at 15 strokes the motion of the piston would be 90 feet per min., and the quantity of air blown by the engine, would have been 1767 cubic feet per min., provided none had been lost by leakage and by compression.

If we assume 1 foot space to have been left at the bottom of the blowing cylinder, when its piston had descended to the lowest, the capacity of that space would be 19·63 cubic feet. Supposing the ordinary elasticity of the atmospheric air to be 14·7 lbs. per square inch, the air which was compressed by the descent of the blowing piston, with a pressure of 2·77 lbs. per square inch, would have its elasticity increased to 17·47 lbs. per sq. inch, in which state, a quantity of air equal to 23·3 cub. ft. of common air, would be crowded into the space of 19·63 cubic feet beneath the bottom of the piston; that is 3·67 cubic feet more than it would contain when the compressure ceased. This quantity which would not be forced out by the piston, would amount at 15 strokes per min. to 55 cubic feet of air lost per min.

Making this deduction, and an allowance of $\frac{1}{15}$ for leakage by the pistons and valves, the quantity of air blown into the furnace would be about 1600 cubic feet per minute, supposing it to have been of the ordinary elasticity of 14·7 lbs. per sq. inch; but when it was compressed to an elasticity of 17·47 lbs. per sq. inch, the volume would be reduced to 1346 cubic feet (c).

The actual weight of this quantity of air would be $120\frac{1}{2}$ lbs.; for the common air is 830 times lighter than water; and as a cubic foot of water weighs $62\frac{1}{2}$ lbs.

(b) Dr. Robison states the diameter of the nose-pipe $1\frac{5}{8}$ inc. which is certainly erroneous.
(c) If the nose-pipe had been $2\frac{5}{8}$ inc. diam. the velocity of the compressed air issuing from it would have been 35 820 ft. per min. or 597 ft. per second.

$(830 \div 62{\cdot}5 =)$ 13·28 cub. ft. of air will weigh a pound; and $1600 \div 13{\cdot}28 =$ 120·5 lbs.

The power exerted by this engine may be thus computed. The weight of the blowing piston $3\frac{1}{2}$ tons ($= 7840$ lbs.) was raised by the steam-piston 6 feet at every stroke; but in returning, some of its power would not be efficaciously applied, on account of the diminution of the volume of the air contained in the blowing cylinder, by the compression; for the first resistance which the common air, in the blowing cylinder, opposed to the motion of its piston commenced at nothing, and increased until that air was compressed from its ordinary elasticity of 14·7 lbs. per sq. inch, to an elasticity of 17·47 lbs.; therefore, out of the whole motion of 6 feet per stroke, only 5·05 feet, would have to overcome the full resistance of 2·77 lbs. per sq. inch; but the first ·95 of a foot, of the motion of the piston, would only be opposed by a variable resistance, beginning at nothing, and increasing to 2·77 lbs. per sq. inc. by the time the piston had descended ·95 of a foot.

On the whole, the power would be equal, to very nearly $5\frac{1}{2}$ feet stroke, with an uniform resistance of 7840 lbs.; and $5\frac{1}{2}$ ft. $\times$ 15 strokes per min. $= 82{\cdot}5$ feet motion per min. $\times$ 7840 lbs. $= 646\,800$ lbs. raised 1 foot per min. $\div 33\,000 = 19{\cdot}6$ horse-power, employed to blow one furnace, with 1600 cubic feet of air per minute when uncompressed, weight $120\frac{1}{2}$ lbs., and which probably issued from the nose-pipe with a velocity of 597 feet per second, in its compressed state (d).

Improved Blowing Machine with a Lifting Pump for Air, 1790.

The construction of blowing engines was afterwards improved, by placing the blowing cylinder in an inverted position, and causing its piston to force out the air during the working stroke of the fire-engine, instead of during the returning stroke. By this arrangement, the heavy load upon the blowing-piston was avoided, the air being blown by the direct action of the atmospheric pressure upon the steam-piston, instead of by the return of the counterweight.

The blowing cylinder was open at bottom, and closed at the top; and the piston, which was fitted into it, was perforated, and fitted with valves opening upwards, so as to admit air to pass through them when the piston descended; but these valves closed, to prevent the escape of that air, when the piston was drawn upwards during the working stroke of the engine. The wind-trunk proceeded from the top of the blowing cylinder, and conveyed the air into the bottom of the regulating cylinder, which was the same as before described, but placed higher than the blowing cylinder; and valves were placed in the wind-trunk, to prevent any air returning from the regulating cylinder.

To communicate motion to the piston of the inverted cylinder, a large rectangular frame of wood was suspended by the chain, from the arch-head of the great lever: and the upright sides of this frame were guided between fixed upright posts, so that it was raised up and down during the motion of the engine, in the same manner as a window-sash rises and falls in its frame. This sliding frame received the blowing cylinder between its two upright sides, the upper cross-piece of the frame extending horizontally across the centre of the close top of the cylinder,

---

(d) This computation corresponds very well with that of the Carron machine (p. 277), which blew 1272 cub. ft. of air, compressed to an elasticity of 17·2 lbs. per squ. inc.; but this, when it was in the ordinary state of the atmospheric air, would have been 1488 cub. ft. or ($\div 13{\cdot}28 =$) 112 lbs. weight. To urge this quantity through the nose-pipe with a velocity of 563 ft. per second required about 15·76 horse-power to be exerted on the water-wheel.

and to the middle of this cross-piece, the main-chain was fastened to suspend and give motion to the frame. The lower cross-piece of the sliding frame extended horizontally across the bottom, or open end of the cylinder, and to the middle of this cross-piece, the upright stem of the piston was fixed, and reached upwards in the centre of the cylinder. The cylinder was suspended between two strong horizontal beams, in the same manner as the cylinder of the fire-engine before described : the ends of these beams being firmly fastened into the side walls of the house, to keep them down.

This cumbrous sliding frame was afterwards laid aside, and the rod of the blowing piston connected immediately with the chain from the great lever ; the close top of the blowing cylinder being perforated in the centre, for the rod to pass through, and the rod being made smooth and true, was fitted in a close collar of leathers, or of hemp stuffing, which prevented any escape of air by the side of the rod. This construction will be readily understood by referring to the figure of the lifting pump for waterworks (see p. 247) ; for the action of the blowing cylinder to compress air, is similar to that by which a lifting pump raises water. The action of the first described blowing cylinder, with a loaded piston, and the cylinder open at top, is similar to that of the forcing pump (see figure, p. 249).

The atmospheric engine, with an inverted blowing cylinder, closed at top, and with a smooth piston-rod passing through a collar of leathers in the top of the cylinder, was a very efficient machine, which was for many years a standard engine for iron furnaces ; and some good specimens still remain in use. The intermitting blast was equalized by a regulating cylinder with a floating piston ; and in some of the best machines, two such regulating cylinders and pistons were used, to give a greater capacity for the air, and thus render the blast more uniform.

Water Regulator for the Blowing Machine, 1794.

The regulating cylinders being found incapable of producing a perfectly equable blast, another plan was at length adopted, which is now become the universal practice in modern blowing machines. This is called the water-regulator : it consists of a very large chest, formed of iron-plates screwed together, and placed in a still larger cistern of water. The chest is open at bottom, but close in all other parts. Its weight is supported upon small piers of brickwork, so as to raise it about a foot above the bottom of the cistern, and therefore the water has free passage into the chest beneath the bottom edge. The wind-trunk, or pipe, from the blowing cylinder, is conducted across the top of this blowing chest, in its way to the furnace, and an upright branch descends from the pipe, to join to the top of the chest. When the air is compressed by the blowing-cylinder, it enters by this branch into the inverted chest, and displaces the water therefrom. The effort which the water makes to return into the chest, produces a constant pressure on the air, so as to force it through the nose-pipe into the furnace with a very uniform velocity.

For instance, when the blowing cylinder makes its stroke, it forces out twice as much air as is required to pass through the nose-pipe into the furnace during that stroke. The surplus of the air therefore enters by the branch-pipe, into the inverted air-chest, and drives out an equal quantity of water therefrom. This water is raised in the external cistern, which contains the chest, and in the regular course of working, the surface of the water in the cistern outside of the chest, is between 5 and 6 feet higher level, than the surface of the water within the chest.

The water having this difference of level, acts as a reacting weight to retain the air in its compressed state, so that the chest forms a vast magazine of compressed air, from which the water will expel as much as is necessary to supply the blast, during the returning stroke of the blowing machine ; but during the working stroke, when the blowing cylinder furnishes more air than is wanted at the moment, the water will recede from the chest, to make room for the superabundant air.

In this manner, there is a continual flux and reflux of the water, out of and into the air-chest ; but the capacity of the chest being very large, in comparison with that of the blowing cylinder, the fluctuations of the surface of the water in the chest are but extremely small, and do not affect the regularity of the blast in any considerable degree. This circumstance, of the great capacity of the inverted air-chest, gives it a decided advantage over the regulating cylinder ; neither can any air be lost by leakage ; and the dilatation and contraction of the capacity of the air-chest, which is produced by the flux and reflux of the water, is not impeded by friction, as is the case with the floating piston.

The inertia of the counteracting weight has a sensible effect on the action of the floating piston, as before stated, because the motion is so considerable. The great weight of the piston causes it to resist the communication of motion at first, and therefore the air must be more compressed than it should be, to make the piston move ; and also, when the piston is put in motion, the energy of the moving mass tends to continue the motion, so that it cannot begin to return and continue the blast at the first instant that the blowing cylinder ceases to force out air. In the water-regulator, the weight of the water, which forms the counteracting weight, is much greater than the weight of the floating piston ; but, on the other hand, the motion given to the water, is so much slower than the motion given to the floating piston, that the energy of the moving mass of water is very small, in comparison with that of the floating piston, and consequently the blast is more equable.

The following are the dimensions of an atmospheric engine and blowing machine, with a water-regulator, which was erected for an iron furnace in Derbyshire, about the year 1794, and which is still in use.

Steam-cylinder 48 inc. diam., the piston making a 6 ft. stroke, and about 12 strokes per minute. The blowing cylinder 72 inc. diam. and 6 ft. stroke. The inverted air-chest is circular, 15 ft. diam., and $7\frac{1}{2}$ ft. deep ; its capacity, when in action, being about six or seven times as great as that of the blowing cylinder. This air-chest is placed in a circular cistern, 26 ft. diam. at top, 10 ft. deep, and 20 ft. diam. at bottom, the sides being a sloping stone wall, and the bottom paved with bricks.

The air-chest is formed of cast-iron plates, screwed together ; the circular top consists of one central plate, with 16 segments surrounding it ; and the upright sides, which form the cylinder of the chest, are also composed of 16 plates $7\frac{1}{2}$ ft. long, and nearly 3 ft. wide, screwed together at the edges. The lower edge of the chest is supported on stone blocks, at $1\frac{1}{2}$ ft. above the bottom of the cistern, to allow the water free passage into and out of the chest. The central plate for the top of the chest, is supported from the bottom of the cistern, by two cast-iron columns. The circular top of the chest is loaded with a solid mass of stone-work, 2 ft. thick and 15 ft. diam. to keep the air-chest down ; for it would float when the water is expelled from it by the air, if it had only its own weight to retain it.

The blowing cylinder is of the kind last mentioned, viz. open at bottom, and closed at top, with a collar of leathers in the centre of the cover, to admit the piston-rod to pass through. An air-pipe of 15 inc. diam. proceeds from the top of the blowing cylinder, towards the furnace, and passes horizontally across the circular top of the air-chest, at 5 ft. above that top, and three vertical branches descend from the pipe, to join to the top of the chest, one in the centre, and two near the circumference.

The communication of the air-pipe with the air-chest is made in this manner, in order that the air which occupies the lower part of the chest may remain unchanged, and not be blown into the furnace, because it is supposed to become charged with moisture by contact with the water, and to be therefore less fit for animating the fire.

The air-pipe, being made of large dimensions, passes in a direct line from the blowing cylinder to the furnace, and connects by its three branches with the top of the air-chest. At every stroke, the surplus of air which the blowing cylinder furnishes more than the furnace requires during that stroke, enters through these branches into the air-chest; but the same air returns immediately afterwards, through the branches into the air-pipe, during the intermission of the action of the blowing cylinder. The air which is thus newly introduced into the upper part of the air-chest, and withdrawn immediately afterwards, will not mix with that air which has been some time contained in the lower part of the chest, near to the water; because when air is suddenly compressed, it generates heat, and its temperature becomes higher than it was before the compression was produced: but that portion of air which has been some time at the lower part of the receiver, becomes cooled by its contact with the water, and is then specifically heavier than the newly-introduced air, which consequently preserves its station at the upper part of the air-chest, near to the entrance branches; so that the same portion of air which is thrown into the chest at each stroke of the blowing piston, is discharged again at the intermission between that and the next stroke: and hence the air cannot acquire any humidity or dampness from the water.

The depth of the water cistern, in which the air-chest is immersed, is so regulated, that when the engine stops, and the compression of the air ceases, the water can never rise so high in the air-chest as to be in any danger of entering the orifices of the three branch-pipes, which conduct the air to the furnace: for if any water were to splash up into those pipes when the engine is stopped, and be thrown into the furnace, an explosion would be produced, by the very sudden rarefaction of the water into steam.

At the first introduction of the water-regulators, great fears and prejudices were entertained by the iron manufacturers, respecting the humidity which the water might communicate to the air, and the danger of water being thrown into the furnace. In some cases, where the apparatus had not been properly constructed, such inconveniences were really felt; but still the great regularity of the blast gave a decided preference to the use of water, in place of the floating piston, and by degrees, the water-regulator has come into general use.

Air-chamber for regulating a Blowing-Machine, 1796.

An attempt was made in 1796, to regulate the blast of a blowing machine, by merely forcing the air into a very capacious air chamber, from which it would issue by its own elasticity, without any reacting weight, of pistons, or of water. This was executed on a large scale, at the Devon Iron Works, near Stirling in Scotland, and an account of the experiment was published by Mr. Roebuck, who had suggested the plan. (See the Edinburgh Philosophical Transactions for 1796, vol. v.)

The two furnaces were not built of masonry, as is the usual practice, but they were excavated in a solid rock; each furnace was 44 ft. high and 13 ft. internal diameter at the largest part. The engine was an atmospheric cylinder, $48\frac{3}{4}$ inches diameter; the piston made $4\frac{2}{3}$ ft. stroke, and usually went 16 strokes per minute, $= 74\frac{1}{2}$ ft. motion per min. The blowing cylinder was 78 inc. diam., and its piston made the same length of stroke as the steam-piston: therefore at 16 strokes per min. it would discharge 2470 cub. ft. of air per min. (the loss by leakage and compression not being taken into account.) This quantity of air was blown into one furnace, through a nose-pipe of $2\frac{3}{4}$ inc. diam.

To regulate the blast, a large vault or air-chamber was excavated in the rock behind the furnaces, and at 16 ft. distance from them. The internal dimensions were 72 ft. long, 14 ft. wide, and 13 ft. high, and its capacity 13 100 cub. ft. or about 84 times the capacity of the blowing cylinder. The air was introduced into one end of this air-vault, by a pipe of 16 inc. diam. leading from the top of the blowing cylinder, and a similar pipe conveyed the air from the other end of the air-vault to the furnaces. This pipe terminated in a cubical box, situated between the two furnaces; and two smaller pipes proceeded in opposite directions from the box to the nose-pipes for each furnace. In the top of the cubical box was an aperture, closed by a safety valve, which was loaded with such a weight as to allow the air to escape, when the compression exceeded what was required; and there was also an inverted syphon-guage, containing mercury, to show the elasticity of the air: this was usually equal to a column of 5 inches of mercury, $= 2\cdot45$ lbs. per squ. inc., and during the returning stroke of the engine the mercury descended half an inch.

The rock in which the air-vault was formed was very close and solid; but where any crevices appeared they were caulked with oakum. The whole of the inside of the vault was plastered, and then covered with brown paper stuck on the plaster with pitch, so as to render it perfectly air-tight.

At the commencement of the operation of this blowing apparatus, drops of water trickled from the rock, within the air-vault, and water was frequently driven through the nose-pipe into the furnace, so as to prevent it from acquiring a proper heat (a).

Mr. Roebuck states, that after the rock had become perfectly dry, by the continual heat of the furnace, all appearance of water at the nose-pipes ceased; but as the furnace did not perform well, and the deficiency was attributed to this new mode of regulating the blast, it was afterwards given up, and the common regulating cylinders applied in lieu of it.

The supposed defect of communicating humidity to the air, might have been easily remedied, by so arranging the pipes that the air in the vault would not be changed, but that the same portion of air which was driven into the vault, should immediately return out of it again, as before explained, in the water regulator (b).

(a) To ascertain the source from which this water proceeded, Mr. Roebuck, with another person, went into the air-vault, and they were shut up in it whilst the engine was at work. The engine had been stopped for about two hours before they went in, and they found the air in the vault damp, and misty; but it cleared up as soon as the engine was set to work. They experienced a pain in their ears, whilst they were confined in the compressed air, but their respiration was not affected. Sounds appeared to be very greatly magnified. The water was found to drip from the rock on the side nearest to the furnace, in consequence of the heat on that side. There was no appearance of damp or mist in the air, whilst the engine continued at work; but the instant after the engine was stopped, the vault became filled with a vapour like a fog; and he concluded that this vapour condensed in the conveyance pipes, and formed drops of water whenever the engine was first set to work.

(b) An air-vault has since been employed for a furnace at Bradley, in Staffordshire, and has answered very well; it is a cylinder of iron plate, 10 or 12 ft. diam., and 50 or 60 ft. long.

MACHINE FOR BORING THE INTERNAL SURFACES OF THE CYLINDERS FOR FIRE-ENGINES.

IN the construction of the earliest steam-engines, the greatest difficulty of execution, was to cast and bore the cylinders; this must have been a serious affair, even for small cylinders of from 22 to 30 inches diameter, such as were used in the first engines, (see p. 155,) although those cylinders were always made of brass, for the facility of boring the inside true and smooth (a).

All the other parts of an engine could be executed, with more or less precision, by such artificers as were to be found in London, and other parts of England, at the beginning of the last century; for instance, the coppersmiths who were in the habit of making large coppers for the brewers and dyers, could readily execute the boiler, which did not require larger dimensions than they were accustomed to; the furnace and brickwork for the boiler would also be the same as the brewing coppers. The dome-top, to keep in the steam, which was the only additional part, was made of lead by the plumbers.

In like manner the pumps and pit-work, and the great lever and chains, would be easily made, by the same workmen who were accustomed to make the pump-work for the water-wheels, by which collieries and mines were then drained; the pump-barrels were made of brass, and the pump-pipes of wood, bored out of the solid, and strengthened by iron hoops on the outside, and from this cause are still called pump-trees by the miners, although they are made of cast-iron.

It is true that the formation of the pump-barrel required all the same operations as would have formed the cylinder, but the latter being on a scale so much larger than any thing which had been then executed, must have presented many difficulties; and to allow the piston to move with freedom in the cylinder, and yet fit air-tight, both the cylinder and piston required greater precision, than the artificers of that day were accustomed to use in any other works.

It is not on record by what means the first brass cylinders were made true, but it appears that they were made near London in 1727 (see p. 230); probably they were cast by the bell-founders. The iron-founders afterwards found out how to cast and bore cylinders of cast-iron, with sufficient accuracy, as is mentioned by Desaguliers (see p. 231).

Towards the middle of the last century, when the demand for steam-engines for collieries became considerable, iron foundries were established in the north of England, particularly after the introduction of the improved method of making iron with coke of pit-coal, instead of charcoal of wood. The quality of cast-iron was so much improved by the new process, that it became a fit material for many new uses, and admitted of being turned and bored with great facility. The processes of moulding and casting iron were studied by many ingenious workmen, and brought to considerable perfection.

(a) The Rev. Mr. Mason, who has recorded some particulars of the state of the Colebrook Dale Iron-works in 1747, (Phil. Trans. vol. XLIV. p. 370), mentions spelter or brass solder as being used for the cylinder of an engine. What spelter is, or what uses are already made of it, Mr. Mason professes not to know; but he believes it was never yet applied to so large a work as the cylinder of a fire-engine, till Mr. Ford, of Colebrook Dale, in Shropshire, did it with success; it ran easier, and cast as true as brass, and bored full as well, or better, when it had been warmed a little; while cold, it is as brittle as glass, but the warmth of his hand soon made it so pliant, that he could wrap a shaving of it round his finger like a bit of paper. This metal never rusts, and therefore works better than iron, the rust of which, on the least intermission of working, resists the motion of the piston.

Most of the large foundries had their boring mill for iron cannon, and pump-barrels, and for steam-engine cylinders : some of these mills on a small scale were turned by horses, but the better sort were moved by a water-wheel ; they were all constructed on the same principle as the common machine used for boring out trees to form wooden pipes for water-works.

The cylinder, or barrel, was bedded with its axis in a horizontal position, on a sledge or sliding carriage, and firmly fastened thereto by chains ; the sledge was moved horizontally, upon a suitable roadway, in order to advance the cylinder towards the borer in the direction of its axis : the borer was a circular iron disk, or wheel, nearly fitting into the interior of the cylinder, in the same manner as its piston would do ; but the circumference of the circular borer was provided with six or eight steel cutters, firmly fastened by wedges, into notches made in the edge of the wheel.

The borer thus furnished, was fixed on the end of a horizontal iron axis, which was turned slowly round by the mill, and the borer being inserted into one end of the cylinder, the cutters, in their revolution, excavated and removed all prominences and roughness, from the interior surface of the cast-iron ; the sledge on which the cylinder was placed, was continually advanced towards the borer, with a very gradual motion, given by a tackle of chains and pulleys, and a windlass, so as to cause the borer to penetrate into it, till by degrees it worked its way quite through, from one end to the other ; but the operation required to be repeated several times, to make the cylinder quite smooth at every part.

The Carron Iron-works had the most complete apparatus for these purposes : the boring mill which they used at the first establishment of their works, was only intended for boring cannon, but they bored barrels and cylinders by it occasionally. In the course of a few years, this mill proved too small to execute the work which their trade required, and in 1769 Mr. Smeaton made them an entire new boring mill for guns, and another for cylinders. Drawings of these mills are engraved in his printed Reports, Vol. I. p. 376, and deserve notice, as being the most powerful machines at that time, and by which many large cylinders were bored for the atmospheric engines already described.

The water-wheel was 18 feet diameter, and 5 feet wide, being of the kind called a breast fall : it had 40 float boards to receive the action of the water, which was poured upon the lower part of the wheel, over the top of a breasting, or wall of stone, built in a cylindrical form, corresponding to the circle described by the extremities of the float boards ; so that about one-fourth part of the wheel worked in a cylindrical pit, to which it fitted in every part as closely as possible, without touching.

The surface of the water in the millpond or reservoir was 2 feet below the level of the centre of the wheel, and therefore the whole fall or descent of the water, to impel the wheel, was 7 feet ; the top of the stone work, or crown of the fall, over which the water poured, was $1\frac{1}{4}$ feet below the level of the surface ; and the water was kept back by an inclined shuttle, or sluice gate, which formed a tangent to the circle of the wheel, and shut down upon the stone work, near the top

This kind of water-wheel was introduced by Mr. Smeaton, in place of what were before called undershot wheels, and was a great improvement in the application of water power. The principal part of the power to impel the wheel, is derived from the weight of the water ; for the curved breast wall together with the side walls, between which the wheel is included, are adapted to the float boards, with the intention that as little water as possible shall leak or pass between the edges of the boards and the walls.

of the fall, so as to stop the water, when the shuttle was down ; but when it was drawn up, a crevice was left between the lower edge of the shuttle and the top of the stone work, through which the water rushed in a thin sheet, and striking upon one of the float boards of the wheel at a time, exerted an effort to turn it round, partly by the impulse with which it struck that float board which was opposite to the issuing stream, but still more by the weight of the water, which rested upon that float board ; and the water being prevented from escaping, by the close fitting of the circular stone wall or breasting, to the extremities of the float boards, it continued to press by its weight against the float board, till, by the rotation of the wheel, it arrived at the lowest part of the circular course, and there it passed off horizontally into the channel of the river.

The axis of the water-wheel was cast-iron, 9 inches diameter, and 7 feet long between the bearings ; upon the middle part of the axis were two large projecting flanges or circular plates, 4 feet diameter, forming the centre pieces to fix the wooden arms of the two rings of the water-wheel ; these arms were applied to the flat surfaces of the circular flanges, and fastened thereto by bolts. The two rings of the water-wheel were made of cast-iron, that they might act as loaded fly-wheels to regulate the motion ; each ring consisted of 8 segments, united together by wrought-iron plates and bolts ; eight mortices were left in each of the cast-iron rings to receive the ends of the wooden arms, and forty other mortices, to receive pieces of wood, to which the float boards were nailed.

On the extreme end of the axis of the water-wheel, within the mill, was fixed a cast-iron cog-wheel of 29 teeth, $3\frac{1}{4}$ feet diameter, 4 inches broad ; and this gave motion to a larger cog-wheel on each side of it, viz. one of 90 teeth, $10\frac{1}{4}$ feet diameter, 2 inches broad ; and the other 60 teeth, $6\frac{3}{4}$ feet diameter, 2 inches broad ; the axes of both these wheels were laid horizontal, and parallel to the axis of the water-wheel, and on a level therewith ; so that all three axes were in the same horizontal plane.

Each cog-wheel was fixed on the middle of its cast-iron axis, which was similar to that of the water-wheel, but shorter, viz. 9 inches diameter, and 2 feet long between the bearings, which were supported on two large horizontal oak beams, laid parallel to each other, and united together by short cross pieces at each end, to form a long narrow frame : these beams were supported upon two parallel walls, and fastened thereto by bolts going down to the foundations ; the space between these walls formed a long narrow pit, in which the lower semicircles of the three cog-wheels worked ; the necks or gudgeons of their axes, which were 8 inches diameter, were supported in bearings fastened on the oak beams, which formed the capping of the two walls, or the margin of the pit.

The extreme end of each axis projected beyond the bearing, and terminated with a square end, or rather with what Mr. Smeaton called a fluted, or citadel head, to couple the end of the boring bar to the axis. The extreme end of the axis of the water-wheel had a similar head upon it ; so that this boring mill contained three parallel horizontal axes, moving with different velocities, to adapt them for boring cylinders of different diameters. The water-wheel was intended to make 7 turns per minute, on an average, and the axis was adapted to give motion to the borers for small pump-barrels. The cog-wheel of 60 teeth, which made rather less than $3\frac{1}{2}$ turns per minute, was employed to bore large pump-barrels, and small cylinders ; the large cog-wheel of 90 teeth, on the opposite side of the water-wheel axis, made about $2\frac{1}{4}$ revolutions per minute, and being the slowest motion, it was intended to bore large cylinders. The motion of the water-wheel was not confined to 7 turns per minute; but it could be regulated to give such a motion to the borers, as was found most suitable for the size of the cylinder which it was required to bore.

The pit containing the three cog-wheels, occupied a narrow space across one end of the mill-house, leaving an extensive platform, or flat floor, opposite to the projecting ends of the three axes, to receive the barrels or cylinders, which were to be bored; a suitable sledge or carriage was provided for each barrel or cylinder, and they were adapted to be advanced towards the borers, with an accurate recti-linear motion, exactly corresponding with the axis or central line of the cylinder; for this purpose a road-way was formed opposite to the end of each axis, by two strong beams laid horizontally on the floor, parallel to each other, and also to the axis, but on a level considerably below the axis; the top surfaces of these beams being worked very exactly straight, level, and parallel, were covered with iron bars, upon which the sledges for the small barrels were adapted to slide, and the carriage for the cylinders had small wheels, to diminish the friction; but every care was taken that the sledge or carriage should fit correctly on its road-way, so as to move freely, but without permitting any sensible deviation from the right line.

The sledges and carriages were flat oblong frames of timber, rather longer and wider than the cylinder; the two side pieces of the carriage were fitted to the road-ways, and upon the cross rails by which the side pieces were united, blocks of wood were fastened, and hollowed out to a suitable form for the cylinder to rest steadily upon them; and by regulating these blocks with wedges, the cylinder could be brought to lie upon the carriage, with its axis exactly horizontal, and coincident with the axis of the boring wheel. The cylinder was fastened down upon its blocks by chains, which passed over it, and were strained very tight, by winding the ends about two windlasses, or small rollers, which extended across the carriage; these rollers were turned by levers, and had ratchet wheels and clicks to retain them from returning. The carriage and cylinder were drawn slowly along the road-way by chains, passing over suitable pulleys, and wound round a roller or wind-lass, which a man turned by a wheel with handles.

The boring head which entered into the cylinder, was a circular wheel of cast-iron, nearly fitting the interior of the cylinder, and it had 6 or 8 notches round its circumference, into which small steel cutters or knives were fixed by wedges; and the cutting edges of all the 6 or 8 cutters were exactly adjusted by wedges, so as to conform to the circumference of a circle, a very little less than the true diameter intended for the cylinder. The boring head, with its cutters, was fixed upon one end of a strong iron axis, of sufficient length to reach quite through the cylinder, and the other end of this axis was connected by a socket, with the projecting citadel head on the extremity of the axis of the cog-wheel, so as to communicate a slow and steady circular motion to the borer, that its cutters might, by turning round within the cylinder, cut away and smooth all irregularities from its interior surface, and form a true cylinder.

The old boring machines, which were made previously to Mr. Smeaton's time, were all defective in strength; and were consequently incompetent to bore large cylinders in a proper manner; for though the boring head had four or six cutters, they were necessarily adjusted to a circle rather smaller than the diameter of the rough cylinder (if it was a large one) and therefore they could not cut or bore all round the interior circumference of the cylinder at the same time. If the cutters for a large cylinder had been adjusted to a circle of the full size, in order to cut all round, then the resistance to the motion of all the cutters at once, would have been greater than the mill could have overcome, consistently with steadiness and regularity of motion ; neither could the cylinder have been retained steadily on its carriage, to have endured such complete boring.

There was no proper support for the weight of the boring head, and its axis, but it was left to rest upon those cutters which were at the lower part of the circumference ; consequently the cutting was almost exclusively confined to the lower part of the circumference of the cylinder ; and after the borer had been worked quite through a large cylinder, it was found to be smoothed at the lower part, but very imperfectly touched, or not touched at all, at the upper part.   To remedy this deficiency, it was necessary, after the first boring, to release the cylinder, and turn it half round on its carriage, then refix it, and bore it through again with the other side upwards ; and for the largest cylinders the operation was repeated four times, to smooth all the four quarters of the interior surface, at successive operations.

Mr. Smeaton intended to give his boring mill a sufficient power in the first mover, and a competent strength in the parts, to cause the six cutters to cut all at once, entirely round in the cylinder ; and to sustain the weight of the boring tool,

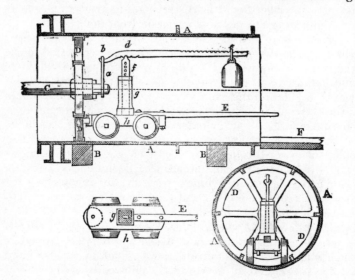

he contrived a small wheel carriage to go into the cylinder, and form a fulcrum or support for a lever or steelyard, from one end of which the end pivot of the axis of the boring head was suspended by a link, and on the other end of the lever, a sufficient weight was applied, to counterbalance the weight of the boring tool, which being by this means undetermined to any side, might be expected to cut equally and truly all round the inside of the cylinder, so as to bore it true at once, and thus obviate the necessity of turning the cylinder on its carriage, and re-boring with different sides upwards.

The construction of this steelyard carriage is sufficiently apparent from the figures : A A is the section of the cylinder ; and B B are the sections of the blocks which are fixed on the cross rails of the carriage, to rest the cylinder upon ; C is the axis of the borer, the extreme end being connected by a coupling, with the projecting end of the axis of the boring wheel, and the boring head, or wheel D, being fastened on the other end.   This wheel is formed with six arms, and furnished with six cutters, as shown in the end section ;   the extreme end of the axis C which projects through the wheel D, is suspended by a link *a*, from the short end *b* of the iron lever or steelyard, *b, d, e*, which has a weight hung on the long end at *e*, and the fulcrum, or centre of motion *d*, is at the top of an iron standard *f*, which slides up or down into a cavity, in a stronger wooden standard *g*, so as to admit of fixing the fulcrum at the required height by inserting a pin through one of the holes in the iron standard *f*. The wooden standard *g* is fixed upon the flat bed plank *h* of the carriage, which has four wheels, adapted to run on the interior surface of the cylinder, or rather to permit the cylinder to move beneath them ; for in fact the carriage should have no motion at all, but it should remain stationary, opposite to the end of the boring axis, in order to sustain its weight, whilst the cylinder and its carriage are progressively advancing, to cause the borer to bore it through from one end to the other.   A roller is fitted into the end of the bed *h* of the carriage, in such a position as to be acted upon by the arms of the boring wheel D, and thus prevent the steelyard carriage being carried along with the cylinder, towards the borer.   E is a long handle or tail projecting out from the end of the carriage, to enable the workman to steer and direct the carriage occasionally, and keep it in its proper place within the cylinder.   At F two planks are fixed on the carriage, beyond the end of the cylinder, to form a continuation of the road for the wheels of the steelyard carriage : these planks sustain the wheels when the borer has nearly passed through the cylinder.   The above sketch is proportioned for a cylinder of 36 inches diameter, but it is obvious that by raising or lowering the iron standard *f,* and altering the fulcrum *d* of the lever, and adjusting the balance weight *e,* the same apparatus will serve different sizes.

In his remarks on this scheme, Mr. Smeaton says, " It has been the opinion of many mechanics, that cylinders can never be truly bored, except in a vertical position ; and if I had not conceived this plan, by which the cylinder is put in the same condition as if it were upright, I should have made a vertical boring mill, although that would have rendered the movements so much more complex, as to have greatly retarded the operation."

It does not appear that Mr. Smeaton's steelyard carriage obtained the favour of the workmen at Carron, or came into use at first ; but some years afterwards, the manager requested Mr. Smeaton to consider, whether it was possible, with ease and accuracy, to take a circular cut quite round a cylinder in the boring ; because the method then practised of quartering the cylinder, did not always form the inside perfectly circular and cylindrical.

Mr. Smeaton, in answer, recommended his former scheme as the best he could devise, to enable a strong built machine to cut all round ; but it appears that the power of the mill, and the means of fastening the cylinder on its carriage, did not prove adequate for boring the largest cylinders at one operation.   To diminish the resistance of boring, he proposed to use only one cutter instead of six, and to fix two smooth pieces of steel, without cutting edges, into those two notches of the boring wheel, which form a triangle with the cutter ; and also to fill up the other three notches with wedges of hard wood, which being driven endways, as they wore away, would always continue to fill the cylinder very exactly, and as tight as the motion would permit.   " The angle that the edge of the cutter makes with the surface of the cylinder, should be such as only just to take a proper cut, and as little of keen cutting as possible ; the cutting edge to be ground very true, and set smooth upon an oil stone ; the edge of the cutter should not stand parallel to the axis of the cylinder, but a little inclined thereto, so as to form a portion of a very long threaded screw, the obliquity being in that direction by which the motion of the cutter will tend to keep the borer back, instead of drawing it into the work. As this method of boring a rough cylinder with one cutter would be tedious, it will be best to prepare the cylinder by boring with six cutters in the usual manner, at four operations, and then to bore it through, for a fifth time, with one cutter, on the system above described ; and with the aid of the steelyard carriage, and good tackle, used with care, I expect that you may take off a clean shaving, and leave a surface quite smooth, and even polished."

It does not appear that Mr. Smeaton's views of a complete boring machine were realized, till Mr. Wilkinson invented a new machine, which is now in universal use, for boring steam-engine cylinders, and which will be described in its proper place.

APPLICATION OF THE ATMOSPHERIC ENGINE TO RAISE WATER FOR WORKING MILLS, BY WATER-WHEELS.

Mr. Savery, in his patent of 1698, and in his Miner's Friend, in 1702, describes the application of his engine, to supply falls of water, for the use of mills, of every form and species, especially where coals are cheap; but this proposal was never acted upon, until after Newcomen's engine had been brought into general use, for draining mines, and supplying towns with water.

About 1752, a fire-engine was used by Mr. Champion, of Bristol, for the first mover in an extensive brass battery work, which he established in that neighbourhood. There were several overshot water-wheels, worked by water, which was pumped up into a large reservoir by the fire-engine, and after flowing on the wheels, to give them motion, it was again raised by the engine from the pool beneath the wheels, into the upper reservoir; so that the same water circulated continually. This work was carried on for about 20 years, but it was then given up, on account of the expense of the coals consumed by the engine.

The manufactories which employed mechanical power and mills, in the early part of the last century, were small, and very few in number, compared with those of the present day; some had horse machines, but most of them were situated upon falls of water, which afforded as much power as they required for a long time after their first establishment; but in time, by the progressive increase of business, manufacturers generally felt a want of greater power.

About 1754, Mr. Smeaton made his improvements upon the mode of applying the natural powers of wind and water, to mills, whereby the performance of almost all the mills then existing, could be doubled with the same supplies of water. He had an extensive practice in directing the construction of such mills during 30 years; and the numerous specimens which he left in all parts of the kingdom, were copied by others in their respective neighbourhoods, till the improved methods of constructing water-mills became general, and was attended with great benefit to trade.

In some instances, where water was scarce in summer, horse machines were constructed to work the mill in dry seasons; and Mr. Smeaton once made a horse machine to return or throw back the water for a mill, in short-water times.

The natural waterfalls being thus employed to advantage, the necessity of employing steam power for mills was not felt until some years afterwards, when manufactories had become greatly extended beyond their former scale; and having no other resource, fire-engines were then erected to return the water, and make up for the deficiency of the natural supplies.

This application was first made, about the year 1760 or 1765, at the great iron-works at Colebrook Dale, or at Carron, where several water-wheels being required for the different operations of their manufacture, they were all worked by the water of a river; but in the summer months, the supply necessarily became short, so as to be insufficient to work them all. It is of the greatest importance in the iron manufacture, to keep the furnaces blown continually without intermission, that the heat may never be abated, and accordingly the first returning engines were erected in order to work the blowing machines for the furnaces ; but the application thus begun from necessity being found to answer very well, the use of engines was afterwards extended for other purposes.

It was not until some time after the returning engines were brought into use to assist mills which had a natural water-fall, that any mills were erected with an entire dependence upon the supply from a fire-engine.   The first attempt of this kind appears to have been made by a Mr. Oxley, in 1762. He constructed a machine to draw coals out of a coal-pit, at Hartley Colliery, near Seaton Delaval in Northumberland : it was originally intended to turn the machine with a continuous circular motion, to be derived from the direct action of the lever of the fire-engine ; but that plan not being found to answer, the engine was afterwards altered to raise water, to supply a water-wheel, by which the machine was worked.   This engine continued to work for some years ; and though it was at length abandoned on account of its defective construction, it proved the practicability of the method, and gave rise to several other attempts to draw coals by water-wheels supplied by a fire-engine ; but these first trials proved abortive, from want of skill in the several projectors, who were the common engine-makers for the collieries, and who were equally ignorant of the principles of the proper application of fire to the engine, and of water to the water-wheel.

In 1776, Mr. Smeaton constructed a small oil-mill at Hull, to be worked by the spare water from the engine for the water-works (see p. 257), and the next year he erected a water-wheel and machine for drawing coals, with a fire-engine to supply it with water ; this was perhaps the second instance of a steam-engine being constructed expressly for the purpose of supplying a water-wheel, or at least which answered the purpose so completely as to become an established plan for real use.

An Atmospheric Engine to raise water, to supply a Water-wheel and Machine for drawing up Coals out of a Coal-mine, designed by Mr. Smeaton, 1777.

Mr. Smeaton had constructed a machine in 1774, to draw coals out of a pit 100 yards deep, by the power of an overshot water-wheel, 37 feet diameter, instead of by horses, as had been the common practice at most collieries. This machine was erected on the estate of Sir Roger Newdigate, at Griff in Warwickshire, upon the same colliery where the first atmospheric engine was erected by Mr. Newcomen in 1712.   The water-wheel was 1 foot wide, and the buckets 10 inches deep ; but it drew up a corf or basket of coals, weighing 5 cwt. from the bottom of the pit, 50 fathoms deep, in a minute and a half.

The success of this machine rendered it desirable to use the same means of drawing coals, in other situations where a natural fall of water could not be obtained.   Mr. S. therefore proposed to construct a fire-engine, to raise the water into a reservoir, from which it should flow into the buckets of an overshot water-wheel ; and this by wheelwork would give motion to a barrel, by which the ropes could be wound up, to draw the baskets of coal out of the mine.   This plan had been practised before at Newcastle, but owing to defective proportions in the engine, and in the machine, such very inadequate effects were obtained, that the method was not made to answer so well as drawing by horses, and had been abandoned.

The first of these fire-engine coal-machines was set up at Long Benton colliery in 1777, to draw coals from a pit 82 fathoms deep.   The cylinder of the

Q Q

engine was 26 inches diameter, and the piston made $5\frac{2}{3}$ feet stroke; $13\frac{3}{4}$ strokes per minute, that is, 78 feet motion per min. : the pump was $18\frac{1}{2}$ inches diameter, and raised the water 34 feet. The weight of the column of water in the pump was 3975 lbs., raised 78 feet per minute, which is equal to $9\frac{4}{10}$ horse power. The quantity of water raised was 146 cubic feet per minute.

The injection water was supplied from the main cistern into which the pump raised the water, so that there was no separate injection pump; the aperture of the injection spout was $\frac{3}{4}$ of an inch square, and the height of the surface of the water in the cistern was $16\frac{1}{2}$ ft. above the level of the orifice (a). Allowing 6 cubic feet per minute to have been expended for the injection, it will leave 140 cubic feet per minute for the supply of the water-wheel, which was 30 feet diameter.

The construction of this engine was different from any of those previously described, because it was intended to be occasionally removed from one pit of the colliery to another, as the convenience of the workings might require. Hence there was no building, or engine-house of masonry, except the setting of the boiler, and the foundation walls or piers; but all the moving parts were supported in a wooden framing, and in place of a great lever, a wheel or large pulley was substituted; the chains from the piston rod being immediately connected with the chains for the pump-rod, were passed round the upper part of the circumference of the wheel, in the manner of a rope and pulley, so that the descent of the piston drew up the pump-bucket, and its column of water, with an equal motion to that of the piston, but in a contrary direction. By this arrangement, the pump was brought much nearer to the cylinder than in lever-engines, and as there was no lever wall between them, the whole engine was brought into less compass. (An engraving and description of this engine is given in Mr. Smeaton's Reports, Vol. II. p. 435).

The boiler was 10 feet diameter; the fire-grate 3 feet square (= 9 square feet). The chimney 19 inches square within, and the top 40 feet high above the fire-grate. The boiler was set in brickwork in the usual manner of Mr. Smeaton's engines, as before described, and formed a cubical mass of about 14 feet square, and about 15 feet high above the foundation, but the upper part of the dome of the boiler rose up above the brickwork; at one side of this cube of brickwork the chimney was placed, and also a wall was carried up on that side, to a height of 4 feet above the top of the brickwork for the boiler, to sustain one end of the timber frame which supported the upper works of the engine, the other end of the frame being placed on the top of a wall, which was built up 19 feet high from the foundation, opposite to one side of the boiler, and 9 feet distant therefrom, so as to leave that space between them for the pump, and the working gear; and the lower part of the space being sunk in the ground, formed a tank or pit for the water, which the pump was to raise.

The wood frame consisted of two isosceles triangles, fixed up in vertical planes parallel to each other, leaving a space of 5 feet clear between them; each triangle was about 23 feet base, and 23 feet perpendicular, and consequently the angle at the vertex was about 60 degrees; the base of each triangle was composed of two beams, so combined as to make a scantling of 24 inches deep, by 12 inches broad, and 27 feet long; the ends of which rested upon the tops of the two foundation walls before mentioned, and which had 23 feet clear space between them. The

(a) The aperture of the injection-cap, and the quantity allowed for injection, are both greater than in Mr. Smeaton's table, p. 183; because the height of the injection column is much less than is prescribed by that table.

axis of the pulley-wheel was sustained by the vertices of the triangles, the gudgeons of the axis being supported in brasses, fitted into cast-iron cap pieces or boxes, which were properly adapted to the upper ends of the beams which formed the sloping sides of each triangle; these beams were 12 inches square.

The two triangular frames were firmly united by several horizontal traverse pieces, framed into them, and proper tyebolts and oblique braces, so as to make a firm frame, which resembled the letter A in its elevation.

The axis of the pulley-wheel was of wood, about 24 inches diameter in the middle, and 18 inches at each end, and 5 feet long, well hooped with iron, and with a cast-iron axis 5 inches diameter passing through the centre of the wood, with gudgeons $4\frac{1}{2}$ inches diameter projecting out from the wood, each end to rest in the brasses at the top of the triangle frames; the wheel which was fixed in the middle of this axis was 10 feet diameter, the circular rim 12 inches broad on the edge, by 5 inches thick, framed with 10 arms mortised into the axis, in the manner of a waggon wheel.

The cylinder stood in a vertical position, within the space between the two triangle frames; and its bottom had strong arms, projecting on each side, from its flange, so as to reach across to the two great horizontal beams, or basis of the two triangles: and the ends of the four arms were bolted to those beams, so as to confine the cylinder very firmly down; it stood over the centre of the dome of the boiler, so that the steam-pipe, which was $4\frac{1}{2}$ inches diameter, rose up to the hemispherical cylinder bottom as usual; but the cylinder was confined down by its bottom, instead of being suspended by its middle, the basis of the triangles serving for cylinder beams.

The piston-shank was of wrought iron, $2\frac{1}{2}$ inches square, connected at the upper end with two parallel chains, which applied upon the circumference of the wheel; these two chains were united by two iron bars, or long curved links, to a similar pair of chains to suspend the pump rod, which was also of iron, and went down into the pump barrel, the bucket being affixed to the lower end. The two long links which connected the two pairs of chains, were curved to fit the circumference of the wheel, and were fastened thereto to prevent slipping.

The upright column of the pump was 35 feet high, composed of 5 lengths, joined by flanges and screw bolts. The lowest piece, called the wind-bore, stood with its flat bottom on a large foundation stone, at the bottom of the pit or tank, which contained about $2\frac{1}{2}$ feet depth of water; and round the lower end were 8 holes or upright slits, $3\frac{1}{2}$ inches wide by 8 inches high, to admit the water freely, but to prevent the entrance of extraneous matters: the wind-bore was 18 inches diameter within, but near the upper part, which was united to the lower end of the working barrel, it was enlarged to form a conical seating, into which the lower clack or fixed valve was jambed, and fastened like a plug; and immediately above this conical seat was an opening 20 inches square in one side, to take the clack in or out when necessary: the opening being closed by a strong iron door, screwed on by 8 bolts, passing through a flange projecting all round the opening.

The working barrel, which was joined to the upper end of the wind-bore, was $18\frac{1}{2}$ inches diameter, and $6\frac{1}{2}$ feet long, truly bored within; and immediately above it was a short length, containing an opening 20 inches square, to give access to the bucket, or to take it out when necessary, and with a door to screw on. Lastly, above this bucket door-piece, were two lengths of pump pipes, $19\frac{1}{2}$ inches diameter, which reached up to the wooden pump-trough or spout, for conveying the water into the reservoir for the water-wheel; the pump-trough was much higher than the top of the cylinder, and reached very nearly up to the pulley wheel, being supported by the pump, which was steadied, by being fastened to one of the horizontal cross pieces between the triangle frames ; part of the pump-rod depending from the chains was of wrought iron, $2\frac{1}{2}$ inches square, but the rest, down to the bucket shank, was made of cast-iron, 4 inches square, made in three lengths of 5 feet each, united together by joints, with pins $1\frac{3}{8}$ inches diameter : the weight of these massive rods served for a counterweight, to balance, and draw up the piston.

The reservoir or trough, which received the water from the spout of the pump and conveyed it to the water-wheel, was 18 feet long, by 5 feet wide, and 2 feet 2 inches deep in water when full ; one end of it, which was next to the engine, rested upon a tall pier of brickwork, and the other end which reached over the water wheel, was supported by two upright posts, erected from the ground, one on each side of the wheel ; the end of the trough over the wheel was made deeper, by an addition of a trough 10 inches deep, beneath the bottom ; and this trough formed the spout, to deliver the water into the buckets of the wheel, and was accordingly fitted with an upright sluice, or drop shuttle, to cut off the flow of water, when the wheel was to be stopped. The bottom of this spout, which was horizontal, was 38 inches beneath the surface of the water in the reservoir when full, and was 2 inches below the top of the water-wheel, which was 30 feet diameter.

The bottom of the wheel was 2 inches above the tail water, in the lower pit or cistern, which received the water from the wheel, and from which the pump of the engine drew its water, as before mentioned ; so that the same water circulated continually, being raised up out of the pit by the pump, into one end of the top reservoir, and running out at the spout at the other end thereof, into the buckets of the wheel ; then descending therein, till it fell out into one end of the lower pit, and running to the other end thereof, it was raised again by the pump.

The upper cistern being 18 feet by 5 feet, contained 180 cubic feet of water, when 2 feet deep, and the expenditure upon the wheel being 140 cubic feet per minute, the cistern would supply the wheel for full $1\frac{1}{4}$ minutes in case the engine stopped by any accident. A large trunk or waste pipe was provided at one side of the cistern, to carry off the superfluous water, whenever it rose to 26 inches deep in the reservoir ; this waste trunk went down by the side of the pier which supported the cistern, and returned the water into the lower pit.

In addition to the drop shuttle before mentioned, there was another shuttle in the spout, which was regulated to permit the water to flow out just as fast as the engine would raise it up again ; and also as the water-wheel required, when working at a proper speed ; this height being found by experiment, the shuttle was nailed fast with that opening ; the other or drop shuttle, was in all cases drawn quite up, to allow the water to run freely, or put quite down to stop it entirely.

The water-wheel was 30 feet diameter outside, with 72 buckets ; the width inside the buckets 16 inches ; and the depth of the buckets 12 inches ; the water was poured from the spout in a horizontal direction, into the highest bucket at the top of the wheel.

The wheel had 12 arms or radii, 6 inches square at the centre, and 4 inches by 6 at the rim. The arms were affixed to a cast-iron axis, 10 feet long between the bearings, 9 inches diameter in the middle, and tapering to $7\frac{1}{2}$ at each end, with gudgeons at each end, 6 inches diameter, and 6 inches length of bearing. The water-wheel was fixed on, near one end of this axis, by means of a circular flange, or flat plate 2 inches thick, and 4 feet diameter, against which the 12 wooden arms were applied, and fastened by two bolts through each ; the axis had also two square boxes, to fasten on the two main cog-wheels; one of these boxes was near the middle of the axis, and the other was near to one end.

Each cog-wheel was $12\frac{1}{2}$ feet diameter, with 88 cogs, or teeth, projecting from the flat face of the wheel, parallel to its axis; the cogs of the two wheels were turned towards each other, and they received a trundle or lantern, of 20 staves, between them, so that it could be turned round either by one wheel, or by the other; but the diameter of the trundle being less than the space between the teeth of the two face wheels, it could not be engaged with both at the same time.

This trundle was fixed on one end of a horizontal axis of fir wood, 37 feet long, and 14 inches square, the other end of which was connected with the axis of the barrel, by which the ropes were wound up out of the pit. The gudgeon, which was fastened into that end of the long wooden axis where the trundle was fixed, was 3 inches diameter, and 5 inches length of bearing, made of wrought iron steeled on the outside, and hardened. This gudgeon was supported in an upright beam or post, which stood up between the two cog-wheels, and the beam was moveable about a centre at the lower end of it, so that by inclining it towards either cog-wheel, the trundle could be thrown into gear, with the teeth of one wheel, or of the other, and would accordingly be turned round, in one direction or in the other, so as to wind up or let down the ropes, although the water-wheel and its cog-wheels, continued always to turn the same way round, by the weight of the water which was poured into its buckets from the spout (a).

The top of the upright beam, which supported the gudgeon of the long axis, was carried up to a considerable height above the cog-wheels, into a small house or cabin which was erected over them, for a man to stand in, and by moving the top end of the upright to one side or the other, the trundle and barrel could be made to turn either way at pleasure; or by keeping it in the middle position, it would remain detached from both wheels; the top of the upright was guided in a groove, or long mortise, in a horizontal beam, and two pairs of pulley-blocks were attached to it, to give the man a purchase, to haul the beam one way or the other, in the same manner as the tiller of a great ship is moved, and when the trundle was engaged with either cog-wheel, the beam could be fastened to that side, by putting a pin through the mortise.

To enable the man to govern the motion of the machine, the great cog-wheel nearest to the water-wheel, was encompassed by a brake or gripe, similar to that for a windmill, and provided with a lever, by which the brake could be drawn tight round the wheel, and thus occasion so much friction, as to stop the motion of the water-wheel; the lever for this brake was placed in the cabin, that the same man might manage it; and he had also a handle like a bell-pull, to raise or let fall the shuttle of the water-wheel, whenever he required to stop or start the machine.

The barrel on which the ropes were wound, was placed at the end of the long horizontal axis, and connected therewith; it was shaped like a barrel, or two cones joined base to base, $5\frac{1}{4}$ feet diameter in the middle, 3 feet diameter at each end, and $10\frac{1}{2}$ feet long; the barrel had a central wooden axis 1 foot square, on which were fastened five wheels of different sizes, and on these the wooden staves of the barrel were nailed; a gudgeon was fixed into each end of the axis to support it; the gudgeon at the extreme end being wrought iron, 3 inches diameter, and $4\frac{1}{4}$ inches length of bearing; the gudgeon at the other end was cast iron, 4 inches diameter, and 6 inches long; it was strongly fixed into the wooden axis, because all

(a) This machine is described in Rees's Cyclopædia, Vol. XXXVIII. article Winding Machine, with an engraving from a drawing made by the author, from Mr. Smeaton's original plans.

the force to turn the barrel was communicated by this gudgeon, the extreme end thereof being a projecting square, which was received into a cast-iron socket, fixed fast on the extreme end of the long wooden axis.

The axis of the barrel was placed horizontal, in a direction at right angles to the axis of the water-wheel, and the gudgeons at each end of the barrel were supported by a suitable frame of timber work; the projecting square and socket, by which the motion was communicated from the long axis, was fitted with sufficient play to allow that small alteration of the direction of the long axis which took place, when the trundle at the other end of the axis was thrown from one cog-wheel to the other, and it was to render this deviation insensible, that the axis was made so long as 37 feet.

The horizontal distance from the axis of the water-wheel to the middle or largest part of the barrel was 50 feet, and the distance from the axis of the barrel to the centre of the pit of the colliery was also 50 feet; the ropes from the barrel were conducted over pulleys, in an inclined direction, to two large pulleys fixed in a frame, over the mouth of the pit, and at 20 feet high above the surface of the ground; after bending over these pulleys, the ropes descended into the pit, which was 82 fathoms perpendicular depth.

There were two separate ropes wound about the barrel, in opposite directions, in order to draw two corves, or baskets, of coals alternately, a full corfe being drawn up by one rope, whilst the empty corfe was let down by the other rope; each half of the length of the conical barrel, was adapted to wind up one of the ropes; and the end of the rope being fastened to the small end of the barrel, every successive coil of the rope wrapped round a larger part of the barrel, till it arrived at the middle or largest part; the surface of each half of the barrel was calculated to hold about 84 fathoms of that rope which wound about it. The diameter of the rope was about $1\frac{5}{8}$ inches, and each half of the barrel being $5\frac{1}{4}$ feet long ($= 63$ inches), it would receive very nearly 39 coils of the rope, supposing them to be laid in close contact side by side; the small ends of the barrel (3 feet diameter) was 9·42 feet circumference, and the largest part ($5\frac{1}{4}$ feet diameter) was 16·49 feet circumference, or 7·07 feet larger than the smallest part; as this increase was gained in 39 coils, the progressive increase of length, wound up at each successive coil, must have been ·18 of a foot.

The intention of giving the barrel a conical form, was to compensate for the unequal resistance occasioned by the weight of the two ropes; for instance, when the full corfe was at the bottom of the pit, the weight of all the 82 fathoms of rope, which was hanging down in the pit, would be an extra weight to be raised by the machine in addition to that of the corfe itself and the coals in it; because the other rope for the empty corfe, being all wound up round the barrel, its weight could not act as a counterpoise. Now, in proportion as the full corfe was drawn up from the bottom of the pit, the weight of the rope by which it was suspended would occasion less and less resistance to the motion; and also the weight of the other rope, by which the empty corfe was let down into the pit, would begin to act as a counterpoise to that resistance, and would balance a greater and greater part of the resistance, as more of that rope was let down, at the same time that the resistance itself diminished, as more of the ascending rope was wound up.

A rope, which is 5 inches circumference, weighs about 5·6 lbs. per fathom, so that 82 fathoms would have weighed nearly 460 lbs. The weight of the empty corfe was about 84 lbs. and the weight of the coals contained in it was about $6\frac{1}{4}$ cwt. or 700 lbs.

The weight to be drawn up, when the full corfe was at the bottom of the pit, was (700 + 84 + 460 =) 1244 lbs. which acted at the smallest part of the barrel, or a radius of 1½ feet, and would be equal to (1244 × 1·5 =) 1866 lbs. acting at a radius of 1 foot ; the counterbalance to this would be merely the weight of the empty corfe, 84 lbs. but as it acted on the largest part of the barrel at a radius of 2⅝ feet, it would be equal to (84 × 2·625 =) 220 lbs. acting at a radius of 1 foot, and the whole resistance to the motion of the barrel would be equal to (1866 — 220 =) 1646 lbs. acting at one foot radius.

When the full corfe was drawn up to the top of the pit, the weight would be (700 + 84 =) 784 lbs. acting at 2⅝ feet radius, or equal to (784 × 2·625 =) 2058 lbs. at one foot radius ; the counterbalance was (84 + 460 =) 544, acting at 1½ feet radius, or (544 × 1·5 =) 816 lbs. at one foot radius, and the whole resistance to the motion of the barrel would be equal to (2058 — 816 =) 1242 lbs. acting at one foot radius.

Here we see that the form of the barrel was not sufficiently conical, to produce a complete uniformity of resistance ; for it was 1646 at the beginning, and only 1242 at the end, or very nearly as 4 to 3, and therefore, if the supply of water to the water-wheel had been uniform, its motion would have accelerated considerably during the drawing up of the corfe.

This machine was adapted to stop the flow of water on the water-wheel by its own motion ; and the moment of stopping was so regulated as to bring the full corfe exactly to the top of the pit and no more : for this purpose a pinion of 15 teeth was fixed on the extreme end of the gudgeon of the long axis, at that end where the trundle was fixed ; this pinion turned a wheel of 60 teeth (or 4 to 1), fixed on the end of a horizontal spindle, which was placed between the two great cog-wheels, in continuation of the direction of the long axis, but higher up ; on the other end of this spindle was a pinion of 8 teeth, to turn a wheel of 80 teeth (or 10 to 1 of the pinion), it was called the count-wheel, and was mounted on a horizontal axis, situated over the first spindle, and parallel thereto ; it is evident that the count-wheel would make one revolution in 40 turns of the rope barrel, and about 38 coils of the rope round the barrel was sufficient to wind up the 82 fathoms of rope.

The count-wheel operated upon the shuttle of the water-wheel, by means of two curved pieces of iron, which were affixed to its rim, and projected from the flat face of the wheel, so as to intercept and lift up a dettent, whenever these pieces were brought beneath the dettent, by the revolving motion of the count-wheel. The dettent was connected with the lever which governed the moveable shuttle, and raised or lowered it, to regulate the flow of water upon the wheel. The projections were screwed against the face of the count-wheel, and their positions were so adjusted, that when the machine had drawn the corfe nearly up to the top of the pit, one of the projections of the count-wheel would intercept the dettent, and put down the shuttle, to stop the water at the proper moment, so that the motion of the water-wheel would expend itself, and the machine would come to rest when the corfe was just at top, without any attention on the part of the attendant. This condition could be easily ensured by regulating the position of the projections on the count-wheel, and when they were once adjusted, they would always operate correctly ; as the count-wheel turned only once for 40 turns of the barrel, the machine would draw up a corfe from the bottom of the pit before the count-wheel made one complete turn.

As the two corves ascended and descended in the same pit with contrary motions, they passed each other at some part of the depth of the pit, near the half way ; and by the swinging motion of the corves they were liable to strike together the moment of meeting, and sometimes part of the coals would be thrown out by the shock. To diminish the evil of such accidents, a smaller projection was fixed on the count-wheel, and it raised the dettent in order to lower the shuttle, and diminish the flow of water on the wheel, that the machine might move slowly whilst the corves were passing each other in the pit; but as soon as they had passed, the projection, having quitted the dettent, the shuttle would be again raised to give a full supply of water, and restore the full motion to the machine, which it continued, till the full corfe nearly reached the top of the pit.

This machine drew up a corfe, containing 20 pecks of coals, weight $6\frac{1}{4}$ cwt. or 700 lbs. from a depth of 82 fathoms, in the space of 2 minutes, which is at the rate of 30 corves per hour; and, including all stoppages, it commonly drew 17 score, or 340 corves in 12 hours, and sometimes 20 score ($=$ 400 corves) in 14 hours.

The engine was worked entirely with waste coals, which would otherwise have been left at the bottom of the pit, but in 12 minutes' time the machine drew up all the waste coals which the engine would consume in 12 hours. The engine and machine were attended by three men.

Sixteen horses had been before required to do the same work, and were only able to continue it during 12 hours per day: they required four men to look after them.

As the machine raised a weight of 700 lbs. through a height of 82 fathoms in 2 minutes, it would be at the rate of 246 feet per min. $\times$ 700 lbs. $=$ 172 200 lbs. raised 1 foot per min. $\div$ 33 000 lbs. $=$ 5·22 horse-power. The quantity of water supplied by the engine to the water-wheel was 140 cubic feet per min. raised 34 feet $=$ 4760 cubic feet, raised 1 foot, and $\div$ 528 cubic feet $=$ 9·02 horse-power; so that the friction and loss of power to move the machine, supposing no coals to have been raised by it, was 3·8 HP out of 9 HP. Part of this loss was occasioned by the loss of the fall of the water, in entering and quitting the buckets of the water-wheel, and may be estimated at 2·25 HP or $\frac{1}{4}$ of the whole power exerted by the engine; at this rate the friction of the several gudgeons and cog-wheels, and of the pulleys for the ropes, and the rigidity of these ropes, must have occasioned a resistance of 1·55 HP.

After this machine was set to work, it was found to answer so much better than the former method of drawing coals by horses, that the proprietors set up another of the same dimensions, which also gave great satisfaction; and then a third was erected: these machines were afterwards copied at other collieries near Newcastle.

Mr. Smeaton's coal machines were found to answer the original intention of taking them to pieces and removing them to other pits. One of the first machines made in 1778 was removed, in 1785, to a new pit 78 fathoms deep, from which it drew up a corfe containing 20 pecks, in one minute; the engine making 14 strokes per minute. This engine being situated on ground which was entirely undermined and laid dry by the subterranean workings, it was difficult to obtain a full supply of water for injection, the utmost quantity which could be brought to it from a distance through a great length of small lead pipe was only 90 gallons per hour; but the walls of the foundation being carefully built with cement, to avoid leakage, and the warm water from the hot-well being conveyed by an open trough into the most distant part of the pit, in which the water-wheel worked, it became so cooled by the time it ran back to the pump, and was raised up into the reservoir, as to work the engine very well (a).

(a) In this case, Mr. Smeaton recommended that the waste water from the overflow of the hot well, should be conveyed in a lead pipe down into the centre of a circular cistern, or pond, in the open

In 1785, a larger coal machine was constructed at Long Benton, on the same plan aṡ the first one, but to draw from a pit 104½ fathoms deep. The cylinder was 32 inches diameter; the piston made 6 feet stroke, and 14 strokes per minute, = 84 feet motion per min. ; the boiler was 11 feet diameter; the pump was 21 inches diameter, and the same stroke as the cylinder : it raised the water about 34 feet high; weight of the column of water 5112 lbs. This was an exertion of 13 horse-power; for 5112 lbs. ✕ 84 feet = (429 408 ÷ 33 000 =) 13 HP.

The water-wheel was 30 feet diameter, and 1½ feet wide, within the buckets ; the iron axis 9¼ inches diameter; the gudgeons 7 inches diameter, and 7 inches length of bearing (b). The main cog-wheels had 96 teeth, and the trundle 20 rounds; the rope barrel was 12 feet long, 3½ feet diameter at the ends, and 6 feet diameter in the middle.

The depth of the pit was 104½ fathoms, and a corfe containing 20 pecks of coals, weight 620 lbs., was drawn up from the bottom in 1¼ minute, including the time of striking, or detaching the full corfe at the top of the pit, and hanging on another empty one to go down again; the empty corfe at the bottom being changed for a full one in the same time. In regular working, 18 score, or 360 corves, were drawn in 12 hours, which is at the rate of 1 corfe in 2 minutes; the motion of the corfe must therefore have been at the rate of 313½ feet per minute.

This is an effect of 5·89 horse-power; for 620 lbs. ✕ 313½ feet = (194 370, ÷ 33 000 =) 5·89 HP; so that out of 13 HP exerted by the engine, nearly 6 HP was realized; and assuming that ¼ of the whole power was lost, in applying the water to the water-wheel, we shall have (¾ of 13 =) 9·75 HP applied to the machine, to produce an effect of 5·89 HP : and at that rate the friction of the machine must have been equal to 3·86 HP.

In 1782, Mr. Smeaton erected another coal machine at Walker Colliery,

---

air, on the level of the water in the wheel pit. The end of the pipe to be turned up with a spout, so as to throw up the water in the air, with a jet or fountain 8 or 10 feet high ; and by falling down again, in a shower of drops through the air, the water would be very effectually cooled ; and still farther by its exposure in the shallow circular basin of the fountain, before it returned to ḍhe wheel pit. This expedient was not found requisite in the above case, but it may be of use in others; it is probable that by this mode of cooling the water, it would become more charged with air than usual, but as it is advantageous to admit some air into the cylinder of the atmospheric engine, this would be no objection.

(b) The cast-iron axes for the water-wheels of these machines, as first executed, proved too weak, they all broke off near the large flange to which the arms of the wheel were fixed; they had been in constant use for eight years without any indications of weakness, and then both failed about the same time, and in frosty weather; they were 9 inches diameter, at the part where they broke off, and the metal was found to be unsound internally; they were replaced by other axes 10¼ inches diameter.

A similar accident happened to the axis of a larger machine at Walker Colliery ; it was replaced, and then it broke again, and was repaired with an axis 10¼ inches diameter.

The motion of the water-wheel being stopped by a brake, applied round the circumference of the great cog-wheel, the energy of the water-wheel occasioned a very great twist upon the axis, particularly when the machine was stopped suddenly; some of these axes broke when the machines were working moderately, but probably they had been previously overstrained.

At the time of these accidents between 25 and 30 large axes of cast-iron were in use for windmills and water-mills, which Mr. Smeaton had executed, without having had a single instance of a failure ; but some of them broke after having been in use five or six years : they all failed nearly at the same places, and the metal was found porous and hollow within. This defect was in consequence of the large flanges being cast in the same piece with the axis; for being 4 feet diameter and 2 inches thick, the contraction of so large a mass of metal at the moment of its congelation must have absorbed or drawn out the fluid metal from the central part of the axis, so as to render it porous and spongy. The flanges were afterwards made separately from the axis, and fastened on it by wedges, and in this way the cast-iron axes are found to be very permanent.

R R

Newcastle; it was on the same construction as that at Long Benton, but on a larger scale, and drew coals from a pit 95 fathoms deep. As this machinery was set up in the expectation of working for 18 years at the same pit, the engine was built in a house with brick walls, on the usual plan, and timber framing was avoided as much as possible.

The cylinder was 36 inches diameter, and the piston made $6\frac{1}{2}$ feet stroke, 13 strokes per minute, on an average $= 84\frac{1}{2}$ feet motion per minute. The pump was 24 inches diameter, and the same length of stroke as the cylinder; it lifted the water 34 feet high; the weight of the column of water was 6677 lbs. which it raised at the rate of $84\frac{1}{2}$ feet per minute, which is an exertion of 17·1 horse-power.

The boiler was 12 feet diameter, and 12 feet deep, the other dimensions of the engine were according to Mr. Smeaton's table (p. 183). The overshot water-wheel was 30 feet diameter, $1\frac{1}{2}$ feet broad, with 72 buckets, it was fixed upon a cast-iron axis, the same as the former machine at Long Benton. The great face wheels upon the axis had 108 cogs, $5\frac{1}{4}$ inches pitch; they worked into an iron pinion of 18 teeth, $2\frac{1}{2}$ feet diameter; the rope barrel, which was connected with the axis of this pinion, was 13 feet long, $5\frac{1}{4}$ feet diameter in the middle, and $2\frac{1}{4}$ feet diameter at each end; the diameter of the rope was $1\frac{3}{4}$ inches, or $5\frac{1}{2}$ inches circumference.

This machine drew 43 corves per hour, each containing 22 pecks of coals, or 682 lbs. weight, from a depth of 95 fathoms; the motion of the corfe was therefore $408\frac{1}{2}$ feet per minute. The useful effect was 8·44 horse-power, out of 17·1 HP exerted.

The consumption of coals by the engine was usually about 36 cwt. in 12 hours, or 360 lbs. per hour, that is 4 bushels; consequently a bushel would have lasted 15 minutes, and in that time the engine must have produced an effect of (6677 lbs. $\times$ $84\frac{1}{2}$ feet $\times$ 15 min. $=$) 8 463 097 lbs. of water raised 1 foot high; that is, 8·46 millions pounds weight raised 1 foot high by the consumption of 1 bushel, or 84 lbs. weight, of Newcastle coals.

Or if we consider the useful mechanical effect produced by the coal machine during 15 minutes, it will be (682 lbs. weight, $\times$ 570 feet depth, $\times$ $10\frac{3}{4}$ corves $=$) 4 178 955 of coals raised 1 foot; that is, 4·18 millions pounds weight raised 1 foot high, by the consumption of 1 bushel of coals (a).

Conclusion respecting Newcomen's Engine.

We may now conclude our account of Newcomen's steam-engine, having traced its progress, from its first origin, through its various stages of perfection, and its different useful applications, until the period when that kind of steam-engine began to be superseded by the improved form given to it by Mr. Watt.

The atmospheric-engine has been farther improved since Mr. Smeaton's time, and has been rendered applicable to many other purposes, in addition to those already described; but as all these improvements have been derived from Mr. Watt's inventions, we will reserve our remarks upon them to follow the account of his engine.

(a) A second machine was afterwards constructed at Walker Colliery, to draw coals from a pit 85 fathoms deep. The engine was the same size as the former, but the pump was $22\frac{3}{4}$ inches diam.; the piston made a stroke of 6 feet 10 inches, and 13 strokes per minute $=$ 89 feet motion per minute. This machine drew 50 corves per hour, each containing 22 pecks of coals, weight 682 lbs.

The great obstacle to the more extensive use of Newcomen's engine was the expense of fuel. An engine having a cylinder of 48 inches in diameter, and making 12 strokes per minute, of 7 feet each, would be equal to 36 horse-power, according to Mr. Smeaton's table, and it would consume $6\frac{1}{2}$ bushels of good Newcastle coals per hour, or, according to the statement, p. 181, it would burn 7 bushels per hour at the least; this is at the rate of 33 chaldrons per week, or near 1700 chaldrons per annum, supposing it to work night and day without intermission.

This is taking the consumption according to an experiment made on the engine, when in its very best condition; but taking the average of circumstances, such an engine might be expected to burn full one-third more, or $9\frac{1}{3}$ bushels per hour $= 44$ chaldrons per week $= 2267$ chaldrons per annum.

This circumstance limited the use of this kind of engine very much. To draw water from coal-pits, where they can be worked with unsaleable small coals, they are still universally employed; and they answered very well for draining valuable mines; or for supplying a great and wealthy city with water; also for some other purposes, where a great expense could be borne; but in a great number of cases, to which the unlimited powers of steam-engines are now applied with advantage, the expense of fuel would have precluded the use of them, if a more economical system than Newcomen's had not been invented.

The attention of engineers was much directed to every thing that could promise a diminution of this expense. Every one had his particular plan for the construction of his furnace, and some were undoubtedly more successful than others; but nothing of importance was done till Mr. Watt established his great invention about 1774. Mr. Smeaton's improvements on Newcomen's engine consisted only in proportioning its parts, but without altering any thing in its principle: it was still Newcomen's, though perfected.

In 1759, Mr. James Brindley, the celebrated engineer who directed the execution of the Duke of Bridgewater's Canal, obtained a patent for improvements in the structure of the fire-engine. The boiler he proposed to be made of wood and stone, with a stove or fire-place of cast-iron, to be placed withinside of it, and surrounded on all sides by water, so as to give heat to the water. The chimney was to be an iron pipe or tube, also immersed in the water of the boiler; this plan he expected would save a considerable portion of fuel. The feeding-pipe for supplying the boiler with water was to be fitted with a valve, and adapted to be opened and shut by a buoy, floating upon the surface of the water in the boiler, so as to keep it constantly full, up to the proper level, without any care on the part of the engine-man. The chains for the arches of the great lever were to be made of wood, and the pumps were also to be made of wooden staves hooped together. These are all the improvements mentioned in the specification of his patent: but in the Biographia Britannica we are informed, that, in 1756, Mr. Brindley undertook to erect a fire-engine, near Newcastle-under-Line, according to his scheme; but it does not appear that he established any real improvement; and he is stated to have been discouraged by the obstacles which were thrown in his way (a).

The greatest cause of the loss of heat, and waste of fuel in Newcomen's engine, is from the condensation of the steam, when it is admitted to fill the cold cylinder

(a) In the Philosophical Transactions for 1741, a proposal is recorded for a new system of raising steam to supply engines; the boiler was to be a metallic vessel placed over a furnace, and heated nearly red-hot; and water was to be thrown into it, by small jets or streams, so as to be dispersed in every part, and it was expected that the water would be instantly converted into steam. The plan was never acted upon.

during the returning stroke, as before explained (p. 143). The quantity is very considerable, and is the greatest objection to this form of the steam-engine.

An attentive observation to the action of an engine will show that there is a waste of steam, but not the quantity which is condensed. The moment the regulator is opened, when the piston is at the bottom of the stroke, the steam may be perceived to issue from the snifting-valve with a strong puff, because the steam is more elastic than the common air, by about one pound per square inch; but as the piston rises, this snift ceases, and no more steam will issue during the whole rise of the piston. It has been calculated, in the case of the engine of 52-inch cylinder (see p. 174), that the loss of steam by condensation is very nearly $\frac{3}{8}$ of the whole quantity of steam; and consequently that $\frac{3}{8}$ of all the fuel is wasted.

The condensation being in proportion to the extent of internal surface to which the steam is exposed, it is obvious that the loss of steam, in proportion to the quantity which must be produced, will be less in large cylinders than in small ones. Here we see the reason for Mr. Smeaton's rule (p. 186), of making the proportion, that the surface of the cylinder bears to its capacity, the ground-work for the calculation for the quantity of coals.

As the 52-inch cylinder is above the medium size, we may safely take the loss of steam at one half of the whole; therefore, in the common atmospheric fire-engines, when they are loaded to seven or eight pounds upon the square inch, and are of a middle size, the quantity of steam which is condensed in restoring to the cylinder that heat of which it had been deprived by the previous injection of cold water, is equal to the full contents of the cylinder, besides what it really required to fill it; so that a quantity of steam equal to twice the content of the cylinder, is required to make it raise a column of water, equal in weight to between seven and eight pounds for each square inch of the piston.

Or, to state it more simply, the condensation of a cubic foot of steam will leave such a vacuous space, that the atmospheric air, in returning to fill it up, will exert a sufficient force to raise a cubic foot of water to a height of about seven or eight feet; besides overcoming the friction of the engine, and the resistance of the water to motion.

For instance, the engine at Long Benton expended 62·95 cubic feet of steam per minute for every horse-power of the useful mechanical effect produced by it (see p. 174). A horse-power is 528 cubic feet of water raised 1 foot high per minute; therefore (528 CF ÷ 62·95 =) 8·39 cubic feet of water, were raised 1 foot high, by each cubic foot of steam expended; or, in other words, 1 cubic foot of steam raised 1 cubic foot of water 8·39 feet high.

As this engine was on a large scale, and well constructed, its performance was greater than most others; and we may safely take between 7 and 8 feet for the height to which steam, when applied in the best manner in Newcomen's engine, will in practice raise its own bulk of water.

CHAPTER V.

Mr. James Watt's Steam-Engine, 1769.

THE principle of this valuable invention will be best explained by a statement of the manner in which it originated in the mind of the ingenious inventor. The steps by which it was brought to perfection are highly interesting, as an instance of very acute philosophical research, terminating in an important practical result (a).

(a) Mr. Cleland, in his History of the Rise and Progress of the City of Glasgow, 1820, has given the following memoirs of Mr. Watt.

James Watt, Esq. was born in Greenock, on 19th Jan. 1736. His father was a respectable blockmaker and ship-chandler in that town. Mr. Watt came to Glasgow in 1752, and resided with his uncle. From an early period, and during the time of receiving his education, Mr. Watt displayed a superior genius for mechanics. In 1754 he went to London, and bound himself apprentice for three years to the profession of a philosophical instrument maker, and in 1738 he returned to Glasgow, and commenced that business on his own account. As the art was then but little known in Scotland, the professors, from a desire to patronize it, requested his acceptance of the use of three apartments within the college, for carrying on his business; shortly after this, Mr. Watt opened a shop in Glasgow for the sale of his instruments. He also amused himself by making musical instruments, and made some improvements in the organ.

In 1763, Mr. John Anderson, professor of natural philosophy in the university, sent to Mr. Watt's workshop, for repair, a small working model of a steam-engine for raising water. In contemplating the principles of this machine with a view to make it work, Mr. Watt thought it capable of improvement, and soon afterwards began a larger model of an improved steam-engine, which he fitted up at the Delft-house in Glasgow: his experiments were not however completely successful, and the machine was laid aside.

Some time after this, he removed to another part of Glasgow, but in 1768 he gave up shop, and the next year commenced business as a civil engineer; he surveyed the river Clyde, and gave an elaborate and luminous report upon its improvements. He was also employed in making a survey of the Monkland Canal. Although busily engaged, he never abandoned his grand object, the improvement of the steam-engine.

Having been accidentally employed as a civil engineer by Dr. Roebuck, of Carron Iron-works, who at that time rented the extensive coal-works at Kinneal, from the Duchess of Hamilton, the Doctor proposed to form a copartnership with him, and to obtain a patent for a steam-engine, on the principle of the model fitted up at the Delft-house. Accordingly, in the following year, Mr. Watt made an engine on a larger scale, with an 18-inch cylinder, and fitted it up in the offices of Kinneal House; this machine at the first offset exceeded his most sanguine expectation, and the patent was obtained in 1769.

The Doctor's affairs, however, soon became embarrassed, and the copartnership was abandoned; but the fame of Mr. Watt's experiments having reached England, he formed a beneficial and happy connexion with Mr. Boulton of Birmingham, who purchased Dr. Roebuck's interest in the patent in 1773, and in the succeeding year, Mr. Watt having sent his first engine to Soho, they obtained an act of parliament to extend the term of the patent, and they then entered with the greatest spirit upon the manufacture of steam-engines of all dimensions.

This concern had the greatest success, and became very lucrative; their engines were continually improved by new inventions of Mr. Watt's, and his application of steam power to turn mills, which he completed about the year 1784, was very successful. He continued to direct the works at Soho with increasing reputation until the year 1800, when the patent expired: he then withdrew from the partnership, and from public business, and spent the remainder of his useful life in retirement.

Independently of his attainments in mechanics, Mr. Watt was a man of great general information. He read much, and recollected what he read; and although his quickness of apprehension was proverbial, he was far from assuming a superiority over men of inferior genius. He was curious as an antiquarian, and well versed in metaphysics, chemistry, and architecture, quite at home in music, and familiar with most of the modern languages.

Origin of Mr. Watt's Invention.

In 1763, Mr. James Watt was a maker of mathematical instruments at Glasgow in Scotland. Being a man of a truly philosophical mind, and conversant with all branches of science, he was in the habit of associating with the most eminent men of that time in Scotland: among these may be named Dr. Robison, then a young man, but afterwards the celebrated professor of philosophy at Edinburgh; Dr. Roebuck, who first began the Carron Iron-works; Dr. Black, the chemical professor, and others.

Mr. Watt first turned his attention to the powers of steam in 1759, in consequence of some speculations by Dr. Robison for propelling wheel carriages by that agent, but nothing was tried. Again, in 1762, Mr. Watt tried some experiments on the force of very hot steam confined in a Papin's digester, and he constructed a simple model, to obtain motion from that power; with this he showed the practicability of what is now called the high pressure engine, but he did not pursue it, on account of the danger of working with such compressed steam as it required.

In 1763, Mr. Watt undertook to repair a small model of Newcomen's steam-engine belonging to the university of Glasgow: the cylinder was only two inches diameter, and six inches stroke. After it was put in complete order, he found that the boiler could not supply it with steam, though much larger in proportion than those of real engines; and to enable it to work, he found it necessary to lower the column of water in the pump, so as to reduce the load on the piston very much below the usual standard for real engines.

In considering the cause of this very great inferiority in the performance of the model, he concluded that it was owing to the small size of the cylinder, and that, being made of brass, it conducted the heat away from the steam, more readily than the large cylinders of cast-iron; for he observed that the small cylinder was so heated, when the steam was admitted into it, that it could not be touched by the hand; but all the heat thus communicated to it, contributed nothing to the performance of the engine, for before a vacuum could be made, the cylinder required to be cooled by the injection, and was then to be heated again by the re-entrance of the steam: this could not happen without the heat was abstracted from the steam, occasioning the condensation and waste of a considerable portion.

Mr. Watt's first attempt to improve the engine was by employing a wooden cylinder, which would transmit the heat more slowly than metal; he made a model with a cylinder of wood, soaked in linseed oil, and baked to dryness. Many

Mr. Watt was elected a member of the Royal Society of Edinburgh in 1784, of the Royal Society of London in 1785, and of the Batavian Society in 1787. In 1806, the honorary degree of Doctor of Laws was conferred on him by the university of the city of Glasgow, and in 1808 he was elected a member of the National Institute of France. At his last visit to Glasgow in 1817 he was full of intellectual vigour, and his conversation cheerful and animated. Since that time, at the age of 83, he applied himself, with all the ardour of youth, to the invention of a machine for mechanically constructing all sorts of seulpture and statuary, and distributed among his friends some of its earliest performances.

This truly great man, who had long been one of his country's greatest ornaments, terminated a lengthened and honourable life on the 25th August, 1819, at his seat at Heathfield, in the neighbourhood of Birmingham, and was buried at Handsworth.

Mr. Watt was twice married; first to Miss Millar, daughter of the chief magistrate of Calton, and latterly to Miss M'Gregor, daughter of James M'Gregor, Esq. an eminent merchant of Glasgow. He left one son, who had long joined him in his business, and two grand-daughters, the Misses Miller of Glasgow.

experiments were made with this, and it was found that by throwing in a small proportion of injection water, much less steam was required; but in that case the force of the engine was greatly impaired, and when more injection water was allowed, with a view of gaining a better vacuum, a very disproportionate waste of steam ensued.

The fact of water boiling in a vacuum, at a low temperature, which had been discovered by Dr. Cullen, was about this time communicated to Mr. Watt, and seemed to offer some explanation of the circumstances of his little engine; for when the heat was kept in by the wooden cylinder, the injection water might become hot enough to boil in the vacuum, and thus afford steam which would in part resist the pressure of the atmosphere. The temperatures at which water boils in vacuo being unknown, or the progression the boiling temperatures observe under different pressures, Mr. Watt began a series of experiments on that subject; but not knowing any simple method of trying the elasticity of steam, when less than that of the atmosphere, he contented himself for the present with trying the elasticity of steam when confined in a Papin's digester, and heated to different temperatures. The results of these experiments he laid down in a series, forming a curve, of which the abscissæ represented the temperatures, and the ordinates the elasticities; the law of increase being thus represented by the curve, he continued it so as to represent the elasticities of steam at lower temperatures than the range of his experiments.

With the knowledge thus obtained he saw what was going on in the cylinder; for it became evident that the injection water thrown into the cylinder to condense the steam, became hot, and being in a vessel exhausted of air, produced a steam or vapour, which in part resisted the pressure of the atmosphere upon the piston, and lessened the power of the engine. This might be remedied by throwing in so much water as would cool the whole vessel below the point at which water boils in vacuo; but that would increase the first-mentioned inconvenience, viz. the destruction of steam, which unavoidably happens upon attempting to fill a cold cylinder with that fluid.

Others, who had constructed steam-engines, had found, that as they rendered the exhaustion more perfect, by making the cylinder colder, they increased the consumption of steam in a greater proportion than they gained power; and though it appears they were ignorant of the cause, they were so sensible of the effect, that they contented themselves with causing the engine to raise a load equal to seven pounds per square inch of the area of the piston; whereas the pressure of the atmosphere would have raised much more, if the cylinder had been perfectly exhausted.

Mr. Watt was then led to inquire what portion of the steam was wasted in giving heat to a cold cylinder, in any particular case; but so very few experiments had been made, even upon the most essential part of the subject, that the real volume of water, when converted into steam of a given temperature, remained unknown, until he determined it by new experiments in the year 1764. The opinions which had been entertained concerning the volume of steam before that time, were exceedingly beyond the truth, which could by no means be deduced from the very inaccurate observations which had been made by Mr. Beighton; and the more exact experiments published by Mr. Payne, in the Philosophical Transactions, were not much known or credited.

Mr. Watt's method of determining the volume of a given weight of steam was, to fill a thin Florence flask with steam of the same elasticity as the atmosphere, and after weighing it accurately, it was filled with water, and weighed again, and then the empty flask was weighed by itself. By comparing the results of the three weighings, he deduced that the weight of steam, equal to the atmosphere in its elasticicity, is about one eighteen-hundredth part of the weight of an equal volume of cold water.

By measuring the quantity of water which was evaporated from the boiler of his model, he was now enabled to calculate the quantity of steam which was supplied to the cylinder. Accordingly he constructed a boiler, which showed by inspection what quantity of water was evaporated from it in a given time, and considering this water to have formed eighteen hundred times its volume of steam, he found that, at every stroke, the cylinder required three or four times as much steam as would fill it, and that the rest was lost, in the condensation occasioned by heating the cold cylinder.

He then proceeded to measure the quantity of cold water requisite for injection into the cylinder of his model, so as to compare it with the quantity of steam which was condensed by it, but finding the quantity of cold water to be very great, compared with the quantity of water in the steam, he was greatly struck with the considerable heat, which so much cold injection water must have received from a very minute quantity of water in the state of steam. Suspecting some mistake in his proportions, he made a direct experiment on the heat of steam, by conveying steam through a small pipe, the end of which was turned down, and immersed in a glass jar containing cold water at 52 degrees of temperature. The steam was therefore mixed with, and condensed in, that water, which received all the heat of the steam, till it became boiling hot, and could condense no more : the water in the jar was then found to have gained about one-sixth part of its weight, by the addition of the condensed steam ; whence it appeared that one pound of water, in the state of steam, can heat six pounds of water from 52 deg. to 212 deg.

Mr. Watt had also remarked, that when a quantity of water is heated several degrees above the boiling point in a close digester, and a hole is opened, the steam rushes out with great violence, and in three or four seconds, the heat of the remaining water is reduced to the boiling temperature. If the steam be condensed, the whole of it will afford but a few drops of water ; yet this small quantity, in the state of steam, can carry off with it all the excess of heat from the water of the digester. From all this it became evident, that an immense quantity of heat is contained in steam.

On mentioning these circumstances to his friend Dr. Black, he was for the first time instructed in the principle recently discovered by that distinguished philosopher, that heat combines with bodies, so as to become dormant or concealed in them, and that this concealed, or, as he called it, latent heat, is the cause of fluidity and of elasticity.

For instance, water boiling in an open vessel, will be of the temperature of 212 deg., and the steam which rises from it will be also 212 deg., and yet a pound of that steam, contains as much more heat, than a pound of the boiling water, as would heat 960 pounds of water 1 deg. say from 211 deg. to 212 deg. And it is this great quantity of internal fire, or concealed heat, which gives the elasticity to steam without augmenting its temperature. Consequently, to condense a pound of steam at 212 deg. and reduce it to a pound of water at 212 deg. as much cold water must be applied, as will absorb all the latent heat of the steam.

Mr. Watt having now become intimate with the principal facts relative to steam, viz. its corresponding temperature, elasticity, weight, and concealed heat, he was enabled to reason correctly on all the circumstances affecting the performance of an engine ; and it became obvious that if so great a quantity of heat is contained in a certain quantity of steam, the economical use of the steam was of more importance to the improvement of the engine, than the construction of the boiler and furnace, which had been the main object of former efforts to improve the engine.   The application of the steam had been neglected after it was first settled by Mr. Beighton in 1719 ; and those projects for improvements, which are recorded in the Philosophical Transactions, and in the patent offices, show how very imperfectly the theory of the steam-engine was understood.

It is worthy of remark, that Mr. Beighton had noticed the fact of the great heat of steam in 1719 ; for he says, " out of a cylinder of an engine 32 inches diameter, there came a gallon of water at every stroke, and it is surprising how that steam, which is made of about 3 cubic inches of water, should heat one gallon of cold water, so as to make it come out scalding hot as it does, and the cylinder and all its parts is but warm when the piston is down."   Desaguliers' Exper. Phil. vol. ii.   In Beighton's time, science was not sufficiently advanced to afford an explanation of this circumstance, but Mr. Watt having the truth unfolded to him, was placed in a new field for the exercise of his inventive talents.

He had now arrived at the conclusion, that to make the best use of the steam, the cylinder must in all cases be maintained as hot as the steam itself, or about 212 deg.; and that, to condense the steam effectually, the water of which it is composed, should be cooled down at least to 100 deg. or lower if possible.   The means of fulfilling these conditions did not immediately present themselves, though it was obvious that such complete heating and cooling could not be performed in the same vessel.

It was not until the next year (1765) that Mr. Watt made his great invention for performing the condensation in a separate vessel from the cylinder.   He then conceived, that if a vessel, which he afterwards called the condenser, were made to communicate with the cylinder by a pipe, so as to be filled with steam at the same time as the cylinder, then an injection being thrown into that condenser, it would condense the steam therein, and cause a vacuum.   Under these circumstances, the elasticity of the steam in the cylinder, would cause it to rush into the condenser to restore the equilibrium : and by the continuance of the injection, this steam would be condensed as soon as it entered the condenser, so that the vacuum still continuing, would draw off the remaining steam from the cylinder, until none was left. In this way the vacuum would be produced in the cylinder, without any necessity for diminishing the temperature of the cylinder below the boiling point.   Having thus obtained the vacuum, to cause the descent of the piston, the subsequent reascent could be effected by cutting off the communication between the cylinder and the condenser, and admitting a fresh supply of steam from the boiler into the cylinder ; but it was not necessary to admit any fresh steam from the boiler into the condenser, because as the vacuum produced therein, still continued, it would be ready to receive and condense the steam from the cylinder, as soon as the piston arrived at the top of the cylinder, ready to make another stroke.

By this arrangement he could therefore have a constant vacuum, or empty space prepared at all times to receive the steam, and empty the cylinder, the instant the communication was opened, in the same manner as the boiler maintained a constant plenum, or supply of steam, always ready to fill the cylinder, the instant the proper communication was opened.

The first difficulty which opposed itself to this beautiful chain of reasoning was, how to continue the action, and prevent the separate condensing vessel from filling up with the injection-water, and also how to get rid of the air.   To snift by blowing steam into the condenser, in the manner of Newcomen's engine, would have caused him as great a waste of steam from condensation, as he could save by all his discovery.   He then thought of condensing without injection, simply by the application of cold water to the outside of the condenser, on Savery's first plan ; and to get rid of the small quantity of water produced by the condensation of the steam, he intended to carry a drain-pipe down from the condenser, to a depth of 34 feet, from the lower end of which, the water would run off by its own gravity. But the air, which is carried over by the steam, and which would accumulate in the condenser by degrees, could not be so easily evacuated ; a small pump must therefore be applied to draw it off, and keep the condenser empty ; and this pump could be worked by the great lever.

Another difficulty was to keep the piston tight.   In the old engine, water was always kept upon the piston, and if any leaked down into the cylinder, it did very little injury, because it went in aid of the injection-water, though it was not so cold as fresh water ; but in the new method, if any water entered into the hot cylinder, it would be converted into steam, and would impair the vacuum, and it would also cool the cylinder.   The essence of the improvement was to keep all water out of the cylinder, that it might be always dry and hot ; he therefore concluded to use wax, oil, and tallow, to keep the piston tight, instead of water.

It next occurred to him that the atmospheric air, entering into the hot cylinder, every time the piston was pressed down into it, would cool the inside of the cylinder, and occasion loss ; the most obvious remedy for this, was to make the doors and windows of the engine-house very close, and keep in the heat, that the air in it might become hot like that of a drying stove.

On further consideration it appeared, that if the house could be entirely filled with steam, instead of air, it would press down the piston equally well, and would be better than even hot air ; for if any leaked into the cylinder, it would do no harm.   Having proceeded thus far, he saw that the effect would be attained by merely enclosing the cylinder within a vessel filled with hot steam, so that it should have a steam atmosphere of its own, quite independent of the external air. This vessel would require to be only a little larger than the contained cylinder, having a close cover at top, the centre of which might be perforated, to admit the shank of the piston to pass through, and that shank being made smooth, and truly fitted to the perforation, could be stuffed round with hemp, to make a close joint without impeding the motion of the piston.

This method promised the advantage of keeping the cylinder and piston always hot, both outside and inside, and Mr. Watt also foresaw that it would be very advantageous, to employ the elastic or expansive force of the steam, to impel the piston in its descent, instead of the pressure of the atmosphere, as it would be more manageable in its intensity ; for by increasing or diminishing the heat of the steam, in which the cylinder was enclosed, its elasticity would be regulated at pleasure, and it would urge the piston with greater or lesser force as was required. The power of the new engine would therefore fall completely under the control of its attendant, and it would be quite independent of all variations in the barometric weight of the atmosphere.

The principle of the invention was now made out, and it only remained to invent the details of the mechanism to carry it into effect, and to establish by experiment, the requisite proportions of the parts.   Mr. Watt says himself, that " when once the idea of the separate condensation was started, all these improvements followed as corollaries, in quick succession ; so that, in the course of one or two days, the invention was thus far complete in my mind."

The first experiment on these new ideas was, to try the effect of the separate condenser; but before he had made the apparatus for the experiment, he resolved to condense the steam by external cold, without injection, and to extract the condensed water from his condenser, by means of the same pump as should draw off the air.

A large brass syringe, $1\frac{3}{4}$ inches diameter, and 10 inches long, was fitted with a cover, and a bottom of tin plate; and to simplify the apparatus, he made this cylinder operate in an inverted position: it had a pipe to convey steam from a small boiler, into the lower end of it, and an additional branch from the same pipe, to the upper end of the cylinder, also an eduction-pipe, to convey the steam away from the upper end, to the new condensing apparatus; a hole was drilled longitudinally up the centre of the rod of the piston, and a valve fitted at its lower end, to drain off the water which was produced by the condensed steam, on first filling the cylinder. The condenser used upon this occasion, consisted of two pipes of thin tin plate, ten or twelve inches long, and about one-sixth of an inch diameter, standing perpendicularly, and communicating at top with a short horizontal pipe of larger diameter, having an aperture on its upper side, which was shut by a valve opening externally. These upright pipes communicated at bottom, with another perpendicular pipe, of about an inch diameter, and fitted with a piston, which served for the air and water-pump, and was moved by the hand. These condensing-pipes, and the air-pump, were placed in a small cistern filled with cold water. The steam was supplied by a small boiler, and conveyed constantly by the steam-pipe to the bottom of the cylinder, and it could be occasionally admitted to the top of the cylinder through a cock in the branch-pipe; the eduction-pipe had also a cock, to intercept the passage of steam through it.

In the first instance, both cocks being open, the steam entered, and expelled the air from the cylinder, through the hollow piston rod, and from the condenser through its external valve; when all the air was cleared, and the whole apparatus filled with hot steam, the eduction cock and the steam cock were shut, and the air-pump piston drawn up, to cause a vacuum within the condenser, the external cold condensing the contained steam; the eduction cock was then opened, and the steam in the upper part of the cylinder immediately rushed away to the vacuous condenser, and being there condensed, by the cold of the surrounding water, the vacuum still continued, till all the steam was exhausted from the top of the cylinder, but the steam from the boiler, entering freely into the lower part of the cylinder, pressed the piston upwards, although a weight of 18 pounds was hung to the lower end of the piston rod. The eduction cock was then closed, and the steam cock opened, to admit the steam to the top of the cylinder, and allow the piston to descend; then the steam cock being shut, and the eduction opened, the operation was repeated, as often as was requisite. The quantity of steam expended to produce a given number of strokes, was inferred from the quantity of water evaporated from the boiler, and the weights the piston could raise, being observed, the invention was found complete, as far as regarded the saving of steam.

A large model was then constructed, with an outer cylinder or steam case to surround the cylinder, and the experiments made with it, fully verified the expectations the ingenious inventor had formed, and left no doubt of the advantages of the new method. The weights raised by the piston, showed that the vacuum in the cylinder was almost perfect; and he found, that when he used water in the boiler, purged of air, by long boiling, the weight raised was not much inferior to the whole pressure of the atmosphere on the piston. This alone was gaining a great deal; for, in the old engines, the remaining elasticity of the steam, arising from the heated injection-water, was never less than one-fourth of the atmospherical pressure, and therefore took away one-fourth of the power of the engine.

In this second model, the cylinder was placed in the usual position, with a working lever, and all the other apparatus employed in the old engines ; the invertion of the cylinder in the first trial being only an expedient to try the new principle, but would be subject to many objections in real engines (a).

Mr. Watt's First Patent, 1769.

Mr. Watt did not apply for his first patent until 1768; it bears date 5th January, 1769, and is for his new invented method of lessening the consumption of steam and fuel in fire-engines.   The specification is as follows :

My method of lessening the consumption of steam, and consequently of fuel, in fire-engines, consists of the following principles.

" First.   That vessels in which the powers of steam are to be employed to work the engine, which is called the cylinder in common fire-engines, and which I call the steam-vessel, must, during the whole time the engine is at work, be kept as hot as the steam that enters it : first, by enclosing it in a case of wood, or any other material that transmits heat slowly ; secondly, by surrounding it with steam, or other heated bodies ; and thirdly, by suffering neither water, nor any other substance colder than the steam, to enter or touch it during that time.

" Secondly.   In engines that are to be worked wholly or partially by condensation of steam, the steam is to be condensed in vessels distinct from the steam-vessels or cylinders, although occasionally communicating with them.   These vessels I call condensers ; and whilst the engines are working, these condensers ought at least to be kept as cold as the air in the neighbourhood of the engines, by application of water, or other cold bodies.

" Thirdly.   Whatever air, or other elastic vapour, is not condensed by the cold of the condenser, and may impede the working of the engine, is to be drawn out of the steam-vessels, or condensers, by means of pumps, wrought by the engines themselves, or otherwise.

" Fourthly. I intend, in many cases, to employ the expansive force of steam, to press on the pistons, or whatever may be used instead of them, in the same manner as the pressure of the atmosphere is now employed in common fire-engines. In cases where cold water cannot be had in plenty, the engines may be wrought by this force of steam only, by discharging the steam into the open air, after it has done its office.

" N. B.   This should not be understood to extend to any engine, where the water to be raised, enters the steam-vessel itself, or any vessels having an open communication with it.

" Fifthly.   Where motions round an axis are required, I make the steam-vessels in form of hollow rings, or circular channels, with proper inlets and outlets for the steam, mounted on horizontal axles, like the wheels of water-mills. Within them are placed a number of valves, that suffer any body to go round the channels in one direction only.   In these steam-vessels are placed weights, so fitted to them as entirely to fill up a part or portion of their channels, yet rendered capable of moving freely in them, by the means hereinafter mentioned or specified.   When the steam is admitted in these engines, between these weights and the valves, it acts equally on both, so as to raise the weights to one side of the wheel, and by

(a) In 1824, Messrs. Boulton and Watt made a large engine in London on the plan of Mr. Watt's first model.

the re-action on the valves, successively, to give a circular motion to the wheel ; the valves opening in the direction in which the weights are pressed, but not in the contrary one ; as the steam-vessel which moves round, it is supplied with steam from the boiler, and that which has performed its office, may either be discharged by means of condensers, or into the open air.

" Sixthly.   I intend, in some cases, to apply a degree of cold, not capable of reducing the steam to water, but of contracting it considerably, so that the engines shall be worked by the alternate expansion and contraction of the steam.

" Lastly.   Instead of using water, to render the piston, or other parts of the engines, air and steam tight, I employ oils, wax, resinous bodies, fat of animals, quicksilver, and other metals, in their fluid state."

In making this specification, Mr. Watt was obliged to define the properties of a machine which had never been tried upon a large scale, and from the general terms in which the specification is worded, it would appear, that, although he had tried a sufficient number of experiments, to assure himself of the value of his invention, and of the truth of his deductions, he was not at all perfect in the executive part of his scheme, and had not settled what might be the best form for the construction of an engine on his principles ; accordingly, his patent was merely for lessening the consumption of steam ; and he describes no machine or form of engine by which it was to be effected, but directs generally that the cylinder is to be kept as hot as the steam ; that the steam is to be condensed in vessels distinct from the cylinder ; that the air is to be discharged therefrom by pumps ; that steam is to be employed to press on the piston, in place of the pressure of the atmosphere, &c. &c.

After Mr. Watt had developed his principle, and made many engines in which it was employed to great advantage, several imitations were made, in the expectation of avoiding his patent-right ; but the very general terms of the specification, enabled him to repel his adversaries, whatever might be their pretensions. He was then assailed on the ground that his specification did not contain such plain directions, as would enable competent workmen to construct his engines, when his patent might expire.   No patent ever underwent more severe ordeals in the courts of law, than this one ; but, fortunately for the proprietors, it withstood all attacks.

About the time he obtained his patent, Mr. Watt became associated with Dr. Roebuck, who had commenced the Carron iron-works in Scotland.   They proposed establishing an extensive manufactory for such engines, under the patent ; and Mr. Watt began his first real engine, at Kinneal-house, with a cylinder of 18 inches diameter.   It was a sort of experimental engine, and was successively altered and improved, till it was brought to a considerable degree of perfection.

In the details of its construction, the greatest difficulty occurred in packing the piston, so as to be steam-tight ; because Mr. Watt's principle did not admit of water being kept upon the piston, to prevent leakage, as in the old engines. He was therefore obliged to have his cylinder very nicely bored, perfectly cylindrical, and finely polished ; and he made trial of many different soft substances for packing the piston, which would make it tight without enormous friction, and which would remain so, in a situation perfectly dry, and hot as boiling water.

It was with great difficulty that he procured a cylinder sufficiently accurate for the first trial, and it then appeared almost hopeless to expect to make large cylinders for the new engines, without an entire new system of boring them.   For any rings or irregularities in the internal surface of the cylinder, soon destroyed the dry packing by the friction, and occasioned leakage.

Fortunately for the progress of Mr. Watt's invention, the performance of his engine is not greatly diminished, by a small want of tightness in the piston.   In the atmospheric engine, if the air enters in any quantity, it impedes the working; but in the new engine, if a considerable quantity of steam escapes past the piston, during its descent, the rapidity of condensation is such, that scarcely any diminution of pressure can be observed, and the loss of fuel, by waste of steam, is the only inconvenience.

After all he could do, the defects in the cylinder, and in the fitting of the piston to it, allowed a considerable portion of steam to escape by the piston, during its descent, or if the hemp packing of the piston was rammed very tight, the friction against the inside of the cylinder was so increased, as to outweigh the saving of steam.

The cover to the cylinder, and the fitting of the piston-rod to slide through it, in a close collar of hemp, was a new contrivance in steam-engines, and required very superior workmanship, to render it effective.   In Newcomen's engine, the piston-shank was rough and square, and required no other property than strength, and the piston could be secured to it by two or four branches if requisite; but Mr. Watt was obliged to form the piston-rod smooth, straight, and cylindrical, with the same precision as the cylinder itself; and it required to be fixed exactly in the centre of the piston, and in the axis of the cylinder; the socket or stuffing box, in the cover of the cylinder, through which the rod passed, was equally difficult of execution in the first instance.

Whilst Mr. Watt was contending with the obstacles which impeded his progress, Dr. Roebuck became embarrassed, from the failure of his undertaking in the Borrowstowness coal and salt-works, and was unable to prosecute the intended manufactory of steam-engines.   In 1773, he therefore disposed of his interest in Mr. Watt's patent, to Mr. Matthew Boulton, whose establishment at Soho, near Birmingham, was already the most complete manufactory for metal work in England, and conducted with the most spirit.   A portion of the works was allotted to Mr. Watt, who soon erected a foundry, and the necessary works to carry his invention into effect, on a grand scale.

In consequence of the great loss of time, and of the expense necessary to bring the engine to perfection, Mr. Watt was unable to produce any large engines, as specimens of his invention, until 1774; and he found, from the difficulty of introducing them, that the fourteen years term of his patent was likely to pass away, before he would be reimbursed his expenses, and recompensed for his invention.   He therefore applied to parliament for an extension of the term of his patent, and being supported by the testimony of Dr. Roebuck, Mr. Boulton, and others, both as to the real merits of his invention, and as to the exertions he had used to get it into use, an act was passed in 1775 to extend the original patent of fourteen years, to twenty-five years from that date; so that, in the whole, the patent was in force more than thirty years (a).

(a) The estimation in which Mr. Watt's invention was held, will best appear by the following extracts from the preamble of this act, which is entitled:

"An Act for vesting in James Watt, engineer, his executors, administrators, and assigns, the sole use and property of certain steam-engines, commonly called fire-engines, of his invention, throughout his majesty's dominions, for a limited time.

".And whereas the said James Watt hath employed many years, and a considerable part of his fortune, in making experiments upon steam-engines, commonly called fire-engines, with a view to improve those very useful machines, by which several very considerable advantages over the common steam-engines are acquired; but upon account of the many difficulties which always arise in the execution of such large and complex machines, and of the long time requisite to make the necessary trials, he could not complete his intention before the end of the year 1774, when he finished some large

With this encouragement, which secured to Mr. Watt the recompense for his labours, he immediately commenced a partnership with Mr. Boulton ; and Messrs. Boulton and Watt carried on the manufacture of patent steam-engines at Smethwick, Soho, near Birmingham, with great success, till the final expiration of the patent in 1800 (a).

This establishment, which was then the only one of the kind, proved a real source of national wealth, at the same time that it realized independent fortunes for the ingenious proprietors.

Mr. Watt was ably seconded by his new associate, in the executive part, and in training workmen for the different departments of the manufacture of the parts at home, whilst Mr. Watt was principally employed in visiting mines and other situations where engines were wanted, to form suitable plans for each particular case, and afterwards to fix the engines, and get them to work.

They had also a corps of workmen at Soho, who were regularly instructed in the management and working of the patent engine, and one of these men was sent out with every new engine which they fixed, to work and keep it in order, until other persons could be properly instructed in that duty.

Thus aided, encouraged, and supported with ample funds, Mr. Watt made rapid progress ; and before 1778, he fixed some capital engines in Staffordshire, Shropshire, Warwickshire, and a small one at Stratford, near London.

These early specimens were doubtless very imperfect in execution, compared with what he afterwards produced ; but the economy of fuel, and complete command of power, was fully realized in all of them.

Mr. Watt had not then made those improvements in the construction of every

---

engines, as specimens of his construction, which have succeeded, so as to demonstrate the utility of the said invention.

" And whereas, in order to manufacture these engines with the necessary accuracy, and so that they may be sold at moderate prices, a considerable sum of money must be previously expended, in erecting mills and other apparatus, and as several years, and repeated proofs, will be required before any considerable part of the public can be fully convinced of the utility of the invention, and of their interest to adopt the same, the whole term granted by the said letters patent may probably elapse before the said James Watt can receive an advantage adequate to his labour and invention.

" And whereas, by furnishing mechanical powers at much less expense, and in more convenient forms than has hitherto been done, his engines may be of great utility in facilitating the operations in many great works and manufactures of this kingdom ; yet, it will not be in the power of the said James Watt to carry his invention into that complete execution which he wishes, and so as to render the same of the highest utility to the public, of which it is capable, unless the term granted by the said letters patent be prolonged, and his property in the said invention secured for such time, as may enable him to obtain an adequate recompense for his labour, time, and expense.

" To the end, therefore, that the said James Watt may be enabled and encouraged, to prosecute and complete his said invention, so that the public may reap all the advantages to be derived therefrom, in their fullest extent, it is enacted,

" That from and after the passing of this act, the sole privilege and advantage of making, constructing, and selling the said engines, hereinbefore particularly described, within the kingdom of Great Britain, and his majesty's colonies and plantations abroad, shall be, and are hereby declared to be, vested in the said James Watt, his executors, administrators, and assigns, for and during the term of twenty-five years," &c. &c.   The act was passed in 1775.

(a) Several years after the death of Mr. Boulton, Mr. Watt, at the advanced age of 74 years, wrote thus of his former associate : " Our partnership, which commenced in 1775, terminated with the exclusive privilege in the year 1800, when I retired from business, but our friendship continued undiminished to the close of his life. As a memorial due to that friendship, I avail myself of this, probably a last, public opportunity of stating, that to his friendly encouragement, to his partiality for scientific improvements, and his ready application of them to the processes of art, to his intimate knowledge of business and manufactures, and to his extended views, and liberal spirit of enterprise, must in a great measure be ascribed whatever success may have attended my exertions."   Vide Mr. Watt's Annotations on Dr. Robison's article Steam-Engine, 1814. Mr. Boulton died in 1809, in his 81st year.

part of the engine, which he afterwards introduced in the course of his practice; but in his first engines he used the same parts as those of the old atmospheric-engines, and only applied his new cylinder, and its steam case, condenser, and air-pump. The steam boiler required only half the capacity of the old one; the steam regulator valve was also new modelled (a).

In his first essays at Kinneal, Mr. Watt employed cocks, and also sliding-valves, such as the steam regulator of the old engine; but he always found them to become leaky after a short time. This is not surprising, when we consider that they were always perfectly dry, and kept continually boiling hot. He was there-fore obliged to change them all, for spindle-valves, which being truly ground and nicely fitted at first, are not found so liable to get out of order. The spindle-valves were then a new application to steam-engines, except for the safety-valves to the boilers; they were also called puppet-clacks, or button-valves.

The spindle-valve is a flat circular plate of bell-metal, with a round iron spindle passing perpendicularly through the centre of it, and projecting above and

(a) Mr. Watt and Mr. Smeaton were upon terms of intimacy, and kept up an occasional corre-spondence, in which Mr. W. communicated the progress of his invention. In one of these letters, April, 1776, Mr. Watt complains of having been " tormented with exceedingly bad health, resulting from the operation of an anxious mind, the natural consequence of staking every thing upon the cast of a die; for in that light I look upon every project, that has not received the sanction of repeated success.

" I have made considerable alterations in our engine lately, particularly in the condenser; that which I used at first, was liable to be impaired by incrustations from bad water, therefore we have substituted one which works by an injection. In pursuing this idea, I have tried several kinds, and have at last come to one, which I am not inclined to alter. It consists of a jack head pump, shut at bottom with a common clack bucket, and a valve in the cover of the pump, to discharge the air and water. The eduction steam-pipe which comes from the cylinder, communicates with this pump, both above and below the bucket, and has valves to prevent any thing from going back, from the pump to the eduction-pipe. The bucket descends by its own weight, and is raised by the engine, when the great piston descends, being hung to the outer end of the great lever; the injection is made both into the upper part of this pump, and into the eduction-pipe, and operates beyond my ideas in point of quickness and perfection.

" The size of the air-pump to our own engine at Soho, is 7 inches diameter, and 20 inches stroke, the vacuum equal to from 27 to 29 inches of mercury, or in general only about 1½ inches below the barometer. When the engine is loaded to above 11 lbs. upon the square inch, it works very well if the steam supports from 1 to 3 inches column of mercury. One cwt. that is 120 lbs. Wednesbury small coals; raises between 20 and 30,000 cubic feet of water to 20 feet high; the cylinder is 18 inches diameter, and raises 7·8 cubic feet of water each stroke, 24 feet high.

" We have now two large engines going, one about 10 miles from Birmingham, the cylinder 50 inches diameter, intended to work a 14¼ inc. working barrel, to lift water from 100 yards deep, but the pit is only sunk to 40 yards at present, they have a good deal of water, and the engine goes con-stantly. Their boiler is 12½ feet diam. and is very bad, and the cylinder is not protected from the cold air. They burn only 25 cwt. of the sweepings of an old coal-hill in 12 hours. I have never seen this engine go, so can tell you nothing more about it.

" The other engine is a 38 inch cylinder, which blows an iron furnace at New Willey, in Shrop-shire; it acts immediately to compress the air in a blowing cylinder of 72 inches diameter and 7 feet stroke; as that cylinder is very rough and unevenly bored, I am uncertain what power it exerts, but it raises a column of water 5½ feet high, in the air-chest of the water regulator, and goes 14 strokes per minute, that is, a column of water 72 inches diameter and 5¼ feet high. When I left it, there were several things unfinished; yet the quantity of fuel used seemed to be very moderate. Both these engines please even the workmen, who are all sufficiently captious. We are going on with several other large engines, one of a 58 inch cylinder in Warwickshire, and our concern wears a bu-siness-like face.

" Mr. Wilkinson has improved the art of boring cylinders; so that I promise upon a 72 inch cy-linder being not further distant from absolute truth, than the thickness of a thin sixpence in the worst part. I am labouring to improve the regulators; my scheme is, to make them acute conical valves, shut by a weight, and opened by the force of the steam. They bid fair for success, and will be tried in a few days."

below it ; this valve is very exactly fitted by grinding, into a circular seat or aperture, which forms the passage that the valve is required to open or shut, and the valve being conical on the edge, fits like a plug into the aperture of the seat, which is also conical, so as to entirely stop the passage when it is let fall ; but when the valve is raised up above, or out of its seat, a free passage is allowed on all sides of it.　The spindle of the valve is supported in two sockets, one above the aperture, and the other below, to retain it in its exact perpendicular direction, and cause it to drop exactly into its place.

Mr. Watt's First Steam-Engine, 1775 to 1778.

The annexed section of the cylinder of one of Mr. Watt's earliest engines, has been made from an imperfect sketch in the 2nd edition of the Encyclopædia Britannica.　It is supposed to be an alteration of the same engine as is represented in plate II. ; the old cylinder being removed, and replaced by a new one, on Mr. Watt's system.

The building, the great lever, and the pump-work in the mine, remain as they were ; but the boiler under the cylinder is removed, the side boiler being sufficient to supply the new cylinder, because it requires only half the steam to perform the same work as the old one.　The beams across the house, to suspend the cylinder, are also removed ; likewise the jack head pump, and its ascending pipe, together with the injection cistern, and its pipe, and all the working gear.

The vacancy left by the removal of the boiler, is filled up by building a strong wall across the house, and it is carried up to a proper height, to form a basement for the new cylinder, which is supported on its bottom, instead of being suspended by the middle.　The new cylinder EE, is made much smaller than the former one, being loaded with a column of water equal to 10 or 11 lbs. per square inch, instead of $7\frac{1}{2}$ lbs.　The cylinder is inclosed within another cylinder, called the steam-case, which is so much larger than the real cylinder, as to leave a small space all round between the two, for the admission of steam.

The real cylinder is open at the top, and is bored very truly within side ; the piston J is accurately fitted into it ; the steam-case is closed at top, by a circular cover, through the centre of which the piston rod n passes, and this rod being perfectly straight, and cylindrical, is fitted very exactly into the perforation through the centre of the cover.　On the top of the cover, a short cylindrical tube is fastened, to contain a stuffing or packing of hemp, which is crammed tight into the space between the piston rod and the internal surface of the cylindrical tube, and surrounds the rod so closely on all sides, as to prevent the passage of any steam ; to retain the hemp, a cover is applied on the top of the tube, and is fastened down by screws ; this cover is perforated in the centre, to admit the piston rod to pass through.　The piston rod is straight and well polished, so as to move freely up and down, through the stuffing box in the centre of the cover, without allowing any steam to escape.

The steam from the boiler, is conveyed by a steam pipe a, into the space between the steam-case and the cylinder ; and as the upper edge of the cylinder does not reach quite up to the cover, there is an open space all round, which admits the steam to flow into the top of the cylinder, so as to press constantly on the piston.　The cylinder may therefore be considered as included in an atmosphere of steam, in the same manner as the cylinder of Newcomen's engine is surrounded by the atmospheric air ; but the effect of the steam around the cylinder

T T

is, to preserve every part of it, at the same uniform temperature as the steam itself, and thereby avoid condensation.

At the lower part of the cylinder, a passage ff communicates with it, for the purpose of alternately conveying steam into it, and of exhausting that steam from it. The steam is admitted from the steam-case, through a valve e, called the steam or the equilibrium valve, which being opened, the steam can enter freely beneath the piston J, and then the steam will press as much beneath the piston, as above, whereby it will be placed in equilibrium, so as to be free to rise up by the action of the counter-weight. When the steam is to be exhausted from beneath the piston, the equilibrium-valve e must be shut, and another valve i, called the exhausting-valve must be opened, to form a communication be-

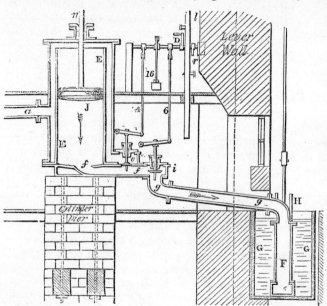

tween the passage f, at the bottom of the cylinder, and the eduction-pipe $g\ g$, which leads to the condenser F, where the steam is cooled.

The condenser F is a close vessel, made of thin copper plate, and of such a form, as to expose a very extensive surface, externally and internally; it is immersed in cold water contained in a cistern G, and as the heat can pass readily through the thin plate of which the vessel is composed, the cold water will absorb the heat from the steam which enters into the condenser. The form of the condenser was sometimes a flat vessel like a book; or several such vessels were united, each being surrounded by the cold water: in other cases, upright cylinders were used, as shown by F; and for large engines, four or five such cylinders were placed in a row, being all united together at top and at bottom, so as to form one vessel, with a great extent of external surface.

The water produced by the condensation of the steam within the condenser, is precipitated in dew against its internal surface, and trickles down to the bottom in drops. This water is extracted from the condenser by a pump H, which is called the air-pump; because it is also required to extract any air, which may gain admission into the condenser by leakage, or with the steam. The construction of the air-pump H is nearly the same as the jack-head, or injection-pump, of the atmospheric engine; and its rod is suspended by a chain from the great lever of the engine. The barrel of the pump communicates at the bottom with the condenser, through a valve which shuts towards the condenser; and there are valves in the moveable bucket of the pump, opening upwards. The operation of the pump is to extract the water and air from the condenser every time that the bucket is drawn up; and when the bucket descends, the lower valve prevents any return of that water or air into the condenser; so that, by the continual working of the air-pump, and by the cooling effect of the surrounding cold water, a perpetual vacuum is maintained within the condenser.

The two valves *e* and *i* are opened and shut alternately by the working gear, which operates nearly on the same principle as that of the old engine. The plug-beam *l* has pins fixed in it, to give motion to two handles D and *r*, which are fixed upon a horizontal axis, placed near the top of the cylinder : on this axis, two short levers are also fixed, which are connected by rods 4 and 6, with two levers which are situated immediately over the two valves *e* and *i*, and are connected with the upright stems of those valves, so as to lift them up and down, and open or shut the passages. The two short levers are so fixed on the horizontal axis of the working gear, as to open one valve and shut the other, whenever the handles D or *r* are moved by the plug, but they will not permit both valves to be open at once (*a*); 16 is a weight suspended by a rod, from a joint at the upper end of a short lever, which is fixed upon the horizontal axis, in such a position, that it inclines to one side of the perpendicular when the axis is turned one way ; but it inclines to the opposite side of the perpendicular when the axis is turned the other way ; so that the weight 16 produces the same effect as the tumbling bob, or hammer, of the old engine, viz. it gives the working gear a decided tendency to rest either one way or the other, according as it is placed by the handles, and thereby always keeps one valve open and the other shut close.

In order to give motion to the valves *e* and *i* from the outside, their spindles are brought up through the covers of the boxes which contain them, and the spindles are packed round with hemp, in the same manner as the stuffing-box of the great piston-rod, so as to prevent the passage of steam or air. The covers of the boxes are so made that they can be taken off, to give access to the valves, when requisite.

The operation of Mr. Watt's engine is as follows. Suppose the steam to be continually supplied from the boiler, through the steam-pipe *a*, to the steam-case, it will fill the space between the steam-case and the cylinder, and also the upper part of the cylinder, above the piston. The cylinder and condenser must first be cleared of the common air (*b*), which is usually done by opening both the valves *e* and *i* at once, and allowing the steam to pass from the steam-case through the valve *e*, and then by the passage *f* through the other valve *i*, and by the exhausting-pipe *g g* into the condenser F. The steam must be sufficiently accumulated in the boiler, to cause it to rush with rapidity through these passages, so as to blow forcibly through the condenser, in order to expel the contained air therefrom

(*a*) The sketch of the working gear is not a very exact representation of that which Mr. Watt used in his early engines, but will serve to give an idea of its properties. In reality, the mechanism for opening one valve, was quite independent of that for opening the other ; for there were two separate horizontal axes in the working gear: one moved by the handle D, and giving motion to the valve *e*, by the rod 4 ; the other axis moved by the handle *r*, and giving motion to the valve *i*, by the rod 6.

The object of making the movements for the two valves independent, was for the convenience of shutting both valves at once, when the engine is to be stopped, and for opening both valves at once when the parts are to be cleared of air, preparatory to starting the engine ; but in the regular course of working, the action of the two valves are simultaneous, reciprocal, and alternate, as above described ; the double mechanism of the working gear, acting just as a single axis would do, if it had been constructed according to the sketch.

(*b*) In some of Mr. Watt's first engines, a smaller air-pump was connected with the condenser, and adapted to be worked with a lever or handle, by the strength of men, in order to exhaust the air, and obtain a vacuum to start the engine; but it was afterwards found, that the air could be very easily expelled by blowing through with steam, as above described, without the trouble of exhausting by a hand-pump.

through the valves of the air-pump (*a*); and this action, which is termed blowing through, must be continued till all the air is expelled from all parts of the engine. If both the valves *e* and *i* are then shut, so as to prevent the great and direct passage of steam from the boiler through the steam-case to the condenser, the external cold water, which is continually absorbing the heat of the contained steam, will very quickly condense it all, so as to leave a vacuum in the condenser, and in the pipe *g*; we may conclude that at this stage, the cylinder E will remain full of steam from the steam-case, both above and below the piston, which is therefore in a state of equilibrium, resulting from an equality of pressure upwards and downwards.

To put the engine in motion, the handle *r* is lifted up, and the working gear turned into that position which will open the exhausting valve *i*, the equilibrium-valve remaining shut; the steam by its own elasticity then rushes with violence out of the cylinder, through the passage *f*, into the pipe *g*, and into the vacuous space of the condenser; but by coming in contact with the cold interior surface of that vessel, the steam is deprived of its heat, and is reduced into drops of warm water, which trickle down to the bottom of the condenser. The condenser is not filled up by the steam which thus flows into it, because, by the condensation of all the steam which enters, the original vacuum is maintained; consequently the steam in the lower part of the cylinder must continue to rush out into the condenser, till none is left, and there is the same vacuum beneath the piston, as in the condenser. This exhaustion is effected almost instantaneously on opening the valve *i*, and then the elastic force of that steam, which occupies the upper part of the cylinder, and presses continually downwards upon the piston, being no longer counteracted by the upward pressure of any steam beneath it, the equilibrium is destroyed, and in consequence the piston is pressed down into the cylinder, by the steam acting above it; the great lever at the same time raising the column of water in the pumps, and also drawing up the bucket of the air-pump. When the piston reaches the bottom of its course, a pin in the plug presses down the handle D, and turns the working gear in that position which will shut the exhausting-valve *i*, and open the equilibrium-valve *e*, with a simultaneous motion; the steam from the steam case is thereby admitted into the cylinder beneath the piston, to counteract that steam which is always admitted above; consequently the equilibrium of pressure is re-established, and the piston is set at liberty, so as to be drawn up by the counterweight, till it arrives at the top of the cylinder, which fills with steam as fast as it rises; the position of the two valves is then reversed by the working gear, and the steam beneath the piston is allowed to pass off to the condenser, in order to make another stroke.

In this way the motion is kept up, and at every stroke, the bucket of the air-pump draws up out of the condenser, all the water which has accumulated there from the condensed steam, and also any air which has been extricated from the water in boiling, or which has gained admission by leakage.

Several of Mr. Watt's first engines were constructed in this form, and, in some cases, where atmospheric engines were altered to this plan, the old cylinder, being inverted, served for the steam-case, or external cylinder of the new engine,

(*a*) A small valve was also provided to the condenser, to permit the air and steam to escape, during the blowing through, with less resistance than would be required to lift up the valves of the air-pump. This is called the *blow-valve*; it opens outwards from the condenser, and its action is very similar to that of the snifting-valve of the atmospheric-engine, but it is only used at first to start the engine.

the real cylinder of which required to be but little more than half the area of the old one, to produce the same effect.

In all the first engines, the condensation was performed by the external cold of the water surrounding the condenser; but in succeeding engines, he found that this mode was not sufficiently rapid in its action, and that a better effect was obtained by admitting a small jet of cold injection water into the condenser at each stroke, and employing an air-pump of sufficient dimensions to extract the injection water, as well as the condensed steam and the air. The action of the engine was so much improved by the injection, as to afford amply for the additional power requisite to work the air-pump, when so enlarged.

The idea of condensing the steam, by injecting into the eduction-pipe, occurred to Mr. Watt as early as the other kinds of condensers, and was tried in the first engine at Kinneal; but owing to its leaks, and defective workmanship, that engine made a bad vacuum, which being attributed to the air which came in with the injection water, Mr. Watt disused the injection into the condenser, until the size and expense of tubulated condensers for large engines, made him resolve to sacrifice a part of the power of the engine to convenience, and to employ larger air-pumps. For he found that the tube condenser, to have exposed sufficient surface for condensing the steam of a large engine, would have been very voluminous, and the bad water with which engines are frequently supplied, would have crusted over the external surfaces of the condenser tubes, so as to thicken their substance, and prevent that quick transmission of heat which is requisite.

In an engine at Bedworth, near Coventry, three air-pumps were used, two below, which were side by side, and worked by chains from each side of the great lever, and a third above these two, in the middle between them: this third one received the hot water lifted up by the other two; and by lessening the surface exposed to the pressure of the atmosphere, the water was extracted with greater ease. In 1778, Mr. Watt only employed two air-pumps for the largest engines, one being double the area of the other, and in succeeding engines used only one, which plan has been followed ever since; it is in fact the air-pump of Smeaton, having a cover or close lid to the barrel, which prevents the atmospheric air pressing on the bucket, and renders the working more easy.

The injection was admitted into the condenser by a short pipe, from the cold water surrounding the condenser, the pipe being stopped with a small valve, which was opened by the plug, every time the engine was required to make a stroke. There was also a cock in the pipe, to regulate the quantity of cold water which should be injected; and this cock was intended to be so much opened, as to reduce the hot water which was drawn out by the air-pump, to blood warm, or 96 degrees of Fahrenheit. This was when the engine was in excellent order, but it was never, in any case, to exceed 110 deg. and at an average would be about 102 deg. In most of Mr. Watt's early engines, the condensing cistern was placed out of doors, between the lever wall, and the great pit; and the air-pump was suspended from the outer end of the great lever, in the same position as the jack-head or injection-pump of the old engines; so that the bucket was drawn up by the descending force of the piston.

He afterwards altered this arrangement, and placed the cistern close to the pier or wall on which the cylinder is erected, and in the space between that pier and the lever wall of the engine; the air-pump was then suspended from the inner end of the great lever, and consequently its bucket was drawn up by the counterweight, which therefore required some increase.

This is now the common practice, and is an advantageous arrangement; because the eduction-pipe *g g* being shortened, the steam has less distance to go, in order to be condensed, and therefore the cylinder is more instantaneously exhausted; also the bucket of the air-pump being raised, when the piston is at the top of the cylinder, the vacuum is prepared in the condenser, before the steam is admitted into it, and then the steam rushing suddenly into the condenser, acts upon that air and rare vapour which still remains in the condenser, and assists to expel some of it, through the lower valve of the air-pump into the empty barrel thereof, so as to be extracted by the next stroke of the pump.

In this way a greater portion of the air will enter into the air-pump, than if it made its stroke during the descent of the piston; for then the air could only pass the lower valve by its own elasticity.

The imperfections of the cylinders, which occasioned so much trouble in the first engine, were removed by degrees, principally in consequence of a new machine for boring the insides of cylinders, which was introduced by Mr. John Wilkinson, at his foundry at Bersham, near Chester, about 1775. In the old method, the borer for cutting the metal, was not guided in its progress, and therefore followed the incorrect form given to the cylinder in casting it (see p. 291); it was scarcely ensured that every part of the cylinder should be circular; and there was no certainty that the cylinder would be straight. This method was thought sufficient for the old engines; but Mr. Watt's engines required greater precision.

Mr. Wilkinson's machine, which is now the common boring machine, has a straight central bar of great strength, which occupies the central axis of the cylinder, during the operation of boring; and the borer, or cutting instrument, is accurately fitted to slide along this bar, which, being made perfectly straight, serves as a sort of ruler, to give a rectilinear direction to the borer in its progress, so as to produce a cylinder equally straight in the length, and circular in the circumference. This method ensures all the accuracy the subject is capable of; for if the cylinder is cast ever so crooked, the machine will bore it straight and true, provided there is metal enough to form the required cylinder, by cutting away the superfluities.

After repeated trials of different kinds of pistons, and of soft materials for stuffing, Mr. Watt ultimately fixed upon hemp for packing, and applied it to the piston in a manner which has since been universally adopted.

The piston has a projecting rim at bottom, which is fitted as accurately into the cylinder as it can be, to leave it at liberty to move freely therein. The part of the piston above this rim, is less than the cylinder, to leave a circular groove or channel, about two inches wide, into which hemp, or soft rope which is called gasket, is rammed, to form the packing; then, to keep the packing in its place, a lid or cover is put over the top of the piston, with a rim or projecting part, which enters into the circular groove and presses upon the packing, the cover being forced down by screws. The lower part of the groove round the piston, is made rounding, with a curve, that the pressure on the packing may force it against the inside surface of the cylinder. The piston must be kept supplied with melted grease; for which purpose a funnel is fixed on the top of the cylinder, with a cock and pipe to let the grease down.

The stuffing-box round the piston-rod, at the top of the cylinder, is packed with hemp in a similar manner, a collar or gland, with a hole through it, for the passage of the rod, being screwed down, to confine the packing in its place, round the rod.   In both the piston and the stuffing-box, when the hemp wears loose, so as to become leaky, it can be rendered tight, by forcing down the screws and compressing the hemp into close contact.

Mr. Watt found very early that most kinds of grease would answer the purpose of keeping the piston tight, but beef or mutton tallow were the most proper, being the least liable to decompose. When cylinders were new and imperfectly bored, the grease soon disappeared, and left the piston dry ; he therefore endeavoured to detain it, by thickening it with some substance which would lubricate the cylinder, and not decompose by heat and exhaustion.   Black-lead dust seemed a proper substance, and was therefore employed, especially when a cylinder or the packing of the piston was new ; but it was found in the sequel, that the black-lead wore the cylinder, though slowly ; and by more perfect workmanship, cylinders were at last made so true, as not to require it, or at least only for a very short time at first using.

When a cylinder is perfectly true and smooth within, there is no difficulty with the piston ; for although the hemp packing is rammed quite solid into the groove round the piston, it will work free, after it has made a few strokes ; there is not in fact any elasticity or softness in the stuffing, but it is to all intents and purposes a solid piston, which, being moulded into the cylinder itself, fits it perfectly, and when it wears loose, the hemp being compressed by the screws, is moulded anew into the cylinder, and thus made to fit again as well as ever.   In the old engines, the packing being kept soft and spongy by the water, was capable in some degree of accommodating itself to small irregularities in the cylinder, and therefore the perfect form of the cylinder, was of less importance than in Mr. Watt's engine, where the packing being always hot and dry, and rammed very hard, cannot yield so as to work at all, in an incorrect cylinder.

The progress of Mr. Watt's improvements, may be traced, by the following dates.   In 1767, and 1770, Mr. Watt was employed in planning the Monkland Canal, in Scotland ; and a bridge over the river Clyde, at Hamilton.   It appears that he had not made any steam engine on his new invention in 1770, his attention having been entirely occupied by other objects ; but in 1774, he removed to Soho near Birmingham, and formed his association with Mr. Boulton, and very soon made his first engine there.

In 1776 Messrs Boulton and Watt had brought their new engines to a considerable degree of perfection, as appears by the following extracts from a proposal made by Mr. Boulton to the Carron Company, to rebuild their returning engine, on the new system.

" Every size of Mr. Watt's engine, from a cylinder of 36 inches diameter to one of 72 inches, will raise water equally cheap ; for the consumption of fuel will be exactly as the steam used, and that will be in proportion to the capacity of the cylinder.   The waste of steam will certainly be greater in very small engines, than in large ones, yet there will be but little difference in the proportion of loss, between a cylinder of 36 inches, and one of 72 inches.

" The cylinder of our Soho engine, is only 18 inches diameter, and the pump raises water 24 feet high ; and has made 3000 strokes, with one hundred weight of Wednesbury slack and lump coal ; each stroke, raising 7.8 cubic feet of water, 24 feet high.

" We have no objection to contract with the Carron Company to direct the making of an engine, to return the water for their mills.   And we will guarantee that each hundred weight of such coals as are commonly used for engines at Carron, shall raise 20 000 cubic feet of water, 24 feet high, (=480 000 cubic feet one foot high) ; if we can do more, our profits will be in proportion to the savings."

" An engine with a cylinder of 50 inches diameter, will work a pump of 52 inches diameter, to lift the water 24 feet high.   We have erected such an engine near Tipton, in Staffordshire, and which is ready to set to work ; the buildings, the engine, with all things relative thereto, will come to about 2000l.

" We do not aim at profits in engine-building, but shall take our profits out of the savings of fuel ; so that if we save nothing, we shall take nothing.   Our terms are as follows : We will make all the necessary plans, sections, and elevations for the building, and for the engine, with its appurtenances, specifying all cast and forged iron work, and every other particular, relative to the engine. We will give all necessary directions to your workmen, which they must implicitly obey. We will execute, for a stipulated price, the valves, and all other parts which may require exact execution, at Soho ; we will see that all the parts are put together, and set to work properly ; we will keep our own work in repair for one year, and we have no other objection to seven years, than the inconvenience of the distance. We will guarantee that the engine so constructed, shall raise at least 20 000 cubic feet of water 24 feet high, with each hundred weight of coals burnt.

"When all this is done, a fair and candid comparison shall be made between it, and your own engine, or any other engine in Scotland, from which comparison the amount of savings in fuel shall be estimated, and that amount being divided into three parts, we shall be entitled to one of those parts, in recompense for our patent license, our drawings, &c. &c. Our own share of savings shall be estimated ; in money according to the value of your coals, delivered under the boiler, and you shall annually pay us that sum, during 25 years, from the day you begin to work ; provided you continue the use of the engine so long. And in case you sell the engine, or remove it to any other place, you must previously give us notice, for we shall then be entitled to our thirds of the savings of fuel according to the value of coals at such new place. This is a necessary condition, otherwise the engine which we make for you at an expense of 2000l. may be sold in Cornwall for 10 000l.

" Such parts of the engine as we execute at Soho we will be paid for at a fair price ; I conclude from all the observations I have had an opportunity of making, that our engines are four times better than the common engines. In boilers, which are a very expensive article, the savings will be in proportion to the savings of coal. If you compare our engine with the common engine (not in size but in power), you will find the original expense of erecting one to be nearly the same."

" Mr. Wilkinson has bored us several cylinders almost without error ; that of 50 inches diameter which we put up at Tipton, does not err the thickness of an old shilling in any part, so that you must either improve your method of boring, or we must furnish the cylinder to you."

In 1777 Mr. Watt constructed a steam engine for blowing an iron furnace at Wilson House, in Lancashire, belonging to Mr. Wilkinson ; both the engine and the furnace were worked with peat for fuel. It had a steam cylinder of about 30 inches diameter, and at the opposite end of the great lever, a blowing cylinder, or air pump, of about 36 inches diameter. This cylinder being closed at both ends, and provided with suitable valves, blew out the air, both in ascending and descending ; the piston rod passing through a collar of leathers in the top or cover of the blowing cylinder. The air from this double blowing cylinder, was received into another regulating cylinder of about the same size, but open at top, and fitted with a floating piston which was loaded with weights, and rose and fell, so as to float, as it were, upon the air contained in the cylinder (see p. 282). This was the first application of the double acting air pump, or blowing cylinder for furnaces : it is now the universal practice.

In 1778, Mr. Watt erected an engine at Ketley Iron Furnace in Shropshire. The cylinder was 58 inches diameter, and it worked two pumps 34 inches diameter, which raised water 36 feet high, to turn the water wheels, by which the blowing machines for the furnaces were worked.

According to Pryce's Mineralogia Cornubiensis, which was written in 1778, Messrs. Boulton and Watt had then erected several engines in Staffordshire, Shropshire, and Warwickshire, and one small one near London.

" An engine erected by Mr. Watt at Hawkesbury Colliery, near Coventry, is supposed to be the most powerful engine in England. The cylinder is 58 inches diameter, the piston moves 8 feet stroke, and 12 strokes per minute ; the pump is 14 inches diameter, 65 fathoms lift. Messrs. Boulton and Watt are now erecting 3 engines in Cornwall, viz. Ting Tang, Owanvean, and Tregurtha-downs. They have lately set to work a small engine, at Huel-Bussy mine ; cylinder 30 inches, 8 feet stroke : it makes 14 per minute, and works two pumps 6½ inches diameter, in two separate shafts 300 feet asunder, 45 fathoms lift in each, or 90 fathoms in all ; the pumps are worked by flat rods, with great friction.

" The engines are built at the expense of the proprietors ; Messrs Boulton and Watt furnish such drawings, directions and attendance, as may be necessary to enable a resident engineer to complete the machine ; and, in lieu of all profits, they take one-third of the annual savings in fuel, which their engine makes, when compared with a common engine of the same dimensions, in the neighbourhood. Mr. Watt's new engine will raise 20 to 24 thousand cubic feet of water to 24 feet high, by 1 cwt. of good pit coal." (= 480 to 576 thousand cubic feet 1 foot high).

The engines constructed by Messrs Boulton and Watt at this period, were nearly after the plan of that which we have described. The elastic force of the steam was in all cases employed to press down the piston, instead of the atmospheric air; and the steam was condensed in a vessel detached from the cylinder, and kept constantly cool, by external cold, aided by an injection of cold water, which was afterwards extracted by the air pump. These two parts of Mr. Watt's improvement are distinct from each other, and a condenser and air pump might have been applied to exhaust the steam from a cylinder, which being open at the top would admit the atmospheric air to act on the piston, in the usual manner of Newcomen's engine; but by avoiding the injection of cold water into the cylinder, a considerable part of the steam, which is commonly lost by condensation, might have been saved.

This partial adoption of Mr. Watt's complete invention, offered great facility of application to existing engines on Newcomen's construction, because the old cylinder, and piston, with all its connecting rods and chains to suspend it from the great lever, would then have remained unaltered; but to apply a close cover to the cylinder, so as to enable the steam to act upon the piston, required an entire new construction of the cylinder and piston, with all its appurtenances; and it became indispensable to have an accuracy of execution, which was so extremely difficult of attainment at that time, as to be nearly impracticable except to the patentees, who had established their new workshops at Soho expressly for the purpose; and even after they had succeeded in getting cylinders properly executed, with covers, pistons with polished rods, and spindle valves, to admit the steam on Mr. Watt's plan, the novelty of that construction, and the supposed difficulty of keeping such new work in good order, was strongly objected to, by those who required the new engines.

To avoid these objections, and yet have some of the benefit of Mr. Watt's invention, it was often proposed to the patentees, to grant licenses for the use of a condenser, to be applied to an atmospheric engine. Mr. Watt's ideas on this subject, are thus stated in a letter to Mr. Smeaton, who had applied to them for such a license in 1778.

" I have several times considered the propriety of the application of my condensers to common engines, and have made experiments with that view, upon our engine at Soho; but have never found such results as would induce me to try it any where else; and in consequence, we refused to make that application to Wheal Virgin engines in Cornwall, and to some others; our reasons were, that though it might have enabled them to have gone deeper with their present engines, yet, the savings of fuel would not have been great, in comparison to the complete machine. By adding condensers to engines that were not in good order, our engine would have been introduced into that country (which we look upon as our richest mine), in an unfavourable point of view, and without such profits as would have been satisfactory, either to us or to the adventurers; and if we had granted the use of condensers to one, we must have done so to all, and thereby have curtailed our profits, and perhaps injured our reputation. Besides, where a new engine is to be erected, and to be equally well executed in point of workmanship, and materials, an engine of the same power cannot be constructed materially cheaper on the old plan than on ours; for our boiler and cylinder are much smaller, and the building the lever, the chains, together with all the pump and pit work, are only the same.

" We have now laid aside the clumsy cast-iron outside cylinder, and have substituted a case of wrought iron plates, which fits within an inch and a half of the cylinder, and extends from within 6 inches of the upper flange, to within 3 inches of the lower flange of the cylinder. This seems to answer as well as the outside cylinder. We have also executed some engines without outside cylinders altogether; but have had no reason to applaud our economy, as the consumption of fuel was considerably greater.

" We charge our profits in proportion to the saving made in fuel by our engine, when compared with a common one which burns the same kind of coals; we ask one-third of these savings to be paid us annually, or half yearly; the payment being redeemable in the option of our employer, at ten years' purchase: and when the coals are low priced, we should also make some charge as engineers. In all these comparisons, our own interest has made us except your (Mr. Smeaton's) improved engines, unless we were allowed a greater proportion of the savings."

" The idea of condensing the steam, by injecting into the eduction pipe, was as early as the other

U U

kinds of condensers, and was tried at large by me at Kinneal; but the other imperfections of that engine, made me attribute a bad vacuum, to the air which entered with the injection; and, consequently I disused the injection, until the size and expense of the tubulated condensers for large engines, made me resolve to sacrifice part of the power to convenience, and to employ large air pumps. At Bedworth we used three air pumps; viz. two below, and one above in the middle: the use of the middle one is to receive the hot water lifted up by the other two, and by lessening the surface which is exposed to the pressure of the atmosphere, the water is extracted with greater ease. The fear of twisting the lever beam, made me apply three pumps at Bedworth; we always made two serve in lesser engines, and now we use only two for the largest engines, but we make one double the area of the other."

In the same correspondence, 1778, Mr. Boulton says, " we shall have four of our engines at work in Cornwall this summer: two of them are cylinders of 63 inches diam., and are capable of working with a load of 11 or 12 pounds on the square inch.

" We are systematizing the business of engine making, as we have done before in the button manufactory; we are training up workmen, and making tools and machines to form the different parts of Mr. Watt's engines, with more accuracy, and at a cheaper rate than can possibly be done by the ordinary methods of working. Our workshop and apparatus will be of sufficient extent to execute all the engines which are likely to be soon wanted in this country; and it will not be worth the expense for any other engineers to erect similar works, for that would be like building a mill to grind a bushel of corn."

" I can assure you from experience, that our small engine at Soho, is capable of raising 500 000 cubic feet of water 1 foot high, with every 112 lbs. of coals, and we are in hopes of doing much more. Mr. Watt's engine has a very great advantage in mines, which are continually working deeper; suppose, for instance, that a mine is 50 fathoms deep, you may have an engine which will be equal to draining the water, when the mine is worked, to 100 fathoms deep, and yet you can constantly adapt the engine to its load, whether it be 50 or 100 fathoms, or any intermediate depth; and the consumption of coals will be less in proportion, when working at the lesser, than at the greater depths; supposing it works, as our engines generally do, at 11 lbs. per square inch, when the mine becomes 100 fathoms deep."

Advantages of Mr. Watt's Engine over Newcomen's.

These are numerous, and very important, both for economizing fuel and augmenting the useful performance of the engine.

1st, the cylinder being surrounded with hot steam from the boiler, it is always kept at the same temperature as the steam itself; and is, therefore, incapable of condensing any part of the hot steam which is admitted into it.

2nd, the condenser being always kept cooled to a temperature of 100 degrees, or colder if possible, the steam will be so completely condensed and exhausted from the cylinder, that the small proportion which remains, cannot greatly oppose the descent of the piston, which is, therefore, pressed down by almost the entire elastic force of the steam from the boiler; and this may be greater or less than that of the atmosphere, as occasion requires.

3rd, The elasticity of the steam from the boiler, being employed to force down the piston instead of the pressure of the atmosphere, the air does not enter the cylinder so as to cool its interior surface; and the engine is not confined to work with its whole force; for if the cylinder is supplied with steam of a suitable elasticity, the power of the engine may be varied very considerably, without making any alteration in its construction.

In calculating the dimensions of an engine (to perform any given task,) on Mr. Watt's principle, the engineer has less to consider than in Newcomen's engine, because the pressure on the piston can be so easily regulated. It was Mr. Watt's practice, to assume 10 lbs. per square inch of the piston's surface for a suitable load; and then in working the engine, if it becomes necessary, the load can be diminished to 7 lbs., or increased to 11 lbs., which is a great latitude for future contingencies.

It has already been shown, that the quantity of steam lost by condensation, when it enters into the cold cylinder of Newcomen's engine, is in most cases nearly as great as that which ultimately produces its power (a); therefore, such engines consume double the quantity of fuel which would be requisite, if the waste could be entirely avoided.

In the improved engine of Mr. Watt, the waste of steam is not more than one-fourth of the quantity which is necessary to fill the cylinder, so that $1\frac{1}{4}$ times the contents of the cylinder must be supplied at each stroke; and it is probable, that a considerable portion of this waste is occasioned by the leakage of the piston, and not much from condensation.

In addition to this advantage of saving steam, which would otherwise be wasted, the power of the engine is so much increased, that Mr. Watt's engine will work as well when loaded, at the rate of 10 lbs. per square inch of the piston, as New-comen's can do with only 7 lbs. For in Newcomen's engine, to work properly, the hot well cannot be cooled below 152 degrees, as already explained (p. 159); and, therefore, the elasticity of the uncondensed steam, which remains in the cylinder during all the working stroke, must be nearly 4 lbs. per square inch; but in Mr. Watt's engine, the hot well may be always kept down to a temperature of about 102 degrees, and then the elasticity of the uncondensed steam in the cylinder, during the working stroke, will not be quite 1 pound per square inch; accordingly it was found in practice, that Mr. Watt's first engines performed the same work with less than half the fuel that the best atmospheric engines consumed.

Before a Newcomen's engine is erected, the engineer must make an accurate estimate of the work to be performed by it, and he must proportion his engine accordingly. He must be careful to make it fully able to execute its task; but the power must not exceed the load in any extravagant degree, for that would produce such a rapid motion, as to make the catch pins of the great lever strike on the spring beams at every stroke, with a force which no building or machinery could withstand. Engines are frequently shattered by the pump drawing air, or by a pump-rod breaking; the steam piston then descends with such force and rapidity, that every thing must give way. In most operations of mining, the task of the engine increases as the mine grows deeper, and it must be so constructed, at first, as to be able to bear this addition, and must be worked partially in the first instance. But with every precaution it is very difficult to manage an atmo-spheric engine, when its power is much greater than its task; the usual methods of moderating the force have been stated (p. 188); viz. to limit the supply of in-jection water, and to admit air into the cylinder by the air-cock; or to work the engine slowly by a cataract; or the engine may stand still for some hours every day, to allow a sufficiency of water to accumulate in the mine, and then the engine may be worked with its full motion, till the collected water is drained.

Mr. Watt's engine has none of these inconveniences, because its power can at all times be exactly adapted to the load of work it has to perform, for as the piston

---

(a) In Mr. Smeaton's small engine, (p. 168), the capacity of the cylinder was but 2·34 cubic feet, and yet it consumed 8·76 cub. feet of steam per stroke, or $3\frac{3}{4}$ times. And in the engine at Long Benton, (p. 174,) the capacity of the cylinder was 133·3 cub. feet, and the steam consumed was 212·5 cub. feet per stroke, or 1·6 times.

These are instances of the very best of Newcomen's engines; but according to the statement in p. 234, the common engines were so much inferior, that we may safely state the waste of steam, at one half of all that is produced.

is pressed down during the working stroke, by the elasticity of the steam which is admitted into the cylinder, it is only requisite to administer steam of a suitable elasticity, to overcome the resistance and produce a moderate motion. The dimensions of the engine may be such, that it would be capable of exerting double the power requisite to drain the mine in the first instance, and yet the engine may be made to work steadily with only a small portion of its force, by very simple means.

One of these methods is, to diminish the passages through which the steam passes into or out of the cylinder; for this purpose, the exhausting valve through which the steam escapes from the cylinder to the condenser, may be regulated so as to be lifted only partially out of its conical aperture, and more or less of the passage may be opened at pleasure, by limiting the height to which the valve shall be raised by the working gear. The degree of opening, thus given to the exhausting valve, determines the rapidity with which the steam will be drawn off from the lower part of the cylinder to the condenser, and consequently the velocity of the motion of the piston during the working stroke. In like manner, the equilibrium valve may be regulated to open only just so much, as to cause the returning stroke to be performed with a steady motion, by limiting the rapidity with which the steam shall be re-admitted into the lower part of the cylinder.

This method of curbing the too violent motion of the engine, was employed by Mr. Watt in his earliest engines, the working gear being adapted to open the valves either partially, or entirely, as the case required; but he afterwards effected the regulation in a better manner, by restraining the flow of steam from the boiler into the cylinder; for this purpose he provided an additional valve, or conical plug, to stop the interior passage of the steam-pipe, which conveyed the steam from the boiler, into the steam-case round the cylinder; this valve, which he called the throttle-valve, is solely for the purpose of regulating the motion of the engine, on the same principle that the shuttle of a water-wheel limits the flow of water upon the wheel, and consequently the rapidity of its motion.

The throttle-valve is not connected with the working gear, so as to be opened and shut at every stroke, but it is always kept partially closed, when the engine is not fully loaded; and it is only opened to such an extent, as will not allow the steam to pass out quite so fast from the boiler, as the descent of the piston makes room for it in the cylinder, consequently the steam which does enter into the top of the cylinder, and into the surrounding space of the steam-case, must expand to fill a greater space than it previously occupied in the boiler, whereby it becomes less elastic, and presses with diminished force upon the piston.

As this mode of regulation only limits the rapidity of the supply of steam to the cylinder, whilst the steam retains its full elasticity in the boiler, a very considerable increase of the load of the engine cannot stop its motion, although it may retard it; because the obstruction occasioned by the throttle-valve, only diminishes the velocity of the motion, but not the force which the engine is capable of exerting, when moving with a still less velocity; for when the load is so much increased as to make the engine move slowly, the steam will flow through the throttle-valve into the steam-case, and into the upper part of the cylinder, faster than the descent of the piston will make room for it; and in consequence it will accumulate, until it has acquired the same elasticity as within the boiler, or a sufficient pressure to overcome the resistance to the piston, and make it descend.

Improved Form of Mr. Watt's Engine, 1778.

The form of the engine already described, was that which Mr. Watt followed in his first engines for pumping water, to drain mines, or to supply towns; he altered a great many atmospheric engines to this plan with success, but he afterwards adopted another arrangement of the parts, in which the steam for the supply of the cylinder does not pass through the steam-case, but enters from the steam-pipe, through a valve immediately into the top of the cylinder; and although the steam-case has a communication with the boiler, the steam which is admitted into it is only for the purpose of keeping up the heat, and preventing any condensation of the steam within the cylinder. In this way the steam-case becomes less essential to the engine; and, about 1778, Messrs. Boulton and Watt began to make the steam-cases of wrought iron-plate, about $1\frac{1}{2}$ inch from the cylinder all round; and, in some cases, they laid them aside altogether; but they found this an ill-judged economy, and returned to it again.

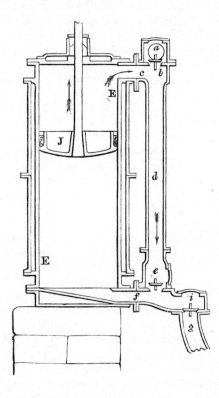

The cylinder E E is fitted with its piston J, and is closed at top by the cover, which is screwed to the top flange of the cylinder itself, instead of the top flange of the steam-case. The steam is brought from the boiler to the cylinder by the pipe a, which appears like a circle, being cut across the direction of its length: b is the regulating or throttle-valve in that pipe, and c the communicating passage into the top of the cylinder, immediately beneath the valve, so that the steam has constant admission through this valve into the top of the cylinder, and presses upon the piston, in such quantity as the constant opening of the regulating-valve b will allow.

d is the steam-pipe, which descends to the bottom of the cylinder, for the purpose of admitting steam into the bottom of the cylinder, when the piston is to ascend; and e is the equilibrium-valve, which opens or shuts that communication at pleasure: i is the exhausting-valve, which being opened when the equilibrium valve, e, is shut, allows the steam to pass off through the eduction-pipe g to the condenser, which is on a similar construction to that already described; but it is provided with an injection-valve, to admit a jet of cold water to flow into the condenser, so as to cool the steam, and produce the vacuum therein.

The operation of this engine is as follows: Suppose the piston J to be drawn up to the top of the cylinder by the counterweight, and all the parts to be filled with steam, except the eduction-pipe *g*, and the condenser and air-pump, which are supposed to be exhausted, the equilibrium-valve *e*, and the exhausting-valve *i*, are shut.   Now, if the exhausting-valve *i* is opened, the steam contained in the lower part of the cylinder, beneath the piston J, will pass by its own elasticity through the valve *i*, and rush into the vacuous space in the eduction-pipe *g*, and condenser ; and being met there by the jet of cold water, the steam is cooled and condensed, so that the vacuum in the condenser is maintained, notwithstanding the rush of steam which is admitted into it ; consequently the steam continues to flow out of the cylinder, till none remains beneath the piston, or only a very rare vapour, which offers scarcely any sensible reaction to the steam above the piston, which is continually pressing it downwards.

The piston is therefore put in motion, and begins to perform its working stroke, as soon as the steam is exhausted from the cylinder, and it continues to descend, till it arrives near to the bottom of the cylinder : the exhausting-valve *i* is then shut, and the equilibrium-valve *e* is opened by the working gear ; this allows the steam from the top of the cylinder to enter by the steam-pipe *d* through *f*, into the bottom of the cylinder, as is shown by the arrows, so as to press upwards beneath the piston J, with the same elasticity as the steam from the boiler presses downwards upon the upper side of the piston ; for there now is an open communication through *c*, *d*, *e*, and *f*, between the top and bottom of the cylinder, whereby the piston is placed in equilibrio, and is consequently drawn up by the counter-weight, until it arrives at the top of the cylinder.   The returning stroke being thus completed, the equilibrium-valve *e* is shut, and the exhausting-valve *i* is opened again, to make another working stroke as before.

This sketch was taken from an engine erected at Hull, in 1779 ; and, with some slight variations in the manner of its action, this is the form of the present standard engine for pumping water : the variation is, that the regulating-valve *b*, is made to open and shut at every stroke, and another regulating-valve is sometimes applied in the steam-pipe, to intercept the steam, just before it arrives at *b*.

Another Form of Mr. Watt's Engine, 1780.

In the two forms of the engine already described, the working stroke is performed by the elastic force of the steam acting upon the upper side of the piston, whilst there is a vacuum beneath it ; and to produce the returning stroke, the piston is placed in equilibrio, by making an open communication between the upper and lower parts of the cylinder ; and provided that the equilibrium-valve opens a sufficient communication, the piston will thus be placed in a much more exact equilibrium than in Newcomen's engine, where the varying pressure of the steam, and of the atmosphere, renders the circumstances in which the piston rises very uncertain.

Mr. Watt's engine may be so arranged, that the ascent of the piston may be performed in vacuo, the vacuum being made, at the same time, both above and below the piston.   This form of the engine is here represented, and the same letters of reference are used as in the preceding sketch. *b* is the steam-valve, which admits the steam to the top of the cylinder E, to press upon the piston J ; this valve is shut when the engine makes its returning stroke. *e* is the equilibrium-valve, placed beneath the steam-valve *b*, instead of being at the bottom of the steam-pipe *d d j g*, which descends direct to the condenser, and turns off with a branch *f*, to the bottom of the cylinder; by this passage, *f j g*, the steam is always drawn off from the bottom of the cylinder, so as to keep a constant vacuum beneath

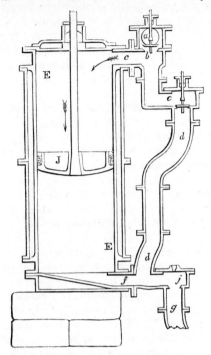

the piston. In this construction the equilibrium-valve *e* becomes also the exhausting-valve, for when it is opened it allows the steam to pass from the top of the cylinder to the condenser. In these engines a valve was sometimes placed in the eduction-pipe at *j*, but it was only used to regulate the rapidity of the flow of the steam through *f* and *g* to the condenser, and was not connected with the working gear; it was always kept partially open.

Suppose that the equilibrium-valve *e* is opened, the steam in the top of the cylinder will pass off to the condenser through the steam-pipe *c e d g*, and leave a vacuum in the top of the cylinder above the piston; in this case the piston will be drawn up by the counter-weight to the top of the cylinder, and being arrived there, the equilibrium-valve *e* is shut, and the steam-valve *b* is opened. The steam from the boiler now enters into the top of the cylinder, and pressing between the cover and the top of the piston, forces the latter down to the bottom of the cylinder; the working stroke being thus performed, the steam-valve *b* is shut, to prevent the farther admission of steam from the boiler; and, at the same instant, the equilibrium-valve *e* being opened, the steam from the top of the cylinder passes off to the condenser; and leaves a vacuum above the piston, the same as is already left beneath it: in consequence, the piston is set at liberty, and rises by the action of the counter-weight, until it arrives at the top: the equilibrium-valve *e* is then shut, and the steam-valve *b* opened, to make another descent.

The intended advantage of this construction is, that the whole time of the ascent of the piston is allowed for the condensation of the steam, and therefore it might be expected to produce a better vacuum, and a more immediate stroke, than in the other constructions, in which the condensation of all the steam in the cylinder, must be made after the piston arrives at the top of the stroke, and before it can begin to return.

But in practice this advantage was not found to be of much importance, because the condensation takes place with such rapidity in the condenser, as to exhaust the cylinder as quick as ever the steam can pass through the exhausting-valve, and eduction-pipe, into the condenser: and all that could be gained by having the cylinder ready exhausted in preparation for the working stroke, might have been equally well attained by making larger passages through the exhausting-valve and eduction-pipe, so as to permit the steam to pass more readily to the condenser.

On the whole, the engines of which the pistons rise in vacuo, were not found to answer so well in practice as others, and have been long disused; one reason was, that the leakage of air was greatly increased at the stuffing-box, and at the joint of the cylinder cover (a).

When the vacuum is only made beneath the piston, the joints at the upper part of the cylinder, if defective, can only allow the escape of some steam into the air, and leaks into the lower part of the cylinder, can only admit air during the descending stroke; but when the piston rises in vacuo, all the defective joints, in every part, are continually admitting air. In engines for mines, which cannot be kept in the best state of repair, this is an objection fully equivalent to all the advantage of condensing during the returning stroke.

An engine is also found to require a greater counterweight to draw up the piston in vacuo, than in steam, or else it will rise slower; this is because the steam contained in the upper part of the cylinder, must have time to pass away through the equilibrium-valve e, and eduction-pipe, d, g, in order to be condensed, so as to produce the vacuum above the piston (or at least a state of equilibrium with the exhaustion below the piston) before it can rise; thence in reality, whatever may be the loss of time, or force, to evacuate the steam through the valve and pipe, from the cylinder to the condenser, it is not entirely removed by the piston rising in vacuo, but is only transferred from the working stroke to the returning stroke.

Another circumstance is also disadvantageous to an engine, when its piston rises in vacuo; viz. that when the cylinder is continually exhausted, the rare steam which always remains in it cannot be above 102 degrees of temperature, and about one pound per square inch elasticity. The metal of the cylinder being kept at 212 deg., by the steam surrounding it in the steam-case, must continually communicate heat to the rare steam; this is a complete waste, and also tends to render that rare steam more elastic, so as to offer a greater resistance to the motion of the piston, than if it had not been heated. Hence the steam-case would be a source of waste; and such engines would probably perform better, without keeping the cylinder hot.

On the actual Performance of Mr. Watt's first Engines in respect to Fuel.

At the first establishment of their engines, Messrs. Boulton and Watt charged their profits, in proportion to the saving of fuel which their engines made, when compared with common engines, burning the same kind of coals. They had one-third of these savings paid to them annually during the term of the patent, or the payment might be redeemed at ten years' purchase.

It should be observed, that Mr. Smeaton's improvements were introduced about the same time as Mr. Watt's, and therefore the comparison was not made with his engines, but with the former ones. It was Mr. Smeaton's rule, judging from some experiments made before him on some of Mr. Watt's early engines, to estimate their consumption of coals at one-half of that of his own engines, when performing the same work; this was in large engines, but the proportion of

---

(a) The above sketch was taken by the author in 1804, from an engine then in use at the Chelsea Waterworks, Pimlico; it had been erected by Messrs. Boulton and Watt about 1782, and was intended to work by the piston rising in vacuo; but it was afterwards altered, by applying a valve at j, and connecting it with the working gear, so as to open and shut it at every stroke, in order to cause the engine to work in the manner previously described. The steam-valve a, of this engine, was also opened and shut by the working gear, in order to produce the expansive action, which will be next explained.

waste of steam, being greater in small atmospheric cylinders than in large ones, the comparison between small engines would be still more favourable to Mr. Watt's, in which the disadvantages of small cylinders is less considerable.   As Mr. Smeaton reckoned his own engines to consume only half as much fuel as the engines in common use when he began his improvements (see p. 234), Mr. Watt's must have consumed only one-fourth as much as the common Newcomen's engines.

In 1778, when Mr. Watt first established his engine, his proposals were to raise 500 000 cubic feet of water 1 foot high, by the consumption of one hundred weight ($=$ 112 lbs.) of Wednesbury coals.   After he began to make engines for the mines in Cornwall, he adopted another term for expressing their performance, this was the number of millions pounds weight of water, which could be raised one foot high, by the consumption of a bushel of coals; the coals in that district are brought by sea from Swansea, in South Wales, and are measured by bushels (a).

Assuming a bushel of coals to weigh three quarters of a hundred weight (or 84 lbs.) the above proposal would be (500 000 $\times$ 62·5 $=$) 31 250 000 lbs. raised 1 foot high by a hundred weight of coals; or ($\frac{3}{4}$ of that number $=$) 23·44 millions pounds raised 1 foot high, by the consumption of a bushel ($=$ 84 lbs.) of coals. Mr. Watt was, at that time, in expectation of making a much greater improvement, by means of his expansive method, which will be next explained.

This proposal was made to Mr. Smeaton, who was desirous of promoting Mr. Watt's discovery; but the improvement so much exceeded his expectations, that he felt some doubt of the accuracy of the statement; and, in consequence, he made an experiment himself, in 1778, on an engine built on the Birmingham

---

(a) The custom of measuring coals by a bushel is rarely practised, except for those coals which are transported by sea, and which are, in consequence, chargeable with a duty.   The great exports of coals by sea are from Newcastle in Northumberland, and Swansea in South Wales, and Workington in Cumberland.   It is usually implied that coals are brought from one or other of those places, when they are reckoned by bushels; and both the Newcastle and Swansea coals are commonly of a better quality than the inland coals of Staffordshire, Yorkshire, or Lancashire, which are reckoned by the hundred weight ($=$ 112 lbs.) and the ton ($=$ 2240 lbs.)

The mode of measuring coals in a bushel is subject to considerable uncertainty, because they are directed to be heaped up in a conical form, above the top edge of the bushel, but the height of the heap is only determined by the eye.   It requires a reference to several acts of parliament to find what the dimensions of the legal coal bushel ought to be.

By an act of parliament, 16th and 17th year of King Charles II. it is enacted, that all coals brought into the river Thames, and sold by the chaldron, shall be at the rate of 36 bushels heaped up.   According to an act of the 12th of Queen Anne, the coal bushel must be round, with an even bottom, and be 19½ inches, from outside to outside, and must contain one Winchester bushel, and one quart of water.   Now, the Winchester bushel (also called the malt bushel) according to an act 13th William III. 1713, is a circular measure, 18½ inc. diam., and 8 inc. deep; this contains 2150·42 cubic inches; and a quart $=$ 67·2 cubic inc.; hence the contents of the coal bushel should be 2217·62 cubic inch.; and a cylindrical measure, to contain this quantity, would be 18·8 inc. diam., and 8 inches deep.   Again, another act, 47th George III. (1807) directs that the coals shall be heaped up above the bushel, in the form of a cone, at least 6 inches high, and of the same size at the base, as the outside of the measure, viz. 19½ inc diam.; this cone will contain 597·3 cubic inches.

The legal coal bushel may, therefore, be stated to contain 2815 cubic inches, or 1·63 cubic feet.   Three bushels of coals are put into a sack, and 12 such sacks ($=$ 36 bushels) make a chaldron. This is the mode of reckoning coals by retail in London, but in the wholesale trade, when 5 chaldrons are measured, an ingrain, or extra allowance of $\frac{1}{70}$ is given, making 189 bushels in 5 chaldrons, which is called the pool measure.

The weight of coals varies considerably in the different sorts, but the chaldron ($=$ 36 bushels) is usually reckoned to weigh 27 hundred weight ($\times$ 112) $=$ 3024 lbs., which is at the rate of $\frac{3}{4}$ of a cwt. (or 84 lbs.) per bushel: this number has been adopted in the present work, because it is the actual weight of the best qualities of the Newcastle coals, though inferior sorts are not above 78 or 80 lbs. per bushel.

Canal, for raising up water into a reservoir, to supply the waste occasioned by the passage of boats through the locks.

The cylinder was 20 inc. diam., and the pump also 20 inc. diam., lifting the water 27 feet; the piston made 11 strokes per minute, of 5 ft. 9 inc. each. It worked for an hour with 65 lbs. of Wednesbury coals.

The load on the piston is equivalent to a column of water 27 feet high, or (× ·434 =) 11·7 pounds per square inch. With this load the piston moved (11 × 5·75 ft. =) 63·25 feet per minute.

The weight of the column of water was (20 × 20 = 400 × ·341 =) 136·4 lbs. for 1 foot high, or (× 27 ft. =) 3683 lbs. in all. This weight was raised 63·25 feet per min. or = 232 937 lbs. raised 1 foot per min. ÷ 33 000 = 7 horse power, nearly.

And 232 937 lbs. × 60 min. 1 397 622 lbs. raised 1 foot per min. with 65 lbs. of coals. This would be very nearly 18 millions pounds raised 1 foot, by the consumption of a bushel (= 84 lbs.) of Wednesbury coals. The coals used in this trial were broken into small pieces, scarcely any being larger than a hen's egg, and none less than a nutmeg.

In 1779, Mr. Smeaton also tried the small engine at Hull, of which a sketch has been already given, p. 333. The cylinder was 22 inc. diam., the piston made 8 feet stroke, and worked at the rate of 11 strokes per minute, = 88 feet motion. The pump was 18 inc. diam., and raised the water 34 feet high. The consumption of coals, when working at the above rate, was 90 lbs. per hour.

Weight of the column of water 18 inc. diam. 34 ft. high, = 3756 lbs.; which, on a piston of 22 inc. diam. is a load of 9·88 lbs. per squ. inc.

The load of 3756 lbs. moving 88 feet per min. is 330 528 lbs., raised 1 foot per min., or very near 10 horse power. As 90 lbs. of coals were consumed in 60 minutes, 84 pounds must have lasted 56 minutes; in which time the engine raised (330 528 × 56 =) 18½ millions pounds raised 1 foot by a bushel.

These performances are considerably less than the proposals, but as both these were small engines, it might be expected that larger ones would have produced a greater effect from the same fuel: we do not meet with any accurate accounts of the performance of larger engines at that period (a).

In 1778, Messrs. Boulton and Watt put up an engine at Hawkesbury Colliery, near Coventry; it had a cylinder 58 inc. diam., 8 ft. stroke, and could make 12 strokes per minute, = 96 ft. motion; the pump was 14 inc. diam., and lifted 65 fathoms (= 390 feet). Load on the piston 9·9 lbs. per squ. inc.; weight of the column of water 26 064 lbs.

This engine must have been nearly 76 horse power; it was considered to be the most powerful in England at that time, for it was capable of performing more work than some of the old atmospheric engines, with cylinders of 70 and 72 inc. diam., and that with less than half their consumption of fuel.

(a) Mr. Boulton laid down the following rule for computing the consumption of fuel in Mr. Watt's engines, 1778.

The space occupied in the cylinder, by the motion of the piston, at each stroke, being calculated in cubic feet, and multiplied by the load upon each square inch of the piston in pounds; the product will be the weight of coals, in pounds, which will be consumed to work the engine 1800 strokes.

Example. A cylinder 20 inc. diam. 5¾ feet stroke, would contain 12·5 cubic feet, which × 11·7 lbs. load per squ. inc. = 146 lbs. of coals should make 1800 strokes. N. B. The engine really consumed 65 lbs. of coals to make 660 strokes, and at the same rate 177 lbs. would be consumed to produce 1800 strokes, instead of 146, as by computation.

This rule is at the rate of nearly 21¾ millions pounds of water, raised one foot by a bushel, or 84 lbs. of coals; for if we imagine a cylinder of one square foot area, and the piston to make 1 foot stroke, its capacity will be 1 cubic foot; and supposing the load to be at the rate of 1 pound per squ. inc., then, by the rule, 1 pound of coals should work it for 1800 strokes; now the effect would be 144 lbs. load, raised 1800 feet, = 259 200 lbs. raised 1 foot, by each pound of coals; or × 84 lbs. = 21 772 800 lbs., raised 1 foot high by a bushel.

It was stated in some of the newspapers of the day, that on a trial of this engine, it raised 99 711 cubic feet of water in 48 hours, from a depth of 130 yards; and consumed 96 cwt. of coals, that is, 2 cwt. per hour. The engine was then regulated by a cataract, so as to work at only about one-third of the full speed.

This is at the rate of 406 000 cubic feet of water, raised one foot high, with each hundred weight of coals; or equivalent to very nearly 19 millions pounds, raised 1 foot by the consumption of a bushel, or 84 lbs. of coals. This is also below the patentees' standard, but it must be expected that a large engine, when working with only part of its speed, would be under a disadvantage, and perform no better than a small engine properly loaded.

MR. WATT'S EXPANSION STEAM-ENGINE, 1778.

This was a most important improvement, of which Mr. Watt had the first idea in 1769, but did not put it in practice until 1778. The general principle is thus stated in a letter from Mr. Watt to his friend Dr. Small, of Birmingham, dated Glasgow, May, 1769:

" I mentioned to you a method of still doubling the effect of the steam, and that tolerably easy, by using the power of steam rushing into a vacuum, at present lost. This would do little more than double the effect, but it would too much enlarge the vessels to use it all; it is peculiarly applicable to wheel-engines, and may supply the want of a condenser, where the force of steam only is used; for open one of the steam valves, and admit steam until one-fourth of the distance between it and the next valve is filled with steam, then shut the valve, and the steam will continue to expand, and to press round the wheel, with a diminishing power, ending in one-fourth of its first exertion. The sum of the series you will find greater than one-half, though only one-fourth of steam was used. The power will indeed be unequal, but this can be remedied by a fly, or by several other means."

Mr. Watt did not pursue this idea till about 1776, when the engine at the Soho Works was fitted up to act on the principle of expanding the steam. It consists in shutting off the farther entrance of steam from the boiler, when the piston has been pressed down in the cylinder for a certain proportion of its total descent, and then leaving the remainder of the descent to be accomplished by the expanding force of that steam which is already introduced into the cylinder.

The economy of this method of using steam will become evident, by considering that if the cylinder, at the conclusion of the stroke, is quite full of steam, equal in elasticity to the atmospheric air, that steam will, when the valve is opened to the condenser, rush out into the vacuum with a very great force and velocity; because it is still capable of exerting more force, after it has depressed the piston; but this extra force which it possesses is entirely lost to the engine, for want of some part to act upon.

If we suppose the further supply of steam to be cut off when the cylinder has received a certain charge of steam, all the force which that steam continues afterwards to exert, in expanding itself, is part of the force which would otherwise have been lost by the steam rushing into the vacuum; and, accordingly, the steam, which has been so expanded, is found to have lost part of its elastic force; but it has not been lost to the engine, because the steam, during its expansion, has exerted a sufficient pressure on the piston to have continued the motion, after the expenditure of steam from the boiler had ceased.

This **method** of working gives the means of regulating the acting force of the engine; because the pins of the plug-beam for the working-gear can be so placed, as to shut the steam-valve, when the piston has descended one-half, one-third, or any other proportion of its course; and so far the cylinder will be charged with steam of the same elasticity as that in the boiler, which is usually a little more than the atmosphere. But to press the piston farther down, the steam so admitted must expand; and though its elasticity will diminish, it may be sufficient to complete the stroke. The power of the engine may thus be modified at pleasure, and according as more or less power is required, the adjustment of the pins in the plug can be varied, so as to allow the steam to act with its full force upon the piston, for a greater or lesser proportion of its total descent.

If this method of working an engine had no other advantage than the regulation of the power, it would not effect that object better than the throttle-valve; but by the application of the expansive principle, a great saving of steam is made. It has been before observed, in describing the action of Newcomen's engine, p. 179, that the motion of the piston is accelerated during its descent, by the continued action of the pressure of the atmosphere, whilst the resistance is constant, or even greatest at the first, considering the inertia of the moving parts. Mr. Watt's engine is the same, but in a less degree, because when it has a throttle-valve, the steam cannot come to the piston, except with a limited velocity. When the top of the cylinder is open to the boiler, or the throttle-valve fully opened, the effect is the same, as when the atmospheric air has free entrance into the top of the cylinder. But, by stopping the further entrance of the steam at a certain portion of the descent, the piston can be made to descend the remainder, with a more uniform velocity; and the stroke is thus performed by the expenditure of only a portion of that quantity of steam, which would be required, if steam of the full density were employed to press it down to the bottom with an accelerated velocity.

When the steam is shut off at a portion of the descent, the subsequent pressure on the piston diminishes as the steam becomes more and more rare; and, consequently, the accelerating force which impels the piston diminishes; so that the motion of the descent will not be accelerated, but it will approach to uniform velocity, or it may even be retarded; because, although the pressure on the piston at the beginning of the stroke may considerably exceed the resistance of the load, yet when the piston arrives near the bottom, the diminution of the pressure may occasion it to fall short of the resistance: in this case the motion can only be continued by the impetus of the moving parts; or some other equivalent contrivance must be used, which will tend to oppose the descent of the piston at the beginning of the stroke, and to assist it towards the conclusion.

For whatever may be the law by which the pressure on the piston decreases, it is possible to contrive the connecting machinery, so that the chain at the outer end of the great lever, shall continually exert the same force to lift the pump-rods, because the machinery may be made to vary in leverage, according to any law which is found most convenient.

This can be done on the same principle as that of the fusee in a watch, which by its peculiar form, transmits a uniform force to the wheel-work, from a very unequal action of the main-spring. In like manner, by making the communication from the piston-rod to the pump-rod, by means of chains, which apply upon arch-heads, formed to portions of proper spirals, instead of circles, the force exerted by the piston upon the lever and pump-rods may be regulated at pleasure, so as to produce a uniform effect in raising water.

This compensation was the subject of Mr. Watt's third patent, March 12, 1782, for certain improvements upon steam-engines, and certain new pieces of mechanism to be added thereto. He had constructed an engine at Shadwell Water-works, London, in 1778, the cylinder of which was nearly on the plan represented in the sketch, p. 333; but the pins in the plug were regulated so as to shut the upper steam-valve *b*, when the piston had descended about two-thirds of the stroke, so that the remaining third of the motion was performed by the expansion of that two-thirds charge of steam. With this moderate degree of expansion, no machinery was required for equalizing the decreasing force, the impetus of the great lever, piston, pump-rods, and other moving parts, being sufficient for the purpose.

Mr. Watt's statement of the operation of his expansion engine, is contained in the following table, which exhibits the pressure of the steam upon the piston, at 20 different stages of its descent: the piston is supposed to make an 8 feet stroke, and the supply of steam from the boiler is assumed to be cut off at one-fourth of the descent, or 2 feet from the top. The pressure of the dense steam in the boiler is supposed to be nearly equal to that of the atmosphere, or 14 lbs. per square inch, and the load of water in the pumps equal to 10 lbs. per square inch of the area of the piston, tending to resist its motion.

Operation of the Steam in Mr. Watt's Expansion Engine.

	Portions of the Descent of the Piston.		Decreasing effort of the Steam on the Piston.	
Top of cylinder	· 00		1·	This is the full elasticity of 14 lbs. per squ. inc. before the supply of dense steam is intercepted.
	· 05	Full supply of dense steam from the boiler.	1·	
	· 1		1·	
	· 15		1·	
	· 2		1·	
One-fourth, or	· 25		0· 833	
	· 3		0· 714	
	· 35		0· 625	
	· 4		0· 555	
	· 45		0· 500	Or half the original elasticity; or 7 lbs. per squ. inc.
One-half, or	· 5	Supply of steam intercepted, the action being produced by the expansion of the steam already admitted into the cylinder.	0· 454	
	· 55		0· 417	
	· 6		0· 385	
	· 65		0· 357	
	· 7		0· 333	Or one-third of the original elasticity, or 4⅓ lbs. per squ. inc.
Three-fourths, or	· 75		0· 312	
	· 8		0· 294	
	· 85		0· 278	
	· 9		0· 263	
	· 95		0· 250	Or one-fourth of the original elasticity; or 3½ lbs. per. squ. inc.
Bottom of cylinder	1.			
			11· 570	

The sum of all the forces of the steam, taken at 20 places, during the descent of the piston, is therefore 11· 57; for the direct and uniform action of the dense steam entering from the boiler produces 5, and the remaining 6· 57, is the sum of the decreasing forces exerted by the expansion of that steam, in proceeding to occupy greater spaces, until it reaches four times the original volume.

If the same cylinder had been fully supplied with dense steam from the boiler, during the whole stroke, the sum of all the forces taken at 20 places, would have been 20 ; but in that case, the expenditure of dense steam would have been 4 times the preceding estimate ; so that one-fourth of the dense steam, necessary to fill the cylinder, will, by the above expansive method, perform 57 hundredths, (or .5785), which is more than half the work, which could be performed by the whole cylinder full of dense steam ; and at this rate, any given quantity of dense steam, being used expansively to the above extent of four times the original volume, will exert .5785 × 4 = 2·314 times the mechanical power which it could do, if it were used only in its entire density to exert a uniform effect, without any expansive action.

Dr. Robison has made a more exact investigation of the action of Mr. Watt's expansion engine ; it is as follows, with some few additions, to render it more explicit to practical men.

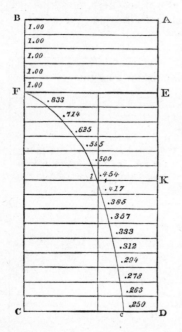

Let A B C D represent a section of the cylinder of a steam-engine, and E F the surface of its piston. Let us suppose that the dense steam was freely admitted from the boiler, whilst E F was in contact with A B, but that as soon as that steam had pressed it down to the situation E F, the steam-valve was shut. The steam will, nevertheless, continue to press down the piston by its own elasticity and force of expansion ; but as the steam expands, its elasticity and pressure must diminish.

We may express the uniform pressure which the dense steam exerts whilst the steam-valve is open, and whilst the piston is moving from A B to E F, by the length of the line E F. And if we assume the elasticity of steam to be proportional to its density (as is nearly the case with air), we may express the varying pressure of the expanding steam on the piston, when in any other position by the lengths of the lines K *l*, and D *c*, which are the ordinates of a rectangular hyperbola F *l c*, whereof A D, A B are the assymptotes, and A the centre.

The accumulated pressure exerted by the steam in expanding, after the valve is shut ; viz. during the motion of the piston from E F to D C, will be represented by the area E F *l c* D E ; and the pressure during the whole motion by the area A B F *l c* D A. Hence by computing these areas, we may find the whole pressure exerted on the piston, by the accumulation of all the varying forces.

Now, it is well known, that the area E F *l c* D E, (which represents the expansion), is equal to the rectangle A B F E, (which represents the full supply of dense steam) multiplied by the hyperbolic logarithm of the number which is obtained by dividing (the whole stroke) A D, by the portion A E during which the steam is freely supplied. For instance, if A D ÷ A E = 4, that is, if the steam is supplied during one-fourth of the stroke ; then the full pressure exerted by the dense steam, whilst the valve is open, being multiplied by the hyperbolic logarithm of 4, the product will be the accumulated pressure exerted by the steam during its expansion (*a*).

To apply this reasoning, suppose the diameter of the piston to be 24 inches, = 452·4 square inches area; and that the full pressure or elasticity of the dense steam in the boiler, is 14 lbs. per square inch, then the uniform pressure of that steam upon the piston, whilst the valve is open, will be (14 × 452 4 =) 6333 lbs. Suppose the length of the stroke A D, to be 6 feet, and that the supply of dense steam is stopped when the piston has descended to E F =1·5 feet, or ¼ of the stroke.

Now 6 ft. ÷ 1·5 feet = 4, the hyperbolic logarithm, of which is 1.386, and × 6333 lbs. uniform pressure = 8777 lbs., is the accumulation of the varying pressure during the expansion; and the total pressure during the whole stroke will be (6333 + 8777 =) 15 110 lbs.

Or the area A B F l c D A, which represents the total pressure, may be found at once, by adding 1 to the hyperbolic logarithm, previously to multiplying it by the uniform pressure A B E F; for instance, 1 + 1·386 = 2·386 × 6333 lbs. = 15 110 lbs. is the accumulated pressure during the whole stroke, as before (*b*).

As few professional engineers are possessed of a table of hyperbolic logarithms, while tables of common logarithms are, or should be, in the hands of every person who is much engaged in mechanical calculations, the following method may be practised.

Take the common logarithm of the quotient which is obtained by dividing the whole length of the stroke A D, by that portion A E, through which the piston descends before the dense steam is cut off; and multiply that common logarithm by 2·3026; the product is the hyperbolic logarithm, to which add 1, and then multiply by the uniform pressure exerted upon the piston, by the dense steam before the steam is stopped; the product will be the accumulated pressure during the whole stroke.

Example, 6 ft. stroke ÷ 1·5 ft. = 4, the common logarithm of which 0·6020 × 2·3026 = 1·386 for the hyperbolic log. to this add 1 = 2·386 and × 6333 lbs. 15 110 lbs. accumulated pressure.

The uniform pressure whilst the piston moves from A B to E F, (that is before the dense steam is shut off) is 6333 lbs.; therefore, that charge of dense steam in the act of expanding to four times the original volume, so as to fill the

---

(*a*) The hyperbolic logarithm of 4 is 1·386: therefore, in the preceding case, stated by Mr. Watt, the uniform force exerted by the dense steam being represented by 1, the sum of the decreasing force exerted by the same steam, during its expansion into a quadruple space, would be 1·386, and the sum total of all the force 2·386. Mr. Watt's method of computation, p. 342, gives 2·314, or nearly the same; the difference arises from the mode of examining the decreasing forces of the steam during its expansion.

Mr. Watt's table states this force at 15 different stages, the sum of which is 6·57 and (÷15=) ·438 is the mean of the decreasing forces, but as they act through 3 times the space that the dense steam acted; they produce a mechanical effect (·438×3=) 1·314 times greater than that of the dense steam.

The computation by any given number of stages, assumes that the force changes all at once at each stage, instead of diminishing gradually; and hence, it is obvious, that the more numerous the stages, the more correct the result will be. The other method of computation, which assimilates the decreasing forces to the ordinates of a curve, is, in effect, the same as taking an infinite number of stages of observation; and by the aid of logarithms, we can calculate the area of such a curve with great precision. F.

(*b*) The above method assumes that the piston occupies the whole capacity of the cylinder in its motion, without leaving any vacancy at the top or bottom, but in practice the effect of those vacant spaces is too great to be neglected.

whole cylinder, exerts an additional pressure of 8777 lbs., making up 15 110 lbs. total.

Supposing that the dense steam had been freely admitted during the whole descent of the piston, the accumulated pressure would have been 6333 × 4, or 25 332 pounds. Mr. Watt observed that the quantity of dense steam expended, in that case, would have been four times greater than when it was stopped at one-fourth, and yet the accumulated pressure would not be twice as great, being nearly five-thirds. So that one-fourth of the steam performs nearly three-fifths of the work; or an equal quantity performs more than twice as much work, when thus admitted, only during one-fourth of the motion (a).

This information is curious and important, and the advantage of this method of working a steam-engine, increases in proportion as the steam is sooner stopped; but the increase is not great after the steam is rarefied four times; for the curve having approached near to the asymptote A D, only small additions are made to the area. The friction of the pistons for such large cylinders as would be requisite, to carry the expansion to a greater extent, would also be very considerable, and would perhaps overbalance the advantage.

The Effects of the Expansive Action of Steam.	
Let the Steam be stopped at	Its Performance is multiplied
One-half	1· 69 ⎫
One-third	2· 10 ⎪
One-fourth	2· 39 ⎪
One-fifth	2· 61 ⎬ times.
One-sixth	2· 79 ⎪
One-seventh	2· 95 ⎪
One-eighth	3· 08 ⎭

(a) The above mode of computing the sum of the decreasing forces of the steam, during its expansion, assumes that the elasticity of steam, is exactly proportioned to its density. The best experiments on air, and other elastic fluids, seem to show that this is the case, when the temperature of the elastic fluid continues to be the same, during all the progress of the expansion; but unless an accession of heat is actually communicated to the steam as it expands, the temperature will subside, of itself, and consequently the elasticity of the expanding steam will diminish by more rapid steps than its density diminishes.

It is probable, that when any change is effected in the volume, and consequently in the elasticity, of steam, it will of itself assume a new temperature, corresponding with the new elasticity, according to the law stated in the table of elasticities for different temperatures. For instance, steam of the same elasticity as the atmospheric air, or 14· 7 lbs. per squ. inch, has a temperature of 212 deg. of Fahrenheit: if this steam is expanded until it possesses only half that elasticity, or 7· 35 lbs. per squ. inc. its temperature will subside to 178· 6 degrees; and when it is further expanded, so that the elasticity falls to one-fourth of an atmosphere, or 3· 675 lbs. per squ. inc. the temperature will sink to 149· 2 deg.

It is true, that in Mr. Watt's engine, the cylinder being surrounded by a steam case, and being thereby always kept as hot as the dense steam, would communicate heat to the expanding steam, and tend to keep it to a uniform temperature; but as such communication of heat would not be sufficiently rapid to have much effect, it is most probable, that the above computation over rates the expanding force considerably; for, in fact, the curve F *l* *c* is not an hyperbola, but a curve which approaches more rapidly towards the asymptote A D.

The following Table contains the Hyperbolic Logarithms for as many numbers as can be required in practice, for calculations of this kind.

Table of Hyperbolic Logarithms.

Numb.	Hyp. Log.	Numb.	Hyp. Log.	Numb.	Hyp. Log.	Numb.	Hyp. Log.	Numb.	Hyp. Log.
1·05	·049	3·05	1·115	5·05	1·619	7·05	1·953	9·05	2·203
1·1	·095	3·1	1·131	5·1	1·629	7·1	1·960	9·1	2·208
1·15	·140	3·15	1·147	5·15	1·639	7·15	1·967	9·15	2·214
1·2	·182	3·2	1·163	5·2	1·649	7·2	1·974	9·2	2·219
1·25	·223	3·25	1·179	5·25	1·658	7·25	1·981	9·25	2·225
1·3	·262	3·3	1·194	5·3	1·668	7·3	1·988	9·3	2·230
1·35	·300	3·35	1·209	5·35	1·677	7·35	1·995	9·35	2·235
1·4	·336	3·4	1·224	5·4	1·686	7·4	2·001	9·4	2·241
1·45	·372	3·45	1·238	5·45	1·696	7·45	2·008	9·45	2·246
1·5	·405	3·5	1·253	5·5	1·705	7·5	2·015	9·5	2·251
1·55	·438	3·55	1·267	5·55	1·714	7·55	2·022	9·55	2·257
1·6	·470	3·6	1·281	5·6	1·723	7·6	2·028	9·6	2·262
1·65	·500	3·65	1·295	5·65	1·732	7·65	2·035	9·65	2·267
1·7	·531	3·7	1·308	5·7	1·740	7·7	2·041	9·7	2·272
1·75	·560	3·75	1·322	5·75	1·749	7·75	2·048	9·75	2·277
1·8	·588	3·8	1·335	5·8	1·758	7·8	2·054	9·8	2·282
1·85	·615	3·85	1·348	5·85	1·766	7·85	2·061	9·85	2·287
1·9	·642	3·9	1·361	5·9	1·775	7·9	2·067	9·9	2·293
1·95	·668	3·95	1·374	5·95	1·783	7·95	2·073	9·95	2·298
2·0	·693	4·0	1·386	6·0	1·792	8·0	2·079	10·	2·303
2·05	·718	4·05	1·399	6·05	1·800	8·05	2·086	15	2·708
2·1	·742	4·1	1·411	6·1	1·808	8·1	2·092	20	2·996
2·15	·765	4·15	1·423	6·15	1·816	8·15	2·098	25	3·219
2·2	·788	4·2	1·435	6·2	1·824	8·2	2·104	30	3·401
2·25	·811	4·25	1·447	6·25	1·833	8·25	2·110	35	3·555
2·3	·833	4·3	1·459	6·3	1·841	8·3	2·116	40	3·689
2·35	·854	4·35	1·470	6·35	1·848	8·35	2·122	45	3·807
2·4	·875	4·4	1·482	6·4	1·856	8·4	2·128	50	3·912
2·45	·896	4·45	1·493	6·45	1·864	8·45	2·134	55	4·007
2·5	·916	4·5	1·504	6·5	1·872	8·5	2·140	60	4·094
2·55	·936	4·55	1·515	6·55	1·879	8·55	2·146	65	4·174
2·6	·956	4·6	1·526	6·6	1·887	8·6	2·152	70	4·248
2·65	·975	4·65	1·537	6·65	1·895	8·65	2·158	75	4·317
2·7	·993	4·7	1·548	6·7	1·902	8·7	2·163	80	4·382
2·75	1·012	4·75	1·558	6·75	1·910	8·75	2·169	85	4·443
2·8	1·030	4·8	1·569	6·8	1·917	8·8	2·175	90	4·500
2·85	1·047	4·85	1·579	6·85	1·924	8·85	2·180	95	4·554
2·9	1·065	4·9	1·589	6·9	1·931	8·9	2·186	100	4·605
2·95	1·082	4·95	1·599	6·95	1·939	8·95	2·192	1000	6·908
3·0	1·099	5·0	1·609	7·0	1·946	9·	2·197	10 000	9·210

Hyperbolic Logarithms are so called, because they express the areas of the asymptotic spaces of a right angled or equilateral hyperbola; that is, a hyperbola whose asymptotes make a right angle with each other; and whose conjugate and transverse axes are equal to each other. The asymptotic spaces are the areas included between the asymptote and the curve; those areas being limited by ordinates drawn at equal distances, parallel to the other asymptote; the lengths of those ordinates decreasing in geometrical progression.

Hyperbolic Logarithms may be found by multiplying common logarithms by 2·30258. For instance, the common logarithm of 4 is 0·6026 and × 2·30258 = 1·38629, which is the hyp. log. of 4.

MR. HORNBLOWER'S PATENT, 1781.

AFTER Mr. Watt had made trial of the expansive method, at Soho and at Shadwell as already stated, Mr. Jonathan Hornblower of Penryn, Cornwall, took out a patent, dated 13th July, 1781, " for a machine or engine for raising water or other liquids by means of fire and steam."

As this project has since been matured into a valuable improvement, its first origin is deserving of notice, although it produced no useful results during the time of Mr. Watt's patent ; for as Mr. Hornblower required all Mr. Watt's invention to begin with, he was prevented by the patent.

The intention of Mr. Hornblower's improvement was, to use the steam twice over, in two successive cylinders ; and thus obtain a greater power from it on the expansive principle, than could be done, in the simple way of using the steam only in its entire density.

The specification of Mr. Hornblower's patent was as follows :

" First, I use two vessels, in which the steam is to act, and which in other engines are generally called cylinders.

" Secondly, I employ the steam, after it has acted in the first vessel, to operate a second time in the other, by permitting it to expand itself, which I do by connecting the vessels together, and forming proper channels and apertures, whereby the steam shall occasionally go in and out of the said vessels.

" Thirdly, I condense the steam, by causing it to pass in contact with metalline surfaces, while water is applied to the opposite side.

" Fourthly, To discharge the engine of the water used to condense the steam, I suspend a column of water in a tube or vessel constructed for that purpose, on the principle of a barometer, the upper end having open communication with the steam-vessels, and the lower end being immersed in a vessel of water.

" Fifthly, To discharge the air which enters the steam vessels with the condensing water, or otherwise, I introduce it into a separate vessel, whence it is protruded by the admission of steam.

" Sixthly, That the condensed vapour shall not remain in the steam-vessel in which the steam is condensed, I collect it into another vessel, which has open communication with the steam-vessels, and the water in the mine, reservoir, or river.

" Lastly, In cases where the atmosphere is to be employed to act on the piston, I use a piston so constructed as to admit steam round its periphery, and in contact with the sides of the steam-vessel, thereby to prevent the external air from passing in between the piston, and the sides of the steam-vessel."

As Mr. Hornblower did not construct any engines according to this patent, till a subsequent period, it is best to pass on to Mr. Watt's next invention.

MR. WATT'S SECOND PATENT, 1781.

This was for " certain new methods of applying the vibrating or reciprocating motion of steam or fire engines, to produce a continued rotative or circular motion round an axis or centre, and thereby to give motion to the wheels of mills or machines."—Dated 25th October, 1781.

This invention has in the end proved a most valuable application ; but at its first origin, as detailed in this patent, it was not complete, being an application of the steam engine, which required greater perfection in the performance of the machine, than had been attained at that time ; in consequence, it was not acted upon immediately.

At the time of obtaining this patent, Mr. Watt was fully occupied in the further improvement of his engine for pumping water, and in perfecting the details of its construction; but after he had succeeded in effecting that object, he was enabled to apply its powers to produce circular motion with the greatest success. The account of this second patent will therefore be deferred, until the date when it was brought into use, and which should be marked as a most important era in the history of mechanical inventions.

MR. WATT'S THIRD PATENT, 1782.

This was for " certain new improvements upon steam or fire engines, for raising water and other mechanical purposes, and certain new pieces of mechanism applicable to the same."—Dated 12th March, 1782.

The specification is voluminous, and is elucidated by many drawings; it contains several different improvements, of which the following extracts and explanations will give a sufficient idea.

" My first new improvement in steam or fire engines consists, in admitting steam into the cylinder of the engine, only during some certain part or portion of the descent, or ascent of the piston, and using the elastic forces, wherewith the said steam expands itself, in proceeding to occupy larger spaces, as the acting powers on the piston, through the other parts or portions of the length of the stroke of the piston.

" Also in applying combinations of levers, or other contrivances, to cause the unequal powers wherewith the steam then acts upon the piston, to produce uniform effects, in working the pumps or other machinery required to be wrought by the said engine : by these improvements certain large proportions of the steam, hitherto found necessary to do the same work, will be saved."

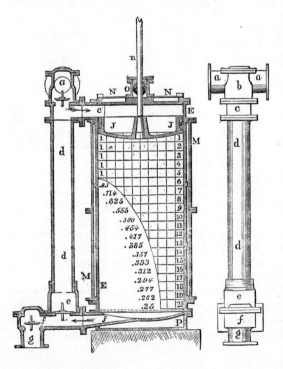

" This principle is explained by the annexed section of a steam cylinder, which is closed at the lower end by its bottom, and also at the upper by its cover. The solid piston is accurately fitted into the cylinder, so that it may slide easily up and down, and yet suffer no steam to pass by it. The piston is suspended by a rod which is capable of sliding through a hole in the cover of the cylinder, and the junction is made air and steam-tight round the rod, by a collar of oakum well greased and rammed into the box round the rod ; near the top of the cylinder there is an opening c to admit steam from the boiler, through a suitable valve a.

" The whole cylinder, or as much of it as possible, is inclosed in a case containing steam, whereby it is maintained at the same heat with the steam from the boiler.

" All things being thus situated, and the piston being near to the top of the cylinder ; suppose the space in the cylinder under the piston to be exhausted or entirely emptied of the steam, and suppose there is a free passage into the space above the piston, for the entry of steam from the boiler ; and that the steam is nearly of the same pressure or elastic force as the atmospheric air, when it supports a column of mercury of 30 inches high in the barometer.

" In such case, the pressure or elastic power of the steam, on every square inch of the area of the upper side of the piston, would be above 14 pounds avoirdupois weight.

" If the said power were employed to act upon the piston through the whole of its stroke, and to work pumps by the piston rod, either directly from that rod, or through the medium of a great lever, as is usual in steam engines ; then such an engine would raise, through the whole length of its stroke, a column of water, equal in weight to 10 pounds, for each square inch of the area of the piston, besides overcoming the friction of all the moving parts of the engine, and the *vis inertiæ* of the water.

" But supposing the whole stroke or motion of the piston, from the top to the bottom of the cylinder, to be eight feet, and that the passage which admitted the steam from the boiler, is perfectly shut when the piston has descended 2 feet, or one-fourth of the length of the stroke. In such case, when the piston had descended 4 feet, or one-half of the length of the stroke, the elastic power of the steam would only be equal to 7 pounds on each square inch of the area of the piston, or one-half of the original power ; and when the piston had descended 6 feet, or three-fourths of its course, the power of the steam would be one-third of the original power, or $4\frac{2}{3}$ pounds per square inch ; and when the piston had arrived at the bottom, or the end of its stroke of 8 feet, the elastic power of the steam would be one-fourth of its original power, or $3\frac{1}{2}$ pounds per square inch. And the elastic power of the steam at all other sub-divisions of the length of the stroke, are represented by the lengths of the horizontal lines, or ordinates of the curve delineated in the figure, and are further expressed in decimal fractions, of the whole original power, by the numbers marked opposite to the said ordinates or horizontal lines.

" The sum of all these powers is greater than fifty-seven hundredth parts of the original power, multiplied by the length of the stroke ; whereby it appears that when only one-fourth of the steam necessary to fill the whole cylinder, is employed, the effect produced, is more than one half of the effect which would have been produced in filling the whole cylinder full of steam, by admitting it to enter freely above the piston during the whole course of its descent.

" Consequently, the said new or expansive engine is capable of easily raising columns of water, whose weights are equal to five pounds on every square inch of the area of its piston, by the expenditure of only one-fourth of the contents of the cylinder of steam at each stroke.

" And though, for example, I have mentioned the admission of one-fourth of the cylinders full of steam, as being the most convenient, yet any other proportion of the content of the cylinder will produce similar effects ; and in practice I actually do vary those proportions as the case requires.

" In some cases I admit the required proportion of steam to enter below the piston, and then pull the piston upwards by some external power, against the elastic force of the steam from the boiler, which at that time communicates freely with the upper part of the cylinder, and this method of working produces similar effects to those described.

" The power which the steam thus exerts during its expansion is unequal, but the weight of the water to be raised, or other work to be done by the engine, being supposed to resist equally, throughout the whole length of the stroke, it is necessary to render the acting power equal, by other means. I perform this in various ways as follows:

" First, by means of two wheels or sectors of circles, one of which is connected with the pump-rod, and the other with the piston rod of the engine, and the two wheels are connected together by means of a rod or chain, pulling obliquely in such direction, that the effective lengths of leverage whereby the two wheels act upon one another, will decrease and increase respectively during the ascent or descent of the piston, nearly in the ratio required to produce a uniform effect.

" My second method for equalizing the varying powers of the steam, is by means of chains, which are wound upon one spiral, and wound off another, as the piston descends ; these spirals are fixed upon two wheels or sectors of circles, to which the chains of the piston and pump rods are attached respectively.

" My third method is by means of a friction wheel attached to or suspended from, one sector or wheel, and acting upon a curved or straight part of another sector, wheel, or working lever, so as to continually change its distance from the centre of motion.

" My fourth method is, by causing the centre of suspension of the working beam, or great lever, to change its place during the time of the stroke, whereby that end of the lever to which the piston is suspended shall become longer, and the other end to which the pump rods are suspended shall become shorter, as the piston descends in the cylinder.

" My fifth method consists in loading the working lever of the steam-engine (or some other wheel, or lever connected with it) with a heavy weight, in such a manner that the weight shall act against the power of the piston, at the commencement of its descent, but that as the piston descends, the weight shall gradually move towards that end of the lever from which the piston is suspended, so as to act in favour of the piston at the latter part of the stroke."

This may be carried into effect, by forming the top of the working lever to a curve, and applying a heavy roller, or weight, to roll along the same, and change its place upon the lever during the motion thereof.

Or the same effect can be produced by a large weight attached to the working lever, at a considerable height above the centre of motion. When the piston begins its descent, this weight opposes itself to the motion, until the lever is moved so much, that the centre of gravity of the weight is perpendicularly over the centre of motion of the lever: the weight then ceases to have any effect on the engine; but after it has passed this position, it tends to aid the effort of the piston to draw up the load of water in the pumps. It is possible to adjust the weight in its position, quantity, and height above the centre of motion, so that it will very nearly equalize the diminishing force of the piston.

The weight of a quantity of water may be very conveniently applied to this purpose, by means of two large cylinders or pumps, open at top and bottom, and each fitted with a solid piston without valves. The pistons of the two cylinders are suspended from the opposite ends of the great lever, so that the descent of one, will produce the ascent of the other. The cylinders are filled with water, and a large trough conducts it from the top of one cylinder to the top of the other, so that the weight of the water is alternately transferred from one piston to the other. When the engine is at the middle of its stroke, and each piston at the middle of its respective cylinder, the water will be equally divided between them, and their pistons will hang in equilibrium on the lever; but when one end of the lever is depressed, the piston suspended from the opposite end will rise, and raise up part of the water which rests upon it, into the trough, by which it will run into the opposite cylinder, and the corresponding descent of that piston having made room for it, the water will become unequally divided. When one piston is at the top of its cylinder, and the other is at the bottom, the latter will have the whole of the water resting upon it.

Suppose the steam-piston to be at the top of its cylinder, the water-cylinder at the opposite end of the lever, will then contain all the water, and the other cylinder none of the water; as the steam presses down the steam-piston, it must raise the water in the cylinder at the outer end of the lever, and the weight of this water will oppose the motion of the piston, because the whole of the water in the water-cylinder must be lifted; but by the time the steam-piston has descended one-third or one-fourth, and the steam is shut off, the pressure on the steam-piston begins to diminish; and the weight of the water on the water-piston begins to diminish at the same moment, because part of the water begins to run through the trough, and enter into the opposite cylinder, where its weight upon the piston tends to aid the steam-piston in descending; and this aid continues to increase as the piston descends further, because the water is regularly transferred from the ascending to the descending piston, until the whole of it rests upon that piston which is at the bottom of its stroke.

" My sixth method, or contrivance, for equalizing the varying powers of the steam, consists in employing that surplus power which the steam exerts upon the piston in the first parts of its motion, to give a proper rotative, or vibratory velocity, to a quantity of matter, which, retaining that velocity, shall act along with the piston, by its energy or impetus, to assist in raising the columns of water during the latter part of the motion, when the powers of the steam become defective."

This may be carried into effect by a toothed-rack on the piston-rod, working into the teeth of a small pinion upon the axis of a fly-wheel, so that the wheel will be turned round one way, when the piston descends, and the other way when it returns: this fly-wheel being at rest when the piston first starts, it will oppose the motion by its inertia; but after that inertia is overcome, and the fly wheel is put in motion, it will aid the piston by its energy, or tendency to continue in motion.

A still better method is, to connect the great lever by a connecting-rod, with a crank upon the axis of the fly-wheel, so as to give a continuous circular motion to the fly-wheel, by the reciprocating motion of the engine, in the same manner as the foot-wheel of a lathe or spinning-wheel, is regularly turned by reiterated treads of the workman's foot. Mr. Watt also combined two equal cog-wheels for this purpose, in such manner that one revolves round the other, as the earth revolves round the sun, and hence they are called sun and planet wheels: this arrangement, which has the effect of giving two complete revolutions of the fly-wheel for each stroke of the engine, afterwards came into general use for turning mill-work with circular motion, but it is proposed in the specification to this patent, as one of the means of equalizing the varying powers of steam in engines working on the expansive method for raising water.

Hence we see that Mr. Watt proposed to equalize the varying force of expanding steam, by very different means; by combinations of levers varying in their effective leverage upon each other; by the gravitation of weights so disposed as to oppose the piston at first, and afterwards to assist it; and also by the impetus of rapidly moving weights, which alternately retard, and aid the piston in its motion.

" My second improvement upon steam-engines consists in employing the elastic power of the steam, to force the piston upwards, and also to press it downwards alternately, by making a vacuum above or below the piston; and, at the same time, employing the steam to act upon the piston in that end, or portion of the cylinder which is not exhausted. An engine constructed in this manner, can perform twice the quantity of work (with a cylinder of the same size) or exert double the power, in the same time, which has hitherto been done by any steam-engine, in which the active force of the steam is exerted upon the piston only in one direction, whether upwards or downwards.

" My third improvement on steam-engines consists in connecting together, by pipes, or other proper channels of communication, the steam-vessels, or condensers of two or more distinct steam-engines, either of which has its separate working lever, and other constituent parts of a steam-engine; or is otherwise so constructed, that it can work pumps, or other machinery, which are either connected with, or are independent of, those wrought by the other engine, and which two engines can take their strokes alternately, or both together, as may be required.

" My fourth improvement on steam-engines consists in applying a toothed rack, and sector of a circle, for suspending or connecting the pump-rod or piston-rod, with the working lever, or other machinery used in place thereof, instead of chains, which have hitherto been used for these purposes.

" My fifth new improvement on steam-engines consists in making the steam cylinders in the form of greater or lesser segments or sectors of cylindrical vessels; and placing in the centre or axis of the circular curvature of such vessels, a round shaft or axle, which passes through, and extends beyond, one or both ends of the steam-vessel; and I shut up the ends of the steam-vessel with smooth flat plates, which have proper apertures for the axle or shaft to pass through; and within the steam-vessel I fix to the axle a plate by way of a piston, extending from the axle to the circular circumference of the steam-vessel, and also extending from one end of the steam-vessel to the other end thereof; and I make such piston steam tight, by surrounding the parts which fit to the steam-vessel, with hemp or other soft substances soaked in grease or oil, or by means of springs made of steel, or other solid and elastic or pliable materials; and within the steam-vessel I fix one or more plates or divisions, extending from the axle, to the circumference of the steam-vessel; and where these plates or divisions join to, or approach to the axle, and also where the said axis passes through the end plates of the steam-vessel, I make such joinings steam and air tight by similar means.

" In the steam-vessel, on each side of the said piston, I make channels or apertures for receiving and discharging the steam, which channels I furnish with proper valves for that purpose: I also apply to the said engine, proper condensers and air-pumps; and the pumps which raise water, or such other machinery as is required to be wrought by the said engine, are put in motion, or worked by wheels fixed upon the external parts of the said axle, or by any other suitable mechanism.

" The engine so constructed, is wrought by admitting the steam between the fixed divisions and the moveable piston, and exhausting or making a vacuum on the other side of the piston, which accordingly, by the force of the steam, moves into the said vacuum, and turns the axle a greater or lesser portion of a circle, according to the structure of the machine. The piston is returned to its former situation, by admitting steam on the other side of the said piston, and drawing the piston back by some external power, or by exhausting the part of the steam-vessel which was first filled with steam (a)."

The different inventions described in this patent, are highly ingenious and creditable to Mr. Watt, and show the minuteness with which he had considered all the possible means of effecting his purpose. The expansive method of using steam has been very generally adopted in engines for mines, and with great advantage, particularly in modern engines; but the contrivances for equalizing the varying forces, have not been practised nor found requisite in practice, for the following reasons.

The application of any spirals, or oblique levers, to large engines employed in pumping, would be attended with great difficulties; and would be inadequate to equalize the action of the expansive principle, when applied in its fullest extent, that is, when the stoppage of the supply of steam is made to take place at a small portion of the descent; for in that case, an excessive strain would be thrown upon the centres of the spirals or levers, if they were sufficiently curved, or oblique, to equalize the action. Lord Stanhope has since applied the principle of Mr. Watt's oblique leverage in a very judicious manner, to the printing-press; but, in so

(a) All the drawings annexed to this specification have been engraved and published in Hall's Encyclopædia, folio, under the article STEAM-ENGINE, vol. ii. The specification was afterwards printed by Mr. Watt, with the same plates.

small a machine, worked only by the strength of one man, the strong cast-iron frame of the press has been frequently broken.

The other principle of shifting weights, would be dangerous and unmanageable on a large scale.

One objection applies to all the above methods, viz. that they would require the steam to be always stopped off, at the same proportion of the stroke, because, if by any of these methods the engine was adapted to equalize the varying forces of the steam, it could only do so, as long as the variations occasioned by the expansion continued to be the same. Again, the action of the equalizing contrivance must be the same in the returning stroke, as in the working stroke, so that these contrivances would only be adapted for the double acting engines, and they are not commonly used for pumping water.

Hence, we may consider all these as first ideas, and projects not reducible to practice ; and some were no doubt introduced into the specification rather for the purpose of guarding against piratical evasions, than as real proposals for practice.

The regulation by means of the inertiæ of moving masses, is more practicable than the other methods. That of a continuously revolving fly wheel is a very useful addition to an expansive engine, but it has been rarely used for the object stated by Mr. Watt in this patent (a).

In short, it has not been found advisable to work Mr. Watt's engines with so great a degree of expansive action, as to require any contrivance for the equalization of the force, because the power exerted by an engine of a given size is greatly diminished, by allowing it only a small supply of steam, at the same time that its performance in respect to fuel is improved. This circumstance restricts the use of the expansion method to small limits.

The advantages of the expansive method of working a steam-engine, are more fully obtained when the steam has an elasticity considerably greater than that of the atmosphere : this is a modern improvement by Mr. Woolf, and it admits of stopping off the supply, at a very small portion of the stroke, and still obtaining a great force from an engine of suitable dimensions.

The practice of Messrs. Boulton and Watt has always been to employ steam a little more elastic than the pressure of the atmosphere, and to cut off the supply at one-half, or two-thirds of the stroke, according to the circumstances under which the engine works. In this case, the decreasing pressure in a large engine, is not much greater than to counteract the acceleration ; and the energy of the heavy working lever, pump-rods, and the rising column of water, produces a nearly uniform motion.

About 1785, Mr. Watt had brought this kind of engine to great perfection, and established a form of construction and a scale of proportions for all its parts, which has undergone no material alteration to the present day.

(a) One of the earliest engines which Mr. Watt put up in London about 1782, was constructed with a crank and fly-wheel, for the purpose of regulating the expansive action ; this was at Pimlico Wharf, for the Chelsea water-works. The author frequently visited this engine in 1803 and 1804, and made sketches of its parts ; it then worked constantly, but the connecting rod of the crank was detached from the great lever, and appeared not to have been used for many years. At first this engine had been intended to work with the piston rising in vacuo, as described in p. 335, but it was afterwards altered, as there mentioned.

DESCRIPTION OF MR. WATT'S SINGLE ENGINE FOR PUMPING WATER FOR
DRAINING MINES, 1788.   (Plate X.)

The drawing is a section of the whole engine, to show as much of all its parts, interior and exterior, as can be done at one view.   The boiler is placed in an outbuilding erected across the end of the engine-house, and covered by its own separate roof.   The engine-house itself is very nearly the same as for the atmospheric engines before described, p. 139.

A is the fire-place, provided with its fire-door in front, to introduce fuel.

B the fire-grate, formed of parallel iron bars on which the fire is made.

C C the boiler, seen in its transverse section ; it is made long and narrow, the upper part being a semi-cylinder, with its axis placed horizontally ; this upper part is reserved for steam ; the lower part, which contains the water, has upright sides, and a concave bottom to receive the action of the fire.   This form of boiler is called a waggon boiler, from its resemblance to a tilted or covered waggon. The fire-grate B, is situated beneath one end of the long boiler, and the flame from it proceeds beneath the concave bottom of the boiler, to its farther end, where it rises up into a channel or flue 9, which is formed all round the outside of the lower part of the boiler, wherein the water is contained ; the flame thus communicates heat to the boiler and its contents, and after making the complete circuit round the boiler, the smoke escapes into the perpendicular chimney D D ; the aperture of the passage from the flue 9, into the chimney, is regulated by an iron damper or slider w, which is adapted to rise and fall, in the manner of a sluice or a sash window, so as to open or close the passage ; this sliding damper w, is suspended by a chain passing over a pulley, with a sufficient balance-weight to retain it wherever it is placed.   By raising or lowering this damper, the rapidity of the draft of air into the chimney, and consequently the combustion of the fuel on the fire-grate is regulated at pleasure, according to the supply of steam that the engine may require.

y are the two guage-pipes and cocks, to show when the water in the boiler stands at its proper height ; and T is the upright feeding-pipe, which introduces the supply of water into the boiler.

a is the steam-pipe, which proceeds from the top of the boiler, and passes through the wall, into the engine-house ; it conveys the steam to the upper nossel or steam-box b, which contains the valve, by which the steam is admitted into the cylinder, to perform the working stroke.

E is the cylinder, very truly bored within, and fitted with its moveable piston J ; the top of the cylinder is closed by the cover, which has an opening through the centre, for the rod of the piston to pass through, and this opening is packed tight round the rod with hemp, to make a close fitting.   The cylinder is enclosed within a cylindrical case, leaving a space all round the outside of the cylinder, which space is filled with steam, to keep the cylinder hot.   This steam is supplied to the steam-case by a small copper-pipe branching from the great steam-pipe a, with a cock to stop it occasionally.   There is also a drain-cock at the lower part of the steam-case, to tap or drain off the water which collects in the steam-case, from the condensation of the steam.

b, is the upper steam-box, containing the steam-valve, which, when opened, admits the steam into the top of the cylinder, through the branch or lateral passage c ; this valve is opened when the engine is to make its down stroke.

z z

d is the upright steam-pipe, which conveys the steam from the top of the cylinder to the bottom, when it is to make its returning stroke.

e, is the box containing the lower steam-valve, or equilibrium-valve, which, when open, admits the steam into the lower part of the cylinder through the branch or lateral passage *f*; this valve by opening a free communication between the top and the bottom of the cylinder, puts the piston in equilibrium, and allows it to rise by the counterweight.

g, the eduction-pipe, which conveys the steam away from the cylinder to the condenser, to exhaust the cylinder, and form a vacuum beneath the piston preparatory to the working stroke.

i is the box containing the exhausting-valve, which opens the communication between the branch *f* at the bottom of the cylinder, and the eduction-pipe *g*.

F is the condenser or receiver; it is a cylindrical vessel immersed in cold water to keep it cool; the eduction-pipe *g* joins to the top of it. At *x*, a small copper-pipe is joined to the condenser, and it extends to the outside of the cistern; the orifice is covered with a small valve opening outwards, and immersed in a small vessel of water; this is called the snift or blow-valve; it serves to discharge the air from the vessels at first starting the engine.

G G the condensing cistern containing the cold water which surrounds the air-pump and condenser; the cistern is kept supplied with cold water by the pump R, and there is a constant surplus or overflow, which runs off by a waste-pipe.

j, the injection-valve and cock, to admit a jet of the cold water from the cistern G, into the condenser, to cool the steam therein. The valve, when shut, stops the passage of the injection-spout, and prevents the entrance of any water during the returning stroke, but it is opened for the descending stroke. The cock is placed in the short pipe, to regulate the flow of the water, after it has passed through the valve.

k, the foot-valve contained in a lateral passage leading from the lowest part of the condenser F to the air-pump H, to extract the water and air therefrom; the valve *k* opens to allow the water and air to pass out from the condenser, but shuts to prevent its return.

H, the air-pump, or discharging-pump, which is truly bored within, and fitted like the cylinder with its moveable piston, or bucket, but this bucket is perforated to allow the air and water to pass up through it, and is provided with valves opening upwards, which prevent any return of water or air downwards. The top of the air-pump is closed by a cover, like that of the cylinder, through the centre of which the air-pump rod *l* passes in a close stuffing box; at the upper part of the air-pump, is a discharging spout *m*, to convey the hot water and air, into the hot well; the opening of this spout is closed by a flap-valve *m* opening outwards to permit the discharge of the water and air, but to prevent any returning.

I is the hot well, which is a part of the condensing cistern G, partitioned off to form a small cistern, for the reception of the hot water which is discharged by the air-pump, and prevent it from mixing with the cold water of the condensing cistern.

J J is the piston of the cylinder, very accurately fitted into it, and surrounded with a packing of hemp rammed hard into a groove round the edge of the piston, and retained therein by a ring or cover which is fitted into the groove over the hemp, and strongly pressed down upon the hemp by screws.

n is the piston-rod, made of iron very straight and true, and well polished, that it may slide freely up and down through the stuffing box, in the centre of the cylinder cover; the lower end of the piston-rod is firmly fastened into the centre of the piston J, and the upper end is connected with the main chains K, which apply upon the arch head at the end of the great lever.

L L the great lever; it is a single beam of oak, moveable about its axis or centre of motion *p*; this axis is affixed to the upper side of the beam, by iron straps; the arch heads at each end of the lever are firmly fastened on, and supported by iron braces, as shown in the figure; and to strengthen the beam, an upright king post *q* is erected upon the axis *p*, and from the top of this king post, iron braces are carried obliquely to each end of the beam, so as to form a truss, similar to that of a roof. Two iron catch-pins are fixed in the arch heads K, at each end of the great lever L, and their ends project out at each side of each arch, so as to strike upon the stop springs of the spring beams U, in order to limit the motion of the piston, and prevent it from exceeding the proper length of stroke.

M, the great pump-rod or spear, suspended from the arch head at the outer end of the lever by its chains, and descending into the mine, to work the pumps therein. The weight of the rods M exceeds that of the piston J, so as to form a preponderating counterweight, which draws up the piston, and moves the whole engine during the working stroke. N. B. the pumps and pit-work down in the mine are the same as already described, (p. 154 and p. 214).

7 is a small arch, affixed to the great lever, and there is another similar one behind it, to receive two chains which suspend the working-plug beam *l*; the rods from the two chains being united, and connected with the upper end of the plug, whilst the rod *l* of the air-pump is affixed to the lower end. To keep the plug steady as it works up and down, a piece of wood is fixed horizontally across it at the lower end, and the two ends of this cross piece slide freely up and down in vertical grooves, made in two upright posts. The plug has three prominent pieces of wood fixed to it, to give motion to the handles of the working-gear; two of them are seen at 1 and 2.

r, *s*, 3, *t*, *u*, 4, is the working-gear, consisting of three horizontal axes placed one above the other, at *t*, *u*, and 4, and each one with a handle or spanner, *r*, *s*, and 3, fixed upon it, to give it motion, either by the hand, when the engine is to be started, or stopped, or regulated, or when the engine is in regular action, the same handles are moved by the chocks 1 and 2 of the plug *l*; each of the axes and handles of the working-gear is connected by suitable levers and rods, with one of the valves *b*, *e*, and *i*. The pivots of the three axes are sustained by two upright posts which are erected upon a beam placed across the top of the condensing cistern, the upper ends of the posts reach up to the spring beams V.

R is the cold water-pump, which raises a constant supply of cold water from a tank or pit at the outside of the engine-house, and delivers it by a trough through the wall V, into the condensing cistern G. N. B. in the figure, the hot well I appears just beneath the end of the trough, but the hot well is behind the trough, and does not receive the water therefrom. The rod of the cold water-pump is suspended from the great lever L, which thus gives it motion.

S, is the hot water-pump, fixed against the outside wall of the house, and worked by a small rod which is suspended from the great lever L; the suction-pipe of this pump draws water out of the hot well I, and it is raised into the cistern at the top of the pump S, whence it is conveyed by a long-pipe 8, to the perpendicular feeding-pipe T, which supplies the boiler with water.

z z 2

T, the perpendicular feeding-pipe; it is open at top, and receives water by the pipe 8 8, through a regulating cock, from the hot water-pump; the lower part of the feeding-pipe T passes through the top of the boiler, and descends into the water nearly to the bottom.

V, the main lever wall which sustains the fulcrum, or centre of motion p of the great lever L. The other walls and the roof of the building, and other parts not already described, are self-evident from the drawing; the floors and timber work of the house are very nearly the same as the atmospheric engine (p. 139). The principal change made by Mr. Watt in the manner of fixing the engine in the building, was in fastening the cylinder down upon its bottom, instead of suspending it between beams, the boiler being removed from beneath the cylinder into a side building, so as to make room for the cylinder pier and the condensing cistern.

X, the pier of masonry on which the cylinder E is placed; it is built upon two very strong cross beams which reach across between the two side walls of the engine house; the cylinder is held down upon the pier X by four strong bolts, which pass down through the pier, and also through the cross beams beneath it, and have cross keys below; the upper ends of the bolts have nuts screwed upon them, by which the flange at the bottom of the cylinder is bound down, so as to keep it very firm upon its basis.

The condensing cistern G, which contains the air pump and condenser, is made of thick planks; it is placed before the cylinder pier, and is supported upon three strong beams, which extend across the house; one of these beams passes immediately beneath the air-pump, which is fastened down thereto by two bolts which go down through the bottom of the cistern.

Mechanism by which the engine is caused to work itself, and perform its reciprocations without assistance. Each valve is a circular plate of brass formed conical on the edge, and accurately fitted into the aperture of a circular brass ring, which forms the passage for the steam, and is situated at the bottom of the square iron box in which the valve is contained. The valve must be lifted up to open the passage for the steam, or let down into its seat to stop the passage; for this purpose a toothed rack is fixed on the stem or spindle of the valve, and it is lifted by a toothed sector, fixed upon the middle of a short horizontal spindle, which passes across the valve box, and one end comes through to the outside, being very accurately fitted into a socket in the side of the box, so that it can turn round freely, but will not permit any leakage; a lever is fixed on the extreme end of the spindle at the outside of the box, and this lever is connected by a rod with one of the levers of the working gear, in order to open or shut the valve within-side its box.

The valves which must be opened and shut in succession, to produce the alternate motions of the engine are four in number; viz. the upper steam valve b called *the expansion-valve;* the lower steam valve e called *the equilibrium-valve;* the lower valve i called *the exhausting-valve;* and the small valve down in the cistern at j called *the injection-valve.*

These valves are opened and shut by means of three handles r, s, and 3, each fixed upon a separate axis t, u, and 4; the handles project outwards from their axis, and are disposed close by the side of the plug beam l, so as to be moved by means of pieces of wood called chocks, which are fastened to the plug as shewn at 1 and 2, and project so much sideways from the plug, as to intercept the handles.

Each axis has a short lever fixed to it, and the end is connected by a small rod, with that lever at the outside of each valve box, which, by means of its axis going into the box and its internal sector and rack, opens or shuts the valve contained in the box.

Each axis has also another short lever fixed to it, with a weight suspended by a rod from its extremity; these levers are so disposed, that the weights will always tend to turn the axes and the handles in that direction which will open the valves respectively; lastly, a suitable catch is provided to each lever, to detain it when the weight is raised, and the valve closed, and then the catch will keep it so closed; but by disengaging the catch, the weight will drop, and open the valve instantaneously.

The upper handle r, for the left hand, is called *the expansion handle;* it is pressed down by the plug, to close the valve; and it is raised up, by the action of its weight, when the valve is to be opened.

The middle handle s, for the right hand, is called *the exhausting handle;* it is also pressed down by the plug, to close the valve; and it is raised up, by the weight so as to open the valve.

The lower handle 3, for the left hand, is called *the equilibrium handle;* it is lifted up by the plug, to close the valve; and is thrown down by the weight in order to open the valve.

The injection-valve j is opened by a strap, which wraps round the middle axis u, and the strap has a wire descending from it to the valve; therefore, when the exhausting handle s rises up suddenly by its weight, and opens the exhausting valve i, the strap is wound up, and opens the injection-valve g also.

The injection-valve j is adapted to close the orifice at the end of a short pipe, which enters into the condenser and turns up; and the pipe has a cock in it, between the valve and the condenser, in order to cut off the communication occasionally, and to regulate the supply of injection whilst the valve is open, during the working stroke. This cock is always open when the engine is in action, but it must be shut when the engine ceases to work, to prevent the condenser filling with water.

OPERATION OF MR. WATT'S SINGLE ENGINE.

Preparation. In the quiescent position of the engine, when it is at rest, the preponderance of the pump-rod M, or counter-weight, always draws the piston up to the top of its cylinder, as in the figure; the air-pump bucket being also at the top of its barrel: and all the valves shut.

Suppose the fire to be lighted at A beneath the boiler C, and that all the valves b, e, i, and j are shut, by pressing down the two upper handles r and s, and lifting up the lower one 3; their respective catches will detain them in those positions, until the steam is sufficiently heated, and the engine is ready to work.

In order to prepare the engine to work, all the three valves b, e, and i must be opened at once. This is done by relieving the handles from their several catches, and the weights immediately open the valves. The steam then enters through the valve b, into the top of the cylinder, and passes by the pipe d, through the lower steam-valve e and branch f, into the bottom of the cylinder; also through the exhausting-valve i, and eduction-pipe g, into the condenser F.

At first the coldness of the metal parts, condenses all the steam which enters, and until all the iron, with which the steam comes in contact, becomes heated nearly to the temperature of boiling water, the steam flows from the boiler in a rapid stream, and is condensed as fast as it arrives at the cylinder and condenser; but when the parts become hot, the steam will acquire nearly the same force in the cylinder and pipes, that it had in the boiler ; it then begins to occupy every cavity of the engine, and in a little while displaces the air, first from the valve boxes, condenser, and pipes, and ultimately from the cylinder also ; this air passes out, and is discharged at the snifting or blow-valve *x*.   The blow-valve is always covered with water, in a small cistern attached to the side of the large one G, for it is difficult to fit it so accurately as to ensure its tightness, unless it is covered with water.

Through this valve the air is discharged by the steam, not at every stroke, as in Newcomen's engine, but only at first setting the engine to work, and this operation is called the *blowing through*, because of the bubbling noise it makes. When the cylinder and other parts are properly heated, and the air discharged from them, a very smart crackling noise is heard at the valve *x* ; this noise is occasioned by the water in the small cistern, producing a sudden and rapid condensation of the issuing steam, when the air is all gone.

The engine being cleared of air by two or three repetitions of this blowing, all the three valves are to be shut, by pressing down the two upper handles *r s*, and lifting up the lower handle 3, in which situations their catches will retain them.   The further supply of steam from the boiler is now intercepted, and the passage of the steam from the cylinder to the condenser is stopped ; as the cold surface of the condenser still continues to condense steam, there will soon be a vacuum formed in the condenser, whilst the cylinder both above and below the piston remains full of steam.   The vacuum in the condenser will become perfect in a few seconds, from the external cold water-alone, though more rapidly when an injection is made into the condenser, to mix with the steam.

Action during the working stroke.   In this state, the engine is prepared for starting at a moment's notice ; by releasing the catches, and letting the two upper handles, *r* and *s*, rise up by their respective weights, the expansion-valve *b*, and the exhausting-valve *i*, and injection-valve *j*, are opened all at once ; the former admits the steam into the top of the cylinder, to press upon the piston ; whilst the latter allows the steam already in the lower part of the cylinder, to flow out therefrom into the vacuous condenser ; the injection-valve was lifted at the same time with the exhausting-valve ; and the injection-cock being opened, admits a jet of cold water into the condenser, which condenses the steam as fast as it arrives from the cylinder, so that in an instant nearly all the steam in the lower part of the cylinder is drawn off and condensed.   The pressure of the steam above the piston being now unbalanced by any sensible pressure of steam beneath, it is forced downwards, and draws up the pump-buckets, and columns of water in the pumps, with a velocity depending upon the descending force of the piston, compared with the resistance of those columns.

The piston having descended about one-half of its stroke, the chock 1 of the plug *l* meets the expansion handle *r*, and, pressing it down, shuts off the steam from the boiler.   That part of the handle against which the chock 1 acts, becomes perpendicular when the valve is so shut, the handle being bent for that purpose, so that the chock will slide against the perpendicular part of the handle, which is straight, without producing any further depression of the handle by the remaining motion of the plug, but the handle is held down to the same position, so as to keep the valve shut.   The piston therefore continues its descent by the expansion of the quantity of steam first admitted into the cylinder ; but having arrived at the bottom of its stroke, another chock on the opposite side of the plug *l* seizes the middle handle *s*, and presses it down, which shuts the exhausting-valve *i*, and also shuts the injection-valve *j* by the strap and rod.

The catch which is to retain the handle s in that position, is so connected with the other catch which retains the handle 3, that, in the act of latching the handle s, the other handle 3 is released from its catch, and then it falls by its weight, and suddenly opens the equilibrium-valve e.

Action during the returning stroke. Let us now consider the position of the engine. The middle handle s will be held down by its catch or detent, so as to keep the exhausting-valve i shut; the handle r of the expansion-valve b, is also kept down by the same chock 1, which kept it shut during the latter half of the descent of the piston. Under these circumstances the piston is at liberty to rise by the action of the counter-weight M, because the opening of the equilibrium-valve e has established a free communication between the top and bottom of the cylinder, so that the steam in the top of the cylinder can flow through the pipe d, and enter into the bottom of the cylinder, as fast as ever the piston is drawn up by the action of the counter-weight.

When the piston has thus risen about half way to the top of the cylinder, the chock 1 in the plug, quits the expansion handle r; but this handle cannot yet be thrown up by its weight, so as to open the upper valve, because its catch keeps it down until the piston arrives nearly at the top of the cylinder, as in the figure; a third chock 2, which is fixed to the plug l, then raises up the lower handle 3, and closes the equilibrium-valve e; and when the piston is quite up, another small pin in the plug releases those catches which detain the other two handles r and s; consequently the weights of those handles cause them to rise up, as in the figure, so as to open the exhausting-valve i, and injection-valve j, and the expansion-valve b, immediately after the equilibrium-valve e is shut.

Repetition of the working stroke. The steam from the boiler is now admitted to press upon the upper surface of the piston, whilst the steam from the lower part of the cylinder, beneath the piston, rushes into the condenser, where, being met by the cold injection water, it is condensed, and a vacuum being formed in the lower part of the cylinder, the steam presses down the piston to make another stroke.

At one-half of the descent, the chock 1 of the plug depresses and holds down the upper handle r, as before, to close and keep the expansion-valve b shut; and when the piston has arrived at the bottom, the plug presses down the middle handle s, to shut the injection and the exhausting-valves; and in catching, it disengages the catch of the lower handle 3, and the weight thereof opens the equilibrium-valve i.

The piston then rises by the counter-weight, and when at the top of its stroke, the plug lifts the lower handle 3, and shuts the equilibrium-valve, and also discharges the two other handles so as to open the expansion-valve, the exhausting-valve, and the injection-valve; this produces another descent of the piston as before, and in this manner the engine continues its action.

The action of the air-pump H does not take place until the engine has made one or two strokes, when the condensed steam, and the injection-water, will have accumulated in some quantity in the condenser, also any air which may have entered, or have been extricated from the water in boiling; then every time the bucket descends to the bottom of its course, it dips into the water contained in the bottom of the barrel, and part of that water passes through the valves in the bucket; when the bucket is drawn up, those valves shut, and all the water which is above them is lifted up to the top of the barrel, and there forced out through the discharging-valve m into the hot-well I.

The same drawing up of the bucket, causes a vacuum in the pump-barrel beneath it ; and if the condenser F contains either air or steam, such vacuum will be more perfect than that in the condenser ; consequently those fluids will press by their elasticity upon the surface of the water contained in the lower part of the condenser, so as to force it through the foot-valve at *k* into the lower part of the barrel H of the air-pump ; and after all the water is thus driven from the condenser, part of the air or elastic vapour in the condenser, will follow that water, and enter into the pump until the vacuous space in the barrel beneath the bucket becomes filled with vapour to nearly the same elasticity as in the condenser.

This takes place whilst the pump-bucket is at the top of its barrel, as in the figure ; and on the descent of the bucket, the space beneath it is diminished, until the rarified vapour is so much compressed, that its elasticity will be first sufficient to close the hanging-foot-valve *k*, and next to lift up the valves in the bucket, so as to pass through them into the space of the barrel above the bucket ; and when the bucket has descended to the very lowest, the water contained in the bottom of the barrel, not being able to escape through *k*, must also pass up through the bucket valves, so as to get above the bucket ; consequently, when the bucket ascends, it will carry up this air and water before it, and the space of the barrel above the bucket diminishing as it rises, the rare vapour or air which it contains is compressed by being crowded into less space, until it becomes equally elastic with the atmospheric air, and then it suddenly makes its way through the discharging-valve *m*, into the hot-well ; and the water follows the air.

The ascent of the bucket left a vacuum beneath it in the barrel, as before ; and, therefore, another portion of the air or vapour from the condenser expands into that vacuum, ready to be drawn out by the bucket at the next stroke. As soon as the bucket begins to descend, the discharge-valve *m* shuts, and prevents the hot water or air from returning into the pump, to press upon its bucket.

By this operation the vapour in the condenser is evacuated, even when it is so rare that the whole contents of the barrel of the pump must be compressed into a few cubic inches, in order to become equal in elasticity to the pressure of the atmosphere ; still this small quantity of air will be effectually expelled through the discharge-valve *m*, because the water resting upon the bucket of the pump, follows the air, so as to drive the whole of it from the top of the pump, and then the water itself will follow.

The action of the cold water pump R, is to supply the condensing cistern G with a sufficient quantity of cold water, to absorb and carry away all the heat of the steam which is condensed, without raising the temperature of the water which remains in the cistern.

The principal part of the heat is absorbed by that portion of the cold water which is admitted into the condenser by injection, and which is discharged by the air-pump into the hot-well, at a temperature of from 96 deg. to 110 deg. ; as much of this hot water as is required to supply the evaporation from the boiler, is drawn from the hot-well by the hot-water pump S, and the surplus is carried off from the hot-well by an overflow pipe and drain, which must be provided to convey it quite away from the engine, without allowing it to mix with the cold water in the condensing cistern, or with the cold water in the pit, from which the pump R draws its supply.

The cold-water pump is adapted to raise considerably more water into the condensing cistern G, than is required for the injection, so that there will be a continual surplus of cold water running waste out of the great cistern G, by its overflow pipe, the same as out of the hot-well ; by thus separating the hot water from the cold, and allowing a surplus of both to overflow, the water which surrounds the condenser, and from which the injection is derived, is kept very cool ; a condition which is essential to the good performance of the engine.

The water that the hot water pump S draws from the hot-well, is conveyed by a long horizontal pipe and cock 8, 8, from the cistern at the top of the pump, to the upright feeding pipe T, by which it is introduced into the boiler near the bottom ; the hot pump is calculated to raise much more water than the boiler requires, and the surplus runs waste from the cistern at the top of the pump, by an overflow pipe, into the common waste drain.   The feeding cock in the horizontal pipe 8, must be adjusted by the engine-man, so as to admit a proper supply of water into the boiler, and keep the water always at the same height therein.

MANAGEMENT OF THE ENGINE.

THE engine-man must keep the fire as regular as possible, by feeding it with fresh coals at intervals, and occasionally raking with a hook between the fire-bars from the underside of the grate, to clear away all impediments to the free entrance of air.   He must also regulate the draft by the damper w, according to the force of fire that the engine requires ; the fire-grate should be in all cases covered with a small thickness of fuel evenly spread over it, but not thrown on in heaps, and the rapidity of the combustion should be governed by the draft of air ; therefore, if the steam accumulates in the boiler, the damper w should be lowered, or if the steam become deficient, the damper must be raised (a).

The principal care of the engine-man must be to keep the engine to a steady and regular motion, so that it may always perform the full length of stroke for which it is intended, and yet never exceed that length so as to strike the catch pins upon the stop springs ; the due performance of this duty will require a constant attention to the fire, and to the working gear.

He must so manage the fire, by supplying the fuel, and regulating the damper, as to keep the steam always to the same elasticity without variation ; and he must so adjust the chocks and pins in the plug l, as to allow the engine the requisite dose of steam to enable it to perform its task ; but he must allow neither more nor less steam than what is requisite ; for all excess will do mischief, and cause the catch pins to strike violent and dangerous blows on the springs ; and any deficiency will occasion the length of the stroke to fall short of what it should be, and thereby diminish the useful performance of the engine.

The necessity of this care will be evident, when we consider that the supply of steam for performing the whole of the working stroke, is furnished to the cylinder during the first half, or two-thirds of that stroke ; so that the manner in which the motion will be completed, is determined by the manner of its commencement.   Whilst the steam acts on the piston with its full force, its tendency is to urge it to a very rapid motion, but the inertiæ of all the moving parts resists any sudden communication of motion, and causes it to take place with moderation.

The motion produced under these circumstances begins and continues by a law of acceleration, which if it were continued throughout the stroke, would give the piston a very great and dangerous velocity ; but the supply of steam being intercepted at the proper stage of the process, puts a stop to the continuance of the

(a) The regulation of the damper has since been effected by means of the force of the steam itself, so as to require no attention from the engine-keeper ; it is much better regulated by that means, as will be explained.

3 A

acceleration; and the further effort which the steam makes upon the piston in the act of expanding itself, decreases continually in intensity, so as to become incapable of continuing the rate of motion, which the piston had acquired at the moment of stopping the supply; consequently the velocity is retarded, until the piston comes to rest.

During this course of retardation, the energy of the moving parts has the same tendency to preserve the continuance of the motion, that it had to resist the commencement of the motion in the first instance; and in coming to rest the moving parts faithfully restore all the power which they had previously absorbed in the act of putting them in motion.

The same principle applies to the regulation of the returning stroke, as to the working stroke. The counter-weight, or preponderance of the pump-rods, is apportioned in the same manner in Mr. Watt's engine, as for the atmospheric-engine, and is usually at the rate of 1 lb. or $1\frac{1}{2}$ lbs. per square inch of the piston, (see pp. 144, 176, 187, and 202); but with this allowance of counter-weight, Mr. Watt's engine will perform its returning stroke with more vivacity, than Newcomen's engine can (a); and this is a considerable advantage, even if it is not required to return in less time, because the inertiæ of the moving parts is more quickly overcome, and the engine is more promptly put in motion to return.

In an engine which has a great length of pump rods, the weight, and consequently the energy of the moving parts, is of necessity very considerable, see p. 175; this is in some measure advantageous to the action of Mr. Watt's expansion engine, because it tends to equalize the variable action of the steam, during its expansion towards the termination of the working stroke. But the energy of the moving parts being nearly as great, in the returning, as in the working stroke, it is requisite to have some provision to check the motion of the piston, when it arrives near the top of its course, so as to bring it quietly to rest without striking the catch pins.

This is effected by shutting the equilibrium valve e, before the piston reaches the top of its course, so as to intercept the passage of any more steam from the top of the cylinder into the bottom. After the valve is so shut, the piston will cease to be in equilibrium, and when it is drawn up any higher, by the preponderance of the counterweight, aided by the impetus of the moving parts, it must compress the steam in the upper part of the cylinder, into a less space, so as to render it more dense; and at the same time, that portion of steam which is contained in the lower part of the cylinder, must be expanded into a greater space, so as to become more rare.

The resistance which the steam opposes to such compression, prevents the continuance of the motion of the piston, and brings it to rest without any shock, in the most easy manner possible; it may in fact be said, that the piston strikes against a pillow of elastic steam contained in the top of the cylinder; an obstacle which is incomparably more elastic than any stop springs which could be applied to receive the stroke of the catch pins.

In this way of stopping the ascent of the piston, no power is lost, because all

(a) The reason of this is, that in Mr. Watt's engine the piston is but slightly resisted during the returning stroke, because it has only to transfer the steam from the top of the cylinder, through the pipe d and equilibrium-valve e, to the bottom of the cylinder; now the quantity of steam to be thus passed through the passage, is not above half as much as must pass through the regulator of a Newcomen's engine of the same size, because none is lost by condensation; hence if the apertures of the valve and of the regulator be equal in the two cases, the returning stroke will be performed with much more ease, in Mr. Watt's, than in Newcomen's (see p. 180).

the impetus which is elicited from the moving parts, in the act of putting them at rest, is rendered available to rarefy the steam beneath the piston, and to compress the steam above it; these are suitable preparations for the return of the piston, because there is less steam beneath the piston to be exhausted by the condenser, and there is more steam contained above the piston ready to press it down again, and therefore less fresh steam will be required from the boiler, than if the equilibrium valve had been kept open, till the piston arrived at the top.

If an engine of this kind has much energy from the weight of its parts, and the celerity of their motion, the equilibrium valve must be shut some space before the piston reaches the top, and in that case the steam in the top of the cylinder may be compressed so much, as to become equal in elasticity to that in the boiler, and then it will lift up the expansion valve b, and return into the boiler; being in fact pumped back by the piston, and going to augment the stock of steam from which the supply for the next stroke is to be derived (a).   Hence, by very simple means, Mr. Watt avoided the principal evils arising from a great mass of matter being put in motion alternately in opposite directions.

Adjustment of the working gear.   The essential conditions are, that at the commencement of the working stroke, such a charge of steam shall be admitted into the cylinder, as to produce so much energy of motion in the moving parts, as will just expend itself, in aiding the expansive action of that steam to continue the motion to the exact termination of the stroke, without either exceeding or falling short of that termination; also, that towards the conclusion of the returning stroke, so much steam shall be retained from entering into the lower part of the cylinder, as will expend all the energy of the moving parts, in aiding the counterweight to compress that retained steam.

It requires but little skill to effect these adjustments by experiment.   The chocks 1 and 2, which act upon the handles of the working gear, are fastened to the flat sides of the plug l, by means of screw bolts, which pass through oblong holes or slits in the chocks, so as to admit of moving the chocks upon the plug at pleasure, by striking them gently with a hammer, and yet the pressure of the bolts will retain the chocks wherever they are placed.

When the engine is first set to work, the valves are very cautiously and gradually opened and shut by hand, until the trim of the engine is ascertained, and the proper places for the chocks roughly determined; and afterwards, if the piston does not come quite down to its intended limit, the chock 1, for the expansion handle r, must be raised higher up upon the plug, so that it may not meet the expansion handle so soon, and by leaving the steam valve longer open, it will then give a greater dose of steam to the cylinder.   On the other hand, if the piston comes down too low, so as to strike the catch pins on the springs, then the chock 1 must be set lower down on the plug, in order to meet the handle, and shut the valve sooner.

In like manner, if the piston does not rise high enough, the chock 2, which is to raise the equilibrium handle, must be set lower down on the plug; or if the piston rises too high, that chock must be put higher up, until it will close the equilibrium valve so much before the piston arrives at the top of the course, that the

---

(a) This method of drawing up the piston by its counterweight in opposition to the elastic force of the steam from the boiler is recommended in Mr. Watt's third patent, 1782 (see p. 348).

N. B. It is not peculiar to Mr. Watt's engine, that the steam passage is shut before the piston arrives at the conclusion of its stroke, and that the parts are brought to rest by expanding the steam. The atmospheric engine is regulated in that manner (see pp. 148, 179, 181, and 210).   In fact it must be so, or else the catch pins would strike the springs every time.

inherent motion of the engine will just carry it to its exact limits, and become expended in doing so.

By these means the motion of the engine is regulated, and its power is adapted to the resistance it has to overcome, and whenever any alteration takes place in that resistance, the engine-man must make a corresponding regulation of the chocks for the expansion and equilibrium handles. If a great and sudden alteration takes place, owing to a pump drawing air, or the failure of a valve in the pumps, the catch-pins will strike; but it is the duty of the engine-man to be constantly on the watch against such accidents, and to go immediately to the handles, to prevent any continuance of violent shocks. As the valves are opened by weights, and shut by the plug, they are at all times under the control of the handles, and they may be shut or opened by hand at any time.

To hand the engine, the engine-man takes the equilibrium handle 3 in his left hand, and the exhausting handle s in his right, and by moving these he may have an entire command of the motion of the engine; as the plug is just before him, he can see by that, when the motion of the engine is too rapid, and if he is expert he may correct the acceleration before a dangerous blow can be struck.

When the water is nearly pumped out of the pit, and it is therefore expected that the pumps will draw air, the engine-man must shorten the length of the stroke of the engine in due time to avoid accidents. For this purpose it is usual to have a pin hole in the plug, just beneath the lower end of the expansion chock 1, and a pin being put into this hole, it will shut the expansion valve so much sooner than usual, that the energy of the piston will be diminished as much as is requisite to avoid danger. This pin can easily be inserted whilst the plug is in motion, and it may be so placed as to paralyse the piston so much, that it will not have energy enough to come quite down, consequently the other chock in the plug will not reach low enough to depress the exhausting handle s; the man may, therefore, depress that handle, or not, as he sees fit, and thus he takes upon himself to shut the exhausting valve, and open the equilibrium valve, whereby he can make the piston return before it has reached to the bottom of the stroke; and if he also holds the equilibrium handle, he can prevent it from opening, or he can open it slowly, or only partially, as the case requires. With this preparation, if the pumps draw air he can prevent any mischief.

It is very common to work Mr. Watt's engine without any catch to keep the equilibrium valve e shut; the weight operates to depress the handle 3, and open the valve e, as already stated, but the pressure of the steam upon the valve is sufficient to keep it shut in opposition to the weight; for instance, when the piston arrives near the top of its course, the chock 2 of the plug, lifts up the equilibrium handle 3, raising the weight, and closing the equilibrium valve e; and when the piston is quite up, a pin in the plug releases the catches of the expansion, and exhausting valves, so that those valves may be opened by their weights, in order to exhaust the cylinder and admit steam to press the piston down. In this state the equilibrium valve will have a vacuum beneath it, and a pressure of steam from the boiler above it, whereby it will be held down very steadily, notwithstanding the effort of the weight to open it; but, as the piston descends, the expansion valve b is shut, and after that, the steam in the top of the cylinder, and in the pipe d, becomes less and less elastic, so as to press with less and less force upon the valve e, until the pressure is no longer equal to the weight, which then opens the valve, and allows the piston to return.

In this way, it is obvious, that the opening of the valve is not regulated by the motion of the piston, but by the pressure which the steam exerts upon it, and by increasing or diminishing the weight which acts against the valve, it may be made to open sooner or later; but it must be so adjusted, that in a regular course of working, the equilibrium valve will not open of itself, until the piston has reached the bottom of its course, and depressed the exhausting handle, so as to shut the exhausting valve in due preparation for the return of the piston.

Whilst the engine continues so regulated, it will work just the same as if the equilibrium valve was kept shut by a catch, as first described; but when the motion of the engine is deranged by any accidental cause, the equilibrium valve will open before the piston reaches the bottom of its course. For instance, if a valve in the bucket of the pump fails, then the resistance being diminished, the piston will descend very rapidly, but as less time will be allowed for the steam to flow through the expansion valve, before it is closed, a less charge of steam will enter into the cylinder than usual; and it follows, that by the expansion of this diminished charge of steam, it will sooner become unequal to resist the weight, which tends to open the equilibrium valve; in this case the equilibrium valve will be opened thereby before the exhausting valve is shut; and, therefore, the steam will be drawn off from the top of the cylinder to the condenser: this will tend to diminish the violence of the decending force of the piston.

Cataract. Mr. Watt's engines for draining mines are very frequently worked by a cataract, which is equally applicable to them as to Newcomen's; it may be constructed in the manner represented in p. 189, and its wire or chain may be connected with those catches of the working-gear, which restrain the weights of the expansion and exhausting valves from opening those valves. In this case, the pin which is provided in the plug, for the purpose of releasing those catches, must be removed; and consequently, when the piston arrives at the top of its course, it will wait there, until the cataract falls over and releases the catches; the engine then makes its working and returning stroke, and waits again, till the falling of the cataract causes it to make another reciprocation.

Another plan is, instead of a cataract, to have a float on the surface of the water in the pit, in the same manner as described in p. 224, but with a strong copper wire, instead of a cord, to reach up into the engine-house; the end of this wire is connected with the catches of the expansion and exhausting handles, in such manner as to cause the engine to make a stroke whenever there is a sufficient accumulation of water in the pit to supply the pumps; for the float then rises up and disengages the catches, and the engine makes its stroke. In this way the motion of the engine always adapts itself very exactly to the quantity of water which is to be drained, without regard to time.

PARTICULARS OF DIFFERENT PARTS OF MR. WATT'S SINGLE ENGINE.

Mr. Watt found, that with the most judiciously constructed furnaces, it required 480 square feet of the surface of the boiler, to be exposed to the action of the fire and flame, to boil off a cubic foot of water in a minute. At that rate one square foot of surface will evaporate 3·6 cubic inches per minute.

He also determined that a bushel (= 84 lbs.) of Newcastle coals so applied, would boil off from 8 to 12 cubic feet of water; that is at the rate of from 6 to 9 lbs. of water evaporated by each pound of coals. A hundred weight (= 112 lbs.) of Wednesbury coals, he found to be equivalent to a bushel of Newcastle; or these two sorts of coals are as 3 to 4 in effect.

Mr. Watt estimated that a cubic inch of water produced a cubic foot of steam of such an elasticity, as was requisite to impel the piston of his engine, at the rate of (12 strokes per min. of 8 feet =) 96 feet motion per min., when loaded with a column of water equal to 8⅔ pounds per squ. inch of the piston (*a*). And he reckoned the actual expenditure of steam, to be one tenth more than the capacity of that space within the cylinder which was occupied by the piston in its motion ; this was to allow for the extra spaces at the top and at the bottom of the cylinder, through which the piston does not pass.

From these data the following rules may be deduced, but they all assume that the supply of steam from the boiler is not shut off, until the piston reaches the bottom of the cylinder, which is therefore quite filled with dense steam.

To find the quantity of water which will be evaporated from the boiler of Mr. Watt's steam engine.
RULE. Multiply the square of the diameter of the cylinder in inches, by the motion of the piston per minute in feet, and divide the product by 288 000; the quotient is the quantity of water evaporated per minute, in cubic feet.
Example. Cylinder 48 inc. diam. squared = 2304 circular inches × (8 feet stroke 12 per min. =) 96 feet motion per min. = 221 184 ÷ 288 000 = ·768 cubic feet evaporated per min.

Sliding Rule. { C Motion ft. per min. Evap. cub. ft. per. min. / D 536·6 Diam. of cyl. inches. } *Ex.* C 96 ft. mo. ·768 cub. ft. per min. / D 536·6 48 inches diamter.

To find the surface of the boiler requisite to be exposed to the fire and flame for Mr. Watt's engine.
RULE. Multiply the square of the diameter of the cylinder in inches, by the motion of the piston per minute in feet, and divide the product by 600; the quotient is the fire surface in square feet.
Example. 48 inc. squared = 2304 × 96 feet motion = 221 184 ÷ 600 = 368· 64 square feet.

Sliding Rule. { C Motion ft. per min. Fire surface squ. ft. / D 24·5 Diam. of cylin. inc. } *Ex.* C 96 ft. motion. 368 squ. ft. / D 24·5 48 inc. dia.

To find the quantity of coals which will be consumed by Mr. Watt's engine. Assuming that one pound of Newcastle coals will boil off 7½ lbs. of water; or that a bushel (= 84 lbs.) of such coals will boil off 10· 08 cubic feet of water.
RULE. Multiply the square of the diameter of the cylinder in inches, by the motion of the piston per minute in feet, and divide the product by 576; the quotient is the quantity of Newcastle coals consumed per hour, in pounds.
Example. 48 inc. squared 2304 × 96 feet motion = 221 184 ÷ 576 = 384 lbs. of coals per hour.

Sliding Rule. { C Motion ft. per min. Coals per hour lbs. / D 24 Diam. cylin. inc. } *Exam.* C 96 ft. motion. 384 lbs. coals. / D 24 48 inc. diam.

To find the power of Mr. Watt's engine, in horse-power. Assuming the load on the piston to be 8· 68 lbs. per square inch, and that the steam is not expanded (*b*).
RULE. Multiply the square of the diameter of the cylinder in inches, by the motion of the piston per minute in feet, and divide the product by 4840; the quotient is the power of the engine in HP.
Example. 48 inc. squared = 2304 × 96 feet motion = 221 184 ÷ 4840 = 45· 6 HP.

Sliding Rule. { C Motion ft. per minute. Horse-power. / D 22 Diam. of cyl. inc. } *Ex.* C 96 ft. motion. 45· 6 HP. / D 22 48 inc. dia.

(*a*) In Mr. Watt's annotations upon Dr. Robison's account of Steam-engines, from which these particulars are taken, the load on the piston is stated to be nearly 11 lbs. per square inch, exclusive of friction ; but to correspond with other facts there mentioned, it must be 8· 68 lbs. instead of 11.
(*b*) In an engine for pumping water, there must of necessity be some expansive action, because the energy or inherent motion of the moving parts, cannot be expended at the termination of the stroke, without striking the catch pins, unless the steam-valve be shut a little before the end of the stroke, so as to cause the motion to finish quietly. Now the less weight of moving parts an engine has, the nearer may its cylinder be filled with dense steam, so as to approach to the above assumption of no expansion ; and rotative engines, which are regulated by a crank and fly wheel, are commonly worked with dense steam, to the very end of the stroke, without any expansion whatever, as will be explained in its proper place.

According to the above data, a cubic foot of steam applied in Mr. Watt's engine will make a sufficient vacuum to raise a cubic foot of water 18· 18 feet high, besides overcoming the friction of the engine, and the resistance of the water to motion.   For a pressure of 8· 68 lbs. per square inch, is equivalent to that of a column of water 20 feet high, and the expenditure of steam being one-tenth more than the space occupied by the piston in its motion, we obtain (20 feet ÷ 1· 1 =) 18· 18 feet effective.

The consumption of a bushel of good Newcastle coals (= 84 lbs.) will raise 19· 8 millions pounds of water 1 foot high ; or one bushel per hour, will supply an engine of 10 horse-power.   And at the same rate, each horse-power will require 8· 4 lbs. of such coals per hour.

Proof. A bushel of coals evaporates 10· 08 cubic feet of water, and each cubic inch of water produces a cubic foot of steam ; therefore a bushel will produce (10· 08 × 1728 =) 17 418 cubic feet of steam ; and as each cubic foot of steam raises one cubic foot of water 18· 18 feet high, a bushel must raise (17 418 × 18· 18 =) 316 666 cubic feet, or (× 62· 5 lbs. =) 19 791 625 lbs. one foot high.

Note. The above statement supposes that the steam-valve is kept open, until the piston reaches the bottom of the stroke, and that the cylinder will then be full of steam a little less than the atmosphere, or 14 lbs. per square inch, elasticity.   But if the engine works expansively to any given extent, we must divide the results given by the preceding rules, by the number of times that the volume of the dense steam is expanded.

To find how many times the volume of the dense steam is enlarged in its expanded state ; or the number which represents the extent of the expansion. Assuming that the vacant space in the top of the cylinder and its passages, which the piston does not occupy when at the top of its stroke, is equal to one-tenth of the capacity of that part of the cylinder which is occupied by the piston in making its stroke.

RULE. Add one-tenth of the whole length of the stroke, to that part of the stroke, during which the dense steam is admitted ; this is for a divisor.   Then add one-tenth of the whole length of the stroke, to that whole length ; and divide the sum by the divisor ; the quotient is the number which represents the expansion (*a*).

Example. Suppose that the engine of 8 feet stroke, is supplied with dense steam, during 5 feet of that stroke.   Now, 8 ft. ÷ 10 = ·8 + 5 ft. = 5· 8 for divisor.   Then, 8 ft. + ·8 = 8· 8, which ÷ 5· 8 = 1· 518 times, that the dense steam is expanded.

Sliding Rule, slide inverted.	A whole stroke.	Dense steam + 1/10 stroke.	*Exam.*	A 8 ft.	5. 8 feet.
	○ 1· 1	Expansion number.		○ 1· 1	1. 518 times.

The results given by the preceding rules, may now be reduced to the following quantities. (·768 ÷ 1· 518 gives) ·506 of a cubic foot of water evaporated per minute.   (368 ÷ 1· 518) = 242 square feet of fire surface (*b*), and (384 ÷ 1· 518 =) 253 lbs. of coals consumed per hour, by the above engine, when working expansively to the extent of 1· 518 times.

(*a*) This rule is in effect to divide the space which the steam occupies when in its expanded state, by the space which it occupied when in its dense state, before the expansion began.   To attain this, we must in both cases include the vacant space which remains unoccupied in the top of the cylinder, and its passages and steam-pipe, when the piston is at the top of its course.   The piston must not approach nearer to the top or bottom of the cylinder, than 3 or 4 inches, to avoid all risk of striking, and the capacity of the passage *c* which leads to the valves must be estimated ; also the capacity of the upright steam-pipe, between the expansion-valve *b* and the equilibrium-valve *e*, is part of the vacant space above the piston ; in general it may be safely assumed at one-tenth of the space occupied by the piston during its motion, for in some cases it will be more, and in others less ; but if the vacant space is known, the above rule may be adapted exactly.

(*b*) An engine which is intended to work expansively, should nevertheless have a boiler of sufficient size to supply it fully with dense steam, when required, and also the hot-water pump and cold-water pump should be calculated to supply the engine properly, when working with dense steam ; for this will occasion no inconvenience when working expansively.

To find the power of Mr. Watt's engine when working expansively to any given extent.

RULE. Having calculated the power which it would exert, if the steam were not expanded, multiply that power by the portion of the stroke through which the piston is moved by the dense steam ; and divide the product by the whole length of the stroke ; the quotient is the power exerted by the dense steam alone. Then to find how much more power is exerted, during the expansive action, take the hyperbolic logarithm of the number representing the expansion, and multiply it by the power exerted by the dense steam ; the product is the additional power gained by the expansion. Lastly, the sum of the two powers, is the whole power of the engine. See table of Hyp. Log. p. 345.

Example. The engine above quoted would be 45· 6 HP without any expansion × 5 feet of the stroke with dense steam, = 228 ÷ 8 feet whole stroke = 28· 5 HP by the dense steam alone.

The expansion is 1· 518 times ; the hyp. log. of which is ·418 and × 28· 5 HP = 11· 9 HP additional by expansion. Lastly, 28·5 + 11. 9 = 40· 4 HP exerted by the engine (*a*).

The vacant spaces at the top and bottom of the cylinder which the piston does not occupy in its motion, occasion a loss of steam, and should therefore be made as small as. is consistent with other circumstances. The space left vacant beneath the piston, when at the bottom of its course, is smaller than that left vacant above the piston, when at its highest, because the upright steam-pipe *d*, forms part of the latter. The actual loss of steam occasioned at every stroke by the vacant space at the bottom of the cylinder, may be found nearly, by dividing the capacity of that space, by the number representing the expansion ; because the vacancy must be exhausted for every working stroke, and then refilled with ex-panded steam for every returning stroke.

For instance, in a cylinder of 48 inches diam. the piston in making an 8 feet stroke will occupy a space of nearly 100 cubic feet by its motion ; and supposing that the piston does not approach nearer than within 4 inches of the top or bottom, the vacant space at the top or bottom of the cylinder would be about 4 cubic feet ; and allowing 1 cubic foot for each of the passages to the valves, we may say 5 cubic feet vacant space beneath the piston when at its lowest ; and taking the capacity of the upright steam-pipe *d*, at 5 cubic feet, the vacant space above the piston when at its highest, will be (5 + 5 =) 10 cubic feet, which is one-tenth of the space occupied by the motion of the piston, as is assumed in the rule.

(*a*) The engine of 48 inc. cylinder above quoted, when working thus expansively, exerts the same power as Mr. Smeaton's engine of 52 inch cylinder (see p. 173), and will serve to compare Mr. Watt's engine with Newcomen's. An estimate is given in p. 181, of perhaps the utmost practi-cable performance of a Newcomen's engine ; and the preceding calculation is probably the utmost practicable performance of Mr. Watt's engine of the same size ; at least there are few situations where the expansion can in practice be carried farther than 1½ or 1¾ times, for the reasons explained at p. 352.

Performance of a steam-Engine of 40 HP.	On Newcomen's prin-ciple ; Atmospheric.	On Mr. Watt's principle.	
		Without expansion.	1· 518 times expansion.
Coals consumed per hour	(7· 56 bush. =) 635 lbs.	(4 bus. =) 336 lbs.	(2· 98 bush. =) 250· 5 lbs.
Coals consumed per hour by each horse-power	15· 87 lbs.	8· 4 lbs.	6· 26 lbs.
Pounds of water raised 1 foot high by each bushel (= 84 lbs.) of coals -	10· 5 million.	19· 8 million.	26· 6 million.
Water evaporated from the boiler, per min. - - -	1· 33 cub. ft.	·674 cub. ft.	·501 cub. ft.
Cold water injected per min. temp. 52 deg. - -	13· 66 cub. ft.	19· 3 cub. ft.	14· 35 cub. ft.
Height to which a cubic foot of water is raised, by each cubic foot of steam	9· 31 feet.	18 ·18 feet.	24· 43 feet.

The space in the cylinder, occupied by the steam at the end of the working stroke, when in its expanded state is (100 + 10 =) 110 cubic feet, and if the expansion-valve is shut, when the piston has descended 5 feet, the quantity of dense steam then in the cylinder would be (62·5 cubic feet + 10 =) 72·5 cub. ft.; and 110 ÷ 72·5 = 1·518 times for the expansion, the same as by the rule. The vacant space beneath the piston, being 5 cubic feet, the actual loss will be about (5 ÷ 1·518 =) 3·29 cubic feet of dense steam at each stroke; or at 12 strokes per minute, 39·5 cubic feet lost per min.

Note. This though sufficiently near is not strictly correct, because the steam is further expanded than above supposed, when it is admitted into the vacant space beneath the piston, for it must then occupy (10 + 100 + 5 =) 115 cubic feet which ÷ 72·5 is 1·58 times expansion (a), and 5 cubic feet vacancy ÷ 1·58 = 3·16 cubic feet of dense steam actually lost at each stroke.

The vacant space above the piston, occasions a loss of as much steam as will give a full density to the expanded steam with which the vacancy is filled during the returning stroke; therefore, if the vacancy be 10 cubic feet, and the expansion 1·58 times, that vacancy will contain (10 ÷ 1·58 =) 6·32 cubic feet of dense steam; and, therefore, (10 — 6·32 =) 3·68 cubic feet of dense steam will be required to fill it up to a full density.

Hence the loss of dense steam, at every stroke of this engine, would be 3·16 cubic feet, by the vacancy beneath the piston, and 3·68 by the vacancy above = 6·84 cubic feet by both. The quantity of dense steam expended at the same time, by the motion of its piston through 5 feet of its stroke, being 62·5 cubic feet, the actual expenditure is, as before assumed, very nearly one-tenth more than the space occupied by the piston.

If an engine were to work entirely with dense steam, without any expansion, then the loss of steam at each stroke would be the full content of the vacant space beneath the piston = 5 cubic feet in the above case; but there would be little or no loss in consequence of the vacancy above the piston, because it would be at all times filled with steam of the entire density, or very nearly so. At the same time the expenditure of dense steam by the motion of the piston through its whole stroke, would be 100 cubic feet, so that the addition in this case would be only one-twentieth.

Apparatus connected with the boiler.—It is one of the first duties of an engine-keeper, to maintain the water in the boiler always at the same height; for if it is too full, the water will be liable to go up the steam-pipe in spray, and pass into the cylinder; or if the surface of the water is allowed to subside beneath the flues 9, which go round the lower part of the boiler, then the metal of the boiler will be worn; and if he should be so negligent as to allow the boiler to boil dry, it may be burned out at once.

The rate of the supply is regulated by turning the cock in the feeding-pipe 8, T; but, in order to know the exact height of the water in the boiler, two gauge-pipes and cocks *y* are employed; one of them descends to within a little of the height or level at which the water should stand, and another goes down a little below that level. If the water is at the desired height, the first-mentioned cock *y* being opened, will give out steam; or the other cock will emit water, in consequence of the pressure of the superincumbent steam on the surface of the water. But if water should issue from both cocks, it will be too high in the boiler; and if steam issues from both, it will be too low. It is the duty of the engine-keeper to make frequent trial of the state of the water by these cocks, and to regulate the feeding-cock accordingly, so as to give the requisite supply to meet the wants of the boiler.

These guage-cocks are the same as were used in Savery's and Newcomen's engines; but, in his first engines, Mr. Watt used another kind of water-guage, which shows the height of the water in the boiler by inspection. It is a small vertical glass tube, which has a communication with the boiler by two short copper pipes, one cemented to each end; thus, one pipe proceeds from the top of the glass tube, and enters the boiler above the intended level of the water; and the other

---

(a) This supposes that the equilibrium-valve is not shut till the piston reaches the top of its course, but if the engine is adjusted as described in p. 362, then the steam remaining in the vacant space beneath the piston will be still further expanded, and the loss will be diminished in consequence.

pipe, from the bottom of the glass tube, enters the boiler below the surface of the water. In this way, it is evident that the glass tube will always be filled with water to the same level as the water in the boiler, and may be graduated with inches, to inform the engine-man how much the surface of the water is above or below its true level. This water-guage was usually placed at the end of the boiler, over the fire-place, at the outside of the brick-work ; the small copper pipes, from each end of the glass tube, passing through the brick-work. It was a very convenient guage, but it has since been laid aside, from the liability of the glass tube to be broken.

Another useful water-guage was sometimes applied in these engines, to give audible notice of the water in the boiler being too low. It is a pipe going down beneath the surface of the water in the boiler, when at its intended level ; and at the upper end of the pipe, at the top of the house, a whistle mouth-piece is formed : then, if the water in the boiler sinks too low, the steam will issue at the pipe ; and, passing through the whistle, will make such a noise as to alarm the neighbourhood, and call the engine-man to his duty, even if he should have fallen asleep. This is a useful precaution against negligence, which ought never to be omitted in any steam-boiler ; it was used by Mr. Smeaton in his engines (see p. 193).

These contrivances for ascertaining the state of the water in the boiler, were rendered less necessary by a subsequent one, wherewith the boiler will always feed itself exactly as fast as its evaporation requires ; the water being admitted by a feeding-valve, which is opened by a floater on the surface of the water in the boiler : this will be described in its proper date.

The boiler has a drain-pipe from the lowest part, coming through the brickwork to the outside ; and the end is furnished with a cock, or a plug, to draw off the water, and empty the boiler occasionally.

In the upper part of the boiler is an oval opening, called the man-hole ; when the engine is at work, this is covered with a lid, fastened on by screws, but which can be easily removed, to admit a man into the boiler, to clear out any sediment which may be deposited therein from the water. The boiler should always be kept clean from all such sediment, as it greatly impedes the transmission of the heat through the metal to the water, and tends to destroy the metal of the boiler.

The sediment in the boilers is a most serious inconvenience in many situations where pure water cannot be obtained for the boiler ; and water which contains mineral substances in solution, frequently deposits a hard stony crust on the bottom of the boiler, which is very difficult to remove. The only remedy, in such cases, is to clean out the boiler frequently, so as to avoid all accumulation of sediment ; and, for this reason, all engines which require to work constantly should have a spare boiler, that one may be cleansed whilst the other is working (see pp. 151 and 240).

Safety valve.—Although the steam in the boiler is not required to be much more elastic than the atmospheric air, there would be danger of bursting the boiler, if the steam should accidentally become too strong ; particularly when the engine is stopped. The boiler is, therefore, furnished with a safety-valve, which is so loaded, that its weight, added to that of the atmosphere on its upper surface, may exceed the elasticity of the steam, and keep the valve shut so long as the steam is only of sufficient strength. But when the force of the steam increases so as to become at all dangerous to the boiler, its pressure will exceed the weight of the valve, and the pressure of the atmosphere, so as to force open the valve, and the steam will escape from the boiler, till its strength is sufficiently diminished ;

and then the safety-valve will shut again, by the predominance of its weight, over the pressure of the interior steam.

The fire should be so regulated by the charge of fuel, and by the opening of the damper, that the steam will never accumulate during the working of the engine ; but, when it stops, the remaining fire will continue to produce steam, and it must therefore be discharged at the safety-valve.

The safety-valve is a circular brass plate, made conical on the edge, and fitted into a conical aperture in a brass ring, or seat, fixed in the top of the boiler ; a perpendicular stem passes through the centre of the valve, and is retained in sockets, which guide the valve into its proper seat, as it rises and falls. The top of the valve is loaded with a sufficient weight to keep it down when the engine is at work ; and to open the valve when the steam is to be let off, a small rectangular lever, with equal arms, is placed at the side of the valve, and connected with the top of its stem. To the other arm a chain is attached, and is conducted into the engine-house, where it passes over a pulley, so as to hang in a vertical direction, like a bell-pull. By pulling it, the valve is opened, when the engine is stopped.

As the steam in the boiler cools, when the fire is extinguished, the air must be admitted into the boiler, or else a vacuum would be formed within it, and the atmospheric pressure would crush the top inwards. If the safety-valve is kept open, the air will be admitted, but if this is neglected it can always gain admission by the feed-pipe T, which is open at top. An additional inverted safety-valve was afterwards applied to admit the air when the engine cools, as will be described.

Note. When the fire is first lighted, the boiler being cold, and the upper part containing air, the safety-valve should be kept quite open, to allow that air to be displaced by the first steam which rises from the water ; and when the steam is observed to issue copiously from the valve, it may be let down again, to retain the steam ready to start the engine. If the air in the boiler were not thus previously discharged at the safety-valve, it would take a very long time to clear it all out through the blow-valve *x*, by blowing through.

Steam-gauge.—Although the load on the safety-valve makes a sufficient regulation of the strength of the steam, to avoid any danger of bursting the boiler, it is not a sufficiently accurate indication, to enable the engine-man to keep up the steam always to the same elasticity. Mr. Watt, therefore, employed a steam-guage, which operates by a column of mercury. This steam-gauge consists of an inverted siphon, or bent tube, of glass or iron ; one leg of which communicates with the boiler, being joined to the steam-pipe, and the other is open to the atmosphere. A quantity of mercury is poured into the tube, to occupy the bent part which joins the two legs ; and the surface of the mercury in one leg being exposed to the pressure of the steam, while the external air acts upon the other, it is evident that the difference of level of the two surfaces, will express the elasticity of the steam above or below the atmospheric pressure, by the height of a column of mercury it will support.

When the tube is of glass, this difference of level may be seen and measured on a scale of inches ; but when an iron tube is used, a small light wooden rod, floating on the surface of the mercury in the open leg, points out the height of the column against a scale of half inches, fixed above the open end of the tube. In this case the divisions, which are numbered for inches, must be only half inches ; because the mercury will descend in one leg as much as it rises in the other, so that the scale must be doubled to show the real difference of level between the two surfaces.

3 B 2

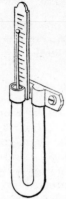

The tube is made of wrought-iron, in the same manner as a gun-barrel, but with the two ends bent parallel like the letter U ; the interior of the tube ought to be bored, in order that both legs may be precisely of the same diameter, otherwise the gauge will not show the pressure correctly, because the mercury will not sink so much in one leg, as it rises in the other.

A steam-gauge of this kind is usually attached by two screws to the steam-pipe *a*, or else to the end of the boiler C, and sometimes to the steam-case of the cylinder E ; or at any part, having open communication with the boiler, and in a convenient place for the engine-keeper to see it, because this should be his constant guide for the regulation of the fire and damper.

Piston.—The piston of the engine is of the same kind as that already described for the first of Mr. Watt's engines (p. 326) ; he was unable to make any progress with his engine, until he had brought the construction of the piston to a considerable degree of perfection, and his method·has not been altered since ; viz. the groove round the edge of the piston is filled with a soft plaited rope of hemp, well supplied with tallow grease, and a ring is applied in the groove over the hemp, in the manner of a cover to the piston ; this cover is pressed down very hard upon the hemp by several screws, which are tightened from time to time, as the packing wears loose, and thus the hemp is pressed forcibly against the inside of the cylinder.

The edge of the piston is turned in a lathe, to make it truly circular, and to fit it exactly to the interior of the cylinder ; the edge of the piston-cover is also turned, so that the two rims, which form the top and bottom of the groove in which the hemp-packing is lodged, are as closely fitted to the cylinder as possible not to actually touch ; by this means the packing cannot get out of the groove.

The piston-rod is also turned true in a lathe, and the lower end, which passes through the centre of the piston, is made of a conical form, with the largest end downwards ; this cone being inserted into a corresponding hole, which is bored very exactly through the centre of the piston, unites it very firmly and correctly to the rod. The upper end of the rod is fitted into a socket, and fastened by a cross-key driven into an opening through the socket, and through the rod : this socket has a proper joint, to connect it with the chains of the arch-head.

The polished piston-rod passes through a socket, or stuffing-box, formed in the centre of the cover of the cylinder ; the lower part of the socket is a hole, which the piston-rod fills very exactly ; but the upper part is enlarged, to leave a space round the rod, into which the hemp-stuffing is rammed ; and to keep it forcibly compressed round the rod, a ring or collar is fitted to the rod above the hemp, and forcibly pressed down upon the same by two screw-bolts.

The cover of the cylinder is a circular plate, which applies to the flange round the top edge of the cylinder, and is fastened down by a number of strong screw-bolts and nuts. The cover drops into the cylinder a little, so as to find its place exactly ; and the joint round the circumference of the cover is made steam tight, by a packing of plaited hemp-rope smeared with white lead ; this is interposed in the joint, and compressed by the bolts.

The cylinder-cover must be lifted up, whenever the piston is packed ; and there is a windlass and tackle fixed up in the roof, to lift up the cover, when the bolts are removed.

Construction of the valves. These are circular brass plates, made conical on their edges, and accurately fitted into corresponding brass rings or seats, which are fixed in the cast iron boxes of the nossels or steam-pipes ; the diameter of the apertures which are closed by the valves should be one-fifth of the diameter of the cylinder, and then the area of the passage will be $\frac{1}{25}$th of that of the piston. Consequently the valves for a 48 inc. cylinder, should be 9·6 inc. diam. of aperture, in the smallest part.

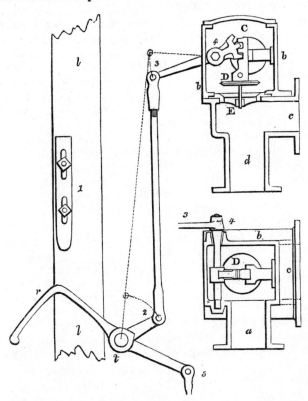

The sketch represents the upper steam-valve. The circle is the interior aperture of the steam-pipe, which introduces the steam from the boiler into the box *b b*; *c* is the branch or passage into the top of the cylinder, and *d* is part of the upright pipe ; E is the seat of the valve, it is a brass ring, fitted with cement into a recess or cell, formed for its reception in the joint of the two flanges, by which the box *b b*, is united to the top part of the pipe *d*; the interior aperture of the seat is formed with an accurate conical aperture.

The valve itself is a circular brass plate D, turned conical on the edge, so as to fit the conical aperture of the seat, and the two cones are very nicely ground to each other with emery. This conical joining is so obtuse, that although it is air-tight, the valve will not adhere to its seat. The valve has a spindle or round axis, which is freely moveable up and down in a hole through a cross-piece which extends across the ring or seat E ; on the upper side of the valve a strong piece of metal, C, is firmly jointed to it, and one side of C is formed into a toothed rack, the teeth of which are acted upon by the teeth of a metal sector, fixed on an iron axle 4, the ends of which are supported in the opposite sides of the valve-box ; one of these ends passes quite through the side of the box, and is nicely fitted into the hole therein by grinding, so as to be air-tight and prevent all leakage.   The extremity of this axis projects outside the box, and has a lever 4 3 fixed on it, which is connected by a rod 3 2, with another shorter lever 2 *t*, fixed on the axis *t* of the handle *r*, which is moved by the projecting chock 1 of the plug *l*.

It is extremely important to make the valves fit tight, and also to open the passages for the steam very quickly, and with a small exertion of force ; this was one reason for adopting circular lifting valves, instead of the old turning regulator; the pressure of the steam acts upon the circular valves, and keeps them shut very close, but they require a considerable force to lift them out of their seats or apertures, this is done by weights, which at first required to be so heavy, as to occasion great inconvenience to manage the engine by hand, until Mr. Watt invented a very simple and effective method of arranging the levers for raising up the valves.

The force which holds down the valve is momentary; and the instant the valve is detached from its seat the pressure nearly ceases, although the valve has not risen half an inch; the force is, therefore, no impediment to the engine, but would be an inconvenient labour to the man who is to start and stop it.

By Mr. Watt's contrivance, the lever by which the valve is to be opened is put in such a position, that when it begins to raise the valve its force is exceedingly great. The figure represents the valve open, and suppose the plug l is descending, and one of its pins or chocks 1 just coming into contact with the handle r, so as to shut the valve, by depressing that handle; the lever t 2, on the same axis, is connected by the rod 2 3, with the lever 3 4, fixed on the axis, 4, of the sector, contained within the valve-box b. Therefore, when the chock 1 of the plug l depresses the handle r, and turns the lever t 2 round upon their axis, the rod 2 3 pushes up the lever 3 4, until the valve is closed. At that time, the rod 3 2, and the arm 2 t, of the lever, are in one straight line, while the lever 3 4, on the axis of the sector, is nearly at right angles to rod 2 3, which moves it, consequently the rod 2 3 is acting with the full length of leverage 3 4 to turn the axis 4, upon which the sector is fixed.

This situation is shown by dotted lines; the valve is kept closed by means of the catch or detent before explained, which holds down the handle r, until the valve requires to be opened; the plug then, by another pin, relieves that catch, and by a weight which is suspended by a rod 5, from the end of a lever t 5, fixed on the axis t of the handle r, that axis, and also the levers, are turned into the position of the figure, whereby the lever t 2 draws down the rod 2 3, and depresses the lever 3 4, so as by the sector and rack within the box b b to lift up the valve D, and open the passage for the steam. By this arrangement, a very great force is exerted by the weight and its lever 5 t, by means of the lever t 2, upon the rod 2 3 during the first moment of opening the valve; but as the levers depart from the dotted position, the lever t 2 acquires an effective leverage to pull down the rod 2 3, and as the motion continues, that leverage increases in length, on the same principle as the action of a crank. So that although the force of the levers to lift the valve is great at first, when the pressure upon the valve is great, yet when the valve is lifted up, and the great force of its resistance overcome, the levers have then assumed a position, in which they act with celerity to lift up the valve to its full height.

The height to which a circular valve should be raised out of its seat, is $\frac{1}{4}$ of the diameter of the aperture, in order that the steam may have an uninterrupted passage through that aperture. At this rate, to open a valve of 9·6 inc. diam. it should be lifted up 2·4 inc.; for the area of a circle 9·6 diam. is 72·4 square inc.; and its circumference 30·15 inc. which $\times$ 2·4 inc. in height is also 72·4 square inches.

Condensing Apparatus.—It is necessary that the parts appropriated to the condensation of the steam should be kept as cold as possible, and those intended for the operation or passage of the steam should be kept as hot as possible; hence the condenser and air pump are placed in the cistern of cold water G, which is kept constantly full by the cold-water pump R, and a little surplus runs away waste, to carry off all the accession of heat; and if the injection-valve is placed low in this cistern, it will take the water in the coldest state.

Injection.—The quantity of injection-water must be regulated according to its temperature, so that the hot water may issue at the discharging valve *m*, at about 96 deg. of Fahrenheit, that is, blood-warm; the engine will then be in excellent order; but the hot well should never exceed 110 deg., or the vacuum will not be good; 102 deg. may be taken for a fair average temperature for the hot well.

It often happens that there will be a deficiency of condensing water, so that it must be used over and over again, without having had sufficient time and exposure to the air, to become quite cold; in this case it is not advisable to admit such a great quantity of water as would be necessary to keep the hot well at its proper temperature, because it will be better to have an imperfect vacuum, than to have such a great quantity of injection water to extract by the air-pump.

Mr. Watt's mode of computing the quantity of water requisite for the injection in his engines is founded on the following principle.

It is to be considered that every cubic foot of steam will produce about a cubic inch of water, when it is condensed; and also that a cubic foot of steam contains as much more heat than a cubic inch of boiling water, as would raise the temperature of 960 cubic inches of water 1 deg. It is by this great quantity of heat becoming latent, or insensibly incorporated with the cubic inch of water, without raising its temperature, that the particles are separated, and removed from each other, to twelve times the distance they were before, and the liquid water having its particles thus separated, becomes aeriform steam.

Now, when this steam is admitted into the vacuous condenser of an engine, the object is to condense it into its original state of fluid water; this is effected by cooling it by the injection, for the cold water coming in contact with the hot steam, abstracts heat therefrom, in raising its own temperature; the heat so abstracted from the steam, being part of that latent heat which occasioned its aeriform state, the greatest part of the steam ceases to continue in that state, and the particles of water returning to each other, resume their liquid form. But the whole of the steam cannot be condensed by this process, because it is relieved from the ordinary pressure of the atmosphere, and therefore the vacuous space in the condenser will remain filled up with rare steam, or uncondensed vapour, possessing a greater or lesser density and elasticity, according to the temperature to which it has been cooled by the injection-water, or according to the quantity of heat which the steam preserves to itself, after all that the injection-water has taken from it.

Mr. Watt ascertained, that if the temperature of the steam and water contained in the condenser of his engines, is reduced to 100 deg., then as the steam remaining in the exhausted space, will only have an elastic force equal to a pressure of not quite one pound on the square inch, it is better to leave this weak steam always in the condenser, and also in the cylinder, to oppose the descent of the piston, than to throw in any more injection-water, which would be requisite to cool it more, and render it more rare and feeble.

Assuming the temperature of the steam in the boiler to be 212 deg., if we add 960 for the latent heat of that steam, the number 1172 may be considered as a representation of the heat contained in the steam, and if from this we deduct 100 deg., for the temperature of the hot well, then 1072 will represent that heat which must be withdrawn from the steam by the cold water, in order to condense and change the state of the greatest part of it, from aeriform steam at 212 deg. of temperature, and the elasticity of atmospheric air, into liquid water at 100 deg.

of temperature; leaving a very small portion of the steam in the aeriform state, at 100 deg. of temperature, and about one pound per sq. inch elasticity.

The quantity of cold water requisite to absorb this heat, will depend upon its temperature when first injected, if this was 50 deg., then the heat which that water will absorb in raising its temperature to 100 deg. will be represented by $(100 - 50 =)$ 50; and as many times as 50 is contained in 1072 (viz. 21·44), so many times must the quantity of injection-water exceed the quantity of water contained in the steam which is to be condensed by it.

Heat of the steam in the boiler (212 sensible temperature + 960 latent heat) = 1172
Deduct the heat which it retains after being condensed, (viz. temp. of hot well) = 100

Heat abstracted from the steam during its condensation   -   - = 1072

Heat of the injection-water, after it has been mixed with the steam, } 100
   viz. the temperature of the hot well   -   -   -
Deduct the heat which it possessed before the injection, viz. the tem- } 50
   perature of the cold water   -   -   -   -

Heat acquired by the injection-water in condensing the steam   - = 50

Now as a cubic foot of steam is assumed to contain a cubic inch of water, $(1072 \div 50 =)$ 21·44 cubic inches (or nearly one-eightieth part of a cubic foot) of cold water, at 50 deg. must be injected into the condenser, for every cubic foot of elastic steam at 212 deg., which is admitted into it; and the mixture of that steam and water will produce 22·44 cubic inches of hot water at 100 deg.

But Mr. Watt remarks, that, as the injection water may not be quite so cold as 50 deg., and as other circumstances may require an extra allowance, he considered that a wine pint (= 28·9 cubic inches) of injection-water to every cubic inch of water evaporated from the boiler was amply sufficient.

This is very nearly one measure of cold water to 60 measures of steam, and greatly exceeds the quantity required in a Newcomen's engine, (which, according to Mr. Smeaton's practice, was about $10\frac{1}{2}$ times, see pp. 180 and 239,) and by showing the more perfect condensation, points out the superiority of the new engine; for Newcomen's, if working to the greatest advantage, should not be loaded to more than 7 lbs. per square inch, whereas Watt's engine will bear a load not much less than 11 lbs., when making 12 strokes per min. of 8 feet = 96 feet motion of the piston per min. :—

"What has now been said is not a matter of mere curiosity, it affords an exact rule for judging of the good working order of the engine; for we can measure with accuracy the water admitted into the boiler during an hour, without allowing its surface to rise or fall, and also the water employed for injection. If the last be above the proportion now given, (adapted to the temperatures of 50 deg. and 100 deg.) we are certain that steam is wasted by leaks or by condensation in some improper place."

It is of great importance to reduce the temperature of the hot well as low as possible, because the cylinder will always remain filled with steam of the same, or rather of a higher temperature than that of the hot well, and possessing an elasticity which opposes part of the pressure on the other side of the piston, so as to diminish the power of the engine. This is shown by the barometer which Mr. Watt applied to his engines to prove the state of the vacuum; and, together with the height of the mercury in the steam-guage, points out the real power the engine is capable of exerting.

To find the quantity of injection-water required for Mr. Watt's engine.   Assuming that it has no expansive action ; that the cold water is at 50 deg. of temperature, and the hot well at 100 deg.

RULE.   Multiply the square of the diameter of the cylinder in inches, by the motion of the piston in feet per minute, and divide the product by 10 000 ; the quotient is the quantity of cold water required per minute in cubic feet.

Example.   48 inch diam. squared = 2304 × 96 feet motion per min. = 221 184 ÷ 10 000 = 22 cubic feet per min.

Sliding Rule. $\begin{cases} \text{C} & \text{Motion ft. per min} & \text{Inject cub. ft. per min.} \\ \text{D} & 1 & \text{Diam. of cylin. inch.} \end{cases}$ *Ex.* $\dfrac{\text{C}\ \ 96\ \text{ft. mot.}\ \ \ 22\ \text{cub. ft. per min.}}{\text{D}\ \ \ \ 1\ \ \ \ \ \ \ \ \ 48\ \text{inches diameter.}}$

Note.   To adapt this result to the case of an engine which works expansively to any given extent, divide it by the number representing that expansion. (See p. 367.)

Example.   Suppose the piston of the above engine is moved 5 feet by the dense steam, and that the expansion number is 1·518, as before shown; then 22 ÷ 1·518 = 14·5 cubic feet of injection per minute.

Capacity of the cold water pump.   This must be calculated to furnish a greater quantity of water than is absolutely requisite for injection, because a portion of the cold water should at all times run waste out of the condensing cistern by its waste pipe, in order to keep down the temperature of the water in the cistern.   The usual proportion for the capacity of the cold water-pump is to make it $\frac{1}{12}$th of that of the cylinder, in engines which do not work expansively.   At this rate, if the bucket of the pump makes half the length of the stroke of the piston, the area of the pump must be $\frac{1}{24}$th of the area of the cylinder.   The proper diameter for the pump may therefore be found, by dividing the diameter of the cylinder by 4·9.

Example.   48 inch cylinder ÷ 4·9 = 9.8 inches for the diam. of the cold pump, its length of stroke being 4 feet, when that of the piston is 8 feet.

Barometer guage.   This is an instrument to indicate the degrees of exhaustion within the condenser ; it is constructed like a weather-glass or barometer, except that the upper end of the glass tube communicates by a small copper pipe with the condenser : the lower end of the glass tube is open, and is immersed in mercury contained in a small cistern or bason, the surface being exposed to the atmospheric pressure.   The glass tube, which is about 32 inches long, is placed in a perpendicular direction, and the copper pipe from the condenser is very closely cemented to the glass at the upper end.   When the engine is at work, and a vacuum is formed in the condenser, the barometer tube is exhausted, so that the surface of the mercury in that tube is relieved from pressure, and the weight of the atmosphere which presses upon the surface of the mercury in the bason, causes a column thereof to mount up in the tube, to a greater or less height according as the vacuum in the condenser is more or less perfect, and according to the weight of the atmosphere, which will be shown by a common Torricellian barometer or weather-glass.

The pipe which leads from the condenser to the top of the barometer tube must be provided with a cock, which should be only very partially opened, or else the variations of the vacuum at every stroke of the engine, would occasion such continual oscillations of the height of the mercury in the tube, as to prevent any accurate observation.   A ruler or scale divided into inches, and numbered upwards from the surface of the mercury in the bason, is fixed behind the tube to measure the height to which the mercury rises therein, in the same manner as the scale of a common weather-glass.

When an engine is in good order, free from leakage and with cold injection-water, the cock of the barometer being opened very gradually, the mercury will ascend in the tube to 28 inches above the level of the surface of the mercury in the bason, the ordinary atmospheric barometer at the same time standing at 30 inches.   This shows that the elasticity of the rare steam which remains uncon-

3 c

densed in the condenser, is equal to the pressure of a column of two inches of mercury. According to the table of elasticities, the temperature of that steam would be near 102 deg. Fahrenheit, supposing that no atmospheric air is present ; but as some air is always mixed with the rare steam, and augments its elasticity, the temperature of the water discharged into the hot well will be about 100 deg.

The mercury will be observed to rise gradually, and fall suddenly, at every stroke of the engine. The vacuum is most perfect towards the end of the return-ing stroke, when the piston has risen to the top of its course, because the greatest cold has been then produced by the injection, and also the air-pump has just taken its stroke, and exhausted part of the air, water, and steam, from the con-denser ; at this period, the mercury will often stand at 28 or $28\frac{1}{2}$ inches, but when the exhausting valve is opened, and the steam is admitted into the con-denser, the mercury will fall rapidly to 27 or $26\frac{1}{2}$ inches ; but before the piston has completed its stroke, the mercury will become stationary, because the cylin-der being exhausted, no more steam is let into the condenser ; the mercury will even begin to rise again slowly, because the injection-water continues to flow and condense the steam. After the injection is shut, and whilst the engine makes its returning stroke, the mercury will continue to rise with more rapidity, in con-sequence of the condenser being further exhausted by the ascent of the air-pump bucket ; and the mercury will have regained its original height of 28 or $28\frac{1}{2}$ inches, by the time the piston reaches the top of the cylinder.

The cock of the barometer tube should be shut when the observation is finished, and should always remain so. It must be shut when the engine is blow-ing through, to avoid throwing the mercury out of the bason. The bason for the mercury must be large enough to contain all the mercury, because when the engine is not at work, the air will leak in, and the mercury will descend into the bason.

As glass tubes are liable to be broken, the barometer tube may be made of iron, in the form of an inverted syphon, like the steam-guage, before described, (p. 372), one leg communicating with the condenser, and the other left open to the air, the rise of the mercury in one leg, will produce a corresponding fall of the mercury in the other ; and the scale being applied to only one leg, the divi-sions must be half inches, that is, provided the two legs are of the same bore ; but if they are different, the scale must not be half inches, but must be divided by trial, in a proper proportion to show the true difference of level between the surfaces of the mercury in the two legs.

The inverted syphon for this barometer may be made of glass ; but both parts of the tube, must be correctly of one diameter, or else the result will be erroneous ; it is difficult to graduate a scale by experiment in an iron tube, be-cause the difference of level of the mercury in the legs cannot be seen. This barometer tube must communicate with the condenser by a small copper pipe, and a stop-cock. The index is a light wooden rod, which is put into the open leg of the tube, the same as in the steam-guage ; the mercury being poured into the tube to within a few inches of the open end, the rod floats on the surface of the mercury, and its upper end points to the divisions on the scale ; the num-bers on that scale must be inverted with respect to those of a single tube.

The engine barometer shows the perfection of the vacuum, or the pressure of the atmosphere to enter into the exhausted space ; the steam-guage shows the pressure of the steam to escape into the air, and the height of these two columns added together, shows the utmost force which the steam is capable of exerting on the piston, supposing that the steam met with no resistance whatever, in passing from the boiler into the cylinder, or in passing out of the cylinder into the condenser.

The Air-Pump. To preserve a good vacuum in the condenser, this pump must be of large dimensions, and the condenser is usually of the same size; therefore the rarefaction of the elastic vapour contained in the condenser will be doubled by the action of the air-pump, for it will extract half the quantity of the elastic vapour at every stroke. If the vacuous space which the bucket of the pump leaves beneath it, when it is drawn up, is equal to the capacity of the condenser, then the vapour in the condenser must expand itself to fill a double space; one half of it will enter into the pump, and will be drawn out at the succeeding stroke, while the other half will remain in the condenser; this is on the assumption that the foot valve *k* offers no obstruction to the passage of the vapour from the condenser into the air-pump.

The usual proportion for the air-pump in Mr. Watt's single engine is to make the pump-bucket half the diameter of the piston, and to move half the length of the stroke of the piston. With these proportions, the area of the air-pump will be one-fourth of that of the cylinder, and the capacity one-eighth. It must not be supposed that as much air as will fill this large pump, is to be drawn out of the condenser at every stroke; for, as before stated, (p. 360), the rare vapour with which the pump is filled, must reduce itself to a very small bulk before it comes to the elasticity of atmospheric air.

As the discharge valve at *m* keeps off the pressure of the atmosphere from the air-pump bucket, it can oppose but little resistance to the motion of the engine, (except by the friction of its packing) until it arrives near the highest limits of the stroke, and the water which rests upon it has actually opened the valve *m*, to discharge itself into the hot well. The resistance during the previous ascent of the bucket, is merely to compress that rare vapour with which the barrel of the pump is filled in the first instance, and to reduce it into such a small bulk, that it will acquire an elasticity equal to that of the atmospheric air, and become able to force its way through the discharge valve *m*. This resistance to compression begins at nothing, and increases very slowly at first, but the rate of increase grows more rapid as the bucket rises, so that the resistance to its motion increases very rapidly, and becomes considerable when it approaches towards the top of its course; but taking the sum of the varying resistances from the commencement of the stroke to the termination, it will be found to be but small in comparison with the power of the great piston.

After the bucket is raised so high as to begin to discharge the water through the valve *m*, the resistance it has to overcome amounts to 13 or 14 lbs. per square inch of the bucket, for it must sustain the whole pressure of the atmosphere on its upper surface, and it is only pressed beneath by the rarefied vapour proceeding through the foot valve *k* from the condenser. As this resistance is opposed to the motion of the engine just at the conclusion of the returning stroke, it tends to destroy the energy of the moving parts, and to bring them quietly to rest without striking the catch pins.

The power required to give motion to the air-pump bucket is little more than is necessary to extract the water which is injected into the condenser, the quantity, according to Mr. Watt's estimate, is about $\frac{1}{60}$th part by measure of the quantity of steam which operates in the cylinder ; now if the effective power of the engine, exclusive of friction, is at the rate of $8\frac{2}{3}$ lbs. per squ. inc. of the piston, and the elasticity of the vapour in the condenser $13\frac{3}{4}$ lbs. per squ. inc. less than the atmospheric air, then the power to work the air-pump would not exceed $\frac{1}{38}$th ; and probably taking the friction and other circumstances into account, the resistance by the air-pump is not above one-thirtieth part of the power of the engine, when it works with dense steam without expansion ; and from one-thirtieth to one thirty-sixth part, in engines working more or less expansively.

The friction occasioned by the air-pump may be estimated at one-fourth or one-fifth of the friction of the piston ; for the circumference of the bucket, and the extent of rubbing surface is nearly one-half of that of the piston, but it moves through only half the space.

The cover and stuffing-box at the top of the air-pump barrel, are constructed in the same manner as for the cylinder, but on a smaller scale. The bucket has a groove round its edge, which is filled with a gasket of hemp, but there is no moveable ring or cover to the groove, as in the piston ; the bucket must therefore be drawn out of the barrel to wind the packing round its groove, and then inserted again with its packing into the barrel.

The valves in the air-pump bucket are flap doors, opening on hinge joints, like butterfly valves, but they are made of brass, without any leather about them ; the top surface of the bucket is also covered with brass for the valves to fit upon. The foot valve k, and the discharging valve m are also made of brass, and hang upon hinge joints at their upper sides ; great freedom of motion, and correctness of fitting, is of importance in all these valves, because they must be opened by the rarefied steam, or vapour, and air contained in the condenser, and all resistance which they oppose to the passage of that vapour causes the exhaustion within the condenser to be less perfect than within the air-pump when the bucket is raised to the top.   For the same reason, it is requisite that the apertures of these valves should be of a sufficient size. The usual proportion is $\frac{1}{4}$ of the area of the air-pump ; at this rate, for an engine of 48 inc. cylinder, the air-pump being 24 inc. diam. ($= 452$ square inches area) the aperture of the foot valve and discharging valves should be $6\frac{1}{2}$ inc. by $17\frac{1}{2}$ inc. $= 113\frac{3}{4}$ sq. inc. ; and the apertures of the two valves in the bucket should be about the same area.

Steam-case for the cylinder.   As the condensing apparatus is immersed in water to be kept cold, so the cylinder should be immersed in steam, to be kept hot ; for this purpose, Mr. Watt, from the first, used a casing to hold steam round the cylinder, and also similar casings for the top and bottom thereof ; this was thought to be attended with very beneficial effects, although it enlarged the steam surface, and therefore exposed the external case to a more rapid condensation than would have taken place from the surface of the cylinder itself. To obtain a good vacuum, it is thought to be necessary to keep the cylinder up to such a temperature, as to prevent any condensation of the steam upon the internal surface, either above or below the piston ; because it is supposed that if the inside of the cylinder were to become wet, as in the atmospheric engine, the vacuum would be vitiated by that rare steam or vapour which would be occasioned by the evaporation of such moisture or wetness from the interior surface, because such evaporation would take place at a low temperature in the vacuum (*a*).

(*a*) This circumstance must be but of very slight importance, because the condenser being

The steam-case being filled with steam from the boiler, keeps the cylinder always as hot as the steam which is admitted into it, and therefore no condensation or wetness can form on the internal surface ; and though heat is lost from the external surface of the casing to the surrounding air, and the steam is in consequence condensed into water, on the internal surface of the casing, still this loss of heat does not impair the operation of the engine.

The steam-case is usually surrounded with a wooden casing, which transmits heat very slowly to the air : if it were possible to cover this casing again with any substance which would entirely prevent the transmission of heat from it, such substance would supersede the use of the steam-case altogether ; for it might be applied with more advantage to the cylinder itself ; but we do not know of any substance which will not permit this transmission more or less.   The best plan is to wrap the cylinder round with hay bands, for a considerable thickness, and then to apply a casing of wooden staves hooped together like a cask ; and if this wood is again covered with a casing of thin sheet copper, well polished on the outside, the loss of heat to the surrounding air will be exceedingly small.   If a cylinder is well covered in this manner, it will make but very little difference whether it has a steam casing or not ; and when a steam-case is used, it ought always to be thus covered, to prevent needless loss of heat to the air, which will be considerable, if the steam-case is not covered.   Cylinders, or their steam-cases, are often built round with bricks, and plastered on the outside ; this is a very good practice.

The steam-case is supplied by a small copper branch-pipe from the main steam-pipe, with a stop-cock to close the passage ; and another cock at the bottom to drain away the hot water which accumulates in the steam-case, from the condensation of the steam therein ; or instead of this cock, a small copper pipe is joined to the lower part of the steam-case, and descends by the side of the condensing cistern, then the end is bent upwards again, nearly as high as the bottom of the cylinder, and the extremity is left open ; the condensed water runs off freely through this inverted syphon-pipe at all times, but the steam cannot follow it, because a sufficient column of water is always retained in the bent part of the tube, beneath the level of the open end.

In large engines, the bottom of the cylinder has a false bottom beneath it, with a space between the two, to receive steam, which is supplied by a small copper pipe from the outside steam-case ; and the cylinder cover is also made hollow to form a steam-case, so that the cylinder is completely surrounded on all sides with steam.

The steam-case is put together round the cylinder in segments, united by bolts, and as it is frequently exposed to be heated and cooled more or less than the cylinder itself, it is joined to the cylinder at the upper part, with a socket joint, packed with hemp and tallow, so as to make a tight joint, but to allow a little drawing out in the height, whenever the cylinder expands by the heat, more than the steam-case.

The upright steam-pipe d in front of the cylinder is united by a similar joint to the square box e, which contains the equilibrium valve, in order to obviate the effects of any unequal expansion between that pipe and the cylinder.

Joints.   In his first engines, Mr. Watt followed the same method of making

kept cool and exhausted, will always draw off any such vapour as could be formed in the cylinder, and prevent any vitiation of the vacuum from that cause.

the joints as had been used in Newcomen's engines for the cylinder bottom,
viz. by interposing between the flanges of the cylinder and its bottom, a ring
of lead covered with glaziers' putty, and compressing the lead very forcibly by
the screw bolts. This method was found sufficient for the atmospheric engines,
as the entrance of a small quantity of air into the cylinder did not materially
affect the performance ; but in Mr. Watt's engine, the joints being very numerous, and the leakage of air very prejudicial, he found it requisite to have better
joints. For some years he followed the plan of making the surfaces of the flanges
very correct, and interposing pasteboard packings, soaked in drying linseed oil,
and smeared with putty ; these being screwed up very tight, made good joints.
The pasteboard being cut out to fit the flanges, was soaked in warm water
till it became quite soft, and was then applied on one of the flanges, and kept
down thereto with small weights, till quite dry. The pasteboards being thus
moulded and adapted to their places, were immersed in a flat pan, containing hot
drying linseed-oil, and remained soaking therein, till the pasteboard ceased to
emit air-bubbles ; the oil was heated to the temperature of boiling water, in
a water-bath. The pasteboards thus prepared, were anointed on both sides with
thin putty, made of very fine and dry whiting, incorporated with the same drying
oil, and then they were screwed very tight between the flanges. White lead was
not found to be so good as whiting. Joints made in this way, though very tight
at first, were not found sufficiently durable, and a better cement was afterwards
found out, called iron cement ; this is now universally employed.

   The great Lever. A single beam of oak was preferred, whenever a tree
of sufficient size could be procured ; and to enable it to resist the strain, it was
trussed on the same principle as the framing for a roof : see the engraving, Plate
X. A strong piece of timber is fixed upon the top of the lever, over its axis or
centre of motion, in a position at right angles to the length of the beam ; and
from the upper end of this piece of timber, strong iron braces are extended
to each end of the lever, being very firmly secured thereto by bolts. The lever
thus trussed or braced forms a triangle, of which the beam forms the base,
the upright post the perpendicular, and the iron braces the sides : and it is plain,
that if the braces are well united to the beam, it cannot bend without
elongating the braces, or compressing the upright post endways.

   Mr. Watt commonly applied the axis p of the great lever, to the upper side
of the beam, and fastened it firmly thereto by iron straps, as represented in
Plate X. By this means, the centre of gravity of the lever was brought below
the centre of motion, as had before been done by Mr. Fitzgerald. (See p. 243.)
Mr. Watt began this practice in his earliest engines, before he adopted the expansive system of working ; but for the expansion engine, if the beam had been
placed above the axis, so as to have raised the centre of gravity above the centre
of motion, it would have tended to equalize the diminishing effort of the steam
during its expansion. This was the fifth method recommended in his specification of 1782. (See p. 349.)

   The difficulty of procuring a tree of sufficient size to form the lever of
a powerful engine in a single piece, rendered it necessary in many cases to
combine four or six beams into one, by placing them in pairs side by side, and
placing two or three such pairs, flat one upon another, screwing them together
upon dowells or joggles, to prevent them from sliding one upon the other ; and
the beam thus formed, was trussed with iron braces, as before mentioned ;

in some cases, the braces were applied to the upper ends of two beams or posts extended from the centre of motion, in the form of the letter K, thus ⋉.

The engine above described continued for several years to be a standard engine for draining mines; and even to the present day no material alteration or improvement has been made in its mode of operation, or in the proportion of its parts, since Mr. Watt brought it to the state represented in Plate X.

Great numbers of these engines were manufactured by Messrs. Boulton and Watt at Soho, and set up in all parts of the kingdom where mining operations are carried on; but the county of Cornwall was their greatest field: between the years 1778 and 1790, they had replaced all the atmospheric-engines in that district by patent engines, which greatly facilitated the mining operations, and produced large emoluments to the patentees; for in addition to their profits as engine-makers, they had their thirds of the savings paid annually; the large size of the engines, their numbers, and the high price of coals, rendered these profits very great (a). The command of power given to the miners by the new engines, enabled them to pursue the veins of ore deeper into the earth, whereby many new and valuable discoveries were made; and then they were emboldened to procure larger engines from Soho, to drain out extensive workings at great depths.

The size of these engines was from 48 to 66 inch cylinders. Respecting their performance Mr. Watt states, " that the burning of one bushel of good Newcastle or Swansea coals in these engines, when working more or less expansively, was found, by the accounts kept at the Cornish mines, to raise from 24 to 32 millions pounds of water one foot high: the greater or lesser effect depending upon the state of the engine, its size, and rate of working, and upon the quality of the coals (b)."

Messrs. Boulton and Watt made a few of these engines for the lead mines in Derbyshire, also in Cumberland, and in Scotland; but the coal mines, except in a very few instances, continued to be drained by Newcomen's engines.

They also made engines of this kind, for most of the water-works in London, in lieu of the atmospheric engines. Two large engines, and a small one, were sent out to Paris, for the public water-works in that city; and others for mines in France, and in Silesia.

Another extensive application of Mr. Watt's single engines was to the iron-works, sometimes as returning engines to supply water-wheels, and in other cases to blow the furnaces by direct action, with double acting blowing cylinders, which Mr. Watt first brought into use about the year 1777. (See p. 328.)

Many of the engines constructed by Messrs. Boulton and Watt 40 and 45 years ago are still in useful operation, except in Cornwall, where they have of late years been either replaced by Mr. Woolf's engines, or else altered to work expansively with high pressure steam, according to Mr. Woolf's system: the high price of fuel in that county, and the great scale of the mining operations, offering a higher premium for the production and adoption of real improvements, there than any where else.

(a) " This was originally their method of agreeing, but afterwards, to avoid disputes and trouble, they fixed certain rates for each sized engine, according to the value and quality of coals in the neighbourhood." See Mr. Watt's Annotations.

(b) The particulars of these performances are not stated, but it is probable that the steam must have been made considerably stronger than the pressure of the atmosphere, to have admitted of such an extent of expansive action, as would be required to raise 32 millions.

MR. JONATHAN HORNBLOWER'S DOUBLE CYLINDER EXPANSION STEAM-ENGINE, 1781.

Mr. Jonathan Hornblower had been an engineer of extensive practice in Cornwall, before the introduction of Mr. Watt's engines into that district; and he had constructed many large atmospheric engines, which were in use at the mines in Cornwall, when Mr. Watt's engines began to be adopted in place of them.

Mr. Jabez Carter Hornblower states, in the first edition of Gregory's Mechanics, vol. ii. p. 385, that Mr. Hornblower took up the subject of his double cylinder engine in the year 1776; and continued it till he had made a large working model, whereof the cylinders were 14 inc. and 11 inc. diam. The following drawing and description of this engine was sent to the editor of the Encyclopædia, Britannica, and was published in Dr. Robison's article STEAM-ENGINE, in the third edition of that work; it is there stated to be a construction near Bristol.

A patent was taken out for this engine, by Mr. Jonathan Hornblower, in 1781, as before stated (p. 346); and about 1790 an engine on this plan was erected by Messrs. Hornblower and Winwood, at Tincroft mine in Cornwall; another was also put up at a mine in the vicinity of Bath.

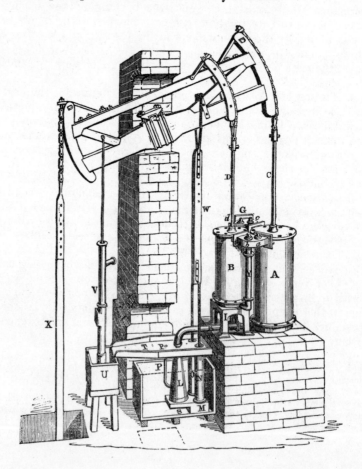

Let A and B represent two cylinders, of which A is the largest; a piston moves in each, their rods, C and D, passing through stuffing-boxes in the covers of each cylinder. The smallest of these cylinders may be supplied with steam from the boiler, by means of the square pipe G, which has a flange to connect it with the steam-pipe proceeding from the boiler. This square part is represented as branching off to both cylinders; but the large cylinder is to be supplied with the waste steam proceeding from the small cylinder; c and d are two cocks, which have handles and tumblers as usual, worked by the plug-beam W. On the side of the cylinders, nearest the spectator, is another communicating pipe, whereof the section is square, or rectangular, having also two cocks, a b. The pipe Y, immediately under the cock b, establishes a communication between the upper and lower parts of the small cylinder B, by opening the cock b. There is a similar pipe on the other side of the cylinder A, immediately under the cock c; and, by opening that cock, the upper and lower parts of the great cylinder are made to communicate with each other.

When the cocks d and a are open, and the cocks b and c are shut, the steam from the boiler has free admission through d into the upper part of the small cylinder B; and the steam from the lower part of the small cylinder B has free admission through Y a into the upper part of the great cylinder A; but the upper part of each cylinder has no communication with its lower part; in this condition the engine is ready to begin its working stroke.

From the bottom of the great cylinder, proceeds the eduction-pipe K, having a valve at its opening into the cylinder; it then bends downward, and is connected with the conical condenser L. The condenser is fixed on a hollow box M, on which stand the pumps N and O, for extracting the air and water, which last runs along the trough T, into a cistern U, from which it is raised by the pump V, for recruiting the boiler, being already hot. Immediately under the condenser there is a spigot-valve, at S, to admit cold water into the condenser L; and over this is a small jet-pipe, reaching to the bend of the eduction-pipe K. The whole of the condensing apparatus is contained in a cistern, P, of cold water; a small pipe, P, comes from the side of the condenser, and terminates on the bottom of the trough T, and is there covered with a valve, p, which is kept tight by the water that is always running over it.

The piston-rods, C and D, of the two cylinders are suspended by chains from arches affixed to one end of the great lever; from the opposite end of which the main spear, or pump-rod X, is suspended by a chain from its arch-head. Therefore, the pistons of the two cylinders will descend at the same time that the pump-spear is drawn up; but the length of the stroke of the small cylinder B is shorter than that of the large cylinder A. The pump-spear X causes the outer end of the great lever to preponderate, so that the quiescent position of the engine is that represented in the figure, both the pistons being at the tops of their cylinders respectively.

Suppose all the cocks opened at the same time, and steam coming in copiously from the boiler, and no condensation going on in L, the steam must drive out all the air, and at last follow it through the valve p, and all the parts will be then filled with hot steam.

Now shut the cocks b and c, and open the injection-valve S of the condenser; the condensation will immediately commence therein, and the steam will be exhausted from the lower part of the large cylinder A. There will then be no pressure on the under side of the piston of the large cylinder; and, at the same time, the communication, Y a, between the lower part of the small cylinder B, and the

3 D

upper part of the large cylinder A, being open, the steam will go from the lower part of B into the top of A; and by its pressure upon the piston thereof, whilst there is a vacuum beneath it, that piston will be pressed down, and the engine thereby put in motion.

At the commencement of that motion, the small cylinder B is filled beneath its piston, with steam remaining from the last stroke, and which is as dense and elastic as that in the boiler; consequently its piston is in equilibrio, and it can have no tendency to descend, although it is pressed upon by the fresh steam from the boiler entering through d, into the top of the small cylinder; but when the motion begins, and as it proceeds, the steam from the lower part of B passes away through Y a into the space left in the top of A, by the descent of its piston. The cylinder A is larger than B, and the arch of the great lever, from which the large piston is suspended, is at a greater distance from the centre of motion, than the arch which suspends the small piston of B; it follows that when the pistons of both cylinders descend into their respective cylinders, the steam between the two pistons must occupy a larger space, than it did when both pistons were at the top of their cylinders. It must therefore, expand in volume, and its elasticity diminishing, it will no longer balance the pressure of the steam which comes from the boiler and presses above the piston of B; consequently that piston will begin to pull the lever down, and assist the large piston to move the engine.

As the pistons continue to descend, the steam beneath the small piston, and above the large one, must continue to diminish in density and elasticity, as the space it occupies continues to enlarge. The steam beneath the small piston is, therefore, less and less able to balance the pressure of the steam on the upper side of the same; and that piston will act to depress the lever with the difference between these pressures, which is continually on the increase from the beginning of the stroke to the end. But the pressure upon the great piston is continually decreasing, from the beginning to the end of the stroke, because it is derived from the elasticity of the steam contained between the two pistons. Therefore, as the effective force of the small piston is continually augmenting, and that of the large one is continually diminishing, the combined effect, to raise the main spear X will approach more nearly to a uniform action, than if it had been moved by either piston alone.

Suppose now that each piston has reached the bottom of its cylinder; shut the cocks d and a, and also the eduction-valve at the bottom of A, and open the cocks b and c. The communication being thereby established between the upper and lower part of each cylinder, their pistons will be in equilibrium, being pressed equally on the upper and lower surfaces, and then the counter-weight will raise the pistons to the tops of their cylinders respectively.

Suppose them to be arrived at the top of their course, the cylinder B is then filled with steam of the ordinary density; and the cylinder A with an equal absolute quantity of steam, but expanded so as to fill a larger space. Shut the cocks b and c, and open the cocks d and a, and also the eduction-valve at the bottom of A; the condensation will again operate to exhaust the large cylinder, and cause the pistons to descend; and thus the operation may be repeated as long as steam is supplied; and one charge of the cylinder B, with ordinary steam from the boiler, is expended during each working stroke.

The cocks of this engine are each composed of two flat circular plates, ground very true to each other, and one of them turns round on a pin inserted through the centres of both plates; each plate is pierced with three sectorial apertures, exactly corresponding with each other, and occupying a little less than one-half of their surfaces.   By turning the moveable plate so that the apertures coincide, a large passage is opened for the steam ; and by turning it so that the solid parts of one plate cover the apertures of the other, the cock is shut.   Such regulators are now very common in the hot-air stoves for warming apartments.

Mr. Hornblower's contrivance for making the stuffing-boxes for the piston-rods air-tight, is by means of two stuffings, placed at a small distance from each other ; and a small pipe, branching off from the steam-pipe, communicates with the space between the collars.   This steam being a little stronger than the pressure of the atmosphere, effectually prevents the air from penetrating through the upper collar ; and though a little steam should get through the lower collar into the cylinder A, it can do no harm.

The manner of making this stuffing-box is as follows: on the top of the cylinder is a nozel or box, to contain a soft plaited rope-yarn, which surrounds the piston-rod, this is rammed down, and occupies about a third of its depth ; upon it a sort of tripod is placed, having a flat ring of brass for its upper, and another for its lower part.   These rings are equal in breadth to the space between the piston-rod, and the inside of the box ; and the two rings are separated to a certain distance, by three small pillars.   The compound ring being put on over the end of the piston-rod, another quantity of rope-yarn is put upon it, and rammed down as before.   The hollow space left between the rings which retain these two packings, is supplied with steam from the boiler ; and thus the packing about the piston-rod will be kept in such a state, as to prevent the air from entering the cylinder, when the elasticity of the steam above the piston is less than that of the atmospheric air.

Performance of Mr. Hornblower's Engine.   It does not appear that Mr. Hornblower was able to obtain any greater effect in practice, from the application of the expansive action in two cylinders, than Mr. Watt did in one cylinder.

In 1791, Messrs. Hornblower and Winwood erected an engine in Cornwall, at Tin Croft mine (*a*), of which the large cylinder was 27 inches diameter, and its piston worked with a stroke of 8 feet long, and the small cylinder was 21 inches diameter, working with a 6 feet stroke.   The column of water in the pumps was said to be 5541 lbs. weight ; and the engine made a little more than 7 strokes per min. of 5 ft. 10 inches, or 41 feet motion per min. ; this would be near 7 horse power. 22 bushels of coals per day was stated as the average consumption, during a month's work ; that is · 92 of a bushel per hour, which would be at the rate of 458 strokes to a bushel, $\times$ 5· 83 ft. stroke $=$ 2670 feet motion $\times$ 5541 lbs. $=$ 14· 8 millions pounds raised 1 foot by the consumption of a bushel of coals.

(*a*) The only account the author has been able to obtain of the performance of this engine, is from a pamphlet published in 1793, by Thomas Wilson, an agent of Messrs. Boulton and Watt, professedly on the part of his employers, to prove the inferiority of Mr. Hornblower's engine, to that of Mr. Watt.   As some angry feelings appear to have been excited in a newspaper controversy on the subject, between Messrs. Boulton and Watt's agents, and Mr. Hornblower and his friends, there is no probability that the performance of the engine would be overrated in the above statement.

The author of the pamphlet above mentioned, offered to bet a sum of money with Mr. Hornblower, that he would erect one of Boulton and Watt's engines on the same mine, with a cylinder 27 inches diameter, and 8 feet stroke, which should have power sufficient to work pumps at a 6 feet stroke, with columns of water weighing 16 623 lbs. ; the number of strokes per min. is not specified ; but with the same fuel it was to raise 50 per cent. more water, than the calculation of Tin Croft engine, by which it appears, that it raised only 14 222 120 lbs. of water one foot high with each bushel of coals.   So that in effect this proposition was, that Boulton and Watt's engine should raise $21\frac{1}{3}$ millions by each bushel.

3 D 2

In Mr. J. C. Hornblower's account of this kind of engine, in Gregory's Mechanics, he states that an engine was erected in the vicinity of Bath, on this principle, and under very disadvantageous circumstances. The engine had its two cylinders 19 inches and 24 inches diameter, with 6 feet and 8 feet lengths of stroke respectively. The area of the large piston was therefore 1·6 times greater than that of the small one ; and the capacity of the large cylinder 2·13 times that of the small cylinder.

The condensing apparatus was a very bad one, through a fear of infringement on Mr. Watt's patent ; and the greatest exhaustion which could be obtained, was 27 inches of mercury. The engine worked four lifts of pumps to the depth of 576 feet, 4500 lbs. 14 strokes in a minute, 6 feet each, with a great deal of inertia and friction in the rods and buckets ; some of the latter of which were not more than $3\frac{1}{2}$ inches diameter ; it consumed 70 lbs. of coal (light coal) per hour.

Now the load 4500 lbs. $\times$ 6 feet stroke $= 27000$ lbs. which the engine raised one foot high at every stroke ; $\times$ 14 strokes per minute $= 378000$ lbs. raised one foot high each minute ; this would be $11\frac{1}{2}$ horse power nearly ; and at the rate of 27·2 millions pounds, raised 1 foot high, with a bushel of coals $=$ 84 lbs. ; if the statement may be depended upon, it was a very good performance for an engine of that size, and more than could be expected from Mr. Watt's engine.

The great lever of the latter engine was framed as shown in the figure, with a view to obtain great stiffness without much weight ; the principal beam is suspended beneath the axis of motion, with four strong straps and nuts, to unite it firmly to the axis ; at each end of this beam the arch heads are joined by mortices and tennons ; the upper ends of the arches are connected by a tye-beam, parallel to the principal beam, and the ends of the tye-beam are firmly united to the arch by iron straps and bolts ; lastly, two strong struts or oblique braces are extended from the middle part of the principal beam, to the ends of the top tye-beam, and are stepped or lodged against shoulders formed at the ends of that top tye.

In Nicholson's Journal, Vol. II. p. 68, 8vo. Mr. J. C. Hornblower has given an account of this lever, which he says continued to work many years under a great load ; the length of the lever was 21 feet, the arms of the lever being as 3 to 4 ; viz. the radius of the arches for the pump-rod, and for the small piston-rod 9 feet, and the radius of the arch for the piston of the large cylinder 12 feet ; the height of the whole when put together 30 inches, and the scantlings of the wood 12 inches by 6. In the figure he there gives of it, the lower or principal beam is represented as being of the same size as the tye-beam above ; but this is probably a mistake ; the weight of the column of water is stated 4800 lbs. in 4 lifts, with 440 fathoms of pump-rods, which with the appendages at the other end of the lever added to the power necessary to overcome the resistance, amounted to about 7 tons.

In this engine, Mr. Hornblower says that two circumstances showed the advantages of this application of the principle : one was, that the man who attended the engine would sometimes detach the small piston from the lever, and work only with the large one, and then the boiler would scarcely raise steam enough to keep the engine going ; but no sooner was the piston-rod of the small cylinder attached to the lever, than the engine resumed its activity, and the steam would blow up the safety-valve.

Also that when the detent, which kept the exhausting-valve shut, happened to miss its action, the piston would be checked, as it were, not being permitted to rise through the whole of the returning stroke, and it would, as by an instinctive nature, come down again and again, until the detent performed its office ; which is a practical argument for the power of the engine at the termination of its stroke.

These accounts are too uncertain to determine the real value of Mr. Hornblower's improvement, and even if we had correct accounts of the performance of these two engines, it would prove but little, because Mr. Hornblower had no chance of bringing his engine to perfection, during the continuance of Mr. Watt's patent, by which he was restrained from proceeding, and could obtain no other opportunities of making his engines, than the two before mentioned.

In 1792, Mr. Jonathan Hornblower made an application to parliament for an act to prolong the term of his patent, of 1781 ; but he was opposed by Messrs. Boulton and Watt, who maintained that it was a plagiarism of Mr. Watt's invention, and no improvement (a).   The prolongation was refused.

(a) The following paper was presented to the members of the parliament.

Short Statement, on the part of Messrs. Boulton and Watt, in opposition to Mr. Jonathan Hornblower's Application to Parliament for an Act to prolong the Term of his Patent.

The steam engine invented by the Marquis of Worcester, brought into use in an imperfect state, by Captain Savery, and almost totally changed for the better by Mr. Newcomen, was still found to consume so great a quantity of coals, that many principal mines in Cornwall, where coals are dear, must have been abandoned, if Mr. James Watt had not invented those improvements which are now well known, and have been brought into very extensive use, throughout the kingdom (and particularly in Cornwall) by the exertions of Mr. Watt, in conjunction with his partner Mr. Boulton.

In 1769 Mr. Watt obtained a patent for his invention ; and in 1775 an act of parliament was passed, whereby his interest in the invention was secured to him for twenty-five years from that period.

The following account *contrasts* Mr. Watt's invention with the state of the engine immediately antecedent to his improvements (b).

1st, The steam is condensed in a distinct vessel, *and not* (as in Newcomen's) *in the body of the cylinder, in which the powers of the steam are exerted.*

2nd, The steam-cylinder, is kept as warm as the steam that enters it, by surrounding it with steam, or with bodies that part with heat slowly, *and not* (as in Newcomen's) *alternately heated and cooled, by admitting steam and cold water by turns into the cylinder.*

3rd, The air that is either mixed with steam, or that enters the cylinder through any defective joint or otherwise, is extracted by the air-pump, *and not* (as in Newcomen's) *blown out by the steam at a snifting clach or valve.*

4th, The piston is pressed down by the expansive power of steam, *and not* (as in Newcomen's) *by the weight of the atmosphere.*

5th, Oil, wax, or other similar substances, are used to keep the piston air-tight, *and not water,* as in Newcomen's.

In 1781, Mr. Jonathan Hornblower obtained a patent for a fire-engine alleged to be his invention. If there had been merit and novelty in that engine, Messrs. Boulton and Watt would have acknowledged the facts.   They were willing that Mr. Hornblower should bring his engine to a fair trial, and therefore have not hitherto interrupted him : but now (1792) after *eleven years' trial*, Mr. Hornblower having erected only two engines, and those upon the same principles as Mr. Watt's in all essential points, and attempting as he is now doing, to procure the sanction of the legislature to his proceedings, it becomes impossible for Messrs. Boulton and Watt longer to continue silent spectators, or to permit an imposition to be practised, so injurious to their just rights ; they, therefore, undertake to prove, by competent evidence, that Mr. Hornblower's engine is a direct and palpable plagiarism of Mr. Watt's invention.   With this view Boulton and Watt propose to prove,

1st, That Mr. Hornblower makes use of the expansive force of steam to press down the pistons of his cylinders, as Mr. Watt does, although he (Mr. Hornblower) has omitted to state in his specification what acting power he proposed to use for that purpose.

2nd, That in regard to the condensation of the steam, by causing it to pass in contact with metalline surfaces, while water is applied to the opposite side, it is only one of the many possible ways

(b) Mr. Smeaton made very important improvements in the construction of Newcomen's engine, and brought it to the utmost pitch of perfection of which perhaps it was capable, whilst its mode of condensation, and its other principles, remained unaltered.

It is stated in the Edinburgh Review, Vol. XIII. p. 327, that "in 1799 Mr. Hornblower's engine became the subject of an action, as an infringement of Mr. Watt's patent; and that the miner who had used the engines of this construction, paid the portion of savings in fuel claimed by Messrs. Boulton and Watt for the use of their invention, rather than risk the event of a law suit." It is also added, that not one engine of the kind has been erected since the expiration of Mr. Watt's patent in 1800.

Professor Robison has given a mathematical investigation of the principles of the action of Mr. Hornblower's engine, to show that it is the same in effect, as Mr. Watt's expansion-engine; but although this is true, there is a difference in the steps by which the effect is attained, and which is an advantage in practice, because the combined effect of the two pistons, approaches more nearly to a uniform action, than could be done by the same extent of expansive action of the steam, when operating in only one cylinder on Mr. Watt's system.

The effect of the steam in expanding itself in the large cylinder of Mr. Hornblower's engine, may be explained in the following manner; assuming that the elasticity of the steam is proportionate to its density, and therefore inversely as the space it occupies, the same as air or other permanently elastic fluids.

It appears that the large cylinder was rather more than double the capacity of the small one; for in the Tin Croft engine, the small cylinder 21 inc. diam. 6 feet stroke, would contain 14·4 cubic feet, and the great one 27 inc. diam. 8 feet stroke would contain 31·7 cubic feet or 2·2 times. And in the other engine, the small cylinder 19 inc. diam. and 6 feet stroke, is 11·8 cubic feet, and the large one 24 inc. diam. 8 feet stroke, is 25·1 cubic feet, or $2\frac{1}{8}$ times. We may, therefore, assume the great cylinder to be twice the capacity of the small one; and to avoid any complication from a difference in the length of the stroke in the two cylinders, we will suppose both pistons to make 8 feet stroke, the great cylinder being 27 inc. diam. = 572 square inches, and the small one 19·1 inc. diam. = 286 squ. inc.

We may also assume that the effective force of the piston is about $8\frac{3}{4}$ lbs. per squ. inc. when the dense steam is admitted from the boiler, to press upon it, the steam beneath being at the same time exhausted by a communication with the condenser; this is the effective force, clear of all deduction for the friction of the moving parts. At this rate the utmost force of the large piston will be 572 × 8·75 = 5000 lbs. and that of the small one, one-fourth as much, or 1250 lbs. As the following computation is only for an example, we need not make any allowance for the vacant spaces at the top and bottom of the cylinders, through which the pistons do not pass.

The following table exhibits the power which is exerted by the large piston, and by the small piston, and also by both pistons combined, taking the pressure of the steam upon them at 16 different periods of the descending stroke.

---

of using Mr. Watt's method, which is expressed in the specification, " condensing by the application of water, or other cold bodies to the condenser."

3rd, That Mr. Hornblower does not use the means expressed in his specification, to discharge the engine of the water, air, and condensed vapour, but in fact employs the very means pointed out, and employed by Mr. Watt, namely, a pump wrought by the engine.

4th, That Mr. Hornblower makes use of oil, wax, or other similar substances, to make his pistons air-tight, in the same manner as expressed in Mr. Watt's specification, and as used by Boulton and Watt.

When these points are proved, Messrs. Boulton and Watt trust, that neither the legislature, nor any individual, will think Mr. Hornblower's pretensions entitled to any countenance or favour.

Operation of the Steam in Mr. Hornblower's Expansion Engine.

Descending Power of the large Piston, area 572 square inches.		Descending Power of the small Piston, area 286 square inches.		Combined Power of both Pistons, area 858 square inches.	
	Pounds.		Pounds.		Pounds.
At top the power will be =	5000	At top the power will be =	000	At top . =	5000
In consequence of the pressure of 2¾ lbs. per square inch upon its upper surface, and no pressure beneath.	4705 4444	because the piston is in equilibrio, having 2500 lbs. pressing upwards, and 2500 lbs. pressing downwards.	147 278	One-eighth .	4852 4722
At one-fourth of the stroke, the power will have diminished to . .	4210 4000		395		4605
		At one-fourth, the power will be . .	500	At one-fourth	4500
because the steam between the two pistons occupies three-fourths of the small cylinder, and one-fourth of the large cylinder, or equal to one and one-fourth of the original space; therefore the spaces will be as five to four; and the pressure on the great piston must be four-fifths of 5000.	3810 3636 3480	for at one-fourth of the descent, the pressure beneath the small piston is reduced by the expansion of the steam between the two pistons, to four-fifths of 2500 = 2000 lbs.; whilst the pressure above the piston continues to be 2500. The power is, therefore, 2500 — 2000 = 500.	595 682 760	Three-eighths	4405 4318 4240
At one-half of the stroke the power will have diminished to	3333	At one-half of the stroke, the power will have increased to	833	At one-half .	4166
for the steam between the pistons occupies one-half of the small cylinder, and one-half of the large one, or equal to one and one-half of the space it filled originally. The spaces will therefore be as six to four, and the pressure on the great piston four sixths, or two-thirds of 5000.	3200 3075 2963	for the pressure beneath is diminished by the increased rarity of the steam, to two-thirds of 2500 = 1666; whilst the downward pressure continues to be 2500. The power is, therefore, 2500 — 1666 = 833.	900 962 1018	Five-eighths .	4100 4037 3981
At three-fourths of the stroke the power will be only .	2857	At three-fourths of the stroke, the power will be .	1072	At three-fourths	3929
for the steam occupies one-fourth of the small cylinder, and three-fourths of the large cylinder, or equal to one and three-fourths of the original space. The spaces will be as seven to four, and the pressure on the great piston four-sevenths of 5000.	2760 2666 2583	for the pressure beneath is reduced by the rarity of the steam, to four-sevenths of 2500 = 1428; therefore the power is 2500 — 1428 = 1072.	1120 1167 1208	Seven-eighths	3880 3833 3791
At the bottom of the stroke the power will be .	2500	At the bottom, the power will be	1250	At the bottom	3750
for the steam occupies the whole of the large cylinder, equal to twice the small cylinder which it at first filled. The pressure will therefore be half of 5000.		for the steam beneath the piston is reduced to one-half of its pressure, or 1250, which deducted from 2500, leaves 1250.			
Power exerted by the large piston } Sum 59222 } Mean 3485		Power exerted by the small piston } Sum 12887 } Mean 758		By both pistons combined } Sum 72109 } Mean 4243	

This table sufficiently explains itself; the two pistons will begin to descend with a force of 5000 lbs. in consequence of the steam contained in the space between the two pistons, pressing on the upper surface of the large piston, and beneath that piston there is nothing to counteract the pressure. At the same time, the small piston, having steam of equal density above and below it, is in equilibrio. As the pistons descend, the steam contained in the space between the two pistons must expand itself, to fill the increasing space occasioned by the equal descent of both pistons in the cylinders, of which one is twice the area of the other; and as the steam becomes rarer, its pressure on the large piston must diminish in proportion; but this diminution occasions the small piston to have a power of descent.

From this computation it appears that the small piston would exert a force of 758 lbs. taking a mean of the whole stroke; and the large piston 4·6 times as much, or 3485 lbs.; hence the combined force of both pistons would be equal to 4243 lbs. acting through a space of 8 feet each stroke. If the pumps worked only a 6 feet stroke, the weight of the columns of water therein might be 5660 lbs. which is a little more than in the engine at Tin Croft. The quantity of dense steam expended from the boiler at each stroke, would be the capacity of the small cylinder = 15·85 cubic feet, and this would be expanded to fill a double space or 31·7 cubic feet, at the conclusion of the stroke, when it is ready to be drawn off to the condenser.

To compare Mr. Watt's system with Mr. Hornblower's, we must suppose the small cylinder to be taken away, and the large one of 27 inc. diam. and 8 feet stroke, to be used by itself on Mr. Watt's expansion method, of intercepting the supply of steam from the boiler when the piston has completed one half of its stroke, or descended four feet; the cylinder will then have received 15·85 cubic feet of dense steam from the boiler, and the remaining 4 feet of the motion will be produced by the expansion of that steam, in proceeding to fill a double space, or 31·7 cub. ft. The power exerted by the piston is represented in the following table, taking it at 16 different periods of the stroke.

The result is nearly the same as that of Mr. Hornblower's engine, and would be precisely the same if both were calculated upon a correct principle; but this method of computation only considers the force that is exerted at a certain number of stages of the stroke, and assumes that the force diminishes all at once, at each of the stages at which it is examined, instead of diminishing gradually. In the table of Mr. Hornblower's engine, the decreasing force is examined at 16 places; but in Mr. Watt's, as the expansion does not begin till half the stroke is completed, the decreasing force is only examined at 8 places, instead of 16, and therefore it is still farther from the truth, which can only be ascertained by computing the force at an infinite number of places. This may be done by the aid of the hyperbolic logarithms, as before explained (see p. 342) (a).

(a) The expansion being into a double space, we must take out the hyperbolic logarithm of 2, which is ·693; then 5000 lbs. the force with which the piston is urged by the dense steam × ·693 = 3465 lbs. is the mean of the decreasing force, which that steam exerts in expanding itself into a double volume. As the original uniform force of 5000 lbs. only acts through half the length of the stroke, we may take 2500 lbs., moving 8 feet, to represent the power exerted by the dense steam; and as the decreasing force of 3465 lbs. acts also through half the stroke, we may take 1733 lbs. moving 8 feet to represent the additional power gained by the expansion. Hence the whole power will be equal to (2500 × 1733 =) 4233 lbs. acting through a space of 8 feet each stroke; this is nearly the same result as the preceding computations.

Operation of the Steam in Mr. Watt's Expansion Engine.

Descending Power of the Piston, area 572 square inches.

		Pounds.
At first the power will be . . . . =		5000
In consequence of the pressure of 8¾ lbs. per square inch upon its upper surface, and no pressure beneath.		5000
At one-eighth of the stroke . . . . =		5000
		5000
At one-fourth of the stroke . . . . . . =		5000
		5000
Three-eighths . . . . =		5000
		5000
At one-half the stroke . (Here the expansion commences.) . =		5000
		4444
Five-eighths . . . . =		4000
		3636
At three-fourths of the stroke the power will be diminished to . . .		3333
for the steam now occupies one-fourth of the cylinder, in addition to that half		3075
of the cylinder which it occupied before the expansion began; or one and a		2857
half times the original space, or as three to two, and the pressure will be two-		2666
thirds of 5000 = 3333.		
At the bottom the pressure will be . . . . . . =		2500
for the steam is expanded to occupy twice the space it filled before.		
Power exerted by the piston . . { Sum		71511
{ Mean		4207

Thus it appears that the total amount of the power exerted by a given quantity of steam, is the same whether it is employed in two cylinders according to Mr. Hornblower's method, or in one cylinder according to Mr. Watt's plan; but the former is more uniform in its action, for it begins at 5000, and ends at 3750; whilst Mr. Watt's begins at 5000, and ends at 2500.

To obtain an expansion of only twice the original volume of the steam, it is not requisite to encounter the complication of two cylinders; for that occasions an increase of friction, and of surface producing condensation, also an extra loss of steam, at the top and bottom of the additional cylinder, and in the pipes and passages which convey the steam from one cylinder to the other; the advantage of the greater uniformity of the force would be overbalanced in this case, and the simple method of Mr. Watt is to be preferred.

The advantage of employing two cylinders is considerable, when the expansive action is carried to an extent of five or six times the original volume of the dense steam, which must, in that case, be raised to an elasticity three or four times greater than the atmosphere: this is a modern improvement introduced by Mr. Woolf, as will be explained in its place.

We may conclude the account of Mr. Hornblower's engine with the remark, that he brought forward a real improvement in his arrangement of appropriating one cylinder and piston to receive the expanding steam, and derive force from it during that expansion, whilst the action of the dense steam is previously performed in another smaller cylinder and piston, exclusively reserved for that purpose. But Mr. H. did not himself apply that improvement in a suitable manner, so as to gain any sensible advantage from it, as his successors have done.

3 E

MR. WATT'S EXPERIMENTS AND INVESTIGATIONS ON THE NATURE AND
PROPERTIES OF STEAM.

This distinguished philosopher, and ingenious inventor, gives the following
account of the state of his knowledge on this subject, previous to inventing his
improved steam-engine (a).

That steam is condensed by coming in contact with cold bodies, and communicates its heat to them.   Witness the common still, &c. &c.

That evaporation causes the cooling of the evaporating fluid, and of other
bodies in contact with it.

That water and other liquids will boil in vacuo at very low temperatures ; viz.
water below 100 degrees of Fahrenheit : this had been shown experimentally by
Dr. Cullen.

That the capacity for heat (since called the specific heat) is much smaller in
mercury and tin than in water : this had been shown by Dr. Irvine ; and Mr.
Watt had himself determined the capacities for heat, of iron, copper, and some
sorts of woods, comparatively with water.

He had also tried some experiments on the volume of steam, compared with
that of an equal weight of water.   And on the elasticity of steam at various temperatures greater than that of boiling water, with an approximation to the law
which the elasticity follows at other temperatures.

The quantity of water which could be evaporated from a certain boiler, by
the combustion of a pound of coals.   And the quantity of water in the form of
steam, which was expended at every stroke, by a small Newcomen's engine, with a
wooden cylinder six inches diameter, and the piston making a stroke one foot long.
Also the quantity of cold water required in every stroke, to condense the steam in
that cylinder, so as to give the piston a working power of nearly 7 lbs. per square
inch.

On comparing the latter quantities, he was unable to comprehend how that
quantity of cold water could be heated so much, by its admixture with so small a
quantity of water in the form of steam ; and after trying a direct experiment on the
heat of steam (as related in p. 312) he consulted his friend Dr. Black, who then
explained to Mr. Watt his discovery of what is called latent heat.

With this stock of knowledge, Mr. Watt began to consider the means of improving the steam-engine, and soon satisfied his own mind that the following are
the requisites for a complete engine ; viz. that as water boils in vacuo, at a temperature below 100 degrees, a good vacuum cannot be obtained, unless the cylinder
and its contents are cooled at every stroke, to below that temperature.   But, at the
same time, to avoid useless condensation, the cylinder ought always to be kept as
hot as the steam itself.

On considering this contradiction between the two important requisites, it
occurred to him that they might be reconciled, by performing the condensation of
the steam in a separate vessel from the cylinder, which vessel might be cooled
down to as low a temperature as is necessary to effect a complete condensation

(a) See Mr. Watt's letter of introduction to his revision of Dr. Robison's Memoir on the Steam-
Engine, dated 1814.   This memoir, with many valuable annotations by Mr. Watt, forms part of an
edition of Dr. Robison's principal writings, edited by Dr. Brewster, and published in four volumes,
8vo. 1821.

therein, without affecting the cylinder ; and it followed as a consequence, that as the air and condensed steam could not be blown out of this separate condensing vessel by the steam, as in Newcomen's engine, they must be extracted by a pump, or other equivalent contrivance.

It also appeared, that to avoid the cooling of the interior of the cylinder, by the contact of the atmospheric air, during the descent of the piston, it would be advisable to apply steam, to press upon the piston, instead of the atmospheric air. And, lastly, that to prevent the cylinder from being cooled by the external air, it would be advantageous to enclose it within a case, containing steam, and again to surround that case with wood, or other substance which would not transmit heat rapidly.

Thus the philosophical principles of the invention were very soon arranged, but it cost Mr. Watt an immense labour to reduce them to practice in an efficient manner. The form of the engine by which these principles were carried into effect has been already described, and also its progressive improvements, until it became a model for established practice. The first experiments on which the principle was founded, were not made with sufficient accuracy to satisfy the mind of the philosopher, though they enabled the mechanician to complete his invention. In the course of years, Mr. Watt repeated and enlarged his experiments in the following manner.

Experiments on the elastic force of steam, when confined and heated to different degrees of temperature. Mr. Watt's first investigations on this subject were made in 1764, by ascertaining only a few facts, and forming from them an approximation for all the others, as already mentioned in p. 311 : it was not until 1773 and 1774 that he found leisure to make a complete series of exact experiments, and at that time no other philosopher had made such experiments. Mr. Watt used two different forms of apparatus ; one which he used at first, was for determining the temperature of steam when its elasticity is greater than that of the atmospheric air ; but he afterwards added another apparatus for smaller elasticities.

To determine the temperature of steam possessing a greater elasticity than the pressure of a column of mercury 30 inches high, he used a small boiler or digester, fitted with a close cover ; a glass tube, 55 inches long, and open at both ends, was inserted through a hole in the cover, and made tight by being lapped round with paper ; the lower end of the tube was immersed in mercury, contained in a small cistern, placed within the digester. A thermometer was also applied at another hole through the cover, its bulb being within the digester, and its stem and scale outside ; the bulb was kept half an inch from the cover, by a wooden collar.

The annexed sketch of an apparatus for this purpose, will serve to explain Mr. Watt's, though the construction is somewhat different. A A is a section of the digester with its cover : *e f g h* the glass-tube, to receive the column of mercury ; the ball *f* forms the cistern, but in Mr. Watt's apparatus it was an open cistern placed within the digester : *a a* is the thermometer. B *b c d* is a safety-valve and piston, not mentioned by Mr. W.

This digester being half filled with water, its cover was screwed on close, and it was placed over a large lamp, to heat the water. When the water boiled, the air contained in the upper part of the digester was allowed to escape as fast as it could be displaced by the steam proceeding from the water, but it was supposed that some air remained to the last. The digester was heated very slowly, and as the steam accumulated, and increased in its temperature, its pressure upon the surface of the mercury in the internal cistern, raised a column of mercury in the tube ; the height

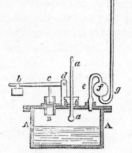

to which this column was raised in inches, was observed by Mr. Irvine, who assisted Mr. Watt, and expressed the elasticity of the confined steam; whilst Mr. Watt himself observed the corresponding degrees of temperature indicated by the thermometer. The heat was kept as much as possible stationary for some time at each observation, and the whole series occupied some hours.

The results are contained in the annexed table, but the column of mercury is there stated as 30 inches higher than it really was, because it is intended in the table to express the absolute elasticity of the steam, whereas the elasticity observed was only the excess above that of the atmosphere. Mr. Watt remarks that he was by no means satisfied with this series of observations, and at a subsequent period he caused them to be repeated by Mr. Southern, as will be described in its proper date.

Mr. Watt's Experiments on the Elastic Force of Steam, 1774.

Temperature. Degrees of Fahrenheit.	Elasticity. Inches of Mercury.	Temperature. Degrees.	Elasticity. Inches.
213	30	240	49
215	31	$242\frac{1}{2}$	50
217	32	$244\frac{1}{2}$	52
219	33	247	54
$220\frac{1}{2}$	34	$248\frac{1}{2}$	56
222	35	$250\frac{1}{2}$	58
$223\frac{1}{2}$	36	$252\frac{1}{2}$	60
225	37	255	62
$226\frac{1}{2}$	38	257	64
228	39	259	66
$229\frac{1}{2}$	40	261	68
231	41	$262\frac{1}{2}$	70
$232\frac{1}{2}$	42	$264\frac{1}{2}$	72
234	43	$266\frac{1}{2}$	74
235	44	268	76
$236\frac{1}{2}$	45	$269\frac{1}{2}$	78
$237\frac{1}{2}$	46	271	80
$238\frac{1}{2}$	47	$272\frac{1}{2}$	82

To determine the elasticity of steam, or the vapour of water, when its temperature is lower than that of boiling water, Mr. Watt used a different apparatus, which is represented in the sketch. The principal part was a straight glass tube, of the size usually employed for barometers, and about 36 inches long, having at the upper end a ball of about $1\frac{1}{2}$ inches diameter, which was capable of containing nearly as much as the rest of the tube: the lower end of the tube was open. The ball of this tube was placed in a circular tin pan, of about 5 inches diameter, and 4

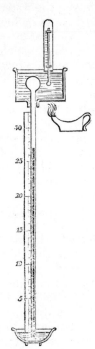

inches deep ; a hole being made in the bottom of the pan, near the circumference, and a socket soldered into the hole, to receive the tube, which being lapped with paper near the ball, was forced tight into the socket, so that the ball remained within the pan, whilst the rest of the tube was beneath it.

To prepare the tube for the experiments it was inverted with the pan, and as much clean mercury poured into the open end, as to fill the ball, then the rest of the tube was quite filled with distilled water, fresh boiled ; the open end of the tube being closed by the finger, the tube was restored to its proper position with the ball and pan uppermost, and then the mercury ran down into the tube, and the water ascended into the ball. The lower end of the tube being then placed in a cistern of mercury, and the finger removed, the mercury and water descended in the tube, in the same manner as in a barometer, leaving the ball void, except from such vapour as might rise from the water (a). In this state a divided scale was applied by the side of the tube, the lower end of the scale floated upon the surface of the mercury in the cistern, and its divisions measured the height at which the mercury and water stood in the tube above that surface.

The tin pan was next filled with water, to surround the ball ; and a thermometer was placed in the water, to determine the temperature, which at the beginning was 55 degrees. After standing some time, in order that the ball, and the vapour within it, might acquire the same temperature as the surrounding water, the

(a) It is essential to the accuracy of these experiments, that the air should be completely excluded from the ball, when it is filled with mercury and water ; for this purpose all the same precautions were used, as should be practised in filling barometers, and as are recommended by M. De Luc, in his treatise on the Modifications of the Atmosphere. After the tube was first filled, and the column allowed to subside, it was agitated to make the column oscillate ; then after allowing it to stand some time, much air became extricated from the water, as was shown by inclining the tube, so as to lower the ball, but without removing the lower end of the tube from the cistern of mercury ; the water then returned into the ball, and filled it so as to show the air in a small bubble. To get rid of this air, the lower end of the tube being stopped with the finger it was again inverted, the air then rose to the open end, and its place was supplied with a drop of boiling water.

The tube being restored to its proper position, water was poured into the tin pan and heated by a lamp till it boiled. The heat converting the water in the ball into steam, caused the mercury to descend in the tube, nearly to the level of the mercury in the cistern. The water in the pan being then suffered to cool, the mercury rose again in the tube, but not so high as before, because more air was liberated from the water by the boiling ; this quantity was increased by agitating the tube, and then leaving it standing erect for some time. The air thus separated, was removed from the tube, in the same manner as before ; and the same series of operations was repeated, till the water became so perfectly cleared, that the air only formed a speck the size of a pin's head, even when expanded by a column of mercury 27 inches high.

When these operations had been successfully conducted, Mr. Watt found that the column of mercury and water would not descend out of the ball into the tube, but it remained suspended by the adhesion of the water to the glass of the ball, although the column of mercury was 34 inches high, and some inches of water above that, and there was no other counterpoise to that column, except the pressure of the atmosphere, on the surface of the mercury in the cistern. The tube required to be violently shaken to make the column subside, and then it fell down suddenly, and settled at 28¾ inches, leaving a void space in the ball. This condition of the apparatus was not obtained, till after much pains had been taken in inverting and reinverting the tube again and again ; it had also been suffered to stand still, after these operations for a long time, in an erect position with the ball exhausted, and then the air which had accumulated in the interval, was discharged.

column of mercury was observed, and found to be 28·75 inches high ; the water above it being 6·5 inches high : the whole column was equivalent to 29·25 inches of mercury. At the same time the common atmospheric barometer stood at 29·4 inc., therefore the difference (= ·15 of an inch) was the column of mercury, which was supported by the elasticity of the vapour within the ball, when at a temperature of 55 degrees.

Mr. Watt's Experiments on the Elastic Force of Steam, 1774.

Temperature. Degrees of Fahrenheit.	Elasticity. Inches of Mercury.	Temperature. Degrees.	Elasticity. Inches.
55	0·15	164	10·10
74	0·65	167	11·07
81	0·80	172	11·95
95	1·30	175	12·88
104	1·75	177½	13·81
118	2·63	180	14·73
128	3·60	182½	15·66
135	4·53	185	16·58
142	5·46	187	17·51
148	6·40	189	18·45
153	7·32	191	19·38
157	8·25	193½	20·34
161	9·18	196½	21·26

Atmospheric barometer 29·4 inches.

The pressure of the atmosphere, on the surface of the mercury in the cistern of the common atmospheric barometer, and on that of the apparatus would be the same. In the barometer, this pressure was totally unbalanced by any addition to the weight of the column of mercury, because the upper part of the tube was a perfect void. But in the apparatus, the compound column of water and mercury, in addition to their own weight, had a slight pressure exerted on the top of the water, by the elasticity of the vapour contained in the ball ; hence, as the height of a column of mercury equivalent to the compound column of water and mercury, fell short of the height of the column of mercury in the common barometer, the deficiency became the measure of the absolute elasticity of the vapour in the ball.

To fill up the series of experiments, a lamp was applied beneath the tin pan, to heat the water within it very slowly, and the water was continually stirred with a feather, to diffuse the heat equally. When the thermometer showed an increase of 10 degrees of temperature, the height of the compound column of water and mercury was again noted ; and in this way the scale was continued nearly to the boiling point, so as to connect with the series of experiments made by the preceding apparatus, on steam of greater elasticity than the atmospheric air. The results of this series of experiments are contained in the annexed Table. The atmospheric barometer, with which the comparison was made, stood at 29·4 inches.

On considering the whole range of experiments on the elasticity of steam, from the temperature of 55 degrees, and 0·15 inches elasticity, up to the temperature of 272½ degrees, and 82 inches elasticity ; we shall find the elasticities to increase much more rapidly than the temperatures, so much so, that the elasticities appear to proceed by a sort of geometrical progression, whilst the temperatures proceed by an arithmetical progression ; this is not exactly the case, but the elasticities approximate to such a law.

This circumstance has given rise to many fallacious notions respecting the mechanical power, which may be exerted by steam of a great elastic force, in preference to steam of the ordinary elasticity; and because the elastic force which steam will exert to burst open any vessel in which it is confined, is found to be very quickly obtained by an increase of temperature, it has been supposed that a much greater mechanical power may be derived from the same heat when accumulated into steam of a high temperature, than when applied in steam of a less elasticity.   In reality, the quantity of heat required to produce a given quantity or bulk of steam, bears no direct proportion to the temperature of that steam, but is nearly according to its elasticity; so that highly elastic steam requires as much more heat for its formation, as it is more elastic; for it is in fact only a greater quantity of heat and water crowded into a smaller space; and hence, any greater power that it possesses, will be obtained at a proportionably greater expenditure of heat.

Mr. Watt also made a series of similar experiments, with the same apparatus, on the elasticity of the vapour proceeding from a saturated solution of common salt, when heated to different temperatures; he found this liquid to be more free from air, than pure water, but it was very difficult to disengage it from the air which it did contain.

In the same manner he tried another set of experiments, on the vapour of spirits of wine.   The results of these two sets of experiments are contained in the following Tables.

Elasticities of the Vapour of the Solution of Salt in Water.

Temperature. Degrees of Fahrenheit.	Elasticity. Inches of Mercury.	Temperature. Degrees.	Elasticity. Inches.
46	0· 01	187	12· 67
76	0· 36	193½	14· 50
85	0· 58	195½	15· 34
92	0· 81	198½	16· 25
113	1.72	201½	17· 16
129	2· 63	203½	18· 10
139	3· 54	205½	19· 03
147	4· 45	207	19· 94
154	5· 36	208	20· 86
160	6· 27	210	21· 80
165	7· 20	212	22· 74
169	8· 12	214	23· 66
172	9· 03	216	24· 60
177	9· 94	218	25· 52
180	10· 85	220	26· 50
183	11· 76		

Atmospheric barometer 29· 5 inches.

Elasticities of the Vapour of Spirits of Wine.

Temperature. Degrees of Fahrenheit.	Elasticity. Inches of Mercury.	Temperature. Degrees.	Elasticity. Inches.
34	0· 22	144	14· 10
40	0· 93	146½	15· 03
67	1· 90	148½	15· 97
84	2· 81	151	16· 91
95	3· 74	152½	17· 85
103	4· 73	155	18· 80
110	5· 63	157	19· 75
114	6· 58	160	20· 71
120	7· 12	162½	21· 65
124½	8· 46	164	22· 59
128	9· 40	166	23· 53
132	10· 34	167	24· 47
135	11· 32	168	25· 40
139	12· 21	169	26· 35
141½	13· 15	171	27· 30

Atmospheric barometer 29· 4 inches.

NOTE. 100 parts of water, at 60 degrees of temperature, will dissolve 35· 42 parts of common salt, and the saturated solution will be 1· 198 times heavier than pure water.

Note. Respecting the correspondence of Mr. Watt's experiments on elasticities as above, with the subsequent repetitions of the same series, by Mr. Southern in 1796, Mr. Watt remarks that his experiments may be reconciled with Mr. Southern's, by adding two-tenths of an inch to all the columns of mercury expressing the elasticities; this is on the supposition that his atmospheric barometer, with which the comparison was made, had its scale placed two-tenths of an inch too low.

Experiments to determine the volume of steam of the same elasticity as the atmospheric air, compared with an equal weight of water. In other words, to measure the quantity of water contained in a given measure of steam, or the relation between the volume and the weight of steam; for, as heat has no effect on the weight of bodies, any quantity of liquid water being rarefied by heat into elastic steam, will weigh just the same when in the state of steam, as it did when in the state of water; and the present question is, how much the volume of the water is expanded in order to effect its rarefaction into steam of the same elasticity as the atmospheric air.

For this purpose, Mr. Watt, by the advice of Dr. Black, made a series of experiments, in which he used a florence flask capable of containing about a pound of water, but only about an ounce of distilled water was put into it; a glass tube was fitted into the neck of the flask, so as to reach down into it nearly to touch the surface of the water; this tube was open at both ends, and was secured to the neck of the flask by lapping it with packthread covered with glaziers' putty. The flask thus prepared, was placed in an upright position in a tin reflecting oven, and placed before a fire, so as to make the contained water boil gradually, but with a heat very little exceeding that of boiling water. The steam which ascended from the water, being lighter than the air contained in the upper part of the flask, displaced and expelled that air through the tube; and it was presumed, that by the time that the last drop of water was evaporated, the flask would become entirely filled with steam. It was then removed from the fire and cooled, by blowing a current of air against the bottom part of the outside of the flask, so as to occasion all the steam to condense, and collect into one drop of water at that place. After the flask had become quite cold, the glass tube was taken out, and the flask with the contained drop of water, accurately weighed.

The flask was then heated again, until the drop of water became entirely evaporated from it, and the inside was then perfectly dried, by making the flask hot, and blowing air into it with a pair of bellows, so that no water or humidity remained within it; in this state it was weighed again, and found to be $4\frac{1}{3}$ grains less than before: this was consequently the weight of the flask full of steam. Lastly, the flask being filled with water, it then weighed about 7492 grains more than when it was empty. Now $7492 \div 4 \cdot 333$ gives 1727 for the volume of steam of the same elasticity as the atmospheric air, and of the temperature of boiling water, the volume of an equal weight of cold water being 1.

The experiment was repeated with nearly the same result, and then a third time, when the flask was supposed to be wholly filled with steam, it was inverted whilst still hot, and the orifice of the tube immersed in cold water, which was drawn up into the flask as it cooled, and nearly filled it, except a small bubble of air about equal to a half ounce measure of water, which showed that the air had not been perfectly displaced by the steam.

In subsequent experiments of this kind, Mr. Watt omitted to apply the glass tube to the flask, as above, but he merely placed the flask on its side in the oven, with the neck inclining downwards, and the mouth closed by a cork, having a notch cut in the lower side, so as to leave a free passage, at the very lowest part, in order that the steam rising from the water in the flask, might displace all the air from the upper part thereof, and fill the whole capacity with steam. Mr. Watt finally concluded, from a review of all his experiments, that *one measure of liquid water rarefied by heat, will produce eighteen hundred measures of*

steam of 212 degrees of temperature, and of a sufficient elasticity to support a column of mercury 30 inches high. Upon this Mr. Watt grounded all his future calculations (*a*); and he remarks, that from his experience in actual practice, he afterwards had reason to think that the expansion is rather more than this computation; but more recent experiments show the expansion to be 1700 times.

These experiments are not to be considered as very accurate, for it is difficult to weigh the vessel with sufficient precision, and the temperature of the steam is very uncertain, nor is it ensured that all the air is displaced by the steam.

Experiments on the quantity of heat required to form steam. It was discovered by the celebrated Dr. Black in 1761, that a much greater quantity of heat must be communicated to water, in order to convert it into steam, than will be afterwards perceptible in that steam by the thermometer, so that the heat becomes concealed, or as he termed it, latent in the steam; and heat when in this latent, or concealed state, is the immediate cause of the elastic form which the water assumes. For instance, when water is boiling in an open vessel, its temperature will be 212 degrees, and the steam which rises from it, will be at the same temperature, of 212 deg.; but nevertheless the steam contains within itself a great quantity of heat, more than an equal weight of water of the same temperature. This excess of heat in the elastic steam beyond that which is contained in an equal weight of water at the same temperature, is called the latent heat of steam, or its heat of elasticity.

The experiments which Dr. Black made himself upon this subject, were sufficient to prove the fact of the existence of a great quantity of heat in a latent state in steam, but not being sufficiently precise to determine its quantity satisfactorily, Mr. Watt made a set of experiments with greater care in 1781; he proceeded by mixing steam and cold water together in known proportions in the following manner. Water was boiled and converted into steam in a common tea-kettle, from the spout of which it was conveyed by a pipe, into a pan of water, the end of the pipe being turned down, and immersed in the water, so that the steam was condensed therein, and communicated all its heat to that water, by mixing with it.

The tea-kettle was filled with water, half way up the entry of the spout, and the lid being fitted on tight, and luted with oatmeal paste, was fastened down by a piece of wood reaching up to the handle. The copper pipe was fitted quite tight to the spout at one end, and the other end was bent down for 3 inches; the pipe was 5 feet long, and $\frac{5}{8}$ of an inch diameter inside, the copper $\frac{1}{50}$ of an inch thick; it was extended from the spout of the kettle, in an inclined direction, the bent end being about 2 feet higher than the spout; the orifice of the extreme end,

(*a*) Mr. Smeaton made several experiments of the same kind as the above; by weighing a thin Florence flask of four inches diameter, first when it was perfectly dry and empty, and afterwards when it was full of water; then pouring out all the water, except a small quantity, he put the flask on the fire, and made that water boil strongly. The neck of the flask being open the steam escaped, until the last drop of water disappeared, and at that instant he stopped up the mouth to retain the steam which was within it. The flask being thus filled with steam of an elastic force equal to that of atmospheric air, was weighed in that state, and deducting the weight of the empty flask from each of these experiments, their comparison gave the proportion of the weight of any bulk of water, to that of an equal bulk of steam; this, by the mean of six different experiments, he determined to be 2459 times. But suspecting that some air was contained in the flask along with the steam, and made it appear lighter than it really was, he inverted the mouth of the flask in water, when it was filled with the hot steam, and found it to draw up the water, in the same manner as described by Mr. Savery (see p. 100); but it did not quite fill with water, for a small bubble of air remained in the flask, and this he estimated to be such a portion of the whole content, as induced him, after a communication with Mr. Watt upon the same subject, to reduce his estimate of the expansion from 2459 times to 1800 times, which was the proportion they both adopted in their subsequent calculations.

3 F

was diminished by fitting a cork into it, and this was perforated with a hole about $\frac{2}{10}$ of an inch diameter, which was kept open by a piece of quill.

The tin pan was 4 inches deep and 6 inches diameter, and had exactly $2\frac{1}{2}$ pounds of cold water poured into it; this quantity filled it nearly $2\frac{1}{2}$ inches deep. The pan was supported upon a piece of flannel folded up; but before the extremity of the bent pipe was immersed in the water, the kettle was made to boil rapidly, to throw out steam, until the pipe becoming quite hot, all water ceased to drop from the quill, and then it emitted a steady current of dry steam. This being the case, the end of the tube was immersed in the water in the pan (the temperature of which was previously noted) and the issuing steam, was condensed in the water with a crackling noise; the water was stirred about with a feather, to distribute the heat which it acquired from the condensation of the steam, and the thermometer was introduced to ascertain the accession of temperature. When it reached between 80 or 90 degrees, the experiment was commonly concluded, and the water in the pan was immediately covered up with a circular disk of strong paper prepared with linseed oil, and dried in a stove; this cover was to prevent evaporation of the water. The pan was then accurately weighed, to find how much the water had gained in quantity, by the condensed steam: the tin pan and the paper disk had been previously counterpoised in the balance when quite dry.

A number of experiments were made in this manner in a room where the air was usually 56 deg. (see the table). The tin pan was afterwards made dry, and placed half an hour in the air of a room at 40 degrees to acquire that temperature, and then 2 pounds of water at 76 deg., being poured into the pan, that water became cooled to $75\frac{1}{2}$; whence it was concluded, that out of every 44 degrees of temperature, in $2\frac{1}{2}$ pounds of water, the pan absorbed $\frac{1}{2}$ a degree.

Mr. Watt's Experiments on the Heat of Steam, 1781.					
Cold water, always 17 500 grains. Temperature. Degrees.	Temperature of the heated water. Deg.	Temperature gained thereby. Deg.	Weight of the condensed steam. Grains.	Sum of the sensible and latent heat.	Latent heat.
43· 5	89· 5	46· 5	760	1159· 5	947· 5
44· 5	98	54	899	1149· 1	937· 1
44· 5	73· 5	29· 5	467· 5	1175· 6	963· 6
44· 5	67· 3	23	369	1158· 0	946
47· 5	87	40	642	1177· 3	965· 3
49	84· 5	36	588· 5	1155· 0	943
47	87· 5	41	675	1150· 5	938· 5
45	86 5	42	680· 5	1166· 5	954· 5
45	85· 5	41	664· 2	1165· 6	953· 6

Dr. Priestley was present with Mr. Watt, during some of these experiments.

Example of one of the experiments; 17500 grains of water in the pan, at $43\frac{1}{2}$ degrees, at the commencement, and at the end 18260 grains at $89\frac{1}{2}$. Therefore the weight of the steam condensed was 760 grains; the accession of heat by the water was 46 degrees, and allowing $\frac{1}{2}$ a degree to be absorbed by the pan = $46\frac{1}{2}$ deg.

The rule for finding the latent heat of steam, from such experiments, is as follows. Multiply the original weight of the cold water, which received the augmentation of heat, by the number of degrees of temperature communicated to it by the steam; this product being divided by the accession to the weight of that water, by the steam which communicated the heat (that is, by the weight of the condensed steam), the quotient will be the degrees of temperature which the steam has lost;

to this add the temperature it retained; viz. that of the hot water at the conclusion of the experiment, and we have the whole heat of the steam, or the sum of its sensible and latent heat. Lastly, by deducting therefrom the sensible heat of the steam, the remainder is the latent heat.

Thus 17 500 grains of water × 46½ degrees of heat gained = 813 750 ÷ 760 grains of steam = 1070 deg. lost. To this add 89½ degrees, the temperature of the mixture, and we have 1159½ for the sum of the sensible and latent heat of the steam; and deducting the sensible heat 212 deg., leaves 947½ for the latent heat of the steam, according to this experiment; and others gave nearly the same result, as is stated in the preceding table.

Two other experiments of this set have been rejected, because they gave a much less result than the others; the mean of those which are retained gives 949·9 for the latent heat. There was no allowance made in the calculations, for the heat which must have been communicated to the surrounding air from the tin-pan, and from the water it contained, during the progress of each experiment, which lasted from four to six minutes; hence this determination is rather below what it should be, and, in consequence, Mr. Watt afterwards adopted 960 for the latent heat of steam, of 212 degrees of temperature, and possessing an elasticity equal to the weight of a column of mercury 30 inches high.

Or in more exact terms; *Any weight of steam of 212 degrees of sensible temperature, contains as much more heat in a latent state, than an equal weight of water, at the same temperature of 212 deg. as would be capable of heating 960 times that weight of water, so much as to raise its temperature one degree.*

Note. If we speak of quantities of heat as being so many degrees, we must specify the substance, and the quantity of matter in the body whose temperature would be raised so many degrees after it has absorbed the quantity of heat in question; without this condition the thermometer affords no measure of quantities of heat; and it would be an improper expression, to say that the latent heat is 960 degrees; for, in fact, that number only expresses the quantity of water, which will be warmed one degree, by a certain quantity of heat, which we designate by its capability to produce that effect.

Mr. Watt also made some experiments on the latent heat of steam of a low temperature in 1765, and repeated them in 1783, but not with sufficient precision to determine the exact quantity. He found that water would distil rapidly in vacuo at the temperature of 146 deg. and he estimated the latent heat of steam in that case to be about 1000. Some other coarse experiments made him suppose that the whole heat of steam (that is, the sum of the latent and sensible heat of steam) would be found to be a constant quantity, at all temperatures and elasticities. This hypothesis has been rather confirmed by succeeding experimenters.

Mr. Watt conducted his experiments on the distillation of water in vacuo in the following manner. He constructed a small still, consisting of a close vessel, or boiler, made of tin plate, and surrounded on the outside with water contained in an external vessel, which was placed upon a chafing-dish, to heat it, and cause the water to boil; the water in the external vessel formed a water-bath, to transmit the heat uniformly to the water contained in the internal vessel, or still. A pipe proceeded from the still, to convey the steam produced therein, to a close refrigerator or condensing vessel, also made of tin plate, in the form of two cones joined base to base. This refrigerator was, at one period of the experiment, wholly immersed in a tin pan containing a known quantity of water, at a known temperature.

The quantity of heat latent in the steam which came over from the still through the pipe, was deduced from the increase of the temperature of this water surrounding the refrigerator. The quantity of the steam was determined by that of the condensed water collected in the refrigerator, and the temperature of that water was noted.

To commence the experiment, a pint of water was poured into the still, and as much into the outside bath; but the refrigerator was not yet immersed in its vessel of water. Heat was then applied, to make the water in the still boil very rapidly, so as to raise steam, and drive out the air, from all parts of the apparatus, through holes left open for that purpose; when it was thought that the air was completely cleared, and all parts of the apparatus made very hot, these holes were stopped, and the apparatus was cooled, by filling up the bath with cold water, so as to quite cover the still. This cooling produced an exhaustion within the apparatus, which was allowed to stand, until it was thought that the tin of the refrigerator, and all the other parts, as well as the rare steam within them, had come to the same temperature as that of the water in the bath surrounding the still. The refrigerator was assumed to contain no condensed water at this period, because it had hitherto been hotter than the steam within it, and could not therefore condense any of that steam.

In this state the refrigerator was completely immersed in the water in the tin pan, which was 6 inches deep, by $8\frac{3}{4}$ inc. diam. and contained 62 800 grains of water, then at $51\frac{3}{4}$ degrees of temperature. The temperature of the water in the bath was intended to be kept up uniformly to 134 deg.; but it rather increased during the experiment, which lasted 9 minutes, so as to become 158 deg. at the conclusion, or about 146 deg. for the mean : with this temperature the water in the still kept boiling all the time, and sending over rare steam, through the pipe into the refrigerator, where it was condensed by the cold of the surrounding water; that is, by communicating its latent heat thereto; so that at the conclusion of the experiment, the temperature of that water had become 61 deg. having gained $9\frac{1}{4}$ deg.

The pan of water was now removed, and the air admitted into the still, so as to put a stop to the distillation; the water arising from the condensed steam was drained out of the refrigerator, and found to be 62 degrees of temperature; and it weighed 534 grains, to which 6 were added, to allow for what would not drain out, making 540 grains.

The refrigerator, which was a double cone made of tin plate, weighed 1000 grains, and being about $7\frac{1}{2}$ times the specific gravity of water, its bulk was equal to $134\frac{1}{2}$ grains of water. The capacity of tin, for heat, being about $\frac{3}{4}$ of that of an equal volume of water, the metal of this cone must have absorbed about as much heat as 101 grains of water would have done in like circumstances. Its temperature, at the beginning, was 134 deg. but on being cooled by the water of the refrigeratory to 62 deg. it must have lost 72 deg.: as this heat was communicated to the water, the quantity of water (101 grains) equivalent to the cone, must be deducted from the 62 800 grains, leaving 62 699 grains of water for refrigeration.

An addition must be made to this as an equivalent for the tin of the pan, in which that water was contained, or at least for so much of the tin as was in contact with the contained water; this weighed 9600 grains, and was equal in bulk to 1280 grains of water, but its capacity for heat, being only $\frac{3}{4}$ of that of an equal bulk of water, it follows that 960 grains of water would have absorbed as much heat as the tin pan did.

Hence the total weight of water heated should be accounted as (62 699+960 =) 63 659 grains; according to the former rule this must be multiplied by 9·25 degrees of temperature acquired = 588 846, and this product being divided by 540 grains of condensed steam, gives 1089; to which adding 62 deg., the temperature of the condensed steam, we have 1151 for the sum of the sensible and latent heat. This is nearly the same as in the former experiments; or, deducting the sensible heat 146 deg. leaves 1005 for the latent heat of the steam.

Mr. Watt remarks, " I am by no means satisfied with the accuracy of this experiment; too many things are taken by estimation, which ought to have been ascertained by experiment, and the degree of exhaustion was not so great as it might have been, or as I have obtained in some other experiments. The vacuum ought to have been made more perfect by repeatedly boiling the water, condensing the steam, and blowing out the air, until the distillation should take place at 70 or 80 degrees.

" The capacity of the refrigerator for heat should have been examined accurately; and it hould have been made to take off from the steam pipe, in order that it might have been weighed before and after each experiment, so as to ascertain the quantity of water which adhered to the inside." M. de Luc attended Mr. Watt during the above experiment in 1783.

It is most probable that the same weight of steam (or of water in the state of steam) always contains the same quantity of heat, or very nearly so, let the temperature and elasticity of the steam be what it may; one part of that heat is sensible to the thermometer, and occasions the elastic force of the steam; and the remainder of the heat being latent in the steam, or insensible to the thermometer, occasions the expansion of volume from the state of liquid water, into elastic steam.

It would follow, that when the expansion of volume is greater, the quantity of heat in a latent state must be greater. Also that the elasticity being less, when the volume is greater, the quantity of heat which is in a sensible state will be less, when that which is in a latent state is greater.

Note. This supposes that the steam is in contact with liquid water, so as to be at liberty to take up as much of that water, as will satisfy the natural disposition of water and steam to receive heat.   The steam may then be said to be saturated with water.

We may conclude Mr. Watt's investigations, respecting steam, with the remark that although the results of his experiments were not fully satisfactory to himself, and therefore he caused them all to be repeated some years afterwards, it appears that his deductions were correct ; and any slight want of precision which there might have been in his quantities, must have been quite insensible in the application of the principles to practice.   All Mr. Watt's experiments as above recorded, are sufficient approximations to the exact truth, to be safe guides to practical engineers ; but, in another part of this work, tables will be calculated according to more exact experiments on steam.

Mr. Watt's experiments were not published till 1821, in his annotations on Dr. Robison's Memoir; nor was there any publication of experiments on steam until 1790, when M. Bettancourt, in France, made a series upon the elasticities of steam at different temperatures.

CHAPTER VI.

On the Application of the Steam Engine to produce continuous Circular Motion, for working Mills and Machinery.

THIS was another grand invention made by Mr. Watt about the year 1784, and which he soon brought to such perfection, that his engines have been found to answer the purpose of working mills, or any kind of machinery, as well as water-wheels; and in most large manufactories, these engines now supply the place of water, wind, or horse-mills, so that, instead of carrying the work to the power, the prime mover is placed wherever it is most convenient to the manufacturer.

The rapid increase of the manufacturing towns of England and Scotland within the last thirty years, and their great prosperity, have been entirely owing to Mr. Watt's invention of the rotative steam-engine; for it would have been impossible, by any other means yet known, to have obtained sufficient power for the performance of the work which is now executed by its assistance, or to have established the present system of manufacturing which is so advantageous to this country.

The general application of steam power, to manufacturing purposes, is with much justice, attributed to Mr. Watt, for although the atmospheric engine is very capable of working mills, and is actually in use for such purposes to a considerable extent, yet without Mr. Watt's improvements, which gave certainty and regularity to its action, and diminished the consumption of fuel, the steam-engine could not have been beneficially employed in a great number of cases where it is now the prime mover. The application of steam power to produce circular motion for mills, is quite an era in the history of mechanical inventions, and it is desirable to record the various attempts which were made, before it was brought to bear.

In Papin's project in 1690, he proposed to employ several of his new cylinders and pistons, with toothed racks formed on the piston rods, in order to act upon pinions so fitted on a common axis as to give that axis a continuous circular motion, by the alternate and successive actions of the pistons of the different cylinders; and he proposed to row boats, by means of paddle wheels fixed upon such an axis, (see p. 98.) Thus we see that the idea of continuous circular motion is coeval with the first notion of the cylinder engine.

Also in 1698, Mr. Savery in his demand for a patent, announces his fire-engine as being of great use for working all sorts of mills; and in his Miner's Friend, 1702, he describes his plan for working mills with water-wheels; to be supplied with artificial falls of water which had been previously raised by his fire-engines, (see pp. 117 and 122.)

This method of raising water to turn the mill, by falling back on a water-wheel, was afterwards frequently put in practice, (see p. 296), before the more simple plan was adopted of applying the force of the engine at once to the mill, by means of a crank fixed on the extremity of the main axis of the mill, in the same manner as a man turns a grindstone by a winch handle.

In the common foot lathe, or in a spinning-wheel, the reciprocating motion given to the treadle by the pressure of the foot, is made to act by a crank and fly-wheel, so as to produce a continuous rotative motion ; this is an old invention of great merit, and would be sufficient to suggest the first idea of a rotative steam-engine, to any person who had considered the action of that simple machine ; for it would seem very easy to apply the piston rod of a Newcomen's engine, to give successive impulses to the treadle, in lieu of the workman's foot. But to carry such an invention into effect, it required a much greater regularity and certainty in the motion of the engine, than was necessary for the simple purpose of raising water ; and although this application appears so obvious, it was not put in practice till a late period, for it was considered as an impracticable scheme, until after many more complicated methods had been attempted and failed ; nor was it brought into any extensive use, until Mr. Watt took up the subject, and applied his improved engine for that purpose.

In 1736, Mr. Jonathan Hulls obtained a patent from King George II., for a machine for towing ships out of, or into harbours or rivers, against wind and tide : he proposed to use revolving paddle-wheels, and to impel them with con-tinuous circular motion, by the force of a Newcomen's cylinder engine. The action of the piston was to be communicated to the paddle-wheels, by a rope pass-ing over a pulley, fixed on their axis, and the motion was to be continued during the returning stroke of the piston, by a counterweight suspended to the end of the rope. It does not appear that this plan was completely made out in a prac-ticable form, or that any thing was ever tried.

In 1759, Mr. Kean Fitzgerald proposed, in the Philosophical Transactions, to give motion to a ventilator, to supply fresh air to mines, where a fire-engine was employed to draw off the water. The proposed ventilator was a fanner or vane revolving circularly within a box, and was intended to blow a constant current of fresh air down into the mine. This was one of Papin's inventions of 1690, and he called it the Hessian bellows. A machine of this kind requires a continuous circular motion in one direction, and the object of Mr. Fitzgerald's contrivance was to obtain such a motion from the great lever of the engine, which works up and down, and, he says, performs at a medium about twelve strokes in a minute ; but by his plan, it was made to turn a wheel constantly one way round, and the number of revolutions was increased to fifty or sixty in a minute.

This was to be effected by a combination of toothed wheels, with ratchet wheels, reversed to each other and mounted on a common axis, in such manner as to be worked by an arc, or toothed sector, connected with the great lever. One of these ratchet wheels carried the axis round with it for about half a turn, during the descent of the piston, and during the ascent of the piston it turned back-wards on the axis, without producing any effect, but then the other ratchet-wheel, which was reversed to the former one, came into action, and by means of interven-ing wheel-work, it turned the axis round another half turn, in the same direction as the former motion ; so that one or other of these reversed ratchet-wheels was always in action, in order to communicate a continued rotative motion to the common axis on which they were both placed. This motion was transmitted by a large toothed wheel, to a smaller pinion, on the axis of which a fly-wheel was fixed, and turned round quick enough to accumulate a sufficient energy to keep up a continual motion, and urge the machinery forward, whilst the engine was per-forming its returning stroke. This latter axis was proposed to turn the ventila-tor ; and it is stated that it may easily be made to turn a mill to grind corn, or to raise coals out of the mine. The machine is fully described, with an engraving, in the 50th volume of the Philosophical Transactions ; but it was never put in practice, and from its extreme complexity, it had no chance of a successful application.

About 1762, a machine was constructed by Mr. Oxley at Hartley colliery, near Seaton Delaval, in Northumberland, for drawing coals out of the pit by a fire-engine ; it was similar in principle to Mr. Fitzgerald's scheme. A toothed sector at the end of the great lever, gave a reciprocating motion to a trundle, which was fitted on an axis, and connected therewith by the intervention of two pinions, with ratchet-wheels and clicks so combined as to communicate a continuous circular motion to the axis, and in the same direction, when urged either by the ascending or descending stroke of the piston. The ratchets and clicks were so contrived that the motion could be reversed at pleasure, in order either to wind up, or to let down the corves in the coal-pit. This engine had no fly-wheel, and it went sluggishly and irregularly ; the machinery was frequently deranged in consequence of the violent jerks to which it was subjected at every change in the direction of the motion. The same engine was afterwards applied to raise water by a pump, to supply a water-wheel, by which the coals were drawn.

In 1769, Mr. Dugald Clarke contrived a plan to derive a continuous rotative motion from a fire engine, and he proposed it to be applied to sugar-mills in Jamaica. A machine was erected in London, and the inventor obtained a patent. The circular motion was produced by means of a toothed rack jointed to the great lever of the engine, and adapted to ascend and descend in a vertical groove, so as to give motion to two rack wheels fixed on the same axis, and this communicated motion to the great horizontal cog wheel of the sugar-mill, by means of trundles. As the rack moved up and down, the axis was alternately carried round with one of the rack wheels, by means of a catch, whilst the other rack wheel was turned back by the rack. The objection to this plan was its complexity, and also that its operation was liable to be affected by any variations in the stroke of the engine.

In 1777, a paper was presented to the Royal Society by Mr. John Stewart, describing a plan for converting the reciprocating motion of the fire-engine into a progressive circular motion, for turning all kinds of mills. It consisted of two endless chains circulating over pulleys fitted into a frame, which was moved up and down by the motion of the engine, in the manner of a window-sash ; the joint pins of the links of the two chains were applied into the teeth at the opposite sides of a main cog-wheel, in such manner as to give it a circular motion, by first one chain, and then the other, acting alternately on the opposite sides of the wheel. For instance, one chain impelled it during the descent of the piston, and the other chain in the ascent ; but one of the endless chains always circulated freely over its pulleys, so as to produce no effect on one side of the cog-wheel, whilst the other chain operated on the opposite side of the wheel to turn it round. For this purpose, each chain was provided with a catch, to prevent its circulating over its pulleys in one direction, but to allow it a free motion in the other direction.

The inventor says, if the main cog-wheel just mentioned be applied to the main axis of a flour-mill, the power of the reciprocating engine will thereby keep it continually going, and the mill-stone will serve as a fly to regulate any such intermissions as will happen at the change of each stroke of the engine. In saw-mills, or any other mills which do not communicate a great velocity to some heavy body, that may serve as a regulating power, a large fly-wheel may be added.

In this paper Mr. Stewart speaks of a crank or winch, as a mode of obtaining the circular motion, which he says naturally occurs in theory, but in practice he thought it would be impossible, from the nature of the motion of the engine, which depends on the force of the steam, and cannot be ascertained in its length; therefore on the first variation, the machine would be either broken to pieces, or turned back; Mr. Smeaton entertained the same idea (a).

In 1779, Mr. Matthew Wasbrough, an engineer at Bristol, took out a patent for improvements in Steam or fire engines; the specifications which is deposited in the Rolls Chapel, is very concise, and gives but an imperfect explanation of the plan: it appears to have been intended to produce a continuous circular motion by ratchet wheels, similar to those tried at Hartley Colliery; but Mr. Wasbrough added a fly wheel, which was then employed for the first time in a steam-engine. The advantage of this auxiliary proved so great in practice, that Mr. Wasbrough was enabled to make a few of these engines, which were used for some time; one was in his own workshop at Bristol for turning lathes; another was set up at Mr. Taylor's saw mills and block manufactory at Southampton; and some others were made for grinding corn. The ratchet work of these engines was found to get continually out of order; and, in 1780, one of them which had been erected by Mr. Wasbrough at Birmingham, was altered by some of the people who had the management of it, the ratchet work was all removed, and a simple crank substituted, the fly wheel only being retained. The engine then answered so much better than any thing which had been tried before, that the same principle has been followed ever since.

In 1780, a patent was taken out by Mr. James Pickard of Birmingham, button-maker, " for a new method of applying steam-engines, commonly called fire-engines, to the turning of wheels; whereby a rotative motion, or motion round an axis, is performed, and the power of the engine is more immediately and fully applied (where motion round an axis is required) than by the intervention of a water-wheel." The patent is dated the 23d of August. The specification, which is deposited at the Petty-bag Office, states that a short lever or crank, is to be fixed on the extreme end of the axis which is to be turned round, and the extremity of that crank is to be united by a pin, to the lower end of a rod, which is jointed to the end of the great working lever (or which may, in some cases, be jointed to the piston itself) but in such manner as to turn the crank once round, at every stroke of the engine : and if a regular motion is required, a cog-wheel is to be fixed on the axis of the crank, in order to turn another cog-wheel of half the size, fixed on a second axis, which will make two revolutions for each stroke of the engine. A heavy balance weight is to be fixed to the circumference of this second wheel, in such a position as will cause the weight to be raised upwards, whenever the crank is at or near the horizontal position, so as to be capable of receiving an effectual impulse from the connecting rod; and consequently the weight will act by its descent, so as to impel the machinery, whenever the crank is at or near the line of the direction of the connecting rod, in which position it is incapable of receiving any effective impulse therefrom. A figure is given of the crank with the two cog-wheels, and the revolving balance weight; but the engine is not represented, nor is any mention made of the application of a fly-wheel.

A revolving balance weight thus applied at the circumference of a wheel

(a) Mr. Stewart's paper was referred by the Council of the Royal Society to Mr. Smeaton, who remarked upon the difficulty arising from the absolute stop of the whole mass of moving parts, as often as the direction of the motion is changed; and though a fly wheel might be applied to regulate the motion, it must be such a large one, as would not be readily controlled by the engine itself; and he considered that the use of such a fly wheel, would be a greater incumbrance to a mill, than a water wheel to be supplied by water pumped up by the engine.

which is made to turn twice for each stroke of the engine, would produce a great uniformity in the force exerted by the circular motion; but it would require a fly-wheel to regulate the velocity. It does not appear that Mr. Pickard's patent was acted upon, although engines with a simple crank and fly-wheel began to be used very soon after.

APPLICATION OF THE ATMOSPHERIC ENGINE TO TURN MILLS WITH CONTINUOUS CIRCULAR MOTION, BY MEANS OF A CRANK AND FLY-WHEEL.

THE atmospheric engine requires no alteration in its boiler, cylinder, or working-gear, from what has been already described; in fact all the parts are the same, except the extremity of the great lever, which, instead of the arch head and chains, is formed with a strong joint to unite it to the connecting rod, or spear M, at the lower end of which another joint is formed, to fit the pin of the crank N.

This crank is a short lever, firmly fixed on the extreme end of the axis P, of the large fly-wheel Q Q, beyond the neck, or part of that axis which rests in the bearing; the length or acting radius of the crank, from the centre of motion to the centre of the crank pin, is half the length of the stroke which is to be made by the piston; the crank projects from the end of the axis of the fly-wheel, and the crank pin projects out from the crank; so that the rod M, not being in the plane of the crank, the latter can be turned completely round, its pin then describing a circle, or moving round in a circular orbit, whereof the diameter is the length of the stroke of the piston. The lower end of the connecting rod M accompanies the crank pin in its circular motion, whilst the upper end thereof moves up and down with the end of the great lever L.

The fly-wheel Q is made heavy in the rim, but balanced on its axis so as to be equally heavy on all sides ; the millwork which the engine is required to turn is connected with the other extremity of the axis P of the fly-wheel.

The action of this engine is obvious from the figure ; during the working stroke, the force of the piston pulling up the crank pin, by the spear M, causes the crank to turn round, and the fly-wheel also ; and the latter, by the weight and rapid motion of its rim, acquires a very considerable energy of motion, whereby it continues to move forwards, when the piston has completed its working stroke, and also during the returning stroke.

The movement of the spear M is compounded of two alternating motions ; for the upper end moves up and down, in an arc of a circle described by the joint at the extremity of the great lever L, whilst the lower end of the spear moves continually forwards in the circle described by the crank pin.   The spear has, therefore, an oscillating motion of its lower end, like that of a pendulum, but compounded with another motion, whereby the centre of suspension of that pendulum rises and falls.   Neither of these are uniform motions, but it is intended that the resulting motion of the crank pin shall be nearly uniform in its circular orbit, so as to turn the fly-wheel steadily round, and give a regular motion to the millwork which is connected with its axis ; but the vertical motion of the piston in its cylinder is very variable, being slow at the top and bottom of its course, and rapid in the middle thereof.

The effort which the piston can exert upon the crank and fly-wheel to turn it round continuously, is extremely variable throughout the different periods of the stroke.   At the first beginning of the stroke, the crank being in a line with the spear or connecting rod, the force of the piston can have no action whatever to turn the crank ; but as soon as the crank begins to make a sensible angle with the connecting rod, the force of the piston begins to operate upon the crank to turn it round, and this operation increases with the angle at which the connecting rod acts upon the crank, until near the middle of the stroke, when they are at right angles to each other, the whole force of the piston becomes operative to turn the crank.   As the motion continues onwards in the circle, the crank begins to approach again towards the line of the connecting rod, and therefore the power which the latter can exert upon the crank to turn it round, diminishes until the end of the stroke, when they come once more into a line, and then there can be no farther action to turn the crank.

The use of the fly-wheel, which is interposed between the engine and the mill, is to reduce all these variations of effort, to a regular and uniform action, or as nearly so as possible ; when the piston is near the middle of its stroke, and is exerting its full power to turn the crank round, the fly-wheel acts by its inertiæ, in aid of the resistance of the mill-work, to retard the piston from descending too rapidly ; but the position of the crank soon becomes altered, so that the whole power which the piston can exert upon it to turn it round, will be less than the resistance of the mill work ; the fly-wheel then gives back part of that excess of power which had been just before communicated to it, and thereby keeps up the motion of the mill-work, even when the engine can exert no power whatever to turn the crank round.

The rapidity with which the piston moves in the cylinder, is continually varying, from the same cause that produces the variations of its effect upon the crank, for when the engine exerts the greatest force on the crank, the piston moves quickest ; and as the power to turn the crank diminishes, the rapidity of the motion of the piston diminishes in the same degree.   Hence the velocity of the piston in the cylinder is to be considered as varying continually; but if the fly-wheel is large enough, and sufficiently heavy, it will be found that the rotative motion resulting from the combination of all the variations, will be nearly regular, whilst the ascending and descending motion of the piston is accelerated from nothing at the top of the cylinder, to its greatest velocity at the middle, or near the middle, and from that point it is retarded, until it comes to nothing at the bottom of the motion.

As the effective force of Newcomen's engine on which the crank and fly-wheel was first tried, is confined to its working stroke, it became advantageous to equalize, as nearly as could be, the power of the working and returning strokes. For this purpose the spear or connecting rod M, was made of cast-iron of large dimensions, and equal in weight to half the unbalanced pressure of the atmosphere upon the piston of the engine; or when the weight was not in the spear, it was placed on the great lever L at that end.

Suppose that by this means the engine is made to exert an equal force to impel the crank pin, in the ascent and in the descent of the connecting rod, still some safe depository of force is required, where a sufficient power may be lodged and kept in readiness to continue the motion, in the interval of the change of the motion from ascending to descending, and *vice versa*. It is with this object that the fly-wheel is applied to the axis of the crank; the weight and motion of the fly-wheel must be sufficient to accumulate in itself the whole force of the engine during its time of action; and therefore it will continue the motion, by its energy, or tendency to preserve the motion it possesses, and it will urge forward the mill-work, whilst the piston is going through its inactive period of changing the stroke. This it may do, provided that the resistance opposed by the mill-work during the whole period of the working and returning stroke of the engine, together with the friction of both, does not exceed the whole power exerted by the piston during its periods of efficient action upon the crank pin; and provided that the energy of the fly-wheel, arising from its weight and velocity, be sufficiently great; so that the resistance of the mill-work, during the change of the stroke, will not make any very sensible diminution of the velocity of the fly-wheel.

This is evidently possible, for the fly-wheel may be made of such magnitude, that the heavy rim or circumference will move with great velocity, and being exactly balanced round its axis, it will soon acquire that velocity which is due to the full motion of the engine. The energy of matter in motion, being as the weight multiplied into the square of the velocity with which it moves (see p. 18), a fly wheel of large dimensions, and having a considerable weight of matter in the rim, will take a great power to put it in motion in the first instance, or to stop or impede it when in motion.

During the working stroke of the engine, the fly-wheel is uniformly accelerated; and by its acquired energy it produces the movement of the mill, whilst the engine changes its action, and until it begins its returning stroke; but in doing this, a small part of the motion which constitutes its energy is taken away, in the motion it has so communicated to the mill work, and consequently its velocity is diminished, but not very considerably. The weight of the connecting rod, by pressing on the crank pin afresh during the returning stroke, restores to the remaining velocity of the fly-wheel, all that it lost during the inactivity of the engine. Hence the fly-wheel receives a part of the power of the engine, whenever it has any power to spare, and keeps that power in reserve, and communicates it faithfully to the mill, whenever the engine is deficient.

The crank and fly-wheel is a very important addition to the steam-engine ; and though it is so obviously an imitation of the foot lathe, that it can scarcely be considered as any invention, still it required considerable skill and address to make it effective. The principal difficulty arose from the slowness of the motion and from the want of regularity in the action of the old engine (a). The condensation of the steam by injecting cold water into the cylinder, takes a sensible time to cool all the steam, and in order to produce a tolerable vacuum, there must be a sensible stop or hesitation at the top of the stroke, before the piston can begin its descent, at least with its full effect. This is a defect in the principle, in consequence of which a regular and rapid motion cannot be acquired, except by a great sacrifice of force.

An engine to act with a crank must be made to work with celerity, and to move at all times exactly the same length of stroke ; and to make it perform well, these strokes should be all performed in an equal period of time. The old engine had very little exactness in either of these particulars ; for several reasons. The boiler was commonly too small, and in consequence, the least alteration in the intensity of the fire, caused the engine to vary its speed ; the alteration of the barometrical weight of the atmosphere, occasioned considerable variations in the descending force of the piston, and also affected the engine, by diminishing or increasing the force, and velocity of the flow, of the injection into the cylinder, for this depends in some measure upon the atmospheric pressure on the surface of the water in the cistern (see p. 180) ; the old working gear was likewise very uncertain and inaccurate in its operation.

Operation of an Atmospheric-engine with a Crank and Fly-wheel.

To explain the application of Newcomen's engine to turn mill-work, we may recur to the case already stated, p. 131, of a cylinder 24 inc. diam. with a 5 feet stroke of the piston, and making 15 strokes per minute ; the unbalanced force on the piston being 4859 lbs. ; the counterweight 662 lbs. ; and the load, or resistance opposed to the motion of the piston 3535 lbs. As this force acts in only one direction, one-half of it (= 1768 lbs.) should be added to the 662 lbs. of counterweight, so as to give a preponderance of 2430 lbs. more weight at the con-

(a) Mr. Smeaton's opinion on this subject, will be found in a report which he made to the Commissioners of the Victualling Office, in 1781, on the best method of constructing a flour mill for the public service, to be worked by a steam-engine, and whether the mill should be turned by the intervention of a water-wheel, or by a direct application of the force of the engine, to a crank and fly-wheel. (See his Reports, vol. ii. p. 378). " I apprehend that no motion communicated from the reciprocating lever of a fire engine, can ever produce a perfect circular motion, like the regular efflux of water in turning a water wheel. All the fire engines that I have seen, are liable to sudden stoppages, and may in the course of a single stroke pass from almost the full power and motion, to a total cessation ; for whenever by inattention of the engine keeper, the elasticity of the steam becomes too low, the engine will be incapable of continuing its motion. The fire engine seems peculiarly adapted for raising water, and then a stoppage for a few strokes, is only a loss of so much time ; but such stoppages in the motion of mill-stones for grinding corn, would greatly confuse the regular operation of the mill, as it would stop with the stones full charged with corn, and before the motion could be renewed, the stones must be raised up from their bearings.

" By the intervention of a water wheel these uncertainties and difficulties are avoided, for the work is in fact a water mill, and if there is a sufficiency of water in the reservoir to work the mill one minute without sensible abatement, it will seldom happen but that the engine may be set going again in less than half a minute after it stops by any common accident, so that the mill will continue regularly at work."

necting rod end of the lever, than at the piston end. When the engine is thus adjusted, the connecting rod will press the crank pin downwards, during the returning stroke, with as much force as it will draw it upwards during the working stroke, and that force will be equal to a weight of 1768 lbs. The radius of the crank will be $2\frac{1}{2}$ feet.

The usual dimensions for a fly-wheel for such an engine as they were constructed at the first commencement, was about 16 feet diameter outside, and about 15 ft. diam. inside ; the rim being made of cast iron, 6 inches by 6 inches ; the arms were commonly made of wood.

To find the weight of the rim, take the area of a circle 16 feet diameter = 201·06 square feet, and deducting the area of a circle 15 feet diam. = 176·71 sq. ft., we have 24·35 sq. ft., × ·5 of a foot thick = 12·18 cubic feet of cast iron, which × 450 lbs. per cubic foot, gives 5481 lbs. for the weight of iron in the rim.

To determine the space through which this mass is moved, and its energy when in motion, or the mechanical power which must be communicated to it, to give it such motion from a state of rest, we must assume for the present, that the fly-wheel revolves with an uniform motion, at the rate of 15 revolutions per minute ; and by the rule given at p. 33, the velocity of the circumference of the rim, taken at the middle of its thickness will be (15 revo. × 15·5 feet diam. ÷ 19·1 =) 12.17 feet per second.

Again by another rule (at p. 23), the height through which a body must fall by gravity to acquire this velocity, would be 2·3 feet; consequently the weight of the moving mass 5481 lbs. × 2·3 feet fall, gives 12 606 lbs. acting through one foot space, for the mechanical power which must be communicated to the rim of the fly-wheel, to urge it from a state of rest into motion (see p. 31), with a rapidity which, if it were continued uniformly, would cause it to make 15 revolutions per minute (a).

The force exerted by the connecting rod upon the crank pin is 1768 lbs. which acting through 5 feet during each half stroke is = 8840 lbs. acting through 1 foot space. The power or energy of the rim of the fly-wheel (= 12 606 lbs. acting through 1 foot, ÷ 8840 lbs.) is 1·428 times the power which is exerted by the engine during one-half stroke.

If the mill-work were detached from the axis of the fly-wheel, so that the engine would have no resistance to overcome, except that which this fly-wheel could oppose by its inertiæ, the power of the engine might be expected to urge the fly-wheel from a state of rest, into motion with the above velocity, by the time it had made rather less than $\frac{3}{4}$ of a complete stroke ; viz. when the piston had moved $7\frac{1}{5}$ feet, having completed its working stroke, and almost half its returning stroke (b).

A fly-wheel of this proportion, would be subject to a very sensible inequality in its motion ; for when the piston is near the middle of its course, the force exerted by the connecting rod upon the crank pin, to turn it round, will greatly exceed the resistance occasioned by the mill-work; and all this excess of force being communicated to the fly-wheel, would cause it for a moment to move considerably quicker than the supposed uniform velocity, 12· 17 feet per second ; but in thus accelerating, it would so much increase its own energy as to take up all the super-

(a) These results may be obtained direct, by the rule given at the bottom of p. 33 ; thus 15.5 ft. diameter × 15 revolutions per rim = 232·5 ÷ 153·2 constant number = 1·518; which squared gives 2· 3 feet, for the height due to the velocity.

Or the calculation may be still more readily performed, by using the slide rule as directed at p. 31 ; viz.

Sliding Rule. $\begin{cases} C & 5481 \text{ lbs. mass.} \quad\quad 12\,696 \text{ lb. acting one foot.} \\ D & 8·02. \quad\quad\quad\quad\quad 12\ 17 \text{ feet per second veloc.} \end{cases}$

(b) The fly-wheel above quoted, would not possess half the energy it ought to have, according to the present practice of engineers; whereby the power which must be communicated to the rim of the fly-wheel, to give it its proper motion from rest, would be from 3 to $3\frac{3}{4}$ times the power which is exerted by the engine, during one half stroke: and with this proportion the motion of an engine becomes sufficiently uniform for most purposes of manufacturing.

fluous power.   When the piston approaches the end of its course, the force exerted on the crank pin to turn the crank round, will become considerably less than the resistance of the mill-work ; the fly-wheel would then make up the deficiency ; and by a momentary diminution of its velocity (which would then become less than the supposed uniform motion of 12· 17 feet per sec.) it would restore to the mill-work all that superfluous power which it had just before received or accumulated in itself, when its velocity was increased.

The velocity of the fly-wheel must, therefore, be irregular to a certain extent, for it must move quicker towards the middle of each stroke of the piston, and slower towards the ends of those strokes ; but the greater the energy of the fly-wheel is, in proportion to the power of the engine, the more its motion will approach to uniformi!y, because the motion will be less deranged by that alternate excess and deficiency of the force exerted on the crank, which the fly-wheel must alternately accumulate in itself, and then restore again, during the progress of each half stroke.

And yet, however great the energy of the fly-wheel of a steam-engine may be, it can never turn the mill-work with a perfectly uniform motion ; and in fact, if the fly-wheel did move uniformly, it would be a useless incumbrance, because it is only by change of its velocity that it can either give out, or take up power.   It would be a difficult problem to calculate by common arithmetic, what are the irregularities of the motion of a fly-wheel which is connected with a steam-engine ; but some idea may be formed of the law of the force and velocity with which the piston acts through the intervention of the crank, to turn the fly-wheel round.

The crank pin describes a semi-circle, during the time that the piston moves in a straight line through a space equal to the diameter of that semi-circle ; therefore, the space passed over by the crank pin, with circular motion, must be (3·1416 ÷ 2 =) 1·5708 times greater than the course described by the piston during the same time in its rectilinear motion ; nevertheless, when the piston is near the middle of its course, the connecting rod draws in a direction at right angles to the crank, and then the motion of the piston must be quite as quick as that of the crank pin ; but the motion of the piston begins to diminish immediately from this point, whilst the crank pin keeps moving on in its circular orbit nearly uniformly ; and the piston moves continually slower and slower, till it comes gradually to a stop at the end of the stroke, and then it begins to return.

The following statement of the rate of the motion of the piston, at different periods of the stroke, will make this subject clear.   Let the crank of an engine be $2\frac{1}{2}$ feet radius, and the piston will make a stroke of 5 feet long ; let the connecting rod be 15 feet long, and the great lever 15 feet long, or $7\frac{1}{2}$ feet radius.   Now suppose the circle of 5 feet diameter, which is described by the crank pin, to be divided into 24 equidistant points ; two of those points being on a level with the centre of motion, and two of them in a vertical line passing through that centre.

The force which the connecting rod can exert to turn the crank pin round in its circular orbit, may be represented by the direct distance from the centre of motion to the length of the connecting rod, measured in a direction at right angles to that length (a).   This distance may be called the effective leverage of the crank,

(a) This direct distance, is in reality the sine of the angle that the crank makes with the connecting rod ; and when that angle is known in degrees, the distance can be found from a table of sines.

because the effect is the same for the moment, as if the connecting rod were applied to a short crank of only that length, but acting in a direction at right angles to its length. For in all cases of levers, pressed by forces which do not act at right angles to the length of the lever, the result of the oblique action is to reduce the effective length of leverage, to that of the direct distance between the centre of motion, and the line of the direction, in which the force does act. (See p. 47.)

The effective leverage of the crank, in each position, is put down in inches, in the following table, as it was ascertained by measuring from a drawing (a); and by summing up these numbers, and taking a mean of them, it appears that the effective leverage of the crank is 19·1 inches, at an average of all its variations.

From these numbers we may proceed to compute the velocity with which the piston must move in its cylinder, when the crank is in each position. If the piston made 15 strokes per min. of 5 feet, as already supposed, it would pass through a space of 150 feet in a minute. But the crank pin making 15 revolutions per min. in a circle 5 feet diam. (= 15·7 feet circumference) must pass through 235·6 feet per minute, and for an instant when the crank is at right angles with the

(a) The different positions of the engine may be denoted by the angle which the crank makes with the vertical line; and then by the following means we may compute the angle that the connecting rod makes with the vertical line, in every one of these positions. The same sine is common to both those angles, and is the horizontal distance of the crank pin from the vertical line; or in other words, the sine of the angles made with the vertical line, is the same for the crank, as for the connecting rod; but the connecting rod is usually six times as long as the crank.

If we take from the tables, the sine of the angle which the crank makes with the vertical line, in any position, (45 degrees for instance, the sine by the tables is ·707) and divide that sine by 6 (if that is the proportion between the lengths of the connecting rod and of the crank) we shall obtain (·1178) the sine of the angle, that the connecting rod makes with the vertical line; and by reference to the table of sines we may find what that angle is (viz. 6 deg. 46 min.).

The angle so found is to be deducted from the angle that the crank makes with the vertical, in all positions when the crank-pin is below the horizontal line; or added to the same, if the crank-pin is above the horizontal; and we then obtain the required angle, that the crank makes with the connecting rod, and whereof the sine is the effective leverage of the crank.

For instance, if the crank is below the horizontal position, and makes an angle of 45 deg. with the vertical line; then, as above shown, the connecting rod will make an angle of 6 deg. 46 min. with the vertical, which being deducted from 45 deg. leaves 38 deg. 14 min. for the angle between the crank and the connecting rod. The sine of this angle = ·6189, is the effective leverage of the crank, when in that position, the full length of the crank being 1; and if that length is 30 inches, then the effective leverage will be (30 inc. × ·6189 =) 18·567 inches.

For another instance, suppose the crank is above the horizontal position, and at an angle of 45 deg. from the vertical; then the angle of the connecting rod from the vertical, = 6 deg. 46 min. must be added to 45 deg. and gives 51 deg. 46 min. for the angle between the crank and connecting rod. The sine of this angle = ·7855 is the effective leverage of the crank in that position, or (× 30 inc. =) 23·565 inches.

Note. This mode of computation supposes that the upper end of the connecting rod never departs from the vertical line which passes through the centre of the crank: this is the case in some steam-engines, and then the above method is exact; but when the upper end of the connecting rod moves in the arch of a circle, as it must do when it is jointed to the great lever, its deviation from the vertical line occasions a minute difference in the angle of the connecting rod with the vertical, from what is given by the preceding mode of calculation; but the difference is so small, as not to be worth the trouble of computing a correction for it.

The effective leverage of the crank is greater, when it makes any given angle above the horizontal position, than when it makes the same angle below the horizontal position; because the obliquity of the connecting rod tends to increase the effective leverage in one case, and to diminish it in the other. And for this reason if the sines of the angles that the crank makes with the vertical line are assumed to represent the effective leverage of the crank, in all its different positions, they will give a correct result for the action during the whole stroke.

connecting rod, the piston must move at the same rate, but that is its maximum velocity.

We may find the velocity with which the piston moves, when the crank is in any other position, by multiplying the velocity of the crank-pin (= 235·6 feet per min.) by the effective leverage of the crank in inches, when in that position, and dividing the product by the actual radius of the crank, also in inches (= 30 inc.). Or the same results may be obtained direct by multiplying the effective leverage in inches by (235·6 ÷ 30 =) 7·854; the numbers so obtained form one of the columns of the table. By summing up all those numbers, and taking the mean, the velocity of the piston appears to be 150 feet per min. at an average of all its variations.

We may then calculate the actual power exerted by the engine, when the crank passes through each of the twenty-four positions, by assuming the force or pressure exerted on the crank-pin to be uniform, and equal to 1768 lbs., acting nearly in a vertical direction, first upwards, and then downwards. This force must be multiplied by the velocity of the piston, at any period of the stroke, and the product divided by 33,000 lbs., then the quotient will give the horse power actually exerted at that moment. The numbers so obtained are put down in one of the columns of the table, and by comparing them we may form some idea of the extent of the irregularities which the fly-wheel is required to equalize, as far as it can do so.

Operation of a Steam-Engine to turn a Crank and Fly-wheel.					
Positions of the Crank.	Effective Leverage: Inches.	Velocity of the Piston: ft. per min.	Horse-power exerted: HP.	Steam expended per min., cub. ft.	Motion of the Piston: inches.
Vertical, down	0·00	0·0	0·00	00·0	at top 1·00
	6·80	53·4	2·86	167·8	2·45
	13·10	102·8	5·51	323·2	4·00
45 degrees, down	18·60	146·1	7·82	459·0	5·45
	23·55	184·9	9·90	581·0	6·60
	27·30	214·4	11·48	673·7	7·45
Horizontal	29·45	231·3	12·38	727·0	7·90
	30·00	235·6	12·62	740·0	7·80
	27·90	219·1	11·74	688·5	7·00
45 degrees, up	23·60	185·5	9·94	583·0	5·65
	17·00	133·5	7·15	419·5	3·45
	8·60	67·6	3·62	212·5	1·25
Vertical, up	0·00	0·0	0·00	00·0	at bot'om
Averages	19·1	150·0	8·00	235	5·00

We see that the power exerted, when the piston is near the middle of the stroke, is 12·62 HP; and then diminishes to nothing at the two extremes; but by summing up all the varying powers, and taking a mean of them, the uniform exertion will be found to be very near 8 HP. The uniform resistance of the mill-work must have been equal to 8 HP, but the power communicated from the piston to the

3 H

fly-wheel exceeded this by 1·57 times, when at the middle of the stroke, and failed totally at each end.

The preceding table also contains a column expressing the extent of the motion of the piston, in passing from one position to the next; this was ascertained by measurement from the drawing: the numbers show, that when the effective leverage of the crank is small (and therefore but little power is communicated to the fly-wheel), the motion of the piston is proportionably slow, so that only a small expenditure of steam takes place. This is stated more completely in the last column of the table, which is formed by multiplying the area of the cylinder, (3·1416 square feet) by the velocity of the piston in feet per minute at each position; the product is the rate at which the steam is expended in cubic feet per minute, and which varies in every position, in the same manner as the effective leverage and the power varies.

Note. As the expenditure of steam is confined to the working stroke, it will only amount to half as much per minute, as the numbers in the table, which merely represent the rate of expenditure, to show how it is perpetually varying. The quantity actually expended is 235 cubic feet per minute; and it will appear, by summing up all the numbers in the last column, and taking the half of their mean, that the expenditure would be nearly at that rate (*a*).

To find by the sliding rule, the effective length of leverage of a crank, when in different positions. We must use the line marked s at the back of the slider of the Soho rule; draw out the slide towards the right hand, until the angle which the connecting rod makes with the crank, in degrees, is brought to the end of the rule, then opposite to 1 on the line A, we shall find the sine of that angle on the line B. Or opposite to the real radius, or full length of the crank on the line A, its effective leverage will be found on B, thus:

Soho Sliding Rule. { Line s behind the slide { A 1 or Length of crank, inches. / Angle of crank with rod { B sine / Effective leverage, inches.

To find the power actually exerted by an engine, when its crank is in different positions. Multiply the power of the engine by 1·57, and having set the rule as before, according to the position of the crank, seek the product on the line A, and opposite to it on B, is the power exerted by the engine, when the crank is in that position.

Soho Sliding Rule. { Line s behind the slide { A 1·57 times the HP. / Angle of crank with rod { B horse power exerted.

To find the rate at which the steam is expended in different positions of the crank, proceed in the same manner, by means of the product obtained by multiplying the quantity of steam expended per minute by 1·57.

Note. The angle that the crank makes with the connecting rod may be found by the means already stated, or instead thereof the angle that the crank makes with the vertical line may be taken, and will give a correct result on an average of all the different positions, which the crank assumes in its rotation.

Many engineers have exercised their ingenuity to contrive some means of obtaining a circular motion from the piston of a steam-engine, whereby the power

(*a*) The numbers above stated, as obtained from summing up the numbers in the table, and taking their mean, are what they ought to be, but are not exactly the average of the numbers in the table; because, in any series of variable quantities, it requires a mean to be taken of an infinite number of steps, to obtain the correct results, such as are above stated.

communicated to the fly-wheel, should be at all times equable and constant, and the velocity of the piston uniform, from one end of the stroke to the other. This was the object aimed at in all the early attempts to produce a rotatory motion, and which was tried, as already mentioned, at Hartley Colliery in 1762, and by Mr. Wasbrough in 1779. Such attempts have been repeatedly renewed since that period, by other mechanicians, but without success. One great defect of all these contrivances for communicating uniform power to the rotative axis is, that the whole force of the piston would come to act upon the axis at once, with a shock, which in course of time must destroy the best constructed mechanism (*a*).

The crank, on the contrary, by the variable motion that it gives to the piston, prepares it for changing the direction of its motion, so that it does not come suddenly to rest, nor is the piston required to start suddenly into motion, in the opposite direction, but the change is effected imperceptibly. This circumstance is highly advantageous to the operation of the steam-engine, because it allows time for the condensation of the steam to be performed. In an engine which is pumping water, the piston may be observed to wait sensibly at the commencement of its stroke, before it begins to move, for although the condensation of steam takes place with extreme rapidity, yet it does require a sensible time for the steam to be condensed, and if a sufficient time is not allowed, then the exhaustion within the cylinder will be more imperfect than usual, and consequently a less mechanical effect will be produced by the engine, with the same expenditure of steam.

Another great advantage of the crank is the accuracy with which it measures the length of the stroke of the piston, which may therefore be safely allowed to move through the whole length of the cylinder, so as to go very near to the top and bottom thereof, because there is no danger of its ever striking. On the contrary, the piston of an engine without a crank, must never be allowed to go too near its total limits, because the least variation of the force, or of the resistance, would occasion the catch pins to strike with dangerous force against the springs which must be provided to limit the length of the stroke ; all the vacant space within the

---

(*a*) It has long been an established opinion amongst practical men, that the crank, in consequence of its perpetual variation of leverage, must occasion an actual loss of a part of that mechanical power, which the engine would be capable of exerting, if by some contrivance it were enabled to operate upon the rotative axis with a constant and unvarying length of leverage. All the different contrivances above alluded to, have originated from this mistaken notion of the action of the crank, and without taking into account, that the expenditure of steam is in all cases proportionate to the effect produced by it; so that when the piston acts on the fly-wheel, through the medium of a short leverage of the crank, the motion of the piston in the cylinder is proportionably slow, and the consumption of steam proportionably small.

It is obvious that the expenditure of steam must be proportional to the motion of the piston in its cylinder, and the semi-circumference of a circle being 11, when its diameter is 7, it follows, that if the force of the piston were applied continually in the direction of a tangent to the circle described by the crank pin (or always with the full length of leverage), the effect to turn the fly-wheel round would be 11, instead of only 7, as it is when the force is transmitted by the common crank. But then, the space described by the piston would be 11 instead of 7; and, therefore, if the effect obtained by applying the force in a tangent to the circle, would be four-sevenths greater than when it is applied by the crank, the reason is not, that any part of the power exerted in the latter case is lost, but that four-sevenths, less power is exerted.

The only difference in favour of a new contrivance would be, that a steam cylinder of any given diameter might produce a greater effect, if the force of its piston were applied at a tangent, with a full length of leverage; but then the cylinder must be longer, and the piston would move quicker in it, than when it acts with a crank; so that the expenditure of steam would be proportionably greater; and there would be no sensible advantage to be derived from the adoption of any such contrivance, whilst the disadvantages of all the plans which have been proposed for this object are very great, as stated above.

cylinder, which the piston does not occupy in its motion, occasions an expenditure of steam without augmenting the power of the engine.

The crank has also another property, which adapts it very completely to the purpose of changing the reciprocating motion of the piston, into a continuous circular motion, viz. that by the gradual diminution and final cessation of the leverage at the conclusion of each reciprocation, all that power is realized which has been necessarily communicated to the reciprocating parts of the engine (such as its piston, great lever, and connecting rod) to put them into motion from a state of rest, and which constitutes their energy whilst in motion (see p. 31). As the motion is gradually taken away from them, and they are brought imperceptibly to rest, by the action of the crank, all their energy is faithfully transmitted by the crank to the fly-wheel, in aid of its motion (*a*).

This is a most important circumstance, for a considerable power must be communicated to the moving parts, to urge them from rest into motion, with the velocity which they must acquire towards the middle of the stroke, and it would occasion a loss of part of that power to bring those parts to a sudden stop ; in fact that could not be done, unless the catch pins were to strike at the end of each half stroke, so as to expend the energy of the moving parts upon the building.

When the reciprocating parts are made to operate upon the fly-wheel, by the intervention of a crank, the motion of the reciprocating mass is regularly accelerated and retarded alternately ; this retardation is gradual, and finally ends in total rest ; after which the acceleration commences again, and continues gradually, until the moving parts acquire their maximum velocity.

During the course of the retardation, the whole of the energy which the reciprocating mass possessed at the time of its greatest velocity, will be gradually transferred to the fly-wheel, for the retardation is occasioned solely by the resistance which the crank pin opposes to the continuance of the motion ; and therefore, it is evident, that whilst it is opposing such resistance, it must receive all the impetus which is elicited from the moving parts in putting them at rest, and the power so communicated from the moving parts, goes to augment the energy of the fly-wheel. In like manner, during the acceleration, this same amount of energy must be renewed again, and communicated from the piston to the reciprocating mass.

It is by this gradual transference of all the energy of the reciprocating mass to the fly-wheel, that the reciprocating motion is checked and caused to cease, and not by the power of the steam urging the piston in an opposite direction. It may be safely affirmed, that no loss of power can be occasioned by the weight and energy of the moving parts, in a reciprocating steam-engine which acts with a crank and fly-wheel, except by the friction of surfaces which are in actual contact, and which move and rub one against the other ; so that the evil arising from the weight of the moving parts will only be felt in the friction that they occasion.

(*a*) This action of the crank has been but very imperfectly understood by practical engineers, who have commonly supposed that the reciprocation of the moving parts in steam-engines, occasions a great loss of power ; because they have assumed that the power of the piston must be exerted to bring the moving masses to rest, as well as to urge them into motion, from rest. In consequence, many plans have been tried to construct steam-engines, with fewer and lighter moving parts, in the expectation of reducing this supposed loss of power from their reciprocation ; but the results of such trials have not shown the expected advantage, which could be but very inconsiderable, as will appear from a careful examination of the manner in which the force exerted by the piston in a right line, is transferred by the connecting rod and crank, to the circumference of the circle, described by the crank pin.

It must be kept in mind, that a reciprocating steam-engine can never produce a perfectly uniform motion of the fly-wheel, by means of a crank; for the power transmitted from the piston varies so much, that at one time it will exceed the resistance opposed to the motion, and then the next moment that resistance will exceed the power exerted. The fly-wheel is interposed as a receptacle of mechanical power or moving force, and is adapted to receive and accumulate in itself, all the superabundant power when it exceeds the resistance, and to restore it again as soon as the resistance exceeds the power. But it is only by changing its velocity that a fly-wheel can either take up, or give out any of its power, and therefore it can have no operation, except when the velocity of the machine is variable; for having naturally a tendency to preserve an uniform velocity in itself, it will tend to equalize the motion of the machine, which is connected with it, by impeding the motion, when any cause operates to accelerate it, and assisting the motion when any thing tends to diminish it. Hence it would be useless to employ a fly-wheel in any machine where the power and resistance are always uniform, or where they always bear the same proportion to each other.

In a steam-engine, however large the fly-wheel is, its motion cannot be rendered perfectly regular; and the effect of the fly-wheel to continue the motion will be greater or less, according as the change produced in its velocity, and consequently in that of the mill-work, is greater or less.

If it is rigorously examined, the movement of any mill or machinery which is worked by a steam-engine, will be found unequal in some degree; but the inequality may be rendered insensible to all common observation, if the fly-wheel is made large enough, and moves with sufficient velocity, and has a sufficient weight of matter in its circumference. By these means its energy may be rendered so great, that the impetus required to be elicited from it, in order to propel the mill-work, during the ineffective position of the crank, may bear a small proportion to the whole energy that the fly-wheel possesses; the diminution of its velocity will then be trifling, and consequently the motion of the mill-work which is connected with it, will be tolerably regular.

The action of fly-wheels was so little understood before they were applied to steam-engines, that it was considered as quite impracticable to give to a fly-wheel a sufficient power to govern and equalize the irregular motion of a steam-engine; and after the common atmospheric engine had been made to work with a fly-wheel, as above described, the motion was found to be irregular, because the fly-wheels were not made sufficiently powerful; for when an engine is worked with such a slow motion as 15 strokes per minute, it will require a very heavy fly-wheel to obtain a tolerably uniform rotative motion (a). When Mr. Watt applied his improved engine to work mills, he made it act with celerity, so as to repeat the impulses on the crank more frequently, by which means the energy of the fly-wheel is greatly increased, in consequence of its quicker motion;

(a) It was not then generally known by practical engineers, or even fully admitted by philosophers, that the mechanical power resident in a moving mass, or its energy, is proportionate to the square of the velocity, (see p. 18). At the commencement of the last century, a keen controversy on the nature of moving force, had been carried on during many years, between two parties in the learned world; and their disputations had left common men in doubt whether they ought to estimate the effect of a fly-wheel according to the simple velocity, or according to the square of the velocity.

Mr. Smeaton made a series of experiments, of which he published an account in the Philosophical Transactions for 1776; whereby he clearly proved, that the power which must be communicated to any mass of matter, to give it different velocities of motion from a state of rest, will be represented by the product of the mass, multiplied by the square of the velocity with which it moves.

and those irregularities of the action of the engine, which it is required to remedy, are lessened.

The advantage of a rapid action of the piston, was soon felt by those who constructed atmospheric engines for mills, and when they were caused to work quicker, they produced a steady motion.   For instance, if the atmospheric engine above described had been adjusted to make 21 strokes per minute, instead of 15, then the motion of the piston would have been 210 feet per minute, and the velocity of the rim of the fly-wheel 17·05 feet per second.   The engine could not have been worked at that rate, without diminishing the resistance in a still greater proportion than the increase of its velocity, and therefore the power of the engine would have been lessened to 7, or perhaps 6-horse power; but in recompense, the effect of the fly-wheel to regulate its motion, would have been more than doubled, and would have produced a regular motion.

About the years 1790 to 1793, when steam-mills began to be introduced into all the large manufacturing towns, with Mr. Watt's improved engines, great numbers of atmospheric engines were also made for turning mills, particularly in the districts where coals were cheap.   The principal makers of these engines were Messrs. Bateman and Sherrat of Manchester, who supplied Lancashire, and also sent some to London: and Mr. Francis Thompson of Ashover, in Derbyshire, who made engines for that district, and for Sheffield, and Leeds.

These engines answered the purposes to which they were applied, and were used for many years; some of them are still in use, and a few have since had condensers and air-pumps applied to them in part of Mr. Watt's improvement.

Both Messrs. Bateman and Sherrat, and Mr. Thompson, had plans for combining two atmospheric cylinders in one engine, so as to produce a double action; and they made a few such engines, which answered better than the single action.

It is still a very common practice, in districts where coals are cheap, to work machinery by Newcomen's engines, by means of a crank and fly-wheel.   The steam in the boiler is made much stronger than formerly, to enable it to enter the cylinder with a very sudden puff, so as to displace the air and water in an instant. The resistance or load in such engines is not above $4\frac{1}{2}$, or 5 lbs. per square inch of the piston; and with this load they will make from 20 to 24 strokes per minute, of 5 feet long.   All these circumstances tend to reduce the performance of an engine with respect to power, and the consumption of fuel is very great in comparison with the work performed.

These atmospheric engines act very well if the work or resistance opposed to them is constantly the same, so that the counterweight of the connecting rod, may be properly adjusted to half the descending force of the piston, in order to make the pressure upon the crank-pin of equal force in ascending as in descending.   But they are not so well adapted for breweries or other works, where the same engine being applied to several different operations, the resistance must be perpetually changing, when any part of the machinery is suddenly disengaged.   If something is not then done to check the engine, it will increase its velocity so as to be in danger of breaking to pieces.   As the only remedy is to diminish the supply of steam at the returning stroke, the discharge of the air will be interrupted, and the motion of the engine will become so irregular, as to be liable to stop, or turn back, when the piston gets nearly down to the bottom of the cylinder, in consequence of the accumulation of air therein; and yet if the fly-wheel can only just carry the crank past the vertical position, the heavy connecting rod then urges it into motion again with violence.

Mr. Watt's Second Patent, 1781, *for applying the Steam-engine to produce continuous Circular Motion.*

FOR some years after the first introduction of his improved engines, Mr. Watt was fully occupied in substituting them for the large atmospheric engines at mines, where the expense of fuel was threatening to put a stop to their proceedings; and he found no leisure for new speculations, although the idea of applying engines on his principle to actuate mill-work and machinery had early occurred to him. He did not seriously set about reducing his ideas to practice until the year 1778 or 1779 (*a*).

In the first model which he made, he employed two cylinders and pistons, in order to equalize the power; they acted upon two cranks formed upon the same axis, at an angle of 120° from each other, and a fly-wheel was fixed upon this axis, with a weight placed upon the circumference of the fly-wheel, at an angle of 120° from each of the cranks; the weight was so adjusted, as to turn the wheel, when neither of the cranks could do so, and consequently the power was rendered nearly equal.

This model performed to satisfaction; but Mr. Watt neglected to take out a patent immediately, and very soon afterwards some of the persons engaged about one of Mr. Wasbrough's engines applied a crank and fly-wheel thereto, as before mentioned, p. 409; and a patent was taken out in 1780, by Mr. Pickard, for the

---

(*a*) Mr. Watt's account of the origin of this part of his invention is given as follows in his notes upon Dr. Robison's article STEAM-ENGINE.

" I had very early turned my mind to the producing of continued motions round an axis, and it will be seen by reference to my first specification in 1769 (see p. 316), that I there described a steam-wheel, moved by the force of steam, acting in a circular channel, against a valve on one side, and against a column of mercury or some other fluid metal, on the other side. This was executed upon a scale of about six feet diameter at Soho, and worked repeatedly, but was given up, as several practical objections were found to operate against it; similar objections lay against other rotative engines, which had been contrived by myself and others, as well as to the engines producing rotatory motions by means of ratchet wheels, mentioned in p. 408.

" Having made my single reciprocating engines very regular in their movements, I considered how to produce rotative motions from them in the best manner; and amongst various schemes which were subjected to trial, or which passed through my mind, none appeared so likely to answer the purpose, as the application of the crank, in the manner of the common turning lathe; but as the rotative motion is produced in that machine, by the impulse given to the crank in the descent of the foot only, it requires to be continued in its ascent by the energy of the wheel which acts as a fly; being unwilling to load my engine with a fly-wheel heavy enough to continue the motion during the ascent of the piston, (or with a fly-wheel heavy enough to equalize the motion, even if a counterweight were employed to act during that ascent) I proposed to employ two engines, acting upon two cranks fixed on the same axis, at an angle of 120 degrees to one another, and a weight placed upon the circumference of the fly-wheel at the same angle to each of the cranks, by which means the motion might be rendered nearly equal, and only a very light fly-wheel would be requisite.

" This had occurred to me very early, but my attention being fully employed in making and erecting engines for raising water, it remained *in petto* until about the year 1778 or 9, when Mr. Wasbrough erected one of his ratchet wheel engines at Birmingham, the frequent breakages and irregularities of which recalled the subject to my mind, and I proceeded to make a model of my method, which answered my expectations; but having neglected to take out a patent, the invention was communicated by a workman employed to make the model, to some of the people about Mr. Wasbrough's engine, and a patent was taken out by them, for the application of the crank to steam-engines. This fact the said workman confessed, and the engineer who directed the works, acknowledged it, but said, nevertheless, that the same idea had occurred to him, prior to his hearing of mine, and that he had even made a model of it before that time: which might be a fact, as the application to a single crank was sufficiently obvious.

application of a crank with cog-wheels, and a balance weight to revolve twice for each stroke of the engine. This did not discourage Mr. Watt, and without troubling himself to dispute a patent which, so long as it continued attached to the common atmospheric engine, could not rival his own, he contrived other means of effecting the same object, and took out a patent for several new methods of producing a continued rotative motion, from the intermitting action of a piston. One of these was a beautiful contrivance of one wheel revolving in an orbit round another, and they were called the sun and planet wheels, from the resemblance to the motion of those luminaries.

Mr. Watt's patent, which is dated the 25th October 1781, is entitled " for certain new methods of applying the vibrating or reciprocating motion of steam or fire-engines, to produce a continued rotative or circular motion, round an axis or centre, and thereby to give motion to the wheels of mills or other machines." The specification describes five different methods, but only two have since been brought into use ; viz. the simple crank, and the sun and planet wheels.

All the methods described in this specification are adapted to be worked by single acting engines ; and a counterweight is to be applied in different ways, according to circumstances, so as to impel the machinery during the returning stroke of the engine.

The first method described by Mr. Watt is, to give the continued motion to a vertical axis, upon which a large horizontal wheel is fixed, and beneath the lower surface of the wheel, a double circular inclined path is formed, in an inverted position ; at one part of the circumference this path descends considerably beneath the horizontal plane of the wheel, and from this point, the path returns in both directions, by gradual inclination, until at the opposite part of the circumference, it coincides with the plane of the wheel ; beneath this wheel a lever is placed so as to form a diameter to the same, and each end of the lever carries a roller ; these rollers act beneath the circular inclined path, at two places diametrically opposite to each other, and the fulcrum or centre of motion of the lever, being just beneath the centre of the wheel, one of the rollers which applies to the inclined path, will ascend, whilst the other descends, and *vice versa*.

The lever is connected by a chain with the great lever of the steam-engine, so as to be moved up and down thereby, and in this motion, the rollers of the lever will rise vertically and act beneath the inclined path, first on one side of the wheel and then on the opposite side in a contrary direction, with an action which will tend to turn the wheel continually round in one direction, in consequence of the obliquity of the circular inclined path to the plane of the motion of the wheel. A fly-wheel must be applied to keep up the motion at the change of the stroke of the engine.

" In these circumstances I thought it better to endeavour to accomplish the same end by other means, than to enter into litigation, and if successful, by demolishing the patent, to lay the matter open to every body. Accordingly, in 1781, I invented and took out a patent for several methods of producing rotative motions from reciprocating ones, amongst which was the method of the sun and planet wheels. This contrivance was applied to many engines, and possesses the great advantage of giving a double velocity to the fly-wheel ; but is perhaps more subject to wear, and to be broken under great strains, than a simple crank, which is now more commonly used, although it requires a fly-wheel of four times the weight, if fixed upon the first axis ; my application of the double engine to these rotative machines (see the patent of 1782, p. 350) rendered the counterweight unnecessary, and produced a more regular motion."

The method of combining two cylinders to act with two cranks formed upon the same axis, has since been brought into use with great advantage in the modern engines for propelling carriages ; and for steam-boats. It is an excellent plan, and has also been applied to turn mills.

Mr. Watt's next method is the same in its operation as a crank ; an excentric circle is fixed upon the axis of the fly-wheel, the distance from the centre of the circle, to the centre of motion, being equal to half the stroke of the piston ; and in order to turn the axis of the fly-wheel round, the centre of the excentric circle must be carried round in a circular orbit. For this purpose it is surrounded by a ring, or open frame, shaped like a stirrup, and provided with three rollers, which enclose the excentric circle, and bear upon its circumference at three different points. The stirrup is formed at the lower end of the connecting rod, which is suspended in place of the pump rod, from the extremity of the great lever of the engine, so as to receive a reciprocating motion, which being communicated by the stirrup to the excentric circle, tends to move it round in its orbit, and turn the axis and fly-wheel continually round.

The action is precisely the same as a crank, for if the connecting rod were attached to a pin placed in the centre of the excentric circle, that pin would describe the same orbit as the excentric circle, which may, in fact, be considered as a mere enlargement of the crank pin, and it is only in consequence of such enlargement that the three rollers become necessary to diminish the friction.

Another plan is to use a crank, with the addition of a balance-weight, applied in such a manner as to produce a regular motion, when a single acting engine is the moving power. This he effected by fixing an iron wheel on the extreme end of the axis of the fly-wheel, in place of a crank, and with a pin projecting from it, to which the connecting rod is jointed : one half of the wheel is made a solid semicircle of cast-iron, so as to be sufficiently heavy on that side in which the crank-pin is fixed, to urge the fly-wheel round during the returning stroke of the engine ; the other half of the wheel is made open with arms, to be light, that it may not oppose this weight.

A similar loaded wheel was proposed with two crank-pins, to combine the action of two different cylinders and pistons, as before mentioned, and the heavy side of this wheel answered the purpose of the weight placed at the circumference of the fly-wheel.

The Sun and Planet-wheels is the last method, and as this was constantly used for many years, it deserves a more particular notice ; it is represented in Plates XI. and XII. The sun-wheel OO, is a toothed wheel fixed fast on the extreme end of the horizontal axis P of the fly-wheel QQ, and the planet-wheel NN is also a toothed wheel, exactly similar, but fixed fast to the lower extremity of the connecting rod M, and adapted to revolve round the sun wheel in a circular orbit with a planetary motion ; to retain the planet-wheel in its orbit, its centre pin is received in a fixed circular groove, or else a link $q\,q$ is fitted on the axis P of the fly-wheel, and also on the centre-pin of the planet-wheel N, so as to keep the centres of the two wheels always at an invariable distance, and prevent their teeth from quitting each other.

The planet-wheel being fixed fast to the connecting rod, it cannot turn round upon its own centre, but can only travel round in its orbit, and its teeth acting in the teeth of the sun-wheel, will turn the same round with a double velocity ; that is, the sun-wheel and fly-wheel make two complete revolutions, whilst the planet-wheel makes one circumvolution in its orbit. Therefore, one-half of a revolution of the planet-wheel (which is made during a half-stroke of the engine) will give the fly-wheel a whole turn.

This method of obtaining the circular motion by the sun and planet-wheels was afterwards generally adopted for steam-engines ; but Mr. Watt did not proceed immediately to apply his single engines to turn mills, because many more

3 I

contrivances became necessary, to enable him to make that application with complete success. The next great step was the double action, by which the piston exerts a continual force, and requires no counterweight ; this important invention was made in 1782, as already stated (p. 350).

DOUBLE-ACTING STEAM-ENGINES.

As the piston of a steam-engine operates by a reciprocating motion, it is advantageous to enable it to act with force in both directions, and thereby exert a continuous effort, instead of expending one-half the time in useless returning motion. It was not easy to produce a double action in a cylinder, until Mr. Watt had invented his engine, in which the piston is urged by the elastic force of the steam itself, independently of the pressure of the atmosphere.

The useful operation of the piston of an atmospheric engine is necessarily limited to one direction, because the air which must be admitted into the cylinder, to press the piston down by its weight, cannot be exhausted or extracted from the cylinder, except by drawing the piston up again by the counterweight ; and the steam which is admitted into the cylinder, merely restores the equilibrium, by refilling the space within the cylinder as fast as the piston rises out of it.

It is true, that if the steam were heated so as to become highly elastic, with an expansive force considerably greater than the pressure of the atmosphere, it might exert a force to lift the piston upwards, when it was admitted into the cylinder. This plan was attempted by Mr. Hindley of York (see p. 257) ; but no advantage could be gained by such a mode of working, because the cylinder would be filled with dense and hot steam, which could not be condensed so suddenly and completely as steam of equal elasticity with the atmosphere ; and hence any power which could be gained from the elasticity of the steam during the ascent of the piston, would be attended with a deduction from the power which the atmospheric pressure would otherwise have exerted during the descent. The consumption of fuel to produce such dense steam, would be excessive, without any adequate advantage.

The only practicable method of obtaining a continual action in the atmospheric engine, is to combine two or more cylinders in one machine, in such manner that whilst one piston is ascending, and therefore inactive, the other shall be descending and exerting its force. This was the scheme of Papin, in 1690, when he gave the first idea of a cylinder engine (see p. 98), and the common double-barrelled air-pump suggests a very good model for the construction.

This kind of double-acting steam-engine was proposed, in 1779, by Dr. Falck, who published an account and description of it ; in which he says, that with the same quantity of fuel, and in an equal space of time, it will raise above double the quantity of water raised by any lever-engine of the same dimensions ; but he does not appear to have proved the assertion, or constructed even a working model of his proposed engine, which was on Newcomen's principle.

The improvement which he suggests is to use two cylinders, into which the steam is admitted alternately by a common regulator-valve, which always opens the communication of the steam to one cylinder, whilst it shuts up the opening to the other.

The piston-rods of these two cylinders are formed with teeth like racks, which work into teeth at the opposite sides of a cog-wheel, situated between the two racks, in such manner that the pistons which are attached to them shall ascend and descend alternately ; but they will always move in opposite directions to each other, in the same manner as the pistons of a common double-barrelled air-pump.

The engine actuates two pumps, which raise water alternately, so as to make a continual stream.   The two pumps are wrought by means of another cog-wheel, similar to that for the cylinders, and fixed on the same axis : this second cog-wheel gives an alternate motion to two perpendicular racks, which are affixed to the two pump-rods ; hence the two pump-buckets will always move in opposite directions to each other, in the same manner as the pistons of the cylinders do.

There is nothing in this engine which could enable it to raise much more water than a common atmospheric engine, with the same fuel, and its performance must have been very nearly the same as two separate engines ; for they would gain nothing by working in concert, except the regularity of the stream of water ; and unless the water was required to be forced through pipes for water-works, that would give no advantage.

For the mere purpose of raising water, a single-acting engine is sufficient, and double engines are rarely employed ; but for giving an uninterrupted motion to mills, a double or continuous action is of the utmost consequence to the regularity of the motion.

Since the improved engines of Mr. Watt have been introduced, Dr. Falck's method of combining the alternate action of two single engines has been applied to work machinery with continuous circular motion by a crank and fly-wheel (a).

Another plan of combining two atmospheric cylinders into one double-acting engine, was also brought forward, in 1793, by Mr. Francis Thompson ; one cylinder was placed over the other, the pistons of both cylinders being affixed to the same rod ; the lower cylinder was open at top, in the usual manner of New-comen's engine, but the upper cylinder was inverted, being suspended with the open end downwards, so that the pressure of the atmosphere forced its piston upwards, and urged the piston-rod in ascending.   Several such engines were con-structed by the inventor, who had a patent for it : they were used for working mills.

MR. WATT'S DOUBLE-ACTING STEAM-ENGINE, 1782.

The double engine, which is now in use, was invented by Mr. Watt, and has but one cylinder ; the force of the steam being applied alternately to press the piston downwards in its cylinder, and then to force it upwards again, so that the counterweight becomes unnecessary, and the action is continuous.

In the specification of his third patent, 1782, he says, " My second improve-ment upon steam or fire-engines consists in employing the elastic power of the steam to force the piston upwards, and also to press it downwards, by alternately making a vacuum above or below the piston respectively; and at the same time employing the steam to act upon the piston, in that end or portion of the cylinder which is not exhausted.   An engine constructed in this manner, can perform twice the quantity of work (with a cylinder of the same size) or exert double the power, in the same time which has hitherto been done by any steam-engine, in which the

() About the year 1790, a large engine of this kind was constructed at Mr. Thackray's Cotton-mill, Manchester, and is still in use (1825). It has two atmospheric cylinders of 36 inches diameter, with a condenser and air-pump to exhaust the cylinders ; the pistons go 4 feet stroke, and formerly each one made 40 strokes per minute, it was then rated at 70 horse power. It has been altered to make 30 strokes per minute.

Several small engines of this kind were also made about the same time, by Messrs. Sherrat of Manchester, for working cotton-mills.

active force of the steam is exerted upon the piston only in one direction, whether upwards or downwards.

Mr. Watt's double engine operates in the same manner as his single engine does when performing its working stroke, and the returning stroke is performed in the same manner as the working stroke, for in fact they are both working strokes. To produce this effect, the valves to admit and exhaust the steam from the cylinder required a new arrangement, and likewise the working gear to open and shut those valves.   It also became necessary to apply additional chains to the arch heads of the great lever, to enable the piston to force the lever upwards, as well as to pull it downwards ; or a better plan was, to apply a toothed rack to the piston rod, to work into a toothed sector formed on the arch head of the great lever, so as to com- municate the double motion of the piston to it.

This forms another item of the specification of his patent of 1782 : " My fourth new improvement on steam or fire-engines consists in applying toothed racks, and toothed sectors of circles, for suspending or connecting the pump-rods or pistons with the working levers, or other machinery used in place of them, instead of the chains and arch heads which have hitherto been used for these purposes."

The engine, thus improved, he called the double engine, as in fact it doubled the power exerted within the same cylinder.   He had long conceived the idea of this improvement in his mind, and had produced a drawing of it to the House of Commons, in 1774, at the time when he procured the act to prolong his original patent.   The first double engine was executed at Soho, in 1782 ; and a large double engine was put up at Wheal Maid copper-mine in Cornwall, in 1787 ; about the same time other double engines were made at Soho, and erected in London, to turn machinery for the breweries, and for a flour-mill.

Mr. Watt says himself, that having encountered much difficulty in teaching others the use and construction of the single engine, and in overcoming prejudices, he did not proceed in the double engines at first ; but finding himself beset with an host of plagiaries and piracies about the year 1782, he thought proper to insert a drawing and description of it in the specification of his patent that year.   It was the representation of an experimental engine, and no others have been made exactly similar.

Mr. Watt's double engines, with some subsequent improvements upon the original project, have been chiefly applied to turn mills with circular motion ; yet in some cases, where mines are very deep, they have been advantageously applied to the working of two sets of pumps in the same pit, by a reciprocating motion ; one set or half of the pump-rods being suspended by means of a sloping rod con- nected with the great lever near the cylinder end of it, and the other half of these rods being suspended directly from the outer end of the lever ; so that the ascending motion of the piston draws up one half of the rods, and works the pumps belonging to them, whilst the descending motion of the piston draws up the other half of the rods, and works their pumps.

A large engine of this construction was erected at Wheal Maid mine, in Corn- wall, in the year 1787.   The cylinder was 63 inches diameter, and the piston made 9 feet stroke ; but the stroke in the pumps, which were 18 inches diameter, was only 7 feet. This engine, which at the time it was made, was the most powerful in existence, worked remarkably well, though, like many others in Cornwall, it was loaded with an enormous weight of dry pump rods.

In other cases, when it has not been convenient to divide the pump rods into two sets, the ascending motion of the piston has been employed to raise a counterweight, equal to half the weight of the column of water in the pumps ; this counter-weight acts in aid of the power of the engine, during the descending stroke of the piston ; but the former method is preferable where there is room in the pit to contain two sets of pumps, and their appurtenances.   There are several of these large double engines still in use at some mines in Cornwall ; but as they have not been found to perform so well as single engines, double engines have not been constructed there of late years.

Parallel motion.   When Mr. Watt began to apply his double-acting engine to turn mills, he found that the toothed rack and sector, which he first used to connect the piston-rod with the great lever, did not move with smoothness, and the teeth were very subject to wear ; and chains were still less adapted for communicating the double action.   It occurred to him, that some mode might possibly be devised, of applying rods or levers, moving upon centres or joints, in such manner as to counteract the circular motion of the extremity of the great lever, and retain the piston-rod in its vertical direction, both in pushing upwards or pulling downwards ; a smoother motion would thus be obtained, and with less liability to wear.   In pursuing this idea, he contrived several methods of effecting the purpose in the course of the year 1783 ; and in the following year he took out a patent for a number of improvements relative to steam-engines, and the applications of their power ; but the parallel motion has proved the most important amongst them : it is one of the most ingenious contrivances in mechanics, and communicates the vertical motion of the piston-rod, with the greatest certainty, to the extremity of the great lever, which moves in the arch of a circle.

MR. WATT'S FOURTH PATENT, 1784.

THIS patent was granted " for certain new improvements upon fire and steam-engines, and upon machines worked or moved by the same, dated 28th April, 1784."   The specification, which is dated 24th August, 1784, is deposited in the Roll's Chapel ; it has never been printed, although it contains several ingenious inventions which are worthy of being recorded.

First, a new rotative engine in which the steam-vessel is turned continually round upon a pivot, by means of a dense fluid contained in the vessel, but which is forced out by the steam, so as to issue with great velocity through a spout properly disposed in the lower part of the revolving vessel, which being wholly immersed in the like dense fluid, the reaction occasioned by the issuing stream, urges the vessel into motion, on the principle whereby a sky-rocket mounts in the air, and in the same manner as those fire-works which are called Catherine-wheels, are caused to turn round.

The steam vessel is shaped like a bottle, or decanter; it is sustained on a pivot in the centre of its bottom, and is supported by its neck, the upper end of which forms the axis to give motion to the machinery intended to be propelled ; this vessel is immersed in the dense liquid up to its neck ; its interior capacity is divided by a vertical partition into two compartments, each provided with a valve in the bottom of the vessel, opening upwards, to permit the dense fluid to enter into, and fill, that compartment.   The steam-pipe from the boiler is so connected with a collar or box, which surrounds the neck or upper end of the axis, as to introduce the steam into one or other of the compartments, through suitable passages formed in the neck or axis ; the steam so admitted, pressing by its elasticity upon the fluid contained in that compartment, forces it out in a horizontal current, through a spout or opening, at the lower part of the revolving vessel, and so disposed that the current will issue in the direction of a tangent to the circular base of the vessel, and this current striking the motionless fluid in which the whole vessel is immersed, causes a re-action at the orifice of the spout, whereby the vessel is put in motion about its vertical axis. Whilst the entering steam is thus emptying the fluid out from one of the compartments, the steam is allowed to escape from the other compartment, that it may be refilled with the fluid, through the valve at bottom, in order to operate in its turn, and so keep up a continual action.

It is proposed either to use steam of a high elasticity, or else to close up the cistern or vessel containing the dense fluid, in which the revolving steam-vessel is immersed, and to draw off that steam which has performed its office, by means of a condenser, so as to exhaust the space above the surface of the dense fluid; and then steam of ordinary elasticity being admitted into the revolving steam-vessel, will force out the fluid through the spouts, with great rapidity.

Second. Methods of causing the piston-rods, pump-rods, and other parts of engines, to move in perpendicular or other straight lines, and to enable the engine to act upon the working lever, both in pushing and in pulling. This is now called the parallel motion, and several varieties are described.

The upper end of the piston-rod is in all cases connected with the extremity of the great lever, by a link jointed to both, so that the upper end may describe an arch of a circle, whilst the lower end moves in a right line, and the upper end of the piston-rod is retained in that right line by various means, as follows.

In one of the varieties two long levers are connected with the piston-rod by joints, and extend out therefrom each way towards the walls of the house; they are formed with arched heads, or sectors of circles at their extremities, to apply with their convex circumferences in contact with straight guides, fixed in an upright position against the walls of the building; viz. one against the lever wall, and the other against the end wall. The circular arches roll up and down, in contact with these guides, and the centres of curvature being the joints by which their levers are united to the piston-rod, those joints are confined to move in straight lines, on the same principle as the wheels of a carriage run upon a road-way, for the arches may be considered as portions of very large rolling wheels.

The weight of the arches are suspended by straps of leather, fastened at their upper ends to the upright guides, and at their lower ends to the lower extremities of the arches; part of the straps being straight, and applying to the upright guides, and part being curved, and applying to the circumferences of the arches.

The other varieties of parallel motions consist of combinations of levers or rods jointed together, and moving about fixed centres, in such manner, that the curvature of one circular motion may be opposed to, and counteracted by, a similar curvature operating in a contrary direction, in order that they may neutralize each other, and produce a rectilinear motion of that point of the link to which the piston-rod is united. The simplest variety is as follows.

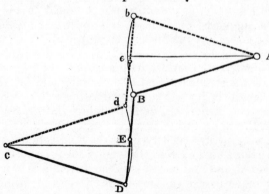

Suppose two equal levers A B and C D to be placed one beneath the other, in the same vertical plane, and to be movable about the fixed centres A and C, so as to be capable of assuming the different positions represented by the dotted lines, and by the strong black lines in the figure; their extremities B and D are connected together by the link B D, so that if one lever is moved up or down, the other must accompany it, and partake equally of its motion.   In such motion the upper end B of the link will describe an arch B b of a circle about the centre A, whilst the lower end D of the same link will describe an arch D d about the centre C ; now as the curvatures of these two arches are equal and similar, but reversed to each other, and acting in opposite directions, the point E in the middle of the link B D will describe a nearly straight and vertical line ; the motion of each of the levers should not much exceed an angle of 40 degrees, about their respective centres A and C.

In applying this to practice, the great lever of the engine forms the upper lever A B ; the link B D is suspended from the end of the great lever, and the upper end of the piston-rod is connected by a joint-pin, with the middle point E of that link, whilst the lower end D, is connected with the extremity of the counteracting lever C D.   In this way the piston-rod always preserves its vertical direction, in moving up and down ; and in so doing it communicates its motion to the extremity B of the great lever, although that point moves in the arch of a circle.

The radius C D of the counteracting lever may be made shorter, as two-thirds or half the length of the radius A B of the great lever, provided that the joint E, which connects the piston-rod with the link, is not applied in the middle of its length between B and D, but at two-fifths, or at one-third, of the length of the link from its upper end B ; and with this condition the joint E will still move in a vertical line, or very nearly so.

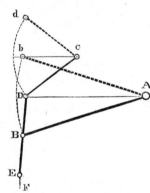

Another modification of the same principle is to connect the piston-rod F with the lower end E of the link E B D, which is connected with the end B, of the great lever A B by the point B, at the middle of its length between E and D ; the upper end D of the link is jointed to the extremity D of the counteracting lever D C, which is movable about the fixed centre C, as is shown by the dotted lines.   The length or radius D C is half the length or radius A B, and consequently the angular motion of the counteracting lever D C, about its centre C, will be twice as great as that of the great lever A B about its centre A.   The lower end E of the link will describe a straight line, or very nearly so, because the middle point of that link moves in the arch B b, described by the long radius A B, whilst its upper end D, being confined to move in the arch D d, described by the shorter radius C D, will be subject to a curvature four times as rapid as that of the other arch ; and although both curvatures are in the same direction, their effects upon each other will be neutralized at the point E, to which the piston-rod is connected.

The complete parallel motion, which has since been universally adopted, is also described in this specification, viz. The piston-rod F is connected with the extremity B of the great lever A B, by the link B E, in order to allow for the deflection of the circular arch B b, described by the upper joint B, from the right line described by the lower joint E.   To retain this point in its right line, it is united by a rod E D with the extremity D of the counteracting lever D C, which is movable about a fixed centre at C, so that the joint D will describe an arch D d of a much smaller circle than the joint B ; but as

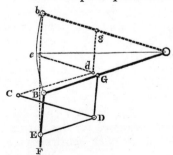

the joint D is connected with the point G of the great lever, by a link D G, the angular motion of the counteracting lever D C will be so much less than that of the great lever A B, as to compensate for the difference of curvature between the two arches B b and D d, which are described by B and by D respectively, and which being thus combined cause the point E to describe a right line.

Third. Improved methods of applying the steam-engine to work pumps, or other alternating machinery, by making the pump-rods balance each other. It is proposed to divide the pumps in a mine into two sets, so apportioned that the rods of one set will be nearly the same weight as those of the other ; and to suspend one rod from each end of a lever, so that one rod will ascend and raise water by one set of pumps, whilst the other rod descends, to perform the returning stroke of that set of pumps. This lever may be the great lever of the engine itself, or else a separate lever connected therewith, by a rod jointed to one end of it, so as to communicate an alternating motion from the great lever. The two rods may be carried down from the opposite ends of this lever, in a sloping direction, so as to approach towards each other, until the two rods are near enough together, to come conveniently into the space of the pit or shaft of the mine ; and the rods may then be continued in a vertical direction down in the mine, being guided by rollers into their vertical direction.

This plan is particularly adapted for the double-acting engine. Another variety of it is also described, with two bended or elbow-levers, moving about their elbows as centres of motion, and having the pump-rods suspended from the extremities of their horizontal arms, which are turned towards each other, so as to bring the rods near together ; the two elbow-levers are connected together, at the lower ends of their upright arms, by a horizontal rod extending from the arm of one lever to that of the other. By this arrangement the two rods will always move in opposite directions, and any suitable part of either of the elbow-levers may be connected with the great lever of the engine.

Fourth. A new method of applying the power of steam-engines to move mills, which have many wheels required to move round in concert. This is applicable to mills for rolling and slitting iron, in which the upper and lower rollers require to turn round in opposite directions to each other, but with the same rapidity. For this purpose the rollers which are to be turned round by the engine are placed immediately beneath the extreme end of the great lever, and are put in motion by means of two distinct pairs of sun and planet-wheels, one pair applied to turn the axis of the upper roller, and the other pair to turn that of the lower roller ; the connecting rods for the two planet-wheels, are suspended from the two ends of a strong wooden beam, which is fixed horizontally across the extreme end of the great lever of the engine, at right angles to its length, or parallel to its axis of motion, and projecting out on each side of the great lever, to a sufficient distance to reach over the two sun-wheels, which are fixed on the extreme ends of the axes of the two rollers. Each of those axes is provided with its own fly-wheel, to regulate the motion ; but these two fly-wheels are turned round in opposite directions to each other, suitable for the motions of the upper and lower rollers respectively. The sun and planet motion will turn a fly-wheel either way round, according to the direction in which it is first put in motion, and therefore to ensure that the two fly-wheels shall always be turned in contrary directions, their axes are connected by toothed wheel-work properly disposed for that purpose.

Fifth. A simplified manner of applying the power of steam-engines to work heavy hammers or stampers. A single-acting engine is proposed for this purpose, with a cylinder and pipes, air-pump and condenser, of a similar construction to those represented in Plate X., but made for a very short stroke ; a cylinder, to work a hammer of five hundred weight, being 15 inches diameter, and the length of the stroke 1 foot.   The piston-rod is connected with one end of the great lever, and the opposite end is connected by a rod, with the helve or handle of a forge hammer, which is constructed in the same manner as if it were to be worked by tappets on a revolving wheel ; but in this method, the hammer is lifted up by the descending force of the piston, when it makes its working stroke, and is let fall at the returning stroke of the piston.

Sixth. A new construction and mode of opening the valves, and an improved working gear : the construction of the parts is very nearly the same as is represented in Plate X., except that the valves are inverted, to give them a tendency to open very quickly, and they are shut in opposition to the pressure of the steam, which tends to force them open, so that no weights are required.   The valves are shut by the working gear, when the handles are pressed by the pins in the plug : and in order to retain the valves when so shut, the levers and rods of the working gear are combined on the principle already described in p. 374 ; viz. when the valve is shut, the short lever on the axis of the handle, is brought into a straight line with the connecting rod, which leads to the other lever for opening the valve.   In this position, the force which the steam exerts to open the valve is effectually resisted, for the action of the connecting rod on the lever, being exactly in the direction of its length, cannot have any tendency to turn the lever round upon its axis ; but the resistance in such a position is of the nature of an unstable equilibrium, for if the handle is moved, so as to turn the lever but a very little out of the line of the direction of its connecting rod, then the force with which the valve tends to open, will be able to turn the lever round so easily, that it will offer little or no resistance to the opening of the valve.

Seventh. A portable steam-engine, and machinery for moving wheel-carriages. The boiler for such an engine is proposed to be made of wood staves, put together in a cylindrical form, in the manner of a cask, and strongly bound with iron hoops.   The furnace to be made of iron plates, and fixed within the wooden boiler, so as to be surrounded with the water ; the flue, or passage for the smoke, being conveyed from the internal fire-place to the outside of the wooden boiler, to join to a perpendicular tube, which serves for a chimney to carry off the smoke, and cause a draught of air.   This boiler is to be mounted like a carriage, upon wheels, which are to be put in motion by the force of a piston fitted into a cylinder, and impelled with double action by the elasticity of the steam generated in the boiler : It is stated that the steam may be discharged into the open air, after it has performed its office, or that it may be cooled, and partially condensed, in condensing vessels exposing a great surface to the open air.   The reciprocating motion of the piston is to be communicated to the wheels of the carriage, by toothed wheels revolving with a sun and planet motion, and a fly-wheel (or by any other equivalent contrivance) with a suitable connection of wheelwork, to give the carriage-wheels a proper velocity : it is also proposed to provide two sets of such connecting wheelwork, of different proportions, and one or the other of them to be brought into action, according to the state of the road on which the machine is to travel, in order that the engine may be adapted in all cases to act upon the wheels with a sufficient power to advance the carriage properly.   It is stated, that a carriage of this kind, for two persons, should have a cylinder of 7 inches diameter, the piston to make a stroke of 1 foot, and to work at the rate of 60 strokes per minute.   It is also stated, that two cylinders may be used for such a portable engine, to act in concert upon its wheels.

The inventions described in this specification are very ingenious, but only a part of them have been brought into extensive use; viz. the parallel motion, and the locomotive engine: the others serve to show the great fertility of Mr. Watt's genius; and we may conclude, that his grand inventions, which have produced such extensive benefits to mankind, were selected from a great number of ideas, which he must have entertained in his mind: he was as remarkable for the soundness of the judgment which he exercised in that selection amongst his own schemes, as for the facility with which he produced them.

First applications of Mr. Watt's rotative engines. Soon after the date of this patent, Messrs. Boulton and Watt began to make rotative steam-engines at Soho, as a regular branch of their manufacture. The first one that they sent out was put up at Mr. Goodwyne's brewery in London, in 1784 (*a*); it was applied to grind malt or to pump up water, &c. and was found to answer the purpose so completely, that the other principal brewers in London were induced to order similar engines from Soho; and the next which was set up, at Mr. Whitbread's brewery, in 1785, completely established their reputation.

Mr. Watt did not adopt his plan of double action for these first rotative engines, but he used a single acting cylinder, and applied a sufficient counter-weight, to impel the machinery during the returning stroke of the piston, as proposed in his second patent (p. 424). He adopted the sun and planet wheels to obtain the rotative motion of the fly-wheel, which by the increased velocity thus given to it, produced a sufficiently steady motion.

In these applications of their engines to turn millwork and machinery, Messrs. Boulton and Watt received great assistance from Mr. John Rennie, then a young man, just entering into active business in London as a millwright and engineer; he had been brought up in Scotland, and at his first arrival in England, he was employed by Messrs. Boulton and Watt to superintend the millwork department of their business, and to set up the new engines which they sent out. Mr. Rennie made great improvements in the construction of the millwork which he executed for them, and commenced that system of accurate execution for powerful machinery which has since proved so beneficial to the manufacturers of this kingdom (*b*).

In these engines, the power and resistance being subject to frequent varia-

(*a*) Extract of a letter from Mr. Watt to Mr. Smeaton, dated Oct. 22, 1784.
" I have lately contrived several methods of getting entirely rid of all the chains and circular arches about the great levers of steam-engines; and nevertheless making the piston-rods ascend and descend perpendicularly, without any sliding motions, or right lined guides, merely by combinations of motions about centres; and with this further advantage, that they answer equally well to push upwards as to pull downwards; so that this method is applicable to our double engines, which act both in the ascent and descent of their pistons.

" A rotative engine of this species with the new motion, which is now at work in our manufactory (but must be sent away very soon), answers admirably. It has cost much brain-work to contrive proper working gear for these double engines, but I have at last done it tolerably well, by means of the circular valves, placed in an inverted position, so as to be opened by the force of the steam; and they are kept shut by the working gear. We have erected an engine at Messrs. Goodwyne and Co.'s brewery, East Smithfield, London."

(*b*) Extract of a letter from Mr. Rennie to a friend in Scotland, dated November 1784; " Mr. Watt's steam-engine has continued without any change of late; but mills are undergoing a total change, both in point of construction, and in the execution of their different parts; cog-wheels are now made with very fine teeth, and the millstones are hung loose upon their spindles, being suspended on their centres of gravity, so that they move very exactly in horizontal planes. Cast-iron shafts to mills have taken place of wooden ones, and fly-wheels are applied to every case where there is the least irregularity of motion. I am now planning a large rolling mill to be worked by one of Mr. Watt's engines in London."

tions, Mr. Watt applied an apparatus, called the throttle-valve and governor, to regulate the admission of steam from the boiler into the cylinder, and preserve a uniform motion of the piston, by administering the exact quantity of steam required to overcome the resistance, but no more. The governor operates by the centrifugal force of two revolving pendulums, on a principle which had been previously used in wind and water-mills; but being applied to the steam-engine, gave the finishing stroke to this great invention, and rendered its motion nearly as uniform as that of a water wheel; so that its mighty powers may now be applied with certainty and safety to actuate the most delicate machines.

The throttle-valve. Mr. Watt used this contrivance in his former engines for pumping water, as has been before described (p. 332); but he now adopted the following construction of the valve.

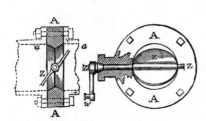

A thin circular plate Z is fitted into the aperture of the steam-pipe, by means of an axis Z Z, which passes across its diameter; and it can be turned by this axis from the outside, so as to present either its thin edge or its flat surface to the steam; in the latter case it quite fills the circular aperture of the steam-pipe, so as nearly to stop the passage of the steam.

The circular turning valve Z, is accurately fitted into a corresponding aperture in a ring of cast iron A A, which ring is of sufficient thickness to admit the axis Z Z of the valve to pass through it edgeways; and the axis is fitted steam tight at the part where it passes through the metal, but with liberty to turn freely. The ring A A is interposed in the joint between those two flanges by which the steam-pipe is united to the valve-box; the aperture in the ring then corresponds with that in the pipe; and the four bolts, which unite the two flanges, also pass through the ring, so as to fasten them all firmly together, as shown by the dotted lines *a a*. One end of the axis Z projects on the outside of the ring A A, and a short lever or handle *w* being fixed upon that end, the valve Z within the passage can thereby be turned more or less edgeways to the steam; and it will allow more or less steam to pass from the boiler into the cylinder, according to the angle at which the valve is placed.

In the first rotative engines, the throttle-valve was regulated at the discretion of the attendant, who turned it by hand from time to time, when requisite; as the axis Z of the circular valve Z passes diametrically across its centre, the pressure of the steam against the valve, acts equally above and below the axis, so that one pressure balances against the other, and so little tendency to turn the valve, or to alter the position which may be given to it, that the friction of its axis is quite sufficient to retain it as it is placed. When a very regular velocity is requisite, Mr. Watt contrived to regulate the throttle-valve by the motion of the engine itself, without any attention from the engine-keeper. For this purpose he tried various methods, but at last resorted to the centrifugal or revolving pendulums, which he called the governor, and which has been used ever since.

The governor. This apparatus consists of a vertical axis D D, which is supported on pivots at its lower and upper ends; and it is connected with the axis of the fly-wheel by wheel-work, or by an endless rope passing round the pulley *d*, so as to turn it round with a continuous circular motion, at a quicker rate than the fly-wheel, but partaking of any irregularity in its motion. To this vertical axis two rods E *e*

3 K 2

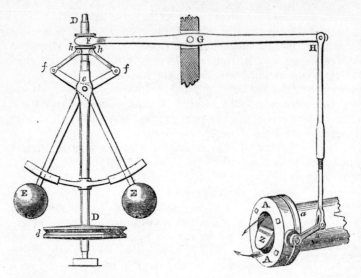

E *e* are attached by a joint at *e*, and they have heavy balls E E fixed at their lower ends, to form two pendulums, which being carried round with the axis, their balls E E will describe a horizontal circle, and their centrifugal force will cause them to recede from the axis D, in opposition to their own weight, which tends to approach them to that axis (see an explanation of this action, p. 42). The quicker the axis is turned round, the farther the balls will expand; but when it revolves slower, the balls will contract; and this motion of the pendulums is communicated to the lever or handle *w* of the throttle-valve Z, by means of a combination of levers *e f f h h* at the upper part of the axis, and other levers F G H *w*.

Thus the rods E *e* or arms of the two pendulums, pass through a mortise in the axis D, and cross each other beyond the joint or centre of motion *e*, in the same manner as the two parts of a pair of scissars, forming two short arms or levers *e f, e f;* to the upper end of each of these arms a link *f h* is jointed, and the upper ends of these links are jointed to a moveable collar *h h*, which slides up and down upon a square part of the axis D; a circular groove is formed round the collar *h h*, to admit a fork F at the extremity of a long lever F G H, which is moveable upon the centre pin G, and the opposite end H is connected by the rod H *w*, with the handle *w* of the throttle-valve Z.

When the motion of the engine is increased, the motion of the governor partakes of that increase, and the balls E E fly farther out, by their increased centrifugal force; the links *f h, f h* draw the collar *h h* lower down upon the axis D, and the lever F G H, and its rod H *w*, communicates this motion to the throttle-valve Z, so as to turn it a little more flatways across the aperture of the steam-pipe A A *a*, and thus diminish the supply of steam to the cylinder; the engine will in consequence move slower, and then the balls E E contract a little, so as to open the valve Z a very little wider, until the motion of the engine comes nearly to its required velocity; and it cannot vary considerably from that velocity, because the alteration of the centrifugal force of the revolving pendulums, will apply an immediate correction, by opening the valve wider whenever the engine slackens in its speed, or closing the valve if the engine moves quicker.

Iron Cement. Another important improvement which Mr. Watt introduced about this time, was the iron cement, for filling up the joints of the parts of the steam-pipes, valve-boxes, cylinder, condenser, &c.: his first method, of applying pasteboards with putty, has been already mentioned (p. 382), but such joints not being found sufficiently durable, he endeavoured to find out some more lasting substance to cement the joints; he observed that the iron founders filled up hollows and flaws in their castings by ramming in borings or filings of iron, moistened with urine, and that the mixture became hard in time.

Mr. Watt improved upon this plan, by mixing the iron borings with a small quantity of sulphur, and a little sal ammoniac; he afterwards used a portion of fine sand from the grindstone troughs, in addition. This mixture being moistened with water, and spread upon the surfaces to be joined, they are screwed together; the cement heats at first and swells a little, so as to fill up the joint very close, and in a short time, it becomes a hard stony substance, like iron pyrites, which will remain good, and keep a joint tight for years.

Mr. Murdock about the same time, and without communication with Mr. Watt, made a cement by mixing iron borings with sal ammoniac, but without sulphur; in this way the cement is equally good, but it takes some time to set, whereas, with the sulphur, it becomes hard almost immediately. The application of this cement has contributed in no small degree to the perfection of steam-engines.

The success which attended the first application of Mr. Watt's rotative engine to the machinery of the breweries, occasioned a great and increasing demand for them; and in the years 1784 and 5, Messrs. Boulton and Watt made several rotative engines at Soho, which were fixed and set to work by Mr. Rennie in the course of those years: one of the first of these was set up at Mr. Whitbread's brewery in Chiswell Street; another at Mr. John Calvert's brewery, and others at Mr. Felix Calvert's, at Messrs. Gyford's, and at Mr. Thrale's; all these were single-acting engines of about 10 horse-power; the cylinders 24 inches diameter, and the pistons making 6 feet stroke, and 18 strokes per minute. Mr. Whitbread's engine was set to work in 1785; it worked four pairs of millstones for breaking malt, and consumed one bushel of coals per hour (a).

In their general appearance these engines were very much like that represented in Plate XI., having the same kind of parallel motion, sun and planet-wheels, and governor; but the cylinder with its nosels or pipes, and the air-pump and condenser, being for single action, were nearly on the same construction as the engine in Plate X., the exhausting valve was inverted, and the working gear was on the plan described in Mr. Watt's patent (p. 433), so as to open the valve very suddenly. The connecting rod was made sufficiently heavy to impel the fly-wheel during the returning stroke of the piston. The waggon form of boilers, represented in Plates X. and XI., had not then been adopted, but the boilers of these first engines were made circular, like those used by Mr. Smeaton (see Plates II. and III.).

Each of the breweries at which these engines were first applied, had previously employed several horses, to walk round in a circular path, and turn a large horizontal wheel to which they were harnessed; and this large wheel, by other connecting wheelwork, gave motion to their millstones for grinding malt, and various pumps, for raising the water and liquor, to different parts of the brewery. Mr. Watt's new steam-engine was applied by means of connecting wheelwork, to turn the large horse-wheel, instead of the horses, leaving all the millwork unaltered, so that in case of any accident happening to the engine, the connecting wheelwork could be disengaged, in order to detach the engine from the

(a) The author has a sketch taken by his father from Mr. Whitbread's engine at the time when it was fixing, in 1785; the valves were arranged in the same manner as the sketch in p. 347. The author examined the two single engines at Messrs. Calverts' breweries in 1805, when both were in use in their original state; but at that time he found Mr. Whitbread's engine had been altered to a double action, such as is represented in Plate XI., so as to be a 20 horse-power, and it is still in use. All the other single engines for the London breweries have since been altered to double action.

horse-wheel, and then horses could be harnessed thereto, and the millwork impelled by the original method, without any interruption to the works. This precaution against a failure of the new power was considered to be very necessary at the first commencement; but Mr. Watt had then attained such a degree of perfection in the execution of his machinery, that very few accidents or stoppages were found to occur, and it became evident that a complete dependence might be placed upon the steam-engine, without the complication of any substitute.

In 1784, a company was formed in London for the purpose of building a very extensive flour mill, called the Albion Mills; they were to be worked solely by the new double-action engines, on a large scale: the direction of the works was given to Mr. Rennie, who superintended the execution of the building and the millwork, near Blackfriars' Bridge, whilst Messrs. Boulton and Watt made the engines at Soho; one of them was set to work at the beginning of the year 1786, with ten pairs of millstones for grinding wheat. This engine was 50 horse-power, of double action, very nearly on the same construction as is represented in Plate XI., except that the first variety of parallel motion was used, on account of its simplicity (see p. 430). As the Albion Mills were begun entirely on the faith of Mr. Watt's new engines, and being on a larger scale than any thing which had been before thought of, Mr. Watt and Mr. Rennie exerted their utmost skill to render the engine and machinery perfect, both in design and execution; their success was such, that the performance of these first steam-mills, in 1786, has never been exceeded by any of the modern engines which operate on the same principle; and Mr. Watt's rotative double-acting engine may be considered as brought to a standard from that time. A second engine of 50 horse-power, with 10 other pairs of millstones, was set to work in the same building in 1789; and space was provided for a third engine, but before it was erected the mill was destroyed by fire.

Estimation of the Force of Steam-engines by Horse-power, 1784.

The only unequivocal mode of expressing the mechanical power exerted by an engine or by an animal, is the weight which can be raised through a certain space, in a given time by that exertion (see pp. 16 and 58); and unless we define what a horse-power is, in those terms, it is a very vague expression, on account of various degrees of strength which different horses possess, and their capability of enduring fatigue.

When Messrs. Boulton and Watt first began to introduce their rotative steam-engines into manufactories, about 1784, they found it necessary to adopt some measure of the power which they were required to exert; this they endeavoured to express in such terms as would be readily understood by the persons who were likely to want such engines. The machinery in the great breweries and distilleries in London was then moved by the strength of horses, and the proprietors of those establishments, who were the first to require Mr. Watt's engines, always inquired what number of horses an intended engine would be equal to.

In consequence, Mr. Watt made some experiments on the strong horses employed by the brewers in London, and found that a horse of that kind, walking at the rate of $2\frac{1}{2}$ miles per hour, could draw 150 pounds avoirdupois, by means of a rope passing over a pulley, so as to raise up that weight, with vertical motion, at the rate of 220 feet per minute. This exertion of mechanical power is equal to 33 000 pounds (or 528 cubic feet of water) raised vertically through a space of one foot per minute, and he denominated it a horse-power, to serve for a measure of the power exerted by his steam-engines; that is, of the resistance actually overcome, in addition to the friction of the engine itself, and the resistance of its air-pump.

In apportioning the dimensions of engines according to this measure, the effective pressure on the piston was taken at rather less than 7 pounds per square inch; and with this load it is found that the piston will move through a space of from 200 to 256 feet per minute, according to the size of the engine.

The expenditure of steam in these engines may be taken at 33 cubic feet per minute for each horse-power, reckoning only by the space actually occupied by the piston in its motion, without considering the waste of steam which is occasioned by the vacant spaces, at the top and bottom of the cylinder, through which the piston does not pass.

Messrs. Boulton and Watt's standard for the horse-power, is very much beyond the actual power of any horses, except the very strongest, and they cannot long endure the exertion of raising 33 000 lbs., at the rate of 1 foot per minute. Mr. Smeaton and other engineers made many observations on the work actually performed by horses when working regularly in mills, and the results seem to show that 22 000 lbs., raised at the rate of 1 foot per minute, may be taken for a real horse-power, or as the exertion that a good horse can overcome with so much ease as to continue to work for 8 hours per day. This is only two-thirds of the estimated horse-power of Boulton and Watt, so that it would require 30 horses to act in concert, to perform the work of a 20 horse-power engine. And as the steam-engine is capable of working incessantly, whilst horses cannot work for more than one-third of the time, it would require three relays, or 90 horses, to be kept, to do as much, as a 20 horse engine is capable of doing.

These circumstances must be kept in mind, whenever the power of steam-engines is considered with reference to that of horses, or else very erroneous conceptions will be formed; and to avoid them, it is best to consider the term horse-power as a determined measure of mechanical power which is sufficient for the purpose to which it is applied, but which bears that title without any direct reference to the actual power of horses (a).

(a) Mr. Olinthus Gregory made the following remark in the first edition of his Mechanics, 1805, vol. ii. " The usual method of estimating the effects of engines, by what are called horse-powers, must inevitably be very fallacious, unless all engineers could agree as to the quantity of work which they would arbitrarily assign to one horse; and in that case the term would manifestly be nugatory."

In the Edinburgh Review, vol. xiii., the following reply is given : " If nothing more definite were said of any engine, than that it did the work of a certain number of horses, it would not convey a sufficiently accurate idea of its power, for purposes of science; but Messrs. Boulton and Watt felt that such a mode of comparison would be very intelligible to common apprehensions. It expressed the power of an engine in numbers, of which the ordinary strength of a horse is the unit. This is not in itself very exact, the unit being large and subject to some variations. But relatively to the purpose for which it was used, it was sufficiently correct; and a more minute measurement would have been less useful. When a historian would express the interval of time between two events, he is satisfied with counting the number of years; and it would be a useless affectation of accuracy to reckon up the number of months, days, hours, and minutes, in order to measure the interval with mathematical exactness. So also, if a man were to ride post from London to York, it would serve his purpose as well to know the distance between these cities in miles, as in feet, inches, and decimals of an inch. Messrs. Boulton and Watt have given to their measure all the accuracy that can be required, when, from the result of experiments made with the strongest horses employed by the brewers in London, they have assumed for the standard of a horse-power, a force able to raise 33 000 lbs. one foot high in a minute. This was intended to include an ample allowance of power to cover the usual variations of the strength of horses."

N. B. This was the first time that Messrs. Boulton and Watt's standard for the horse-power had been stated in print; but Mr. Watt afterwards repeated it himself, in his annotation on Dr. Robison.

Mr. Watt established the following proportions for his rotative engines.

The first engines which were made for the breweries in London, were only single-acting cylinders, 24 inches diameter, the length of stroke 6 feet, and they made 18 strokes per minute; these were about 10 horse-power.

A 10 horse double rotative engine, was made with a $17\frac{1}{2}$ inch cylinder, the length of its stroke 4 feet, and to make 25 strokes per minute.

The most standard size of the double-acting rotative engines was called a 20 horse-power, with a cylinder $23\frac{3}{4}$ inc. diam., the piston making a 5 feet stroke, and working at the rate of $21\frac{1}{2}$ strokes per minute.

For a 40 horse double engine, the cylinder was $31\frac{1}{2}$ inches diameter, the piston made a 7 feet stroke, and went at the rate of $17\frac{1}{2}$ strokes per minute.

To calculate the power of any machine in horse-power. RULE.—Multiply the force exerted, in pounds, by the space, in feet, through which that force acts during one minute; and divide the product by 33 000 lbs. The quotient will be the horse-power exerted by the machine.

Example. Suppose the piston of an engine to be impelled by a force, or unbalanced pressure, of 3070 lbs. and that it moves at the rate of $21\frac{1}{2}$ double strokes of 5 feet length, or 215 feet motion per minute. Then 3070 lbs. load, × 215 ft. per min. = 660 050 lbs. acting through 1 foot per min., and this ÷ 33 000 lbs. = 20 horse-power.

Sliding Rule. { A Force in pounds.   33 000 lbs.   *Examp.*   A 3070 lbs. 33 000.
slide inverted. { Motion ft. per min.   Horse-power.   215 ft.   20 HP.

To calculate the power of a steam-engine in horse-power. RULE. Multiply the square of the diameter of the cylinder, in inches, by the motion of the piston, in feet, per minute; multiply the product, by the effective pressure upon the piston, in pounds, per square inch (exclusive of friction); and divide the latter product, by 42 017; the quotient is the horse-power exerted by the engine.

Example. Suppose a cylinder $23\frac{3}{4}$ inches diameter; the piston urged with a force of 6·94 lbs. per square inch, and to move 215 feet per minute. Then, 23·75 squared = 564 circular inches area of the piston × 215 ft. = 121 260 × 6·94 lbs. per squ. inc. = 842 029 ÷ 42 000 = 20 horse power.

Sliding { C Mot. ft. per min. × pres. lbs. per sq. inc.   Horse-power.   *Ex.*   C 145·9   20 HP.
Rule.   { D   205 gage point.   Diam. of cyl. inc.   D 205   23·75 inc.

Note. If the effective pressure on the piston had been stated in pounds per circular inch, then the divisor would have been 33 000; but the pressure being stated per square inch, it is (33 000 ÷ ·7854 =) 42 017 as above. If an engine expends 33 cubic feet of steam per minute for each horse-power, the effective pressure on the piston will be 6· 944 lbs. per square inch.

Mr. Watt's fifth and last Patent, 1785.

This was for " new methods of constructing furnaces or fire-places, for heating, boiling, or evaporating water; which methods are applicable to steam-engines; and whereby greater effects may be produced from the fuel, and the smoke in a great measure prevented, or consumed." Dated 14 June, 1785. The specification of this patent is printed in the Repertory of Arts, First Series, vol. iv. p. 226.

The specification states, that the smoke which proceeds from the raw coal when it is first thrown into the furnace, must be supplied with a current of fresh air, so as to become mixed therewith; and that mixture of smoke and air, must be caused to pass through, or over that part of the burning fuel which has become intensely red hot, and has already ceased to smoke, by being converted into ignited coaks or cinders. The smoke and grosser part of the flame, being thus brought into close contact with the intensely hot fuel, and being already mixed with a due proportion of fresh or unburnt air, will ignite and burn with a pure flame, free from smoke.

To put this in practice, every passage must be stopped up where air might gain admission into the furnace or flue, and pass into the chimney, leaving only the interstices between the pieces of fresh coals, for the admission of the air; the fresh raw coals must always be supplied at the outside part of the furnace, nearer to the external air, than that part where the intense heat and ignition takes place. In consequence the smoke proceeding from the raw coals will be mixed with, and carried along with, the fresh air which insinuates itself through the interstices between the coals, and will pass, in mixture with that air, through the hottest part of the burning fuel, so as to be consumed; but as a sufficient supply of fresh air would not always find its way through the interstices of the pieces of coals, an opening must be provided to admit a current of air, at discretion, into the burning part of the fuel, with a stopper to regulate the quantity of air which shall be admitted, according as the combustion is found to require.

Mr. Watt proposed for this plan, to place the furnace in front of the boiler, and without any fire-grate, the fuel being thrown into an upright funnel or hopper, through which the draught and current of air passed downwards, so that the hottest part of the fire was at the bottom. The fresh coals were thrown in at the top of the funnel, which was left open to the air, and therefore the funnel could not be placed beneath the boiler, but the flame and heat was conveyed from the lower part of the funnel through a passage which rose up again, and then passed horizontally beneath the boiler. Or it was proposed to place that funnel or hopper, in an inclined or nearly horizontal direction, so that it might extend beneath the bottom of the boiler, with only its mouth or open end on the outside; and still preserving the principle that the raw fuel shall be supplied at the part nearest to the supply of fresh air, in order that the smoke may be mixed with that air, and be carried into the hot fire with it.

In other cases Mr. Watt proposed to place the fresh fuel, in the usual manner, on a fire-grate situated beneath the boiler, at one end of it, and to fix another small fire-grate beneath the opposite end of the boiler, near the place where the flame and smoke passes from under the boiler, to turn up into the external flue around the same; a fire is to be maintained on this second grate with cokes or cinders which have been previously burnt until they have ceased to emit smoke; this fire will give an intense heat to the smoke of the first fire which is conveyed through or over it, and a suitable supply of fresh air being admitted to mix with that smoke, it will ignite and be consumed.

The plan of introducing the coals into a funnel, or furnace without a fire-grate, was not found to answer well in practice, because the coals being subject to cake together in masses, would not always permit a sufficient passage of air through the interstices; and it required too much attention and discretion to regulate the fresh air which was admitted at the air passage, in order to supply that deficiency; when that fresh air entered in too great a body, it would be only partially intermixed with the smoke, which could not therefore be completely consumed. After many trials, Mr. W. returned to the old method of placing a fire-grate beneath the bottom of the boiler, but he altered the construction of the furnace so as to attain in part the object stated in the specification.

This method, which he afterwards adopted generally, he called the smokeless furnace; it is founded upon the same principle as Argand's lamps. The fire-grate and mouth piece of the furnace are laid sloping downwards from the fire-door, at an angle of about 25 to 30 degrees to the horizon, and the mouth-piece is adapted to receive and contain a quantity of coals. The fire is lighted on the sloping grate in the usual way, and a small quantity of air must be admitted through one or two apertures in the fire-door, so as to blow down exactly on the blazing part of the fire, which at first burns principally at the upper part of the grate, and the fresh coals with which the fire is to be supplied when requisite, are laid upon the sloping bottom plate of the mouth-piece, close to the burning fuel, but not upon it. When the fire requires fresh coals, they must not be thrown upon those already on fire, as that would greatly increase the smoke; but the burning coals on the grate, and the unconsumed coals already on the dead plate of the mouth-piece, must be pushed further down the slope on to the grate. With this precaution the fresh coals are gradually dried by lying in the mouth-piece, in the vicinity of the fire; and the current of air which proceeds from the door, mixes with the smoke from the fresh coals, and causes it to pass over the bright burning fuel so as to be consumed. The apertures in the door are so constructed as to admit that quantity of air which is requisite for the consumption of the smoke, but too much air would be prejudicial.

The original plan was somewhat different from this description, which is of the most improved form; it answers the purpose perfectly with free burning coals when properly attended to; but it is difficult to manage with coals which melt and cake together.

The First Establishment of Steam-mills, 1785 to 1793.

MESSRS. Boulton and Watt executed their first and greatest work of this kind in London in 1785 and 1788, at the Albion Mills, which was a new establishment for grinding corn, entirely by the power of Mr. Watt's new rotative engines. The mills were contained in an elegant and commodious building erected under the direction of Mr. Samuel Wyatt, by the river side, at the foot of Blackfriars Bridge; and two engines were made for them at Soho, by Messrs. Boulton and Watt, each rated at 50 horse-power, or equal in all to the power of 150 horses working in concert. Each engine worked 10 pairs of millstones, or 20 in the whole mill, of which 12 or more were generally kept at work with the requisite machinery for dressing the flour, and other purposes (a).

The Albion Mills were established by a company, who purchased wheat in the London market, and manufactured it immediately into flour for sale; whereby they avoided the loss of time, and the expense and damage of carrying the corn up the river Thames, to be ground by water-mills, and then bringing back the flour. It was expected that the saving of expense, in time and carriage, would compensate for the cost of fuel consumed by the engines, and afford a profit to the proprietors.

These mills deserve particular notice, as being the first of those numerous establishments which have since arisen in all the manufacturing districts of

---

(a) The following account of this work is given in a letter to Mr. Smeaton by a friend, in August, 1785.

"The new steam flour-mill now building near Blackfriars Bridge, is an immense undertaking, which is carried on with great spirit and judgment; they expect to begin to work before Christmas, with one engine, and 10 pair of stones, 6 of them to work at once. The wheels and axes are all cast iron; there are two fly-wheels of 18 feet diameter, to be turned by a spear suspended from the great lever of the engine, and each of the fly-wheels is to work 5 pair of millstones of 4 feet 4 inches diameter."

By another letter dated 17th March 1788: "I have this day been through the Albion Mills with Mr. Boulton. Five bushels of Newcastle coals are said to work the engine an hour, and those 5 bushels will grind 50 bushels of wheat per hour, upon 6 pair of stones. It is reckoned that one horse-power is equal to the grinding of one bushel of wheat per hour, and therefore the engine is equal to 50 horse-power. The steam cylinder is 34 inches diameter."

England, Scotland, and Wales, entirely out of the advantages of Mr. Watt's engines; the unlimited command of power thus attained, enables a manufactory to be placed at once in the vicinity of the market for the purchase of its materials, and for the sale of its produce, instead of carrying the materials to a water-fall.

The Albion Mills were said to be the means of sensibly reducing the price of flour in the metropolis, whilst they continued at work, for they occasioned a greater competition amongst the millers and meal-men than had ever existed before; but nevertheless great prejudices were excited against the company amongst the lower class of people, to whom it was represented as a monopoly highly injurious to the public. These mills were destroyed by fire in 1791 (a), and it has been suspected that it was not occasioned by accident; the satisfaction of the populace was afterwards expressed by songs in the streets of London.

The engines at the Albion Mills were constructed with great care, and contained all Mr. Watt's improvements combined into the most advantageous form; the invention of the double acting steam-engine for turning mills may be considered as completed at that time, nor has any important improvement in the operation of the steam-engine been made since, except Mr. Woolf's method of working with less fuel, by high pressure steam used expansively. All the modern improvements of Mr. Watt's engine have been confined to the methods of putting the parts together, in more substantial and durable forms.

The mill-work and machinery of the Albion Mills was executed under the direction of Mr. John Rennie, who had then commenced business in London, as a millwright and engineer, and who finally succeeded Mr. Smeaton in most of his public appointments. In place of wooden wheels, which are always subject to change of figure, wheels and axes of cast iron were employed, with their teeth, accurately formed, and proportioned in strength to the strain they had to bear; many other parts of the machinery, which had been usually made of wood before, were then made of cast iron, in improved forms.

Mr. Smeaton was the first who began to use cast iron in mill-work, several years before (see p. 275), and he made larger and more powerful mills than his predecessors; but the founder's art was so imperfect in his time, that he was obliged to proceed with the utmost caution in the use of cast iron, and experienced some vexatious failures (see p. 305). Mr. Rennie had greater advantages, and was enabled to apply cast iron in machinery with more success; and as the treatment of that material has become more known, it has been more and more used, up to the present time, when nothing but stone and metal is used in the construction of mills and engines, either for the fixed framing, or for the moving parts.

The practicability and advantages of Mr. Watt's double acting rotative-steam-engines, being completely established by the performance of those at the Albion

(a) Extract from the Annual Register, vol. 33. "On the 3d of March, 1791, soon after 6 o'clock in the morning, a fire broke out in the Albion Mills, and raged with such fury, that in about half an hour, that extensive edifice with an immense stock of flour and grain, was reduced to ashes; the house and offices of the superintendant only escaping by the thickness of the party wall. It was low water when the fire was first discovered, and before the fire-engines were collected, their assistance was ineffectual, for the flames burst out on every side with such force, and heat, that it was impossible to approach any part, until the roof fell in. The wind blew across the wide street, or rather place, at the foot of Blackfriars Bridge; and the heat of the fire was so great as to scorch the houses on the opposite side of that street. The insurances in 5 different fire offices amounted to 26 000l. on the premises, and 35 000l. on the stock; 4 000 sacks of corn were destroyed."

Mills, they began to be generally adopted, and large manufactories were in a few years founded in towns, to work entirely by the new power (*a*).

After the rotative engine had thus been brought to a standard form, Mr. Watt made no further alterations during the remainder of the term of his patent, except such modifications as became necessary in manufacturing engines of all sizes from 4 horse-power to 100 horse-power. A great number of rotative-engines were made by Messrs. Boulton and Watt at their manufactory at Soho near Birmingham, and were fixed up in every part of the kingdom, and many were exported. They were called patent engines, and were looked upon as standards of perfection in mechanical works : many of them are still in use, and their performance is fully equal to that of any modern engines which operate on the same principle. It is now time to describe the construction of the complete patent engine in detail.

DESCRIPTION OF MR. WATT'S ROTATIVE STEAM ENGINE FOR TURNING MILLS, CON-
STRUCTED BY MESSRS. BOULTON AND WATT, AT SOHO, FROM 1787 TO 1800.

Plate XI. is a general elevation of the whole engine, and its boiler.

Plate XII. contains a cross section, and an end elevation of the engine ; also a plan, and a longitudinal section of the boiler : all these are drawn to a scale of a quarter of an inch to a foot, and represent an engine of 10 horse power.

Plate XIII. contains a front view, sections and plan of the cylinder, with its valves and working gear, and condenser and air pump : all on a scale of half an inch to a foot, and some enlarged sections of the valves. The same letters and characters of reference, are employed in all the different figures.

(*a*) At Manchester a cotton-mill was built on Shude Hill in 1783, by Messrs. Arkwright and Simpson, with an atmospheric-engine ; and some years afterwards it was replaced by one of Messrs. Boulton and Watt's patent engines.

At Glasgow Messrs. Scott and Stevenson began spinning cotton, by steam-power, in 1792.

At Leeds a large woollen manufactory was built by Mr. Gott in 1793, with a 40 horse-engine by Boulton and Watt ; it continued in use till 1825, when it was replaced by an 80 horse-engine.

Mr. Marshall began the spinning of flax by steam-power at Leeds, about 1793 ; an engine on Savery's principle was first tried (see p. 123), and then a 28 horse-engine of Boulton and Watt. This establishment has since been extended by other engines, until it has become the largest manufacturing establishment in the kingdom by steam-power.

The spinning of worsted was begun by steam, at Arnold near Nottingham, by Messrs. Davison and Hawksley ; they first had an engine on Savery's principle, and afterwards a large atmospheric-engine, with a crank and fly-wheel, constructed by Mr. Thompson, with two cylinders combined to produce a double action.

At Warwick a worsted spinning-mill was begun about 1793 by Mr. Parkes, with a 26 horse-engine by Messrs. Boulton and Watt.

At Birmingham a large rolling-mill was set up with steam-power, about 1792 ; and another at Liverpool, with an atmospheric-engine.

At Sheffield a steam-rolling-mill was built, about the same time : and soon after, a mill for grinding cutlery ; with two engines of 40 horse-power each.

In London all the large breweries and distilleries were in a few years furnished with engines, from Soho, except one or two with atmospheric-engines from Manchester. An iron forge was built at Rotherhithe, about 1787, for making up scraps of old iron into bars. Also a paper-mill with an atmospheric-engine. A fulling-mill and logwood-mill was set up in the Borough in 1792, at a large dyehouse for woollen cloth, with a 20 horse-engine. An oil mill was begun in the Borough soon after ; and then a mill for grinding apothecaries' drugs ; also a mill for callendering, glazing, and packing cloths for exportation.

These first establishments in each place, were greatly multiplied and extended in the course of 10 years, particularly the steam-cotton-mills at Manchester and Glasgow, and steam corn mills in every large town ; the extension has been still more rapid since the expiration of Mr. Watt's patent in 1800, when several new establishments were formed for making steam-engines.

General Description of the Engine.

A is the furnace or fire-place, with an opening in front closed by a small iron door, which opens on hinges, to give access to the fire, and to introduce fresh fuel when necessary; but the door is shut close at all other times, to exclude the air.

B, the fire-grate, composed of several iron bars, upon which the fire is made; the bars lie parallel to each other, but do not touch (see fig. 2.), so that interstices remain between them, to admit a free current of air to the burning coals, which are spread upon the grate; the space beneath the grate, which is called the ash-pit, is open in front, to admit the fresh air freely to the underside of the grate.

C C the boiler, made of thin iron plates, with a semi-cylindrical top, so as to resemble the shape of a covered waggon; it is enclosed in brick-work, and the fire-place is fixed beneath one end of it, so that the bottom of the boiler may receive the direct radiant heat of the fire, and cause the water contained in the boiler to evaporate in steam. The water in the boiler occupies about two-thirds of its capacity, leaving the other third for steam. The flame and smoke from the fire-place, is carried beneath the whole length of the bottom of the boiler, and then rises up at the extreme end into a long passage or flue 9, which returns and encompasses all the lower part of the boiler (as shown by the arrows, fig. 2.): the flame by thus circulating round the outside of the boiler, gives the utmost effect of the fire to the contained water; the flue terminates with a damper or sluice-door w, which rises and falls in the manner of a window sash, in order to increase or diminish the passage for the heated air, and smoke, and thus regulate the draft through the flue 9, according as more or less fire is required; beyond this damper the flue joins to the base of the chimney D.

D the perpendicular stack or chimney, which conveys the smoke and heated air upwards to a great height; and from the lightness of this heated smoke, compared with the surrounding external air, it acquires a sufficient power of ascent in the chimney D, to cause the flame and heated air to draw through the long flue 9, with a rapid current; and the cool fresh air pressing upwards through the interstices of the grate-bars, animates the fire.

a, the steam-pipe which conveys the steam from the interior of the boiler, to the steam boxes of the cylinder; the throttle-valve is fitted into this pipe at z, figs. 4. and 6. (also fig. 12. and p. 435.), it is a turning vane placed across the steam passage so as to fill the circular aperture; but being mounted on a spindle, it can be turned edgeways in the passage, and will not then intercept the steam sensibly. The throttle valve being turned by its spindle, the passage is opened or shut as may be required to increase or diminish the flow of steam from the boiler to the cylinder.

E the cylinder surrounded by an external steam case, containing hot steam; the cylinder is bored truly cylindrical within, and a piston or moveable stopper is accurately fitted into it, being packed with hemp round the circumference, so that no steam can pass by the piston, although the latter moves freely up or down in the cylinder. The top of the cylinder is closed by a cover, through the centre of which the polished piston rod n passes, and the opening is stuffed with hemp round the rod, to prevent the passage of steam, or air, but the piston rod can move freely up and down, when the piston (which is fastened to the lower end of the rod) is impelled by the steam.

b the upper steam-box, containing the upper steam-valve, which being opened admits the steam to enter into the upper part of the cylinder, through

c, a branch or lateral passage into the upper part of the cylinder; it communicates with the space beneath the upper steam-box *b*, so that the steam which passes through the upper steam valve (when it is opened), will flow through *c* into the upper part of the cylinder, and press on the piston to force it downwards.

d d, the descending steam-pipe, which conveys the steam from the upper steam box *b* to the lower one *e*.

e, the lower steam-box, containing the lower steam-valve, which, being opened, admits the steam to enter into the lower part of the cylinder, through

f, a branch or lateral passage, into the lower part of the cylinder; it communicates with the space beneath the lower steam-box *e*, so as to receive steam through the lower steam valve, when open, and admit it into the lower part of the cylinder, where it presses beneath the piston, to force it upwards.

g g the eduction-pipe, which conveys the steam from either of the lower valve boxes *h* and *i* of the cylinder, down into the condenser F.

h the box beneath the upper exhausting-valve, which is situated in that part of the upper box which communicates with the upper branch *c*, of the cylinder; so that the valve being opened, will allow the steam to pass out of the cylinder by the passage *c*, into the box *h*, which communicates with the upper end of the eduction-pipe *g*, to convey the steam down to the condenser F.

i the box beneath the lower exhausting-valve, which is situated in the middle part of the lower box, and communicates with the lower branch *f* of the cylinder, so that this valve being opened, allows the steam to pass out of the cylinder by the passage *f*, into the box *i*; which communicates by a short branch with the eduction-pipe *g*, and thereby with the condenser.

The boxes, *b c h* and *e f i*, which contain the four valves, are called the nossels, and their pipes *d* and *g* are called the side-pipes; the upper-nossel is shown on a larger scale in fig. 10 and 11, Plate XIII., to explain the two valves within it.

Each nossel is divided by two horizontal partitions into three compartments, *b c* and *h*, and in each partition a brass ring or valve seat is cemented, having a circular aperture into which the conical-valve is very accurately fitted by grinding, so as to stop the aperture entirely, when the valve is down, but when it is lifted up, there is a free passage for the steam; the upper compartment of each nossel, is always filled with steam from the boiler, and the lower one is always exhausted, having an open communication to the condenser; and the middle compartment, or space between the two valves, is the branch or passage communicating with the cylinder.

F the condenser or receiver, placed in a cistern of cold water G : it receives the steam, in order to cool and condense it, being the reverse of the boiler, where the water is so much heated by fire as to expand and be converted into steam, but in the condenser the excess of heat is taken away from the steam by cold water, in order to contract or reduce it again to water.

G the condensing cistern, containing the cold water, in which the condenser and air-pump are immersed.

j the injection-pipe and cock, which admits a jet of cold water from the cistern into the condenser, to assist in cooling the steam; the plug of the injection-cock is turned by a spindle which rises up by the side of the eduction-pipe *g*, (see fig. 6.), and has a handle fixed on at the top, in a convenient position for the engine-man to turn it; a divided plate and index is fixed beneath the handle to show how much the cock is opened.

k a lateral branch from the very lowest part of the condenser, to drain off the condensed steam, and injected water therefrom, and convey it into the air-pump H. In this passage is a hanging, or flap-valve *k*, called the foot-valve : it opens outwards from the condenser, to allow the water and air to pass out of the condenser into the air-pump, but shuts to prevent any return that way.

H the air-pump, so called because it exhausts air, as well as water, from the condenser, in order to maintain a perpetual exhaustion therein, or as great a rarefaction and approach to a vacuum as can be obtained. The air-pump, like the cylinder, is very truly bored within, and accurately fitted with a piston, which is called its bucket, because it is perforated, and has valves in it opening upwards, in order that the bucket in descending may not displace or disturb either the air or the water in the lower part of the pump; but in ascending the valves close, so as to prevent the water and air returning, and the bucket draws out both, leaving a void space beneath it in the air-pump, into which the water from the condenser runs through the foot-valve k by its gravity, and then the air follows by its elasticity. The top of the air-pump is closed by a cover, through which the polished rod of the bucket passes; and the opening is stuffed with hemp round the rod, to exclude the atmospheric air from the upper part of the pump.

l the rod of the air-pump by which its bucket is moved up and down to work the pump; the upper end of this rod is connected with the great lever of the engine; it has two pins or clamps fastened to it at 1 and 2, and projecting from it, to give motion to the handles r and s of the working gear.

m is the discharging branch or spout, to convey the water and air from the upper part of the air-pump, into the hot-well I; the orifice of this spout is covered by a hanging-valve, called the discharging-valve, which permits the water and air to pass out from the air-pump, but prevents its return.

I the hot-well, which is a part of the condensing cistern G divided off, to receive the hot water discharged by the air-pump, through the discharging-valve m, and prevent it from mixing with the cold water in the cistern (a).

J J fig. 8. Plate XIII. is the piston, fitted into the cylinder, with a packing of hemp contained in a groove round the edge of it, so as to apply to the interior surface of the cylinder, and fill it exactly, but leaving the piston at liberty to move up and down freely in the cylinder.

Action of the Piston. The piston may be considered as a moveable partition, dividing the interior capacity of the cylinder into two close compartments, which have no communication with each other, but which are capable, by the motion of the piston, of enlarging and diminishing their respective capacities; and the enlargement of one compartment, is always attended with a corresponding diminution of the other. The motion of the piston is produced by admitting the elastic steam into one or other of these compartments, and the effort which that steam makes to increase its volume, tends to enlarge that compartment which it occupies, but such enlargement cannot take place without pushing the piston before it. At the same time that steam is thus admitted into one of the compartments of the cylinder, the other compartment is so much exhausted as to leave little or no steam therein to resist the motion of the piston, or press it in an opposite direction to that in which the steam contained in the full compartment, tends to impel it.

Hence there is in all cases a plenum of steam in one compartment, and a

---

(a) The construction and operation of the condensing apparatus, and of all the parts contained in the condensing cistern G, are very similar to the corresponding parts in Mr. Watt's single engine, Plate X. already described at pp. 354, and 375, excepting that there is no injection-valve; but the injection-cock is always kept open, in order to admit a continual jet of cold water into the condenser, whenever the engine is at work. There is no necessity for an injection-valve in the double engine, and the use of the injection-cock is to regulate the quantity of the injection, or to stop it when the engine is not at work.

vacuum, or exhaustion in the other; under these circumstances the piston will be moved up or down in its cylinder, provided that the friction of the packing against the inside of the cylinder, and the resistance opposed to its motion by the machinery which is connected with the piston rod, are less than the impellent force of the steam. The piston will move, with greater or less celerity, according as the force exceeds the resistance in a greater or lesser degree.

Action of the steam. The steam is rarefied from the state of liquid water, by the heat of the fire beneath the boiler C, and of the flame around it. Any quantity of water thus converted into elastic steam, by heat, will fill 1700 times the space that it occupied when in the liquid state. This steam passes from the boiler through the steam-pipe *a*, and is admitted by the throttle-valve *z* to the upper steam-valve in the box *b*; and also through the descending steam-pipe *d*, to the lower steam-valve in the box *e*. These valves being thus constantly supplied with steam from the boiler, we may, by opening one or other of them, admit the steam into the cylinder, either above or beneath the piston, according as we require to impel the piston upwards or downwards.

After having exerted its force in the cylinder, the steam becomes useless and must be evacuated from it, through one or other of the two exhausting-valves; the upper exhausting-valve is situated between the passage *c* and the box *h*; and the lower one between the passage *f* and the box *i*; and both the boxes, *h* and *i*, communicate with the condenser F by the eduction pipe *g*; so that by opening one or other of the exhausting-valves, the useless steam will pass out of the cylinder, either from above or below the piston, and descend through the eduction-pipe *g*, into the vacuous condenser F, where it will meet the jet of cold injection water, which by absorbing the heat from the steam takes away its elastic property, and converts it again into a small quantity of hot water; and this mixing with the cold water, they both acquire a medium temperature, leaving a void exhausted space, or nearly a vacuum, in all that great space, which the elastic steam recently occupied within the cylinder.

The hot water proceeding from the mixture of the steam and cold water cannot accumulate considerably in the condenser, before it is drawn out therefrom, through the foot-valve in the passage *k*, by the air pump H; and after passing through the valves in the air-pump bucket, it is discharged through the valve of the spout *m* into the hot well I.

The moving parts of the Engine. The piston J being moved up and down within the cylinder, by the steam, as already mentioned, the piston rod *n* communicates that motion to the external mechanism.

n is the piston rod made of iron, very straight and true, and highly polished, that it may slide freely up and down through the stuffing-box in the cover of the cylinder; the lower end of the piston rod is made fast to the piston J J, and the upper end is connected with the parallel motion by

o the cross pin, which is firmly fastened to the upper end of the piston rod; and the two ends of the pin are cylindrical pivots which form joints for

K K the main links of the parallel motion; they are double, being placed one on each side of the piston rod, and fitted at their lower ends, on the two pivots of the cross pin *o*; and at their upper ends, they are fitted on the pivots at the ends of a similar pin, which is fastened beneath the end of the great lever, by two bolts which pass up through the wood.

L L the great lever, also called the working beam, because it is made from a beam of oak: it is poised on a centre or axis *p* in the middle of its length, so as to be capable of librating like a scale beam; the piston rod *n* is suspended from one end of it, and the connecting rod M from the other.

p the fulcrum, or centre of motion of the great lever; it is a short iron axis firmly fastened across, beneath the beam, and formed with cylindrical pivots at each end ; the pivots are received in fixed sockets or bearings, which are firmly secured to the framing.

M M the connecting rod or spear, made of cast iron: the upper end is connected with the end of the great lever L by an universal joint ; and the planet-wheel N, which gives a circular motion to the fly-wheel, is fastened to the lower end. The weight of the connecting rod and planet-wheel, is sufficient to balance the weight of the piston, and piston-rod, at the other end of the lever.

N the planet wheel, it is firmly bolted to the connecting rod, so that it cannot turn round on its own centre.

O the sun-wheel, it is fastened on the extreme end of the axis of the fly-wheel, in order to give motion thereto.

q q, Pl. XII. fig. 5, is a short connecting link, which retains the centres of the sun-wheel, and of the planet-wheel, always at the same distance from each other. One end of this link is fitted on a round part of the axis of the fly-wheel, behind the sun-wheel, and the other end is fitted on a round pin, which projects backwards, from the centre of the planet-wheel. The link causes the centre of the planet-wheel to travel in a circular orbit (as shown by the dotted circle in Pl. XI.), when it revolves round the sun-wheel, by the motion given to it by the connecting rod.

P P the axis of the fly-wheel ; it is made of cast iron, square in the middle part, but with cylindrical necks near each end, which necks are received in sockets or bearings supported by the framing, and in these sockets the axis can turn freely round ; one end of the axis projects considerably beyond the bearing, and has the link *q* fitted on the cylindrical part, and then the sun-wheel, is fixed fast on a square at the extreme end. The other end of the axis also projects beyond the bearing, and is made square, in order to fasten on a cog-wheel, or a coupling, to connect the axis with the mill-work which the engine is intended to propel.

Q Q the fly-wheel, made of cast iron, and heavy in the rim, that it may tend by its energy or impetus to preserve an uniform rotative motion of the machinery.

Action of the sun and planet-wheels. The alternate reciprocating motion of the piston in a straight line, is made to produce continuous circular motion of the fly-wheel in the following manner. The piston-rod *n*, is connected to the extremity of the great lever L, by the links K, and by the aid of the parallel motion or combination of jointed rods 5, 6, and 7, the rectilinear motion of the piston-rod is preserved, although the extremity of the great lever describes the arc of a circle, when it librates upon its centre of motion *p* (see p. 431.).

The connecting rod M, and planet-wheel N, being suspended from the opposite end of the great lever, the planet-wheel must always move at the same time with the piston, but in an opposite direction; rising when the piston descends, and *vice versa*. The link *q* behind the sun and planet-wheels confines the centre of the planet-wheel to move in the circular orbit around the centre of the sun-wheel. During the descent of the piston, the planet-wheel is drawn up from its position shown in Plate XI. *i. e.* beneath the sun-wheel ; and is raised to the top thereof, in which motion the centre of the planet-wheel describes a semicircle. When the piston returns, the planet-wheel descends, and completes the circuit in its orbit round the sun-wheel, so as to regain the position shown in Plate XI.

The planet-wheel being fastened to the connecting-rod, so that it cannot turn round upon its own centre, the action of its teeth in those of the sun-wheel, will cause the latter to make two revolutions about its own centre, for each circuit which the planet-wheel makes in its orbit: the effect of the planet-wheel, circulating round the sun-wheel, is to make the latter, and the fly-wheel, turn twice, during each complete stroke of the engine.

An idea of the action of this ingenious contrivance, may be formed by tracing the motion of the planet-wheel, beginning at the position represented in Plate XI. The piston is then at the top of its course, and the teeth at the upper part of the planet-wheel are engaged with those at the lower part of the sun-wheel; suppose that the piston descends to about half way down in the cylinder, it will draw up the planet-wheel in its circular orbit, until its centre is on a level with that of the sun-wheel; the planet-wheel will then have made one-fourth of its circuit, and it is plain that one-fourth of the whole number of the teeth in the planet-wheel, must have passed in contact with one-fourth of those of the sun-wheel; so that the two wheels will have made one-fourth of a revolution in respect to each other; but, in addition to this, the planet-wheel will also have made one-fourth of the circuit in its orbit round the sun-wheel; and will have carried the sun-wheel with it, so as to have caused it to complete half a revolution.

In the same manner, when the piston gets quite down, and has raised the planet-wheel to the summit of the sun-wheel, the two will have made half a revolution in respect to each other; and in addition, the planet-wheel having made half the circuit in its orbit round the sun-wheel, will have produced a complete revolution of the latter.

The return of the piston depresses the planet-wheel, which continues its motion round the sun-wheel to the first mentioned position, when it will have completed its circuit, and caused the sun-wheel, and fly-wheel, to have made two complete revolutions (a).

The action of the fly-wheel, is to equalize the force of the circular motion, and to impel the machinery in the intervals when the planet-wheel is above or below the sun-wheel; for the centres of the two wheels being then in a line with the connecting-rod, the force of the piston can have no tendency to turn the planet-wheel round in its orbit; but when the centre of the planet-wheel arrives at the same horizontal line with the centre of the sun-wheel, then the whole force of the piston is exerted upon the sun-wheel, in a favourable direction to turn it round. The fly-wheel is urged into an accelerating motion at these favourable positions, and when the position of the planet-wheel is less advantageous, it will continue to turn all the machinery, by the impetus of its accumulated motion.

The force which the planet-wheel can exert to turn the sun-wheel and fly-wheel varies continually, being nothing when the piston is at the top or bottom of its course, and then it increases until the whole force of the piston is active at the middle of the course, but afterwards it decreases again to nothing at the end of the course. This variation is owing to the changeable direction in which the connecting rod acts, in respect to the line, which joins the centres of the sun and planet wheels. The action in this respect is the same as that of the simple crank already stated, p. 417. The force of the piston (and consequently the ascending or descending force of the planet-wheel) in a perpendicular line, is nearly uniform;

(a) Although the sun-wheel makes two revolutions for one circuit of the planet-wheel, the wheels only turn round once in respect to each other; for any two of their teeth will meet but once in a stroke, and the double velocity of the sun-wheel is a consequence of the circulation of the planet-wheel around it. If the planet-wheel were twice the size of the sun-wheel, the latter would make three revolutions for one circuit; or if the planet-wheel were only half the size of the sun-wheel, the latter would make one and a half revolutions in each circuit of the planet-wheel.

but this force is applied with a greater or lesser efficacy to turn the fly-wheel round, according as the planet-wheel, occupies a more or less advantageous position, in respect to the sun-wheel.

Construction of the working gear, or hand gear for opening and shutting the four valves in due succession, in order to distribute the steam so as to impel the piston upwards and downwards alternately.

Motion is given to the working gear by two round pins, 1 and 2, which project from the rod *l* of the air-pump and move up and down with it; the pins act in succession upon two handles *r* and *s*, which are affixed to two horizontal axes *t* and *u*, and by a system of levers and rods, two of the four valves are connected with one handle, and the other two valves with the other handle.

The four valves are thus connected with the handles *r* and *s* in two pairs; each pair consisting of a steam-valve and an opposite exhausting-valve; the two valves of the same pair always move together, one pair of valves being opened produces the ascent of the piston, and the other pair produces the descent.

This system is called the hand-gear, because the two handles are adapted to be moved by the attendant, when he wishes to govern the engine; though in the ordinary course of working, the handles are moved by the pins 1 and 2 on the air-pump rod, every time that the piston arrives nearly at the top or bottom of its course.

The valves are lifted by small spindles, which pass horizontally across the boxes (as shown in figs. 10 and 11); on each spindle a small lever or sector of three teeth is affixed, and they act in corresponding teeth on a short rack, which is connected to the stem of the valve; one end of the spindle passes through the side of the box, being exactly fitted into a socket by grinding, and the extremity which projects on the outside, has a lever fixed on it, which is connected by a rod with another lever on one of the axes *t* or *v* of the handles. By this arrangement the motion is communicated from the handles to the valves within the boxes, so as to raise them up, or let them fall at pleasure (see p. 373.).

The internal racks and sectors, for the two steam-valves, are placed at *b* and *e* above the valves; and the racks and sectors for the two exhausting-valves are placed at *h* and *i* beneath the valves, as shown in the sections fig. 8 and 9; but all the four valves open upwards.

r is the upper handle for the right hand; it governs the upper steam-valve *b*, and the lower exhausting-valve *i*, opening them both, when it is raised up as in fig. 8, or shutting them both when it is put down as in fig. 9.

t is the upper axis of the hand-gear, upon which the upper handle *r* is fixed; a lever is also fixed upon it with two arms, from one of which arms a rod 10 goes up to the lever of the upper steam-valve *b*, and from the other arm a rod 11 goes down to the lever of the lower exhausting-valve *i*. On the extreme end of the axis *t* a lever is fixed, and a rod 4 hangs down from it into the cistern, being loaded with a weight, which always tends to throw the handle *r* up, into the position of fig. 8, so as to open both the valves *b* and *i*. A short lever 12 is also fixed on the axis *t*, as shown in fig. 10, to act in the upper hook of the catch 3, and keep the valves shut, as in fig. 9, in opposition to the weight 4; but the instant the catch 3 is removed, that weight will open both valves, as in fig. 8.

s is the lower handle for the left hand: it governs the lower steam-valve *e*, and the upper exhausting-valve *h;* opening them both, when it is put down, as in fig. 9; and shutting them both, when it is raised up, as in fig. 8.

u is the lower axis of the hand-gear, with the lower handle *s* fixed upon it; also a lever with two arms, from one of which a rod 13 goes up to the lever of the upper exhausting-valve *h;* and from the other a rod 14 goes down to the lever of

3 м 2

the lower steam-valve *e*. On the extreme end of the axis *u* a lever is fixed, and a rod 15 which descends from it, into the cistern, is loaded with a weight, tending always to pull down the handle *s* and open the valves *e* and *h*, as in fig. 9. A short lever 16 is also fixed on this axis, to act with the lower end of the catch 3, as shown in fig. 10, and keep the valves shut, as in fig. 8, notwithstanding the weight 15; but when the catch 3 is removed, that weight opens both valves, as in fig. 9.

The pins 1 and 2 on the air-pump rod act on the handles *r* and *s* to raise up the weights which are connected with those handles, and this action closes the valves. The levers 12 and 16, and the catch 3, then hold up those weights, so as to prevent them from opening the valves, until the catch 3 is removed, and then that weight which is released, opens its pair of valves suddenly.

3, fig. 10, is the catch for the levers 12 and 16, which retain the handles of the working gear in the proper positions to keep the valves shut. The catch is poised upon a centre pin, which passes through it, near the middle of its length, and it has a hook at the upper end, to detain the lever 12 of the upper handle *r*, and a tooth at the lower end, to stop the lever 16 of the lower handle *s*; the parts are so arranged, that when either of the levers 12 or 16 pushes back one end of the catch, in order to pass under its hook or tooth, that motion of the catch will cause it to release the other lever from the opposite end of the catch. Therefore the act of moving one handle to shut one pair of valves, releases the catch of the opposite pair, and then they are opened immediately by their weight.

Position of the valves and working gear, during the descending stroke, see fig. 8. The lower handle *s* is raised, and the weight 15 is held up by the catch, consequently the lower steam-valve *e*, and the upper exhausting-valve *h*, are both kept shut. The upper handle *r* is raised, and its weight 4 is down, whereby the upper steam-valve *b*, and lower exhausting-valve *i*, are opened, the former admits steam above the piston, and the latter exhausts the steam from beneath it, as shown by the arrows, therefore the piston is pressed down; and the air-pump rod *l* descends also.

When the piston arrives near to the bottom of its course, the pin 1 on the air-pump rod *l* meets the handle *r*, and presses it down; raising its weight 4, and shutting the upper steam-valve *b*, and the lower exhausting-valve *i*, as shown in fig. 9. When they are shut the lever 12, fig. 10, catches the upper hook of the catch 3, which will prevent the return of the weight 4. The catch 3 being poised on a centre pin, as shown in fig. 10, the lever 12, in removing the upper end, in order to catch under the hook thereof, removes the lower end from the lever 16 of the lower axis and handle by the same action, so as to release that lever, and then its weight 15 suddenly opens the lower steam-valve *e*, and the upper exhausting valve *h*, as shown in fig. 9.

Position of the valves and working gear during the ascending stroke, see fig. 9. The action is now reversed, for the steam is admitted beneath the piston, and is exhausted from above it, so that it is impelled upwards; and the air-pump rod moves with it. When the piston gets near to the top of its course, the pin 2 on the air-pump rod, meets the lower handle *s*, and lifts it up, raising its weight 15 and closing the lower steam-valve *e* and the upper exhausting-valve *h*; and when its lever 16 passes the lower tooth of the catch 3, it disengages the hook at the upper end of the catch, so as to release the lever 12 of the upper axis and handle; consequently its weight 4 suddenly opens the upper steam-valve *e* and the lower exhausting-valve *i*, as represented in fig. 9, and then the piston is impelled downward again, as above-mentioned.

In this manner the alternating motion is kept up without any interruption; for as soon as the piston has run its course, one of the pins of the air-pump rod acts upon one of the handles, so as to close that pair of valves which allowed the steam to impel the piston ; and as soon as they are closed (but not before) the catch of the other handle being set free, its weight suddenly opens the other pair of valves, and they allow the steam to act on the piston with a returning impulse in the opposite direction.

Description of Particular Parts of the Engine.

It now remains to describe such other parts of the engine as are essential to the continuance and regularity of its operation ; and also the structure of such parts as are not already explained.

R is the cold water or house water pump, it is worked by a rod jointed to the great lever L of the engine ; the coldness of the water in the condensing cistern G is maintained, by pouring in a constant supply of cold water by this pump, which is a common sucking-pump, with valves in the bucket, and others at the bottom of the barrel ; the suction-pipe goes down into a well or pit, beneath the engine. The motion of the engine works the bucket of the pump, to draw cold water from the well, into the top of the barrel, from whence it runs by a pipe into the condensing cistern G.   *Note.* It appears in Plate XI. as if this spout would discharge into the hot-well I, but the spout is at the side of the hot-well, and discharges the cold water into the cistern G.

This pump raises more water into the cistern, than is required for the injection into the condenser, and the surplus runs off from the top of the cistern G by a waste pipe at one side, as is shown by the dotted lines 20 in the cross section fig. 4: this pipe conveys the water away into a drain ; the water in the cistern being thus changed, it is always kept cold, and assists to condense the steam in the condenser and air-pump by the external cold, as well as by the injection.   When a continual supply of fresh cold water cannot be obtained, the waste water from the engine is conveyed by a drain, to a large shallow pond in the open air, where the water cools by exposure to the air, and is then drawn back again, by the cold water pump.

S is the hot water pump to supply the boiler; it is a small sucking-pump, and its rod is moved by the great lever L ; the suction-pipe of this pump, draws its water from the hot-well I, and having raised it to the proper level, it runs from the top of the pump, by a horizontal pipe 8 into the feeding cistern T.

T is the feeding cistern ; it is placed high up over the boiler, being fixed on the upper end of the upright feeding-pipe 20, which goes down through the top of the boiler, into the water.   The passage of the water through this pipe, is regulated by a valve at the bottom of the cistern T, which closes the aperture of the pipe 20: the valve is suspended by a wire, from a lever 21, the centre of which is supported by a bracket fixed to the side of the cistern T, and from the opposite end of the lever, a wire descends through a small stuffing-box in the top of the boiler, and has a stone 22 affixed to the lower end of it, within the water in the boiler ; see fig. 3.

This stone is so balanced by a weight suspended at the opposite end of the lever 21, that it will act as a float on the surface of the water, and open the feeding-valve at T whenever the boiler requires a supply of water, or shut the valve as soon as the boiler is replenished ; for when a heavy body is suspended in a fluid, it will lose as much of its own weight as the weight of that quantity of the fluid which it displaces.   When the water in the boiler diminishes by the conversion of a part of it into steam, the upper surface of the stone will remain above the water, and consequently a less proportion of its weight will be supported by the fluid, because it ceases to bear up that part of the stone that is not immersed. By the action of this additional portion of its weight, the stone will overcome the balance-weight at the end of the lever 21, so as to elevate that weight, and open the valve at the top of the pipe 20, to admit water into the boiler in sufficient quantity to replace that which is carried off by evaporation. Too much water can never be introduced, for as soon as the surface of the water in the boiler coincides with the surface of the stone, the balance weight recovers its former preponderance, and the valve at the bottom of the cistern, T, shuts the top of the pipe 20, so as to prevent any more water entering the boiler, until the float or stone 22 descends again by the diminution of the water.

When the engine is steadily at work, the stone subsides and opens the valve until it will admit a regular stream of water, which is equivalent to the waste by evaporation; and then the operation will go on regularly, without any action of the float, until something is altered. The pipe 20 has a cock in it, to regulate the rapidity with which the water shall descend into the boiler.

The hot-water pump S is adapted to raise more water than the boiler ever can require, and the surplus is carried off from the cistern T by a waste pipe, to avoid overflowing; this waste pipe leads to the common waste water drain.

The feeding cistern T being placed at about 8 feet above the surface of the water in the boiler, will give the water a sufficient column to run into the boiler at all times; because the steam is never intended to be raised to an elasticity of more than 4 or 6 feet column of water.

y are the two gauge cocks, to ascertain the height of the water in the boiler; they close the upper ends of two pipes, which pass down through the top of the boiler, and the lower ends are open; one pipe descends lower than the intended level of the water, but the other does not reach down so low; in consequence, when the water stands at its intended level, one cock being opened will emit water, and the other steam. These cocks are tried occasionally to ascertain that the feeding apparatus operates correctly; but in the regular course, the water will always preserve the same level, without any attention from the attendant.

The boiler is provided with a drain pipe which proceeds from the lowest part of it, and comes out through the brick work; it has a stop cock at the end (as shown in Plates XI. and XII.) to draw off the water, and empty the boiler occasionally in order to cleanse it.

v is the safety-valve; it is contained within a box or chamber fixed upon the top of the boiler, with a cover, through which a rod passes in a stuffing-box, to open the valve within, and discharge the steam, when the engine is not to be worked any longer. When it is at work, the valve is pressed down by a weight. At any time, whether the engine is at work or not, if the steam should accumulate, so as to be in any danger of bursting the boiler, the force of the steam beneath the valve will lift up the weight, and allow the steam to escape through a branch pipe which opens into the chimney; see the dotted lines fig. 3.

There is also another smaller safety-valve to the boiler, for a contrary object to that of the principal safety-valve v; it is inverted and opens internally, and is balanced by a small lever, with a sufficient weight to keep the valve shut, until the pressure of the steam within the boiler, becomes less than the external air, which then forces open the valve, and enters into the boiler, till the equilibrium is restored. This valve never operates whilst the engine is at work; but it prevents the sides of the boiler being crushed in, by the pressure of the air, when the steam within it cools and condenses. This small safety-valve may be fitted upon any part of the boiler: it is represented at &, fig. 3, fitted upon the lid of the man hole.

& is the man hole; it is a short cast iron tube of an oval shape, and large enough within, for a man to pass through ; it is fastened to the top of the boiler by means of a flange round its lower edge, and at the top is a similar projecting flange, to fasten on the lid or cover, by means of screw bolts round the edge ; a packing of pasteboard and white lead being interposed to make a tight joint.   This cover being removed, a man can go down into the boiler, and the water being previously drawn off, he scrapes away any dirt or sediment which may adhere to the bottom.

The cylindrical top of the boiler is covered over with brick-work, to retain the heat ; and the several parts of the apparatus connected with the boiler, are fitted to pipes, which rise up above the brick-work.

There are openings with moveable covers, to give access to all the valves in the engine, in order to repair them when necessary: viz. the four principal steam valves, can be examined through square openings in front of the nossels at $c\ h$ and $f\ i$, as shown in fig. 6 ; and also two other openings in the tops of the steam boxes b and e.   Each opening is closed by a cover fitted exactly to it, and fastened on by two screw bolts to each, as shown in fig. 10.   The foot valve at k, between the condenser and the air-pump, is accessible by removing a cover immediately over it.

The covers of the cylinder, and of the air-pump, are fastened on by screw bolts passing through projecting flanges ; they can be removed to gain access to the piston, and air-pump bucket.   The buckets of the cold water-pump R, and of the hot water-pump S, can be drawn up out of their respective barrels, to repair their valves ; and each pump barrel has an opening at the lower part, with a moveable cover, to give access to the lower valve of each pump.   All these covers are fastened to their places by strong screw bolts, and packings of pasteboard soaked in oil, and covered with whiting or white-lead, being interposed in the joints, will make them tight, when the covers are drawn forcibly into their places by the screw bolts.

x is the blow-valve, to allow the air to escape from the condenser, when the steam is first admitted into it, previous to setting the engine to work ; but this valve does not act whilst the engine is working.   It is a small light brass valve, fitted into a brass seat, at the orifice of a pipe, which proceeds from the lower part of the condenser F, and passes through the side of the cistern G to the outside. The blow-valve is immersed beneath the surface of water, contained in a small cistern, which is fixed to the side of the great cistern G ; the small cistern is kept constantly full, by a dropping out of a small hole in the side of the large one; this water keeps the valve tight, but it can open freely to allow the air to pass.

The fixed framing to support the parts of the engine is chiefly wood.   The building or engine-house, consists of two side walls, and two end walls, covered with the roof.   The boiler is placed at the outside of the walls, either across the end of the house, as shown in Plate XI. or else along the side wall, according to convenience ; the chimney is carried up against the wall of the house.

The framing is erected upon two long beams, or ground sills, which extend horizontally all the length of the house, parallel to each other ; the ends are worked into the end walls, and they are supported by foundation walls, built across the house, one beneath the cylinder, another under the centre of the great lever, and a third, beneath the centre of the fly-wheel.   Upon these sills a cross beam V is laid, and posts V V are erected on that cross beam, to support another cross beam W, upon which the spring beams U U rest ; the two cross beams V and W, are firmly bound together by long iron bolts which are carried up outside the posts V V; and the ends of both beams are worked into the side walls of the house.

The spring beams U U extend parallel to each other all the length of the house, and are fastened into the end walls ; the bearings for the axis p of the great lever are fixed upon the spring beams, and the ends of the lever, move up and down between the spring beams. A strong piece of wood is fixed across each end of the great lever, at its upper side, by iron straps, and the ends which project out each way over the spring beams UU, serve for catch pins, to limit the motion of the lever, in case the link of the sun and planet-wheel, or the connecting rod, should break, but in the regular course of working catch pins are not necessary.

The cylinder is supported upon two cross beams XX, which are worked into the side walls at each end, and are supported by two upright posts, from another cross beam X, which lies upon the two long ground sills; the cylinder is fastened down upon the cross beams XX, by two bolts, which go down through the two long sills, and are fastened beneath. The condensing cistern G is placed upon these sills, and by its weight keeps them steady (a).

The air pump stands upon a block of wood placed on the bottom of the cistern, and it is fastened down by two bolts, which go down through the long sills, and are fastened beneath.

The bearings for the axis of the fly-wheel are supported by beams, which extend horizontally from one of the posts V, to the end wall, and are borne up in the middle by posts erected upon the long ground sills; the bearings are bound down by two iron bolts, which pass down through those sills and are fastened beneath. The cold water pump R is fixed against one of the fly-wheel beams.

The engine-house is lighted by large windows in the end walls, or else by one large skylight in the roof.

A floor is laid over the condensing cistern G, on the level of the cylinder beams X to give access to the working gear. The entrance door into the engine-house usually leads to this floor, either through the end wall or the side wall. A small staircase passes behind the cylinder, under the steam-pipe a, to a stage fixed around the top of the cylinder, for the convenience of packing the piston, and thence another small staircase ascends to the floor, which is laid upon the spring beams UU.

The construction of the remaining parts of the double engine are the same as the corresponding parts of the single engine already described, viz. The steam gauge, to indicate the elasticity of the steam within the boiler, which is usually kept up so as to sustain a column of mercury of four or five inches high; one of these steam gauges is attached to the steam case of the cylinder, to show the state of the steam therein, and another to some convenient part of the steam-pipe a or d d; the construction of these gauges has been already described, p. 372.

The cylinder, with its cover, and its steam case, together with the piston and piston-rod, are made the same as for the single engine, see pp. 372 and 380. The structure of the valves, and that powerful arrangement of the levers of the working gear, whereby each valve is lifted, is the same as described in p. 373; but the nossels or boxes, to contain the valves and the working gear for them, are different in the double engine, as already described, because there are four valves. The condensing apparatus, in the condensing cistern, consisting of the condenser, the air-pump, with its bucket and valves, the hot well, and barometer, are the same as for the single engine (see p. 374 to 379), except in the proportions, the effective capacity of the air-pump, for the double engine, being one-ninth part of that of the cylinder.

The structure of the working-gear has been already sufficiently described to

(a) The cistern G is represented shorter in Pl. XI. than it was usually made; it ought to extend quite beneath the cylinder, and stand upon the lower cross beam X so as to keep it down; the bolts to fasten the cylinder, then go down through the cistern. A cistern of the dimensions represented in the plate would be sufficient, but a larger one by containing more water would be better.

give an idea of its action. The four short levers which are fixed upon the two axes *t* and *u* for the handles *r* and *s*, are so arranged that when the valves are all shut, these levers will assume the same directions as their rods 10, 11 and 13, 14, by which they are connected with the other longer levers, on the axis of the sectors for the different valves. The levers 12 and 16 are so regulated to the catch 3, as to detain the handles, when the levers are in those positions, as is shown in fig. 10. The chocks 1 and 2 on the plug rod *l*, are adjusted so that they will just move the handles *r* and *s* into those positions, in order that the catch 3 may retain them, where the chocks leave them. When either of the handles is released from the catch 3, the weight 4 or 15 belonging to that handle will fall, and will open the corresponding pair of valves; in that action, the weight has a great advantage of leverage to lift up the valves, as is stated p. 374.

The valves must be opened and shut by hand, in order to stop or start the engine, or to regulate its motion; and for this purpose, the attendant always retains a full control over all the four valves, by means of the two handles *r* and *s*, either to open or to shut either pair of the valves, whatever position the chocks 1 or 2 of the plug may be in; for even when either of those chocks has moved the handle belonging to it, so as to shut one pair of valves, they may nevertheless be opened again immediately, by the same handle, without waiting for the return or removal of the chock from that handle. This is done by moving the handle still further in the same direction as the chock has just moved it; for when the valves are shut, the short levers on the axes of the handles, being in the lines of their connecting rods, those levers will have the effect of opening the valves, when they are turned in either direction from that position. In the regular action of the working-gear when moved by the chocks, the valves are closed by the motion which those chocks give to the handles; and are opened again by the fall of the weights 4 and 15; nevertheless, when the engine is regulated by hand, that action may be reversed if required; for the same effects may be produced by the contrary motions of the handles; and then the weights will tend to close the valves, instead of to open them.

This property of Mr. Watt's working-gear is very important, for although the four valves are moved with great certainty by the self-action of the engine, still they always remain under the control of the attendant, who can at any time give either two of the four valves, or all the four, an opposite position to that which the self-action would give them; or he can give them the same position, or any intermediate position, as the case may require. *Note.* The working-gear of the single engine has the same property; but it was omitted to be mentioned in the description of the sketch, p. 374.

Regulation of the Motion of the Engine.

z the throttle-valve, which regulates the passage of steam from the boiler, through the steam-pipe *a*, to supply the cylinder; see fig. 12, (see also p. 435). It is a circular valve or vane of metal, mounted on an axis or spindle, which passes across its centre, and forms a diameter to the circle; this vane is fitted into the aperture of a circular ring of cast iron, which is fixed between the flange of the steam-pipe *a*, and that of the steam box *b*, and is fastened to them by four screwbolts, so that the circular aperture through the ring coincides with the internal passage of the steam-pipe. The spindle of the throttle valve passes across the ring, and one end, which is accurately fitted into a hole through the metal ring, projects on the outside, and has a short lever or handle fixed on it, to turn the spindle and the valve within the pipe.

When the throttle-valve is set, so that its plane is perpendicular to the axis of the pipe, it fills the circular passage, and allows no steam to pass by it; but when the vane is turned edgeways, it presents a very small surface, and the steam passes by it without obstruction to the steam-boxes *b* and *e*, and thence into the cylinder. By the handle of the throttle-valve, the engine-man can turn the valve at any required angle in order to obstruct the passage, and regulate the speed of the engine; the friction of the axis is sufficient to retain the valve as it is placed.

This method of regulation is sufficient for many engines; but when the steam engine is employed to drive machinery, in which the resistance is variable, and where a determinate velocity cannot be dispensed with, Mr. Watt applied the revolving pendulum or governor, in order to regulate the opening of the throttle-valve according to the velocity with which the engine is intended to move.

Z is the governor or revolving pendulum, for procuring uniform velocity, (see also p. 435); it consists of two pendulums, attached by a joint pin, to a vertical axis, which is turned round by the engine, so that the balls of the pendulums revolve in a horizontal circle, and the velocity of their motion is sufficient, to make them expand or recede from the axis, by their centrifugal force, notwithstanding that the weight of the balls being suspended by the joint, they incline to collapse or approach towards the axis.

The governor is turned round by means of an endless strap, which encompasses a pulley fixed on the main axis P of the fly-wheel, and also another pulley, fixed on a short horizontal axis, which carries a bevelled wheel Y, to turn a similar wheel fixed on the lower end of the upright axis Z of the pendulums, which is by this means carried round. The two rods, or arms of the pendulums, which have the balls appended to their lower ends, cross each other at the upper ends, through an opening in the vertical axis, and are connected therewith by the joint pin, in the manner of a pair of scissars; after crossing each other, the upper ends of the two rods are bent outwards from each other, in contrary directions, and are connected by two short links, with a collar, which is so fitted on the upper part of the axis, as to be capable of sliding up and down upon it; the sliding collar has a groove formed round it, to receive the prongs of a fork, formed at the end of a long lever, which is poised nearly horizontally on a centre, and the opposite end is connected with one end of a second lever, the other end of which reaches over the handle of the throttle-valve, and is joined to it by a wire. The result of this arrangement is, that whenever the balls of the pendulums expand, or fly out by their centrifugal force, the short links from the upper ends of the pendulums, will draw the sliding collar downwards on the axis; and this motion being transmitted by the two levers, and the wire, to the throttle-valve, turns that valve a little upon its axis, so as to present more surface, and close or diminish the passage for the steam. Or on the contrary, when the balls contract, or approach the axis by their own weight, that motion will open the throttle-valve a little wider, to supply more steam to the cylinder.

As the governor derives its motion from the fly-wheel, it must partake of every increase or diminution of velocity which takes place; but the pendulums being freely suspended, will fly out from the vertical axis to a different distance corresponding to every different velocity, so as to assume that position in which the weight of the balls, and their centrifugal force, will exactly balance each other (see p. 42). As every alteration in the position of the pendulums produces a corresponding alteration in the opening of the throttle-valve, the governor will keep the engine nearly to one regular motion. For whenever the engine exceeds that regular motion, the pendulums of the governor by expanding, will close the throttle-valve, and diminish the supply of steam, until the motion of the engine is reduced to proper speed. Or if the motion becomes slower than the regulated speed, then the balls, by collapsing, will open the throttle-valve, and give a better supply of steam, in order to quicken the motion of the engine.

OPERATION OF MR. WATT'S ROTATIVE STEAM-ENGINE. *Plates XI. XII. and XIII.*

Suppose the engine is at rest in the position represented in Plate XI.; except that the upper handle *r* should be depressed, whilst the lower handle *s* is raised; and both handles being detained in those positions by the catch 3, all the four valves *b i* and *h e* will be kept shut. Suppose the cylinder to be quite cold, and the condensing cistern G to be full of cold water. The fire being made upon the fire-grate, will heat the water in the boiler C, and when it first begins to boil, and produce steam, the safety-valve must be set open, to allow that steam to drive out all the air contained in the upper part of the boiler; but when the steam is observed to issue at the valve, it may be shut, and the steam retained in the boiler until so much is accumulated therein, as to acquire a temperature of about 220 degrees, and an elastic force of about $2\frac{1}{2}$ pounds per square inch, and then it will rush forcibly out of the boiler, if it is permitted to escape into the open air.

Preparation. The attendant must admit the hot steam into the space between the cylinder and the steam case, by opening the cock in the small pipe which branches off from the main steam pipe *a ;* this steam will expel the air contained in that space (and also between the two bottoms of the cylinder) through another small cock fixed in the hollow bottom, and which is previously opened for that purpose; this cock is not seen in the figures. When all the air is expelled, and the metal of the cylinder thoroughly warmed by the steam, the latter cock is to be shut, and all the water which will afterwards collect in these spaces, by the condensation of the steam therein, during the working of the engine, will drain off through the inverted syphon pipe, which leads from the lowest part of the steam case, as is shown beneath the cylinder near *x* in Plate XI.

Blowing through. The engine being in this situation, the first operation in preparation for a commencement of motion, is to discharge the air from all the steam-vessels by blowing through. To do this, the attendant must release the catch 3, and then the weights 4 and 15, will raise the upper handle *r*, and depress the lower handle *s*, so as to open all the four valves *b e* and *i h* together; the injection cock being shut. The steam thus admitted into all parts of the engine, will drive out the air through the blow-valve *x*, so as to displace it from the valve boxes *b* and *e*, from the steam-pipe *d*, and eduction pipe *g*, and the condenser F, but not from the cylinder E E.

As soon as all the air is gone through the blow-valve *x* into the open air, it will be succeeded by some steam, which will occasion a sharp cracking noise in the water of the cistern, wherein the valve *x* is immersed, owing to the sudden condensation of the issuing steam, by that water. All the four valves are then to be shut, by raising the lower handle *s* and depressing the upper one *r*, and they must be kept shut, until it is thought that the steam which had entered the condenser, is condensed, and a partial exhaustion formed in the condenser; then on opening first one pair of valves, and then the other, part of the air in the cylinder E E will pass by its own elasticity through the exhausting valves *h i* and the eduction pipe *g* into the condenser.

The same operation of blowing through is then to be repeated, by opening all

3 N 2

the four valves at once, whereby that part of the air which was drawn out of the cylinder E E into the condenser, will be discharged from it, through the blow-valve *x*. All the valves are then to be shut, and a longer time should be allowed, for the steam which is introduced into the condenser to cool, between the times of blowing; when it is found that upon opening the injection cock, some water will enter into the condenser, and that the barometer will then show some degree of exhaustion in the condenser, another repetition of the blowing should be made, so as to clear the cylinder of air, and its place being supplied with steam, the engine is ready for work.

Starting. If the piston is quite at the top of its stroke, as in Plate XI. the upper handle *r* must be released from the catch 3, and allowed to rise up, so as to open the upper steam-valve *b*, and the lower exhaustion-valve *i*, both at once, as in fig. 8, the other two valves remaining closed. The fly-wheel Q Q must be turned round by hand, for about one eighth of a revolution or more, in that direction in which it is intended to move; and the injection cock *j* must be opened by its handle.

At the moment of opening the two valves *b* and *i*, the steam which was then in the lower part of the cylinder, beneath the piston, will begin to pass through the lower exhausting-valve *i*, and the eduction pipe *g*, into the condenser F, where the jet of cold water, which is admitted through the injection cock *j*, coming in contact with the steam, will cool and condense it into water, so as to form a vacuum in the condenser. The steam which remains in the lower part of the cylinder, beneath the piston J, will continue to expand itself, and rush rapidly through the lower exhaustion-valve *i*, and eduction-pipe *g*, into the exhausted condenser, leaving the cylinder also exhausted beneath the piston.

Action during the descending stroke. The steam from the boiler entering freely through the upper steam valve *b*, into the upper part of the cylinder, above the piston, and pressing continually upon it, will now cause the piston to descend, and by its conneixon with the great lever L, through the piston rod *n*, and parallel motion K, the piston will pull down the cylinder end of the lever, and raise up the opposite end, together with the connecting rod M, and the planet-wheel N; whereby the latter begins to revolve in its orbit round the sun-wheel O. The planet-wheel N being fixed fast to the lower end of the connecting rod M, so that it cannot turn upon its own centre, and its teeth being engaged in those of the sun-wheel O, the latter, and the fly-wheel Q, which is fixed upon their common axis P, are made to revolve in the desired direction, and give motion to the mill-work.

As the piston descends, the plug or air-pump rod *l* descends also, and when they are near the end of their course, the pin 1 which is clamped upon the rod, presses down the upper handle *r*, and thereby shuts the upper steam-valve *b*, and the lower exhausting-valve *i*; the same movement, by disengaging the catch 3, permits the weight 15, which is connected with the axis *u* of the lower handle *s*, to depress that handle, and thereby to suddenly open the lower steam-valve *e*, and the upper exhausting-valve *h*, as shown in fig. 9.

Action during the ascending stroke. At the moment of opening these valves, the piston will have very nearly reached the lowest position of its stroke; the steam which filled the upper part of the cylinder, above the piston, rushes with extreme rapidity by its own elasticity, through the upper exhausting-valve *h*, and the eduction pipe *g*, into the vacuous condenser F, so that the upper part of the cylinder above the piston, becomes exhausted in a very short space of time. The steam from the boiler entering freely, through the lower steam-valve *e*, into the lower part of the cylinder, and acting beneath the piston, now forces it upwards until it returns to the top of the cylinder.

Repetition of the descending stroke. When the piston arrives near the top of its stroke, the other pin 2, which is clamped upon the air-pump rod *l*, raises up the lower handle *s*, so as to close the lower steam-valve *e*, and the upper exhausting-valve *h*; in so doing the catch 3 is disengaged, and then the weight 4 throws up the upper handle *r*, and opens the upper steam-valve *b* and the lower exhausting-valve *i*, as in fig. 8; the downward stroke is then repeated in the same manner as has been already described.

At the first commencement of the action, the piston moves but slowly, because the inertiæ of the fly-wheel, as well as the resistance of the mill-work, is opposed to the motion; but as the action goes on, each succeeding half stroke is performed in less and less time, giving the fly-wheel an accelerating velocity, until, in about three or four strokes, the full velocity will usually be attained, and then the motion of the fly-wheel continues to be nearly uniform. In a 10-horse engine this motion is at the rate of 25 strokes per minute, and the fly-wheel makes 50 revolutions per minute; a 20-horse engine makes $21\frac{1}{2}$ strokes per minute, and the fly-wheel 43 revolutions per minute; and a 40-horse engine makes $17\frac{1}{2}$ strokes per minute, and the fly-wheel 35 revolutions per minute.

Action of the Air Pump. When the piston descends, the bucket of the air pump H descends also; the water which is contained in the lower part of the pump, passes through the valves in its bucket, which being drawn up again, when the piston ascends, the water above the bucket of the air-pump is drawn up, and is discharged through the discharging valve *m*, into the hot well I. The ascent of the bucket leaving a vacuum beneath it in the barrel of the pump, the air and elastic vapour contained in the condenser, pressing by its elastic force upon the surface of the water contained in the condenser, forces that water to pass through the foot-valve *k* into the air-pump, and when all the water has passed, part of the air or vapour will follow it; then at the next descent of the bucket, the air and water contained in the barrel will pass through its valves, and get above the bucket, so as to be drawn out and discharged at the subsequent ascent of the bucket. The action of the air-pump is exactly the same as that of the single engine already described at p. 360; and also the action of the cold water pump R, and of the hot water pump S, by which part of the hot water is raised from the hot well I, and conveyed through the pipe 8 8 and T 20, for the supply of the boiler.

The admission of the feed water into the boiler is regulated according to the height of the surface of the water therein, by means of the float 22, which by the lever 21 opens the feeding valve, whenever the surface sinks, and its balance weight closes that valve whenever the surface rises. The valve is thus regulated to such an extent of opening, as will admit a sufficient stream of water into the boiler, to supply the waste of evaporation; there is also a cock in the feeding pipe 20, whereby the attendant can render the adjustment more complete, see p. 369. This feeding apparatus was first brought into use in the rotative engines, but is equally applicable to all steam-boilers. The management of the fire and the regulation of the sliding damper *w* are conducted in the same manner as already described, p. 361.

Regulation of the Motion. A rotative engine requires much less care and attention, while it is working, than the simple engine for pumping; because the length of stroke of the piston, being regulated by the link *q q* behind the sun and planet wheels, the catch pins can never strike on the spring beams. When any casual or slight alteration takes place in the resistance, or in the force of the steam, so as to cause the engine to begin to move faster or slower, the governor Z will immediately regulate the throttle-valve, and supply less or more steam to the cylinder, according as the actual resistance requires, in order to preserve nearly the same velocity of the fly-wheel, without any attention on the part of the attendant, who has little to do in ordinary cases, except to take care of the fire, and keep up the steam to a steady force.

Oiling. He must also apply oil and grease occasionally, to all the moving parts of the engine, to diminish the friction ; to introduce the grease to the piston, a funnel is provided in the cover of the cylinder, with a stop cock and a short pipe passing down through the double cover. A small depression or bowl is formed in the cover, to contain tallow, which becomes melted by the heat, and is then filled into the funnel with a spoon ; the cock being shut. To let this melted grease down into the cylinder, the attendant must open the cock, whilst the piston is ascending, and the cylinder is exhausted; the pressure of the atmosphere will then force the grease in the funnel, through the cock and pipe, into the cylinder, where it falls upon the piston, and runs over the edges thereof, to lubricate the hemp packing, which rubs up and down within side the cylinder. The joints of the parallel motion and connecting rod, and the bearings for the axis of the great lever, and of the fly-wheel, must also be kept supplied with oil or tallow grease ; for this purpose all the sockets are perforated with small holes from the upper sides, to admit the oil or melted grease to run down into the joints, when it is poured in from a small pot with a spout. If the oiling is neglected, the friction of the metallic surfaces will increase so much, as to produce a considerable heat, which will dry up all the remaining grease, whereby the surfaces will come into absolute contact, and then the friction and heat will increase beyond all bounds, so as to make the metal of the axis and the socket hot, even to redness, if the motion were continued, and the metals will cut and wear each other so that it is impossible to proceed. In the usual course of working, the friction occasions a gentle warmth of all the rubbing parts, which are in contact. This is advantageous by melting the grease and rendering it more fluid ; the heat will not increase, so long as the surfaces are kept well supplied with grease, but if that is neglected, and the heating begins, it will go on increasing, until it becomes difficult to correct it ; for after the rubbing surfaces have become hot, and worn themselves into such close contact as to exclude the grease, it is difficult to get it between them again ; and the engine must be stopped, till the heated parts become cool ; or the cooling may be effected by throwing on cold water, but the work cannot proceed whilst the heating continues.

To Stop the Motion of the Engine; this must be done gradually, because the energy of the fly-wheel is too great to permit it to be suddenly checked ; the attendant should first turn the handle of the injection-cock, so as to diminish or cut off the supply of cold water, and thereby impair the vacuum in the condenser ; he then takes hold of the upper handle *r* with the right hand, and the lower handle *s* with the left hand (*a*), and endeavours to counteract, or diminish the tendency of the weights 4 and 15 to open the valves ; thus every time that either handle is urged by its weight, he restrains the valves from opening so suddenly, as in the usual course of working ; and he also prevents them from opening to their full extent ; the apertures through which the steam is to pass

(*a*) It is usual to apply the right hand to the handle of the upper axis *t*, and this commands the two valves which are to be opened to produce the descending stroke ; whilst the left hand is applied to the handle of the lower axis *u*, which opens or shuts the two valves for the ascending stroke. In figures 6 and 7 the handles are by mistake represented reversed to this arrangement, the upper handle being on the left hand side.

into, or out of the cylinder, being thus contracted, the cylinder is imperfectly exhausted, and the force of the piston being diminished, the motion of the fly-wheel soon expends itself.

When the great velocity is thus checked, and the fly-wheel is moving slowly round, he proceeds to stop its motion entirely, leaving the parts in that position which will be most favourable for starting the next time; this is when the line between the centre of the planet-wheel, and that of the sun-wheel, has passed the vertical line to an angle of about 45 degrees.; the piston being near the top of the cylinder. To accomplish this he takes the opportunity of the piston arriving at the top of its course, and when the lower handle s is raised, so as to close the lower steam-valve e, and the upper exhausting-valve i, he holds down the right hand handle r, so as to prevent the upper steam-valve and the lower exhausting-valve from opening by the weight 4; consequently all the four valves will then remain shut at the same time, and the double catch 3 will retain them so; in this situation the cylinder remains full of steam beneath the piston, and as it cannot escape, the further motion of the piston is opposed by all the resistance which that steam can offer to compression. This resistance very soon expends the remaining energy of the fly-wheel, so as to stop the motion by the time the piston has descended a little, and after that, it would recoil by the action of the compressed steam; but to prevent this, the upper handle r is depressed, and the lower handle s elevated, so as to open all the four valves at once, for a moment, in order to establish a free communication, and consequently an equilibrium of the steam on both sides of the piston, which then loses all further tendency to motion; both handles being allowed to return till the catch 3 retains them, and in this situation the engine will remain at rest, with all the four valves shut, but ready for starting at any time.

A little dexterity and practice will enable a man to stop the engine exactly at the proper position, without making any recoil; or turning the fly-wheel at all backwards, which in some kinds of mill-work cannot be allowed. In all cases the speed must be reduced as before described, until the remaining energy of the fly-wheel, is only so much as may be checked at once, by retaining the steam within the cylinder.

In a large engine which has a great energy in the moving parts, it is usual to detach the throttle-valve from the governor, previous to stopping the engine; this is done by removing the joint-pin which unites the connecting wire to the handle of the throttle-valve; then the throttle-valve being closed, or very nearly so, at the same time that the injection-cock is closed, the piston is at once deprived of almost all its force, and the motion of the fly-wheel is soon expended in turning the mill-work, and moving the engine. This precaution is unnecessary in a small engine, although the governor tends to keep up the motion, by opening the throttle-valve wider as the motion becomes slower.

It is obvious, that at any time, if either of the handles were retained from moving by its weight, so as to keep all the four valves shut, the steam would be retained in the cylinder, and would stop the engine at once, even in its fullest speed; this does sometimes happen by accident, but the energy of the fly-wheel is very likely to break some part of the engine, as the teeth of the sun or planet-wheels, or some of the joints or axes; and, therefore, such an occurrence ought to be carefully avoided, particularly in large engines.

In cases when it is required to stop the engine very promptly, the action of the steam may be reversed, so as to act in opposition to the motion of the piston, and produce a rapid retardation of the motion of the fly-wheel; for this purpose the working gear must be placed in a contrary position to that which is given to it, in the usual course of working.

Suppose, for instance, that the piston has reached the top of the cylinder, and that the chock 2 of the plug rod *l* has raised the lower handle *s*, so as to close the lower steam-valve *e*, and the upper exhausting-valve *h*; as in fig. 8. The upper handle *r* is then to be held fast, to prevent the weight 4 from lifting it up; and the lower handle *s* is to be lifted up, higher than the chock 2 has lifted it, whereby (as already stated, p. 457) the lower steam-valve *e* and the upper exhausting-valve *h* will be opened again. In this state the steam will continue to press upwards beneath the piston, whilst there is a vacuum above it, and the steam will oppose the descent of the piston, because the steam must be driven back before the piston, into the boiler again, by an action similar to that of a pump. The motion of the fly-wheel will be greatly retarded by this resistance, though not absolutely stopped, as when all the valves are kept shut, to retain the steam from escaping out of the cylinder. When the piston reaches the bottom of its course, then the lower handle *s* must be let down again, until it is stopped by the catch 3, so as to close the upper exhausting valve *h*, and the lower steam-valve *e*; and the upper handle *r* is depressed to open the upper steam-valve *b*, and the lower exhausting-valve *i*, whereby fresh steam from the boiler will be admitted again to fill the vacuous part of the cylinder, above the piston, and this steam will oppose its ascent, for there is a vacuum beneath the piston. This resistance will so reduce the remaining motion of the fly-wheel, by the time the piston gets again to the top of the cylinder, that it may then be completely stopped, by keeping all the four valves shut together, to retain the steam in the manner before described. This method of retarding the motion of the fly-wheel, by reversing the action of the steam, occasions a considerable strain on the parts of the engine, which small engines may very well endure, but which it is best to avoid in large engines, except in cases when it is requisite to stop quickly.

The fly-wheel may be turned round in either direction by the engine, without any alteration of the working gear; for the direction of the motion of the fly-wheel depends entirely upon that which is given to it at the first starting; but when once in motion it will continue to turn in the same direction, by its own energy, which carries on the motion during the inactivity of the piston at every change of its stroke. In starting the engine the position of the planet-wheel in respect to the sun-wheel, must always be attended to, in order that the motion may be begun in the right direction.

When rotative engines first came into use, it was thought that the planet-wheel should move away from the cylinder, when passing over the top centre, that is, over the summit of the sun-wheel; but this circumstance being quite immaterial, it was never attended to; but an engine which turns its fly-wheel in the opposite direction to that above stated, is commonly said to go backwards.

As soon as the engine is stopped, if the working is over for the day, the safety-valve should be opened to discharge the steam from the boiler. The fire should be raked out, and the grate cleared of cinders, ready for the next day's work. When an engine is stopped with all its four valves shut, the steam contained in the condenser, and in the pipes and the cylinder, will be gradually condensed, leaving all those parts exhausted; but by degrees the air will leak in, so as to fill up the vacuous spaces, particularly when the metal grows cool. If it is required to start the engine again in a short time after stopping it, the steam may be retained and its heat kept up, by leaving the fire-door open, so as to admit a current of cold air into the furnace, over the fire upon the grate, and beneath the bottom of the boiler; this will impair the draught of air through the fire bars, and cause the fire to burn so moderately, as to maintain the heat of the steam in the boiler, without producing more steam; and yet, by stirring up the fire, and shutting the door, the steam will at any time begin to be generated, ready for starting the engine again.   If all the joints are tight and good, the condenser will remain exhausted for an hour or two, as can be seen by the barometer, and the engine may be set to work again without blowing through.   In a common state, as much air will enter by leakage in the course of an hour, as to require to be cleared by blowing through, in the manner first described; and to perform this with effect, the steam must be accumulated in the boiler till its elasticity is at least equal to 5 inches of mercury by the steam-guage (see p. 372), that is, about $2\frac{1}{2}$ pounds per square inch more elastic than the atmospheric air.   When the engine is steadily at work, the steam should be kept nearly to that elasticity, at least if the engine is fully loaded by the resistance that it is to overcome; but if the engine is not fully loaded, the steam will be sufficient at 2 or 3 inches; it ought in all cases to be kept regular.

Action of the governor and throttle valve (see the sketch, p. 435).   The effect of this apparatus to regulate the velocity of an engine is not quite perfect, although it is a sufficient approximation for practice.   The revolving pendulums cannot operate beforehand to prevent alterations of velocity, but only to correct the alterations after they have taken place; and such corrections will always be in excess; for whenever the governor has altered the throttle-valve to a sufficient extent to correct any sudden change in the motion of the engine, it must retain the valve in its altered position, until the motion of the engine is affected thereby, and the centrifugal force of the governor being also affected, it will produce an alteration of the throttle-valve, in a contrary direction to the first one; another alteration of speed must therefore ensue, and then a consequent correction will be made by the governor and throttle-valve.   As each succeeding correction becomes more and more minute, the motion of the engine is not perceptibly deranged thereby, in an ordinary course of working, provided that the governor is properly proportioned; but whenever a great and sudden change of the motion of the engine takes place, the governor will make a correspondingly great and sudden alteration of the opening of the throttle-valve, so as not merely to correct the change of the motion, but it will cause an alteration of an opposite character to that of the original one, though to a less extent.   The governor will afterwards apply a remedy to that second alteration, and proportionably less active, though still in excess; and thus, after a few fluctuations the governor will bring the valve to its proper adjustment.   To prevent any evil from this property of the governor, it must be made to act delicately on the throttle-valve, so that a considerable expansion or contraction of the pendulums will not produce too much alteration in the opening of the valve.   The steam in the boiler should have at least 2 lbs. per square inch more elasticity than it is required to have, after it has passed the throttle-valve, and entered into the cylinder, to impel the piston.   The complete action of the governor and throttle-valve depends greatly upon this circumstance.

The revolving pendulums are very sensitive of any changes in their motion; they are usually adjusted to make 36 revolutions per minute, but if the motion is increased to the rate of 39 revolutions, the balls will fly out to the utmost extent which is allowed by the iron guides or grooves, in which the arms E *e* E *e* move; or if the motion is diminished to the rate of 34 revolutions per minute, the balls will collapse as much as the grooves will permit; hence the power of regulation by the governor is confined to narrow limits, and when great irregularities of the motion take place, they must be corrected by other means.

The office of the governor is to correct small and casual derangements of the dynamic equilibrium of the forces of the machine, which may arise either from fluctuations of the resistance which is opposed by the mill-work, or from variations in the force of the engine. If the velocity of the engine is not greatly deranged from these causes, the corrective action of the governor will reduce the alterations until they become so small between one stroke and another, that the average motion of the engine, or the number of strokes it makes, during each minute, will be quite exact to the proposed standard for which the governor is adjusted. The governor, therefore, serves for an exact standard measure or regulator of the going of the engine, for it diminishes the variations which would take place without it, and keeps them within certain small limits, alternately above and below some average speed, which the engine will preserve very exactly.

Adjustment of the governor. The governor and the throttle-valve must be properly adjusted to each other, according to the resistance that the engine is to overcome, so that when the governor (and consequently the fly-wheel) is moving with the intended velocity, the throttle-valve shall be sufficiently opened to supply the cylinder with as much steam, as will produce that velocity; and then the governor will adjust that supply, so as to meet any casual alterations of velocity, and produce a correct average speed. A different quantity of steam will be required for every permanent alteration of the resistance, but the governor can make no change in the opening of the throttle-valve, except in consequence of a change of the velocity of the engine; and that ought to be avoided, by adjusting the connexion between the governor and the throttle-valve according to every permanent change of resistance. This is done by lengthening or shortening the wire or rod H w (p. 436) which connects the lever HF of the governor, with the handle w of the throttle-valve, a screw socket being provided in the rod for that purpose. Suppose that the resistance is permanently increased by applying more work to the mill, the engine will move slower, and then the governor will open the valve wider, and will keep it so, as long as the engine continues to move slow. In this case, the screw socket in the rod H w must be turned so as to lengthen the rod, and open the valve wider, until the motion of the engine is accelerated to its proper velocity, and then the governor will tend to keep it at that velocity. In like manner if the resistance is diminished, the engine will move quicker, and the governor will close the valve a little; but to retain the valve in that position, the engine must continue to move quicker, unless it is corrected, by turning the screw-socket so as to shorten the rod H w, and close the valve, until the proper speed is restored.

There are some purposes to which rotative engines are applied, whereby they are exposed to extreme variations of resistance; for instance, the machinery for breweries and distilleries, and some other manufactories, is required to perform a variety of operations at different times of the day; sometimes several pumps and machines are required to be actuated at once, but when that work is done, only one machine may be required to continue working. In mills for rolling iron, or lead, or copper, the changes of resistance are still greater, and more sudden, owing to the frequent application and cessation of nearly all the resistance at once. The keeper of such an engine must always be prepared to regulate its motion by the handles, so as to limit the opening of the valves, and prevent any dangerous acceleration of the velocity. In case any of the connecting axes or wheels of the mill-work should break, an engine may be suddenly relieved from the whole, or a great part of its usual resistance. In such circumstances the governor will not produce an adequate regulation, because the fly-wheel and the governor must be turned round very quickly, to make its pendulums expand sufficiently to close the throttle-valve as much as would restrain the velocity of the engine, so as to avoid the danger of breaking to pieces. Whenever the attendant observes that the engine is not sufficiently controlled by the governor, he must be in readiness to take the handles.

Small engines, whereof the pistons work with a short stroke of 3 or 4 feet, will bear to accelerate their motion very considerably, without much danger of breaking; but large engines, with 7 and 8 feet stroke of the pistons, having great masses of matter in the reciprocating parts, are very subject to accident, if they greatly exceed their usual velocity; for the energy which these parts acquire by their rapid motion, must be checked by the fly-wheel at every stroke, in order to bring them to rest, ready to return in an opposite direction; a very great strain is thus thrown upon the axis of the fly-wheel, the teeth of the sun and planet wheels, and the link which connects those wheels; also upon the connecting rod, and the great lever, with all their centres and connecting joints; and unless all the parts are excessively strong, some of them will be liable to give way, by the jerks to which they are exposed.

Engines which are applied to mills for grinding corn, or to spinning mills, and other similar purposes, are not exposed to great changes of resistance, and then the governor will produce a very complete regulation, and uniform motion.

Regulation of the valves and working gear. It is of the utmost importance to the performance of the engine, to exhaust the steam very promptly from the cylinder, at each turn of the stroke; for the strokes succeed each other with such rapidity, that no interval of rest can be allowed between them for the evacuation of the steam to be effected; but the time whilst that exhaustion is going on, is a period of imperfect action, for the motion of the piston, which takes place during that period, will be attended with only a partial exertion of the force of the steam, which is expended during such motion.

The force which puts the steam in motion, when the exhausting-valve is opened, is that of its own elasticity, which exerts itself to press equally in all directions against all the interior surfaces of boundary to the space which it occupies (see p. 53). The steam presses with a force of more than 15 pounds, against every square inch of the inside surface of the cylinder and of the piston, and whilst the exhausting-valve is shut, it must sustain that pressure against its surface; but when the valve is raised, the aperture which it opens presents scarcely any thing to resist the pressure, for the eduction-pipe and condenser only contain a vapour or rare steam of an elasticity of $1\frac{1}{2}$ lbs. per square inch, consequently the steam exerts an unbalanced force of $13\frac{1}{2}$ lbs. per square inch, which becomes active to urge the steam itself into motion, through the valve and eduction-pipe. The rush of the elastic steam through the valve, and along the pipe, is prodigiously rapid, so that it reaches the condenser in an exceedingly short space of time. The steam which is thus introduced into the condenser, being cooled and condensed, by the injection, almost as quickly as it arrives there, the exhaustion within the condenser is maintained without any material diminution; and the tendency for the steam to flow out of the cylinder into the condenser, continues nearly unabated, until scarcely any is left. For although the elastic force of the steam in the cylinder becomes less, yet the steam becomes lighter and rarer, at the same time, so as to require as much less force in proportion to give it motion. Hence the steam continues to flow through the pipe without any great diminution of velocity, from the time that the valve is opened, until the cylinder becomes exhausted to very nearly the same degree of rarefaction as exists in the condenser.

3 o 2

To obtain the full effect of the steam which is expended by an engine, its cylinder ought to be exhausted before the piston begins its motion ; and in an engine for pumping water, if it is sufficiently loaded, so as to act with a slow motion, the piston will wait a sensible time after the exhausting-valve is opened, and will not begin its motion until the steam has passed out of the cylinder, so as to leave it very nearly exhausted ; but an engine which is applied to turn a fly-wheel requires to move quickly, in order that it may produce a regular motion, and therefore less time can be allowed to effect the exhaustion.

It is a great advantage, that the action of the crank (or of the sun and planet wheels) causes the piston to move slowly, when it is at the beginning and end of its course, and whilst it is in the act of changing the direction of its motion ; because this period of slow motion is favourable for exhausting the cylinder, and to employ it to the greatest advantage, the exhausting-valve should be opened a little before the centre of the planet-wheel arrives at the vertical line, in order that the exhaustion may be commenced during the slow motion with which the piston terminates its stroke, and completed during the slow motion with which it begins to return for the succeeding stroke. If the working gear is properly adjusted, the exhaustion may be rendered very complete, before the piston has moved far, or acquired any considerable velocity ; but so long as any steam remains in the cylinder it must offer a resistance to the motion of the piston, even though that steam is in the act of quitting the cylinder ; for as the steam rushes out through the valve, by its only elasticity, that elasticity must exert itself against the piston, as much as it does at the aperture of the valve, where the steam urges itself into motion.

During the progress of the evacuation of the steam from the cylinder, the elastic force, with which it presses against the piston, will diminish from the full pressure of more than $15\frac{1}{2}$ lbs. per square inch, by a very rapid rate of decrease at first, and then with a more and more gradual rate of decrease, until the elasticity becomes only about $1\frac{1}{2}$ lbs. or nearly the same as in the condenser. It is obvious that the time in which the exhaustion can be effected, must depend upon the size of the aperture of the exhausting-valve and pipe, compared with the quantity of steam which is to pass through it.

Mr. Watt proportioned the sizes of the valves of his rotative engines, by the same rule that he had before adopted for the single engines (see p. 373). The diameter of the aperture opened by each valve, being one-fifth of the diameter of the cylinder, and consequently its area one twenty-fifth ; but allowing for the space occupied by the spindle of the valve, and by the cross bar which guides it, the real area of the passage, would be about one twenty-seventh. The single engines for pumping have the same proportion, but the pistons of those engines move with a slower motion than the double rotative engines, and the small intervals of rest, which take place between the strokes, allow more time for the steam to escape from the cylinder to the condenser ; hence, one twenty-seventh of the area of the cylinder is found to be a sufficient passage for the single engine, and the same has been adopted for the double engine, though it would have been better to have allowed a greater proportion.

Adjustment of the working gear. The length of the levers upon the axes *t u* of the handles, *r, s,* must be properly proportioned to that of the levers upon the axes of the spindles which enter into the valve boxes, and also to the radii of the sectors within the boxes ; in order that the valves may be lifted up the proper height to open the full passages for the steam, when the axes of the handles are turned round about one-fifth of a circle. The height that a valve should be lifted up, is one-fourth of the diameter of its aperture (as already stated, p. 374) ; and that diameter being one-fifth of the diameter of the cylinder, the valves should be lifted at least $\frac{1}{20}$ of the diameter of the cylinder.

The chocks 1 and 2 are fastened on the plug or air-pump rod l by screws, which pinch them upon the rod, but admit of placing the chocks higher or lower on the rod at pleasure. Their positions should be so adjusted, that when the centre of the planet-wheel is one-twelfth of the length of the stroke of the piston, distant from the vertical line of the centre of the sun-wheel, the acting chock should then have given the full motion to the acting handle, so as to have closed one pair of valves, and disengaged the catch to allow the weight of the other handle to begin to open the other pair of valves. For instance, if the piston makes a 4 feet stroke, then the working gear should be adjusted so as to close one pair of valves, and release the other pair, by the time that the centre of the planet-wheel has approached within 4 inches of the vertical line, through the centre of the sun-wheel. This is equal to an angular motion of $9\frac{1}{2}$ degrees, or allowing for a little looseness of the teeth, we may say that the valves are changed when the planet-wheel has still to describe 10 degrees of angular motion in its orbit, to arrive at the vertical line. Supposing the wheel to circulate in that orbit with uniform motion, this will be $\frac{1}{18}$ of the time occupied in performing half a stroke, which, if the engine makes 25 strokes per minute, will be $\frac{1}{900}$ part of a minute, or $\frac{1}{15}$ of a second. This is a very short interval of time, but it is of consequence to the performance of the engine that it should be well employed in opening the valves, and allowing part of the steam to pass out of the cylinder, that it may become exhausted before the piston has moved far on its course. The motion of the piston during this $\frac{1}{18}$ part of the time of half a stroke is imperceptible, being only $\frac{1}{132}$ part of the length of the stroke.

When the working gear is adjusted in this manner very little effect of the steam will be lost, for the exhaustion being begun before the piston arrives at the termination of its course, will be completed very soon after the piston begins to return, and before it has performed any considerable portion of its whole stroke, so that the motion of the piston is not materially impeded by any steam remaining in the exhausted end of the cylinder. On the other hand, if the exhausting-valves are not opened suddenly, and at the proper time, and to their full extent, or if the apertures are too small, or if the passage through the eduction-pipe is contracted, then the steam will take more time to get away out of the cylinder, so that the piston will begin its returning stroke with a considerable resistance from the un-exhausted steam which remains in the cylinder, though in the act of escaping from it (a). If the piston is thus allowed to overtake the escaping steam, a portion of the stroke will be performed with diminished effect, before the resistance of the unexhausted steam is removed; and the force of the fresh steam, which is entering from the boiler into one end of the cylinder, to impel the piston before it, will be partly expended, in assisting to drive out the steam from the other end of the cylinder, through the contracted valve and passages, with a quicker motion than

(a) This deficiency, or want of rapidity of exhaustion, is one of the most common defects of steam-engines, and the cause of it is very imperfectly understood by the common makers of those machines, who esteem an engine to be in good order if the barometer shows a good vacuum or exhaustion in the condenser, without considering whether the exhaustion is good and quick in the cylinder. This may be tried by an instrument invented by Mr. Watt, and called the Indicator.

The modern steam-engines are by their construction less capable of speedy exhaustion of the cylinder than the original construction, with four valves and hand-gear. Mr. Watt was very particular in attending to this circumstance in his engines, and hence his plan of inverted exhausting-valves (see p. 433); and it is from this cause, that the old engines which he made and adjusted under his own superintendence, perform better than modern engines usually do.

the impulse of its own elasticity can give to it ; whereas the steam ought to escape, and exhaust the cylinder, merely by its own elasticity, before any sensible motion of the piston takes place, and therefore without occasioning any sensible opposition to its motion.

Although it is so important to open the exhaustion-valves suddenly, and to their full extent, in order to exhaust the cylinder promptly, the steam-valves do not require to be opened so fully, because a much smaller aperture than one twenty-seventh of the area of the piston will admit the steam into the cylinder quick enough to follow the piston, so as to fill up the space left by its motion, and that is all that can be required ; in fact, the steam has all the time of performing the stroke to enter into and fill the cylinder, but there is scarcely any time which can be allowed for exhausting that steam from the cylinder, without a deduction from the performance as before explained, and hence the necessity of using every precaution which can expedite the exhaustion.   It is usual in Mr. Watt's engines to make the steam-valves of the same size as the exhausting-valves, and to open them to the same height ; but the flow of steam through the steam-valves is continually impeded by the throttle-valve, whilst there is nothing to retard the rush of the steam through the exhaustion-valves (*a*).

All those parts of the double engine which sustain any rubbing or friction, are made adjustable at pleasure, so as to make up for the wear of the rubbing surfaces, and it is also provided, that when those rubbing parts are greatly worn, they may be taken out and renewed by fresh ones.   For instance, the piston, and the air-pump bucket, are packed with hemp rammed very hard into grooves round their circumferences, at the parts where they rub against the interior surfaces of the cylinder and pump ; this hemp can be renewed when it is worn, and at any time when the packing of the piston becomes loose, it can be compressed and moulded anew to the size of the cylinder, by screwing down the piston cover, so as to keep the packing always very solid, and extended to the full size of the cylinder.   The

---

(*a*) It would have been better to have made the apertures of the steam-valves rather smaller than they were, or about one thirty-sixth of the area of the cylinder, and the exhausting-valves larger, or about one-eighteenth of that area, the passage through the eduction-pipe being in every part of the same size, or rather larger.   With these proportions the exhaustion of the cylinder would be more quickly performed, and the force requisite to lift each part of valves would be no greater than usual, because the area of the steam-valve would be diminished, as much as that of the exhausting-valve would be increased.   The apertures of the exhausting-valves can never be too large.

The promptitude of the exhaustion of the cylinder depends greatly upon the velocity of the motion of the piston, and upon the force of resistance it has to overcome ; for that size of the exhausting apertures, which would be fully sufficient for an engine which acts with a slow motion, or which is lightly loaded with work, would be too small to obtain a good performance, with a quick motion, and with a full load of resistance.   Because there is less time for evacuation when the motion is quicker ; and when a greater resistance is to be overcome, a greater quantity of steam must be admitted into the cylinder, so as to fill it with denser and more elastic steam ; and then larger passages will be requisite to discharge that steam ; or else more time will be spent in the discharge.

As the steam passes out of the cylinder into the condenser by its own elastic force, it will move with nearly the same velocity, whatever may be its density ; for when the elastic force is greater, the density, or weight of steam to be urged into motion by that force, will be greater in nearly the same proportion.   The apertures of the valves and passages should be proportioned to the quantity of steam which is to pass through them in a given time, and that will be in proportion to the power exerted by the engine.   Mr. Watt's proportion of making the area of the aperture one twenty-seventh of that of the piston, gives about 8 tenths of a square inch of exhausting aperture for each horse power exerted by an engine, whereof the cylinder is so proportioned as to expend 33 cubic feet of steam per minute, for each horse-power, according to Mr. Watt's practice, as stated in p. 440.   The above proposal of one thirty-sixth for the steam passages, would be about 6 tenths of a square inch to a horse-power ; and one eighteenth, for the exhausting passages, would give about 1·2 square inches to each horse power, taking an average of engines of different sizes.

stuffing boxes have the same property. The valves which are fitted into their seats by grinding with emery, can be repaired by the same means when worn; and when they are too much worn to be repaired, the seats can be removed out of the boxes of the nossels, and new ones put into their places with new valves.

It has been already stated (p. 437), that the first engines which Mr. Watt made for the breweries in London, in 1784 and 5, were single engines; it does not appear why he adopted the single action, after having made out the invention of the double action, unless it was that patterns were prepared for forming the nossels and valves of single engines, and that there was not time to get double valves made for that set of engines. All these first single engines were altered in the course of a few years to double, except one, which worked in its original state for 40 years, and then it was altered like the others.

In adapting the steam-engine to work with double action, it became requisite to fit all the connecting joints with great accuracy into their sockets; in the single action the joints being always pulled in one direction, chains answer the purpose very well, for the pins of the joints, however loosely they may be fitted into their sockets, are certain to come to a bearing. In the double engine the action is alternately in both directions, and the most exact fitting is requisite, for if there is any sensible looseness in the fitting of each joint, a violent jerk will take place in the sun and planet wheels, and all other parts, at every change of the direction of the motion, so as to endanger the breaking of the work to pieces. This evil is very great, even when there is only a trifling looseness in each joint of the parallel motion, the great lever, the connecting rod, and the sun and planet wheels, because it will be all accumulated into one place, so as to have the effect of a considerable play, at the change of the stroke; whereby the piston will move a sensible space before it is opposed by the resistance which is applied to the axis of the fly-wheel, and then it will act with violence and accumulated force, to urge the machinery forwards. The piston will in such case exert a much greater force to strain and break the parts by jerks, than can happen by its proper action, if no play is allowed in the joints.

To attain this accuracy of fitting Mr. Watt was obliged to contrive every joint in his double engine, so as to admit of exactly adapting the size of the socket to that of the pin, and to make a good fitting in all cases, by closing the socket about the pin, whenever any diminution has taken place by wearing. For this purpose, all the connecting joints between the piston rod and the fly-wheel, viz. those of the parallel motion, the great lever, the connecting rod, the sun and planet wheels, the link which connects their centres, and the bearings for the axis of the fly-wheel, are fitted into brass sockets, capable of adjustment; each socket is composed of two pieces of hard brass or gun metal, accurately fitted into a cell or groove, in which one of the brasses can be slided or moved towards the other, by means of a wedge, or of screws; each brass has a semicylindrical cavity in it, and the two cavities being opposite to each other, they form a sufficient socket to receive the cylindrical joint pin, which is always made of iron, either wrought iron, or cast iron; and by means of the wedge or screws, one of the two brasses can at any time be tightened round the pin, so as to enclose it quite tight between the two cavities, and prevent any shake or looseness, which, in an engine that acts both in ascending and descending, would soon be destructive of its parts.

The two great links K of the parallel motion, are each composed of a strap or bar of iron, bent so as to make a fork or double link; the upper part of the bend or loop, receives the two brasses which form the socket for the joint pin to connect the links with the end of the great lever; and the lower end of each forked link receives two other brasses to form the sockets for the ends of the cross pin o which is fastened on the top of the piston rod. All these four brasses are accurately fitted into the space between the two prongs of the forked link, and the brasses have projections which apply to each side of those prongs, to prevent any motion sideways.

The lower brass for the upper joint, and the upper brass for the lower joint, are kept extended to their proper distance asunder, by a piece of wood which is fitted in between the prongs; and all the four brasses, for both joints, are retained in the fork, by means of a wedge, or cross key, put through mortices made in the two prongs at their lower ends, so that by driving the wedge farther into the mortices, the brasses can be drawn tight round both the joint pins at once : for by means of the piece of wood, the wedge closes both brasses, so as to make both joints tight at the same time. To confine the lower ends of the two prongs of the forked link at their proper distance, and prevent them from springing outwards from each other, a short piece of iron called a *keeper*, is inserted into the lower ends of the mortices, beneath the wedge; this keeper has a hook at each end, and they project out beyond the outside of the prongs, so as to hook over the edges of the mortices, and bind the prongs together, with the brasses tight between them; the mortices are long enough, up and down, to allow the hooked ends of the keeper to be introduced through them, but when the wedge is put in, over the top of the keeper it nearly fills up the mortices, and also presses the four brasses towards each other. The link q at the back of the sun and planet wheel, which connects the centre pin of the planet wheel, with the axis of the sun wheel, is a double or forked link, with two pair of brasses and a cross key and keeper of the same kind as the links of the parallel motion.

The sockets for the pivots at each end of the axis of the great lever, and also those for the necks of the axis of the fly-wheel, are each composed of two brasses fitted into a cell or box of cast-iron, with a cover, or cap, fastened down upon the box by two screws, and pressing upon the upper brass, so that by turning the screws, the brasses can be advanced towards each other, in order to confine the pivot. The sockets for the joint by which the connecting rod is united to the end of the great lever, have brasses confined by caps, with two screw bolts and nuts, to each cap.

Mr. Watt also contrived the main joints at the ends of the great lever, so as to form universal joints, having liberty of motion laterally, as well as in the direction of the vertical plane of the great lever. This was a provision against any insensible deviation of the axis of the great lever, or of the fly-wheel, from the horizontal position, owing to the wearing away of its brasses, or from any settlement or partial sinking of the frame-work; or for any trifling deviation of the cylinder from the vertical. The universal joint is made by means of a joint pin, formed in one piece of iron, like a cross, with four pivots, their axes crossing each other at right angles; this cross is fastened beneath the underside of the great lever, by two sockets, which receive two of the pivots, and hold them parallel to the length of the lever, whilst the other two pivots project out sideways, parallel to the axis p of the lever, and upon these pivots the upper end of the cast-iron connecting rod is suspended, being forked as shown in figure 5, Plate XII., in order to reach to both the pivots. By this means the two latter joints allow the requisite motion of the connecting rod in its regular course of working, whilst the other two joints of the cross, allow for any small lateral flexure which may become necessary from derangement of the parallelism of the axis of the great lever. The connecting rod joint was always made thus universal, by Messrs. Boulton and Watt; and, in some engines, the upper joint pin of the links K of the parallel motion was fitted in the same manner to the underside of the great lever; but in general that pin was straight and fastened to the underside of the lever by a socket, with two bolts which bind the two halves of the socket together round the middle of the pin, by nuts screwed on the upper ends of the bolts, as is shown in figure 4, plate XII.

In all these details of execution, and the several provisions to guard against the effects of wearing and derangement, Mr. Watt showed no less judgment than in the great outlines of his invention.   We may compare his rotative engine, in the state above described, with Newcomen's engine in its greatest perfection, as made by Mr. Smeaton, and considering the degree of invention requisite to produce all the new properties of the improved machine, from its original state, and also taking into account the great perfection with which it is adapted to its purpose, the rotative steam-engine will be found to be the greatest step of useful invention ever made by one individual, at any time in the history of the arts.

Messrs. Boulton and Watt carried on a very extensive business at Soho, during many years, in manufacturing their patent rotative steam-engines of the kind before described.   These engines were found to answer so well for impelling the machinery of mills and manufactories, that they were set up in all parts of the kingdom, and they contributed very much to that great extension of trade and commerce which took place at the time of their introduction, and which has been going on ever since with a more rapid progression.   Many engines were sent abroad by the patentees, and were brought into successful use in foreign countries.

It is a circumstance highly creditable to the character of Mr. Watt as an inventor and as an engineer, that his first double rotative engine which he made in 1787 at the Albion Mills, performed quite as well as any engine which has since been constructed, to employ steam on the same principle.   Some important improvements have been made in the construction of modern engines, by substituting cast-iron, and stone work, in place of wood ; and by putting the parts together in more substantial modes ; but all those essential forms and proportions which affect the performance of the machine were so ascertained by the first inventor, that no improvement has been since made in them, and every departure from those forms and proportions has impaired the performance, in a greater or lesser degree (a).

(a) The best modern steam-engines have nearly the same proportions and dimensions that Mr. Watt established for the furnace, the boiler, the cylinder and piston, the air-pump and condenser, and the pipes to convey the steam, from one of these vessels to another.   The parallel motion, great lever, and connecting rod, are used to transmit the reciprocating motion of the piston, to the fly-wheel, and turn round it with continuous rotative motion.   The sun and planet wheels, which occasioned the fly-wheel to make two revolutions for each complete stroke of the piston, have been laid aside, in favour of a simple crank, which produces only one revolution of the fly-wheel for each stroke ; the fly-wheel then requires to be four times the weight of that of the original plan, to give it the same power of regulating the motion.   It is a common fault to make the fly-wheel lighter than four times the weight of Mr. Watt's fly-wheel, and then the motion of the engine is rendered less uniform.   The modern engines are thus rendered simpler, more durable and more easily kept in order, but there is rather a deficiency, than an improvement in the performance.

The four independent circular valves for distributing the steam to the cylinder, have been changed for two combined sliding valves : this change has been attended with some advantage, as to the simplicity of the action, ease of management, durability, and above all in the symmetry of external appearance ; but the evacuation of the steam from the cylinder is not effected so promptly by the modern sliding valves, as by the original plan of four lifting valves ; and hence the useful effect of the modern engines is sensibly less, than that of the old ones of the same size, whilst the consumption of fuel is sensibly greater, so that the mechanical effect produced by the consumption of a given quantity of fuel, is considerably less in modern engines, than in the old ones.

The substitution of iron in place of wood, for the great lever and for the fixed framing, are great improvements as to the permanency of the machinery, but do not affect the performance ; a wooden-framed engine being just as good as an iron one, for so long as the wood continues firm and solid.

3 P

In Mr. Watt's large rotative engines, the great lever was supported upon a wall of solid masonry, which extended across the house, and divided it into two apartments, one containing the cylinder, and the other the fly-wheel. Engines of 36 horse-power and upwards were constructed in this way, but smaller engines were set up in wood framing, in the manner represented in Plate XI. and already described, p. 455. The 20 horse engine, which was usually referred to as a standard size, was always mounted in such a wooden framing; and the following sizes were also made in that manner, 4, 6, 8, 10, 12, 14, 16, 20, 24, 26, and 30 horse-power, the lengths of the stroke of the pistons being from 3 to 5 and 6 feet.

It was common to place the wood framed engine in a house of more than double the width of that represented in the engravings, in order to have the boiler in the same house, at the side of the engine, instead of outside the walls of the house. Only one boiler was used, and in the first engines, the boiler was made of copper in the same circular form as was used by Mr. Smeaton : this boiler occupied ᴜᴇ corner of the wide engine-house opposite to the cylinder of the engine; and the space before the boiler, opposite to the sun and planet wheels, gave access to the furnace. The engine was placed along one of the side walls of the house, and occupied all its length. The main cross-beam W, beneath the centre of the great lever, extended all across the house, from one side wall to the other; and the other cross-beams V, and the beams X X to support the cylinder, were worked at their ends into the walls which enclosed the boiler.

The larger sized engines of 36, 40, 45, 50, 60, 80, and 100 horse-power, having the lengths of stroke 6, 7, 8, or 9 feet, were supported by the masonry of the building, in the same manner as the engine represented in Plate X., and described p. 356; viz. the cylinder is mounted upon a massive pier of stone-work, and fastened down thereto, by three or four bolts, which pass down through the whole depth of the pier, to its foundation. The fulcrum of the great lever is sustained upon a lever wall, built very solidly, and the bearings for the axis of the lever, are held down by four long bolts, which pass down in the thickness of the wall from top to bottom, so as to prevent the fulcrum from rising up, unless it lifts all the weight of the wall; nor can the fulcrum sink, unless the masonry of the wall could yield to compression. The spring beams are worked into the end walls of the house, in the manner before described, and extend all the length of the building, being sustained upon the lever wall in the middle.

The house for the large rotative engine is long enough to contain all the engine, in the manner represented in Plate XI., but the lever wall being built across the middle of the house, in place of the posts V V, it divides the house into two spaces; the lever wall does not rise higher than the spring beams U U, so that the upper part of the house is all one room. A floor is fixed in the house around the cylinder, at a convenient height for packing the piston, and another at the level of the cylinder pier, to give access to the working gear : this floor extends over the space, in which the condensing cistern is placed, between the lever wall and the cylinder pier. The building for large house built engines, is never wider than is requisite for the engine, the boiler being always placed outside of the walls of the house.

Management of the piston. To keep an engine in good order, the state of the hemp packing of the piston must be carefully attended to, and it must be so managed as to prevent any looseness taking place in the fitting between the piston and the cylinder, for that would occasion loss of steam by leakage; and on the other hand, the hemp packing must never be fitted too tight into the cylinder, or it would occasion a loss of force by friction. The hemp should be rammed very hard and solidly into the groove round the piston, at first; and if it is not too much confined by the piston cover, the motion and friction against the inside of the cylinder will loosen the packing sufficiently, after the piston has made a few strokes, so that it will work properly during the first two or three days; in that time, the hemp will be consolidated into a mass, so as to become rather loose in the groove, and then the piston cover should be screwed down as much closer as is necessary to confine the hemp properly in its groove. It will be requisite occasionally to repeat this screwing down of the piston cover, in order to make good for the diminution of the hemp, by the wearing away of its outside surface, after it has become settled into a solid mass, so as not to be susceptible of further consolidation by the friction to which it is subjected.

In a regular course of working, if the cylinder is true and smooth withinside, and if the piston is properly supplied with grease, the piston cover will require to be screwed down about once every fortnight, in order to keep the packing solid and make up for its diminution by wearing at the outside. If the cylinder is at all untrue, or tapering, or if the metal is porous and spongy, so as to have any small pin holes, or cavities withinside, then the hemp packing will wear away more rapidly, and the piston cover will require to be screwed down frequently. The hemp packing will require to be renewed about every two or three months, if the cylinder is a good one, and if careful management has been used; but in a rough cylinder the hemp may be cut away so fast, as to occasion great trouble, and to render it impracticable to keep the piston in any tolerable order; hence, it is of the greatest importance to have a very true and smoothly polished cylinder.

To give access to the piston, the cylinder cover must be unscrewed, and lifted up by the windlass tackles; for this purpose two eye-bolts are screwed into the upper side of the cylinder cover, near its circumference, at opposite sides of the piston rod; these eyes are to receive hooks at the ends of two ropes, which pass over two pulleys, suspended from a cross-beam in the roof of the engine-house, over the cylinder, and the other ends of both ropes are wound round the same roller or barrel of the windlass, one on each end of it. This roller is placed horizontally and across, opposite to the cylinder end of the great lever, the pivots of the roller being supported by two upright posts of wood, which are erected upon the spring beams U U, and affixed at their upper ends to a cross-beam in the roof. A large cog-wheel is fixed on one end of the roller, and it is turned round by a small pinion, fixed on a second axis, which is mounted between the two posts, and which has a winch handle fitted on each end, for two or more men to turn it by. The two ropes which descend from the pulleys at the roof, to the eye-bolts in the cylinder cover, pass clear down on each side of the great lever, and parallel motion, but within the spring beams. When the winch handles are turned round, by two or more men, a slow motion is communicated to the roller, and it winds up the two ropes, with a proportionable increase of force, so as to lift the weight of the cylinder cover, and slide it upwards upon the piston-rod; if the two ropes are properly adjusted to one length, they will take up the cover in a horizontal position.

The gasket and white-lead which is interposed in the joint between the cylinder cover, and the flange at the top of the cylinder, becoming dried by the heat, will then adhere to the iron, and cement the cover so very firmly to the flange, that in a large engine it would be difficult to detach the cover by the force of the windlass alone, although all the screws are removed from the flange. The usual practice is to take up the cylinder cover immediately after the engine has ceased working, and to keep up the steam to its full elasticity, in order that the steam may be applied in aid of the windlass to lift the cover. For this purpose the engine being stopped with the piston very near the top of the cylinder, the fly-wheel must be turned round so much, as to place the piston at the very highest part of its course. All the four valves being shut, the nuts of the screws around the edge of the cylinder cover are to be removed, the ropes of the windlass tackles hooked to the eyes in the cylinder cover, and the force of the windlass applied to lift it ; then by opening the upper steam-valve, the steam is admitted into the upper part of the cylinder, where it acts by its elastic force, between the upper surface of the piston, and the under surface of the cylinder cover ; and as the piston cannot descend (because the centre of the planet-wheel is exactly under the centre of the sun-wheel), the steam will rarely fail to force up the cylinder cover, and detach it from its place, with little or no effort at the windlass.

The cylinder cover may then be hoisted up upon the piston-rod, by the force of the windlass, until it is raised high enough to give access to the piston beneath it ; but before the men begin to work at the piston, the cylinder cover should be propped up very securely, to avoid any chance of its falling down upon the men. The prop is a square trunk of wood, which is applied round the piston-rod, in an upright position ; it consists of three boards, nailed together like a trough, of a suitable size to fit round the piston-rod, and a fourth side is attached to it, with hinges like the lid of a box, and it fastens with hasps. When the cylinder cover is raised high enough up on the piston-rod, this trough is applied round the rod, so as to inclose it within the trough, which is fastened by the hasps of its lid ; and then the cylinder cover is lowered down, until its weight rests upon the upper end of the trough, the lower end of which rests upon the central part of the piston ; this makes a very secure prop, and being in the centre of the cylinder it is out of the way of the men.

If it is not required to pack the piston, but merely to screw its cover down, it is done by turning the screws round, by a wrench applied to their square heads ; the piston cover is thus pressed down, with great force upon the hemp which is lodged in the groove round the edge of the piston, and the form of the top and bottom of that groove being adapted to cause the packing to spread outwards by the pressure, it will expand against the interior surface of the cylinder, so as to make a close fitting between them, but care must be taken to avoid compressing the packing any more than is requisite for that purpose, lest it should occasion the piston to move with unnecessary friction.

When the packing of the piston requires to be renewed, the piston cover must be lifted out of its place by the windlass. The piston cover is a flat circular plate, or broad ring of cast-iron, which is applied upon the piston, and extends over all the circumference of its upper surface ; there is a large circular hole in the centre of the cover, which reduces it to a broad flat ring, and the underside of this ring has a deep rim projecting downwards, all round its circumference, so as to fit round over the circumference of the upper part of the piston, in order to fill up, and cover over, the annular groove, or space between the edge of the piston and the inside of the cylinder, where the hemp packing is to be lodged (see a section of the piston and its cover at J J in figure 8, Plate XIII.). The lower part of the piston is formed with a projecting rim, which is large enough to fill the cylinder, and form the bottom of the annular groove, whilst the projecting rim at the under side of the piston cover, which also fills the cylinder, forms the top of that groove. The circular edge of the projecting rim at the bottom of the piston, as well as that of the piston cover, are turned in a lathe and accurately fitted into the cylinder, so as to prevent the hemp passing those edges and getting out of its groove. The piston cover is confined by means of four, six, or eight screws, according to the size of the engine, which pass down through the flat plate of the cover, and screw into nuts which are fixed into the cast-iron piston; these screws have square heads to turn them with a wrench, and then they force the projecting rim of the cover down upon the hemp which is contained in the groove, so as to compress it very forcibly; and the upper and lower parts of that groove being made rounding or inclined, the pressure has a tendency to spread the packing outwards, against the inside of the cylinder. To take up the piston cover after the cylinder cover is raised and propped up, two temporary eye-bolts are screwed into two holes tapped into it, at opposite sides of its circumference; these eyes receive the hooks of the windlass tackles by which the ring can be lifted up out of the groove, in order to take out the old packing, and put in new.

It is not necessary to have windlasses to small engines of ten or twelve horse-power, such as is represented in Plate XI., because the cylinder cover, after it has been detached by the force of the steam, may be lifted up by turning the fly-wheel round. The piston being first placed near to the bottom of the cylinder, the cover may be fastened by a rope to the joint-pin *o* of the parallel motion, and then by turning the fly-wheel round, the cover may be raised up, along with the piston, and afterwards let down again by the same means. The piston cover being light, may be lifted out by two men, without tackle.

Manner of packing the piston. The hemp which is used for packing the piston, is prepared by first spinning it into large threads, or yarns, about $\frac{1}{8}$ of an inch diameter; about 30 or 36 of these yarns are then twisted together, to form a round strand of about $1\frac{1}{8}$ inch diameter, in the same manner as if it were intended to form a rope, except that the hemp is not twisted so hard as for a rope; lastly, four or five of these strands are plaited together, and form what is called a gasket, or soft flat plaited band, about $1\frac{1}{4}$ inch thick, by $2\frac{1}{4}$ inches broad, which is adapted to fill the annular groove round the piston; this groove being about 2 inches wide between the cylinder and the piston, the gasket when bent circularly, and rammed down flat ways into the groove, will fill the same very completely.

The gasket is coiled spirally round the piston, each coil lying upon those beneath it, until the depth of the groove is filled up. The end of the gasket which is first put in, to fill the narrow bottom of the groove, is made of a smaller size, by reducing the number of yarns in each strand; and when one coil of the small part of the gasket is rammed in, at the bottom of the groove, a second coil of the same gasket, but which is of a larger size, is put in and rammed down, and then a third coil which is of the full size. Every coil that is put in, is well rammed down before another coil is laid upon it. The best plan of ramming down the packing for a small engine, is with a drift of hard wood inserted into the groove, over the hemp, and driven by a mallet; but as this is tedious for a large engine, which requires a great deal of ramming, it is usual to employ an iron rammer with a long handle, which the man holds in both hands, and applying it in an upright position against the inside of the cylinder, he rams it down upon the hemp with hard blows, taking care to strike vertically, so as to avoid battering the inside of the cylinder, or the edge of the piston. The gasket should be rammed down very solid, and it is usual to sprinkle black-lead powder, and to smear tallow-grease upon the different coils of the gasket, to fill up the interstices, and more effectually prevent the penetration of the steam or melted grease through the packing.

If the gasket happens to be in two or three lengths, the ends where the lengths join, should be neatly tapered away, and overlapped, so as to make up the same thickness as the other part of the gasket. In filling and ramming the gasket, particular care must be taken, to insert an equal quantity at every part of the circumference of the piston, so as not to force it against any one side of the cylinder; for it is not intended that the iron rims at the top and bottom of the piston should touch the cylinder at all, but only the hemp which is included between those rims; and this may be done if the packing is carefully applied. An iron gauge should be fitted into the width of the groove, when the piston occupies the exact centre of the cylinder; and by occasionally inserting this gauge into the groove at different places, during the operation, it will be observed if the packing is becoming too full at any one part, and it may be corrected accordingly.

When the groove is thus filled with coils of gasket, the piston cover may be lowered down into its place, and fastened down by its screws, with a sufficient pressure to confine the hemp. These screws are liable to work loose, and turn back, in a double engine, by the continual reciprocating motion of the piston; but to prevent this, a circular ring of iron plate is applied over the heads of all the screws; and this ring has a suitable number of square holes cut through it, to fit round and encompass the square heads of the screws, and thus confine them, so that they cannot turn round, until the ring is lifted up clear above the heads.

It has been before remarked, p. 327, that the hemp packing is not soft or elastic, but that it forms a solid plug of a very hard ligneous substance, which is accurately fitted into the cylinder by moulding it therein, and which can at all times be restored to its original size, by moulding it anew into its place. Hence the exactness of the fitting depends chiefly upon the accuracy and polish of the cylinder.

At first starting the piston with fresh packing, it moves exceedingly stiff, but it soon works itself free, because the hardest contact at first, is in the prominent parts of the folds or plaits of the gasket, but the friction soon produces a uniformity of bearing. By the above method of applying the hemp around the piston, all the fibres are bent or doubled into loops, or small spiral coils; and the convex folded parts of those loops, form the contact with the internal surface of the cylinder, so that the rubbing action which they must endure, when the piston moves up and down, is in a direction obliquely across the lengths of the fibres. The hemp is found to resist wearing for a long time in this way, and when the first or outside fibres are worn through at the folds, the other fibres, which are behind them, make good the fitting, whilst the stumps or ends of the worn fibres apply to the cylinder, so as to make a still better fitting between the folds of the plaits, and at the edges of the different coils of gasket. For this reason, the fitting of the piston is improved by the wearing of the hemp to a certain extent, provided that a suitable compression is always preserved, by screwing down the piston cover as often as is necessary.

The outside of the hemp packing, where it applies to the inside of the cylinder, being composed of successive coils of gasket, it will be most solid and hard at the middle part of the thickness of each coil, and less dense at the edges and in the jointings between the coils; this difference of solidity diminishes as the packing wears away; but after a certain proportion of the breadth of the gasket is worn away from the outer edge, the pressure of the piston cover, will no longer be able to expand the hemp outwards, sufficiently to fill the cylinder, and then the packing must be renewed.

The pistons of engines which work with a short stroke, and make a great number of reciprocations, are observed to wear out their packing sooner than engines with a long stroke, which make fewer reciprocations; although the pistons of the latter pass through the greatest space in a given time. The packing of single engines is found to last longer than that of double engines, because the strain being always in one direction in a single engine, the packing settles down more securely in its groove, and does not require so much confinement by the piston cover; but in double engines, if the packing is in the least degree loose, the steam will alternately press it upwards and downwards, towards the top or the bottom of the groove, so as to unsettle it, and facilitate the wearing of the hemp. This is shown by the necessity of confining the screws of the piston cover from turning back, but such a provision is not essential in single engines.

Manner of packing the air-pump bucket. The edge of the bucket has a groove formed round it, which is filled by winding a small gasket spirally round in the groove, until it is quite filled; this is done when the bucket is drawn out of the barrel: the hemp is consolidated by striking it with a mallet as it is wound on; the ends of the gasket are secured by inserting them into holes in the edges of the groove, and fastening them in, by driving wooden pegs into those holes at the side of the gasket. The gasket is also secured in other parts of the groove, by driving wooden pegs through it, into holes in the iron bucket. When the packing is finished, the bucket is lowered down into the barrel of the air-pump, which is made rather larger at the upper part, above the part where the bucket works, and that enlarged part joins to the working part of the barrel with a bell mouth, in order to facilitate the entrance of the bucket into the barrel.

This method of packing the air-pump bucket for a large engine is very troublesome, particularly when the inside of the barrel becomes worn by long use, so as to be a little larger in the working part, than at the top; it is then impracticable to insert the bucket with a sufficiency of packing about it, to make it fit tight. It is a much better method to apply a packing ring to the air-pump bucket, with screws in the same manner as to the piston, and then the bucket may be packed without drawing it out; but this was not the practice of Messrs. Boulton and Watt in their engines, and has only been very lately adopted.

Method of proving the piston and valves. It is necessary to make occasional trials of the piston and valves, to ascertain whether they fit tight, or whether they will allow steam to pass. For this purpose, the steam must be kept up to its full elasticity, after the engine is stopped; and the doors or covers for the two exhausting boxes h and i must be removed, by unscrewing the nuts from the two bolts which confine each cover in its place. Two pieces of wood may be applied across the framing, between the arms of the fly-wheel, to prevent it from turning round; and then by moving the handles r and s, so as to alternately open either pair of the steam and exhausting valves, it will be seen if the fittings are tight; for if they are not, the steam will issue through the open doors h or i.

For instance, if both the handles r and s are raised, as in fig. 8, Plate XIII., the steam will be admitted through the upper steam-valve b into the top of the cylinder, so as to press downwards upon the piston, and also upon the upper exhausting valve at h; then if the packing of the piston is leaky, the steam which passes by it will issue from the lower opening at i, after passing through the lower exhausting valve i, which is open. Also if the upper exhausting valve h is leaky, the steam which escapes through it will, at the same time, blow out through the upper opening at h. If no steam whatever escapes through the lower opening i, the same trial also proves that the lower steam valve e is tight, as well as the piston.

The two handles r and s may then be put down, as in fig. 9, and the steam will be admitted through the lower steam-valve e into the bottom of the cylinder, so as to press upwards beneath the piston, and also upon the lower exhausting-valve i: if that valve is leaky, the steam will issue at the lower opening i; and if the piston or the upper steam-valve b are leaky, the steam will escape at the upper opening h.

If steam does escape during these trials, it may still remain uncertain whether the leakage is through the steam-valves, or by the piston. If the leakage is considerable, it will most probably be through the piston, because the valves can scarcely ever be so leaky as to allow a very considerable quantity of steam to pass. Also if the leakage is found by repeated trials to be as great when the valves are placed in the position, fig. 9, for making the up stroke, as when they are set like fig. 8 for the down stroke, then it is most likely that the leakage is in the piston. The sound of the steam which escapes, will also afford some indication of the place of the leakage.

Under these circumstances the cylinder cover should be taken up, to expose the piston, and then the valves being placed properly for making the upward stroke, as in fig. 9, the state of the packing will be apparent, and the defective part of the circumference will show itself, if there is any leakage. At the same time, the tightness of the upper steam-valve b will be proved; for if any steam leaks through that valve, it will issue through the passage c, and show itself in the open cylinder.

When it is required to ascertain the state of the steam-valves, without reference to that of the piston, the two covers to the middle compartments c and f of the nossels must be removed, and then both the steam-valves being shut, the steam will issue at the openings, if there is any leakage, and it may be seen where it proceeds from. The tightness of the throttle-valve may be tried at the same time, by closing it, and then by opening one or both of the steam-valves b or e, the quantity of steam which passes by the throttle-valve may be observed. It is not expected that the throttle-valve will fit quite tight; for it is of no consequence if it does allow a little steam to pass by it; but it ought to move very freely, without any danger of sticking when it is shut.

In making these trials of the valves, they should be repeatedly opened to their full extent, and then shut again, with a similar motion to that of their regular action in working; this is to prove that they will always drop correctly into their seats, for if the spindles of the valves are worn very loose in their sockets, the valves may be subject to drop on one side of their seats occasionally, so as to leak; although they may fit tolerably, when they happen to fall correctly into their places. Whenever this appears to be the case, the sockets for the spindles of the valves ought to be refitted. The state of the fitting of the spindles of the valves should be occasionally examined, by actual inspection of the valves through the doors of the nossels; for if the spindles wear loose, the fittings of the circumferences of the valves must soon be destroyed, owing to the valves frequently dropping to one side of their seats, and afterwards sliding and settling into their places by the pressure of the steam, the conical fitting will thus be worn out of the true circular form. It is of essential importance to the good management of an engine, to prevent any looseness taking place in the spindles of the valves, for that is the part where their defective fitting commences.

If an engine is required to be in good order, there should not be any sensible leakage through the valves, or the piston, during these trials; for it must be kept in mind, that when the engine is at work, the pressure with which the steam will be urged through any leaks, will be five or six times as great as takes place during the trials, because there is then no exhaustion or vacuum. To make the trials as efficacious as possible in that respect, an extra weight may be applied on the stem of the safety-valve V, and the steam should be kept up to the greatest elasticity above the atmospherical pressure, that the height of the feeding cistern T of the boiler will permit.

Also, in trying the piston, if there is any doubt of the correctness of the cylinder, the piston should be placed successively at different parts of its course, by turning the fly-wheel round; and the trials should be repeated at each place, viz.

at the top of the cylinder, in the middle, and at the bottom.   The friction of the packing of the piston may be tried, by a suitable number of men applying their strength to the rim and arms of the fly-wheel, to heave it slowly round; and they can judge whether it requires the same exertion to turn it, as it did on former occasions, when the piston was known to be in good working condition.

The above methods of trial will prove all the joints of the cylinder and nossels, as well as the fittings of the piston and valves.   The state of the joints of the condensing apparatus may be known by means of the barometer which is connected with the condenser, see p. 377, and Plate XI.   When the engine is about to stop working, the height at which the column of mercury stands in the barometer tube should be noted; and after the engine is stopped, the rapidity with which the mercury subsides, will show the extent of the leakage.   To make this trial most effective, the water should be allowed to run out of the condensing cistern G, as soon as the engine is stopped, by opening a plug, which is provided in the bottom of the cistern for that purpose.   The external air will then be at liberty to enter into the condenser through the defective joints, so as to let the mercury fall more rapidly than when water entered.   A tolerable judgment may at all times be formed of the state of the condensing apparatus, by the facility with which the engine may be set to work again, after it has been stopped, and remained some time at rest.   If the joints are good it may start very well, without blowing through, after standing two hours; but if there are considerable leaks, the condenser may require clearing by blowing through, when the engine has only stood half an hour, or still less. Even though the leakage may be to that extent, it will not materially impair the performance of the engine, because all the joints ought to be immersed in the water in the condensing cistern G, and the leakage of water into the condenser does no harm, for it is only in lieu of a part of the injection.

Mr. Watt's Indicator, or Instrument to show the Force actually exerted by the Piston of his Engine, when it is at work.

It has been already stated (p. 468) that the evacuation of the steam from the cylinder is progressive, and that the degree of exhaustion within the cylinder varies very considerably during the motion of the piston, so that it is only towards the termination of its course, that the cylinder becomes exhausted, to nearly the same degree of rarefaction as is continually maintained within the condenser.   The barometer, which is connected with the condenser, is well adapted to show the degree of exhaustion within the condenser, where the variations of that exhaustion, during each stroke of the engine, are inconsiderable and gradual, see p. 378; but if the same kind of barometer were to be connected with the cylinder, the great and rapid changes of elasticity of the steam, whilst in the act of escaping from the cylinder during its evacuation, would cause such great oscillations of the column of mercury, at different periods of the stroke, as to prevent any correct observations.

Mr. Watt was, therefore, led to contrive a different instrument, or barometer gauge, which should be less subject to vibration than a mercurial column, and which should be capable of showing with tolerable accuracy, the degree of exhaustion within the cylinder, at all periods of the course of the piston.   This instrument, which he called the Indicator, is found to answer the end sufficiently.

Mr. Watt's indicator is a small cylinder like a syringe, about an inch diameter, and six inches long, very accurately bored withinside, and fitted with a solid piston, which slides very easily in the cylinder, and fits tolerably tight, by the aid of oil

3 Q

poured upon the piston. The stem of the piston is guided and retained in the direction of the axis of the cylinder, so that it will not be liable to jam, and cause unnecessary friction in its motion. The bottom of this little cylinder terminates with a cock, and a small projecting pipe, the end of which is of a conical form, that it may be inserted into a hole drilled through the metal of the engine cylinder, near one of its ends, or else through the cover of the cylinder, or through the bottom: and then, by opening the cock, a free communication will be formed between the inside of the cylinder of the engine, and that of the indicator. The upper end of the small cylinder is open, to admit the atmospheric air to act freely upon its piston.

The cylinder of the indicator is fixed to a frame of metal or wood, which is more than twice the length of the cylinder, and which extends beyond its top, or open end; a spiral worm spring, which is coiled into a cylindrical form, like that of a spring steelyard, is fastened with one end to the upper part of the frame, and the other end of the same spring is fastened to the upper end of the piston rod of the indicator. This spring is made of such a strength, that when the indicator cylinder is quite exhausted, the atmospheric pressure acting upon the surface of the small piston of the indicator, will extend the spiral spring so much, as to force the piston down into the cylinder, to within an inch of its bottom: or when the cylinder is filled with steam of a greater elasticity than that of the atmospheric air, the pressure of that steam will force the piston up in the cylinder, and compress the spiral spring.

An index is fixed to the top of the piston rod of the indicator, and points to a scale which is formed upon the frame; this scale is divided in the following manner. A point is marked where that index stands, when the cylinder is open at both ends, so as to admit the air to act freely upon both sides of the piston, the spring then assumes its quiescent position, and this forms the zero of the scale; other points are next marked, by connecting the pipe of the indicator with the receiver of a pneumatic air-pump, which has a mercurial barometer attached to it; the cylinder being exhausted by the air-pump, to different degrees of rarefaction (which are determined by the barometer), corresponding divisions are marked on the scale, at the places to which the index points in each case. A number of principal points for the scale being thus determined, the rest of the divisions are filled up, so as to indicate what degree of exhaustion is made within the small cylinder; the pressure which is exerted against the small piston being expressed in pounds per square inch of its surface. The scale is also continued beyond the zero, or quiescent position of the spring, in order to measure the elasticity of any steam, stronger than the atmosphere, with which the cylinder may be filled. These divisions are likewise adapted to express the pressure which is exerted against the small piston, in pounds per square inch.

When the power of a steam-engine is to be tried by this indicator, its pipe is driven tight into the hole in the cylinder of the engine, and oil is poured upon the small piston, to make it tight; then the engine being set to work, and the cock opened, the interior of the cylinder of the indicator, communicates with that of the cylinder of the engine. At every stroke, when the steam is admitted into the top of the cylinder, to impel the piston downwards, the index of the indicator will point out on the scale what is the elasticity of that steam; whether it is greater or less than that of the atmosphere, and how much greater or less, in pounds per square inch. And as soon as the upper exhausting-valve of the engine is opened, to exhaust the upper part of the cylinder, and make the upwards stroke, the indicator cylinder partakes of that exhaustion, and the atmospheric pressure upon its piston extends the spring, and forces the piston down into the cylinder; the index shows on the scale what the extent of the exhaustion is, and at what rate it proceeds during the motion of the piston of the engine, from the plenum, or full charge of steam, to the vacuum, or greatest exhaustion.   When the steam is readmitted into the engine cylinder, to return its piston, then the spring restores the little piston suddenly to its former position.

The small cylinder and piston of the indicator, therefore, forms a correct working model of the cylinder and piston of the engine, and the force which is exerted by the steam, upon both the pistons, is actually measured at all times by the indicator spring ; and the pressure is pointed out by the index, in pounds per square inch. The motion of the indicator piston, and of the index, along the scale, is very rapid at first ; and it continues to move slowly till the piston of the engine has finished its course, so that it requires an attentive examination to determine the extreme exhaustion ; and what the mean state of that exhaustion is during the whole of the stroke, must be guessed at, but after some practice it may be estimated with tolerable precision.

When the exhaustion begins, the index moves quickly along the scale, from near its zero ; and by the time that the piston of the engine has run through about one-third of its course, the index becomes almost stationary, having reached nearly the lowest division to which it will descend.   The quickness of the motion with which the index passes over the divisions at first, shows the promptitude of the exhaustion, and whether the working gear of the engine is properly adjusted according to the principles explained at p. 469.   When the steam is readmitted into the cylinder of the engine, the index of the indicator shoots up quite suddenly along the scale, and then remains very nearly stationary during all the stroke of the engine.

The observations which are thus taken by the indicator are proper data for calculating the force which is exerted by the engine, when it is at work, for they show the degree of the plenum of steam, and also of the exhaustion of that steam which is produced alternately in one end of the engine cylinder ; it may be assumed, that the same effect takes place in the other end of the cylinder ; or if there is any doubt, the indicator may be also applied to that other end, and similar trials made with it.

By observing the index of the indicator, the mean state of exhaustion within the engine cylinder, during all the motion of the piston, may be estimated in pounds per square inch, less than the atmospheric pressure ; and the mean elasticity of the steam, with which the cylinder is filled, during the plenum, to impel the piston, being also observed ; if that elasticity is less than the atmospheric pressure, it must be deducted from the exhaustion ; or if the elasticity of the steam, during the plenum, is greater than the atmospheric pressure, then it must be added to the exhaustion.   The sum of the exhaustion and the plenum, or their difference, as the case may be, will represent the unbalanced pressure actually exerted by the steam, against each square inch of the surface of the piston in pounds.

From this actual pressure a suitable deduction must be made, on account of the force which is necessarily expended in overcoming the friction of the moving parts of the engine itself, and in working the air-pump and cold water-pump, &c.

This force has been determined, by many trials of Mr. Watt's engines, to be between $2\frac{1}{2}$ pounds per square inch of the piston, for the smallest and worst conditioned engines ; and $1\frac{1}{2}$ pounds in the largest, and best managed engines.   When this deduction is made, the remainder will represent the effective force exerted in pounds, by each square inch of the surface of the piston, that is, the force which the engine will realize, and actually exert to impel the machinery, which is connected with the extreme end of the axis of its fly-wheel.

The calculation of the power of the engine in horse power, may then be made, according to the second rule given in p. 440.   The result of the observations made with the indicator, and summed up as above, with the proper deduction for friction, must be used as the effective pressure upon the piston in pounds per square inch (exclusive of friction), as is directed in that rule.

Example of the use of Mr. Watt's Indicator.   Suppose that the indicator is applied to a hole in the cover of the cylinder of a double acting rotative engine ; that cylinder being 24 inches diameter, and its piston being adapted to make a stroke of 5 feet. Suppose that the trial is made when the piston is working at the rate of 20 double strokes per minute, so as to move through 200 feet per minute ; and that the indicator then shows the extreme exhaustion within the cylinder at the termination of the stroke, to be $12\frac{1}{4}$ pounds per square inch, less than the pressure of the atmosphere. Suppose also, that the mean state of the exhaustion during all the course of the piston, is estimated (by observing the indicator in the manner above described) at $9\frac{3}{4}$ lbs. per square inch less than the atmosphere, so that the mean state of exhaustion is about ·8 tenths of the final or greatest exhaustion. Lastly, suppose that the indicator shows that the elasticity of the steam with which the cylinder is filled, during the plenum, to impel the piston, is half a pound per square inch less than the elasticity of the atmospheric air.

Upon these data the effective force, and the mechanical power exerted by the engine, may be computed in the following manner. The unbalanced pressure actually exerted upon each square inch of the piston is ($9\cdot75$ lbs. − ·5 lbs. =) $9\cdot25$ lbs ; from which deduct $1\cdot75$ lbs. on account of friction, and it leaves $7\cdot5$ lbs. per square inch effective pressure, exclusive of friction ; that is, the engine will overcome a resistance which is equal to $7\frac{1}{2}$ pounds to each square inch of the surface of the piston. Then, by the rule (p. 440) 24 inches diameter, squared is 576 circular inches area, × 200 feet motion per minute = 115 200 cylindrical inch feet of steam expended per minute, by the motion of the piston, × $7\cdot5$ lbs. per square inch effective pressure (exclusive of friction) = 864 000 ÷ 42 000 = $20\cdot57$ horse power exerted by the engine to turn machinery. Or by the sliding rule.

Calculation, $\left\{ \begin{array}{l} \text{Sliding} \\ \text{Rule} \end{array} \right\}$ $\dfrac{\text{C}\ (200\ \text{ft. per min.} \times 7\cdot5\ \text{lbs. per squ. inc.} =)\ 1500}{\text{D} \qquad\qquad\qquad\qquad\qquad\qquad \text{gauge point}\quad 205}$ $\begin{array}{l} 20\ 6\ \text{horse power} \\ 24\ \text{inches diam.} \end{array}$

The indicator is a most valuable instrument in the hands of an intelligent engineer, for independently of its use to ascertain the power exerted by an engine, it will give him a correct knowledge of the state of an engine, and of the promptitude with which its cylinder is exhausted. He may thereby adjust the working gear, as is directed in p. 469, in the most advantageous manner for the good performance of the engine. This state of the engine will be shown, by the index of the indicator moving very rapidly over the scale, so as to arrive in the quickest possible manner, at the greatest degree of exhaustion that the condensing apparatus is capable of producing within the cylinder. There is no method of adjusting the action of an engine with so much accuracy, as by the result of trials made with the indicator (*a*).

Dr. Robison proposed the following kind of indicator in his article Steam-Engine, in the third edition of the Encyclopædia Britannica. " It would be very desirable to get an exact knowledge of the elasticity of the steam in the cylinder of a steam-engine, and this is by no means difficult. Take a long glass barometer tube exactly calibered, and closed at one end. Put a small drop of some coloured fluid into it, so as to stand nearly at the middle. Let the tube be placed in a long box filled with water, to keep it at a constant temperature ; and let the open end of the tube communicate with the cylinder, with a cock between."

" The moment that the steam-valve of the engine is opened, to admit steam into the cylinder, open the cock of this instrument ; and the coloured drop will be pushed by the steam towards the close end of the tube, provided that the steam in the cylinder is more elastic than the air with which that end of the tube is filled ; or the drop will be drawn the other way, when that steam is less elastic than the air. By a divided scale properly adapted to the tube, the elasticity of the steam which is contained within the cylinder at every position of the piston, may be discovered. The same thing may be done more accurately, by a barometer properly constructed, so as to prevent the oscillations of the mercury."

(*a*) The indicator is an instrument but little known or understood by the common makers of steam-engines ; but it is of the utmost importance to the perfecting of their productions, that they should accustom themselves to frequent trials of the performance of their engines by that means ; and then they would often find defects and deficiencies in their operations, which are at present unsuspected, but to which the engines of Mr. Watt's original construction were not so liable as the modern engines are.

" It is equally necessary to know the state of exhaustion, or the elasticity of the steam which remains in the cylinder, during the descent of the steam-piston ; and this question may be decided in the same way as the other, by a tube containing air, and a coloured drop, or by a barometer connected with the interior of the cylinder.   And thus we shall learn what is the state of the moving force at every moment of the performance, so as to obtain a correct mean of all the variations, whereby the machine will become as open to our examination, and as subject to calculation, as any water-mill, or horse-mill.   Until this be done, or something equivalent, we can only guess at what an engine is actually performing, and we cannot tell in what particulars we can lend it a helping hand."

" We are informed that Messrs. Boulton and Watt have made this addition to some of their engines ; and we are persuaded, that from the information which they have derived from it, they have been enabled to make the curious improvements from which they have acquired so much reputation and profit.   Mr. Watt's observations, by means of the barometer, must have given him much valuable information in this particular, and we hope that he will not always withhold them from the public."

Mr. Watt's original indicator has been greatly improved, by applying a blacklead pencil to the top of its piston rod, in such a manner as to trace a line upon a card, or piece of paper, fastened flat against a small moveable tablet, which is placed vertically, or edgeways upwards, in a grooved frame, and the tablet is moved backwards and forwards, by means of a string, which is connected with the great lever of the engine ; the motion of the tablet is edgeways, and in a direction at right angles to the motion of the piston rod of the indicator.   The object of adding the pencil and moving tablet is, to obtain an exact record of the progressive exhaustion of the steam from the cylinder, at every stage of the motion of the piston ; for the motion of the piston of the indicator being compounded with that of the motion of the tablet (which is simultaneous with the motion of the piston of the engine) the pencil is caused to describe a curved line upon the card, and this curved line returns into itself, so as to include a space or area within it.   The area so included within the curve, compared with the area of a parallelogram, which circumscribes or includes the curve within it, will represent the mean state of the exhaustion within the cylinder, throughout the stroke of the piston, compared with the final or greatest degree of exhaustion, at the termination of that stroke.

For instance, in the case before stated, if the area of the parallelogram had been represented by 11·75, that of the included curve would have been represented by 9·25 ; showing that the mean of all the variations of unbalanced pressure which is exerted by the steam against the piston during the whole stroke, is (9·25 ÷ 11·75 =) ·787 of the ultimate exhaustion at the end of the stroke (a).

This method of recording and computing the result of the trials made by the indicator, is truly scientific, and highly useful, for it gives a correct mean of all the variations, without trusting to any estimation by the eye, which was a source of some uncertainty in the original instrument.   The construction of a complete indicator will be described in the proper place with figures.

(a) The author has not been able to learn who was the inventor of this ingenious device, of tracing a card by the indicator ; but he has seen cards made in that way which are said to have been taken 25 and 30 years ago.   Mr. Watt has briefly described the indicator, in his Appendix to Dr. Robison's article Steam-Engine ; but as he makes no mention of the card, it may be concluded that it was added to his indicator by some other person, and did not, therefore, come within the account of his own inventions.

Performance of Mr. Watt's double rotative engines, according to the Indicator. By a series of observations made with the indicator, upon steam-engines of all kinds, it appears that the original plan of Mr. Watt's rotative double engine, with four separate valves, and hand gear, is one of the best systems which has yet been devised for that part of a steam-engine; for engines of that kind perform better than the modern engines, such as are now usually constructed by the best makers (*a*); the reason of this superiority has been already pointed out, p. 470, note.

Many of the engines which were made under Mr. Watt's own direction, 35 and 40 years ago, are still in use; and if they are in tolerable order, they are capable of working effectually, and at their full speed, when they are loaded with a resistance which requires half as much more as their nominal horse-power to overcome it, and they will work continually under such a load, without inconvenience; or if they are urged to their utmost, and fully supplied with steam, and with cold condensing water, they will do even more than that.

Such engines, when thus loaded, being tried by the indicator, it will often be found that the exhaustion within the cylinder, at the end of the stroke, is $11\frac{1}{2}$ pounds per square inch, less than the pressure of the atmosphere; and that the mean state of exhaustion during the whole stroke, is ·78 hundredths of the complete exhaustion, or 9 pounds per square inch. The mean elasticity of the steam with which the cylinder is filled during the plenum, to impel the piston, being at the same time $3\frac{1}{2}$ pounds per square inch, greater than the pressure of the atmosphere, the unbalanced pressure against the piston will be $(9 + 3\frac{1}{2} =)$ $12\frac{1}{2}$ pounds per square inch; and deducting 2 pounds for the friction of the engine, there is full $10\frac{1}{2}$ lbs. effective pressure, to impel the mill-work, exclusive of friction. If the steam in the boiler is kept up to 4, or $4\frac{1}{2}$ pounds per square inch, then that effective pressure may be increased to 11, or $11\frac{1}{2}$ pounds per square inch.

The standard force which Mr. Watt assumed for calculating the sizes of the cylinders for his rotative engines, was rather less than 7 lbs. per square inch (or according to the rule given in p. 440) 6·944 pounds per square inch, the expenditure of steam being at the rate of 33 cubic feet per minute, for each horse-power, reckoning only the space which is actually occupied by the piston in its motion, without any allowance for the steam which is expended to fill the vacant spaces at the top and bottom of the cylinder, where the piston does not pass.

Hence, when an engine is working as above stated, and overcoming a resistance equal to $10\frac{1}{2}$ (or $11\frac{1}{2}$) pounds per square inch of the piston, provided that it keeps up to its proper velocity, it will exert one and a half (or one and two-thirds) times, the mechanical power at which it is rated in horse-power, according to the standard of 0·94 lbs. per square inch.

When an engine on Mr. Watt's original construction, is not overloaded, if it is in good order, and is properly supplied with cold condensing water, at a temperature of not more than 50 degrees of Fahrenheit's thermometer, its performance will usually appear by the indicator to be nearly as follows. The temperature of the hot water which is discharged by the air-pump, will be about 100 degrees of Fahrenheit. The exhaustion within the condenser will be such, as to raise a column

---

(*a*) It is well known to engineers and manufacturers, who still have those old engines in use, and also engines of the modern construction employed in the same kind of work, that the old ones perform better than the new ones, supposing both to be in equally good order. The old engines are capable of exerting a greater power without becoming overloaded, and diminishing in their speed; and also, when they are moderately loaded, they consume less fuel in proportion to the power they exert, than the modern engines, as they are usually constructed.

This retrograding in the perfection of engine making, would have been avoided if the engineers of the present time had been as well acquainted with the internal operation of the machine as Mr. Watt was; but as they may easily acquire that knowledge, by the aid of the indicator, it is inexcusable that any engines now made, should be inferior to those which were made forty years ago.

of mercury 27 inches high in the barometer, which is connected with the condenser.   The greatest exhaustion within the cylinder at the termination of the
stroke of the piston, will be about 13 lbs. per square inch less than the atmospheric
pressure ; and the mean state of the exhaustion throughout the stroke about $10\frac{3}{4}$
pounds per square inch, less than the atmospheric pressure ; this is ·827 of the final
exhaustion in the cylinder.   The mean elasticity of the steam with which the
cylinder is filled during the plenum, will be about 2 lbs. per square inch less than
the atmospheric pressure.   Hence, the unbalanced pressure against the piston will
be $(10\frac{3}{4} - 2 =)$ $8\frac{3}{4}$ lbs., and deducting $1\frac{3}{4}$ lbs. for the friction of the engine,
(which is sufficient when it is in good order, and lightly loaded), leaves 7 lbs. per
square inch effective pressure to impel the machinery, which is according to Mr.
Watt's standard.

When an engine is working in this manner, the governor must be adjusted
as directed in p. 466, so as to keep the throttle-valve very much closed, in order
to restrain the flow of steam from the boiler into the cylinder.   The elasticity of
the steam in the boiler, must, in the first instance, be raised to about $2\frac{1}{2}$ pounds per
square inch greater than the atmosphere, in order to clear the condenser by blowing
through, previous to starting the engine ; but that being done, it is needless to
keep up the steam to more than 1 or $1\frac{1}{2}$ lb. per square inch above the pressure of
the atmosphere ; because it is only required to be admitted at such a rate through
the throttle-valve into the cylinder, as to possess an elasticity of 2 pounds per
square inch less than the atmosphere, when it forms the plenum in the cylinder, to
act against the piston.   The cylinder is, therefore, supplied with steam of a diminished density, which is a very convenient manner of working an engine, because,
by opening the throttle-valve wider, the power of the engine can be instantly increased to a very considerable extent, so as to meet any increase of resistance which
may be applied to the engine.   Under these circumstances, if the governor is properly adjusted, it will act very promptly to produce a proper regulation of the motion, as is already explained, p. 465, because there is always a considerable reserve of
power at the disposal of the governor, ready to be called into action, whenever it
is required.

The latter statement of the operation of Mr. Watt's rotative engine, is that
state in which it will produce the very best effect that he was able to attain by these
engines (a), but it requires an engine to be in very good order to perform in that
manner.

The expenditure of steam for each horse-power, is then at the rate of 33 cubic feet per minute,
the steam being 2 pounds per square inch less elastic than the pressure of the atmosphere, which
being taken as equal to a column of mercury 30 inches high, or 14·7 lbs. per square inch, the actual
expenditure of steam equal to the atmospheric pressure, would be $(14·7 - 2 =)$ 12·7 lbs. × 33
cub. ft. = 419 ÷ 14·7 lbs. = 28· 5 cubic feet per minute for each horse power.

And allowing one-tenth of that quantity to be wasted, in filling the vacant spaces at the top and
bottom of the cylinder, and by condensation and leakage, the actual expenditure of steam of the same
elasticity as the atmospheric air, would be $31\frac{1}{3}$ cubic feet per minute, to produce one horse-power.

As a horse-power is capable of raising 528 cubic feet of water one foot high per minute, each
cubic foot of steam will raise $(528 ÷ 31·3 =)$ 16· 85 cubic feet of water one foot high.

(a) Mr. Watt did not apply his principle of expanding the steam, in his rotative engines, or he
might have obtained an increased performance, in the same manner as he had before done in his single
engines for pumping (see p. 367).   For by shutting the steam-valve, so as to intercept the supply of
steam to the cylinder, when the piston had moved through a portion of its stroke, the piston might
have been impelled through the remainder of its course, without any farther expenditure of steam, by
the elastic force which the steam already admitted into the cylinder would exert during its expansion,
in the act of proceeding to occupy a larger space, on the principle stated at p. 339.

According to this data, a cubic foot of steam of the same elasticity as the atmospheric air, being applied in Mr. Watt's rotative engine of the best construction, under the most favourable circumstances, will cause a sufficient vacuum, to raise a cubic foot of water 16·9 feet high, besides overcoming the friction of the parts of the engine itself, and the resistance of the water to motion. This is rather less than the performance of Mr. Watt's single engine for pumping water (see p. 367).

The consumption of a bushel of good Newcastle coals (= 84 lbs.) will raise 18·34 millions pounds of water one foot high.   Or 1·08 bushels per hour will supply an engine of 10 horse-power.   And at the same rate each horse-power will require 9·07 lbs. of such coals per hour.

For a bushel of coals (= 84 lbs.) may be assumed to evaporate 10·08 cubic feet of water into steam, and every cubic inch of water so evaporated, will produce a cubic foot of steam, of the same elasticity as the atmospheric air ; hence a bushel of coals will produce (10·08 × 1728=) 17 418 cubic feet of steam ; and if each cubic foot of steam raises one cubic foot of water 16·86 feet high, then the consumption of a bushel of coals must raise (17 418 × 16·85 =) 293 500 cubic feet of water, or (× 62·5 =) 18 340 000 lbs. of water one foot high.

The above is the utmost that can be done by Mr. Watt's rotative engines ; but if an engine of that kind is not in the very best order, although it may be properly adjusted, and may be in the best condition that can be maintained under ordinary management for a long continuance, the performance at an average will commonly be found to be about 15·84 millions pounds, raised one foot high, by each bushel of good Newcastle coals that is consumed.   In that case one bushel per hour, will supply an engine of 8 horse-power ; and at the same rate, each horse-power will require (84 ÷ 8 HP =) $10\frac{1}{2}$ lbs. of coals per hour ; this supposes the engine to be a good one, and that it is properly loaded, to be exactly to the standard of 8 horse-power.

Nevertheless it is common to find rotative engines of 6 horse-power, consuming at the rate of a bushel of coals per hour ; this is about 11·88 millions pounds raised one foot high, by the consumption of a bushel of coals ; and then each horse-power occasions a consumption of (84 lbs. ÷ 6 HP =) 14 lbs. of coals per hour.   Such engines must be in very bad condition as to leakage by the piston and valves ; or the working gear may be out of adjustment, or the exhausting passages too small to exhaust the cylinder properly ; the vacancies at the top and bottom of the cylinder may be needlessly large ; or there may be a deficiency of condensing water, or the coals may be of inferior quality.

The resistance with which engines are loaded, is rarely known, for the notions of the people who usually have the care of them, are very fallacious ; and even reputed good judges of engines are frequently deceived in their opinion of the resistance that they are actually overcoming ; for it is only by frequent trials of different engines with an indicator, so as to obtain an actual knowledge of their performance under different circumstances, and an accurate observation of all the apparent symptoms during such trials, that an engineer can acquire the tact of judging with tolerable accuracy what load an engine is working against.

The different means of trying the state of an engine when it is at work, so as to form a probable estimate of its performance, will be fully explained in its proper place ; but the indicator is the only mode of trial which can be relied upon.

The friction of the small piston of the indicator occasions it to represent the pressure of the steam against the piston of the engine rather less than it really is.   Hence, it is of importance to make the indicator piston move with the least friction possible ; and with that view it is better to make the piston of a larger diameter than Mr. Watt states; a convenient size is 1·95 inches diameter, which is 3 square inches area. The deduction which is made from the unbalanced pressure against the piston of the engine, on account of the friction of its parts, and its air-pump, ought to be apportioned in order to compensate for so much deficiency in the pressure which is shown by the indicator, as may arise from the necessary and unavoidable friction of the indicator piston; and then, if that piston is kept in good order, it will have the same amount of friction in every trial, and we shall obtain exact results.

Having now described the structure and operation of Mr. Watt's rotative engine, and stated some particulars of its performance, we may proceed to give the actual dimensions and proportions of some of the best specimens of his engines of different powers.

DIMENSIONS OF A 10 HORSE-POWER PATENT ROTATIVE ENGINE MADE BY MESSRS. BOULTON AND WATT, 1795.

THE drawings in Plates XI, XII, and XIII, are on too small a scale to admit of sufficiently exact measurements of the dimensions of the parts. The engine from which those drawings were taken, was not made at Soho, and some of the dimensions differ from the standard followed by Messrs. Boulton and Watt. The following are the dimensions of a ten horse-engine, which was put up by them in 1795, at a manufactory of starch at Lambeth, in London. The dimensions were taken in 1805, by the author, who finds that the same engine remains still in use, after having been constantly worked for more than 30 years. The construction is the same as is represented in the engravings, with a few exceptions.

The cylinder E is $17\frac{1}{2}$ inches diameter, $= 240·5$ square inches area, or 306·25 circular inches. The length of the stroke of the piston J is 4 feet; and it makes 25 double strokes per minute, so that it moves through a space of 200 feet per minute. The quantity of steam expended by the motion of the piston, without reckoning the quantity wasted by the vacant spaces at top and bottom of the cylinder, is 334 cubic feet per minute; that is at the rate of 33·4 cubic feet of steam per minute for each horse-power. When the cylinder had been in use about 18 years, it became worn and was bored out anew; it is now $18\frac{1}{4}$ inches diameter.

The effective force of the piston, when the engine is exerting 10 horse-power, is at the rate of 6·86 pounds upon each square inch of the piston, exclusive of the friction of the engine; and 6·86 multiplied by 240·5 square inches area, gives 1650 pounds for the force exerted by the piston, taking the mean of all the variations of force during every part of the stroke; for 1650 lbs. force, acting through a space of 200 feet per minute, is equal to 330 000 lbs. raised one foot per minute, which ($\div$33 000 lbs.) amounts to ten horse-power. The utmost force that the piston is ever expected to exert, is about twice as much as the uniform force, or at the rate of 14 pounds per square inch, or 11 pounds per circular inch, which in this engine would be 3369 pounds, or 1·5 tons. The force will very rarely amount to this, yet it will sometimes be as great; and, therefore, the strength of all the machinery must be adapted to bear that force without any danger of breaking.

The boiler C is made of copper, in a circular form, similar to the boilers used by Mr. Smeaton, see Plate II, and pp.173, and 191. The boiler is $8\frac{1}{2}$ feet diameter, and about $7\frac{1}{2}$ feet deep. The bottom is $7\frac{1}{2}$ feet diameter, and rises 8 inches in the middle.

The fire-grate B is three feet square, or 9 square feet of surface, which is at the rate of ·9 of a square foot of fire-grate, to each horse-power.

3 R

The horizontal surface of the water in the boiler, is 57 square feet. The concave bottom of the boiler exposes about 38 square feet of surface to the direct action of the fire; and the surface exposed in the flue around the lower part of the boiler is (25 feet mean circumference × 3 feet high =) 75 square feet. Hence 113 square feet of surface is exposed in the whole; or at the rate of 11·3 square feet to each horse-power (a).

The air-pump H is 12 inches diameter = 144 circular inches area. The area of the cylinder is 2·13 times the area of the air-pump. The bucket makes rather less than 2 feet stroke; the effective capacity of the air-pump is rather less than one-eighth part of that of the cylinder. The condenser F is the same size as the air-pump.

The steam and exhausting valves b, c, h, i, are $3\frac{5}{8}$ inches diameter in the apertures which they open for the passage of the steam; and the uninterrupted areas of those apertures are 8·9 square inches, or about $\frac{1}{27}$ of the area of the cylinder, this is at the rate of ·89 square inch of passage to each horse-power. The steam and eduction-pipes are $3\frac{3}{4}$ inches diameter inside at the smallest parts, so as to have an area of 11 square inches, or about $\frac{1}{22}$ of the area of the cylinder.

As the piston moves through 200 feet per minute with a reciprocating motion, and with a varying velocity, its greatest velocity when it is near the middle of its stroke, will be at the rate of (1·57 times 200 =) 314 feet per minute, for a very short space of time, as is already explained at p. 415. When the steam is admitted into the cylinder, it must pass through the apertures of the steam-valves b or c, with 27 times the velocity of the piston, viz. (200 × 27 =) 5400 feet per minute, or 90 feet per second, taking the mean of all the variations of velocity throughout all the stroke; or the greatest velocity of the steam, when the piston is near the middle of the stroke is (314 × 27 =) 8482 feet per minute, or 141 feet per second.

When the steam escapes from the cylinder through the exhausting-valves h or i into the condenser, its velocity is not regulated by that of the piston, but it depends upon its own elasticity, as is already explained at p. 469. Whereas the steam in entering into the cylinder, must follow the piston; and the velocity with which it will flow through the aperture of the steam-valve b or c, so as to fill up the space left by the motion of the piston, will bear the same proportion to the velocity of the piston, as the area of the piston bears to that of the aperture of the valve, on the principle stated in p. 53.

The piston-rod n is $1\frac{3}{4}$ inc. diameter. The joint pins o of the main links K of the parallel motion, which connect the piston-rod with the great lever, are made of wrought iron $1\frac{7}{8}$ inc. diameter where they fit into the brass sockets, and 2 inc. length of fitting in those sockets. The wrought iron straps which form the links K, are $1\frac{3}{8}$ inc. by $\frac{5}{8}$ inc. thick, four of which, amounting in all to 3·44 square inches of wrought iron, bear all the strain of the piston. The length of the main links K K is 20 inches from one joint to the other. The upper joint pin of the main links K is fastened to the under side of the wooden beam L, by two iron bolts which pass up through the wood with nuts at top; the screws are $\frac{7}{8}\frac{1}{16}$ inc. diameter outside the threads, or about 1 square inch of solid iron in both screws.

(a) According to the rule given in p. 366, the surface exposed to the heat in the boiler of this engine, should have been (cylinder $17\frac{1}{2}$ inc. diam. × $17\frac{1}{2}$ = 306·25 circular inches area × 200 feet per min. = 6125 cylindrical inch feet of steam expended per minute ÷ 600 =) 102·1 square feet. The boiler actually has 113 square feet. That rule is formed from Mr. Watt's data for his single engines for pumping, and not for the rotative engines, which from their construction cannot exhaust the cylinder so completely, and have no advantage of the expansive action of the steam; hence, they require rather more steam in proportion to the size of the cylinder, than the single engines for pumping. There are many instances of rotative engines which have worked for years, with boilers apportioned according to the above rule, which gives very nearly ten square feet of boiler surface exposed to the heat for each horse-power; but it is found better to allow 12 square feet of heating surface for every horse-power. In which case the divisor used in the above rule should be 500, instead of 600, as before directed.

The great lever L is an oak beam, $13\frac{1}{3}$ feet long from the joint at one end to that at the other, the axis p being in the middle of its length. The wood is $16\frac{1}{2}$ inches deep, and 12 inc. broad. The axis p is cast iron, $4\frac{3}{8}$ inc. square, and 22 inc. length between the bearings; the cylindrical pivots at each end are $3\frac{1}{8}$ inc. diameter, and 4 inc. length of bearing in their brass sockets. The beam is fastened to the axis p, by four bolts, with nuts at top; the screws are $1\frac{1}{4}$ inc. diameter outside the threads. The centre of motion of the axis of the great lever is in one straight line with the joints at each end of the lever.

The connecting-rod M is jointed to the end of the great lever, by a joint-pin 2 inches diameter, and $2\frac{1}{2}$ inches length of fitting in the sockets. The connecting-rod is made of cast iron $5\frac{3}{8}$ inches by $3\frac{3}{4}$ inches, with the angles taken off; it is about $12\frac{1}{4}$ feet long from the upper joint to the centre of the planet wheel (a).

The planet-wheel N is made of cast iron, rather less than 2 feet geometrical diameter, with 31 teeth; the teeth are 2·36 inches pitch from the centre of one tooth to the centre of the next tooth, and 4 inches broad. The sun-wheel O is rather more than 2 feet geometrical diameter, with 33 wooden teeth $4\frac{1}{4}$ inches broad, fixed into mortices in the iron wheel. The axis P of the sun-wheel and fly-wheel, is made of cast iron, $5\frac{1}{2}$ inches square, and $4\frac{1}{2}$ feet long between the bearings; its necks or pivots on which it turns, are 5 inches diameter, and $5\frac{1}{2}$ inches length of fitting in their brass sockets.

The fly-wheel Q is 12 feet diameter outside, and the rim is $5\frac{1}{2}$ inches by $2\frac{1}{2}$ inches, $= 13\frac{3}{4}$ square inches; the wheel has 8 cast iron arms, of about 5 inches, by $1\frac{3}{8}$ inches thick, with ribs $\frac{3}{4}$ thick, projecting 7 inches at the centre, and tapering away, so as not to project at all at the run.

The rim of this fly-wheel contains 3·46 cubic feet of cast iron ($\times$ 450 lbs. =) 1557 lbs. weight. The fly-wheel makes $48\frac{1}{2}$ revolutions per minute when the piston makes 25 strokes per minute.

From these data we may compute the effect of the fly-wheel in the manner explained in p. 414. The velocity of the middle of the rim of the fly-wheel is 29·3 feet per second. The height through which a body must fall by gravity to acquire this velocity is 13·26 feet, which × 1557 pounds weight of iron in the rim, gives 20 650 lbs. (=9·22 tons) raised one foot high, for the mechanical power which must be communicated to the rim of this fly-wheel, to urge it from rest into motion with the velocity of 29·3 feet per second. The power exerted by the engine during each half stroke is (1650 lbs. pressure on the piston × 4 feet stroke =) 6600 lbs. raised one foot high; therefore the energy of the rim of the fly-wheel is (20 650 lbs. ÷ 6600 lbs. =) 3·13 times the power exerted by the engine during one half stroke. This fly-wheel is found sufficient to render the motion of the engine very regular.

This engine is applied to turn two pair of mill-stones 4 feet diameter for grinding corn. They make about 120 revolutions per minute, or $2\frac{1}{2}$ turns for one of the fly-wheel. The axis of the fly-wheel is prolonged by a horizontal axis of cast iron, which is connected with the extreme end of it; at the farther end of this second axis, a bevelled cog-wheel is fixed, in order to turn another bevelled cog-wheel, which is mounted on a vertical axis, and the same axis has a large spur cog-wheel fixed upon it, above the bevelled wheel. The spur-wheel turns pinions at the lower ends of the upright spindles of four running mill-stones, two of which are commonly worked at once.

(a) The weight of the connecting-rod and planet-wheel, being greater than that of the parallel motion and piston-rod, the catch-pin at the cylinder end of the great lever is made of cast iron, and hollow within, like a box, in order that as much lead may be put into it, as will bring the lever to a true balance; the other catch-pin at the fly-wheel end of the lever is of wood. Both these catch-pins are fastened across the top of the wooden beam by four iron bolts for each pin; the bolts being applied on the outside of the wood, in the same manner as those which fasten the beam to its axis p, so as to avoid piercing and diminishing the strength of the timber; but the bolts bind the wood, and prevent it from splitting and cracking by the heat to which it is exposed.

3 R 2

The engine is supposed to be loaded rather above 10 horse-power by this mill-work, and the consumption of fuel is from 18 to 20 bushels of ordinary Newcastle coals, such as are usually employed for engines, burned in each day's work. It is commonly worked during 12 hours per day, but some part of the coals are consumed in getting up the steam in the morning. On an average it may be stated to burn $1\frac{1}{2}$ bushels per hour, when in good working order.

Messrs. Boulton and Watt continued to make their rotative engines upon the plan of that represented in Plate XI, until near the expiration of Mr. Watt's original patent in 1800, when they frequently made engines without the sun and planet-wheels. In some cases they applied a single crank to the extremity of the axis of the fly-wheel, in the manner of the atmospheric engine, already described at p. 410; but as this construction requires the rim of the fly-wheel to be four times as heavy as it is, when sun and planet-wheels are used (a), they more commonly employed a multiplying-wheel and pinion for the fly-wheel, so as to give it the same, or even a greater velocity, than with sun and planet-wheels.

Multiplying-wheel and Pinion for the Fly-wheel, instead of Sun and Planet-wheels.

For this purpose a large cog-wheel is fixed upon the axis of the crank, in the place of the fly-wheel Q, p. 410; and it turns a small cog-wheel or pinion upon a second axis, upon which the fly-wheel is also fixed, in order to be turned round twice or three times, for each double stroke of the piston, and thus give the loaded rim a great velocity. This construction is represented in its more perfected state in Plate XV.; the plan was thought to be so much preferable to that of the sun and planet-wheels, that Messrs. Boulton and Watt gave up making the latter entirely.

One convenience of the multiplying-wheel and pinion is, that the fly-wheel being mounted upon a separate axis from that of the crank, it may be placed nearer towards the cylinder, so as to include the engine within a less space than with sun and planet-wheels, or with a simple crank. It was found that the multiplying-wheel and pinion would be less liable to be worn out and broken, than the sun and planet-wheels, because they admit of being made very strong, without inconvenience from their weight, which is truly balanced upon the axis of the crank, and that of the fly-wheel, and they move continuously in one direction; whereas the greater part of the weight of the planet-wheel must be balanced by applying an equivalent weight on the great lever, at the cylinder end; and this counterpoise occasions a still greater accession to the weight of the reciprocating parts of the engine.

The multiplying-wheel being of at least twice the diameter of the planet-wheel, and having a double number of teeth, a greater number of teeth are brought to act upon the teeth of the pinion at once, so as to resist wearing for a longer time. As the multiplying-wheel may be made of any required diameter, without reference to the length of the stroke of the piston, the pressure upon the teeth of the wheel, and of the pinion, may be rendered as much less than the whole force exerted by the piston, as the diameter of the multiplying-wheel is made greater than the length of the stroke.

With this advantage of diminishing the pressure upon the teeth, and the large

(a) To have rendered the fly-wheel equally capable of regulating the motion of the engine, incase it had been fixed upon the axis of the crank, the rim of the fly-wheel being 12 feet diameter outside, it would have required to have been 7 inches by 8 inches = 56 square inches area in its cross section; the effect would be the same as that of the fly-wheel with a rim of $13\frac{3}{4}$ square inches, turned with sun and planet-wheels. The rim of 56 square inches would contain 13·84 cubic feet of cast iron × 450lbs. = 6228 pounds weight, or 2·7 tons.

size of the wheel allowing a greater number of its teeth to be interlocked with those of the pinion at the same time ; and also, as any required weight and substance may be given to the wheel and pinion, they can be rendered so strong and durable, as to avoid any chance of breaking ; and the teeth will wear a very long time. Also, when the teeth do become worn, the drop, or jerk, occasioned by the looseness of the teeth, will not be half as great as would take place with the sun and planet-wheels, when their teeth are worn equally loose ; this advantage results from the large size of the multiplying-wheel.

Dimensions of a 10 *horse-engine, with multiplying wheel and pinion.* This was made by Messrs. Boulton and Watt, in 1800, for a Dye-house in London, where it is still in use. The arrangement of the crank, multiplying-wheel, and fly-wheel, is the same as that of the engine in Plate XV. ; but the construction of the cylinder, the valves and working gear, air-pump and condenser, parallel motion and great lever, and the fixed framing, is the same as that of the engine represented in Plate XI. ; and the dimensions of all those parts are nearly the same as above stated.

The axis of the crank is situated in the place of the axis of the sun-wheel and fly-wheel in Plate XI. ; this axis is made of cast iron, 6 inches square, and only $2\frac{1}{4}$ feet long between the bearings; the neck of the axis where it rests in the bearing, is $5\frac{1}{2}$ inches diameter, and 6 inc. long in that bearing. The crank is fixed on the extreme end of the axis, which projects out beyond the bearing ; this part, which is $6\frac{3}{4}$ inc. diam., is turned true, and a circular hole is bored out in the central part of the crank, to fit very tight upon the extremity of the axis. The crank is prevented from turning round upon the axis, by three round steady pins, about one inch diameter, which are driven into holes bored into the circular joint, in the direction parallel to the axis, so that one half of each pin is inserted into the solid metal of the axis, and the other half into the metal of the crank (*a*).

The multiplying wheel which is fixed upon this axis is about 4 feet diameter, with 64 teeth, which are 2·36 inches pitch, and 4 inches broad, or the same size as the teeth of the sun and planet-wheels of the former engine. The wheel has eight arms, and is all cast in one piece; it has a square hole through the centre, to receive the axis, and it is fastened on by wedges (*b*). The pinion, which is turned by the multiplying wheel, has 33 wooden teeth, $4\frac{1}{8}$ inc. broad, fixed into mortises in the cast iron pinion ; the axis of the pinion, which also carries the fly-wheel, is $4\frac{1}{4}$ inc. square, and about 4 feet long between the bearings.

(*a*) The original method of fastening the sun-wheel upon the projecting square at the extremity of the axis of the fly-wheel, was by means of eight iron wedges driven very tight into the joint between the square hole through the centre of the sun-wheel, around all the four sides of the square axis which entered into that hole ; two wedges being inserted at each side of the square, near to its angles, in a direction parallel to the length of the axis. This is a very good method, if the wedges are accurately fitted into their places, and driven tight. The same plan was used, in the first instance, for fastening the crank on the end of the axis ; but afterwards Messrs. Boulton and Watt adopted the above method of turning the end of the axis, and boring out the centre of the crank to fit it, and fastening the crank on the axis by three steady pins: this, which is now the universal practice, was also used for fixing the sun-wheel of some of the latest engines, which they made with sun and planet-wheels.

(*b*) The multiplying wheel of this engine being of the same diameter as the length of the stroke made by the piston, the strain upon the teeth of the wheel and pinion will be equal to the whole force of the piston, the same as if the sun and planet-wheels had been used ; and accordingly, the strength given to the teeth is the same as to those of the sun and planet-wheel before described.

It would have been better to have put the wooden teeth into the multiplying-wheel, and to have made the pinion with solid cast iron teeth ; and then the wooden teeth in the multiplying-wheel being in greater number, would have endured longer wearing than they can do, when they are in the pinion, because as the pinion turns nearly twice, for once of the multiplying-wheel, every one of its teeth must come into action twice in each stroke of the engine, whereas each tooth of the multiplying-wheel comes into action only once in each stroke.

The fly-wheel is 12 feet diameter outside, and the rim is 6 inches by $2\frac{1}{2}$ inches, so as to be heavier than the fly-wheel before mentioned for the 10 horse-engine, with sun and planet-wheels; both these fly-wheels make $48\frac{1}{2}$ revolutions per minute, and therefore the heaviest must have the greatest effect, to regulate the motion of the engine (a). The axis of the pinion, and fly-wheel, is placed on the same level as the axis of the crank and multiplying wheel; but the fly-wheel being 3 feet nearer to the cylinder, than is represented in Plate XI. the engine occupies a space 3 feet shorter than if it had been made with sun and planet-wheels; in some situations this is a convenience.

The bearing for the front end of the axis of the fly-wheel, is fixed upon the same wooden frame that sustains the bearing for the axis of the crank; that framing is the same as is represented in Plate XI., and the fly-wheel is placed behind the back frame, which supports the farther end of the axis of the crank; hence it occupies the space between that frame and the wall. The farther end of the axis of the fly-wheel is supported in a bearing fixed in the wall; and the extreme end of that axis projects out through the wall, beyond the bearing, in order to connect the axis with the machinery, which it is intended to turn.

This engine works six pair of stocks, of a similar construction to those of a fulling-mill; but these are lighter, being used for washing woollen cloth after it has been dyed; each pair of stocks has two hammers, which are lifted alternately by means of tappets projecting from a horizontal axis, placed underground, beneath three pairs of the stocks, which stand side by side; and there is another similar horizontal axis placed parallel to the former, to work the other three pairs of stocks in another row. The two axes underground are turned by means of large spur cog-wheels, fixed on each of them, at one end; and these wheels are both turned by means of a large intermediate spur-wheel, which is placed between the two, so as to cause them both to turn the same way round. The centre of the large intermediate wheel is placed on a higher level, than those of the two axes underground, so that the upper part of the intermediate wheel is at a suitable height to be turned by a small cog-wheel, which is fixed on the axis of the fly-wheel of the engine. The proportions of the wheel-work is such, that when the fly-wheel makes $28\frac{1}{2}$ revolutions per minute, each hammer will strike about 30 blows per minute, and the six pair of stocks contain 12 such hammers; but when they are all in action at once, the resistance or load upon the engine is more than ten horse-power.

A large pump is also connected with the great lever of the engine, to be worked occasionally, to throw up water from the river into a large cistern, which is fixed over the dye-house. The engine also turns a pair of rolling stones, called runners upon edge, to crush dyeing materials; and a colour mill, for mixing and grinding indigo. It is capable of turning all the machinery at once as its utmost exertion, but commonly some of the stocks are disengaged, when the other works are applied.

The boiler of this engine is made of iron plate, of a waggon shape, such as is represented in Plates XI. and XII.; the upper part is a half cylinder $4\frac{1}{4}$ feet diameter, and $9\frac{1}{3}$ feet long.

The surface which is exposed to the direct action of the fire and flame, beneath the bottom of the boiler, is about $(3\frac{3}{4} \times 9\frac{1}{3} =)$ 35 square feet; and the surface exposed in the flue, which passes round the sides of the boiler at the lower part, is about 26 feet circumference, by $2\frac{1}{2}$ feet high = 65 square feet. The whole of the surface exposed to the action of the heat is, therefore, 100 square feet (b), which is

---

(a) The rim of the above fly-wheel contains 3·76 cubic feet of cast iron, and weighs 1693 pounds. Its energy is equal to 3·4 times the power exerted by the piston at each half stroke.

(b) The surface exposed in this boiler corresponds very nearly with the calculation, p. 490, according to the rule in p. 336.

at the rate of 10 square feet to each horse power.   The horizontal surface of the water in the boiler is ($4\frac{1}{4}$ × $9\frac{1}{3}$ =) $39\frac{2}{3}$ square feet.

The fire grate is 3 feet long by 3 feet wide, having 16 grate bars, $2\frac{1}{4}$ each inches wide : this is 9 square feet area, or at the rate of ·9 tenths of a square foot of each horse power.

This boiler is found capable of supplying the engine with steam, when it is not overloaded.   Two boilers are placed one by the side of the other, and they are used alternately, one being cleaned, or repaired, whilst the other is at work.   One of these boilers has been worn out since the original erection, and has been replaced by another boiler of a larger size (a), which is found to answer better than the original one, because it is more easy to keep up the steam to a regular elasticity, when the engine is working with only its proper load of 10 horse power ; and also this larger boiler will supply the engine with steam when it is loaded with all the work, and exerting its utmost power.   The consumption of coals, by this engine, is supposed to be on an average $1\frac{1}{2}$ bushels per hour, but as both boilers are frequently worked at once, and then steam is taken from the same boilers to heat water for the dye-house, this is not very exactly ascertained.

The cold water pump R is $5\frac{1}{2}$ inches diameter, = $30\frac{1}{4}$ circular inches area, and its bucket makes a stroke of 2 feet, or one half of that of the piston ; but the bucket only raises water during the descending stroke of the piston, whilst the piston expends steam equally in the up stroke, and in the down stroke.

The effective capacity of the steam cylinder is 40·5 times the effective capacity of the cold water pump, reckoning merely by the spaces which are occupied by the piston, and by the pump bucket, without considering the extra spaces at the top and bottom of the cylinder, through which the piston does not pass ; but allowing these spaces to be $\frac{1}{10}$th of the space occupied by the piston in its motion, and also allowing that the water raised by the pump is $\frac{19}{20}$ths of the space occupied by the pump bucket in its motion, the actual quantity of steam which passes through the cylinder, will be about 47 times the quantity of cold water which is raised by the pump into the condensing cistern, G.   Mr. Watt's allowance of injection water for his single engines was 60 times (see p. 376.)

DIMENSIONS OF A 20 HORSE-POWER PATENT ROTATIVE ENGINE, MADE BY MESSRS. BOULTON AND WATT, 1792.

This engine is on the same construction as that represented in Plates XI. and XII. in almost every respect, except that the dimensions are larger.   It was set up at a dye-house in the borough of Southwark, near London Bridge, in 1792 ; and the following dimensions were taken in 1805, by the author, who has frequently examined the engine since that time, in order to observe the effect of its wearing during 20 years.   It is still in use, without having undergone any alteration from its original plan.

The cylinder is 24 inches diameter, = 452·4 square inches area, or 576 circular inches.   The length of the stroke of the piston is 5 feet ; and it makes 21 double strokes per minute, so that the piston passes through 210 feet per minute. The expenditure of steam is 660 cubic feet per minute, or at the rate of 33 cubic feet per minute, to each horse power.   The effective pressure upon the piston is

(a) The boiler which is represented in the drawings, Plates XI. and XII., was not made by Messrs. Boulton and Watt; it is $9\frac{3}{4}$ feet long by $5\frac{1}{4}$ feet diameter, so that the horizontal surface of the water in the boiler is 51·2 square feet, and the surface which is exposed is ($9\frac{3}{4}$ feet by $4\frac{3}{4}$ =) 46·3 square feet beneath the bottom, and (29 feet × 3 feet =) 87 square feet in the flue. Total 133 square feet to ten horse-power, or $13\frac{1}{3}$ square to each horse power.   This is a more modern proportion.

6·944 lbs. per square inch, or ($\times$ 452·4 squ. inc. =) 3142 lbs. for the effective force of the piston, taking a mean of all its variations throughout the stroke; and this force acting through a space of 210 feet per minute, is equivalent to 660 000 lbs. raised one foot per minute, or ÷ 33 000 lbs. = 20 horse power. The utmost force that the piston can ever exert may be computed at 11 lbs. per circular inch, or ($\times$ 576 circ. inc. =) 6336 lbs. = 2·83 tons; and the strength of all the parts must be adapted to resist this force, without any danger of breaking.

The boiler is made of iron plate, of a waggon shape; the upper part is a half cylinder 12$\frac{2}{3}$ feet long, and 4 feet diameter, with its axis horizontal : the total depth of the boiler is 7$\frac{1}{2}$ feet.

The bottom of the boiler exposes (3$\frac{2}{3}$ by 12$\frac{1}{4}$ =) 44 square feet to the fire and flame. An internal tube is fixed horizontally in the boiler, and extends all the length, through the mass of the contained water, so as to be entirely surrounded on all sides by the water; and the flame, smoke, and heated air, after passing from the fire grate at the front end of the boiler, and spreading beneath all the surface of its bottom, rises up, and is conducted back again to the front, through this internal flue, in order to communicate its heat to the water. The tube is of an elliptical figure, flattened at the top and bottom; its height 2$\frac{1}{4}$ feet, and its breadth 1$\frac{2}{3}$ feet. The sides and the top part of this flue may be reckoned as effective boiling surface, so that it exposes (5$\frac{1}{2}$ × 12$\frac{2}{3}$ =) 69·7 square feet, to the action of the heat. The flame and smoke is afterwards conveyed all round the outside of the boiler, at the lower part, in the same manner as the boiler already described Plate XII. and acts upon a surface of (31 × 3 =) 93 square feet. The whole surface exposed to the heat is 206·7 square feet for 20 horse-power, which is at the rate of 10·33 square feet of surface to each horse power. The horizontal surface of the water in the boiler is (12$\frac{2}{3}$ × 4 =) 50$\frac{2}{3}$ square feet.

The fire grate is 3$\frac{1}{2}$ feet wide by 4 feet long, or 14 square feet of horizontal surface, which is at the rate of ·7 of a square foot for each horse power.

When the engine was first erected, it had two such boilers, to be used alternately, one of them working, whilst the other was cleaning. One of these boilers has since been replaced by a larger one of 16 feet long, but of the same diameter and depth : this large boiler is found to be a better proportion than the original, which requires hard firing to supply the engine sufficiently with steam when it is loaded to its utmost, although it is sufficient when the engine is not overloaded. The surface exposed to the heat in the 16 feet boiler is 261 square feet, or at the rate of 13 square feet to each horse power.

The air-pump is 15 inches diameter, = 225 circular inches area; therefore the area of the cylinder is 2·56 times the area of the air-pump : the bucket makes rather more than 2$\frac{1}{2}$ feet stroke; the effective capacity of the pump, is about one-tenth of that of the cylinder.

The apertures of the steam, and exhausting-valve seats, are 4$\frac{1}{8}$ inches diameter, or about 17 square inches of uninterrupted passage, that is, ·85 of a square inch to each horse power. The steam and eduction pipes are 5$\frac{1}{4}$ inches diameter withinside : the area of the steam passages bear very nearly the same proportion to that of the cylinder, as those of the 10 horse-engine, p. 490; but as the motion of the piston is 210 feet per minute, instead of 200 feet, the velocity with which the steam must pass through the apertures is greater than is there stated.

The exhausting-valves of this engine are inverted, on the principle stated at p. 433, in order that they may open very quickly by the pressure of the steam. The construction may be easily understood, by supposing the bell-metal seats of the two exhausting-valves h and i, fig. 8 and 9, Plate XIII., to be fixed in the same place as is there represented, but inverted, so that the valves are below the seats instead of above them; and therefore the valves open the passages when they are let downwards into the boxes h and i; or the valves shut the passages when they are raised upwards; consequently the pressure of the steam tends to open the valves, and the working gear is adapted to resist this force, and to retain the valves so close shut as to prevent any leakage. The racks and sectors, to give motion to these inverted valves, are like those represented in figures 8 and 9, except that the racks are connected with the valves themselves, in the same manner as the steam-valves are, instead of being affixed to the spindles of the valves.

The working gear is made without any catches, to retain the handles in that position which occasions the valves to be shut, but the levers of the working gear are so arranged, that they will be turned into a particular position when the valves are closed, and they will retain that position, in the manner stated at p. 433, in opposition to the greatest force that the exhausting-valves can ever exert to open them. Each of the handles for this working gear requires to be moved both ways, by the chocks of the plug or air-pump rod, in order to lift up each handle, as well as to put it down; consequently there is no risk that all the four valves can ever remain shut at once, which may happen by accident, in an engine whose valves are opened by the mere action of weights, in case a careless person takes the handles, and confines them from moving; the motion of the engine would then be stopped all at once, as is stated in p. 463, and some parts may be broken by the strain.

The piston is 8 inches thick, and its cylindrical edge, which applies to the interior surface of the cylinder, is $7\frac{1}{2}$ inches deep; of which $4\frac{1}{2}$ inches is composed of the hemp packing, and the cast iron rims above and below that packing are each $1\frac{1}{2}$ inc. deep; the latter are accurately fitted into the cylinder, to prevent the hemp from escaping past them; but it is not intended that the iron rims of the piston shall be pressed against the interior of the cylinder with any force, so as to occasion friction; the real fitting of the piston is intended to be confined to the zone or belt of hemp, $4\frac{1}{2}$ inc. deep, which encompasses its outer edge. The upper cast iron rim, is that of the moveable cover of the piston, which is screwed down by six screws, as already described (p. 476.)

The piston rod is $2\frac{3}{8}$ inches diameter (a). The joint pins of the main links of the parallel motion, and also those of the connecting rod, are $2\frac{3}{8}$ inches diameter, and $2\frac{1}{2}$ inches length of bearing in their brass sockets. The links K of the parallel motion, are not made in the manner described at p. 472, but each link is a solid iron bar, $2\frac{1}{2}$ inches by $1\frac{1}{4}$ inches; the section of the two is $6\frac{1}{4}$ square inches of iron, to endure the whole strain of the piston.

The great lever is an oak beam 16 feet long, from centre to centre, and 19 inches deep, by $16\frac{1}{2}$ inc. broad. Its axis is cast iron, $4\frac{3}{4}$ inches square, and $2\frac{1}{2}$ feet long between the bearings; the pivots at each end of the axis are $4\frac{1}{8}$ inches diameter, and $5\frac{3}{4}$ inches long in the bearing parts. The axis is fastened to the wood beam by four wrought iron bolts, with nuts on the tops; the screw parts are $1\frac{5}{8}$ inches diameter, outside of the threads; and the section of the solid iron in each screw is $1\frac{1}{2}$ square inches, or 6 square inches in the four.

The joint pin, which connects the main links K of the parallel motion with the end of the great lever, is fastened beneath the beam by two bolts, which pass up

(a) The original piston rod was $2\frac{3}{8}$ inches diameter, and when the author first took the dimensions, in 1805 he found it reduced in the working part to about $2\frac{1}{4}$ inc. by wearing; it had then been working for 13 years. In 1818, after 26 years' constant service, it had become worn to $2\frac{1}{8}$ inches diameter in the part most exposed to friction; and it was then taken out, and replaced by a new piston and rod. The two iron rims of the piston were also worn so much smaller than the cylinder, in the course of the first 20 years, as to require to be surrounded by two thin hoops of brass, which were fastened round the circumference of the rims with small screws; these hoops were almost worn away, when the piston was laid aside.

3 s

through the wood, and are fastened by nuts screwed on the upper ends; the screw parts are $1\frac{1}{2}$ inches diameter outside the threads, and the section of the solid iron in each screw is $1\frac{1}{4}$ square inches, or $2\frac{1}{2}$ square inches in both; these bear all the force of the piston. The joint pins of the links, by which the rod of the air-pump is connected with the great lever, are $1\frac{5}{8}$ inches diameter.

The connecting rod is of cast iron, about 15 feet long, and 5 inches square, with the angles off. The sun and planet-wheels are $2\frac{1}{2}$ feet geometrical diameter, with 42 teeth in each; the teeth are $2\frac{1}{4}$ inches pitch, and 5 inches broad. The planet-wheel with all its teeth, is in one piece of cast iron. The sun wheel is also made of cast iron, but with wooden teeth or cogs driven into mortises in the iron. The cogs are $2\frac{1}{4}$ inches pitch, and $5\frac{1}{4}$ inches broad; they are made of live oak, and they last about two years, when they become so worn, as to require to be renewed.

The axis of the sun-wheel and fly-wheel is cast iron, $6\frac{3}{4}$ inches square, and about 5 feet long between the bearings. The neck is 6 inches diameter, and 6 inches long, in the part which rests in the brass bearings: the pivot, at the other end of the axis, is $4\frac{1}{8}$ inches diameter, and $5\frac{3}{4}$ inches long in the bearing.

The fly-wheel is 14 feet diameter outside, and the rim is 6 inches square $= 36$ square inches area. The fly-wheel was originally that of a 12 horse engine, and the rim has been made up to its present dimensions, by fastening segments of cast iron against the original rim of the wheel.

The rim of the fly-wheel contains $10\cdot6$ cubic feet of cast iron, or ($\times 450$ lbs. =) 4770 lbs. weight; it makes 42 revolutions per minute, and the velocity of the central part of the rim is $29\cdot7$ feet per second. The height through which a body must fall to acquire that velocity, is $13\cdot7$ feet; which $\times 4770$ lbs. weight, gives 65 349 lbs. (or $19\cdot17$ tons) acting through a space of one foot, for the mechanical power which constitutes the energy of the rim of the fly-wheel. The mechanical power exerted by the piston in a half stroke, is (3142 lbs. $\times 5$ feet stroke =) 15 710 lbs. acting through a space of one foot. Hence the mechanical power which must be communicated to the rim of the fly-wheel, to produce its motion from rest, is (65 349 $\div$ 15 710 =) $4\cdot16$ times the power that the engine exerts during one half stroke. This is a greater proportion of fly-wheel than the 10 horse engines before stated.

The moving parts of this engine are supported in a wooden framing, of the same form as is shown in Plate XI. The main posts V V, and the cross beams V and W, to support the centre of the great lever, are of oak, 13 inches by 14 inches; and the frame for the fly-wheel, is composed of beams of the same size. The engine-house is very wide, and the two boilers are placed within the house, one at each side of the engine. The cross beam W extends all across the wide house. The condensing cistern is of a large size, and extends quite under the cylinder; the cross beams XX, on which the cylinder stands, are supported at each end by two sloping oak posts, of 9 inches by 10 inches, which stand up at each side of the cylinder, from the lower cross beam X, to the spring beams U U.

The machinery which is impelled by this engine is as follows: A forcing pump of $11\frac{1}{4}$ inches diameter, and its piston making a $2\frac{1}{2}$ feet stroke, is worked by the great lever of the engine, in place of a cold water pump R, in order to force up water, about 36 feet high, from the level of the Thames, into a cistern at the top of the dye-house. A logwood-mill, or machine for rasping dye-wood into powder: it is a cylindrical rasp, or roller of about $2\frac{1}{2}$ feet diameter, and $2\frac{1}{2}$ feet long, with 60 knives, or cutters, placed round its circumference: the axis is horizontal, and it is turned round about 20 times per minute. The pieces of logwood are presented to the cylinder, and pressed up very hard, by means of rack-work, against the revolving knives, which cut and rasp away the wood into powder. A chipping wheel, for cutting up logwood into chips: it is a large vertical wheel of cast iron, with four sharp knives fixed into openings in it, so that their cutting edges protrude a very little beyond the smooth flat surface of the wheel, in a suitable manner to cut away thin chips from the ends of pieces of logwood, which are presented to the revolving knives ; the action of the knives is similar to that of a carpenter's plane.   Two pairs of mill-stones, or runners upon edge, to grind the logwood chips to powder.   A pair of small mill-stones, like those of a flour-mill ; they are used to grind indigo to powder.   Also some other small machines for grinding and mixing indigo.   Eight pairs of stocks for washing woollen cloth, similar to those worked by the 10 horse engine before described.   Also some small machines for agitating cloth in water to wash it, and machines for brushing the dyed cloth when finished.

The engine would be loaded to its very utmost, to impel all this machinery at once ; and some part or other is always disengaged ; but the engine usually works with a greater load than 20 horse power.   The consumption of coals is from 2 to 3 bushels per hour, or about 2½ bushels per hour on an average.

A 20 horse engine without sun and planet-wheels.   The following are the dimensions of some parts of a 20 horse engine, which was put up by Messrs. Boulton and Watt, about the year 1801, to turn machinery at a public work in London. The engine is similar to that in Plate XI. except that the fly-wheel is turned by a simple crank, in place of the sun and planet-wheels, and therefore it makes only one revolution for each stroke of the engine.   The dimensions of this engine are nearly the same as those of the 20 horse engine already quoted, except the follow-ing.   It has an iron boiler of the waggon shape, 14 feet long, 5 feet wide, and about 7 feet deep ; the horizontal surface of the water is, therefore, 70 square feet : this boiler is without any internal tube for the flue.

The surface exposed to the fire beneath the bottom is about (4½ × 14 =) 63 square feet, and the surface exposed in the external flue is about (4 × 37 =) 148 square feet : total 211 square feet for 20 horse power, or at the rate of 10·5 square feet to each horse power.   The fire grate is 4 feet square, or 16 square feet, for 20 horse power ; which is at the rate of ·8 of a square foot for each horse power.

The cold water pump is 7 inches diameter = 49 circular inches, and the bucket makes a 2½ feet stroke, or half the length of the stroke made by the piston. The effective capacity of the steam cylinder is, therefore, 47 times that of the cold water pump, reckoning merely by the spaces occupied by the piston, and by the pump bucket, without any allowances for loss by the vacant spaces.   With due allowances for those vacancies, the quantity of steam which passes through the cylinder is about 54 times the quantity of the water, which is thrown into the con-densing cistern by the cold water pump.

The connecting rod is made of cast iron in one piece ; the pin which is fixed in the crank, and projects out, to joint the connecting rod to, is 4¼ inches diameter, and 5½ inches length of bearing in its socket.   The crank is made of cast iron, and is fastened by wedges, upon the square extremity of the projecting end of the cast iron axis of the fly-wheel.   This axis is 7¾ inches square, and 5 feet long between the bearings ; the neck of the axis is 7½ inches diameter, and 8½ inches length of bearing in its brass socket.   The fly-wheel, which is fixed upon the middle of this axis, is 18 feet diameter outside, and the rim is 7 inches by 6 inches, = 42 square inches area.

The rim of this fly-wheel contains 15·95 cubic feet of cast iron, or (× 450 lbs. =) 7177 pounds weight, independently of the arms.   It makes 21 revolutions per minute, and the velocity of the middle part of the rim is then 19·15 feet per second.   The height due to this velocity is 5·7 feet, and the power which is resident in the rim, to produce its motion from rest is (5·7 × 7177 lbs. =) 40 909 lbs. (or 18·27 tons) acting through a space of one foot.   The force exerted by the piston of the engine in each half stroke is a force of (3142 lbs. acting through 5 feet =) 15 710 lbs. acting through a space of one foot.   Therefore the energy of the rim of this fly-wheel is (40 909 ÷ 15 710 =) 2·6 times the power exerted in each half stroke.

This engine was employed to turn 8 pairs of rolling stones, called runners upon edge, for grinding and mixing up the cement for the mortar used in building public docks. It was but lightly loaded, the resistance of the machinery being less than 20 horse power: the engine consumed 2 bushels of coals per hour upon an average.

An engine of the same dimensions was also set up, about the same time, at another of the docks which were then building in London. And for a smaller work, a 10 horse engine was erected, of nearly the same dimensions as that already quoted in p. 489, and it turned four runners upon edge, for grinding cement or mortar in the wet state, in the same manner as the other engines.

DIMENSIONS OF A 30 HORSE POWER PATENT ROTATIVE ENGINE MADE BY MESSRS. BOULTON AND WATT, 1798.

This engine was erected for a cotton spinning mill at Manchester; it is on the same construction as that represented in Plates XI. XII. and XIII. in almost every respect, except the dimensions, which are as follows (a).   The cylinder is $28\frac{1}{4}$ inches diameter, = 626·8 square inches area, or 798 circular inches.   Length of the stroke of the piston 6 feet, and it makes 19 double strokes per minute, so as to pass through a space of 228 feet per minute.   The expenditure of steam is 993 cubic feet per minute, or at the rate of 33·1 cubic feet per minute for each horse power.

The effective pressure on the piston, taking a mean of all its variations, is 6·93 lbs. per square inch, and (× 626·8 squ. inc. =) 4344 lbs. is the mean effective force of the piston; this force acting through 228 feet per minute, is equal to 990 000 lbs. raised one foot per min., or (÷ 33 000 lbs.) = 30 horse power.

The utmost force that the piston can ever exert, being at the rate of 11 pounds per circular inch, is (× 798 cir. inc. =) 8778 lbs.; and the strength of all the parts of the engine must be adapted to endure that force without any risk of breaking.

The boiler is made of iron plate, of a waggon shape, 5 feet diameter, 16 feet long, and $7\frac{1}{2}$ feet deep, with an internal flue to convey the flame through the water.   The horizontal surface of the water in the boiler is (16 × 5 =) 80 square feet.   The fire grate is $4\frac{1}{2}$ feet wide, by 5 feet long, = $22\frac{1}{2}$ squ. feet; or at the rate of $\frac{3}{4}$ of a square foot to each horse power.

The surface exposed to the flame is ($4\frac{1}{2}$ ft. by $15\frac{1}{2}$ ft. =) 69·8 square feet beneath the bottom of the boiler.   The internal flue, for the passage of the flame and smoke, is of an elliptical figure flattened at top; it is $2\frac{1}{4}$ feet deep, and 2 feet wide.   The surface of the sides, and the upper part of the internal flue, is about ($6\frac{1}{2}$ × 16 =) 104 square feet. The surface exposed in the flue which passes round the lower part of the boiler, is (38 × $3\frac{3}{4}$ =) 142·5 square feet. The whole surface exposed to the heat is 316·3 square feet for 30 horse power; this is at the rate of 10·52 square feet of boiler surface to each horse power.

The apertures of the valves are $5\frac{3}{4}$ inches diameter, and the areas of the uninterrupted passages through them are about 24 square inches, which is at the rate of ·8 of a square inch to a horse power, the same as the other engines before stated.

The air-pump is 19 inches diameter, = 361 circular inches area; therefore

(a) Messrs. Boulton and Watt's 30 horse engines were sometimes built with a lever wall, and sometimes they were mounted in wooden framing; their 36 horse engines were always built with lever walls.

the area of the cylinder is 2·21 times that of the air-pump. The air-pump bucket makes a stroke of 3 feet long, and the effective capacity of the cylinder is 8·84 times that of the air-pump.

The piston rod is $2\frac{3}{4}$ inches diameter. The joint pins of the main links of the parallel motion, and of the main joint for the connecting rod, are $3\frac{1}{4}$ inches diameter, and $3\frac{1}{2}$ inches length of bearing, in their sockets. The iron bands of which the head links are composed, are $2\frac{1}{2}$ inches wide, by $\frac{5}{8}$ thick ($= 1\cdot56$ square inches area), or the united areas of all the four bands is $6\frac{1}{4}$ square inches, to sustain all the force of the piston.

The great lever is $18\frac{1}{2}$ feet long; the depth of the beam 24 inches, and the breadth 19 inches. The axis of the great lever is $5\frac{1}{2}$ inches by $6\frac{1}{2}$ inches deep, and $2\frac{1}{2}$ feet long between the bearings. The pivots at the ends of the axis are $4\frac{1}{2}$ inches diameter, and 6 inches length of bearing in their sockets.

The upper part of the connecting rod is an oak beam, about 10 inches broad, by 9 inches thick, and the whole length of the rod is 18 feet from the upper joints to the centre of the planet-wheel. The joint at the upper end of the connecting rod is attached to the wood by two long iron straps, which apply at each side of the wood, and are fastened to it by several bolts which pass through the wood, and also through both straps. At the lower end of the wooden connecting rod, a short length of cast iron is fastened by several bolts, in order to prolong the rod sufficiently to attach the planet-wheel to the lower end. This piece of cast iron is fastened to the outside face of the wooden rod, so as to project beyond the central plane of the engine; and the planet-wheel being fixed at the back or inner face of the cast iron rod, the wheel comes in the central plane of the engine; for by the manner of combining the wood and the iron parts of the rod, the whole rod is of a crooked form, such as is represented in fig. 5, Plate XII. so that the lower, or cast iron part of the rod, passes clear in front of the sun-wheel, although the upper part of the rod is exactly in the plane of the great lever, and of the sun and planet-wheels.

The sun and planet-wheels are 3 feet geometrical diameter, with 39 teeth, of 2·9 inches pitch, and $5\frac{1}{2}$ inches broad; both wheels are made of cast iron, with solid teeth.

The axis of the fly-wheel is of cast iron, 8 inches square, and about 5 feet long between the bearings; the necks are $7\frac{1}{4}$ inches diameter, and $7\frac{1}{2}$ inches length of bearing in their sockets. The centre pin, or axis of the planet-wheel, projects outwards from it on both sides; the front end of the pin passes through the connecting rod, and the extremity, which projects through in front of the wheel, is connected by a radius link, with a fixed pin, which is situated exactly in the line of the axis of the sun-wheel and fly-wheel: this link assists the back link q, fig. 5, Plate XII. to retain the centre of the planet-wheel very steadily in its circular orbit, because, on this plan, there are two radial links, one applied in front of the sun and planet-wheels, and the other behind them; the back link is fitted upon the neck of the axis of the fly-wheel, as before described, and as is represented at $q\ q$ in the figure; the front link is fitted upon the central pin, which is fixed very fast in front of the end of the axis of the sun and planet-wheels, upon a wood framing, similar to that which supports the axis of the fly-wheel.

The fly-wheel is 16 feet diameter outside, and the rim is 8 inches by $4 = 32$ square inches area of the cross section. The fly-wheel has 6 arms, and makes 38 revolutions per minute.

The rim of this fly-wheel contains 10·7 cubic feet of cast iron, or ($\times$ 450 lbs. $=$) 4815 pounds weight. The middle part of the rim moves with a velocity of 30·5 feet per second, and the height due to that velocity is 14·46 feet. The mechanical power which must be communicated to the rim of this fly-wheel, to give it motion from rest, is equal to that of a force of (4815 lbs. × 14·46 ft. =) 69 600 pounds, (or 31·1 tons) acting through a space of one foot.

The power exerted by the engine during one half stroke is (4344 lbs. × 6 ft. =) 26 064 pounds acting through a space of one foot, hence the energy of the rim of the fly-wheel is (69 600 lbs. ÷ 26 064 lbs. =) 2·67 times the power exerted by the engine in each half stroke. The energy of this fly-wheel is smaller in proportion to the power of the engine, than that of the fly-wheels of some of the other engines before quoted; and the motion of this engine and its machinery is sensibly variable at each stroke of the engine, though not so irregular as to be a great defect.

The cold water pump is 10 inches diameter = 100 circular inches area, and its bucket makes a stroke of nearly 2 feet, or about one-third of that of the piston. The effective capacity of the steam cylinder is 47·88 times the effective capacity of the cold water pump, without making any allowance for the steam which is wasted in filling the vacant spaces; or when due allowances are made, the quantity of steam which is expended by the cylinder will be found to be 55·6 times the quantity of the cold water that is raised by the pump.

This engine turned all the machinery of one of the largest cotton spinning mills in Manchester for about ten years; the mill then contained 28 000 mule spindles for spinning the very finest yarn; the engine was fully loaded, so as to exert more than 30 horse power, and its furnace consumed at the rate of from 3½ to 4 hundred weight of coals per hour. After this engine had been in use about ten years, a large addition was built to the cotton mill, and a new engine was put up in the new part, having a sufficient power to turn the whole mill. The 30 horse engine was then removed, and put up at another manufactory, for a different purpose, where it remains still in use, but some alterations were made in the valves and working gear when it was removed.

Several other 30 horse engines, of the same dimensions as the above, were set up by Messrs. Boulton and Watt for cotton mills, a few years before the expiration of the prolonged term of Mr. Watt's original patent; most of these were at Manchester, and the neighbourhood, one at Liverpool, another in Scotland. One of these 30 horse engines gave motion to 18 000 mule spindles, for spinning ordinary cotton yarn. A 30 horse engine was the power requisite to impel the machinery of a first rate cotton mill, as those mills were built at that time; most of these engines are still in use, though other and larger engines have been since added to the same establishments, suitable to the enlargements which have been made in the buildings and machinery; several of them have had great levers of cast iron applied in place of the wooden ones, and also the valves and working gear have, in some instances, been removed, and others of a more modern fashion substituted.

DIMENSIONS OF A 40 HORSE POWER PATENT ROTATIVE-ENGINE, MADE BY MESSRS. BOULTON AND WATT, 1792.

The construction of the moving parts of this engine is nearly the same as that represented in Plates XI. and XII., but those moving parts are not supported in a wooden framing; the bearings for the axis p, which forms the centre of motion for the great lever, are fixed upon the top of a strong lever wall, which is built of solid masonry across the engine-house, between its two side walls, in the same manner as the engine for pumping water, Plate X. This lever wall divides the engine-house into two compartments, one containing the cylinder and its working gear, air-pump and condenser; and the other containing the connecting rod, sun and planet-wheels, and the fly-wheel.

The cylinder is fastened down upon a strong pier of solid masonry, in the same manner as the single engine in Plate X. The wood frames to sustain the bearings of the axis of the fly-wheel, are fastened at one end into the great lever wall, and at the other end into the end wall of the house, so that they are very solidly supported; the spring beams are supported across the lever wall in the middle, and their ends are worked into the end walls of the engine-house. In other respects the description already given of the 10 horse engine, will answer to this 40 horse engine, and its dimensions are as follows.

The cylinder is $31\frac{1}{2}$ inches diameter = 779·3 square inches area, or 992·25 circular inches. The length of the stroke of the piston is 7 feet, and it makes $17\frac{1}{2}$ strokes per minute, so as to pass through 245 feet per minute.

(a) The expenditure of steam by the motion of the piston is 1326 cubic feet per minute, or at the rate of 33·15 cubic feet per minute for each horse power. The mean effective pressure upon the piston is 6·92 pounds per square inch, and the mean effective force of the piston is 5390 pounds, which acting through 245 feet per minute is = 1 321 000 lbs. acting through one foot per minute, or ÷ 33 000 = 40 horse power. The utmost pressure or force exerted by the piston, being estimated at the rate of 11 pounds per circular inch, is 10 915 pounds (= 4·87 tons), and the strength of all the parts of the machinery must be calculated to resist that strain, without any danger of breaking.

The engine has two boilers, each one is $5\frac{1}{2}$ feet diameter, and 20 feet long, with an internal flue through it; one of these boilers is sufficient to supply the engine with steam.

The surface exposed beneath the bottom is ($5 \times 19\frac{1}{2}$ =) 97·5 square feet. The surface exposed at the top and sides of the internal flue is ($6\frac{1}{2} \times 20$ =) 130 square feet, and the surface exposed in the external flue, around the lower part of the boiler, is (48×4 =) 192 square feet. Total of the surface exposed 419·5 square feet for 40 horse power, or at the rate of 10·05 square feet of surface for each horse power.

The air-pump is 21 inches diameter = 441 circular inches, so that the area of the piston is $2\frac{1}{4}$ times that of the air-pump bucket; and the length of the stroke made by the bucket being 3 feet, or half of that of the piston, the effective capacity of the cylinder is 9 times that of the air-pump.

The cold water pump is 10 inches diameter = 100 circular inches area, and its bucket makes a stroke of $2\frac{1}{2}$ feet. The effective capacity of the cylinder is 47·6 times that of the cylinder, or making due allowance for the steam wasted by the vacant spaces, the quantity of steam expended by the cylinder is about 55·1 times the quantity of cold water raised by the pump.

The diameter of the piston rod is $3\frac{1}{8}$ inches. The joint pins of the main links of the parallel motion, and of the connecting rod, are $3\frac{1}{4}$ inches diameter, and $3\frac{1}{2}$ inches length of fitting in their sockets. The iron of the straps for the main links is 3 inches broad, by $\frac{5}{8}$ of an inch thick = $1\frac{7}{8}$ of a square inch area, or there is $7\frac{1}{2}$ square inches of iron to endure the whole force of the piston.

The great lever is an oak beam, $21\frac{1}{4}$ feet long between the centres of the joints at each end; the depth of the timber is 28 inches, by 20 inches broad. The axis of the great lever is 6 inches by 7 inches deep, and $2\frac{3}{4}$ feet long between the bearings; the pivots at each end are $5\frac{1}{4}$ inches diameter, and 7 inches length of bearing in their sockets. The axis is bound to the wood by four bolts, with screws of $1\frac{3}{4}$ inches diameter outside the threads.

The connecting rod is an oak beam, 11 inches by 10, with a cast iron stem

(a) Messrs. Boulton and Watt afterwards made some of their 40 horse engines with a 6 feet stroke, to make 19 strokes per minute = 228 feet motion per minute, instead of 245; and to make up for the deficiency of speed, the cylinder was made $32\frac{1}{2}$ inches diameter, instead of $31\frac{1}{2}$ inches. The expenditure of steam by the motion of the piston is 1314 cubic feet per minute, or 32·8 cubic feet per minute to each horse power.

bolted to it at the lower end, to fasten the planet-wheel to : the length of the connecting rod, from the joint at the upper end to the centre of the planet-wheel, is about 20 feet.

The sun and planet-wheels are both made of cast iron, with solid teeth ; each wheel is $3\frac{1}{2}$ geometrical feet diameter, with 44 teeth of 3 inches pitch, and 6 inches broad. The centre pin of the planet-wheel is retained in its circular orbit by two radial links, one applied in front of the wheels, and the other at the back, in the same manner as already mentioned (p. 501) for the 30 horse engine. The sun-wheel is fastened upon the square end of the axis of the fly-wheel by wedges ; that axis is $9\frac{1}{2}$ inches square, and about 6 feet long between the bearings : the necks of the axis are $8\frac{1}{4}$ inches diameter, and 10 inches long in their bearings.

The fly-wheel is 18 feet diameter outside, and the rim is 9 inches by 5, or 45 square inches in the cross section ; the wheel has 8 arms.

The rim of the fly-wheel contains 16·94 cubic feet of cast iron, or (× 450 lbs. =) 7623 pounds weight ; it makes 35 revolutions per minute, and the middle of the rim then moves with a velocity of 31·6 feet per second. The height due to this velocity is 15·53 feet, × 7623 lbs. weight, gives 118 400 lbs. (= nearly 53 tons) acting through a space of one foot for the energy of the rim of this fly-wheel. The power exerted by the piston in each half stroke is 5390 pounds, acting through 7 feet, which is equivalent to 37 737 lbs. (=16·85 tons) acting through one foot. And, lastly, the energy of the rim of the fly-wheel is (118 400 lbs. ÷ 37 737 lbs. =) 3·14 times the power exerted by the piston in each half stroke.

This engine was put up at an iron works in Staffordshire, to turn a rolling mill for rolling out iron into bars, and it continues still in use, but some of the principal parts have been renewed, and the plan of the wheelwork has been altered. A spur cog-wheel, of about 6 feet diameter, was originally fixed upon the axis of the fly-wheel ; and it turned another spur-wheel, of about 8 feet diameter, to the axis of which the rollers for laminating iron bars were connected, so that they made about 26 turns per minute ; and the rollers being about 13 inches diameter (= 3·4 feet circumference), the iron was rolled through between them at the rate of about 88 feet per minute, which was considered a sufficient speed for rolling iron, at the time when this engine was put up.

After this engine had been in use for some years, an additional fly-wheel was applied to the millwork, in order to accumulate a greater energy in the moving parts, and thus enable the rollers to draw through a large piece of metal, without stopping or sensibly impeding the motion of the engine. For this purpose, a large spur cog-wheel was fixed upon the axis of the sun-wheel and fly-wheel, in place of the original spur-wheel, of 6 feet diameter ; this large wheel turns another spur-wheel, of about half the size, so that the latter makes about 60 revolutions per minute. On the axis of this is fixed the second fly-wheel, of 16 feet diameter, its rim being 10 inches by 7 inches, = 70 square inches of cross section ; and to the same axis the rollers are connected, so that they turn nearly 60 times per minute, and the metal passes through the rollers at the rate of 188 feet per minute. The engine, in this state, does considerably more work than it could do formerly, with only one fly-wheel.

It does happen occasionally that the accumulated power of both fly-wheels, in addition to that which the planet-wheel exerts to turn the fly-wheel round, is insufficient to overcome the resistance of laminating a large piece of metal, if it is put in when the rollers are too near together ; the whole of the machinery is then brought very suddenly to rest, from its state of rapid motion ; and, in such cases, the parts are very liable to be broken by the strain, which is prodigiously great.

The rim of this additional fly-wheel contains 23·15 cubic feet of cast iron, or (× 450 lbs. =) 10417 pounds (or 4·65 tons) weight. The mean velocity of the rim is 47·65 feet per second, and the height due to that velocity is 35·3 feet; which × 10 417 lbs. gives 367 720 lbs. (= 164 tons) acting through a space of one foot, for the energy of this rim; and added to (118 400 lbs.) the energy of the rim of the other fly-wheel, gives 486 120 lbs. (= 217 tons) acting through one foot, for the united energy of both rims. This (÷ 37 737 lbs.) is 12·88 times the power exerted by the engine in each half stroke of the piston.

The consumption of coals by this engine is at the rate of about 4½ hundred weight of coals per hour. The resistance occasioned by the rolling of metals, is subject to such great variations, that it is difficult to estimate what power the engine does exert; and though it is continually made to exert its very utmost force, the resistance is probably less than 40 horse power, upon an average of all the circumstances, because there are so many intervals when there is scarcely any resistance, and then the engine has but little to do, except to accumulate energy in the fly-wheels.

Messrs. Boulton and Watt made several engines similar to the above for rolling iron. One of the first rotative engines which they made in 1785, was set up by Mr. Rennie for a rolling mill at Rotherhithe in London (see p. 434); it was a 20 horse engine, and answered the purpose very well. The business of the manufactory increased so much, in consequence of the facility of making iron bars by this mill, that, in 1790, a large engine was put up in place of the original one. This engine is still in use, its dimensions are the same as those of the 40 horse engine above stated, except that the length of the stroke is only 6 feet, instead of 7 feet; and it makes about 20 strokes per minute, so as to be about 39 horse power.

The wooden connecting rod of this engine was removed after it had been in use about 12 years, and a cast iron rod, 6¾ inches square, was substituted, with a universal joint, to connect it to the extremity of the great lever (see p. 472). The sun and planet-wheels were afterwards laid aside, and a simple crank substituted, with a multiplying cog-wheel fixed upon the axis of the crank, to give a rapid motion, by a pinion, to a very large fly-wheel, which is fixed on the axis of the pinion.

The axis of the crank is 12¼ inches square, the crank is fastened upon the extreme end of the square axis by wedges. The multiplying-wheel, upon the axis of the crank, is 16½ feet diameter, with teeth of 5 inches pitch, and 10 inches broad: the rim of this cog-wheel is 3 inches thick, by 10 inches broad, and together with the teeth it contains nearly 15·9 cubic feet of cast iron, or 7160 pounds (= 3·2 tons) weight, moving with a velocity of about 17 feet per second, so that it serves as a considerable fly-wheel, its energy being 32 100 lbs. (= 14⅓ tons) acting through a space of one foot. The teeth of this large multiplying-wheel gave motion to a pinion of about 6 feet diameter, so as to turn it round about 55 times per minute: on the axis of this wheel is a fly-wheel of 19 feet diameter at the outside, and its rim is 10 inches by 12 inches, = 120 square inches area.

The rim contains 47·1 cubic feet of cast iron, or (× 450 lbs. =) 21 195 pounds weight (9·46 tons) moving with a velocity of 51·8 feet per second. This fly-wheel has 8 arms made of oak wood, 9½ inches by 8 inches; its axis is connected by wheel-work with the axis of the rollers, so that they made about 85 turns per minute.

The energy of the rim of this great fly-wheel is 885 000 pounds (= 395 tons) acting through one foot; and the united energies of the rims of the fly-wheel and of the multiplying-wheel is 917 100 lbs. (= 409 tons) acting through one foot, which is 28·3 times the force exerted by the piston in each half stroke, viz. (5390 lbs. force, × 6 ft. stroke =) 32 340 lbs. acting through one foot, and 917 100 lbs. ÷ 32 340 lbs. = 28·3 times.

This engine performed so much better with this great augmentation of energy in the moving parts, that a further alteration was afterwards made, by fixing a larger rim of teeth upon the multiplying-wheel, around the former rim of teeth, which is become useless, except by its weight in aid of the fly-wheel; a smaller pinion is fixed on the axis of the great fly-wheel, so that it is turned at the rate of 72 times per minute; and its axis is connected with the rollers by wheel-work, to turn them about 110 times per minute.

Three rollers are also applied one over the other, according to a modern improvement, in order to roll out the iron bars both ways, viz. the bar is passed between the lower roller, and the middle roller, in going through in one direction; and then the same bar is returned back again between the middle roller and the top roller, so that very little time is lost, and the operation is completed before the metal loses its heat (a). The additional rim to the multiplying-wheel is 18 feet diameter. The teeth are 4 inches pitch, and 10 inches broad; the rim is $2\frac{1}{2}$ inches thick, and together with the teeth it contains 14·5 cubic feet of cast iron, = 6525 pounds weight, or 2·91 tons.

This additional mass moves at the rate of 18·5 feet per second. Its energy is equal to 34 720 lbs. (= 15·5 tons) acting through a space of one foot. The velocity of the rim of the great fly-wheel, when making 72 turns per minute, is 67·82 feet per second, and its energy is equal to 1 522 000 lbs. (= 679 tons) acting through one foot. The united energy of all the three rims is 1 588 820 pounds (= 709 tons) acting through a space of one foot, which is 49·1 times the power exerted by the piston in one half stroke.

This prodigious energy is not the whole that the machinery possesses when in motion, because no notice has been taken of the weight of the arms of the wheels, nor of that of the great lever, connecting-rod, and piston, of the engine. It requires nearly two minutes exertion of the power of the engine, to get the fly-wheel to its full speed of 72 turns per minute, but which being attained, the iron is presented between the rollers, and the resistance of rolling it out into long bars occasions but very little retardation of the motion.

This engine is chiefly employed in the manufacture of bars from scrap iron, which consists of small pieces, and useless fragments, of all sorts of old articles of wrought iron, which are collected in London; these scraps of old iron are sorted into different qualities of metal, and are packed up into suitable parcels, which are heated to a good welding heat in a reverberatory furnace, and then welded into solid masses, by repeated blows of a very heavy forge hammer, which is worked by another steam-engine of 14 horse power. These masses are wrought by the forge hammer, at one heat, into the form of short thick bars, called blooms, which are then taken to the rolling-mill, and heated in another reverberatory furnace, to a proper heat, to be rolled out between the rollers, at one heat, into long smooth bars.

The metal is usually rolled five times backwards, and five times forwards, so that each mass is passed ten times through between the rollers, to reduce it from the form of the rough bloom, or short thick bar, of a very irregular form, into that of a smooth long bar. The engine is capable of rolling a ton of iron in about 3 hours, or 4 tons per day of 12 hours. There are two large reverberatory furnaces for heating the blooms, and each receives a number of them; and whilst the contents of one furnace is acquiring the proper heat, the other hot blooms are worked

(a) This case forms, in some degree, an exception to the general rule, that whatever is gained in time, must be by virtue of an increased exertion of force, because the lamination of the iron is so easily performed whilst the metal retains its heat, that the quicker it can be passed through the roller, the less resistance it opposes to the required change of form; the practice of rolling iron with a rapid motion is in pursuance of the old injunction, to strike whilst the iron is hot.

through the rollers, one at a time, till they are all finished ; and by that time the charge of the other furnace will be ready for rolling, so that the engine is kept at work, with very few intermissions ; or if three furnaces are employed, and additional men to relieve each other, the work may be kept on continually ; the great heat and weight of the iron bars, and the quick motion with which they must be handled, renders the labour too severe to be continued, without intervals of rest.

An engine of the same dimensions as the above was set up at Sheffield, about 1793, for a rolling-mill ; and after it had been three or four years in use, the sun and planet-wheels were removed, and a large multiplying-wheel and pinion were substituted, to turn a very heavy fly-wheel upon a second axis, with a rapid motion, in the manner already described. Another engine, of the same size, was then added to the establishment ; the wheel-work for this second mill was made after the same plan as the first engine in its improved state ; both these engines are still in use.

Messrs. Boulton and Watt's 36 horse, and 40 horse rotative-engines became a sort of standard for rolling-mills at iron works, and a great number of the kind already described, were erected in Staffordshire, Shropshire, and Yorkshire, between the years 1790 and 1800. The engine-house is built with four massive walls, of a proper size to inclose the cylinder and working gear ; one of these walls supports the centre of the great lever, but the axes of the fly-wheel, and of the crank and multiplying-wheel, as well as of the other wheel-work, and of the rollers, are supported upon very solid foundation walls, which do not rise above the level of the ground, so that the wheel-work is very securely fixed ; but as it is not inclosed by the walls, it is more accessible for repairing.

The whole of the space occupied by the mill-work, and by the rollers, and their furnaces, is covered over by a series of large roofs, to form a vast shed, which is supported upon pillars of brick-work, placed at suitable intervals apart. The boilers for the engine are placed at the end of the engine-house, beyond the cylinder end of the great lever, and the furnaces for the rolling-mill are situated at each side of the engine-house.

It is usual to apply two series of rollers, by suitable connexions of wheel-work, at each end of the axis of the great fly-wheel ; each set of rollers being adapted for a particular purpose, such as for rolling bars, or for plate, or for round bolts, or for nail rods. A 40 horse engine is not sufficiently powerful to impel both sets of rollers at once, except for rolling light work ; the rollers are therefore used alternately, to manufacture the different kinds of iron, for which they are intended.

At some iron-works rolling-mills were erected with large atmospheric engines, with cylinders of 48 and 50 inches diameter ; they acted with cranks, and multiplying-wheels and pinion, for the fly-wheel ; these performed very well, but it was found better to add condensers and air-pumps to them. Mr. Watt's complete single engine, such as is represented in Plate X., was in some instances applied to turn a rolling-mill, by means of a connecting rod and crank, with a multiplying-wheel and pinion, to turn the fly-wheel with a sufficiently rapid motion.

As it is requisite to employ such very powerful fly-wheels in rolling-mills, a single engine will impel them very well, though a double action is much better for that purpose, as well as for all other rotative engines. A rolling-mill was erected, about 1794, at an iron works in Scotland, with a single engine, having a cylinder 52 inches diameter, and the piston making an 8 feet stroke ; it exerted about 56 horse power, but the strength of all the parts of the engine, required to be twice as great as if it had been a 56 horse double engine, with a cylinder of only 37 inches diameter.

The 40 horse rotative-engine was a larger size than was commonly required for the use of manufactories, at the first establishment of steam-mills ; but, in 1793, Messrs. Boulton and Watt made a 40 horse engine of the kind already described, and of nearly the same dimensions, for a large woollen manufactory, which was founded at that time by Mr. Gott at Leeds, in Yorkshire.   This was the first complete manufactory of that kind, which was established with an entire dependence upon steam power for its first mover ; and from the manner of its execution, this work reflects great credit upon all the persons concerned in the planning and execution of the building, the engine, the mill-work, and the machinery, for the sound judgment displayed in the general arrangement of the whole, as well as in the contrivance of the details : the works were erected under the superintendence of Mr. Peter Ewart, who had been educated under Mr. Rennie.

The principal building of Mr. Gott's manufactory is five stories high, with 24 windows in a line, making a front of more than 200 feet long : this was a larger size than any manufactory which had been built at that time, although of late years the common standard for the buildings of first rate cotton-mills at Manchester has reached to nearly that size.   It required no ordinary spirit of enterprise in the proprietor, to adventure so large a capital at once, upon a new foundation, with new machinery and new processes, at a period when they had not been extensively used in that district, and when they were still considered, by the labouring part of the population, as unjust and injurious innovations upon their rights and interests ; and when the better informed viewed such establishments as hazardous and desperate speculations.

The engine and mill-work were so well planned and executed, that every part answered its intended purpose very completely, and scarcely any material alterations have been made in it, to the present time.   After the engine had been in use about 16 years, the valves and working gear were renewed upon a more modern plan ; and the great lever was strengthened, by fastening a strong plate of cast iron upon the upper side of the wooden beam ; the engine continued in constant use, exerting between 50 and 60 horse power, until 1825, when a large addition was made to the mill, and a new engine of 80 horse power, was erected in place of the old one, which was removed to another manufactory, being still in good condition.

A very large mill was built at Sheffield, in Yorkshire, in 1797, for grinding cutlery, with two 40 horse engines of Messrs. Boulton and Watt's plan, placed in the same building ; they were executed by Mr. Francis Thompson, under licence from the patentees, who supplied some of the parts from their manufactory.

DIMENSIONS OF MR. WATT'S 50 HORSE POWER, PATENT ROTATIVE-ENGINES, AT THE ALBION FLOUR-MILLS, 1786 AND 1789.

This was the first establishment which was founded to be actuated entirely by Mr. Watt's rotative-engines, as has been already stated, pp. 438 and 442.   The flour-mills were worked by two engines of 50 horse power each, and it was intended to have added a third engine, but the mills were burned before it was made. Mr. Watt took great pains to proportion the parts of these engines correctly, according to his previous experience from his single engines for pumping water ; and he succeeded so well, that the first engine at the Albion Mills became, in a great degree, a standard for the proportions of all his other engines, and these proportions have not been materially altered to the present time.   The construction of the first of these engines, which was made in 1785, was similar to that of the engine represented in Plate XI., except in the following particulars.

The engine was not placed in a wooden framing, but the great lever was supported upon a wall 3 feet thick, in the same manner as the 40 horse engine described at p. 502. The parallel motion was not the same as is represented in Plate XI., but it was on the construction described by the sketch, p. 430. The exhausting-valves were inverted, on the plan described at p. 497; and a second air-pump, or large hot-water pump, was applied to this engine, to take the hot water away from the delivering-valve *m*, of the principal air-pump H, in order to relieve that valve from the pressure of the atmosphere, as is mentioned p. 325 and 330.

The engine had two fly-wheels, and two pairs of sun and planet-wheels, one situated at each side of the connecting rod; these fly-wheels were fixed upon two separate axes, which were placed exactly in the same line, one at the end of the other, so that the two sun-wheels, which were fixed upon the extreme ends of the two axes respectively, were exactly opposite to each other, but leaving a clear space between them for the connecting rod to pass through, in the motion which it had when the planet-wheels circulated round the sun-wheels.

Each of the axes of these two sun-wheels and fly-wheels, gave motion to a separate flour-mill, containing 5 pairs of mill-stones. To compel the two fly-wheels to turn round, with an exactly corresponding motion, a large spur-wheel, of about $7\frac{1}{2}$ feet diameter, was fixed upon the axis of each fly-wheel; and these two wheels worked into two other spur-wheels, of the same size, which were fixed upon a strong horizontal axis, 7 inches diameter; this axis extended from one wheel to the other, and formed the connexion between the axes of two fly-wheels. The centre of this connecting axis was nearly on the same level with the axes of the fly-wheels, and being $7\frac{1}{2}$ feet distant therefrom, room was left for the planet-wheels to go round clear of it; the bearings for the necks of this axis were supported by the same wood framing as the bearings for the axes of the fly-wheels. In all other respects the construction of this engine was the same as is represented in Plate XI.

The second engine, which was put up in 1789, was the same size as the first; the parallel motion represented in Plate XI. was used, and the valves and working gear were constructed as is shown in Plate XIII. This engine had also two air-pumps, and two pairs of sun and planet-wheels, and two fly-wheels, to turn two separate flour-mills; but the four spur-wheels, and their connecting axis, not being found necessary, were omitted. This second engine is described by Mr. Watt, in his Appendix to Dr. Robison's article Steam-engine, with four engravings, from which some of the following dimensions have been measured; and others have been obtained from different persons who had examined these engines, when they were first brought into use, and when they were an object of general interest.

The cylinder was 34 inches diameter, = 908 square inches, or 1156 circular inches area, or 6·3 square feet. The length of the stroke of the piston was 8 feet, and in regular working it was intended to make 16 strokes per minute, and then the piston passed through a space of 256 feet per minute.

The space occupied by the piston in its motion occasioned an expenditure of 50·4 cubic feet of steam at each half stroke, or (× 32 =) 1614 cubic feet of steam expended per minute, without any allowance for loss, by the vacant spaces at the top and bottom of the cylinder; this quantity divided by 50 horse power, gives 32·28 cubic feet per minute for each horse power. This would produce an effective force of the piston of 7·1 pounds per square inch, independently of friction.

The space left vacant at the top of the cylinder, when the piston was at its highest position, contained about 2·85 cubic feet; viz. the clear space above the piston was about 4 inches, = 2·1 cubic feet; and the passage *c*, fig. 8, Plate XIII., to the upper valves, was 5 inches by 11 inches, and nearly 2 feet long, so that it contained about ·75 of a cubic foot. The piston rod was 3·4 inches diameter, and the capacity of 8 feet length, was ·5 of a cubic foot. The quantity of steam expended in making the descending stroke was (50·4 + 2·85 − ·5 =) 52·75 cubic feet.

The space left vacant at the bottom of the cylinder, when the piston reached its lowest position, was about 3 inches; and including the passage (*f*, fig. 8) to the lower valves, the vacancy was about 2·85 cubic feet, so that about 53·25 cubic feet of steam was expended to produce the ascending stroke.

To make a complete stroke (52·75 + 53·25 =) 106 cubic feet of steam was expended, or × 16 strokes per minute, = 1696 cubic feet expended per minute (*a*).

The effective force of the piston, when the engine was exerting 50 horse power, was at the rate of 7·1 pounds per square inch of the piston, exclusive of friction; for the efficient force of the piston to impel the mill-work was (7·1 lbs. per squ. inc. ×908 squ. inc. =) 6446 pounds, taking the mean of all its variations. As this force acted through a space of 256 feet per minute, it was equivalent to a force of 1 650 176 lbs. acting through one foot per minute, or ÷ 33 000 = 50 horse power.

The additional force requisite to overcome the friction of the parts of the engine, and to give motion to its air-pump and cold water-pump, is commonly about 1⅔ pounds per square inch of the piston, for an engine of this size, so that the actual force, or unbalanced pressure against the piston, must have been about 8¾ pounds per square inch, taking the mean of all its variations, through the whole stroke.

The utmost force which can ever be exerted by the piston may be taken at 11 pounds per circular inch, or (× 1156 circ. inc. =) 12 716 pounds = 5·68 tons; and all the parts must be adapted to endure this force, without any risk of breaking.

When an engine of this kind is at work, if it is in excellent order, the barometer, which is attached to the condenser, will usually raise the column of mercury to 27 inches, or about 3 inc. lower than the atmospheric barometer, or weather-glass; thus indicating that the absolute elasticity of the steam which remains uncondensed in the condenser, is able to support a column of about 3 inches of mercury, or 1½ lbs. per square inch. The temperature of the hot well, in such case, is usually about 100 to 107 degrees of Fahrenheit's thermometer.

The steam remaining in the cylinder, when it is exhausted to its utmost, will be very nearly the same in elasticity as that in the condenser; but as the cylinder cannot be exhausted all at once, a considerable resistance is opposed to the motion of the piston at the commencement of each stroke, by the steam which has not then had time to get away, through the exhausting-valve and eduction-pipe, to the condenser, (see p. 468 and 486). On this account, the mean of the resistance of the unexhausted steam, taken through the whole stroke, will commonly be about 4 lbs. per square inch, in an engine which is loaded so as to exert its intended power; and the elasticity of the steam with which the cylinder is filled to form the plenum, and impel the piston, will be about 12¾ pounds per square inch absolute elasticity, or 2½ pounds less than the ordinary pressure of the atmosphere.

This engine had two waggon boilers, one placed at each side of the engine-house, the length of the boilers being parallel to the great lever of the engine; only one boiler was used at once, and was sufficient to supply the engine. The first engine had only one boiler, which was placed at the side nearest to the other engine. The third engine was also intended to have had one boiler; so that four boilers being provided for three engines, one would always have been cold for cleaning or repairing.

Each boiler was 16 feet long, and 6 feet wide; so that the horizontal surface of the water was 96 square feet. The top of the boiler, which was a half cylinder with the axis horizontal, received the steam; and the water in the lower part stood about 6½ feet deep at a mean.

The fire was made upon two grates beneath one end of the boiler, each grate was 3½ feet square, or 24½ square feet of surface in the two, which is at the rate of only ·49 of a square foot of grate to each horse power. The flame from the grates, after passing beneath all the bottom of the boiler, rose up through that bottom, near one end, into an internal tube, or flue, which was carried horizontally through the water all the length of the boiler, and passed through the front end of the boiler, over the fire, to communicate with an external flue which was formed in the brick-work, and passed all round the outside of the lower part of the boiler, before it

(*a*) Mr. Watt's estimate of the quantity of steam wasted by the vacant spaces at the top and bottom of the cylinder of his single engines, was one-tenth of the quantity expended by the motion of the piston, see p. 366; but, in the above, the waste was very little more than one-twentieth; for 1614 cubic feet per minute, ÷ 20 = 80·7 + 1614 = 1694·7 cubic feet per minute, instead of 1696. This diminished waste in the rotative engine is owing to the circumstance that the motion of the piston is accurately measured by the sun and planet-wheels, and therefore it can be allowed to pass nearer to the top and bottom of the cylinder, than the piston of a pumping engine could do, on account of the inconveniency of striking the stops of the catch pins too frequently.

entered into the perpendicular chimney.   By this means the flame was conveyed through the centre of the mass of the water in the boiler.

The two fire-grates were situated side by side, with the length of the bars crosswise beneath one end of the boiler; and their fire-doors opened at one side of the boiler.   The front ends of the fire-grates were 2 feet beneath the bottom of the boiler, but they sloped downwards from the fire-doors, at an angle of about 18 degrees from the horizontal; and the far ends of the grates were $3\frac{1}{2}$ feet below the bottom of the boiler, which was concave or arched beneath, to receive a greater action from the fire.

The flame from the fires acted beneath all the bottom of the boiler, against a surface of $5\frac{1}{2}$ feet wide, by $12\frac{3}{4}$ feet long, $= 70$ square feet, and from this surface the greatest quantity of steam was raised; near to the other end of the boiler, farthest from the fire-grates, a flue or ascending passage for the flame, rose up from the bottom of the boiler within the water, so as to be surrounded with water on all sides: this rising flue was of a pyramidical form, the base being nearly the whole width of the bottom of the boiler, or $5\frac{1}{2}$ feet wide, by $2\frac{1}{4}$ feet in the other direction; and it stood up in the boiler about $5\frac{1}{2}$ feet high, or within 1 foot of the surface of the water; the pyramid diminished to $2\frac{1}{4}$ feet wide at the top, and this part joined to the horizontal tube or flue, which passed through all the length of the boiler.

The surface exposed to the heat in the ascending flue was about $64\frac{1}{2}$ square feet, and the horizontal flue which proceeded from it was $2\frac{1}{2}$ feet wide, by $4\frac{1}{2}$ feet high, with the angles rounded off. The circumference of this tube, or internal flue, was $12\frac{1}{3}$ feet, and its length $12\frac{1}{4}$ feet, to reach from the top part of the ascending flue, to the other end of the boiler over the fire-places.   The surface of this flue was $150\frac{3}{4}$ square feet, but as heat is transmitted very slowly downwards, the bottom or lower part of the flue, being about $30\frac{3}{4}$ square feet, may be considered as contributing very little to the useful effect of the boiler.

The flame, smoke, and heated air which proceeded from the fire-grates, passed beneath all the length of the bottom of the boiler, and was carried up by the internal ascending flue at the far end, within the centre of the mass of water, and was then conducted back again, horizontally through all the length of the boiler, and conveyed through that end which was over the furnaces, into the external flue, which passed all round the outside of the boiler at the lower part.   The height of that flue was $5\frac{1}{2}$ feet, and the circumference of the boiler 43 feet, so that the surface exposed to the heat was $236\frac{1}{2}$ square feet.

The total surface exposed to the heat, in an efficient manner for raising steam, was 491 square feet, for 50 horse power, $= 9\cdot82$ square feet to a horse power.

The capacity of the water part of the boiler was 16 ft. $\times$ 6 $\times$ $6\frac{1}{2}$ $= 624$ cubic feet, of which the internal flue occupied about 187 cubic feet, so that the boiler contained 437 cubic feet of water; the capacity of the half cylinder top of the boiler was 226 cubic feet, and this space was reserved for the steam.   The whole capacity of the boiler was 663 cubic feet.

The quantity of water evaporated from the boiler, in the usual course of working, was nearly one cubic foot per minute (a); therefore the whole quantity of water contained in the boiler was equal to 437 minutes, or $7\frac{1}{4}$ hours evaporation; and the horizontal surface of the water being 96 square feet, the evaporation of one cubic foot would sink it $\frac{1}{8}$ of an inch in a minute, supposing that no water was supplied.   The depth of water over the internal flue being one foot, the evaporation of 96 minutes, or about $1\frac{1}{2}$ hour, without feeding, would have laid the flue bare.

The transverse section of the internal flue was about $10\frac{1}{4}$ square feet, for the passage of the flame, heated air, and smoke.   The area of the passage, through the external flue around the outside of the boiler, was $7\frac{1}{3}$ square feet.   The perpendicular chimney was 2 feet square withinside, $= 4$ square feet, and about 50 feet high.

(a) This boiler evaporated one cubic foot of water per minute from 491 square feet of surface, which rather exceeds Mr. Watt's estimate of 480 square feet to evaporate one cubic foot per minute, stated at p. 365.   The rule given in p. 366, for finding the fire surface of the boiler, is founded upon Mr. Watt's statement; and, according to that, the engine at the Albion Mill should have had $493\cdot2$ square feet of surface exposed in the boiler, instead of 491.

Thus cylinder, 34 inches diameter, squared $= 1156$ circular inches area, $\times$ 256 feet motion of the piston per minute, $= 295\,936$ cylindrical inch feet of steam expended per minute, $\div$ 600, $= 493\cdot2$ square feet of fire surface.

Also, according to another rule given in p. 366, to find the evaporation from the boiler, this engine should have evaporated $1\cdot03$ cubic feet of water per minute.   Thus cylinder 34 inches diameter, $= 1156$ circular inches area, $\times 256 = 295\,936$ cylindrical inch feet, $\div 288\,000 = 1\cdot028$ cubic feet per minute.

The steam and exhausting-valves were 7 inches diameter in the apertures, and allowing for the spindles and cross-bars, left about 35 square inches of uninterrupted area for the passage of the steam. This is very nearly one-26th of the area of the cylinder, or (35 squ. inc. ÷ 50 HP =) ·7 of a square inch of aperture to each horse power.

The air-pump was 18 inches diameter, = 324 circular inches, so that the area of the cylinder was 3·57 times as great; the bucket of the pump made 4 feet stroke, or half as much as the piston, and the effective capacity of the cylinder was 14·28 times the effective capacity of the cylinder. The secondary air-pump, or hot water-pump, was also 18 inches diameter; but its bucket was suspended by a rod from the great lever, at one-fourth of its length, from the centre of motion to the suspension of the great piston, so that it made only 2 feet stroke, or one-fourth of that of the piston; and the effective capacity of the cylinder was 28·56 times the capacity of this secondary air-pump.

The construction of the secondary air-pump was similar to that of a common sucking-pump, with valves in the bucket, and the barrel open at top; the pump stood in the condensing cistern, in place of the hot well I; and the lower part of the barrel communicated by a square trunk with the delivering spout m, fig. 8, Plate XIII. at the top of the air-pump, the discharging-valve m being contained in that trunk, in the same manner as the foot-valve k, is contained in the passage between the condenser and the air-pump. The discharging-valve served for the lower valve of the secondary-pump, and formed the communication between that pump and the upper part of the air-pump, in order to allow the air and hot water to pass out from the air-pump, but to prevent any return; a cistern or trough was fixed on the open top of the secondary-pump, to serve for a hot well to receive the hot water which that pump drew out of the air-pump: this trough was situated close beneath the floor, which covered over the space between the cylinder pier and the lever wall; the trough passed through the lever wall, and the suction-pipe of the hot water-pump S, was immersed in it.

The hot water-pump, to supply the boiler, was 6 inches diameter, and its bucket made a stroke of 1⅔ feet; it was suspended at the outer end of the great lever, outside of the lever wall, in the manner shown in Plate X.

The cold water-pump was 12 inches diameter, = 144 circular inches, and its bucket was suspended from the great lever, at one-third of its length from the centre of motion, so as to make a stroke of 2⅔ feet. The capacity of the cylinder was 48 times the capacity of the pump, or allowing for the vacant spaces, the quantity of steam expended by the cylinder, was about 52 times the quantity of cold water that the pump raised into the condensing cistern.

The great lever was an oak beam, 25 feet long, between the joints at each end; and 29 inches deep, by 24 inches wide. The connecting rod was also an oak beam, 13 inches by 12 inches, with a cast-iron stem fixed to the lower end of it, to fasten the planet-wheels to: the length of the connecting rod, from the joint at the upper end to the centre of the planet-wheels, was 22 feet.

The wood framing, to sustain the axes of the fly-wheels, was composed of oak beams, 15 inches deep, by 13 inches wide: there were two of these frames, to support the necks of the axes, close to the two sun-wheels, and the other ends of the axes were supported on the framing of the flour-mills.

The sun and planet-wheels were each 4 feet geometrical diameter, with teeth of 3 inches pitch, and 5 inches broad; this would give 50 teeth in each wheel; the wheels were cast iron, with solid teeth. The axes of the sun-wheels and fly-wheels were 9 inches square, and 8⅔ feet long between the bearings; the necks of the axes were 8 inches diameter, and 8 inches length of bearing in their sockets.

The fly-wheels, which were fixed on these axes, were 18 feet diameter outside; and the rims were 8 inches by 4 inc. = 32 square inches of transverse section. These fly-wheels made 32 revolutions per minute each.

The rim of each fly-wheel contained 12·1 cubic feet of cast iron, or 24·2 cubic feet for both, × 450 lbs. = 10 890 pounds weight, (or 4·86 tons) ; this mass moved with a velocity of 29 feet per second, and the height due to that velocity is 13·1 feet, × 10 890 lbs. = 142 659 lbs. (or 6· 37 tons) acting through a space of one foot, is the energy of the rims of these fly-wheels. The force exerted by the piston during each half stroke was (6446 lbs. × 8 feet stroke =) 51 568 lbs. acting through one foot ; and the energy of the rims of the two fly-wheels was (142 659 lbs. ÷ 51 568 =) 2·77 times the power exerted by the piston in each half stroke.

Each of the axes of the fly-wheels worked five pairs of mill-stones, for grinding wheat into flour ; the mill-stones were arranged in a circle around a large horizontal spur-wheel, which gave motion to all the five. A system of mill-stones thus driven by one horizontal wheel is called a hirst, and this engine was said to work two separate hirsts, of five pairs of stones each. The large spur-wheel of each hirst was $8\frac{1}{4}$ feet geometrical diameter, and was made of cast iron, with wooden teeth, fastened into mortices formed round the edge ; the teeth were about $2\frac{3}{4}$ inches pitch, and 5 inches broad ; there would be 113 teeth in the wheel. This wheel was fixed nearly upon the middle of an upright axis of cast iron, which was 8 inches square, and 18 feet high, from the pivot at the bottom, to the bearing by which it was supported at top.

On the lower part of the upright axis was fixed a bevelled cog-wheel of $5\frac{1}{4}$ feet diameter, and it was turned by a similar wheel of nearly the same size, which was fixed upon the end of the fly-wheel axis : these bevelled wheels, being equal to each other, are called mitre-wheels ; they were made of cast iron, with cast iron teeth, of about $2\frac{3}{4}$ inches pitch, and 6 inches broad ; this would give 72 teeth to each wheel. The upright axis and great spur-wheel made 32 revolutions per minute, the same as the fly-wheels.

The five pinions, by which the great wheel was surrounded, were each $2\frac{1}{2}$ feet diameter, with solid iron teeth of about $2\frac{3}{4}$ inches pitch, and 5 inches broad, to correspond with those of the great wheel. At this rate each pinion must have had 34 teeth, and must have made (125 ÷ 38 =) 3·3 turns for each turn of the fly-wheel ; which being at the rate of 32 revolutions per minute, the mill-stones must have made about $105\frac{1}{2}$ revolutions per minute.

The mill-stones, which were suspended upon the upper ends of the upright axes of the pinions, were $4\frac{1}{2}$ feet diameter, and about $1\frac{1}{4}$ feet thick when new, but they became thinner as they were chipped away by repeatedly dressing or picking the under surfaces, in order to sharpen the edges of the furrows, by which the corn was cut or ground. The engine was capable of impelling all the grinding machinery at once, but two or more of the mill-stones would be at all times under the repairing operation ; therefore the engine usually worked 8 pairs of mill-stones for grinding wheat into meal, and also a corresponding system of machinery for dressing the meal into flour ; this flour machinery was turned by wheel-work from the upper end of the upright axis of the great horizontal spur-wheel, and the motion was communicated to each of the bolting-mills by endless straps.

The energy of all the mill-stones of these mills was very considerable ; thus each mill-stone, $4\frac{1}{3}$ feet diameter, and $1\frac{1}{4}$ feet thick, would contain $19\frac{7}{8}$ cubic feet of stone, of which the specific gravity is about 2·5 ; so that a cubic foot would weigh (62· 5 lbs. a cubic foot of water × 2· 5 times =) $156\frac{1}{4}$ pounds × 19· 87 cubic feet = 3105 pounds weight of each stone. The velocity of the circumference when the stone was making $105\frac{1}{4}$ revolutions per minute, was 24· 87 feet per second ; and the height due to that velocity is 9· 62 feet.

3 U

The energy of a solid circular wheel, like a mill-stone, is equal to that of one half its weight of matter, moving with the same velocity as the outside circumference (see p. 36) ; therefore the energy of each mill-stone is (1552 lbs. × 9·62 feet =) 14935 pounds acting through one foot ; or × 10 stones = 149 350 lbs. or 6·66 tons, which is more than the energy of the rims of the two fly-wheels of the engine ; for 149 350 lbs. ÷ 51 568 lbs., is 2·9 times the power exerted by the engine in each half stroke.

The energy of the ten mill-stones, together with that of the rims of the two fly-wheels, was 5·67 times the power exerted in each half stroke of the piston.

The building of the Albion Mills was adapted to contain three 50 horse engines side by side, with four boilers in the two spaces between the three engines ; each engine was to have had two hirsts of 5 pairs of mill-stones connected with it. Two engines were put up, but the third was not begun when the whole was burned. The building is situated at the end of Blackfriars Bridge, on the South-wark side of the river ; and presents a front of 160 feet long, to the street which leads to the bridge. The end of the building along the river side, was nearly as long. A basin or dock was carried beneath the centre of the building, all its length, to admit barges with corn from the river ; and tackle was provided to draw up the sacks of corn out of the barges into the upper rooms of the mill, by the power of the engines. The front external walls of the building still remain.

The engines were placed across the length of the building side by side, and the three would have occupied all its length. They were situated at the back of the building, and occupied its width, from the wall of the back front to the dock. At the opposite or front side of the dock, was a road way for carts and waggons to enter and be loaded under cover. The upper rooms which extended over the dock and the cart way, contained the mill-stones, and formed a vast warehouse for the corn and flour.

Each engine-house consisted of four very massive walls to enclose the cylinder, and to afford a support to the centre of the great lever, in the same manner as the building for the single engine represented in Plate X. The lever wall was built 4 feet thick from the base, and up as high as the level of the bottom of the cylinder ; then $3\frac{1}{2}$ feet thick, up to the level of the cylinder cover ; and the upper part, as high as the spring beams, was 3 feet thick. The two side walls of the engine-house were $2\frac{1}{2}$ feet thick at the bottom, then 2 feet, and the upper part $1\frac{2}{3}$ feet. The offsets of the masonry and brick work, were at the outside of the building, so as to leave the inside space of the house 13 feet wide, by $16\frac{1}{2}$ feet long. The end wall of all the three engine-houses formed part of the outside wall of the back front of the building.

The two boilers of the middle engine were placed at each side of the engine-house, the length of the boilers being across the building ; the brick work which enclosed one side of each boiler, was built in contact with the side walls of the engine-house, and the brick work at one end of each boiler joined to the outside wall of the build-ing. The chimneys were carried up against that wall withinside of the building, in the spaces between the engine-houses, and there was a separate flue for each boiler, or two chimneys with two flues in each. The fire-doors of the furnaces opened beneath the side of the boiler near one end, as before mentioned, and close to the chimney. The brick work of each boiler occupied a space of $11\frac{1}{2}$ feet wide by 22 feet long, at the side of the engine-house.

The two hirsts of the flour-mills of each engine were situated opposite to the ends of its two boilers, the centres of the upright axis of each hirst being $12\frac{3}{4}$ feet from the central plane of the engine, or $25\frac{1}{2}$ feet apart. The outer end of the great lever, and the connecting-rod, the sun and planet-wheels, and the two fly-wheels, were not enclosed between brick walls, but by a wooden framing, forming an apart-ment of about $10\frac{1}{2}$ feet wide inside, to include those parts. The upright posts of this framing also sustained the frame of the hirsts for the mill-stones, and the different floors of the building were also supported by the same framing. The centres of the five mill-stones in each hirst were arranged at equal distances of $6\frac{1}{3}$ feet apart, in a circle of $10\frac{3}{4}$ feet diameter; and the five mill-stones occupied all the space of that angle which was left at the ends of the boiler, and at the outside of the framing which enclosed the fly-wheels.

The building for the Albion Mills was erected upon a very soft soil, consisting of the made ground at the abutment of Blackfriars Bridge; to avoid the danger of settlements in the walls, or the necessity of going to a very unusual depth with the foundations, Mr. Rennie adopted the plan of forming inverted arches upon the ground, over the whole space on which the building was to stand, and for the bottom of the dock.

For this purpose the ground upon which all the several walls were to be erected, was rendered as solid as is usual for building, by driving piles where neces-sary, and then several courses of large flat stones were laid, to form the founda-tions of the several walls; but to prevent any chance of these foundations being pressed down, in case of the soft earth yielding to the incumbent weight, strong inverted arches were built upon the ground, between the foundation courses of all the walls, so as to cover over the whole surface included between the walls; and the abutments or springings of the inverted arches being built solid into the lower courses of the foundations, they could not sink, unless all the ground beneath the arches had yielded to compression, as well as the ground immediately beneath the foundations of the walls.

By this method the foundations of all the walls were joined together, so as to form one immense base, which would have been very capable of bearing the required weight, even if the ground had been of the consistency of mud; for the whole building would have floated upon it, as a ship floats in water; and whatever sinking might have taken place, would have affected the whole building equally, so as to have avoided any partial depressions or derangements of the walls; but the ground being made tolerably hard in addition to this expedient of augmenting the bases by inverted arches, the building stood quite firm.

Performance of Mr. Watt's Engine at the Albion Mills.

Soon after this engine was set to work, several exact trials of its performance were made by Mr. Watt and Mr. Rennie; the results of one of these trials, which was continued during ten hours, and in which the greatest performance was attained, are as follows.

In 10 hours continual working, a counter which was fixed on the great lever showed that the engine had made 10 735 strokes, which is at the rate of $17 \cdot 89$ strokes per min. ✕ twice 8 feet stroke of the piston = 286 feet motion per minute.

The furnaces consumed 48 bushels of the best Newcastle coals, which is at the rate of $4 \cdot 8$ bushels per hour, or (✕ 84 lbs. per bushel =) 403 pounds of coals consumed per hour, or (÷ 50 horse-power =) $8 \cdot 06$ pounds of coals per hour for each horse-power.

The boiler evaporated 556 cubic feet of water into steam; this is at the rate of $55 \cdot 6$ cubic feet per hour, or $\cdot927$ cubic feet per minute; hence, (556 cubic feet

$\div$ 48 bush. $=$) 11·58 cubic feet of water was evaporated by each bushel, or 8·62 pounds of water was evaporated by each pound of coals.

The engine worked 8 pair of mill-stones, and ground 124 sacks of wheat, each containing 4 bushels, or 496 bushels in ten hours, or 49·6 bushels of wheat ground per hour (*a*); which is at the rate of (496 bush. wheat $\div$ 48 bush. coals $=$) 10⅓ bushels of wheat ground, for each bushel of coals consumed; and (49·6 bush. $\div$ 50 HP) $=$ ·992, or very nearly 1 bushel of wheat was ground per hour by each horse-power. At this rate the expenditure of mechanical power requisite to grind one bushel of wheat, is equal to that of a force of (33 000 lbs. $\times$ 60 minutes $=$ 1980 000 lbs. or), nearly 2 millions pounds acting through a space of one foot.

The flour machines also dressed 70 sacks of flour, of 5 bushels each $=$ 350 bushels, or 35 bush. per hour; which is at the rate of (350 $\div$ 48 $=$) 7·3 bushels of flour dressed, and 10⅓ bushels of wheat ground, by each bushel consumed.

This was a very great performance, and to attain it the cutting edges of the furrows in the mill-stones, required to be in the most excellent order, so as to render the resistance of the grinding as small as possible; the coals must also have been of the best quality, as is shown by the quantity of water evaporated, which greatly exceeds the estimate given at p. 366. The engine must also have been in the best condition, as to the adjustment of its working gear, the tightness of the fittings, and the packing of the piston.

Assuming, according to Mr. Watt's estimate, that a cubic inch of water produces a cubic foot of steam (see p. 366), the quantity of steam supplied to the engine during the above trial, must have been (·927 of a cub. ft. of water evaporated $\times$ 1728 cub. inc. in one cub. ft. $=$) 1602 cubic feet per minute of steam having the same elasticity as the atmospheric air, or 14·7 pounds pressure per square inch. The total expenditure of steam by the cylinder, when the piston made 17·89 strokes per minute, being at the rate of 106 cubic feet per stroke, must have been 1896 cubic feet per minute, including the waste in the vacancies. Hence, the elasticity of the steam with which the cylinder was filled to make the plenum to impel the piston, must have been (14·7 lbs. $\times$ 1602 cub. ft. $\div$ 1896 cub. ft. $=$) 12·42 pounds per square inch; that is 2·28 pounds less elastic than the atmospheric air, or making a further allowance for leakage, and for the condensation of steam in the steam case, we may say 2¾ pounds below the atmospheric pressure.

When the piston was moving through a space of 286 feet per minute, the effective force of the piston would be 5770 pounds, in order to exert 50 horse-power upon the mill-work; that is ($\div$ 908 square inches $=$) 6·36 pounds effective force per square inch, and allowing the friction of the engine itself to have been 1·75 pounds, the actual force exerted, or the unbalanced pressure of the steam against the piston, must have been 8·11 pounds, at the mean of the whole stroke.

(*a*) Corn is measured by the Winchester bushel, which, according to an act of parliament passed in 1701 (13th of William III.), should be 18½ inc. diameter, and 8 inc. deep. This would contain 2150·42 cubic inches. A bushel of the best wheat will weigh 60 lbs. or more, but according to an act of the 29th of George III., it is to be reckoned as 57 pounds; 4 bushels of wheat are put into a sack, and 2 such sacks, or 8 bushels, make a quarter; and 5 quarters, or 80 bushels, make a load, which is contained in 10 sacks.

Meal or flour is reckoned by bushels and sacks, but instead of measuring the meal, it is weighed, and half a hundred weight, or 56 lbs., is to be called a bushel, according to an act of 31st George III., because it is supposed to be the produce of a bushel, or 57 lbs. of wheat, one pound being accounted as lost and wasted in grinding; 5 bushels of flour are put into a sack, which weighs 2½ cwt., or 280 lbs., according to the act of 31 George III.

An act of parliament was passed in 1824 (5 George IV.), to regulate weights and measures, and establish a new standard called Imperial measure. According to this the Imperial bushel is to contain 8 Imperial gallons, each containing 10 pounds avoirdupois of pure rain water, at the temperature of 62 degrees of Fahrenheit's thermometer, or 277·274 cubic inches; hence, the Imperial bushel contains 80 pounds of water, or 2218·192 cubic inches $=$ 1·284 cubic feet. By this measure all corn is now to be measured; and 8 such bushels make a quarter.

The old coal bushel which is to be heaped up, as stated at p. 337, is so very nearly according to this standard, that it is not altered by this act; for the heaped bushel of Imperial measure would contain 2815·48 cubic inches.

It has been already inferred from the quantity of water evaporated, that the steam with which the cylinder was filled, was 2·75 pounds per square inch, less than the atmospheric pressure, and at that rate to have produced an effective force of 8·11 lbs., the mean state of exhaustion within the cylinder during the whole stroke, must have been 10·86 pounds per square inch, less than the atmospheric pressure. It is proved by observation with the indicator, that this degree of exhaustion may be obtained in engines similarly situated (see p. 487).

The greater or lesser extent of the mean state of exhaustion in the cylinder, depends very much upon the size of the apertures which are opened by the exhausting-valves to discharge the steam from the cylinder, compared with the quantity of steam which must pass through them. When the engine was working at 17·89 strokes per minute, as above, the quantity of steam expended by the motion of the piston without allowance for waste, must have been 1802 cubic feet per minute ; as this quantity had to pass through an aperture of 35 square inches, it required to move with a prodigious velocity.

The area of the apertures in this engine was at the rate of ·7 of a square inch to each horse-power, which is a less proportion than in the smaller engines, because Mr. Watt proportioned the area of the apertures according to that of the cylinder; but the power of the engine depends upon the area of the cylinder and the velocity of its piston; and as the pistons of large engines move with the greatest velocity, they exert more power in proportion to the size of the cylinder (and of the apertures) than smaller engines.

Assuming the expenditure of steam to have been 1602 cubic feet per minute for 50 horse-power, as above, it would be 32·04 cubic feet per minute to each horse-power, including waste. As a horse-power will raise 528 cubic feet of water one foot high per minute, each cubic foot of steam must have raised (528 ÷ 32·04 =) 16·48 cubic feet of water one foot high.

Assuming the engine to have exerted 50 horse-power, when it consumed 4·8 bushels of coals per hour, it would be at the rate of 8·06 pounds of coals per hour to each horse-power, or 20·63 millions pounds weight raised one foot high, by the consumption of each bushel of coals; for 50 HP × 33 000 lbs. = 1 650 000 × 60 minutes = 99 millions pounds per hour raised one foot ÷ 4·8 bushels consumed per hour = 20·63 millions raised one foot by each bushel consumed. This is a greater performance than the estimates of Mr. Watt's single engine (p. 367), or of his double engine (p. 488), because the quantity of steam produced by a given consumption of fuel at the Albion Mills, was greater than is assumed at p. 366.

In February, 1789, an exact trial was made of the quantity of coals consumed by both engines, when they were working at once without intermission for 26⅓ hours, and it was found to be 8 chaldrons. The furnace of each engine burned 3¼ chaldrons of Newcastle coals, of the kind called Warebottle, and about ¼ of a chaldron of small coal.

The consumption of each engine was very nearly the same, but the first engine went at the rate of 16 strokes per minute, and the second engine 14½ strokes per minute. Eight chaldrons of coals × 36 bushels per chaldron = 288 bushels ÷ 26⅓ hours = nearly 11 bushels per hour for both engines, or 5½ bushels for each engine of 50 horse-power.

Dimensions of a 50 horse-power patent Rotative-engine made by Messrs. Boulton and Watt, 1797.

This engine was set up in London, at a large distillery on the bank of the river Thames, in order to pump liquors, and to turn machinery for grinding malt, and other work for the service of the distillery. The author had constant opportunities of examining the construction and operation of this engine, in the years 1804 and 1805, when he first turned his attention to the subject of steam-engines, and being the largest specimen of Mr. Watt's rotative-engine, that he had then any opportunity of studying minutely, he selected it as a standard for the proportions of that kind of steam-engine.

The construction of this engine was in every respect the same as that at the Albion Mills, with two fly-wheels, and two pairs of sun and planet-wheels, each giving motion to a separate hirst of mill-stones.

This distillery was destroyed by fire in 1806, and has since been rebuilt, but the engine has been replaced by a 30 horse engine, as the former engine was found to be larger than necessary for the work required to be done in the distillery.

The diameter of the cylinder was 36 inches = 1017· 9 square inches area, or 1296 circular inches. The length of the stroke of the piston was 7 feet, and it made 17 strokes per minute, so as to pass through a space 238 feet per minute. The expenditure of steam was 1683 cubic feet per minute, or at the rate of 33· 65 cubic feet of steam per minute, for each horse-power (a).

The mean of the effective pressure upon the piston when the engine was exerting 50 horse-power, was at the rate 6· 81 pounds upon each square inch of the piston, without allowance for friction; this multiplied by 1018 square inches gives 6933 pounds for the mean effective force of the piston; and this force acting through a space of 238 feet per minute, would be equal to 1 650 000 lbs. acting through one foot per minute, or ÷ 33 000 lbs. = 50 horse-power. The utmost force that the piston could ever be expected to exert, being taken at the rate of 11 pounds per circular inch, would be (× 1296 circular inches =) 14 256 pounds = 6· 36 tons, and the strength of all the parts must have been adapted to endure that force without any risk of breaking.

The engine had three iron boilers, and two of them were worked at once, leaving the third cold, for cleaning; they were 4 feet diameter, and 16 feet long, with internal flues through the centres of them. The dimensions of these boilers being the same as those of the larger boiler of the 20 horse engine mentioned at p. 496, the surface exposed to the heat was 261 square feet in each boiler, or in the two boilers 522 square feet for 50 horse-power; which is at the rate of 10· 45 square feet to each horse-power.

The air-pump was 23½ inches diameter = 552 circular inches area, or the area of the cylinder was 2· 35 times the area of the piston. The air-pump bucket made a stroke of 3½ feet, or half as long as that of the piston, so that the effective capacity of the cylinder was 9· 4 times that of the air-pump. The condenser was the same size as the air-pump. The aperture through the foot valve, to communicate from the condenser to the air-pump, was 17½ inches wide, by 6½ inches high, = 113¾ square inches area; so that the area of the air-pump was 3· 81 times that of the aperture of the foot valve. The area of the delivering valve at the top of the air-pump was 20 inches wide, by 6 inches high; or 120 square inches area. The hemp packing around the bucket was 3½ inches deep, and the iron rims above and below that packing were 1¼ inches each, so as to make the whole depth of the bucket 6 inches. The rod of the air-pump was 2¼ inches diameter.

The cold water pump was 11¼ inches diameter, = 126½ circular inches, and its bucket made a 3 feet stroke; hence, the effective capacity of the cylinder was 47· 8 times that of the cylinder; or allowing for the waste of steam, and of water, the expenditure of steam was about 55· 35 times the quantity of cold water which was raised by the cold water pump, into the condensing cistern.

The piston-rod was 3½ inches diameter. The joint pins of the main links of the parallel motion were 3⅞ inches diameter, and 4 inches length of fitting in their

(a) Messrs. Boulton and Watt afterwards called an engine of nearly the same dimensions as the above, 53 horse-power, instead of 50 horse; the cylinder being 36⅛ inches diameter = 1025 square inches area, and the piston making 17½ double strokes per minute of 7 feet = 245 feet motion per minute. This is an expenditure of 1744 cubic feet of steam per minute, or (÷ 53 HP =) 32· 9 cubic feet per minute by each horse-power.

Mr. Watt's double engines at the Albion Mills were called 50 horse-power each. The cylinder was 34 inches diameter, and the piston made an 8 feet stroke; one of the engines worked at the rate of nearly 18 strokes per minute = 286 feet motion of the piston per minute; and the expenditure of steam was 1802 cubic feet per minute, which is at the rate of 36· 04 cubic feet per minute for each horse-power. The other engine, which was made afterwards, made 16 strokes per minute (see p. 510).

sockets. The iron bars of which the main links were formed, were 3 inches broad, by $\frac{3}{4}$ of an inch thick, or $2\frac{1}{4}$ square inches area, and the four bars contained 9 square inches, which sustained all the force of the piston. The smaller joint pins of the parallel motion were $1\frac{3}{4}$ inches diameter, and 2 inches length of fitting in their sockets ; the rods of the parallel motion were $1\frac{1}{4}$ inches diameter ; the pins of the joints which suspended the rod of the air-pump, were $2\frac{3}{4}$ inches diameter, and $2\frac{3}{4}$ length of fitting in their sockets ; the iron bars of which these links were composed, were 2 inches broad, by $\frac{1}{2}$ an inch thick.

The great lever was an oak beam $21\frac{1}{2}$ feet long, between the main joints at each end, and 29 inches deep, by 22 inches broad ; the axis of the great lever was $7\frac{1}{2}$ inches broad, by $8\frac{1}{2}$ inches deep, and 3 feet length between the bearings ; the pivots at each end of it were $5\frac{3}{4}$ inches diameter, and 7 inches length of bearing in its sockets. The connecting-rod was of oak, 12 inches by 10 inches, with an iron stem fastened to it at the lower end, to fix the planet wheels to ; for this engine had two pairs of sun and planet-wheels, one at each side of the connecting-rod, to give motion to two fly-wheels, one on each side of the engine-house ; the length of the connecting-rod from the joint at the upper end, to the centres of the planet-wheels, was about 20 feet long.

The planet-wheels were $3\frac{1}{2}$ feet diameter, with 48 teeth, $2\frac{3}{4}$ inches pitch, and 5 inches broad. The sun-wheels the same ; they were all made of cast-iron, with solid teeth. The axes of the two sun-wheels and fly-wheels were placed exactly in the same line, the end of one being opposite to the end of the other, and the two sun-wheels which were fixed upon the extreme ends of the two axes respectively, were exactly opposite to each other, with a small space between them.

The axes of the fly-wheel were $8\frac{1}{4}$ inches square, and about 7 feet long between the bearings ; the necks of the axes were $7\frac{1}{2}$ inches diameter, and 8 inches length of bearing in their sockets. The fly-wheels were 20 feet diameter outside, and the rims were 7 inches by 3 inches = 21 square inches of cross section. One of these fly-wheels was fixed upon each axis near to the side walls of the engine-house, and their axes were prolonged each way through those walls, to give motion to the machinery at each side of the house. The fly-wheels made 34 turns per minute.

The rims of these two fly-wheels together contained $17 \cdot 8$ cubic feet of cast-iron $\times$ 450 lbs. = 8010 lbs. (or $3 \cdot 57$ tons). The rims moved with a velocity of $34 \cdot 55$ feet per second ; and the height due to that velocity is $18 \cdot 56$ feet, which $\times$ 8010 lbs. weight of the rim = 148 700 lbs. (or $6 \cdot 64$ tons) acting through one foot, is the energy of the two rims. The power exerted by the piston in one half stroke is (6933 lbs. $\times$ 7 feet =) 48 530 pounds acting through one foot. And lastly, the combined energy of the two fly-wheels is (148 700 lbs. $\div$ 48 530 lbs. =) $3 \cdot 06$ times the power exerted by the piston in each half stroke.

This engine gave motion to 10 pairs of mill-stones for grinding malt ; they were arranged in two separate hirsts, of 5 pairs of stones, and each set received motion from the axis of one of the fly-wheels, in the same manner as the engines at the Albion Mills (see p. 513). The different pumps which this engine was required to work for the use of the distillery, were actuated by a second great lever of oak, which was placed at the end of the engine-house, in the same line with the great lever of the engine, but on a higher level. One end of this second lever extended over the great lever, at the cylinder end, and was connected therewith by two links similar to those of the parallel motion. The rods of six pumps were suspended from beneath this second lever.

In 1798, Messrs. Boulton and Watt made a 50 horse engine of the same construction as the above, and it was set up in the Citadel, at Petersburgh, to turn a rolling mill for the Mint, and part of the coining machinery (*a*). The sun and planet-wheels have since been removed, and a crank substituted, with a multiplying-wheel and pinion, to turn a large fly-wheel upon the second axis with a great velocity, in the same manner as the 40 horse engine mentioned at p. 505. The boiler of this engine is 17 feet long, by 7½ feet wide; fir wood is used for fuel. The author examined this engine in 1819, at which time a new one of the same size was making in Petersburgh, to be put up in the place of the old engine.

Having now stated the dimensions and proportions of five standard sizes of Mr. Watt's rotative engines, we may proceed in the next chapter to give rules by which the principal parts of those engines ought to be proportioned. The author has not met with any rotative engines of greater power than 50 horse, which were made by Messrs. Boulton and Watt during the term of the patent. At the expiration of that patent, Mr. Watt retired from the business of the Soho manufactory, and alterations were afterwards introduced in the construction of their engines. Other engine makers, who began business at the expiration of the patent, followed the plan of the patent engines, and some were made of 60, 80, and 100 horse-power, which the author has examined, and their dimensions have been taken into account in forming the rules contained in the next chapter.

Mr. Watt's Counter to ascertain the number of strokes made by a Steam-engine in a given time.

This is a small machine something like a clock, which is placed on the top of the great lever, and contains a short pendulum, which by the inclination of the great lever from the horizontal position, is caused to vibrate, every time that the engine makes a stroke; and an accurate count of all the vibrations it has made, is kept by means of wheel-work, and indexes pointing to numbers upon dials like those of watches (*b*).

The counter contains a train of small wheels and pinions, mounted in a frame, similar to that of a clock, and about the same size; but there is no spring barrel or weight. An escapement-wheel of ten teeth, is fixed upon the last axis of the train, and the teeth act against a pair of pallets, similar to those of a clock, but the pallets are so formed and adapted to the teeth of the wheel, that every vibratory movement of the pallets (and of a short pendulum which is connected with them) will turn the escapement-wheel round half a tooth; and each complete vibration of the pendulum will turn the wheel round a whole tooth.

(*a*) The self-acting coining machinery now used in the Royal Mint, was invented by Mr. Boulton, who set up a mint for copper money at Soho, about 1788. The rolling mill for laminating the copper, was worked by a water-wheel, which was supplied by a large single returning-engine of Mr. Watt's, such as is represented in Plate X. An extensive business was carried on at Soho for some years in coining copper for provincial tokens, and for the East India Company; and in 1797 they began a new copper coinage for the British government.

The advantage of this system of coinage induced the Empress Catharine to order a complete set of machines from Mr. Boulton, which were made at Soho, and set up at Petersburgh, in 1798. There is a 20 horse engine to work the coining presses, as well as the 50 horse engine for the rolling mill. Similar machinery was afterwards made at Soho for the Danish government, and another set was sent out to the Brazils; and lastly, in 1810, a new Royal Mint was established in London, on Mr. Boulton's plan, with four steam-engines.

(*b*) The counter should have been mentioned in describing Mr. Watt's single engine for draining mines (in p. 365), because it was first applied to those engines at the mines in Cornwall; and by the accounts thus kept, of the actual number of strokes made by each engine, Messrs. Boulton and Watt charged the proprietors of the mines with their premium for the portion of savings made in fuel, as has been stated at pp. 329 and 383.

Upon the axis of the escapement-wheel, is a pinion, to turn another wheel; and on the axis of that wheel, another pinion, to turn another wheel, and so on. Each wheel has ten times as many teeth, as the pinion by which it is turned, and the extreme end of the axis of each wheel carries a small index, to point out to ten numbers which are arranged around a circle, like the dial-plate of a watch, but it is only divided into ten. There are seven or eight of these indexes; the first index is fixed on the axis of the escapement-wheel, and indicates tens on a dial; for the index passes once round in ten complete vibrations of the pendulum. The next indicates hundreds, then thousands, and so on, up to millions, or to tens of millions.

The counter is inclosed in a small wooden box, which is fastened by screws upon the upper side of the great lever, at any convenient part of its length. The pendulum, which is short, and loaded with a considerable weight, is not suspended in a vertical position in the manner of that of a clock; but it lies in a horizontal position, with its length across the length of the great lever; the axis of the pendulum, and the axes of the several wheels and pinions, stand in a vertical position, when the great lever is horizontal, and when the piston is at the middle of its course. But when the great lever inclines from the horizontal position alternately to one side, and then to the other, the weight at the end of the pendulum will move across the width of the box, from one side to the other, every time that the piston arrives at the top or bottom of its stroke, and one of the pallets of the escapement, will turn the escapement-wheel round half a tooth. By the indexes and dials, an account of each half stroke is kept, and when ten strokes are made, they will be shown by the index of the second dial, and so on of all greater numbers.

The box which contains the counter is shut up by a door, which is kept locked, so that no person can gain access to it, except by means of the key; and the heads of the screws by which the box is fastened upon the great lever, being within the box, it cannot be taken off, without first opening the box. A counter which has eight dials, is sufficient to keep an account of the going of an engine which works constantly day and night, during a whole year. This is an accurate mode of determining the number of strokes actually made by an engine.

Construction of the Nossels which contain the four valves of Mr. Watt's Double Rotative-Engines.

The construction of these parts, and their manner of action, has been already explained, pp. 451, 457, and 467; but there are some further particulars which remain to be stated, and some varieties in the construction of the working gear which deserve notice.

The nossels or boxes which contain the four valves b, h, and e, i, and their side pipes d d and g g, were always arranged in the manner shown in fig. 6, Plate XIII.; but in all the earliest rotative-engines which were made by Messrs. Boulton and Watt, the nossels were made of several pieces, and screwed together. The middle compartment of each box, which joins to the passage c or f, at the top or bottom of the cylinder, was a separate piece from the two boxes, b and h, or e and i, which are situated above and below the passages c or f, to contain the racks and sectors to lift the valves; hence there were six separate boxes of cast iron to compose the nossels, and they were united by means of 16 nuts and screw bolts put through projecting flanges, which were formed at each side of each of the six boxes.

3 x

The four bell-metal valve seats were interposed between the joinings of these six boxes, being fitted with cement into circular recesses or rebates, which were formed in the tops and bottoms of the middle boxes, or passages c and f, so that the seats became firmly fixed in their places, when the flanges round the edges of the other four boxes b and h, e and i, were screwed to those passages or middle boxes c and f, above and below the same.   See the sketch, p. 373.

The two side pipes $d\ d$, and $g\ g$, were joined to the boxes by means of flanges, in the manner shown in the sketch, p. 524; for this purpose, a short length of pipe projected out from each of the four boxes $b\ h$, and $e\ i$, and terminated with a circular flange; and these flanges were fastened by screw bolts to the corresponding flanges at the curved ends of the upright pipes $d\ d$ and $g\ g$. Each of these upright pipes was made in two lengths, which were united together by a socket joint, to admit of the expansion or contraction of the metal, by the heating and cooling of the steam; these joints were near the lower ends of the pipes, and were made tight with stuffings of hemp rammed into the sockets, around the lower ends of the upper lengths of the pipes.

In addition to all these joints were six moveable doors or covers, for the openings in front of the boxes, which gave access to the different valves.

The nossels, on this plan, required 21 joints to be made tight, by screwing flanges together, viz. : Two vertical joints, and 4 bolts and nuts, to connect the two middle boxes of the nossels with the branches c and f, at the top and bottom of the cylinder.   Four horizontal joints, and 16 bolts and nuts, to unite the six boxes b $c\ h$, and $e\ f\ i$, together, as before mentioned.   Six vertical joints, and 20 bolts and nuts, to join the two upright pipes to the boxes; and to join the throttle-valve and steam-pipe thereto.   Also one horizontal flange, with 4 bolts and nuts, to join the eduction-pipe $g\ g$ to the condenser, and the two expansion stuffed joints in the upright side-pipes.   Two of the six moveable doors, in front of the nossels, are horizontal, to cover the tops of the steam boxes b and e; and the other four doors are vertical at c, h and f, i.   Each door is fastened by two bolts and nuts.

With this great number of joints it was very difficult to keep the nossels free from leakage.   The joints were made close by interposing pasteboards between the flanges, with putty made of lintseed oil and whiting, as directed in p. 382; these packings being compressed very hard by the screws, made good joints at first, but by the heat of the steam, and the continual working of the engine, they became loose in time, and required to be renewed.   The bell-metal seats for the valves were commonly taken out, and new ones put in, at the same time that the joints between the boxes were refitted; but these seats could not be removed without taking the boxes apart.

In course of time, as the founders became more expert in moulding intricate forms in cast iron, the top box $b\ c\ h$, and the bottom box $e\ f\ i$, for the nossels, were each cast in one piece, with two internal partitions, to divide each box into three compartments; and these compound boxes were united to the upright side pipes by flanges as before: this method was used for large engines, and the side pipes had socket joints, to allow for the expansion.   For small engines, the whole of the nossels and side pipes were cast in one piece, in the manner represented at figures 6, 7, and 8, Plate XIII.

When the boxes are cast in one piece, the bell-metal seats for the valves are fitted into round apertures in the partitions; and there is a square flange, or projecting border, round each seat, to fit down upon the flat surface of the partition, with cement beneath, to make a tight joint.   These square flanges, or borders round the valve seats, for the steam-valves b and e, are fastened down by a screw at each angle ; and the heads of those screws are accessible, when the covers at the tops of the boxes, *b* and *e*, are taken off.   The seats for the two exhausting-valves, *h* and *i*, must be introduced through the openings in front ; and they are fastened down in their places by means of two broad thin wedges of iron, which are driven in, at each side of the box edgeways upwards ; and they fit between the border of the valve seat, and the top of the box, as is seen at *c*, fig. 11.   When these wedges are withdrawn, the valve seats can be removed from the boxes, in order to replace them by others, in case the old ones are worn out.

The circular aperture of each valve seat has a bar extending diametrically across it, with a hole through the centre of the bar, to receive the upright spindle of the circular valve.   The racks, by which the valves are lifted, are united to the spindles of the valves by cross pins, as is shown in figures 10 and 11.   The racks for lifting the two steam valves, are applied to the upper ends of the spindles above the valves ; but the racks for the exhausting-valves are applied to the lower ends of the spindles, below the cross bars which guide them. To retain the racks steadily in a vertical position, brackets are fixed to the insides of the boxes, and project into the centre ; the end of each bracket is formed with a groove, to receive the back of the rack, and hold it upright.   See the sketch, p. 373.

Each of the sectors is made with three teeth ; it is fitted very tight upon the middle part of the spindle, and is prevented from turning round upon the spindle by a cross pin.   The sector must be fastened on the spindle, after the spindle is inserted into its place across the box. The pivot, at the extreme end of the spindle, is lodged in a socket formed in the inside of the box ; and the conical part, at the other end of the spindle, passes through an opening in the side of the box ; a brass bush is driven tight into that opening, and the spindle is accurately fitted into the bush, by grinding, so as to make a tight fitting, at the same time that it can turn round freely, in the same manner as the turning plug of a cock.

The lever at the outside of the box, to give motion to the spindle, and lift the valve, is fixed fast upon a square at the outer end of the spindle, beyond the conical part.   To keep the conical end of the spindle confined in its socket, a light iron clamp is applied at the outside of the box, in the manner shown at figures 6 and 7 : at one end of this clamp is a sharp point, which is lodged in a small puncture, at the end of the projecting boss which contains the socket for the pivot at the end of the spindle, inside the box ; but the puncture does not pass through the metal.   At the opposite end of the clamp, is a small screw, the point of which is inserted into a hole at the end of the spindle ; and by this screw the conical part of the spindle is forced into its socket, with just so much force as is requisite to make it fit tight, but without jambing fast.

In the course of time, the racks and sectors were laid aside by Messrs. Boulton and Watt ; and instead of the sector, a simple short lever, or lifting finger, was fixed upon the middle of each spindle ; and the end of the lifter entered into an opening, or stirrup, which was formed in the spindle of the valve. This construc-tion is shown at the lower part of Plate XIX. ; the operation of lifting the valve, on this plan, is exactly the same as the racks and sectors.

The upper box *b*, for the nossels, has a short branch pipe projecting from it on each side, and terminated with a flange, as is shown in fig. 6, in order to connect the steam pipe from the boiler with either side of the nossels, according to convenience ; the other short pipe being stopped up, by screwing a flat plate against the flange. When two boilers are used, they are sometimes placed on each side of the engine, and then the steam pipes are connected with the nossels on both sides. The throttle-valve *z*, fig. 12, is fixed in the joint between the flange of the short pipe which pro-

3 x 2

jects from the nossels, and that of the steam-pipe, as is shown at z, figs. 6 and 7. If the engine has two boilers, each steam pipe must be provided with a conical stop-valve, to shut off the communication from that boiler to the engine, at pleasure. This valve must always be quite open when the engine is at work, but it is useful to cut off the supply of steam, in order to stop the engine.

Construction of the working gear in Mr. Watt's double rotative-engine. The working gear, which has been already described, p. 451, and which is represented in Plate XIII., was the plan which Messrs. Boulton and Watt adopted for all their engines, after having had some years' experience in making them. Mr. Watt's original plan, which he used for the first engine at the Albion Mills, in 1785, and in several subsequent engines, deserves to be recorded.

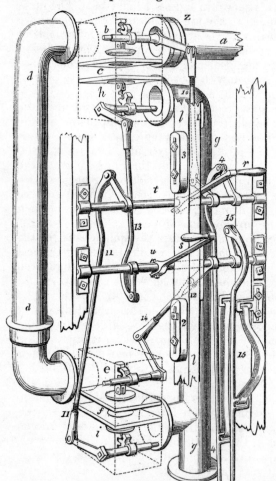

The bell-metal seats for the two exhausting-valves h and i are inverted, and the valves are fitted into the under sides of the seats, so as to open when they are drawn downwards, or to shut when they are pushed upwards; hence the pressure of the steam with which the cylinder is filled during the plenum, tends to force the exhausting-valves open, and the working gear is adapted to keep them shut, until the end of the stroke, and then they are opened very suddenly by the working gear (or rather they are permitted to open, by the pressure of the steam against them), in order to exhaust the cylinder as quickly as possible. The principle is laid down in Mr. Watt's fourth patent, 1784, (see p. 433); but he had formed the idea of this plan in 1776 (see note, p. 320.)

The axis t of the upper handle r, gives motion to the upper steam-valve b, and the lower exhausting-valve i, by means of two short levers which are fixed upon that axis, and which are connected by the rods 10 and 11, with the levers which open those two valves respectively. When the upper handle r is raised up, as in the figure, it opens the two valves b and i, in order to make the down stroke of the piston; or when the handle is put down, it shuts the two valves, in preparation for the upwards stroke. A short lever is also fixed upon the upper axis, and a weight is applied to it by the rod 4, in order to keep the valves open, after they have been opened by raising the handle r.

The axis u of the lower handle s gives motion to the lower steam-valve c, and the upper exhausting-valve h, by means of two short levers, and the rods, 13 and

14, which connect them with the levers, to open those valves respectively.   When the lower handle *s* is put down, it opens both valves *e* and *h*, in order to make the upward stroke of the piston ; or when that handle is raised up, as in the figure, it shuts both the valves, in preparation for the downward stroke.   A weight is also made to operate upon the axis of the lower handle *s*, by means of a rod 15, which is jointed to the end of a short lever fixed upon that axis ; this weight keeps the valves open, after they have been opened, by pressing down the handle *s*.

l l is the plug rod, which in these old engines was a large piece of wood suspended by its upper end from the great lever ; and the air-pump rod was affixed to it at the lower end, in the manner represented in Plate X.   On each side of the rod plug, two projecting chocks were fastened, by two bolts and nuts to each chock, in the manner shown in the figure.   The chocks 2 and 3 give motion to the lower handle *s* ; and the chocks 1 and 12 actuate the upper handle *r*, at the proper intervals, when the piston arrives near the end of its stroke.

This working gear requires no catches to retain the handles in those positions, which occasion the valves to be shut ; because the short levers, on which the weights operate, are reversed ; for they are fixed on the upper sides of the axes *t* and *u*, and the rods 4 4, and 15 15, by which the weights are connected to those levers, are bent at the upper ends, as is shown in the figure, in order to allow the levers to pass a little beyond the vertical position, when the valves are shut. This is shown at 15, where it is apparent that the action of the weight which is applied by the rod 15, will tend to raise the handle *s*, and shut the valves *h* and *e*, rather than to open them ; and yet if the handle *s* is put down, so as to open the valves a little, then the short inverted lever will depart so much from the vertical position, as to give the weight 15 a suitable action to throw down the handle, and open the valves completely, and also to keep them open, until the handle is raised again.   The action of the other handle *r* is the same, except that it is raised to open the valves *h* and *i*, and put down to shut them.   The proper extent of opening for each pair of valves is determined by two check straps, which are applied to the rods 4 and 15, and their upper ends are attached to fixed supports, so as to prevent the weights descending farther than is intended ; but when the valves are shut, the check straps become slack, as is shown at 15.

The two short levers upon the axes *t* and *u*, which are connected by the rods 11 and 13, with the two exhausting-valves, are reversed, as is shown in the figure ; and the ends of those rods are bent so much, as to allow the levers to come very nearly into the direction of the rods 11 and 13, when the valves are shut, and then those levers point away from the valves.   The levers for the two steam-valves are also arranged, so as to correspond with the directions of their connecting-rods respectively, when the valves are shut ; but the levers point towards the valves.

In any of these levers, when the valve is shut, the straight line which may be supposed to exist between the joints at each end of the connecting-rod, will pass very nearly through the centre line of the axis of the handle, as is shown at 12 or 13 ; the length of each lever therefore coincides with that of its connecting-rod, and consequently the effort which the steam exerts against the exhausting-valves to open them, is effectually resisted, for it can produce no sensible effect to turn the levers and their axes round ; but when the handle is moved, it turns the axis with all the levers round, and causes them to depart from their coincidences with their respective connecting-rods ; so that they will acquire the properties of levers, and will move their connecting-rods endways, in order to open the steam-valves, and to allow the exhausting-valves to open by the pressure of the steam.

The object of this system of combining the levers and their connecting-rods has been already stated at p. 374; when the valves are fully open, the levers make a considerable angle with their connecting-rods, and then any motion which is given to the handle will be transmitted by the levers and connecting-rods, so as to give a considerable motion to the valves; but as the valves become more closed, the levers approach towards the directions of their respective connecting-rods, and therefore their effective lengths of leverage diminish, on the principle stated at p. 46, whereby they acquire an augmented force to act upon the valves. The levers which actuate the inverted exhausting-valves, by coming very nearly into the lines of their respective connecting-rods, exert a great force to press those valves very close into their seats; and also to keep them shut close during all the course of the piston. This effect is attained by regulating the lengths of the connecting-rods 11 and 13, in a suitable manner, so that the levers will come very nearly into the lines of their connecting-rods when the valves are shut (*a*).

The engines which were made with this kind of valves and working gear perform extremely well, when they are kept in good order, with the working gear accurately adjusted, so as to shut the valves quite close, and to hold them with sufficient firmness to keep them shut very close; but as any neglect in this particular will allow the steam to leak continually through the inverted valves, this working gear requires careful management, and therefore it was thought best for general use to alter the plan, and to make the exhausting-valves shut downwards by the pressure of the steam, as is represented in Plate XIII., because they will then shut tight, independently of the accuracy of the working gear, which is only required to open them, and they will shut of themselves when permitted.

One of the advantages of this working gear without catches, is that the handles are moved both upwards and downwards by the chocks on the plug-rod; consequently one pair of the four valves must in all cases be opened, at the same time when the other pair are shut; and there is no danger that they can all four remain shut at once, so as to confine the steam in the cylinder and stop the engine suddenly, in the manner stated at p. 463. In the other kinds of working gear the valves are opened by weights, which may be obstructed from falling, and then all the four valves will remain closed at once.

When the attendant takes hold of the handles, to stop the motion of the engine, the inverted exhausting-valves are very easy to manage, from the facility with which they open; and if he is so careless or inexpert as to keep all the four valves shut at once, and confine the steam in the cylinder, when the engine is moving with its full speed, the inverted valves would probably be forced open by the pressure of the steam, when it became so much compressed as to endanger the breaking of the parts of the engine; whereas the other kind of exhausting-valves would shut closer when the steam was compressed, and would therefore render it difficult to open them by the handles, to allow it to escape.

The most dangerous accident to which a rotative-engine is liable, is when the working gear fails to open one pair of valves as soon as the other pair is shut; for the steam with which the cylinder is filled will then be confined in it, and therefore the motion of the piston must be stopped from passing through all its

---

(*a*) In some of these engines the connecting-rods 11 and 13 were made with springs in them, to admit of a little yielding in their length, and make them more certain to shut the valves close. They were spiral-worm springs, similar to those of a spring steelyard for weighing, and were applied in the middle of the length of each rod. In other cases the rods themselves were made of flat steel bars, and bent into waves or zig-zag folds of a serpentine form, so as to render the rods capable of a little elongation, when they were forcibly strained. These were expedients to cause the exhausting-valves to shut close, even if the working gear should not be quite exactly adjusted.

stroke by that confined steam, which has no escape, except the little that can leak upwards through the steam-valves *b* or *e*, and return to the boiler.   If the fly-wheel is a powerful one, the strain of thus checking its motion all at once is prodigious, and must injure the parts of the engine, even if they are not actually broken by it at the time.   Almost all cases of breaking down in steam-engines may be traced to this source, for the strength of the parts of engines is such, that the utmost force of the unconfined steam can never produce any injury (*a*).   The certainty of the valves being duly opened, depends upon the free action of the weight by which they are to be opened, and also upon the catch which is provided to retain them shut, being properly released ; it should be observed, that if the weight fails to open the valves at first, it will not be able to open them after the compression has begun, because the compression tends to keep the exhausting-valves shut.

Another variety of working gear for Mr. Watt's double engine.   The catch which Mr. Watt used in most of his rotative-engines, during the first ten years, is different from the diagonal catch which is described in p. 452, and represented at 3, 3 fig. 10, Plate XIII.   The catch is poised on a small horizontal axis, which is placed behind the axis *t u* of the two handles, as is shown at 19, fig. 13.   The catch 3, 3 stands in an upright position, and the end of the lever 12, of the axis *t* of the upper handle, is adapted to lodge upon the upper end of it ; or the end of the lever 12, of the axis *u* of the lower handle, is adapted to lodge beneath the lower end of the catch, as is shown in the figure ; but both levers cannot catch at the same time.

This catch is situated quite at one side of the working gear, in the same relative position as the diagonal catch 3 in fig. 6, and the axis 19 of the catch extends all across the working gear ; a lever 20 is fixed to that axis, at the middle of its length, and extends forwards horizontally, so that the extremity is close behind the plug-rod *l*.   Two small chocks which are fixed at the back of the rod project out so much as to intercept the extremity of the lever 20 ; one of these chocks lifts the end of the lever 20 a little up, every time that the piston reaches the top of its course, and that removes the upper end 3 of the catch backwards, so as to release the lever 12 of the upper handle, in order to allow the upper steam-valve, and the lower exhausting-valve, to be opened by their weight 4 as in figure 8. The same motion of the catch 3 brings the lower end of it forwards, in order to catch the end of the lever 16, which has been previously placed in the position shown in the figure, by the operation of the chock 2 of the plug-rod *l*, when it

(*a*) Some security against this complete confinement of the steam may be had, by a proper adjustment of those two short levers upon the axis of the handles which act upon the two steam-valves *b* and *c*, by the rods 10 and 14.   For if those levers pass a little beyond the lines of the direction of their rods respectively, when the steam-valves are shut, and when the handles are resting upon the catch, then, in case of a very violent compression of the steam within the cylinder, the steam-valve for that end of the cylinder would be pressed upwards with so much force, as would turn the lever and the axis of the handle round, and open the steam-valve sufficiently to allow the steam to escape to the boiler.

This security would become quite complete, if the connecting-rods 10 and 14 were made with socket joints in the middle of their length, and with spiral-worm springs, so applied to those joints, as to allow the rods to shorten when forced, but not to lengthen. In that case the steam-valves could never be held down into their seats with any greater force than the strength of those springs could occasion, and the springs might be regulated so as to cause a sufficient confinement of the steam to stop the motion of the engine in a proper manner as directed at p. 463, but no more, because the steam-valve would lift up and allow the steam to escape into the boiler.

The author has never seen this plan adopted, but it would be a certain remedy for the evil, because the two steam-valves would become regulated safety-valves for each end of the cylinder, to prevent any dangerous compression of steam within it.

raises the lower handle *s*, in order to close the lower steam-valve *e*, and the upper exhausting-valve *h*.

As the piston descends, the chock 2 quits the lower handle *s*, but the lower end of the catch 3 will detain the lever 16, so as to prevent the handle following the chock, until the piston arrives at the bottom of its course; the chock 1 of the plug then depresses the upper handle *r*, so as to close the upper steam-valve *b* and the lower exhausting-valve *l*; and, after that, the small chock at the back of the rod *l* depresses the end of the lever 20 so much as to move the lower end of the catch 3 backwards, and then the lever 16 becomes released, in order that the lower steam-valve *e* and the upper exhausting-valve *h* may be opened by their weight 15, as in fig. 9, in order to produce the upwards stroke of the piston. At the same time, the upper end of the catch 3 being brought forward beneath the end of the lever 12, will retain that lever in its position after the chock 1 rises from the upper handle *r*.

The operation of the working gear with this catch is precisely the same as with the diagonal self-acting catch, fig. 10, in which the action of latching one lever, discharges the other, as already described, p. 452; but on the plan of fig. 13, the catch is governed by the plug itself, independently of any other part of the working gear. This kind of catch insures that one pair of valves shall be completely closed before the others are opened, because the different chocks on the plug-rod must of necessity be so regulated, as that one of the handles of the working gear shall be moved into their required positions, before the catch is moved; and if this is not the case, the levers 12 or 16 will lock the catch fast, and prevent its motion, whereby the lever 20 will be strained, or some part of the catch work will be broken. This kind of catch is incapable of retaining all the four valves shut at the same time, which is a great security against accidents; but when the engine is required to stand still, the two handles must be held fast, by means of a short iron rod, with a hook at each end, and the hooks are applied to the two handles so as to keep the lower handle up, and the upper handle down. This was thought an inconvenience by the engine-keepers, and also if they suffered the chocks to get loose, and out of adjustment, the catch work would be broken; hence Messrs. Boulton and Watt adopted the diagonal self-acting catch, fig. 13, which is free from those minor inconveniences; but it is liable to a much more objectionable accident; for it is capable of retaining all the four valves closed, whereby the steam may by accident be confined in the cylinder, when the engine is working at full speed (*a*).

A different kind of catch for the working gear. The levers 12 and 16, which are fixed upon the upper and lower axis *t* and *u*, of the working gear, to act with the catch which is to retain one pair of valves closed, might be formed as in the sketch, so as to catch each other alternately, and effect the required purpose without any separate catch. For instance, the figure shows the upper

---

(*a*) The author happened once to be present when an engine, which had a diagonal self-acting catch, fig. 10, to the working gear, was suddenly stopped, in consequence of a careless person inadvertently holding one of the handles from moving by the action of the weight, when the engine was going at its full speed. Being struck with the danger of this accident, he was led to contrive a different catch, which cannot retain all the valves at once, and would not be liable to derangement, but would be very certain in its action. The author constructed a model of this catch many years ago, but it has never been put in actual practice. He has wished, in one or two cases since, to have applied it to old engines with hand gear, when they required repairing; but as it is necessary to place the axes of the two handles nearer together, than they usually are for the common catch work, this alteration would have involved several others, and would have occasioned more trouble than the proposed advantage of the plan would have been worth.

handle c depressed, and, consequently, the upper steam-valve b and the lower exhausting-valve i are closed. The lower handle s is, at the same time, depressed by the action of its weight 15, and, therefore, the lower steam-valve e and the upper exhausting-valve h are open. In this position of the valves (see fig. 9, Plate XIII.) the piston will make its upwards stroke.

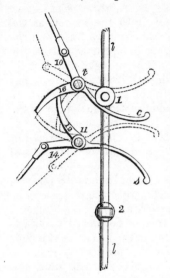

The lever 12 of the lower axis 11 then serves as a prop or catch, to retain the upper lever 16 and prevent it from descending, until the piston arrives at the top of its stroke. The chock 2 of the plug will then raise the lower handle s, in order to close the lower steam-valve c, and the upper exhausting-valve h; in which motion the lower lever 12 will move so far backwards from beneath the upper lever 16, as to release it; and then, by the action of the weight 4, the upper handle is suddenly raised, in order to open the upper steam-valve b, and the lower exhausting-valve i (see fig. 8), which position of the valves will produce the downward stroke. In this state the levers 12 and 16, and the handles r and s, assume the positions of the dotted lines in the sketch, and the upper lever 16 serves as the catch to prevent the lower lever 12 from rising, and thus that pair of valves are kept shut.

The working-gear with this kind of catch could never open one pair of valves before the other pair were closed, nor could it retain all the four valves closed at once.

Note. To make this plan most effective, the two axes t and 11 should be placed nearer together than is represented in the sketch, in order that they may have a sufficient angular motion about their respective centres, so as to open and shut the valves completely. This motion is usually between one-fourth and one-fifth of a circle.

Weights to the working-gear for opening the valves. In Mr. Watt's first engines, the rods 4 and 15, to which the weights of the working-gear are applied, were connected at their lower ends, with the ends of horizontal pieces of wood, the other ends of which rested upon centre pins, in the manner of horizontal levers, or like the treadles of a loom; and large iron weights were applied upon the treadles, to aid their own weight, in drawing down the rods 4 and 15, in order to open the valves; these weights could be moved along the treadles, to a greater or lesser distance from their centres of motion, so as to act with any required degree of force upon the working-gear.

These treadles were placed beneath the floor over the condensing cistern, and check straps were applied to them, to limit the distance to which they should fall, and consequently to regulate the extent of the opening of the valves. The lower ends of these check straps were passed through loops, which were fastened to the treadles; and the upper ends of the straps were passed through similar loops, which were fixed to the posts for the working-gear, as is shown in the sketch, p. 524; the ends of the straps were united together by buckles, by which the lengths of the check straps could be regulated at pleasure, in the same manner as the stirrups-leathers of a saddle are adjusted. The check straps were intended to stop the motion of the falling weights, and avoid noise or concussion; but a better plan, called the plunger weights, was afterwards adopted.

3 Y

Plunger weights for opening the valves, are cylindrical weights of cast iron, which are fastened upon the rods 4 and 15 of the working-gear, so that the central line of the weight corresponds with that of the rod. Each plunger is fitted into a short hollow cylinder, like that of a pump barrel, which is fixed down in the condensing cistern, beneath the water. The lower end of each barrel is closed, but the bottom has a hole through the centre of it, which is covered by a leather clack-valve opening upwards. This clack will admit the water freely into the barrel, as the plunger weight is raised upwards in it, whilst the valves are closing ; but when the plunger is left to fall suddenly by its own weight, in order to open the valves, the contained water must make its escape out of the barrel. The plunger does not fill the barrel very exactly, but a sufficient space is left, to allow the water to squeeze out around the plunger on all sides as it descends ; and so much resistance is thus opposed to the descent of the plunger, as will give it a suitable motion for opening the valves quickly, and yet without noise or concussion. This method allows the plunger weights to be made so heavy, that they can have no chance of sticking, or failing to open the valves ; and yet they will act quietly, and without shaking and deranging the working-gear.

Joints of the working-gear. The connecting-rods 10 and 11, and 13 and 14, require to be very exactly adjusted in their lengths, in order that the valves may be properly closed, when the short levers on the axes of the working-gear are so placed, as to correspond with the directions of the rods respectively. To effect this adjustment, those ends of the connecting-rods which are jointed to the levers on the ends of the spindles of the valves, are formed into screws, which are tapped into the joint pieces for those levers. Consequently, by taking out the pin of the joint, and turning the joint piece round upon the screw at the end of the connecting-rod, the effective length of the rod can be altered at pleasure. In the best engines, the ends of all the levers, and of the rods, are made of steel, or else they are case-hardened, that the joints may resist wearing ; the joint-pins are made of hard wood, which cannot wear the steel, and do not make a noise in working ; these pins can easily be replaced when they become worn.

The several levers which are fixed upon the axes *t* and *u*, are bored out with round holes, to fit upon those axes ; and they are fastened at their required places, and in the proper positions, by wedges, which are driven tight in between the axis and the lever, in a direction parallel to the axis ; the wedges are contained in notches, which are cut out at one side of the circular openings through the levers, and also flat places are filed upon the round axes, for these wedges to bed upon (*a*).

In the first engines the pivots for the axes *t* and *u* of the working-gear, were sustained by upright wooden posts, as is represented in the single engine, Plate X. ; but in subsequent engines, cast iron brackets were affixed to the side pipes *d d* and *g g*, as is represented in Plate XI., and the brackets projected out to a sufficient distance to support the axis of the handles in their proper situations.

(*a*) *Note.* The engine from which figure 6 was drawn, was not made by Messrs. Boulton and Watt. The short levers for the rods 10 and 11, or for the rods 13 and 14, are made two in one piece, so that one fixing served for both levers. This is the best plan ; but Messrs. Boulton and Watt always made the four short levers separate, and each one was fixed separately upon the axis, as is shown in the sketch, p. 524.

CHAPTER VII.

Application of the sliding rule for calculating the dimensions for the parts of Steam-Engines.

MR. WATT proportioned all the parts of his patent rotative-engines so judiciously, that after a few years' practice in making those engines, he ascertained the proper proportions for every part, and established standards for the dimensions of engines of all sizes; these dimensions have been followed ever since, by the best engineers and makers of steam-engines, with very few deviations, because long experience has proved that those standards were extremely well proportioned.

In this part of his subject, Mr. Watt was greatly assisted by several ingenious workmen and operative engineers, who had been educated under his own eye, in the manufactory at Soho, and who had acquired a stock of experience in the course of practice. The calculations which were required for proportioning the dimensions of engines, were commonly intrusted to Mr. Southern, who was a skilful mathematician, and to whom Messrs. Boulton and Watt were induced to give an interest in their manufactory chiefly on that account. Mr. Watt, with the assistance of Mr. Southern, investigated all the circumstances which can affect the proportions of each part of a steam engine; and thence formulæ were deduced by which the dimensions could be calculated for each individual case. The dimensions so ascertained were communicated to the workmen for their guidance, but the rules themselves, or the principles of calculation which were followed, are very little known.

At the same time Mr. Watt employed logarithmic scales, on a sliding rule, for performing calculations relative to steam-engines and machinery. These instruments had been long in use amongst gaugers and officers of the excise, and were also used by carpenters; but they were very coarsely and inaccurately divided, and required some improvements to render them serviceable to engineers. Mr. Watt and Mr. Southern arranged a series of logarithmic lines upon a sliding rule, in a very judicious form, and they employed the most skilful artists to graduate the original patterns, from which the sliding rules themselves were to be copied.

The Soho sliding rules are made of box-wood, 10½ inches long, with one slider, and four logarithmic lines on the front face; and at the back are tables of useful numbers, divisors and factors, for a variety of calculations. Sliding rules of this kind are still called the Soho rules, and they are so correctly divided by some of the best makers of mathematical instruments in London, that they are capable of performing ordinary calculations with sufficient accuracy for practice; and by means of the tables at the back of the rule, most questions in mensuration may be very readily solved.

These sliding rules were put into the hands of all the foremen and superior workmen of the Soho manufactory, and through them, the advantage of calculating by means of the sliding rule has become known amongst other engineers, and some do employ it for all computations of ordinary mensuration; but the habit of using it upon all occasions, is almost confined to those who have been educated at Soho. To apply the sliding rule extensively for the calculation of the dimensions of the parts of steam-engines and machinery, particular formulæ are required, which

3 Y 2

were confined to a very few of the principal engineers at Soho, and have not been at all disseminated in the profession.

The great number of patent engines which Messrs. Boulton and Watt sent out, and fixed in every part of the kingdom, furnished a sufficient number of models, from which other engineers could easily ascertain the proper dimensions for the parts of an engine of any particular size which they might be required to construct. The knowledge of these dimensions has also been made generally known by workmen who were brought up at Soho, and who have by degrees become dispersed over the kingdom; but they have rarely acquired, or been able to communicate, much knowledge of the principles by which the dimensions are regulated (*a*).

From the great experience which engineers have acquired since Mr. Watt's time, it may be presumed, that if any considerable deficiencies or errors had existed in the dimensions of his standard engines, they would have been corrected in the modern engines. This has actually been the case, in some few instances; but in almost all essential particulars, the practice of the most skilful engineers of the present day, is very nearly the same as that of Mr. Watt himself; and in those few instances where they differ, the modern practice is for the most part inferior to the original, which ought to be studied with care by all engineers, as the fountain-head for that kind of knowledge. In that view the information contained in this and the next chapter will be very useful to the profession, and accordingly the author has taken great pains to verify all the proportions and rules which he has formed, by a continual reference to Mr. Watt's own practice, so as to be assured of their correctness.

The properties of the logarithmic lines upon a sliding rule are not very generally known to practical engineers, and there is no complete treatise extant upon the subject (*b*), hence it becomes necessary for the instruction of students, to supply

(*a*) The author, at his first entrance into business, made it his particular study to acquire a complete knowledge of the structure of Mr. Watt's steam-engines, and of the proportions and dimensions of all their parts; as being in every respect the very best course of instruction for a practical mechanician. With this view, in the years 1804 and 1805, he examined and took exact drawings of a number of those engines of all sizes, with their dimensions; and after having accumulated a sufficient collection of observations, they were arranged and compared, to find out the proportions that the different dimensions bear to each other; which being ascertained, corresponding rules were formed for calculating the dimensions, in every case, either by common arithmetic, or by the sliding rule.

The author is not aware, whether the rules which he thus made himself, are exactly the same as those which Messrs. Boulton and Watt followed; but the rules in question have been proved in the course of several years' practice, and corrected when necessary, so as to give results which upon an average, correspond very nearly with the practice of the most experienced engineers, who have all taken their proportions from the established models of Messrs. Boulton and Watt's standard engines.

(*b*) The author procured a Soho sliding rule at his first commencement in business; but not being able at that time to obtain any instructions for the mode of using it, and having observed the facility with which the Soho workmen performed their ordinary calculations by it, he was induced to investigate the properties of the instrument very fully, and thence deduced formulæ for its application upon all occasions.

The same course had been begun a little earlier by Mr. Benjamin Bevan, civil engineer and architect, who has since published a practical treatise on the sliding rule, in octavo, 1822; this is a valuable little work, which contains a collection of useful theorems for performing all kinds of calculations; but as it is intended for the instruction of all professions who require calculations, and not particularly for any one class, it does not contain many of those specific formulæ which render the instrument particularly valuable to mechanicians.

A treatise on the sliding rule was published some years ago, by Dr. Mackay, for the particular instruction of navigators, to enable them to perform nautical calculations by it. A small book has also been printed by Mr. Routledge, engineer, of Leeds, to explain the Soho sliding rule, and promote the use of it amongst engineers: it is sold by the makers of those rules.

that deficiency, and to explain the construction of the various formulæ which are given in this work, for calculating particular quantities, by the aid of the sliding rule.

METHOD OF PERFORMING CALCULATIONS BY THE SLIDING RULE.

THIS instrument is a mechanical application of logarithms; and to have a correct idea of its principle of action, we must consider the operation of logarithms, whereby they perform the multiplication and division of numbers.

LOGARITHMS are a series of artificial numbers, adapted in a particular manner to a series of real numbers, and arranged in a table, wherein every real number has its corresponding logarithm; so that by inspection of such a table, any number can be converted into its logarithmic representative; and conversely any logarithm can be converted into the real number which it represents (*a*).

To multiply any two numbers together, by the aid of a table of logarithms, we must substitute the logarithm of each number, for the number itself, and then add the two logarithms together; their sum will be another logarithm, which being reconverted by the table, into a real number, that number will be the product of the two original numbers, which have in effect been multiplied together, by this addition of their logarithmic representatives.

And conversely, the division of one number by another, can be effected by subtracting the logarithm of the divisor, from the logarithm of the dividend, and the remainder is the logarithm of the quotient.

Hence, logarithms tend to facilitate computations, by substituting the operations of addition and substraction, for those of multiplication and division, which are more tedious and difficult to be performed.

Logarithms were first invented by John Napier, Baron of Merchiston, in Scotland, who published an account of his discovery in Latin, intitled *Mirifici Logarithmorum Canonis descriptio*, 1614. The numbers given in the inventor's table were Hyperbolic Logarithms, see p. 345; but the tables of common loga-

---

(*a*) *The logarithm of a number is the index of that power of ten which will produce the number.* Hence the logarithm is one more, than the number of times that ten must be multiplied by itself, to produce the number which the logarithm represents.

Example. The logarithm of 100 is 2, because 100 is the 2nd power (or square) of 10; for, as 10 must be multiplied once by 10 to produce 100, the logarithm of 100 is $1 + 1 = 2$. And in like manner, 1000 being the third power (or cube) of 10, its logarithm is 3; for ten must be multiplied twice by itself to produce 1000, and $2 + 1 = 3$.

It follows from this construction of logarithms, that they will form a series which increases in arithmetical progression, whilst the numbers which they represent, form a series which increases in geometrical progression; and the two series are so adapted to each other, that 0 in the arithmetical series corresponds to 1 in the geometrical series.

Thus $\begin{cases} 0 & 1 & 2 & 3 & 4 & 5 & \&c. \quad \text{Logarithms forming an arithmetical series.} \\ 1 & 10 & 100 & 1000 & 10000 & 100\,000 & \&c. \quad \text{Numbers forming a geometrical series.} \end{cases}$

This is the skeleton of a table of logarithms, for all the intermediate number between 1 and 10; 10 and 100, &c., in the geometrical series, may be filled up, and may have logarithms properly proportioned to them, to fill up the corresponding intervals between 0 and 1; 1 and 2, &c., in the arithmetical series.

It is a consequence of this construction, that the logarithms of all numbers which are less than 10 will be decimal fractions; for instance, the log. of 5 is 0·69897. And the logarithms of the numbers between 10 and 100, will be 1, with certain decimal fractions in addition; example, the log. of 50 is 1·69897. And the logarithms of numbers from 100 to 1000, will be 2, with the addition of suitable decimals; for instance, the log. of 500 is 2·69897.

rithms now generally used, were deduced from Napier's, by Mr. Henry Briggs in 1615, and for a long time they were called Briggs's logarithms.

The following table contains the logarithms of every number from 1 to 100, but the decimal fractions of the logarithmic numbers are only carried to three places of figures, because the divisions of sliding rules are not commonly made to represent more minute quantities than thousandth parts of the whole scale.

Num.	Logar.	Num.	Logar.	Num.	Logar.	Num.	Logar.	Num.	Logar.
1	0·000	21	1·322	41	1·613	61	1·785	81	1·908
2	0·301	22	1·342	42	1·623	62	1·792	82	1·914
3	0·477	23	1·362	43	1·633	63	1·799	83	1·919
4	0·602	24	1·380	44	1·643	64	1·806	84	1·924
5	0·699	25	1·398	45	1·653	65	1·813	85	1·929
6	0·778	26	1·415	46	1·663	66	1·820	86	1·934
7	0·845	27	1·431	47	1·672	67	1·826	87	1·940
8	0·903	28	1·447	48	1·681	68	1·833	88	1·944
9	0·954	29	1·462	49	1·690	69	1·839	89	1·949
10	1·000	30	1·477	50	1·699	70	1·845	90	1 954
11	1·041	31	1·491	51	1·708	71	1·851	91	1 959
12	1·079	32	1·505	52	1·716	72	1·857	92	1 964
13	1·114	33	1·519	53	1·724	73	1·863	93	1·968
14	1·146	34	1·531	54	1·732	74	1·869	94	1·973
15	1·176	35	1·544	55	1·740	75	1·875	95	1·978
16	1·204	36	1·556	56	1·748	76	1·881	96	1·982
17	1·230	37	1·568	57	1·756	77	1·886	97	1·987
18	1·255	38	1·580	58	1·763	78	1·892	98	1·991
19	1·279	39	1·591	59	1·771	79	1·898	99	1·996
20	1·301	40	1·602	60	1·778	80	1 903	100	2·000

Examples of the use of logarithms. To multiply 16 by 4. The log. of 16 is per table 1·204; add thereto the log. of 4, which is 0·602, and we have 1·806 for the sum of the two logarithms. If we seek this amongst the logarithms in the table, we find its corresponding number is 64; which is the product of the two numbers 16 × 4.

To divide 96 by 8. Take the log. of 96, which is 1·982; deduct from it the log. of 8, which is 0·903, and the logarithm remainder is 1·079; according to the table, this difference of the two logarithms is the logarithm of the number 12; which is the quotient of 96 ÷ 8.

The most important use of logarithms is to abridge those complicated multiplications and divisions, which are required to perform involution, or the raising of the powers of numbers; and evolution, or the extraction of their roots. For instance, to obtain the square of a number, it must be multiplied by itself; but if we multiply its logarithm by 2, the product will be the logarithm of its square. And conversely, if we divide the logarithm of any number by 2, the quotient will be the logarithm of its square root. In like manner, by multiplying or dividing the logarithm of a number by 3, we obtain the logarithm of its cube, or of its cube root. And any other power, or root, of a number may be raised or extracted by multiplying, or dividing, the logarithm of the number by the index of the power in question.

Examples. To extract the cube root of 64. Its log. is 1·806, which being divided by 3, gives 0·602, which is the log. of 4, the root required. Again, to extract the square root of 81. Divide its log. 1·908 by 2, and we have 0·954, which is the log. of 9, the root sought. Or to obtain the square of 5, multiply its log. 0·699 by 2, and we obtain 1·398, which is the log. of 25, the square demanded.

The above is a sufficient statement of the properties of logarithmic numbers, to explain the construction and operation of the sliding rule, which consists of a combination of straight lines, engraved on the edges of rulers, and graduated with unequal divisions, in such manner, that the magnitudes of the spaces or distances of each of the divisions, from the first division of the scale, will represent the series of logarithmic quantities, not by numbers, as in the table, but by spaces. And the divisions are figured, so as to denote the real numbers, corresponding to the logarithms which are represented by those spaces.

Hence in the ordinary use of logarithms, we substitute for the real numbers, certain artificial logarithmic numbers, as their representatives; and in the sliding rule we make a still farther substitution, viz. that of spaces, to represent these logarithmic numbers, which are themselves only artificial quantities.

At first sight this would appear to be a complication of the logarithmic method of computation; but in effect it will be found to be a great simplification, because it divests that system of all idea of number, in reference to the logarithms; for those numbers, which as logarithms are purely artificial quantities, become realities as spaces; also when the logarithmic quantities are represented by spaces, we can very conveniently perform the addition or subtraction of the logarithms, by applying those spaces in actual contact with each other, and joining one space to another, so as to obtain their sum, in order to perform multiplication; or cutting off from one space, a quantity equal to another space, so as to obtain their difference, in order to perform division.

To divide a line into logarithmic spaces, we must provide a plane scale of any suitable length, which is accurately divided into 10 equal parts, with decimal subdivisions at one end, and with diagonal lines, to obtain such minute subdivisions as will represent thousandths of the whole length. From this scale we must measure off with a pair of compasses, such a space from the zero of the scale as will represent any logarithm in the table; and the space so measured must be transferred to the line which we intend to divide logarithmetically, by placing one point of the compasses upon the first division or commencement of that line, and marking a division upon it, at the proper place, with the other point of the compasses.

Note. In thus forming a logarithmic scale, we must disregard the whole numbers which constitute a part of the logarithms in the table, and only consider the decimals; whereby we assume that all the logarithmic numbers are decimal fractions of unity; and therefore to represent them, we take corresponding decimal fractions of the whole space or length on the plain scale, from 0 to 10 (*a*).

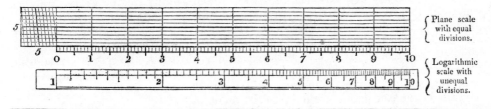

(*a*) This method of expunging the whole numbers of the logarithms is not peculiar to the sliding rule, but it is done in all modern tables of logarithms; those whole numbers are called the indices of the logarithms to which they belong; and the indices are omitted in the tables, because they can be very readily supplied, being in all cases one less than the number of places of figures contained in the whole number which is to be represented by the logarithm, when that index is prefixed to it.

For instance, the logarithm of 4 is 0·60206; and that of 40 is 1·60206; or of 400, 2·60206; or of 4000, 3·60206. In all modern tables the logarithm ·60206 is given for 4 every time it recurs, as at 40, 400, &c. without regard to the number of 0's which follow it; and the index, or whole number, which is to be prefixed to the logarithm, is omitted in the table; but it must in all cases be supplied, to the logarithms of any numbers which are taken out from the table for use.

For example, to set out the primary or figured divisions of the logarithmic line, we must successively transfer from the plane scale spaces of 301, 477, 602, 699, &c. as per table; that is to say, the logarithms of 2, 3, 4, &c. must be represented by 301, 477, 602, &c. thousandths of the whole length of the plane scale from 0 to 10; and those spaces must be marked off respectively from the first division of the logarithmic line, which is numbered 1.

The primary divisions thus obtained, may be numbered with figures 1, 2, 3, 4, 5, &c. to 10, and these figures will denote real numbers (either whole numbers or decimal fractions) whose logarithms are expressed by the spaces or distances at which they are placed respectively from the first division marked 1; and all those logarithms are considered as decimal fractions of the whole space from 1 to 10, which is taken for the unity of the system, and is called its radius. By continuing the same operation of measuring off the intermediate logarithms from the plane scale, and transferring those measures to the logarithmic scale, all the intervals between the primary divisions may be filled up, and the scale completed, as in the figure. Or, by reference to a more extensive table of logarithms, the scale may be filled up with as many other subdivisions as it will admit of, without becoming too crowded with minute divisions, to leave them distinguishable by the eye.

The simple scale, or line, with logarithmic divisions, is called Gunter's scale, from the name of the celebrated mathematician who invented it about 1623. The line contains two series of the logarithmic divisions from 1 to 10, placed one at the end of the other in continuation, to form one line, which is used with the compasses, to perform the multiplication and division of numbers, and the evolution or involution of powers, and roots of numbers. Thus the space from the first division 1 of the scale, to the division representing any number which is to be multiplied (16 for instance) being taken in the compasses; we can set off that space from the division representing the other number, or multiplier (4 for instance) measuring forwards in the direction from 1 towards 10; and thus prolonging one space by adding the other to it, we shall find the other point of the compasses will reach to some further division (64 in this case) which represents the product of the multiplication (of 16 by 4).

Or to divide any number (as 96) by another number (as 8), take the space in the compasses from 1 to the division representing the divisor (8); and set off that space from the number to be divided (96) measuring backwards from 10 towards 1; then the other point of the compasses will fall on a division (12) representing the quotient obtained by dividing the large number (96) by the other (8).

The simple Gunter's scale was improved by Mr. Wingate in 1627, who formed it on two separate rulers, which were applied one against the other, to avoid using compasses. This was modified into the present sliding rule by a Mr. Milburne in 1650, and by Seth Partridge in 1657, it was for a long time called the sliding Gunter. This instrument consists of a ruler, having a moveable slider fitted into a groove along the middle of the ruler; and the adjacent edges of both the groove and the slider being graduated with logarithmic divisions, the spaces representing the logarithms can be added to, or subtracted from each other, by comparison of contact, without using compasses; and when the slider is properly placed, the results of the calculation are obtained by mere inspection.

The lines upon sliding rules have been combined in various forms, to suit the purposes of particular calculators; but that which has been found most convenient for the use of engineers, was arranged by Mr. Southern, under the direction of Mr. Watt, expressly for the use of the engineers of Messrs. Boulton and Watt's manufactory at Soho, and is in consequence called the Soho rule. The original divisions for these instruments were made with the greatest accuracy, the radius of the scale being 10 inches long; but similar lines have since been divided, with equal accuracy, on a larger scale; and the divisions being less minute, they are more easily read off, and admit of more minute subdivisions.

DESCRIPTION OF THE SOHO SLIDING RULE. The rule itself is made of hard box-wood, 10¼ inches long, $\frac{8}{10}$ broad, and about $\frac{2}{10}$ thick. A groove is formed along the middle of one side of it; and a slider of the same wood, $\frac{3}{10}$ wide by $\frac{1}{10}$ thick, is fitted into the groove, so as to slide freely endways, backwards or forwards therein.

The face of the slider, and that of the ruler itself, are reduced to one flat surface, upon which the divisions and figures are engraved, as is represented in the sketch. There are four lines, divided logarithmically, viz. two on the fixed ruler, at the margins of the groove, and two others on the margins of the moveable slider; so that there are two pairs of lines, each pair consisting of one moveable, and one fixed line of divisions.

The several lines are designated by the four first letters of the alphabet; the two upper lines are marked A and B, and the two lower lines are marked C and D. The lines A, B and C are all alike, being fac similes of each other, but the lower line D has divisions of exactly twice the magnitude of the others; this lower line, which is called a line of single radius, is formed as before described, and the space from 1 to 10, which is called the radius, occupies 10 inches of the length of the rule. Each of the upper lines contains two repetitions of the series of smaller logarithmic divisions, each series being 5 inches radius; these are called lines of double radius, but it would be more correct to say lines of two radii, and to call line D a line of double radius.

NUMERATION, OR NOTATION, ON THE SLIDING RULE. The first step to learning the use of this valuable instrument, is to acquire a facility in reading its divisions, so as to find that division which represents any required number. It should be premised, that the value of the divisions on the logarithmic lines is not permanently fixed; but that an arbitrary value is given to them by the calculator in each particular operation, so that the ten primary figured divisions of the scale may equally represent 1, 2, 3, 4, &c.; or 10, 20, 30, 40, &c.; or 100, 200, 300, 400, &c.; or ·1, ·2, ·3, ·4, &c., according to the assumption made in each case.

Hence the sliding rule does not determine the numbers which result from the calculations which are made by it, but only the figures by which the numbers are to be denoted, and the order in which those figures are to stand; but their actual value must be determined by other means.

The reason of this deficiency will be evident, when we reflect that in forming the logarithmic scale, all the whole numbers appertaining to the logarithms were rejected, and only the decimal portion of each logarithm retained, to be represented by the spaces of the divisions. These whole numbers indicate the exact value of the numbers represented by the logarithms; and hence the whole numbers are called the indexes or characteristics of the logarithms to which they are prefixed.

By altering the index of the logarithm of any number, it will become the logarithm of any other number which can result from multiplying or dividing the original number, by 10, or 100, or 1000, &c. For instance, the log. of 2·5 is 0·39794; and of 25, 1·39794; and of 250, 2·39794; and of 2500, 3·39794; so that, by applying a suitable index, the same decimal portion of the logarithm will serve for all numbers which can be denoted by 2, followed by 5. The sliding rule, as commonly constructed, has nothing to represent the index or whole number of the logarithms, and, consequently, its divisions cannot have a determinate value.

In general, whatever value is assumed for the first division of a logarithmic line, whether ·1 or 1·, or 10 or 100, the other primary or figured divisions of the same line should preserve an uniform progressive value, according to their respective places in the scale; so that if the first division is accounted ·1, the next will be ·2, then ·3 and ·4, &c.; but the first division being called 1, then the others will be 2, 3, 4, &c. Or if the first is assumed to be 10, the others will be 20, 30, 40, &c.; or the first being 100, we shall have 200, 300, 400, &c. for the others.

All intermediate subdivisions must be reckoned according to the value that has been assumed for the primary divisions between which they are situated.

Note. At the same time that we ought thus to preserve an uniformity of value, in the progressive divisions of the same line, we may, at pleasure, assume different values for the divisions of the different lines on the same rule, even when they are used in concert for one operation.

We may now explain the notation by some examples, which the reader must perform with the rule itself, as the sketches being small and immovable, are not sufficiently explanatory to a learner (*a*). It is easy to find the division which will represent any number which can be expressed by one figure with any number of 0's (as ·6, or 6, or 60, or 600,) because the primary divisions of the scale are figured suitably for that purpose.

To find numbers of more than one place of figures on the sliding rule, we must keep the following facts in mind : The primary figured divisions, represent the first or left hand figures of the numbers. The subdivisions between the primary divisions, represent the figures in the second place from the left towards the right hand. The intermediate divisions between these subdivisions, represent the figures in the third place from the left. In some long rules the intermediate divisions are again partially subdivided, to represent some of the figures in the fourth place. Lastly, in all cases when the subdivisions and intermediate divisions required to express any number, are not engraved on the scale, they must be imagined, by estimating the intermediate spaces by the eye.

Example. To find any number of two figures (for instance 64, or the line A), we must begin with the first or left hand figure (6), and select that primary division which is marked with the same figure (6). Then we must take the second figure towards the left hand (4), and find its corresponding subdivision, between the primary division last found (6) and the next beyond it (7). There are always ten of these subdivisions engraved between every two primary divisions, in the shortest rules, and the middle subdivision is drawn longer than the others, in order to distinguish it for five (and in some long rules it is figured with a small 5) ; taking this long subdivision for a guide, it is easy to acquire the habit of counting all the ten subdivisions, as quickly as if each one were figured ; thus 64 will be found at the short subdivision adjacent to 65, on the side towards 60. For numbers which are denoted by only two figures, we shall always find a subdivision engraved to represent the second or last figure, and we shall have no occasion to imagine intermediate divisions, or to estimate any spaces by the eye. The sketch of the rule shows its slider drawn out so far, at the left hand, that the middle division of the line B (figured 10) is opposite to 64 on A, and serves as a pointer thereto. Also 5, on line B, points out 32 on line A ; again, 4 on line D points out 25 on line C.

Example. To find a number of three figures (for instance 1·56 on the line B). The first figure (1) is represented by the primary division figured 1, at the commencement of the line B ; then for the second figure (5), we must look amongst the subdivisions between the primary divisions (1 and 2). The fifth of those subdivisions is marked by a long stroke, to represent 1·5, and the short subdivision which follows it is 1·6 ; therefore the number we seek must be between those two ; and if there are no intermediate graduations between the subdivisions, we must estimate $\frac{6}{10}$, or a little more than half the space between the subdivisions 1·5 and 1·6 to fix the place of 1·56. In the sketch of the rule 1·56,

---

(*a*) Mr. Bate, optician in the Poultry, London, has taken great pains in dividing correct logarithmic scales for sliding rules, and has brought the manufacture of those instruments to the highest perfection ; they are made by him in ivory and in box-wood, of various dimensions, suitable for different purposes. The original Soho rule, of 10 inches radius, is a very convenient size for the pocket ; but they are also made of 24 inches radius, which is preferable for an office.

The calculations in this work have been made with a sliding rule, of nearly 28 inches radius, which was made by Mr. Bate for the author, and is divided with extreme precision. In correcting the impression, the same calculations have been repeated with another of Mr. Bate's sliding rules, of double the length of the former, or 56 inches radius. All engineers ought to be provided with such instruments, and with a little previous study, to become familiar with the notation ; they will be enabled to calculate the proportions of machines and engines with the utmost facility, by the aid of the formulæ laid down in this work.

upon the line B, is opposite to 1 on line A, whereby the latter division becomes a pointer to the former. Also 10, at the middle of the line C, will point out to the student 2·53, or 253 on line D.

Note. Learners are liable to mistakes in reading off numbers of three or more figures, when the second figure is an 0 (such as 105 or 401), and it requires some attention to avoid taking 150 or 410 instead. It may be kept in mind, as a guide in these cases, that the divisions representing numbers which have an 0 in the second place of figures, will in all cases be found close to, or very near to, a primary figured division.

In some parts of the scales of short lines, the subdivisions which represent the second place of figures, are too near together to admit of engraving ten intermediate divisions between each of them ; in such places five minute divisions are sometimes inserted, to represent the third place of figures; and therefore in reading such intermediate divisions, each one must be counted as two units of the third place of figures; the five divisions will therefore represent the even units, viz. 2, 4, 6, and 8 ; and consequently when the odd units 1, 3, 5, 7, or 9, occur in the third place of figures, the spaces between the five minute divisions must be subdivided by estimation.

Again, in other parts of the scale, the spaces between the subdivisions, which repreesnt the second place of figures, will only admit one intermediate division to be inserted; it will therefore represent five units of the third place of figures, and care must be taken to count it as such.

Hence, on the same rule, we shall find, at different parts of its scale, three kinds of minute intermediate divisions engraved between the larger subdivisions, in order to fill up and represent the third place of figures. At the commencement of the scale the interval between the subdivisions is filled up complete, with ten intermediate divisions; each division will then represent an unit in the third place of figures. In the middle parts of the scale only five intermediate divisions are inserted; and then each one will represent two units in the third place of figures. And in the higher parts of the scale, only one intermediate division is engraved; and it will represent five units in the third place of figures.

It is needless to give any other examples of this kind, as nothing but practice with the rule itself, can give a facility in reading off its divisions, and in estimating the exact places for those small intermediate divisions, for the third place of figures which are not engraved ; and a considerable practice is requisite to acquire the habit of reading quick and correct. It is best to begin with a small Soho rule of 10 inches radius, and first practise as above, with the aid of the sketch and directions, to find out numbers of only two figures ; then proceed to seek for numbers of three figures, which will require some estimation for the third place of figures, except at the commencement of the lower line D, which being a 10 inch radius, the intermediate divisions for each number, in the third place of figures, are inserted.

Having acquired the habit of reading the small rule, the learner should then proceed to practise with a larger one, in which the intermediate divisions are more numerous, and the spaces larger, so as to admit of more accuracy in estimating the last place of figures by the eye. Two kinds of long sliding rules have been divided by Mr. Bate, one being 28 inches radius, and the other 56 inches radius. The latter has divisions at the commencement of its scale, which represent every other number in the fourth place of figures, without estimation ; and the middle part contains every division for the third place of figures ; and the upper part of the scale every other division for the third place of figures.

MULTIPLICATION BY THE SLIDING RULE. This is performed by the two similar lines marked A and B. The division representing one of the factors, or numbers to be multiplied, being found on the line B, the slider must be drawn out until that division is brought opposite to 1 on the line A. Then the division representing the other factor, or multiplier, being found on the line A ; the product of the multiplication, will be found opposite to it, upon the line B ; see the sketch, and also the following precept.

Sliding Rule. $\left\{ \begin{array}{lll} \text{A} & 1 & \text{Multiplier.} \\ \text{B} & \text{Multiplicand.} & \text{Product.} \end{array} \right.$ *Example.* $\dfrac{\text{A} \quad 1 \quad 45}{\text{B} \quad 1·56 \quad 70·2}$

This mode of stating the precept and example, is an exact representation of the manner of placing the slider, and of seeking the coincidences of the divisions upon the adjacent lines. The two letters at the commencement, denote the lines which are to be used. Mr. Watt and Mr. Southern used this form of stating theorems, and it is followed by the engineers of their school; it has also been adopted by Mr. Bevan in his treatise on the sliding rule.

When the slider is thus set, the rule forms a complete table of products of the multiplicand, into any multiplier, we choose to select on the line B, thus

Exam. A	1	2	3	4	5	6	7	8	9
B	1·56	3·12	4·68	6·24	7·80	9·36	10·92	12·48	14·04

Note. As the lines A and B are divided exactly alike, it is immaterial upon which line we choose the multiplicand, and on which the multiplier, only observing that the two numbers to be multiplied, must in all cases be chosen on two different lines, and never both on the same line: also, that the product must be sought on the opposite line to that whereof the first division or 1 is used. A constant attention to this circumstance is indispensable, for if we depart from the precept, and read off from the wrong line, the operation of the rule will be to divide instead of to multiply, so as to give us a quotient where we required a product. The following precept gives the same results as the preceding:

$$\text{Sliding Rule.} \left\{ \begin{array}{lll} \text{A} & \text{Multiplicand.} & \text{Product.} \\ \hline \text{B} & 1 & \text{Multiplier.} \end{array} \right. \quad \textit{Example.} \quad \begin{array}{lll} \text{A} & 3·14 & 22 \\ \hline \text{B} & 1 & 7 \end{array} \text{ or } \begin{array}{l} 11 \\ \hline 3·5 \end{array}$$

When we wish to use the sliding rule as a table, to obtain the various products of the multiplicand by different multipliers, we must place the multiplicand or factor, which is not to be altered, opposite to 1, and seek the different multipliers on the same line as the 1; then the products will be found opposite to those multipliers, on the same line as the multiplicand.

The sliding rule may have its slide inverted, and, in many cases, this is the most convenient mode of performing calculations by it. Thus draw the slider quite out of its groove, then turn it end for end, and slide it back into the groove, so that it will be inverted and reversed, as in this sketch, and the divisions on the slider must be counted backwards, in a reversed order to those of the direct line upon the ruler against which they apply.

This condition of the sliding rule is always marked *slide inverted* in the precepts.

Multiplication by the lines A *and* ℧ *with the slide inverted.* Move the slider, so as to bring the two factors, or numbers which are to be multiplied together, opposite to each other, and then the product of their multiplication will be found opposite to 1 or 10; thus—

$$\begin{array}{l} \text{Sliding Rule,} \\ \text{slide inverted.} \end{array} \left\{ \begin{array}{lll} \text{A} & \text{Multiplier.} & 1 \\ \hline \text{℧} & \text{Multiplicand.} & \text{Product.} \end{array} \right. \quad \textit{Example.} \quad \begin{array}{lll} \text{A} & 3·14 & 1 \\ \hline \text{℧} & 7 & 22 \end{array} \text{ or } \begin{array}{ll} 9 & 1 \\ \hline 3 & 27 \end{array}$$

It requires some practice to acquire the habit of reading off the divisions of the slide when inverted, but no new instructions can be wanted. On this plan, we must move the slider, so as to set the rule properly for each multiplication, and therefore it cannot form a table for the products of different multipliers, but the lines A and ℧ form a table of all the different factors or pairs of numbers, which, being multiplied together, will give the same product.

In most cases of multiplication, the inverted slider will be preferable to the direct slider, because it will not admit of mistakes in reading off the results, and it is immaterial which of the two lines we use; for wherever we can find the two numbers which are to be multiplied together, we may place them opposite to each other, by moving the slider one way or the other; and the rule being so set, then wherever we can find 1 or 10, the product will be opposite to it.   There will be more than one coincidence of the same numbers; but as they will all give the same result, we may examine them all, so as to obtain a correct result, independently of any small inaccuracy in the divisions of the rule.

Suppose, for instance, that 23 is to be multiplied by 25; the inverted slider must be set as in the sketch, with 23 on one line, opposite to 25 on the other line. On examination of the two lines, we shall find four different places where 23 on one line corresponds to 25 on the other line, and if the rule is accurately divided, those coincidences will be exact; or if there is any inaccuracy, it will be apparent, and the slider may be adjusted so as to obtain a mean of the errors, by making the coincidence at one place as exact as that at another.   The slide being thus set, the product 575 will be found opposite 1, and opposite 10, so that there will be four places on the two lines A and C where this result may be found, and which may be compared together, to obtain a mean of the errors of the divisions.

THE INVERTED SLIDING RULE has two logarithmic lines, and is constructed expressly to be used in the manner before described. It may be formed by providing an extra slider to put into the groove of the Soho rule; such new slider being divided with a line of single radius, answering to that on the line D; but inverted or reversed in respect to it, and so arranged that the divisions begin at 3 instead of 1, and proceed to 10, which is situated near the middle of the length, and thence the divisions are continued onwards to 3, which forms the other end of the slider.   This is called a broken inverted line of single radius. The figures to the divisions being engraved erect, it will be easier to read than the preceding.

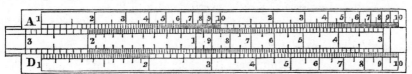

In this way, two lines of single radius are brought into use instead of two of double radius, and, consequently, in a rule of given length, the divisions are twice as large.   In this form of the inverted slider, we must use the two lower lines for simple multiplications and divisions.   In the sketch, the slider is represented as set with 3 opposite 9, and we may trace two such coincidences; the product of this multiplication 27 is found opposite to 1, and there are two of those coincidences, both giving the same result.

The inverted rule is very convenient for performing all the ordinary processes of arithmetic, and is more accurate than the direct sliding rule, both by avoiding mistakes in reading, and by having divisions of double size, upon a rule of the same length.   The inverted rule of 56 inches radius, which is made by Mr. Bate, gives very accurate results, and the sub-divisions being figured with small figures, in addition to the large figures which mark the primary divisions, it is extremely easy to find any number upon it.

In reading off the divisions of the inverted line on the slider, it must always be kept in mind, that the order of the numbers is reversed; so that they proceed from the right hand towards the left, contrary to the order of the direct line which counts from left to right.   It would be more proper to call it the reversed slider, instead of the inverted slider; but the term has been adopted from the common sliding rule, when its slide is inverted as before described. The object of arranging the inverted line on the slider, with 1 near the middle of the slider, and 3 at each end, is to avoid the necessity of ever drawing the slider any farther out of its groove, than is absolutely necessary; and it will be found, on trial, that it is never required to draw out the slider much more than half its length at one end, or the other end, for then 1 at the middle of the slide comes opposite to 1 at either end of the rule; and this extent of motion brings the whole range of numbers into action (*a*).

Directions for moving the slider of the inverted sliding rule. In all cases of setting a slide rule, two numbers are given, one of which must be found upon the divisions of the rule, and the other upon those of the slider; and the slider must be placed so that those two numbers will correspond with each other. As there is only one series of numbers upon the slider of the inverted rule, and another series upon the rule, without any repetitions of the same numbers on each line, it is necessary in every case to draw out the slider, either at one end of the rule or at the other end, according to the following directions; for if the slider is drawn out at the wrong end, the required coincidences will not appear amongst the divisions of the rule and those of the slider.

In sliding rules which have the broken inverted line of divisions engraved upon the rule (*b*): Multiply the first figure of each of the two numbers together, and if the first figure of their product is more than 3, then draw out the slider at the left hand end of the rule; this case will occur most frequently. Or if the first figure of the product is less than 3, then draw out the slider at the right hand end of the rule. For instance: to set 7 to 9, the first figure of their product (63) is 6, which being more than 3, the slider must be drawn out at the left hand end, until 7 on one line of divisions is observed to correspond with 9 on the other. Again: to set 10 to 12, the first figure of their product (120) is 1, which is less than 3, therefore the slider must be drawn out at the right hand end, until 10 on one line of divisions is brought to correspond with 12 on the other line. These directions should be preserved in the memory, until it becomes a habit to move the slide the right way without any thought.

In order to set the slider with expedition, when it is decided which way to move it, both the required numbers (or their nearest whole numbers) should be glanced at with the eye, one number on the rule and the other on the slide, so as to contemplate the distance between them, and then the slider may be moved one way or other over that distance with rapidity, in order to bring the two numbers near enough together, that they may both come within the field of vision, and be contemplated at once. The attention should then be steadily fixed upon the division or space between two divisions upon the rule, which represents one of the given numbers, and at the same time observing the divisions of the slider as they pass by it in succession, when the slider is moved slowly along in its groove. The motion given to the slider should not be more rapid than will allow the eye to recognize the different divisions as they pass by; and when the required di-

(*a*) The use of the sliding rule, with its slide inverted, was first proposed by the Rev. Wm. Pearson, in 1797, in Nicholson's Journal, quarto, vol. i., p. 450; and the author adopted the method from reading that paper. The complete inverted, or more properly, the reversed slide rule, was first arranged by Dr. Wollaston, for the purpose of calculating chemical equivalents, and other similar calculations. An inverted sliding rule, of a short length for the pocket, was afterwards arranged by Mr. John Taylor, with a small printed table of directions for the use of it; but it is not yet brought into extensive use, nor is it explained in any book on the sliding rule.

(*b*) In sliding rules which have the broken inverted line of divisions engraved upon the slider, as is shown in p. 541, then the directions must be reverse to the above; for right hand read left hand, and vice versa.

vision, or space between two divisions, on the slider comes in sight, the motion should be so much retarded, as to enable the eye to count the several subdivisions which arrive in succession opposite to the number which has been chosen on the rule; and when the required coincidence is obtained, the slider is properly set.

Note. To attain this final adjustment of the subdivisions, the slider should always be moved in that direction which will cause the numbers to count upwards, as they pass in succession, and not downwards: if the inverted line of divisions is upon the rule, then the slider must be moved from right to left for that purpose; and therefore in cases when it is necessary to draw out the slider at the right hand end, it should be moved a little beyond the required coincidence in the first instance, and should then be returned slowly from left to right, to establish the exact coincidence.

DIVISION BY THE SLIDING RULE. This may be performed either with the direct slider, or with the inverted slider; and in either case it is the reverse of multiplication.

If the slider is direct, set it so that the divisor is opposite to 1; then seek the dividend on the same line as the divisor is upon, and the quotient will be found opposite to the dividend, upon the same line as the 1; thus,

Sliding Rule. $\left\{\begin{array}{l} \text{A} \quad\quad 1 \quad\quad\quad \text{Quotient.} \\ \hline \text{B} \quad \text{Divisor.} \quad\quad \text{Dividend.} \end{array}\right.$ *Example.* $\dfrac{\text{A} \quad 1 \quad\quad 50}{\text{B} \quad 1\text{·}56 \quad 78}$

Note. The rule when thus set forms a table of the several quotients which may be obtained by dividing different numbers by the same divisor.

With the slide inverted, the operation of division is very simple, thus; set the slider so that the number to be divided is opposite to 1 or 10, and then the quotient will be found opposite to the divisor. There are two coincidences for each reading, and they cannot be mistaken, for we may take either of the numbers on either of the lines, wherever they can be found.

Sliding Rule, slide inverted. $\left\{\begin{array}{l} \text{A} \quad\quad 1 \quad\quad\quad \text{Quotient.} \\ \hline \text{Ɔ} \quad \text{Dividend.} \quad \text{Divisor.} \end{array}\right.$ *Example.* $\dfrac{\text{A} \quad 1 \quad 50}{\text{Ɔ} \quad 780 \quad 156}$ or $\dfrac{1 \quad 4\text{·}5}{27 \quad 6}$

Note. The rule when thus set forms a table of the several quotients which may be obtained by dividing the same number by different divisors.

Multiplication and division may be performed at one operation by the sliding rule. For the product which would result from the multiplying one number by another, may be divided by a third number, and the resulting quotient may be obtained, by inspection, with very great facility, because it is not necessary to observe the intermediate product. It is one of the great advantages of calculating by the sliding rule, that it is capable of abridging the arithmetical operations which it performs, by concentrating two or three operations into one, and exhibiting the final result at once, whereby the risk of errors is avoided, as well as the trouble of forming and recording intermediate products or quantities, which are commonly of no use in themselves, but only as stages of the process which must be gone through, in the usual mode of numerical computation.

When the slider is direct, one of the factors must be found upon one line A or B, and the divisor being found upon the other line, the slider must be set so as to bring them to correspond with each other; then the other factor being sought upon the same line as the divisor, the result will be found opposite to it, upon the same line as the factor first found.

Sliding Rule. $\left\{\begin{array}{l} \text{A} \quad \text{Factor.} \quad \text{Result.} \quad \text{Quotient.} \\ \hline \text{B} \quad \text{Divisor.} \quad \text{Factor.} \quad\quad 1 \end{array}\right.$ *Examp.* $\dfrac{\text{A} \quad 16 \text{ mltplier.} \quad\quad 28\text{·}8 \text{ result.}}{\text{B} \quad 25 \text{ divisor.} \quad\quad 45 \text{ mltplcand.}}$

The product of the multiplication cannot be found upon the rule, when the slide is direct, because it operates first to divide one of the factors by the divisor, and then to multiply the other factors by the quotient obtained by that division; thus $16 \div 25 = \cdot64 \times 45 = 28\cdot8$, which is the result, the same as above. The intermediate quotient may, if required, be found upon the same line as the result, opposite to 1 upon the same line as the divisor.

When the slider is inverted it must be set so as to bring the two factors, or numbers, which are to be multiplied together, to correspond with each other; and then the divisor being sought upon one line, the result will be found opposite thereto, upon the other line. *Example*. To multiply 16 by 45, and then to divide their product by 25. Thus $45 \times 16 = 720 \div 25 = 28\cdot8$ result.

Sliding Rule, slide inverted.	A Factor.	Divisor.	*Example*.	A 45 factor.	25 divisor.
	Ɔ Factor.	Result.		Ɔ 16 factor.	28·8 result.

The intermediate product of the multiplication need not be observed or attended to; but, if it is wanted, it may be found, by inspection, upon either of the lines opposite to 1 upon the other line.

The above operation is the same as that by which the rule of three or proportion is performed. For an example of the use of this property of the sliding rule, suppose that the dimensions of the two sides of a rectangular parallelogram are given in inches, and that it is required to find its area in square feet; the product obtained by multiplying the two sides of the rectangle together, will represent its area in square inches, which being divided by 144, the quotient will give the area in square feet. The rule must be set thus:

Sliding Rule, slide inverted.	A Side of rectangle inc.	Divisor 144	*Examp.*	A 8 inc.	div. 144
	Ɔ Side of rectangle inc.	Area sq. feet.		Ɔ 9 inc.	·5 sq. ft.

In like manner to obtain the area of a circle or an ellipsis in square feet, having the diameter, or the two diameters given in inches. The divisor for such cases will be 183·34, for that number of circular inches are equal to a square foot.

Sliding Rule, slide inverted.	A Diameter inches.	Divisor 183·3	*Exam.*	A 43 inc. diam.	div. 183·3
	Ɔ Diameter inches.	Area sq. feet.		Ɔ 04 inc. diam.	15 sq. ft.

PROPORTION, OR THE RULE OF THREE BY THE SLIDING RULE. It is necessary to have a very precise idea of the method on which proportionate numbers are to be found by the rule of three, which is so called because three numbers are given, and the object of the calculation is to find such a fourth unknown number, as shall bear the same ratio to one of the three given numbers, as that which already exists between the other two given numbers. The two corresponding numbers between which the ratio is established, may be called *the terms of the ratio*, and, in all cases, they must both express different quantities of the same kind of thing. The other or third number, to which the unknown number is to be adapted, may be called the *known number*, and it will express a certain quantity, the same kind of thing, as that which the unknown number is required to express.

The problem is to find such a number for the fourth term, as will bear a certain ratio to the known number; that ratio is expressed by the two terms of the ratio, considered merely as relative or comparative numbers, without regard to their individual values as quantities. If one of the terms of the ratio is divided by the other, the quotient will show how many times one term is contained in the other, so as to represent the ratio which exists between them in the simplest manner; and then if the known number is multiplied by that quotient, the product will be the fourth number required; and which, from the manner by which it is obtained, will be so adapted to the known number, as to bear the same ratio to it, as that term of the ratio which was divided, bears to the other term of the ratio which was used for the divisor.

Hence, when the fourth number is obtained, there will be two pairs of numbers; the two numbers of each pair expressing the same kind of thing, and the ratio between them being the same in both pairs; for if the largest number of each pair is divided by the smallest, the quotient so obtained from each pair, will be the same number.

Note. It is immaterial whether the division or the multiplication is made first, for the known number may be multiplied by one of the terms of the ratio, and then the product being divided by the other term, the quotient will be the same number as would be obtained, by first dividing one term of the ratio by the other, and then multiplying the quotient by the known number, in the manner above stated.

The circumstances on which the proportion depends must be considered in each case, and it will be obvious, from the nature of the question, whether the unknown number ought to be greater, or less, than the known number to which it is to be adapted; that circumstance will determine whether it is a case of increasing or of diminishing proportion; and, accordingly, the smallest or the largest of the terms of the ratio must be taken for the divisor.

Rule of three questions may be considered as if the two terms of the ratio constituted the numerator and denominator of a vulgar fraction, by which the known number is to be multiplied, and the result will be the fourth or unknown number. Thus, instead of the usual expression, as 16 is to 12, so is 8 to 6; we may say ($\frac{12}{16}$ths of 8 is 6) twelve sixteenths of eight is six. Or instead of saying as 12 is to 16, so is 8 to $10\frac{2}{3}$; we may say ($\frac{16}{12}$ths of 8 is $10\frac{2}{3}$) sixteen twelfths of eight, is ten and two thirds.

Increasing proportion is when the unknown number ought to be larger than the known number to which it is to be adapted; for instance, as 12 is to 16, so is 8 to $10\frac{2}{3}$; or $\frac{16}{12}$ of 8 is $10\frac{2}{3}$. In such cases the smallest of the two terms of the ratio must be taken for the divisor, and the largest for the multiplier.

Diminishing proportion is when the unknown number ought to be smaller than the known number to which it is to be adapted; for instance, as 16 is to 12, so is 8 to 6; or $\frac{12}{16}$ths of 8 is 6. In such cases the largest of the two terms of the ratio must be taken for the divisor, and the smallest for the multiplier.

Example of increasing proportion. Suppose that the multiplying-wheel of a steam-engine has 72 teeth, and is turned round 19 times per minute, and that it actuates a pinion of 38 teeth upon the axes of the fly-wheel; query how many revolutions will that pinion and fly-wheel make per minute?

The terms of the ratio are 38 and 72, for those two numbers form one pair, and both represent the same kind of thing, viz. teeth; and the given ratio is that which exists between those two numbers. The known number is 19; it shows itself, because it has no correspondent, for want of the unknown number which is to be adapted to it, to complete the second pair of proportionate numbers: that unknown number will be the same kind of thing as the known number, viz. revolutions per minute. It is obvious that as the wheel is larger than the pinion, the pinion must turn round quicker than the wheel, therefore the unknown number will be greater than the known number 19, and this is a case of increasing proportion; consequently the smallest number of the two terms of the ratio (viz. 38) must be taken for the divisor.

The calculation may be as follows. 72 teeth ÷ 38 teeth = 1·895 times as many teeth are contained in the wheel as in the pinion; and therefore whilst the wheel makes 19 turns, the pinion must make $19 \times 1·895 = 36$ turns per minute, which is the fourth number required. Or else the calculation may be, 72 teeth $\times$ 19 turns per minute = 1368 teeth of the multiplying-wheel will act on those of the pinion in a minute; and as every 38 teeth which act, will cause one turn of the pinion, the latter must make (1368 ÷ 38 =) 36 turns per minute, as before.

If we consider the two terms of the ratio as constituting the numerator and denominator of a vulgar fraction, by which the known number is to be multiplied, the unknown number which is sought, would be seventy-two thirty-eighths of 19, which is 36; for as $\frac{72}{38}$ths is equal to 1·895, it will be 1·895 times 19, which is = 36 turns per minute.

Proportion may be performed by the sliding rule, either with the slider direct or inverted.

When the slider is direct, the largest term of the ratio must be found on one line, and the slider must be set with the smallest term upon the other line, in correspondence with the largest; then in cases of increasing proportion, the known number must be found upon the same line with the smallest term of the ratio, and the unknown number will be found opposite to it, on the same line with the largest term of the ratio. The known number must be sought upon the same line with that term of the ratio which is to be used as the divisor.

| Sliding Rule. | A Largest term of ratio. | Unknown number. | *Example.* | A | 72 teeth. | 36 turns. |
| | B Smallest term of ratio. | Known number. | | B | 38 teeth. | 19 turns. |

The rule when thus set forms a complete table of all the possible pairs of numbers which have a common relation or analogy to each other; in the example, it is the ratio that 38 bears to 72; hence any two numbers which correspond on the lines A and B, would be proper for the number of

4 A

teeth in the wheel, and in the pinion respectively; for instance, 108 teeth for the wheel, to 57 teeth in the pinion. Also any two corresponding numbers on the two lines, will represent the number of turns which the multiplying-wheel and the pinion must make in the same time; for instance, if the wheel made 24 turns per minute, the fly-wheel would make $45\frac{1}{2}$ turns per minute.

When the slider is inverted, the known number must be found upon one of the lines; and in cases of increasing proportion, the slider must be set with the largest of the two terms of the ratio in correspondence with the known number; then the smallest term of the ratio being found upon either of the lines, the unknown number will be opposite to it. The result or unknown number will be found opposite to that term of the ratio which is to be used for a divisor.

Sliding Rule, slide inverted.	A	Known number.	Smallest term of ratio.	Exam.	A 19 turns.	38 teeth.
	Ɔ	Largest term of ratio.	Unknown number.		Ɔ 72 teeth.	36 turns.

Example of diminishing proportion. Suppose that the multiplying-wheel has 72 teeth, and makes 19 turns per minute, and that the fly-wheel and pinion make 36 revolutions per minute; query how many teeth must there be in the pinion?

The terms of the ratio are 19 and 36; for those two numbers form one pair, and both represent revolutions per minute. The known number is 72 teeth, and requires the unknown number, which must represent the same kind of thing, viz. teeth. It is obvious that as the pinion revolves quicker than the wheel, the pinion must have fewer teeth than the wheel; so that the unknown number will be less than the known number; hence this is a case of diminishing proportion, and the largest term of the ratio (36) must be taken for the divisor.

Thus, 19 revolutions ÷ 36 revolutions = ·5278 of a revolution of the wheel is made, whilst the pinion makes one turn, and as the wheel has 72 teeth, the pinion must have 72 teeth × ·5278 = 38 teeth.

Or 72 teeth × 19 revolutions = 1368 teeth of the wheel act in a minute, and as they cause the pinion to make 36 turns per minute, it must have 1368 ÷ 36 = 38 teeth.

Or we may say that the number of teeth in the pinion must be ($\frac{19}{36}$ of 72 = 38) nineteen thirty-sixths of seventy-two, which is thirty-eight.

Sliding Rule.	A	Smallest term of ratio.	Unknown number.	Exam.	A	19 turns.	38 teeth.
	B	Largest term of ratio.	Known number.		B	36 turns.	72 teeth.

The known number must be sought upon the same line with that term of the ratio which is to be used as the divisor. The rule when thus set forms a complete table of all the pairs of numbers which have the same ratio, as before explained.

Sliding Rule, slide inverted.	A	Known number.	Unknown number.	Ex.	A 72 teeth.	38 teeth.
	Ɔ	Smallest term of ratio.	Largest term of ratio.		Ɔ 19 turns.	36 turns.

When the slider is inverted, the result or unknown number, will be found opposite to that term of the ratio which is to be used for a divisor.

To convert vulgar fractions into decimal fractions for the sliding rule. All numbers which are to be used on the sliding rule must be according to the decimal notation, because the divisions and subdivisions of the logarithmic lines proceed by tens (a). According to the established customs of practical artists, quantities are subdivided into halves, thirds, quarters, eighths, twelfths, and sixteenths, and but few instances occur of fifths or tenths. For instance, a fathom is divided into half, to make a yard; the yard is divided into three, for feet; the foot into twelve, for inches; and each inch into halves, quarters, eighths, and sixteenths. The pound avoirdupois is divided into halves, quarters, and sixteenths.

This want of uniformity in the subdivision of weights and measures, and the want of correspondence in any of those subdivisions, with the established system

(a) Sliding rules have been made with the intermediate spaces between the primary divisions divided into twelfths, and those subdivided into quarters and eighths, conformably to the division of linear measures into feet and inches, and eighths of inches; others have been divided into eighths and sixteenths. Such rules are convenient for particular purposes, but are necessarily limited to those purposes; and none but decimal divisions can be recommended to engineers.

A series of tables of logarithms, in which the numbers proceed by fractions of twelfths, and by sixteenths, was published in 1817 by Mr. Thomas Preston, in a small octavo volume, entitled; a New System of Commercial Arithmetic, of such construction as to obviate all the inconveniences arising from the irregularity in the division of our moneys, weights, and measures. This is a useful book, and contains very explicit directions for the application of logarithms to common business.

of arithmetical notation, is a source of continual trouble in calculation; and to acquire a facility in the use of the sliding rule, it becomes necessary to retain in the memory, the decimal value of all those vulgar fractions which are in most common use, in the same manner as the multiplication table is learned.

The following table exhibits all these fractions. The twelfths in the table serve for converting inches into decimals of a foot; and the sixteenths for converting sixteenths of an inch into decimals of an inch; or ounces avoirdupois into decimals of a pound.

$\frac{1}{2}$ = ·5							
$\frac{1}{3}$ = ·3333	$\frac{2}{3}$ ·6666						
$\frac{1}{4}$ = ·25	$\frac{2}{4}$ ·5	$\frac{3}{4}$ ·75					
$\frac{1}{5}$ = ·2	$\frac{2}{5}$ ·4	$\frac{3}{5}$ ·6	$\frac{4}{5}$ ·8				
$\frac{1}{6}$ = ·1666	$\frac{2}{6}$ ·3333	$\frac{3}{6}$ ·5	$\frac{4}{6}$ ·6666	$\frac{5}{6}$ ·8333			
$\frac{1}{8}$ = ·125	$\frac{2}{8}$ 25	$\frac{3}{8}$ ·375	$\frac{4}{8}$ ·5	$\frac{5}{8}$ ·625	$\frac{6}{8}$ ·75	$\frac{7}{8}$ ·875	
$\frac{1}{12}$ = ·0833	$\frac{2}{12}$ ·1666	$\frac{3}{12}$ ·25	$\frac{4}{12}$ ·3333	$\frac{5}{12}$ ·4166	$\frac{6}{12}$ ·5	$\frac{7}{12}$ ·5833	$\frac{8}{12}$ ·6666
$\frac{1}{16}$ = ·0625	$\frac{2}{16}$ ·125	$\frac{3}{16}$ ·1875	$\frac{4}{16}$ ·25	$\frac{5}{16}$ ·3125	$\frac{6}{16}$ ·375	$\frac{7}{16}$ ·4375	$\frac{8}{16}$ 5
$\frac{9}{16}$ = ·5625	$\frac{10}{16}$ ·625	$\frac{11}{16}$ ·6875	$\frac{12}{16}$ ·75	$\frac{13}{16}$ ·8125	$\frac{14}{16}$ 875	$\frac{15}{16}$ ·9375	

Separate (twelfths): $\frac{9}{12}$ ·75 | $\frac{10}{12}$ ·8333 | $\frac{11}{12}$ ·9166

When the decimal value of such vulgar fractions as occur, is not retained in the memory, they may be converted into decimals by the lines A and B of the sliding rule, either with the slider inverted or direct.

When the slider is direct, the numerator must be found on the line A, and the denominator on the line B, and the slider must be set so that they will correspond, and stand one over the other, in the same manner as fractions are usually written; the equivalent decimal may then be found on the line A, opposite to 1 on the line B; viz. on the same line with the denominator.

Sliding Rule. $\begin{cases} \text{A} & \text{Numerator.} & \text{Decimal.} \\ \text{B} & \text{Denominator.} & 1 \end{cases}$ *Examples.* $\begin{array}{cc} \text{A} & 3 & ·375 \\ \text{B} & 8 & 1 \end{array}$ or $\begin{array}{cc} 13 & ·812 \\ 16 & 1 \end{array}$

The rule when thus set forms a complete table, of all the possible fractions that are equivalent to the decimal number, which is opposite to 1 upon the same line with the denominators of the fractions.

When the slider is inverted, it must be set so that the numerator will correspond with 1, thus;

Sliding Rule, slide inverted. $\begin{cases} \text{A} & \text{Numerator.} & \text{Decimal.} \\ \text{Ɔ} & 1 & \text{Denominator.} \end{cases}$ *Examples.* $\begin{array}{ccc} \text{A} & 3 & ·375 \\ \text{Ɔ} & 1 & 8 \end{array}$ or $\begin{array}{ccc} 13 & ·812 \\ 1 & 16 \end{array}$

It is so frequently required to express inches, and parts of inches, in decimals of a foot, that the following tables will be found to save much time.

Parts of an Inch.	Decimals of an Inch.	Decimals of a Foot.
$\frac{1}{32}$	·0312	·0026
$\frac{1}{16}$	·0625	·0052
$\frac{1}{8}$	·125	·0104
$\frac{1}{8}$ & $\frac{1}{16}$	·1875	·0156
$\frac{1}{4}$	·25	·0208
$\frac{1}{4}$ & $\frac{1}{16}$	·3125	·0260
$\frac{3}{8}$	·375	·0312
$\frac{3}{8}$ & $\frac{1}{16}$	·4375	·0365
$\frac{1}{2}$	·5	·0417
$\frac{1}{2}$ & $\frac{1}{16}$	·5625	·0469
$\frac{5}{8}$	·625	·0521
$\frac{5}{8}$ & $\frac{1}{16}$	·6875	·0573
$\frac{3}{4}$	·75	·0625
$\frac{3}{4}$ & $\frac{1}{16}$	·8125	·0677
$\frac{7}{8}$	·875	·0729
$\frac{7}{8}$ & $\frac{1}{16}$	·9375	·0781

Table of Inches and fractional Parts, expressed in Decimals of a Foot.

Inches and parts.	Decimals of a Foot.	Inches and parts.	Decimals of a Foot.	Inches and parts.	Decimals of a Foot.
$\frac{1}{8}$	·0104	4	·3333	8	·6667
$\frac{1}{4}$	·0208	$4\frac{1}{4}$	·3542	$8\frac{1}{4}$	·6875
$\frac{1}{2}$	·0417	$4\frac{1}{2}$	·375	$8\frac{1}{2}$	·7083
$\frac{3}{4}$	·0625	$4\frac{3}{4}$	·3958	$8\frac{3}{4}$	·7292
1	·0833	5	·4167	9	·75
$1\frac{1}{4}$	·1042	$5\frac{1}{4}$	·4375	$9\frac{1}{4}$	·7708
$1\frac{1}{2}$	·125	$5\frac{1}{2}$	·4583	$9\frac{1}{2}$	·7917
$1\frac{3}{4}$	·1458	$5\frac{3}{4}$	·4792	$9\frac{3}{4}$	·8125
2	·1667	6	·5	10	·8333
$2\frac{1}{4}$	·1875	$6\frac{1}{4}$	·5208	$10\frac{1}{4}$	·8542
$2\frac{1}{2}$	·2083	$6\frac{1}{2}$	·5417	$10\frac{1}{2}$	·875
$2\frac{3}{4}$	·2292	$6\frac{3}{4}$	·5625	$10\frac{3}{4}$	·8959
3	·25	7	·5833	11	·9167
$3\frac{1}{4}$	·2708	$7\frac{1}{4}$	·6042	$11\frac{1}{4}$	·9375
$3\frac{1}{2}$	·2917	$7\frac{1}{2}$	·625	$11\frac{1}{2}$	·9583
$3\frac{3}{4}$	·3125	$7\frac{3}{4}$	·6458	$11\frac{3}{4}$	9792

To find the square of a number by the sliding rule. For this purpose the number must be multiplied by itself, so that it is only a case of multiplication, which may be performed either with the slide direct or inverted ; thus,

Sliding Rule. $\left\{\rule{0pt}{20pt}\right.$

A	1	Number.
B	Number.	Square.

Example.

A	1	8
B	8	64

Sliding Rule, slide inverted. $\left\{\rule{0pt}{20pt}\right.$

A	Number.	1
Ɔ	Number.	Square.

Example.

A	8	1
Ɔ	8	64

It is most convenient to use the slide inverted to obtain the squares of numbers, and particularly in cases where the square of a number is required to be divided by another number.

For instance, the rule given at the bottom of p. 31, to find the height that a body must fall to acquire a given velocity ; Divide the square of the velocity in feet per second by $64\frac{1}{3}$; the quotient is the height fallen in feet. This and other similar cases may be performed by the sliding rule.

Sliding Rule, slide inverted. $\left\{\rule{0pt}{18pt}\right.$

A	Veloc. ft. per second.	64·33 feet.
Ɔ	Veloc. ft. per second.	Height fallen ft.

Ex.

A	30 ft. per sec.	64·33 feet.
Ɔ	30 ft. per sec.	14 ft. faln.

Sliding Rule, slide inverted. $\left\{\rule{0pt}{18pt}\right.$

A	Veloc. ft. per minute.	231600
Ɔ	Veloc. ft. per minute.	Height fallen ft.

Ex.

A	1800 ft. per min.	·2316
Ɔ	1800 ft. per min.	14 ft. faln.

Sliding Rule, slide inverted. $\left\{\rule{0pt}{18pt}\right.$

A	Veloc. miles per hour.	29·9
Ɔ	Veloc. miles per hour.	Height fallen ft.

Ex.

A	20·45 mil. per hr.	29·9
Ɔ	20·45 mil. per hr.	14 ft. faln.

To find the square root of any number by the sliding rule. This is a case of division, and the rule may be set with the slider inverted, so as to form a complete table of all the quotients which can be obtained by dividing the required number by different divisors ; hence we can select, by inspection, such a divisor as will produce a quotient equal to itself ; and that divisor and quotient will represent the square root of the number.

The slide being inverted, set it so that 1 upon one line will point to the required number upon the other line ; the divisions of the two lines are then to be examined, to find two coincident numbers, which are both of the same value, and those numbers will be the square root required. *Note.* As the numbers on one line proceed in a contrary direction to those upon the other line, it is easy, by counting along them, to find when the same number on both lines meet ; or if it happens that two divisions representing the same number do not meet, the coincident point must be within the space between those two divisions which are nearest to a coincidence ; and as there is only one such point, there can be no danger of mistake in finding it.

Sliding Rule, slide inverted. $\left\{\rule{0pt}{18pt}\right.$

A	Number.	Root.
Ɔ	1	Root.

This is where the same numbers on both lines meet.

Ex.

A	64 numb.	8 root.
Ɔ	1	8 root.

The rule thus set, forms a table of all the quotients which can be obtained by different divisors.

The above cases of squares and square roots, may be more conveniently performed by means of the lines marked C and D, which are laid down on the sliding rule, expressly for such purposes as will be explained ; but the inverted rule deserves the preference in all cases to which it is applicable, because its divisions are double the size of those on the line C, upon a rule of the same length ; and that is a great advantage, both for accuracy, and for the facility of reading off the quantities.

To find the reciprocals of numbers by the sliding rule. When 1 is divided by any number, the quotient will be a decimal fraction, which is termed the reciprocal of the number. The chief use of reciprocals is to enable us to substitute the operation of multiplication for that of division, or vice versa ; because the same results may be obtained by multiplying a number by a decimal fraction, as by dividing it by the whole number to which that decimal fraction is reciprocal.

For instance, a cubic foot of water weighs 62·5 pounds; and to find the weight of any number of cubic feet of water in pounds, that number must be multiplied by 62·5. The reciprocal of 62·5 is $(1 \div 62·5 = ·016$; and if the number in question is divided by ·016, the quotient will be the same, as the product which would be obtained by multiplying it by 62·5.

Example. 14 cubic feet $\times$ 62·5 lbs. = 8750 lbs. Or 14 cubic feet $\div$ ·016 = 8750.

In calculating by the sliding rule, it is very often advantageous to employ the reciprocals of numbers for divisors, instead of the numbers themselves for multipliers; or the reciprocals for multipliers, instead of the numbers for divisors. Suppose for instance, that two multiplications or divisions are required to be made by the sliding rule at one operation; it is capable of performing multiplication and division at one operation, as before explained; and therefore, by substituting the reciprocal of one of the multiplying numbers, and using that reciprocal as a divisor, or vice versa, the same result may be obtained by one operation of the rule, as by two multiplications or divisions.

Reciprocals are most conveniently found with the slide inverted; for if it is set so that 1 on one line, corresponds with one on the other line, then the two lines will form a complete table of numbers and their reciprocals opposite to them.

$$\text{Sliding Rule,} \left\{ \begin{array}{lll} \text{A} & 1 & \text{Numbers.} \\ \hline \text{ↄ} & 1 & \text{Reciprocal.} \end{array} \right. \quad Examples. \frac{\text{A} \quad 62·5 \text{ numb.}}{\text{ↄ} \quad ·016 \text{ recipro.}} \text{ or } \frac{16}{·0625} \text{ or } \frac{3}{·333}.$$

To find the reciprocals of numbers by the sliding rule with the slide direct, it must be set as follows:

$$\text{Sliding Rule.} \left\{ \begin{array}{lll} \text{A} & 1 & \text{Reciprocal.} \\ \hline \text{B} & \text{Number.} & 1 \end{array} \right. \quad Examples. \frac{\text{A} \quad 1 \quad ·016}{\text{B} \quad 62·5 \quad 1} \text{ or } \frac{1 \quad ·333}{3· \quad 1}.$$

USE OF THE LINE OF SINGLE RADIUS MARKED D ON THE SLIDING RULE. This line is used in concert with the line of double radius marked C, for performing such calculations as involve the squares of numbers, or their square roots; for by means of these two lines, a number may be either multiplied or divided by the square of another number, or by its square root.

It has been already stated, p. 534, that when the logarithm of any number is multiplied by 2, the product will be the logarithm of the square of that number; and conversely, if the logarithm of any number is divided by 2, the quotient will be the logarithm of the square root of that number.

The divisions upon the logarithmic line marked C, upon the slide rule, called the line of double radius, or the line of squares, are exactly half the size of the divisions upon the line D, which is called the line of single radius, or the line of roots. If the first divisions of each of those two lines are placed in correspondence, then every number upon the line C will have its square root opposite to it upon the line D; and conversely, every number upon the line D, will have its square opposite to it upon the line C; thus,

$$\text{Sliding Rule.} \left\{ \begin{array}{lll} \text{C} & 1 & \text{Square} \\ \hline \text{D} & 1 & \text{Number.} \end{array} \right. \text{ or } \frac{\text{Number.}}{\text{Root.}} \quad Examples. \frac{\text{C} \quad 1 \quad 2 \text{ numb.}}{\text{D} \quad 1 \quad 1·414 \text{ rt.}} \text{ or } \frac{16 \text{ squ.}}{4 \text{ numb.}}$$

Note. The rule being thus set forms a complete table of the squares and square roots of numbers.

To multiply the square of a number by another number. The number which is to be used for the multiplier being found on the line C, and placed opposite to 1 on the line D, the product of the multiplication will be found on C, opposite to the number which is to be squared for the multiplicand on D, thus,

$$\text{Sliding Rule.} \left\{ \begin{array}{lll} \text{C} & \text{Multiplier.} & \text{Product of multiplication.} \\ \hline \text{D} & 1 & \text{Factor to be squared.} \end{array} \right. \quad Example. \frac{\text{C} \quad 8 \text{ mult.} \quad 72 \text{ prod.}}{\text{D} \quad 1 \quad 3 \text{ squ.}}$$

Or the same result may be obtained with the slide inverted, when it is set as follows:

$$\text{Sliding Rule,} \left\{ \begin{array}{lll} \text{B} & \text{Multiplier.} & \text{Product.} \\ \hline \text{D} & \text{Factor to be squared.} & 1 \end{array} \right. \quad Example. \frac{\text{B} \quad 8 \text{ mult.} \quad 72 \text{ pro·}}{\text{D} \quad 3 \text{ squ.} \quad 1}$$

This property of the lines C and D renders the sliding rule extremely convenient for computing the solidities of all prismatic, cylindrical, or spherical bodies. For the side of a square prism, or the diameter of a cylinder, being taken on the line D, as that factor which is to be squared, and the length of the prism or cylinder being taken for the other factor on the line C or ꝗ, then the solidity may be found on the line C or ꝗ, as follows, according as the slider is direct or inverted.

Sliding Rule.	C	Length of cylinder.	Solidity of cylinder.	Exam.	C	4 ft. long.	16 cyl. ft.
	D	1	Diameter of cylinder.		D	1	2 ft. dia.

Sliding Rule, slide inverted.	ꝗ	Length of cylinder.	Solidity of cylinder.	Exam.	ꝗ	4 ft. long.	16 cyl. ft.
	D	Diameter of cylinder.	1		D	2 ft. diam.	1

The solidity thus obtained is merely the product of the three dimensions which become multiplied together into one product by the above operations of the rule; consequently those products will express the solidity in various terms, either cubic, or cylindrical feet; cubic or cylindrical inches; square inch feet, or cylindrical inch feet, &c. according to the different terms in which the given dimensions of the solid are expressed. But by dividing these different expressions of the solidities, by suitable numbers, they may be converted into any other required measures of solidities.

The necessary division of the product may be effected in the same operation of the rule, by taking some other number upon the line D, instead of 1, as is directed in the above precepts. The number to be taken for that purpose is called a gauge point, and it must be the square root of the required divisor, as will be more fully explained in another article; it is only intended here to show how the rule at one operation, multiplies the three dimensions of the solid together into one product, and also divides that product by any required divisor.

To find how many times the area of one square, or circle, is greater than that of another square or circle. This is best done by the lines C and D, thus.

Find the side of the largest square (or the diameter of largest circle) upon D, and set the slider so that 1 upon C corresponds therewith; then find the side of the smallest square (or the diameter of the smallest circle) upon D, and opposite to it on C will be the number of times that the area of the smaller is contained in that of the larger.

Sliding Rule.	C	1	Proportion of areas.	Exam.	C	1	4 times.
	D	Diam. of large circle	Diam. of small circle.		D	3 ft. dia.	6 ft. dia.

To find the corresponding diameters and lengths, of a number of square prisms or cylinders which will have the same solidity. This can be done by the lines ꝗ and D, with the slider inverted thus.

The side of any square prism, or the diameter of any cylinder, being found upon D, the slider must be set so that the length of that prism or cylinder upon ꝗ corresponds therewith; the rule then forms a table, which shows the lengths on ꝗ, and on D the sides of square prisms, or diameters of cylinders, corresponding to those lengths, so as to exhibit the dimensions of a number of square prisms or cylinders having the same solidity.

Sliding Rule, slide inverted.	ꝗ	Length of cylinder.	Length of cylinder.	Exam.	ꝗ	5 ft. long.	1·25 ft. long.
	D	Diam. of cylinder.	Diam. of cylinder.		D	2 ft. diam.	4 ft. diam.

If the solidity of these cylinders is required, it may be found at the same set of the rule, upon the line ꝗ opposite to some particular point upon the line D, which is called a gauge point; and there are different gauge points for different cases, as will be explained.

In cases when the dimensions of the cylinder are given in feet, and the solidity is required in cubic feet, the gauge point will be at 113 or 357 upon D; by this means we may find the dimensions of all kinds of cylinders, which will contain a given number of cubic feet, the slider being set so that the given number of cubic feet on line ꝗ, corresponds with 113 or 357 on C, thus.

Sliding Rule,   ꟼ   Solidity cubic ft.   Lengths of cylinders ft.   *Exa.*   ꟼ   15·71 cub. ft.   5 ft. long.
slide inverted.   D   113 or 357.   Diam. of cylinders in ft.   D   113 or 357.   2 ft. dia.

In cases of square prisms, when all the dimensions are given in feet, and the solidity in cubic feet, then the gauge point on D will be 1 or 10. If the sides of the square prisms are required in inches, and their lengths in feet, the solidity being given in cubic feet; then the gauge point will be 12 or 379 upon D. Or in cases of cylinders, when their diameters are required in inches, and their lengths in feet, the solidity being given in cubic feet; the gauge point will be 135 or 423 on D. The nature of these gauge points will be further explained in the proper place.

To find the areas of circles. The areas and circumferences of circles are so frequently required by engineers, that it is very desirable to have tables ready calculated to show them by inspection(*a*); but when a table is not at hand, the sliding rule is a good substitute; it may be set either for a particular case, or it may be set so as to form a table.

To find the area of a circle in square inches, having given its diameter in inches.
RULE. Multiply the square of the diameter in inches, by the decimal ·785398; the product is the area in square inches. *Note*, dividing the square by 1·273239, which is the reciprocal of the decimal number, will give the same result as multiplying by the decimal number.

Example. 25 inches diameter, squared = 625 circular inches area × ·7854 = 490·87 square inches. Or 625 circular inches ÷ 1·273 = 490·87 square inches.

Sliding Rule,   A   Diam. of circle inches.   1·273   *Exam.*   A   25 inc. dia.   1·273
slide inverted.   Ɔ   Diam. of circle inches.   Area squ. inc.   Ɔ   25 inc. dia.   491 squ. inc.

This requires the rule to be set for each operation, but the lines C and D being placed as follows, will form a complete table of the areas of circles in square inches.

Sliding Rule.   C   95   Area square inches.   *Exam.*   95   49 squ. inc.   or   77
              D   11   Diameter in inches.   11   7·9 inches.   9·9

(*a*) The author felt the want of a table of the areas of circles, at the commencement of his professional studies, and being unable to find such a table in any book, he was induced to calculate the areas and circumferences of circles, from one inch in diameter to one hundred, with every intermediate half inch, to form a table, which he has had in use for twenty years past.

A more extensive table, containing every quarter of an inch of diameter, has since been calculated by the late Mr. Goodwyn, the proprietor of an extensive brewery in London. This gentleman had always shown a great spirit for improvements, he was the first person who adopted Mr Watt's rotative steam-engine in 1784 (see p. 434); and after his retirement from business, he devoted much time to researches into the properties of numbers, and methods of computation, some of which were published during his life. The table in question has been printed by Dr. Gregory, in his Mathematics for Practical Men, 8vo., 1825. During the progress of the present work, the author has also found a table of the areas of circles for every inch in diameter, in Sir Samuel Morland's Elevation des Eaux, 1685, (see p. 92), and the same table is printed in Ozanam's Cours de Mathematique, 1697, Vol. III.

The author's table has been verified from both these sources, so that every inch in diameter, in the above table, is the result of three independent calculations; and every half inch is derived from two independent calculations; the intermediate quarters of inches are Mr. Goodwyn's calculations; and the author has proved as many of them as could be done by comparison with each other. It has been thought sufficient, for all practical purposes, to give the numbers to five places of figures, but they have been calculated to seven places of decimals, and the last figure of decimals in the table is the nearest which could be chosen to the truth. The foundation of the calculations in this table are as follows.

The circumference of any circle is 3·14159265 times its diameter. Or, reciprocally, the diameter of any circle is ·31830989 of its circumference.

The area of any circle is ·78539816 of the area of its circumscribing square. Or, reciprocally, the area of any square is 1·2732395 times the area of the circle which may be inscribed within it.

A square foot contains 144 square inches; or (144 × 1·2732=) 183·346 circular inches.

TABLE OF THE CIRCUMFERENCES AND AREAS OF CIRCLES OF ALL DIAMETERS, FROM ONE QUARTER OF AN INCH TO ONE HUNDRED INCHES.

The diameters are expressed in inches and quarters; the circumferences in inches; and the areas in circular inches, in square inches, and in square feet.

Circumf.	Area.	Diam.	Area.		Circumf.	Area.	Diam.	Area.	
Inches.	Circular inc.	Inches.	Square inc.	Square feet.	Inches.	Circular inc.	Inches.	Square inc.	Square feet.
·7854	·0625	0 25	·04909	·00034	38·484	150·06	12·25	117·86	·81847
1·5708	·25	0·5	·19635	·00136	39·270	156·25	12·5	122·72	·85221
2 3562	·5625	0·75	44179	·00307	40·055	162 56	12·75	127·68	·88664
3·1416	1·	1·	·7854	·00545	40·841	169·	13·	132 73	·92175
3·927	1·5625	1·25	1·2272	·00852	41·626	175·56	13·25	137 89	·95754
4·7124	2·25	1·5	1·7671	·01227	42·411	182·25	13·5	143·14	·99402
5·4978	3·0625	1·75	2·4053	·01670	43 197	189·06	13·75	148·49	1·0312
6·2832	4·	2·	3·1416	·02182	43·982	196·	14·	153 94	1·0690
7 0686	5 0625	2 25	3·9761	·02761	44·768	203·06	14·25	159·48	1·1075
7 854	6·25	2·5	4·9087	·03409	45·553	210·25	14·5	165·13	1·1467
8 6394	7·5625	2·75	5·9396	·04125	46·338	217·56	14·75	170·87	1·1866
9·4248	9·	3·	7·0686	·04909	47·124	225·	15·	176·71	1·2272
10·210	10·562	3·25	8·2958	·05761	47 909	232·56	15·25	182·65	1·2684
10·996	12 25	3·5	9·6211	·06681	48 695	240·25	15 5	188·69	1 3104
11 781	14·062	3 75	11·045	·07670	49·480	248·06	15·75	194·83	1·3530
12 566	16·	4·	12·566	·08727	50·265	256·	16·	201·06	1·3963
13·352	18·062	4·25	14·186	09851	51·051	264·06	16·25	207·39	1·4402
14·137	20 25	4·5	15·904	·11045	51·836	272·25	16·5	213 82	1·4849
14 923	22·562	4·75	17·721	·12306	52·622	280·56	16·75	220·35	1·5302
15·708	25·	5·	19·635	·13635	53·407	289	17·	226·98	1 5762
16·493	27·562	5 25	21 648	·15033	54·192	297·56	17·25	233·71	1·6230
17·279	30·25	5·5	23 758	·16499	54 978	306·25	17 5	240·53	1·6703
18·064	33·062	5·75	25 967	·18033	55·763	315·06	17·75	247·45	1·7184
18·85	36·	6·	28·274	·19635	56·549	324·	18·	254·47	1·7671
19·635	39·062	6·25	30·680	·21305	57·334	333·06	18·25	261·59	1·8166
20·420	42·25	6·5	33·183	·23044	58·119	342·25	18 5	268·80	1·8667
21·206	45 562	6·75	35·785	·24850	58 905	351·56	18·75	276·12	1·9175
21·991	49·	7·	38·485	·26725	59·690	361·	19·	283·53	1·9689
22·777	52·562	7·25	41·282	·28668	60·476	370·56	19 25	291·04	2·0211
23·562	56·25	7·5	44·179	·30680	61·261	380·25	19·5	298·65	2·0740
24·347	60·062	7·75	47·173	·32759	62·046	390·06	19·75	306·35	2·1275
25·133	64·	8·	50·265	·34907	62·832	40)·	20·	314·16	2·1817
25·918	68·062	8·25	53·456	·37122	63·617	410·06	20·25	322·06	2 2366
26·704	72 25	8·5	56·745	·39406	64·403	420 25	20 5	330·06	2·2921
27·489	76 562	8 75	60·132	·41758	65·188	430·56	20·75	338·16	2 3484
28·274	81·	9·	63·617	·44179	65·973	441·	21·	346 36	2·4053
29 060	85·562	9·25	67 201	·46667	66·759	451·56	21 25	354·66	2 4629
29 845	90·25	9·5	70·882	·49224	67·544	462 25	21·5	363 05	2·5212
30·631	95 062	9·75	74 662	·51849	68 33	473 06	21 75	371·54	2 5802
31·416	100·	10·	78·540	·54542	69·115	484·	22·	380·13	2·6398
32 201	105·06	10 25	82·516	57303	69·900	495·06	22 25	388·82	2·7001
32 987	110·25	10·5	86 590	·60132	70·686	506·25	22·5	397·61	2·7612
33·772	115 56	10·75	90 762	63029	71·471	517·56	22·75	406·49	2 8229
34 558	121·	11·	95·033	·65995	72·257	529·	23	415·48	2 8852
35 343	126·56	11·25	99 402	·69029	73·042	540·56	23·25	424·56	2 9483
36 128	132 25	11·5	103 87	·72131	73·827	552·25	23·5	433·74	3·0121
36 914	138·06	11·75	108·43	·75301	74 613	564·06	23·75	443·01	3 0765
37 699	144·	12	113 10	·78540	75 398	576	24·	452 39	3 1416

Circumf.	Area.	Diam.	Area.		Circumf.	Area.	Diam.	Area.	
Inches.	Circular inc.	Inches.	Square inc.	Square feet.	Inches.	Circular inc.	Inches.	Square inc.	Square feet.
76·184	588·06	24·25	461·86	3·2074	113·88	1314·06	36·25	1032·06	7·1671
76·969	600·25	24·5	471·44	3·2739	114·67	1332·25	36·5	1046·35	7·2663
77·754	612·56	24·75	481·11	3·3410	115·45	1350·56	36·75	1060·73	7·3662
78·540	625·	25·	490·87	3·4088	116·24	1369·	37·	1075·21	7·4667
79·325	637·56	25·25	500·74	3·4774	117·02	1387·56	37·25	1089·79	7·5680
80·111	650·25	25·5	510·71	3·5466	117·81	1406·25	37·5	1104·47	7·6699
80·896	663·06	25·75	520·77	3·6164	118·60	1425·06	37·75	1119·24	7·7725
81·681	676·	26·	530·93	3·6870	119·38	1444·	38·	1134·11	7·8758
82·467	689·06	26·25	541·19	3·7582	120·17	1463·06	38·25	1149·09	7·9798
83·252	702·25	26·5	551·55	3·8302	120·95	1482·25	38·5	1164·16	8·0844
84·038	715·56	26·75	562·00	3·9028	121·74	1501·56	38·75	1179·32	8·1897
84·823	729·	27·	572·56	3·9761	122·52	1521·	39·	1194·59	8·2958
85·608	742·56	27·25	583·21	4·0501	123·31	1540·56	39·25	1209·95	8·4025
86·394	756·25	27·5	593·96	4·1247	124·09	1560·25	39·5	1225·42	8·5098
87·179	770·06	27·75	604·81	4·2000	124·88	1580·06	39·75	1240·98	8·6179
87·964	784·	28·	615·75	4·2761	125·66	1600·	40·	1256·64	8·7266
88·750	798·06	28·25	626·80	4·3528	126·45	1620·06	40·25	1272·39	8·8361
89·535	812·25	28·5	637·94	4·4301	127·23	1640·25	40·5	1288·25	8·9462
90·321	826·56	28·75	649·18	4·5082	128·02	1660·56	40·75	1304·20	9·0569
91·106	841·	29·	660·52	4·5870	128·81	1681·	41·	1320·25	9·1684
91·892	855·56	29·25	671·96	4·6664	129·59	1701·56	41·25	1336·40	9·2806
92·677	870·25	29·5	683·49	4·7465	130·38	1722·25	41·5	1352·65	9·3934
93·462	885·06	29·75	695·13	4·8273	131·16	1743·06	41·75	1369·00	9·5069
94·248	900·	30·	706·86	4·9087	131·95	1764·	42·	1385·44	9·6211
95·033	915·06	30·25	718·69	4·9909	132·73	1785·06	42·25	1401·98	9·7360
95·819	930·25	30·5	730·62	5·0737	133·52	1806·25	42·5	1418·63	9·8516
96·604	945·56	30·75	742·64	5·1572	134·30	1827·56	42·75	1435·36	9·9678
97·389	961·	31·	754·77	5·2414	135·08	1849·	43·	1452·20	10·0847
98·175	976·56	31·25	766·99	5·3263	135·87	1870·56	43·25	1469·14	10·2023
98·960	992·25	31·5	779·31	5·4119	136·66	1892·25	43·5	1486·17	10·3206
99·746	1008·06	31·75	791·73	5·4981	137·44	1914·06	43·75	1503·30	10·4396
100·531	1024·	32·	804·25	5·5851	138·23	1936·	44·	1520·53	10·5592
101·316	1040·06	32·25	816·86	5·6727	139·02	1958·06	44·25	1537·86	10·6796
102·102	1056·25	32·5	829·58	5·7609	139·80	1980·25	44·5	1556·28	10·8075
102·887	1072·56	32·75	842·39	5·8499	140·59	2002·56	44·75	1572·81	10·9223
103·673	1089·	33·	855·30	5·9396	141·37	2025·	45·	1590·43	11·0447
104·458	1105·56	33·25	868·31	6·0299	142·16	2047·56	45·25	1608·15	11·1677
105·243	1122·25	33·5	881·41	6·1209	142·94	2070·25	45·5	1625·97	11·2914
106·029	1139·06	33·75	894·62	6·2126	143·73	2093·06	45·75	1643·89	11·4159
106·814	1156·	34·	907·92	6·3050	144·51	2116·	46·	1661·90	11·5409
107·600	1173·06	34·25	921·32	6·3981	145·30	2139·06	46·25	1680·02	11·6667
108·385	1190·25	34·5	934·82	6·4918	146·08	2162·25	46·5	1698·23	11·7932
109·170	1207·56	34·75	948·42	6·5862	146·87	2185·56	46·75	1716·54	11·9204
109·956	1225·	35·	962·11	6·6813	147·65	2209·	47·	1734·94	12·0482
110·741	1242·56	35·25	975·91	6·7771	148·44	2232·56	47·25	1753·45	12·1767
111·527	1260·25	35·5	989·80	6·8736	149·23	2256·25	47·5	1772·05	12·3059
112·312	1278·06	35·75	1003·79	6·9707	150·01	2280·06	47·75	1790·76	12·4358
113·097	1296·	36·	1017·88	7·0686	150·80	2304·	48·	1809·56	12·5664

To find the area of a circle in square feet; having given, its diameter in inches.

RULE. Divide the square of the diameter in inches, by 183·346. The quotient is the area in square feet.

Example. 25 inc. diameter squared = 625 circular inches area, ÷ 183·346 = 3·4088 square feet.

Sliding Rule. { A Diam. of circle inches. 183·34
slide inverted. { ꓛ Diam. of circle inches. Area square feet.   *Example.* A 25 inc. dia. 183·34
ꓛ 25 inc. dia. 3·409

Or the lines C and D will form a complete table of diameters and areas when the slider is set thus :

Sliding Rule. { C 3·14 Area square feet. *Example.* C 3·14 15 squ. ft. or 22
{ D 24 Diameter inches. D 24 52·5 inc. 63·5.

4 B

Circumf.	Area.	Diam.	Area.		Circumf.	Area.	Diam.	Area.	
Inches.	Circular inc.	Inches.	Square inc.	Square feet.	Inches.	Circular inc.	Inches.	Square inc.	Square feet.
151·58	2328·06	48·25	1828·46	12·698	189·28	3630·06	60·25	2851·04	19·799
152·37	2352·25	48·5	1847·45	12·829	190·07	3660·25	60·5	2874·75	19·964
153·15	2376·56	48·75	1866·55	12·962	190·85	3690·56	60·75	2898·56	20·129
153·94	2401·	49·	1885·74	13·095	191·64	3721·	61·	2922·47	20·295
154·72	2425·56	49·25	1905·83	13·235	192·42	3751·56	61·25	2946·47	20·462
155·51	2450·25	49·5	1924·42	13·364	193·21	3782·25	61·5	2970·57	20·629
156·29	2475·06	49·75	1943·91	13·499	193·99	3813·06	61·75	2994·77	20·797
157·08	2500·	50·	1963·49	13·635	194·78	3844·	62·	3019·07	20·966
157·96	2525·06	50·25	1983·18	13·772	195·56	3875·06	62·25	3043·47	21·135
158·65	2550·25	50·5	2002·96	13·909	196·35	3906·25	62·5	3067·96	21·305
159·44	2575·56	50·75	2022·84	14·047	197·13	3937·56	62·75	3092·55	21·476
160·22	2601·	51·	2042·82	14·186	197·92	3969·	63·	3117·25	21·648
161·01	2626·56	51·25	2062·90	14·326	198·71	4000·56	63·25	3142·03	21·820
161·79	2652·25	51·5	2083·07	14·466	199·49	4032·25	63·5	3166·92	21·993
162·58	2678·06	51·75	2103·34	14·607	200·28	4064·06	63·75	3191·91	22·166
163·36	2704·	52·	2123·72	14·748	201·06	4096·	64·	3216·99	22·340
164·15	2730·06	52·25	2144·19	14·890	201·85	4128·06	64·25	3242·17	22·515
164·93	2756·25	52·5	2164·75	15·033	202·63	4160·25	64·5	3267·45	22·691
165·72	2782·56	52·75	2185·42	15·176	203·42	4192·56	64·75	3292·83	22·867
166·50	2809·	53·	2206·18	15·321	204·20	4225·	65·	3318·31	23·044
167·29	2835·56	53·25	2227·04	15·466	204·99	4257·56	65·25	3343·88	23·221
168·07	2362·25	53·5	2248·01	15·611	205·77	4290·25	65·5	3369·55	23·400
168·86	2889·06	53·75	2269·06	15·757	206·56	4323·06	65·75	3395·33	23·579
169·65	2916·	54·	2290·22	15·904	207·35	4356·	66·	3421·19	23·758
170·43	2943·06	54·25	2311·48	16·052	208·13	4389·06	66·25	3447·16	23·939
171·22	2970·25	54·5	2332·83	16·200	208·92	4422·25	66·5	3473·23	24·120
172·00	2997·56	54·75	2354·28	16·349	209·70	4455·56	66·75	3499·39	24·301
172·79	3025·	55·	2375·83	16·499	210·49	4489·	67·	3525·65	24·484
173·57	3052·56	55·25	2397·48	16·649	211·27	4522·56	67·25	3552·01	24·667
174·36	3080·25	55·5	2419·22	16·800	212·06	4556·25	67·5	3578·47	24·850
175·14	3108·06	55·75	2441·07	16·952	212·84	4590·06	67·75	3605·03	25·035
175·93	3136·	56·	2463·01	17·104	213·63	4624·	68·	3631·68	25·220
176·71	3164·06	56·25	2485·05	17·257	214·41	4658·06	68·25	3658·43	25·406
177·50	3192·25	56·5	2507·19	17·411	215·20	4692·25	68·5	3685·28	25·592
178·29	3220·56	56·75	2520·42	17·503	215·98	4726·56	68·75	3712·23	25·779
179·07	3249·	57·	2551·76	17·721	216·77	4761·	69·	3739·28	25·967
179·86	3277·56	57·25	2574·19	17·876	217·56	4795·56	69·25	3766·43	26·156
180·64	3306·25	57·5	2596·72	18·033	218·34	4830·25	69·5	3793·67	26·345
181·43	3335·06	57·75	2619·35	18·190	219·13	4865·06	69·75	3821·01	26·535
182·21	3364·	58·	2642·08	18·348	219·91	4900·	70·	3848·45	26·725
183·00	3393·06	58·25	2664·90	18·506	220·70	4935·06	70·25	3875·99	26·917
183·78	3422·25	58·5	2687·83	18·665	221·48	4970·25	70·5	3903·63	27·109
184·57	3451·56	58·75	2710·85	18·825	222·27	5005·56	70·75	3931·36	27·301
185·35	3481·	59·	2733·97	18·986	223·05	5041·	71·	3959·19	27·494
186·14	3510·56	59·25	2757·19	19·147	223·84	5076·56	71·25	3987·12	27·688
186·92	3540·25	59·5	2780·51	19·309	224·62	5112·25	71·5	4015·15	27·883
187·71	3570·06	59·75	2803·92	19·472	225·41	5148·06	71·75	4043·28	28·078
188·50	3600·	60	2827·43	19·635	226·19	5184·	72·	4071·50	28·274

To find the side of a square which will have the same area as a circle of a given diameter.

RULE. Multiply the diameter of the given circle by the constant decimal ·8862269. The product will be the side of the equivalent square.

Sliding rule. $\begin{cases} A & 39 & \text{Side of equivalent square.} \\ B & 44 & \text{Diameter of the circle.} \end{cases}$ *Exam.* $\dfrac{A \quad 39 \quad \text{square 70.}}{B \quad 44 \quad \text{circle 79 diam.}}$

The rule being thus set, the lines A and B will form a table showing the diameters of all circles, and the sides of their equivalent squares. Or it may be done by the lines C and D, thus:

Sliding rule. $\begin{cases} C & 70 & 55 \\ D & \text{Diam. of circle.} & \text{Side of equivalent square.} \end{cases}$ *Exam.* $\dfrac{C \quad 70 \quad 55.}{D \quad \text{circle 79 diam.} \quad \text{square 70.}}$

Circumf.	Area.	Diam.	Area.		Circumf.	Area.	Diam.	Area.	
Inches.	Circular inc.	Inches.	Square inc.	Square feet.	Inches.	Circular inc.	Inches.	Square inc.	Square feet.
226·98	5220·06	72·25	4099·83	28·471	270·96	7439·06	86·25	5842·63	40·574
227·76	5256·25	72·5	4128·25	28·668	271·75	7482·25	86·5	5876·55	40·809
228·55	5292·56	72·75	4156·77	28·866	272·53	7525·56	86·75	5910·56	41·046
229·34	5329·	73·	4185·39	29·065	273·32	7569·	87·	5944·68	41·282
230·12	5365·56	73·25	4214·10	29·265	274·10	7612·56	87·25	5978·89	41·520
230·91	5402·25	73·5	4242·92	29·465	274·89	7656·25	87·5	6013·20	41·758
231·69	5439·06	73·75	4271·83	29·665	275·67	7700·06	87·75	6047·61	41·997
232·48	5476·	74·	4300·84	29·867	276·46	7744·	88·	6082·12	42·237
233·26	5513·06	74·25	4329·95	30·069	277·25	7788·06	88·25	6116·73	42·477
234·05	5550·25	74·5	4359·16	30·272	278·03	7832·25	88·5	6151·43	42·718
234·83	5587·56	74·75	4388·46	30·475	278·82	7876·56	88·75	6186·24	42·960
235·62	5625·	75·	4417·86	30·680	279·60	7921·	89·	6221·14	43·202
236·40	5662·56	75·25	4447·37	30·884	280·39	7965·56	89·25	6256·14	43·445
237·19	5700·25	75·5	4476·97	31·090	281·17	8010·25	89·5	6291·24	43·689
237·98	5738·06	75·75	4506·66	31·296	281·96	8055·06	89·75	6326·43	43·934
238·76	5776·	76·	4536·46	31·503	282·74	8100·	90·	6361·73	44·179
239·55	5814·06	76·25	4566·35	31·711	283·53	8145·06	90·25	6397·12	44·424
240·33	5852·25	76·5	4596·35	31·919	284·31	8190·25	90·5	6432·61	44·671
241·12	5890·56	76·75	4626·44	32·128	285·10	8235·56	90·75	6468·20	44·918
241·90	5929·	77·	4656·63	32·338	285·88	8281·	91·	6503·88	45·166
242·69	5967·56	77·25	4686·91	32·548	286·67	8326·56	91·25	6539·67	45·414
243·47	6006·25	77·5	4717·30	32·759	287·46	8372·25	91·5	6575·55	45·664
244·26	6045·06	77·75	4747·78	32·971	288·24	8418·06	91·75	6611·53	45·913
245·04	6084·	78·	4778·36	33·183	289·03	8464·	92·	6647·61	46·164
245·83	6123·06	78·25	4809·04	33·396	289·81	8510·06	92·25	6683·79	46·415
246·62	6162·25	78·5	4839·82	33·610	290·60	8556·25	92·5	6720·06	46·667
247·40	6201·56	78·75	4870·70	33·825	291·38	8602·56	92·75	6756·44	46·920
248·19	6241·	79·	4901·67	34·039	292·17	8649·	93·	6792·91	47·173
248·97	6280·56	79·25	4932·74	34·255	292·95	8695·56	93·25	6829·48	47·427
249·76	6320·25	79·5	4963·91	34·472	293·74	8742·25	93·5	6866·15	47·682
250·34	6360·06	79·75	4995·18	34·689	294·52	8789·06	93·75	6902·91	47·937
251·33	6400·	80·	5026·55	34·907	295·31	8836·	94·	6939·78	48·193
252·11	6440·06	80·25	5058·01	35·125	296·10	8883·06	94·25	6976·74	48·449
252·90	6480·25	80·5	5089·58	35·344	296·88	8930·25	94·5	7013·80	48·707
253·68	6520·56	80·75	5121·24	35·564	297·67	8977·56	94·75	7050·96	48·965
254·47	6561·	81·	5153·00	35·785	298·45	9025·	95·	7088·22	49·224
255·25	6601·56	81·25	5184·85	36·006	299·24	9072·56	95·25	7125·58	49·483
256·04	6642·25	81·5	5216·81	36·228	300·02	9120·25	95·5	7163·03	49·743
256·83	6683·06	81·75	5248·86	36·450	300·81	9168·06	95·75	7200·58	50·004
257·61	6724·	82·	5281·02	36·674	301·59	9216·	96·	7238·23	50·265
258·40	6765·06	82·25	5313·27	36·898	302·38	9264·06	96·25	7275·98	50·528
259·18	6806·25	82·5	5345·62	37·122	303·16	9312·25	96·5	7313·82	50·790
259·97	6847·56	82·75	5378·06	37·348	303·95	9360·56	96·75	7351·77	51·054
260·75	6889·	83·	5410·61	37·574	304·73	9409·	97·	7389·81	51·318
261·54	6930·56	83·25	5443·25	37·800	305·52	9457·56	97·25	7427·95	51·583
262·32	6972·25	83·5	5475·99	38·028	306·31	9506·25	97·5	7466·19	51·849
263·11	7014·06	83·75	5508·82	38·256	307·09	9555·06	97·75	7504·53	52·115
263·89	7056·	84·	5541·77	38·485	307·88	9604·	98·	7542·96	52·382
264·68	7098·06	84·25	5574·81	38·714	308·66	9653·06	98·25	7581·50	52·649
265·46	7140·25	84·5	5607·94	38·944	309·45	9702·25	98·5	7620·13	52·917
266·25	7182·56	84·75	5641·17	39·175	310·23	9751·56	98·75	7658·86	53·186
267·04	7225·	85·	5674·50	39·406	311·02	9801·	99·	7697·69	53·456
267·82	7267·56	85·25	5707·93	39·638	311·80	9850·56	99·25	7736·61	53·726
268·61	7310·25	85·5	5741·46	39·871	312·59	9900·25	99·5	7775·64	53·997
269·39	7353·06	85·75	5775·08	40·105	313·37	9950·06	99·75	7814·76	54·269
270·18	7396·	86·	5808·80	40·339	314·16	10000·	100·	7853·98	54·542

4 B 2

TABLE OF DIVISORS FOR CALCULATING THE QUANTITIES OF MATTER IN
SQUARE PRISMS, CYLINDERS, OR SPHERES, BY THE SLIDING RULE.

Mr. Watt and Mr. Southern calculated a series of numbers to form a concise table, which is engraved at the back of the Soho rule, to serve as theorems for the mensuration of solid bodies. The solidities of regular bodies will be represented by the products which are obtained by multiplying three of their principal dimensions together, viz., length, breadth, and thickness; and then, by dividing those products by suitable divisors, the quotients will express the solidities in any terms that may be required (see p. 550).

For instance, suppose a cylinder to be 2 feet diameter, and 5 feet long; its capacity will be (2 ft. diam. squared = 4 circular feet area, × 5 ft. long =) 20 cylindrical feet. If it is required to express that capacity in cubic feet, we must divide the 20 cylindrical feet by 1·2732, (because one cubic foot is equal to that number of cylindrical feet,) and we have 15·708 cubic feet. Or, instead of dividing, multiplying by the decimal number ·7854 (which is the reciprocal of 1·2732) will give the same result in cubic feet; because one cylindrical foot is equal to that fraction of a square foot.

By means of the lines of squares and roots, C and D, on the sliding rule, we can, at one operation, multiply the square of a number by any other number, and then divide the product by a third number, so as to obtain the quotient without any necessity for observing or recording, either the square or the intermediate product. This property has been already explained, p. 550, and it is a great convenience in saving time, and avoiding errors.

To perform the calculation, all the four lines of the Soho sliding rule are used at once, thus; The proper divisor being found on the line A, the slider must be set, so that the length of the cylinder on B, will correspond with that divisor on A; then the diameter of the cylinder being found on the line D, the contents of the cylinder will stand opposite to it on the line C.

Soho Sliding Rule, all four lines,	A	Divisor 1·273	Example.	A	Divisor 1·273
	B	Length of cylinder feet.		B	Length 5 feet.
	C	Content of cylinder cubic feet.		C	15·708 cubic feet.
	D	Diameter of cylinder feet.		D	2 feet diameter.

The rule being thus set, it forms a complete table of the contents of all cylinders which are of the same length, but of different diameters; for opposite to any diameter on D, the cubic contents will be found on C. Or the rule may be set as follows, and then it will form a complete table of the contents of all cylinders which are of the same diameter, but of different lengths, for opposite to any length on the line B, will be the content on the line A.

Soho Sliding Rule, all four lines,	A	Content of cylinder cubic feet.	Example.	A	15·708 cubic feet.
	B	Length of cylinder in feet.		B	5 feet long.
	C	Divisor 1·273		C	Divisor 1·273
	D	Diameter of cylinder feet.		D	Diameter 2 feet.

The operation of the divisors being now explained, we may proceed to state how they are arranged in the table which is engraved at the back of the rule. The different titles, Cubic feet, Cubic inches, Water pounds, Cast iron lbs., &c., along the top of the table, denote the terms in which the contents of solids will be expressed, when they are calculated by means of the divisors which are arranged under each title.

The different columns marked FF, FI, and II, under each title, denote the particular divisors which are to be used in each case, according to the terms in which the dimensions of the solids are given, as the data for the calculation. Thus FF denotes that all the dimensions are given in feet; FI, that the length is given in feet, and the diameter in inches; and II denotes that all the dimensions are given in inches.

Table of Divisors for calculating the solidities or weights of square prisms, cylinders, or globes.

Solids.	Solidity in cubic feet.			Solidity in cubic inches.		Water pounds, a cubic foot 62·5 lbs.			Cast iron lbs. a cubic foot 450 lbs.			Wrought iron lbs. a cubic foot 485 lbs.		
	FF	FI	II	FI	II	FF	FI	II	FF	FI	II	FF	FI	II
Squares ..	1·	144	1728	·08333	1·	·016	2·304	27·648	00222	·32	3·84	·002062	·2969	3·563
Cylinders .	1·273	183·3	2200	·1061	1·273	·02037	2·933	35·203	00283	·4074	4·889	·002625	·378	4·536
Globes ...	1·91	—	3300	—	1·91	·03056	—	52·804	00424	—	7·334	·003938	—	6·805
						Specific gravity 1			Specific grav. 7·2			Specific grav. 7·76		

Solids.	Lead lbs. a cubic foot 710 lbs.			Copper lbs. a cubic foot 555 lbs.			Brass lbs. a cubic foot 525 lbs.			Stone lbs. a cubic foot 155 lbs.			Brick lbs. a cubic foot 125 lbs.		
	FF	FI	II	FF	FI	II	FF	FI	II	FF	FI	II	FF	FI	II
Squares ..	·00141	·2028	2·434	·00180	·2595	3·114	·0019	·2743	3·291	·00645	·929	11·148	·008	1·152	13·824
Cylinders.	·00179	·2582	3·099	·00229	·3304	3·964	·00243	·3492	4·191	·00821	1·182	14·195	·01019	1·466	17·601
Globes ...	·00269	—	4·648	·00344	—	5·946	·00364	—	6·286	·00123	—	21·292	·01528	—	26·402
	Spec.grav.11·36			Spec. grav. 8·88			Specific grav. 8·4			Specific grav. 2·48			Specific gravity 2·		

The three different lines of the table are marked at the beginning with the words Squares, Cylinders, and Globes, to denote what kind of solid each divisor is applicable to. The principal divisors in the above table are obtained in the following manner (*a*):

The three divisors in the column marked FF, under the title of cubic feet, are to be used in cases when all the dimensions of the solids are given in feet, and their solidities are required in cubic feet. The product which is obtained in such cases by multiplying the three dimensions together, viz., in cubic feet, if the solid is a square prism; or in cylindrical feet, if it is a cylinder; or in spherical feet, if it is a globe. The several divisors must be adapted to reduce all those different denominations to cubic feet; and they are the number of cubic feet, cylindrical feet, or spherical feet, which are equal to one cubic foot. Thus, The divisor for square prisms is 1, because no reduction is required. The divisor for cylinders is 1·27324, for that number of cylindrical feet are contained in one cubic foot. The divisor for spheres is 1·90986, for that number of spherical feet make one cubic foot.

The three divisors in the column marked II, under the title of Cubic inches, are the same numbers as the above, being the number of cubic inches, cylindrical inches, or spherical inches, which are equal to one cubic inch.

The two divisors marked FI, under the title of Cubic feet, are to be used for square prisms or cylinders, when their lengths being given in feet, and their other dimensions in inches, their solidities are required in cubic feet. In such cases the products which are obtained by multiplying the three dimensions together, will express the solidities in different terms, viz. The solidities of square prisms will be expressed in square inch feet, that is, square prisms one inch square and one foot long. And the solidities of cylinders, in cylindrical inch feet, that is, cylinders one inch diameter, and one foot long. The divisors are 144, which is the number of square inch feet in one cubic foot. And 183·346, which is the number of cylindrical inch feet in one cubic foot.

The three divisors marked II under the title of Cubic feet, are to be used when all the dimensions are given in inches, and the solidities are required in cubic feet. The product of the multiplication of the three dimensions will then give the solidities in cubic inches, cylindrical inches, or spherical inches. The divisor for square prisms is 1728 cubic inches in a cubic foot. For cylinders 2200·15 cylindrical inches in a cubic foot. And for globes 3300·23 spherical inches in a cubic foot.

The two dvisors marked FI, under the title Cubic inches, are to be used when the lengths are given in feet, and the other dimensions in inches, and the solidities are required in cubic inches. The divisor for square prisms is ·08333, because that decimal portion of a square inch foot, is equal to one

(*a*) The above table of divisors was calculated by the author, and it is more complete and exact than the table usually engraved at the back of the sliding rule.

cubic inch. And the divisor for cylinders is ·1061, for that portion o.̈ a cylindrical inch foot is equal to one cubic inch.

The divisors for cylinders are 1·27324 times the corresponding divisors for square prisms, under the same title and denomination; and the divisors for globes are 1·90986 times the corresponding divisors for square prisms. Or the divisors for globes are 1·5 times the corresponding divisors for cylinders. The divisors which are marked FI under each title, are 144 times the corresponding divisor marked FF under the same title; and the divisors II are 12 times the corresponding divisors FI; consequently those marked II are 1728 times those marked FF.

The divisors which are placed under the different titles of Water, Cast iron, &c. are to find the weight, in pounds avoirdupois, of different solids composed of those substances. Those divisors which are marked FF, are the number of cubic feet, cylindrical feet, or spherical feet of each substance, that will weigh one pound. The divisors marked FI are the number of square inch feet, or cylin-drical inch feet of each substance, that will weigh one pound. And the divisors marked II, are the number of cubic inches, cylindrical inches, or spherical inches, of each substance, that will weigh one pound.

For instance, a cubic foot of water weighs 62·5 pounds; and under the title Water, the divisors marked FF are as follows: ·016 of a cubic foot of water weighs one pound (that is, a pound of water is = ·016 of a cubic foot); ·02037 of a cylindrical foot weighs one pound; and ·03056 of a spherical foot weighs one pound. The divisors marked FI are as follows; 2·304 square inch feet of water weigh one pound; or 2·9335 cylindrical inch feet of water weigh one pound. The divisors marked II are, 27·648 cubic inches of water weigh one pound; or 35·2025 cylindrical inches of water weigh one pound; or 52·8038 spherical inches of water weigh one pound.

The divisors for calculating the weight of any other substance in pounds, may be obtained by dividing the proper divisors for water by the specific gravity of the substance in question. Or else by dividing the proper divisors for cubic feet, or cubic inches, by the weight in pounds, of a cubic foot, or a cubic inch, of that substance.

For instance, the specific gravity of cast iron is 7·2; that is, any bulk of cast iron is 7·2 times the weight of an equal bulk of water. And ·016 of a cubic foot of water weigh one pound; therefore (·016 ÷ 7·2 =) ·00222 of a cubic foot of cast iron will weigh one pound; and that number is the proper divisor for finding the weight of square prisms of cast iron in pounds, when all the dimensions are given in feet: accordingly it is marked in the table. Cast iron lbs. FF, Squares ·00222.

In like manner the divisor cast iron lbs. FI cylinders is (2·9335 ÷ 7·2 =) ·40743. This division may be performed by the sliding rule, with the slide inverted, thus:

Sliding rule,   { A   1   Divisor for that sub.   A   1   ·407 divis.
slide inverted. {   ꓳ Divis. for wat. lbs.   Specific grav. of sub.   Ex.   ꓳ 2·93 wat. FI   7·2 sp. gr.

When the weight of a cubic foot, or of a cubic inch, of any substance is given in pounds, divisors for calculating the weights of solids of that substance may be found by dividing the divisors under the titles of cubic feet or cubic inches, by the weight in pounds of a cubic foot, or a cubic inch of that substance. For instance, the weight of a cubic foot of cast iron is (7·2 × 62·5 =) 450 pounds. The divisor for cubic feet, cylinder, FI, is 183·346 which ÷ 450 gives ·40743, which is the proper divisor for cast iron cylinders, FI, as before. Or by the sliding rule with the slide inverted.

Sliding rule,   { A   1   Divisor for that substance.   A   1   ·407
slide inverted. {   ꓳ Divisor for cubic feet.   Weight of a cubic foot, lbs.   Ex.   ꓳ 183·3   450lbs.

To explain the mode of finding divisors for new cases, we may suppose that divisors are wanted to calculate what weight of water in tons (of 2240 pounds) will be contained in different vessels, all the dimensions of those vessels being given in feet. The products obtained by the multiplications of the three dimensions, will express the solidities in cubic feet, or cylindrical feet, or spherical feet, and the divisors must therefore be the number of cubic feet, cylindrical feet, or spherical feet, which will weigh one ton, or 2240 pounds. As a cubic foot of water weighs 62·5 pounds, we have (2240 lbs. ÷ 62·5 lbs. =) 35·84 cubic feet of water weigh a ton, for the divisor for square prisms. Or a cylindrical foot of water weighs 49·087 pounds; therefore (2240 lbs. ÷ 49·087 lbs. =) 45·633 cylindrical feet of water weigh one ton: this is the divisor for cylinders. And a spherical foot of water weighs 32·725 pounds; hence (2240 lbs. ÷ 32·725 lbs. =) 68·449 spherical feet of water weigh a ton: this is the divisor for globes.

It is obvious that the divisors which will give the weight of bodies in tons, must be 2240 times those divisors which will give their weight in pounds; or for hundred weights 112 times, &c. Hence to obtain divisors for calculating the weights of bodies in tons, or in hundred weights, we have only to multiply the numbers in the table by 2240 lbs. or by 112 lbs. *Example.* The divisor for the weight of cylinders of water FF in pounds, is ·016; this multiplied by 2240 lbs. is = 35·84 for the divisor for the weight of cylinders of water, FF, in tons, as before.

Examples of the use of the table of divisors. A cylindrical piston-rod of wrought iron being 3½ inches diameter and 9 feet long, how many pounds will it weigh? The divisor for this case is ·378 according to the table; for the solid being a cylinder, the required divisor must be in the middle line, of the three horizontal lines of the table; and it will be under the head of Wrought iron, in the column FI, because the length of the cylinder is given in feet, whilst its diameter is in inches. The divisor may be taken on the line A thus, and the rule will form a table showing the weight of every different diameter, when the length is 9 feet.

Soho Sliding Rule, all four lines,
{
A   ·378 divisor.
B   9 feet long.
C   292 pounds weight.
D   3· 5 inches diameter.
}

(without altering the slider.)

or { 214 pounds. / 3 inc. diam. }   or   251 pounds. / 3· 25 inc. diam.

Or the divisor may be taken on the line C thus, and the rule will form a table, showing the weight corresponding to every different length, the diameter being always 3½ inches.

Soho Sliding Rule, all four lines,
{
A   292 pounds weight.
B   9 feet long.
C   ·378 divisor.
D   3· 5 inches diameter.
}

or { 324 pounds. / 10 feet long. }   or   259 pounds. / 8 feet long.

(without altering the slider.)

Gage points for the sliding rule. If the square root of any divisor is taken upon the line D, it will point out the same result upon the line C or q, as the divisor itself does upon the lines A or C or q. The square roots of divisors are called gage points; for instance, the square root of 1·273 is 1·128, which is the gage point corresponding to the divisor 1·273, and may be used thus on the line D.

Sliding Rule.
{
C Length of cylinder ft.   Content of cylinder cubic ft.
D Gage point 1·13 or 3·57. Diameter of cylinder in feet.
}
Ex.
{
C 5 ft. long.   15·708 cub. ft.
D 1·13 g. p.   2 feet diam.
}

When the rule is thus set, it forms a table for all cylinders which are of different diameters, but of the same length; for opposite to any diameter on D, the content will be found on C. If the slider is inverted and set as follows, then the rule will form a table of the corresponding lengths and diameters of a number of cylinders, which will have the same content that is pointed out by the gage point; for opposite to any diameter on D will be the requisite length on q.

Sliding Rule, slide inverted.
{
q Content of cylinder cub.ft.   Length of cylinder in ft.
D Gage point 1·13 or 3·57   Diameter of cylinder ft.
}
Ex.
{
q 15·708 cub.ft.   5 ft.long.
D 1·13 g. p.   2 ft.diam.
}

For instance, in the above example, a cylinder 1 foot diameter must be 20 feet long, in order to contain 15·708 cubic feet; or a cylinder 3 feet diameter would require to be 2·22 feet long.

If it is required to find the product which results from the multiplication of the three dimensions of the solid, it may be found on the line q, opposite to 1 on the line D; in the above case, that product is the solidity of the cylinder expressed in cylindrical feet.

Note. Instead of taking the square root of the divisor for a gage point on the line D, we may take the square root of ten times that divisor, and it will give the same results; this will be apparent by trial with the rule; the square root of 12·73 is 3·568, and opposite to that number on the line D, we shall find 15·708 on C or q, which is the same result as is opposite to the other gage point 1·128; because the series of numbers on the line C or q, are twice repeated. (See table, p. 569.)

To calculate the solidities of rectangular prisms, such as planks of wood, or flat bars of metal, by the sliding rule with the table of divisors. In these cases we must first calculate the proper size for a square prism, whereof the sectional area would be the same as that of the rectangular prism in question, and then the calculation of its solidity or weight may be made, in the same manner as if it were a square prism. The two sides of the rectangle must be multiplied together to represent the sectional area by their product, and the square root of that product will be the side of a square which will have the same area as the rectangle.

For instance, a flat bar 2 inches thick, by 4½ broad (2 × 4½) = 9 square inches, the square root of which is 3 inches, and a square bar of that size would be equivalent to the flat bar. This may be done by the sliding rule thus:

Sliding rule, { A Large side of rectangle.   Side of equivalent square. } This is where the same numslide inverted. } ○ Small side of rectangle.   Side of equivalent square. { bers on both lines correspond.

Or still more conveniently by the lines C and D. The number representing one of the sides of the rectangle (either the largest or the smallest) being found upon one of the lines C or D, the slider must be placed with that number corresponding to the same number upon the other line. Then the number representing the other side, being found upon the line C, the side of the equivalent square will be opposite to it, upon the line D, thus:

Sliding rule. { C Large side of rectan.   Small side of rectangle.   *Ex.* C 4½ broad   2 thick.
              { D Large side of rectan.   Side of equivalent square.   D 4½ broad   3 square.

By the same process we may find the diameter of a circle which shall have the same area as that of any given ellipsis; whereby an elliptical prism may be assimilated to a cylinder thus:

Sliding rule. { C Conjugate dia. of ellips.   Transverse diameter.   *Ex.* C 16 con. dia.   9 trans. dia.
              { D Conjugate dia. of ellips.   Diam. of equal circle.   D 16 con. dia.   12 dia. of cir.

To calculate the solidities of pyramids or cones by the sliding rule, with the table of divisors. The solidity of any pyramid, or cone, is one-third of that of a prism, or cylinder, of the same base and vertical height as the pyramid, or cone; hence we must calculate the solidity, or weight, of a prism, or cylinder, of the same base and vertical height as the pyramid, or cone, by the sliding rule, with the proper divisor, or gage point, taken from the table; and then one-third of the result, will be the solidity, or weight, of the pyramid, or cone.

The solidities of similar pyramids, or cones, are as the cubes of the sides, or diameters, of their bases respectively; similar pyramids, or cones, are those which have the same angle at the vertex, and therefore the dimensions of their bases, bear some constant proportion to their vertical heights.

To find the solidities of frustums of square pyramids, or cones; having given, the vertical height of the frustum, its diameter at the base, and its diameter at the top; all the dimensions being in the same terms, either feet or inches, &c.

RULE. Divide the difference between the cubes of the sides, or diameters, of the two ends, by the difference between those sides, or diameters; and multiply the quotient by one third of the vertical height of the frustum. The product is the solidity, which will be expressed in cubic feet, or cubic inches, if it is the frustum of a square prism; or in cylindrical feet, or cylindrical inches, if it is a conical frustum.

Note. The square root of the quotient, which is obtained by dividing the difference between the cubes of the sides, or diameters, by the difference between those sides, or diameters, will be the diameter of a square prism, or of a cylinder, whereof the height being equal to that of the frustum, it will have three times the solidity: the following rule is deduced from this fact.

To find the diameter of a cylinder which will have the same solidity as that of a given frustum of a pyramid or cone.

Rule. Divide the difference between the cubes of the sides, or diameters, by three times the difference between the sides, or diameters, and then extract the square root of the quotient. That root will be the diameter of a cylinder, which, being of the same height as the frustum, will have the same solidity. The solidity or weight of such equivalent prism, or cylinder, may then be calculated with the sliding rule, by means of a suitable divisor, or gage point, selected from the tables.

Example. Suppose a conical frustum to be 6 feet vertical height; 5 feet diameter at the base; and 3 feet diameter at the top. The cube of 5 is 125; from which deduct 27 (the cube of 3), and the remainder is 98, for the difference between the cubes; divide this by (three times 2, the difference between 5 ft. diam. and 3 ft. diam. =) 6, and the quotient is $16\frac{1}{3}$; the square root of this is 4·04 feet for the diameter of a cylinder, which being 6 feet high, will have the same content as the frustum in question. Thus 4·04 ft. dia. squared = $16\frac{1}{3}$ circular feet area, × 6 ft. high = 98 cylindrical feet, or × ·7854 = 76·96 cubic feet. If it had been a frustum of a square prism, the content would have been 98 cubic feet.

This calculation will require three operations by the sliding rule; first, two operations to find the cube of the side, or diameter, of the base, and the cube of the side, or diameter, of the top, thus:

Sliding rule, $\left\{\begin{array}{l}$ ꟼ Side or diameter.   Cube of side or diam. \\ D Side or diameter.   1 \end{array}\right.$ $Ex.$ $\begin{array}{l}$ ꟼ 5 ft. dia.   125 cube. \\ D 5 ft. dia.   1 \end{array}$
slide inverted.

The difference between the two cubes cannot be taken by the sliding rule; but it must be found by subtracting the cube of the greatest side, or diameter, from the cube of the smallest side or diameter (thus, 125 − 27 = 98); and also the difference between those sides, or diameters, must be taken; (thus, 5 ft. dia. — 3 ft. dia. = 2). We may then find the side of an equivalent square prism, or the diameter of an equivalent cylinder, of the same height as the frustum, thus:

Sliding rule, $\left\{\begin{array}{l}$ ꟼ Dif. of cubes.   3 times dif. of dia. \\ D   1   Diam. of cylinder. \end{array}\right.$ $Ex.$ $\begin{array}{l}$ ꟼ 98 dif. cub.   (3 × 2 =) 6. \\ D   1   4·04 diam. \end{array}$

Note. As the same numbers are twice repeated on the line ꟼ, there are two different results which may be obtained in each case, by following the above precept; and care must be taken to choose the right one, which may be known by its being rather greater than a mean between the two diameters of the frustum. For instance, in the above example, one 6 on the line ꟼ will point out 4·04 ft. diam.; and the other 6 will point out 1·28 ft. diam.; so that there is no danger of mistaking them in this case, if we are aware of the fact that there are two different results.

In some cases of frustums it is desirable to know what would be the solidity of the whole pyramid or cone, if it were completed; and then the solidity of the upper part, or small pyramid, which is wanting, being also calculated, the difference between the two solidities will be the solidity of the frustum.

To find the vertical height of the complete pyramid, or cone, having given, the vertical height of the frustum, the side or diameter of its base, and the side or diameter of its top.

Rule. Multiply the vertical height of the frustum, by the side, or diameter of its base; and divide the product by the difference between the side, or diameter, of the base, and the side, or diameter, of the top. The quotient will be the whole height of the complete pyramid, or cone.

Example. Suppose a conical frustum to be 6 feet vertical height, 5 feet diameter at the base, and 3 feet diameter at the top. Then, 5 feet diameter at base, × 6 feet high = 30 ÷ by (5 ft. — 3 ft. =) 2 feet difference of diameters = 15 feet, would be the height of the complete cone.

Sliding Rule, $\left\{\begin{array}{l}$ A Diameter of base.   Height of complete cone. \\ Ɔ Height of frustum.   Difference of diameters. \end{array}\right.$ $Ex.$ $\begin{array}{l}$ A 5 ft. diam.   15 ft. high. \\ Ɔ 6 ft. high.   2 ft. diff. \end{array}$

4 c

The solidity of a cone 5 feet diameter at the base, and 15 feet high, is as follows : 5 feet diameter squared = 25 circular feet area, × (one-third of 15 feet high =) 5 = 125 cylindrical feet ; or, if it had been a square prism, the solidity would have been 125 cubic feet.   The small cone which is wanting, is 3 feet diameter at the base, and 9 feet high.   Therefore, 3 squared = 9 × (one-third of 9 ft. high =) 3 = 27 cylindrical feet.   Hence, the solidity of the frustum must be 125 — 27 = 98 cylindrical feet ; or 98 cubic feet if it had been a square prism.

To find the solidity of an ellipsoid, or spheroid.   The solidity of an ellipsoid is two-thirds of that of its circumscribing cylinder; hence, we may calculate the solidity or weight of such a cylinder by the sliding rule, with the aid of a suitable divisor or gage point from the table ; and then take two-thirds of the result, for the solidity or weight of the ellipsoid in question.

As the divisors for globes are $1\frac{1}{2}$ times those for cylinders, they will enable us to calculate the true result for ellipsoids, as well as for spheres, at one operation.   The length of the axis of the ellipsoid is to be taken on the line B or C, and the diameter of the great or equatorial circle of the ellipsoid on the line D.   Thus,

Sliding Rule, all four lines.				
	A	Divisor for a globe.	A	Divis. for cub. ft. FF. 1·91
	B	Length of axis of ellipsoid.	B	9 ft. length of axis.
	C	Solidity or weight of ellipsoid.	C	230 cubic feet.
	D	Equatorial diam. of ellipsoid.	D	7 feet diameter.

(with *Exam.* between the two halves, opposite rows B and C)

Or, if the square root of the divisor is taken as a gage point on the line D, instead of the divisor itself upon A, then the rule may be set thus,   (See the table of gage points, p. 569.)

Sliding Rule, slide inverted.				
	ꟼ	Solidity or weight of ellipsoid. Length of axis.	ꟼ	230 cb. ft. 9 ft. lon.
	D	Gage point for a globe. Diam. of equator.	D	138 g. p. 7 ft. dia.

(with *Ex.* between the halves)

To find the cube of any given number by the lines C *and* D, *on the sliding rule.*

Sliding Rule.				
	C	Number to be cubed. Cube of the number.	C	4 numb. 64 cube.
	D	1 Number to be cubed.	D	1 4 numb.

(with *Ex.* between the halves)

Or thus, by the lines ꟼ and D when the slide is inverted, which is the best method.

Sliding Rule, slide inverted.				
	ꟼ	Number to be cubed. Cube of the number.	ꟼ	4 numb. 64 cube.
	D	Number to be cubed. 1	D	4 numb. 1

(with *Ex.* between the halves)

To extract the cube root of any number by the lines ꟼ and D on the sliding rule, with the slide inverted.   This is done according to the last precept, with the slider inverted, by finding the number whose root is to be extracted, upon the line ꟼ, and placing it opposite to 1 on the line D ; then seeking along the lines ꟼ and D, for the place where the divisions representing the same numbers on both lines, meet together, those numbers are the cube root required.   Thus,

Sliding Rule, slide inverted.					
	ꟼ	Numb. Cube root.	This is where the same num-	ꟼ	64 cube. 4 root.
	D	1 Cube root.	bers on both lines correspond.	D	1 4 root.

(with *Exa.* between the halves)

Note. If two divisions which are of the same value, upon the lines ꟼ and D, do not exactly coincide, the coincident point will be within the space between those two divisions of the same value, which are nearest to a coincidence.   There are three such points of coincidence of similar numbers along the lines, one denoting the cube root required; the others the cube roots of 10 times, and of 100 times the number; care must be taken to choose the proper root of the three, but there is nothing on the rule to point it out.

The following precepts show how the sliding rule can perform successive multiplications, and divisions, of numbers with the squares, and cubes, of other numbers, by one operation of the lines C and D ; or by all the four lines of the Soho rule.   These precepts will be very useful guides to those calculators who require to adapt the sliding rule to new cases.   Two precepts are given for each case, to show the process which must be gone through, either with the slider direct, or with the slider inverted ; and the calculator can select that which is most suitable to his particular purpose.

To square a given number, and multiply that square by another given number, at one operation. For instance, multiply the square of 6 by 4.

The square root of the multiplicand is 6 ; the square of which is (6 × 6 =) 36 for the multiplicand itself. The multiplier is 4. And 144 is the resulting product.

Sliding Rule.	C Multiplier	Resulting product.	*Ex.*	C 4 mult.	144 result.
	D 1	Sq. root of multiplicand.		D 1	6 sq. rt.

Sliding Rule, slide inverted.	Ↄ	Multiplier.	Resulting product.	*Ex.*	Ↄ 4 mult.	144 result.
	D Sq. root of multiplicand.		1		D 6 sq. rt.	1

To square a given number, and divide that square by another given number, at one operation. For instance, divide the square of 12 by 6.

The square root of the dividend is 12 ; the square of which is (12 × 12 =) 144 for the dividend itself. The divisor is 6. And 24 is the resulting quotient.

Sliding Rule.	C Resulting quotient.	Divisor.	*Ex.*	C 24 result.	6 divis.
	D Square root of dividend.	Divisor.		D 12 sq. rt.	6 divis.

Sliding Rule, slide inverted.	Ↄ Divisor.	Resulting quotient.	*Ex.*	Ↄ 6 divis.	24 result.
	D Square root of dividend.	Divisor.		D 12 sq. rt.	6 divis.

To cube a given number, and divide that cube by the square of another given number, at one operation. For instance, divide the cube of 8 by the square of 4.

The cube root of the dividend is 8 ; the cube of which is (8 × 8 × 8 =) 512 for the dividend itself. The square root of the divisor is 4 ; the square of which is (4 × 4 =) 16 for the divisor itself. And 32 is the resulting quotient.

Sliding Rule.	C Cube root of dividend.	Resulting quotient.	*Ex.*	C 8 cube rt.	32 result.
	D Square root of divisor.	Cube root of dividend.		D 4 sq. root.	8 cube rt.

Sliding Rule, slide inverted.	Ↄ Cube root of dividend.	Resulting quotient.	*Ex.*	Ↄ 8 cube rt.	32 result.
	D Cube root of dividend.	Sq. root of divisor.		D 8 cube rt.	4 sq. root.

To multiply the square of a given number, by another given number, and divide the product by the square of a third given number, at one operation. For instance, multiply the square of 8 by 12, and divide the product by the square of 4.

The square root of the multiplicand is 8, the square of which is (8 × 8 =) 64, the multiplicand itself. The multiplier is 12. And 768 is the product, which, being divided twice successively by the square root of the divisor (thus, 768 ÷ 4 = 192 ÷ 4 =) gives 48 for the resulting quotient.

Sliding Rule.	C Multiplier.	Resulting quotient.	*Ex.*	C 12 mult.	48 result.
	D Sq. root of divisor.	Sq. root of multiplicand.		D 4 sq. rt. div.	8 sq.rt.mult.

Sliding Rule, slide inverted.	Ↄ Multiplier.	Result. quotient.	*Ex.*	Ↄ 12 mult.	48 result.
	D Sq. root of multiplicand.	Sq. root of divisor.		D 8 sq.rt.mult.	4 sq. rt. div.

Note. This is the same operation as that by which the solidities, or weights, of square prisms, or cylinders, or spheres, are calculated by the aid of gage points ; which are the square roots of the divisors given in the table at the back of the Soho rule. See p. 559 and 569.

To cube a number, and divide that cube by another given number, at one operation.   For instance, divide the cube of 8 by 16.

The cube root of the dividend is 8; the cube of which is 512 for the dividend itself.   The divisor is 16.   And 32 is the resulting quotient.   This operation cannot be performed by the lines C and D alone, but it requires all the four lines of the Soho rule, thus,

Soho Rule, all four lines.

A	Divisor.		A	16 divisor.
B	Cube root of dividend.	*Example.*	B	8 cube root.
C	Resulting quotient.		C	32 result.
D	Cube root of dividend.		D	8 cube root.

Soho Rule, all four lines, slide inverted.

A	Divisor.		A	16 divisor.
Ɔ	Resulting quotient.	*Example.*	Ɔ	32 result.
ꓭ	Cube root of dividend.		ꓭ	8 cube root.
D	Cube root of dividend.		D	8 cube root.

To multiply the square of a given number by another given number, and divide the product by a third given number, at one operation.   For instance, multiply the square of 8 by 16, and divide the product by 8.

The square root of the multiplicand is 8; the square of which is (8 × 8 =) 64 for the multiplicand itself.   The multiplier is 16.   And 1024 is the product; which, divided by the divisor 8, gives 128 for the resulting quotient.   This operation requires all the four lines of the Soho rule.

Soho Rule, all four lines.

A	Divisor.		A	8 divisor.
B	Multiplier.	*Example.*	B	16 multiplier.
C	Resulting quotient.		C	128 resulting quotient.
D	Square root of multiplicand.		D	8 sq. root of multiplicand.

Soho Rule, slide inverted, all four lines.

A	Divisor.		A	8 divisor.
Ɔ	Resulting quotient.	*Example.*	Ɔ	128 resulting quotient.
ꓭ	Multiplier.		ꓭ	16 multiplier.
D	Square root of multiplicand.		D	8 sq. root of multiplicand.

Note.   This is the same operation as that by which the solidities, or weights, of square prisms or cylinders, are calculated by means of the divisors in the table p. 557, at the back of the Soho rule.

To multiply two given numbers together, and divide their product by the square of a third given number, at one operation.   For instance, multiply 16 by 9, and divide the product by the square of 6.

The multiplicand is 16.   And 9 is the multiplier.   Their product is 144; which divided twice successively by 6 the square root of the divisor (thus, 144 ÷ 6 = 24 ÷ 6 =), gives 4 for the resulting quotient.   This operation requires all the four lines of the Soho rule.

Soho Rule, all four lines.

A	Multiplicand.		A	16 multiplicand.
B	Resulting quotient.	*Example.*	B	4 result.
C	Multiplier.		C	9 multiplier.
D	Square root of divisor.		D	6 sq. root of divisor.

Soho Rule, slide inverted, all four lines.

A	Multiplicand.		A	16 multiplicand.
Ɔ	Multiplier.	*Example.*	Ɔ	9 multiplier.
ꓭ	Resulting quotient.		ꓭ	4 result.
D	Square root of divisor.		D	6 sq. root of divisor.

To divide a given number by the square of another given number, at one operation. For instance, divide 288 by the square of 6.

The dividend is 288; which divided twice successively by 6, the square root of the divisor, (thus, $288 \div 6 = 48 \div 6 =$) gives 8, for the resulting quotient.

Sliding rule. $\left\{\begin{array}{l} \text{C} \quad\quad\quad \text{Dividend.} \quad\quad\quad \text{Resulting quotient.} \\ \overline{\text{D} \quad \text{Square root of divisor.} \quad\quad\quad\quad 1} \end{array}\right.$ *Exa.* $\begin{array}{l} \text{C} \quad 288 \text{ div.} \quad 8, \text{ result.} \\ \overline{\text{D} \quad 6 \text{ sq. rt.} \quad\quad\quad 1} \end{array}$

Sliding rule, $\left\{\begin{array}{l} \text{g} \quad \text{Dividend.} \quad\quad\quad \text{Resulting quotient.} \\ \overline{\text{D} \quad 1 \quad\quad\quad\quad \text{Square root of divisor.}} \end{array}\right.$ *Exa.* $\begin{array}{l} \text{g} \quad 288 \text{ div.} \quad 8, \text{ result.} \\ \overline{\text{D} \quad 1 \quad\quad 6 \text{ sq. root.}} \end{array}$
slide inverted.

To find such an unknown number as will bear the same proportion to a given number, as that which exists between the squares of two other given numbers. For instance, as the square of 4, is to the square of 6: so is 12, to a fourth unknown number. The terms of the ratio are $(4 \times 4) = 16$, and $(6 \times 6 =) 36$; it is an increasing proportion, and therefore the smallest of those terms (16) must be used for the divisor of the product which is obtained by the multiplication of the largest term 36, by the known number 12.

Thus, $\frac{36}{16}$ths of $12 = 27$ is the fourth unknown number; for 12 the known number $\times 36 = 432 \div 16 = 27$. The arithmetical process is in reality, $12 \times 6 = 72 \times 6 = 432$ product, which, $\div 4 = 108$ for the first quotient, and $\div 4 = 27$, resulting quotient, which is the fourth unknown number.

Sliding rule. $\left\{\begin{array}{l} \text{C} \quad \text{Known number.} \quad\quad \text{Unknown number.} \\ \overline{\text{D} \quad \text{Sq. rt. small. term.} \quad \text{Sq. rt. large term.}} \end{array}\right.$ *Ex.* $\begin{array}{l} \text{C} \quad \text{known 12.} \quad \text{unknown 27.} \\ \overline{\text{D} \quad \text{sq. rt. 4.} \quad\quad \text{sq. rt. 6.}} \end{array}$

Sliding rule, $\left\{\begin{array}{l} \text{g} \quad \text{Unknown number.} \quad\quad \text{Known number.} \\ \overline{\text{D} \quad \text{Sq. rt. small term.} \quad \text{Sq. rt. large term.}} \end{array}\right.$ *Ex.* $\begin{array}{l} \text{g} \quad \text{unknown 27.} \quad \text{known 12.} \\ \overline{\text{D} \quad \text{sq. rt. 4.} \quad\quad \text{sq. rt. 6.}} \end{array}$
slide inverted.

To divide a number by the square root of another number at one operation. For instance, divide 18 by the square root of 9.

The dividend is 18. The square of the divisor is 9; the square root of which is 3, for the real divisor. The resulting quotient is $(18 \div 3 =) 6$.

Sliding Rule. $\left\{\begin{array}{l} \text{C} \quad \text{Square of divisor.} \quad\quad 1 \\ \overline{\text{D} \quad \text{Dividend.} \quad\quad \text{Result. quotient.}} \end{array}\right.$ *Exa.* $\begin{array}{l} \text{C} \quad 9 \text{ sq. of divis.} \quad 1 \\ \overline{\text{D} \quad 18 \text{ dividend.} \quad 6 \text{ result.}} \end{array}$

Sliding Rule, $\left\{\begin{array}{l} \text{g} \quad 1 \quad\quad\quad \text{Square of divisor.} \\ \overline{\text{D} \quad \text{Dividend.} \quad \text{Result. quotient.}} \end{array}\right.$ *Exa.* $\begin{array}{l} \text{g} \quad 1 \quad\quad 9 \text{ sq. of divis.} \\ \overline{\text{D} \quad 18 \text{ divid.} \quad 6 \text{ result.}} \end{array}$
slide inverted.

To extract the square root of a number, and multiply that root by another number. For instance, extract the square root of 16, and multiply that root by 7.

The square of the multiplicand is 16; the square root of which is 4, for the real multiplicand. The multiplier is 7. The resulting product is $(4 \times 7 =) 28$.

Sliding Rule. $\left\{\begin{array}{l} \text{C} \quad 1 \quad\quad\quad \text{Sq. of multiplicand.} \\ \overline{\text{D} \quad \text{Multiplier.} \quad \text{Resulting product.}} \end{array}\right.$ *Exa.* $\begin{array}{l} \text{C} \quad 1 \quad\quad 16 \text{ sq. of mult.} \\ \overline{\text{D} \quad 7 \text{ mult.} \quad 28 \text{ result.}} \end{array}$

Sliding Rule, $\left\{\begin{array}{l} \text{g} \quad 1 \quad\quad\quad \text{Sq. of multiplicand.} \\ \overline{\text{D} \quad \text{Resulting prod.} \quad \text{Multiplier.}} \end{array}\right.$ *Exa.* $\begin{array}{l} \text{g} \quad 1 \quad\quad 16 \text{ sq. of mult.} \\ \overline{\text{D} \quad 28 \text{ result.} \quad 7 \text{ multiplier.}} \end{array}$
slide inverted.

To divide the square root of one number, by the square root of another number. For instance, divide the square root of 64, by the square root of 16.

The square of the dividend is 64; the square root of which is 8, for the real dividend. The square of the divisor is 16, the square root of which is 4, for the real divisor. The resulting quotient is $(8 \div 4 =) 2$.

Sliding Rule.	$\dfrac{C}{D}$	Squ. of dividend.	1								

Sliding Rule. $\dfrac{\text{C}}{\text{D}}$　$\dfrac{\text{Squ. of dividend.}}{\text{Squ. of divisor.}}$　$\dfrac{1}{\text{Result. quotient.}}$　*Exa.*　$\dfrac{\text{C}}{\text{D}}$　$\dfrac{\text{64 sq. of divid.}}{\text{16 sq. of divis.}}$　$\dfrac{1}{\text{2 result.}}$

Sliding Rule, slide inverted. $\left\{\dfrac{\text{C}}{\text{D}}\right.$　$\dfrac{\text{Squ. of dividend.}}{1}$　$\dfrac{\text{Sq. of divisor.}}{\text{Result. quotient.}}$　*Exa.*　$\dfrac{\text{C}}{\text{D}}$　$\dfrac{\text{64 sq. of divid.}}{1}$　$\dfrac{\text{16 sq. of divis.}}{\text{2 result.}}$

Precepts of this kind might be greatly extended, but the above will serve for those operations which are of the most common occurrence. When a long calculation has been made by a series of arithmetical operations in the usual manner, the sliding rule by the aid of one or other of the above precepts, will, in most cases, be found capable of performing several of the succeeding operations at once, so as to abridge the process very considerably.

To find the hypothenuse of a right-angled triangle, by one operation with the sliding rule; having given the dimensions of the base, and of the perpendicular, both in the same terms, as feet or inches, &c. For instance, suppose the base to be 4 feet, and the perpendicular 3 feet, what will be the hypothenuse in feet?

The usual arithmetical process is to add the square of the base, to the square of the perpendicular; viz. $(4 \times 4 =)16 + (3 \times 3 =)9 = 25$; and then the square root of their sum being extracted; viz. 5, that root is the hypothenuse required.

The same result may be obtained thus: Divide (9) the square of the perpendicular, by (16) the square of the base; the quotient will be a decimal fraction, (·5625), to which prefix 1·, then multiply (1·5625) the number so obtained, by (16) the square of the base; the product will be the square of the hypothenuse; and the square root of that product, viz. (5), will be the hypothenuse required.

The latter process, though it appears complicated in figures, can be very readily performed by the lines C and D of the sliding rule, with the slider direct, thus: Find the base on the line D, and set the slider so that 1 upon the line C, corresponds to that base; then opposite to the perpendicular on D, is the required decimal fraction, to which 1 is to be prefixed. The number formed by that addition being sought upon C, the required hypothenuse will be found opposite to it on D, thus,

Sliding rule. $\left\{\dfrac{\text{C}}{\text{D}}\right.$　$\dfrac{1}{\text{Base.}}$　$\dfrac{\text{Decim. fract.}}{\text{Perpendicular.}}$　&　$\dfrac{1 \cdot + \text{Dec. frac.}}{\text{Hypothenuse.}}$　*Ex.*　$\dfrac{\text{C}}{\text{D}}$　$\dfrac{1}{4 \text{ base.}}$　$\dfrac{\cdot 5625 \text{ dec.}}{3 \text{ perpend.}}$　&　$\dfrac{1.5625}{5 \text{ hypoth.}}$

Note. This is one of the theorems invented by Mr. Watt, or Mr. Southern.

Directions to Engineers for the choice of a Sliding Rule.

The slider of a sliding rule should fit very accurately into its groove, and must slide very freely in it, without being so loose as to drop, or move by its own weight. The face of the slider should form a very even surface with the face of the rule, when it is put into the groove either way, direct or reversed. Sliding rules are commonly made of box wood; but short ones of 11 or 13 inches should be made of very white ivory, and highly polished. In all cases the rules should be made a long time before the divisions are engraved upon them, that the wood or ivory may become seasoned, and shrink as much as it is disposed to do. The larger rules, from 18 to 30 inches, can only be made of box wood; and the very long ones of lance wood. It is of great importance to have a light-coloured surface for the ground of the divisions, as they can be read so much more easily, and with less fatigue to the eye. No substance is so good as ivory for this reason, but on account of the expense of that substance, and the small demand for excellent sliding rules, the instrument makers do not construct them of ivory, except when expressly ordered.

The goodness of the divisions of sliding rules must depend upon the accuracy of the original patterns, and the skill of the engraver. The rules made by Mr. Bate are more correctly divided than any others that the author has examined.

The accuracy of the principal divisions may be tried by comparing the coincidences of the divisions representing numbers which are multiples of each other, and if these coincidences are examined with the slider in several positions, both when direct and when inverted, the errors will discover themselves if they are of a sensible magnitude. The regularity of the intermediate subdivisions which fill up between the primary divisions, may be judged of by the eye with considerable accuracy.

Engineers are usually provided with a Soho sliding rule, and they should also have an inverted sliding rule; these have hitherto been made as separate instruments, which is not so convenient for use, as if they were combined into one. Persons who are already accustomed to use the Soho rule, and who have acquired facility in calculating by it, will wish to retain an instrument to which they are habituated; but the author has found the advantages of the inverted rule to be so great in his own practice, that he most strongly recommends it to the notice of engineers, and, consequently, in forming the precepts which are given in this work, a preference has been given to the slide inverted, except in cases where there is some good reason for using the slide direct. The advantages of the inverted method can only be attained in part, when the slide of the Soho rule is inverted, because the divisions are only half the size that they would be, upon a proper inverted rule of the same length; and therefore such an inverted rule should always be used when the precepts are entitled slide inverted, and have the letters A and Ɔ prefixed.

A very convenient sliding rule for the use of engineers who are accustomed to the Soho rule, may be made with two sliders, one in the back, and the other in the front face of the rule; one face being engraved to form a complete Soho rule, such as is represented in the sketch, p. 537, and the other face being engraved with an inverted broken line upon the slider, as is represented in the sketch, p. 541. The latter should be used in all cases which are directed to be performed with the slide inverted, and the lines A and Ɔ. The divisions on the rule will be the same at both sides. The table of divisors may be engraved at the back of the extra slider, and in the bottom of its groove.

This mode of combining the Soho rule with the inverted rule, the author has found to answer the intended purpose very well; but such a rule has a double quantity of divisions, and he afterwards found that the same advantages may be attained, by combining four lines upon a rule with one slider, leaving all the back of the rule to receive tables of gage points, and specific gravities. This new arrangement he recommends as that which will be found the most convenient for engineers, who have not already become familiar with the use of the Soho rule; for it has no more divisions upon it than the Soho rule, and, with a very few exceptions, it will perform all the calculations for which precepts are given in this work; and it has some new properties.

A new arrangement of Logarithmic Lines upon a Sliding Rule, by the Author.

The inverted slide rules which have been hitherto made, contain only two lines of single radius; viz. one placed direct on the rule, and the other placed inverted on the slider, and broken into two parts, as already described, p. 541. This rule is very convenient for performing multiplication and division, and all cases of proportion or rule of three, the advantages of certainty in reading off the results, and of accuracy, from the large size of its divisions, have been already stated.

For calculations which involve the square roots, or the squares of numbers, it is requisite to have a line of double radius acting against a line of single radius ; but the slide rules which are usually constructed (such as the Soho rule) require three lines of double radius and one line of single radius, which occupy the entire face of the rule, and leave no room for an inverted line, without having another additional slider, with an inverted line to fit the groove of the Soho rule, as has been already described.

To avoid the inconvenience of changing the slider, and to make a complete sliding rule for engineers, with only one slider, the author has made the following arrangement of four logarithmic lines ; (*a*)

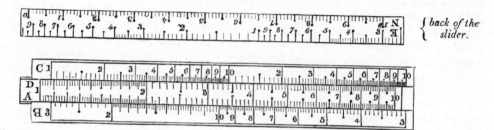

viz., three of single radius and one of double radius, which are so disposed upon one face of a sliding rule, as to perform multiplication, division, and simple proportion, on the inverted method, by means of the line of single radius A, at the lower part of the slider, acting against the inverted broken line ꓭ, at the lower part of the rule.   And all cases of squares, or square roots, may be solved by means of the line of single radius D, at the upper part of the slider, acting against the line of double radius C, at the upper part of the rule.

By this means the properties of the inverted rule, and also those of the Soho rule, are attained by the new sliding rule, with only one slider, and with no more divisions than the Soho rule has upon it ; and three of the lines being single radius instead of double radius, their divisions are twice as large, on a rule of a given length, so as to be more exact, and less fatiguing to the eye to read off.

This new rule is composed of the two lower lines of the Soho rule, disposed over the two lines of the inverted rule, and the only change made in the action of either of them is, that those lines which were formerly engraved upon the rule, are now put upon the slider, and *vice versa*.   The two upper lines of the Soho rule are omitted, because their place is more advantageously supplied by the inverted lines at the lower part of the new rule.

The divisors in the table which is engraved on the back of the Soho rule, for calculating the contents, or the weights of prisms, cylinders, or spheres, of different substances, are adapted to be used on the upper line A of that rule (see p. 556), and are not applicable to this new rule ; but the following table of gage points is engraved on the back of the new rule, to show the gage points which must be used on the line D for such calculations.   The numbers in this table are the square roots of the numbers in the former table of divisors (p. 559), or the square roots of ten times these divisors, as is explained at p. 560.

(*a*) Mr. Bate has undertaken to make sliding rules of this kind, according to the author's directions, with very accurate divisions; he expects the new rules will be found more suitable for the use of engineers than any similar instruments which are now to be procured.

Table of Gage Points for calculating the solidities or weights of square prisms, cylinders, or globes, by the sliding rule.

SOLIDS.	Cubic Feet.			Cubic Inch.		Water, lbs.			Tons.		Cast Iron, lbs.		
Dimensions.	FF	FI	II	FI	II	FF	FI	II	FF	FI	FF	FI	II
Squares {	1 / 1	12 / 379	416 / 131	289 / 913	1 / 1	4 / 126	48 / 152	166 / 526	599 / 189	72 / 227	149 / 471	179 / 566	62 / 196
Cylinders {	113 / 357	135 / 428	469 / 148	103 / 326	113 / 357	45 / 143	54 / 171	188 / 593	214 / 675	811 / 256	168 / 532	202 / 638	70 / 221
Globes {	138 / 437	— / —	574 / 182	— / —	138 / 437	175 / 553	— / —	23 / 727	262 / 827	993 / 314	206 / 652	— / —	271 / 856

SOLIDS.	Bar Irn. lbs.		Brass, lbs.		Lead, lbs.		Copper, lbs.		Stones, lbs.		Bricks, lbs.	
Dimensions.	FI	II	FI	II	FI	II	FI	II	FF	F I	FF	FI
Squares {	172 / 545	189 / 597	166 / .524	181 / 574	45 / 142	156 / 493	161 / 509	56 / 176	254 / 803	305 / 964	283 / 894	339 / 107
Cylinders {	194 / 615	213 / 673	187 / 591	205 / 647	51 / 161	176 / 557	182 / 575	63 / 199	287 / 906	344 / 109	319 / 101	383 / 121
Globes {	— / —	261 / 825	— / —	251 / 793	— / —	216 / 682	— / —	77 / 244	111 / 351	— / —	391 / 124	— / —

The gage points in the above table are to be taken upon the line D of the sliding rule, in the manner stated in page 559. The lengths of square prisms, cylinders, globes, or ellipsoids, are to be taken upon the line C, if the slider is direct, or on ꬶ if it is inverted; and their sides or diameters upon the line D. Their solidities or weights will be found upon the line C, if the slider is direct, or on ꬶ if it is inverted.

If the slider of the rule is direct, then the length upon C must be set opposite to either of the gage points upon D: and the solidity or weight will be found upon C, opposite to the side or the diameter upon D; thus,

Sliding rule. { $\dfrac{\text{C Length of cylinder. Solidity or weight.}}{\text{D Gage points. Diam. of cylinder.}}$ *Ex.* $\dfrac{\text{C 210 ft long. 660 cub. feet.}}{\text{D 135 or 428 24 inc. diam.}}$

If the slider is inverted, then the length upon ꬶ must be set opposite to the diameter upon D: and the solidity or weight will be found upon ꬶ, opposite to either of the gage points upon D; thus,

Sliding rule, slide inverted. { $\dfrac{\text{ꬶ Length of cylinder. Solidity or weight.}}{\text{D Diam. of cylinder. Gage points.}}$ *Ex.* $\dfrac{\text{ꬶ 210 ft. long. 660 cub. ft.}}{\text{D 24 inc. diam. 135 or 428.}}$

The above table contains two gage points for each case, but as either of them will give the same result, they may be taken indifferently upon the line D; the number which stands at top is to be preferred for common use, as being the most convenient and exact. In the table which is engraved at the back of the new sliding rule, only those upper numbers are inserted, and those which stand beneath are omitted on account of room; but on the long rules the complete table may be engraved as above.

4 D

In addition to the above table of gage points, a table of specific gravities is engraved at the back of the sliding rule, for the purpose of calculating the weight of solids of substances for which no gage points are given. Having found by means of the gage points for water, what would be the weight of the solid in water, the result may be multiplied by the specific gravity of the substance in question. Or else we may calculate the quantity of matter in cubic feet, and then multiply the result by the weight of a cubic foot of the substance in question.

A Table of the Specific Gravities of different Substances, and the weight of a cubic foot in pounds; adapted to be engraved at the back of a sliding rule.

Platina	21·04	710	Lead	11·36	Glass	2·88	163	Purbeck	2·61	Ivory	1·92
Gold	19·36	555	Copper	8·88	Marble	2·72	160	Portland	2·56	Horn	1·84
Mercury	13·55	525	Brass	8·40	Slate	2·67	155	Millstone	2·48	Bone	1.65
Silver	10·51	485	Iron	7·76	Granite	2·64	152	Paving	2·43	Box	1·33
Steel	7·84	450	Cast Iron	7·20	China	2·38	125	Brick	2·00	Coal	1·28

60	Wax	·96	44	Ash	·70	Carbonic acid gas 544	
58	Oil	·93	40	Fir	·64	Oxygen gas . 747	Times lighter than water.
55	Oak	·88	37	Elm	·59	Azotic gas . 854	
47	Beech	·75	35	Deal	·56	Common air . 830	Barom. 30 inches.
45	Mahogany	·72	15	Cork	·24	Hydrogen gas 11960	Thermom. 60 degr.

The divisors and gage points in the preceding tables (p. 557 and 569), are adapted to the specific gravities here marked, and which have been adopted from the best authorities.

The malleable metals are supposed to be hammered as much as they usually are, when they are wrought into shape.

The different kinds of stone have been selected amongst those which are in most common use; and the specific gravities marked above are the averages of different specimens. The gage points marked stone, are adapted to a specific gravity of 2·48, or 155 lbs. to a cubic foot.

The specific gravity of wood is taken at an intermediate state between unseasoned wood, and extremely dry wood, being the state in which it is fit for carpenters' use; but if the wood is cut into thin planks and dried sufficiently for joiners and cabinet makers, it will become lighter. Different specimens of the same tree, will differ very considerably in specific gravity.

The weight of a cubic foot of such substances as are in most common use, is given in pounds, to facilitate calculations; it may be obtained in other cases by multiplying the specific gravity of the substance by 62½ lbs., which is the weight of a cubic foot of rain water; or dividing the specific gravity by ·016 (the reciprocal of 62·5) will give the same result.

There is nothing peculiar in the use of this new sliding rule, for ordinary purposes, which can require any specific directions. Those precepts which are marked C and D, are to be performed with the two upper lines of the new rule which are marked C and D. The slide may be inverted, when necessary, for the precepts marked ꟼ and D, but which will be C and ʌ on the new rule. When the precepts are marked slide inverted A and ꓛ, the two lower lines A and ꟼ of the new rule are to be used without inverting the slider, because it is made inverted. For cases which are marked A and B simply, the slide must be inverted, and the two lower lines employed as direct lines, but they are not so well adapted for this as for the other cases.

To enable all the four lines of the new rule to be used in concert, in the same manner as those of the Soho rule, the 10 in the middle of the broken inverted line ꟼ, at the bottom of the rule, is placed exactly opposite to the 10 at the middle of the line C, at the top of the rule. This is not correctly represented in the sketch, but it may be easily imagined, and it enables the rule to perform some compound operations which cannot be done on other sliding rules : for instance,

To multiply two numbers together, and multiply their product by the square root of a third number. For instance, multiply 8 by 4, and multiply their product by the square root of 16.

Thus, the two factors are 8 and 4 ; and the square of the multiplier is 16, the root of which is 4 for the real multiplier. The resulting product is $(8 \times 4 = 32 \times 4 =)$ 128.

The author's Sliding Rule, all four lines.	C	Square of multiplier.	*Example.*	C	16 square of mult.
	D	Resulting product.		D	128 result.
	A	One of the factors.		A	4 one factor
	ꟼ	Other factor.		ꟼ	8 other factor.

Note. If the intermediate product of the multiplication is required, it may be found upon either of the lines A or ꟼ opposite to 1 upon the other line. The series of numbers on the line C is twice repeated, and a different resulting product will be found opposite to each of the repetitions of the number which represents the square of the multiplier ; the rule does not point out which of these results is the true one, but it can be known by other means.

Use of the line of one and a half radius on the Author's New Sliding Rule.

The line of divisions marked ꟼ, at the back of the slider of the author's new sliding rule, is a logarithmic scale like the others ; but the length of its radius, or the distance from 1 to 10, is only two-thirds of the length of the radius, or the distance from 1 to 10, upon the line of single radius, ꟼ.

The use of this additional line is to enable us to raise the 1·5th power of any number ; that is a fractional power which is less than the square of the number ; being equal to the square root of the cube of the number.

This fractional power is sometimes called the power of three halves : it may be raised by multiplying the logarithm of the number by 1·5, instead of by 2, or by 3, as must be done if it were required to raise the square or cube of the number.

Example. To raise the three-half power of 4. The cube of 4 is 64, the square root of which is 8, for the 1·5th power of 4, as required. Or by logarithms ; the log. of 4 is $0 \cdot 602 \times 1 \cdot 5 = 0 \cdot 903$, which is the log. of 8.

To use the line ꟼ, the slider must be taken out of its groove and turned upside down, and then replaced in the groove, with the line ꟼ applied against the lower line ꟼ of the rule ; the numbers of both lines will then count the same way, because both are inverted lines ; and if the slider is set so that 1 (or 31·623) upon the line ꟼ corresponds with 1 (or 10) upon the line B, then any number being chosen upon the lower line ꟼ, the 1·5th power of that number will be found opposite to it on the line ꟼ, thus,

The author's Sliding Rule.
$$\begin{cases} \text{ꟼ} & 1 \text{ or } 31·6 & 1·5\text{th power of numb.} \\ \text{ꟼ} & 1 \text{ or } 10 & \text{Number.} \end{cases} \text{or } \frac{\text{Number.}}{1·5\text{th root.}} \quad Ex. \quad \frac{\text{ꟼ } 31·6 \quad 8 \text{ power.}}{\text{ꟼ } 10 \quad 4 \text{ numb.}} \text{ or } \frac{64 \text{ num.}}{10 \text{ root.}}$$

And conversely the lines ꟼ and ꟼ, serve to extract the 1·5 root of a number ; which is the same thing as the square of the cube root of the number.

Example. To extract the three half root of 64. The cube root of 64 is 4 ; the square of which is 16, for the 1·5th root of 64, as required. Or the logarithm of 64 is $1·806 \div 1·5 = 1·204$, which is the log. of 16.

The principal use of the line ꟼ is for facilitating calculations relative to the discharge of water through apertures, such as sluices, or in cascades over weirs, or mill-dams, thus.

4 D 2

To calculate the quantity of water which will flow in a cascade over the edge of a weir, or through an aperture or notch in the edge of a board, such notch or aperture being open at top. Having given the depth in inches, from the level surface of the water in the reservoir, down to the edge over which the cascade of water flows; to find the quantity of water in cubic feet, which will be discharged per minute, over one foot wide of such cascade.

RULE. Multiply the square root of the cube of the depth in inches, by 5, and the product will be the number of cubic feet which will be discharged per minute over every foot in width; and therefore by multiplying the product by the width of the cascade in feet, the number of cubic feet discharged per minute will be obtained.

Example. What quantity of water will cascade over a weir 20 feet wide, when the level of the water in the reservoir is 4 inches above the edge of the weir. 4 inc. deep cubed = 64, the square root of which is 8 × 5 = 40 cubic feet per minute will be discharged over each foot wide; and × 20 feet wide = 800 cubic feet will be discharged per minute.

The author's new Sliding Rule. $\dfrac{\text{Ǝ} \quad \text{40 cubic ft. per min. over 1 ft. wide.}}{\text{Ḃ} \quad \text{4 inches deep below level surface.}}$ *Example.* $\dfrac{\text{Ǝ} \quad 40 \quad \text{158 cub. ft.}}{\text{Ḃ} \quad 4 \quad \text{10 inches.}}$

The rule being thus set, the two lines form a table; the numbers on the line Ǝ showing how many cubic feet which will be discharged per minute over a cascade one foot wide, at that depth in inches, which is denoted by the corresponding numbers on the line Ḃ.

Note. This table only extends from 3 inches deep, to 30 inches deep; it will comprise most cases in practice, but to make the rule serve as a table from ·3 tenths to 3 inches deep, it must be set as follows.

The author's new Sliding Rule. $\left\{ \dfrac{\text{Ǝ} \quad \text{5 cubic ft. per min. over 1 ft. wide.}}{\text{Ḃ} \quad \text{1 inch deep below level surface.}} \right.$ *Exam.* $\dfrac{\text{Ǝ} \quad 1 \quad \text{1·77 cub. ft.}}{\text{Ḃ} \quad 5 \quad \text{·5 of an inc.}}$

The rule when thus set will also serve as a table from 30 inches deep to 300 inches deep, if the value of the divisions is estimated as follows.

The author's new Sliding Rule. $\left\{ \dfrac{\text{Ǝ} \quad \text{5000 cubic ft. per min. over 1 ft. wide.}}{\text{Ḃ} \quad \text{100 inches deep below level surface.}} \right.$ *Exam.* $\dfrac{\text{Ǝ} \quad 5000 \quad \text{1560 cub ft.}}{\text{Ḃ} \quad 100 \quad \text{46 inches.}}$

The following table will point out the proper value for the divisions in these computations.

Depth inches.	Cubic feet per minute.	Depth inches.	Cubic feet per minute.	Depth inches.	Cubic feet per minute.	Depth inches.	Cubic feet per minute.	Depth inches.	Cubic feet per mnute.
$\frac{1}{4}$	625	4	40	13	234·36	22	515·94	40	1264·9
$\frac{1}{2}$	1·767	5	55·9	14	261·91	23	551·52	50	1768
$\frac{3}{4}$	3·248	6	73·48	15	290·47	24	587·87	60	2323·8
1	5	7	92·60	16	320	25	625	70	2928·3
$1\frac{1}{2}$	9·185	8	113·14	17	350·46	26	664	80	3577·7
2	14·14	9	135·	18	381·83	27	701·57	90	4269
$2\frac{1}{2}$	19·76	10	158·11	19	414·1	28	740·8	100	5000
3	25·98	11	182·41	20	447·22	29	780·85	200	14142
$3\frac{1}{2}$	32·74	12	207·85	21	481·17	30	821·58	300	25980

Note. The above rule will give results which correspond very nearly with the best experiments on the discharge of water through apertures. The quantity discharged is assumed to be ·65 of that which might be expected, according to the theory of bodies falling freely in vacuo (see p. 23 and 272); because the flowing water cannot acquire its full velocity all at once. Whilst the water is actually passing through the aperture, it usually moves with about ·65 of the velocity that a body would acquire by falling through a height equal to the depth of the water; but the motion being accelerated, the stream acquires very nearly that velocity after it has quitted the aperture.

The method of calculation best adapted for the purposes of engineers being now explained, we may proceed, in the next chapter, to state a number of specific rules by which the dimensions of every important part of a steam-engine of any size, may be calculated with great facility, so as to obtain the same proportions as those which were given by Mr. Watt to his engines.

CHAPTER VIII.

Rules for calculating the Proportions and Dimensions for the Parts of Mr. Watt's Rotative Steam-Engines of different Powers.

THE power of these steam-engines is measured by what is termed the horse-power (see p. 438).   That is, such an exertion of mechanical power, as is equivalent to a force of 33 000 pounds, acting through a space of one foot per minute, without making any deduction on account of the friction of the moving parts of the engine itself.   It is supposed that the denomination by horse-power represents the ulti-mate effect of the engine, or the power which is exerted at the extremity of the axis of its fly-wheel, and therefore the power actually exerted by the piston must be greater, by all that which is expended to overcome the friction of the moving parts of the engine, to work its air pump, cold water pump, &c. (*a*).

Mr. Watt's engines when in good order are capable of exerting a force of $10\frac{1}{2}$ pounds for each square inch of the piston, when it acts with a sufficient celerity of motion (see p. 486); but it was judged advisable in calculating the size of the engines, to make a very ample allowance for the probability of their not being kept in the best order, and therefore the load was only assumed at about 7 pounds to each square inch of the piston.

At this rate about 33 cubic feet of steam per minute, must be allowed to pass through the cylinder for every horse-power that the engine exerts.   This expendi-ture of steam is independent of the loss that is occasioned by the vacant spaces at the top and bottom of the cylinder through which the piston does not pass, and also any loss which may arise from leakage or condensation.

Mr. Watt gave the following account of the dimensions and performance of his rotative engines, in his annotations upon Dr. Robison's article Steam-Engine.

An engine upon the rotative double construction, having a cylinder of $31\frac{1}{2}$ inches diameter, and making $17\frac{1}{2}$ strokes per minute, of 7 feet long (=245 feet motion of the piston per minute) was called 40 horse-power (*b*), meaning the constant exertion of 40 horses; for which purpose (supposing the work to go on night and day) three relays, or at least 120 horses must be kept.   This engine con-sumed about 4 bushels of good Newcastle coals per hour (= 336 lbs.); or four hundred weight of good Wednesbury coal.

A rotative double engine, with a cylinder of $23\frac{3}{4}$ inches diameter, making $21\frac{1}{2}$ strokes per minute, of 5 feet long, was called 20 horse-power. And an engine, with a cylinder of $17\frac{1}{2}$ inches diameter, making 25 strokes per minute, of 4 feet long, was called 10 horse-power.   The consumption of coals by these engines was nearly proportional to that of the 40 horse-power, the greater or lesser effect depending upon the state of the engine, and upon the quality of the coals.

(*a*) The power required to work the cold water pump varies in different engines, according to the depth from which the cold water must be drawn out of the well or reservoir, in order to raise it into the condensing cistern of the engine.   In all cases, when the water is merely raised for the service of the engine, the power expended for that purpose should not in strictness be considered as forming any part of that power which is expressed when the engine is stated to exert so many horse-powers. But in practice, if the cold water pump raises the water more than 10 or 12 feet, all the remaining depth of the column is stated as an extra resistance which the engine must overcome, in addition to the usual and fair expenditure of power to give motion to its own parts.

(*b*) A strong mill-horse going at the rate of $2\frac{1}{2}$ miles an hour, is assumed to be able to raise a weight of 150 pounds, by a rope passing over a pulley; the power thus exerted is equal to raising 33 000 pounds one foot high in a minute.   Ordinary mill horses cannot do above two-thirds of this.

The following table has been formed from the three standard engines of 10, 20, and 40 horse-power mentioned by Mr. Watt; the other sizes being calculated according to the same proportion. The author has met with instances of engines made by Messrs. Boulton and Watt during their patent, which correspond with all those which are marked * in the table below. These form a considerable portion of the whole, and being proportionate to the others, the table may be taken for the exact scale which Mr. Watt established, and followed in his practice.

Dimensions of the Cylinders and Velocities of the Pistons, in Messrs. Boulton and Watt's Rotative Engines of different powers.

Horse-power.	Cylinders.		Stroke of Pistons.			Cubic ft. of steam per min.	Effective pressure per squ. inc. lbs.	Cubic ft. of steam per min. by 1 HP.	Effective load on the piston pounds.
	Diam. inches.	Area squ. inc.	Length feet.	Number per min.	Feet per min.				
4	12	113·1	3	29	174	135	6·8	33·7	759
* 6	14	153·9	$3\frac{1}{2}$	27	189	202	6·82	33·6	1048
* 8	16	201·1	4	24	192	268	6·84	33·5	1375
*10	$17\frac{1}{2}$	240·5	4	25	200	334	6·86	33·4	1650
*12	19	283·5	4	25	200	400	6·88	33·3	1980
14	$20\frac{5}{8}$	334·1	4	25	200	465	6·91	33·2	2310
16	$21\frac{3}{4}$	371·5	$4\frac{1}{2}$	23	207	531	6·91	33·2	2550
18	23	415·5	$4\frac{1}{2}$	23	207	598	6·91	33·2	2870
*20	$23\frac{3}{4}$	443·0	5	$21\frac{1}{2}$	215	662	6·92	33·1	3070
22	25	490·9	5	$21\frac{1}{2}$	215	728	6·92	33·1	3376
*24	26	530·9	5	$21\frac{1}{2}$	215	794	6·92	33·1	3684
*26	$26\frac{3}{4}$	562·0	$5\frac{1}{2}$	20	220	861	6·92	33·1	3900
28	$27\frac{7}{8}$	610·3	$5\frac{1}{2}$	20	220	927	6·92	33·1	4200
*30	$28\frac{1}{4}$	626·8	6	19	228	993	6·92	33·1	4344
*36	$30\frac{1}{8}$	748·7	6	19	228	1192	6·92	33·1	5210
*40	$31\frac{1}{2}$	779·3	7	$17\frac{1}{2}$	245	1324	6·92	33·1	5390
*45	$33\frac{1}{3}$	875·4	7	$17\frac{1}{2}$	245	1490	6·92	33·1	6060
50	$35\frac{1}{8}$	969·0	7	$17\frac{1}{2}$	245	1650	6·94	33·0	6737
60	$38\frac{1}{2}$	1164·2	7	$17\frac{1}{2}$	245	1980	6·94	33·0	8082
70	$40\frac{3}{4}$	1304·2	8	16	256	2310	6·94	33·0	9023
80	$43\frac{1}{2}$	1486·2	8	16	256	2640	6·94	33·0	1032
90	$46\frac{1}{8}$	1670·9	8	16	256	2970	6·94	33·0	1160
100	$48\frac{5}{8}$	1857·0	8	16	256	3300	6·94	33·0	1289

The 50 horse engine was at first made with a cylinder 34 inches diameter, and 8 feet stroke (see p. 509); and afterwards with a cylinder 36 inches diameter, and 7 feet stroke; but ultimately an engine of the last dimensions was rated at 53 horse-power (see Note, p. 518); this accords very nearly with the proportion of the other engines in the above table.

It appears from this table that Mr. Watt assumed the expenditure of steam by each size of engine, to be very nearly proportionate to the power exerted by it;

for in the smallest engine 33·7 cubic feet of steam per minute is assumed to exert a horse power, and in the largest 33 cubic feet.   It will be sufficiently near for practice to take the expenditure of steam at 33 cubic feet per minute, for each horse power, or ($33 \times 183·346 =$) 6050 cylindrical inch feet, and then the effective pressure upon each square inch of the piston will be 6·944 pounds, or ($\times$ ·7854 =) 5·454 pounds per circular inch.   And each cubic foot of steam will raise 16 cubic feet of water one foot high.   The following rule corresponds to this allowance.

To find the proper diameter for the cylinder of Mr. Watt's rotative steam-engine, to exert a given power; having given the space in feet through which the piston is intended to move per minute.

RULE. Multiply the number of horse power by the constant number 6050; divide the product by the motion of the piston, in feet per minute, and the square root of the quotient is the proper diameter for the cylinder of the engine in inches.

Example; for a 20 horse engine, if the piston is to make 21½ double strokes per minute of 5 feet long = 215 feet motion per minute.   Then 20 horse power × 6050 cylindrical inch feet of steam per minute for each horse power = 121 000 cylindrical inch feet must be expended per minute by this engine ÷ 215 feet motion per minute = 562·8 circular inches for the area of the piston, the square root of which is 23·72 inches for the diameter of the cylinder.

Sliding Rule. $\begin{cases} \text{C Motion of piston ft. per min. Horse power of eng.} \\ \text{D Gage point 246 or 778. Diam. of cylind. inc.} \end{cases}$ *Ex.* $\begin{array}{ccc} \text{C 215 ft. per m.} & \text{20 HP.} \\ \text{D 246.} & 23·72. \end{array}$

Note. This rule will give the diameters of the cylinders for engines above 20 horse power, a very little less than Mr. Watt's scale.   For engines less than 20 horse power, the cylinders should be rather larger than this rule would give.   The proper sizes may be found by using the following gage points for the sliding rule, viz. for 20 horse and upwards, 246 as above.   For 14 horse to 20 horse, 247. For 8 horse to 14 horse, 248.   And for 4 horse to 8 horse, 249, for the gage point.

The engines made at the Soho manufactory, for some years after Mr. Watt retired from the business, continued to be proportioned by his scale, although the structure of the engines was considerably altered, as will be described.   Of late years it has become customary amongst the makers of engines, to proportion the pistons to act with a shorter stroke than the above scale.   For instance, Messrs. Boulton, Watt, and Co.'s 40 horse engine is now made with a cylinder 32½ inches diameter, and 6 feet stroke (see note, p. 503).   Their 80 horse engine, with a cylinder 44⅛ inches diameter, and 7 feet stroke; and their small sized engines, under 20 horse power, act with very short strokes.   This departure from Mr. Watt's scale enables an engine to be brought into less compass, and it requires less materials for its construction, so as to cost less to the maker; but it is the general opinion of the most experienced engineers, that the best performance will be attained with as great a length of stroke for the piston, as convenience will allow; and they follow Mr. Watt's scale, in cases when they wish to produce an excellent engine.

After the expiration of Mr. Watt's patent in 1800, several establishments for making steam-engines were founded in Lancashire, Yorkshire, and Stafford-shire, in Scotland, and in Wales; but owing to defective workmanship, and want of knowledge of the true proportions, it was generally found that the engines first executed by these new makers, fell very short of the performance of the patent engines; and to make up for the deficiency, after it had become known, the makers were required by their customers to put in larger cylinders, than had been usually sent from Soho.   The steam cases to the cylinders were sometimes omitted; and as this was supposed to impair the effect of the engines, it was alleged as a reason for making the cylinders larger.   In this manner a 20 horse engine acquired a cylinder 25 inches diameter; a 30 horse, 30 inches; a 40 horse, 33 inches; a 50 horse, 37 inches; a 60 horse, 40 inches; an 80 horse, 45 inches; and a 100 horse engine, a cylinder of 50 inches diameter.

Engines made upon such a scale, are very common in Yorkshire and Lancashire; they ought to be capable of exerting considerably more power than Mr. Watt's own engines of the same nominal power; and that is the case with those now made by the best makers; but many engines, even with that increased size of the cylinder, are not capable of doing more work than if they were made in a proper manner, according to Mr. Watt's scale.

In all cases it must be understood that the above calculation of Mr. Watt's engines by horse power, does not express the utmost power that they are capable of exerting; because all good engines, with a suitable allowance of fuel, are capable of exerting half as much more power, as that at which they are rated; for instance, a 20 horse engine can exert 30 horse power; a 40 horse engine 60 horse power, and so on: see p. 486. The nominal horse power by which each engine is rated in the table, is that exertion which it is competent to overcome in the most advantageous manner for a continuance, considering all the attendant circumstances, such as expense of fuel, wear of the machinery, and consequent stoppage and expense for repairs, and the first cost of the engine. It is found that Mr. Watt's engines never answer so well, when they are loaded with a resistance which is considerably greater or considerably less, than the nominal horse power of the engine as expressed in the table.

To find how many cubic feet of steam, must be expended per minute, for each horse-power that is exerted by a steam-engine; having given the effective pressure exerted by the steam on the piston, in pounds per square inch.

Suppose a piston of one square foot, or 144 square inches area, to move through one foot per minute, it would expend one cubic foot per minute; and if this piston were loaded with 33 000 pounds, it would exert one horse power; the pressure in that case would be (33 000 ÷ 144 =) 229·167 pounds per square inch; or (33 000 ÷ 183·346 =) 180 pounds per circular inch; hence the following rule.

RULE. Divide the constant number 229·17, by the effective pressure of the steam upon the piston, in pounds per square inch; the quotient is the number of cubic feet which must be expended per minute, to produce one horse power.

Example. If the effective pressure on the piston is 6·994 pounds per square inch, then 229·17 ÷ 6·944 = 33 cubic feet of steam per minute to each horse power.

| Sliding Rule, slide inverted. | $\begin{cases} A & 1 \\ \cap & 229 \end{cases}$ | Pressure lbs. per circ. inc. / Cub. ft. per min. to 1 HP. | *Exam.* | $\begin{array}{cc} A & 1 \\ \cap & 229 \end{array}$ | 6·94 lbs. / 33 cubic feet. | or | $\dfrac{6}{38·2}$ |

If the pressure is given in pounds per circular inch, then the constant number to be divided must be 180. Thus for 5·45 lbs. per circular inch; 180 ÷ 5·45 = 33 cubic feet per minute.

| Sliding Rule, slide inverted. | $\begin{cases} A & 1 \\ \cap & 180 \end{cases}$ | Pressure lbs. per circ. inc. / Cubic feet per min. to 1 HP. | *Ex.* | $\begin{array}{cc} A & 1 \\ \cap & 180 \end{array}$ | 5·45 lbs. per circ. inc. / 33 cubic ft. per min. |

On the actual Expenditure of Steam in Mr. Watt's Rotative Engines.

When an engine is fairly loaded so as to exert the power at which it is rated, the steam with which the cylinder is filled during the plenum, to impel the piston, will commonly be less elastic than the atmospheric air. It has been already stated, p. 487, that if an engine is in excellent order it may exert its proper power, when the steam in the cylinder is as low as two pounds per square inch, less elastic than the atmospheric air. Or (14·7 − 2 =) 12·7 lbs. per square inch actual elasticity.

The above is an extreme case of good performance, but we may assume for an average of the best engines when in good order, that the elasticity of the steam with which the cylinder is filled during the plenum, is about $1\frac{1}{3}$ pounds per square inch less elastic than the atmospheric air, or (14·7 − 1·34 =) 13·36 lbs. per square inch.

Assuming the expenditure of steam of that elasticity to be at the rate of $(33 \times 1\cdot1 =)$ 36 ·3 cubic feet per minute for each horse-power, including waste, then the expenditure of steam equal to the atmosphere, would be $(36\cdot3 \times 13\cdot36 = 485 \div 14\cdot7 =)$ 33 cubic feet per minute for each horse-power.

This will be found to correspond nearly with the performance of the best engines, if they are in good order, so as to be free from leakage, and it is a convenient proportion for computations, because the expenditure of steam equal in elasticity to the atmosphere, is assumed to be as much as will fill the space that is occupied by the piston in its motion; and the extra quantity of steam which is wasted in the vacant spaces, is supposed to be equivalent to the deficiency that is occasioned by the steam in the cylinder being less elastic than the atmospheric air.

At this rate each cubic foot of steam equal to the atmosphere, will exert a sufficient power to raise 16 cubic feet of water one foot high.   For a horse-power is 528 cubic feet of water raised one foot high per minute; and $528 \div 33$ cubic feet of steam $= 16$ cubic feet of water (see p. 575).

On the quantities of water evaporated from the boilers of Mr. Watt's Rotative Engines.

If the expenditure of steam equal to the atmosphere, is assumed to be as much as would fill the space that is occupied by the piston in its motion; and if the volume of steam equal to the atmosphere, is taken to be 1700 times the volume of an equal weight of cold water, then the quantity of water evaporated from the boiler would be equal to fill one 1700th part of the space that is occupied by the piston in its motion.

This supposes that the engine is in good order, so as to lose no steam by leakage, and that the surface of the steam case to the cylinder is well clothed, so as to avoid loss of steam by condensation.   *Note*, the water which is produced from the steam which is condensed in the top of the boiler, and in the steam-pipe, will run back into the boiler; and hence, that condensation will not occasion any increase of the quantity of water which must be supplied to feed the boiler.

The expenditure of steam equal to the atmosphere being taken at 33 cubic feet per minute, for each horse-power, including waste (but supposing that there is no loss by leakage) and the volume of that steam being 1700; then the evaporation of water from the boiler will be at the rate of $(33 \div 1700 =)$ ·0194 of a cubic foot per minute for each horse-power; or one cubic foot evaporated in 51 ·5 minutes; that is, 1 ·165 cubic feet evaporated per hour for each horse-power.

An engine must be in good condition to enable it to exert its full power, with no greater evaporation than as above.   The engine at the Albion Mill was 50 horse-power, and by this proportion should have evaporated $(50 \times \cdot0194 =)$ ·97 of a cubic foot per minute.   The evaporation really was only ·927 of a cubic foot (see p. 515); hence, the above allowance may be sufficient, though in general the evaporation will be found greater, owing to loss of steam by leakage·and condensation (a).

The rule already given at p. 366, to find the quantity of water which will be evaporated from the boiler of Mr. Watt's single engine, is founded on the assumption, that the steam with which the cylinder is filled during the plenum, is of the same elasticity as the atmospheric air, but that the expenditure of such steam is one-tenth more than what would fill the space occupied by the piston in its motion; also, that one cubic inch of water produces one cubic foot of steam.

(a) It is a common saying amongst workmen, that Mr. Watt reckoned the boilers of his engines to evaporate 10 ale gallons (of 282 cubic inches each) per hour for each horse-power; that is, 1 ·632 cubic feet per hour.   If this is correct, it is probable that it was intended as a sufficient allowance for an engine when loaded to the utmost, but adapted to give a considerable surplus at other times.

4 E

According to that rule the evaporation would be at the rate of $(33 \times 1\cdot1 = 36\cdot3 \div 1728 =)\ \cdot021$ of a cubic foot of water per minute for each horse-power; or $1\cdot26$ cubic feet per hour for each horse-power. As this will be found near to the quantity evaporated from the boilers of engines in actual practice, the rule in p. 366 may be followed, and need not be repeated. It may be considered as allowing 33 cubic feet per minute of steam equal to the atmosphere, for each horse-power, with a farther allowance of one-tenth for leakage and condensation.

Note, to adapt that rule to the above statement of $\cdot0194$ of a cubic foot per minute, or $1\cdot165$ cubic feet per hour for each horse-power, the constant divisor must be $(183\cdot346 \times 1700 =)\ 311\,688$, instead of $288\,000$. And the gage point for the line D of the sliding rule, must be $558\cdot3$, instead of $536\cdot6$.

Dimensions of hot water pumps for Mr. Watt's Rotative Engines.

The hot water pump must be adapted to deliver a much greater quantity of water into the feeding cistern, than can be evaporated from the boiler in the same time, and thus afford a very ample surplus, to run off by the waste pipe; because the feeding of the boiler may be interrupted by accident at times, and the water will subside in the boiler; when this is discovered, the deficiency ought to be replenished as quickly as possible, but if the hot water pump could not deliver more water than to supply the evaporation, it would be incapable of making up such an occasional deficiency, which can only be replaced out of the surplus water; and the greater that surplus is, the quicker such deficiency may be made up.

It is advisable to make the hot water pump raise at least three and a half times the quantity of water that is evaporated; and taking the evaporation at one 1700th part of the space occupied by the piston, the capacity of the barrel of the hot water pump should be about one 240th part of the capacity of the steam cylinder; supposing that it is a double engine, which expends steam continually, and that the hot water pump only acts during the ascent of the bucket.

To find the proper diameter for the hot water pump of Mr. Watt's Rotative Engine. Having given, the diameter of the cylinder in inches; the radius or distance, in inches, from the centre of the great lever, to the main joint, by which the piston rod is suspended; and the radius, in inches, of the joint by which the rod of the hot water pump is suspended.

RULE. Divide the square of the diameter of the cylinder in inches, by 240; multiply the quotient by the radius, in inches, of the main joint for the piston rod; and divide the product by the radius, in inches, of the joint for the hot water pump; the square root of the quotient is the proper diameter for the hot water pump in inches.

Example. Suppose the cylinder to be 33 inches diameter; the square is 1089 circular inches area, $\div 240 = 4\cdot538 \times 120$ inches radius of the great lever $= 544\cdot5 \div 30$ inches radius of the hot water pump $= 18\cdot15$ circular inches in the area for the hot water pump; and its square root $4\cdot26$ inches is the proper diameter for the pump.

To perform the calculation by the sliding rule, the slide must be inverted, and set with the radius of the joint for the piston rod on $\mathbb{q}$, corresponding to the diameter of the cylinder on D, and then 24 times the radius of joint for the hot water pump, being found on $\mathbb{q}$, the proper diameter for the pump barrel will be opposite to it; thus,

Sliding Rule, $\left\{ \begin{array}{ll} \mathbb{q} & \text{Rad. of pist. joint inc.} \\ D & \text{Diam. of cyl. inches.} \end{array} \right.$ $\dfrac{24 \times \text{rad. of pum. joint inc.}}{\text{Diameter of hot pump, inc.}}$ *Ex.* $\dfrac{\mathbb{q}\ 120\ \text{rad.}\ \ (24 \times 30 =)\ 72}{D\ 33\ \text{dia.} \qquad 4\cdot26\ \text{diam.}}$

Modes of expressing the performance of Engines in respect to fuel.

It is customary to estimate the performance of the steam-engines which are employed in draining mines, according to the number of millions pounds weight of water, which is raised one foot high, by the consumption of one bushel of coals. This mode of expression was introduced by Mr. Watt, at his first commencement in Cornwall, where coals are sold by measure of bushels and chaldrons: see p. 337. The power of rotative engines which are employed in manufactories, is always estimated by the horse power, and coals being sold by weight, in most of the manufacturing districts, it is usual to express the performance of such engines, by the number of pounds of coals which are consumed per hour, for each horse power. It is requisite to have a rule of comparison to convert these two modes of estimation from one to the other.

To find how many millions pounds weight are raised one foot high, by the consumption of one bushel of coals (or 84 pounds weight) in any steam-engine; having given the number of pounds weight of coals consumed per hour, by each horse power.

RULE. Divide the constant number 166·32 by the number of pounds of coals which are consumed per hour by each horse power; the quotient is the number of millions pounds weight, which are raised one foot high, by the consumption of one bushel of coals.

Example. If an engine of 40 horse power consumed 4 bushels (= 336 lbs.) of coals per hour, it would be at the rate of 8·4 pounds per hour, for each horse power. Then 166·3 ÷ 8·4 = 19·8 millions pounds are raised one foot high, by the consumption of each bushel of coals.

| Sliding Rule, slide inverted. | A | 19·8 | Millions lbs. raised 1 foot per bush. | *Exam.* | A | 19·8 | 17·5 mill. |
| | ၁ | 8·4 | Pounds per hour, per horse power. | | ၁ | 8·4 | 9·5 lbs. |

Note. The constant number 166·32 is thus obtained; a horse power is 33 000 lbs. raised one foot per min. × 60 min. = 1·98 millions lbs. raised one foot per hour × 84 lbs. of coals in a bushel = 166·32 millions pounds would be raised one foot high, by one pound of coals, if that quantity could supply one horse power for an hour.

Assuming a bushel of coals to weigh 84 lbs., if the number of millions pounds which are raised one foot high by its combustion, is divided by 84, the quotient will be the weight in pounds which is raised one foot high, by the consumption of one pound of coals. This would be a more convenient mode of expression than either of the preceding, because the weight raised, and that of the coals consumed, would be in the same terms. For instance, if 19·8 millions pounds weight, is raised one foot high by the consumption of a bushel of coals, then 19 800 000 ÷ 84 = 235 700 times the weight of coals consumed, will be raised one foot by their combustion.

To find how many times the weight of the coals, which are consumed by any steam-engine, is raised one foot high by their combustion. Having given the horse power exerted by the engine, and the weight of coals it consumes per hour in pounds,

RULE. Multiply the horse power exerted by the engine, by 1 980 000 lbs.; and divide the product by the pounds weight of coals consumed per hour. The quotient is the number of times the weight of the coals, that is raised one foot high, by their combustion.

Example. If a 40 horse engine burns 336 pounds of coals per hour. Then 40 HP × 1 980 000 lbs. = 79 200 000 ÷ 336 = 235 700 times the weight of coals.

| Sliding Rule, slide inverted. | A | Horse power | No. of times the weight of coals. | *Ex.* | A | 40 HP. | 235 700 times· |
| | ၁ | 198 | Coals consumed per hour, pounds. | | ၁ | 198 | 336 lbs. per hr. |

DIMENSIONS OF THE BOILERS FOR MR. WATT'S ROTATIVE ENGINES.

The quantity of steam which a boiler is capable of producing, is limited in practice by the extent of surface which is exposed to the action of the fire; for if a very intense degree of heat is applied to a boiler, in order to obtain the utmost quantity of steam from it, a great portion of the heat will escape with the current of heated air and gas which passes up the chimney; and the metal of the boiler will be worn away by rapid oxydation. A boiler must, therefore, be provided with a sufficient heating surface in proportion to the quantity of steam it is required to produce, so that it will not be necessary to apply a greater force of fire to the boiler, than the metal can endure for a long time without injury.

Mr. Watt states in his annotations on Dr. Robison's article on the Steam-engine, that he found " that with the most judiciously constructed furnaces, it required 8 square feet of surface of the boiler, to be exposed to the action of the fire and flame, to boil off a cubic foot of water in an hour." This is at the rate of 480 square feet of surface, to boil off one cubic foot of water per minute, as has been already stated in p. 365; and the rule in p. 366 has been deduced from this data, to apportion the surface of the boiler, according to the diameter of the cylinder, and the rapidity of the motion of its piston, that is according to the expenditure of steam.

Mr. Watt assumed that a cubic inch of water will produce a cubic foot of steam, then each square foot of the surface exposed would produce $(1728 \div 480 =) 3\cdot6$ cubic feet of steam per minute, equal to the atmosphere. If it is assumed that for every horse power exerted, there must be 33 cubic feet of space occupied in the cylinder per minute, by the motion of the piston, and allowing one-tenth additional quantity for waste, by leakage and condensation, each horse power will expend $36\cdot3$ cubic feet of steam per minute, or $(36\cdot3 \times 183\cdot346 =) 6655$ cylindrical inch feet. Hence for each horse power there must be $(36\cdot3 \div 3\cdot6 =)$ very nearly 10 square feet of surface of the boiler exposed to the heat of the fire.

This proportion appears to have been followed by Mr. Watt for the boilers of his earliest rotative engines: see p. 489 to 518. A boiler is very capable of producing steam at this rate (a), and if it is made of copper-plate, as most of these boilers were, it will last a long time. But it has been found adviseable, in the course of practice, to allow a greater surface, and the increased expense of the first construction is amply compensated by the greater durability of the boiler; this is more particularly the case when the boiler is made of iron plate, because that metal is cheap, but it is very subject to oxydation. In many cases, where engines are required to exert their utmost power, it has been found expedient to replace the original boilers with others of larger dimensions; and the increase of surface has been found to save fuel, as well as to prolong the duration of the boilers: see p. 496.

The most adviseable dimensions for iron boilers for rotative steam-engines has been ascertained by long experience since Mr. Watt's time, and from a comparison of the surfaces exposed to the heat, with the quantities of steam produced from a great number of boilers, in actual use in modern engines, it appears that each square foot of surface which is exposed to the heat, will produce about 3 cubic feet of steam per minute equal to the atmosphere, or $(3 \times 183\cdot24 =) 550$ cylindrical inch feet; and allowing that $36\cdot3$ cubic feet of such steam will produce one horse-power (the waste by condensation and leakage being included), the heating surface

(a) The following instances show that Mr. Watt's proportion will produce a sufficiency of steam.

The two boilers of Mr. Smeaton's engine at Long Benton, see p. 173, exposed 459 square feet of surface to the action of the fire and flame; viz. 142 square feet of horizontal surface beneath the boiler bottom, and 317 square feet of vertical surface round the outside of the boiler. These boilers evaporated $1\frac{1}{2}$ cubic feet of water per minute, which is at the rate of 306 square feet of surface to evaporate one cubic foot of water per minute; this is only ·638 of Mr. Watt's standard of 480 square feet.

The surface of the two fire-grates was 35 square feet, to evaporate $1\frac{1}{2}$ cubic feet per minute, that is, $23\frac{1}{3}$ square feet to evaporate 1 cubic foot per minute.

The engine which Mr. Watt made for Whitbread's brewery, see p. 437, has a circular copper boiler, which was put in new in 1795, at the time when the engine was altered to work with double action. This boiler is still in use; it is $10\frac{1}{2}$ feet diameter at the largest part, and 8 feet diameter at the bottom. The total height is about $7\frac{1}{2}$ feet. The surface exposed beneath the bottom of the boiler, is a circle of more than $7\frac{1}{2}$ feet diameter, or about 45 square feet; and the surface exposed in the flue is about $(29 \text{ ft.} \times 3\frac{1}{2} \text{ ft.} =) 102$ square feet. In the whole, 147 square feet of surface is exposed to the heat, to produce steam for a 20 horse engine. This is only $7\cdot35$ square feet to each horse power, instead of 10 square feet according to Mr. Watt's standard. The fire-grate is $4\frac{1}{2}$ feet long, $3\frac{1}{3}$ feet wide, with 14 bars in the width, = 15 square feet.

of the boiler should be at the rate of (36·3 ÷ 3 =) 12·1 square feet to each horse-power that the engine is required to exert; say 12 square feet.

This is the proportion which is now most commonly followed, and it will give a sufficiency of steam, without any unreasonable wear of the boiler; but it is the opinion of the most experienced engineers, that a greater quantity of steam will be produced from a given consumption of fuel, if more surface is given to the boiler, even as much as 15 or 18 square feet to each horse power (a).

Assuming that 12 square feet of the heating surface of the boiler is allowed to one horse-power; and that each square foot will yield 3 cubic feet of steam per minute, then if the steam is 1700 times the volume of an equal weight of cold water, it will require (1700 ÷ 3 =) 567 square feet of heating surface to evaporate one cubic foot of water per minute. At the same rate the thickness of the film of water, which would be evaporated every minute from the heating surface of the boiler, would be (3 feet or 36 inches ÷ 1700 =) ·0212 of an inch; or one 47th of an inch thick.

It must be kept in mind, that the different parts of the heating surface of a boiler yield very different quantities of steam. The greatest production of steam is from the horizontal bottom of the boiler, where the fire is applied directly beneath it, and the water above it; the surface exposed in the flues is much less efficacious in producing steam. The above statement is adapted to an average of all the heating surface, which is exposed in a boiler; and the extent of the horizontal surface which is exposed beneath the bottom, is commonly about three-sevenths of the vertical surface which is exposed in the flue around the outside of the boiler.

To find the surface that the boiler of Mr. Watt's rotative engine ought to expose to the fire and flame.

RULE. Multiply the square of the diameter of the cylinder in inches, by the motion of the piston per minute, in feet; and divide the product by 500. The quotient will be the proper surface in square feet which should be exposed to the heat.

Example. The cylinder for a 20 horse engine, being 24 inches diameter; and the piston making 21 strokes per minute, of 5 feet long = 210 feet per minute. Then 24 inches squared = 576 circular inches × 210 feet = 120 960 cylindrical inch feet of steam is required per minute, without allowance for waste, and ÷ 500 = 241·9 square feet is the proper extent of heating surface for the boiler. This is very nearly 12 square feet to each horse power (b).

Sliding Rule.	C	Motion ft. per min.	Fire surface squ. ft.	*Exam.*	C	210 ft. mo.	242 squ. ft.
	D	22·36 or 707	Diam. of cylind. inc.		D	22·36	24 inc. diam.

Note. This rule supposes the engine to exert its proper power, as rated in horse-power, and then the boiler will work to the greatest advantage. If the engine is overloaded, the boiler will be quite sufficient to supply steam; but it will require more firing than it is adviseable to apply to a boiler, particularly if it is made of iron plate.

The proportionate extent of the horizontal and vertical parts of the heating surface, varies in different boilers according to their structure; the vertical surface being twice as great as the horizontal in some extreme cases, and in others three times as great; but from an examination of a number of boilers of the best construction, the usual proportion appears to be 3 of horizontal surface, to 7 of vertical surface.

(a) Messrs. Rothwell, Hick, and Rothwell, of Bolton in Lancashire, now have a most extensive practice in making Mr. Watt's rotative-engines for manufactories; they have always applied very large boilers to their engines, the surfaces exposed being nearly at the rate of 15 square feet to each horse-power. This is found to be an advantageous proportion for boilers, and renders them less expensive in a course of years, than if they were made on a smaller scale.

(b) To give 15 square feet of heating surface to each horse-power, the divisor in the above rule should be 400 instead of 500; and the gage point for the line D of the sliding rule should be 20 instead of 22·36. At the same rate 709 square feet are allowed to evaporate one cubic foot of water per minute.

The proper extent of heating surface being found by the above rule, it may be multiplied by ·3, and the product will be the extent of horizontal surface beneath which the heat should be applied, the water being applied over that surface. Thus 242 square feet ·3 = 72·6 square feet of horizontal heating surface for the 20 horse engine. Or multiplying the whole extent of heating surface by ·7, we shall obtain the extent of vertical heating surface, against one side of which the heat is to be applied, and the water on the opposite side. Thus 242 squ. feet × ·7 = 169·4 square feet of vertical surface.

The extent of heating surface which may be given to a boiler, admits of considerable latitude, without occasioning any material difference in the effect, but it is advantageous to have too much surface rather than too little, particularly in cases when the water with which the boiler is supplied, is subject to deposit a sediment, or form an earthy incrustation upon the internal surface of the boiler, for such sediment or incrustation greatly impedes the transmission of the heat through the metal to the water. If a large surface is allowed, the deposition will be spread over a greater extent, so as to form a thinner crust, and as less heat is required to be transmitted through that crust, it will prove a less impediment. This circumstance is of the greatest importance, and is the most effectual means of obviating, or diminishing the evil of bad water.

It has never been ascertained what is the proportionate effect of the different parts of the heating surface in a boiler, to produce steam; but it is certain that the horizontal surface must be much more efficacious than the vertical surface, because of the tendency which the heat has to ascend. Some engineers are of opinion that the horizontal surface of a boiler will produce twice as much steam, as the same extent of its vertical surface, but this opinion is not grounded upon experiment; and it is obvious that there must be a very great difference in the quantities of steam which are produced from the various parts of the horizontal surface, as well as from the various parts of the vertical surface, according to the degrees of force with which the fire acts against the different parts of the boiler.

The greatest heat is communicated to the bottom of the boiler, over the back part of the fire-grate, and above the fire-bridge, for there the bottom receives the radiant heat of the burning fuel, as well as the current of flame which proceeds from it. That part of the bottom of the boiler which is over the front end of the grate, does not receive so much heat, because the current of the flame and heated gas draws backwards, and the heat does not ascend vertically, but in a sloping direction, so as to strike beneath the bottom of the boiler, at the back part of the furnace, near to the fire-bridge. The front end of the fire-grate is always placed beneath the front end of the boiler, and it is thought advantageous to place the grate on a slope of about 20 degrees from the horizontal; the front end being highest, and nearest to the bottom of the boiler, and the farther end being lower down, and farther from the bottom, so as to cause the fire to act more uniformly beneath the bottom.

The front face of the fire-bridge, or back end of the furnace, is usually built vertical if the fire-grate is horizontal; but if the fire-grate is on a slope, the front of the fire-bridge is inclined backwards at the top edge, that it may lead the fire and flame backwards beneath the bottom of the boiler, without striking so abruptly upwards beneath the bottom, over the top of the fire-bridge.

That part of the bottom of the boiler which is beyond the fire-bridge, receives the action of the flame beneath it, and produces a great quantity of steam. In some boilers, the brickwork beneath the bottom of the boiler is arched upwards in conformity with the arched bottom of the boiler, so as to leave an equal height or space, for the passage of the flame beneath every part of the bottom of the boiler; it is supposed that the proximity of the brickwork to the bottom of the boiler, will compel the flame to apply more closely to the bottom.   In other boilers the brick-work beneath the bottom of the boiler is laid flat, and a considerable space is left between them for the flame ; the disposition of the flame to ascend will cause it to apply to the bottom of the boiler, and when an ample passage is left for the current of flame, it will move with less rapidity, and consequently it will remain longer in contact with the bottom, so as to communicate more heat thereto, than if the passage were narrower, because then the current would then be more rapid.

The angle at the farther end of the boiler, where the bottom joins to the upright end, receives a considerable action of the flame, but the lateral action of the heat, against the vertical surfaces at the ends and sides of the boiler, is necessarily much less than beneath the horizontal surface at the bottom ; it is pro-bable that the heat acts more powerfully at the upper part of the vertical surface than at the lower part, because the heated current will ascend.   The passage through the external flue, or the space which is left between the sides and ends of the boiler and the surrounding brickwork, must not be too small, otherwise the current through that external flue will be so rapid, that the flame and heated gas will not remain long enough in contact with the boiler, to communicate its heat to the contained water.

All the different parts of the heating surface of a boiler, ought to be adapted to receive and absorb as much heat as possible, from the flame and heated gas, during its passage through every part of its course, from the furnace to the extreme end of the flue, where it quits the boiler, to escape into the chimney.   The bottom part of the boiler receiving the first and direct action of the flame, will produce the principal part of the supply of steam, but as a very considerable degree of heat will pass on from beneath the bottom, into the surrounding flue, the boiler should expose such an extent of surface in that flue, as will cause almost all the heat to be communicated to the water, by the time that the current reaches the end of the flue, in order that it may pass into the chimney, with only as much remaining heat as is necessary to produce a sufficient draft in the chimney, and through the flues, to cause a vivid combustion of the fire on the grate.

The passage through the flues should not be too narrow, because the draft will be impaired by the resistance that the current of heated gas must encounter, in passing through contracted flues.   Or if a sufficient draft is obtained, it can only be occasioned by an unnecessary loss of heat up the chimney, and even if that sacrifice is made, the motion of the current through the flue being rapid, will not allow time for the proper communication of heat to the boiler.

Dimensions for the fire-grate.   The area of the fire-grate should be propor-tioned to the quantity of coals that is required to be consumed upon it, and also to their quality.   Coals of a bituminous quality, which melt and cake together into masses, and burn with much flame and smoke, must be spread thinner upon the fire-grate, than stone coals which do not melt, and which burn with less flame ; because greater interstices will remain between the pieces of stone coal, for the passage of the air into the body of the fire.   On the other hand, the bituminous coals commonly produce more heat by their combustion than the stone coals, so that a less quantity is required to produce the same effect.

Some kinds of coals are very subject to choke up the fire-grate with masses of half melted matter called clinkers, arising from a semi-vitrification of the ashes or earthy matter of the coals, which remains after combustion ; but which, instead of dropping through the grate as ashes, becomes partially melted by the intense heat of the fire, and forms lumps of coarse glass or slag, which stop up the inter-stices between the bars of the grate, so as to prevent the ashes of the coals from falling through as coals are burnt, and then those ashes also run into clinkers, and increase the evil, until the air cannot obtain sufficient admission through the grate, to cause the fire to burn properly; these kinds of coals are very troublesome to manage, and the spaces between the bars require to be frequently raked out from the underside of the grate with an iron hook. The grate for such coals should be a larger size, than for clear burning coals, in order to diminish the tendency to run into clinkers, and to compensate for the interstices being partly clogged up.

The area of the fire-grate in different cases of Mr. Watt's rotative-engine, will be found to vary between half a square foot, and one square foot to each horse-power; but about two-thirds, or three-fourths of a square foot of fire-grate, to each horse-power, will be found to answer in most cases, if the coals are of a good quality.

If we assume that the consumption of good Newcastle coals, is at the rate of $10\frac{1}{2}$ pounds per hour for each horse-power (see p. 488); and that the fire-grate is three-fourths of a square foot to each horse-power; then the consumption of coals will be at the rate of 14 pounds per hour, upon each square foot of the surface of the grate; that is, one-sixth of a bushel of coals per hour to each square foot of the fire-grate. Hence, a fire-grate of 6 square feet area, which is a proper size for an 8 horse engine, will burn one bushel of coals per hour.

And if 12 square feet of heating surface is allowed to the boiler, for each horse-power, the area of the fire-grate will be one-sixteenth of that of the heating surface of the boiler.

The bars of the fire-grate are usually made of cast-iron about 2 or $2\frac{1}{4}$ inches broad upon the upper edge, and intervals of about $\frac{3}{8}$, or half an inch wide, are left between the adjacent bars, for the passage of the air to the fire; the area of the air-passage is commonly about one-seventh of the whole area of the grate. The under edges of the bars are much narrower than their upper edges, so as to leave the spaces wider below, in order that the ashes may more readily fall through. The ends of the bars are directed towards the fire-door, to enable the fireman to rake out the spaces between the bars. If the grate is shorter than $4\frac{1}{2}$ feet long, it is usual to make the bars in one length; but for grates of more than 5 feet long, two lengths of bars are commonly applied, their ends being supported on a bearing bar which is fixed across beneath the middle of the grate.

Dimensions for the furnace which receives the burning fuel. The clear height from the bars of the grate, to the bottom of the boiler, is usually between 20 and 24 inches at the middle of the arched bottom, and about 11 to 14 inches at each side of the boiler. This is when the grate is horizontal, but the sloping bars are frequently 36 inches beneath the bottom of the boiler, at the back end. The whole width of the arched bottom is exposed to the action of the fire, except 3 or 4 inches at each side, where the boiler rests upon the brickwork. The width of the concave bottom which is exposed to the heat, is commonly about nine-tenths of the whole width of the boiler. The fire-grate is usually the same width as that part of the bottom which is exposed, but if not, the brickwork is carried up sloping above each side of the grate, so as to spread outwards to the proper width.

Dimensions of the flues. The back end of the furnace, or fire-bridge, over which the flame is to pass, is commonly carried up above the grate, to within from 9 to 15 inches of the bottom of the boiler, and is usually made to correspond with the arched bottom. The area of the passage which is thus left for the flame, between the top of the bridge, and the bottom of the boiler, is between one-fourth and one-fifth of the area of the fire-grate, or nearly one-72nd part of the area of the heating surface of the boiler; that is, one-sixth of a square foot to each horse-power.

The area of the passage for the flame beneath the bottom of the boiler, beyond the fire-bridge, is about one-third of the area of the fire-grate, or one-48th of the area of the heating surface of the boiler; that is, one-fourth of a square foot to each horse-power.

The area of the passage through the external flue, around the lower part of the boiler, is usually about one-fifth of the area of the fire-grate; or one-84th of the area of the heating surface; that is, one-seventh of a square foot to each horse-power.

Dimensions of the vertical stack or chimney. The area of the internal passage of the chimney, must be proportioned to the quantity of heated air and gas which is to ascend through it: the area is usually at the rate of one-sixth of the fire-grate, or one-96th of the heating surface of the boiler; that is, one-eighth of a square foot to each horse-power. The chimney for a 50 horse-engine is $2\frac{1}{2}$ feet square = 6·25 square feet × 8 = 50 HP (a). The vertical height of the chimney should be from 60 to 120 feet, between the level of the fire-grate, and the top where the smoke issues. Large boilers which have a great length of passage, for the current of heated air and gas to pass through, require a greater height of chimney to produce an efficient draft through those flues, than smaller boilers.

The above proportions for boilers, fire-grates, furnaces, and flues, are not very precise, and in practice very considerable deviations are made from them; but they are, perhaps, the best standard, and the nearest to Mr. Watt's practice.

Dimensions of a waggon boiler for a 30 horse-engine (b). This boiler, which was constructed by very experienced engineers, will serve as an example of the different proportions which have been already laid down, for the dimensions of boilers.

Diameter of the boiler $5\frac{1}{2}$ feet; length 17 feet; horizontal surface of the water $93\frac{1}{2}$ square feet. The whole height of the boiler is $7\frac{1}{2}$ feet. The arched bottom rises 10 inches, in $4\frac{3}{4}$ feet wide, being the width of that part of the bottom which is exposed to the heat. The seating or part of the bottom which rests upon the brickwork is $4\frac{1}{2}$ inches wide at each side. The sides of the boiler are not vertical planes, but are curved inwards 3 inches from the vertical, at the middle of their height; hence, the width of the boiler across the narrowest part is 5 feet. The water line is level with the axis of the cylinder, and the space reserved for steam is a half cylinder, $5\frac{1}{2}$ feet diameter, and 17 feet long; content 202 cubic feet.

The boiler has a tube through all its length, which serves for an internal flue to convey the flame and heat through the body of the water; the top and bottom of this flue are horizontal flat surfaces, 20 inches wide, and 17 feet long = $28\frac{1}{3}$ square feet. The vertical height of the flue, between its top and bottom, is $2\frac{1}{2}$ feet; the water stands 11 inches deep over the top of the flue; and the bottom of the flue is 6 inches above the crown of the arched bottom of the boiler. The sides of the internal flue are curved outwards, so as to make it 28 inches wide at the middle of its height; and the clear spaces left for water, between the sides of the flue and the sides of the boiler, are 16 inches wide at each side in the narrowest part. The area of the passage for the flame through the internal flue is $4\frac{3}{4}$ square feet.

(a) The chimney of Mr. Watt's 50 horse-engine at the Albion Mills was only 2 feet square. See p. 511.

(b) Messrs. Boulton, Watt, and Co. have made some 36 horse engines, with boilers of the same dimensions as above, being proportioned at the rate of 10 square feet of heating surface to each horse-power. According to the proportion of 12 square feet, which is here adopted, a boiler of that size would be suitable for a 30 horse engine, as stated above. In one instance, a 36 horse engine has only one such boiler, which is made of copper plate; it has been found sufficient to supply the engine during several years, but it has not been overloaded, nor is the water with which the boiler is supplied, liable to deposit any hard incrustation, but only a slight deposit of muddy slime.

Another 36 horse engine of the same dimensions and construction (see Plate XXI.) has two iron boilers of the above dimensions, except that the length is $18\frac{3}{4}$ feet, instead of 17 feet. In all these boilers the fire-grates are 6 feet long, instead of 5 feet, so as to give 27 square feet of grate, instead of $22\frac{1}{2}$; the fire-grates are horizontal.

4 F

The horizontal surface exposed to the heat beneath the arched bottom of the boiler, is about 5 feet wide, by $16\frac{1}{2}$ feet long, $= 82\frac{1}{2}$ square feet; which added to $28\frac{1}{3}$ square feet for the horizontal top of the internal flue, makes $110\frac{5}{6}$ square feet of horizontal surface exposed; the flat bottom of the internal flue is not accounted as heating surface.   The two sides of the internal flue expose about $5\frac{1}{3}$ feet by 17 feet long $= 90\frac{2}{3}$ square feet; and the outside of the boiler exposes a surface of 46 inches high in the external flue, by 44 feet long, round the sides and ends of the boiler $= 168\frac{2}{3}$ square feet, from which deduct $9\frac{1}{2}$ square feet for the area of the two open ends of the internal flue, and it leaves $159\frac{1}{6}$ square feet of vertical surface exposed in the external flue; or $249\frac{5}{6}$ square feet of vertical surface exposed in the whole boiler.

The total surface exposed to the heat, in the whole boiler, is $360\frac{2}{3}$ square feet. The proportion of the horizontal surface is ·307, and of the vertical surface ·693 of the whole surface; this is very near to the proportion of 3 to 7, as before stated.

The fire-grate is $4\frac{1}{2}$ feet wide, with 17 fire bars in the width, each occupying about $3\frac{1}{8}$ inches; the width of the top edges of the bars is $2\frac{1}{2}$ inches, and the intervals between them $\frac{5}{8}$ of an inch wide.   The grate is 5 feet long, in two lengths of bars, of $2\frac{1}{2}$ feet each.   Area of the fire-grate $22\frac{1}{2}$ square feet, which is at the rate of $\frac{3}{4}$ of a square foot to each horse-power.   The front end of the fire-grate is 14 inches below the bottom of the boiler at the sides, and 24 inches at the middle of the arch; the bars slope down backwards at about 20 degrees from the horizontal, so that the far end of the grate is 30 inches below the bottom of the boiler at the sides, and 40 inches at the middle of the arch.

The brick wall at the back end of the grate, which forms the end of the furnace, is inclined backwards, nearly at right angles to the grate, and it rises up to within 13 inches of the bottom of the boiler, to form the fire bridge over which the flame is to pass; the top of the bridge is curved to correspond with the arched bottom of the boiler, leaving a passage of about $5\frac{1}{8}$ square feet for the flame to pass; this is at the rate of one-70th of the whole heating surface of the boiler.   The space beneath the bottom of the boiler beyond the bridge, is $4\frac{3}{4}$ feet wide by 15 inches high at the sides, and 25 inches high at the middle; and the area of the passage thus left for the flame, is about $7\frac{1}{2}$ square feet, or one-48th of the heating surface. The area of the passage through the internal flue is nearly $4\frac{3}{4}$ square feet, which is one-76th of the heating surface.

The area of the passage through the external flue is $4\frac{1}{4}$ square feet, or nearly one-85th of the heating surface.   The space between the sides of the boiler and the brickwork, is about 14 inches wide, to form the external flue; the whole height of that flue is 49 inches, its bottom being 3 inches below the seating of the boiler, so that any soot or ashes which may collect, will not touch the boiler.   The height of the side of the boiler which is exposed in the flue is, as before stated, 46 inches, which is seven-tenths of the width of the boiler.   The upper line of that exposed surface is level with the top of the internal flue, and the surface of the water is 11 inches above the top of the external flue, as well as of the internal flue.

The evaporation from the boiler to produce 30 horse-power, would be about ·6-tenths of a cubic foot per minute; and the horizontal surface of the water being 93·5 square feet, that surface would sink at the rate of ·077 (or one-13th) of an inch in a minute, if no water were supplied to the boiler; hence, it would take (11 inc. × 13 min. =) 143 minutes (or 2 hours 23 minutes) evaporation without feeding, to boil away as much water as would lay the top of the flue quite dry (a).

The flame and heated gas, after passing horizontally beneath the bottom of the boiler, rises up at the end of the boiler, and returns horizontally through the

(a) If this boiler supplied steam for 36 horse-power, it would evaporate ·75 of a cubic foot per minute, whereby the surface would sink at the rate of ·096 of an inch per minute without feeding; or nearly one inch in $10\frac{1}{2}$ minutes, and then it would take 115 minutes, or rather less than 2 hours, to lay the top of the flue dry.

internal flue to the front end of the boiler; then the current divides, and one
half passes along the flue at one side of the boiler, and the other half along the
flue at the other side.   The two currents join again into one, beyond the end of
the boiler; the sliding damper is fitted into the single flue which is formed by the
junction, and that single flue is continued until it reaches the vertical chimney,
which is 2 feet square withinside (= 4 square feet) and 80 feet high.

The current of heated gas is not thus divided into two, in all boilers, but it is
more commonly continued round the boiler in one direction, from the internal flue
at the front end, along one side of the boiler and round beyond the end, so as to
return along the other side of the boiler.   As the current must then pass through
a longer course, it is supposed to communicate more of its heat to the water within
the boiler; but the long passage is an impediment to the draft, and if the chimney
is situated at the farther end of the boiler from the furnace, another returning
flue must be made, in the brickwork, all the length of the boiler to lead to the
chimney.   On the other hand, when the current is divided between the two flues,
it moves slower through them, so as to have more time to impart its heat to the
boiler; and the passage being shorter, as well as more open, the draft is not impaired.   It is not decided which of the two methods answers best, but both plans
are in use; and when the chimney is situated at the front end of the boiler near to
the furnace, so that there is no unnecessary length of flue, the continuous flue
answers as well as the divided flue.

Boilers are frequently made according to the above proportions and dimensions, but without internal flues through the water; such a boiler is shown in
Plates XI and XII, see p. 495; and supposing the same extent of surface to be
exposed to the heat, the effect of a simple boiler is very nearly the same as that
of a boiler with an internal flue.   In the simple boiler, the current through the
external flue is always carried round the outside of the boiler in one direction, and
never divided into two currents, to pass one on each side of the boiler.

Simple boilers without internal flues are the least expensive in construction,
and the most durable, because there is a considerable depth of water over those
parts of the surface which are most exposed to the heat, and that surface must
be always well covered with water; but the internal flue having only a small depth
of water over the top of it, the weight or columnar pressure of that water is less
able to keep it in close contact with the heated surface; and from the small depth
of water it is more liable to be injured by negligence in feeding the boiler.   The
internal flues through boilers commonly wear out before the other parts.

When any part of the metal surface of the boiler is strongly heated beneath,
and covered with water above, so as to produce steam very rapidly; the steam is
formed in large bubbles, which rise up from the heated surface, through the whole
depth of the water; the steam must necessarily displace the water from contact
with the heated surface, every time that a bubble of steam is evolved; but when the
bubble rises, the water returns upon the surface, by its own weight or columnar
pressure, and the greater that pressure is, the quicker the water will return.   This
circumstance is of some consequence, when the evolution of steam is very rapid,
for then the bubbles of steam follow each other so quickly, that the water is driven
off from contact with the surface, during a considerable portion of the whole time,
and consequently, the oxydation of the metal must be more rapid than it ought to
be; this is more particularly the case in iron boilers.

As boilers with internal flues contain less water than simple boilers, in proportion to the surface exposed, less time and fuel is consumed to heat the water,
and get up the steam in the first instance; and in those cases, when the fire

must be lighted every morning, in order to perform a short day's work, the saving of a part of the loss in heating the water, may be an advantage; but when the boiler is regularly working, it is an immaterial circumstance how much water the boiler contains, provided that it is full to the proper level.   It is an advantage to have a great mass of heated water in a boiler, because that water serves as a recipient to take up heat, by increasing its temperature when the fire burns more rapidly than ordinary; and the water so heated, by lowering its temperature again, will give out that heat to the steam, when the fire slackens.   If the mass of water is small, it cannot absorb so much heat, and consequently a greater part will escape up the chimney, whenever the fire glows strongly.

In waggon boilers, which are constructed according to the above proportions, the surface which is exposed to the heat may be computed in square feet, by the following rule.

RULE.   Multiply the length of the boiler in feet, by the breadth in feet, and the product is the horizontal surface of the water in square feet; this being multiplied by 2·3, if it is a simple boiler without a tube (or by 3·5, if it is a boiler with an internal tube), the product will be the surface exposed by the bottom and sides in square feet. To this add the surface exposed by the two ends, which may be found in square feet, by multiplying the square of the breadth of the boiler in feet by 1·4; the sum will be the whole of the heating surface; but if there is an internal tube, one-39th of that sum must be deducted, to allow for the area of the two open ends of the internal tube, which is wanting at each end of the boiler.

Example.   A waggon boiler 17 feet long, by $5\frac{1}{2}$ feet diameter, with an internal flue through it; thus 17 ft. × 5·5 ft. = 93·5 square feet of water surface × 3·5 = 327·25 squ. ft.; to which add (5·5 squared, or 30·25 × 1·4 =) 42·35 square feet for the two ends = 369·6 squ. ft.; from which, if we deduct one-39th part, or $9\frac{1}{2}$ squ. ft., for twice the area of the internal flue, we have 360·1 square feet of surface exposed, the same as by the former computation.

If the same boiler had been without the internal flue; then 17 ft. × 5·5 ft. = 93·5 square feet × 2·3 = 215·05 square feet + (5·5 × 5·5 = 30·25 × 1·4 =) 42·35 square feet for the ends = 257·4 square feet of surface exposed.   Or (÷ 12 squ. feet =) 21·45 horse-power.

Note.   Instead of multiplying by 3·5, the same result will be obtained by dividing by its reciprocal ·2857; and this can be conveniently done by the sliding rule.   The reciprocal of 1·4 is ·7143. And for simple boilers without tubes, the reciprocal of 2·3, viz. ·4348. may be used as a divisor, thus.

Sliding Rule, $\left\{ \begin{array}{l} \text{A} \\ \text{Ɔ} \end{array} \right.$ Width of boiler ft.   ·286 $\quad$ Ex. $\dfrac{\text{A } 5\cdot5 \text{ ft. wide.} \qquad ·286}{\text{Ɔ } 17 \text{ ft. long.} \quad 327\,\text{sq. ft.}}$
slide inverted.   Length of boiler ft.   Surf. bott. and sides sq. ft.

Note.   When the slider is thus set, the horizontal surface of the water in square feet may be found opposite to 1 on either of the lines; viz. in the above example it is 93·5.   The surface of both the ends of the boilers may be computed in a similar manner, as follows, and the result added to that of the former calculation; then the sum being multiplied by 39, and the product divided by 40, the quotient will be the correct surface exposed.

Sliding Rule, $\left\{ \begin{array}{l} \text{A} \\ \text{Ɔ} \end{array} \right.$ Width of boiler ft.   ·714 $\quad$ Ex. $\dfrac{\text{A } 5\cdot5 \text{ ft. wide.} \qquad ·714.}{\text{Ɔ } 5\cdot5 \text{ ft. wide.} \quad 42\cdot4 \text{ sq. ft.}}$
slide inverted.   Width of boiler ft.   Surf. of ends sq. ft.

Note.   These rules can only apply to boilers constructed according to the above proportions; the width of the bottom exposed to the heat being ·9 tenths of the whole width of the boiler; the height of the boiler exposed in the external flue, being ·7 tenths of the whole width; and the proportions of the internal flue being as above.   In simple boilers constructed on this proportion without internal flues, the heating surface of the boiler is nearly 2·75 times the horizontal surface of the water; and allowing 12 square feet of heating surface to a horse-power, that horizontal surface of water must be at the rate of (12 square feet ÷ 2·75 =) 4·36 square feet to each horse-power.   The proportion of the horizontal surface of water in the boiler is used as a mode of computation by some engineers, but as it is only comparative, it can only be applicable when the length, breadth, and depth of the boilers for different engines are made upon the same proportion (*a*).

(*a*) Mr. Hick of Bolton, is accustomed to proportion his boilers at the rate of $5\frac{1}{7}$ square feet of the horizontal surface of water to each horse-power.   The boilers being without internal tubes, and nearly of the proportion above stated, the heating surface is nearly 15 square feet to each horse-power, instead of 12, see Note, p. 581. For 5·5 squ. ft. × 2·75 times = 15·1 square feet.

Note. If a large boiler is applied to an engine, in order to raise steam with great facility from an extended surface, or if a smaller boiler is applied, in order to save room and expense, in either case, the dimensions of the fire-grate, the furnace, the flues, and the chimney, ought to be nearly the same; because those parts should be proportioned to the power that the engine is required to exert, and not according to the heating surface of the boiler; in fact, they should be according to the quantity of fuel that is to be consumed, and the object of giving a large surface to the boiler, is not to consume a greater quantity of fuel, but to raise the requisite quantity of steam with less fuel; and if a boiler with a small surface is applied to an engine, it will require rather more fuel instead of less.

The preceding rules and proportions respecting the furnaces and flues of boilers, are all adapted to an allowance of 12 square feet of heating surface to each horse-power (see p. 581), and whenever that proportion is departed from, the dimensions of the fire-grate, flues, and chimney, should be proportioned to the horse-power, and not to the heating surface of the boiler, because that admits of a greater or lesser proportion, which the other parts do not, in the same degree.

The external surface of the upper parts of all boilers should be covered over with a non-conducting substance, to prevent waste of heat. It is usual to cover boilers with brickwork, but some proprietors of engines are of opinion, that boilers decay more rapidly when they are thus covered; and they prefer to leave the metal exposed to the air, because, if any steam escapes through minute leaks in the upper part of the boiler, it becomes condensed into water, which is retained by the brickwork so as to oxydate the metal. If boilers are well made they are not liable to such leaks; and the economy of a covering is very sensible.

An excellent plan of covering boilers is to place semicircular hoops of cast iron, over the semicylindrical top of the boiler, at distances of about four feet apart, but they need not touch the boiler; a planking of fir-wood is applied upon the iron semicircles, so as to form a wooden semicylinder, which incloses the boiler, leaving a vacant space of three or four inches around the metal for air. If the wood is covered over with brickwork, or paving stones, scarcely any heat will be lost, and such boilers may be exposed to the open air, without any roof over them, excepting a shed over the fire-place, to shelter the fireman.

The steam-pipe should be joined to the boiler at the highest part of it, so that the orifice of the pipe will be situated at the greatest height above the surface of the water, in order to avoid any chance of the water being carried into the pipe by the violent agitation of the boiling; and, for the same reason, the pipe should rise up perpendicular to some height, before it turns horizontally to go to the cylinder. It is usual to place the steam-pipe with a slope of about one inch in a foot, from the horizontal; the lowest end being towards the boiler, that the water, which is condensed from the steam, may drain back to the boiler. The steam-pipe should be clothed by winding haybands very tight round it, and then covering those bands with canvas, which may be painted white on the outside, and it will retain the heat very completely.

Dimensions of the steam-pipe. Its internal diameter is usually rather more than one-fifth of the diameter of the steam cylinder, so that its area is about one-23rd of that of the cylinder. The size of the steam-pipe ought to be proportioned to the quantity of steam that is required to pass through it.

Taking a 20 horse-engine for a standard, the cylinder being (24 inches diameter =) 576 circular inches ÷ 20 HP, gives 28·8 circular inches of the piston to a horse-power. The steam-pipe is 5 inches diameter; its area being one-23rd of 576 = 25 circular inches. This is at the rate of $1\frac{1}{4}$ circular inches (= ·982 of a square inch) to each horse-power. Or reciprocally ·8 tenths of a horse-power to each circular inch of the steam-pipe. And allowing 12 square feet of heating surface of the boiler to each horse-power, then the area of the steam-pipe should be at the rate of one circular inch to (12 ÷ 1·25 =) 9·6 square feet of heating surface.

To find the internal diameter for the steam-pipe to convey the steam from the boiler. Having given, either the power that the steam is required to exert in horse-power. Or the extent of surface that the boiler exposes to the heat, in square feet. Or else the number of cylindrical inch feet of space that is occupied per minute by the motion of the piston; the latter is obtained by multiplying the square of the diameter of the cylinder in inches, by the motion of the piston per minute, in feet.

RULE. Divide the horse-power that the steam is required to exert, by ·8 (or the square feet of heating surface in the boiler by 9·6); or the cylindrical inch feet of space per minute in the cylinder, by 4840. The quotient is the proper area of the passage through the steam-pipe in circular inches; and the square root of that quotient, is the proper internal diameter for the steam-pipe in inches.

Example. For 20 horse-power ÷ ·8 = 25 circular inches for the area of the steam-pipe; (Or 240 square feet of heating surface ÷ 9·6 = 25 circular inches,) the square root of which is 5 inches, for the internal diameter of the pipe. Or, by the other method: the cylinder being 23·72 inches diameter, which squared is 562·8 circular inches, × 215 feet motion of the piston per minute = 121 000 cylindrical inch feet of space occupied per minute; this divided by 4840, as above directed, gives 25 circular inches, for the area of the steam-pipe, the same as before.

Sliding rule. $\begin{cases} \dfrac{C \quad 80 \quad \text{Horse-power.}}{D \quad 10 \quad \text{Diam. of steam-pipe inc.}} \end{cases}$ *Exa.* $\dfrac{C \quad 80 \quad 20 \text{ HP}}{D \quad 10 \quad 5 \text{ inc. dia.}}$ or $\dfrac{50 \text{ HP}}{7·9 \text{ inc. dia.}}$

The rule thus set forms a complete table of the steam-pipes of any horse-power. Or, if the rule is set as follows, it will become a table of the steam-pipes for any extent of heating surface.

Sliding Rule. $\begin{cases} \dfrac{C \quad 960 \quad \text{Heating surface sq. ft.}}{D \quad 10 \quad \text{Dia. of steam-pipe inc.}} \end{cases}$ *Ex.* $\dfrac{C \quad 960 \quad 240 \text{ sq. ft.}}{D \quad 10 \quad 5 \text{ inc. dia.}}$ or $\dfrac{600 \text{ sq. ft.}}{7·9 \text{ inc. dia.}}$

Sliding Rule, slide inverted. $\begin{cases} \dfrac{\text{q Mot. of pist. ft. per min.} \quad \text{Divisor 4840.}}{D \quad \text{Dia. of cylind. inc.} \quad \text{Dia. of steam-pipe inc.}} \end{cases}$ *Ex.* $\dfrac{\text{q } 215 \text{ ft. per min.} \quad 484}{D \quad 23.72 \text{ inc. dia.} \quad 5 \text{ inc. dia.}}$

Allowing 33 cubic feet of steam, equal to the atmosphere, per minute for each horse-power, or 6050 cylindrical inch feet; and 1¼ circular inches of area in the steam-pipe, being allowed for each horse-power, the velocity with which the steam must pass through the pipe would be (6050 ÷ 1·25 =) 4840 feet per minute, or (÷ 60 =) 80·66 feet per second, taking a mean of all the variations of velocity throughout all the stroke of the piston. When the piston is near the middle of its stroke, the velocity must be (80·66 × 1·57 =) 126·6 feet per second, on the principle already stated p. 416 and p. 490.

When an engine has more than one boiler, a stop-valve must be provided at the top of the upright steam-pipe which rises from each boiler, in order to shut off the communication from that boiler with the sloping steam-pipe at pleasure. The aperture of such stop-valve should be fully equal to that of the upright steam-pipe, which is to convey the steam from the same boiler.

If two or three boilers are to be worked at once, the upright steam-pipe from each boiler, and the stop-valve at the top of each pipe, must be proportioned to the heating surface of each boiler respectively; and the area of the common steam-pipe, which is to convey the steam from all the boilers to the engine, must be proportioned to the sum of the heating surfaces of all those boilers.

The safety-valve, or discharge valve. As all the steam which the boiler produces must be occasionally discharged through this valve, it would appear reasonable to make it the same size as the other steam valves of the engine. In practice, safety-valves are not made above half that size, and are thought to be sufficient for the intended purpose.

There is no rule established for the size of safety-valves, but it will be sufficient to allow ·8 tenths of a circular inch to each horse-power; that is 1·25 horse-power to each circular inch; or 15 square feet of heating surface to each circular inch of the aperture of the safety valve.

Sliding Rule. $\begin{cases} \dfrac{C \quad 1·25 \quad \text{Horse-power.}}{D \quad 1 \quad \text{Diam. of safety-valve inc.}} \end{cases}$ *Ex.* $\dfrac{C \quad 1·25 \quad 20 \text{ HP}}{D \quad 1 \quad 4 \text{ inc. dia.}}$ or $\dfrac{50 \text{ HP}}{6·33 \text{ inc. dia.}}$

Sliding Rule. $\begin{cases} \dfrac{C \quad 15 \quad \text{Heating surface squ. ft.}}{D \quad 1 \quad \text{Diam. of safety-valve inc.}} \end{cases}$ *Ex.* $\dfrac{C \quad 15 \quad 240 \text{ sq. ft.}}{D \quad 1 \quad 4 \text{ inc. dia.}}$ or $\dfrac{600 \text{ sq. ft.}}{6·33 \text{ inc. dia,}}$

Sliding Rule, slide inverted. $\begin{cases} \dfrac{\text{q Motion of pist. ft. per min.} \quad \text{Divisor 7562.}}{D \quad \text{Dia. of cylind. inches.} \quad \text{Dia. of safety-valve inc.}} \end{cases}$ *Ex.* $\dfrac{\text{q } 215 \text{ feet.} \quad 756}{D \quad 23.72 \text{ dia.} \quad 4 \text{ inc. dia,}}$

The load upon the safety-valve should be at the rate of about $2\frac{1}{2}$ pounds to each circular inch of the area of the passage which is closed by the valve. Hence, if the square of the diameter of that passage in inches, is multiplied by 2·5, the product will be the proper weight in pounds, for the valve, together with the load which is to be applied upon it.

The surface of the water in the feeding cistern, usually stands 8 feet above the level of the surface of the water in the boiler; and consequently, if the steam were to accumulate in the boiler to a greater elasticity than is equal to the pressure of a column of water 8 feet high, then the water would be forced out at the top of the feeding cistern. The safety-valve should be regulated so as to open and discharge the steam, before it can acquire that force, which it will do when proportioned as above, for $2\frac{1}{2}$ pounds per circular inch is (1·273 =) 3·183 pounds per square inch; or equal to a column of mercury $6\frac{1}{2}$ inches high; or to a column of water $7\frac{1}{3}$ feet high.

The safety valve is accurately ground, and fitted very correctly into its seat, so that the border or edge of the valve is in close contact with the seat; when the surfaces so in contact, are quite dry, without any moisture between them, the pressure of the steam will only act against that part of the under surface of the valve which is exposed in the open aperture beneath. But after the same valve has been lifted up, to allow steam to pass by it, an exceedingly thin film of condensed water, or moisture, will remain interposed between the surfaces, when the valve is shut down again; and this film will be sufficient to transmit the pressure of the steam throughout all the surface in contact, so as to cause the pressure of the steam to act beneath the whole surface of the valve, including its margin or edge, which is in contact with the seat. From this cause a safety-valve will remain closed, as though it adhered to its seat, until the steam acquires sufficient strength to force it open, by acting beneath its under surface, or merely against the interior circle within the fitting part; but after the valve has once been lifted, it appears to rise more readily; for it will not retain the steam to so great a strength as in the first instance, because the steam acts against all the surface of the exterior circle, to the outside of the fitting part.

Dimensions of the working valves. Mr. Watt's practice was to make the apertures of the circular lifting valves one-fifth of the diameter of the cylinder; the area of the passage being about one-27th of that of the cylinder, as already stated, p. 468. This is not a correct mode of proportioning the valves, because the size of their passages ought to be according to the quantity of steam which is to pass through them, without regard to the size of the cylinder. It appears, from an examination of several of Mr. Watt's own engines, that, on an average, the area of the passages through the valves is nearly ·8-tenths of a square inch to each horse power, see p. 470. This is a larger passage than is requisite to admit steam into the cylinder; but it is not so large as is desirable, for the passage through the exhausting valves.

The induction passages to admit the steam into the cylinder will be almost as large as Mr. Watt's proportion, if 1 circular inch (= ·785 of a circular inch) is allowed to each horse-power, then the square root of the number of horse-power that the engine is to exert, will be the proper diameter for the aperture of the steam-valves in inches.

Sliding Rule. $\left\{\begin{array}{ll} \text{C} & 4 \\ \text{D} & 2 \end{array}\right.$ $\dfrac{\text{Horse-power.}}{\text{Diam. of steam passage inc.}}$ *Exa.* $\begin{array}{ll} \text{C} & 4 \\ \text{D} & 2 \end{array}$ $\dfrac{20\text{ HP.}}{4\cdot47\text{ inc. dia.}}$ or $\dfrac{50\text{ HP.}}{7\cdot07\text{ inc. dia.}}$

Sliding Rule, $\left\{\begin{array}{ll} \text{日} \\ \text{D} \end{array}\right.$ slide inverted. $\dfrac{\text{Motion of pist. ft. per min.}}{\text{Diam. of cylinder inc.}}$ $\dfrac{\text{Divisor 6050.}}{\text{Dia. of steam passage inc.}}$ *Ex.* $\begin{array}{l} \text{日} \\ \text{D} \end{array}$ $\dfrac{215\text{ feet.}}{23\cdot72\text{ dia.}}$ $\dfrac{605.}{4\cdot47\text{ dia.}}$

It was Mr. Watt's practice to make the eduction passages of the same size as the induction passages, but it would be better to make the former a larger proportion, as follows.

The eduction passages to exhaust the steam from the cylinder should be at the rate of 1·515 circular inches (= 1·19 square inches) to each horse-power. Or reciprocally, ·66 of a horse-power to each circular inch. The latter number may be used as a divisor for the horse-power, to find the area of the exhausting apertures, in circular inches. Thus 20 HP. ÷ ·66 = 30·3 circular inches; the square root of which is 5·5 inches diameter.

$$\text{Sliding Rule.} \begin{cases} \text{C } 66 \quad\quad \text{Horse-power.} \\ \text{D } 10 \quad\quad \text{Dia. of eduction passage inc.} \end{cases} Ex \quad \frac{\text{C } 66 \quad 20 \text{ HP.}}{\text{D } 10 \quad 5\cdot5 \text{ inc. dia.}} \text{ or } \frac{50 \text{ HP.}}{8\cdot7 \text{ inc. dia.}}$$

$$\begin{matrix}\text{Sliding Rule,} \\ \text{slide inverted.}\end{matrix} \begin{cases} \text{ꓭ Motion of pist. ft. per min.} & \text{Divisor 4000.} \\ \text{D } \text{Diam. of cylinder inc.} & \text{Dia. of eduction passage inc.} \end{cases} Ex. \frac{\text{ꓭ } 215 \text{ feet.} \quad\quad 4.}{\text{D } 23\cdot72 \text{ dia.} \quad 5\cdot5 \text{ dia.}}$$

The steam-pipes and the eduction-pipes should be rather larger than the above proportions, which are for the uninterrupted apertures through the valves. The valves should be lifted up rather more than one-fourth of their diameters, in order to open the passage fully. See p. 374.

Note. The preceding computation for the heating surface of the boiler, and the sizes of the different passages for the steam, are proportioned to the quantity of steam; and when the term horse-power is used in those rules, it is supposed that it has been correctly calculated, according to the rule of 33 cubic feet of space in the cylinder, to each horse-power, see p. 575. In cases where there is any doubt of the horse-power being correctly stated, the rules which proceed by the diameter of the cylinder, and the motion of the piston, should be preferred.

Dimensions of the air-pump. The usual proportion in Mr. Watt's double-acting engines is, for the diameter of the air-pump to be two-thirds of the diameter of the cylinder, that is, four-ninths of the area; but the motion of the air-pump bucket being only half as great as that of the piston, the capacity of the air-pump will be only four-18ths of that of the cylinder: and again, as the cylinder receives steam both in the ascent and descent of the piston, but the air-pump only exhausts when its bucket is drawn up, the effective capacity of the pump will be only four-36ths, or one-ninth of that of the cylinder.

The area of the passage through the foot-valve is about one-fourth of the area of the air-pump, and the apertures through the two valves in the air-pump bucket, are nearly the same proportion. The discharge-valve is usually made rather larger than one-fourth of the area of the pump. The structure of the air-pump is exactly the same for the double engine as for the single engine. See p. 379.

Dimensions of the cold water-pump. Mr. Watt's practice was, to make the actual capacity of the barrel of the cold water-pump one-24th part of the capacity of the steam cylinder; reckoning those capacities according to the motion of the piston, and of the pump-bucket, respectively, without allowing for any vacant spaces. The cylinder of the double engine expends steam continually, when its piston is descending, as well as in ascending, but the pump only raises water during the ascent of its bucket, hence the effective capacity of the pump to raise water is one-48th of the effective capacity of the cylinder to expend steam.

To find the proper diameter for the barrel of the cold water-pump of Mr. Watt's double engine. Having given, the diameter of the steam cylinder in inches; the distance in inches from the centre of the great lever, to the joint by which the piston is suspended; and the distance in inches, from that centre, to the joint by which the bucket of the pump is suspended.

RULE. Divide the square of the diameter of the cylinder in inches by 24; multiply the quotient by the distance in inches at which the piston is suspended; and divide the product by the distance in inches, at which the pump-bucket is suspended. The square root of the last quotient is the proper diameter for the pump barrel in inches.

Example. Suppose a 20 horse-engine, with a cylinder 24 inches diameter. The piston being suspended at 96 inches from the centre of the great lever, and the pump-bucket being suspended at 40 inches from the same centre. Then 24 inc. squared = 576 circular inches ÷ 24 = 24 × 96 inches radius = 2304 ÷ 40 inches radius, gives 57·6 circular inches for the area of the pump barrel; and the square root of 57·6 is 7·59 inches for the diameter of the pump barrel.

To perform this calculation by the sliding rule, the radius or distance at which the piston is suspended, must be previously divided by 24. Thus, for the above example, 96 inc. ÷ 24 = 4.

Sliding Rule, $\left\{\begin{matrix} \text{g} & \text{one-24th of rad. of pist. inc. Rad. of cold pump inc.} \\ \text{D} & \text{Diam. of cylinder inc. Dia. of cold pump inc.} \end{matrix}\right.$ $Ex.$ $\dfrac{\text{g} \quad (96 \div 24 =) 4 \quad \text{rad. 40 inc.}}{\text{D} \quad \text{Diam. 24} \quad \text{dia. 7.59 in.}}$

Allowing the vacant spaces at the top and bottom of the cylinder, through which the piston does not pass, to be one-tenth of the capacity of the space which is occupied by the piston in its motion; then the quantity of steam would be $(48 + \frac{1}{10} = 4{\cdot}8 =)$ 52·8 times the quantity of cold water; also allowing that the pump loses one-20th of the capacity of its barrel, by the return of the water before the valves can shut to retain it, we shall have $(52{\cdot}8 + \frac{1}{20} = 2{\cdot}64 =)$ 55·44 measures of steam to one measure of cold water.

According to Mr. Watt's allowance, (see p. 376), the quantity of cold water to be injected into the condenser would be one-60th part, by measure, of the quantity of steam actually expended; and as the pump raises a little more than that proportion into the condensing cistern, a small surplus of water will run waste from it.

The cold water-pump being made to the above proportion, will be adapted to supply the engine properly, when it is exerting its standard power, and neither more nor less than that. When the engine is underloaded, the injection cock will be partially closed, so as not to inject all the water that the pump raises. But if the engine is overloaded, then the pump will not supply as much water as would be necessary to keep the condenser properly cool.

In cases where a sufficiency of cold water can be obtained at a moderate depth, it is expedient to make the cold water-pump larger than the above proportion, in anticipation of supplying the engine properly when it becomes overloaded (*a*); but if the cold water must be pumped up from a considerable depth, it would not be advisable to make the pump larger than necessary at first, but rather exchange the pump barrel and bucket for a larger size, when the overload is applied to the engine. The above rule will give the size for the cold water pump according to any required proportion, by using a proper divisor to express that proportion instead of the divisor 24.

Quantity of cold water required to supply Mr. Watt's double engine.

The capacity of the cold water-pump being one-48th of the capacity of the steam cylinder, and assuming the loss of water through the pump valves to be one-20th, then the quantity of cold water raised by the pump would fill $(48 + \frac{1}{20} = 2{\cdot}4 =)$ one-50·4th part of the space occupied by the piston in its motion. That space being at the rate of 33 cubic feet per minute to each horse-power, the quantity of cold water would be $(33 \div 50{\cdot}4 =)$ ·655, or, we may say, two-thirds of a cubic foot per minute for each horse-power (*b*); that is, 40 cubic feet per hour.

The aperture through the injection cock must be adapted to admit as much cold water into the condenser, as is equal to one-60th, by measure, of the quantity of steam actually expended by the engine. According to Mr. Watt's proportion,

(*a*) Mr. Hick, of Bolton, usually makes the cold water-pumps for small engines, under 20 horse-power, at the rate of one-36th of the effective capacity of the cylinder instead of one-48th; this enables an engine to work with advantage when it is considerably overloaded.

(*b*) It is said that Mr. Watt's allowance for cold water to his engines was at the rate of $3\frac{1}{2}$ ale gallons (of 282 cubic inches each) per minute to each horse-power, that is, ·571 of a cubic foot; this is rather less than the above computation, but probably it is the quantity of cold water to be actually injected into the condenser.

see p. 376, the quantity of cold water injected into the condenser must be 28·9 times as much as is evaporated from the boiler; and that quantity being taken at ·021 of a cubic foot per minute, for each horse-power (see p. 578), the quantity of cold water injected ought to be (28·9 × ·021 =) ·607 of a cubic foot per minute to each horse-power actually exerted by the engine; that is, equal to (·607 × 144 =) 87·4 square inch feet, or prisms of water one inch square and one foot long.

The cold water is forced through the injection cock into the condenser, by the pressure of the atmosphere, which exceeds the elasticity of the steam, or vapour, within the condenser, so much as to be capable of sustaining a column of mercury, of at least 26 inches high, in the barometer, which is connected with the condenser. Taking 26¼ inches of mercury for the usual state of exhaustion, it is equal to a column of water (26·5 × 1·129 =) 29·92 feet high; and that may be assumed for the pressure which is to urge the cold water through the passage of the injection cock.

A heavy body falling through a height of 29·92 feet, would acquire a velocity of 43·9 feet per second (see p. 23); thus the square root of 29·92 = 5·47 × 8·021 = 43·87. It has been determined, by experiments on the actual efflux of water through narrow apertures, that it commonly moves with a velocity of about ·65 of the velocity that a body would acquire by falling through the depth that the aperture is beneath the surface of the water (see p. 572). Whence we may conclude, that the water actually passes through the aperture of the injection cock with a velocity of (43·9 × ·65 =) 28·53 feet per second, or 1712 feet per minute; and the quantity of water injected will be at the rate of (1712 ÷ 144 =) 11·89, say 12 cubic feet per minute, through each square inch of the aperture. The injection cock should be set, so as to open a passage equal to (1712 ÷ 87·4 =) one-19·6th, or, we may say, one-20th of a square inch for each horse-power that the engine is actually exerting. *Proof.* If each square inch admits 11·89 cubic feet per minute to be injected, and if one-19·6th of a square inch is allowed to each horse-power; then the quantity injected would be (11·89 ÷ 19·6 =) ·607 of a cubic foot per minute for each horse-power.

When the injection cock is set quite open, the area of its aperture should be at the rate of one-15th of a square inch for each horse-power that the engine is rated at; that is supposing the capacity of the cold water-pump to be one-48th of that of the cylinder; but if the pump is larger than that proportion, then the cock should also be larger, because it should in all cases be capable of injecting very nearly as much water as the pump can supply (a).

PROPORTIONS FOR THE LENGTHS OF THE PRINCIPAL MOVING PARTS OF MR. WATT'S ROTATIVE ENGINE.

The length of the stroke for the piston, and the rapidity of its motion. There is no settled proportion between the diameter of the cylinder and the length of the stroke, as may be observed by referring to the table, p. 574. In the 20 horse-engine, which is usually considered as a standard, the length of the stroke (5 feet) is 2½ times the diameter of the cylinder, 24 inches. The 12 horse, 16 horse, 26 horse, 30 horse, and 45 horse-engines in the table, are nearly after the same proportion; but in others it must necessarily be different, because there are three or more sizes with the same length of stroke; in the large engines, above 50 horse-power, the stroke is shorter than 2½ times the diameter of the cylinder.

The proportion between the diameter of the cylinder, and the length of the stroke, is not of much consequence, for if the piston makes a shorter stroke, it is expected to make a greater number of strokes in the same time, in order to compensate for that deficiency; but according to the established practice, the space through which the piston passes in a given time, is greater in those engines which make a long stroke, than in those which make a shorter stroke.

It is advantageous to cause the piston to act with a rapid motion, because in

(a) The author is not aware what size Mr. Watt gave to the apertures for the injection cock of his engines. The above proportion is merely deduced from the investigation, and not from observation. The author has found the apertures in some engines considerably less than is directed by the above rule, but as those engines are deficient in injection, they are not proper examples to form a rule from. For instance, an engine exerting nearly 80 horse-power has a circular aperture of two inches diameter; this is only one-20th of a circular inch per horse-power.

order to realize the same mechanical power in a given time, a smaller force will be required to be exerted by the piston, in proportion as it acts more rapidly, and consequently all the moving parts which are to transmit that force may be made slighter, so as to be more easily moved, having less friction, and they will be less liable to be broken by accident.   To explain this, we may suppose that the 10 horse-engine, which makes 25 strokes per minute, of four feet long (see p. 489), were to be worked at the rate of 30 strokes per minute, it would then exert 12 horse-power, without any extra force, or strain on the parts; and the rotatory motion of the fly-wheel would be more regular, because it would have more energy in proportion as the square of 25 ($= 625$) is to the square of 30 ($= 900$), that is 1·44 times, whilst the power exerted by the engine during each half stroke would remain the same.

The rapidity with which the piston can be made to act advantageously in practice, is limited by the promptitude with which the steam can be exhausted from the cylinder; for if the steam cannot make its escape in time, to exhaust the cylinder very completely, before the piston has made much progress in its stroke, then the effective force exerted by the piston, will be lessened by the negative pressure of the unexhausted steam; and this evil will be greater, as the piston moves with a greater speed (see p. 469).

The quickness with which the cylinder can be exhausted, depends upon the size of the apertures through which the steam must pass, compared with the quantity of steam which is to be exhausted.   If these apertures were made sufficiently large (see p. 470 and 592), and if the valves were opened very suddenly, the piston might be made to act with a much greater velocity than is now practised, without occasioning any deficiency in the effective force exerted by the piston.   In that case, the performance of the engine would be improved, by giving such an increased speed of the piston, because a smaller sized engine would exert the same power. If the exhausting-valves are proportioned according to Mr. Watt's scale, of about one-27th of the area of the cylinder, see p. 468, then the motion of the piston cannot advantageously be made greater than about 220 feet per minute (a); because the steam cannot be exhausted from the cylinder through such valves, with sufficient promptitude to admit of a quicker motion of the piston, without sensibly impairing its effective force, in consequence of the negative pressure of the unexhausted steam.

It is advantageous to proportion an engine with as great a length of stroke for the piston as can be conveniently used, because a greater speed of the piston can be attained, with fewer reciprocations, and the waste of steam will be less; because the spaces which must be left vacant at the top and at the bottom of the cylinder, where the piston does not pass, and for the passages to the valves, must be nearly the same whether the stroke is shorter, or longer, provided that the cylinder is of the same diameter; but that waste will be more frequently repeated when the engine makes a greater number of strokes in a given time.

On the other hand, if an engine is proportioned with a short length of stroke for the piston, it will occupy less room, and will be less expensive in the first

(a) This is the only point of importance, in which Mr. Watt's proportions appear to admit of improvement; he fixed the size for the valves, according to his experience, with his single engines for pumping water, see p. 373; and he afterwards adopted the same proportion for the valves of his double rotative engine, see p. 468.   It would have been better to have given them a much larger proportion, because the piston of a rotative engine must act with a much quicker motion, than that of a pumping engine, in order to produce a regular motion of the fly-wheel, and it would be advantageous to make the motion still more rapid than the present practice, for the reason above stated, provided that it can be done without diminishing the effective force of the piston.

construction; because the magnitude of many of the parts must be proportioned according to the length of the stroke, and will therefore require less materials for their construction when the stroke is shorter. Another advantage of a short stroke, and a quick repetition is, that the rotatory motion will have a less tendency to irregularity, in consequence of the impulse upon the crank being more frequently repeated, so that a less power in the fly-wheel will render the motion sufficiently regular for turning mills.

Taking all these different circumstances into consideration, it will be found that the lengths of the strokes for the pistons of cylinders of different diameters, has been very judiciously chosen by Mr. Watt, as laid down in the table, p. 574, except that it might be better to give the 80 horse and 100 horse-engines a stroke of nine feet. It would be an advantageous improvement of Mr. Watt's scale, to apply larger exhausting passages, as recommended p. 592, and to cause the pistons to make a greater number of strokes per minute, so that each engine would become more powerful, in proportion to its increased speed of action, and its motion more regular, even with a less power of fly-wheel.

The length of the great lever should be rather more than three times the length of the stroke of the piston; the centre of motion being in the middle of the length between the centres of the joints at each end of the lever, so that both those joints may have an equal motion. It is usual to make the horizontal distance between the vertical centre line of the cylinder (or of the piston rod) and that of the centre of the axis of the crank, exactly three times the length of the stroke of the piston, which is usually a whole number of feet; because as that distance must be set out by the masons, in building the engine-house, before the engine is set up, it will be more likely to be correctly measured, if it is some certain number of feet, than if there were fractions. For instance, if the piston makes a six feet stroke, the vertical centre line of the crank axis, should be 18 feet horizontal distance from the vertical centre line of the piston rod.

The length of the great lever, or the distance between the centres of the joints at each end of it, must be greater than the horizontal distance between the vertical centre lines of the piston rod, and of the crank; because as the joints describe an arch of a circle in their motion, they will deviate sensibly from those vertical lines. The length of the stroke, is the chord of that arch, and the deviation is its versed sine; the deviation is termed the *vibration* of the great lever. The vertical centre line of the piston rod must bisect that versed sine or vibration; consequently, when the piston is at the middle of its course, and the lever is horizontal, the centre of the joint at the end of the lever, will deviate from the centre line of the piston rod beyond that line, or farther away from the centre of the lever; but when the piston is at the top, or bottom of its course, and the lever is at its greatest inclination, the centre of the joint will deviate as much from the centre line of the piston rod on the opposite side, or within that line, nearer towards the centre of the great lever.

The horizontal distance between the centre of the great lever and the centre line of the piston rod being $1\frac{1}{2}$ times the length of the stroke of the piston, as above directed, the centre of the main joint by which the main links for the parallel motion, are suspended from the great lever, will deviate from the vertical line, one-12th of the length of the stroke, during that stroke; that is, one inch of vibration, or deviation, for every foot of the stroke. The radius of the great lever, or the distance between the centre of the above joint, and the centre of motion, must then be $1\frac{13}{24}$ of the length of the stroke, that is, $18\frac{1}{2}$ inches, for every foot of the stroke. And the centre of the lever being in the middle of the length, or distance between the centres of the joints at each end, that length will be $3\frac{1}{12}$ times the length of the stroke; or 3 feet 1 inch for every foot of the stroke.

For instance: for a 4 feet stroke, the distance between the vertical centre lines should be 12 feet, and length of the great lever $12\frac{1}{3}$ feet. For a 6 feet stroke, the distance between the vertical centre lines should be 18 feet, and the length of the great lever $18\frac{1}{2}$ feet.

The angular motion of the great lever about its centre will be 37·85 degrees, during the whole stroke, that is nearly one-19th part of the whole circle during the half stroke.

The above proportion was followed by Mr. Watt, and is the most convenient; but as circumstances will sometimes require different proportions to be established between the length of the stroke, and the length of the great lever, the following rule will be useful.

To find the vibration of the main joint, at the extremity of the great lever; that is, how much the centre of the joint will deviate from a vertical line, during its motion in the circular arch; having given half the length of the stroke in inches; and the radius of the main joint in inches, that is the distance between its centre, and the centre of motion.

RULE. From the square of the radius of the joint in inches, deduct the square of half the length of the stroke in inches; extract the square root of the remainder; and deduct that root from the above radius in inches. The remainder is the vibration in inches. Also, to find the proper horizontal distance between the centre of the lever, and the vertical centre line of the piston rod; add the above square root, to the radius of the lever in inches, and half their sum will be that horizontal distance in inches.

Example. Suppose half the length of the stroke to be 36 inches; and that the radius of the lever is 111 inches. The square of that radius is 12 321, from which deduct 1296 (the square of half the stroke), the remainder is 11 025, and its square root is 105 inches; which being deducted from the radius 111 inches, leaves 6 inches for the vibration. Also, adding 105 to 111, their sum is 216, the half of which is 108 inches, (=9 feet) for the horizontal distance between the centre of the lever, and the centre line of the piston rod.

Note. The same rule will apply, to find the vibration for the joints which suspend the rods of the air-pump, and the cold water-pump, or any other cases of levers to which rods are jointed.

Proportions for the parallel motion. The main links (marked K, Plate XI.) are jointed at their lower ends, to the cross pin *o*, which is fastened upon the upper end of the piston rod, *n*; and the upper ends of the links, K, are united by another joint pin, to the extremity of the great lever, L. The latter joint must move in the arch of a circle, about the centre of motion, *p*; but the joint *o* must be constrained to move in a vertical straight line, corresponding with the centre line of the piston rod; for this purpose the parallel rods, 5, are connected with the cross pin, *o*, at one end, and at the opposite ends, they are jointed to the lower ends of the back links, 7, which are suspended from the great lever. The main links, K and the back links 7, together with the parallel rods, 5, below, and the great lever above, form the four sides of a parallelogram, all the angles of which are moveable joints, and therefore that parallelogram may be placed in the form of a rhombus.

The joint which unites the lower ends of the back links, 7, to the parallel rods, 5, is also connected with one end of the bridles, or radius rods, 6, the other ends of which are moveable about a fixed centre, being fastened to a horizontal axis, which is supported at each end on pivots, resting in bearings, fixed to the under sides of the spring beams, U. Consequently, the joint which unites the bridles 6, to the back links, 7, must in its motion describe an arch of a circle, which will deviate from a vertical line, in an opposite direction to the deviation of the joint at the end of the great lever, but only about half as much deviation, and that being transmitted by the parallel rods, 5, to the joint, *o*, which connects the lower ends of the main links, K, with the top of the piston rod, it will have the effect of retaining that joint in a vertical line.

The accuracy and straightness of the vertical line in which the parallel motion will retain the piston rod, during its motion, depends chiefly upon the length of the bridles, 6, which must be adapted to produce such an extent of contrary deviation, to that of the joint at the end of the lever, as will just neutralize and counteract that deviation.

To obtain a clear idea of the principle by which the lengths of the different 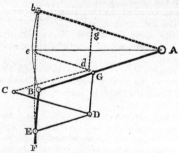 rods of the parallel motion are proportioned; we must consider how much the joint D, at the end of the bridle-rod C D, deviates from a vertical line, during its motion, in describing an arch of a circle from *d* to D, about its centre C. The joint D is connected with the upper end of the piston-rod E, by the parallel-rod D E, and that rod inclines from the horizontal position, during its motion, in the same degree as the great lever does, because it is retained parallel thereto by the main links B E, and the back links G D. In consequence the inner end D of the parallel-rod, moves in an arch of a circle described about the centre C, and the problem requires that the other end E, shall be retained to move in a vertical right line F.

The vertical ascent of the outer end E of the parallel rod D E, must necessarily be less than the vertical ascent of the inner end D of that rod, in the same proportion as the radius A B of the main joint, is greater than the radius A G of the joint for the back links G D. In consequence of the inclination of the parallel rod D E from the horizontal position, its inner end D must deviate from a vertical line in its motion, when the outer end E moves in the vertical line F; the amount of that deviation may be found by calculation, and the length of the bridle rod C D must be adapted, so that it will in its motion, produce that amount of deviation which the inner end D of the parallel rod requires, in order that its outer end E may move in the vertical line.

The inner end D of the parallel rod D E will deviate from a vertical line, less than the main joint B of the great lever deviates from a vertical line, in the same proportion as the length of the parallel rod D E is less than the radius A B of the great lever; because the angular motion of the parallel rod, being the same as that of the great lever, their deviations respectively will be as their lengths.

To find the length of the bridle rods C D *of the parallel motion,* by a method communicated to the author by Mr. Benjamin Hick of Bolton; it is attributed to Mr. Stevenson of Newcastle.

RULE. Square the distance A G in inches, from the centre A of the great lever, to the joint G, by which the back links G D are suspended from the great lever; and divide that square by the length D E of the parallel rods in inches. The quotient is the proper radius C D of the bridle rods in inches, that is the distance from their centre of motion C, to the joint D, by which they are united to the back links G D, and to the parallel rods D E.

Example. Suppose the radius A G of the joint G for the back links G D, to be 68 inches, and the length D E of the parallel rods to be 63 inches. Then 68 squared = 4624 ÷ 63 = 73·397 inches, is the required radius C D of the bridle rods.

Sliding Rule, slide inverted.	A Rad. of joint for back lin.	Length of paral. rod.		A 68 inc. rad.	63 inc.
	○ Rad. of joint for back lin.	Rad. of bridle rod.	*Ex.*	○ 68 inc. rad.	73·4 inc.

The best proportions for the parallel motion, and that which is now universally followed, is to suspend the back links G D from the great lever, at half way between its centre of motion A, and the centre of the main joint B, at the end of the lever; in that case the bridle rod C D must be of the same length as the parallel rod D E, and also equal to half the radius A B, of the great lever. The centre of motion C for the bridle rod C D, will then coincide with the centre line F of the piston rod; and therefore that centre cannot be a horizontal axis extending across between the spring beams, as was the practice in Mr. Watt's original engines, such as is represented in Plate XI.; but the bridle rods must move about two studs or centre pins, which are fixed to the spring beams, one at each side of the piston rod.

The length of the main links B E, and of the back links G D, are not important, but the longer the better; it would be a good proportion to make them equal to half the length of the stroke, but Mr. Watt's practice was to make them about three-sevenths of the length of the stroke.

When the parallel motion is proportioned as above, which is the modern practice, the air-pump rod may be suspended from a joint pin, which is adapted to the back links G D, at the middle of their length, and then it will move in a vertical line, on the principle of Mr. Watt's simplest form of the parallel motion (see p. 430), and the length of the stroke made by the air-pump bucket, will be exactly half that of the piston.

The length of the connecting rod is not important, but the longer it is, the better it will act on the crank; it is often made the same length as the great lever, or three times the length of the stroke of the piston, but sometimes a little shorter.

The diameter of the fly-wheel may be made according to convenience, provided that the weight of the rim is proportioned accordingly. The fly-wheel in engines with sun and planet wheels, was commonly made rather less than three times the length of the stroke of the piston. When the fly-wheel was placed on a second axis, to be turned by a multiplying wheel and pinion, it was frequently made still smaller in diameter, and the wheel and pinion so proportioned as to turn it round more than twice, for each stroke of the piston. When the fly-wheel is placed on the axis of the crank, so as to make only one revolution for each stroke of the piston, the diameter of the fly-wheel is usually between three and four times the length of the stroke of the piston; four times is a very suitable proportion in such cases.

RULES FOR FINDING THE DIMENSIONS OF THE MOVING PARTS, WHICH TRANSMIT THE FORCE OF THE PISTON OF MR. WATT'S DOUBLE ROTATIVE ENGINE.

All the different parts which transmit the force of the piston, must possess sufficient strength to enable them to resist the utmost strain that the piston can ever exert upon them, without any danger of breaking or bending. The proper dimensions for each of the moving parts were determined with great care by Mr. Watt, with the assistance of Mr. Southern, and his scale has been followed ever since, with very little deviation. For it has been found by long experience, that engines constructed after his models, are not liable to break by any continuance of fair working; and on the other hand, their moving parts are as light as it is advisable to make them, considering all the circumstances.

The makers of steam-engines have acquired great experience, in the strength which is required in the different parts, for sustaining the pressure to which they are exposed; and their practice affords the best information which can at present be procured, respecting the strength of materials, or the strain that they are capable of enduring continually, without any risk of receiving injury. A rotative steam-engine contains in its different parts, examples of all the different kinds of strains, to which materials can be subjected. On this account the following rules are very important, not merely in their application to the construction of steam-engines, but as data for proportioning the strength of the parts of any other machines, or edifices; and with that view the strain to which each part of a steam-engine is subjected, is computed in pounds, as a measure for the application of the rules in new cases, and also for proportioning the strength of the parts of high pressure engines, whereof the pistons exert a greater force in proportion to their size, than Mr. Watt's engines.

In Mr. Watt's steam-engines, when in good order, the greatest exhaustion within the cylinder, is about 13 pounds per square inch, less than the atmospheric pressure; this is usually the case at the latter part of the stroke of the piston when time has been allowed for exhausting the cylinder completely (see p. 487); and at times, when the engine is fully loaded, the steam with which the other end of the cylinder is filled to form the plenum, and impel the piston, is frequently as much as one or two pounds per square inch, more elastic than the atmospheric air; hence, we may safely assume that the piston of an engine, when fully loaded, exerts a force of 14 pounds for each square inch of its surface, or 11 pounds for each circular inch, at some part of every stroke; and accordingly the parts must all be adapted to sustain that degree of force, with perfect safety.

Note.   In calculating the power exerted by an engine, the effective pressure upon the piston is usually taken at rather less than 7 pounds per square inch, and from that to $10\frac{1}{2}$ pounds per square inch (see p. 486); for that is the average force exerted by the piston, throughout the whole length of the stroke, when a suitable deduction has been made on account of the friction of the moving parts; but in order to proportion the strength of the parts, we must consider the utmost force which the piston can ever exert upon them in fair working; and that is assumed to be 11 pounds per circular inch, which is a convenient number for computation, because the square of the diameter of the cylinder in inches, being multiplied by 11, the product will represent the utmost force exerted by the piston in pounds.

The size of the piston-rod for Mr. Watt's engine.   This is always made of wrought iron; and for a double acting engine its diameter is made nearly one-tenth of the diameter of the cylinder, so that the transverse section of the iron rod, is one-hundredth part of the area of the piston.   In large engines the piston-rod is rather less than one-tenth of the cylinder, because the length is less in proportion to the diameter.   For instance, a cylinder 48 inches diameter, the piston making 8 feet stroke, has a piston-rod $4\frac{1}{2}$ inches diameter.

The utmost force exerted by the piston being 11 pounds per circular inch, and allowing the weight of metal in the piston to be one pound per circular inch in addition, the strain to which the piston-rod must be subjected, is at the rate of 1200 pounds for each circular inch of its transverse section.   This proportion may be followed for other cases of wrought iron props or standards, where the length of the bar does not exceed 30 times its diameter.   For instance, a wrought iron bar one inch diameter, and 30 inches long, being used as a prop to sustain a load of 1200 pounds weight, would be loaded in the same degree, as the piston-rod of Mr. Watt's double engine during the ascending stroke.

For single engines, the piston-rod having only to resist a pulling action, it is usually made lighter, or one-12th of the diameter of the cylinder; and in some of Mr. Watt's old single engines, in Cornwall, the piston-rod is only one-14th of the diameter of the cylinder.   For instance, a single engine with a cylinder 70 inches diameter, has a piston-rod 5 inches diameter.

The wrought iron bands which form the main links of the parallel motion, to connect the piston-rod with the end of the great lever, and also those bands which suspend the connecting-rod from its joints, are usually made of such dimensions, that the area of the transverse section of all the iron in them, is about one-113th part of the area of the piston; consequently they are not quite so strong as the piston-rod which is one-100th part of that area.

To find the proper dimensions for the wrought iron links which suspend the piston-rod of Mr. Watt's steam-engine.

RULE. Divide the square of the diameter of the cylinder in inches by 144; the quotient is the proper sectional area in square inches, for the wrought iron in the links.

Example. For a 36 inch cylinder squared = 1296 circular inches ÷ 144 = 9 square inches of wrought iron should be allowed in the main links.

Sliding Rule,	A Dia. of cylind. inc.	144	*Exam.*	A 36 inc. dia.	144
slide inverted.	O Dia. of cylind. inc.	Area of iron links squ. inc.		O 36 inc. dia.	9 sq. in.

It is usual to make the main links, of two open links or straps, in the manner described at p. 472, so that there are four flat iron bars, to sustain the force of the piston. The breadth of each of these bars is usually one-12th of the diameter of the cylinder; and the thickness one-4th of the breadth, or one-48th of the diameter of the cylinder. This proportion corresponds to a divisor of 144 in the above rule; for instance, for a cylinder 36 inches diameter, the main links of the parallel motion should be made of iron bars, 3 inches broad, and $\frac{3}{4}$ of an inch thick. The transverse section of four such bars, will contain 9 square inches.

From the results of experiments which have been made on the absolute strength of wrought iron (*a*), it appears that a bar of sound English wrought iron, of an average quality, one inch square, may be expected to be torn asunder by a force of about 60 000 pounds, pulling in the direction of its length. The absolute strength of a circular inch of wrought iron is therefore (60 000 × ·7854 =) 47 124 pounds.

Assuming that the utmost force exerted by the piston, is 11 pounds per circular inch, and that the weight of metal in the piston, and piston-rod, amounts to one pound more per circular inch; then if the section of the iron in the main links is taken at one-113th part of the area of the cylinder, each square inch of wrought iron will be loaded with (144 × 12 =) 1728 pounds, when in service; but its absolute strength is 60 000 pounds, or 34·7 times the force to which it is subjected. Hence, there can be no chance of the main links being broken by the force of the piston. The piston-rod is still stronger, as 113 is to 100.

It is not necessary to have such a great strength of iron in the main links, for the force of the piston may be sustained by a much smaller section of iron with perfect safety. The bolts which are inserted through the wood of the great lever, to fasten the joint pins for the main links, and for the connecting-rod, to the ends of the great lever, sustain all the force of the piston, in the same manner as the main links do; in some instances, the area of the transverse section of these bolts, in the solid part of their screws, was one-260th part of the area of the piston; we may assume it at one-261·8th part to found the following rule.

(*a*) Mr. Barlow, in his excellent Essay on the Strength of Timber (at p. 236), has recorded a series of experiments, which were made at Messrs. Brunton's Chain Cable Manufactory, London, by Mr. Telford, when he was forming his plan for the suspension bridge at Runcorn near Liverpool. And two other experiments, made by Captain Brown, at his Patent Iron Cable Manufactory. From the results of these experiments, Mr. Barlow concludes, that the medium strength of a bar of British wrought iron, one inch square, is 27 tons = 60 480 pounds. Messrs. Price also made a series of experiments at Neath Abbey in South Wales, which give the same result. Mr. George Rennie made some experiments on a smaller scale, which give 55 872 lbs. for English iron, and 72 064 for Swedish.

Similar experiments were tried at Neath Abbey, and by Mr. Rennie, on the strength of cast iron, which agree very nearly. The mean is 19 157 pounds, to tear asunder a bar of cast iron one inch square. Some other experiments which have been communicated to the author, being upon better specimens of cast iron, were as high as 24 000 lbs.; but for an average we may take 20 000 pounds for the absolute strength of a square inch of cast iron; that is one-third of the strength of wrought iron.

4 H

To find the proper dimensions for the wrought iron bolts, which will be sufficient to sustain the force of the piston of Mr. Watt's engine.

RULE. Multiply the square of the diameter of the cylinder in inches by ·003; and the product will be the sectional area of the bolts in square inches. *Note.* Dividing by $333\frac{1}{3}$ would give the same result, as multiplying by ·003.

Example. If the cylinder is 48 inches diameter, the square is 2304 circular inches, × ·003 = 6·912 square inches of wrought iron, will be sufficient to bear the strain of the piston (*a*).

According to this proportion each square inch of wrought iron is subjected to a strain of $(333\frac{1}{3} \times 12 =)$ 4000 pounds when in service; that is one-15th part of (60 000 lbs.) the absolute strength of the iron; hence, the rule makes an ample allowance to avoid any risk of accidents.

To find the proper dimensions for bolts or bars of wrought iron, to sustain a given force without any danger of breaking.

RULE. Divide the given strain, in pounds, by 4000; the quotient will be the proper section in square inches. Extract the square root of the quotient, and that root is the proper dimensions, in inches, for the side of a square wrought iron bar.

Example. The piston of a cylinder 48 inches diameter = 2304 circular inches, at 12 lbs. per circular inch, would exert a force of 27 648 lbs. to separate the piston rod from the main links of the parallel motion. Then 27 648 ÷ 4000 = 6·912 square inches of iron; the square root of which is 2·629 inches square, would be the proper size for a single wrought iron bar, to sustain the force.

Sliding Rule,	C	4000	Force in pounds that the bar is to bear.	*Ex.*	C	4000	27 648 lbs.
	D	1	Side of square bar of wrought iron, inc.		D	1	2·63 in. sq.

ON THE STRENGTH OF WROUGHT IRON SCREW BOLTS, WHICH ARE USED IN STEAM-ENGINES FOR FASTENING THE PARTS TOGETHER.

THE absolute strength of screw bolts must depend upon the sectional area of the cylinder of solid iron, around which the spiral threads project; and the size and strength, of as many of those threads as are contained within the nut, must be at least equal to that of the solid cylinder; so that the threads may not strip off, with any less force, than that which would tear the solid of the screw in two.

It is usual amongst workmen in speaking of the size of screw bolts, to express the diameter that the bolt had, before the spiral groove was cut round it, to form the thread, that is the diameter to the outside of the thread of the screw. According to the proportion of the best specimens of screws, the spiral groove penetrates into the cylinder, about one-12th of its diameter, consequently the diameter of the solid cylinder which remains, when the screw is formed, is five-sixths of the diameter, taken at the outside of the threads according to the customary mode.

The area of that solid cylinder will be twenty-five-36ths, or ·694 of the area of the cylinder which is formed by the outside of the threads; we may assume, that the strain to which wrought iron may be safely subjected in machinery is, one-15th of its utmost strength, or 4000 lbs. to a square inch, or (× ·7854 =) 3141·6 lbs. to a circular inch; hence (3141·6 × ·694 =) 2180 lbs. may be borne by a screw one inch diameter, at the outside of the threads.

(*a*) In one of Messrs. Boulton, Watt, and Co's engines with a 48 inch cylinder, the joint for the connecting rod is suspended by four wrought iron bolts, $1\frac{3}{4}$ inches diameter, outside of the threads; or about 1·46 inches diameter of solid iron, = 1·675 square inches in each bolt, × 4 = 6·7 square inches of solid iron in all four, or a little less than the above rule would give. As this engine has been working under a heavy load for several years, without any failure, it proves that this strength of iron is quite sufficient.

To find the proper diameter for screw bolts of wrought iron, to enable them to bear a given strain without any danger of breaking.

RULE.   Divide the given strain in pounds by 2200, and extract the square root of the quotient. That root is the proper diameter for the screw, at the outside of the threads.

Example.   Suppose that the piston of a 48 inch cylinder, exerts a force of 27 648 lbs.   Then 27 648 ÷ 2200 = 12· 56, the square root of which is 3· 544 inches, for the diameter, outside the threads, of a single screw bolt, which would be sufficient to sustain the force.   Or if four bolts are used, then each one would bear 6912 lbs. ÷ 2200 = 3· 14 ; the square root of which is 1· 772 inches, for the diameter of each screw, outside of its threads.   In practice they are $1\frac{3}{4}$ inches diameter.

$$\text{Sliding rule.} \begin{cases} \text{C} & 22 \\ \hline \text{D} & 1 \end{cases} \frac{\text{Force that the screw sustains, lbs.}}{\text{Diam. of screw outside the thread, inc.}} \quad Ex. \quad \frac{\text{C} \quad 22}{\text{D} \quad 1} \quad \frac{6912 \text{ lbs.}}{1 \cdot 77 \text{ inc. dia.}}$$

The proportions of the threads of screws of different sizes, vary considerably, but the following proportions may be recommended, for all screws which are to be used as holding bolts, and not as mechanical powers; these proportions are taken from an examination of several large screws, which have been formed with care, by the best makers, for suspending the arch head chains for the pistons of large steam-engines in Cornwall, and the screws for the pillars of rolling mills; and also from an average of some good specimens of screw bolts, used for engine work.

The diameter of the screw, at the outside of its threads, being multiplied by 5, and the product divided by 6, the quotient will be the diameter of the solid cylinder within the threads.

The diameter outside of the threads being divided by 6, the quotient will be the pitch, or distance between the threads, from the middle of one thread to the middle of the next.   In that case, the obliquity of the thread (see p. 50), or the rise of the inclined plane, will be 1 in 17· 28 of its length. For taking the mean between the outside of the thread, and the inside, it will be eleven-12ths of the outside diameter; therefore (·917 × 3· 1416 =) 2· 88 times that diameter, is the circumference in each turn; and 2·88 × 6 turns is 17· 28 times that diameter, for the whole length of thread within the nut.

The depth or thickness of the nut, should be equal to the diameter outside of the threads, and then the nut will contain 6 turns of the threads; this is proper for nuts which are not required to be frequently unscrewed.   The strength of the threads will then be $1\frac{7}{8}$ times the strength of the solid cylinder of the screw; for assuming that the thickness of the threads is equal to the spaces between them, the threads remain adhering to one half of the surface of the solid cylinder; the length of that cylinder, which is contained in the nut, is 1· 2 of its diameter, and that surface is (1·2 × 3· 1416 =) 3· 77 ; one half of which, is 1· 88 times the area of the solid cylinder.

This is the strength of the threads of screws with square threads, but the sharp or angular threads have more adherence to the solid cylinder, so as to have $2\frac{1}{2}$ times the strength of the solid cylinder, which is not too great an allowance for wearing, and for any unequal bearing in the threads.   If the nuts are intended to be frequently unscrewed, their thickness should be $1\frac{1}{3}$ of the outside diameter of the threads, and the nut will contain 8 turns of the thread.

Dimensions of the connecting-rod.   The upper part of the rod was made of oak, in the old engines (see p. 504), and the area of the transverse section of the wood was about one-6th part of the area of the cylinder; the lower part was made of cast iron, its transverse section being about one-20th part of the area of the cylinder.   In succeeding engines the connecting-rod was a square bar of cast iron, in one piece, and its transverse section was about one-18th of the area of the cylinder.

The connecting-rods for modern engines are also made of cast iron in one piece, but much lighter; the form of the middle part is that of four ribs adhering together, so that the transverse section of the bar resembles a ✦.   This form gives the rod stiffness to resist lateral flexure, when it transmits the thrust of the engine. The area of the transverse section of the cross, at the middle part of the rod, is about one-28th part of the area of the cylinder; and the breadth across the two opposite arms of the cross, is about one-20th of the whole length of the rod.   The lower end of the rod, which acts near to the crank, is shaped like a flattened oval bar, its area in the smallest part, being one-35th of the area of the cylinder.

The connecting-rods of large engines are subject to be broken sometimes, and it would be an improvement to make them of wrought iron, instead of cast iron.

The lower end of the connecting-rod has only the force of the piston to bear; that being at the rate of 14 pounds per square inch, and the cylinder being 35 times the area of the rod, each square inch of cast iron must be strained with a force of (14×35=) 490 pounds.

In single engines the strain upon the pump-rod is always in one direction; if the rod is made of cast iron, its area is about one-45th of the area of the cylinder; therefore each square inch of cast iron must bear a strain of (14 $\times$ 45 =) 630 pounds. The best form for a pump-rod in cast iron, is to make it of a cylindrical figure, and hollow within, like a tube.

The experiments which have been made upon the absolute strength of cast iron, show that a bar of sound English cast iron, of an average quality, one inch square, will require a force of 20 000 pounds to pull it asunder in the direction of its length. (See Note, p. 601.)

The cast iron connecting-rods for double engines being pulled with 490 lbs. per square inch, it would probably require 40 8 times that strain, to break the cast iron. And the cast iron pump-rods of single engines, being loaded with 630 lbs. per square inch, might be expected to endure 31·75 times that strain, before they would break.

Dimensions of the main joint pins for the parallel motion, and for the connecting-rod. These joints sustain the whole force of the piston, and the weight of the piston and piston-rod, in addition. The joint pins were made of wrought iron in the first engines, and then the diameter was one-tenth of the diameter of the cylinder, or the same size as the piston-rod. In modern engines the joint pins are made of cast iron, and the diameter is one-ninth of that of the cylinder, consequently the area of each pin, is one-81st part of the area of the cylinder. As there are two pins to bear the strain, and it is equally divided between them, the strain to which each pin must be subjected in working, is at the rate of (81 times 6 lbs. =) 486 pounds for every circular inch of its area. The length of the pin in the bearing part, is equal to its diameter.

The dimensions for the joint pins, and rod, by which the air-pump bucket is suspended; these are proportioned according to the diameter of the air-pump, by the same rule as that by which the piston-rod, and main joints, are proportioned to the diameter of the cylinder.

The axis of the great lever was made square in the middle part which applies beneath the lever, and cylindrical at each end, where it rests in the bearings or sockets. For small engines up to 20 horse-power, the axes were usually made of wrought iron, but the larger sizes are made of cast iron. In large engines the depth of the axis is greater than its breadth. The length of the axis between the bearings is between one-seventh and one-eighth of the length of the great lever.

The diameter of the cylindrical pivots at each end of the axis, is $\frac{5}{80}$ths of the diameter of the cylinder, and may be found by multiplying the diameter of the cylinder by ·16; or dividing by 6· 25 will give the same result. The length of each cylindrical pivot in the bearing part, is about $1\frac{1}{4}$ times the diameter.

The pivots of the great lever have to bear double the force of the piston, and the resistance of the air-pump, when at its greatest; and in addition to that, they must bear the whole weight of the great lever, the parallel motion, piston-rod and piston, the connecting-rod and its planet-wheel; also the air pump-rod and bucket, the cold water pump-rod and bucket, and as much balance weight at the cylinder end of the lever, as will make up a counterpoise to the weight of the connecting-rod at the other end.

The absolute weight of all these parts does not bear any constant proportion to the force of the piston, in engines of different sizes, but in most cases it will amount to as much as the force of the piston; and we may safely assume, that the pivots of the great lever are loaded with at least three times the utmost force of the piston, or at the rate of 33 pounds for each circular inch of the piston. The diameter of the cylinder being 6·25 times the diameter of the pivots, the area of the cylinder will be (6·25 squared =) 39 times the area of the pivots, and as there are two pivots to bear the strain between them, each one must be loaded with (39 × 33 = 1287 ÷ 2 =) 643·5 pounds for every circular inch of its area. The length of bearing is 1·25 times the diameter; but if the length of the pivot had been only equal to its own diameter, then it would have been equally well able to have borne 804·4 pounds per circular inch.

The dimensions of the crank pin. The pin which projects out from the crank, to form the joint with the lower end of the connecting rod, must endure all the force of the piston. The crank pin is usually made of cast iron, and its diameter is nearly the same as that of the pivots for the axis of the great lever; the diameter may be found by dividing the diameter of the cylinder by $6\frac{1}{3}$. The area of the piston is therefore 40 times the area of the crank pin; and the force of the piston being 11 pounds per circular inch, the strain on the crank pin will be at the rate of 11 × 40 = 440 pounds for every circular inch of its area. The length of bearing is about $1\frac{1}{2}$ times the diameter; but if it had been only equal to the diameter, then the pin would have been equally able to have borne 660 pounds per circular inch. It should be remarked, that the crank pin has to bear a greater strain than that of the mere force of the piston, in consequence of the energy of the moving parts of the engine being brought to rest at the termination of each stroke, see p. 420; all the reciprocating parts partake of this strain, but the crank pin, in a greater degree than the others. There are sufficient instances of the crank pins having been broken in working, to prove that they are not stronger than necessary, but are even weaker than other parts of engines.

The different joint pins or gudgeons, used in Mr. Watt's rotative engines, are not loaded in the same proportion to their dimensions; but as the main pivots of the great lever are never known to break, we may conclude that in cases where the motion is slow, so that sudden jerks are not likely to happen, a cast iron pivot or gudgeon, whereof the diameter is equal to its own length, may be safely loaded at the rate of 800 pounds for every circular inch of its transverse section.

According to the dimensions of the gudgeons for the axes of water-wheels, as they are made by the most experienced millwrights, they are loaded with about 500 pounds to each circular inch, which is only a little more than the strain that the main joint pins for the parallel motion are loaded with; this forms the basis of the following general rule for the strength of cast iron gudgeons.

ON THE STRENGTH OF CYLINDRICAL GUDGEONS, OR PIVOTS, FOR THE AXES OR SHAFTS OF MACHINERY.

The strength of different gudgeons is proportionate to the quotient which is obtained by dividing the cubes of their diameters by their lengths of bearing. If the lengths of the bearing parts of the gudgeons, bear a constant proportion to their diameters, then the strength of the gudgeons will be proportionate to the areas of their cross sections.   In practice, if the pivot is to turn round with a slow motion, it is usual to make the length of bearing about $1\frac{1}{4}$ of the diameter; but for rapid motions it is better to make the length $1\frac{1}{3}$ or $1\frac{1}{2}$ times the diameter.

FOR CAST IRON GUDGEONS.

To find the proper size for a cast iron gudgeon to sustain a given weight, that it may have sufficient strength to avoid any danger of breaking, in working with a slow and regular motion.

RULE. Multiply the weight that the gudgeon is to sustain, in pounds, by so much of the length of the gudgeon, as is received in the socket or bearing, measuring from the shoulder; divide the product by 500, and extract the cube root of the product: that root is the proper diameter for the gudgeon in inches.

Example. Suppose a fly-wheel, with its arms, centre piece, and axis to weigh 28 800 pounds, and that the wheel is placed nearly at the middle of its axis, between the two bearings, so that the weight on each gudgeon of the axis is 14 400 pounds.   The length of bearing being 10 inches, then $14\,400 \times 10 = 144\,000 \div 500 = 288$, the cube root of which is 6·6 inches diameter, for the proper size for the cast iron gudgeon.   This rule supposes that it has no twist to endure, but merely to support the weight of one end of the axis (*a*).

To perform this calculation with the sliding rule, we must take double the weight in pounds that the gudgeon is required to bear, which will, in most cases, be the total weight of the axis and wheels, and using that number on either of the upper lines of the Soho rule, with the slide inverted, it will have the effect of dividing by 500; for, in fact, we may suppose that we have multiplied by its reciprocal ·002.

Soho Rule, slide inverted, all four lines.	A Double the weight pounds. C Length of bearing inches. B Diam. of gudgeon inches. D Diam. of gudgeon inches.	} This is where the same numbers meet on both.	*Exam.* A 28 800 lbs. C 10 inch. bearing. B 6·6 inches diam. D 6·6 inches diam.

The following is a more convenient form of the same rule, because it fixes a proper proportion for the length of the gudgeon.   Both these rules give a little less strength to the gudgeon than that of the main joint-pins, for the parallel motion of Mr. Watt's engine, when their diameters are one-ninth of the diameter of the cylinder, but more strength than the pivots of the axis of the great lever (*b*).

(*a*) The weight poised upon the axis of Mr. Smeaton's atmospheric engine at Long Benton (see p. 182), was 60 386 lbs., or 30 193 lbs. on each pivot; they were seven inches diameter, and of the same length.   The section being 49 circular inches, the load was 616 pounds upon each circular inch; the length of bearing was equal to the diameter.   This is considerably less in proportion, than the strain upon the same part of Mr. Watt's engine, which is 804 lbs.; Mr. Smeaton's proportion is nearer to the above rule for the strength of gudgeons, which is at the rate of 500 lbs.

The weight borne by the main arch head chains of the same engine, was 27 056 lbs. upon six square inches of wrought iron; this is at the rate of 4509 pounds to each square inch, and is rather greater than the rule in p. 602, which is 4000 lbs.   These chains were considered to be so weak, that after they had been some time at work, it was proposed to Mr. Smeaton to apply a third additional chain to the arch head; but as the original chains continued to work without breaking, this example may give us confidence in the sufficiency of the rule.

(*b*) If the divisor in the above rule were 800 instead of 500, it would then proportion the gudgeons with the same strength as the main pivots for the great lever of Mr. Watt's engine possess, when the diameter of the cylinder is $6\frac{1}{4}$ times the diameter of the pivot.   That would be a very sufficient strength to avoid the danger of breaking the gudgeon by mere pressure; but it is requisite to have a certain size of gudgeon, independently of the consideration of strength, in order that there may be a sufficient surface of bearing, to avoid excessive friction, and cutting and heating of the gudgeon (see p. 61).   It is injudicious to make cast iron gudgeons too long, in order to increase their surface of bearing, because they are so much more liable to be broken, and particularly if they ever become hot by the friction.   The above rules are perhaps the best proportion, for general practice.

RULE. Multiply the weight in pounds by the intended length of bearing, expressed in terms of the diameter; divide the product by 500, and extract the square root of the quotient; that root is the diameter in inches. Also multiply the diameter thus found, by the proportion that the length is to bear to the diameter, and the product is the length of bearing in inches.

Example. Mr. Robertson Buchanan, in his Essay on the Shafts of Mills, mentions an overshot water-wheel of cast iron, 16 feet diameter, and 8 feet broad, which weighed 330 hundred weight; each pivot of its axis sustained 18 480 pounds; the length of bearing is not stated, but we will suppose it to have been 1·25 diameters. Then 18 480 lbs. × 1·25 = 23 100 ÷ 500 = 46·2; the square root of which is 6·797 inches for the diameter. And 6·797 × 1·25 = 8·5 inches length of bearing. The gudgeon really was 6·625 inches diameter.

According to another example given by the same author, a wooden overshot wheel of the same dimensions as the above, would weigh about 10 500 lbs. on each pivot, × 1·25 = 13 250 ÷ 500 = 26·5, the square root is 5·148 inches for the diameter. And 5·15 diameter × 1·25 = 6·44 inches length of bearing. The gudgeon really was 6 inches diameter.

Sliding Rule, slide inverted.
$\left\{\begin{array}{l}\text{A Double the weight pounds. \quad Dia. of gudg. inc.}\\ \text{ɔ Lengt. of bearing in diams. \quad Dia. of gudg. inc.}\end{array}\right.$
Exam.
$\begin{array}{l}\text{A 36 960 lbs. \quad 6·8 inc. dia.}\\ \text{ɔ 1·25 bear. \quad 6·8 inc. dia.}\end{array}$

FOR WROUGHT IRON GUDGEONS.

Wrought iron gudgeons may be safely proportioned by the preceding rules, if 1000 is used for a divisor, instead of 500; they will then be loaded with double the strain to which cast iron gudgeons of the same dimensions are subjected, according to the above rules. This proportion appears to correspond with the established practice in the axletrees of wheel carriages.

To find the proper diameter for a wrought iron gudgeon to support a given weight; or for the axletrees for a wheel carriage.

RULE. Multiply the weight which the gudgeon or axletree is to sustain in pounds, by the length of the bearing from the shoulder in inches; divide the product by 1000, and extract the cube root of the product. That root is the proper diameter for the axletree at the shoulder in inches.

Example. The weight on the axletree of a Landau being 8 cwt. = 896 lbs., and the length of bearing from the shoulder 9 inches; then 896 × 9 = 8064 ÷ 1000 = 8·064, the cube root of which is 2·005 inches, for the proper diameter of the axletree at the shoulder.

Soho Rule, slide inverted, all four lines.
$\left\{\begin{array}{l}\text{A \quad\quad Weight in pounds.}\\ \text{ɔ \quad Length of bearing inches.}\\ \text{ᙠ \quad Diam. of axletree inches.}\\ \text{D \quad Diam. of axletree inches.}\end{array}\right.$
$\left.\begin{array}{l}\\ \end{array}\right\}$ This is where the same numbers meet on both.
Exam.
$\begin{array}{l}\text{A \quad 896 lbs. weight.}\\ \text{ɔ \quad 9 inc. bearing.}\\ \text{ᙠ \quad 2·005 inches dia.}\\ \text{D \quad 2·005 inches dia.}\end{array}$

The dimensions of the axletrees for wheel carriages are very correctly proportioned to the weights they are intended to carry; for it has been determined by the result of experience, what strength they require, to avoid any danger of breaking, and yet no more strength is given than is consistent with safety, because it would occasion unnecessary weight and friction.

The following are the dimensions of the axletrees of different descriptions of travelling carriages, made by Messrs. Hobson and Co. of Long-acre, London, whose extensive practice places them at the head of the coach-making business.

The lightest two-wheeled carriage, for one horse, is called a Dennet; when complete it weighs 4 cwt. = 448 pounds, and the two wheels weigh 102 pounds. The dennet is adapted to carry two persons, and including luggage, it may occasionally be loaded with about 336 lbs., and then the weight resting upon each axletree is 341 lbs. The axletrees are 8 inches length of bearing, from the shoulders; the diameters at the small ends, where the linch-pins go through, 1¼ inch, and at the shoulders 1½ inch diameter. By the above rule, 341 lbs. × 8 inc. = 2728 ÷ 1000 = 2·728, the cube root of which is 1·4 inches for the diameter, instead of 1·5 inches.

A strong two-wheeled carriage, for one horse, called a Cabriolet, weighs 9 cwt. when complete; the wheels weigh 150 lbs. It is adapted to carry two persons inside, and a servant boy behind, and may be loaded with about 448 lbs; the load upon each axletree will then be 653 lbs. The length of bearing is 9 inches. The diameter at the linch-pin $1\frac{3}{8}$ inch, and at the shoulder $1\frac{5}{8}$ inches diameter. By the above rule it should be 1·8 inches.

A four-wheeled carriage for two horses, called a Britzschka, weighs 12 cwt. when complete; the four wheels $2\frac{1}{4}$ cwt. It is adapted to carry six persons, and luggage, or about 9 cwt. This would be 518 lbs. load on each axletree. They are 8 inches length of bearing, $1\frac{3}{8}$ inches diameter at the linch-pins, and $1\frac{5}{8}$ inches at the shoulders. By the rule they should be 1·61 inches diameter.

A four-wheeled Chariot for two horses, weighs 12 cwt.; its four wheels 3 cwt. It is adapted to carry seven persons, and luggage, or about $10\frac{1}{2}$ cwt.; the weight on each axletree is 602 lbs. They are 9 inches length of bearing, $1\frac{1}{2}$ inches diameter at the linch-pins, and $1\frac{3}{4}$ inches at the shoulders. By the rule they should be 1·756 inches diameter.

A four-wheeled carriage for two or four horses, called a Landau, weighs about 20 cwt.; its four wheels $3\frac{1}{8}$ cwt. It is adapted to carry ten persons, and luggage 15 cwt.; the load on each axletree is then 896 lbs. Their length of bearing is 9 inches, their diameters at the linch-pins $1\frac{3}{4}$ inches, and at the shoulders 2 inches. The rule gives 2·095 inches diameter.

The Mail coaches are manufactured by Mr. Vidler, in London, who has taken great pains to make them as light as is consistent with safety. A mail coach is drawn by four horses, and when complete weighs about 17 cwt.; its four wheels 5 cwt. It is adapted to carry eight persons, weighing about 12 cwt., luggage about 4 cwt., and the mail about 3 cwt. The utmost weight in service does not exceed 40 cwt., and then the weight on each axletree is about 980 lbs. Their length of bearing is 11 inches, their diameters at the small end is $1\frac{5}{8}$ inches, and at the shoulder $2\frac{1}{4}$ inches. By the rule it should be 2·21 inches diameter.

A two-wheeled cart used in London, with two horses, to carry about 20 cwt. of lime, the weight of the body being 4 cwt. The load upon each axletree 1344 lbs., their length of bearing $11\frac{1}{2}$ inches, their diameters at the linch-pins $1\frac{3}{4}$ inches, and at the shoulders $2\frac{1}{2}$ inches diameter. By the rule it should be 2·49 inches diameter.

A four-wheel waggon, used in London with three horses, to carry $2\frac{1}{2}$ chaldrons of coals, or about 60 cwt.; weighs complete 37 cwt.; its four wheels about 17 cwt.; the rims of the wheels are 6 inches broad upon the edge; the weight borne by each axletree is 2576 lbs.; their length of bearing 13 inches, their diameters at the linch-pin $2\frac{1}{2}$ inches, and at the shoulder 3 inches diameter. By the rule it should be 3·08 inches diameter.

A heavy stage waggon, drawn by seven horses, weighs 40 cwt.; its four wheels about 20 cwt.; the rims of the wheels are 9 inches broad; and the total weight allowed by law for the waggon and its load is 6 tons; the weight on each axletree would then be 2800 lbs. The length of bearing is 13 inches, the diameter at the linch-pin $2\frac{3}{4}$ inches, and at the shoulder $3\frac{3}{8}$ inches diameter. By the rule it should be 3·31 inches diameter.

These different examples agree so nearly with the rule, as to give us great confidence in its accuracy, particularly when we consider that the above proportions have not been formed by any rule whatever; but the proper size for the axletree of each kind of carriage has been found out by the makers from experience, without reference to that of the others. Hence we may follow the rule with safety for proportioning any cases of wrought iron gudgeons in millwork and machinery. It should be observed, that an axletree which does not turn round is less liable to be broken, than a gudgeon which revolves in its bearing; because the axletree is always bent in the same direction by the force; but a gudgeon is made to bend successively in every direction in each revolution, and, therefore, if it is bent so far that the elasticity of the iron will not cause it to return to its proper form, it will very soon be broken.

DIMENSIONS OF THE GREAT LEVERS FOR MR. WATT'S ROTATIVE ENGINES.

The great lever, LL, is made of one beam of English oak, well seasoned. A hole of about $1\frac{1}{2}$ inch diameter was sometimes bored through the centre of the log, from one end to the other, in order to admit the air into the wood, to dry it, and thus avoid cracking. The beam is placed across the axis of motion p, and fastened to it by four bolts, as is represented in Plate XI.; these bolts being on the outside of the wood, the whole substance is preserved without any perforation, and as the bolts tend to bind the wood together, they prevent it from splitting. The two catch pins are fastened on the top of the lever in a similar manner, by four bolts to each pin. The joint to suspend the link for the air-pump rod *l*, is formed in the head of a bolt, which passes through the beam and has a nut screwed on the top. The socket for the upper joint of the main links, K, of the parallel motion, is fastened to the end of the lever by two or four bolts, which pass up through the wood, and have nuts at top. The joint for the connecting rod is fastened to the lever in the same manner.

To find the size of the oak beam for the great lever of Mr. Watt's rotative steam-engine.

RULE. Divide the diameter of the cylinder in inches by 1·2; the quotient is the proper depth for the beam, when the breadth is one-12th of the length of the lever, from centre to centre.

Example. Cylinder 24 inches diameter ÷ 1·2 = 20 inches for the depth of the beam; and if the length between the centres is 16 feet, then its breadth must be 16 inches.

Note. If the size of the timber is such that it is not convenient to make the breadth and length in this proportion, then the proper dimensions may be found by the following

RULE. Divide the diameter of the cylinder in inches by 1·2, then square the quotient, and multiply that square by the length between the centres in feet. *If the breadth is given, then to find the depth*; divide the last product by that breadth in inches, and extract the square root of the quotient; that root will be the proper depth in inches. *Or if the depth is given, then to find the breadth*; divide the above product by the square of that depth in inches, and the quotient will be the proper breadth in inches.

Example. The cylinder of the engine at the Albion Mills (p. 512) was 34 inches diameter ÷ 1·2 = 28·33 which squared is 802·6 × by 25 feet length between centres = 20 065. Supposing the breadth to be 24 inches, then 20 065 ÷ 24 = 836, the square root of which is 28·9 inches, for the proper depth; it was 29 inches. Or, supposing the depth to be 28 inches, its square is 784; and (20 065 ÷ 784 =) 25·6 inches would be the corresponding breadth.

Sliding Rule, slide inverted. $\left\{ \dfrac{\text{ᗺ Length of lever ft.   One-7th of breadth inc.}}{\text{D Dia. of cylind. inc.   Depth of beam inches.}} \right.$ *Ex.* $\dfrac{\text{ᗺ 25 ft. long.   3·43 (×7=24)}}{\text{D 34 inc. dia.   29 inc. deep.}}$

The number thus obtained on the line ᗺ of the sliding rule must be multiplied by 7, and the product will be the breadth in inches; thus 3·43 × 7 = 24 inches breadth. When the breadth in inches is given, it must be divided by 7, and the quotient being found upon the line ᗺ, the proper depth in inches will be opposite to it upon line D.

In Newcomen's atmospheric fire-engine, the cylinder being less perfectly exhausted than in Mr. Watt's, the utmost force which can be exerted by the piston must be proportionably less; and as that force depends upon the atmospheric pressure, it cannot be increased, as in Mr. Watt's engine, by accumulating a greater elasticity of steam in the boiler. Taking the utmost force exerted by the piston of Mr. Watt's engine to be 11 pounds per circular inch, or 14 pounds per square inch, we may safely assume, that the utmost force of the piston of the atmospheric engine is four-fifths of Mr. Watt's, viz. 8·8 pounds per circular inch, or 11·2 pounds per square inch. (See p. 175.)

To adapt the preceding rules for the strength of the great lever, to the atmospheric engine, we should use for a divisor (five-4ths of 1·2 =) 1·5 instead of 1·2. Calculations thus made will be found to agree very well with the practice of the best engineers.

Example. In Mr. Curr's table (see p. 207), the oak beam for the great lever of an atmospheric cylinder, 60 inches diameter, is directed to be 35 inches deep, by 32 inches broad at the middle, and 25 feet long. According to the above modification of the rule, the calculation would be as follows: 60 inches diam. ÷ 1·5 for the divisor = 40, which squared is 1600, × 25 feet long = 40 000, and ÷ 32 inches broad = 1250, the square root of which is 35·36 inches for the depth; instead of 35 inches, as directed by Mr. Curr.

4 I

The above rules assume that the strength of a prismatic beam may be represented by multiplying the square of the depth into the breadth, and dividing the product by the length. The product thus obtained is a correct representation of the absolute strength of the beam to resist fracture, but does not represent its stiffness, or the resistance it can make to flexure. For instance, a beam, which is of a considerable depth, and narrow in breadth, will not bend so much with a given force, as another beam of the same length, which has a less depth, but so much more breadth as will give it an equal strength, according to the above rule. The deep beam cannot bend so far without breaking, as the shallower beam can; but it takes as much more force to bend it, so that they will both break with the same force. This circumstance must be kept in mind in applying the second of the above rules, and the proportion of the length and breadth, should be kept as nearly to the first rule as the timber will admit, and then the lever will have a suitable stiffness to resist bending, as well as strength to resist breaking.

The above rules give such dimensions, as have been proved by long experience to be sufficient to sustain the force of the piston of Mr. Watt's rotative engine; and we may deduce from them a general rule for the weight with which oak beams may be loaded in other situations. The preceding rule is intended for rotative engines; but the levers of engines for draining mines, usually have a great weight of pump rods to sustain, in addition to the force exerted by the piston, and a suitable allowance must be made for such additional weight; this may be done by adopting the general rule.

Each end of the great lever of a rotative engine must sustain a force equal to the whole force of the piston, together with the weight of the connecting rod and planet wheel; all which may be assumed to be 14 pounds for each circular inch of the piston. Or we may suppose the beam to be supported at each end, and the load applied in the middle of its length, in which case that load will be at the rate of 28 lbs. per circular inch of the piston.

According to the preceding rules, the diameter of the piston is 1·2 times the depth of the lever; and the length of the lever is 12 times the breadth; hence (1·2 squared = 1·44 $\times$ 28 lbs. =) 40.32 lbs. is the weight which may be loaded on the middle of an oak bar, 1 inch square, and 1 foot long between the supports; the following rule is according to that proportion.

To find the proper dimensions for an oak beam, to sustain a given weight on the middle, when it is supported at each end.

RULE. Divide the weight in pounds, which is to be applied at the middle of the beam (a), by 40, and multiply the quotient by the distance in feet between the supports, then; *If the depth of the beam is given, to find the breadth,* divide the above product by the square of that depth in inches; the quotient will be the required depth in inches. *Or if the breadth is given, to find the depth,* divide the same product by that breadth in inches, and extract the square root of the quotient; that root will be the required depth in inches.

Example. The strain occasioned by the piston of a cylinder 34 inches diameter (= 1156 circular inches), at 28 lbs. per circular inch, would amount to 32 368 pounds; suppose that weight is to be applied on the middle of an oak beam, 25 feet long between the supports. Then 32 368 lbs. $\div$ 40

(a) In the above rule the weight of the beam is not taken into account, but, in strictness, half the weight of the beam ought to be added to the weight which is applied on the middle of it; because a load distributed uniformly along the length of a beam, will have the same effect to break it, as half that weight would have, if it were applied at the middle of its length.

= 809·2 × 25 feet long = 20 230. Supposing the depth to be 29 inches, the square is 841 ; and 20 230 ÷ 841 = 24·05 inches, for the breadth of the beam. Or supposing the breadth to be 24 inches, then 20 230 ÷ 24 = 843, the square root of which is 29·03 inches, for the depth of the beam.

The calculation may be made by the Soho sliding Rule, with the slide inverted, using all the four lines ; and in place of dividing the weight by 40, we may use one-fourth of the length of the beam in feet, on either of the upper lines A or ꓛ, thus:

Soho Rule, Slide inverted, all four lines.	A	One-4th length of beam feet.	_Exam._	A (25 ft. ÷ 4 =) 6·25	
	ꓛ	Weight on the middle pounds.		ꓛ Weight 32 368 lbs.	21
	Я	Breadth of the beam inches.		Я Breadth 24 inc.	or ──
	◻	Depth of the beam inches.		D Depth 29 inc.	31

The rule being thus set, the two lower lines, я and D, form a table, showing all the corresponding depths and breadths, which will give the requisite strength. For instance, if the breadth were 21 inches, the depth must be 31 inches ; or for 14 inches broad 38 inches deep.

Mr. Smeaton made a series of experiments on the strength of different woods, and found that a bar of good English oak, 1 inch square, and 2 feet long between the supports, was broken by 280 pounds hung on the middle. That is at the rate of 560 lbs. for the absolute strength of a bar of English oak, 1 inch square and 1 foot long.

Mr. Barlow, in his Essay on the Strength of Timber (p. 181), states, that a bar of English oak, 2 inches square, and 7 feet long between the supports, was broken by 637 lbs. hung on the middle. At that rate, the absolute strength of a bar of oak, 1 inch square and 1 foot long, is 557 lbs. The bar was bent down at the rate of 1·28 inches, by every succeeding 200 lbs. which was loaded upon it.

The preceding rule is so proportioned, that a bar 1 inch square and 1 foot long would only be loaded with 40 lbs. ; and, according to the above experiments, it would require 560 or 557 lbs. to break it ; that is, 14 times, hence there can be no danger of the beam being broken by the strain.

The flexibility of beams follows a different law from that of their absolute strength ; the lateral stiffness of a prismatic beam to resist flexure, may be represented by multiplying the cube of the depth by the breadth, and dividing the product by the cube of the length. This fact is demonstrated theoretical and experimentally by Mr. Barlow, and the following rule is deduced from his experiments and investigations.

To find how much a beam of English oak will bend, when it is supported at each end, and loaded in the middle with a given weight.

RULE. Multiply the cube of the depth of the beam in inches, by the breadth in inches, and divide the product by the cube of the length in feet ; multiply the quotient by the constant number 3360, and reserve the product for a divisor. Divide the weight in pounds, which is applied at the middle of the beam, by that divisor, and the quotient is the deflexion of the beam in inches.

Example. Suppose an oak beam, 29 inches deep, by 24 inches wide, and 25 feet long between the supports, to have a weight of (1156 circ. inc. × 22 lbs. =) 25 432 pounds applied on the middle. Then 29 inc. deep cubed = 24 389 × 24 inc. wide = 585 336 ÷ (25 feet long cubed =) 15 625 gives 37·46 to represent the stiffness of this beam × 3360 = 125 860 lbs. is the weight that would cause it to bend an inch. Lastly, the load 25 432 lbs. ÷ 125 860 gives ·202 of an inch for the deflexion of the beam ; that is one-1485th part of its length.

From this calculation it would appear, that in the engine at the Albion Mills, the great lever must have been bent down from its natural form, 2 tenths of an inch at each end, during the descending stroke of the piston ; and as it must have been bent as much upwards, during the ascending stroke, the flexure of the lever was more than 4 tenths of an inch in the whole.

The beams for great levers being proportioned according to the former rule, the flexure of long levers, for large engines, would be rather greater in proportion to their length, than for the smaller ones ; for instance, the flexure of the lever for the 20 horse engine would be only ·12 of an inch.

4 I 2

Another Example. An oak beam 20 inches deep, by 16 inches wide, and 16 feet long between the supports, having (576 circ. inc. × 22 lbs. =) 12 672 lbs. applied at the middle. Then 20 inc. deep cubed = 8000 × 16 inc. broad = 128 000 ÷ (16 feet long cubed) 4096 = 31·24 for stiffness × 3360 = 10 500. And 12 672 lbs. load ÷ 10 500 = ·1206 of an inch deflexion; that is, one-1590th part of the length of the lever (*a*).

CAST IRON LEVERS FOR STEAM-ENGINES.

In modern engines the great lever is always made of cast iron, instead of wood. The depth of the lever in the middle is made equal to the diameter of the cylinder, or nearly so; and for rotative engines, the breadth of the iron is one-108th part of the length of the lever; or the breadth in inches, is one-ninth part of the length in feet. The lever is a curved plate of cast iron placed edgeways upwards, and the depth at each end of the lever, is usually about one-third of the depth in the middle: see Plates XVIII and XXI. The axis is inserted through the middle of the lever, in a socket suitably formed for its reception; and as the breadth of the beam is so small in proportion to its depth, that it might twist sideways, it is strengthened by a plate, which projects out horizontally at each side of the lever, and extends all its length, along the middle of its depth, so as to make the transverse section of the beam like a cross, thus ✚; this plate is so thin edgeways, as not to occasion any great increase of weight.

The lever is farther strengthened by projecting borders, which extend all round the exterior margin on both sides, so as to make the top and bottom edges about 1½ times broader than the middle part. The depth of each of these projecting margins is about one-20th of the whole depth of the lever, at the middle. The metal contained in them, is included in the estimated breadth of the lever, when it is stated as being 108th part of the length; because the breadth is greater than that proportion, at the top and bottom edges, but less in the middle part.

The lever is cast in one piece, except the largest sizes, for engines of 7 and 8 feet stroke, and these are made of two separate plates, united together parallel to each other; the axis, and also the pins for the main joints, passes through both plates, and assists to unite them into one lever.

To find the dimensions for the cast iron levers for Mr. Watt's steam-engine.
RULE. Divide the square of the diameter of the cylinder in inches by 9, and multiply the quotient by the length of the lever in feet. *If the depth is fixed, then to find the breadth.* Square that depth in inches, and divide the above product by that square; the quotient will be the proper

---

(*a*) To find the actual deflexion of the beam from its natural form, we should take five 8ths of its own weight, as a part of the load which is applied to the middle of it, because a load which is distributed uniformly along the length of a beam has the same effect to bend it, as five 8ths of that load would have if it were applied at the middle of its length. We should also make an allowance of 3 lbs. per circular inch for the weight of the piston parallel motion and connecting rod, air-pump rod and bucket, in the same manner as is done for ascertaining the strength of the beam.

The object of the above examples is only to find how much the great lever is bent by the force exerted by the piston, therefore the load is taken at only 11 pounds per circular inch, instead of 14 lbs.; and the weight of the beam and other parts is omitted, because all the flexure which is occasioned by that weight will remain when the piston exerts no force, and the object of inquiry is to find how much that flexure is increased by the force which is exerted by the piston during the descending stroke, and how much it is diminished during the ascending stroke.

breadth in inches. *Or if the breadth is fixed, then to find the depth.* Divide the above product by the breadth in inches, and extract the square root of the quotient; that root will be the proper depth in inches.

Example. Suppose the lever for a cylinder, 44 inches diameter, is 21·6 feet long, and that it is 34 inches deep in the middle. Then 44 squared = 1936 circular inches area, ÷ 9 = 215·1 × 21·6 feet long = 4646 product, ÷ (34 inc. deep squared =) 1156 = 4·02 inches is the proper breadth for the great lever (*a*).

If the above lever had been 44 inches deep, which is the proper proportion, then the breadth of the lever would have been (4646 ÷ (44 squared =) 1936 =) 2·4 inches. *Note.* The breadth given by this rule is greater than the breadth of the middle part, and less than that of the projecting margins at top and bottom (*b*).

To perform the calculation by the sliding rule, we must divide the length of the lever in feet by 9, and use the product on the line ꝗ, thus:

Sliding Rule,   { ꝗ   One-9th length of lev. ft.   Bread. of lev. inc.   *Ex.* ꝗ (21·6ft.÷9=)2·4   4 inc.
slide inverted. { D   Diam. of the cylinder inc.   Depth of lev. inc.   D Dia. of cylin. 44   34 inc.

The above rule is intended for the levers of rotative engines, but the levers of engines for draining mines, frequently have a great weight of pump rods in addition to the force of the piston, such cases must be calculated by another rule, which proceeds according to the weight with which the lever is loaded.

We may estimate the total force which the great lever of a rotative engine sustains at each end, to be 14 lbs. per circular inch of the piston; or if the lever is supposed to be supported at each end, and the load applied in the middle, it may be assumed to be equal to 28 lbs. per circular inch of the piston. According to the preceding rule the depth of the lever is equal to the diameter of the cylinder, and the breadth is equal to one-108th part of its length; hence (108 inc. = 9 feet × 28 lbs. =) 252 pounds, is the weight which may be safely loaded on the middle of a cast iron bar, 1 inch square, and 1 foot long; the following rule is very nearly according to that proportion.

To find the proper dimensions for the middle part of a cast iron beam, which is required to sustain a given weight in the middle, when supported at each end; the depth of the beam at each end, being one-third of the depth at the middle.

RULE. Divide the weight in pounds, which is to be applied on the middle of the beam, by 250 (*c*), and multiply the quotient by the distance in feet, between the supports. *If the depth of the beam is given, then to find the breadth;* divide the above product by the square of that depth in inches; the quotient will be the proper breadth in inches. *Or if the breadth of the beam is given, then to find the depth;* divide the same product by that breadth in inches, and extract the square root of the quotient, that root will be the required depth in inches.

Example. The strain occasioned by a cylinder of 36 inches diameter = 1296 circular inches area, at 28 lbs. per circ. inc., is 36 288 lbs. ÷ 250 = 145 ·15 × 21·6 feet long = 3135. Then if the beam is to be 36 inches deep at the middle, 3135 ÷ (36 squared =) 1296 = 2·42 inches, is the proper breadth of the beam. Or if the breadth is to be 3 inches, then 3135 ÷ 3 = 1045, the square root of which is 32·33 inches, for the depth of the beam at the middle.

(*a*) The author has had constant opportunities of observing two engines of the above dimensions, in which depth of the lever is 34 inches, and the breadth of the plate is only 2 inches; as these engines have worked for some years, loaded to their utmost, without any signs of failure in the levers, they prove that the rule will give a very sufficient strength.

(*b*) The cast iron levers of the steam-engines made by Messrs. Boulton, Watt, and Co. are, in many cases, stronger than the above rule would give, the breadths of the levers being between one-108th, and one-84th part of their length. The slightest levers that they make, correspond to the above proportion. For instance, their 36 inch cylinder, has a great lever 36 inches deep at the middle, 21·58 feet long, and 2 4 inches broad at a mean, including the broad margins at the top and bottom edges. The rule would give (21·58 ft. ÷ 9 =) 2·396 inches.

(*c*) To make this rule strictly correct, one half of the weight of the beam itself, should be added to the weight which is supposed to be applied at the middle of it.

The calculation may be made with the Soho slide rule, with the slide inverted, using all the four lines; and in place of dividing by 250, we may take four times the length of the beam in feet, on either of the upper lines A or Ɔ.

Soho Rule, slide inverted, all four lines.	A	Four times the length of beam feet.		A	(21·6 ft. × 4 =)86·4	
	Ɔ	Weight applied on middle pounds.	Ex.	Ɔ	Weight lbs. 36 288	
	ᗺ	Breadth of the beam inches.		ᗺ	2 42 inc. broad	3 in. broad.
	D	Depth at the middle inches.		D	36 inc. deep.	or 32·3 in. deep.

The above rules, for cast iron beams, and for oak beams, assume that the strength of a beam of cast iron is 6·25 times stronger than a beam of good English oak of the same dimensions. But by the results of direct experiments, it appears to be only $4\frac{5}{8}$ times. The average of a number of experiments made by Mr. Banks, shows that a bar of cast iron one inch square, and one foot long between the supports, will be broken by 2560 pounds applied on the middle; and similar experiments by Mr. George Rennie give more than 2700 pounds. The above rule assumes the load to be at the rate of 250 lbs., and therefore it gives the beam a sufficient strength to bear between 10 and 11 times the weight with which it is loaded.

SUN AND PLANET-WHEELS FOR MR. WATT'S ROTATIVE ENGINE.

The sun and planet-wheels usually have the same number of teeth in each; and assuming the main joints at each end of the great lever, to be equally distant from the centre of motion, then the geometrical diameters of the wheels (that is the diameters of their pitch circles) must be equal to half the length of the stroke made by the piston.

The sun-wheel is wedged fast upon a projecting square, at the extreme end of the axis of the fly-wheel. For small engines, and up to 20 horse-power, the sun-wheels were made with wooden teeth, driven into mortices in the cast iron rim. The planet-wheels were always made with solid teeth of cast iron, and for large engines the sun-wheels were the same. The planet-wheel is fastened against the connecting-rod, by its centre pin, which passes through both, and it is secured to the rod by three staple bolts, which pass round three of the arms of the wheel; and by nuts screwed on the ends of the bolts in front of the connecting-rod, the wheel is held very fast, and prevented from turning round upon its own centre pin.

The teeth of the sun and planet-wheels must be sufficiently strong to enable them to endure the whole force of the piston, without danger of breaking. From an examination of some of Messrs. Boulton and Watt's earliest engines, it does not appear that they had a settled rule for the proportions of the teeth, at their first commencement, but in the course of their practice they determined the dimensions of the teeth, for the sun and planet-wheels of each size of engine; and those dimensions will accord nearly with the following rule, which assumes that the strength of the teeth of cog-wheels, may be represented by the product obtained by multiplying the square of the pitch of the teeth in inches, by the breadth of the teeth in inches (a). Also that teeth of one inch pitch, and one inch broad, are competent to transmit a force of 176 pounds, without any risk of breaking.

(a) The pitch of the teeth of wheels, is the distance from the centre of one tooth, to the centre of the next, measured on the pitch line, or geometrical circle of the wheel. The breadth of the teeth is measured across the edge of the wheel, in a direction parallel to its axis. The length of the teeth is the distance that they project out from the circular rim of the wheel. The thickness of the teeth is their substance measured in the same direction as the pitch, and is usually rather less than half the pitch.

To find the pitch of the teeth of any cog-wheel. Having given, the geometrical diameter of the wheel (that is the diameter of the pitch circle) in feet, and the number of the teeth.

RULE. Multiply the diameter of the pitch circle in feet, by 37·7 inches, and divide the product by the number of teeth; the quotient is the pitch of the teeth in inches.

Example. Suppose a wheel 2½ feet diameter to have 37 teeth. Then 2·5 ft. dia. × 37·7 = 94·25 inches circumference of the pitch circle ÷ 37 teeth = 2·546 inches pitch of teeth.

Sliding Rule,	A	37·7 inches.	Number of teeth.	Exam.	A	37·7 inc.	37 teeth.
slide inverted.	C	Pitch dia. feet.	Pitch of teeth inc.		C	2·5 ft. dia.	2·55 pitch.

To find the breadth for the teeth of the sun and planet-wheels of Mr. Watt's Rotative Steam-engine. Having given the diameter of the steam cylinder in inches, and the pitch of the intended teeth in inches.

RULE. Divide the square of the diameter of the cylinder in inches, by 16 times the square of the pitch of the teeth, in inches. The quotient is the proper breadth for the teeth in inches.

Example. Suppose the cylinder is 24 inc. diameter, and the sun-wheel 2½ feet diameter, with 37 teeth; their pitch will be 2·55 inches, the square of which is 6·5, and × 16 = 104 for the divisor. And 24 inc. diam. squared = 576 circular inches ÷ 104 = 5·54 inches for the breadth.

Sliding Rule,	B	625 gage point.	Breadth of teeth inc.	Example.	B	625	5·54 brd.
slide inverted.	D	Dia. of cylind. inc.	Pitch of teeth inches.		D	24 dia.	2·55 pit.

Note. This gage point is really ·0625, the reciprocal of 16, used as a multiplier.

The sliding rule when thus set, forms a table for that size of cylinder; so that any other pitch for the teeth being chosen on the lower line D, the proper breadth, to give the teeth the same strength, will be opposite to it on the line B. For instance, if the above sun-wheel had 42 teeth, then the pitch would be 2·244 inc., and according to the sliding rule, its breadth ought then to be 7·15 inches (a).

The dimensions for the sun and planet-wheels for the standard engines, would be as follows according to this rule.

For a 10 horse-engine, 17½ inc. cylinder. The wheels 2 feet diameter, 32 teeth, 2·356 inc. pitch, 3·45 inches broad. N.B. They were usually made 4 inc. broad (see p. 491), but the same wheels served for a 12 horse-engine.

For a 20 horse-engine, 23¾ inc. cylinder. The wheels 2½ feet diameter, 36 teeth, 2·618 inc. pitch, 5·15 inches broad. They were usually 5 inches broad.

For a 30 horse-engine, 28¼ inc. cylinder. The wheels 3 feet diameter, 40 teeth, 2·828 inc. pitch, 6·23 inches broad. They were usually 6 inches broad.

For a 40 horse-engine, 31½ inc. cylinder. The wheels 3½ feet diameter, 44 teeth, 3 inc. pitch, 6·9 inches broad.

For a 50 horse-engine, 34 inc. cylinder. The wheels 4 ft. diameter, 50 teeth, 3·015 inc. pitch, 8 inches broad. N.B. This size of engine was made with two sun and two planet-wheels, to turn two fly-wheels (see p. 512); the breadth of the teeth of each wheel was 5 inches; or equal to 10 inches broad, which is much stronger than the calculation.

The modern steam-engines which are made with multiplying wheels and pinions to turn the fly-wheel, have 1·6 times greater strength in the teeth, than if they were proportioned by the above rule; but as the sun and planet-wheels were very rarely broken, the modern proportions may be considered as giving a great

(a) The author has had continual opportunities during 20 years past, of observing the condition of a pair of sun and planet-wheels in an engine of 20 horse-power (see p. 498), with a cylinder 24 inc. diam. and 5 ft. stroke, which was constructed by Messrs. Boulton and Watt in 1792, and has continued in constant use to the present time, 1827. These wheels have 42 teeth (= 2·24 inc. pitch), and are only 5 inches broad, so that they are but seven 10ths of the strength given by the calculation. The sun-wheel has wooden teeth, which require to be renewed every second year. The planet-wheel has iron teeth, and they become worn, so as to require a new wheel, once in 8 or 10 years. The engine is fully loaded, and the teeth have been broken once or twice by accident.

The first rotative-engines which Mr. Watt made for the Breweries in London (see p. 437), are 24 inch cylinders, and 6 feet stroke; the sun and planet-wheels being 3 feet diam., have 44 and 45 teeth = 2·54 inc. pitch, and were 4 inches broad; the preceding rule would give 5·57. Several of these wheels have been in constant use for 36 years. The wooden teeth of the sun-wheels are made of live oak, and they require to be renewed every third or fourth year. Some of these engines are now working with cylinders of 26¼ inc. diam., and sun and planet-wheels with 44 and 45 teeth, but 5¼ inches broad; the rule would give 6·67 inc. broad. These instances show that the strength given by the rule is sufficient to avoid breaking.

excess of strength to the teeth, so that they can scarcely ever be broken by any accident to which steam-engines are liable.

It is a good practice to give the planet-wheel one more tooth than the sun-wheel, in order that the same teeth may not meet each other frequently, but that each tooth of the planet-wheel may in succession act with every one of the teeth in the sun-wheel. It was formerly a rule amongst mill-wrights to put fractional numbers of teeth into all wheels which are to work together, in order to compensate for any unequality amongst the teeth; for if any one tooth of one wheel is larger, or rougher, or harder, than the rest, it will by this expedient be caused to act in succession amongst all the teeth of the other wheel, so as to wear them all equally; the extra tooth which is put into the large wheel for this purpose, is called *the hunting cog*. This custom has fallen into disuse since workmen have become expert in forming all the teeth of a pair of wheels very accurately alike; but it is a correct principle, and produces a good effect in the sun and planet-wheels, or the multiplying wheels of steam-engines, because the piston exerts a much greater strain upon the teeth, when it is near the middle of its course, than when it is at the top or bottom, consequently those teeth which are then in contact, will be subjected to greater wear than the others.

If the planet-wheel has an extra tooth, all the teeth of the sun-wheel must successively be subjected to this greatest stress of the piston, so as to become equally worn by it. Those teeth of the planet-wheel which are at its two sides, near to its horizontal diameter, will wear away more rapidly than those which are nearer to the top and bottom of the wheel; and the former will be worn at the fronts, or driving sides of the teeth, whilst the latter are worn at the backs, or following sides of the teeth, for when the piston is near the top or bottom of the stroke, the sun-wheel must push the planet-wheel round in its orbit, in order to pass it through the inactive period of its course; but during the greater part of the stroke, the planet-wheel must impel the sun-wheel.

If there is any play or looseness between the teeth, an unpleasant jerk will take place, every time that the planet-wheel passes over the top, or under the bottom, of the sun-wheel, in consequence of the contact of the teeth changing from the leading side, to the following side, and vice versa. This is called *back lashing*, and to avoid it, the teeth of the two wheels must be made to fit each other very accurately in the first instance; and when the teeth at the sides of the planet-wheel are at all diminished by wearing, it should be loosened from the lower end of the connecting-rod, and turned round one-sixth on its centre pin, and fastened again in the new position; this being done every 12 months, the teeth of the planet-wheel will wear away very equally, as well as those of the sun-wheel.

When the teeth of the sun and planet-wheels become worn away, so as to render them loose, the planet-wheel should be replaced by a new one, having rather thicker teeth than those of the original wheel, and fitting exactly between the reduced teeth of the sun-wheel; when these teeth are worn loose, a new sun-wheel should be made, with thick teeth, to fit the reduced teeth of the original planet-wheel, which should be kept in use until the teeth of the new sun-wheel become worn, and then it should be exchanged for the planet-wheel with thick teeth. By thus having two spare wheels, and changing them occasionally, they may always be kept in excellent order.

For small engines up to 20 horse-power, the sun-wheels were made with wooden teeth, driven into mortices formed in the cast iron rim, and fastened in by cross pins; this is an excellent method, as the wooden teeth receive the action of the iron ones of the planet-wheel, without any noise or rattle. The wood called live oak is found to answer best for teeth, but crab-tree or beech is used. The wooden teeth wear away, so as to require to be renewed every two or three years, but the iron teeth of the planet-wheel do not wear sensibly; and by having two planet-wheels, one with thin teeth, to be used when the wood teeth of the sun-wheel are new and full, and the other with thicker teeth, to be put in when the wood teeth are worn, the teeth may be always made to fit each other.

The sun-wheel is fastened upon the projecting end of the axis of the fly-wheel; that part is square, and the wheel having a square hole through its centre, is fastened upon the axis by eight taper keys, or long narrow wedges, which are driven very fast into the space between the axis, and the hole in the wheel, the length of the wedges being parallel to the axis (see p. 493); two wedges are inserted near each angle; these wedges require to be filed true, and accurately fitted to their places, so as to jamb very fast when they are driven in tight; if this is not well executed, they are liable to become loose in working. As a security, a round flat plate of iron is usually applied over the end of the axis, to cover the ends of all the eight wedges, and conceal them; this plate is fastened to the axis by a screw tapped into the centre of it, and the ends of the wedges being left rather prominent, the plate always bears against them, so as to press them into their places, and prevent them coming out, if they do work loose.

When the sun and planet-wheels have the same number of teeth in each, the fly-wheel will make two revolutions for each complete stroke of the piston. If the sun-wheel is smaller than the planet-wheel, then the fly-wheel will make more than two revolutions at every stroke, according to the following rule (see p. 450).

To find what number of revolutions the fly-wheel will make per minute, when the sun and planet-wheels have different numbers of teeth.

RULE. Multiply the number of strokes made per minute by the piston, by the number of teeth in the planet-wheel, divide the product by the number of teeth in the sun-wheel, the quotient is the additional number of revolutions per minute that the fly-wheel will make; therefore add the number of strokes made per minute by the piston, and the sum is the number of revolutions the fly-wheel will make per minute.

Example. The 10 horse-engine, p. 491, makes 25 strokes per minute. The planet-wheel had 31 teeth, and the sun-wheel 33 teeth. Then 25 strokes × 31 teeth = 775 ÷ 33 teeth = 23·48 additional revolutions + 25 strokes = 48·48 revolutions of the fly-wheel per minute; it is stated at 48½.

Sliding Rule, slide inverted.	A Strokes of pist. per min.	Teeth in sun-wheel.	Ex.	A 25 strokes.	33 teeth.
	◯ Teeth in planet-wheel.	Additional revolut.		◯ 31 teeth.	23·48 rev.

To find the pitch of the teeth for the sun and planet-wheels, when they have different numbers of teeth.

RULE. Add the diameters in feet of both wheels together; the sum will, in most cases, be the length of the stroke of the piston; multiply that sum by 37·7 inches, and divide the product by the sum of the numbers of teeth in both wheels; the quotient is the pitch in inches.

Example. The stroke of the piston being 4 feet; the sun-wheel having 33 teeth, and the planet-wheel 31 teeth. If the two ends of the great lever are of equal length, then the diameters of the two wheels together must be 4 feet, × 37·7 inc. = 150·8 inches circumference of both wheels ÷ (33 teeth + 31 teeth =) 64 teeth = 2·356 inches pitch.

The radial link, q, behind the sun and planet-wheels, which retains the centre of the planet-wheel in its circular orbit (see p. 449), is constructed in the same manner as the main links for the parallel motion (see p. 472); and the same size of iron is used for making the radial link, as is used for the main links. The centre pin of the planet-wheel, which projects out behind the wheel, and upon which the radial link is fitted, is about one-eighth of the diameter of the cylinder, and its length equal to its diameter; this pin is made of wrought iron. The

4 K

other end of the radial link is fitted upon the neck of the axis of the fly-wheel, which is made of cast iron, and its diameter is proportioned according to the following rule.

DIMENSIONS FOR THE ROTATIVE AXES OF MR. WATT'S STEAM-ENGINES.

In axes which transmit circular motion, the strength of the cylindrical necks to resist twisting off, is in proportion to the cubes of their diameters; the lengths of the necks being supposed to bear some constant proportion to their diameters.

To find the proper diameter for the neck of the cast iron axis of the fly-wheel, for Mr. Watt's rotative steam-engine, with sun and planet-wheels.

RULE. Multiply the square of the diameter of the cylinder in inches, by the diameter of the sun-wheel in feet; multiply the product by the constant decimal ·15, and extract the cube root of the last product. That root will be the proper diameter for the neck in inches.

Example. Suppose the cylinder to be 24 inches diameter, and the sun-wheel $2\frac{1}{2}$ feet diameter. Then 24 inc. squared = 576 circular inches × 2·5 ft. diam. = 1440 × ·15 = 216; the cube root of which is 6 inches, for the diameter of the neck (a).

To calculate this by the sliding rule two operations are required on the lines q and D, thus:

Sliding Rule, slide inverted. { q Diam. of sun-wheel ft.   Cube of diam. of neck inc.   *Ex.*   q 2·5 ft. dia.   216 cube

{ D Diam. of cylinder inc.   258 gage point.   D 24 inc. dia.   258 g. pt.

Then to extract the cube root of the number thus obtained, proceed as follows:

Sliding Rule, slide inverted. { q Cube   Diam. of neck inches. } This is where the same   *Ex.*   q 216   6 inc.

{ D 10   Diam. of neck inches. } numbers meet on both.   D 10   6 inc.

In the old engines the length of bearing was the same as the diameter; but in more modern engines the length was increased to $1\frac{1}{5}$ times the diameter.

Dimensions of the necks of the fly-wheel axes, for the standard engines, according to the above rule. They are made of cast iron.

10 horse ..	$17\frac{1}{2}$ inc. cylinder.	Sun-wheel 2 ft. diam.	Neck 4·5 inc. diam. by 5·4 inc. long.
20 p. 498 ..	$23\frac{3}{4}$	.... $2\frac{1}{2}$	.. 5·96 ...... 7·15 ....
30 p. 501 ..	$28\frac{1}{4}$	.... 3	.. 7·11 ...... 8·53 ....
40 p. 504 ..	$31\frac{1}{2}$	.... $3\frac{1}{2}$	.. 8·06 ...... 9·67 ....
50 p. 519 ..	34	.... 4 .... Two necks 7·03	...... 8·43 ....

The modern engines, which are made with a crank, instead of sun and planet-wheels, require the neck of the axis of the crank to be twice as strong as is required for the neck of the axis of the sun-wheel; because the crank acts to twist off the neck, with a leverage twice as long as that of the sun-wheel. As the fly-wheel is fixed upon the axis of the sun-wheel, the neck requires strength to bear the weight of the fly-wheel, as well as to resist the twist to turn the axis round.

In engines which act with a crank, and have only a multiplying-wheel upon the axis of that crank, to turn the fly-wheel by a pinion on a second axis, the neck

---

(a) The earliest rotative engines made by Mr. Watt for the London Breweries with 24 inch cylinders, being for a 6 feet stroke, instead of 5 feet, the planet-wheels were 3 feet diameter (see p. 437); the necks of the axes were 6 inches diameter, or only $\frac{5}{6}$ of the strength given by the above rule. Some of these engines have been in constant use for 40 years past, without any failure.

of the axis of the crank will have only the strain of torsion to bear; in such cases the following rule may be used.

To find the proper diameter for the cast iron neck of the axis of the crank for Mr. Watt's rotative steam-engine, when the fly-wheel is not upon that axis, but only a multiplying-wheel.

RULE. Multiply the square of the diameter of the cylinder in inches, by the length of the stroke in feet; divide the product by 8, and extract the cube root of the quotient. That root is the proper diameter for the crank neck in inches; and that diameter, multiplied by 1·2, will give the proper length of bearing for the neck, in inches.

Example. Cylinder 36 inches diameter squared = 1296 circular inches × 7 feet stroke = 9072 ÷ 8 = 1134, the cube root of which is 10·43 inches for the diameter of neck. And (10·43 × 1·2 =) 12·5 inches is the proper length of bearing.

Sliding Rule, ⌐ B	Length of stroke feet.	Cube of dia. of neck inc.	*Ex.* B	7 feet.	1134
slide inverted. ⌐ D	Diam. of cylind. inc.	283 gage point, or 895.	D	36 inc.	283

Then to extract the cube root of the number so obtained, proceed as follows:

Sliding Rule, ⌐ B	Cube.	Diam. of crank neck inches.	*Exa.* B	1134	10·43 inc.
slide inverted. ⌐ D	10	Diam. of crank neck inches.	D	10	10·43 inc.

According to this rule, a small engine with a cylinder 2·83 inches diameter = 8 circular inches, and the piston making 1 foot stroke, would require the crank neck to be one inch diameter, and 1⅕ inches length of bearing.

When the fly-wheel is fixed on the axis of the crank, as is the universal practice in modern engines, the neck of that axis should be rather stronger than the last rule, to enable it to bear the weight of the fly-wheel with safety. The same rule, as is before given for engines with sun and planet-wheels, may be used, only substituting the length of the stroke, for the diameter of the sun-wheel.

Example. Cylinder 36 inches diameter = 1296 circular inches × 7 feet stroke = 9072 × ·15 = 1361, the cube root of which is 11·08 inches for the diameter of the neck. And (11·08 × 1·2 =) 13·3 inches, is the proper length of bearing.

The utmost force exerted by the piston, at the most favourable period of its action, being assumed to be at the rate of 11 pounds per circular inch of the piston, then a crank neck, proportioned by the rule with 8 for a divisor, would be twisted with a strain, which may be found by multiplying the cube of its diameter in inches, by 88 pounds; the product will represent the force of torsion in pounds, acting with a leverage of half a foot from the centre. Hence a cast iron neck one inch diameter, and 1⅕ inch length of bearing, may be safely subjected to a twisting action of 88 pounds, acting at half a foot from the centre.

Example. A cast iron neck 10·43 inc. diam. cubed = 1134 × 88 lbs. gives 99 792 lbs. acting at six inches radius from the centre, for the force of torsion which this neck should endure, according to the rule.

Proof. 36 inches diameter of cylinder, squared is 1296 circular inches × 11 lbs. = 14 256 lbs. utmost force of the piston; as this acts at 3½ feet radius, it is equal to (14 256 × 7 =) 99 792 lbs. force of torsion, acting at 6 inches radius.

Mr. Dunlop, of Glasgow, made a series of experiments upon the force required to twist off cast iron necks of different sizes, from 2 inches diameter to 4¼ inc. diameter; the lengths of bearing being 1¼ times the diameter. It appears from these experiments, that if the cube of the diameter of the neck in inches, is multiplied by 880, the product will be nearly the force of torsion in pounds, which acting at half a foot radius, will actually twist them off. Hence the necks,

4 K 2

proportioned by the above rule, will only be twisted with one-tenth of the force which would break them off.

According to the preceding statement, a cast iron neck of 1 inch diameter, and $1\frac{1}{5}$ inches long, is capable of sustaining a force of torsion, amounting to 88 lbs. acting at half a foot distance from the centre of motion, without any risk of injury, for it will require ten times that force, or 880 pounds to twist off the neck. If such a neck made one turn per minute, the mechanical power it would transmit, may be represented by a force of 88 lbs., acting uniformly through the circumference of a circle one foot diameter, or a space of 3·1416 feet per minute; which is equal to 276·46 lbs., acting through a space of one foot per minute. This, divided by 33 000 lbs. for a horse-power, gives ·00838 of a horse-power, which may be transmitted by a cast iron neck one inch diameter, and $1\frac{1}{5}$ inc. long, when making one turn per minute. Or conversely, such a neck must make 119·37 revolutions per minute, in order to transmit one horse-power.

From the practice of the most experienced millwrights, it would appear that they do not follow any rule to proportion the size of the necks of shafts, according to the velocity of their rotation, and to the power they are to transmit, for different cases, will be found to vary very greatly; and there are but few instances in steam-mills, of necks which correspond with the above proportion, although there can be no doubt of its sufficiency, because it is derived from the strain to which the crank necks of steam-engines are subjected.

The following rule, which may be taken as an average of the millwrights' practice, assumes that a neck of cast iron, 1 inch diameter and $1\frac{1}{5}$ inches long, must make 300 turns per minute, in order to transmit one horse-power; that is $2\frac{1}{2}$ times the strength allowed for the crank necks of Mr. Watt's engines; and the force which would be required to twist off the necks, would be 25 times the force to which they are subjected, when they are proportioned according to the following rule.

To find the dimensions for the cast iron neck of an axis, which is to transmit any given number of horse-power, when it makes a given number of revolutions per minute.

RULE. Multiply the number of horse-power that the neck is to transmit, by the constant multiplier 300; divide the product by the number of revolutions that the shaft is to make per minute, and extract the cube root of the quotient. That root is the proper diameter for the cast iron neck in inches; supposing its length of bearing to be $1\frac{1}{5}$ of that diameter.

Example. The whole force of an engine which exerts 60 horse-power is transmitted by the neck of an axis, which makes 50 revolutions per minute. Then 60 HP × 300 = 18 000 ÷ 50 revolutions = 360, the cube root of which is 7·11 inches for the diameter of the neck. And (7·11 × 1·2 =) 8·53 inches is the proper length of bearing. The neck really is 7 inches diameter, and has worked many years.

To perform this calculation by the slide rule, with the slide inverted, we must first divide the number of horse-power that is to be transmitted by the neck, by the number of revolutions it is to make per minute, and set the quotient on the line ʚ, opposite to the gage point 5·48 (the square root of 30) upon D, thus:

Sliding Rule, slide inverted. { ʚ HP ÷ rev. per mi. Dia. of neck inc. / D Gage pt. 547 or 173. Dia. of neck inc. } *Exa.* ʚ (60 HP ÷ 50 rev. =) 1·2 7·11 inc. dia. / D 173 or 547 7·11 inc. dia.

Wrought iron axes for steam-engines. In some of Mr. Watt's earliest engines, the neck for the axis for the fly-wheel was made of wrought iron fitted into the end of the axis, which was a hollow tube of cast iron, and the neck was fastened in, by a cross pin put through both. In modern engines for steam-boats, the axes for the cranks are usually made of wrought iron, and it would be a good practice to use wrought iron, for the main axes of steam-engines for many other purposes.

To find the diameter for the wrought iron neck, for the crank axis of Mr. Watt's rotative steam-engine.

The rule already given (p. 619) will serve, by using 9·6 or 10 for a divisor instead of 8, thus:

Example. An engine with a cylinder 32 inches diameter (squared) = 1024 circular inches × 3 feet stroke of the piston = 3072 ÷ 9·6 = 320, the cube root of which is 6·84 inches for the diameter of the neck. And (6·84 × 1·2 =) 8·2 inches is the length of bearing.

To perform the calculation for wrought iron necks by the sliding rule, we must proceed as before directed in the precept for cast iron necks (p. 619), only using the gage point 31 or 458, instead of 283 or 895.

According to this rule a wrought iron neck, one inch diameter and $1\frac{1}{5}$ inches long, will be able to transmit one horse-power, when it makes (119·37 × 8 = 954·9 ÷ 9·6 =) 99·5 ; say 100 revolutions per minute.

The main axis of the crank is usually of a square form, and the side of that square is about one-tenth larger than the diameter of the crank neck, as given by the preceding rules. Therefore, multiply the diameter of the neck by 1·1, and the product is the side of the square for the axis.

Example. For a 36 inch cylinder, 7 feet stroke, the crank neck being 10·43 inches diameter. Then 10·43 × 1·1 = 11·47 inches square, is the proper size for the axis. Its length is about 5 feet, between the bearings.

Note. The strength of a neck, or of an axis, to resist twisting, is less when the length exposed to the twist is greater ; hence the above rules for necks must be considered as adapted to some certain proportion between their diameters and their lengths. The main axes of the cranks are, in all cases, so short between the bearings, in proportion to the size of the axis, that if the side of the square is one-tenth larger than the diameter of the neck, the axis can never suffer from the twisting action ; for it can only require that size, to enable it to carry the weight of the multiplying-wheel, or of the fly-wheel, without risk of breaking by a sudden concussion, to which it may occasionally be subjected, from a slight lifting up of the axis from its bearings, if they are allowed to get loose.

STRENGTH OF THE MULTIPLYING-WHEELS AND PINIONS, TO TURN THE FLY-WHEEL.

The multiplying-wheels of modern steam-engines are usually made considerably larger in diameter, than the circle described by the crank pin, and hence the pressure to which the teeth are subjected is reduced, so as to be less than the whole force exerted by the piston. The size of the multiplying-wheel is regulated by convenience, and according to the purpose to which the engine is to be applied.

When an engine is required to turn machinery with a rapid motion, as in spinning-mills, it is advantageous to use a large multiplying-wheel, in order to gain a considerable increase of velocity by the first motion. In such cases the diameter of the wheel may be twice the length of the stroke of the piston, and then the pressure to be transmitted by the teeth of the wheel will be only half as much as the force exerted by the piston. In general, the diameter of the wheel is about one and a half times the length of the stroke, or between that and twice. So that for a 4 feet stroke, the wheel will be 6 feet diameter ; for a 5 feet stroke, $7\frac{1}{2}$ feet diameter ; for a 6 feet stroke, 9 feet diameter ; for a 7 feet stroke, $10\frac{1}{2}$ feet diameter ; and 8 feet stroke, 12 feet diameter. The pinions may be from one-half, to one-fourth, of the size of the multiplying-wheel, so as to turn the fly-wheel from twice to four times, at each stroke of the piston.

The teeth of the multiplying-wheels used in modern engines, are made much stronger than those of the sun and planet-wheels used by Mr. Watt, or according to the rule in p. 615. The following rule will proportion the teeth for the multiplying-wheels of steam-engines on Mr. Watt's principle, so as to correspond nearly with the practice of the best engineers.

To find the dimensions for the cast iron teeth of the multiplying-wheel of Mr. Watt's rotative engine.

RULE. Multiply the square of the diameter of the cylinder in inches, by the length of the stroke in feet; divide the product by the diameter of the pitch circle of the wheel in feet, and take one-tenth of the quotient to represent the proper strength for the teeth; that is, to represent the product of the square of the pitch of the teeth in inches, into their breadth in inches. Therefore, *If the pitch for the teeth is given, then to find their breadth;* divide the above number representing the strength, by the square of the pitch in inches; the quotient is the proper breadth for the teeth in inches. *Or if the breadth for the teeth is given, then to find their pitch;* divide the number representing the strength, by the breadth in inches, and extract the square root of the quotient; that root is the proper pitch for the teeth in inches.

Example. Suppose the diameter of the cylinder to be 36 inches, the piston to make a 7 feet stroke, and the multiplying-wheel to be 12 feet diameter. Then 36 squared = 1296 circular inches × 7 feet stroke = 9072 ÷ 12 feet diameter = 756; and one-tenth of this is = 75·6 for the strength of the teeth. Suppose that the wheel is required to have 151 teeth; then 12 ft. dia. × 37·7 = 452·4 inc. circumference, ÷ 151 teeth = 3 inc. pitch, the square of which is 9. And 75·6 strength ÷ 9 = 8·4 inches is the proper breadth for the teeth. They are 8¼ inches broad.

Or suppose that the wheel is to be 9 inches broad; then 75·6 ÷ 9 = 8·4, the square root of which is 2·9 inches for the pitch of the teeth; 2·9 inc. pitch would give 156 teeth to the wheel.

To perform the calculation by the sliding rule, with the slide inverted, the length of the stroke of the piston in feet, must be previously divided by the diameter of the pitch circle of the wheel in feet; this may be done by the lines A and ꙅ, with the slide inverted. For instance, 7 feet stroke ÷ 12 feet diam. = ·583; this quotient being found on the line ꙅ, the slider must be set, to bring it opposite to the diameter of the cylinder in inches, on the line D, thus:

Sliding Rule, { ꙅ Stroke ÷ dia. of wheel.   Breadth of teeth inc.   *Exa.*   ꙅ (7 ÷ 12 =) ·583.   8·4 in. br.
slide inverted. { D Diam. of cylin. inches.   Pitch of teeth inc.   D   36 inc. diam.   3 inc. pit.

The rule being thus set, forms a table of all the different pitches and corresponding breadths of teeth, which will have the proper strength. For instance, opposite to 10 inches broad on ꙅ is 2·75 inc. pitch on D, or for 8 inches broad 3·07 inc. pitch.

The teeth of the multiplying-wheel being proportioned by the above rule, will have a sufficient strength to sustain the utmost force that the piston can exert upon them, when it is near the middle of its course, and consequently they will be much stronger than is necessary, to sustain the force which the piston exerts when it is passing through other parts of its course. The greatest force is 3·14 times the mean or average of the force exerted through the whole stroke. See p. 416.

According to the proportion of the above rule, if we suppose a small engine with a cylinder 3·16 inches diameter = 10 circular inches, and with a multiplying-wheel one foot diameter (the length of the stroke of the piston being the same), then, according to the rule, the size of the teeth should be one inch pitch, and one inch broad. The teeth would have to bear a force of (10 circ. inc. × 11 lbs. =) 110 lbs., and that they can sustain without any risk of injury. If we suppose the pitch circle of such a wheel to be one foot diameter, and that the wheel makes one turn per minute, then the mechanical power that it would be capable of transmitting, may be represented by 110 lbs. moved through the circumference of a circle one foot diameter, or 3·1416 feet per minute; that is equal to 345·576 lbs. acting through a space of one foot per minute, which divided by 33 000 lbs., for a horse-power, gives ·01047 of a horse-power, which may be transmitted by the teeth of a cast iron cog-wheel, one foot diameter, with teeth of one inch pitch, and one inch broad, when it makes one turn per minute. Or reciprocally, such a wheel should make 95·5 revolutions per minute, in order to transmit one horse-power by its teeth.

As this proportion is obtained from the strain to which the teeth of the multiplying-wheels of steam-engines are actually subjected, every time that the piston arrives near the middle of its course, there can be no doubt of its sufficiency; but according to the practice of the best mill-wrights, the teeth of cog-wheels are made much stronger than the above proportion.

The following rule is formed from the average of a number of good examples; it assumes that a cast iron wheel one foot diameter, with teeth one inch pitch, and one inch broad, must make 240 revolutions per minute, in order to transmit one horse-power by its teeth; hence it will give $2\frac{1}{2}$ times the strength which is allowed for the teeth of the multiplying-wheels of Mr. Watt's engines, according to the preceding rules, and 4 times the strength allowed to the teeth of the sun and planet-wheels, according to the rule given in page 615.

To find the proper dimensions for the teeth of a cast iron cog-wheel, which is required to transmit a given number of horse-power, when it makes a given number of revolutions per minute.

RULE. Multiply the diameter of the pitch circle of the wheel in feet, by the number of revolutions it is required to make per minute, and reserve the product for a divisor. Multiply the number of horse-power to be transmitted, by 240, and divide the product by the above divisor; the quotient represents the proper strength for the teeth. *If the pitch is given, then to find their breadth;* divide the above strength, by the square of the pitch in inches; the quotient is the proper breadth for the teeth in inches. *Or if the breadth of the teeth is fixed, then to find their pitch;* divide the above strength, by the breadth in inches, and extract the square root of the quotient; that root is the proper pitch for the teeth in inches.

Example. The whole power of an 80 horse-engine, is transmitted by the teeth of a cog wheel 9 feet diameter, which makes 17 revolutions per minute. Then 9 ft. dia. × 17 revolutions = 153 for a divisor. And 80 HP × 240 multiplier = 19 200, which ÷ 153 divisor, gives 125·5 for the strength of the teeth. If the breadth were $12\frac{1}{3}$ inches, then 125·5 ÷ 12·33 = 10·18, the square root of which is 3·19 inches, for the pitch of the teeth. Or if the pitch is $3\frac{1}{4}$, the square is 10·56, and 125·5 ÷ 10·56 = 11·87 inches for the breadth of the teeth. The pitch of the teeth really is 3·14 inches, and the breadth 12 inches. The wheel has 108 teeth.

To perform the calculation by the sliding rule, with the slide inverted, the diameter of the wheel in feet, must be previously multiplied by the number of revolutions it is to make per minute, and after the slider has been set as follows, the product being sought upon either of the lines, A or ꓳ, the number representing the strength of the teeth will be opposite.

Sliding Rule, { A HP exerted.	Dia. wheel ft. × rev. per min.	*Ex.*	A 80 HP	9 × 17 = 153
slide inverted. { ꓳ Multiplier 240.	Strength of the teeth.		ꓳ 240	125·5 strength.

Then set the slider with that number on the line ꟼ, opposite to 1 or 10 on the line D:

Sliding Rule, { ꟼ Strength of teeth.	Breadth of teeth inc.	*Ex.*	ꟼ 125·5 strong	12·33 inc.
slide inverted. { D 10	Pitch of teeth inc.		D 10	3·19 inc.

The slider being thus set, the lines ꟼ and D form a table of breadths and corresponding pitches.

In all cases where the stress upon the teeth will be constant and uniform, the teeth will not require so much strength as the above rule will give; but it should be used for wheels which are so connected with a steam-engine, that they may occasionally be subjected to the utmost strain which the piston can exert when it is near the middle of its course, although in the usual manner of working the stress upon the teeth may be uniform and constant.

From the preceding statement, it appears that the practice of millwrights is to give a very great excess of strength, to the teeth of cog-wheels, and to the necks of shafts. The reason of this custom is, that they have adopted the dimensions of the wheels and axes used in Mr. Watt's steam-engines, according to the horse-power exerted, assuming it to be produced by a uniform pressure, instead of forming a correct estimate of the strain to which those parts are actually subjected in working, and which varies so greatly, that at the most effective part of every stroke of the piston, the strain is 3·17 times greater than the force which would produce the given number of horse-power, if it acted uniformly and continuously.

The action of the piston upon the neck of the axis of the crank, and upon the teeth of the multiplying-wheel, is a continual alternation of very powerful impulses which are given every time that the piston arrives near the middle of its course, with intervals of very diminished force, and even of total cessation, at the end of each stroke ; the strength of the parts must be adapted to bear the utmost momentary force. In axes and wheelwork which are urged by a water-wheel, or by the axis of the fly-wheel of a steam-engine, the strain is uniform, and then less than one-third of the greatest strain to which crank necks and multiplying-wheels, are subjected, would produce the same power.

The horse-power in steam-engines is calculated at 6·944 pounds effective pressure, acting continually and uniformly upon each square inch of the piston (see p. 575) ; but the utmost force exerted by the piston at its most favourable moment, is 14 lbs. per square inch (see p. 600), or more than double the supposed uniform pressure. The leverage of the crank when it is at right angles to the connecting-rod, is at the rate of 1·57 times the leverage which would produce the same effect with a constant action ; because the crank pin describes the circumference of a semicircle, whilst the piston passes through its diameter (see p. 415). Hence, the utmost effort of the piston, when transmitted by the crank at its most favourable position, is (14 lbs. ÷ 6·944 lbs. = 2·017 × 1·57 times =) 3·17 times the average of all the variations of the force exerted throughout the whole stroke.

According to this principle, if a multiplying-wheel is fixed upon the axis of the crank, to give motion to the fly-wheel upon a second axis, the teeth of the wheel should be three times stronger, than if the fly-wheel is placed upon the axis of the crank, although the multiplying-wheel is still upon the same axis; because in the latter case the wheel is only employed to communicate the uniform force of the engine to other machinery, and its teeth cannot be strained with more than the uniform resistance of that machinery ; but in the former case the teeth of the wheel will be subjected to the utmost force that the piston and crank can exert, and must communicate that force (which is three times the uniform resistance) to the fly-wheel.

For the same reason, if the fly-wheel is fixed upon the axis of the crank, and the machinery is driven from the extreme end of that axis, the neck at that end of the axis on which the crank is fixed, should have three times as much strength as the neck at the opposite end of the same axis, which is to transmit the same power, because the crank neck must endure the utmost force of the piston and crank, once at every stroke, in order to communicate it to the fly-wheel, but the other neck has only to endure the uniform force which will overcome the resistance of the machinery which is to be driven by the engine.

This great difference is produced by the intervention of the fly-wheel, which receives all the successive violent impulses, but will not allow them to be transmitted to the machinery, except with such a uniform and regular intensity as can be kept up continually, so as to avoid all the intermissions which take place when the piston changes its stroke.

This circumstance is not attended to in practice, for the best engine makers and millwrights are accustomed to apportion nearly the same strength to wheels and axes, when they are only subjected to the uniform action of the fly-wheel, as

Mr. Watt's engines are found to require for the crank necks, and multiplying-wheels, which must transmit the utmost force of the piston and crank to the fly-wheel. The reason for this practice is, that in case the wheelwork of the machinery should be forcibly detained by some obstruction, which would greatly increase the resistance, then the crank in moving slowly round, might transmit the utmost force of the piston to all the parts of the machinery which are so detained, and thus subject them all to the same strain as if there were no fly-wheel; because when it moves slowly it can have no effect, to equalise the force.

The liability to such obstructions must depend upon the kind of machinery to which the engine is applied; when an engine is connected with machinery containing very heavy masses of matter, which are to be driven with a great velocity, such as grinding mills (see p. 505 and 514), the inertiæ of those masses will oppose such a resistance to motion, whenever the engine is started, as will subject all the connecting wheels and axes, to the utmost strain of the piston and crank; and therefore all those parts should have the same proportion of strength, as the neck of the crank, axis and the teeth of the multiplying wheel.

For such cases the multiplier in the rule for finding the diameter of neck (p. 620), should be ($119 \cdot 37 \times 3 \cdot 17 = 378 \cdot 42$, or say) 380 instead of 300. And the multiplier in the rule for the teeth of wheels (p. 623), should be ($95 \cdot 5 \times 3 \cdot 17 = 302 \cdot 7$ say) 300, instead of 240 (a).

Example for the neck of an axis on the strongest proportion. The axis of the crank of an engine of 53 horse-power, which has a 36 inch cylinder (see note, p. 518), makes $17\frac{1}{2}$ revolutions per minute. Then 53 HP $\times$ 380 multiplier $= 20140 \div 17 \cdot 5$ revolutions $= 1150$, the cube root of which is $10 \cdot 47$ inches for the diameter of the neck; this is nearly the same as was given by a former rule (p. 619).

Example for the teeth of a cog-wheel on the strongest proportion. A wheel 12 feet diameter transmits 53 horse-power when it makes $17\frac{1}{2}$ revolutions per minute. Then 12 ft. dia. $\times$ $17 \cdot 5$ revolutions $= 210$ for a divisor; and 53 HP $\times$ 300 multiplier $= 15\,900$, which $\div$ 210, gives $= 75 \cdot 8$ for the strength of the teeth; if they were 3 inches pitch, they should be $75 \cdot 8 \div$ (3 squared $=$) 9, $= 8 \cdot 4$ inches broad. Or if the teeth were to be 9 inches broad, then $75 \cdot 8 \div 9 = 8 \cdot 42$; the square root of which is $2 \cdot 9$ inches for the pitch of the teeth. They really are 3 inches pitch, and $8\frac{1}{4}$ inches broad in one of Messrs. Boulton, Watt, and Co.'s 53 horse-engines.

These rules may be considered as giving sufficient strength for the wheels and axles of any kind of machinery, excepting rolling mills, in which extra fly-wheels are introduced (see p. 505). In machinery moved by water wheels, the impulse is quite uniform, and then a much less strength is sufficient; there are constant instances in practice of necks, and wheels, which correspond with calculations made

---

(a) Mr. Hick of Bolton has communicated to the author, the following method of computing the power that the teeth of wheels are capable of transmitting.

RULE. Multiply one-fourth of the square of the pitch in inches, by the breadth of the teeth in inches; the product is the number of horse-power that the teeth will transmit, when the pitch line passes through 4 feet per second.

Example. To transmit the power of a 20 horse-engine, the multiplying wheel which is applied on the axis of the crank, is 7 feet 1 inch diameter, with 108 teeth, $2\frac{1}{2}$ inch pitch, $6\frac{1}{2}$ inches broad; it makes 22 turns per minute, and the pitch circle moves $8 \cdot 16$ feet per second. For $7 \cdot 083$ ft. diam. $\times$ 22 revolutions $= 155 \cdot 8 \div 19 \cdot 1$ divisor $= 8 \cdot 16$ (see rule, p. 33). Then $2 \cdot 5$ inc. pitch squared $= 6 \cdot 25$, $\div 4 = 1 \cdot 562 \times 6 \cdot 5$ inc. broad $= 10 \cdot 156$ horse-power, is what the wheel ought to transmit if its teeth moved 4 feet per second, but as they move $8 \cdot 16$ feet per second; then $10 \cdot 156$ HP $\times$ $8 \cdot 16 = 82 \cdot 87 \div 4$ ft. per sec. $= 20 \cdot 72$ horse-power, is what they may transmit.

According to this proportion, which Mr. Hick follows in his practice, if the square of the pitch in inches is multiplied by the breadth in inches, the product will represent the number of horse-power that the wheel is capable of transmitting when its teeth move through 16 feet per second, that is ($\times 60 =$) 960 feet per minute. A wheel one foot diameter must make ($960 \div 3 \cdot 1416 =$) 305 3 revolutions per minute to have that velocity. According to the author's rules it should be $302 \cdot 7$, which is so near, that the two proportions may be considered as identical.

with 120 and 96 for multipliers, according to the preceding deduction from the necks and multiplying wheels of Mr. Watt's engines.

Examples for the strengths of cast iron necks, and for the teeth of wheels, which are subjected to a uniform strain, with slow motion, by a water wheel. A water wheel which is 16 feet diameter, and 18 feet broad, with a breast fall of about 7 feet effective fall of water, exerts 48 horse-power, when making 4 revolutions per minute. This wheel communicates its power by means of a ring of cast iron teeth, which is fixed on the rim of the wheel, and turns a cast iron pinion, 4 feet diameter, with 48 teeth, 3·14 inches pitch, and 10 inches broad. This pinion is fixed on the extreme end of a cast iron axis, the neck of which is $7\frac{1}{2}$ inches diameter, and $9\frac{3}{4}$ inches long in the bearing part. This neck, and the teeth, transmit all the power of the water wheel, when they make 14 revolutions per minute.

Calculation for the neck. 48 HP × 120 constant multiplier = 5760 ÷ 14 revolutions per min. = 411·43; the cube root of which is 7·44 inches for the diameter of the neck. And (7·44 × 1·2 =) 8·93 inches for the length of bearing; instead of 7·5 inc. diameter, and 9 75 inc. long.

For the teeth. 48 HP × 96 constant multiplier = 4608 ÷ (4 ft. dia. × 14 revol. =) 56=82·29 for the strength of the teeth, ÷ 10 inches broad = 8·229; the square root of which is 2·87 inches for the pitch; instead of 3·14 inc.

Another breast water wheel 18 feet diameter, and 14 feet wide, with nearly 6 feet effective fall of water, exerts about 25 horse-power when it makes 4 revolutions per minute. It transmits that power by means of the neck of its axis, which neck is $9\frac{1}{8}$ inches diameter; and also by the wooden teeth of a cast iron pit wheel, which is fixed on the extreme end of the axis, beyond the neck; that wheel is 13 feet diameter, and its teeth are $2\frac{1}{2}$ inches pitch, and 7 inches broad.

Calculation for the neck. 25 HP × 120 constant multiplier = 3000 ÷ 4 revolut. = 750; the cube root of which is 9·09 inches for the diameter of the neck; instead of 9·125 inc.

For the teeth. 25 HP × 96 constant multiplier = 2400 ÷ (13 ft. dia. × 4 revol. =) 52 = 46·15 for the strength of the teeth ÷ 7 inches broad = 6·59; the square root of which is 2·57 inches for the pitch; instead of 2·5.

An overshot water wheel 22 feet diameter, and $6\frac{1}{2}$ feet wide, exerts about 26 horse-power, when it makes 6 revolutions per minute. It transmits that power by the neck of its axis, which neck is $7\frac{7}{8}$ inches diameter, and 8 inches length of bearing; also by the teeth of a cast iron pit wheel, which is fixed on the end of the axis beyond the neck; that wheel is 6 feet diameter, with 76 teeth, of 2·98 inches pitch, and $8\frac{1}{4}$ inches broad.

Calculation for the neck. 26 HP × 120 multiplier = 3120 ÷ 6 revol. = 520, the cube root of which is 8·04 inches for the diameter of the neck, and (×1·2 =) 9·65 inches for the length; instead of 7·875 inc. diameter, and 8 inches long.

For the teeth. 26 HP × 96 =2496 ÷ (6 ft. dia. × 6 revol. =) 36 = 69·3 for the strength of the teeth ÷ 8 25 inches broad = 8·4; its square root is 2·9 inches for the pitch, instead of 2·98.

The overshot water wheel of Mr. Smeaton's blowing machine at Carron (see p. 278), exerted $15\frac{3}{4}$ horse-power, when making 6 revolutions per minute; it transmits that power by the neck of the axis, which is 7 inches diameter, and 7 inches length of bearing.

Calculation for the neck. 15·75 HP × 120 multiplier = 1890 ÷ 6 revol. = 315; the cube root of which is 6 8 inches for the diameter of the neck, instead of 7 inches.

The above examples are taken from mills which have performed extremely well during a long time, and they prove the sufficiency of the standard which is here adopted, although the ordinary practice of millwrights is to give so much more strength, in millwork driven by steam-engines. The preceding calculations are so proportioned, that a cast iron neck one inch diameter, will not be twisted with a greater force than 88 pounds, acting at half a foot distance from the centre ; and the teeth of a wheel, being 1 inch pitch, and 1 inch broad, will not be pressed with a greater force than 110 pounds (*a*). When the necks of axes, and the teeth of wheels, are merely subjected to a steady pressure, acting with a very slow motion, so as to be secure from jolts and shocks, they will bear from ten to fifteen times that force, without danger of breaking ; this is shown by the following examples.

Examples of the strengths of wrought iron necks, and of the teeth of cast iron wheels, which are used in cranes for lifting weights, with a very slow motion, and a uniform pressure. The strongest crane at the London Docks, is used for lifting blocks of marble, and is capable of lifting a weight 20 tons, by a chain made of iron, $1\frac{1}{4}$ inches diameter, which is wound up on a cylindrical barrel 18 inches diameter. The axis of this barrel is made of wrought iron, $4\frac{1}{4}$ inches square ; the neck is $4\frac{1}{2}$ inches diameter, and 4 inches long ; the barrel is turned round by this neck, to wind up the chain ; a cast iron cog-wheel of 6 feet diameter, is fixed on the extreme end of the axis, beyond the neck to turn it round ; this wheel has 150 teeth, which are 1·51 inches pitch, and $4\frac{1}{2}$ inches broad.

Calculation for the neck. 20 tons is ($\times$ 2240 lbs. =) 44 800 pounds $\times$ 1·5 ft. diam. of barrel = 67 200 lbs. acting at half a foot radius, $\div$ (the cube of $4\frac{1}{2}$ inc. diam. =) 91·12, gives 738 pounds, acting at half a foot radius, for the proportionate strain of torsion upon a wrought iron neck 1 inch

---

(*a*) The following instances will show that the above proportion for the strength of teeth will apply in very extreme cases, of small machinery acting with a very slow and gradual motion, so as to be quite secure against sudden shocks.

The great wheel of a common kitchen jack, or turn-spit, for roasting meat, is 5·46 inches diameter, with 48 teeth, which are ·357 of an inch pitch, and ·4 of an inch broad. The strength of these teeth is (·357 inc. pitch squared =) ·127 $\times$ ·4 inc. broad = ·0508. The wheel is turned by means of a cord, wound round a barrel upon the axis of the wheel, and the cord is reeved through a system of pullies, to which a heavy weight is applied. The cord is drawn with a force of 12 pounds, and the pressure transmitted by the teeth of the wheel is 6 pounds, which $\div$ ·0508 strength of the teeth, gives 118 pounds for the proportionate strength of teeth of one inch pitch, and one inch broad ; instead of 110, which is assumed for a standard. This wheel has been in daily use for more than 30 years, it is made of wrought iron, and its teeth give motion to a pinion of 12 teeth ; those teeth are very much worn.

The great wheel of a very excellent pocket watch, made by Mr. Pennington, is ·764 of an inch diameter, with 60 teeth, which are ·04 of an inch pitch, and ·04 of an inch broad. The strength of these teeth is therefore (·04 squared =) ·016 $\times$ ·04 broad = ·0064. To find the force which these teeth transmit, a wheel was affixed to the watch key, and a sufficient weight applied by a silk thread, to the circumference of the wheel, to wind up the watch. It appeared that the constant pressure upon the teeth is ·83 of a pound, which $\div$ ·0064 strength of the teeth, gives 130 pounds for the proportionate strength of teeth, one inch pitch and one inch broad, instead of 110. This watch has been in use for 10 years past. The wheel is made of fine yellow brass, hammered very hard ; it is gilt, but the gilding is not yet worn off the teeth.

The axis of the same wheel is made of steel, hardened and tempered ; the neck is ·07 of an inch diameter ; the force of torsion upon it is (·83 of a pound $\times$ ·764 of an inch diameter of wheel =) ·634 $\div$ 12 inches diameter = ·0528 of a pound, acting at half a foot radius from the centre, this $\div$ by (·07 inc. dia. cubed =) ·000343, gives 154 pounds, acting at half a foot radius, for the proportionate strain on a steel neck, one inch diameter : the rule previously given for wrought iron necks, is at the rate of 105·5 pounds, and we may expect steel to be half as much stronger than wrought iron.

diameter, and eight-9ths of an inch long.   This is 7 times our assumed standard of 105·5 pounds for a wrought iron neck.

For the teeth.   44 800 lbs. × 1·5 ft. dia. of barrel = 67 200 ÷ 6 ft. diam. of wheel = 11 200 lbs. pressure, exerted upon the teeth of the wheel.   Their strength is 1·51 inc. pitch squared, = 2·28 × 4·5 inc. broad = 10·25.   And 11 200 lbs. ÷ 10·25 strength = 1092 pound pressure, is the proportion for cast iron teeth of 1 inch pitch, and 1 inch broad.   This is 9·93 times our assumed standard of 110 lbs.

Mr. Loyd (formerly Loyd and Ostell) of London, has had a great practice in making strong cranes.   The largest size is adapted to lift 20 tons weight; the chain is made of iron, 1¼ inch diameter, and winds round a barrel 22 inches diameter; a spiral groove is formed round the barrel to guide the chain in winding up, and to receive the lower halves of those links which stand edgeways upwards.   The axis of this barrel is made of wrought iron, 4 inches square; the gudgeon at one end is 3 inches diameter, and 3½ inches long; the neck at the other end is 4¾ inc. diameter, 3½ inches long.   The barrel is turned round by this neck, to wind up the chain; and the cast iron cog-wheel, which is fixed on the extreme end of the axis, beyond the neck to turn it round, is 6 feet diameter, with 152 teeth, which are 1·49 inches pitch, and 3¼ inches broad.

The teeth are rounded at the ends nearly to a semicircle, and the spaces between the teeth are also hollowed out nearly to a semicircle; this form gives the teeth very great strength at the roots, where they join to the rim of the wheel. The wheel is turned by a cast iron pinion of 17 teeth, which are of a corresponding form to those of the wheel, but the pinion has a circular rim, or flange, at each end of the teeth, which greatly strengthens them, being cast in one piece with the teeth.   With this form of teeth, only one pair of the teeth of the wheel and pinion, are in actual contact at the same time, so that one tooth bears all the strain.

Calculation for the neck.   44 800 lbs. × 22 inc. diam. of barrel = 985 000 ÷ 12 inc. diam. = 82 133 lbs., acting at half a foot from the centre, is the force of torsion for the neck; and ÷ (4·75 inc. dia. cubed =) 107·17, gives 756 pounds, acting at half a foot radius, for the proportionate torsion of a wrought iron neck 1 inch diameter, and three-fourths of an inch long.   This is 7·16 times our assumed standard of 105·5 lbs. for a wrought iron neck.

For the teeth.   44 800 lbs. ÷ 22 inc. dia. of barrel = 985 600 ÷ 72 inc. dia. of wheel = 13 689 lbs. is the pressure upon the teeth of the wheel and the pinion.   Their strength is (1·49 inc. pitch squared, = 2·22 × 3·25 inc. broad =) 7·22.   And 13 689 ÷ 7·22 strength = 1898 pounds, is the proportionate pressure for cast iron teeth, of 1 inch pitch and 1 inch broad.   This is 17¼ times our assumed standard of 110 lbs.

Mr. Loyd's crane to lift 15 tons, has a chain made of iron 1⅛ inches diameter; the barrel is 22 inches diameter; the neck is 4¼ inches diameter; the cog-wheel is 4½ feet diameter, the pitch and the breadth the same as above, and the strain upon the teeth is also the same.

These cranes are always proved by lifting the full weight that they are intended to bear; but it is only occasionally that they are loaded so heavily, when in use; and the motion being exceedingly slow, the wearing is quite inconsiderable.   It scarcely ever happens that the wheels or necks of these cranes are broken, even when fully loaded; and although by a very severe strain the necks have been sometimes twisted round considerably, yet they have not broken.

According to the standards which we have assumed, for the strength of necks, and of the teeth of wheels, the following rule is formed, to apportion the size of the neck of an axis to that of the teeth of a wheel which is fixed upon the same axis, so that they will have corresponding strength, and be competent to transmit the same force, with equal security against breaking.

To find the proper size of a cast iron neck, to transmit such a force of torsion as is proper to be communicated to it, by the teeth of a given cog-wheel, fixed upon the same axis.

RULE. Multiply the square of the pitch of the teeth in inches, by the breadth of the teeth in inches; multiply the product by the diameter of the wheel in feet; multiply the product by 1·2, and extract the cube root of the quotient. That root will be the required diameter for the neck in inches; the length of the neck being 1·2 times the diameter so found.

Example. The multiplying-wheel for a steam-engine of 24 inch cylinder (see p. 625) being 7$\frac{1}{12}$ feet diameter, with teeth of 2$\frac{1}{2}$ inches pitch, and 6$\frac{1}{2}$ inches broad. Then 2·5 inc. pitch squared, = 6·25 × 6·5 inc. broad = 40·6 for the strength of the teeth, × 7·08 feet diam. of wheel = 287·3 × 1·2 constant multiplier = 344·76 for the strength of the neck; the cube root is 7·012 inches for the required diameter of the neck. And (× 1·2 =) 8·41 inches is the proper length of bearing.

Note. The constant multiplier should be (110 lbs. ÷ 88 lbs. =) 1·25 instead of 1·2; or dividing by ·8, instead of multiplying, will give a correct result, corresponding with the assumed standards for the strengths of necks, and the teeth of wheels. The multiplier 1·2 is most convenient for calculation, and sufficiently near the truth; it is particularly convenient for the sliding rule, because the diameter of the wheel being taken in inches, instead of feet, the multiplication by 1·2 will be unnecessary, thus.

To perform this calculation with the sliding rule, with the slider inverted, we must previously multiply the breadth of the teeth in inches, by the diameter of the wheel in inches; and set the slider with that product on the line ꟼ, opposite to the pitch in inches, on the line D; then the diameter of the neck in inches, will be shown where the same numbers meet on both the lines ꟼ and D. For instance, 6·5 inc. broad, × 85 inc. diam. = 552·5, is the number to be found on the line ꟼ, and set opposite to 2·5 inches pitch on the line D, thus,

Sliding rule, slide inverted.	ꟼ Bread. of teeth inc. × dia. wheel inc.	Dia. neck inc.	Ex.	ꟼ 6·5 ×85 =552	7 inc. dia.
	D Pitch of the teeth of the wheel inc.	Dia. neck inc.		D Pitch inc. 2 5	7 inc. dia.

The examples above given, show that in different cases, there is a great disproportion in the strength which is allowed to the necks of axes, and the teeth of wheels, according to the strain they are to endure. The greatest strength is given in cases of steam-engines, where the action is communicated with a rapid motion, and subject occasionally to violent concussions, and sudden jolts; the excessive strength is requisite to resist those shocks. Also in rapid motions, it is requisite to have a certain surface of contact, in the bearing parts, in order to endure the pressure, without excessive friction, and cutting and wearing of the metal. All these circumstances must be taken into account, in choosing multipliers to calculate the proper strengths of necks, or of the teeth of wheels for new cases, but the preceding examples are sufficiently numerous to afford the requisite information, and to direct that choice.

Wooden teeth for cog-wheels. It is an excellent practice in wheelwork to transmit quick motions, to form the teeth of the larger wheel of hard wood, such as crab tree, horn-beam, live oak, or beech. The wooden cogs are driven tight into mortices, formed in the cast iron rim of the wheel, and are fastened in their places by cross pins. The teeth of the smaller wheels or pinions, which are to work with the wooden teeth, are made of cast iron, filed very true and smooth. Iron and wood teeth are found to work together much better than if both were wood, or both iron.

In the action of the teeth of wheels, each succeeding pair of teeth which come into contact, must meet each other with a slight blow, and if the motion is rapid, and the pressure considerable, these successive blows will make a violent noise, and tremor, which is much greater when both sets of teeth are made of iron, than if one set is of wood, for they will yield a little, and ease that tremor. The largest of the two wheels should be furnished with the wooden teeth, and the smallest with the iron teeth, that they may wear each other more regularly (see note, p. 493). The thickness of the iron teeth should be rather less than the spaces between them, in order to allow the wooden teeth to be rather thicker than the iron ones, and thus give strength to the wooden teeth.

It is found by experience, that wooden teeth are not more liable to be broken, than solid iron teeth of the same size, but the wooden teeth wear out sooner than the iron teeth. This arises from the wooden teeth yielding a little to sudden concussions, so as to avoid the effects of jolts or blows upon the teeth ; and also, that in consequence of the wooden teeth yielding a very little when forcibly pressed, all the different pairs of teeth which are in contact at the same time, are caused to take their share of the pressure, whereby the surface in contact is greatly increased. Cast iron teeth which are formed of one solid piece, with the rims of the wheels, cannot yield sensibly, and however exactly the teeth may be formed, the greatest pressure will be thrown upon that pair of teeth which is nearest to the line between the centres of the two wheels, and very little pressure will be borne by any other pair of teeth. If both the wheels are large enough to have a sufficient number of teeth, two or three pairs of teeth will appear to touch each other at once ; but in solid cast iron teeth, that contact will not be so perfect as to divide the pressure fairly amongst them all. When one of the wheels has wooden teeth, the contact will be more perfect, and it may be safely assumed that in wood and iron teeth, the efficient bearing surface is twice as great as in solid iron and iron teeth of the same dimensions, and transmitting the same force.

Wooden teeth are not so proper for slow motions, with great pressure, as iron teeth, but for quick motions, they should always be used. In corn mills which are turned by a water wheel, for four pairs of millstones, the first motion, or bevelled pit wheel, on the axis of the water wheel, is usually wooden teeth, $2\frac{1}{2}$ inches pitch, and 7 inches broad. The large horizontal wheel which turns the four pinions for the mill stones, has also wooden teeth, 2 inches pitch, and 6 inches broad. The diameter of this wheel is about four times the diameter of the bevelled pinion upon the same axis, so that the stress upon the teeth of the second motion is only one fourth of the stress upon the teeth of the first motion ; and this stress being divided amongst all the three pinions which are worked at once, is reduced to one-12th of the pressure, acting with a quadruple velocity, and three times the frequency of action. It is observed in these mills, that the teeth of the first motion usually wear out as soon as those of the second motion. In such cases the first motion would act best with iron and iron teeth.

It is the opinion of the most experienced millwrights, that when the motion of the pitch line exceeds $3\frac{1}{2}$ feet per second, (= 210 feet per minute) the largest wheel will work better with wooden teeth ; but for that speed, or slower, iron teeth will answer best.

On proportioning wheels and pinions for millwork. The diameters of toothed wheels should always be made sufficiently large, to allow the smallest of the two wheels which are to work together to have a considerable number of teeth. Pinions with 12 to 18 teeth are frequently used to drive a large wheel, and if the motion is slow, they answer tolerably well for crane work ; but such small pinions are quite unfit to be driven with a rapid motion, by a large wheel for millwork ; because the teeth of the pinions move in a small circle, and advance so suddenly towards the teeth of the wheel, as to meet abruptly ; hence they come in contact with violence, and only one pair of teeth can be in contact at a time. When the number of teeth in the two wheels is larger, the first contact of the meeting teeth will be performed more quietly, and the teeth will have better hold of each other.

For quick motions in millwork, the pinion which is driven by a wheel should never have less than 30 or 40 teeth, to enable them to work in a proper manner, and allow a sufficient number of teeth to be in action at the same time. If the requisite velocity of motion cannot be attained by one wheel, working into such a large pinion, it will be advisable to apply an intermediate multiplying wheel and pinion; the motion may then be attained with wheels of a smaller diameter, and yet the driven wheels may have a sufficient number of teeth, to perform well. The friction arising from the intermediate axis, and the extra pair of wheels thus introduced, will commonly prove less, than the friction of the teeth of a single wheel and pinion, when they must work to a disadvantage, in consequence of the pinion having too small a number of teeth; and as the single wheel must be of a larger diameter, it will be heavier than when a small intermediate wheel is used.

Bevelled wheels act better than spur wheels of the same proportions, because the teeth do not meet each other so abruptly, but they take and leave more gradually. And rings of internal teeth, with the pinion within them, act best of all, because the curvature of the wheel and pinion being in the same direction, the advance of the teeth towards each other is more progressive, and they meet less abruptly than in any other kind of teeth (a). In some instances it is a great convenience to have the centres of the wheel and pinion as near to each other as they can be; a wheel with internal teeth admits of this, and yet allows a much larger size to the pinion, than could be given to a common spur wheel and pinion. Spur wheels act best when both are of the same diameter; and for millwork, the difference between the driving wheel and the driven pinion, should never be excessive.

The pitch of the teeth of wheels should always be chosen as fine as is consistent with strength and durability, keeping in view the rapidity of the motion with which the teeth are to act upon each other, and apportioning the strength of the teeth according to such one of the preceding examples, as is most similarly circumstanced; preferring in all cases to have the driving wheels of as small diameter as can be, and the driven pinions with as great a number of teeth; and introducing intermediate motions, where they are necessary to get up the required velocity, rather than making a great disproportion between the wheel and pinion.

It is a good practice to give the largest of two wheels which are to work together, one more tooth (or one less) than a number which can be exactly divided by the number of teeth in the other wheel; hence, the quotient resulting from that division will be a fractional number. For instance, if the pinion has 38 teeth, and it is required to give it about four revolutions to one of the wheel, then that wheel should have ($38 \times 4 = 152 + 1 =$) 153 teeth (or else 151 teeth). The difference that this extra, or deficient, tooth will make in the motion, will be quite unimportant in most cases, but the teeth will wear each other more equally in consequence, because the same pair of teeth will not meet each other often. For instance, each tooth in the wheel will meet the same tooth in the pinion, in the course of 38 revolutions of the great wheel, or 153 revolutions of the pinion; and during this interval, each tooth of the pinion will act successively with each tooth of the pinion, whereby if there is any difference amongst the teeth they will tend to wear themselves to a uniform size; or if the teeth are correctly of a size at first, they will have the best chance to continue so, as they wear away. The extra tooth

(a) Mr. James Kennedy has applied a wheel with internal teeth, for the multiplying wheel of a steam-engine of 53 horse-power (see note, p. 518), which drives his Cotton Mill at Manchester. The wheel is 14 feet diameter, with 168 teeth, of 3·14 inches pitch, and 8 inches broad. The internal pinion on the end of the axis of the fly-wheel, is $4\frac{1}{2}$ feet diameter; the distance between the centres is only $4\frac{3}{4}$ feet.

which is termed by millwrights *a hunting tooth*, was formerly introduced in all millwheels, but it has fallen into disuse in modern millwork, because wheels are now executed with such accuracy, that it is supposed to be unnecessary to take such a precaution; but this expedient has no inconvenience, and ought to be practised, although it is less necessary than formerly.

The proportion between the pitch of the teeth, and their breadth. In practice, the breadth of the teeth varies in different cases, from $1\frac{1}{2}$ to $3\frac{3}{4}$ times the pitch. The strength of teeth depends more upon the pitch, and less upon the breadth, but the wearing depends principally upon the breadth. It is necessary to give so much breadth to the teeth as will enable them to endure the friction that they have to bear, without excessive cutting and wearing; but the rate of wearing depends in a great measure upon the rapidity of the action. If the breadth of the teeth in inches, is multiplied by half the pitch in inches, the product may be assumed to be the extent of surface, in square inches, which is in contact to communicate the motion from one wheel to the other. This supposes that only one pair of teeth is in absolute contact at the same time. From an examination of a number of cases of the multiplying wheels of steam-engines, which act with a quick motion, it appears that the force transmitted by their teeth, is at the rate of 550 pounds for each square inch of the surface so assumed. With this proportion the teeth are found to work properly, and last a long time; they are commonly made with iron and iron teeth. Wood and iron teeth would bear as great a proportion of pressure, and work more pleasantly; but the wooden teeth will wear out, and require to be renewed occasionally.

Proportion between the pitch of the teeth and their length. According to the practice of the best millwrights, the length, or projection of the teeth from the rim of the wheel, is five-8ths of the pitch. The longest teeth which are ever made, are $\frac{3}{4}$ of the pitch, but that length renders them more liable to be broken by jerks. The projection of the ends of the teeth beyond the pitch circle, should be one-fourth of the pitch, and then the surface in contact between the teeth of the two wheels will be half the pitch, leaving one-8th of the pitch unoccupied at the bottom of the teeth, for clearance. When the teeth of wheels are made on this proportion, the outside diameter of the wheel, taken to the extremities of its teeth, will be half the pitch more than the geometrical diameter, or diameter of the pitch circle; and the inside diameter, or that of the rim from which the teeth project, will be three-fourths of the pitch less than the geometrical diameter.

Form of the teeth for wheels. Several ingenious mathematicians have investigated the subject of the teeth of wheels, and have recommended various forms, which would cause the teeth to transmit a uniform motion, without any irregularities of impulsive action. None of these forms have been brought into use, and they are objectionable in a greater or lesser degree, because they require the teeth to be brought to rounded points at their extremities; to attain this form, the teeth must be of a greater length than usual, in proportion to their pitch. The sides of the teeth within the pitch circles are also required to converge towards the centre of the wheel, and therefore the teeth are thinner where they join to the rim of the wheel, than they are at the pitch circle. Teeth formed on these principles would be more subject to be broken than the common teeth, on account of their greater length, and from want of substance at the roots. Or if the teeth were made of a larger and coarser pitch than usual, in order to give them sufficient strength, they would not work so regularly (whatever their form might be) as the common form of teeth, which being short, and strong at the roots, may be of a fine pitch; but the extremities cannot be pointed, for they must have a considerable thickness at the ends, and those ends are formed to correspond with the circle by which the extremities of the teeth are bounded. The curved sides of each tooth, which form the acting surfaces to transmit the force, are portions of circular arches, described from the centres of the adjacent teeth; and the curvature must be such, that the ends of the teeth will be sufficiently rounded off, to allow them to enter properly into the spaces between the teeth of the other wheel, but no more curvature should be given, than is requisite for that purpose.

Another difficulty in the application of any of the forms of teeth recommended by mathematical writers is, that the teeth must be shaped according to the relative sizes of the two wheels which are to work together. In practice this cannot be observed, because a wheel must be cast from the same model, whether it is to work with a larger or a smaller wheel; and in many cases, the same wheel is required to work into two other wheels of different sizes.

The great improvements which have been made in modern wheelwork, compared with what was in use before Mr. Watt and Mr. Rennie executed the Albion Mills, have been attained by accurate execution, and judicious proportioning of the pitch and breadth of the teeth, according to the force they are required to transmit, and the rapidity of their motion. The teeth being short, and interlocking only a small depth into each other, with spreading bases where they join to the rim, they have great strength, with only a fine pitch, and the breadth being considerable, there is a sufficient surface in contact to endure the pressure. These conditions being observed, the most important point is accurate execution, to make all the teeth precisely alike, and equidistant from each other, and from the common centre. It is also indispensable that teeth of the two wheels, which are to work together, shall be of exactly the same pitch.

In iron wheels, where the teeth are cast solid with the rim, the mathematical accuracy of the teeth, and the identity of the pitch in both wheels, is of the greatest importance, in order that the pressure may be divided as fairly as possible amongst those two or three pairs of the teeth which are in contact at the same time; for if there is the smallest difference between the pitch of the two sets of teeth, then only one pair of teeth can be in contact at a time, and their surfaces must bear all the pressure, whereby they are worn and cut away very rapidly. When one of the wheels has wooden teeth, the wearing will be more uniform, because that wooden tooth which receives the principal strain, will compress and yield a very little under the pressure, so as to allow the adjacent pairs of teeth to come into close contact, and take their full share of the strain; but solid metal teeth cannot yield at all, and if there is the least want of exactitude in the pitch of the teeth, they will be worn very rapidly; although iron teeth which are truly pitched will last a very long time.

Method of finding the true geometrical diameters of the pitch circles for cogwheels. The circumference of a circle being 3·14159 times its diameter, if the geometrical diameter of the wheel in inches, is multiplied by that number, it will give the circumference in inches, and that circumference being divided by the number of teeth in the wheel, the quotient will be the pitch of the teeth, nearly.

The rule given in p. 615 is founded on this principle, and is a sufficient approximation for common use, but it is not mathematically correct, because it supposes the pitch, or distance from the centre of one tooth to the centre of the next, to be measured on the curved circumference of the pitch circle, whereas the pitch should be measured on the straight line drawn from one tooth to the next. In strictness, the pitch lines of cog-wheels are not circles, but regular polygons having as many sides as there are teeth in the wheel. In any two wheels which are to work together, the sides of the supposed polygons, must be precisely the same in both wheels; but if the two polygons have different numbers of sides, then the diameters of their circumscribing circles will not be exactly proportionate to their number of sides respectively.

In order that the pitch of the teeth may be exactly alike, in two wheels of different sizes, the diameters of their pitch circles must be correctly proportioned to each other; but that will not be exactly the same proportion as exists between the numbers of teeth. The pitch of the teeth is the chord of the arch subtending the angle which is made at the centre of the wheel, by the interval between two adjacent teeth. The chord of any angle is double the sine of half that angle; and by finding half the angle that the teeth subtend in degrees, we may make the calculation by the aid of a table of sines, in the following manner.

To find the true geometrical diameter for the pitch circle of any cog-wheel or pinion, which is to work with another cog-wheel or pinion. Having given the number of teeth in the wheel, and in the pinion, and the diameter for one of the pitch circles (either of the wheel, or of the pinion), to find the true diameter for the other pitch circle.

RULE. Divide 180 degrees, by the given number of teeth, in the wheel, and in the pinion; the respective quotients will be the halves of the angles which are subtended by two adjacent teeth of each wheel, in degrees. Take out the natural sines for those halves from the table of sines.

Multiply the given diameter of the pitch circle of one wheel in inches, by the sine previously found for half the angle subtended by its teeth, and divide the product (which is the true pitch in inches) by the sine of half the angle subtended by the teeth of the other wheel, whereof the diameter of the pitch circle is required. The quotient is that diameter in inches.

Example. The multiplying-wheel for a 30 horse-engine being $6\frac{1}{8}$ feet diameter at the pitch circle, and having 77 teeth; if the pinion has 38 teeth, what will be the true diameter of its pitch circle. Thus 180 deg. ÷ 77 teeth = 2·337 degrees; or 2 degrees 20·26 minutes, is half the angle subtended by two teeth; the natural sine of that half angle is ·04079 of the radius. And 180 ÷ 38 = 4·737 degrees, or 4 degrees 44·21 minutes; the natural sine of this angle is ·08258 of the radius. Then 73·5 inc. diam. of wheel, × ·04079 = 2·988 inc. is the true pitch; and ÷ ·08258 gives 36·304 inches for the required diameter of the pitch circle of the small wheel of 38 teeth.

Note. If these two wheels were proportioned according to their respective numbers of teeth, the diameter of the small wheel would be (73·5 inc. diam. × 38 teeth = 2793 ÷ 77 teeth =) 36·273 inches; so that in this case the approximate method is very near to the truth.

The following table was calculated by Mr. Donkin, and published in 1803, to show the exact diameters of the pitch circles of cog-wheels, of any number of teeth less than 300; the diameter is expressed in terms of the pitch of the teeth, measured on a straight line from the centre of one tooth, to the centre of the next adjacent tooth.

The first column expresses the number of teeth in the wheel, and the second column contains the diameter of the pitch circle in inches, supposing the pitch of the teeth to be one inch. Hence, by multiplying the numbers in the second column by the intended pitch, the proper diameter for the pitch circle in inches, may be obtained in any case.

Examples of the use of the table of the diameters of the pitch circles for wheels of one inch pitch.

The pitch circle of a wheel of 77 teeth, being 73·5 inches diameter, what must be the diameter of a pitch circle of a wheel of 38 teeth, to work therewith. In the table, 77 teeth is 24·52 inches diameter, if the pitch had been 1 inch; and 73·5 inc. dia. ÷ 24·52 = 2·998 inches, is the true pitch. Also, 38 teeth in the table is 12·11 inc. diam., if the pitch had been 1 inch; and 12·11 × 2·998 = 36·304 inches, is the required diameter of the pitch circle for the wheel of 38 teeth.

Suppose two wheels of 77 teeth and 38 teeth, are required to work together; the distance between their centres being fixed at 54·9 inches. According to the table, a wheel of 77 teeth would be 24·52 inches diameter; and of 38 teeth 12·11 inches diameter, if the pitch had been 1 inch. The sum of these two numbers is 36·63; and half that sum is 18·315, for the distance of the centres, if the pitch had been 1 inch. Then 54·9 ÷ 18·315 = 2·998 inches, is the true pitch; and 24·52 inc. dia. × 2·998 inc. pitch = 73·5 inc. is the true diameter of the large wheel. Or 12·11 inc. dia. × 2·998 inc. pitch = 36·304 inc. is the true diameter of the small wheel.

Geometrical diameters, in inches, of cog-wheels with teeth of 1 inch pitch.

Number of teeth 1 inc. pit.	Diameter of pitch line in inches.	Number.	Diameter.	Number.	Diameter.	Number.	Diameter.	Number.	Diameter.
		61	19·42	121	38·52	181	57·62	241	76·72
		62	19·74	122	38·84	182	57 94	242	77·03
3	1·155	63	20·06	123	39·16	183	58·25	243	77·35
4	1·414	64	20·38	124	39·48	184	58·57	244	77·67
5	1·701	65	20·70	125	39·79	185	58·89	245	77·99
6	2·000	66	21·02	126	40·11	186	59·21	246	78·31
7	2·305	67	21·34	127	40·43	187	59·53	247	78·63
8	2·613	68	21·65	128	40·75	188	59·85	248	78 94
9	2·924	69	21·97	129	41·07	189	60·16	249	79·27
10	3·236	70	22·29	130	41·38	190	60·48	250	79 59
11	3·549	71	22·61	131	41·70	191	60·80	251	79·90
12	3·864	72	22·93	132	42·02	192	61·12	252	80·22
13	4·179	73	23·24	133	42·34	193	61·44	253	80·53
14	4·494	74	23·56	134	42·66	194	61·76	254	80·85
15	4·810	75	23 88	135	42·98	195	62·07	255	81·17
16	5·126	76	24·20	136	43·30	196	62·40	256	81·50
17	5·442	77	24·52	137	43·61	197	62·71	257	81·81
18	5·759	78	24·84	138	43·93	198	62 03	258	82·13
19	6·076	79	25·15	139	44·25	199	63·35	259	82·44
20	6·392	80	25·47	140	44·57	200	63·66	260	82·76
21	6·710	81	25·79	141	44·88	201	63 98	261	83·08
22	7·027	82	26·11	142	45·20	202	64·30	262	83·40
23	7·344	83	26·43	143	45·52	203	64·62	263	83·72
24	7·661	84	26·74	144	45·84	204	64·94	264	84·04
25	7·979	85	27·06	145	46·16	205	65·26	265	84·35
26	8·296	86	27·38	146	46·48	206	65·57	266	84·67
27	8·614	87	27·70	147	46·80	207	65·89	267	84 99
28	8·931	88	28·02	148	47·11	208	66·21	268	85·31
29	9·250	89	28·34	149	47·43	209	66·53	269	85·63
30	9 567	90	28·65	150	47·75	210	66·85	270	85·95
31	9·885	91	28·97	151	48·07	211	67·17	271	86·26
32	10·20	92	29·29	152	48·39	212	67·48	272	86·58
33	10·52	93	29·61	153	48·71	213	67·80	273	86 90
34	10·84	94	29·93	154	49·02	214	68·12	274	87·22
35	11·16	95	30·25	155	49·34	215	68·44	275	87·54
36	11·47	96	30·56	156	49 66	216	68·76	276	87·85
37	11·79	97	30·88	157	49·98	217	69·08	277	88·17
38	12·11	98	31·20	158	50·30	218	69·39	278	88·49
39	12·43	99	31·52	159	50·62	219	69·71	279	88·81
40	12·75	100	31·84	160	50 93	220	70·03	280	89·13
41	13·06	101	32·16	161	51·25	221	70·35	281	89·45
42	13·38	102	32·48	162	51·57	222	70·67	282	89·77
43	13·70	103	32·79	163	51 89	223	70·99	283	90·08
44	14·02	104	33·11	164	52·21	224	71 31	284	90·40
45	14·34	105	33·43	165	52·52	225	71·62	285	90·72
46	14·65	106	33·75	166	52·84	226	71·94	286	91·04
47	14·97	107	34·06	167	53·16	227	72·26	287	91 36
48	15·29	108	34·38	168	53·48	228	72·58	288	91·68
49	15·61	109	34·70	169	53·80	229	72·89	289	91·99
50	15·93	110	35·02	170	54·12	230	73·21	290	92·31
51	16·24	111	35·34	171	54·43	231	73·53	291	92·63
52	16·56	112	35·66	172	54·75	232	73·85	292	92·95
53	16·88	113	35·97	173	55·07	233	74·17	293	93·27
54	17·20	114	36·29	174	55·39	234	74·49	294	93·58
55	17·52	115	36·61	175	55·71	235	74·81	295	93·90
56	17·84	116	36·93	176	56·03	236	75·12	296	94·22
57	18 15	117	37·25	177	56·34	237	75·44	297	94·54
58	18·47	118	37·57	178	56 66	238	75·76	298	94·86
59	18·79	119	37·88	179	56·98	239	76·08	299	95·18
60	19·11	120	38·20	180	57·30	240	76·40	300	95·49

4 M 2

Mr. Joseph Brewer, of Preston, in Lancashire, published a small tract, in 1816, on a method of calculating the diameters of wheels which are to turn each other by teeth; his method of calculation is by the tables of sines, on the principle above directed, and he gives a similar table to the above. As Mr. Brewer makes no mention of Mr. Donkin's table, we may conclude that he was not acquainted with it, and formed his own table from original calculation. The table here given is derived from a comparison of both Mr. Donkin's and Mr. Brewer's tables, and where they differ, the correct numbers have been found by a new calculation; from which it appears that Mr. Donkin's table is very correct, but Mr. Brewer's contains many errors.

ON THE PROPORTIONS FOR THE FLY-WHEELS OF STEAM-ENGINES.

The principles from which these proportions must be deduced, are explained in the mechanical definitions at the commencement of this work, under the heads Inertiæ, Force, Mechanical Power, Energy, and Impetus (see pp. 15 to 20). The law of the motion produced in bodies which fall freely by their own gravity (see p. 23) is the standard by which all cases must be illustrated and proved. The application of these principles to the most simple cases of communicating motion to masses of matter, is explained generally in p. 31, and more particularly in the cases of bodies moving in circles, in p. 33 to 37. It is unnecessary to repeat any of these principles, for the reader is supposed to be fully informed on all these points; and also on several of their applications to the subject of fly-wheels, which have been shown at p. 414; and pp. 491 to 519.

The force which the piston of a steam-engine exerts to turn the crank and fly-wheel round, is most active when the piston is near the middle of its course, but it diminishes gradually, until it comes to nothing at the end of that course; then beginning again at nothing, it increases until it comes to its maximum, about the middle of the stroke. Hence, in the course of each half stroke, the impulse of the piston upon the fly-wheel ceases totally, and is renewed again; this takes place twice in each complete stroke. The office of the fly-wheel is to fill up the intervals between the successive efforts which are exerted upon the crank during each half stroke, and to reduce the whole action to one continuous impulse on the millwork, with as much uniformity in velocity as can be obtained (see p. 415).

To produce this effect, the fly-wheel must receive and accumulate in itself all the surplus force which is exerted by the piston, whenever that force exceeds the resistance occasioned by the millwork; and by transmitting that surplus regularly to the millwork, whenever the power falls short of the resistance, the action on the millwork will be continually kept up, without intermission, notwithstanding the intermitting action of the piston upon the crank. As the fly-wheel must undergo continual additions to, and abstractions from, its own energy or inherent power of motion, the amount of that energy must be very considerable, in order that no very notable change may be occasioned in its velocity, by all the additional energy which is communicated to it, when the piston is at the middle of its stroke, or by all the impetus which is elicited from it, to continue the motion, when the piston is at either end of its stroke.

It must be recollected that a fly-wheel cannot in any case transmit a perfectly uniform motion to the millwork, but only an approximation thereto (see p. 415), because it is only by accelerating its velocity, that it can receive any addition to its energy; and only by diminishing its velocity that it can exert any impetus, or bring into activity any of that energy with which it is endowed (see p. 17).

The energy or inherent force which is resident, or accumulated, in a fly-wheel, is in fact all the mechanical power which has been communicated to it, to urge it from a state of rest, into that of motion with the velocity which it actually possesses. The amount may be determined, by multiplying the mass of moving matter by the height due to the velocity with which it moves, and the product so obtained is the mechanical power which must have been communicated to it to produce its motion (see p. 31 and p. 414). Or the energy may be represented by the product which is obtained by multiplying the moving mass, by the square of the velocity with which it moves (see p. 17).

The energy required to be accumulated in the fly-wheel of a steam-engine, should be proportionate to the power which is exerted by the piston during one half stroke; for the operation of the fly-wheel is repeated, at every successive impulse which the piston exerts upon the crank; hence it must receive the surplus force that is exerted by the piston, when at the middle of its course, and it must transmit that force to the millwork, when the piston arrives near to the end of its course, and exerts but little impelling action.

The office of the fly-wheel is, to carry on the operation of the millwork, from one favourable period of action upon the crank, to the succeeding favourable period; and the energy requisite in the fly-wheel to enable it to produce this effect, must depend upon the magnitude of the interval which elapses between the successive impulses, and upon the amount of the force which is to be received, and transmitted by it, during each of those intervals.

For instance: supposing an engine which exerts 4 horse-power, makes only 20 strokes per minute, and that another engine which exerts 8 horse-power, makes 40 strokes per minute. In this case the same energy would be requisite in the fly-wheels of both, although one engine exerts twice as much power as the other; because in the small engine the intervals between the successive impulses on the crank are twice as long as those of the larger engine; and although the effect to be produced by the fly-wheel of the small engine, is only half the force of the larger one, it requires to be exerted through double the space.

Supposing the fly-wheels of the two engines to be of the same diameter, so that the rim of the 8 horse fly-wheel, would move with double the velocity of that of the 4 horse; then the rim of the 4 horse would require to be $(2 \times 2 =)$ 4 times the weight of that of 8 horse, in order to give both wheels the same energy; so as to produce an equal regulation of both engines.

Or supposing the 4 horse engine to act with sun and planet-wheels, and the 8 horse fly-wheel to be fixed on the crank axis; if both wheels were of the same diameter, they would move with the same velocity, and should therefore be of the same weight.

If we suppose the case reversed, viz. that the 8 horse engine acted with sun and planet-wheels, so that its fly-wheel would make 80 revolutions per minute, whilst the 4 horse engine had its fly-wheel on the crank axis, and made only 20 revolutions per minute; then the fly-wheels being the same diameter, the rim of the 8 horse fly-wheel would move with four times the velocity of that of the 4 horse; in that case, the rim of the 4 horse fly-wheel would require to be $(4 \times 4 =)$ 16 times the weight of that of the 8 horse fly-wheel.

These examples show the great advantage of celerity of motion, in a steam-engine to turn mill-work. If the 4 horse engine made the same number of strokes per minute as the 8 horse engine (say 30 per minute), then the energy required for the fly-wheel of the 4 horse engine would be only half as great as is required for the 8 horse; and if both rims moved with the same velocity, the 4 horse fly-wheel would require to have half the weight of matter in the rim that the other must have.

The proportion between the power which is exerted by an engine during one half stroke, and the power which must be resident in its fly-wheel, to constitute its energy, ought to be the same in engines of all sizes, in order to produce the same approximation to an uniform motion of the millwork. The dimensions of fly-wheels which will have energy enough to produce a sufficiently regular motion for the purposes of manufactories, has been determined by repeated practice, probably without any previous theory.

By a series of observations on engines of all sizes, from 2 horse-power to 100 horse-power, it appears that the energy of the rim of the fly-wheel is usually between 3 and $3\frac{3}{4}$ times the power exerted during half a stroke; hence if the engine had no other resistance opposed to its motion, than that of giving motion to the rim of the fly-wheel, it would be capable of urging that rim from rest to motion, with its full velocity, in 3 or $3\frac{3}{4}$ half strokes of the piston.   The energy of the arms of the fly-wheel, and the crank, and the main cog-wheel by which the millwork is moved, is not considered in this estimate, but only the rim of the fly-wheel, because the weight of that rim, in some measure, regulates all the rest.

The different specimens of Messrs. Boulton and Watt's patent engines,

Page 491, 10 HP. 3·13
494, 10 HP. 3·40
498, 20 HP. 4·16
500, 20 HP. 2·60
502, 30 HP. 2·67
504, 40 HP. 3·14
513, 50 HP. 2·77
519, 50 HP. 3·06

which have been already quoted as standard examples (see p. 489 to 520), show that there may be a considerable variation in the proportion between the energy of the rim of the fly-wheel (that is the power which must be communicated to it, to produce its motion from rest) and the power exerted by the piston in one half stroke.   The least proportion is 2·6 times, and the greatest is 4·16 times; the mean of all these examples is $3\frac{1}{6}$ times.

From similar computations of the fly-wheels of a number of more modern engines, a proportion of $3\frac{1}{4}$ times appears to answer extremely well, and the established practice of the best engineers corresponds very nearly with it.

The method of computing the proportion between the energy of the rim of the fly-wheel, and the power exerted by the engine in each half stroke, has been already given at p. 414, and also in the different examples above mentioned; but the following rule will facilitate such computations.

The mechanical power exerted by the piston in each half stroke, may be assumed to be represented by the capacity of the cylinder.   This may be found, in cylindrical inch feet, by multiplying the square of the diameter of the cylinder in inches, by the length of the stroke of the piston in feet.

If the expenditure of steam by the engine, is taken at the rate of 33 cubic feet (= 6050 cylindrical inch feet) per minute, for each horse-power, then the effective force of the piston will be at the rate of 6·944 pounds per square inch of the piston; or 5·454 pounds per circular inch (see p. 575). If the capacity of the cylinder in cylindrical inch feet, is multiplied by 5·454 lbs., the product will be the force in pounds, with which the piston would be capable of acting through a space of one foot, during each half stroke.

The mass of matter in the rim of the fly-wheel, may be assumed to be represented by the product obtained by multiplying the mean diameter of the rim, by the area of its transverse section in square inches; this product is the number of masses of cast iron one inch square, and 3·1416 feet long, which are contained in the rim.   Each such mass weighs 9·817 pounds; and therefore by multiplying that product by the constant number 9·82, we may obtain the weight of the rim in pounds.

The velocity with which the rim of the fly-wheel moves, may be assumed to be represented by the product obtained by multiplying the diameter of the middle of the rim, by the number of revolutions which it makes per minute; that product is the number of spaces, of 3·1416 feet each, through which it passes in a minute; and being divided by the constant number 153·2 (as directed in p. 33), the square of the quotient is the height in feet, through which a body must fall, to acquire the velocity with which the rim moves.   The product obtained in multiplying this height by the weight of the rim in pounds, will be its energy, expressed by the number of pounds which by acting through one foot, will exert the power that must be communicated to it, to produce its motion from rest.

Lastly, that energy, or power, being divided by the number representing the power exerted by the piston in each half stroke, the quotient will show the proportion between them.

The three constant numbers which are stated above, may be reduced to one in the following manner.   The multiplier 5·454 must be employed to obtain the force exerted by the piston during each half stroke, in pounds acting through one foot; this force is to be used for a divisor. And to find the mass of cast iron in pounds (which is the dividend), the multiplier 9·817 is required.   By substituting a constant multiplier (9·817 ÷ 5·454 =) 1·8, instead of the multiplier 9·817, the previous multiplication by 5·454 may be omitted, and the same result will be obtained. Also, by dividing the other constant number 153·2 by 1·342 (which is the square root of 1·8) we obtain 114·16 for a constant divisor, which being used instead of 153·2, will answer the purpose of all the three constant numbers, whence the following rule.

To find the proportion between the power which must be communicated to the rim of a fly-wheel to produce its motion from rest, and the power which is exerted by the piston of the engine in each half stroke.

RULE *First.* Multiply the square of the diameter of the cylinder in inches, by the length of the stroke of the piston in feet; the product may be assumed to represent the power exerted during each half stroke of the piston, and it must be reserved for a divisor.

Second. Multiply the mean diameter of the rim in feet, by its section in square inches, and the product will represent the mass of cast iron in the rim. Multiply the mean diameter of the rim, by the number of turns it makes per minute, the product will represent the velocity with which the rim moves; then divide that product by the constant number 114·16, and square the quotient; this square will represent the height due to the velocity with which the rim moves; and being multiplied by the second product (representing the mass of cast iron), the resulting product will represent the power communicated to the rim, which is to be used for a dividend.

Lastly. Divide this power by the divisor reserved at first (representing the power of the piston); the quotient will be the number of times that the power exerted by the piston in each half stroke, is contained in the power which constitutes the energy of the fly-wheel.

Example. The 10 horse-engine (p. 494). Cylinder 17½ inches diam. squared = 306·25 circular inches area, × 4 feet stroke = 1225 for the power exerted by the piston in each half stroke. (*Note.* If this is multiplied by 5·454 lbs., it gives 6681 pounds acting through one foot.) The mean diameter of the fly-wheel 11·5 feet × 15 square inches area of the cross section of its rim = 172·5 for the mass of cast iron in the rim. (*Note.* If this were multiplied by 9·817 lbs., it would give 1693·4 pounds for the weight of the rim.)

The mean diameter 11·5 feet × 48·5 revolutions per min. = 557·75 for the velocity, ÷ 114·16 constant divisor = 4·885, the square of which is 23·86 for the height due to that velocity, and × 172·5 the mass of cast iron = 4116, to represent the energy of the rim. Lastly, 4116 energy ÷ 1225 power of the piston, gives 3·358 times, that the energy of the fly-wheel exceeds the power of the piston in each half stroke (*a*).

The proportion that the energy of the rim of the fly-wheel should bear to the power that the engine exerts in each half stroke, may be varied according to circumstances; the principal circumstance to be considered is the kind of machinery to which the engine is to be applied. If the millwork contains heavy masses of matter moving rapidly, such as the mill-stones of grinding mills, the energy of those masses will act in aid of the energy of the fly-wheel, and therefore it may be made on the least proportion of 2¾ times, without any inconvenience (see p. 513). When the machinery opposes a uniform and constant resistance, without having any considerable energy in itself, to regulate its own motion, the fly-wheel must produce the whole regulation, and should not be less than the proportion of 3¼ or 3½ times, in order to produce a sufficiently uniform motion for manufacturing purposes, such as spinning-mills (*b*).

For steam rolling-mills a very great energy is advantageous, and an additional fly-wheel is usually employed; the energy of the whole being from 30 to near 50 times the power exerted in half a stroke (see p. 506); but those are extreme cases, in which the fly-wheels are not introduced merely for the purpose of regulating the motion, but in order to accumulate the power of the engine, so as to enable it to overcome a very great and sudden resistance; and for this reason very powerful fly-wheels are applied to rolling-mills, when they are

(*a*) The statement in the note, p. 494, is 3·4 times, because it was calculated by assuming the engine to exert exactly 10 horse-power (see p. 489); and then the power exerted in each half stroke would be only 6600 lbs. raised one foot high (see p. 491); but the above rule makes it 6681 lbs. And 3·4 times × 6600 lbs. = 22 440 ÷ 6681 lbs. = 3·358 times.

(*b*) In the engines made by Mr. Hick, the energy of the fly-wheel is usually 3¾ times the power exerted in each half stroke.

driven by water-wheels, although a uniform power is exerted, and no regulation or fly-wheel would be required for other purposes.

When it is determined how many times the energy of the fly-wheel should exceed the power exerted by the piston in each half stroke, that number may be multiplied by 858 000, and the product will form a constant multiplier, to be used in the following rule. Or the square root of one-tenth of that product, may be used for a gage point on the line D of sliding rule. For instance, suppose the energy of the fly-wheel is fixed at 3·2 times, which appears to be a sort of standard in Messrs. Boulton, Watt, and Co.'s engines; then $(3 \cdot 2 \times 858\,000 =)\ 2\,745\,600$, or for a convenient number say 2 760 000; that is, two millions seven hundred and sixty thousand for the multiplier; and the square root of 276 000 (one-tenth of the above number) is 525, for the gage point for the sliding rule.

To find the quantity of cast iron which should be contained in the rim of the fly-wheel of a steam-engine. Having given the power exerted by the engine in horse-power; the number of strokes the piston makes per minute; the diameter of the intended fly-wheel, measured to the centre of the rim in feet; and the number of revolutions that the wheel is to make per minute (*a*).

RULE. Multiply the mean diameter of the rim of the fly-wheel in feet, by the number of revolutions it is to make per minute, and square the product for a divisor. Divide the number of horse-power exerted by the engine, by the number of strokes the piston makes per minute; multiply the quotient by the constant number 2 760 000, and divide the product by the divisor found as above. The quotient is the requisite quantity of cast iron in cubic feet, to form the rim of the fly-wheel.

Example. An engine of 30 horse-power (see table, p. 574) makes 19 strokes per minute; the fly-wheel is $17\frac{1}{3}$ feet diameter to the middle of its rim, and it is turned by a multiplying-wheel of 77 teeth, and a pinion of 38 teeth, so as to make $38\frac{1}{2}$ revolutions per minute. Then 17·33 ft. dia. × 38·5 revol. per min. = 667·3, the square of which is 445 289 for the divisor; and 30 HP ÷ 19 strokes = 1·58 × by the constant multiplier 2 760 000 = 4 360 000, which product divided by 445 289, gives 9·79 cubic feet of cast iron for the rim of the fly-wheel. The rim really is $8\frac{1}{4}$ inches, by $3\frac{1}{8}$ inches = 25·78 square inches of cross section; and it contains 9·74 cubic feet of cast iron.

Sliding Rule. $\left\{\dfrac{\text{C Cast iron in rim cub. ft.}}{\text{Gage point 525.}} \quad \dfrac{\text{HP} \div \text{stro. per min.}}{\text{Dia. ft.} \times \text{rev. per min.}}\right.$ *Exa.* $\dfrac{\text{C } 9\cdot77 \text{ c.f.}}{\text{D } 525.} \quad \dfrac{(30\,\text{HP} \div 19\,\text{stro.} =)1\cdot58}{(17\cdot3\,\text{ft.} \times 38\cdot5 =)\ 667}$

Note. The same fly-wheel is used for an engine of 53 horse-power, making $17\frac{1}{2}$ strokes per minute (see note, p. 518), but the fly-wheel is turned by a multiplying wheel of 144 teeth, and a pinion of 48 teeth, so as to make $52\frac{1}{2}$ revolutions per minute. The calculation is 17·33 ft. diam. × 52·5 revol. per min. = 910, the square of which is 828 100 for the divisor; and 53 HP ÷ 17·5 strokes = 3·028 × 2 760 000 = 8 360 000, which ÷ 828 100 = 10·09 cubic feet of cast iron for the rim. It really was 9·74 cubic feet as above stated.

Sliding Rule, Example. $\left\{\dfrac{\text{C} \quad \text{Rim } 10\cdot04 \text{ cub. ft. of cast iron.}}{\text{D} \qquad\quad 525.} \quad \dfrac{53\,\text{HP} \div 17\cdot5 \text{ strokes per min.} = 3\cdot028.}{17\cdot33\,\text{ft. dia.} \times 52\cdot5 \text{ revol.} = 910.}\right.$

Another example for a small engine which makes a rapid succession of strokes, and therefore requires less energy in the fly-wheel, in proportion to the power exerted. A 10 horse-engine makes 25 strokes per minute (see table, p. 574). The fly-wheel is $11\frac{1}{3}$ feet diameter to the middle of its rim, and being fixed on the axis of the crank, it makes 25 revolutions per minute. Then 11·33 ft. diam. × 25 revolutions = 283·25, which squared is 80 230 for the divisor. And 10 HP ÷ 25 strokes = ·4 × 2 760 000 = 1 104 000 and ÷ 80 230 = 13·75 cubic feet of cast iron for the rim of the fly-wheel. The wheel is 12 feet diameter outside, and its rim is 8 inches by 7 = 56 square inches of cross section; it contains 13·84 cubic feet. If this fly-wheel had been turned by sun and planet-wheels, its rim would have only required one-fourth of the above weight, or 6 inches by $2\frac{1}{3}$ inches = 14 square inches of cross section (see p. 494).

Sliding Rule, Example. $\left\{\dfrac{\text{C} \quad \text{Rim } 13\cdot75 \text{ cub. ft. of cast iron.}}{\text{D} \qquad\quad 525.} \quad \dfrac{(10\,\text{HP} \div 25 \text{ strokes} =)\ \cdot4.}{(11\cdot33\,\text{ft. dia.} \times 25 \text{ rev.} =)\ 283\cdot3.}\right.$

(*a*) It is assumed that the power of the engine is correctly stated, according to Mr. Watt's scale (see p. 574), which allows 33 cubic feet of steam per minute to each horse-power, and is equivalent to an effective pressure of 6·944 lbs. per square inch, exclusive of friction; that is equal to the pressure of a column of water 16 feet high.

The rim and arms of the fly-wheel are made of cast iron, and in Mr. Watt's original engine, the weight of the arms is about two-thirds of the weight of the rim; the fly-wheel above mentioned has 6 arms, of $7\frac{3}{8}$ inches broad at the rim, and $10\frac{1}{8}$ inches broad at the centre, by 2 inches thick; they contain about $6\frac{1}{2}$ cubic feet of cast iron. The energy of a lever of uniform size, which moves about one of its extremities as a centre of motion, is equal to the energy that a mass, one-third of the weight, would have, if it moved with the same velocity as the outer extremity of the lever. The arms of fly-wheels being thicker, and heavier at their central parts than at the rim, the addition they make to the energy of the rim may be computed by supposing one-fourth of their weight to be added to that of the rim. Assuming the whole weight of the arms to be two-thirds of that of the rim, the energy of all the arms will be two-twelfths, or one-sixth, of that of the rim (*a*).

In Mr. Watt's original engines, the arms of the fly-wheel were cast in two halves, each containing three or four arms; the two parts join at the centre, and are united together by bolts. The central part is fastened upon the square axis by wedges. The rim of the wheel is formed in 6 or 8 segments, which are fastened to the arms, and to each other by bolts and nuts.

The operation of the sun and planet-wheels is advantageous to a steam-engine, by giving a double velocity to the fly-wheel; for the energy of moving masses is as the squares of their velocities, and hence a fly-wheel, which is turned by sun and planet-wheels, requires to be only one-fourth the weight that it must be, to produce the same effect in regulating the motion, if it were turned by a simple crank. When a multiplying wheel and pinion is used, the fly-wheel is often turned

---

(*a*) Mr. Robertson Buchanan, in his Treatise on propelling Vessels by Steam, 1816, gives the following rule, which, he says, Messrs. Fenton, Murray, and Wood follow in proportioning the weights of the fly-wheels for their steam-engines.

RULE. Multiply the number of horse-power of the engine by 2000, and divide the product by the square of the velocity of the circumference of the fly-wheel, in feet per second. The quotient will be the proper weight for the fly-wheel, in hundred weights of 112 lbs. each.

Example. To find the proper weight for the fly-wheel of an engine of 20 horse-power; supposing the fly-wheel to be 18 feet diameter, and to make 22 revolutions per minute.

A circle of 18 feet diameter is = 56 feet circumference × 22 revolutions per minute = 1232 feet motion per minute ÷ 60 = $20\frac{1}{2}$ feet per second, is the velocity of the circumference of the fly-wheel; and $20\frac{1}{2}$ squared is = $420\frac{1}{4}$. Then 20 HP × 2000 = 40 000; which ÷ $420\frac{1}{4}$, gives 95·2 cwt. for the required weight of the fly-wheel.

The fly-wheel of a 20 horse-engine, made by Messrs. Rothwell, Hick, and.Co., of Bolton, is 18 feet diameter; the rim is ten inches by four inches, and contains 15 cubic feet of cast iron, which would weigh 60 cwt.; and the 8 arms weigh 33·75 cwt.; making the total weight of this fly-wheel 93·75 cwt., when the above rule would give 95·2 cwt.

Another Example. A 30 horse engine, made by Messrs. Fenton, Murray, and Co., has a fly-wheel 20 feet diameter at the outside; the piston makes a 6 feet stroke, and the fly-wheel makes 19 revolutions per minute. The velocity of the circumference would be 19·9 feet per second, the square of which is 396. And 30 HP × 2000 = 60 000 ÷ 396 = 151·5 cwt. for the weight of this fly-wheel, by the rule.

The rim of the wheel is 11 inches by 6 inches, it contains 27·5 cubic feet of cast iron = 110 cwt.; and allowing 41·5 cwt. for the arms of the wheel, it would come to the calculation.

The preceding rule appears to answer for engines which make about 19 strokes per minute, but it is defective in principle, because it does not take into account the frequency of the intervals at which the strokes, or successive impulses on the crank and fly-wheel are repeated, and this is an essential consideration. The rule would consequently direct that the fly-wheels of small engines, which make their strokes quick, should be much heavier than they are usually made, according to the practice of the best engineers. And it would give large engines, which move slower, smaller fly-wheels than they ought to have, to cause them to move steadily.

three times round for each stroke of the engine, and in some cases four times; a moderate fly-wheel is then sufficient to regulate a very large engine; this is shown in the above examples, where the same fly-wheel is used for a 30 horse-engine or for a 53 horse-engine, by only varying the proportion between the multiplying-wheel and pinion. If the same fly-wheel were fixed on the axis of the crank, it would serve for a 15 horse-engine, making 23 strokes per minute.

To find the quantity of matter in the rim of the fly-wheel in cubic feet. Having given the diameter of the outside of the rim in feet; the depth of the rim in inches, measured in a direction from the outside towards the centre; and also the breadth or thickness of the rim in inches.

The depth of the rim in feet must first be deducted from the diameter, in feet, and the remainder may be taken for the mean diameter of the mass of the rim in feet (*a*). Also the depth of the rim in inches must be multiplied by its breadth in inches, to obtain the area of the transverse section of the rim in square inches.

RULE. Multiply the mean diameter of the rim in feet, by the area of its transverse section in square inches, and divide the product by 45·837; the quotient is the solid content of the rim in cubic feet.

Example. The fly-wheel mentioned at p. 491, is 12 feet diameter outside, and the rim is 6 inches deep by $2\frac{1}{2}$ inches thick. The mean diameter is (12 ft. — ·5 ft. =) 11·5 feet. And the area of the transverse section is (6 inc. × $2\frac{1}{2}$ inc. =) 15 square inches. Then 11·5 feet mean diam. × 15 square inches = 172·5 ÷ 45·84 = 3·763 cubic feet of cast iron in the rim.

Sliding Rule, slide inverted.	A Mean diameter feet.	Content cubic feet.	Exam.	A 11·5 feet,	3·76 cu. ft.
	ᙍ Area square inches.	45·84 divisor.		ᙍ 15 squ. inc.	45·84 div.

Note. The product obtained by multiplying the mean diameter of the rim in feet, by the area of its transverse section in square inches, represents the number of masses of one inch square and 3·1416 feet long, that the rim contains. The divisor 45·84 is the number of such masses that are contained in a cubic foot: thus 144 (square inches in a square foot) ÷ 3·1416 (the circumference of a circle whose diameter is 1) = 45·8367.

Again, 228·16 of such masses of cast iron weigh one ton, of 2240 pounds avoirdupois; because a cubic foot of cast iron weighs 450 pounds; and (2240 lbs. ÷ 450 lbs. =) 4·978 cubic feet of cast iron weigh a ton; whence 4·978 × 45·84 = 228·16.

Lastly, each mass of cast iron of one inch square, and 3·1416 feet long, weighs 9·817 pounds; for a square inch foot of cast iron weighs (450 lbs. ÷ 144 =) $3\frac{1}{8}$ pounds; and 3·1416 × 3·125 = 9·8175. From these data the following rules are derived.

To find the weight of cast iron in the rim of a fly-wheel, in tons of 2240 pounds each.

RULE. Multiply the mean diameter of the rim in feet, by the area of its transverse section in square inches, and divide the product by 228·16; the quotient is the weight of the rim in tons.

Example. Mean diameter of rim $11\frac{1}{2}$ feet × 15 squ. inc. area = 172·5 ÷ 228·16 = ·756 of a ton, is the weight of the rim.

Sliding Rule, slide inverted.	A Mean diameter feet.	Weight in tons.	Ex.	A 11·5 feet.	·756 tons.
	ᙍ Area square inches.	228 divisor.		ᙍ 15 sq. in.	228 divi.

To find the weight of cast iron in the rim of a fly-wheel, in pounds avoirdupois.

RULE. Multiply the mean diameter of the rim in feet, by the area of its transverse section in square inches, and multiply the product by 9·817 lbs.; the product is the weight of the rim in pounds.

Example. Mean diameter of the rim $11\frac{1}{2}$ feet × 15 square inches area = 172·5 × 9·817 = 1693·4 pounds weight of iron in the rim.

Sliding Rule, rule inverted.	A Mean diameter feet.	Weight in pounds.	Exam.	A 11·5 feet.	1693. lbs.
	ᙍ Area square inches.	·1018 divisor.		ᙍ 15 squ. inc.	·1018. dw.

(*a*) This is not mathematically correct, but it is sufficiently exact for all cases of fly-wheels, because the depth of the rim is only a small proportion of the outside diameter. The true method would be to find the area of the internal circle of the rim in square feet, and to deduct it from the area of the external circle in square feet; the difference between them would be the exact area of the rim in square feet, which being multiplied by the thickness of the rim, in decimals of a foot, the product would give the true content of the rim in cubic feet (see an example, p. 414). The areas of the circles may be found by inspection in the table for that purpose, p. 552.

Note. The sliding rule requires a divisor instead of a multiplier, therefore we must take the reciprocal of 9·8175, viz. ·10185 for that divisor.

Proof of the above examples. As the rim contains 3·763 cubic feet, it must weigh (3·763 × 450 lbs. =) 1693·35 pounds, which (÷2240 lbs. in a ton) is equal to ·756 of a ton.

In modern steam-engines the fly-wheel is fixed upon the axis of the crank, and will commonly be found to have rather less energy than is directed by the preceding rule (see p. 640). The energy is usually three times the power exerted in each half stroke, therefore (3 × 858 000 =) 2 574 000 is the multiplier which must be used in that rule, to correspond with the modern practice; or, in even numbers, 2 600 000, and then the gage point for the sliding rule will be 510. The former rule is preferable for all cases, but the difference is not considerable.

Example. When the fly-wheel of the 30 horse-engine is fixed upon the axis of the crank, it must make 19 revolutions per minute; its diameter to the centre of the rim is 20 feet. Then 20 ft. dia. × 19 revol. = 380, which squared is 144 400 for the divisor. And 30 HP ÷ 19 strokes = 1·579 × 2 600 000 = 4 105 400; which, divided by 144 400, gives 28·42 cubic feet of cast iron for the rim of the fly-wheel. The wheel is 21 feet diameter outside, and the rim 12 inches by 5½ inches = 66 square inches section, and contains 28·78 cubic feet.

Sliding Rule, Example. $\begin{cases} C \\ D \end{cases}$ $\dfrac{\text{Rim 28·4 cub. ft. cast iron}}{51}$ $\dfrac{(30\ \text{HP} \div 19\ \text{strokes} =)\ 1·58.}{(20\ \text{ft. dia.} \times 19\ \text{revol.} =)\ 380.}$

Another Example. If the fly-wheel of the 53 horse-engine is fixed on the axis of the crank, it must make 17½ revolutions per minute; the diameter to the middle of the rim is 23 feet. Then 23 × 17·5 = 492·5, the square of which is 162 006 for the divisor. And 53 HP ÷ 17·5 strokes = 3 028 × 2 600 000 = 7 872 800, which ÷ 162 006 = 48·61 cubic feet of cast iron in the rim. The rim really is 12 inches by 8 inches = 96 square inches of cross section, and contains 48·17 cubic feet.

Sliding Rule, Example. $\begin{cases} C \\ D \end{cases}$ $\dfrac{\text{rim 48·6 cub. ft. of cast iron}}{51}$ $\dfrac{(53\ \text{HP} \div 17·5\ \text{strokes} =)\ 3·03.}{(23\ \text{ft. dia.} \times 17·5\ \text{revol.} =)\ 402·5.}$

When the fly-wheel is placed upon the axis of the crank, the rim requires to be very heavy, and the weight of the arms is not above one half of the weight of the rim, or sometimes only one third. In such cases the energy of the arms will be one eighth, or one twelfth of the energy of the rim (see p. 641).

In modern engines the fly-wheel is made with a strong centre piece, or circular plate like a wheel, which is fixed upon the axis, and it has six or eight cells, formed in its flat face, to receive the ends of as many arms, which are well fitted into those cells or sockets, and are held fast in them by three or four strong screw bolts, which pass through each arm, and through the flat circular plate. The extremities of the arms are made with projecting palms, formed to fit into recesses in the rim, which is made in six or eight segments, and they are united together by the projecting parts at the end of each arm, which cross the joints of the segments, and are fastened to them by three screw bolts through each end of each segment.

General observation respecting the Rules, contained in Chapter VIII., for calculating the Dimensions for the parts of Mr. Watt's Rotative Steam-Engines.

When the power of the engine is made a term of the calculation, it is supposed that it is correctly stated in horse-power, according to Mr. Watt's scale, see p. 574, which allows 33 cubic feet of steam per minute to each horse-power. To avoid any uncertainty on this head, some of the rules proceed according to the diameter of the cylinder, and the rapidity of the motion of its piston, instead of the horse-power.

When these rules are to be applied to engines for pumping water, the load upon the piston must be taken into account, and the horse-power calculated according to the rule in p. 440. If the engine operates by expanding the steam in the cylinder, then the power exerted during the expansion (see p. 367) must not be included in that expression of the power which is used as a term for calculation in the rules; because that power is obtained without any additional expenditure of steam.

The rules are adapted to give the proper proportions to all the different parts, when the engine is loaded with its proper resistance of 6·94 lbs. per square inch of the piston, and moves with its proper speed, so as actually to exert the power at which it is rated. If the engine is more or less loaded, then those proportions will not be quite exact. In those rules which relate to the performance of the engine, the power actually exerted by it may be substituted in the calculations, for the power at which the engine is rated. But those rules which give the proportion of the parts of the machine, should be calculated according to the power at which the engine is rated, because that is the power with which it will work to the greatest advantage, and which it is expected that it will exert, on an average of all the circumstances (a).

The cast and wrought iron used in steam-engines, should be of the best quality, and carefully manufactured; materials of an inferior quality have been sometimes used in engine-work, and an increase of dimensions allowed beyond the established proportions, to compensate for the deficiency; but that is a bad system, for it loads the working parts with unnecessary weight, and yet they may be liable to break, because inferior metal is subject to internal defects, of which the extent cannot be known.

It is a common practice with some engineers, to make the parts of steam-engines much stronger than Mr. Watt's dimensions, but if the materials and workmanship are good, this is not at all necessary; for the force exerted by the piston can never be sufficient to occasion injurious strains on the parts, if they are proportioned according to the preceding rules, except in the single instance of the engine being suddenly stopped, when it is in rapid motion, by retaining the steam within the cylinder accidentally (see p. 463). It is better to apply a prevention to guard against such an occurrence, as is recommended in p. 527, than to give increased strength with a view of resisting the strain.

(a) The mode by which the rules contained in this chapter have been formed, has been already stated, note, p. 532, and it is only necessary to add, that, in giving them to the profession, the author feels confidence that they are substantially the same as those which Mr. Watt followed in his practice, because they have been formed from a very extensive series of observations made during a course of twenty years, upon steam-engines of which many were constructed under Mr. Watt's own superintendence; and the rules have been adapted so as to give results which correspond equally well with all sizes of engines, from the largest to the smallest.

The notes and calculations which have been used in the formation of these rules, are very voluminous, but it was thought unnecessary to give more than one example for each rule. Those examples are in all cases taken from some real specimen of Messrs. Boulton and Watt's practice, which was found to correspond nearly with the rule. A great number of similar examples of different sizes, have been used to obtain the rule in every case, and although many of those examples vary from the rule, the variations are not so great as to be of any importance in practice, and they are as often above as below the result given by the rule; they are such variations as might be expected to arise from the rules not being rigorously observed in practice.

The principles of all the most important proportions upon which the rules depend, have been the subject of frequent communication from Mr. Watt himself to the author, and are really those which he followed. At the time the author first became acquainted with Mr. Watt, he had been several years in retirement from business, and minute details were not his favourite subjects of conversation, but only general principles; he himself states, in the introduction to his revision of Dr. Robison's article, "Steam Engine," that the subject of steam and steam-engines had been almost dismissed from his mind, for many years previous to his undertaking that revision (in 1814); in which, on account of his advanced age, he did not attempt to render it a complete history of the steam-engine, or even to give a detailed account of his own improvements upon it, but he only intended to make a commentary upon his friend's work.

Engines for steam vessels which navigate the sea, being exposed to great violence in rough weather, are provided with loaded safety-valves at the top and at the bottom of the cylinder, to permit the escape of any water which may boil over into the cylinder, and they will not allow the steam to be retained within the cylinder, beyond a certain elasticity. These safety-valves were brought into use by Mr. Field ; and if they were generally applied to all engines which have fly-wheels, the breaking of engines would become a very rare occurrence.

If, for any particular reasons, it is judged expedient to give greater strength to the parts of any steam-engine, the dimensions may be calculated with the preceding rules, by adapting them to a larger cylinder than that which is actually used. For instance, if the strength of the parts for a 30 horse-engine are calculated for a cylinder $30\frac{1}{8}$ inches diameter, and the engine is set to work with a cylinder of $28\frac{1}{4}$ inches diameter, then those parts will have one fifth more strength than is proposed to be given by the preceding rules. This method may be followed in cases where it is thought probable that it will be required, at some future period, to apply a larger cylinder to the engine.

CONCLUSION RESPECTING MR. WATT'S INVENTIONS.

We have now traced the history of Mr. Watt's steam-engine through various progressive stages of improvement, during a period of about thirty-five years, from 1765 to 1800 ; in all which time he appears to have been attaining increased perfection in his engines. The first nine years, from 1765 to 1774, were spent in completing the invention of the single engine for pumping water, before he left Scotland, (see p. 310 to 318); he then joined Mr. Boulton, and passed the next ten years, from 1774 to 1784, in perfecting the details of the construction of that kind of engine, and in applying it to use for mines, on a very extensive scale, and in the application of the expansion principle, (see p. 320 to 383).

It appears that, during the last mentioned period, he had also settled the invention of the double rotative engine for mills, in his own mind ; he reduced it to practice, and applied it to a very great extent, and perfected the details of its construction during the next ten years, from 1784 to 1798, (see p. 434 to 530). The remaining seven years of this distinguished career, was occupied in active superintendence of the very extensive business which had then arisen in manufacturing his steam-engines at Soho, and in systematizing that business (a) by forming

---

(a) As a part of that system of business, in addition to the application of the sliding rule, (see chap. vii.) it is proper to mention the method which Mr. Watt invented, of copying manuscript writings, by transferring or printing off a part of the ink upon wet paper ; this was first adopted about 1780, by Messrs. Boulton and Watt, for their office at Soho, and it has since become a very general practice in merchants' counting-houses. Mr. Watt took out a patent in 1780, for his new method of copying letters and other writings. The specification is printed in the first series of the Repertory of Arts, Vol. I. p. 13, and the following extract will explain the method very clearly :

" The letter or other writing that is intended to be copied, is written with a kind of ink which is partly soluble in water, and then a copy of that writing is taken or printed off upon thin paper which contains no size, or not so much as would make it fit or writing upon ; this thin paper being cut into leaves of the same size and shape as the original writing, is wetted with water, or with a weak solution of gall nuts, by means of a sponge, or brush, or by dipping ; and the wetted paper is folded between two thick unsized spongy papers, or two cloths, which are pressed lightly together by laying a flat board upon them, in order that the superfluous moisture may be absorbed from the thin paper.

a scale for the proper dimensions of the parts of engines of all sizes, and also in applying the rotative engines to new purposes in manufactories, (see p. 531 to 645).

In all these points Mr. Watt was most ably seconded by his associate, Mr. Boulton, who, with his assistants, undertook the details of directing the manufacture of a number of copies after every original engine which Mr. Watt produced, leaving him a great share of leisure to devise further improvements. It was their practice to continue to construct their engines according to the most approved model for a long time together, without introducing any of those alterations or improvements upon which Mr. Watt was continually occupied, until he had very maturely settled the details of an improved engine; it was then adopted for a new model, which, in its turn, was followed, until something better was quite arranged. By this means they secured themselves against failures in the application of their new plans; for, in reality, they were not new to themselves, as they had arranged them for a long time in their own minds, and proved them by experiments, before they attempted to put them in practice.

"A leaf of the thin paper, after being thus moistened and pressed, is spread over the surface of the paper containing the writing which is to be copied, so that one side of the thin wet paper will be in contact with the ink of that writing, and a leaf of clean writing paper is applied to the other side of the thin paper. The three papers are then placed upon the board of a common rolling press, such as is used for printing impressions from engraved copper-plates; and the papers are forcibly pressed together, by rolling them through between the rollers of that press, twice or more times, in the same manner as is practised for copper-plate printing. That pressure will cause part of the ink of the original writing to be absorbed by the thin wet paper, so as to penetrate through the same; and consequently, when the thin paper is removed from the original writing, an exact copy of that writing, more or less faint, will appear upon both sides of the thin paper; the side which was in contact with the original writing will exhibit the copy in a reverse order, but on the opposite side the copy will be in the proper order and direction of the lines, exactly the same as the original writing; because the wet paper, by its thinness and spongy quality, allows the ink which it absorbs from the original writing, to penetrate quite through its substance. Instead of a rolling press, a screw press may be used to take the impression.

"If the original writing is in strong characters, with a sufficiency of ink, three or four successive copies may be taken off, upon as many thin papers, and will all be very legible; but when only one copy is taken, the writing upon it will appear very black; and the original will not be at all injured by the operation. Copies of writings thus taken on thin paper, are even more authentic documents than the originals, because it is more difficult to make any alteration in the writing; for as the ink penetrates quite through the thickness of the paper, it cannot be erased; and as the unsized paper will not bear ink, no additions can be made. Hence it would be a good practice to make transferred copies on thin paper, for bills of exchange, and letters of credit, the writer keeping or destroying the original; also for wills, and other writings of importance.

"The ink is prepared as follows:—To four quarts, ale measure ($70\frac{1}{2}$ cubic inches each) of spring water, put one pound and a half (avoirdupois) of Aleppo gall nuts, half a pound of green vitriol or copperas (sulphate of iron), half a pound of gum-arabic, and a quarter of a pound of roach allum. Pound the solid ingredients, and infuse them in the water for six weeks or two months, during which time the liquor should be frequently shaken, then strain it through a linen cloth, and keep it in bottles closely corked, till wanted for use.

"The astringent liquor which is recommended instead of water, to moisten the thin paper, is prepared as follows: In two pounds of distilled vinegar, dissolve one ounce of the sedative salt of borax, and add four ounces of calcined oyster-shells, carefully separated from their brown crust, so as to leave only the white part; shake the mixture frequently, during twenty-four hours, and then allow it to stand and deposit its sediment; after this, decant the clear liquor, and filter it through unsized paper into a glass vessel. To this solution add two ounces of the best blue Aleppo gall nuts bruised, and keep the liquor in a warm place during twenty-four hours, shaking it frequently; then filter the liquor again through unsized paper, and afterwards add one quart, ale measure, of pure water; it must then stand quiet for twenty-four hours, and if it becomes turbid, it must be filtered again, after which it will be fit for use.

In consequence of this cautious mode of procedure, Messrs. Boulton and Watt so completely acquired the confidence of manufacturers and others who required steam-engines, that whatever they proposed to undertake was commonly adopted by their employers, without farther inquiry, in the full assurance of attaining the proposed object. Mr. Rennie was continually associated with them in all applications of steam-engines to public works, and his assistance and concurrence in their plans, tended to increase their reputation.

The monopoly which was secured to Messrs. Boulton and Watt for so long a period, by the prolongation of Mr. Watt's original patent, proved a great advantage, independently of the pecuniary profit that it secured to them ; for by concentrating the business in their hands, they always had orders for a sufficient number of engines of the same kind, to enable them to arrange their manufactory upon system, with a great division of labour, whereby all their workmen acquired dexterity from continual repetition of the same work ; and the work was better executed, at a less expense than if they had made fewer engines.

At the same time by the prohibition of the patent, Messrs. Boulton and Watt were subjected to much jealousy from a number of persons whose situation led them to think of making steam-engines, but who were restricted by the patent from making any engines on Mr. Watt's principle ; their practice was thus confined to the construction of Newcomen's engines, whilst the great superiority of Mr. Watt's engines for most purposes of manufactories, occasioned a greater demand for them, than for atmospheric engines ; hence, continual attempts were made to evade the patent right, and many engines were constructed with part of Mr. Watt's improvements combined with the common form of Newcomen's. As Messrs. Boulton and Watt never neglected to institute legal proceedings against the makers of all such engines, they were commonly stopped at the commencement of their career, without venturing the result of a trial at law. In the two instances their right was tried, and they obtained verdicts in their favour ; their patent right was afterwards considered as valid, and all those who had made engines containing any part of Mr. Watt's improvements, were compelled by the patentees to pay them for licenses to work such engines.

In 1795, Messrs. Boulton and Watt brought an action against a Mr. Bull, for an infringement of the patent, and it was tried in the Court of Common Pleas at Westminster. The engine which formed the subject of this action, was set

---

" The thin paper may be prepared by impregnating it with this liquid, and drying it again ; when it is afterwards wanted for copying, it may be wetted with water. Or the thin paper may be wetted with this liquid instead of water, and it will render the transferred writing blacker than when water alone is used; although in most cases water will be sufficient."

A great number of Mr. Watt's copying presses were made at Soho, and sold under this patent ; they answer the purpose of letter-copying extremely well. They are of two kinds, one a strong rolling press with large rollers proper for an office, and sufficiently large to copy drawings and plans. The other is a compact rolling press and apparatus for copying letters, included within a portable writing desk, containing every other convenience for writing, and which folds up into a moderate compass for travelling.

This method of copying was brought to such perfection in Messrs. Boulton and Watt's office at Soho, that it was universally employed to retain exact copies of all letters, plans, and drawings, which were sent out as directions to their workmen for fixing their steam-engines. This system of business enabled them with very little trouble, and without any delay, to have a precise knowledge of every particular of all the distant works which were going on ; and thus tended to avoid mistakes in their orders, and in fitting the work exactly to the places. For drawings, thick drawing paper properly sized, may be used to receive the copies, instead of thin paper; because it is not material whether a drawing is reversed or not. Mr. Smeaton had one of Mr. Watt's copying presses, which he used to copy his letters and drawings during the latter years of his practice.

up by Mr. Bull, at Oatfield Copper mine in Cornwall. The cylinder, valves, and working gear, were on the usual construction of Mr. Watt's single engine (see Plate X.); but the cylinder was inverted, and placed over the pit, the pump-rod being attached immediately to the piston-rod. The air-pump and condenser were the same as Mr. Watt's, but were not inverted.

This was such a complete adoption of Mr. Watt's apparatus, as to admit of no defence, on the ground of dissimilarity; and accordingly, the validity of his patent was contested on the ground of informality in the terms of the patent, which states the invention to be a method of working, and not a new invented machine; but the act of parliament for prolonging the term, states the patent to be for certain engines. It was also objected, that the specification does not contain sufficient instructions to enable any person to construct an improved engine.

The case of Messrs. Boulton and Watt was supported by the testimony of many distinguished artists, and men of science; who stated that they understood the specification, and could have made an improved engine without any farther instructions. The jury gave a verdict for the patentees; but it was reserved for the judges of the Court of King's Bench, to decide, whether the patent was good in law, and continued in force by the act of parliament, and whether the specification was sufficient in point of law, to support the patent. The case was argued twice before the judges. Exceptions were made to the title of the patent, which is for a new invented method of lessening the consumption of steam and fuel in fire engines; also to the specification, because no particular engine is described in it, but only a bare enunciation of principles. It was contended that mere principles cannot be the subject of a patent; and that the specification ought to have described the method whereby the engine could be constructed, so as to save the greatest quantity of fuel which it is known to be capable of saving, and which in fact it does save, when used by the inventor.

The Lord Chief Justice Eyre, and Judge Rooke, considered the objections as merely formal, and that they did not affect the substantial merits of the patentee, who ought to be protected on account of the great importance of his improvements; but two other Judges, Buller and Heath, gave weight to the formal objections, and the court being equally divided, no judgment was given. This circumstance was considered by their adversaries as a sort of defeat, and in consequence several other infringements on the patent were begun in various quarters, and the patentees commenced as many proceedings against them (a).

In 1799, Messrs. Boulton and Watt brought an action against Messrs. Hornblower and Maberly for infringement of Mr. Watt's patent; it was tried in the Court of Common Pleas, and a verdict given in favour of the patentees. The engines which were the ground of this complaint, were made by Mr. Hornblower for different manufactories, and for winding up coals at collieries. The principal instance was an engine set up at Messrs. Meux's brewery in London, in 1796; it had two single acting cylinders, with their separate condensing apparatus; but they were so combined as to produce a double action to turn the crank and fly-wheel. The proceedings on this trial were nearly the same as on the former, and a similar case was reserved for the Court of King's Bench; but in this instance the Lord Chief Justice Kenyon, and the Judges, Ashurst, Grose, and Lawrence, were unanimous in supporting the patent against the formal objections (b).

(a) It was stated in the proceedings on these trials, that on the 5th September, 1777, Mr. Watt assigned two-thirds of his patent right to Mr. Boulton, for the remainder of the prolonged term.
(b) Judge Ashurst gave the following opinion. Every new invention is of importance to the

This was a more liberal construction of patent rights than has been usually evinced in courts of law, where the judges have been accustomed to exercise a very rigid severity towards patentees. There is no specific law to secure to inventors a property in their inventions, but patents are granted according to a remnant of an ancient prerogative of the crown, whereby the king formerly gave exclusive privileges and monopolies to individuals, to exercise particular trades. These grants became so multiplied at the time when commerce began to extend itself, in the middle part of the reign of queen Elizabeth, that they proved an intolerable grievance to her subjects, and complaints were made in parliament ; in consequence the queen revoked the most obnoxious of the patents then existing ; but as the evil still continued, several successive acts of parliament were passed for the regulation of patent grants, and finally they were all declared void by the statute of monopolies, which was enacted in the 21st year of king James the First.

In this act an exception is made, of letters patent granted for the term of fourteen years, " for the sole working or making of any manner of new manufactures " within this realm, to the true and first inventor and inventors of such manu- " factures, which others at the time of making such letters patent, and grants, shall " not use, so as also they be not contrary to law, nor mischievous to the state, by " raising the prices of commodities at home, or hurt of trade, or generally incon- " venient."

Hence, the rights of inventors are only held by sufferance, and the terms of this exception leave them subject to inquiry into the merits and public utility of their inventions. In the early stages of a new invention, this must be a matter of opinion, and courts of law, from an imperfect knowledge of such subjects, are liable to receive very erroneous impressions respecting inventions, from the contending testimony of witnesses prejudiced against a patent, and that of sanguine enthusiasts in its favour ; independently of interested evidence on either side. Many decisions upon patent rights have been very unjust from this cause, and a well defined law, of the right of property that an inventor may have in his own productions, is exceedingly wanted. Patents should not be granted at all, except a strong case of novelty, merit, and public utility, is previously made out before a tribunal of competent judges, and the results should be ascertained by experiment. The patentee should also be called upon to publish exact specifications of all the improvements he may make in the course of his practice, during the term of his patent, with detailed instructions for practising the invention in its greatest perfection. If this had been done in the case of Mr. Watt, much valuable knowledge would have been disseminated, which would have greatly facilitated the extension of his improvements after the expiration of his patent.

Mr. Watt's original specification was necessarily very imperfect, because it was made at the time when he had not perfected his engine ; but when parliament granted him an extension of the term, it ought to have been on condition, that he should instruct the public as far as he was then able to do. According to the ordinary practice of the courts of law in other cases, Mr. Watt's patent ought to have been annulled, for the insufficiency of the specification, which is a series of definitions of principles of action, without any description of the means of carrying them into effect. And it is certain that if this specification had not been supported by

wealth and convenience of the public; and when they are enjoying the fruits of a useful discovery, it would be hard on the inventor to deprive him of his reward. The jury have found that the inventor has sufficiently described the nature of his invention by his specification, and I think he is, in point of law, as well as justice, entitled to the benefit which was intended to be conferred on him by the patent, and by the act of parliament.

the testimony of many scientific artists, who stated that it was sufficient in their opinion, and if the merit of Mr. Watt's engine had not been so universally allowed at the time of those trials, as to have obtained a leaning in his favour, his right could not have been established, as a mere question of law, according to the usual practice of the courts in other similar cases.

Extent of Mr. Watt's invention in perfecting the Steam-engine.

The first notion of a steam-engine with a cylinder and piston, which was given by Papin in 1690 (see p. 97), was very imperfect and impracticable; he proposed to produce the steam from water which was contained within the lower part of the cylinder, by applying fire at the outside, beneath the bottom of it; and then he expected to obtain the condensation of that steam, by the coldness of the surrounding air; hence his cylinder was to answer the purpose of a boiler, as well as to receive the piston, and the steam was also to be condensed in the same cylinder.

This project was realised by Newcomen in 1712, who succeeded in making a useful engine, chiefly by adding a separate boiler to Papin's cylinder, in order to produce the steam to fill that cylinder when necessary; and then, by injecting cold water into the cylinder, the contained steam was condensed and cooled, so as to occasion a sufficient vacuum to give an efficient action to the piston, by the pressure of the external air upon it.

This system was rendered complete by Mr. Watt in 1769, who added a separate vessel for the condensation of the steam, so as to have a vacuous space constantly exhausted, in preparation to receive the steam from the cylinder, whenever it is required to be exhausted; and the steam-boiler is constantly kept full of steam, in readiness to fill the cylinder when required. Hence the plenum and the vacuum are always maintained in their respective places; and the cylinder can be filled and emptied instantaneously, so as to give a rapid action to the piston, as well as an efficient force. The piston being impelled by the steam from the boiler, instead of the external air, its force is capable of exact regulation, according to the quantity of steam which is admitted into the cylinder.

Mr. Watt's character as an inventor.

In summing up Mr. Watt's character as a mechanical inventor, he will be found to occupy the same supereminent station amongst that class, as the illustrious Newton held amongst philosophers of a higher order; and Mr. Watt was not less remarkable for that fertility of genius which produces new ideas and combinations, than for the sound judgment which he exercised in the arrangement of his plans, before putting them in execution. With a command of inventive faculty, far greater than any of his cotemporaries, he does not appear to have been elated with that distinction, or to have depended upon it, so as to have neglected any precautions to insure success; but he left as little as possible to the chance of after thoughts for the complete accomplishment of his views. This feeling is well expressed in a passage in one of his letters to Mr. Smeaton (see p. 320), and offers a striking contrast to the sanguine expectations and confident assurance of many other inventors, when they have been at a similar stage of their progress.

This union of genius and judgment is very rare, and it is reasonable to expect that it should be so, from the very different operation of mind that those two faculties require. The inventive faculty, when sufficiently powerful to produce original and extensive combinations, must be attended with a sort of spontaneous operation of the mind, which being excited to action by the will, is capable of continuing a course of thought in some measure independently of the will, which to a certain extent remains dormant, for it only exercises a slight direction of that spontaneous thought.

Every person of the least imagination must be conscious of the existence of spontaneous thought, in that scene of fleeting ideas and new combinations which is continually passing through the mind, whenever it is not actively employed under the direction of the will ; and also in dreaming.   The composition of music, poetry, and painting, depends very greatly upon such a spontaneous action of mind ; and the works of different individuals will be found more or less original, in proportion as that action is stronger and less controlled ; or, on the other hand, their works will be more classical, correct, and according to rules, in proportion as that spontaneous action is guided and regulated by the will.

The judgment depends upon a faculty of distinguishing between ideas ; decomposing compound ideas into more simple elements ; arranging them into classes, and comparing them together ; estimating quantities and qualities of things ; the judgment also depends upon the power of forming new combinations of ideas, and foreseeing the results of such combinations, so as to correct and supply deficiencies in them.   All these operations of mind are very distinct from that which produces new ideas, and may as well be exercised upon a stock of ideas which have originated with others, and been communicated to the mind, as upon new ideas which have arisen spontaneously in the mind.

In the present state of knowledge, physicians, architects, and engineers, only require the faculty of judgment, and a very limited degree of invention is sufficient for them, because almost all the ideas which they require to employ in their combinations, have been already produced and recorded, so that they may be acquired from others by study ; and those individuals may be expected to produce the best results who have accumulated the greatest fund of the best ideas, in a retentive methodical memory, and who possess the greatest power of analysis and combination, with discrimination to select judiciously from that fund (*a*).

There is an intermediate class of pursuits in which to attain eminence, men must have a greater power of conceiving new ideas, in order to fill up and complete the stock of those which can be acquired from others, so as to form a complete series in their own minds ; this is requisite for subjects which have been only imperfectly cultivated hitherto, or which are not sufficiently recorded, or readily communicated.   All arts and sciences must, at some period of their progress, have been so circumstanced ; but their advancement and perfection, is the result of long application of mind by a number of individuals, whose ideas, by constant interchange, become a common stock, and in time form a tolerably complete and uniform series ; having undergone a rigid selection, by which the useless supernumeraries have been discarded, and the best have been preserved.   This selection of the ideas being made by the method of trial and error, according to the results of experience, that common stock of ideas which are current amongst communities and professions, will generally prove to be of a better quality than the average of those new ideas, which can be produced by any individual from the operation of

---

(*a*) In many arts and sciences, maxims have been laid down, and rules prescribed, for the guidance of the judgment ; such rules were more esteemed formerly, than they are now ; for it has been found that they generally impede the judgment, instead of assisting it.   This arises from the maxims and rules being imperfect in themselves, and limited in their application.

his own mind, without assistance from others. The most useful additions to that common stock, usually proceed from those individuals who are well acquainted with the whole series. Such are the gradual improvements and perfectionments of details which practical men make; but extensive improvements, which affect the whole series of ideas, are commonly the work of more active geniuses; because practical men cannot receive ideas which are wholly new to them, with sufficient impartiality; and are poor reasoners when they are taken out of that course, or chain of ideas which they have been accustomed to contemplate.

It would appear that the possession of genius, or the power of creating new ideas, ought, in all cases, to give any individual a great advantage over those who possess only a fund of acquired ideas, with judgment to employ them; because the common stock being accessible to the man of genius, his own ideas would form a useful addition; but it is found that such men have generally an undue partiality for their own ideas, a sort of parental affection, which does not allow them to make fair selections from their joint stock of acquired and created ideas. They have also a habit of changing their ideas frequently, and are not accustomed to preserve a complete and uniform series, but frequently leave vacancies in their minds; hence their combinations want harmony, justness of proportion, and completeness. The consciousness of being able to create ideas at pleasure, always takes away very much from the desire of acquiring ideas by communication or observation; and hence men capable of great originality of thought are rarely studious, and not always sufficiently observant of what passes before them.

The practice of oratory, military command, mechanical invention and combination, and of discovery in the arts and sciences, requires a considerable extent of spontaneous action of mind; but to produce good results it must be very greatly under the control of the will. The imagination, when excited to action, will present a number of ideas to the mind, which must be examined, compared, and combined, according to just principles of reasoning, whereby many will be found defective and unsuitable, and must be rejected, and the imagination set to work to produce others to replace them. The power of producing important results will depend upon the fertility of the imagination to conceive rapidly a great number of ideas, and upon an impartial and accurate choice and selection from them, of such as are suitable, and which will combine harmoniously together.

Mr. Watt may be quoted as a most eminent instance of the latter description of mind. He had a most methodical imagination, which could be directed by the will to produce a number of ideas, answering so far to description, that amongst them it was not difficult to select that which would answer the intended purpose. He had also the habit of incorporating his new ideas with those which he acquired by communication or observation, and of arranging the whole into one uniform series, from which he could choose those which were suitable for his combinations; and he appears to have been very free from undue partiality to his own ideas in such selections.

Mr. Smeaton must be mentioned as a most eminent example of sound judgment in mechanical combinations, without great invention or power of creating original ideas; his works and mode of reasoning are upon record, in the numerous reports and opinions which he left behind him (a). Mr. Watt, in addition to his great powers of original invention, was not inferior to Mr. Smeaton as a practical engineer; at least in that branch which they both cultivated with so much success.

(a) Many of these were published during his lifetime, and others since his death, in his Reports, Estimates, and Miscellaneous Papers, in four volumes quarto, London, 1810.

In many cases of great and complicated inventions, the original inventor has not been able to apply his ideas to practice with so much success, as other men of judgment and experience in similar pursuits, who have taken up the new invention; this arises from the undue partiality which most men of genius have for their own ideas, in preference to others which would be more suitable, and hence there are incongruities in the parts of their combinations. Mr. Watt was very free from these defects of mind, but it is certain that he must have thought very deeply upon all the possible, as well as all the practicable, means of obtaining the results he had in view; and he appears to have adopted the best of those means, for, to the present day, nothing of importance has been brought into use for the improvement of the steam-engine, of which the rudiments, or the principle, cannot be found in Mr. Watt's various specifications, or in his practice (a).

It is to be regretted that this great man was not more accustomed to record his ideas in writing, than he appears to have been. The writings he may have left would be a valuable accession to the stock of knowledge possessed by engineers, and it is due to his fame that they should be made public. It does not appear that Mr. Watt ever published even the shortest memoir respecting the subjects with which he was so conversant, until 1814, when he revised the article Steam-engine, which had been composed by his friend Professor Robison many years before, and published in the third edition of the Encyclopædia Britannica; this had been written without any communication with Mr. Watt, and it contained some errors and omissions, which Mr. Watt corrected and supplied by notes. This revision was published in 1822, in a new edition of Dr. Robison's Philosophical Papers, by Dr. Brewster, in four volumes octavo; it contains a cursory review of Mr. Watt's progress during more than twenty years, from the commencement of his invention in 1764 (or half a century before he wrote), until the completion of his invention, about 1787. The great number of years that Mr. Watt lived in an active exercise of mind, is an unusual circumstance for men of his character; and his mental powers did not appear to be greatly impaired in his latter years. He made his inventions between the age of 29 and 48 years; and he lived to nearly 84 years of age (see p. 309) (b).

We may conclude this account of Mr. Watt by remarking, that the great reputution which he so justly acquired by his improvements in the steam-engine, has by almost universal consent been extended to the engine itself, and for a long time it was a received opinion that no better application of steam could be made

(a) Mr. Woolf's engine, which is the greatest modern improvement, is an extension of the principle of expanding the steam, which Mr. Watt invented and reduced to practice with steam of the ordinary elasticity (see p. 339); leaving for Mr. Woolf to make the same application of high pressure steam, and thus realise greater advantages from the same principle.

(b) The author first became acquainted with Mr. Watt in 1814, about the same time that he had made this revision; and having long before formed the design, and collected materials for the present work, Mr. Watt gave him the particulars of all that he had done in that revision before it was printed, in order to guide the author in the sketch of the invention of the steam-engine which he was then engaged to write for Dr. Rees's Cyclopædia.

It was Mr. Watt's custom to visit London every year, in order, as he said, " to follow the world as fast as he was then able to do." The author being much engaged in preparing the specifications and plans of new inventions, for which patents are to be taken out, Mr. Watt always visited him during his stay in town, and took great interest in examining the drawings of all that was new, and worthy of notice; his remarks and judgments upon them, and his references to his own observations and ideas, formed quite a course of instruction, from which the author feels that he has derived much advantage. Mr. Watt's conversation, on these and other occasions, turned very frequently upon that operation of mind by which inventions are produced, and decisions formed on plans for execution. His ordinary language was so forcible, and well arranged and illustrated, that it would have been very desirable to have recorded it exactly; but it is doubtful if he ever wrote any thing upon that subject, although more eminently qualified than any other man.

beyond what Mr. Watt had done himself. This notion has operated, and may still operate, to retard further improvement. Men of superior intellect, who might have been induced to investigate the subject, have been led to suppose that nothing remains to be perfected ; and when practical engineers have attempted improvements, they have experienced strong opposition, and found it difficult to gain any attention to their proposals, or to obtain opportunities of putting them to the test in practice. Mr. Woolf lost some years before he could obtain credit for the reality of his improvement, although it is an extension of Mr. Watt's principle of expanding the steam, during a part of its action on the piston.

On the extent to which Messrs. Boulton and Watt's engines were adopted during the patent. In the year 1805, there were 112 steam-engines at work in London, and the immediate neighbourhood ; they exerted about 1355 horse-power, and were employed in the following trades (*a*).

	Engines.	Horse-power.		Engines.	Horse-power.
Public waterworks . .	10	270	Sail-cloth weaving . .	1	14
Docks for shipping . .	6	150	Flour mills	6	52
Temporary public works .	7	68	Oil mills	3	20
Public baths . . .	2	8	Mustard mills . . .	2	12
			Drug mills . . .	3	28
Power for pumping water .	25	496	Paper mills . . .	2	20
Breweries . . .	17	250	Colour makers . . .	3	26
Distilleries . . .	8	114	Starch maker . . .	1	10
Vinegar makers . .	3	20	Roperies . . .	3	24
Dyehouses . . .	8	80	Iron forge . . .	1	40
Woollen cloth dresser . .	1	24	Foundries and machine makers	10	42
Tanneries . . .	2	8	Cutlers . . .	2	3
Cotton mills . . .	2	12	Glass cutter . . .	1	6
Calenders and packers .	2	22	Diamond cutter . . .	1	4
Calico printers . .	4	20	Silversmiths . . .	1	8

This list was made five years after the expiration of the patent, when a number of persons had begun to make steam-engines, and had fixed a considerable number of small engines in London ; those which were made at Soho during the patent, before the year 1800, amounted to 46 engines, and exerted 648 horse-power.

The number of engines in use at Manchester, before the year 1800, was probably 32, and their power 430 horse. And at Leeds, 20 engines, 270 horse-power (*b*).

Application of Steam-engines in France.

Newcomen's atmospheric engine was introduced into France a few years after it was brought to perfection in England, but the use of them was not greatly extended, as they did not attempt to make them in France ; but the few engines they had, were all made and managed by English people (see p. 212).

Mr. Watt's steam-engine for pumping water by a single action, was first in-

(*a*) The author first turned his attention to the subject of steam-engines in 1804 and 1805, and he visited all the establishments in London and the immediate neighbourhood, where steam-power was then used, in order to obtain drawings and dimensions of engines of all sizes, and the particulars of their performance (as already mentioned, p. 532). These studies formed his entire occupation during that period, and he believes that the above list comprises all the engines which were then in use in London.

(*b*) The author made similar inquiries at Manchester and Leeds, some years afterwards, and the above estimate was put down, by selecting the old engines which were made during the patent.

troduced into France about 1780, when Messrs. Boulton and Watt sent out the parts for a large engine; and some time after another was set up in the same building, on the banks of the river Seine, at Chaillot in Paris: they pump up water for the supply of the public fountains in the town. About the same time Messrs. Perrier established an iron foundry and workshops for making engines on the same spot; they fixed these engines, and executed other parts of the water-works. These engines are described by M. Prony, in the second part of his Nouvelle Architecture Hydraulique, with large engravings. From an expression used by M. Prony, it would appear that one of the engines at Chaillot was constructed by M. Perrier; and that they also made a smaller engine for pumping water, which was fixed up on the opposite bank of the river, at Gros Caillou. These engines, which are still in use, are nearly on the same construction as the engine in Plate XI., except that the air-pump and condenser are at the pump end of the great lever, as described (p. 321). They do not act on the expansion system, nor are there any steam cases to the cylinders.

In 1780, a patent was granted in France to M. Darnal, for working flour mills by the power of a steam-engine: his plan was to pump up water into a large reservoir by two steam-engines; and to give motion to ten small overshot water wheels, by the water flowing out of that reservoir; each water wheel to turn one mill stone. The water was pumped up from a pond beneath the water wheels, into the upper reservoir, to act again upon the wheels, in the same manner as had been often practised in England during some years before (see p. 296). An establishment of this kind was attempted by M. Darnal at Nimes, in 1781. The plan is described with engravings, in the publication of Les Brevets d'Invention, in quarto, vol. i. p. 198.

In 1789, Mr. Watt's double rotative engine was introduced in France, by M. Betancourt, a Spaniard, who had been some time in London, and had examined the engines at the Albion Mills; he afterwards went to Paris and made a working model, from which, according to M. Prony, two large engines were made in 1790, and set up in the Isle des Cygnes at Paris, for grinding corn. The construction of these engines is described in great detail by M. Prony, but they are not very good specimens of Mr. Watt's engines (a). A rotative engine was sent out by Messrs. Boulton and Watt about the same time, and fixed at Nantes.

(a) It is not quite clear whether such engines were really executed or not; there are no such mills now at the Isle des Cygnes; but there is a large building which has chimneys to it, and is called La Triperie: if the mills really were erected in that building, the establishment must have failed soon after its commencement.

The French have never acquired a proper knowledge of the use of steam-engines, and the best of those they have attempted to make have rarely answered so well as the most inferior engines made in England. It is only of late years, and after a great number of engines had been imported from England into France, that they have been able to manage them, so as to succeed well in their application to manufactures; and but very few of the attempts to make steam-engines in France were successful, until English workmen had set up workshops in France for that purpose. The steam-engine is completely an English production, and is so difficult of execution by new hands, that it is not easily transplanted into other countries.

In Russia there are some workshops for making steam-engines, which have been established with success by engineers belonging to a colony of ingenious artists, who were taken from Scotland to Russia, about 1786, by Mr. Gascoigne; he had been manager at the Carron Iron Works (see p. 274), and was invited by the empress Catherine to establish himself in Russia, where he was liberally patronised, and he and his successors have since executed very considerable public works with great success.

CHAPTER IX.

IMPROVEMENTS MADE IN THE STEAM-ENGINE DURING THE TERM OF MR. WATT'S PATENT, AND FOR A FEW YEARS AFTER ITS EXPIRATION.

Messrs. Boulton and Watt's great success in business stimulated several other persons to attempt to improve the steam-engine, but they were all superseded by Mr. Watt's patent, because no more efficient engine than Newcomen's could be made, without using Mr. Watt's system of condensing in a vessel separate from the cylinder. Mr. Hornblower's engine has been already described (see p. 384), and the following short notices of the most feasible of the other attempts will be sufficient to give some idea of such of them as were actually put in practice.

In 1784, Mr. Robert Cameron of London, took out a patent for making or constructing a steam-engine. His specification, which is recorded in the Rolls Chapel, describes what he calls a helical rotative engine, acting by a sort of piston or leaf affixed to an axis, and winding round within a cylinder, which contains a sort of path, winding helically about its axis, something in the manner of a spiral staircase; the piston-rod or axis, by applying to this spiral, is moved endways, at the same time that it is turned partly round with circular motion. There is another plan for impelling a piston continuously round in a circular channel which returns into itself; and also a new method of discharging condensed steam.

These schemes do not appear to be very practicable, but soon after the patent, the same Mr. Cameron made some kind of steam-engine and machine for Mr. Morrison at Mountmoor, in the north of England, to draw coals out of a pit without a water wheel (a). And he afterwards erected an engine to turn millwork at Battersea, near London. The success of these engines was not so great as to bring them into any farther use.

About 1792, Mr. Symington proposed a modification of the atmospheric engine, which it was supposed would possess the properties of Mr. Watt's mode of condensation, and yet would keep clear from his patent, because it was pretended that the injection of cold water was thrown into the cylinder. To accomplish this, the cylinder of an atmospheric engine was made considerably longer than usual; and all the additional length was beneath the horizontal passage, or

---

(a) The following account of this engine was given by an eye-witness in 1786. The fire engine has a boiler of six feet diameter, and a cylinder of 20 inches diameter; the piston gives motion to a horizontal iron axis, by means of a crank on one end, and at the other end is a pinion to turn two spur wheels, which are thrown out of gear alternately; one of these wheels turns another spur wheel, which is fixed on one end of a long horizontal axis, that has a large drum at the other end, on which the rope is wound, to draw up the corfe of coals.

The cylinder is divided into two parts by a partition in the middle, so as to form two distinct cylinders, one above the other; and there are two pistons fixed to an iron rod finely polished, and passing through the centre of the partition. The two pistons move both together, upwards and downwards; each piston makes a stroke of 28 inches, and they are said to make 45 strokes per minute. The under part of the cylinder only has a communication with steam, and the upper with air; one of the pistons acts as an air-pump, making a complete vacuum every stroke; the under part of the cylinder which receives the stream of injection is very hot, but the upper part is cool.

There is some difficulty in striking the corve properly, and in stopping the machine in case of accident. The whole expense, by contract, is not to exceed 500l.

entrance, through which the steam entered from the steam valve into the cylinder.   The whole length of this long cylinder was bored, and another piston was
fitted into the lower part of it, so as to be capable of working up and down therein
with a stroke of about one eighth of the length of the stroke of the working piston.
The piston rod of this extra piston passed down through a stuffing box in the
bottom of the cylinder, and the rod was attached to a lever situated beneath the
cylinder, and connected, by another rod, with the great lever of the engine, so as
to cause the lower piston to move up and down in the lower part of the cylinder, at
the same time, and in the same direction as the working piston moved up and
down in the upper part of the same cylinder; but the motion of the lower piston
was only about one eighth of that of the working piston; and it never rose quite
up to the passage which admitted the steam into the cylinder, nor did the working
piston ever descend quite so low as that passage.

This extra piston, and the space beneath it, in the bottom of the cylinder,
performed the office of Mr. Watt's air pump; and in lieu of his condenser a large
wide passage extended out horizontally from the bottom of the cylinder, but it
was formed in one piece with the cylinder, and a foot valve was inserted into that
passage near to the cylinder, to separate the space beneath the lower piston, from
that other space within the passage, which may be called the condenser; an injection
of cold water was admitted into this condenser, and the eduction pipe, which descended from beneath the exhausting valve of the cylinder, also joined to the condenser, to convey the steam out of the cylinder into it, when that valve was
opened; there was also a passage out of the lower part of the cylinder to the
open air, and it was closed by a valve opening outwards, to act as the discharge
valve of the air pump.   The whole of this additional length at the lower part of
the cylinder, and the projecting part from it, which served for the condenser, was
immersed in a cistern of cold water.

The operation of Mr. Symington's engine was the same as that of Mr. Watt's
single engine.   Suppose the working piston to be at the top of its cylinder, and
the lower piston to be raised to its highest position, so as to be on a level with the
passage leading into the lower part of the cylinder; suppose the cylinder to be
full of steam, and the space beneath the lower piston to be exhausted, as well as
the condenser adjoining to it.   The valve by which steam is admitted from the
boiler into the cylinder being shut, let the exhausting valve be opened, and also a
small valve to admit an injection of cold water into the condenser at the bottom
of the cylinder; the steam will then by its own elasticity pass out from the cylinder, through the exhausting valve and eduction pipe, into the condenser, where
it will be condensed by the injection as fast as it arrives. The space in the cylinder,
beneath the working piston, being thus exhausted, the pressure of the atmosphere
on that piston will force it down into the cylinder, and make the working stroke.
The lower piston, by its connexion with the great lever, is forced down at the
same time, with one eighth of the motion of the working piston, whereby any
water or air, which is contained in the bottom of the cylinder, beneath the lower
piston, will be expelled out of that space, and discharged through the discharge
valve into the open air, or into a hot well.

The working stroke being thus completed, the exhausting valve and the
injection valve are shut, and the steam valve is opened, to admit fresh steam from
the boiler into the cylinder above the lower piston, and beneath the working
piston; this allows the working piston to be raised up again by its counterweight,
and the lower piston is raised at the same time, by its lever; in so doing it leaves
an exhausted space beneath it, in the bottom of the cylinder, into which the water

recently injected into the condenser, drains through the foot valve into the bottom of the cylinder; and the air follows it, so as to preserve the exhaustion within that condenser.

When the piston arrives at the top of its course, the steam valve is shut, and the exhausting valve and the injection valve opened again, to make another stroke; during which the lower piston, by its descent, expels all the water and air which is contained in the space beneath it, and drives it through the discharge valve, into the hot well.

This operation is just the same as that of Mr. Watt's single engine; and the expected evasion of his patent was only by assuming, that the large passage which projected out at the bottom of the cylinder, in the same manner as the foot part of a boot projects out from the leg part, was, in terms of law, a part of the cylinder; but in fact that passage acted in every respect like Mr. Watt's separate condenser, being a space separated by the lower piston, and by the exhausting valve, from that space within the cylinder where the working piston acted. Mr. Symington's engines were never brought to bear, so as to work with good effect; and after a few had been attempted, they were stopped by Messrs. Boulton and Watt; the proprietors not being inclined to come into a court of law upon the strength of this ingenious perversion of terms (a).

DOUBLE ACTING ATMOSPHERIC STEAM-ENGINES FOR TURNING MILL-WORK BY CRANKS AND FLY-WHEELS, 1793.

The application of Newcomen's engine to turn a crank and fly-wheel, was first made about 1780, as already related, p. 410, and in the course of the next ten years a few of those engines were brought into use in the North of England, for turning mill-work. They answered tolerably well for some purposes, which do not require a very regular motion, such as drawing coals out of mines, grinding corn, crushing seeds, &c.

When Mr. Watt had completed his double rotative engine about 1784, it was rendered so extremely correct in its movements, as to be suitable for turning any kind of machinery in manufactories, as well as a water-wheel. As this regularity of action was chiefly to be attributed to the double action, the makers of atmospheric engines were led to consider the means of combining two cylinders together in one engine, and causing their pistons to act alternately upon the same crank and fly-wheel. The following sketches represent two kinds of double atmospheric engines, each with two cylinders, which were made about 1793.

A double-acting Atmospheric Engine to turn Mill-work, executed by Mr. Thompson in 1793.

This plan of combining two atmospheric cylinders in one engine, so as to produce a double action, was arranged by Mr. Francis Thompson, engineer of Ashover, in Derbyshire, who took out a patent for it in 1793. Mr. Thompson

(a) The same Mr. Symington afterwards turned his attention to a better object, and deserves honourable mention for his attempts to apply the steam-engine to row boats; he made the first step towards a useful result in steam-boats about the year 1801, as will be mentioned in its place.

was a practical engine builder, who had rather an extensive business in erecting Newcomen's engines for the mines in Derbyshire, at a period when those mines were in their greatest activity (see p. 238); this was about the time when Mr. Watt first brought forward his improvements, and Mr. Thompson was sometimes employed to make and fix the pit work and pumps, to Messrs. Boulton and Watt's engines, at mines where he had been previously accustomed to make and repair the machinery.

A few years afterwards, when steam-engines first began to be applied to turn machinery for manufactories, Mr. Thompson made several atmospheric engines, with cranks and fly-wheels, which answered very well, and were set up at a small expense (see p. 410). When Mr. Watt had brought his rotative engines to a standard form, about 1786, their performance proved so much superior to that of the atmospheric engines, that the former have since been generally adopted for all considerable manufactories, notwithstanding the great price of their first erection ; hence the makers of atmospheric engines could not obtain orders for them, except at mines, and in the vicinity of collieries, where coals are very cheap ; and they were induced to attempt improvements in the construction of the atmospheric engine, in order to adapt it to turn machinery in a more steady and regular manner, than could be done by the most simple application, see p. 426.

Mr. Thompson's method was to apply an additional cylinder F in an inverted position, or with the open end downwards, over the open top of the ordinary cylinder E of the atmospheric engine, which cylinder is placed in the usual position ; the piston of the upper inverted cylinder F is fixed to the same piston rod *n* as the piston for the lower cylinder E ; and the upper part of this piston rod is

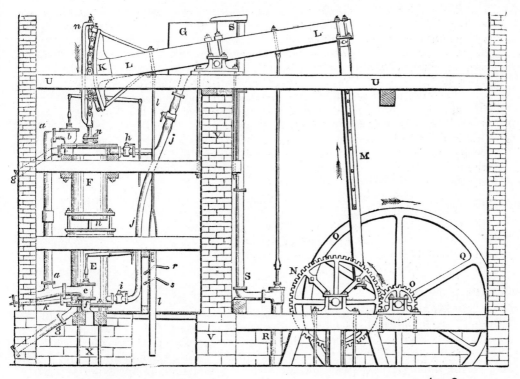

straight and polished, to work through a stuffing box, in the centre of the close end of the inverted cylinder F.

The open ends of both cylinders F and E are near together, and a sufficient space is left between them at *n* to admit the atmospheric air to act freely upon both pistons. The two cylinders are exhausted alternately, by injecting the cold water into one and the other in turn. The unbalanced pressure of the atmosphere upon the piston of the lower cylinder E produces the downward stroke, and then the unbalanced pressure of the atmosphere, beneath the piston of the inverted cylinder F, urges the piston rod *n* upwards, and makes the upwards stroke. By these means the working stroke of one piston produces the returning stroke of the other piston, and no counterweight is required for either of them. The action upon the great lever L is double, or in both directions, so that the connecting rod M will turn the crank or multiplying wheel N continually round, and that by its teeth, turns the small pinion O on the axis of the fly-wheel to Q, which is thus turned with great rapidity, in order to produce a regular continuous circular motion.

The construction of this engine is aparent from the figure, and it is needless to enter into minute particulars. The lower cylinder E is fastened down by four bolts, upon two beams which extend across the house beneath the bottom of the cylinder, and rest upon a solid pier of masonry X; the upper part of the cylinder E is received between two beams which retain it firmly in a vertical position; and the upper cylinder F is suspended between two similar beams. The ends of all these beams are worked into the walls of the house, and the wall which supports the great lever. The cylinders are secured together by four upright bolts.

The steam is brought from the boiler by the pipe *a*; it passes through the end wall, and joins to the box containing the lower steam valve *e*, which, when open, admits the steam through the passage *f*, into the lower part of the lower cylinder E; an upright branch from the steam pipe *a*, conveys the steam up to the box containing the upper steam valve *b*, which, when open, admits the steam into the upper part of the upper cylinder F. The valves are lifted at the proper intervals, by rods and levers, which communicate with the working gear.

R is the cold water pump, which, by the pipe S, forces a supply of cold water up into the cistern G, and it descends again through the injection pipe *j*, which has a branch leading to the upper part of the upper cylinder F, with a cock *h* to admit the injection into that cylinder; and at the lower end of the pipe *j* is another cock *i*, to admit the injection into the lower cylinder E. A cock is also provided in the upper part of the injection pipe *j*, to regulate the flow of water through it, or to stop it when required.

l is the plug rod to give motion to the two handles of the working gear; which is so contructed, that the lower handle being raised by a pin in the plug, whenever the pistons arrive at the top of their course, will shut the lower steam valve *e*, and the upper injection cock *h*; and at the same instant a catch which detained the upper handle is released, and then a weight, which is connected with that handle, throws it upwards, and suddenly opens the upper steam valve *b*, and the lower injection cock *i*, to exhaust the lower cylinder, and produce the descending stroke. When the pistons arrive at the bottom of their course, the action is reversed; the upper handle is depressed, to close the upper steam valve *b* and the lower injection cock *i*, and the catch of the lower handle is disengaged, to allow it to fall by its weight and open the lower steam valve *e*, and the upper injection cock *h*, in order to exhaust the upper cylinder, and produce the ascending stroke.

A sinking pipe *g* passes off, in a sloping direction, from the lowest part of the lower cylinder E, to drain away the hot water from it, when the steam is admitted into that cylinder; and there is a similar pipe, marked g, which communicates with the upper part of the upper cylinder F.   The ends of both these pipes are closed by horse-shoe valves, which allow the water to drain out of the cylinders, but prevent any return. k Is a snifting clack at the lower part of the lower cylinder, to discharge the air from that cylinder; and there is a similar clack to the upper cylinder, but it is not represented.

The piston-rod has three chains connected with it, viz. two which are fastened to the top of the arch head, at the end of the great lever L, and suspend the pistons; these serve to pull down the end of the lever to make the descending stroke; a strong iron rod n is affixed to the top of the piston-rod, and rises up above the top of the arch head; to the top of this rod the third chain is fastened, and the lower end of it is fastened to the lower end of the arch head; the latter chain serves to pull the end of the lever up again, for the ascending stroke; it occupies the space between the other two chains. All the three chains are braced up, by nuts upon the screws by which they are connected to the rod, or to the arch head, so as to be quite tight, and allow no shake or looseness, at the change of the stroke of the pistons.   The arch head is made of cast iron, with a socket to receive the end of the wooden lever L; the arch is fastened on by wedging, and it is steadied by iron braces, as shown in the figure.

The bearings for the fulcrum of the great lever, are firmly fastened down by four bolts, which pass down at each side of the lever wall; their lower ends pass through the ends of the beams which sustain the upper cylinder, and they are fastened beneath those beams by cross keys; these bolts prevent the bearings rising up, unless they could lift the masonry of the lever wall.

The connecting-rod M is made of wood, and strengthened by straps of iron (a); the upper end is jointed to the end of the great lever, and the lower end to the crank pin, which is fixed into one of the arms of the multiplying cog-wheel N.   That wheel is fixed upon the projecting end of its axis, where it extends beyond the bearing, and the crank pin projects out from one of the arms, beyond the plane of the wheel, so that the connecting-rod passes against the wheel in the same manner as if a crank were used; but the pin in the arm of the wheel is better than a crank, because it does not occasion any twist upon the axis of the wheel.   The teeth of the multiplying-wheel turn the small pinion O, on the axis of the fly-wheel Q, which is connected with the millwork in the usual manner.   The bearings for the axis of the fly-wheel, and those for the axis of the multiplying-wheel, are supported on the same wood framing.

The largest engine which Mr. Thompson made on this plan, was applied to turn the first steam-mill for spinning worsted, which was established by Messrs. Davison and Hawksley, at Arnold, near Nottingham.   The cylinders were 40 inches diameter; the pistons made 6 feet stroke, and went 18 strokes per minute; they were lightly loaded, to enable them to work so quick, and the engine exerted about 48 horse-power.   The great lever was 24 feet long; the connecting-rod 22 feet long; the multiplying-wheel was $8\frac{1}{4}$ feet diameter, and had 96 teeth; the pinion being 38 teeth, the fly-wheel went $45\frac{1}{2}$ revolutions per minute, it was 18 feet diameter outside, and its rim $8\frac{1}{4}$ inches by $3\frac{1}{2}$.   This engine worked for some years, and drove the spinning machinery very steadily, but the consumption of fuel was very considerable.   The whole of this establishment was broken up in 1810.

Another engine of the same size was put up for a cotton mill at Maccles-

---

(a) *Note,* the connecting-rod of the atmospheric engine represented in page 410, is supposed to be made of oak wood strengthened with iron straps, though it is not stated so in the description.

field; Mr. Thompson also made a few small engines, with 27 inch cylinders. Two of the cotton mills in Manchester which are now on the largest scale, were commenced in 1794, with those engines, but being found insufficient for the increasing work, they were replaced by larger engines by Messrs. Boulton and Watt. Mr. Thompson's engines were found to be as difficult to keep in order as Mr. Watt's, and very inferior in performance; hence, they were only adopted in a few instances, and have long since been laid aside.

Another double-acting atmospheric Engine to turn Millwork, 1794.

The idea which had been suggested by Dr. Falck, in 1779 (see page 426),

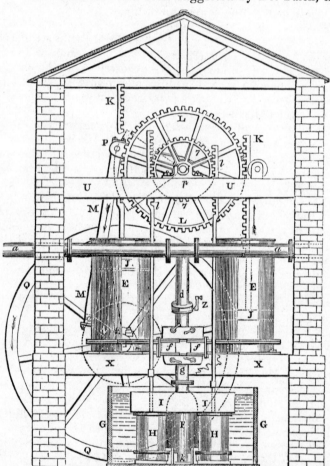

was brought into use about 1794, by some engineers at Manchester, who erected such an engine for Mr. Thackray's cotton mill, at Garrat, in that town (a), and it continued in constant use for more than 30 years. Some small engines had been previously made with two cylinders combined in the manner of the accompanying sketch; but the condensation was effected by injecting cold water into the cylinder. In the present instance a condenser and air-pumps were borrowed from Mr. Watt's engine (b).

E E are the two cylinders, each fitted with its piston, as shown by the dotted lines J J; the two piston-rods K K are toothed racks which are adapted to work into the teeth of a

(a) This was the first cotton mill which was established in Manchester. The spinning machinery was on Sir Richard Arkwright's plan, and was worked by water wheels on the river Medlock; but the water power being found insufficient, two engines on Savery's system were put up by Mr. Joshua Rigley, in 1784 (see p. 125), to return the water of the river for the water wheels, and afterwards those engines were replaced by the above rotative engine.

(b) The proprietors were prosecuted by Messrs. Boulton and Watt for an infringement of Mr. Watt's patent; and the action was compromised, by paying the patentees for a license to work.

large toothed wheel L, and being at opposite sides of that wheel, the two pistons are caused to move in opposite directions, one ascending when the other descends ; hence they balance each other, and no counterweight is required.   On the axis p of the large wheel L, a strong lever is fixed, as shown at P, the connecting-rod M is jointed to the end of the lever P, and the lower end of the rod is jointed to the pin of the crank N, which is fixed on the end of the axis of the fly-wheel Q.   By this arrangement the alternate action of the pistons of the two cylinders is caused to operate with double action on the crank to turn it round.

A smaller toothed wheel, 7 7, is fixed on the end of the axis p of the great wheel L, and at the opposite sides of the wheel, 7, two racks, $l\,l$, are applied, to suspend the rods of two air-pumps H H, which, by that means, act reciprocally and alternately, to keep up the exhaustion within the condenser F, which is common to both cylinders, and to both air-pumps.

$a\,a$ Is the steam-pipe, which brings steam both ways, from boilers which are situated outside of the walls of the engine-house ; from the middle part of this pipe a branch d descends in an inclined direction to convey the steam into a box containing the two steam valves ; those valves are situated over the two passages $f f$, which lead into the lower parts of the two cylinders, and either of these valves being opened, will admit steam into the corresponding cylinder.   Beneath the passages, $f f$, is another box, which communicates by the eduction pipe g, with the condenser F ; the two exhausting valves are situated in the lower parts of the passages $f f$, and those valves being opened, will allow the steam to pass out of either cylinder to the condenser ; Z is the throttle valve to regulate the flow of steam through the descending branch d of the steam-pipe.

The handles of the working gear are moved by chocks on the two rods $l\,l$, and every time that the pistons arrive at the ends of their course, that gear opens and shuts the four valves in alternate pairs ; each pair consisting of one steam valve and one exhausting valve.

The operation of the working gear and valves is the same as Mr. Watt's double engine.   The cylinders being open at top the atmospheric air always acts freely upon the pistons, and when either of the steam valves is opened, to admit steam into the cylinder beneath either of the pistons, that piston is allowed to rise, in equilibrium ; but having risen, the steam is exhausted from that cylinder by opening its exhausting valve (its steam valve being first shut), and then the pressure of the atmosphere will force that piston down into its cylinder, and by the racks KK, the wheel LL, and the lever P, the action of either piston is transmitted to the connecting rod M, one piston urging it upwards, and then the other in turn urging it downwards ; a double action is thus exerted on the crank pin to turn it round, in the circle shown by the dotted line ; the fly-wheel keeps up and regulates the rotative motion.

The cylinders are supported on two very strong beams, XX, which extend across the engine-house, the ends being worked into the side walls.   The bearings for the axis p of the great wheel LL, are supported by similar beams, UU, and the bearing for the neck of the axis of the crank, is fixed on another beam behind the beams X.   The condensing cistern GG is placed on the ground, beneath the beams XX.

The two air-pumps HH are open at top ; they have each a foot valve at the bottom near k ; and they have valves in their buckets like common sucking pumps ; the hot water which they extract from the condenser they discharge into the hot well, II.   These pumps act alternately, and thus keep the condenser constantly exhausted ; in that respect they are better than the air-pump of Mr. Watt's double engine, which is a single action, and exhausts only once in each complete stroke; but the two pumps exhaust twice in each complete stroke. The buckets of the two pumps are always subjected to the pressure of the atmosphere which acts upon the surface of the water in the hot well; but as one bucket is in part balanced against the other, the motion of the engine is not deranged by that pressure.

The engine at Mr. Thackray's mill has two cylinders of 36 inches diameter, and the pistons make a stroke of 4 feet; when it was first erected they were worked at the rate of 40 strokes per minute; this rapid action gave a great regularity to the motion of the fly-wheel, although it was but a small size for the power of the engine, which was rated at 70 horse-power; and it must certainly have been 60 horse-power. After the engine had worked many years, some repairs were made, and it was then regulated to make 30 strokes per minute, at which rate it will exert about 45 horse power; it has two large boilers.

Messrs. Sherratts, of Manchester, who made the principal parts of the work for this engine, afterwards adopted the plan, and made several smaller engines, which answered a good purpose, and some are still in use. They also made similar engines without air-pumps and condensers, but with injection into the cylinders, so as to be clear of Mr. Watt's patent. This is the best method which has yet been proposed for combining two single acting cylinders into one double engine. The weight of the pistons and rods, and those of the air-pumps, are caused to balance each other without any extra weight; and the pressure upon the teeth of the racks K, and upon the pivots p of the axis of the great wheel L, is always downwards, the same as in a single engine; and none of the joints, except those of the connecting rod and crank, are urged in both directions. The construction brings an engine into a small compass.

In 1796, Mr. Hornblower adopted this combination of two single cylinders, in an engine which he made for Messrs. Meux's brewery; but he applied covers to the cylinders, with stuffing boxes for the piston rods to pass through, and they were suspended by chains, applied upon the circumference of the wheel L, which was a plain wheel without teeth, and the chains were fastened to it at the upper part; the horizontal part of the steam-pipe, aa, was joined to both cylinders at the upper part, near their covers, in order to introduce the steam from the boiler into both cylinders at all times, and therefore it pressed constantly upon both pistons. The action of each cylinder was precisely the same as that of Mr. Watt's single engine (see p. 333); but by their combination a constant double action was kept up. The valves and the two air-pumps, and all the other parts, were constructed as in the preceding sketch, except that their rods were suspended from the small wheel, 7, by chains (a).

Mr. Hornblower has stated, in Gregory's Mechanics, vol. ii., that his reason for adopting this mode of combining two cylinders, was to enable the pistons to act upon the crank pin with greater force towards each end of the stroke, than in the usual mode of working with a great lever; for the force of first one and then the other of the two pistons is applied on the circumference of the wheel L, at a constant distance from its centre p; but the end of the lever P to which the connecting rod M is jointed, moves in an arch of nearly one-third of the circumference of a circle; during this motion the connecting rod M advances to, and recedes from, the centre of motion p, so as to cause a continual variation in the effective length of leverage with which the lever P acts upon the rod M; this principle is stated at p. 47 and at p. 416.

(a) Messrs. Hornblower and Maberley were prosecuted by Messrs. Boulton and Watt, in 1796, for making this engine, and some others of the same kind, of a smaller size, as mentioned at p. 648. The proprietors were obliged to pay the patentees for license to use these engines.

When the pistons are at the top or bottom of their stroke, the effective radius of the lever P is considerably shortened, so that it acts with greater power upon the connecting rod, than it does when the pistons are near the middle of their course, for then the lever P is nearly at right angles to the connecting rod. This circumstance tends to compensate for the diminishing leverage of the crank, whereby its power of turning the fly-wheel round decreases so rapidly towards each end of the stroke, as is stated p. 417 ; for the connecting rod being enabled to exert a greater force upon the crank pin, at the time when the crank acts less advantageously to turn the fly-wheel, a more uniform force of rotation is obtained.

It has been already explained, (p. 419), that the slow motion of the piston, at each end of its course, is very advantageous to the action of the engine, by allowing more time for a complete exhaustion of the cylinder ; and the variations of the force transmitted to the fly-wheel can be sufficiently equalised by giving the fly-wheel a proper weight and velocity. The property which Mr. Hornblower points out, is also unfavourable for the permanency of the work, because the connecting rod and crank, with their joints, are subjected to an excessive strain at every change of the stroke.

The double engines, with two single acting cylinders combined, are not so good as Mr. Watt's double engine, because of the increased friction, and weight, of two pistons instead of one, and the increased surface for the radiation of heat ; and also the expense of construction is greater, if the work is equally well executed.

Mr. Cartwright's Steam-engine, 1797.

Mr. Edmund Cartwright took out a patent, in 1797, for improvements in the construction, working, and application of steam-engines ; his specification, which is printed in the first series of the Repertory of Arts, vol. x. p. 1, states the following improvements :

1st, The water, or other liquid, which is used to produce the steam to work the engine, is to circulate continually through the engine, without having any communication with the external air, and without mixing with any other fluid.

2nd, The lower part of the cylinder is to be always open to the condenser, and the condensation is to be performed during the returning stroke of the piston, which has only a single action. There are to be only two valves, one at the top of the cylinder, and the other in the piston ; and these valves are to be opened and shut at the proper intervals, by the motion of the piston itself.

3rd, The condensation is to be performed in a close vessel, by the application of cold water to the external surface of that vessel, on the common principle of distillation, without any injection of cold water into it. The condenser is to be composed of two cylinders of very thin metal, one fixed within the other, so as to leave a very narrow space between them for the steam. This condenser is to be immersed in cold water, which is to surround the outside cylinder, and to fill the internal cylinder ; whereby a very thin stratum of steam will be exposed to a large surface of cold metal. The condensed liquid which drains to the bottom of the condenser, is to be drawn out by a suitable pump, and returned into the boiler ; any air which may get into the condenser being separated from the water before it reaches the boiler, and discharged into the open air.

4th, The piston is to be made entirely of metal, without any hemp packing.

Lastly, Instead of water, the boiler is to be filled with distillers' wash, to produce the steam with which the engine is to be worked; and the engine is to be made to operate as a still and as a moving power at the same time.

A few small single engines were made under this patent, with self-acting valves, according to the following sketch, which is copied from the first volume of the Philosophical Magazine.

EE is the cylinder, and JJ its piston, which is going down. *a* Is the steam-pipe which brings the steam from the boiler; and *d* is the box containing the steam valve, which admits the steam through the cover C of the cylinder to act upon the piston. *e* Is a valve in the piston to admit the steam to pass from the upper part of the cylinder into the lower part.

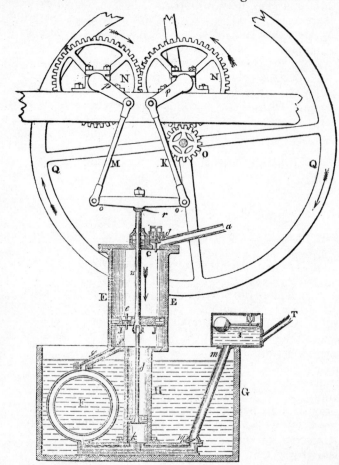

f g Is the eduction-pipe to conduct the steam from the bottom of the cylinder to the condenser F, which consists of two thin cylinders, one within the other, leaving a small space between them, into which the steam is admitted. The inner cylinder is filled with cold water, and the external cylinder is surrounded by the same; by which means the steam is exposed to a large surface of thin metal, which is kept cold by the water; though no water is suffered to come in actual contact with the steam.

To the lower end of the piston rod *n* a smaller rod *d* is attached by a joint; this is the rod of another piston working in the barrel H, which acts as the air-pump, and accordingly it has a passage *k* leading from the lowest part of the condenser F, with a foot valve at *k*, which will allow the water and air to pass from the condenser into the pump, but will prevent its return; there is also a passage proceeding from the lower part of the air-pump, with a discharge valve at *m*, and a pipe *m*, to convey away the hot water and air, which is expelled from the barrel H of the pump, when the piston descends. The piston of the air-pump has no valves; it discharges through *m* when it is forced down, and exhausts from the condenser through *k* when it is drawn upwards.

The piston rod n passes through a stuffing box in the cover C of the cylinder; a cross bar, oo, is fixed on the top of the piston rod n, and to the two ends of this bar two connecting rods, M and K, are jointed; the upper ends of those rods act upon the pins of two equal cranks pp, formed at the extreme ends of two parallel horizontal axes, upon which the two equal cog-wheels, NN, are fixed, and their teeth working together, cause the two cranks to turn round with equal motion; but they move in contrary directions. The consequence of this arrangement is, that the two connecting rods M and K will, in all positions, be inclined equally from the vertical, and their inclinations will be in opposite directions; so that their respective tendencies to draw the piston rod from the vertical, will be corrected one by the other, and consequently that rod will preserve a vertical motion when it works up and down, and turns the cranks round.

The two wheels N are loaded with heavy weights on their rims, at the sides opposite to the cranks pp, in order to counterbalance the weight of the pistons, and the rods d, n, o, M, and K, and the cranks pp.

One of the wheels N gives motion to a pinion O, on the axis of the fly-wheel Q Q, which is turned round with great rapidity; and it is made heavy enough in the rim to regulate and keep up the motion during the ascent of the piston, when it exerts no force.

The action of this engine is the same as that of Mr. Watt's single engine, when the piston rises in vacuo (see p. 334). The equilibrium valve e which is fitted into the piston J J, is opened for the returning stroke, and allows the steam to pass through it during the ascent of the piston. The valve e is opened when the piston is at the bottom of its course, by the lower end of its spindle striking the bottom of the cylinder; and the valve is shut, when the piston reaches the top of its course, by the top of the spindle striking a spring at the underside of the cover C of the cylinder. The steam-valve d, which is fitted into the cover C of the cylinder, is opened when the piston arrives at the top of its course, by the piston striking the lower end of the spindle of the valve; and the valve is shut when the piston reaches the bottom of its course, by a spring r, which projects out from the piston rod, and presses upon the upper end of the spindle of the valve. This spindle passes up through a hole in the top of the box, which contains the valve; and to prevent the escape of the steam, an inverted cup is fixed on the top of the spindle, and the border or lower edge of the cup is immersed in mercury, which is contained in a deep circular groove formed at the top of the box, for the reception of the edge of the cup. The spindles of the valves are pressed laterally by springs, which cause so much friction and impediment to the motion of the valves, that they will remain either open or shut, according as they are placed.

The sketch represents the piston going down: when it reaches the bottom of its course, the spring r shuts the steam valve d to prevent the further admission of steam from the boiler, and the equilibrium valve e in the piston is opened by touching the bottom of the cylinder; the communication is thus opened from above the piston to below it, and the lower part of the cylinder being always open to the condenser F, by the eduction pipe fg, the steam from above the piston passes through the valve e, and is gradually drawn on to the condenser F. The piston is placed in equilibrium by thus opening the valve e, for that allows the steam to pass beneath the piston, which is then drawn up in the cylinder by the action of the fly-wheel; all the time that the piston is making this returning stroke, the condensation of the steam from the cylinder is performed in the condenser. The piston of the air pump being drawn up at the same time as the piston, leaves a vacuum beneath it in its barrel H, and the condensed water and air in the condenser F drains through the pipe and foot valve *k*, into the lower part of the barrel H, leaving the condenser exhausted.

When the piston reaches the top of its course, the equilibrium valve *e* in the piston is closed, by touching a spring under the cover C of the cylinder, and at the same instant the steam valve *d* is opened by the piston, to admit fresh steam from the boiler into the upper part of the cylinder above the piston, and as the space beneath it is already exhausted, the pressure of the steam causes the piston to descend with force through its working stroke, during which it acts on the cranks *p p* ; and the wheels, N N and O, give an impulse to the fly-wheel Q. The piston *d* of the air pump being forced down at the same time, it displaces all the water and air which had entered into the barrel H from the condenser, and forces that water and air through the valve *m* and the pipe *m*, into the hot well I. When the piston arrives again at the bottom of its course, its spring *r* shuts the steam valve *d*, and the equilibrium valve *e* is opened, to allow the steam above the piston to pass beneath the same, and thence it is drawn off to the condenser, leaving the piston at liberty to rise in the partially exhausted cylinder, to make its returning stroke by the impetus of the fly wheel as before.

As no water is admitted into the condenser, except that which is condensed from the steam, the quantity discharged into the hot well I is no more than is sufficient to supply the boiler ; the hot well I is closed with a tight cover, and a feeding pipe T proceeds from the lower part of it, to the boiler. All that air which is disengaged from the water, and which gains admission by leakage, rises to the upper part of the close hot well I, and does not enter the pipe T ; a small inverted air valve is fitted to an aperture in the cover of the hot well I, and is kept shut by a lever, with a hollow copper ball at the end, which floats upon the surface of the water, and bears the valve upwards ; but as the air accumulates in the hot well, the surface of the water is depressed, and the ball descending with it, opens the valve, and allows the confined air to escape.

Mr. Cartwright's Metallic Piston. The piston J J is not packed with hemp, but by a number of loose segments of brass or gun metal, which are accurately fitted into the groove round the edge of the piston, in place of hemp, and are forced outwards by springs, which are so disposed, as to keep the segments in close contact with the interior surface of the cylinder, and prevent the escape of steam ; the different segments are made to overlap each other, so that the middle part of one segment will cover the joints between the ends of the other segments.

The base of the piston, which is fastened to the lower end of the piston rod, is made rather less than the bore of the cylinder ; and its surface being made quite flat and smooth, two circular rings of gun metal are applied upon it, one over the other ; the external circumferences of these two rings are fitted very accurately into the cylinder, and their flat surfaces are accurately fitted to each other, and to the base of the piston. There are also two other rings which are fitted into the interior of the large rings. These four rings are covered over at top by the piston cover, which is very nearly the size of the cylinder, and is firmly fastened to the base of the piston by screws, and the piston forms a complete shell, within which the four rings are contained, and they fit very accurately therein, so as to be capable of moving horizontally, but not at all up or down.

To give the four rings a power of expansion, each one is divided into three segments, and a sufficiency of metal is removed from the ends, to leave room for a spring between the adjacent ends of two of the segments of each ring. This spring is in form of the letter V, the open end of which is placed outwards, near to the circumference ; the two segments against which the two ends of the spring act are pressed, in the direction of the circumference, against the ends of the third segment, and all the three segments are thus kept uniformly in contact with the cylinder. To prevent steam passing through between the ends of the segments, the solid parts of the upper ones are made to fall upon the divisions and springs of the under ones, so as to form break joints ; and the segments of the internal rings are behind the joints or ends of the segments of the outer rings, in order to make a complete fitting.

The stuffing-box round the piston-rod is made in a similar manner.

Mr. Cartwright had acquired some reputation as a mechanical inventor, before he took out this patent, and it was in consequence received very favourably by the public ; but his engines were not found to perform well, because the condensation by external cold, without injection, is not sufficiently rapid to produce a good effect. Mr. Watt had given that method a full trial in his first engines (see p. 322), and had abandoned it in favour of an injection.

Mr. Cartwright's proposal was confined to single engines, which are not so good for turning mills as double engines. Some years after this patent, Mr. Cartwright made a double engine, of about six horse-power, on his plan, for a mill at Wisbeach ; but it could not perform the work it was appointed to do, and was removed ; it was afterwards set up for the Duke of Bedford, at his farm at Woburn, but it proved a very defective engine, and has been broken up. The metallic piston is the only part of Mr. Cartwright's scheme which has been brought into use ; its construction has been improved, and it is found to answer for high pressure steam better than a hemp piston, as will be explained under its proper head.

Mr. Sadler's Steam-engine, 1798.

Mr. Sadler took out a patent, in 1798, for improvements in steam-engines : his object was to combine, in the same engine, two cylinders ; one acting in the same manner as Mr. Watt's single engine, when the piston rises in vacuo, and the other acting in the same manner as Newcomen's atmospheric engine, with an injection into the cylinder. The same steam is used successively to fill both cylinders, and the atmospheric cylinder serves as a condenser and air pump, to exhaust the principal cylinder.

A description and engraving of this engine is given in Nicholson's Philosophical Journal, quarto, Vol. II. p. 231, from which it appears, that the principal cylinder is on the same construction as Mr. Cartwright's cylinder, viz. it is closed at the top, and its piston rod passes through a stuffing-box in the cover ; the steam from the boiler is introduced above the piston, through a self-acting valve in the cover ; and there is another self-acting valve in the piston, to cause the equilibrium for the returning stroke. The lower part of the cylinder communicates by a horizontal passage, with the bottom of the atmospheric cylinder, which is of the same diameter as the other, but its piston makes only half the stroke ; the passage which communicates between the two cylinders, is made like the passage from the condenser to the air pump of Mr. Watt's engine, and has a hanging foot valve in it, which shuts towards the principal cylinder. The atmospheric cylinder has a pipe entering through the bottom, with a valve to admit an injection of cold water ; and there is a valve in the piston opening upwards like the bucket of a pump ; also a drain-pipe leads from the bottom of the cylinder, to convey away the injected water and air from it into the open air, through a valve which opens outwards.

The engine is intended to turn machinery by means of a crank and fly-wheel, but without a great lever ; a strong bar is fixed horizontally across the upper end of the piston rod, like the letter T, and at each end a connecting rod is jointed ; these rods descend at each side of the cylinder, to act upon two cranks formed upon a common horizontal axis, which is situated beneath the bottom of the cylinder, and the fly-wheel is fixed upon it.   The ends of the cross bar of the piston rod have small wheels fitted upon them, and they run up and down between upright guides, by which the piston rod is retained in its vertical direction.   The piston rod of the atmospheric cylinder is suspended by links from the middle of a lever, one end of which is connected by links, with the piston rod of the principal cylinder, and the other end moves about a fixed centre ; the consequence is, that the atmospheric piston moves through only half the length of stroke made by the other piston.

The action of the principal cylinder is precisely the same as that of Mr. Cartwright's, but the condensation of the steam is effected by injecting cold water into the second atmospheric cylinder ; this takes place when both pistons are at the top of their course, and are filled with steam beneath their pistons ; that steam being condensed by the injection, both pistons are pressed down, the principal one by the steam coming from the boiler through the steam valve, and the other by the pressure of the atmosphere.   When they arrive at the bottom, the supply of steam from the boiler is shut off, and the equilibrium valve in the principal piston is opened, to admit the steam to pass from the principal cylinder, through the horizontal passage and its hanging valve, into the space beneath the piston of the atmospheric cylinder ; that steam being at first more elastic than the atmospheric air, displaces the useless injection water from that space, into the open air, partly through the discharge valve, and partly through the valve in the piston.   When the pistons are raised up by the fly-wheel, to make their returning stroke, the steam expands itself, during their ascent, into half as much more volume as that which it had at the commencement of the returning stroke ; for it must fill the atmospheric cylinder, as well as the other.   The two pistons having arrived at the top of their course, the equilibrium valve in the principal piston is shut, and the steam valve opened ; the injection valve is also opened, to admit the cold water into the atmospheric cylinder, so as to condense the steam that it contains, and to exhaust the steam from the other cylinder, in order to produce another working stroke.

The piston of the atmospheric cylinder opposes a very considerable resistance to the motion of the fly-wheel, during the returning stroke ; because the steam beneath that piston, by its expansion, becomes less elastic than the atmospheric air ; and although this force is restored again to the fly-wheel during the subsequent descent of the piston, yet the regularity of the motion must be greatly disturbed.   The second cylinder is in effect the air pump of Mr. Watt's engine, made of double the usual capacity, and left open at top to the atmosphere ; an arrangement from which no advantage can be expected to arise.

A few engines were made on this plan, but did not answer so well as Mr. Watt's engines ; one of them of 8 or 10 horse-power was erected at Portsmouth Dock Yard, and was the first steam-engine which was applied in the public naval establishments.   It was used for a few years, and was then replaced by a 30 horse engine by Messrs. Boulton and Watt.   Another 30 horse engine, which was erected for the same machinery, has connecting rods at each side of the cylinder, and cranks under the cylinder, according to Mr. Sadler's arrangement ; but the cylinder, with its valves, air pump, and condenser, are the same as Messrs. Boulton and Watt's engine ; these engines are used alternately to turn the machine of the wood mill, or to drain the dry docks, in which ships are placed for repairing.

Atmospheric Engine, with Condenser and Air Pump.

In some districts of the north of England, the atmospheric engine was very commonly applied to turn machinery by a crank and fly-wheel (see page 410) : this became a standard engine in that district during the term of Mr. Watt's patent, and is still very extensively used. These engines were rendered much more efficacious, when a condenser and air pump on Mr. Watt's construction were added, to exhaust the cylinder instead of injecting cold water into it ; but this was not allowed by Messrs. Boulton and Watt, who prosecuted many persons for making such engines, and obliged the proprietors to pay for licenses. Engines of this kind being lightly loaded, and worked at a quick speed, perform very well ; and as they are cheaper, and simpler to keep in order, than Mr. Watt's complete engines, they are still used where coals are cheap.

A simpler kind of air pump was adopted for these engines towards the expiration of Mr. Watt's patent, and answered very well. The air pump is nearly as large as the cylinder, and its rod is suspended from the crank end of the great lever, so as to make a stroke half as long as that of the piston. The air pump is open at top to the atmosphere. During the working stroke, the bucket of this large air pump is drawn up, and exhausts the condenser ; the atmospheric pressure upon it, causes a resistance nearly equal to half the force of the piston, and during the returning stroke, the greatest part of that pressure acts in aid of the counterweight, to draw up the piston, and continue the action upon the fly-wheel.

A condenser was sometimes used, but in other cases the cold water was injected into the air pump. The operation of these engines is nearly the same as that of Mr. Sadler's, except that the bucket of the air pump being drawn up during the working stroke, instead of the returning stroke, its resistance tends to equalize the two strokes, instead of to increase the irregularity of the impulse on the fly-wheel, as is the case in Mr. Sadler's original construction.

Attempts to obtain a continuous circular Motion by the Action of Steam, without reciprocating Pistons.

The machines which are required to be impelled by steam-engines, are almost all dependent upon a continuous circular motion, hence it has been thought very desirable to obtain such a motion by more direct means than can be done by a piston moving backwards and forwards in a cylinder. Some of Mr. Watt's speculations upon this subject have been already mentioned, in his first specification, in 1769 (see p. 316) ; and his third specification, in 1782, explains his ideas on this subject more fully (see his fifth improvement, p. 351), and he added a figure of a rotative engine, which is copied in Plate XV, figure 1.

EE is the hollow cylinder, within which the piston A is caused to revolve by the action of the steam, so as to give a continuous circular motion to its axis D, in order to turn any machinery which may be connected with that axis. The cylinder EE has a flange at each end, by which the two ends or covers are fastened to it, and each cover has an opening through the centre to admit the neck of the axis D to pass through, with a proper stuffing box to confine hemp around the neck, and make a close fitting.

The revolving piston A, which is fixed to the axis D, projects out from it, to reach to the interior surface of the cylinder E, and the ends of the piston apply closely to the ends of the cylinder; hemp packing is inserted in a groove round the edge or thickness of the revolving piston A, to make a close fitting, and prevent any steam passing between the edges of the piston and the interior surface of the cylinder.

B is a partition which extends from the circumference of the cylinder E, to the central part D of the piston, around the axis. The partition B is provided with packing, in a groove round its edges, in the same manner as the piston, in order to prevent any steam passing between those edges, and the interior surface of the cylinder, or the central part D of the revolving piston.

The steam from the boiler is introduced into the cylinder by the pipe H, and it is exhausted from the cylinder by another pipe K, so that the space marked *vacuum* in the figure will be exhausted, whilst the space marked *steam* is supplied with fresh steam from the boiler, which, by its elasticity, will force the piston A round into the vacuous space in the direction of the arrow, and thus turn the axis D, and the machinery which is connected with it. When the piston A has nearly completed a revolution by this means, and the cylinder is quite filled with steam, the piston comes in contact with the back of the partition B, which is capable of moving about an axis situated at the circumference of the cylinder, in the same manner as a door moves about its hinges; in order that the partition B may be turned out of the way of the piston A, and then it retreats into a recess, which is formed for its reception, in the circumference of the cylinder E. The partition B, when turned back, exactly fills that recess, and its surface conforms precisely with the interior surface of the cylinder, so that the edge of the piston A will fit to it, as it turns round.

When the partition B is thus turned back into its recess, it closes the passage of the steam-pipe H, and prevents any more steam entering for the present; and that steam which has impelled the piston before it, and which then fills all the interior of the cylinder, passes away to the condenser through the eduction pipe K, as soon as the edge of the piston A has passed beyond the opening of that pipe.

The steam has no power to impel the piston beyond the position where it comes in contact with the back of the partition B, and therefore the motion must be continued by the fly-wheel, during the time that the piston is moving and passing by the partition; but the piston having so passed by, the partition is suddenly turned down out of its recess, into the position represented in the figure, by the pressure of the steam from the pipe H acting behind it, and then the operation above described is repeated, to make another revolution of the piston.

The defect of this rotary engine is, that the piston must act very abruptly against the partition B, when it comes in contact with it, to remove it out of its way; particularly if the motion is sufficiently rapid to produce a proper effect. And it is extremely difficult to keep the packing round the edges of the piston A, and the partition B, in proper order, so as to avoid immoderate leakage on the one hand, or excessive friction on the other. Also the action of the piston is not continual, because it must be suspended whilst the piston passes by the partition, which occupies about one-sixth of the whole circle. The latter defect might be remedied by having two or more pistons A affixed to the same axis, to act in the same cylinder, which must then have as many moveable partitions, B; but that would aggravate the other defects of violent action against the partitions, and their friction.

Mr. Watt abandoned this kind of steam engine after several trials, from the results of which it appeared to be inferior to his reciprocating engine. The plan has been revived, and several small engines made by Mr. Routledge, who took out a patent in 1818, for some improvements in the mode of putting the parts together; they have not proved good engines, though they are better than any other rotary engines yet produced. (See Repertory, 2nd series, vol. xxxiii. p. 129.)

Another of Mr. Watt's schemes for a rotary engine to revolve in mercury or fluid metal, is described in his specification of 1784 (see p. 429). It was never brought to bear, although many trials of that plan were made by Mr. Watt at Soho.

Mr. Hornblower's Rotative Steam-engine, 1798.

Mr. Hornblower took out a patent for this engine in 1798; his specification is printed in the first series of the Repertory of Arts, vol. ix. p. 289, with engravings, from which the figures 4 to 8, plate XV, are taken. Figure 5 is an elevation, and figures 6 and 7 sections corresponding thereto; figures 4 and 8 are horizontal sections.

The action of the steam is exerted between two movable rectangular revolving leaves A and B, figs. 4 and 6, which are placed within a hollow cylinder E E, and extend from its centre to its circumference, so as to divide it into two compartments; into one of which the steam from the boiler is introduced, whilst the other compartment is exhausted, to occasion a vacuum. The elasticity of the steam, by urging the two leaves apart from each other, causes one or other of them to turn round in the direction of the arrow; but only one leaf moves at a time, because they are detained from moving backwards, in the contrary direction to the arrow; consequently one leaf stands still to receive the reaction of the steam, whilst the other is propelled forward, and by its axis C D, fig. 6, gives motion to the fly-wheel, and the machinery that the engine is intended to turn.

When one of the leaves has made about five-sixths of a revolution, and approached nearly in contact with the other leaf which remained stationary, the distribution of the steam is changed; that compartment, or space between the leaves, which was before supplied with steam, is now exhausted, by opening a communication with the condenser; and the other compartment, which was before exhausted, is now filled with steam from the boiler. The pressure is consequently exerted on the contrary sides of the leaves, and as they are not allowed to move backwards, that leaf which has just run through its course stands still, whilst the steam drives forward the other leaf, which was stationary; and by its axis that leaf continues the motion of the machinery for another five-sixths of a turn in the same direction as before; after which the leaves change characters again.

In this manner a continual action is kept up; the moving leaf always pursues the stationary leaf until it overtakes the same, and then the moving leaf becomes stationary, whilst the other leaf moves forwards; that leaf which moves, always carries the rotative axis round with it, so as to keep it in continual motion; and the motion is regulated by a fly-wheel.

The hollow cylinder E E is firmly fixed in a vertical position; it has a flange at each end, by which the ends *e e* and *f f* are fastened on, and about one-fourth of its circumference can be removed, to give access to the moveable leaves A B within it. At the top of the cover *e e*, a hollow box H is fixed, to receive the steam from the boiler, by the pipe *h*; and K is a similar box fixed beneath the lower cover *f f*, to receive the steam, and conduct it away to the condenser, through the exhausting pipe *k*. The entrance of the steam from the box H into the

4 R

interior of the cylinder E, and the subsequent evacuation of that steam from the cylinder into the box K, is effected through the central axes D C of the two leaves A B; which axes, as well as the two leaves themselves, are made hollow for that purpose. There are two steam-valves b, and two exhausting valves a, in the flat surfaces of the leaves A and B, to distribute the steam.

The two leaves A B are put together in the manner of the leaves of a pair of hinges (see fig. 7), but the axis of each leaf is quite independent of that of the other leaf; l C is the hollow axis of the leaf A, and it is prolonged upwards by a solid central axis e c, which passes through all the length of the hollow axis d d of the other leaf B; but as the solid axis e does not fill the hollow axis d, there is room between them for steam to pass. The steam enters from the box H, through a square hole in the axis near r, fig. 6, and passes round the solid axis e, into the hollow leaf B, and thence it passes out through one of the steam valves b, into one or other of the compartments into which the cylinder is divided.

In like manner the steam passes through one of the exhausting valves a, from one or other of those compartments, into the hollow of the leaf A, and its central part C, and passes out at a square hole in its hollow axis l, into the box K, which is always kept exhausted by its communication with the condenser; that condenser and its air-pump are on the usual construction of Mr. Watt's engine.

The two leaves A B are allowed to move, or turn round forwards, but never backwards; for this purpose, pieces M M, fig. 8, called clamps, are lodged in circular channels w w, fig. 6, within the boxes H and K, and are connected by links p p, with the arms of short levers N N, figs. 7 and 8, which are fixed upon the axes d and l of the two levers A B respectively. When the axes are urged round in the direction of the arrow, fig. 8, the links p p arrange the clamps M M in the middle of their channel w w, so that they are quite loose therein, and in that position the links push the loosened clamps freely round in the channel; because the links p p have tails beyond the joints by which they are united to the levers N N, which tails stop the motion of the links about those joints in one direction, so that the links p p become prolongations of the levers N. But when any retrograde motion of the leaves, and their levers N N commences, then the links p p draw in those ends of the clamps M M which are jointed to the links, and contract them towards the axis, whereby the other thick ends of the clamps are caused to jamb fast in the circular channel w w, and thus prevent any farther motion of the clamps, or of the links and the levers N, so as to retain the leaves firmly from retreating by the re-action of the steam. These clamps produce the same effect as ratchet wheels and clicks would do, if they were applied as in the winding up part of a watch.

Similar clamps are applied within circular channels formed at x x and y y, fig. 6, within-side of a revolving box I I, which is fixed on the extreme end of the axis F F of the fly-wheel; the links of those clamps are jointed to the ends of levers x x and y y, fig. 7, one of which is fixed on the extreme end of the central axis e c of the leaf A, and the other on the extreme end of the hollow axis d d of the leaf B. By means of these two pairs of clamps, that leaf which is propelled forwards by the steam always becomes connected with the axis F, to urge it forward; but the other leaf which stands still, is in all cases detached from the axis, which then turns round independently of that leaf.

The borders of the leaves A and B are furnished with hemp packing which is driven tight into grooves formed round their edges, so as to prevent any steam passing by the edges of the leaves, from one compartment of the cylinder to the other; and all the interior surface of the cylinder is made true and smooth for the hemp to move against it; the packing is confined by plates screwed against the surfaces of the leaves. A stuffing of hemp is applied round the hollow axis *d d* at L, fig. 6, and another stuffing is applied in a groove round the central axis *c*, within the hollow axis *d d* (see fig. 7). The weight of the upright axis and the leaves, rests on the point of a pivot *m* at the lower end, in a brass socket, which is borne up by a regulating screw beneath.

The steam which enters constantly from the boiler into the box H, and passes through the axis *d* into the hollow within the leaf B, is throttled by a loose tube *r*, fig. 6, which slides up or down upon the outside of the hollow axis *d d*, so as to contract the square hole through which the steam enters. The tube *r* has rings on the outside of it, to enter between the teeth of a small pinion *s*; and the end of the axis of that pinion which comes through to the outside, being turned, will raise or lower the ring *r*, to regulate the passage of the steam, and consequently the motion of the engine.

The two steam-valves *b* in the leaf B, open inwards towards each other, into the hollow of that leaf, which is always filled with steam; the spindle of each valve projects out so much, that it will press against the other leaf, and open the valve before the leaves actually touch each other. The two opposite valves *b* are connected, so that when one is opened, it will shut the other. The two exhausting valves *a a* in the leaf A, open outwards from each other, to admit the steam into the hollow of that leaf, which is always exhausted, in consequence of its open communication with the condenser; the centres of these valves are prominent, so that they will be pressed by the other leaf B, to close the valve before the two leaves come in contact; and the two valves *a a*, are so connected, that when one is shut, it must open the other.

By this arrangement the four valves are opened and shut by the action of the leaves, whenever the moving leaf overtakes the other, so that they approach close together; and in all cases the steam-valve *b* on that side of the leaf B, which is close to the other leaf, will be opened, and thus the steam will be admitted between the two leaves. Also, the exhausting valve on that side of the leaf A, which is close to the other leaf B, will be shut; but the same motion opens the opposite exhausting valve, in order to take away the steam contained in the rest of the cylinder.

Mr. Hornblower made many attempts to bring this ingenious project to bear, but the engine could not be made to answer, on account of the friction and leakage of the packing, which is more than twice as much as is required in a common engine exerting the same force, because the leaf which is immovable, and which answers to the bottom of the cylinder, is packed and exposed to leakage as well as the acting leaf. The stopping and starting of the pistons was always attended with a violent jerk, and the clamps to prevent the retrograde motion, were very subject to wear out of order, and then to break or run back.

This project was revived by Mr. Carter, who took out another patent in 1821, with some alterations of Mr. Hornblower's original plan. The clamps he used are shown at M M, in fig. 9; they are applied within a circular box H H, opposite to the ends of the lever N N, which is fixed on the axis of one of the leaves, in the manner shown in fig. 7. The ends of the lever N N, against which the clamps M M act, are formed to portions of a spiral curve; consequently, whenever the lever N begins to move backwards, contrary to the direction of the arrows, the spiral curves at its ends, expand the clamps M M outwards, and jamb them so fast against the inside of the box H, as to prevent any motion. But whenever the lever N is turned round in the direction shown by the arrows, the spiral curves at its ends then release the clamps M M, so as to let them loose; and they are easily

4 R 2

pushed round in the box H, by small springs *p p*, which are fixed to the lever N N, and bear behind the clamps.

Mr. Murdock's Project for a Rotative Engine, 1799.

This was one article introduced into the specification of a patent which Mr. Murdock took out in 1799; it is printed in the first series of the Repertory of Arts, vol. xiii. p. 10. And figures 2 and 3 of plate XV., which are copied from that specification, are sufficient to explain the idea.

E E is a steam vessel which is in the shape of two cylinders joined together, with their axes parallel to each other, and their circumferences intersecting each other. The centre of each cylinder is occupied by a revolving axis C D, and upon each axis a roller is mounted; it is formed like a pinion with six teeth, and the teeth of the two rollers fit into each other, so that one pinion will turn the other round. The extremities of the teeth of each roller fit to the interior of their respective cylinders E E, and are provided with packing, which will prevent any passage of steam; the teeth which act together are also exactly fitted to each other, and the packing at the ends of the teeth makes a close junction between the two rollers, at the place where their teeth interlock with each other. The ends of the revolving rollers are closely fitted to the ends *e e* and *f f* of the cylinder; and the ends of the axes are received in stuffing boxes, as is shown in fig. 2.

h Is the pipe by which the steam is brought from the boiler, and enters into the engine at H, over the junction between the teeth of the two rollers; *r*, fig. 3, is a stop-valve to stop the engine, or to regulate its velocity, by restraining the flow of steam. K *k* is the eduction-pipe, leading to the condenser Q; and R is the air-pump, which is worked by a crank Z, fig. 2, fixed upon the end of the axis of one of the rollers. The machinery which is to be turned by the engine, is connected with the ends of either of the axes C or D of either of the rollers.

The operation of this engine is extremely simple. The steam which enters at H fills the spaces between the teeth of both rollers at the upper part, and at the same time a vacuum being formed in the condenser, the steam is exhausted from the spaces between the teeth at the lower part of both rollers, which, by the pressure of the steam against their teeth, are turned constantly round in opposite directions to each other, with a continuous circular motion, without any intermission. In that respect Mr. Murdock's engine would be free from the defect which is so great an objection to all the preceding schemes for rotative engines, viz. that their action is interrupted or suspended once in every revolution; but the objections of excessive friction or leakage, apply in full force to Mr. Murdock's plan, and preclude any chance of a successful application.

A machine of this kind was proposed more than a century ago, as a rotative pump for raising water, and is described with figures in the *Description du Cabinet de M. Grollier de Serviere*, p. 48, *quarto, Lyon*, 1719. See also *Leupold's Theatrum Machinarum Hydraulicarum*, vol. i. p. 123; he calls it Machina Pappenheimiana.

Having now enumerated the best of those projects for improving the steam-engine, which were brought forward by Mr. Watt's cotemporaries, we may proceed to describe some other modifications of his engine, which were made after the expiration of the patent right in 1800, when Mr. Watt retired from business, and the legal restriction to making his engines was removed. The demand for steam-engines being continually on the increase, a number of persons then began to make them, in all the manufacturing districts of the kingdom ; but from deficient execution, and incorrect proportions, the engines which these new makers produced, proved very inferior to those which were made at Soho ; and the deficiencies were so great, that in many instances, the makers were obliged to give up the pursuit, after having made a few engines.   Others who had better means of execution, and who took care to study Mr. Watt's models very closely, succeeded so far as to establish themselves in the business.

Messrs. Bateman and Sherratt, of Manchester, who had long been in the habit of making atmospheric engines, had the greatest success in this new business, and as engines were very much wanted at that time, for the increasing manufactories in Lancashire, their trade became very extensive in a few years.   Messrs. Bowman and Galloway also began at Manchester, and made good engines ; Mr. Bowman having been brought up by Mr. Rennie.   The manufactory of Messrs. Rothwell, Hick, and Co., at Bolton in Lancashire, was commenced by Messrs. Smalley and Thwaites.   Messrs. Fenton, Murray, and Wood, began making steam-engines at Leeds, in Yorkshire, and they attempted to improve the construction of Mr. Watt's engines, but after some years' trial they were obliged to follow the models of the engines made at Soho.

When Mr. Watt retired from the direction of the manufactory at Soho, he was succeeded by Mr. William Murdock, an experienced engineer, who had been stationed in Cornwall for many years, as an agent for Messrs. Boulton and Watt, to fix their engines, and direct reparations.   Mr. Southern continued in the management of the manufactory, and the business was carried on by the sons of the original proprietors, under the firm of Messrs. Boulton, Watt, and Co.

Mr. Murdock's patent of 1799, above quoted, was for new methods of manufacturing and constructing steam-engines ; and the specification contains contrivances which he introduced by degrees into the engines made at Soho.   The principal of these new plans is the sliding valve, which produces the distribution of the steam, to and from the cylinder, in a more simple manner than the four valves and working gear of Mr. Watt's original engine.   The construction of Mr. Murdock's sliding valve has since been improved, and is now universally adopted.

As the applications of steam-engines became more general for manufacturing purposes, a considerable demand arose for small engines of 2, 4, 6, and 8 horse-power ; and to diminish the expense and trouble of fixing such small engines, Mr. Murdock and Mr. Southern began to make what are now called independent engines, all the parts of which are supported one by another, or else by framing, which stands from the foundation, quite independently of the building.   The condensing cistern is made of cast iron, and forms a basement for all the other parts.   The first independent engines were called bell-crank, from the form of the great lever, which was a bent or elbow lever, similar to those used by bell hangers.

BELL-CRANK STEAM-ENGINE FOR TURNING MACHINERY,
(SEE THE LOWER PART OF PLATE XVI.)

This is a very compact form of Mr. Watt's double rotative engine, which was arranged by Mr. Murdock and Mr. Southern, for steam-engines of small power, to enable them to be placed within a less space than Mr. Watt's original construction, and also that they may be set up in any workshop or apartment of a manufactory, without requiring a particular building for the engine.   For this purpose they contrived to support one part of the engine by another, so as to require scarcely any parts for fixed framing : and the whole engine is sustained upon its foundation without any attachment to the walls of the building.   Figures 1, 2, and 3 of plate XVI. are an elevation, end view, and plan, of this kind of engine.

The condensing cistern G G forms the basis of the whole engine ; it is made of cast iron, in one piece, and is firmly supported upon two pieces of oak timber, which are laid upon brick walls, and the cistern is made fast upon the beams by four bolts.   The cylinder E is placed over one end of the condensing cistern, being mounted on brackets X, which are erected from the sides of the cistern G G, and support the bottom of the cylinder.   On the upper end of the piston-rod n, a horizontal cross rod $o o$, figs. 2 and 3, is fixed, and a vertical side-rod K K is jointed to each end of that rod, in order to give motion to the great lever L L, which is formed like a right-angled triangle ; the centre of motion p is at the right angle of the triangle ; the base L forms one of the arms of the lever ; the perpendicular the other arm ; and the hypothenuse of the triangle acts as a brace to strengthen the lever.

There are two of the bell-crank levers L L, one situated at each side of the cistern G G ; they are fixed on the two extreme ends of a horizontal axis p, which forms the centre of motion ; it extends across beneath the cistern from side to side, and is supported in bearings fastened beneath the bottom of the cistern.   The two levers L L are united at the upper ends of their perpendicular arms, by a bar, which extends across, over the top of the cistern, as is shown in figs. 2 and 3, and the ends project out beyond the levers, to form the joint pins for the connecting rods M M ; the other ends of those rods are fitted upon the crank-pins of the two cranks N N ; the cranks are formed on the opposite ends of a horizontal axis P, which is supported in bearings fixed on the angles of the cistern G G.

The axis P of the fly-wheel, is situated exactly in a line with the axis P of the cranks, and is supported in bearings fixed on brickwork.   The fly-wheel Q Q is wedged on the extreme end of its axis P, beyond the bearing, and a pin, which is fixed into one of the arms near the centre, projects out far enough to be connected by a short link, with the extremity of the crank-pin, which, for that purpose, projects out through the joint of the connecting rod ; by this means the fly-wheel is turned round, at the same time that the cranks are turned round by the action of the connecting rods M M, alternately pushing forwards, and drawing backwards, the crank-pins, with the force which is transmitted to them by the great levers L L, from the side-rods K K, and the piston-rod n.

In this engine no parallel motion is required for the piston-rod, because the horizontal arm of the great lever L, being of a considerable length, deviates less from the vertical than usual, and by the great length of the side-rods K K, they have only a very trifling deviation from the vertical, during their motion.   The perpendicular arm of the great lever L is considerably shorter than the horizontal arm, and therefore the radius of the cranks N N, is less than half the stroke of the piston.

The air-pump is worked by two rods $l l$, which are jointed to the great lever L at their lower ends, and at their upper ends to a cross rod, which is fixed on the top of the air-pump rod.   The air-pump and condenser are immersed beneath the cold water, which is contained in the cistern G G.   The condenser is situated beneath the cylinder E, and the eduction-pipe g, which conveys the steam away from the cylinder, passes in an inclined direction, to reach to the condenser. The lower part of the condenser communicates with the lower part of the air-pump, and the foot valve is situated in the bottom of the pump barrel.   The hot well I is placed on the top of the air-pump, the discharge valves being fitted over apertures in the cover of the pump.   A spout *m* conveys the hot water away, into a small box 20 at the end of the condensing cistern G, and from that box it runs off by a pipe.

The hot-water pump, and the cold-water pump are not shown in the sketches, but the rods of those pumps are usually suspended from the ends of two short arms, which project out from the axis *p* of the great lever L, so as to reach beyond the end of the cistern G, and the pumps are fixed down in a well, beneath the cistern.

The cylinder is enclosed within a steam-case, which is cast in one piece with the cylinder (see the section, figure 5) ; they join together at the upper flange, by which the cylinder cover is fastened on, and the bottom of the space between the cylinder and the steam-case, is filled up by a ring of metal, fitted in with cement ; the junction is made quite close by the bottom of the cylinder (*a*).

The construction of the sliding valve, which distributes the steam to and from the cylinder, is explained by the three enlarged figures, 4, 5, and 6. The two passages *c* and *f*, which communicate with the top and bottom of the cylinder, open at the outside of the steam-case, in front of the cylinder, and the parts around the openings are flat surfaces, formed extremely true and smooth ; the margin of these two flat surfaces and all the space between them, is a projecting flange, adapted to receive the projecting flange of an upright steam-pipe *b b*, which is a half cylinder, and which, when fastened in its place against the flat side, in front of the steam-case, forms a suitable cavity to receive the sliding eduction-pipe *d d*, which is likewise a half cylinder, but small and shorter than the upright steam-pipe; the sliding-pipe *d d* is open at both ends, and those ends are adapted to fill the cavity of the steam-pipe *b* exactly, as is shown in the section, fig. 5 ; but the middle part of the sliding-pipe *d d* is smaller, so as to leave a space *e* all round it, within the steam-pipe *b*, for the passage of the steam, which is introduced from the boiler into the steam-pipe *b*, by the steam-pipe *a* ; and the throttle-valve for the regulation of the steam is fitted into some part of the pipe *a*.

The lower end of the upright steam-pipe *b* is joined to the inclined eduction-pipe *g*, fig. 5, and there is a constant communication from the interior of the sliding eduction-pipe *d d* to the condenser, so as to keep that space exhausted, and also those spaces within the steam-pipe *b b*, above and below the fitting parts at the ends of the sliding exhausting-pipe, but at the same time all the space *e* around the outside of the sliding-pipe, between that and the interior of the steam-pipe, is supplied with steam from the boiler.   The interior surface of the upright side-pipe *b b* is made very smooth and true, at the upper and lower ends, where the enlarged parts at the ends of the sliding eduction-pipe *d d* fit into it, in order to allow that pipe to move up and down within the steam-pipe *b b* ; the semicircular part of the sliding-pipe *d d* has two grooves round its enlargement at the top and bottom, and packings are fitted tight into those grooves, in order to make the junction quite tight, between the ends of the sliding-pipe and the interior of the steam-pipe *b*, so as to prevent any steam passing up or down, by the ends of the sliding-pipe.   These packings also press the flat sides at the ends of the sliding-pipe *d d*, firmly against the flat surfaces in which the apertures *c* and *f* open into the top and bottom of the cylinder ; and the surfaces thus in contact are accurately fitted together, so that no steam can pass between them.

When the sliding eduction-pipe *d* is pushed down in the steam-pipe *b*, as shown in figure 5, the fitting parts at the top and bottom of it, will be below the

---

(*a*) This plan of making the cylinder and the steam-case in one piece, was one of the improvements proposed by Mr. Murdock in the specification of his patent of 1799.

two passages *c* and *f* respectively; consequently any steam that is contained in the upper part of the cylinder above the piston J, can pass out freely into the open top of the sliding eduction-pipe *d*, and descend through that pipe to the condenser, so as to exhaust the cylinder. And at the same time the steam from the boiler can enter from the space *e* between the sliding-pipe and the steam-pipe *b*, into the passage *f* at the lower part of the cylinder, so as to act beneath the piston, to impel it upwards.

On the other hand, when the sliding eduction-pipe *d* is raised up, so that its fitting parts will be above the passages *c* and *f*, then the communication will be opened from the condenser to the lower part of the cylinder, to exhaust the same through *f* and *g*, and the steam will be admitted into the upper part of the cylinder, above the piston, to press it down, and make the descending stroke. When the sliding eduction-pipe *d* is in its middle position, the fitting parts at the top and bottom of it, will stand exactly opposite to the apertures of the passages *c* and *f*, so as to close them completely, and allow no passage into or out of either end of the cylinder; and in this state the engine must remain at rest.

The upper end of the upright steam-pipe *b b* is closed by a semicircular cover, which is fastened down by screws, and can be removed to draw out the sliding-pipe when necessary. To give the necessary motion to the sliding eduction-pipe *d*, within the steam-pipe *b*, a small polished rod 4 passes through a stuffing box in the cover; the lower end of the rod is connected with the upper end of the sliding-pipe *d* by three small upright screws; and the upper end of the rod is fastened to an arm projecting out from the upper end of a square bar *s*, which is situated in front of the upright steam-pipe *b*, and is fitted into sockets, which allow it to slide freely up or down, and it is moved by a lever *u*, the end of which enters into a square opening in the upright bar *s*. The lever *u* is fixed on the end of an horizontal axis *t*, which is supported on bearings affixed to the steam-case, and on the other end of the same axis *t*, a longer lever *v* is fixed, and has a fork at the end to embrace the projecting edge of an excentric rim *r r*, which is fixed on the arms of the fly-wheel Q Q.

The excentric rim *r r*, is not a regular circle, but it is composed of portions of two concentric circles, one of a larger radius than the other, and their ends are joined by easy spiral curves, leading from one into the other. The consequence of this arrangement is, that as the fly-wheel turns round, the excentric *r r*, acting in the fork at the end of the lever *v*, will alternately raise and depress that lever, and by the connexion of the axis *t*, the lever *u*, the upright bar *s*, and the rod 4, which passes through the stuffing box in the cover of the steam-pipe *b*, the sliding eduction-pipe *d d* will be moved up and down within the upright steam-pipe *b*, at the proper intervals, to produce the required distribution of the steam.

When the piston reaches near to the top of its course, the sliding-pipe is raised up with a quick motion, so as to admit the steam into the top of the cylinder; and being so raised, it is kept nearly motionless, whilst the fly-wheel makes about one-third of a revolution, because that part of the excentric rim *r r*, which is then in action upon the end of the lever *v*, is nearly concentric with the axis of the fly-wheel. When the piston arrives nearly at the bottom of its course, one of the spiral parts of the curve of the excentric rim presents itself to the end of the lever *v*, and gathers it in towards the centre of the fly-wheel, so as to depress the sliding-pipe *d d* into the situation of the section, figure 5, and then the small circular part of the excentric curve *r r*, coming into action, retains the sliding-pipe nearly at rest, whilst the piston makes its ascending stroke; after which the sliding-pipe is raised up again as before.

A stop valve is applied in the steam-pipe *a*, to shut off the supply of steam, and stop the motion of the engine, without disengaging the sliding-valve from its connexion with the excentric. The blowing through is effected by a small pipe leading from the lower part of the steam-case to the condenser, with a valve in it to be opened by a handle when it is required to clear the condenser of air and water.

A considerable number of these bell-crank engines were made at Soho, during five or six years after the expiration of Mr. Watt's patent; they were of three different sizes, viz. 4, 6, and 8 horse-power; the sketches in Plate XVI. were taken from a 4 horse engine, of which the principal dimensions are as follows :—

The boiler is of the waggon shape, $3\frac{1}{3}$ feet wide, and 6 feet long, so that the horizontal surface of the water is 20 square feet; and, according to the rule given at p. 588, for a boiler without an internal tube, the surface exposed to the heat is 61·5 square feet; or 15·37 square feet to each horse-power.

The cylinder is $12\frac{3}{4}$ inches diameter; the piston makes a 2 feet stroke, and works at the rate of 40 strokes per minute, or 160 feet motion per minute. The expenditure of steam is 141·8 cubic feet per minute; or at the rate of 35·46 cubic feet to each horse-power; and, according to the rule p. 576, the effective pressure on the piston is 6·46 pounds per square inch, at an average.

The air-pump is $6\frac{3}{4}$ inches diameter, and its bucket makes a stroke of $1\frac{1}{6}$ feet; the valve in the bucket is a circular plate 5 inches diameter, fitted round the rod in the manner shown at H, fig. 8, Plate XIII.

The condensing cistern G is $5\frac{1}{2}$ feet long, 25 inches deep, and 20 inches wide, outside. The upright side rods K K, are $8\frac{1}{2}$ feet long, and $1\frac{3}{4}$ inches diameter; the two side-rods are 25 inches apart. The horizontal arm of the great lever L, is $3\frac{1}{2}$ feet radius, and the perpendicular arm 35 inches. The connecting-rods M M, are $4\frac{1}{2}$ feet long, and $1\frac{1}{2}$ inches diameter; the two connecting-rods are 33 inches apart. The cranks are 10 inches radius, and the axis of the cranks is 4 inches diameter. The fly-wheel is 8 feet diameter outside; the rim is 5 inches by $3\frac{1}{2}$ inches = $17\frac{1}{2}$ square inches in the cross section, and it contains 2·89 cubic feet of cast iron; according to the rule given in p. 643, it should be 2·83 cubic feet; thus,

Sliding Rule, Calculation. $\left\{\begin{array}{l} \text{C} \quad \text{2·83 cubic feet.} \qquad \text{(4 HP ÷ 40 strokes =)} \quad \text{·1} \\ \overline{\text{D} \quad \text{Gage point 51.} \quad \text{(7·58 ft. dia. × 40 revol. =) 303}} \end{array}\right.$

The bell-crank engine has the disadvantage of being out of balance, for the weight of the piston, piston-rod, cross-rod, side-rods, and the horizontal arms of great lever, as well as the corresponding parts for the air-pump, are all unbalanced; so that their weight acts to increase the force of the piston during its descending stroke, and to resist the motion of the piston during its ascending stroke. A balance weight has been sometimes attached to the rim of the fly-wheel, but the centrifugal force of such a weight has a tendency to shake the supports for the bearings of the axis of the fly-wheel.

The bell-crank engines made at Soho were of a small size, and this inconvenience was not considerable; but some engines were made in London of a larger size, which did not perform well in consequence; and one large bell-crank engine, of 80 horse-power, which was made in 1807, for a rolling mill, failed totally. These instances showed the defect in a striking light, and such engines have not been made for many years.

About the year 1806, Messrs. Boulton, Watt, and Co. began to make their small independent engines with the great lever over the cylinder, and the parallel

4 s

motion, connecting-rod, crank, and fly-wheel, arranged in the same manner as in Mr. Watt's large engines, without the sun and planet-wheels; these moving parts are supported in a framing of iron, erected upon the cast iron condensing cistern, which serves as the basement for the whole engine. The distribution of the steam is performed by the sliding eduction pipe, moved by an excentric circle on the axis of the crank. This arrangement has since been generally adopted for independent engines.

Mr. Murdock's sliding eduction pipe, as it was first constructed, did not answer very well in practice, because the packings in the semicircular grooves at the back of the pipe cannot be kept in good order. The pipe requires to be drawn up out of its place in the steam-pipe, in order to examine or renew the packing, which is a difficult and troublesome operation. When the packing becomes loose a great leakage of steam takes place, and if the hemp is not applied uniformly all round the semicircle, it occasions an unequal pressure on the sliding surfaces, which causes them to wear away more at one side than at the other, so as to impair their fitting. The apertures for the passage of the steam were very small in most of these bell-crank engines, and this circumstance rendered their performance deficient.

In course of time the causes of these defects were observed, and they were remedied by a few alterations in the construction, whereby the sliding eduction pipe is made to answer very well, and it is now brought into very general use. The two ends of the sliding-pipe are made smooth at the semicircular parts opposite to the passages, and the hemp packings are lodged behind those parts, in suitable grooves formed withinside of the semicylindrical steam-pipe; the hemp is compressed into the grooves by screws, which can be tightened occasionally, to keep the packings quite tight, without immoderate friction. The apertures for the passage of the steam are now made of a larger size than in the first bell-crank engines; and the excentric by which the sliding-pipe is moved up and down, is made a complete circle, so that it acts on the same principle as a crank; viz. it keeps the pipe continually in motion up and down, and moves it very slowly when the direction of the motion changes. This occasions the motion of the pipe to be very easy, without any abrupt jerks, but it is no improvement to the performance of the engine, because the apertures for the passage of the steam are opened gradually; and unless those apertures are made of a larger size than would otherwise be necessary, the steam will not be exhausted promptly from the cylinder, as it should be to enable the engine to perform well, and with a rapid motion.

DESCRIPTION OF PLATE XVII, A ROTATIVE STEAM-ENGINE OF FOUR HORSE POWER, MADE BY MESSRS. FENTON, MURRAY, AND WOOD, OF LEEDS, 1802.

This engine was erected at Mr. Brewin's tannery, in Bermondsey, in 1802 (a), and was applied to turn a mill for grinding bark, as is shown in the side elevation.

This was an arrangement of the parts of Mr. Watt's double engine, by which it was intended to bring them within a small space, and to diminish the weight of the moving parts. The cylinder E is fastened upon a large flat plate of cast iron X X, which is bedded upon brickwork. The condensing cistern G G, which is also made of cast iron, is placed beneath the foundation plate X X, and is fastened to it, so that the weight of the cistern assists to keep that plate steady.

(a) A sketch of this engine in perspective, and a description of it, was made by the author, in 1805, for Dr. Olinthus Gregory, who inserted it in his " Treatise on Mechanics," vol. ii. p. 390.

The piston-rod *n* is jointed to a pin *o o*, which projects out from the circumference of a cog-wheel N of 36 teeth ; that wheel is adapted to roll round withinside of a fixed ring *q q*, having 72 internal teeth, to engage with the teeth of the revolving wheel N.   The wheel is suspended by its centre, upon the pin of the crank O O, at the end of the axis P P of the fly-wheel Q Q, so that when the crank pin moves round in its circular orbit (which is shown by the dotted line in the front elevation) the wheel N will circulate round within the fixed circle *q q*, in the manner of a planet-wheel.   The wheel N being half the size of the fixed ring, and having only half as many teeth, it will be turned once round about its own centre, for every time that it performs the circuit in its orbit ; and the motion round its own centre, will be in a contrary direction to the motion in its orbit.

The effect of this arrangement is, that the pin *o* to which the piston-rod *n* is jointed, will always move up and down in a vertical line, opposite to the centre of the fixed ring *q q* ; and the force of the piston being applied to urge that pin alternately upwards and downwards in its vertical line, will cause the axis P P of the fly-wheel Q Q, to turn round continuously, in the same direction as the planet-wheel N circulates round in its orbit.

The axis P of the fly-wheel is supported in bearings, at the upper ends of two columns of cast iron V V, which are fixed upon a large block of stone ; and that block rests upon the foundation plate X X, to keep it steady by its weight.   The fixed ring of teeth *q q*, is supported by two inclined pillars of cast iron U U, which are erected from the foundation plate, and are fastened to the ring *q q* at each side ; the ring is secured by two curved braces from behind it, which are united to the upper end of one of the columns V.   There is also a horizontal bar which extends across behind the plane of the ring *q q*, from one side to the other ; and a bearing is formed in the middle of the cross bar, to steady the axis P of the crank ; that bearing being exactly in the centre of the ring, causes the teeth of the wheel N to apply correctly to the internal teeth of the ring *q q*, when the wheel circulates within it.

The crank is a flat circular plate O O, which is fixed fast on the extreme end of the axis P, beyond the bearing, and behind the plane of the fixed ring *q q*. The crank pin is fixed fast into this plate, and projects out from it ; the distance from the centre of the axis P, to the centre of the crank pin, is one-fourth of the length of the stroke that the piston is to make.   The centre hole through the planet-wheel N is fitted upon the crank pin, so that the wheel can turn freely round upon it, and the teeth of the wheel engage in those of the fixed ring ; the geometrical diameter of the ring is just equal to the length of stroke that the piston is to make, and the geometrical diameter of the planet-wheel N is equal to half that stroke.   The pin *o*, to which the piston rod is suspended, is firmly fastened to the wheel N, and projects out at right angles to its plane ; the centre of this pin corresponds with the pitch circle of the teeth of the wheel ; and those teeth are applied to the teeth of the fixed ring *q q*, so that during each half circuit that the wheel N makes within the ring, the pin *o* shall move from the highest point of the circumference of the ring to the lowest (or vice versa) in a straight line across its centre.

This parallel motion may be considered as consisting of two short cranks, one of which is mounted upon the pin of the other for its centre of motion, which therefore travels round in a circular orbit, and these cranks are turned round both together in opposite directions, so as to combine the vertical ascent and descent of both into one vertical motion of the crank pin of the second crank, with a stroke of four times the radius of either crank, and without any lateral deviation from a vertical line ; therefore the piston-rod may be applied immediately to that crank pin. The first short crank is fastened on the end of the axis P, to which the rotative motion is to be given (see O O in the front elevation); the crank pin which projects out from that crank, describes the circle shown by the dotted line around the centre of the axis P. The other short crank is the planet-wheel N, which is fitted upon the crank pin of the first short crank, and turns round upon that pin as its centre of motion, with just the same rapidity that the first crank O turns round with the axis P, but in a contrary direction. The pin *o* of the second crank is fixed to the circumference of the planet-wheel N, and the piston rod is suspended from it.

The turning round of the second crank, with equal and contrary motion to that of the first one, is produced by the teeth of the planet-wheel N, rolling round within those of the fixed ring *q q;* and the result of that motion is, that the pin *o* of the second crank combines the vertical ascending or descending motion of both cranks, without the lateral deviation of either; because the lateral deviations of the two crank pins from the vertical lines which pass across their respective centres of motion, will in all cases be equal one to the other, but in opposite directions; so that those deviations will counteract and neutralise each other, and thus the pin *o* is always retained in the same vertical line.

The pin *o* projects out to a considerable distance from the wheel N, and the rod *l l* for the air-pump H is suspended from it, at the extreme end; the bucket of the air-pump, therefore, makes the same length of stroke as the piston, and accordingly the air-pump is of a smaller diameter than is usual in Mr. Watt's engines; in other respects, the condensing apparatus is the same as Mr. Watt's engine, viz. the condenser F, the injection cock *j*, the air-pump H, with its foot valve at *k*, and its discharge valve at *m*, to deliver the hot water into the hot well I.

The steam is brought from the boiler by the steam-pipe *a* into the box *b*, which contains the upper steam valve, and thence it passes by the upright side pipe *d d* into the box for the lower steam-valve *e*. The eduction-pipe *g g* proceeds from the box which contains the upper exhausting-valve, and descends to the condenser F, and that pipe has a short branch to join to the box *i* for the lower exhausting-valve; the passages *c* and *f*, which convey the steam into, and out of, the top and bottom of the cylinder, join to the valve boxes, at the back part between the pairs of valves. This arrangement of the nossels for the four valves, is very similar to that of Mr. Watt's engine, except that the two boxes *b e* and *h i*, which contain the four valves, are of a cylindrical form; and instead of internal racks and sectors to lift the valves, the spindles of the two steam-valves pass up through the covers of those boxes; and the spindles of the two exhausting-valves pass up through the centres of the spindles of the steam-valves, which are for that purpose made hollow like tubes; all the four spindles are packed with hemp, which is confined around the spindles in suitable stuffing boxes, at the parts where they come through to the open air, so as to prevent leakage.

The four valves are opened and shut at the proper intervals, by two upright lifting-rods, 3 and 4, placed in front of the nossels, and guided in sockets, through projections from the covers of the boxes which contain the valves; short arms, 1, 2 and 5, 6, are fixed at the upper and lower ends of each of these rods, and they reach out to the spindles of the valves, which are so connected with those arms, that when either of the lifting rods 3 or 4 is lifted up, it will open one pair of the valves, viz.: the rod 3 is connected at top by its arm 1, with the spindle of the upper exhausting-valve, and by its arm 6 with the hollow spindle of the lower steam-valve; therefore, if the rod 3 is lifted up, it will open that pair of valves which occasion the ascending stroke of the piston. The other lifting rod 4 is connected by its arm 2 with the hollow spindle of the upper steam-valve, and by its arm 5 with the spindle of the lower exhausting-valve ; and therefore the rod 4 being raised, it will open those two valves which produce the descending stroke of the piston (*a*).

The lifting rods are moved by two revolving tappets, which project out from a horizontal axis, *t t*, that passes through large semicircular openings formed in the middle of each of the lifting rods 3 and 4, and the tappets act within those openings. The axis *t t* receives its motion by a pair of equal bevelled wheels *v* from an inclined axis, which extends up towards the main axis of the fly-wheel, and is turned by another pair of equal bevelled wheels *u*, and by a pair of equal spur wheels *p s* from that main axis. By this wheelwork the axis *t t* is turned round continuously, with a corresponding motion to that of the fly-wheel. The tappets project out from the axis on opposite sides of it, so that when one tappet points upwards the other will point downwards. The flat sides of the semicircular openings through the lifting rods, within which the tappets act, are at the upper sides of those openings, above the axis, and when either of the tappets begins to point upwards, above the level of the axis, it acts beneath the flat side of the semicircular opening, and raises up the lifting rod, so as to open that pair of valves which belongs to it. But when the axis has made about one-third of a revolution, and the same tappet begins to approach towards the level of the axis, it ceases to act beneath the flat side of the semicircular opening, and allows the lifting rod to descend by its own weight, so as to shut the pair of valves. The revolving axis *t t* is supported in brackets, which project out from the side pipes *d d* and *g g*.

r are two small handles, which hang loose upon a centre pin projecting out of the front of the valve box ; when either of these handles is turned about sideways upon its centre pin, and raised into a horizontal position, it acts beneath a pin which projects out from the front edge of the corresponding lifting rod, so as to lift up that rod, and open the pair of valves belonging to it. By using both these handles together, all the four valves can be opened at once, to blow steam through the condenser, and then the engine can be started ; it is also stopped by these handles.

The operation of the four valves is the same as that of Mr. Watt's engine ; when the piston arrives very near to the top or bottom of its course, the two opposite tappets on the axis *t t* point out horizontally therefrom, and neither of them act, so that both the lifting rods 3 and 4 are let down by their own weight, and all the four valves will be shut ; but they only remain so for an imperceptible space of time, for the instant that one of the rods is let down by one tappet, the opposite tappet begins to raise the other lifting rod, in order to open

---

(*a*) The plan of the nossels and working gear used in this engine, were introduced by Messrs. Boulton, Watt, and Co. Mr. Watt states, that Mr. Murdock contrived the steam-valves with hollow spindles, like tubes, in order to admit the spindles of the exhausting-valves to pass through them, and thus enable either valve to be lifted from above independently of the other, by levers situated at the outside of the boxes, instead of withinside as was Mr. Watt's plan. Mr. Murdock also first contrived to open and shut the valves, by a circular motion derived from the rotative axis; his first application of that contrivance has been shown in his bell-crank engine, and he afterwards applied the excentric circle on the axis of the crank, as is now commonly used in almost all engines.

Mr. Mathew Murray took out a patent in 1801, for a method of constructing the air-pump, and other parts of steam-engines. The specification is printed in the Repertory of Arts, first series, vol. xvi. p. 298. An arrangement of nossels and working gear is there described, which is very nearly the same as that represented in Plate XVII. This patent was set aside in 1803, by a writ of scire facias, at the instance of Messrs. Boulton and Watt, who had made engines on that plan previous to the date of the patent. See Repertory, second series, vol. iii. p. 235.

its pair of valves.   When the piston arrives near the top of its course, the rod 3 is let down, and the lifting rod 4 is raised ; and, vice versa, when the piston nearly reaches the bottom of its course, the rod 4 is let down, and the rod 3 is raised again.

The parallel motion used in this engine was invented by the late Mr. James White, an ingenious mechanician, who spent several years in France ; it is described in his New Century of Inventions, quarto, p. 30, and he says that, in 1801, it procured him a medal from Bonaparte, then First Consul of the French republic.   The practical defect of the contrivance is, that the socket in the centre of the planet-wheel N, cannot conveniently be made long enough in its bearing upon the crank pin, to enable it to endure the strain to which it is subjected without wearing loose in a short time ; for that strain is more than twice as great as the whole force exerted by the piston ; and the strain upon the teeth of the wheel, and of the fixed ring q q, is equal to all the force exerted by the piston.

The crank pin upon which the planet-wheel N is fitted, ought to project out to a considerable distance from the face of the crank O O, so as to allow the socket in the centre of the wheel N to fit upon it, in the same manner as the centre box of a coach wheel is fitted upon its axletree ; and the length of bearing in that socket, ought not be less than the diameter of the wheel, to give it a fair chance of wearing well ; because the socket on the pin o to which the piston rod n is suspended, must project out beyond the face of the planet-wheel N, in the manner of a crank pin ; and therefore, the force of the piston acts with an increased power to twist off the crank pin, and to wear the socket in the centre of the wheel N on one side.   If such a length were given to the crank pin, then it would cause all the work to project out so far beyond the bearing for the neck of the central axis P, as to occasion a very great increase of the strain upon that neck, tending to break it off laterally.

In the engine above described this objection is greatly aggravated by the manner in which the pin o o is prolonged, to suspend the rod l of the air-pump ; it was attempted to counteract this strain, by applying a balance weight by means of a link K jointed in the middle of the long pin o o, between the joints for the piston rod n, and the air-pump rod l.   This link went up to one end of a long wooden lever, which was situated in a room over the engine, and suspended on a centre like a scale beam, the balance weight being fastened to the opposite end of it.   With this addition the engine continued to work for a few years, but the joints became very loose by wearing, and then the wheelwork went unpleasantly. For these reasons, this kind of parallel motion was laid aside, after a few engines had been made with it, and some of those engines have since been altered to work with other parallel motions.

Mr. White, the inventor, proposed in his Century, p. 338, to double the parts of this parallel motion for large engines, in order to give it stability, and avoid the twisting strain.   Two of the fixed rings q q, with their planet-wheels N within them, are to be placed exactly opposite to each other, with only a sufficient space between them for the piston-rod to work in ; and the pin o o to which it is suspended, being firmly fixed to both the planet-wheels N, so as to unite them together, will cause them both to revolve on their own centres like one wheel, although they are at opposite sides of the piston-rod.

The main horizontal axes P P of the two cranks are exactly in a line, one opposite to the end of the other ; they have no direct connexion, except by the pin o, which will oblige the two planet-wheels N to turn round together, and they by their teeth acting in the two fixed rings q q, will compel the two cranks and their axes to turn together.   To attain certainty in this respect, Mr. White proposed to fix a large spur-wheel upon each of the axis of the cranks, to act as multiplying-wheels, and work into two pinions fixed upon a horizontal axis, which is common to both pinions, and extends over the two fixed rings, so as to receive the united action of both parallel motions to turn it round with an increased velocity; the fly-wheel is to be fixed upon this common axis. This plan would give the parallel motion sufficient strength; but the complication of wheelwork would be very objectionable, for there would be eight toothed wheels; and the weight of the piston and the planet-wheels would not be balanced.

Description of the Bark Mill represented in Plate XVII.

The bark mill operates on the same principle as the small hand mills which are used to grind coffee for domestic use. J is the revolving part of the bark mill; it is a cone of cast iron fixed upon an upright axis, which is turned round by means of the bevelled wheels M M and L, from the axis P of the fly-wheel. The lower part of the cone J is formed into sharp teeth or cutting edges, which are disposed around its circumference obliquely to the axis of the cone; some of the teeth are longer and more prominent than the others, as is shown in the figure.

The revolving cone J is surrounded by a cylinder of cast iron W W, which is open at top and bottom, but the cone J fills up the opening at bottom; the lower part of the cylinder W is enlarged to a conical form, and the interior surface of that conical part is formed with sharp cutting edges, corresponding to those of the revolving cone J. The upper edge of the cylinder W W is surmounted by a conical hopper of iron plate, into which the bark is thrown, and the revolving cone J being in continual motion, the pieces of bark settle down by their own weight, into the narrow space which is left between the outside of the cone and the inside of the cylinder, and are first broken by the large teeth into small pieces, and are then ground into a coarse powder, between the revolving teeth of the cone, and the fixed teeth of the cylinder; the ground bark falls out upon the floor, round the circumference of the cone J.

The fixed cylinder W is supported upon a strong wooden frame or stool, and the pivot at the lower end of the upright axis of the cone J turns in a socket, which is fixed into a cross bar of the stool; the end of the pivot is borne upon the top of a plug which is fitted into the socket, and the lower end of the plug rests upon the top of a screw z. By turning that screw, the axis can be raised or lowered, so as to make the cone J enter more or less into the cylinder W W, in order to cause the teeth to cut at a proper distance from each other, and grind the bark into powder of the desired fineness.

At the upper end of the upright axis is a pivot, which is supported in a bearing fastened against a beam of wood extended horizontally across the building. The horizontal bevelled wheel M M is fastened on the upper end of the axis, and is turned by another bevelled wheel L, of about half the size, fixed on the end of a horizontal axis which extends from the end of the axis P of the fly-wheel of the steam-engine, and is connected with that axis, so as to be a prolongation of it. The pivot at the end of the axis of the wheel L is supported in a bearing bolted to the beam that supports the bearing for the upper end of the vertical axis. This additional axis is omitted in the drawing, in order to bring it into the compass of the plate, and the wheel L is represented as if it were fixed on the end of the axis P of the fly-wheel, which would not answer in practice, because the bark mill would be too near the engine, and the dust raised in grinding the bark would injure the engine. The bark mill is placed in the middle of a large room adjoining to that in which the engine is fixed, and the dust, which is the best part of the bark, is kept in, and preserved from flying about.

This kind of bark mill was brought into use by Mr. James Weldon, who had two patents for its construction in 1797 and 1801. (See the first series of the Repertory of Arts, vol. x. p. 77; and vol. xv. p. 90.) It is very well adapted for the purpose of grinding bark, for it is not liable to clog up, and will take in very large pieces as well as small pieces, and it will grind a great quantity. The long teeth, which are called breaking teeth, are large and prominent, and they extend a considerable distance up the cone J; the other teeth, which are called the grinding teeth, are finer and sharper, they are inserted in the spaces between the large ones. The breaking teeth catch the large pieces of bark, and reduce them small enough to allow them to pass down between the grinding teeth, by which they are ground to powder.

Bark mills of this kind are often turned by a horse, harnessed to a long lever, which is fixed on the upright axis of the cone J, in place of the wheel M M. One horse working at a time, in a mill of this kind, will grind a load of bark of 45 hundred weight, in a day of 12 hours, but the labour is severe, and two horses must be kept to relieve each other every hour; so that each horse works 6 hours per day. The above engine was rated at four horse power, the cylinder was 12 inches diameter, and the piston made a stroke of 3 feet; it went at the rate of 32 strokes per minute. The bark mill made about 16 turns per minute, and usually ground about 4 tons of bark in a day of 12 hours.

An independent Steam-engine, with the great Lever beneath the condensing Cistern, made by Messrs. Fenton, Murray, and Wood, 1806.

The perspective view at the upper part of Plate XVI, is taken from a print in Nicholson's Philosophical Journal, vol. ix. p. 93. The condensing cistern G G, which is of cast iron, in one piece, is supported upon foundation walls at each side, which leave a sufficient space between them for the great lever L L, which is situated beneath the cistern; the two bearings for the pivots of its axis p, are mounted in brackets V, which are bolted to the underside of the bottom of the cistern, and which project down low enough to allow the lever to librate about its axis of motion.

The piston rod n is connected with the extreme end of the great lever by the two side-rods K K, which are jointed at their upper ends, to the two ends of the cross-rod $o\, o$ of the piston-rod n, and their lower ends are jointed to a corresponding cross-rod 6, which is affixed to the end of the great lever, and projects out horizontally from it each way, in a direction parallel to the axis p. The bucket of the air pump is connected with the great lever by the rods $l\, l$ and l in a similar manner to the piston; but being at the opposite side of the centre of motion p, the weight of the parts tend to balance each other.

The connecting-rod M is jointed to the end of the great lever L, and acts upon the pin of a simple crank N, which is fixed on the extreme end of the axis P of the fly-wheel Q Q; the neck for the axis of the crank is lodged in a bearing formed in the upper part of a frame, which has two legs U U; one is a short leg which is bolted to the end of the cistern G, and the other is a long leg, which stands upon the foundation wall. The opposite end of the axis of the fly-wheel is supported by a bearing, fixed in the wall of the building in which the engine is placed.

The condenser is made of thin copper in a globular form, as is shown by the dotted lines F, and the foot-valve is at k in the passage which leads from the condenser to the lower part of the air-pump. The hot well I is placed on the top of the air-pump H, in the same manner as in the bell-crank engine.

The distribution of the steam to and from the cylinder, is performed by one sliding valve, which is contained within the steam-box *b*; the steam from the boiler is introduced into this box by the steam-pipe *a*, in which a throttle-valve is placed. The bottom of the steam-box *b* is a flat and perfectly smooth surface of metal, with three oblong apertures or slits in it, situated in a row side by side, parallel to each other, the spaces between them being nearly equal to their breadth the narrowest way. The middle one of these three apertures is exactly over the eduction-pipe *g*, and is the orifice of the passage into that pipe; one of the side apertures is the orifice of the passage which leads into the bottom of the cylinder; and the other aperture is the orifice of the passage which leads to the top of the cylinder, the latter passage is formed in casting the cylinder, and passes up in front of it, being a portion of the space between the cylinder and its steam-case.

The sliding-valve is a small shallow pan or dish which is placed on the flat bottom of the steam-box *b*, in an inverted position, with its open part downwards, over the three apertures, and the edges of the dish are accurately fitted to the flat surface in which the apertures are formed, so that steam which fills the steam-box above the inverted valve, cannot pass under the edges of the dish, or gain admission into those apertures which are covered by the sliding-valve (*a*). The valve is adapted to cover two of the three apertures, leaving the third uncovered, to admit the steam in the box freely into it. The middle or eduction aperture is in all cases covered by the sliding-valve, so as to prevent the entrance of any steam from the upper part of the box, and one or other of the adjacent passages at the side of the middle one, is also covered by the valve, so as to form a communication with that passage and the eduction-pipe, in order to exhaust the steam either from the top, or from the bottom of the cylinder; for when the valve covers one of the side passages, and the middle passage, the steam can pass up through that side passage into the space within the inverted dish or sliding-valve, and thence it can pass down again through the middle passage to the condenser.

By moving the sliding-valve horizontally, one way or the other, over these three passages, the communication with the condenser is opened, with either of the side passages, so as to exhaust either from the top of the cylinder, or from the bottom. The breadth of the valve is only sufficient to cover one of the side apertures and the middle one at the same time, therefore the opposite aperture to that which is so connected with the condenser, is in all cases left uncovered by the valve, and admits the steam from the boiler to pass out of the steam-box, in which the valve is placed, either into the top of the cylinder or the bottom, in order to cause the plenum therein, and impel the piston, when the opposite end of the cylinder is exhausted, in consequence of the communication that the valve has opened from the same to the condenser.

This single sliding-valve performs the office of the four valves of Mr. Watt's engine, in the same manner as Mr. Murdock's sliding eduction-pipe, from which it originated. The steam in the box *b* always presses upon the top of the sliding-valve, and as the space beneath it is exhausted, it is forcibly pressed upon the flat surface. The sliding-valve is moved backwards and forwards horizontally within the box *b*, by an internal rack and sector, in the same manner as the valves of Mr. Watt's engine are lifted (see p. 523): viz. a toothed rack is formed on the top

---

(*a*) The construction of the valve is nearly the same as that of the engine represented in Plate XVIII, p. 692, except that the valve slides up and down vertically, instead of sliding horizontally.

4 T

of the sliding-valve, and a sector, with corresponding teeth, is fixed upon a horizontal spindle, which passes across the steam-box *b*, and comes through a socket to
the outside, where a handle *v* is fixed on, to give motion to it, and to the internal
sector, so as to slide the valve within the box.

The handle *v* is moved by hand, in order to start or stop the engine; but
when it is regularly at work, the handle is moved by a rod *u*, having a notch in it
near one end, to fit over a pin which projects out sideways from the handle *v*.
The rod *u* extends to the axis of the fly-wheel, and the extreme end is guided
through a fixed socket *t*; near this end a square opening like a stirrup is formed
in the rod *u*, and a revolving tumbler *r* acts within that opening, so as to move
the rod backwards and forwards alternately, at the proper intervals to produce
the requisite distribution of the steam.

The revolving tumbler *r* is fixed upon a small axis, which is situated in the
line of the main axis P of the crank and fly-wheel; that axis rests in bearings fixed
on a frame which is connected with the bearing for the neck of the axis P of the
crank N; and the frame is sustained by a bracket U from the end of the cistern G.

The small axis is turned round at the same rate with the main axis P, by
means of an arm or second crank, which is fixed on the extreme end of the small
axis, opposite to the great crank N, and the end of the main crank pin, which projects
out through the connecting-rod M, enters into a hole in the end of the arm, so as
to turn it round, and thus give a constant motion to the tumbler *r*, which revolves
within the opening in the rod *u*, and acts alternately between the two upright sides
of that opening, to push the rod, with a sudden motion, first one way and then the
other, through a sufficient space, to give the requisite motion to the sliding-valve
within the steam-box *b*. After having so moved the rod either way, the tumbler
retains it at rest, during the remainder of that half stroke of the piston, and then
the tumbler moves the rod suddenly back again, in order to cause the opposite
half stroke of the piston.

To produce such an intermitting motion, the form of the tumbler is an
irregular spherical triangle, something like the sector of a circle; its outline
consists of two circular arcs, one of a larger radius than the other, but both arcs
are concentric with the axis on which the tumbler is fixed, and are opposite to each
other across the centre of that axis; these circular arcs are joined together by two
curved boundary lines, which diverge from each other at an angle. The upright
sides of the opening in the rod *u* form tangents to the two concentric circular arcs
respectively, and the space between those uprights is so adapted, that they will
always be in contact with the two concentric arcs; the chord of that arch which
has the greatest radius, is also equal to the space between those upright sides.
The tumbler being of this form, the rod *u* is never left at liberty, for the tumbler
retains the rod at rest, in the intervals between the motions which it gives to it.

The rod *u* terminates with a handle at the end nearest the cylinder, that it may
be lifted up, in order to disengage it from the pin in the handle *v*, which then becomes
detached, and the handle can be moved backwards or forwards by hand, to stop or
start the engine. When the sliding-valve is allowed to rest in the middle of its
course, its edges, or the border of the inverted dish, will exactly cover over, and
close, both those outside passages which lead to the top and bottom of the cylinder, so
as to prevent any entrance of steam from the boiler into either of them, and then
the engine will remain at rest. The cover of the steam-box *b* can be removed to
give access to the sliding-valve within it when necessary.

This kind of steam-engine, with the great lever beneath the cistern, was not
found so convenient in practice, as those with the lever above, according to the plan

previously adopted by Messrs. Boulton, Watt, and Co. (see p. 682), because it is not so readily accessible to keep the joints in order; the plan of the engine in Plate XVI. was therefore laid aside, after a small number had been made.

DESCRIPTION OF PLATE XVIII, AN INDEPENDENT ROTATIVE STEAM-ENGINE OF SIX HORSE POWER, CONSTRUCTED BY MESSRS. FENTON, MURRAY, AND WOOD, 1810.

The whole engine is erected upon a large flat plate of cast iron X X, the form of which is sufficiently shown in the plan; it is firmly bedded upon four foundation walls of masonry, upon which it is fastened by ten bolts, which have heads on the upper ends, countersunk flush into the upper surface of the plate; their stems pass down through openings left in the thickness of the walls, and the lower ends of the bolts are bound down by keys driven tight through holes across them. Square openings are left out in the walls, to give access to these keys, as is shown at $y\,y$ in the elevation; and cast iron plates are applied around the bolts in the upper part of these openings, for the keys to bear against when they are driven tight.

The space between the four foundation walls, leaves room for the condensing cistern G G, and the cold water pump R; the foundation plate X has three large openings through it, to render it lighter, and also to give entrance into the space beneath. The condensing cistern is made of cast iron, and is suspended to the underside of the foundation plate X, that its weight may assist to keep that plate steady. The cylinder E is mounted upon a square pedestal at one end of the foundation plate, and is fastened to that plate, by four holding-down bolts; one applied at each angle of the square.

The fulcrum of the great lever is sustained by four pillars V V, placed in an inclined direction; the lower ends of the pillars are inserted into corresponding sockets formed in the foundation plate X X, and are fastened in by iron bolts, with cross keys driven very tight through them; their upper ends are inserted and fastened into sockets in a strong plate W W situated across beneath the great lever L L, and on each end of this plate a bearing p is fixed, to sustain the pivots of the axis of the great lever. The four pillars W incline towards each other at the upper ends, so as to form a sort of pyramidical frame, which was intended to be strong enough to stand firm from its own base X, without any bracing or spring beams. The axis which forms the centre of motion for the bridle rods 6 of the parallel motion, is supported by a small pillar U, erected upon a bracket, which projects out from the upper flange of the cylinder.

The piston J, and its rod n, with the parallel motion K, the great lever L L, and connecting-rod M, the crank N, and fly-wheel Q Q, are the same as in Mr. Watt's engine without the sun and planet-wheels, or multiplying-wheels and pinion (see p. 499); but the great lever L L is made of cast iron, deep at the centre, and tapering towards each end, as is shown in the elevation. The rod l of the air-pump H, and the rod 7 of the cold water pump R, are suspended from pins which project out sideways from the great lever L, without any parallel motion.

The bearing for the neck of the axis P of the crank, is formed with two inclined legs or pillars, spreading out at bottom, with broad feet, which are firmly fixed to the foundation plate by four bolts, as is seen in the plan. The fly-wheel Q Q, is mounted upon the same axis P as the crank, and the lower part of its rim descends into a cavity which is formed for it, in the thickness of one of the foundation walls. The other end of the axis of the fly-wheel is supported by the wall of the apartment in which the engine is placed; and the machinery that the engine is required to turn, is connected with the extreme end of the axis P, where it projects through that wall.

The condensing apparatus is nearly the same as in Mr. Watt's engine, excepting the form of the condenser, which is not a cylinder, but it is a flat box placed edgeways upwards (see F in the elevation); in the other direction it is of a considerable breadth; this form exposes a greater surface to the cold water, and leaves more room within the cistern, to gain access to the foot-valve at k, in the passage from the bottom of the condenser, to the bottom of the air-pump. The hot well I is a round cistern fixed on the top of the air-pump H; and the discharge-valves (see $m\,m$ in the plan) are two butterfly-valves at the bottom of that cistern, fitted over openings through the cover of the pump. The surplus hot water overflows from the hot well, by a spout which is not shown in the drawing; nor is the hot water pump represented; it draws hot water from the hot well by a suction-pipe, and forces it by another pipe, up to the feeding cistern over the boiler. x is the blow-valve, to discharge the air from the condenser. j is the stem and handle by which the injection cock is opened or shut.

The distribution of the steam to and from the cylinder, is effected by a cup sliding-valve of the same kind as that of the engine last described, but the valve is applied in front of the cylinder at b, and slides up and down vertically, instead of horizontally; its construction is described by the separate section of the cylinder, and by the three detached figures at the side of the plate. a is the steam-pipe, which brings the steam from the boiler; and z is the throttle-valve to regulate its passage into the steam-box b in front of the cylinder, within which the sliding-valve h is situated.

c and f is a square upright pipe situated in front of the cylinder; the upper end is connected with the branch which leads into the top of the cylinder, and the lower end with the branch leading into the bottom of the cylinder. The middle part of the pipe $c\,f$ is an interrupted passage, or rather each of the passages c and f terminates by turning outwards from the cylinder into the steam-box b, which is fastened against the vertical surface, formed by the front of the upright pipe; this surface forms the bottom or back of the steam-box, and it is made very flat and smooth, for the sliding-valve h to fit against it. The cover in front of the steam-box b can be taken off, to get at the valve within it.

The orifices of the passages c and f which communicate with the top and bottom of the cylinder, are of an oblong form, where they come out in the flat vertical surface against which the sliding-valve h acts; and there is a third orifice in that surface, which is situated between the other two, and communicates with a passage g, extending horizontally across behind the upright pipe $c\,f$, and joining at each side, to the upper ends of two upright eduction-pipes $g\,g$, which descend to the condenser F at each side, as is shown in the elevation, and in the plan. The two eduction-pipes g act as one, and form a constant open communication between the condenser, and the middle of the three oblong apertures in the vertical flat surface against which the sliding-valve acts.

The valve itself is shaped like a square dish or pan, with a broad margin which is accurately fitted against the vertical flat surface. The space within the valve is equal to two of the three orifices in that surface, together with the space between them, so that the valve is capable of covering two of the apertures at once, in order to connect them together. The middle aperture which leads to the condenser, is in all cases one of the two apertures which are covered by the valve, and whenever two of the apertures are thus connected, the third aperture will be left uncovered by the valve, and will admit steam from the box.

This construction has been already described (p. 689), and it is made quite clear by the three separate figures. In the upper small section, the valve is placed in the middle position, and the edges of its border cover and stop the orifices of both the passages *c* and *f* which lead to the top, and to the bottom, of the cylinder; in this situation no steam can gain admission from the steam-box *b* into the cylinder, and the engine must remain motionless.

If the valve is slided upwards in the box, into the position shown by the middle figure, then the interior space within the valve establishes a free communication between the orifices of the eduction-pipe *g*, and of the passage *c*, which leads to the top of the cylinder, so as to exhaust the steam from above the piston; for after passing down from the top of the cylinder through *c*, the steam comes forwards, out at the orifice in the flat surface, into the hollow of the valve, and thence returns backwards again through the middle orifice into the eduction-pipe *g*, and to the condenser. The same motion of the valve, raises the lower edge of its border above the lower orifice of the passage *f* which leads to the bottom of the cylinder, so as to uncover that orifice, and admit the steam from the boiler, to act beneath the piston, and force it upwards.

When the piston arrives near the top of its course, the valve is suddenly slided down, into the position shown in the lower figure, and then the course of the communication is reversed; for the valve connects the passage to the eduction-pipe *g*, with the lower passage *f* from the bottom of the cylinder, so as to exhaust the steam from beneath the piston; whilst the upper passage *c* to the top of the cylinder, is uncovered by the valve, and admits steam from the box *b*, to act above the piston, and force it down.

The sliding-valve *h* is moved up and down within the steam-box *b*, by means of a rack upon the valve, and a toothed sector *d* within the box, mounted upon an horizontal axis, which comes through to the outside; and a handle lever *s* is fixed upon the end of it, to move the valve by. The lever *s* is connected by a rod *r*, with a lever fixed upon a horizontal axis *t t*, which is supported in brackets standing up from the foundation plate X; another small lever *v* is fixed near the end of the axis *t*, and it is connected by the compound rod *u u* with the excentric ring *w*, which encompasses an excentric circle fixed fast upon the main axis P of the crank.

The excentric, which is seen in the elevation, is a circular wheel with a groove round its edge; it has a circular hole through it, which is considerably out of its centre, and by this hole it is fixed fast upon the axis P, so as to turn round with it. The excentric ring *w* is made in two semicircles, which are put together in the groove round the edge of the excentric circle, and are united by screws, as shown in the elevation, so as to form a complete circular ring, which fits truly upon the excentric *w*, but allows that excentric to turn round freely within it. The two rods *u u* are connected with the concentric ring *w* by pins projecting at top and at bottom. When the axis P of the crank is turned round, the excentric turns with it, and moves the ring *w* and the rod *u u* backwards and forwards at each revolution, through a space equal to twice the excentricity, or distance from the centre of the axis P to that of the excentric circle.

The progress of this alternate motion is the same as would be produced by a crank pin, if it were fixed to the extreme end of the axis P at that distance from the centre, and projected out, so as to be jointed to the connecting rod *u*, in the same manner as the pin of the great crank N is applied to the connecting rod M; consequently the motion given to the rod *u u* is slow, when it arrives at either end of its alternate course, and quicker at the middle parts of that course. (See p. 416.)

This slow action would be unfavourable for the motion of the sliding-valve, for it ought to be moved suddenly; to effect this, the hole for the joint pin by which the end of the rod *u u* is united to the lever *v* of the axis *t t*, should be an oblong slit of some length, to allow a considerable play in the joint, so that when the excentric *w* first begins to move the rod *u u*, with its slow and gradual motion, that motion will not be transmitted to the sliding-valve, because the oblong hole will draw over the joint pin, without moving it; but by the time that the rod *u u* has moved through about one-fourth or one-third of its course, and acquired a more rapid motion, the end of the slit will reach the pin, and will then begin to move it, together with the levers and rods *v*, *t*, *r*, and *s*, so as to move the sliding-valve quickly up or down within the steam-box *b* (*a*).

The sliding-valve requires to be moved down into the position shown in the lower figure, when the piston nearly reaches the top of its course; and it must be moved up into the position of the middle figure, when the piston arrives near the bottom of its course. The excentric is fixed on the axis of the crank, in a proper position to produce that effect, when the fly-wheel turns round in the direction of the arrows shown in the elevation; but the engine cannot turn the fly-wheel in the opposite direction. In that respect, an engine which opens its valves by the rotative motion, differs from Mr. Watt's original engine with hand gear, which will turn the fly-wheel either way round, according as it is started. (See p. 464.)

The excentric, and the levers which communicate its motion to the sliding-valve, must be so adjusted, as to have moved it half way, and closed both the passages, before the crank comes into a line with the connecting rod; in order that when the piston arrives at the end of its course, the valve may be already opened in some degree. By this means the evacuation of the steam will take place during the period whilst the piston is moving slowly, at the change of its course; the reasons for this have been already explained at p. 469.

The framing of the engine in Plate XVIII, is found to require horizontal braces, or spring beams, to extend from the top of the column U, to the plate W W, beneath the centre of the great lever, in order to render that column more firm; because the bridle rods 6 of the parallel motion exert a considerable thrust to move their centre of motion, and if its supports yield sensibly, the piston rod will deviate from its vertical motion, so as to cause all the parts to wear rapidly. The rod *l* of the air-pump does not wear well, without a parallel motion.

The box sliding-valve being always forced up towards the surface against which it slides, by the pressure of the steam in the steam-box *b*, it is caused to move with great friction, and the surfaces wear each other away rapidly; another disadvantage is, that the passages *c* and *f*, extending from the top and bottom of the cylinder to the valve, must be filled with steam, and exhausted again, at every stroke, without contributing to the power of the engine; the capacity of these passages is a needless addition to the vacant spaces at the top and bottom of the

---

(*a*) The above method of opening and shutting the valves by means of an excentric circle fixed upon the axis of the crank, was introduced by Messrs. Boulton, Watt, and Co.; and also the expedient of allowing the connecting rod from the ring that surrounds the excentric, to move through a part of its course, before it communicates any of its motion to the valve; in order that the slow progressive motion with which the connecting rod begins its motion may be expended, and that it may communicate its quick motion to the valve so as to move it suddenly. This they adopted in their large rotative engines with four valves, which will be described in the next article (see p. 697).

cylinder, which ought to be rendered as small as possible.   Mr. Murdock's sliding eduction-pipe, being free from all these inconveniences, has obtained the preference, and in its improved form it is now used in almost all independent engines.   The plan of the engine in Plate XVIII was given up after a few years, and Messrs. Boulton, Watt, and Co.'s model for independent engines has been followed very generally by most other engine makers.

DESCRIPTION OF PLATES XIX AND XX, MR. WATT'S ROTATIVE ENGINE OF THIRTY-SIX HORSE-POWER, BY MESSRS. BOULTON, WATT, AND CO., 1808.

This is the construction which Mr. Murdock adopted for large house-built steam-engines for manufactories, during the first ten years that he directed the manufactory at Soho, after Mr. Watt retired from the business.   The action of the steam in this engine is precisely the same as in Mr. Watt's original rotative engine (see p. 459); the chief differences in the construction are in the nossels, and in the working gear for opening the valves.   The steam-valves are made with tubular spindles, to admit the spindles of the exhausting-valves through them, on the plan already described, p. 684; and the valves are opened and shut by the motion derived from an excentric circle fixed upon the axis of the crank.   The foundations are entirely of stone without any timber; and all the other parts, except the spring beams, are of metal.   The great lever is made of cast iron, and its centre of motion is supported by a cast iron column.   The rotative motion is obtained by a simple crank, in lieu of sun and planet-wheels, and a multiplying-wheel is fixed on the axis of the crank, to turn a pinion on the axis of the fly-wheel, and thus give it an increased velocity.

The boiler C, Plate XIX, its furnace A, and flues 9, 9 and 16, and its chimney D, also the sliding damper w in the flue, are on the same construction as before described, p. 445; but this boiler has an internal flue, 16, passing through it from one end to the other, to convey the flame and heated air through the mass of the water, in its passage from beneath the end of the boiler till it enters into the external flue 9 9, which passes all round the lower part of the boiler.   The boiler is $5\frac{1}{2}$ feet diameter, and $18\frac{3}{4}$ feet long; the surface it exposes to the heat is 393 square feet $= 10.9$ square feet to each horse-power.   The fire grate is $4\frac{1}{2}$ feet wide by 6 feet long $= 27$ square feet.   The other dimensions are nearly the same as the boiler for the 30 horse engine, p. 585, note *b*.   & is the man-hole to the boiler. v is the waste-pipe to convey the steam which is discharged from the safety-valve, into the chimney D.   T is the feeding cistern to supply the boiler with water as it evaporates.   The feeding-valve in the bottom of this cistern is the same as described at p. 453; also the lever, with the wire and balanced stone which floats upon the surface of the water in the boiler, and opens the valve whenever the water subsides in the boiler.

Self-regulating damper.   In the present engine the height of the water in the upright feeding pipe T is made to regulate the sliding damper w in the flue, so as to give more or less draft to the fire, according to the strength of the steam in the boiler.   For this purpose the pipe T is made sufficiently large to receive a hollow iron box or float, which can move freely up and down in the pipe; this float is suspended by one end of a chain, which passes up through an open tube in the centre of the feeding cistern, the top of the tube being so nearly on a level with the top of the cistern, that the water in the cistern cannot run down by the side of the chain, which is conducted over two pullies, and the sliding damper w is suspended from the other end of it ; the weight of the damper balances the weight of the float in the pipe T, when one half of its bulk is immersed in the water in the pipe, and then neither the damper nor the float will preponderate.

The height to which the water will rise in the feeding pipe T, depends upon the elasticity of the steam in the boiler ; because the steam presses upon the surface of the water in the boiler, whilst the atmospheric air presses upon the surface of the water in the feeding pipe, and consequently, inasmuch as the steam in the boiler is more elastic than the external air, so the surface of the water in the upright pipe T, will stand on a higher lever than the surface of the water in the boiler ; and whenever the elasticity of the steam increases, the water will be forced higher up in the pipe, and by rising about the float in that pipe, will immerse a greater portion of it, so as to bear up more of its weight ; the weight of the damper will then preponderate, and cause it to descend, in order to diminish the passage from the flue into the chimney, and impair the draft of the fire, whereby the production of steam is diminished.   On the other end, whenever the elasticity of the steam in the boiler diminishes, the water in the pipe T will subside, and the float will follow it, so as to raise up the damper, and increase the draft, in order to cause the fire to burn more rapidly, and produce more steam. This self-regulating damper was introduced about the time that Mr. Watt retired from business ; it is a very useful contrivance, and when well adjusted, it will regulate the draft according to the elasticity of the steam, so as to preserve a tolerably uniform degree of elasticity, during the intervals between the feeding of the fire with fresh coals.

The engine has two boilers placed side by side ; the steam-pipe a passes over both boilers, and an upright pipe ascends from each boiler, to join to the steam-pipe ; a stop-valve is fitted into the upper part of each upright pipe, to shut off the communication from either boiler at pleasure ; only one boiler is used at once, the other being left cool, in order to clean it.   The steam-pipe rises gradually from the boilers towards the cylinder, that the water which is condensed in it, may run back to the boilers.

The construction of the nossels and the four valves. This is explained in figures 3 and 4 of Plate XX : a is the steam-pipe ; z is the throttle-valve placed in the joint which unites the steam-pipe a to one of the branches of the upper steam-box b ; $d\,d$ is the upright steam-pipe, to convey the steam to the lower steam-box e. $g\,g$ is the upright eduction-pipe, joined at top to the upper exhausting-box h, and also to the lower exhausting-box i by a side branch ; the lower end of the pipe g joins to the top of the condenser F ; fig. 4 shows the several flanges by which the different pieces are bolted together, and each upright pipe d and g has a socket joint near the lower end, to allow the expansion and contraction of the metal to take place in heating and cooling.

The four valves $b\,h$ and $e\,i$ are fitted into brass seats fixed within the valve-boxes, as shown in fig. 3, so as to form partitions, which divide each box into three compartments, and the middle compartment c and f of each box is connected with the branches which lead into the top and bottom of the cylinder E.   The spindles of the two exhausting-valves h and i, pass up through the spindles of the two steam-valves b and e, which are for that purpose made hollow like tubes, and they pass through stuffing boxes in the centres of the covers of the two steam-boxes b and e, the junctions being made tight by hemp confined in the stuffing boxes round those tubes ; there are also similar stuffing boxes at the upper ends of the tubular spindles of the steam-valves, to confine hemp around the central spindles of the exhausting-valves. By these stuffings the escape of steam, or the entrance of air is prevented, and the valves can be raised by their spindles on the

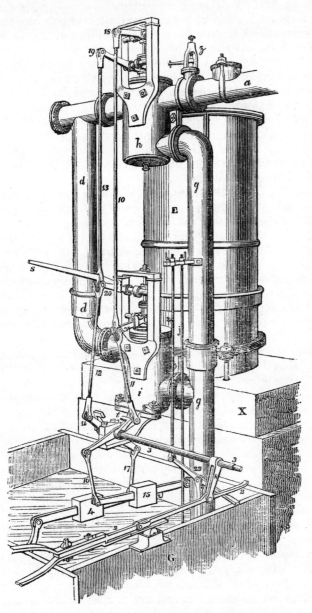

outside of the boxes, in order to open the passages through the valve seats; the two valves of each pair can be opened or shut quite independently of each other.

The valves are raised by suitable short levers fixed upon horizontal axes, which are mounted in brackets fixed to the front of each box in which the valves are contained, and rising up above the cover of that box. The spindle of each of the exhausting-valves has a square opening through it near its upper end, to receive the extremity of the lever by which the valve is to be raised. And the hollow spindle of each steam-valve is included between the extremities of two levers, which are affixed upon the same axis parallel to each other, and those extremities enter into a circular groove formed round the outside of the stuffing box at the top of the hollow spindle, so as to enable those levers to lift up the stuffing box and the valve. The upper ends of the spindles of the exhausting-valves are received in fixed guides which retain them in their vertical position, and the lower ends of the same spindles are retained in sockets in the centres of the valve seats. The spindles being thus guided at their upper and lower ends, the valves will rise and fall very correctly into their seats. The steam-valves are guided by the same spindles which pass through them, and also by the stuffing boxes through which their hollow spindles pass in the centres of the covers. The utmost accuracy is required in the execution of these nossels to make the seats for the two valves exactly concentric with each other, and also with the upper edges of the boxes where the covers are fitted on, and with their stuffing boxes.

The hand gear. The horizontal axes of the levers by which the valves are

4 U

lifted, have other levers 18, 19, and 20, 21, fixed upon them, and projecting out in an opposite direction in front ; the ends of these levers are connected by the upright rods 10 and 13, so as to unite the levers for the upper and lower valves r, s, in two pairs.   The levers 20 and 21 can be prolonged, as shown by the dotted lines in fig. 3, by inserting the ends of bars r and s into sockets formed on the top of each lever ; these bars serve as lever handles, and by depressing them the engine-man can open either pair of valves, in order to start or stop the engine.   The rod 10 connects the end of the lever 20 for the lower exhausting-valve i, with the lever 19, of the upper steam-valve b ; and when the handle r is depressed (as in fig. 3), it opens both these valves in order to produce the descending stroke of the piston. The rod 13 unites the lever 21 of the lower steam-valve e, with the lever 18 of the upper exhausting-valve h ; and when the handle s is depressed, it opens both these valves, to make the descending stroke of the piston.

Weights to close the valves.   The valves tend to shut by their own weight, but that would not be quite sufficient ; therefore two rods 11, 12, are jointed to the ends of the levers 20, 21, to connect them with the upper ends t, u, of two in-clined levers t 16, and u 17, which are poised upon a cylindrical part of a horizontal axis 3, 3, and can turn freely round upon that axis, as a centre of motion ; these levers are inclined in opposite directions, and cross each other like the two blades of a pair of scissors.   The opposite or lower ends of the inclined levers are con-nected by links 16, 17, with balance treadles 15, 4, which are levers loaded with weights, tending to draw down the lower ends of the inclined levers, so as to cause their upper ends t, u, to bear upwards ; this action is transmitted by the rods 11, 12, and 10, 13, to the outer ends of the several levers of the valves, in order to press down the opposite ends of these levers, which act upon the valves, and thus cause the valves to shut down close into their seats, whenever they are allowed to descend.   In opening either pair of the valves, whether it is done by a man with the handles r or s, or by the working-gear of the engine itself, the weight 4 or 15 belonging to that pair of valves must be overcome and raised ; and the effort made by the weight to return, tends to close that pair of valves.

The working gear is moved by an excentric circle 1, which is fixed upon the main axis P of the crank ; it is a circular ring, with a groove formed round the edge like a pulley ; and it has a round hole through it, the centre of which is about six inches out of the centre of the circle.   This excentric is fixed fast upon the square axis P by four wedge pieces, which are applied upon the four flat sides of the square axis, and the outsides of the pieces are formed convex, so as to fill up the round hole through the excentric ; these pieces are driven tight into the excentric, so as to jamb it fast upon the axis, that it cannot slip round upon it.

A circular brass ring is applied in the groove round the edge of the excentric ; it is made in two semicircles put together, and firmly united by nuts screwed upon bolts, which pass through ears projecting out from each semicircle ; these bolts are the extreme ends of the two branches of the compound-rod 2, 2, which conveys the motion of the excentric ring to the working gear ; the two branches are joined together into one rod at the end nearest to the cylinder, but they diverge from each other, so as to include the excentric ring between the screws which are formed upon the other ends of them ; one of those screws being over the top of the ring, and the other beneath the bottom of it.

When the axis P and the excentric turns round, it will move the rod 2, 2, back-wards and forwards through a space of about 12 inches at each revolution.   The ex-centric projects out from the axis P on the same side as the crank N projects out therefrom, consequently when the crank pin, in moving round in its circle, arrives nearly on a level with the axis, and the piston is near the middle of its stroke, the rod 2 will have been moved by the excentric, to the full limit of its course, and will be setting out on its return. When the crank is nearly in a line with the connecting-rod, and the piston is near the top or bottom of its course, the rod 2 will have arrived at the middle of its course.

The end of the compound-rod 2, 2, where the two branches are joined into one rod, rests upon a small roller which is mounted in a box fixed near the edge of the condensing cistern; this roller bears the weight of that end of the rod, and allows it to move backwards and forwards; the end of the rod is prolonged beyond the bearing on the roller, by another short rod which is jointed to it, and which acts on a pin at the lower end of a lever 8, fixed upon the horizontal axis 3, 3, near to one end of it. By this means, at every complete stroke of the piston, the axis 3, 3, is turned backwards and forwards about its own centre line; it is supported in two bearings, one at the extreme end fixed to the side wall, and the other is supported by a bracket, which projects out from beneath the lower box i of the nossels, and is fastened by three bolts, to ears which project out from the lower part of that box to receive the screws. The inclined levers t 16 and u 17 are fitted freely upon a cylindrical part of the axis 3, close to each side of the latter bearing.

To open the valves, two short levers 22, 23 are fixed fast upon the axis 3, close at the sides of the loose inclined levers t 16 and u 17, the ends of the short levers are like hooks or claws, which turn sideways over the upper ends t and u of the loose levers, in such manner that when, by the reciprocating motion of the axis 3, either of the claws 22 or 23 is depressed below a certain position, the hooked part of that claw will press upon that lever t or u which belongs to it, and will force that lever down, so as to lift up and open one pair of valves by the connexion of the rods and levers 11, 20, and 10, 19, or 12, 21, and 13, 18. That pair of valves which is thus opened will be kept open, until the claw 22 or 23 returns, and lets them down again by their own weight, aided by that of the loaded lever 4 or 15; but when the valves are shut, the hook at the end of the claw 22 or 23 rises up, and quits the lever t or u, which then remains stationary without following the claw.

The working gear is so adjusted, that the instant after one of the claws 22 or 23 has allowed one pair of valves to close themselves, and is beginning to quit its inclined lever t or u, the opposite claw 22 or 23 will come in contact with the other inclined lever, and will begin to open the other pair of valves; for the two levers t and u being inclined in opposite directions, they are acted upon alternately by the two claws 22 and 23; one pair of valves being opened when they move backwards, and the other pair when they move forwards; but one pair of valves is always closed before the other is opened.

When the engine is properly at work the axis 3, 3 is moved continually backwards and forwards by the connexion of its lever 8 with the rod 2 of the excentric; but by lifting up the end piece of that rod, it can be disengaged from the joint pin at the end of the lever 8, because the connexion is only made by a notch in the under side of the rod, which fits over the joint pin of the lever 8, so as to cause the lever, to partake of the motion which is given to the rod, by the ring of the excentric, so long as the end rod rests by its weight upon that joint pin, but if the end rod is raised up, its notch will quit the joint pin of the lever 8, and leave that lever at liberty; in which case the weights 4 or 15 will instantly close that pair of valves, which may be open, unless they are held open by pressing down the handles r and s by the hand.

The end piece of the rod 2 can be lifted up by a lever 23, and screw 24 (see fig. 4, and the perspective sketch); this lever moves about a fixed centre at the upper end of an iron standard, which rises up from the top of the condenser; one end of the lever 23 extends beneath the extremity of the end piece of the rod 2, and the other end is jointed to a vertical rod 24, the upper end of which is a screw, and is received in a nut fitted into a small bracket, which projects out from the side-pipe *g*.   The nut has a handle by which it can be turned round, in order to depress the screw, and raise the other end of the lever, so as to lift up the extremity of the rod which is jointed to the end of the rod 2, and thus disengage that rod from the joint pin of the lever 8.   The lever 8 is made double, to receive the rod in the interval between its two sides, and when the extremity is borne up by the lever 23, the rod can move freely backwards and forwards within the double lever 8, without communicating any motion to it.

The cylinder, with its piston, cover, and bottom, and its steam-case (see fig. 3), are the same as in Mr. Watt's engines (see p. 380).   *The steam-case* is put together round the cylinder in six pieces, joined together by small screw-bolts; there are three segments to form the circle, and two sets of such segments in height; one set is fastened by a flange at top, to the underside of the flange round the top of the cylinder, and the other set is fastened in a similar manner to the lower flange of the cylinder.   The flange of the bottom of the cylinder, and that of the base or false bottom beneath it, which is to contain steam to keep up the heat, are all united by the same bolts which fasten the steam-case by passing through all the four flanges.   The large bolts which fasten the cylinder cover, pass through the upper flange of the steam-case, and as many smaller bolts are inserted in the spaces between the other large bolts, they pass through the flanges of the steam-case and of the cylinder, their heads being countersunk into the upper surface of the flange of the cylinder, and the nuts beneath.   The lower set of segments of the steam-case are enlarged at their upper edges, to form a circular socket for the reception of the lower edges of the upper set of segments, and the joint is made tight by ramming in hemp with tallow; this joint allows the steam-case to accommodate itself to the effect of the expansion or contraction of the cylinder by the heat.

The cover of the cylinder is hollow within, to prevent the transmission of heat.   In larger engines the hollow space is filled with steam, by a small pipe of communication from the steam-case.   The false bottom beneath the cylinder is supplied from the steam-case in the same manner.   The steam-case is supplied with steam by a small branch from the steam-pipe *a*, with a regulating cock; and there is also a syphon pipe, proceeding from the lowest part of the steam-case, to drain away the water which condenses therein (see p. 381, and Plate XXIV).

The cylinder is $30\frac{7}{8}$ inches diameter; the piston makes a 6 feet stroke, and it works at the rate of 19 strokes per minute, according to Mr. Watt's standard (see table, p. 574).

The condensing apparatus within the cistern G is the same as in all Mr. Watt's engines (see p. 374).   The lower end of the eduction-pipe *g* is joined by a socket joint to the top of the condenser F.   The foot-valve is at *k*, in the passage between the bottom of the condenser and the bottom of the air-pump H; and the discharge-valve *m*, from the top of the air-pump, is within the hot well I. The blow-valve to discharge the air from the condenser is shown at *x* in the plan; it is fitted at the end of a small pipe, which passes through the side of the condensing cistern G, and the blow-valve is covered with water contained in a small cistern, fixed on the outside of the large cistern G (see p. 455).   The condensing cistern is made of wood; but, in most of these engines, when the cistern has become decayed, it has been renewed by a cast iron cistern.

The injection is admitted into the condenser through a small sluice-valve, the opening of which is regulated by turning the nut of a screw formed on the upper end of an upright rod j, which rises up from that valve (see the sketch, p. 697). The nut has a handle to turn it round, and is supported in the same bracket as the nut of the screw 24 for disengaging the working gear. The sluice-valve is a square brass plate, the flat surface of which is fitted against a corresponding plate fastened to the outside of the condenser, and having an oblong aperture through it, to admit the water whenever the sluice-valve is drawn up so high by its screw j, as to uncover a part of the hole; if the sluice is screwed quite down, it covers and stops the hole entirely. The sluice-valve is retained in its place against the fixed perforated plate by ledges, which are fastened to the latter at each side, so as to form a groove which includes the edges of the valve. The aperture is $1\frac{1}{4}$ inches wide, by 3 inches high; and when the engine is working properly, the sluice is commonly drawn up, so as to open $1\frac{1}{2}$ inches high, making an aperture of $1\frac{7}{8}$ square inches for 36 horse-power, which is nearly according to the rule, p. 594.

The foundations upon which the engine is fixed are built of solid masonry; a thick wall X extends across the house, to form the pier for the cylinder; another wall extends nearly across the house, beneath the centre of the great lever; and from this a wall proceeds to the end of the house, to sustain the bearing for the neck of the main axis of the crank. A horizontal passage, called the tunnel, is left beneath these foundations to give access to the lower ends of the foundation bolts, by which the different parts are held down. The tunnel extends beneath all the length of the building, as is shown in Plate XIX, where the scale of feet is placed; and it has cross branches to pass beneath the bearings for the main axes. The passage is large enough for a man to pass conveniently through it.

The top of the tunnel consists of very large flat stones, laid across the space between two low parallel walls which form the sides of the tunnel; these walls are built upon a firm foundation of two or three courses of broad flat stones; and if the earth is not very hard, similar stones are laid beneath the whole area of the building. The several walls for the engine are built upon the large stones over the tunnel, and vertical holes are left in the thickness of these walls, as shown by the dotted lines, to receive the different foundation bolts, which are about $1\frac{1}{2}$ inch square, with eyes through their lower ends, to receive cross keys, which are put through in the tunnel to prevent the bolts drawing up; at the upper ends of the bolts are screws to receive square nuts, by which the parts of the engine are firmly bound down upon the walls. The holes are considerably larger than the bolts, and large cast iron plates, or washers, are applied upon the bolts beneath the stone work at the top of the tunnel, to form the bearings for the cross keys, which are inserted through the eyes of the bolts beneath these washers.

There are 14 foundation bolts in all, viz. three bolts to hold down the cylinder; two which pass through the bottom of the condensing cistern, to fasten the air-pump; one bolt for the column V, which sustains the fulcrum of the great lever; two for each bearing at each end of the main axis P P of the crank; and two for each bearing at each end of the axis of 26, 26 of the fly-wheel. The upper course of the stone work X beneath the cylinder is made very flat and level to receive the basis of the cylinder, and the screws on the upper ends of the foundation bolts are inserted through holes in three ears which project out from those four flanges at the lower part of the cylinder, by which it is united to the bottom and to the false bottom beneath it, and to the lower flange of the steam-case; the foundation screws pass through the ears of all these four flanges, and the nuts being screwed down upon the top of them, they gain very secure hold of the cylinder.

The square base of the column V is made hollow, like a small box, which is left open at two opposite sides, to introduce a nut upon the screw at the upper end of the foundation bolt, as is shown by the dotted lines in Plate XX, and sufficient room is left within the box, to admit a wrench to turn the nut round, in order to screw it tight and fasten the column down on the stone work V.

The bearings for the axis of the crank require to be firmly secured; for this purpose a cast iron plate is let into the stone work at the top of the wall, and secured by running lead round it, so that it cannot move sideways; the plummer block or bearing for the axis is placed upon this plate, which has two projecting ledges rising up from it at each end, to form a cell for the reception of the lower part of the bearing, and it is held down upon the plate by a foundation bolt at each end. The bearing is not so long as the cell in which it is contained, and an iron wedge key is driven in at each end, to fill up the space, and confine the bearing from moving in the cell; by changing these wedge keys for others of a different breadth, the position of the bearing can be altered as much as is requisite to adjust its position, so that the teeth of the multiplying-wheel and pinion may gear, or work properly together. The bearing for the extreme end of the axis of the crank, and those for each end of the axis 26 of the fly-wheel, are supported on the side walls of the engine-house, and they are fastened in the manner above described.

The upper end of the column V is keyed fast into a socket formed for its reception in the under side of a strong cast iron beam W, which is called the entablature plate; it extends horizontally across the house, and its ends are secured by letting them into large stones, built in each of the side walls at the proper places; the form of the entablature plate is that of a flat horizontal plate, with two parallel vertical plates proceeding from it at the under side, so that it resembles an inverted trough, which is all cast in one piece. The two spring beams U U, which are of wood, are laid across the entablature plate W, and extend all the length of the engine-house, their extremities being fastened into the end walls. The bearings for the axis of the great lever are placed upon the spring beams, and fastened by bolts, four of which go down through the projecting edges of the entablature plate W, and bind the spring beams down upon it.

The parallel motion is proportioned as stated at p. 599, the air-pump rod l is suspended from the middle of the back links 7, which are jointed to the great lever L, at exactly half way between the centre of motion and the joint by which the main links K for the piston rod are suspended; consequently, the bucket of the air-pump makes a stroke half as long as that of the piston. The bridle rods 6 are the same length as the parallel rods 5; and the fixed centre of motion for the bridle rods are two short studs, projecting out from brackets which are fixed beneath the spring beams U. The centre line of these studs intersects the centre line of the piston rod n. The pin which forms the common joint to unite the ends of the parallel rods 5, the bridle rods 6, and of the back links 7, extends across from one side to the other, and has a large opening through the middle of it, to admit the air-pump rod l to pass through without touching it.

The great lever L L is one piece of cast iron, its form is that of a flat plate placed edgeway upwards, and nearly three times as deep in the middle as at the ends; it is strengthened by a projecting border all round it on each side, and also by a projection along the middle of it at each side. A suitable socket is formed at the centre of it, to receive the axis, which is fastened into the socket by wedges (see p. 612).

The main joints at each end of the great lever, which unite it to the connecting-rod M, and to the main links K of the parallel motion, are universal joints

capable of lateral motion (see p. 472).   For this purpose the ends of the lever are
made circular, and terminate with cylindrical pins, which project out like pivots at
the end of an axis; they are truly turned in a lathe to fit them into two collars
which have trunnions projecting out on each side, to form the joint pins for suspending the connecting-rod and the main links.   These collars are retained upon the
pivots at the ends of the lever, by caps, which are fitted upon the ends of the pivots
beyond the sockets, and are fastened by cross pins; but they leave the sockets at
liberty to turn about upon the pivots, in order to allow a lateral motion to the
joints, and render them universal joints.

The connecting-rod M is one piece of cast iron; its form is that of two broad
thin plates intersecting each other at right angles, so that the cross section of the
rod resembles a star with four points (see ✦, p. 603).   The upper end of the rod is
shaped like the top of a crutch, and straps of wrought iron are fastened on each
end of the cross piece, to hold the brasses which form the joints to suspend the
rod from the two projecting trunnions of the socket at the end of the great lever;
hence the upper end of the connecting-rod is forked, and has double joints to
unite it to the trunions of the socket; and that socket gives it universal motion.
The lower part of the connecting-rod where it passes against the crank, is a
flattened bar with its angles removed; the lower end terminates in an enlargement having an opening through it, to receive the two brasses which form the
socket for the crank pin; the upper of these brasses is held down by a cross key,
which can be driven in, to tighten the brasses about the crank pin.

The crank N, fig. 2, is fixed fast on the extreme end of the main axis P; that
end which is $9\frac{3}{4}$ inches diameter, is turned and truly fitted to a corresponding hole
which is bored through the crank; when the crank is driven very tight on the end of
the axis, it is secured from turning round, by three cylindrical steady pins of steel
which are inserted into corresponding holes, formed half in the axis, and half in
the crank; the length of the pins being horizontal, and parallel to that of the axis
(see p. 493).   The outside of the circular part of the crank around the centre is
16 inches diameter, and $7\frac{1}{2}$ inches broad.   The pin of the crank is fitted very
truly into a hole bored through the end of the crank, and fastened in by a cross
pin driven through both.   The main axis P P is 10 inches square, and four feet
long between the bearings; the neck is 9 inches diameter, and 12 inches long in
the bearing part; the rule in p. 619, would give 8·94 inches diameter.

The multiplying-wheel O is cast in one piece, and fastened upon the square
axis P by wedges; it is $6\frac{1}{8}$ feet diameter at the pitch line, with 77 teeth, which are
3 inches pitch, and $7\frac{1}{2}$ inches broad.   These teeth are stronger than those of
the sun and planet-wheels, according to the rule given at p. 615, which would
give 6·6 inches for the breadth of the teeth; but they have less strength than the
more modern multiplying-wheels, according to the rule, p. 622, which would give
10·6 inches for the breadth (*a*).

(*a*) A 30 horse-engine of the same construction as the above, and of the dimensions given in the
table, p. 574, has a multiplying-wheel of 77 teeth, and a pinion of 38 teeth, of the same size as above
(see p. 634), except that the breadth of the teeth is 6 inches.   The strength of these teeth is rather
more than usual for sun and planet-wheels, according to the rule, p. 615, which would give 5·5
inches for the breadth of the teeth instead of 6 inches.   Or according to the rule for multiplying-
wheels, p. 622, the breadth of its teeth should be 8·85 inches.   As this wheel has worked well for
many years, it forms an additional proof of the sufficiency of the strength of the teeth of Mr. Watt's
old sun and planet-wheels.
    The fly-wheel of this 30 horse-engine is quoted for an example of the rule for fly-wheels, p. 640.
Most of the other dimensions of this engine are the same as Mr. Watt's 30 horse-engine, p. 500

The pinion q is 3 feet diameter, and has 38 teeth, corresponding with those of the multiplying-wheel. The axis 26 of the fly-wheel is 8 inches square, and $10\frac{1}{2}$ feet long between the bearings; the necks at each end of it are 7 inches diameter, and $8\frac{1}{2}$ inches long in the bearing part. The ends of the axis project out beyond the bearings, and the machinery that the engine is intended to drive is connected with one of those projecting ends.

The fly-wheel Q Q is fixed on the axis 26, close to the wall at the opposite side of the house, to the multiplying-wheel and pinion; it is 18 feet diameter outside; the rim is $8\frac{1}{4}$ inches, by $3\frac{3}{4}$ inches, and it contains 11·7 cubic feet of cast iron. The fly-wheel makes $38\frac{1}{2}$ revolutions per minute when the engine makes 19 strokes per minute, which is the proper speed.

The velocity of the middle of the rim of the fly-wheel is 34·93 feet per second. According to the rule in p. 640, the rim should contain 11·7 cubic feet of cast iron; thus,

Sliding Rule, Calculation.	$\Big\{$	C 11·7 cubic feet.	(36 HP ÷ 19 strokes =) 1·895
		D 525 gage point.	(17·3 ft. dia. × 38·5 rev. =) 667

The governor Z is placed over the axis 26 of the fly-wheel, and the pivot at the lower end of its upright axis is supported on the vertex of an iron arch which stands astride over the axis, as is shown in the plan, fig. 2. The governor is turned round by a bevelled wheel of 40 teeth fixed on the axis of the fly-wheel, and working another wheel of 44 teeth, fixed on the lower end of the upright axis of the governor, which therefore makes 35 revolutions per minute. The balls of the pendulums are 10 inches diameter, and the centres of the balls are $31\frac{5}{8}$ inches from the joint by which they are suspended from the axis. The sliding collar at the top of the governor acts upon the short arm of a bent lever, the long arm of which hangs down, and the lower end communicates by a long horizontal wire, with the handle of the throttle-valve *z*, as is shown in plate XIX.; the spindle of that valve is placed upright. The upper end of the axis of the governor is retained in a socket fixed to the underside of the spring beam; and the middle part of the axis is steadied by a bearing at the end of a bracket which projects out horizontally from the side wall of the house.

The operation of the engine is the same as is described in p. 459, except the action of the working gear, which is as follows; whenever the engine is standing still, the rod 2 is disengaged from the lever 8, by raising the extremity of the rod by the screw 24 and lever 23; all the four valves then remain closed by the action of their weights 4 and 15. To start the engine, the attendant inserts the lever handles *r* and *s* into their sockets on the tops of the levers 20 and 21, and applying one hand to each lever, he presses them both down at once, so as to open all the four valves together, to allow the steam to blow through, from the boiler to the condenser, and clear it of air. This being done repeatedly, as described at p. 459, and the injection-valve opened, the handle *s* is pressed down whenever the piston is required to be impelled upwards; and the handle *r* is depressed when the piston is to be impelled downwards; therefore every time that the piston arrives near the top or bottom of its course, the position of the handles must be reversed, but in all cases the handle which was depressed, must be allowed to rise up before the other is put down. In thus starting the engine, one or other of the handles must be put down first, according to the position in which the crank stands, and according to the direction in which it is required to be turned round.

The engine being put slowly in motion by thus opening and shutting the valves by hand, the rod 2 is moved backwards and forwards by the excentric at every stroke of the piston without producing any effect; but by turning back the nut of the screw 24, the end of the rod 2 is let down, so as to allow it to rest upon the joint pin of the lever 8; and when the notch in that rod comes over the pin, it will drop upon it, so as to engage the rod with the lever, and move the latter backwards and forwards; the claws 22, 23 upon its axis then act upon the inclined loose levers *t* 16 and *u* 17, and open and shut the valves in alternate pairs, at the proper intervals, to produce the requisite distribution of the steam.

For instance, when the parts are in the position of figure 3, the upper steam-valve *b* and the lower exhausting-valve *i* are open, consequently the piston is making its downward stroke; when it is about the middle of its course (the crank being horizontal, and pointing towards the cylinder) the excentric 1 will have moved the rod 2, and the lever 8, to the extent of its course towards the cylinder, and the claw 22 will have depressed the lever *t*, so as to have opened the two valves *b* and *i* to their utmost by the connexions 11, 20, and 10, 19. As the piston descends, and the connecting rod M turns, the crank N turns round, the excentric 1 will begin to draw the rod 2 back again, away from the cylinder, and the motion thus given to the axis 3, and its claw 22, will allow the lever *t* to return, and begin to let down the valves *b* and *i*, but the motion given by the excentric being very slow at first, those valves are closed with a gradual motion, which becomes quicker and quicker as the excentric proceeds, so that they will be closed with a rapid motion by the time that the piston is nearly at the bottom of its course, and a little before the crank reaches the vertical position. The rod 2 will then have acquired its quickest motion, being at the middle of its course, and at the same instant that the valves *b* and *i* are quite closed; the opposite claw 23 comes in contact with the lever *u*, and depresses it with a sudden motion, which by the rods and levers 12, 21, and 13, 18, is communicated to the lower steam-valve *e*, and the upper exhausting-valve *h*, to open them, and cause the ascent of the piston.

The valves are thus raised very quickly to a sufficient height to open their passages, and as the motion continues they are raised higher than is absolutely necessary, but with a motion which becomes slower and slower, until the piston reaches the middle of its course, and the crank becomes horizontal (pointing away from the cylinder); the rod 2 will then have reached the extent of its motion away from the cylinder, and the valves *e* and *h* will be opened to their utmost; after which they begin to close again with a gradual motion, which quickens as it proceeds, and the valves are completely closed by the time that the piston is near to the top of its course; at the same instant the claw 22 overtakes its lever *t*, and opens the other pair of valves *b* and *i* with a sudden motion at first, and which becomes more gradual and slow as the piston descends, until it reaches the middle of its course, when the valves become fully opened; after which they begin slowly to close again, as already described.

The regulation of this working gear depends upon the positions of the levers *t* and *u*, which must be adapted to the claws 22 and 23 which act upon them, so that the instant that one pair of valves is closed the other pair shall begin to open, and not before. For this purpose the lengths of the rods 11 and 12 can be regulated at pleasure, the lower ends of them being formed into screws, which are screwed into the joint pieces at the ends of the levers *t* and *u*. The joints at the upper ends of the rods 10 and 13 are capable of a similar regulation, in order that the two valves of each pair may close exactly together.

The excentric 1 must be fixed on the main axis P, in such a position that the excentric rod 2 will be exactly in the middle of its course, when the centre of the crank pin arrives within one-twelfth part of the length of the stroke of the

4 x

piston, from a line passing through the centre of the main axis P, and the joint at the extremity of the great lever by which the connecting-rod is suspended. The lengths of the rods 11 and 10, 12 and 13, must be adjusted so as to allow all the four valves to be exactly closed by their respective weights 4 and 15, when the engine is in the above position, the levers $t\ u$ being then barely in contact with the claws 22, 23 without pressing against them. The working gear being thus adjusted, the engine will perform its functions properly, in the manner already described, p. 468, so as to evacuate the steam promptly from the cylinder.

Mr. Murdock's working gear has no decided advantage over Mr. Watt's original plan with handles (see p. 524); the chief difference of action is, that in Mr. Murdock's, the weights are applied to close the valves instead of to open them, and the valves are opened by a sudden motion communicated from the main rotative axis. This kind of working gear is not so convenient for the attendant as Mr. Watt's, because the rod 2 of the excentric must be disengaged before he can govern the valves by the lever handles r and s; and also because the force that he must exert upon those levers is much greater than Mr. Watt's handles require; for the former have none of that powerful combination of leverage which has been described at p. 374.

Mr. Murdock's valves are secure against the accident of retaining steam within the cylinder, if all the four valves are closed when the engine is moving rapidly (see p. 463 and 527), because the valves are only kept closed by the weights 4 and 15, therefore the compressed steam would lift up the steam-valves, and escape by returning to the boiler, and could not be confined sufficiently to break the parts of the engine. The pressure that the weights 4 and 15 exert to close the valves, can be regulated at pleasure by sliding them along the levers to which they are attached, and fixing them at any required distance from the centres of motion.

The construction of the valves with hollow spindles requires great accuracy of execution, to ensure that the seats of the valves, and the stuffing boxes through the covers of the boxes, shall be truly concentric with each other; for this purpose the boxes are bored withinside, at the parts where the brass-valve seats are fitted into the iron partitions, and also at the upper edges of the boxes where the covers are to fit on; and all these parts are bored at one operation with the same tool, to render them very exact. The bottoms of the valve-boxes are open to allow the borer to pass through, but bottoms are fastened in with wedges and cement after the work is fitted together.

The brass seats for the valves are accurately turned on the outsides to fit into the apertures in the partitions, and they are put in with cement. The covers of the boxes, which are of the same form as the cover for the cylinder in miniature, are also turned very true at the parts where they fit upon the top edges of the boxes, and they have projecting rims at the undersides to drop into the boxes. The seats for the lower valves have bars across them to form sockets which guide the lower ends of the spindles. The apertures through these valves are $6\frac{1}{4}$ inches diameter at the smallest parts; the valves are lifted up $2\frac{1}{2}$ inches out of their seats, which is more than is necessary to open their passages (see p. 374), but it occasions the valves to open the requisite passage suddenly. The seats for the upper valve are merely brass rings into which the valves are fitted. The hollow spindles for the upper valves are made of wrought iron like gun barrels; they are truly bored withinside, and turned on the outside to $1\frac{5}{8}$ inches diameter; the brass valves are firmly fastened on the lower ends by screws, and the boxes which contain the hemp packings round the spindles of the lower valves are screwed on their upper ends.

The covers of the boxes are fastened on by three screw bolts to each ; by removing these, the covers can be lifted up to take out the valves ; in replacing the covers, it is requisite to make the packings beneath them very even, and to screw them down equally on all sides, to avoid throwing the stuffing boxes out of the vertical position. The guides for the upper ends of the spindles of the exhausting-valves are fastened by screws on the tops of the upright standards for the axes of the levers which lift the valves ; the pivots of these axes rest in brasses, which are fitted into grooves within the upright standards, and are fastened in those grooves by cross keys ; by withdrawing these keys, and removing the guides, the brasses can be drawn up out of the grooves, so as to take out the axes with their levers. The nossels are fastened to the cylinder by bolts passing through flanges, which project from the valve boxes at the back, and join to corresponding flanges around the edges of the branches *c* and *f* at the top and bottom of the cylinder.

In some of these engines the covers for the valve boxes are made small enough to enter into grooves or circular rebates which are bored out withinside the boxes, so as to enlarge them at the upper edges. The circumference of the cover is fitted very exactly into the groove, and when it is lodged therein, the upper surface of the cover forms a level with the top edge of the box, and then the stuffing box will stand truly vertical and concentric with the axis of the valve seats. The cover is fastened, and its joint made good, by a circular ring which is applied upon it, and screwed down with a pasteboard beneath, to cover over the circular joint or circumference of the cover and of the box ; as the pasteboard is not applied in the joint between the box and the cover, its thickness cannot occasion any derangement of the position of the cover.

A great number of large engines were made on the plan of Plate XIX. and XX., at Soho, between the years 1800 and 1810 ; they were all nearly alike, and performed very well, but not better than Mr. Watt's engines. In the earliest of these engines the diameter of the multiplying-wheels was made equal to the length of the stroke of the piston, as is shown in Plate XIX. ; but in subsequent erections larger multiplying-wheels were used, and the teeth were made stronger, according to the rule, p. 622. For instance, engines of 45 horse-power of this construction were made at Soho in 1808, with cylinders of $33\frac{1}{3}$ inches diameter, and the pistons making strokes of 7 feet, and $17\frac{1}{2}$ strokes per minute (see table, p. 574) ; the multiplying-wheels are 12 feet diameter, with 152 teeth of very nearly 3 inches pitch, and 8 inches broad ; the pinions on the axis of the fly-wheel are $4\frac{1}{4}$ feet diameter, with 54 teeth, and the fly-wheels make $49\frac{1}{4}$ revolutions per minute ; they are 18 feet diameter, and the rims $8\frac{1}{4}$ inches by $3\frac{1}{8}$ inches ; this is the same size as the fly-wheels for the 30 horse-engine, and the 53 horse-engine, quoted at p. 640, and the proportion to the power of the engine is very nearly according to the rule there given.

The dimensions and proportions of the essential part of these engines correspond very nearly with the various rules given in chapter VIII. ; a portion of the observations from which those rules have been deduced, were taken from engines of the above kind. It appears that Messrs. Boulton, Watt, and Co., have continued to follow the standards established by Mr. Watt for the dimensions of all the parts of their large engines to the present time, with but very few alterations. A great number of engines on the plan of Plate XIX. were also made by other makers in imitation of those sent out from Soho, and some of these copies, which were made at Manchester and at Glasgow, proved equal to the originals.

About the year 1810, Mr. Murdock's sliding-valves having been brought to great perfection for small engines, were modified so as to suit large engines ; they were arranged in a very symmetrical form for the nossels, and are called D valves with architectural side pipes; these have since been universally adopted by Messrs. Boulton, Watt, and Co. for their rotative engines, and have also come into general use amongst other engineers.

DESCRIPTION OF PLATE XXI.; A 20 HORSE STEAM-ENGINE APPLIED TO A MILL FOR ROLLING SHEET LEAD, CONSTRUCTED BY MESSRS. LOYD AND OSTELL, 1810.

The construction of the engine is nearly the same as that last described, and most of its dimensions are according to the rules given in chapter VIII. The cylinder is 24 inches diameter; the piston makes a stroke of 5 feet, and usually makes 22 strokes per minute. The boiler is 5 feet diameter, 15 feet long, and 7 feet high, with a flue through the middle of it. The fire-grate is 4 feet square (see p. 496). The following are the dimensions of the millwork.

The axis P, fig. 2 of the crank N, is 7 inches square, and $4\frac{3}{4}$ feet long between the bearings; the neck of the axis is 7 inches diameter, and $7\frac{1}{2}$ inches length of bearing (a). The excentric which gives the motion for opening and shutting the valves is fixed on the axis P of the crank. The multiplying-wheel O, which is also fixed upon that axis, is $4\frac{3}{8}$ feet diameter, with 66 teeth of $2\frac{1}{2}$ inches pitch, and 6 inches broad; it turns the pinion q on the axis 26 of the fly-wheel Q Q, at the rate of $32\frac{1}{4}$ revolutions per minute; that pinion is $3\frac{1}{6}$ feet diameter, with 45 teeth. The axis 26 is $6\frac{1}{2}$ inches square, and $7\frac{1}{2}$ feet long between the bearings; the pivots at each end of it are 6 inches diameter, and $6\frac{1}{2}$ inches long. The fly-wheel is $17\frac{1}{2}$ feet diameter to the outside, and the rim is 9 inches by 4; it has 6 arms.

To give motion to the rollers between which the lead is laminated, the multiplying-pinion q turns a large spur-wheel D of $10\frac{1}{12}$ feet diameter, with 152 teeth of $2\frac{1}{2}$ inches pitch, and 6 inches broad; the centre hole of this wheel, which is 11 inches diameter, and 15 inches through, is fitted upon a cylindrical part of its axis, so that the wheel can turn round freely upon the axis; but it can be engaged with it by means of a sliding clutch box E which is fitted upon the axis. There is another large spur-wheel F, also fitted loose upon the same axis as the wheel D; it is $9\frac{1}{12}$ feet diameter, with 137 teeth of $2\frac{1}{2}$ inches pitch, and 6 inches broad. This is turned from a corresponding wheel G upon the axis of the fly-wheel, $2\frac{2}{3}$ feet diameter with 40 teeth, by means of an intermediate wheel H, fig. 1, of about 4 feet diameter, so that the wheel F is turned round upon its axis in a contrary direction to that in which the wheel D is turned.

The sliding clutch box E has an enlargement at each end of it, forming two circular flanges of 22 inches diameter, in the flat faces of which are projecting teeth adapted to interlock with corresponding teeth at the central parts of the wheels D and F. The clutch box E is prevented from turning round upon the axis by means of two feathers of steel, which are fixed into grooves chisseled out in the solid metal of the axis, so as to project out therefrom; and the projecting parts are received in corresponding notches withinside of the cylindrical hole through the box; but the box is at liberty to slide endways upon the axis, in order to engage its teeth with those of either of the wheels D or F, so as to connect that wheel with the axis, and turn the axis round. The clutch box E is shorter than the space between

(a) The neck of the axis for the crank is rather less than would be given by the rule, p. 619, which would be 7·11 inches diameter instead of 7 inches.

the centres of the two wheels F and D, and when it is placed in the middle of that space, its teeth will remain detached from those of both wheels, which will therefore turn round upon their common axis in opposite directions to each other without communicating any motion to it.   The clutch box is moved endways by means of a horizontal lever e, with a fork at the end, which embraces the small part of the box between its two end flanges; this lever is fixed upon the upper end of an upright axis f, which is sustained by brackets from the end wall of the engine-house, and a longer horizontal lever g is fixed on the lower end of it; this long lever g passes under the wheel D, and extends out beyond the wall of the engine-house; the extremity is connected with a handle, which is situated in a convenient position for a man to move it one way or the other, in order to slide the clutch box E along its axis, so as to engage either of the wheels D or F with the axis, and thus turn it round in either direction at pleasure.

The necks of the axis of the wheels D and F are 9 inches diameter, and 9 inches long; the end of the axis beyond the bearing is connected by a square box i, with one end of an axis I 9 inches square, which is connected at the other end by another box k, with a square on the end of the neck of the lower roller L, which is 18 inches diameter, and 7 feet long; the upper roller M, which is placed exactly over it, as shown in fig. 3, is of the same size, and is turned round by the action of the lead when passing between the two rollers.   The rollers are made of cast iron, very truly turned and polished on the outside surfaces, that they may impart a smooth surface to the lead.   The necks at the ends of the rollers are 9 inches diameter, and 10 inches long; the necks of the lower roller rest in brasses fitted into two massive beds of cast iron R, which are placed upon a strong framing of timber, and held down by two foundation bolts for each bed (see fig. 4).   The necks of the upper roller M, are kept down by two caps r r, which are fitted upon four upright pillars s s, fig. 4, of wrought iron $4\frac{3}{4}$ inches diameter, and the caps are retained by nuts d d, screwed upon screws formed at the upper ends of the standards s s.

The weight of the upper roller M, with the caps r r, and the nuts d d, is counterbalanced by a loaded lever S, situated under ground, beneath the rollers; the necks of the upper roller rest upon brasses in two pieces t t, which are fitted upon the upright standards s s, so as to slide up or down upon them. The pieces t t are supported on the upper ends of two upright bolts v v, which pass down on each side of the necks of the lower rollers, through holes in the beds R R, and the timbers which sustain the same; and their lower ends have nuts screwed upon them to rest upon cross bars w w, which are jointed upon two horizontal levers x y, situated one opposite to the end of the other, beneath the rollers; the fulcrums x x of these levers rest in bearings on the foundations, and the opposite ends y y, which are close together, are connected by links with the end z of the loaded lever z S, the fulcrum of which rests on a bar fixed across between the foundation walls, and a heavy weight is applied at the other end S.   This weight having a double leverage is more than sufficient to balance the weight of the upper roller M and its dependences, and it lifts up those parts, so as to cause the necks of the upper roller M to bear upwards with force beneath the brasses in the caps r r which are over them; and those caps and their nuts d d bear upwards upon the threads of the screws s s, by which means the space between the rollers is always kept open to its full extent for the admission of the lead between them.

By turning the nuts d d round upon their screws s s, the upper roller M can be brought nearer to, or farther from, the lower roller L, in order to regulate the space between them, according to the thickness to which it is required to laminate the lead. The nuts *d d* are turned round by cog-wheels fixed upon them ; and a pinion is applied between each adjacent pair of wheels *d d*, as is shown at *b*, fig. 4, to engage with the teeth of both, and turn both the nuts round together ; on the upper ends of the axis of these pinions, small bevelled wheels are fixed, to be turned round by corresponding pinions upon a horizontal axis *b b*, fig. 3, which extends over all the length of the top roller M, and has a handle A at one end, for a man to turn the axis round by, and then its bevelled wheels and pinions *b b*, and the wheels upon the nuts *d d*, cause all the four nuts *d d* to turn slowly round together, with a corresponding motion, in order to raise or lower the upper roller, and adjust the space between the two rollers. The axis *b b*, and the bevelled wheels and pinions, are mounted in a frame, which is fastened at each end upon the two caps *r r* (see fig. 3), so as to rise and fall with them. The axis *b b* is divided in the middle, and the two parts are joined by a toothed clutch box, which can be disengaged, when it is necessary to screw down the nuts *d d* at one end of the rollers, independently of those at the other end, in order to adjust the two rollers parallel to each other.

On each side of the rollers L M is a long horizontal table B B, fig. 2, to receive the sheet of lead ; it is 72 feet long and $8\frac{3}{4}$ feet wide ; its upper surface consists of a great number of horizontal rollers which facilitate the motion of the lead upon the table. To support this table three parallel beams of oak, 9 inches wide and 10 inches deep, are laid horizontally upon corresponding walls, at right angles to the length of the rollers ; and upon these beams upright iron standards *p*, fig. 4, are placed, to support three corresponding beams of 6 inches wide by 9 deep, which sustain the pivots of the small rollers. The tops of these rollers form the surface of the table, and are on a level with the top of the lower roller L, suitable to bear the sheet of lead ; the small rollers are made of wood, 7 inches diameter and $3\frac{1}{4}$ feet long, each one with an iron hoop at each end, and a projecting pivot to rest in suitable bearings lodged in the horizontal beams ; the rollers are 11 inches apart from centre to centre.

The foundations for this engine are nearly the same as those of the engine in Plate XIX, but the engine-house is wider and longer, in order to allow space for the wheelwork D, F, and the large fly-wheel Q Q. The wall V beneath the column which sustains the centre of the great lever, extends all across the house, and two parallel walls W W extend from it to the end wall of the house ; these walls support the bearings for the three parallel axes of the crank, the fly-wheel, and the large wheels D and F. Beams of oak, 13 inches broad, by 14 inches deep, are laid upon the walls W, and the several bearings are held down upon those beams by foundation bolts, which go down through the thickness of the walls, and are keyed beneath (see fig. 1). The bearings for each end of the axis of the intermediate wheel H are fastened within cast iron frames or boxes, which are built in amongst the stones of the same walls. The fly-wheel occupies the space between the side wall of the engine-house, and one of the walls W for the wheelwork. The ends of the spring beams are bolted down upon a strong beam, which extends all across the house.

The foundations for the rollers are three long parallel walls beneath the table, and two walls built across between them, beneath the rollers ; large beams of oak are laid upon these cross walls (see fig. 3), and they are united by strong cross pieces, upon which the iron beds R for the rollers are laid, and they are held down upon them by foundation bolts, which go down through the walls, and are keyed beneath ; the space between these cross walls under the rollers is left open for the balance levers before described.

The operation of laminating lead by this mill is very simple. The lead is melted in a cast iron caldron, and run out, when in a fluid state, upon a large flat table, so as to form a very thick rectangular sheet, or plate of lead 6½ feet square, and usually about 3 inches thick ; it weighs more than 3 tons. This plate is lifted upon the table B B of the rolling mill by a crane, and, by means of the rollers with which that table is provided, the lead is easily moved along the same, in order to be presented between the laminating rollers L M.

The clutch-box E is placed in the middle position, by the handle of its lever g, so as to disengage the axis of the rollers from both the wheels D and F, until the engine is started, and its heavy fly-wheel Q put in full motion. In the mean time the regulating handle A of the rollers is turned round one way or the other, as may be required, in order to set the rollers at such a distance apart as will just allow the thick plate of lead to be introduced between them, and yet will compress it moderately : by habit and experience the workmen learn to know the proper opening between the rollers when they are to begin upon the lead. The handle of the lever g is then moved sideways, in that direction in which the plate of lead is required to pass through between the rollers ; the sliding clutch-box is thereby engaged with that wheel D or F, which will cause the lower roller L to turn round in the proper direction. The plate of lead being then pushed up between the rollers, with one of its sides parallel to the length of the rollers, they take hold of the edge, and by the motion of the lower roller, the plate is drawn through between the lower and the upper rollers ; the latter receiving its motion from the lead itself. The metal is compressed very forcibly in its passage between the rollers, so as to reduce its thickness, and extend the length of the plate.

When the plate has passed through, the regulating handle A of the wheel-work is turned round a certain space, by one man, to set the rollers a little nearer together, whilst another man pushes back the handle of the lever g in the other direction, and thus, without stopping the engine, the lower rollers are caused to turn round in an opposite direction, and the lead being pushed up to the rollers, will be taken again between them and still farther reduced. The rollers are then set nearer together by the handle A, and their motion reversed by the handle g, in order to roll the lead through them again. In this manner the lamination is repeated, over and over again a great many times, until the plate is reduced to the required thickness.

The lead extends in length, but not in breadth, during the operation, and it becomes warm by the compression. When the lead is rolled out to 35 or 40 feet in length, if it is not then as thin as is required, the sheet is cut in two, and each half is rolled separately until it is thin enough. The surface of the lead is rendered perfectly smooth and even by the lamination, with a fine polish, similar to that of the surfaces of the rollers, which must be carefully preserved, and restored if they are injured by any accident. The thickness of the sheet of lead must necessarily be uniform at all parts. The rollers make about 9½ revolutions per minute, and their surfaces move through about 45 feet per minute. The sheets are passed through the rollers a great number of times to reduce them as much as is required, for the thickness is only diminished a very little at each time. The 20 horse engine is capable of rolling two plates in a day of 14 hours.

This mill and engine was constructed by Messrs. Loyd and Ostell, in 1810, and was put up in London. It was one of the first steam-mills of the kind, horse-mills having been commonly used before in London for rolling lead (*a*). The same engineers made another mill with a 36 horse engine to turn two pairs of rollers; all the cog-wheels are the same as the above, but there are double the number. The engine is situated between the two mills, and has a crank on each side of the connecting rod; the axes of the two cranks are exactly in a line, one opposite to the end of the other, and the same crank pin is common to both cranks, so that they are turned round together. The axis of each crank has its separate multiplying-wheel and fly-wheel, with double cog-wheels, to turn a pair of rollers either way round, in the manner already described. In more modern mills of this kind it is found better to have a greater power: an engine of 25 horse-power is applied to work one pair of rollers, or a 50 horse-engine to turn a double mill with two pairs of rollers.

DESCRIPTION OF PLATE XXII, AN INDEPENDENT ENGINE OF TEN HORSE-POWER APPLIED TO A HORIZONTAL SUGAR-MILL.

The construction and action of the moving parts of this engine are nearly the same as those of the 36 horse-engine, Plate XIX, and require very little description. The perspective view shows the whole of the engine and mill at one view. The cylinder is 18 inches diameter; the piston makes a stroke of 3 feet, and works at the rate of 32 strokes per minute. The boiler C is 12 feet long and 5 feet diameter; its ends are round, and above the brickwork those ends form quarters of spheres. A is the furnace, and D the chimney. T is the feeding cistern at the top of the upright feeding pipe; the pulley and chain to suspend the float in that pipe, with the balance weight to it, and the other pulley fixed on the same axis with the chain *w* to suspend the damper in the flue, act in the same manner as those described in p. 695. The steam-pipe, the nossels in front of the cylinder, the parallel motion, the great lever, the connecting-rod and crank N, are sufficiently apparent in the drawing. The great lever is $10\frac{1}{4}$ feet long, and 18 inches deep in the middle. The fly-wheel Q Q is fixed upon the same axis as the crank N; that axis is 6 inches square, and $3\frac{1}{2}$ feet long between the bearings; the fly-wheel is 9 feet diameter outside, but the rim is represented much smaller than it should be (*b*).

The fixed framing of the engine is on the plan of the independent engines made by Messrs. Boulton, Watt, and Co., as mentioned at p. 682. The condensing cistern G G is made of cast iron, 15 feet long, 4 feet broad, and $4\frac{1}{3}$ feet high outside; it forms the basement for the whole engine, and is placed on foundation

---

(*a*) The method of laminating lead between rollers turned by millwork, first one way round and then back again, has been practised for more than a century past in England, and milled sheet lead is in very general use. The first mills were driven by water-wheels, but some were afterwards made to be turned by horses. A rolling mill of this kind was sent from England to France in 1727, and it was set up in Paris; the machine is fully described in Les Machines approuvées par l'Academie, 1728, vol. v. p. 43. Also in Descriptions des Arts et Métiers, Bertrand's edition, vol. xiii. p. 465; and in the Encyclopédie Méthodique, Arts et Métiers, vol. iv. article Laminage: the two latter articles contain a very detailed account of the process of casting the plates.

(*b*) According to the rule, p. 643, the rim of the fly-wheel would require to be about 8 inches by $7\frac{7}{9}$ inches, or 63 square inches of cross section, for it should contain 11·4 cubic feet, thus:

$$\textit{Sliding Rule Calculation.} \begin{cases} \text{C} & 11\cdot4 \text{ cubic feet.} & (10 \text{ HP} \div 32 \text{ strokes} =) \cdot 312. \\ \text{D} & \text{Gage point } 51. & (8\cdot33 \text{ ft. dia.} \times 32 \text{ revolv.}) \ 266. \end{cases}$$

The fly-wheel of Mr. Hick's 10 horse-engine for the same kind of sugar-mill as the above, is 14 feet diameter outside, and the rim is 9 inches by $4\frac{1}{2}$ inches.

walls, upon which it stands sufficiently steady by its own weight. From each angle of the cistern, a vertical column W W rises up, to sustain the spring beam U U, which is a rectangular open frame of cast iron, in one piece; it is further supported beneath the fulcrum of the great lever, by two other columns V V, which stand up from the middle of the cistern.

The cylinder bottom is made large enough to reach across the width of the cistern, and is bolted down upon the edges of it. The bearing for the neck of the axis of the crank N is also bolted down upon the edge of the cistern, and the pivot at the other end of the axis is supported in an iron standard erected upon the same foundation walls which support the cistern, leaving a sufficient space between the cistern and that standard to receive the fly-wheel, and the excentric by which the working gear is moved, and also the small pinion by which the sugar-mill is turned. This pinion, of which only a very small part can be seen in the drawing, is one foot diameter, with 15 teeth of $2\frac{1}{2}$ inches pitch and 6 inches broad; the teeth are strengthened by solid flanges at each side, which connect all the teeth together (a).

The great wheel O O, which is turned by the pinion, is $7\frac{1}{6}$ feet diameter, with 108 teeth of $2\frac{1}{2}$ inches pitch and 5 inches broad; it is fixed on the extreme end of its axis P, which rests in two bearings fastened on the edges of the cistern G; the necks of this axis are 6 inches diameter, the extreme end projects beyond the bearing over the edge of the cistern with a square of 6 inches, which is connected by a square box k with one end of a round axis I, which is connected at the other end with the projecting end of the axis of the upper roller F F of the sugar-mill.

The two lower rollers H H are placed side by side beneath the upper roller F, so that their centres form the angles of an isoceles triangle. The rollers are 22 inches diameter and 4 feet long, turned truly cylindrical on their outsides, and cut with small grooves parallel to their axes; the rollers are hollow within, and have centre sockets at their ends to fit upon their axes, and they are fastened by wedges; the axes are of wrought iron, 6 inches square, and $4\frac{1}{2}$ feet long between the bearings; the necks at one end of each of these axes are 6 inches diameter, and $6\frac{1}{2}$ long in the bearings; the pivots r, m, m, at the opposite ends, 5 inches diameter and $6\frac{1}{2}$ inches long.

The rollers are supported in a very massive frame of cast iron, consisting of two upright standards X X erected upon a foundation plate B B, which is bedded upon masonry, and is formed with a border projecting upon all sides, to give it strength and also to form a pan for the reception of the juice which is expressed from the sugar-canes in their passage between the rollers. The ends of the foundation plate B are adapted to receive the bases of the upright standards X X, which have feet spreading out from them sideways, and the foundation bolts which fasten the plate down upon the masonry pass through these feet. The form of the standards is sufficiently shown in the figure; the lower part of each standard consists of several upright pillars with openings between them; above these are two cells to receive the brasses for the pivots m m of the two lower rollers H H, with strong regulating screws to set those brasses horizontally towards each other, in order to adjust their distance from the upper roller; between these two cells a very strong

(a) The teeth of the pinion are not so strong as they should be, but the motion being slow, the teeth are capable of enduring a much greater strain than if they acted with more rapidity (see p. 627). The strain upon these teeth is at the rate of 342 pounds for teeth of one inch pitch and one inch broad, which is more than double the strain upon the teeth of sun and planet-wheels, according to the rule in p. 614.

4 Y

upright standard rises up with a cell in the top to receive the brasses for the pivots
r of the upper roller, and those brasses are kept down by a cap which is fastened
over them by two strong bolts and nuts at top.

The ends of the axes of the rollers project out beyond their bearings at the
side of the frame towards the engine, and each one has a pinion fixed upon it;
only the upper one of these pinions can be seen at u; the pinions of both the
lower rollers H H are turned round by the teeth of the pinion u on the axis of
the upper roller, but the horizontal distance between the centres of the lower
rollers is rather greater than the oblique distance from the centre of the upper
roller to the centres of either the two lower ones, and for the same reason there is
a considerable opening between the surfaces of the two lower rollers; but both
the lower rollers are very close to the upper roller, and only leave spaces of about
$\frac{1}{10}$ of an inch between them, in order that the canes may be squeezed into that
thickness in passing between the rollers. These pinions are nearly the same diameter as the rollers, and have each 22 teeth of 3·14 inches pitch, and 5 inches
broad.

The axis I is connected with the extreme end of the axis of the upper roller
F by means of a circular box u, which is fixed on the end of the axis of the upper
roller; the interior of the box is bored rather conical, to receive a circular plug,
which is fixed upon the extreme end of the axis I, so that the plug fits into the box
in the same manner as the plug of a cock fits into its socket, and the plug is pressed
into the box, by means of three or four screws, with so much force, that the friction
of the plug in the box will suffice to turn the rollers round and perform their
work properly; but if by accident the rollers become clogged up with canes, or
with stones, or other solid matter, the plug will slip round within the box, so as to
avoid breaking the teeth of the wheels, or stopping the engine, as might otherwise
happen.

The sugar-canes are presented to the rollers by a man, who spreads them out
upon an inclined table K, which is made of iron plate; the ends of the canes pass
down the inclined table, and are drawn in between the front lower roller H and
the upper roller F, by which they are subjected to a very severe pressure, which
squeezes out the juice or natural sap of the canes, and it trickles down from the
rollers into the pan, which is formed in the middle of the foundation plate B. The
ends of the canes which have passed downwards between the two rollers then meet
with the back lower roller H, which by its motion turns those ends up again, so
as to introduce them between the upper roller F and the back lower roller H,
whereby the canes receive a second pressure, and the latter rollers being set nearer
together than the former, the remainder of the juice is expressed; the canes then
pass from the roller H down another inclined table L, and fall upon the ground.

q is a spout in the side of the pan B, with a small shuttle to run off the collected juice into a trough, which conveys it away to the boiling-house. The inclined
tables K and L are supported by upright pillars, as is seen in the drawing. The
canes, after being pressed, are called cane trash, and are used as fuel for boiling the
sugar, and for working the engine. As this fuel produces a less intense heat than
coals, the fire-grate is of a large size, being $3\frac{1}{2}$ feet square $= 12\frac{1}{4}$ square feet.

When the engine works at its proper speed of 32 strokes per minute, the
rollers make 4·44 turns per minute, and their surfaces move at the rate of 25·6
feet per minute. The cold water pump R, and the hot water pump S, are
placed within the cistern G, as is shown in the figure, and 8, 8, is the feeding
pipe which conveys the hot water for the supply of the boiler, from the hot water

pump, to the feeding cistern T.   There is also a waste pipe to carry away the surplus water, which is not represented.

This engine had no governor, but a governor may be placed at the end of the engine, as shown at Z, above the spring beams U, being fixed on the upper end of its upright axis Y, which has a bevelled wheel upon the lower end of it within the cistern G, to be turned by another bevelled wheel fixed upon a horizontal axis, which passes through the cistern to the outside, and has a spur wheel upon the end of it, which is turned from a similar wheel fixed upon the axis of the crank, with an intermediate wheel to convey the motion.   When the governor is applied in this manner, the pannel at the end of the cistern G must be extended out in a circular form like a bow window, so as to leave a space for the bevelled wheels within the cistern.   Or else the axis of the governor may be placed farther away from the engine, and the bevelled wheels may be at the end of the cistern on the outside.

The valves of this engine are lifted by means of upright rods on the plan already described, p. 685, and plate XVII.; and those lifting rods are raised up alternately by two tappets, which project out in opposite directions from a horizontal axis, and acting beneath feet at the lower ends of the upright rods; this axis does not turn round continuously, as described at p. 685, but it is moved backwards and forwards by means of a lever 4 at the extreme end of it, which is connected with the compound rod 2, extending from the excentric circle, which is fixed upon the axis of the fly-wheel.   This mechanism is the same as that described at pages 693 and 698.

Steam-engines are found to answer extremely well for turning sugar-mills, where water-mills cannot be had; and great numbers of steam sugar-mills have been exported from England to the West Indies within the last twenty years.   The first application was made by Mr. Rennie, about the year 1801, to be worked by an 8 horse-engine made by Boulton and Watt; it was sent to Demerara, and performed very well.   It is advantageous to express the juice from the sugar-canes as soon as ever they are gathered, and before they have time to ferment.   Cattle-mills are deficient in power to despatch the work for a large plantation, and extra cattle must be kept on purpose to turn the mills, because they are wanted at the most busy season of the year, when there are other employments for all the cattle that can be procured.

The common sugar-mills have three rollers placed in a vertical position, because that is most conveniently turned by cattle, without any wheelwork.   Steam-engines are frequently applied to turn vertical sugar-mills, but the horizontal position of the rollers is a great improvement; they were first proposed by Mr. Smeaton to be turned by a water-mill in 1757, and they now are brought into very general use.

DESCRIPTION OF PLATES XXIII. AND XXIV.; MR. WATT'S SINGLE ENGINE FOR PUMPING WATER FOR THE SUPPLY OF TOWNS, BY MESSRS. BOULTON, WATT, AND CO., 1803.

The structure and operation of this engine is so nearly the same as has been already described at p. 353 to 383, that it is unnecessary to enter into a detailed explanation of all the parts; it is sufficient to state the alterations which were introduced in the form of the engine by Mr. Murdock soon after Mr. Watt had retired from business.

The cylinder of this engine is 48 inches diameter; the piston makes a stroke of 8 feet, and about 14 strokes per minute; its principal dimensions are according to the rules given at p. 366. It works a lifting pump of 17½ inches diameter, the bucket making an 8 feet stroke, and raising the water 126 feet.

The boiler C, plate XXIII., is made of iron plate of the waggon form, with an internal flue through it, and is nearly the same as the boiler already described, p. 695, but of a larger size. It is 20⅓ feet long, and 5⅔ feet wide; the horizontal surface of the water is 115 square feet; and according to the rule given in p. 588, the surface that the boiler exposes to the heat should be 436 square feet. The fire grate B is 4⅔ feet square = 21¾ square feet, which is nearly one-20th of the heating surface. The height from the grate to the bottom of the boiler is 24 inches in the centre, and 15 inches at the sides.

The feeding-pipe T, with the balanced stone float within the boiler and feeding-valve 21, to regulate the admission of the water into it, according to the level of its surface, is the same as already described, p. 453. The sliding damper *w* to regulate the passage through the flue, is self-acting on the plan described at p. 695; the damper *w* being suspended at one end of a chain which is conducted over two pulleys 22, 22, and a float is suspended from the opposite end of it within the upright feeding-pipe T, so that it will be moved up or down at every variation of the height of the water in that pipe, and the chain will communicate an opposite motion to the damper, in order to regulate the draft of the furnace according to the strength of the steam in the boiler; 9, 9, are doors to give an entrance into the external flue round the outside of the boiler.

The steam-pipe *a* rises with a slope towards the cylinder E; it has a throttle-valve in it at *z*, and joins sideways to the steam-box for the steam-valve *b*. The nossels, valves, and working-gear, are represented on a larger scale in plate XXIV.; these parts are the same as already described, p. 356, except that, instead of the internal racks and sectors, the valves are lifted by the extremities of simple levers fixed on the horizontal axis, which pass across the valve-boxes; the ends of those levers enter into openings in the stems of the valves. This plan is sufficiently shown by the figures at the bottom of the plate (see also p. 523).

The exhausting-valve i being of a large size, the steam presses down upon it with great force, and it would require a great strain upon the hand gear to overcome that pressure and lift up the valve, if some counterbalance were not applied; the spindle of the valve *v* is therefore prolonged downwards, and a piston 23 is fastened on the lower end of it; this piston is packed with hemp, and is fitted into a cylindrical chamber at the bottom of the box beneath the valve; the chamber is accurately bored for the reception of the piston, so that it can move freely up and down when the valve is opened and shut; the lower end of the cylindric chamber 23 is closed by a cover screwed on, and a small copper-pipe 24 forms an open communication from the passage *f* above the exhausting-valve *i*, to the lower part of the chamber beneath its piston 23. The water which collects from the condensed steam fills the lower part of the chamber in time, and that water transmits the pressure of the steam which acts above the top of the valve through the copper-pipe 24 to the under surface of the piston 23, so as to press it upwards, and assist to open the exhausting-valve *i*. The piston 23 is 8 inches diameter, and the aperture through the exhausting-valve *i* is 9 inches diameter; therefore the upwards pressure beneath the piston is so much less than the downwards pressure above the valve, as to leave it a sufficient preponderance to shut close. The education-pipe *g* proceeds sideways out of the box beneath the exhausting-valve *i*, and turns down; the lower end of the pipe *g* enters the side of the condenser F by a similar turn, and by this means the eduction-pipe passes clear of the cylindric chamber 23.

The nossels b, c, d, e, f, i, are put together in several pieces as described at p. 522, and the cylinder E, with its surrounding steam-case, is constructed as described at p. 699.   Water is kept in the socket joint round the middle of the steam-case to prevent leakage; and the overplus runs down by a small pipe into a copper bason which surrounds the socket joint by which the lower end of the upright pipe *d* is united to the box containing the equilibrium-valve.   25 is a short curved pipe by which steam is conveyed from the steam-case into the hollow bottom beneath the cylinder; and 26 is the syphon-pipe to drain off the condensed water which collects from the steam-case (see p. 381.)   27 is the hollow above the cover of the cylinder; it is supplied with steam by a small crooked pipe similar to 25, which must be removed every time that the cylinder cover is to be taken up.

The piston is the same as is already described p. 372 and 476; the edge of the piston which applies to the cylinder is about 8 inches deep, 5 inches of which is the depth of the hemp packing; the iron rim round the bottom edge of the piston beneath the packing is $1\frac{1}{4}$ inches deep; and the edge that the piston-cover or ring above the packing presents to the cylinder is $1\frac{3}{4}$ inches deep.   This ring is pressed down upon the packing by four clamps which bear upon the ring at their outer ends, and the inner ends rest upon the top of the piston near to its centre, and screws pass through these clamps to screw into the piston (see the figure at the bottom of plate XXIV.)   The thickness of the packing is about $1\frac{3}{4}$ inches.

The condensing apparatus is exactly the same as described at p. 374.   *x* is the blow-valve through which the air is discharged from the condenser previous to starting the engine; this valve is fitted at the end of a small pipe which passes through the side of the condensing cistern G, and the valve is covered with water contained in a small cistern on the outside of the large cistern (see p. 455), so as to prevent leakage, but the water is not deep enough to sensibly impede the passage of the air through the valve.   The manner of joining the condenser F to the air-pump H is apparent, and the foot-valve *k* is placed in the passage between them; the aperture which is closed by that valve is $6\frac{1}{2}$ inches high, by $17\frac{1}{2}$ inches wide (see p. 379).

The air-pump H is 24 inches diameter, and its bucket makes a stroke of 4 feet, so that the effective capacity of the air-pump is one-eighth of that of the cylinder. The butterfly-valves in the bucket of the air-pump are clearly shown in figure 1; also the cover of the pump, and the stuffing-box in the centre of it for the polished rod *l* to pass through.   It is of a similar construction to the cover of the cylinder. The discharge-valve *m*, which delivers the hot water into the hot well I, is a rather larger aperture than the foot-valve *k*.

The suction-pipe of the water-pump S draws its water out of the hot well; the suction-valve is contained in a box in that pipe, with a cover to take off, and give access to the valve; from this box the pipe proceeds horizontally through the lever-wall V to the top of the barrel of the pump S; the forcing-pipe proceeds from the opposite side of the same barrel, and turns up against the wall; it then extends horizontally to the feeding-cistern T over the boiler, as shown by the dotted lines 8, 8; the forcing-valve is contained in a box at the lower part of this pipe, near the pump, with a cover to take off.   The piston of the pump is solid, without any valves in it; the barrel is $4\frac{1}{2}$ inches diameter, and the rod being suspended from the great lever at 3 feet from its centre of motion makes a stroke of nearly 2 feet.

The injection-valve and cock are shown at *j* in fig. 1.   The valve is fitted into a seat, at the upper end of an upright pipe, from the lower end of which a curved pipe descends, and rises up again to join to the underside of the socket of the cock, which is fixed against the side of the condenser F; the conical plug of the cock is fitted into this socket; it can be turned round by a handle, which rises up near the working gear, and regulates the aperture which admits the water into the condenser. The valve *j* is lifted up by the working gear, every time that the piston reaches the top of its course, and is let down again when it reaches the bottom of its course.

The working gear is the same as described at p. 356 and 363, but the structure is better explained in Plate XXIV. There are three horizontal axes *t u* and 4 placed one above the other; they are supported by pivots at their ends, which rest in bearings fastened to two strong upright posts. Each axis has a handle *r s* and 3 fixed on it, to enable a man to govern the engine, and these handles are worked by chocks 1, 7 and 2, which are fastened to the plug *l*, upon the rod of the air-pump.

The axis *t* of the upper or expansion handle *r* carries a short lever which is connected by the rod 10 with the external lever of the steam-valve *b*, and there is also a short lever from which a weight is suspended by the rod 17; this weight acts to throw the expansion handle *r* upwards, and then it opens the steam-valve *b* as in fig. 1. When the handle *r* is pressed down either by the hand or by the chock 1 on the plug, so that the short lever comes into a line with the rod 10, the valve *b* will be shut. In like manner the axis *u* of the middle or exhausting handles has a lever which, by the rod 11, is connected with the external lever of the exhausting-valve *i*; and a weight 4 is applied to throw up the exhausting-handle *s*, and open the valve *i*. The axis 4 of the lower or equilibrium handle *s* is connected by the rod 14, with the external lever of the equilibrium-valve *e*; and a weight 15 is applied to the axis 4 so as to throw down the equilibrium handle *s* and open the valve *e*.

5 is a double or diagonal catch, such as is described at p. 452. It has a hook at each end to catch the ends of the two levers 12 and 16, which are fixed on the ends of the axes *u* and 4, so as to retain those axes in the positions which will cause the exhausting and equilibrium valves *i* and *e* to remain shut, and the catches prevent the weights 4 and 15 from opening the valves, until they are disengaged.

The catch 5, see p. 3 and 4, has a small spherical weight fixed to the extreme end of it, to give it a tendency to catch the levers 12 and 16, and that end is supported on the point of an upright adjusting screw which is screwed into a bracket fixed to one of the upright posts of the working gear.

The axis of the expansion handle *r* has no catch, but it is detained in that position which will keep the expansion-valve *b* shut, by means of a short link 6, the lower end of which is jointed to that lever of the middle axis *u*, from which the weight 4 is suspended, and the upper end is a long loop or slit to receive a joint pin at the end of a short lever on the upper axis *t*. The effect of connecting the two axes *t* and *u* together by the looped link 6, is, that the expansion handle *r* will be kept up as shown in fig. 3, so as to keep the expansion-valve shut, as long as ever the lever 12 is detained by the upper hook of the catch 5, to keep the exhausting-valve *i* shut; but when the catch 5 is disengaged from the lever 12, and allows the weight 4 to throw up the handle *s*, and open the exhausting-valve *i*, then the looped link 6 will no longer detain the upper axis, but its weight 17 will throw up the expansion-handle *r*, and open the expansion-valve *b*; nevertheless the loop in the link 6, allows the expansion-handle *r* to be put down at any time, either by the hand or by the chock 1, in order to close the expansion-valve, independently of any motion being given to the exhausting-handle *s*.

The injection-valve *j* is lifted up by a lever 19 to which its rod is jointed, and the end of the lever is suspended by a rod and a strap 18 which is coiled round the middle axis *u*, so as to be wound up a little whenever the exhausting-handle *s* is thrown up by its weight to open the exhausting-valve *i*.

The piston-rod n is $3\frac{5}{8}$ inches diameter, the lower end of it is a cone which is fitted through the centre of the piston in a conical socket, so that the rod cannot draw up, and it is fastened in by a cross key; the upper end of the rod is fastened into a socket, which is attached to the lower joint-pin of the parallel motion by a cross key.   The main joint-pins of the links K of the parallel motion are 5 inches diameter, and $5\frac{1}{2}$ inches length of bearing in their sockets.   The iron of which the main links K are made, is $4\frac{1}{2}$ inches by $1\frac{1}{8}$ inch; the four contain $20\frac{1}{4}$ square inches of iron to endure the force of the piston.

The great lever L L is of cast iron made in two plates, the form of which is shown in Plate XXIII.; the two plates are put together side by side upon the axis of the lever in parallel vertical planes, as is shown at the bottom of Plate XXV. The plates are each 48 inches deep in the middle, and 16 inches deep at the ends. The length between the centres is $24\frac{2}{3}$ feet.   The breadth or thickness of each plate is $1\frac{3}{4}$ inches; a prominent border, which extends all round the plates, renders the breadth of the edges $2\frac{1}{2}$ inches; this border is three inches deep. The adjacent surfaces of the two plates are flat, and leave a space of 12 inches between them.   They are fastened together by the axis which passes through sockets in the middle of both plates, and is wedged fast; the axis is $10\frac{3}{4}$ inches diameter, and four feet long between the bearings; the pivots at each end of it are 7 inches diameter, and $8\frac{1}{2}$ inches length of bearing.   The main joint-pins at each end of the lever pass through both plates, and are wedged fast; these pins are 6 inches diameter, the ends which project out through the plates form the joint pins for the parallel motion.   The joint pins for the back links 7 of the parallel motion, are fixed in a similar manner through both plates; they are 2 inches diameter in the joints; the parallel rods 5 and the bridle rods 6 are $1\frac{1}{2}$ inches diameter.   The two plates of the great lever are further united by a short pillar at each end, and four bolts which pass through flanges at each end of each pillar.   The catch pins at each end of the great lever, to limit its motion, are fixed across on the upper ends of strong pieces of wood, which are fastened in between the two plates of the great lever, and are strongly connected with the lever by bolts, which pass through the wood and are steadied by oblique bracing bolts.   When the piston moves too far, the ends of the catch pins are stopped by the spring beams U.

The rod l of the air-pump is suspended from a pin, 3 inches diameter, by a short link contained in the space between the two plates of the great lever; the lower end of this link is connected by a very short link, with a joint pin at the middle of the back link 7, and by this means the rod *l* is caused to move vertically; but to keep it steady it is guided through a socket, which rises up from the floor opposite to the steam-box *b*.   In some engines the socket is steadied by a brace projecting out from the cover of that box.   The rod *l* is prolonged by the wooden plug rod *l*, against which the chocks for working the handles are screwed, as shown in Plate XXIV; and to the lower end of the wooden plug the rod *l* of the air-pump is connected by a joint.   The lower end of the plug rod is kept steady by means of a horizontal piece of wood 27 (fig. 2), which is fixed fast across it, and the ends slide up and down in grooves in the upright posts which sustain the axes for the handles of the working gear.

The building is the same as described p. 356.   The centre of the great lever is supported by the lever wall V.   The cylinder is fastened down upon its pier X by four foundation bolts, as described p. 701.   The condensing cistern is placed in the space between the cylinder pier X and the lever wall V, and the air-pump is fastened down by two bolts which pass through the bottom of the cistern. The great pump N is placed in a well formed by the space between the end wall of the engine-house, and a wall which is built across the house opposite to the lever-wall V.   The cold water pump R is placed in the space between the latter walls ; thé rod of this pump is suspended from the great lever at $4\frac{2}{3}$ feet from its centre of motion, so as to make a stroke of 3 feet long ; the barrel is $11\frac{1}{4}$ inches diameter inside.   A pier is built at the base of the end walls to receive the large air-vessel Q, and strong oak beams are fixed horizontally from this pier to the cross wall to receive the great pump N between them ; the projecting flanges of the pump rest upon these beams, and are very firmly fastened down upon them by bolts ; two of these beams extend beneath the air-vessel Q, that its weight may assist to keep them steady.

The parallel motion by which the spear M is suspended, is exactly the same as that at the other end of the great lever for the piston-rod.   The spear M is made of two strong bars of iron, united into one at the upper end, which is suspended from the parallel motion, and separated into two branches below, to receive two heavy plates of cast iron between them, which form the counterweight to draw up the piston during its returning stroke.   The pump-rod is connected to the lower end of the spear by a cross key ; the rod is polished like the piston-rod, in order to pass through a stuffing-box in the cover of the pump, and the bucket of the pump, which is fastened at the lower end of the rod, has two openings through it, covered by double butterfly-valves shutting downwards ; the circumference of the bucket is surrounded with leather to fit exactly into the cylindrical bore of the working barrel, so as to work freely up and down in it.   At the bottom of the working barrel the lower valve, or fixed clack, is placed ; it cannot be seen in the figure ; it has double butterfly-valves shutting downwards, the same as those of the bucket.

The working barrel bucket and clack of the great pump are the same as those represented in Plate IV, fig. 6, 7, and 8 (see p. 215 ; see also the figure p. 247).   The flange at the top of the working barrel serves to join it to a pump head, which has a side flange to join to a branch o, which conveys the water into the lower part of the air-vessel Q, and another flange at top to fasten on the cover of the pump, which has a stuffing-box in the centre of it for the polished pump-rod to pass through, the construction of which is the same as that in the cover of the cylinder.   The pump is $17\frac{1}{2}$ inches diameter, and the motion of its bucket is the same as that of the piston.

The air-vessel Q receives the water which the pump delivers at every stroke through the branch o, and discharges that water again through the conduit pipe P at the opposite side, by the elasticity of the compressed air, in a regular stream. The conduit pipe P is conducted under ground for a great distance up a gradual ascent, to convey the water into a reservoir or pond situated on elevated ground, the level of the surface of the water in that reservoir being usually 126 feet above the level of the surface of the water in the pump well of the engine.   The upper part of the vessel Q is filled with air, in such a state of compression, that its elastic force equals the pressure of the column of water that the pump is intended to raise. The inertiæ of the mass of water contained in the long pipe P is so very considerable, that it requires a great exertion of power to put it in motion, and when once set in motion, it has a corresponding tendency to continue moving; it also resists any tendency to sudden increase or diminution of its velocity. Consequently, every time that the piston makes its working stroke, and the pump delivers the contents of the working barrel into the air vessel, the whole of that water is not forced at once into the conduit pipe P, but the air in the upper part is compressed into a less space, so as to make room for a portion of the water forced in; and that portion is reserved in the air-vessel until the pump ceases to deliver more water; but during the returning stroke, when the pump delivers no water, the compressed air, by its elastic pressure, drives out that portion of water which was reserved, and forces it through the conduit pipe P, so as to keep up the stream through that pipe without any intermission, and with only a slight pulsation or variation of velocity, although the pump only delivers water during the working stroke, and is quite inactive during the returning stroke.

The air-vessel is 4 feet diameter withinside, and $9\frac{1}{2}$ feet high, with a hemispherical dome at top; the pump being $17\frac{2}{3}$ inches diameter, the area of the air-vessel is $7\frac{1}{2}$ times greater than the area of the pump; therefore, if the bucket of the pump makes a stroke of $7\frac{2}{3}$ feet, it will deliver as much water into the air-vessel as would fill it about one foot high, supposing no water to pass out during the working stroke; but if one-half of that water passes away through the conduit pipe P, during the working stroke, then the rise and fall of the surface of the water in the air-vessel would be about six inches. If the height of the upper part of the vessel, which contains air, is assumed to vary from $7\frac{1}{2}$ to 8 feet above that fluctuating surface, the proportion between the greatest and least elasticity of the compressed air would be as 16 is to 15.

To keep the air-vessel properly filled with air, a small cock is fixed in the chamber at the bottom of the working barrel of the great pump N, just above the lower clack; this cock is always open, and admits a small stream of air into the pump at every stroke; the air mixing with the water, tends by its elasticity to diminish the violence of the shocks which take place when the valves in the pump-bucket, and in the clack, shut suddenly at the beginning and end of every stroke. When the piston makes its working-stroke, the air thus introduced is forced out with the water through the branch O, from the top of the pump-barrel N, into the air-vessel Q, where the air separates from the water, and rises up into the upper part of that vessel, so as to keep it as full of air as is necessary.

The surplus air is allowed to escape through an inverted valve contained in a box Z which is fixed at the side of the air-vessel Q; the interior of this box is connected by a small branch from the lower part of it, with the interior of the air-vessel at that part of its height where the surface of the water in the air-vessel is intended to stand in it; consequently, if the level of the water is below the connexion, air will enter into the box; but if the water stands above the intended level, then the box will be filled with water. The inverted air-valve within the box is fitted into an aperture through the cover of the box, and a cylindrical float, which is fixed to the stem of the valve, floats upon the surface of the water in the box, so as to keep the inverted valve shut, whenever the water stands at a proper height in the box; but when the surplus of the accumulated air passes out of the air-vessel into the box, and displaces the water therefrom, the float subsides, and opens the valve, so as to discharge that air; as soon as the surplus has escaped, the water returns again into the box, and raises the float, so as to close the valve. The stem of the valve passes up through the cover of the box, and is connected with a balance lever, which carries a sufficient weight to counterpoise the weight of the float within the box, and keep the valve closed whenever the water stands high enough in the box; but when the water subsides, the float will preponderate, and open the valve.

The operation of this engine is precisely the same as is described at p. 357, and it requires the same management as is directed at p. 361. The adjustment of the chock 1 for the expansion-handle r is effected by two nuts applied upon an upright screw 13, which is fixed to that chock in front of the plug, and the screw passes through a socket fixed to the plug; one of the nuts is above this socket, and the other below it, and they have projecting knobs to turn them round with the fingers, and thus set the chock higher or lower on the plug, as is required; and it is firmly retained wherever it is placed.

This engine was erected for the Chelsea Waterworks, at Pimlico Wharf, on the same foundation as the old atmospheric engine, mentioned at p. 255, and in the same building with Mr. Watt's original engine, p. 352. The water was drawn from Chelsea Creek, which communicated with the Thames at high water; and the engine forced it, a distance of 1831 yards, through the conduit-pipe, into a reservoir situated in Hyde Park, at 126 feet perpendicular height above the water in the well, at high water; the water in the creek ebbed a few feet in the course of the tide, and the average column was about 130 feet; the pressure of which is 56·4 pounds per square inch.

The pump being $17\frac{1}{2}$ inches diameter, the weight of the column was 13 570 pounds; this load on the piston of 48 inches diameter, was about $7\frac{1}{2}$ pounds per square inch. The piston was adapted to make an 8 feet stroke, when the catch-pins struck the spring beams, but on average it made $7\frac{2}{3}$ feet stroke, and went at the rate of $13\frac{1}{2}$ or 14 strokes per minute, or 105 feet effective motion of the piston per minute. This is an exertion of 43·2 horse-power. It delivered 175 cubic feet of water per minute into the reservoir in Hyde Park.

The consumption of coals was on an average about $3\frac{3}{4}$ bushels per hour = 315 pounds; that is 7·3 pounds per hour for each horse-power, or 22·8 millions pounds raised one foot high by each bushel of coals consumed.

The conduit-pipe is double for 1069 yards, as far as Hyde Park-corner, consisting of two pipes, each of 12 inches bore, acting as one conduit. The remaining distance of 762 yards to the reservoir is only one pipe of 12 inches bore. The velocity with which the water moved through the double pipe was 112 feet per minute, and through the single pipe 224 feet per minute.

Regulator. The main conduit-pipe P had some branches from it, to distribute water, and when these were opened so as to draw off part of the water short of its destination, the load upon the engine was considerably diminished. To accommodate the power to these variations of resistance, a regulator or governor was applied to act upon the throttle-valve z, according to the pressure of the column of water which the pump was required to raise. For this purpose a small pipe was conducted from the lower part of the air-vessel Q, to the lower part of a small upright cylinder, into which a piston was fitted; the cylinder was open at top, and the rod of the piston was loaded with a heavy chain, which acted as a weight, and was sufficiently heavy to counteract the pressure of the water beneath the piston. When the column of water increased, this piston would be raised higher up in its cylinder by the extra pressure; and when the column diminished, the piston would descend again. This motion was communicated by suitable levers and rods to the handle of the throttle-valve z, so as to open it wider when the regulating piston rose, and to close it when the piston descended, and thus regulate the opening of the throttle-valve and the supply of steam, according to the resistance to be over-come.

The lower end of the heavy chain which formed the load on the small piston rested upon the ground, and as the piston descended, more of its links came to rest, so as to relieve the piston from their weight; but when the piston rose up, it lifted more of the links up from the ground; by this means the load upon the piston increased continually, in proportion as it ascended higher; and vice versa; consequently the piston would always settle itself in its cylinder at some height, where the varying load of the chain would exactly balance the pressure of the water beneath the piston. The levers and rods by which the piston acted upon the throttle-valve, were so adjusted as to supply the proper quantity of steam which the engine required, to enable it to work at a proper speed, when overcoming that pressure of water in the pump and air-vessel. This contrivance has since been improved, by adapting it to act upon the regulating screw 13 of the chock 1 of the expansion-handle r, instead of the throttle-valve, so as to adjust the position of that chock according to the pressure of the water in the air-vessel. This is now called the *banging apparatus*, because it is intended to prevent the engine striking its catch-pins upon the spring beams.

Safety catch. A contrivance was also introduced into the above engine, to stop its motion whenever the catch-pin at the cylinder end of the great lever struck upon the spring beams, in consequence of the power or resistance being suddenly altered. A short upright prop was erected on the floor near the cylinder cover, to stand up beneath the external lever of the expansion-valve b, so as to sustain that lever near to the end where the rod 10 is jointed to it, and prevent the valve opening. This prop was mounted on a centre pin at its lower end, and the upper end was moveable, being fitted between guides, with a spring applied behind, to press the upper end of the prop continually towards the lever of the valve; the prop was held back out of the way of that lever by a latch similar to that of a door, which was connected by an upright wire, with the tail of a lever placed on one of the spring beams U, and the opposite end of that lever was situated at the place where the catch-pin came down upon the spring beam; consequently when the catch-pin struck, it depressed the end of that lever, and lifted the catch by the upright wire, so as to disengage the upright prop, the upper end of which was then instantly forced by its spring beneath the lever of the expansion-valve, and prevented that valve from being opened by the hand gear. A repetition of the stroke was thus prevented, and the engine was stopped until the attendant came to the handles to start it again, with due precaution to regulate the steam according to the resistance.

In 1810 a new engine was made by Messrs. Boulton, Watt and Co. for these waterworks; it is put up in a new building at Chelsea, near to the river, and draws its water therefrom; a level conduit-pipe of 1287 yards long, and 18 inches diameter, extends from this new engine, to convey the water from it into the above double-pipe of the old engine. The cylinder of the new engine is 50 inches diameter, and the pump $17\frac{3}{4}$ inches diameter. The old engines were pulled down and removed some years afterwards.

DESCRIPTION OF PLATE XXV.   MR. WATT'S DOUBLE STEAM-ENGINE FOR BLOWING
FURNACES BY A DOUBLE FORCING PUMP FOR AIR.

The application of the atmospheric engine for blowing furnaces by a direct action with an air-pump has been already explained, p. 281; and Mr. Watt applied his single engine to work a double forcing pump for blowing furnaces, at a very early period of his career in 1777 (see p. 328) ; he afterwards applied his double engine to that purpose with great success.   The engine represented in Plate XXV. was constructed about 1807, for blowing two furnaces for smelting iron from the ore.   Almost all its parts are the same as those of other engines already described, and a very brief explanation of the engine will be sufficient.

The cylinder is 48 inches diameter, and the piston makes a stroke of 8 feet ; the cylinder E and its steam case, the great lever L, and its parallel motions K at each end, are the same as in the engine last described, p. 719 (Plate XXIII), and are of the same dimensions.   The condensing apparatus is the same, except the dimensions, the air-pump being $31\frac{1}{2}$ inches diameter, and its bucket making a 4 feet stroke.   The foundations and the building are very nearly the same as those of the last engine, Plate XXIII.   The nossels and valves are the same as those of the engine in Plate XX, with hollow spindles to the steam-valves, see p. 697.   The working gear has two handles r and s actuated by chocks on the air-pump rod l, the same as Mr. Watt's original double engine, Plate XIII (see pp. 451 and 457).   All these parts are so obvious from the drawing, that they require no further explanation.   The piston-rod is $4\frac{3}{8}$ inches diameter. The apertures through the steam and exhausting-valves are 9 inches diameter.

This engine has three waggon boilers of $18\frac{1}{3}$ feet long, and $7\frac{1}{6}$ feet wide ; they are placed side by side, and have internal tubes through them for flues ; two of these boilers are worked at once, and the surface exposed to the heat is about 500 square feet in each boiler, or 1000 square feet in the two boilers, which supply steam to the engine.

The cold water pump and hot water pump are not represented in the figure, but they are the same as those in Plate XXIII, except in dimensions. The diameter of the cold water pump is 14 inches, and its rod is suspended from the great lever at $6\frac{1}{6}$ feet distance from its centre of motion, so as to make a stroke of 4 feet. The hot water pump is $6\frac{1}{4}$ inches diameter, its rod is suspended from the great lever at $3\frac{1}{12}$ feet from its centre of motion, so as to make a stroke of 2 feet.

The blowing cylinder N, is a double forcing-pump of 84 inches diameter, and its piston makes a stroke of 8 feet, the same as the steam piston.   The pis-ton-rod is $5\frac{1}{8}$ inches diameter, its parallel motion is similar to that at the other end of the great lever.   The blowing cylinder is formed in two lengths N N, which are united by flanges in the middle of its height ; a bottom is fastened on by a flange at the lower end, and a cover similar to that of the steam cylinder is screwed on the top flange, with a stuffing-box in the centre for the piston-rod to pass through.   The blowing piston is solid, having no valves in it, and it is made tight round the edge by double leathers applied in a similar manner to those of a pump-bucket, in order that there may be as little friction as is consistent with preventing the escape of the air.

At the bottom of the blowing cylinder two large branches, 22 and 25, proceed from it at opposite sides ; and there are two similar branches 20 and 21 from the upper part of the cylinder, immediately beneath its upper flange.   The air valves are placed within these four branches, viz. a set of suction valves in the branches 20 and 25, to admit the external air into the cylinder ; they are hanging valves, like doors suspended by their upper sides, and opening inwards towards the cylinder ; the forcing valves are at the ends of the branches 21 and 22, and they open out-wards from the cylinder into the chambers 23, 24, which join to the branches 21, 22 ; these chambers communicate together by the upright pipe O O, and the

large conduit-pipe P proceeds from the lowest chamber 24, to convey the compressed air to the receiver or water-regulator which supplies the furnaces.

The action of the blowing cylinder is the same as that of a double forcing-
pump. When the steam-piston makes its descending stroke, and draws the
blowing piston up in its cylinder, the suction valves at 20 close, and prevent
the escape of the air which is contained in the upper part of the cylinder
above the piston; that air is therefore compressed until it acquires a sufficient
elasticity to open the forcing valves at 21, and then it is forced out from the
cylinder by the piston, into the chamber 23, and passes through the pipe O to the
lower chamber 24, from which it is conveyed through the conduit-pipe P to the
furnaces. At the same time, the piston as it rises exhausts the lower part of
the cylinder, and the lower forcing valves at 22 close themselves, so as to prevent
any air returning from the chamber 24 into the cylinder; but the lower suction
valves at 25 open, and admit the atmospheric air freely into the cylinder, to fill it
as the piston rises, ready to be blown out again at the returning stroke.

When the steam piston arrives near the bottom of its course, the steam and
exhausting valves are reversed by the action of the working gear, so as to impel
the piston up again for its ascending stroke; the blowing piston is thereby forced
downwards, and the lower suction valves at 25 close themselves, so as to prevent
the escape of the air beneath the piston, it is therefore compressed, and when its
elasticity is sufficient, it opens the lower forcing valves at 22, and enters into the
chamber 24, from which it passes away to the furnaces by the conduit-pipe P.
During this descent of the blowing piston the upper suction valves, 20, are opened
by the external air, which enters into the upper part of the cylinder, and fills up
the space which is left therein by the descent of the piston; the air so admitted
will be blown out through the upper forcing valves 21, during the next stroke,
when the blowing piston is drawn up again.

The air valves of the blowing cylinder are made of thick leather, fortified by
iron plates riveted upon the leather; the valves hang in a vertical position, so as
to close the apertures through which the air is to pass; the apertures are made
through vertical iron plates, which are fixed by bolts to the flanges of the branches
20, 21, 22, and 25. The leather at the upper side of each valve is fastened to
the plate at the upper side of the aperture that the valve is intended to close, and
the leather bends to form the hinge upon which the valve opens. To give a more
ready entrance to the air into the blowing cylinder, other suction valves are applied
in two boxes, which are fixed on the top or cover of the cylinder, one on each
side of the stuffing-box through which the piston rod passes; these are also hanging valves, of the same construction as the others. It is of importance to the performance of a blowing machine, that the passages through the valves should be
sufficiently large to allow the air to pass very freely into, and out of, the blowing
cylinder, as fast as the motion of the piston will propel it.

The blowing cylinder is bored to 84 inches diameter, and the conduit-pipe P
is 21 inches diameter, or one-fourth of the diameter of the cylinder; consequently
the area of the pipe is one-16th of the area of the cylinder. The apertures which
are closed by the valves are each 8 inches high by $14\frac{1}{2}$ inches wide $= 116$ square
inches area in each aperture; there are three such apertures side by side in each
of the branches 20, 21, 22, and 25, covered by three valves which open and shut
all together; hence the area of the passage for the air is 348 square inches, which
is a little more than the area of the conduit-pipe P. The engine usually makes
12 strokes per minute of 8 feet length, and then the motion of the piston is 192 feet

per minute; the motion of the air through the apertures of the suction-valves is about 16 times as much, that is, 3072 feet per minute, or 51 feet per second.

The blowing piston is a circular plate of cast iron, with a rim round the outer edge, which is fitted correctly into the blowing cylinder; the circular edge which applies to the interior surface of the cylinder is about 5 inches broad. The piston-rod is fitted with a cone, into the centre of the piston, and is fastened by a cross key; the flat surface of the piston is strengthened by a number of projecting ribs at the upper and under sides, which radiate from the centre like the arms of a wheel. To prevent the passage of any air by the piston, its edge is furnished with leathers formed into circular rings, which are fastened upon the flat surfaces around the circumference of the piston at its upper and under sides; the circumferences these leathers apply against the interior surface of the cylinder in the same manner as what are called cup-leathers for pump-buckets; the circular border of the leather is turned up all round the piston, into the form of a saucer, and the extreme edge has a tendency to expand itself to a larger size than the cylinder, so as to press lightly against the interior surface of the cylinder, and fit correctly thereto, to prevent any passage for the air, although the piston is capable of moving freely up and down in the cylinder, with but little friction. The edge of the leather, which is fixed on the upper side of the piston, turns upwards, and prevents the escape of the air when the piston is drawn up; the leather which is fixed beneath the piston turns downwards, and prevents the escape of the air when the piston is forced down. The leathers are fastened to the piston by a number of segments of cast iron, which are applied round the edge of the piston, and form complete rings; they are fixed to the piston by screw bolts with nuts, so as to confine the leather between those rings and the piston; each of the circular leathers is made of a number of segments properly moulded and put together, to form a complete ring, which turns up all round like the edge of a saucer.

Water regulator. Although the blowing cylinder is a double acting pump, the discharge of air through the conduit-pipe P is not continuous, because there is an instant at the termination of each half stroke, when no air is blown by the cylinder. The blast of air which is blown through the nose-pipes into the furnaces is required to be as constant and as uniform as possible. To attain a regularity of blast, a water-regulator is applied to the engine, to receive the air from the blowing engine, and deliver it regularly to the furnaces. The water-regulator is a large inverted chest, made of cast iron, and immersed, with its open part downwards, in a cistern of water (see p. 286); the conduit-pipe P, in its passage from the engine to the furnaces, is conducted over this chest, and communicates with the top of it, by perpendicular branches, which form an open communication between the conduit-pipe and the interior of the inverted chest; consequently, when the engine is set to work, and the air in the conduit-pipe is compressed, it will displace the water from the inverted chest, and raise the water in the cistern in which it is immersed; the surface of the water within the inverted chest is thus depressed below the level of the surrounding water in the cistern, until the pressure that is exerted upon the air, by the column of water so raised, will cause all the air that the engine blows, to pass through the nose-pipes into the furnaces with a regular efflux, independently of the slight intermissions in the action of the blowing cylinder at the end and beginning of each half stroke. The operation of the water-regulator has been fully explained already, at p. 286.

The inverted chest is 36 feet long by 14 feet wide, and 8 feet high; it is made of plates of cast iron screwed together, and its weight is nearly 30 tons.

The water-cistern in which the inverted chest is immersed is sunk in the ground, and lined with brick or stone walls : it is 45 feet long by 23 feet wide at the top, and 41 feet by 19 feet at bottom ; the walls are sloping ; the depth is 10 feet. The inverted chest is supported on stone blocks, so that the lower edge is raised $1\frac{1}{2}$ feet above the bottom of the cistern, in order to allow the water to pass freely in and out.   The cistern is filled with water about $5\frac{1}{2}$ feet deep, when the engine is at rest ; and when it is set to work, the air which is forced into the inverted chest displaces the water therefrom, until it stands only about 2 feet deep in the space beneath the inverted chest ; the water so displaced rises in the space around the outside of the cistern, until it is about 9 feet deep therein ; and then the surface of the water within the inverted chest will be depressed by the condensed air, to about 7 feet below the level of the surface of the water in the cistern at the outside of the chest.   The pressure occasioned by a column of 7 feet of water is very nearly 3 pounds per square inch, and the compressed air within the air-chest must be thus much more elastic than the atmospheric air ; or assuming the absolute elasticity of the atmospheric air to be 14·7 pounds per square inch, that of the compressed air will be 17·7 pounds per square inch ; and the volume of the compressed air will be $\frac{147}{177}$ of the volume of the same absolute quantity of common air by weight.

The area of the surface of the water in the inverted chest is (36 ft. $\times$ 14 ft. $=$) 504 square feet, and the area of the surface of the water which is raised in the space between the walls of the cistern, and the outside of the chest, is nearly the same ; consequently, the surface of the water at the outside of the inverted chest, will be raised as much as the surface of the water within it is depressed by the compressed air, and vice versa.   The area of the blowing cylinder (84 inches diameter) is 38·48 square feet, and the horizontal area of the inverted chest is 13·1 times the area of the cylinder.   When the water is displaced from the inverted chest as above stated, so as to raise a column of water 7 feet high, that part which contains the compressed air above the water, is $7\frac{1}{2}$ feet high, and its capacity is 3780 cubic feet ; that is, about 12·3 times the capacity of the blowing cylinder.

A blowing engine of the above dimensions is capable of blowing three furnaces for smelting iron from the ore, if it is worked at its full speed of 16 strokes per minute, which may be done when the ore is of such a quality that it can be reduced with a moderate blast ; but when the operation of the furnaces requires the compressure of the air to be 3 pounds per square inch, the engine usually makes 12 strokes per minute, and blows two furnaces.   The piston is adapted to make a stroke of 8 feet, but it usually works with a stroke of $7\frac{3}{4}$ feet ; and the blowing cylinder being 38·48 square feet in area, it takes in 298·2 cubic feet of common air at every half stroke.   The vacant spaces left at the top and bottom of the cylinder into which the piston does not enter, are each about 12 cubic feet capacity, and when the air is compressed, an additional quantity of ($\frac{30}{147}$ths of 12 $=$) 2·45 cubic feet of common air will be crowded into those spaces, so as to escape being displaced from the cylinder by the piston, hence the quantity of common air discharged by the blowing cylinder would be (298·2 $-$ 2·45 $=$) 295·75 cubic feet at each half stroke, from which, if we deduct $\frac{1}{30}$th to allow for the loss by leakage through the piston and valves, it will be about 286 cubic feet actually discharged at each half stroke, or $\times$ 24 $=$ 6864 cubic feet of common air blown per minute into the two furnaces.   As 13·28 cubic feet of common air weigh 1 pound (see p. 284), the actual weight of 6864 cubic feet of common air is 517 pounds of air blown per minute by this engine.

The volume of this air is reduced by the compression to ($\frac{144}{177}$ths of 6864 =) 5700 cubic feet per minute of air having an elasticity of 17·7 pounds per square inch. It is usual to blow the air into the great smelting furnaces through nose pipes of about 2½ inches diameter; each furnace has two such pipes, and the area of the four apertures is 19·64 square inches. In most iron works a small furnace called a cupola for remelting cast iron, is also blown during the daytime with two nose pipes of 1¼ inches diameter = 2·45 square inches area. We may conclude that 5700 cubic feet of compressed air is forced per minute through 22·09 square inches of aperture, and in that case the velocity with which the condensed air would issue from the orifices would be about 37 200 feet per minute, or 620 feet per second. The resistance or load on the steam piston is at the rate of 9·2 pounds per square inch, when the resistance to the blowing piston is 3 lbs. per square inch.

The mechanical power exerted for this purpose may be thus computed. The motion of the piston is 12 double strokes per minute of 7¾ feet long = 186 feet motion per minute. The area of the blowing cylinder is 5542 square inches × 3 pounds pressure per square inch = 16 625 lbs. for the resistance to the motion when the blowing piston is driving the compressed air through the conduit-pipe; but when the piston first begins to act upon the air which is taken into the cylinder, that air not being compressed, it can only resist the piston by degrees, as it becomes compressed and increases in elasticity; the air does not acquire the elasticity of 17·7 lbs. per square inch until the piston has moved through a space of ($\frac{30}{177}$ of 8⅓ feet =) 1·4 feet; and during this motion the resistance to the motion of the piston increases regularly from nothing at the commencement, to 3 lbs. per square inch at the conclusion; after the piston has moved thus far, the air will become sufficiently elastic to open the forcing valves, and pass away through the conduit-pipe to the furnaces; if we assume for an average, that the resistance during the compression of the air is only half as much as it is afterwards, then the resistance will be equal to (7·75 ft. + 6·35 ft. =) 14·1 feet at each stroke, or × 12 = 169·2 feet per minute × 16 625 lbs. = 2 813 000 pounds raised one foot per minute ÷ 33 000 = 85·2 horse-power is exerted by the engine when it is blowing two smelting furnaces and a cupola.

THE END.

LONDON :
Printed by A. & R. Spottiswoode,
New-Street-Square.

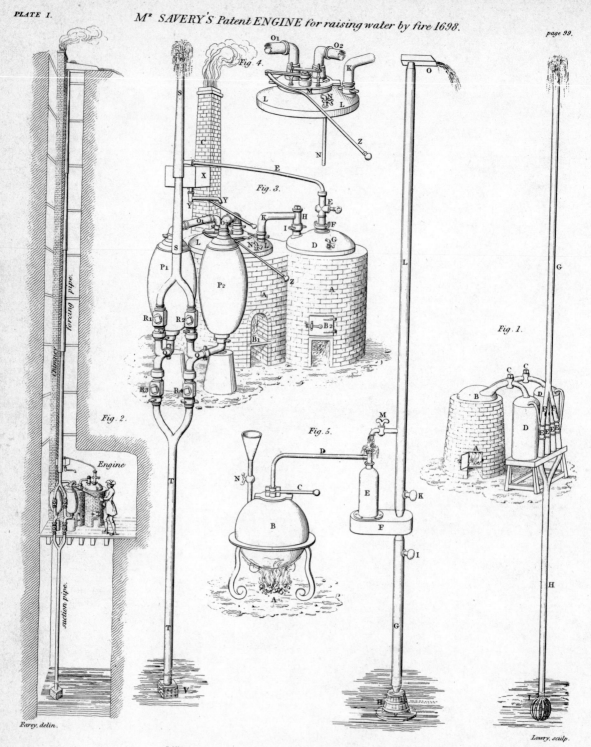

Fig. 4.

Fig. 3.

Fig. 2.

Fig. 1.

Fig. 5.

Engine

Farey, delin.

Lowry, sculp.

Published as the Act directs, 1826, by Longman, Rees, Orme, Brown & Green, Paternoster Row.

PLATE II.

NEWCOMEN'S, *Atmospheric* STEAM ENGINE *for draining Mines,*
constructed by Mr. Smeaton. 1772.

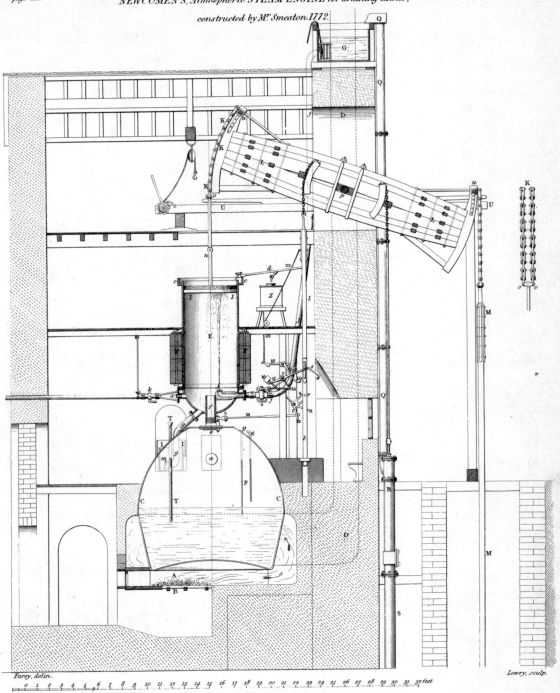

Farey, delin.

Lowry, sculp.

0 1 2 3 4 5 6 7 8 9 10 11 12 13 14 15 16 17 18 19 20 21 22 23 24 25 26 27 28 29 30 31 32 feet

NEWCOMEN'S *Atmospheric* STEAM ENGINE *as constructed by* Mr. *Smeaton.* 1772.

Plate III.

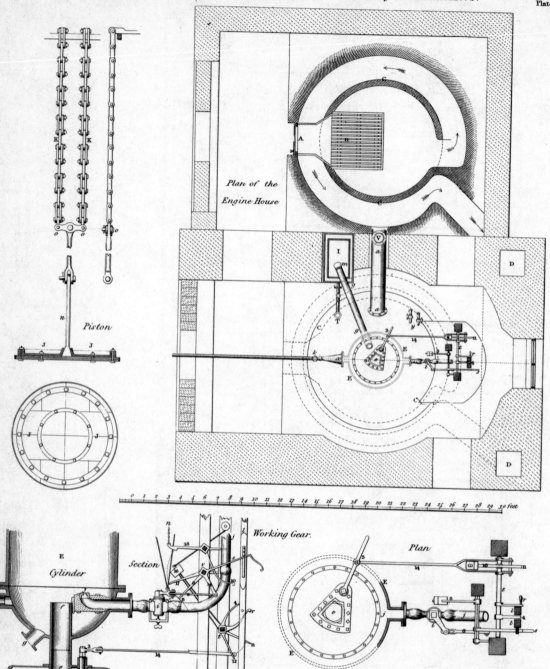

Plan of the Engine House

Piston

Working Gear.

Cylinder

Section

Plan

E

Farey. delin.

Lowry. sculp.

Published as the Act directs, 1826, by Longman, Rees, Orme, Brown & Green, Paternoster Row.

PIT WORK and PUMPS of a STEAM ENGINE for a Colliery.

PLATE IV.

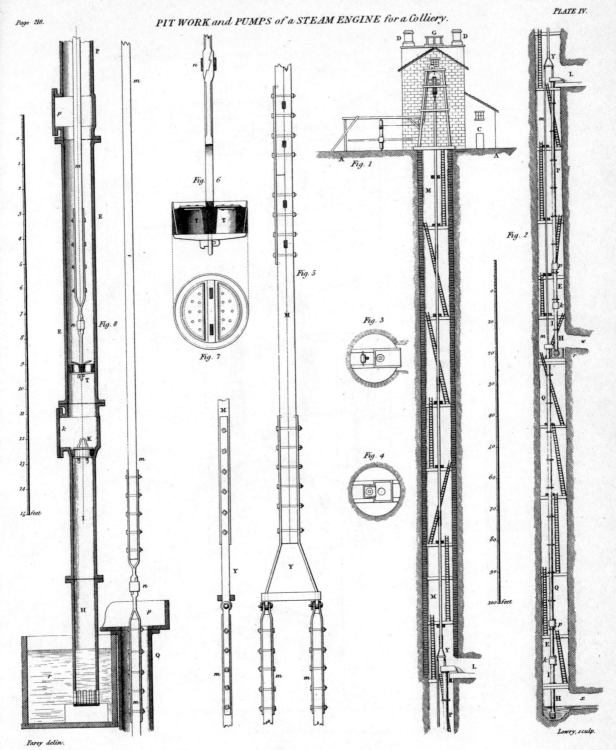

Fig. 1

Fig. 2

Fig. 3

Fig. 4

Fig. 5

Fig. 6

Fig. 7

Fig. 8

Farey delin.

Lowry, sculp.

Published as the Act directs, Dec.r 1.st 1826, by Longman & C.o Paternoster Row.

PLATE V.

page 191.

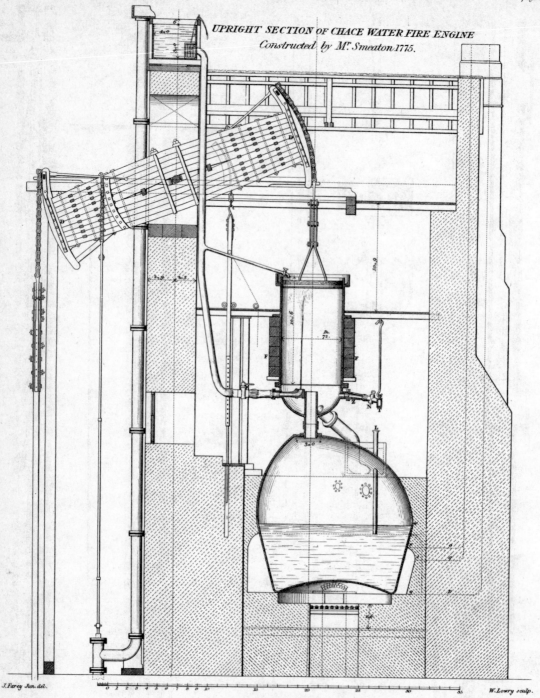

UPRIGHT SECTION OF CHACE WATER FIRE ENGINE
Constructed by Mr. Smeaton 1775.

J.Farey Jun. del.

W.Lowry sculp.

Published as the Act directs, 1826, by Longman, Rees, Orme, Brown & Green, Paternoster Row.

PLATE VI.

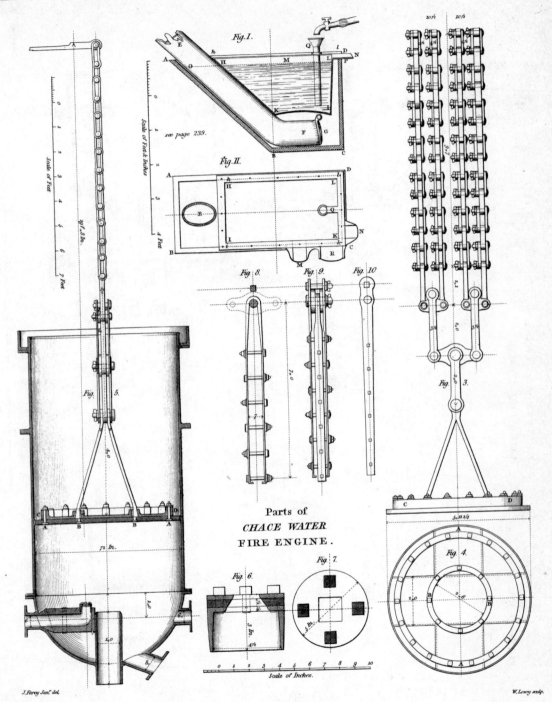

Fig. I.

see page 239.

Fig. II.

Fig. 8. Fig. 9. Fig. 10.

Fig. 3.

Fig. 5.

Parts of
CHACE WATER
FIRE ENGINE.

Fig. 4.

Fig. 6. Fig. 7.

0 1 2 3 4 5 6 7 8 9 10
Scale of Inches.

J.Farey Jun.ʳ del.

W.Lowry sculp.

Published as the Act directs 1826 by Longman, Rees, Orme, Brown, & Green, Paternoster Row.

PLATE VII.

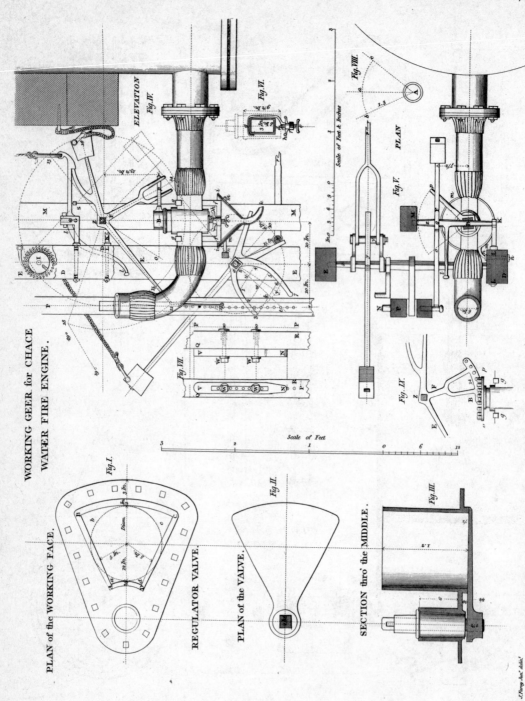

W.Lowry sculp.

WORKING GEER for CHACE
WATER FIRE ENGINE.

ELEVATION
Fig.IV.

Fig.VI.

Fig.VIII.

PLAN

Fig.V.

Fig.VII.

Fig.IX.

Scale of Feet & Inches

Scale of Feet

PLAN of the WORKING FACE.
Fig.I.

REGULATOR VALVE.

PLAN of the VALVE.
Fig.II.

SECTION thro' the MIDDLE.
Fig.III.

Published as the Act directs 1816 by Longman, Hurst, Rees, Orme, Brown, & Green, Paternoster Row.

J.Farey Arch.t delin.t

PLATE VIII.

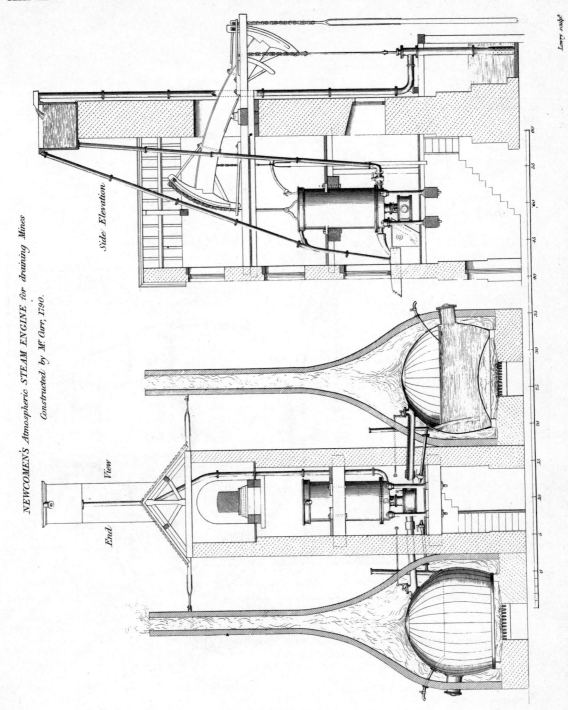

NEWCOMEN'S Atmospheric STEAM ENGINE for draining Mines.

Constructed by Mr Curr, 1790.

Side Elevation.

End View.

Lowry sculp^t

London, Published as the Act directs, by Longman, Rees, Orme, Brown, & Green, Paternoster Row.

Farey delin.

PLATE IX.

Engraved by Wilson Lowry.

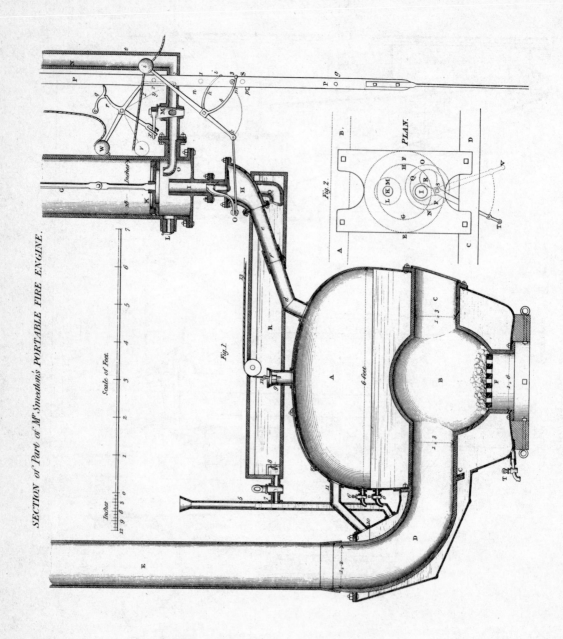

SECTION of Part of M.ʳ Smeaton's PORTABLE FIRE ENGINE.

Fig.1.

Fig.2

PLAN.

Scale of Feet.

Published as the Act directs, 1820 by Longman, Rees, Orme, Brown & Green, Paternoster Row.

Reduced by J.Farey.

Plate X

Mr. WATT'S single ENGINE for pumping water,
for draining Mines, 1788.

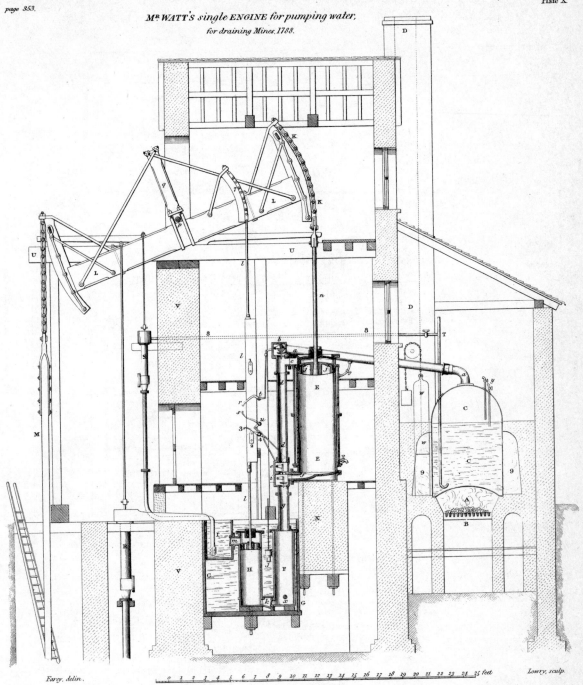

Farey, delin.

Lowry, sculp.

0 1 2 3 4 5 6 7 8 9 10 11 12 13 14 15 16 17 18 19 20 21 22 23 24 25 feet

Published as the Act directs, 1826, by Longman, Rees, Orme, Brown & Green, Paternoster Row.

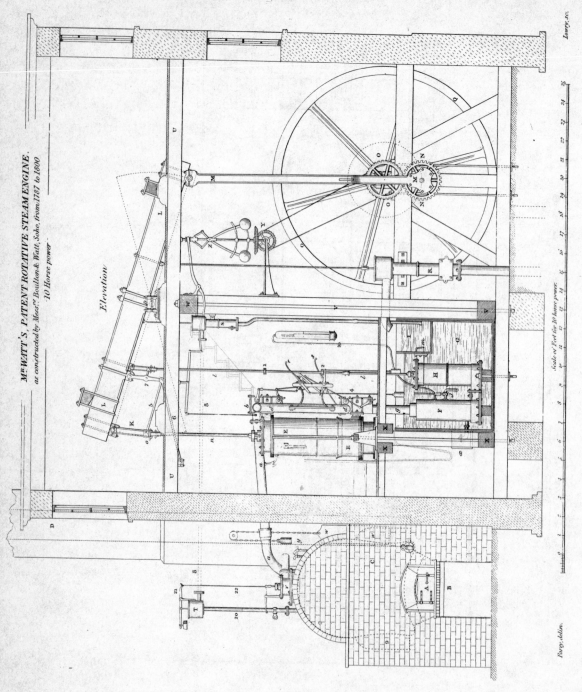

PLATE XI.

page 444.

Lowry, sc.

MR. WATT'S PATENT ROTATIVE STEAM ENGINE.
as constructed by Messrs. Boulton & Watt, Soho, from 1787 to 1800.
10 Horse power.

Elevation

Scale of Feet for 10 horse power.

Published as the Act directs, 1826 by Longman, Rees, Orme, Brown & Green, Paternoster Row.

Farey, delin.

PLATE XII.

Mr. WATT'S Patent Rotative STEAM ENGINE.
10 Horse power.

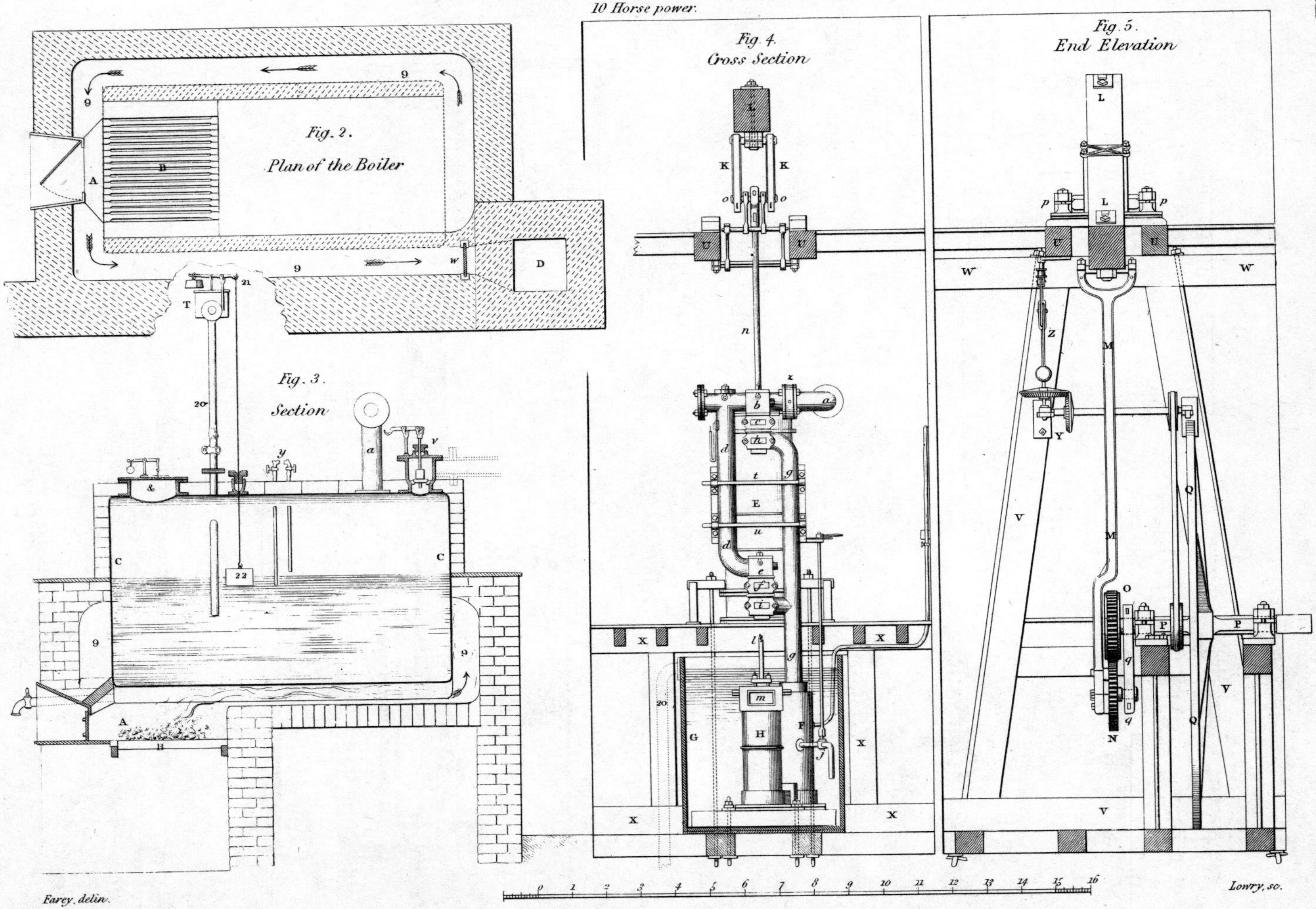

Fig. 5.
End Elevation

Fig. 4.
Cross Section

Fig. 2.
Plan of the Boiler

Fig. 3.
Section

Lowry, sc.

Published as the Act directs. 1816 by Longman, Rees, Orme, Brown & Green, Paternoster Row, London.

Emery, delin.

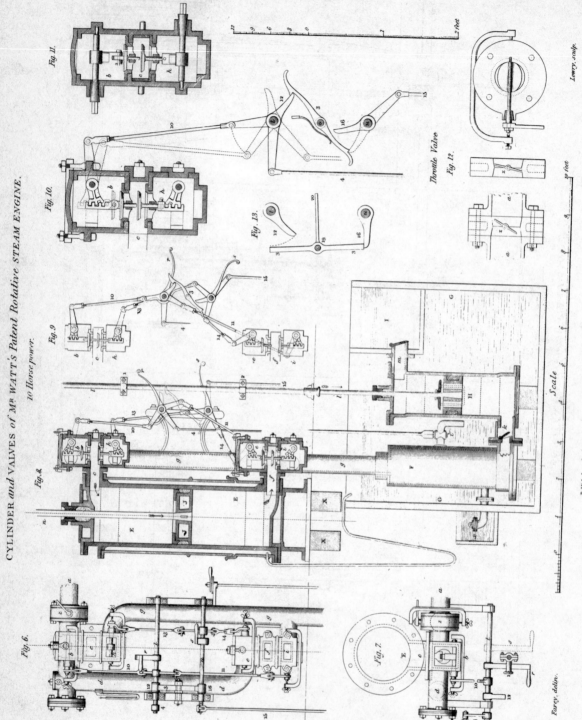

PLATE XIII.

CYLINDER and VALVES of Mr WATT's Patent Rotative STEAM ENGINE.

10 Horse power.

Fig. 11.

Fig. 10.

Fig. 9.

Fig. 8.

Fig. 6.

Fig. 7.

Fig. 13.

Throttle Valve

Fig. 12.

Scale

2 feet

10 feet

Lowry, sculp.

Farey, delin.

Published as the Act directs, 1818 by Longman, Hurst, Rees, Orme, Brown & Green, Paternoster Row, London.

PLATE XIV.

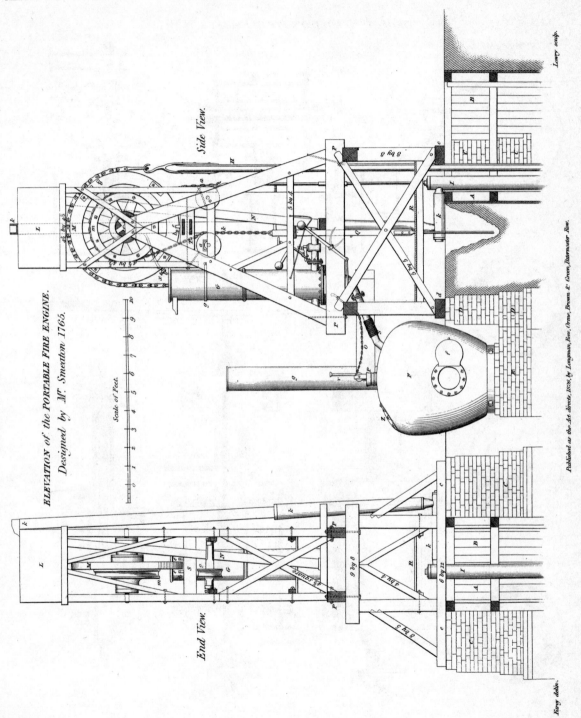

ELEVATION of the PORTABLE FIRE ENGINE.
Designed by M. Smeaton 1765.

Side View.

End View.

Scale of Feet

Lowry sculp.

Furey delin.

Published as the Act directs, 1836, by Longman, Rees, Orme, Brown & Green, Paternoster Row.

PLATE XV.

PROJECTS for ROTATIVE STEAM ENGINES.

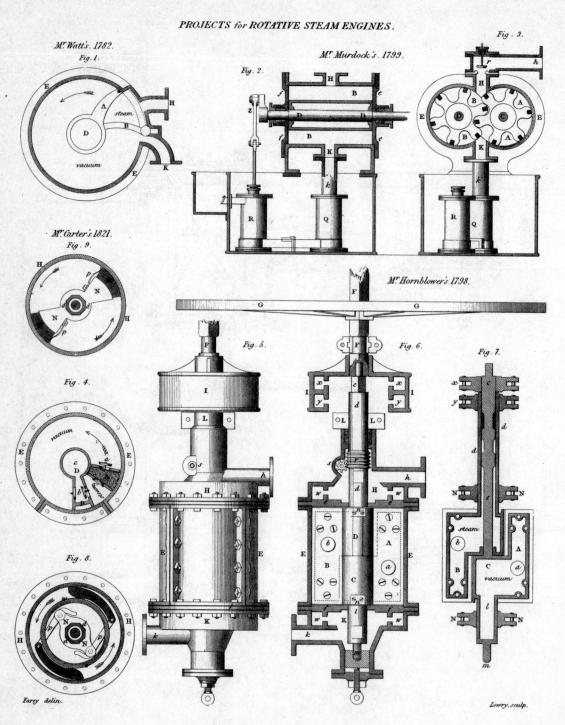

M.ʳ Watt's. 1782.
Fig. 1.

M.ʳ Murdock's. 1799.
Fig. 2.

Fig. 3.

M.ʳ Carter's. 1821.
Fig. 9.

M.ʳ Hornblower's. 1798.

Fig. 5.

Fig. 6.

Fig. 7.

Fig. 4.

Fig. 8.

Farey delin.

Lowry, sculp.

Published as the Act directs, by Longman, Rees, Orme, Brown & Green, Paternoster Row.

PLATE XVI.

STEAM ENGINE of 6 Horses Power
by Messᵣˢ. Fenton, Murray & Wood.
1806.

Fig.1.

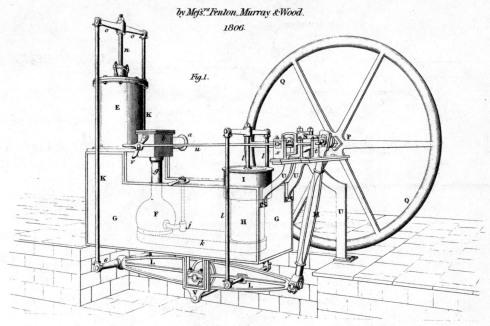

BELL CRANK STEAM ENGINE
of 4 Horses Power
constructed by
Messᵣˢ. Boulton & Watt.
1802.

Fig.4.

Fig.3.

Scale of Feet

Fig.2.

Fig.5.

Fig.6.

Fig.1.

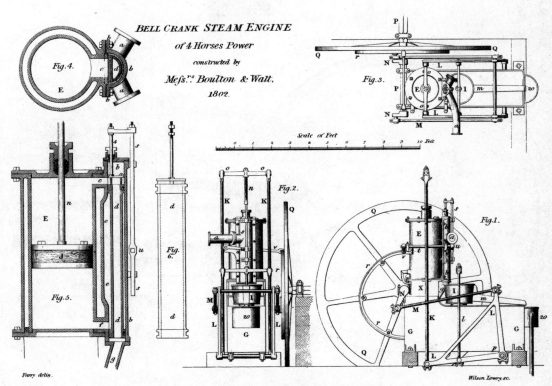

Furry delin.

Wilson Lowry sc.

Published as the Act directs, 1816, by Longman, Rees, Orme, Brown & Green, Paternoster Row, London.

PLATE XVII.

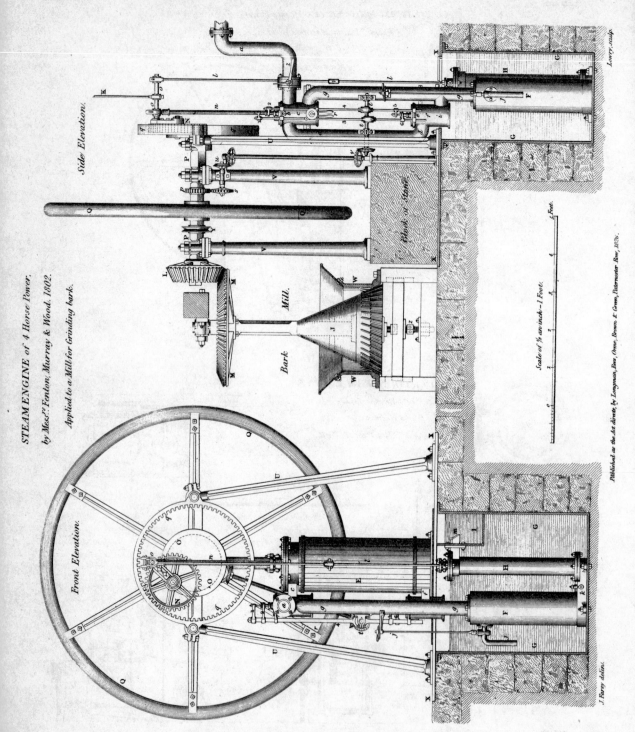

STEAM ENGINE of 4 Horse Power.

by Mess.rs Fenton, Murray & Wood. 1802.

Applied to a Mill for Grinding bark.

Side Elevation.

Front Elevation.

Bark Mill.

Bark

Block of Stone.

Scale of ⅛ an inch=1 Foot.

Published as the Act directs, by Longman, Hurst, Orme, Brown, & Green, Paternoster Row, 1820.

J. Farey delin.

Lowry sculp.

PLATE XVIII

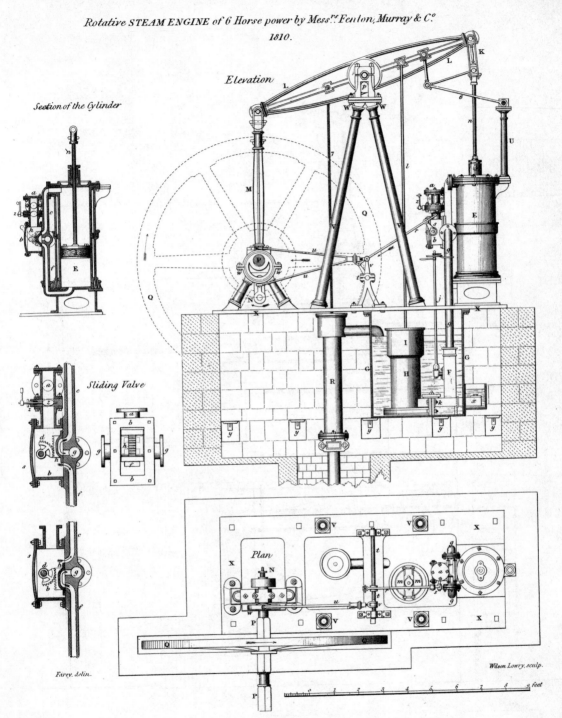

Rotative *STEAM ENGINE* of 6 Horse power by Mess.rs Fenton, Murray & C.o
1810.

Section of the Cylinder

Elevation

Sliding Valve

Plan

Farey. delin.

Wilson Lowry, sculp.

London, Published as the Act directs by Longman, Rees, Orme, Brown & Green, Paternoster Row, 1826.

PLATE XIX.

Mr Watt's Rotative STEAM ENGINE of 36 Horse Power;
by Mes.rs Boulton & Watt, Soho, 1808.

General Elevation. Fig. 1.

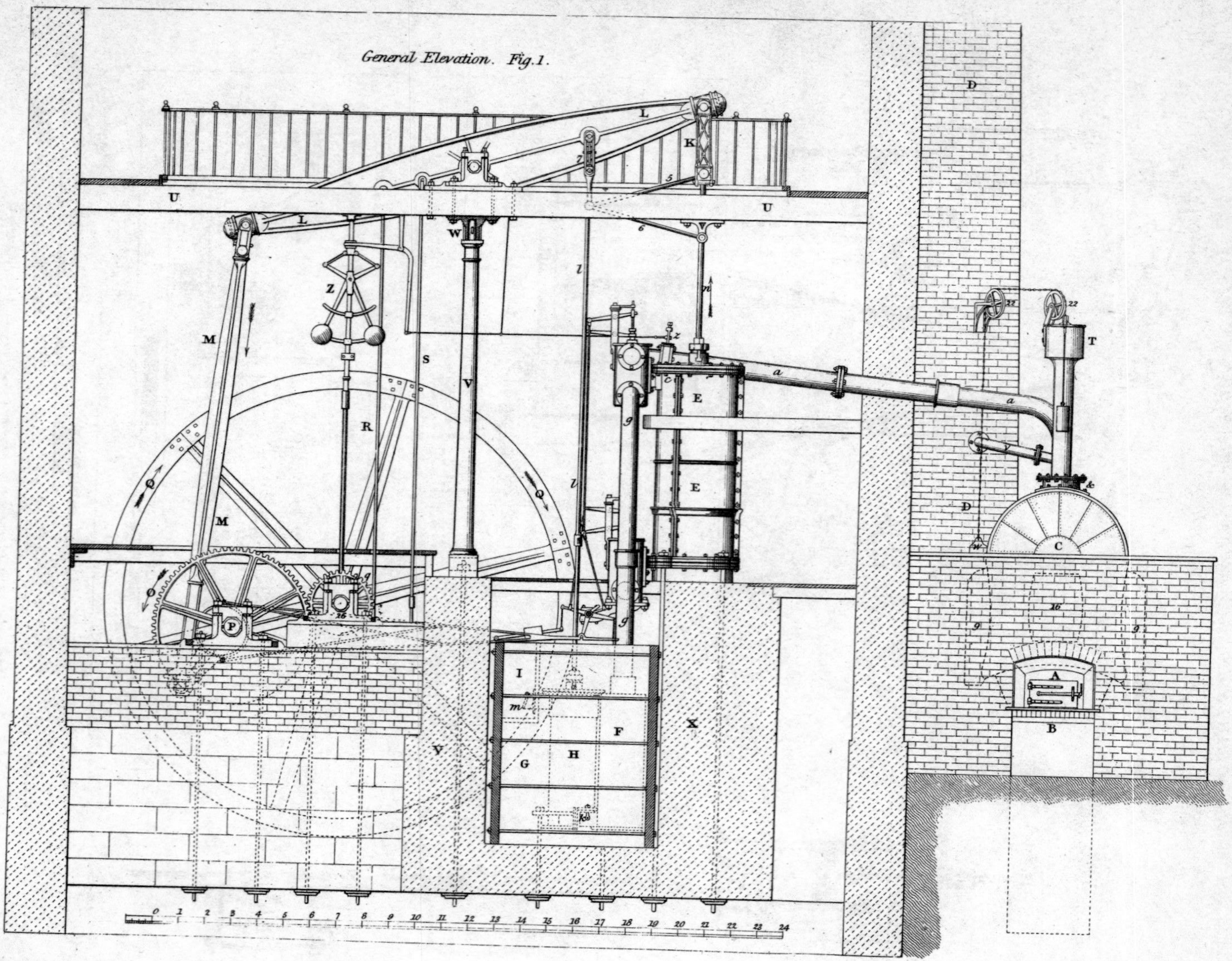

J. Farey delin.t

Wilson, Lowry sculp.t

Published as the Act directs, 1826, by Longman, Rees, Orme, Brown, & Green, Paternoster Row.

PLATE XX.

Cylinder & Valves of Mr Watt's Rotative STEAM ENGINE
of 36 Horse Power, 1808.

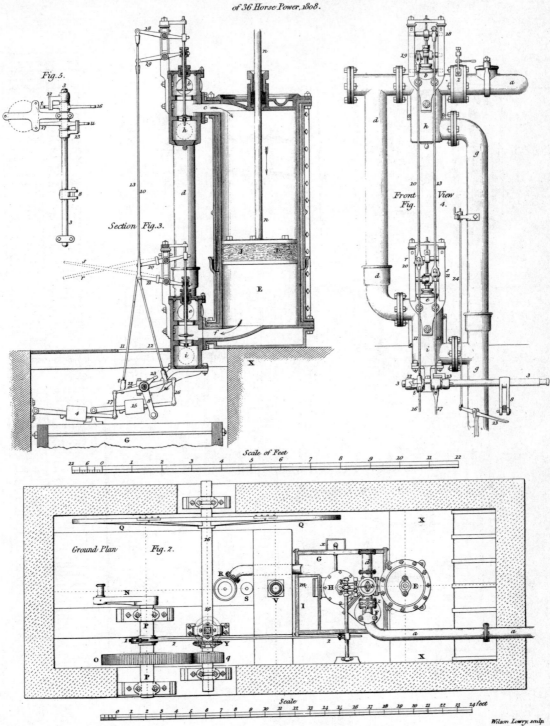

Fig. 5.

Section Fig. 3.

Front
View
Fig.
4.

Scale of Feet

Ground Plan   Fig. 2.

Scale

J. Farey, del.

Wilson Lowry, sculp.

London Published as the Act directs by Longman, Rees, Orme, Brown, & Green, Paternoster Row. 1826.

PLATE XXI.

STEAM ENGINE and MILL for ROLLING SHEET LEAD.

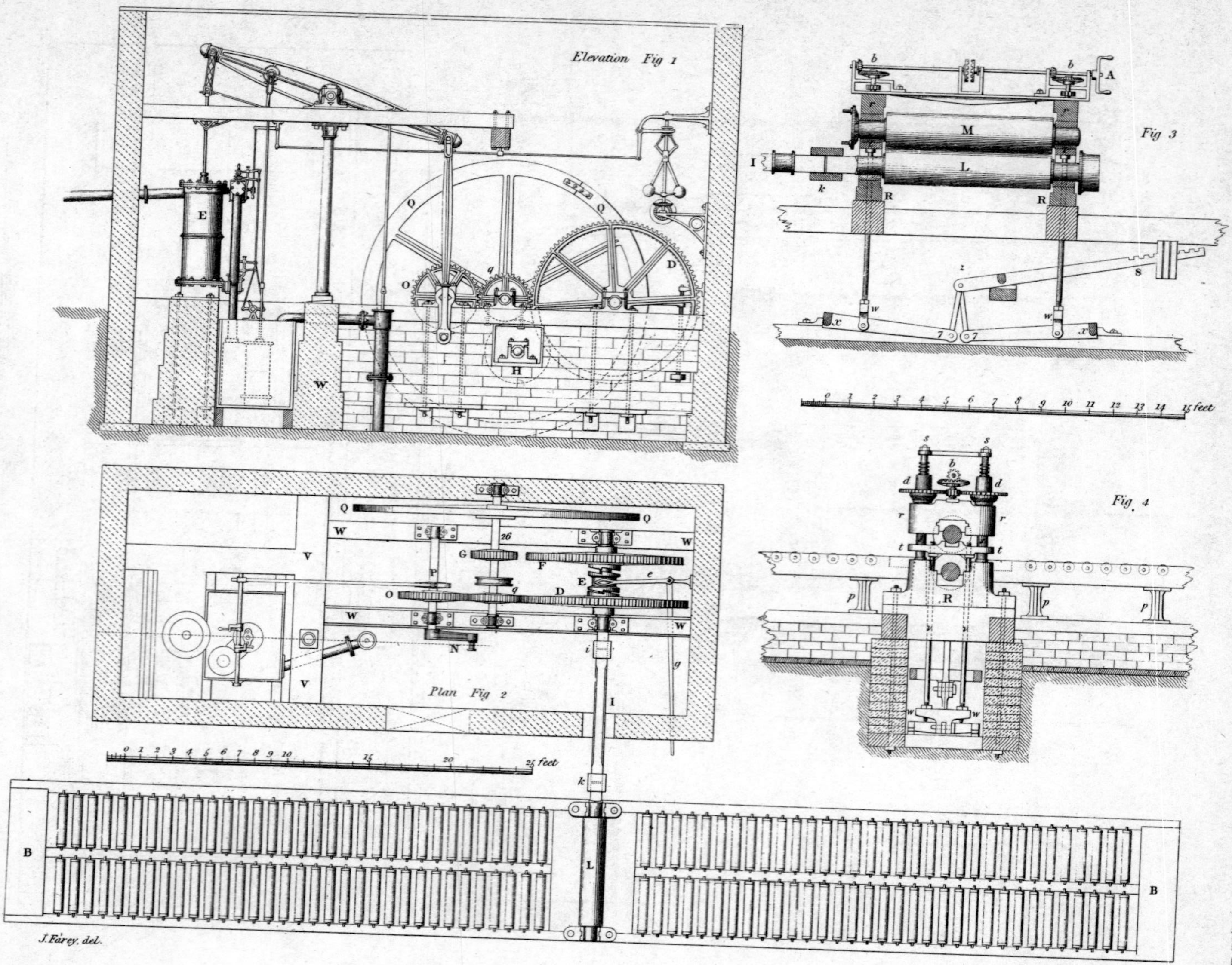

Fig 3

Fig. 4

Elevation Fig 1

Plan Fig 2

Wilson Lowry, sculp.

London, Published as the Act directs by Longman, Rees, Orme, Brown & Green, Paternoster Row, 1822.

J. Farey, del.

PLATE XXII

PERSPECTIVE VIEW of a STEAM ENGINE and HORIZONTAL SUGAR MILL,

as constructed by Jukes Coulson & Co London.

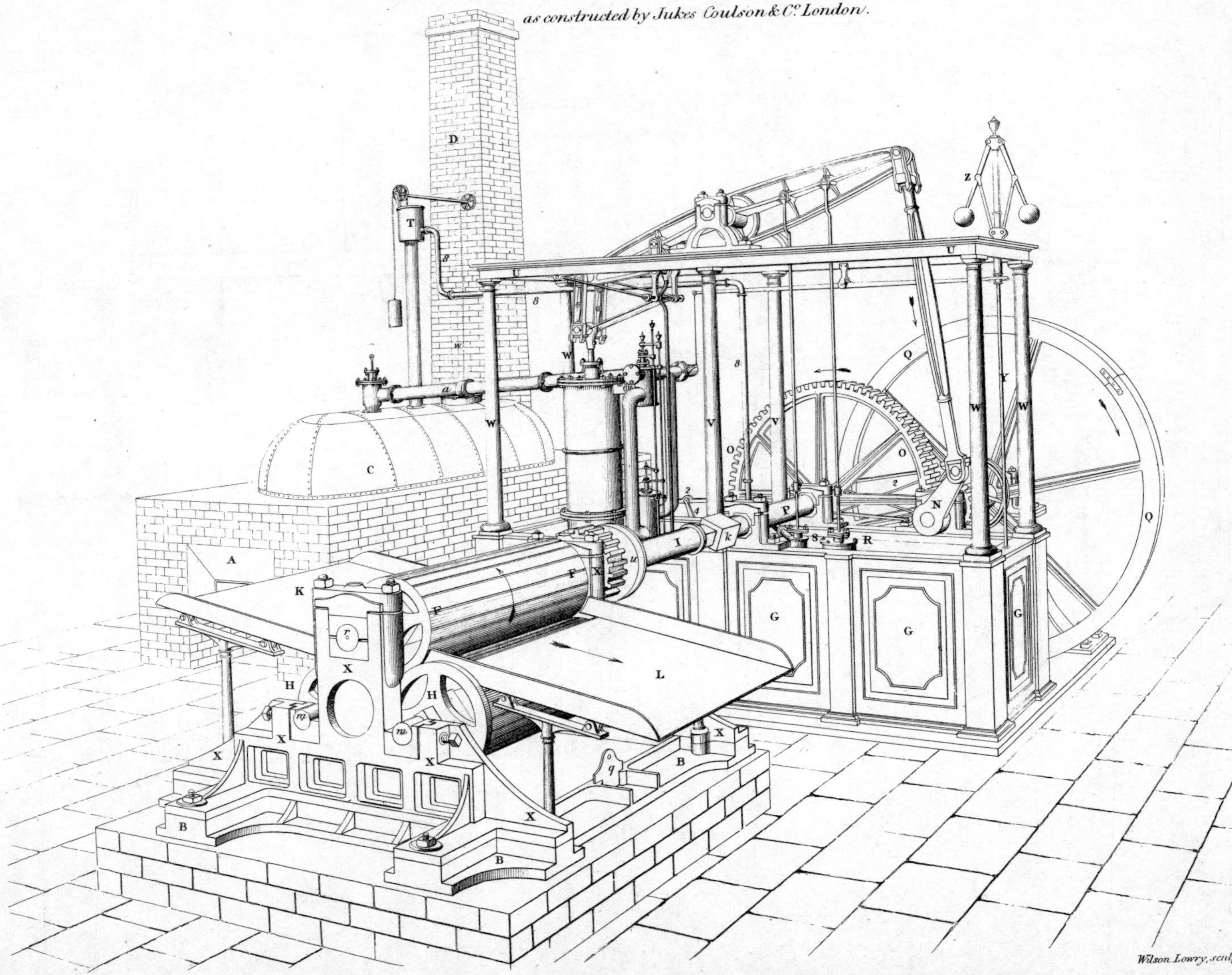

Wilson Lowry, sculp.

J. Farey, del.

London, Published as the Act directs, by Longman, Rees, Orme, Brown, & Green, Paternoster Row, 1827.

PLATE XXIII.

Mr. WATT'S STEAM ENGINE for pumping water for the supply of Towns,

constructed by Mess.rs Boulton & Watt, 1807.

Wilson Lowry, sculp.

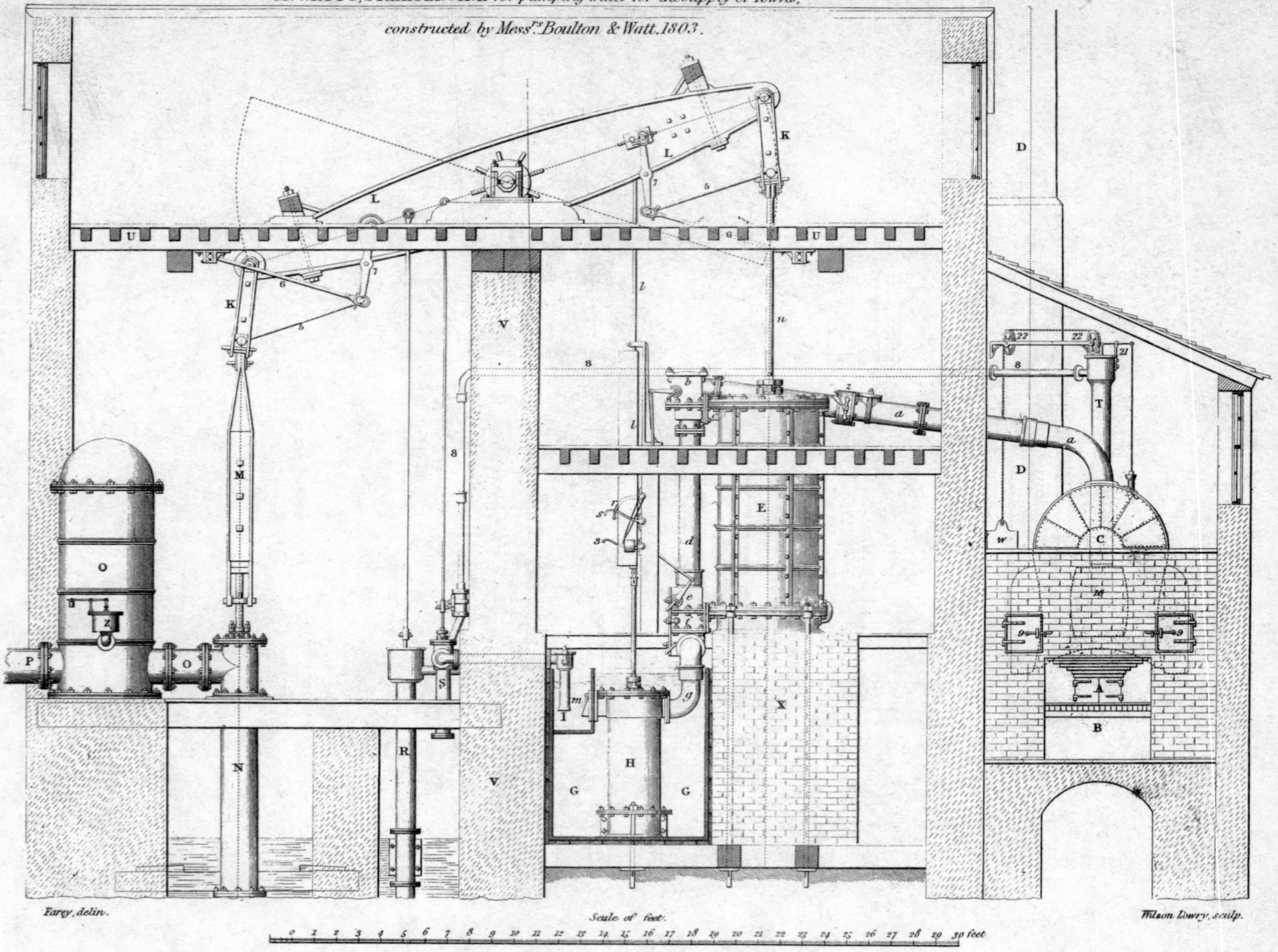

Farey, delin.

Scale of feet.

Published as the Act directs, 1828, by Longman, Rees, Orme, Brown & Green, Paternoster Row.

PLATE XXIV.

Parts of Mr. Watt's, STEAM ENGINE for Pumping water 1803.

Section of the Cylinder and Valves
Air Pump and Condenser.

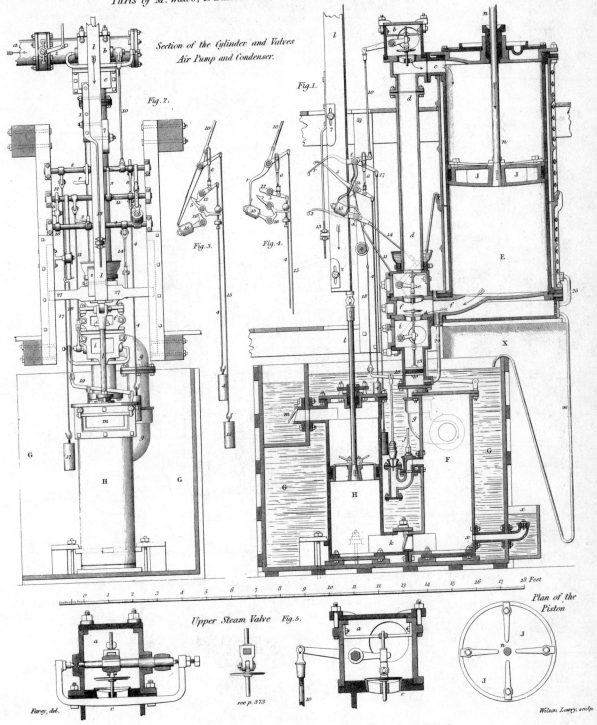

Fig.2.

Fig.1.

Fig.3.

Fig.4.

Upper Steam Valve Fig.5.

Plan of the Piston

see p. 373

Farey, del.

Wilson Lowry, sculp.

London, Published as the Act directs, by Longman, Rees, Orme, Brown & Green, Paternoster Row, 1826.

PLATE XXV.

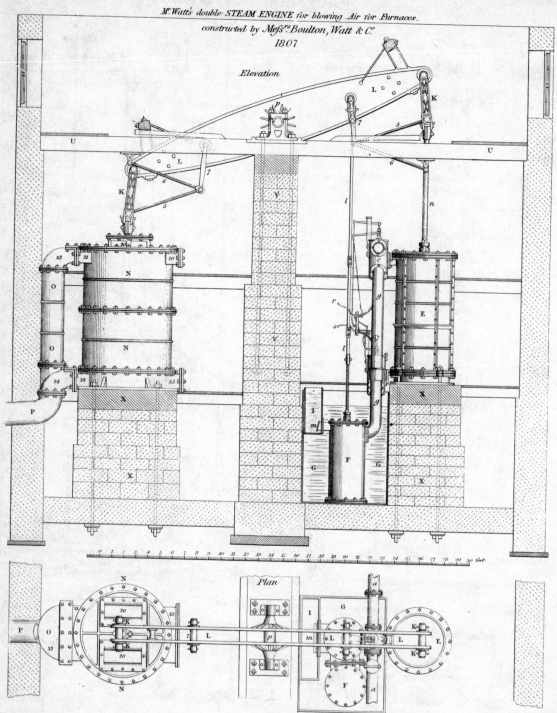

Mr Watt's double STEAM ENGINE for blowing Air for Furnaces.
constructed by Messrs Boulton, Watt & Co.
1807

Elevation

Plan

Farey, delin.

Wilson Lowry, sculp.

Published as the Act directs, 1836, by Longman, Rees, Orme, Brown & Green, Paternoster Row.